SWITCHING THEORY AND LOGIC DESIGN

SWITCHING THEORY AND LOGIC DESIGN

By

M. V. Subramanyam

M. Tech., Ph.D.

Associate Professor and I/C Head of the Department of Electronics and Communication Engineering
R.G.M. College of Engineering and Technology
Nandyal, Kurnool Distt. (A.P.)

UNIVERSITY SCIENCE PRESS

(An Imprint of Laxmi Publications Pvt. Ltd.)

An ISO 9001:2015 Company

BENGALURU • CHENNAI • GUWAHATI • HYDERABAD
KOCHI • KOLKATA • LUCKNOW • RANCHI
NEW DELHI

SWITCHING THEORY AND LOGIC DESIGN

Printed and bound in India

Typeset at : Kalyani Computer Services, New Delhi.

First Edition : 2006; *Second Edition* : 2011; *Reprint* : 2013, 2016, 2019, 2022; *Edition* : 2025

ISBN : 978-93-80386-90-4

UST-9333-999-SWITCH THEO & LOGIC DES-SUB

Price: ₹ 999.00

Branches

Bengaluru	080-26 75 69 30
Chennai	044-24 34 47 26
Guwahati	0361-254 36 69
Hyderabad	040-27 55 53 83
Kochi	0484-405 13 03
Kolkata	033-40 04 77 79
Lucknow	0522-430 36 13
Ranchi	0651-224 24 64

PUBLISHED IN INDIA BY

UNIVERSITY SCIENCE PRESS
(An Imprint of Laxmi Publications Pvt. Ltd.)

An ISO 9001:2015 Company
113, GOLDEN HOUSE, GURUDWARA ROAD, DARYAGANJ,
NEW DELHI - 110002, INDIA

Corporate Office : B-33, Sector-5, Noida,
Gautambuddha Nagar (U.P.) - 201 301, INDIA
Telephone : +91-120-44 46 900
www.laxmipublications.com
info@laxmipublications.com

C—R/025/08
Printed at : Ajit Printing Press, Delhi.

CONTENTS

	Pages
Syllabus	(*xvi*)
1. NUMBER SYSTEM AND CODES	**1—107**
1.0. Introduction	1
1.1. Review of Number Systems	1
1.2. Decimal Number System	2
1.3. Binary Number Systems	3
1.4. Octal Number System	5
1.5. Hexadecimal Number System	6
1.6. Other Number Systems	7
1.7. Number System Conversion	8
1.7.1. Binary-to-Decimal Conversion	8
1.7.2. Decimal-to-Binary Conversion	9
1.7.3. Binary-to-Octal Conversion	12
1.7.4. Octal-to-Binary Conversion	12
1.7.5. Binary-to-Hexadecimal Conversion	13
1.7.6. Hexadecimal-to-Binary Conversion	13
1.7.7. Decimal-to-Octal Conversion	14
1.7.8. Octal-to-Decimal Conversion	15
1.7.9. Decimal-to-Hexadecimal Conversion	16
1.7.10. Hexadecimal-to-Decimal Conversion	16
1.7.11. Octal-to-Hexadecimal Conversion	17
1.7.12. Hexadecimal-to-Octal Conversion	18
1.7.13. Other Radix (r) System-to-Decimal Conversion	18
1.7.14. Decimal System-to-Other Radix (r) System	20
1.7.15. Summary of Conversions	22
1.8. Binary Arithmetic	22
1.8.1. Binary Addition	22
1.8.2. Binary Subtraction	24
1.8.3. Binary Multiplication	34

	1.8.4. Binary Division	36
1.9.	The 9's and 10's Complements	37
1.10.	Octal Arithmetic	41
	1.10.1. Addition	41
	1.10.2. Subtraction	42
1.11.	Hexadecimal Arithmetic	44
	1.11.1. Addition	44
	1.11.2. Subtraction	45
1.12.	Binary Codes	46
	1.12.1. Classification of Binary Codes	46
	1.12.2. BCD Codes or 8421 Code	48
	1.12.3. Other 4-bit BCD Codes	51
	1.12.4. Non-Weighted Codes	52
1.13.	5-Bit Codes	59
1.14.	Codes with More than 5-Bits	60
1.15.	Alpha Numeric Codes	61
1.16.	Error Correcting and Detecting Codes	65
	1.16.1. Types of Errors	65
	1.16.2. Error Detection	65
	1.16.3. Error Correction	73
1.17.	Solved Problems	79
1.18.	Review Questions	98
1.19.	Short Answer Questions	99
1.20.	Multiple Choice Questions	106
2.	**BOOLEAN ALGEBRA**	**108—226**
2.1.	Introduction	108
2.2.	Boolean Postulates	108
2.3.	Huntington Postulates	109
2.4.	Boolean Algebra Symbols	110
2.5.	Two-Valued Four-Element Boolean Algebra	111
2.6.	Basic Theorems and Proofs	112
2.7.	Switching Algebra	114
	2.7.1. Generalized De Morgan's Theorem	115
	2.7.2. Shannon's Theorem	115
	2.7.3. Expansion Theorem	116
	2.7.4. Theorem	116
2.8.	De Morgan's Theorems	116

CONTENTS

	Pages
Syllabus	(*xvi*)
1. NUMBER SYSTEM AND CODES	**1—107**
1.0. Introduction	1
1.1. Review of Number Systems	1
1.2. Decimal Number System	2
1.3. Binary Number Systems	3
1.4. Octal Number System	5
1.5. Hexadecimal Number System	6
1.6. Other Number Systems	7
1.7. Number System Conversion	8
1.7.1. Binary-to-Decimal Conversion	8
1.7.2. Decimal-to-Binary Conversion	9
1.7.3. Binary-to-Octal Conversion	12
1.7.4. Octal-to-Binary Conversion	12
1.7.5. Binary-to-Hexadecimal Conversion	13
1.7.6. Hexadecimal-to-Binary Conversion	13
1.7.7. Decimal-to-Octal Conversion	14
1.7.8. Octal-to-Decimal Conversion	15
1.7.9. Decimal-to-Hexadecimal Conversion	16
1.7.10. Hexadecimal-to-Decimal Conversion	16
1.7.11. Octal-to-Hexadecimal Conversion	17
1.7.12. Hexadecimal-to-Octal Conversion	18
1.7.13. Other Radix (r) System-to-Decimal Conversion	18
1.7.14. Decimal System-to-Other Radix (r) System	20
1.7.15. Summary of Conversions	22
1.8. Binary Arithmetic	22
1.8.1. Binary Addition	22
1.8.2. Binary Subtraction	24
1.8.3. Binary Multiplication	34

1.8.4. Binary Division	36
1.9. The 9's and 10's Complements	37
1.10. Octal Arithmetic	41
1.10.1. Addition	41
1.10.2. Subtraction	42
1.11. Hexadecimal Arithmetic	44
1.11.1. Addition	44
1.11.2. Subtraction	45
1.12. Binary Codes	46
1.12.1. Classification of Binary Codes	46
1.12.2. BCD Codes or 8421 Code	48
1.12.3. Other 4-bit BCD Codes	51
1.12.4. Non-Weighted Codes	52
1.13. 5-Bit Codes	59
1.14. Codes with More than 5-Bits	60
1.15. Alpha Numeric Codes	61
1.16. Error Correcting and Detecting Codes	65
1.16.1. Types of Errors	65
1.16.2. Error Detection	65
1.16.3. Error Correction	73
1.17. Solved Problems	79
1.18. Review Questions	98
1.19. Short Answer Questions	99
1.20. Multiple Choice Questions	106
2. BOOLEAN ALGEBRA	**108—226**
2.1. Introduction	108
2.2. Boolean Postulates	108
2.3. Huntington Postulates	109
2.4. Boolean Algebra Symbols	110
2.5. Two-Valued Four-Element Boolean Algebra	111
2.6. Basic Theorems and Proofs	112
2.7. Switching Algebra	114
2.7.1. Generalized De Morgan's Theorem	115
2.7.2. Shannon's Theorem	115
2.7.3. Expansion Theorem	116
2.7.4. Theorem	116
2.8. De Morgan's Theorems	116

2.9. Duality Theorem 117
2.9.1. Self Dual Function 118
2.10. Switching Functions 118
2.10.1. Representation of Switching Functions 120
2.11. Sum-Of-Products (SOP) Form 121
2.12. Product-Of-Sums (POS) Form 121
2.13. Canonical Forms 121
2.13.1. Canonical SOP 121
2.13.2. Canonical POS 122
2.13.3. Converting SOP Expressions into Canonical SOP Form 122
2.13.4. Canonical SOP Expression from Truth Table 122
2.13.5. Conversion of POS Expression into Canonical POS Form 123
2.13.6. CPOS Expression from a Truth Table 124
2.13.7. Truth Table from SOP Expression 126
2.13.8. Truth Table from a POS Expression 126
2.13.9. Conversion between Canonical Forms 126
2.14. Standard Forms 131
2.14.1. Non Standard Form 131
2.15. Simplification of Boolean Expressions 132
2.16. Karnaugh Map Simplification 141
2.16.1. The Karnaugh Map 142
2.16.2. Plotting a Karnaugh Map 144
2.16.3. Grouping of Cells for Simplification 147
2.16.4. Simplification of SOP Expressions 152
2.16.5. Simplification of POS Expressions 159
2.16.6. Don't Care Conditions/Cannot Happen States 161
2.16.7. Five and Six Variable K-maps 163
2.16.8. Converting from SOP to POS form Using K-map 169
2.16.9. Converting from POS to SOP form Using K-map 170
2.17. Quine-McCluskey or Tabulation Procedure 170
2.17.1. Minimal Functions and their Properties 171
2.17.2. The Binary Representation 173
2.17.3. The Decimal Representation 181
2.18. Reduced-Dimension Map 195
2.19. Solved Problems 200
2.20. Review Questions 215
2.21. Short Answer Questions 217
2.21.1. Multiple Choice Questions 218

2.21.2. True or False 224
2.21.3. Fill in the Blanks 225

3. DESIGN OF COMBINATIONAL CIRCUITS 227—428

3.1. Logical Operators 227
3.1.1. Logical Operator Not/Invert 227
3.1.2. Logical Operator AND 227
3.1.3. Logical Operator OR 228
3.2. Logic Gates 228
3.2.1. AND Gate 229
3.2.2. OR Gate 231
3.2.3. NOT Gate 234
3.2.4. Buffer (OR) Transfer Gate 235
3.2.5. NAND Gate 236
3.2.6. NOR Gate 238
3.2.7. Exclusive–OR Gate 241
3.2.8. Exclusive–NOR Gate 242
3.3. Gates with Inverted Inputs 244
3.3.1. AND Gate with Inverted Inputs 244
3.3.2. NAND Gate with Inverted Inputs 245
3.3.3. OR Gate with Inverted Inputs 246
3.3.4. NOR Gate with Inverted Inputs 246
3.4. Logic Circuit From Boolean Expressions 248
3.5. Analysis of Circuits 253
3.6. Multiple Input Gates 260
3.7. Single Output and Multi-Output Functions 262
3.8. Universal Gates 264
3.8.1. NAND Gate 265
3.8.2. Multilevel Implementation of NOR Gate 267
3.9. Design of Combinational Circuits 269
3.9.1. Multiple Output Functions 271
3.9.2. Which Gate Representation to Use 273
3.9.3. Which Circuit Diagram should be Used 275
3.10. Arithmetic Circuits 276
3.10.1. Adders 276
3.10.2. Subtractors 295
3.11. Code Converters 307
3.11.1. Binary–to–BCD Converter 308
3.11.2. BCD–to–Binary Code Converter 309

3.11.3. Binary–to–Gray Code Converter 312
3.11.4. Gray Code–to–Binary Code Converter 314
3.11.5. BCD–to–Excess–3 Code Converter 316
3.12. Magnitude Comparator 317
3.13. MSI and LSI Circuits 323
3.13.1. Decoders 323
3.13.2. Encoders 331
3.13.3. Multiplexers 339
3.13.4. Demultiplexer 354
3.14. Logic implementation Using PLD's 357
3.14.1. Read Only Memory (ROM) 358
3.14.2. Programmable Logic Array (PLA) 364
3.14.3. Programmable Array Logic (PAL) 371
3.14.4. Comparison between PROM, PLA and PAL 375
3.15. Disadvantages in Asynchronous Circuits 376
3.15.1. Cycles 376
3.15.2. Races 376
3.15.3. Hazards 377
3.16. Solved Problems 381
3.17 Short Answer Questions 420
3.18. Review Questions 427
4. INTRODUCTION TO SEQUENTIAL CIRCUITS 429—598
4.1. Introduction 429
4.2. Flip-Flops 431
4.2.1. Basic Building Block 431
4.2.2. RS-Flip-Flop 432
4.2.3. Clocked RS-Flip-Flop 435
4.2.4. SR-Flip-Flop 438
4.2.5. D Flip-Flop 438
4.2.6. JK Flip-Flop 440
4.2.7. T Flip-Flop 443
4.3. Triggering of Flip-Flops 444
4.3.1. Edge-Triggered Flip-Flops 444
4.3.2. Level Triggered Flip-Flop 446
4.3.3. Race Around Condition 446
4.3.4. Master-Slave Flip-Flop 447
4.4. Asynchronous or Direct Inputs 450
4.5. Flip-Flop Conversions 452
4.5.1. Realization of Delay (D) Flip-Flop using SR Flip-Flop 452

4.5.2. Realization of JK Flip-Flop using SR Flip-Flop 453
4.5.3. Realization of Trigger (T) Flip-Flop using SR Flip-Flop 455
4.5.4. Realization of JK Flip-Flop using D Flip-Flop 456
4.6. Flip-Flop Excitation Tables 457
4.6.1. RS Flip-Flop 457
4.6.2. JK Flip-Flop 458
4.6.3. D Flip-Flop 458
4.6.4. T Flip-Flop 459
4.7. Applications of Flip-Flops 459
4.8. Analysis of Synchronous Sequential Circuits 459
4.8.1. State Table 459
4.8.2. State Diagram 461
4.8.3. State Equations 462
4.8.4. State Reduction 466
4.8.5. State Assignment 468
4.9. Design Procedure 472
4.9.1. Design with Unused States 479
4.9.2. Lockout Condition 484
4.10. Sequence Generator 487
4.10.1. Sequence Generators Using Counters 487
4.10.2. Sequence Generator Using Shift-Registers 491
4.11. Sequence Detector 493
4.12. Counters 508
4.12.1. Asynchronous or Ripple Counters 508
4.13. Design of Divide-By-N Ripple Counter 514
4.13.1. Design of Decade (or Devide-by-10 or Modulo-10) Ripple Counter 515
4.13.2. Design of MOD-6 Ripple Counter 516
4.13.3. MOD-10 Ripple Counter/Decade Counter (or) BCD Counter 517
4.13.4. IC Asynchronous Counters 519
4.13.5. 74LS93/293 4-bit Binary Counter 519
4.13.6. 74LS90/290 Decade Counter 522
4.13.7. CMOS Asynchronous Counters 523
4.13.8. Asynchronous Down Counter 523
4.13.9. Propagation Delay in Ripple Counters 525
4.13.10. Advantages and Disadvantages of Asynchronous Counters 526
4.13.11. MOD Counters Using Reset Input 526
4.14. Synchronous Counters 527
4.14.1. 2-bit Synchronous Counter 527
4.14.2. A 3-bit Synchronous Binary Counter 528

4.14.3. 4-bit Synchronous Binary Counter 529
4.14.4. Maximum Frequency Calculations 531
4.14.5. Design of Synchronous Counters 533
4.14.6. Modulus-N Synchronous Counter 534
4.14.7. Up/Down Counter 539
4.15. Shift Counters (or) Shift Register Counters 543
4.15.1. Johnson Counter 545
4.15.2. IC Synchronous Counters 548
4.16. Presettable Counters 549
4.17. Counter Applications 550
4.18. Registers 554
4.18.1. Serial-In-Serial Out Shift Register 556
4.18.2. Serial-In-Parallel Out Shift Register 556
4.18.3. Parallel-In-Serial Out Shift Register 557
4.18.4. Parallel-In-Parallel Out Shift Register 558
4.19. Shift Registers 558
4.19.1. Shift Right Register 558
4.19.2. Shift Left Register 560
4.19.3. Controlled Shift Register 562
4.19.4. Bidirectional Shift Register 563
4.19.5. Universal Shift Register 563
4.20. Solved Problems 565
4.21. Review Questions 591
4.22. Short Answer Questions 595
5. CAPABILITIES AND MINIMIZATION OF SEQUENTIAL MACHINES 599—640
5.1. Finite State Machine 599
5.1.1. The Finite State Model – Basic Definitions 599
5.1.2. Deterministic Machine 600
5.2. Input and Output Transformations 603
5.2.1. Important Function of a Sequential Machine 604
5.3. Capabilities and Limitations of Finite-State Machines 605
5.4. State Equivalence and Machine Minimization 606
5.4.1. K-equivalence and K-distinguishable 606
5.4.2. The Minimization Procedure 608
5.4.3. Machine Equivalence 609
5.5. Transition Diagrams 610
5.6. Comparison between Moore Machine and Mealy Machine 615
5.7. Simplification of Incompletely Specified Machines 616

5.7.1. Merger Graphs 618
5.7.2. Merger Table 620
5.8. Solved Problems 622
5.9. Review Questions 632
5.10. Exercise 633
5.11. Multiple Choice Questions 635
5.11.1. True or False 638
5.11.2. Fill in the Blanks 639
6. ALGORITHMIC STATE MACHINES 641—732
6.1. Introduction 641
6.1.1. Difference between Conventional Flow Charts and ASM Charts 643
6.2. ASM Charts 643
6.2.1. Comparison between ASM Charts and State Diagrams 644
6.3. Register Operations 648
6.4. Timing Considerations 649
6.5. Circuit Realization 655
6.6. Control Implementation 658
6.6.1. State Table Method 659
6.6.2. Design with Multiplexers Method 662
6.6.3. Design Using ROM and D–Flip-Flop 664
6.6.4. Design Using Programmable Devices 669
6.7. Mnemonics Used in ASM Charts 674
6.8. Threshold Logic 678
6.8.1. Introduction 679
6.8.2. Construction of Threshold Element 681
6.8.3. Capabilities and Limitations of Threshold Logic 682
6.8.4. Elementary Properties 687
6.9. Synthesis of Threshold Networks 688
6.9.1. Unate Functions 688
6.9.2. Geometrical Representation of a Unate Function 689
6.9.3. Linear Separability 690
6.10. Identification and Realization of Threshold Functions 692
6.11. Synthesis of Threshold Networks Using Map as a Tool 695
6.12. Solved Problems 697
6.13. Short Answer Questions 729
6.14. Review Questions 731
Annexure 733—774
INDEX 775—783

PREFACE

The book "Switching Theory and Logic Design" is an outcome after several years of teaching. It is presented in simple and easy-to-understand manner. This book is useful to the undergraduate students of JNT University and almost all other universities in India. This book contains six chapters and each chapter is explained in a plain, lucid and necessary background to each topic with essential examples. Once again this book is useful not only to the students but also to the subject teachers. I have tried my level best to bring this book error-free, but by misfortune some errors might have crept in. I will be very thankful if such errors are brought to my notice. Constructive criticism and positive suggestions for the improvement of the quality of the book are welcome and will be highly appreciated.

—M.V. Subramanyam

KEY FEATURES

★ *Numerous worked-out examples highlight the need for intelligent approximation to achieve more accuracy in lesser time.*

★ *Short answer questions at the end of each chapter helps to easy understands the subject.*

★ *Large number of review questions and unsolved problems to develop a clear understanding of basic principles.*

★ *Previous GATE paper solutions are the unique feature of this book.*

ACKNOWLEDGMENTS

At the very outset, the author solemnly offers his severe gratitude to the Almighty, the most beneficent and merciful for bestowing the opportunity to write this book entitled "Switching Theory and Logic Design".

The author is greatly indebted to his parents Sri. M. Sankaranagaiah, and Late Smt. M. Punyavathamma, without their blessings the author could not have this day to see. Also thankful to his wife Smt. R. Rajyalakshmi Devi and children M.V.S. Sai Snehitha and M.V.S. Sai Abhinay, for their sustained patience during the preparation of this book.

The author is extremely thankful to the chairman, R.G.M. Group of educational institutions Sri. M. Santhiramudu Garu for establishing very good library with many numbers of Journals related to this book. The author is also thankful to the Dr. K. Satyaprasad, Vice-Principal, JNTU Engineering, Kakinada, Principal of this college Dr. P.V. Ramanamurthy, and A.O. Sri. R. Thyaga Raju for their blessings to complete this book.

The author is also thankful to all the Teaching and Non-Teaching Staff of the ECE Department for their support in completion of this book directly or indirectly.

SYLLABUS

JAWAHARLAL NEHRU TECHNOLOGICAL UNIVERSITY, HYDERABAD

II YEAR B.TECH. ECE I SEMESTER

4—0—4 EC 2125

SWITCHING THEORY AND LOGIC DESIGN

(Common for ECE, EEE, EIE, EContE, CSE, CSIT)

UNIT I : NUMBER SYSTEMS AND CODES

Review of number systems—binary arithmetic—binary weighted and non-weighted codes—error detecting and error correcting codes.

BOOLEAN ALGEBRA

Postulates and theorems; representation of switching functions—SOP and POS forms—Karnaugh map representation—minimization using K-maps.

UNIT II : DESIGN OF COMBINATIONAL CIRCUITS

Tabular minimization—design of single output and multi output functions—design using conventional AND, OR, NOT, NAND, NOR and EX–OR gates. Design using MSI and LSI devices—digital multiplexer/selector decoder, demultiplexer—design of 4 bit adder, carry look-ahead adder, magnitude comparator—BCD converter, logic implementations using ROM, PAL and PLA.

UNIT III : INTRODUCTION TO SEQUENTIAL CIRCUITS

Combinational versus sequential circuits, asynchronous versus synchronous circuits—state table and state diagram—state assignment—memory elements and their excitation functions—T flip-flop, D flip-flop, RS flip-flop, JK flip-flops and their excitation requirements—design of synchronous sequential circuits like sequential detectors and binary counters.

UNIT IV : CAPABILITIES AND MINIMIZATION OF SEQUENTIAL MACHINES

Mealy and Moore machines—capabilities and limitations of finite state machine—state equivalence and machine minimization.

UNIT V : ALGORITHMIC STATE MACHINES

ASM chart, timing considerations, control implementation, design with multiplexers and PLA control. Introduction to unate functions and threshold logic.

This book is dedicated to
my mother
Late Smt. M. Punyavathamma

1

Number Systems and Codes

The objective of this chapter is three fold. One is introduction to the different number systems and conversion from one system to another and arithmetic operations on binary numbers.

The second objective is presentation of switching functions and representation of boolean functions in two different forms.

The third objective is minimization of boolean functions using Karnaugh map and tabular method.

1.0 INTRODUCTION

Number systems are useful in digital computers and the knowledge of these systems is necessary to perform reliable, economic and easily understandable arithmetic operations. The term digital implies a system that counts descrete units. A physical system whose behaviour is described by mathematical equations is simulated in a digital computer by means of *"number systems"*. Many digital systems are required with various *'digital codes'* to handle the data which is in the form of numbers, alphabets and symbols.

1.1 REVIEW OF NUMBER SYSTEMS

A number system is nothing more than a code that uses symbols to refer to a number of items. In general, in any number system, there is an ordered set of symbols known as *'digits'*.

Basically, two types of number systems are present. They are

(1) Positional number systems

(2) Non-positional number systems.

In a *positional number system,* the position of each digit of a number indicates the significance to be attached to that digit.

In a *non-positional number system*, a digit of a number does not indicate any significance of its position and weight.

An example of a non-positional number system is *roman numerical system*. In this number system, zero is not present. So, it is difficult to use.

The most widely used number system is positional number system. In this system, the digits in a number are placed side by side and each position in the number is assigned a *weight* or *index* of importance by some predesigned rule.

A number is made up of a collection of digits and it has two parts – *integer* and *fraction,* both are separated by a radix point (.). A positive number N can be represented by

$$N = a_{n-1}\, a_{n-2} \cdots a_1 a_0 . a_{-1}\, a_{-2} \ldots a_{-m}.$$

where N = a number

n = number of digits in integer part

m = number of digits in fractional part.

All positional number systems must contain a *radix* or *base.* It is defined as the weight of a digit which depends on its relative position within the number.

Alternatively, radix of a number system is the number of symbols used in that number system. So any positive number in the positional number system with base r can be represented by

$$N_b = a_{n-1}\, b^{n-1} + a_{n-2}\, b^{n-2} + \ldots + a_1\, b^1 + a_0\, b^0 + a_{-1} b^{-1} + a_{-2}\, b^{-2} + \ldots + a_{-m}\, b^{-m}$$

$$= \sum_{i=-m}^{n-1} a_i\, b^i \text{ where } a = \text{integer in the range } 0 \le a_i \le (b-1)$$

Where b is the base or radix of the number system and its value should be an integer greater than one (1).

The sequence of numbers $a_{n-1}, a_{n-2}, \ldots, a_0$ constitutes the integer part of N, Whereas the sequence of numbers $a_{-1}, a_{-2}, \ldots, a_{-m}$ constitutes the 'fractional part of N'. Thus n and m designate the number of digits in the integer and fractional parts respectively.

The integer part and fractional part are separated by *radix point.*

The digit a_{-m} is referred to as the Least Significant Digit (LSD) and

a_{n-1} is called the Most Significant Digit (MSD).

1.2 DECIMAL NUMBER SYSTEM

The most familiar number system is the decimal system. In day to day life, we deal with decimal number system. It is a positional number system. The base or radix of this system is 10 (deci). This system contains 10 unique symbols to represent a number, known as digits. They are 0, 1, 2, 3, 4, 5, 6, 7, 8 and 9. *i.e.,* the minimum value digit is 0 (zero) and the maximum value digit is 9. All the numbers in this system are represented using these digits. The absolute value of each digit is fixed but its positional value is determined by its relative position or place in the given number. This positional value is known as *weight.* In the decimal number system, we can express any decimal number in units, tens, hundreds, thousands and so on. After reaching 9 in the decimal number system, we form combinations of decimal digits to get 10, 11, 12 etc. In other words, the next decimal number after 9 is obtained by using *the second digit followed by the first* to get 10. The decimal number after 10 is obtained by using second digit followed by the second to get 11 and so forth.

For example, When we write a decimal number say

456.7, we know that it can be represented by $(456.7)_{10}$

$$400 + 50 + 6 + 0.7 = (456.7)_{10}$$

Here 10 indicates the base of the system, default base is 10. The above decimal number can be represented in powers of 10,

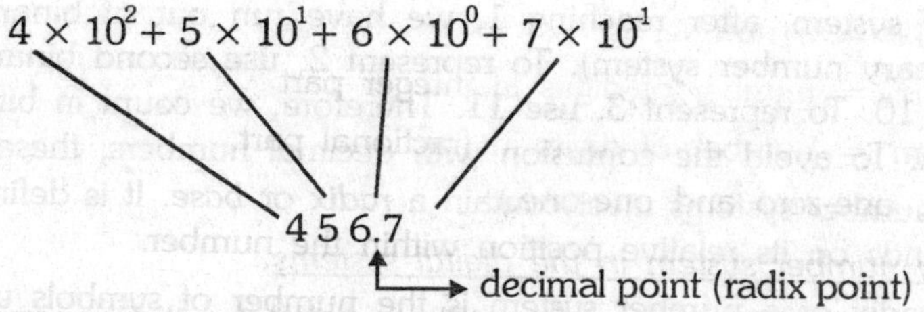

This says that the position of the digit with reference to the decimal point determines its value/weight.

The sum of all weighted digits represents the number.

The left most digit, which has the greatest weight is called the *MSD* and the right most digit, which has the least weight, is called the *LSD*. Fig. 1.1 shows the digits and their weights represented as a power of 10 or in decimal number system.

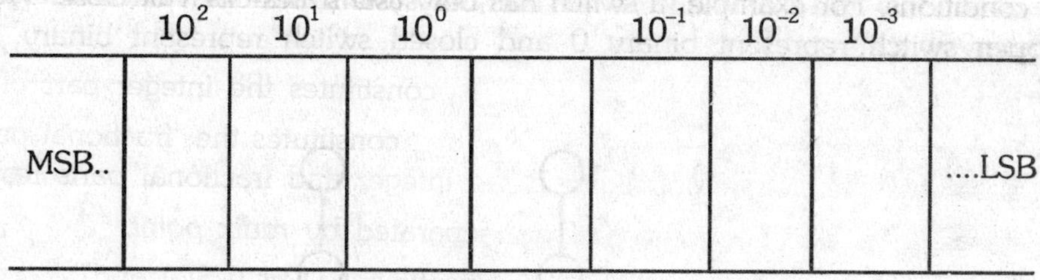

Fig. 1.1 Decimal positional values as powers of 10.

Example 1.1. *Represent a number 456.78 in powers of 10.*

Solution. $N = 4 \times 10^2 + 5 \times 10^1 + 6 \times 10^0 + 7 \times 10^{-1} + 8 \times 10^{-2}$.

The digit 4 has a weight of 100, the digit 5 has a weight of 10, the digit 6 has a weight of 1, the digit 7 has a weight of 1/10 and the digit 8 has a weight of 1/100.

1.3 BINARY NUMBER SYSTEM

The binary number system is employed in digital electronics, in digital computers and microprocessors, since it requires only two digits, 1 and 0. This allows the use of electronic devices with just two states, either HIGH or LOW (ON or OFF).

The number system with base (radix) two is known as *'binary number system'* (Here *bi* means two). This system uses only two symbols *i.e.,* 0 and 1 known as **Binary digits.** The binary numbers are strings of 0's and 1's. The position of 1 or 0 in a binary number indicates left) in a binary number is an increasing *power of two.*

The highest value bit in a binary number is known as the "Most-Significant Digit/Bit" (MSD/B), and the lowest value number is called the 'Least-Significant Digit/Bit' (LSD/B).

The integer part and fractional part of a number in the binary number system are separated by a radix (binary) point.

In the binary system, after reaching 1, we have run out of binary digits (there is no 2, 3, ... in the binary number system). To represent 2, use second binary digit (bit) followed by the first to get 10. To represent 3, use 11. Therefore, we count in binary as follows: 0, 1, 10, 11 and so on. To avoid the confusion with decimal numbers, these binary numbers are read as zero, one, one-zero and one-one.)

Use of binary number system in the digital systems

Digital systems accept binary numbers (0'*s* and 1'*s*) for their operation because

1. Active devices employed are operated as switches and have two stable states, ON and OFF. These conditions can be represented easily in binary number system with two symbols 0 and 1.

2. It is very easy to maintain noise margin during the data processing in digital systems using binary operation.

Binary quantities can be represented by any device that has only two operating states or possible conditions. For example, a switch has only two states-open or close. We can arbitrarily let an open switch represent binary 0 and closed switch represent binary 1 as shown in Fig. 1.2.

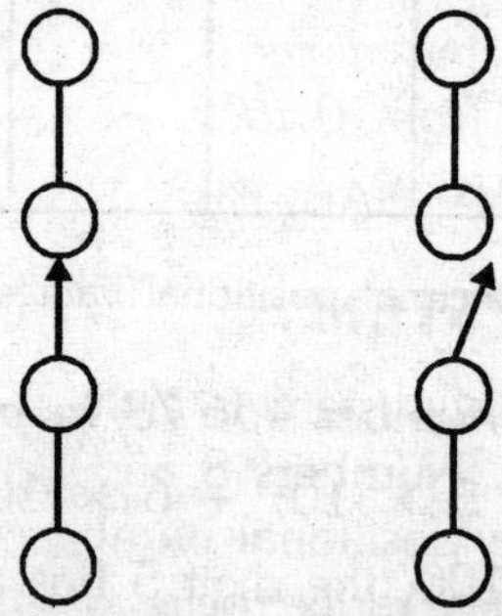

Fig. 1.2 Binary representation using switch.

The group of 4-bits is called a *'nibble'*, group of 8-bits is called a *'byte'* and a group of 16-bits is called a "word"

The weight of each bit in a binary number system is expressed as a power of 2, as shown in Fig. 1.3.

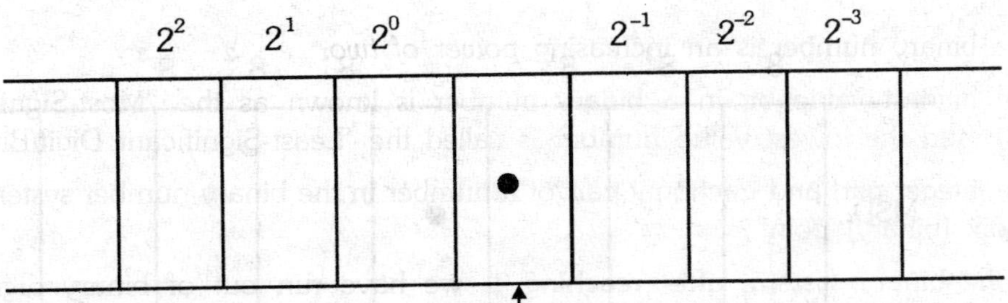

Fig. 1.3 Binary position values as a power of 2.

An example for representation of a binary number is

	MSB							LSB
Binary	.	.	.	.	.		.	.	.
Digit position	4	3	2	1	0 .	−1	−2	−3
Weight in powers of two	2^{+4}	2^{+3}	2^{+2}	2^{+1}	2^{+0} .	2^{-1}	2^{-2}	2^{-3}
Value	16	8	4	2	1 .	0.5	0.25	0.125
Example	10110.1101							

Although binary numbers are advantageous for digital circuitry and systems, they are not easy to be followed by people. For example, the number 1628_{10} is easy to read but its equivalent in binary, $(1101011100)_2$ is not. For this reason, binary numbers are usually converted into any one of other number systems, such as decimal, octal or hexadecimal equivalents before being read out of the system.

Example 1.2. *Instead of using digits 0 and 1 for the binary numbers, use A and B, respectively. Show how to count from 0 to 7.*

Solution. A, B, BA, BB, BAA, BAB, BBA, BBB.

1.4 OCTAL NUMBER SYSTEM

The base or radix of the octal system is eight, hence the digits 0 through 7 are used as symbols to represent any number. Numbers 8 and 9 never appear and can be represented by 10 and 11 and so on. It is also a positional number system and has in general two parts *i.e.*, integer and fraction, set apart by a radix (octal) point.

Advantages

1. *The octal number system* is sometimes employed as a more convenient way of grouping binary numbers to make them easier to read.

2. Electronically, it is easier and hence cheaper, to convert from binary to octal than it is to convert from binary to decimal.

In the octal number system, each number weight is expressed as a power of 8, as shown in Fig. 1.4.

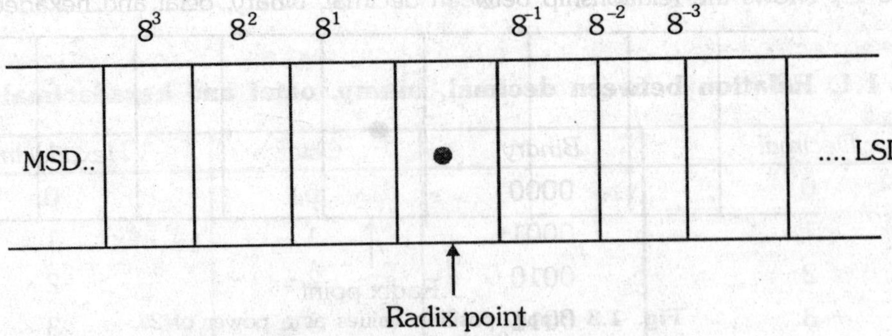

Fig. 1.4 Octal position values as a power of 8.

The minimum digit value in this system is 0 and maximum digit value is 7. So $(74.32)_8$ is a valid number in the octal number system but (78.32) is not.

1.5 HEXADECIMAL NUMBER SYSTEM

Hexadecimal number system has a base of 16 having 16 distinct symbols : {0, 1, 2, 3, 4, 5, 6, 7, 8, 9, A(10), B(11), C(12), D(13), E(14), F (15)}.

Here the minimum digit value is 0 and maximum digit value is 15 and is represented by F. The numbers 0–9 and alphabets A to F are used to represent the digits in hexadecimal number system, hence it is also known as *alpha numeric number system*. A hexadecimal number is usually indicated by H following the number. Thus 16_{10} = 10H, 17_{10} = 11H and so on. It is also a positional number system, and the value of each digit depends on its position. Every positional value is represented by the powers of 16 as shown in Fig. 1.5.

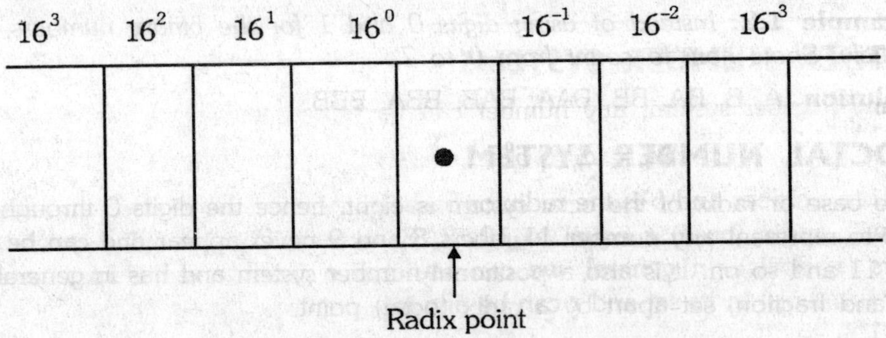

Fig. 1.5 Hexadecimal positional values as a power of 16.

The number in this system has two parts integer and fraction, set apart by a radix (hexadecimal) point.

The right most digit in this system is called LSD and the left most digit is called MSD.

Microprocessors use address and data buses that carry bits in groups of 4, or multiples of 4 such as 8, 16 and 32. Because of this it is generally more useful to employ the 4-bit *hexadecimal number system* which has a base of 16, rather than octal system.

Hexadecimal number system is more convenient to represent a number when compared to octal and binary number systems.

Table 1.1 shows the relationship between decimal, binary, octal and hexadecimal number systems.

Table 1.1. Relation between decimal, binary, octal and hexadecimal numbers

Decimal	*Binary*	*Octal*	*Hexadecimal*
0	0000	0	0
1	0001	1	1
2	0010	2	2
3	0011	3	3
4	0100	4	4
5	0101	5	5
6	0110	6	6
7	0111	7	7
8	1000	10	8
9	1001	11	9
10	1010	12	A
11	1011	13	B
12	1100	14	C
13	1101	15	D
14	1110	16	E
15	1111	17	F

1.6 OTHER NUMBER SYSTEMS

In any number system, any number can be represented as

$$N_b = a_{n-1}\, b^{n-1} + \ldots a_0\, b^0 + a_{-1}\, b^{-1} \ldots a_{-m}\, b^{-m}$$

where N is a number and b is the radix or base. In the previous section, we have seen number systems with radix (base) r equal to 10, 2, 8 and 16. Each number system has 'r' set of characters (symbols). In general, we can say that a number represented in radix r, has r character (symbols) in its set and r can be any integer value greater than one. This is illustrated in Table 1.2

Table 1.2. Radix and symbol set

Radix (base) r	*Character (symbols) in set*
2 (binary)	0, 1
3 (trinary)	0, 1, 2
4 (quinary)	0, 1, 2, 3
5	0, 1, 2, 3, 4
6	0, 1, 2, 3, 4, 5
7	0, 1, 2, 3, 4, 5, 6

8 (octal)	0, 1, 2, 3, 4, 5, 6, 7
9	
10 (decimal)	0, 1, 2, 3, 4, 5, 6, 7, 8, 9
11	0, 1, 2, 3, 4, 5, 6, 7, 8, 9, α
12	0, 1, 2, 3, 4, 5, 6, 7, 8, 9, α, β
16 (hexadecimal)	0, 1, 2, 3, 4, 5, 6, 7, 8, 9, A, B, C, D, E, F

Example 1.3. *Show how to count a number system with a base 3, use the digits 0, 1, 2.*

Solution. First 3 numbers are 0, 1, 2.

2-digit combinations are 10, 11, 12, 20, 21, 22

3-digit combinations are 100, 101, 102, 110, 111, 112, 120, 121, 122, 200, 201, 202, 210, 211, 212, 220, 221, 222

4-digit combinations are 1000, 1001 and so on.

1.7 NUMBER SYSTEM CONVERSION

The binary number system is the most important one in digital systems, but several others are also important. The decimal system is important because it is universally used to represent quantities outside a digital system. Therefore, it is necessary to convert decimal number into its equivalent binary while feeding number into the digital system (computer) and to convert the binary number into its decimal equivalent while displaying the result of operation on the output devices. While dealing with the large quantity of binary numbers of many bits, it is inconvenient to read or write. For this reason, octal and hexadecimal numbers are used as a shorthand means of expressing large binary numbers. So, it is necessary to have the conversion between binary to octal, binary to hexadecimal and vice-versa.

1.7.1 Binary-to-Decimal Conversion

Any given binary number can be converted into its decimal equivalent by adding the products of each bit and its weight.

The general representation of a binary number is

$$N_{10} = A_{n-1}\, 2^{n-1} + A_{n-2}\, 2^{n-2} + \ldots A_1.2^1 + A_0 2^0$$
$$+ A_{-1}\, 2^{-1} + A_{-2}\, 2^{-2} + \ldots A_{-m}\, 2^{-m}$$

where N_{10} = number in decimal

A = digit (bit) value (either 0 or 1)

2 is the base, n = number of digits in the integer portion of a number

m = number of digits in the fractional portion of a number.

Procedure

(1) Write the binary number.

(2) (*a*) Directly under the binary number write 1 (2^0), 2 (2^1), 4(2^2), 8 (2^3), 16 (2^4), ... working from right to left.

(*b*) In case of fractions, the weights of digit positions to the right of binary point are given by 0.5 (2^{-1}), 0.25 (2^{-2}), 0.125 (2^{-3}) ... and so on.

(3) If a zero appears in a digit position, cross out the decimal weight for that position.

(4) Add the remaining weights to get the decimal equivalent. The following example will give a clear idea of the conversion.

Example 1.4. *Convert the binary number 10110 into decimal.*

Solution. Step 1. Given binary number

$= 10\ 110$

Step 2. Corresponding weights

$= 2^4\ 2^3\ 2^2\ 2^1\ 2^0$

Step 3. $\Rightarrow\ = 2^4 \times 1 + 2^3 \times 0 + 2^2 \times 1 + 2^1 \times 1 + 2^0 \times 1$

$= 16 + \not{0} + 4 + 2 + \not{0}$

Step 4. $(22)_{10}$

Example 1.5. *Convert the binary fraction 0.1101 into decimal.*

Solution. $1 \times 2^{-1} + 1 \times 2^{-2} + 0 \times 2^{-3} + 1 \times 2^{-4}$

$\Rightarrow 0.5 + 0.25 + 0 + 0.0625$

$\Rightarrow (0.8125)_{10}$

Example 1.6. *Convert the binary number 1101.011 into decimal equivalent.*

Solution.

Step 1. 1101.011

Step 2. $2^3\ 2^2\ 2^1\ 2^0\ .\ 2^{-1}\ 2^{-2}\ 2^{-3}$

Step 3. $2^3 \times 1 + 2^2 \times 1 + 2^1 \times 0 + 2^0 \times 1 + 0 \times 2^{-1} + 1 \times 2^{-2} + 1 \times 2^{-3}$

$= 8 + 4 + 0 + 1 + 0 + 0.25 + 0.125$

Step 4. $(13.375)_{10}$

1.7.2 Decimal to Binary Conversion

(The Double-dabble method) or (Dibble-dabble method)

There are many ways to convert a decimal number to a binary number, but most popular is the double-dabble method. In this method, successive division or multiplication by decimal number 2 is done for integer part or fractional part respectively.

1.7.2.1 Successive division for integer part conversion

In this method, we repeatedly divide the integer part of the decimal number by 2 until the quotient is zero. The remainder of each division becomes the number in the binary system. The remainders are taken in the reverse order to form a number in base 2. That is, the first remainder is the LSD and the last remainder is the MSD in the binary number system.

The Fig. 1.6 shows the flowchart of conversion.

The procedure is illustrated in the following examples.

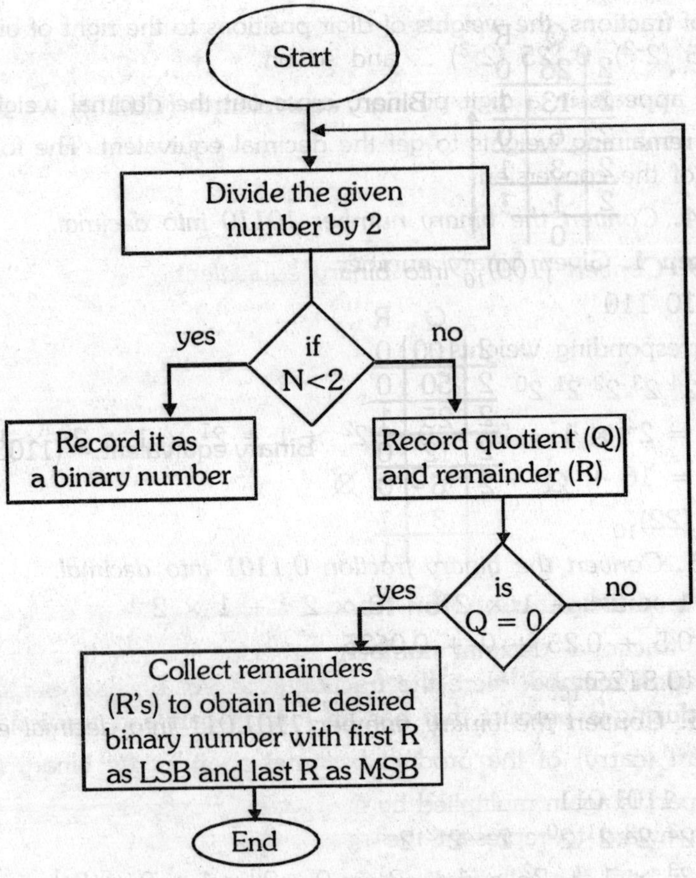

Fig. 1.6 Flow chart for integer part conversion.

Example 1.7. *Convert decimal number $(26)_{10}$ to its binary equivalent.*

Solution.

		Quotient	*Remainder*	
divide 26 by 2	$\frac{26}{2}$	13	0	LSB
divide 13 by 2	$\frac{13}{2}$	6	1	
divide 6 by 2	$\frac{6}{2}$	3	0	
divide 3 by 2	$\frac{3}{2}$	1	1	
divide 1 by 2	$\frac{1}{2}$	0	1	MSB.

(OR)

	Q	R
2	26	0
2	13	1
2	6	0
2	3	1
2	1	1
	0	

Binary equivalent = $(11010)_2$

Example 1.8. *Convert* $(100)_{10}$ *into binary equivalent.*

Solution.

	Q	R
2	100	0
2	50	0
2	25	1
2	12	0
2	6	0
	3	1
	1	1

Binary equivalent = $(1100100)_2$

1.7.2.2 Successive multiplication for fractional part conversion

Conversion of fractional decimal numbers to binary numbers is accomplished using a successive multiplication method. Here the fractional part of decimal number is multiplied by the decimal 2, producing a product that has an integer part and fractional part.

The integer part (carry) of the product becomes a bit in the binary number.

The fractional part is again multiplied by the decimal 2 and the integer part (carry) of each product is read downward to represent the new binary number.

This is illustrated with the following examples.

Example 1.9. *Convert* $(0.65625)_{10}$ *into base – 2.*

Solution.

0.65625×2	0.3125×2	0.625×2	0.250×2	0.5×2
1.3125	0.625	1.250	0.5	1.0
↓ 1	↓ 0	↓ 1	↓ 0	↓ 1

Binary equivalent = $(0.10101)_2$

Example 1.10. *Convert the decimal number* $(0.85)_{10}$ *into binary form.*

Solution.

Fraction base result				*Recorded copy*
0.85×2	= 1.7	→	0.7 with a carry of	1
0.7×2	= 1.4	→	0.4 with a carry of	1
0.4×2	= 0.8	→	0.8 with a carry of	0
0.8×2	= 1.6	→	0.6 with a carry of	1 read
0.6×2	= 1.2	→	0.2 with a carry of	1 down
0.2×2	= 0.4	→	0.4 with a carry of	0 ↓
0.4×2	= 0.8	→	0.8 with a carry of	0

Binary equivalent = $(0.1101100)_2$.

While converting a decimal number containing integer and fractional part, mixed methods have to be used.

Example 1.11. *Convert decimal number $(35.45)_{10}$ to base-2 number.*

Solution. *For integer part*

	Q	R
2	35	1
2	17	1
2	8	0
2	4	0
2	2	0
2	1	1
	0	

For fractional part

$0.45 \times 2 \rightarrow 0.9 \rightarrow 0.9$ with a carry of 0

$0.9 \times 2 \rightarrow 1.8 \rightarrow 0.8$ with a carry of 1

$0.8 \times 2 \rightarrow 1.6 \rightarrow 0.6$ with a carry of 1

$0.6 \times 2 \rightarrow 1.2 \rightarrow 0.2$ with a carry of 1

$0.2 \times 2 \rightarrow 0.4 \rightarrow 0.4$ with a carry of 0

Integer part binary equivalent = 100011

Fraction part binary equivalent = 01110

Therefore binary equivalent for decimal number $(35.45)_{10}$ is $(100011.01110)_2$.

1.7.3 Binary-to-Octal Conversion

The base of the octal numbers is 8 and the base for binary numbers is 2. Thus the base for octal number is the third power of the base for binary number.

For integer part: While converting, divide the binary number into group of three starting from the LSB *i.e.,* binary point and moving towards MSB. Then replace each group of three bits by its octal equivalent.

For fractional part: The grouping of three bits is made starting from the binary point and moving towards right.

Example 1.12. *Convert $(110011)_2$ to base-8 equivalent.*

Solution. To convert binary to octal, each 3-bit binary group is replaced by the equivalent octal symbol.

$$\underbrace{110}_{6}\ \underbrace{011}_{3}$$

Octal number = 63.

Example 1.13. *Convert $(101.10100110)_2$ to octal number.*

Solution. To convert binary to octal, each 3-bit binary group is replaced by the equivalent octal symbol.

$$\underbrace{101}.\underbrace{101}\ \underbrace{001}\ \underbrace{10}$$

Octal number = $(5.514)_8$

1.7.4 Octal-to-Binary Conversion

Conversion from octal to binary is a reversal of the process explained in the previous section. Octal numbers are converted into equivalent binary numbers by replacing each octal digit by its 3-bit equivalent binary.

Example 1.14. *Change $(65)_8$ into binary.*

Solution. To convert the octal to binary, each octal symbol is replaced by a 3-bit binary.

$$(65)_8 = (110\ 101)_2 = (110101)_2$$

Example 1.15. *Convert $(72.43)_8$ into binary equivalent.*

Solution. To convert the octal to binary, each octal symbol is replaced by a 3-bit binary.

```
      7 2. 4 3
    ↙  ↓   ↓  ↘
111 010 100 011
```

Hence binary equivalent $= (111010.100011)_2$

Advantages of converting octal-to-binary and vice-versa

(1) Getting information into and out of digital system requires less circuitry, because it is easier to read, record and print.

(2) • Large decimal numbers are more easily converted to binary, if we first convert to octal and then to binary. The reason is, direct decimal-to-binary conversion requires more divisions than a decimal-to-octal-to-binary conversion.

1.7.5 Binary-to-Hexadecimal Conversion

We know that the base for hexadecimal numbers is 16 and the base for binary numbers is 2. The base for hexadecimal numbers is the fourth power of the base for binary numbers. So, the given binary number can be converted into its equivalent hexadecimal value by grouping four binary bits together starting from LSB and moving towards MSB for integer part. (All the binary groups are converted into their equivalent hexadecimal values).

For the fractional part, grouping must start from MSB of the fractional part *i.e.*, the bit just after the binary point and moving towards right.

Example 1.16. *Convert $(1101.10110011)_2$ to hexadecimal equivalent.*

Solution.

$$\underbrace{1101}_{D}\ \underbrace{1011}_{B}\ \underbrace{0011}_{3}$$

∴ Hexadecimal number $= (DB3)_{16}$ = DB3H.

Example 1.17. *Convert 10100110100101. 10001100 into base-16 number.*

Solution.

```
10 1001 1010 0101 . 1000 1100
 2    9    A    5 .    8    C
```

∴ Hexadecimal number $= (29A5.8C)_{16}$ = 29A5.8CH.

1.7.6 Hexadecimal-to-Binary Conversion

Conversion from hexadecimal to binary is a reversal of the process explained in the previous section. Here every hexadecimal number is converted into equivalent binary number by replacing each 'hex' digit by its equivalent 4-bit binary number. All the binary values grouped together will give the equivalent value.

Example 1.18. *Convert $(2\ B\ D)_{16}$ to binary.*

Solution.

```
2       B       D
0010    1011    1101
```

Binary number $= (10\ 1011\ 1101)_2$

Example 1.19. *Convert $(5A3.CE2)_{16}$ hexadecimal number into binary form.*

Solution. 5 A 3 C E 2

0101 1010 0011 1100 1110 0010

Binary equivalent = $(10110100011\ .\ 11001110001)_2$

In order to perform the conversion between hex and binary, it is necessary to know the 4-bit binary numbers (0000 through 1111) and their equivalent hexdigits. Generally hexadecimal conversion is useful in representing the large binary numbers with less digital circuit complexity.

1.7.7 Decimal-to-Octal Conversion

'Octal dabble' (similar to double dabble) is the method used with octal numbers, while converting the decimal numbers into octal numbers.

Here the integer part of the decimal number can be converted into its equivalent octal number by successively dividing the given number by eight till the final remainder becomes less than eight or equal to zero. The octal equivalent can be obtained by counting all the remainders of each division from bottom to top.

For fractional part of the decimal number, the conversion is affected by continuous multiplication by eight and keeping the track of the integers carry generated.

The above procedure is illustrated with the following examples.

Example 1.20. *Convert the decimal $(160)_{10}$ into octal.*

Solution.

8	160	0 LSD
8	20	4 ↑
	2 →	2 MSD

$\therefore$ $(160)_{10}$ equivalent octal number = $(240)_8$

Example 1.21. *Convert decimal $(266)_{10}$ to octal number.*

Solution.

	Q	R	
8	266	2	LSD
8	33	1	↑
8	4	4	MSD
	0		

$(266)_{10} = (412)_8$

Note Q : Quotient, R : Remainder.

Example 1.22. *Convert decimal 0.52 to octal.*

Solution.

$0.52 \times 8 = 4.16$ → 0.16 with a carry of 4

$0.16 \times 8 = 1.28$ → 0.28 with a carry of 1

$0.28 \times 8 = 2.24$ → 0.24 with a carry of 2

$\therefore$ $(0.52)_{10} = (0.412)_8$

Example 1.23. *Convert $(10.20)_{10}$ to octal.*

Solution.

Integer part

	Q	R
8	10	2
8	1	1
	0	

$(12)_8$

Fractional part

$$\frac{0.20 \times 8}{1.60} \rightarrow \frac{0.60 \times 8}{4.80} \rightarrow \frac{0.80 \times 8}{6.40} \rightarrow \frac{0.80 \times 1}{3.20}$$

$$\downarrow 1 \quad \downarrow 4 \quad \downarrow 6 \quad \downarrow 3$$

$$= (0.1463)_8$$

$$\therefore \quad (10.20)_{10} = (12.1463)_8$$

1.7.8 Octal-to-Decimal Conversion

In the octal number system, each digit corresponds to a power of 8. The weights of the digit positions in an octal number are as follows

etc...	8^3	8^2	8^1	8^0	.	8^{-1}	8^{-2}	8^{-3}	... etc

Therefore, to convert from octal to decimal, multiply each octal digit by its positional weight and add the resulting products.

Example 1.24. *Convert $(475.25)_8$ to its decimal equivalent.*

Solution.
$$(N)_{10} = 4 \times 8^2 + 7 \times 8^1 + 5 \times 8^0 + 2 \times 8^{-1} + 5 \times 8^{-2}$$
$$= 256 + 56 + 5 + 0.25 + 0.078125$$
$$= (317.32825)_{10}$$

Example 1.25. *Convert the given octal number $(362)_8$ into its equivalent decimal value.*

Solution. 3 6 2

Step 1. 8^2 8^1 8^0

Step 2. 64 8 1

Step 3. $3 \times 64 + 6 \times 8 + 2 \times 1$

Step 4. $192 + 48 + 2$

$\Rightarrow (242)_{10}$

$\therefore \quad (362)_8 = (242)_{10}$

Example 1.26. *Convert the octal number $(432.763)_8$ into the decimal form.*

Solution. 4 3 2 . 7 6 3

Step 1. 8^2 8^1 8^0 . 8^{-1} 8^{-2} 8^{-3}

Step 2. 64 8 1 . $\frac{1}{8}$ $\frac{1}{64}$ $\frac{1}{512}$

Step 3. $64 \times 4 + 8 \times 3 + 1 \times 2$ $\quad \frac{1}{8} \times 7 + \frac{1}{64} \times 6 + \frac{1}{512} \times 3$

Step 4. $256 + 24 + 2$ $\quad 0.875 + 0.093 + 0.0058$

$\Rightarrow (282.9738)_{10}$

$\therefore \quad (432.763)_8 = (282.9738)_{10}$

1.7.9 Decimal-to-Hexadecimal Conversion

The method used to convert the decimal to hexadecimal is called "hex dabble".

Here the integer part of the decimal number can be converted into its equivalent hexadecimal number by successively dividing the given decimal number by sixteen (16) till the final remainder becomes less than 16 or equal to zero.

For integer part : The hexadecimal equivalent can be obtained by writing all the remainders of each division from bottom remainder (MSD) to top and first remainder as (LSD). When the remainder is greater than 9, it is represented by alphabet.

For fraction part of the decimal number : The conversion is done by successively multiplying by 16 and keeping the track of the integers carry generated. When the carry is greater than 9, replace it by an appropriate hexadecimal alphabet.

The above procedure is illustrate in the following examples.

Example 1.27. *Convert the decimal number $(3509)_{10}$ to its equivalent hexadecimal.*

Solution.

	Q	R	
16	3509	5	LSD
16	219	–11 – B	
16	13	–13 – D	MSD
	0		

$\therefore \quad (3509)_{10} = (DB5)_{16} = DB5H$

Example 1.28. *Convert the $(0.15)_{10}$ decimal number to its hex equivalent.*

Solution.

$$\frac{0.15 \times 16}{2.50} \rightarrow 2 \qquad \frac{0.50 \times 16}{8.0} \rightarrow 8$$

$\therefore \quad (0.15)_{10} = (0.28)_{16} = 0.28H$

Example 1.29. *Convert $(95.5)_{10} = (?)_{16}$.*

Solution.

	Q	R	
16	95	15	→ F
16	5	5	
	0		= 5FH

$$\frac{0.5 \times 16}{8.0} \rightarrow 8$$

$\therefore \quad (95.5)_{10} = (5F.8)H$

Alternate method for decimal-to-hex conversion is first convert the given decimal number into binary and then group the 4-bits of binary, and assign equivalent hex value.

1.7.10 Hexadecimal-to-Decimal Conversion

In hexadecimal number system, each digit position corresponds to a power of 16. To convert the hexadecimal number to its equivalent decimal, multiply each hexadecimal digit by its positional weight and add the resulting products.

The weights of the digit positions in hexadecimal number are as follows.

etc...	16^3	16^2	16^1	16^0	.	16^{-1}	16^{-2}	16^{-3}	... etc

Example 1.30. *Convert hexadecimal number* $(3A.2F)_{16}$ *to its decimal equivalent.*

Solution. 3 A . 2 F

Step 1. 16^1 16^0 16^{-1} 16^{-2}

Step 2. 16 1 $\frac{1}{16}$ $\frac{1}{256}$

Step 3. $16 \times 3 + 1 \times A . \frac{1}{16} \times 2 + \frac{1}{256} \times F$

$$\Rightarrow 16 \times 3 + 1 \times 10 . \frac{1}{16} \times 2 + \frac{15}{256}$$

$$(3A.2F)_{16} = (58.1836)_{10}$$

Example 1.31. *Convert (9B2.1A) H to its decimal equivalent.*

Solution. $(N)_{10} = 9 \times 16^2 + B \times 16^1 + 2 \times 16^0 + 1 \times 16^{-1} + A \times 16^{-2}$

$$= 9 \times 256 + 11 \times 16 + 2 \times 1 + \frac{1}{16} + \frac{10}{256}$$

$$= 2304 + 176 + 2 + 0.0625 + 0.039$$

$$\therefore (9B2.1A)_{16} = (2482.101)_{10}$$

1.7.11 Octal-to-Hexadecimal Conversion

The easiest way to convert octal number to hexadecimal number which is done in two phases, is given below.

1. Convert octal number to its binary equivalent (octal-to-binary conversion).
2. Convert the binary number to its hexadecimal equivalent (binary-to-hexadecimal conversion).

Other method is to first convert the octal number to decimal and then to hexadecimal.

Example 1.32. *Convert* $(1234)_8$ *into hexadecimal.*

Solution.

Step 1. 1 2 3 4

001 010 011 100

Binary equivalent = (001010011100)

Step 2. 0010 1001 1100

2 9 C

$$= (1234)_8 = (29C)_{16}$$

Example 1.33. *Convert* $(675.42)_8$ *into base-16 number.*

Solution.

Step 1. 6 7 5 . 4 2

110 111 101 . 100 010

Binary equivalent = $(110111101.100010)_2$

Step 2. 0001 1011 1101 . 1000 1000

1 B D . 8 8

= $(1BD.88)_{16}$ = 1BD.88H

1.7.12 Hexadecimal-to-Octal Conversion

The simplest way to convert hexadecimal number to octal number contains two phases.

1. Convert the hexadecimal number to its binary equivalent.
2. Convert binary number to its octal equivalent.

Another method is to convert the given hexadecimal number into decimal and then decimal to the octal number.

The simplest procedure for conversion is illustrated in the following examples.

Example 1.34. *Convert* $(AB6.13)_{16}$ *into octal equivalent.*

Solution.

Step 1. AB6.13

1010 1011 0110 . 0001 0011

Binary equivalent = $(101010110110\ .\ 00010011)_2$

Step 2. 101 010 110 110 . 000 100 110

5 2 6 6 . 0 4 6

⇒ $(5266.046)_8$

Example 1.35. *Convert* $(457D.F8)_{16}$ *into base-8 number.*

Solution. 4 5 7 D . F 8

Step 1. 0100 0101 0111 1101 . 1111 1000

Binary equivalent = $(1000101\ 01111101.11111)_2$

Step 2. 100 010 101 111 101 . 111 110

4 2 5 7 5 . 7 6

Octal equivalent = $(42575.76)_8$

1.7.13 Other Radix (*r*) System-to-Decimal Conversion

Any radix system number can be converted into decimal system, by multiplying the each radix digit by its positional weight and adding the resulting products.

$$N_b = a_{n-1}\, b^{n-1} + a_{n-2}\, b^{n-2} + \ldots + a_0\, b^0 + a_{-1}\, b^{-1} + a_{-2}\, b^{-2} + \ldots a_{-m}\, b^{-m}$$

where b = base of the system.

Let us examine some other number systems with bases 4, 6 and 12.

Number system with base (*b*) = 4

The number in this system is represented by

$$(N)_{10} = a_{n-1}\, 4^{n-1} + a_{n-2}\, 4^{n-2} + \ldots + a_0\, 4^0 + a_{-1}\, 4^{-1} + a_{-2}\, 4^{-2} + \ldots a_{-m}\, 4^{-m}$$

It contains 4 symbols, 0, 1, 2 and 3.

Example 1.36. *Convert $(234)_4$ into decimal system.*

Solution.

	2	3	4
Step 1.	4^2	4^1	4^0
Step 2.	16	4	1

Step 3. $= 16 \times 2 + 4 \times 3 + 1 \times 4$

$= 32 + 12 + 4$

$\Rightarrow (48)_{10}$

Example 1.37. *Convert $(12.23)_4$ into decimal system.*

Solution.

	1	2	.	2	3
Step 1.	4^1	4^0	.	4^{-1}	4^{-2}
Step 2.	4	1	.	$\frac{1}{4}$	$\frac{1}{16}$

Step 3. $4 \times 1 + 2 \times 1 + \dfrac{2 \times 1}{4} + \dfrac{3 \times 1}{16}$

$= 4 + 2 + 2 \times \dfrac{1}{4} + \dfrac{3}{16}$

$\Rightarrow (6.68)_{10}$

Number system with base (*b*) = 6

This system contains 6 symbols, *i.e.*, 0, 1, 2, 3, 4, and 5. In this system, any number can be represented by $a_{n-1}\, 6^{n-1} + a_{n-2}\, 6^{n-2} + \ldots a_0\, 6^0 + a_{-1} 6^{-1} + a_{-2} 6^{-2} + \ldots + a_{-m}\, 6^{-m}$.

Example 1.38. *Convert $(345.12)_6$ into decimal.*

Solution.

3	4	5	.	1	2
6^{+2}	6^{+1}	6^0	.	6^{-1}	6^{-2}

$= 6^2 \times 3 + 6 \times 4 + 1 \times 5 + 1 \times \dfrac{1}{6} + 2 \times \dfrac{1}{36}$

$= 36 \times 3 + 6 \times 4 + 1 \times 5 + 0.16 + 0.06$

$= 108 + 24 + 5 + 0.16 + 0.06$

$\Rightarrow (137.22)_{10}$

Example 1.39. *Convert $(54.23)_6$ into base-10 system.*

Solution. 5 4 2 3

Step 1. 6^1 6^0 6^{-1} 6^{-2}

Step 2. $6 \times 5 + 4 \times 1 + \frac{2}{6} + \frac{3}{36}$

Step 3. $30 + 4 + \frac{2}{6} + \frac{3}{36}$

Step 4. 34.416

$\therefore$ $(54.23)_6 = (34.416)_{10}$

Number system with base (b) = 12

This number system contains 12 symbols, *i.e.*, 0, 1, 2 ... 9, α and β. In this system, any number can be represented as

$$a_{n-1}\ 12^{n-1} + a_{n-2}\ 12^{n-2} + \ldots\ a_0\ 12^0 + a_{-1}\ 12^{-1} + a_{-2}\ 12^{-2} + \ldots\ a_{-m}\ 12^{-m}$$

This number system is called as alpha-numeric number system.

Example 1.40. *Convert $(\alpha\ 3.6)_{12}$ into decimal.*

Solution. α 3 6

Step 1. 12^1 12^0 12^{-1}

Step 2. 12 1 $\frac{1}{12}$

Step 3. $10 \times 12 + 3 \times 1 + \frac{6}{12}$

Step 4. 120 + 3 + 0.5

$\Rightarrow$ 123.5

$\therefore$ $(\alpha 3.6)_{12} = (123.5)_{10}$

Example 1.41. *Convert $(72.3)_{12}$ into base-10 number.*

Solution. 7 2 3

Step 1. 12^1 12^0 12^{-1}

Step 2. 12 1 $\frac{1}{12}$

Step 3. $7 \times 12 + 2 \times 1 + \frac{3 \times 1}{12}$

Step 4. 84 + 2 + 0.25

$\Rightarrow$ $(86.25)_{10}$

1.7.14 Decimal System-to-Other Radix (*r*) System

The conversion from decimal to any other radix (*r*) system is done differently for integer part and fractional part.

The integer part of a decimal number can be converted into its equivalent radix (*r*) number system by successively dividing by *r*, till the final remainder becomes less than *r* or equal to zero.

The number in *r*-number system can be obtained by writing all the remainders of each division from bottom to top.

For fractional part of the decimal number, the conversion is done successively multiplying by r and keeping the track of the integers (carry) generated.

Example 1.42. *Convert the* $(724.63)_{10}$ *into a number system whose base b is*

(a) 4 *(b) 6* *(c) 12*

Solution.

(a) $(724.63)_{10} = (?)_4$

	Q	R	
4	724	0	LSD
4	181	1	
4	45	1	
4	11	3	↑
4	2	2	MSD
	0		

$= (23110)_4$

$0.63 \times 4 = 2.52 \rightarrow 2$; $0.52 \times 4 = 2.08 \rightarrow 2$; $0.8 \times 4 = 0.32 \rightarrow 0$

$(220)_4$

$\therefore \quad (724.63)_{10} = (23110.220)_4$

(b) $(724.63)_{10} = (?)_6$

	Q	R
6	724	4
6	120	0
6	20	2
6	3	3
	0	

$(3204)_6$

$0.63 \times 6 = 3.78 \rightarrow 3$; $0.78 \times 6 = 4.68 \rightarrow 4$; $0.68 \times 6 = 4.08 \rightarrow 4$

$(344)_6$

$\therefore \quad (724.63)_{10} = (3204.344)_6$

(c) $(724.63)_{10} = (?)_{12}$

	Q	R
12	724	4
12	60	0
12	5	5
	0	

$(504)_{12}$

$0.63 \times 12 = 7.56 \rightarrow 7$; $0.56 \times 12 = 6.72 \rightarrow 6$; $0.72 \times 12 = 8.64 \rightarrow 8$

$= (0.768)_{12}$

$\therefore \quad (724.63)_{10} = (504.768)_{12}$

The relation between base-4, base-6, base-12 and decimal system is shown in Table 1.3.

Table 1.3. Representation of integers

Decimal number	*Base-4 system*	*Base-6 system*	*Base-12 system*
0	0	0	0
1	1	1	1
2	2	2	2
3	3	3	3
4	10	4	4

5	11	5	5
6	12	10	6
7	13	11	7
8	20	12	8
9	21	13	9
10	22	14	α
11	23	15	β
12	30	20	10

1.7.15 Summary of Conversions

The following summary will help in doing the different conversions :

1. When converting from any radix (*r*) to decimal, use the method of taking the weighted sum of each digit position.

2. When converting from decimal to any radix (*r*), use the method of repeatedly dividing by *r* and collecting remainders.

3. When converting from binary to other radix (*r*) system, group the number of bits according to the power (power of 2), and convert each group into the correct digit.

4. When converting from radix *r* to binary, convert each digit into its 3-bit, 4-bit etc. according to the power value of 2.

5. When converting from radix *r* (4, 6, 8, 12, 16 etc.) to another radix *r*, first convert to binary and convert the binary into the desired number system.

1.8 BINARY ARITHMETIC

Digital systems do not process the decimal numbers. They process only binary numbers. Therefore, it is necessary to learn the *binary arithmetic i.e.*, arithmetic operations such as addition, subtraction, multiplication and division using binary numbers only (0 and 1).

1.8.1 Binary Addition

The rules for addition of binary numbers are given in Table 1.4.

Table 1.4. Binary addition rules

A (*Augend*)	*B* (*Addend*)	*SUM* *S*	*CARRY* *C*
0	0	0	0
0	1	1	0
1	0	1	0
1	1	0	1

Here, A and B are the two one bit binary numbers and SUM and CARRY are the results after addition.

If A and B are more than one bit numbers, for LSD there are only two bits A and B to consider, the second significant bit, and the following bits, may have a third bit to consider, which is the carry bit, which may come from the previous column. If carry is produced, it must be added to the next more significant column as a carry-in. This gives eight possible combinations in binary addition as shown is Table 1.5.

Table 1.5. 3-bit binary addition rules

Carry-in	*A*	*B*	*Sum*	*Carry-out*
0	0	0	0	0
0	0	1	1	0
0	1	0	1	0
0	1	1	0	1
1	0	0	1	0
1	0	1	0	1
1	1	0	0	1
1	1	1	1	1

Note: (1) If the number of 1's to be added in a column is even, then the sum bit = 0 and if the number of 1's to be added is odd, then the sum bit = 1.

(2) Carry = 1 if number of 1's more than or equal to 2.

Example 1.43. *Add $(1010)_2$ and $(0101)_2$.*

Solution.

$$\begin{array}{r} 1010 \\ 0101 \\ \hline 1111 \end{array} = \begin{array}{l} \text{decimal } 10 \\ \text{decimal } 5 \\ \hline \text{decimal } 15 \end{array}$$

$\Rightarrow$ $(1111)_2$

Example 1.44. *Add $(10)_{10}$ and $(11)_{10}$ in binary.*

Solution.

$$\begin{array}{rr} 10 = & 1010 \\ 11 = & 1011 \\ \hline & 10101 \end{array}$$

$(10101)_2 = (21)_{10}$

Example 1.45. *Add 26 and 13 in binary.*

Solution. decimal 26 = $(11010)_2$

decimal 13 = $(01101)_2$

$$= \begin{array}{r} 11010 \\ 01101 \\ \hline 100111 \end{array} \begin{array}{l} -26 \\ -13 \\ = (39)_{10} \end{array}$$

$\Rightarrow$ $(100111)_2$

Example 1.46. *Add 1011.011 and 110.1.*

Solution.

$$\begin{array}{r} 1011.011 \\ 110.1 \\ \hline 10001.111 \end{array} = \begin{array}{r} 11.375 \\ 6.5 \\ \hline 17.875 \end{array}$$

$\Rightarrow$ $(10001.111)_2$

Example 1.47. *Add 33 3/8 and 22 5/8 in binary.*

Solution.

$$33\ 3/8 = 10001.011$$
$$22\ 5/8 = 10110.101$$
$$111000.000 = (56)_{10}$$

$\Rightarrow$ $(111000)_2$

Example 1.48. *Add 101.11, 1101.01 and 10000.001.*

Solution.

$$101.11 \text{ — } (5.75)_{10}$$
$$1101.01 \text{ — } (13.25)_{10}$$
$$10000.001 \text{ — } (16.125)_{10}$$
$$100011.001 = (35.125)_{10}$$

$\Rightarrow$ $(100011.001)_2$

Example 1.49. *Add 4.25, 7.75 and 8.0 in binary.*

Solution.

$$4.25 \text{ — } 100.01$$
$$7.75 \text{ — } 111.11$$
$$8.0 \text{ — } 1000.00$$
$$(20.00)_{10} = 10100.00$$

$\Rightarrow$ $(10100)_2$

1.8.2 Binary Subtraction

Binary subtraction can be carried out in either of the two different ways.

(1) Direct subtraction.

(2) Complement subtraction.

1.8.2.1 Direct subtraction

The rules for performing the binary subtraction are given in the Table 1.6.

Table 1.6. Binary subtraction

A *(Minuend)*	*B* *(Subtrahend)*	*Difference* *D*	*Borrow* *B*
0	0	0	0
0	1	1	1
1	0	1	0
1	1	0	0

Example 1.50. *Subtract 10101 from 11011.*

Solution.

$$11011 = 27$$
$$-\ 10101 = 21$$
$$00110 = 6$$

$\Rightarrow$ $(110)_2$

Example 1.51. *Use binary arithmetic to subtract 10 from (a) 16 (b) 22.*

Solution.

(*a*)

10000	—	16
1010	—	10
00110	=	6

⇒ $(110)_2$

(*b*)

10110	—	22
1010	—	10
01100	—	12

⇒ $(1100)_2$

Example 1.52. *Subtract the following by using binary arithmetic.*

(*a*) *7.25 from 9.50* (*b*) *6.25 from 10.75*

Solution. (*a*)

1001.10	—	9.50
–0111.01	—	7.25
0010.01		2.25

⇒ $(10.01)_2$

(*b*)

1010.11	—	10.75
0110.01	—	6.25
0100.10		4.50

⇒ $(100.10)_2$

Signed–magnitude representation or signed binary numbers

Until now, we have considered only unsigned numbers, numbers without any positive or negative sign. These unsigned numbers represent only magnitude. To perform the subtraction, representation of negative number is needed. In the decimal system, a + sign is used to denote positive numbers, and – sign is used to denote negative numbers. This type of representation of numbers is known as 'signed numbers'. To indicate this sign we must use a symbol, and in a digital circuit either 0 or 1, generally. 0 is used as Most Significant Bit (MSB) to represent a positive (+) number and 1 is used as Most Significant Bit (MSB) to represent a negative (–) number.

These numbers are represented by the signed magnitude format. Fig. 1.7 shows the signed magnitude format for 8-bit signed number.

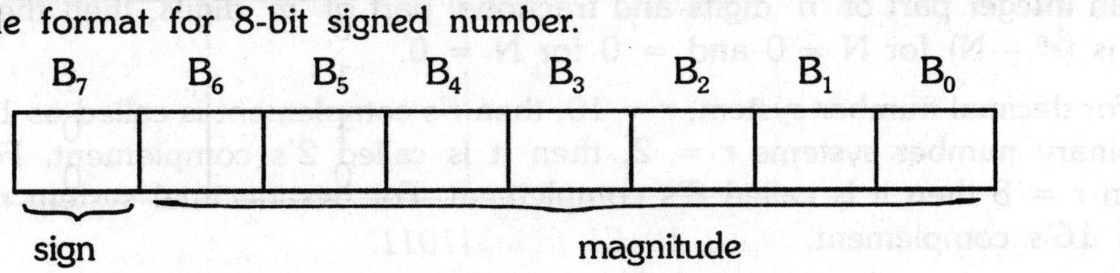

Fig. 1.7 Signed magnitude representation.

Here, the most significant bit (MSB) represents sign of the number. If MSB is 1, number is negative and if MSB is 0, number is positive. The remaining (seven) bits represent magnitude of the number.

Some examples for signed-magnitude numbers.

Example. (1) 01000100 is a $(+68)_{10}$

(2) 11000100 is a $(-68)_{10}$

(3) 10001110 is a $(-14)_{10}$

(4) 00011001 is a $(+25)_{10}$

In case of unsigned 8-bit binary numbers, the decimal range is 0 to 255. For signed magnitude 8-bit binary numbers, the largest magnitude is reduced from 255 to 127. Sign-magnitude numbers require much hardware for addition and subtraction.

1.8.2.2 Complement subtraction

Complements are used in digital computers for simplifying the subtraction operation and for logical manipulation.

There are two types of complements for each 'base-*r*' system.

(1) radix (*r*) complement (or) *r*'s complement

(or)

(2) Diminished radix complement (*r*–1)'s complement.

(1) **(r–1)'s complement.** Given positive number 'N' in base-*r* having an integer part of '*n*' digits and fractional part of '*m*' digits, then the (*r*–1)'s complement of 'N' is defined as

$$(r^n - r^{-m} - N)$$

for only integer part *i.e.*, $m = 0$

The (*r*–1)'s complement of N

$$= \left[(r^n - 1) - N\right]$$

For decimal number system, $r = 10$, then (*r*–1)'s *i.e.*, (10–1)'s complement is called as 9's complement and for binary number system, $r = 2$, then (*r*–1)'s *i.e.*, (2–1)'s complement is called 1's complement. For octal system, $r = 8$, then (*r*–1)'s complement is called 7's complement, for hex system, $r = 16$, then $(r - 1)$'s complement *i.e.*, $(16 - 1)$'s complement is called 15's complement.

(2) **Radix complement (or) *r*'s complement.** For a given number N, with base '*r*', with an integer part of '*n*' digits and fractional part of '*m*' digits, then the *r*'s complement of N is $(r^n - N)$ for $N \neq 0$ and $= 0$ for $N = 0$.

For decimal number system, $r = 10$, then *r*'s complement is called as 10's complement. For binary number systems $r = 2$, then it is called 2's complement. For octal number system $r = 8$ then it is called 8's complement. For hexadecimal system $r = 16$ then it is called 16's complement.

The 1's complement and 2's complement of a binary number represent a negative number and are more useful in binary subtraction.

Note: (1) *r*'s complement of a number exists for any base '*r*' if $r > 1$ and an integer.

(2) *r*'s complement is obtained by adding 1 to the (*r*–1)'s complement

since $r^n - N = [(r^n - 1) - N] + 1$

(3) Complement of complement = original number.

$$r^n - N = r^n - (r^n - N) = N$$

or

$$(r^n - 1 - N) = r^n - 1 - (r^n - 1 - N) = N$$

1's complement system (representation)

The 1's complement of any binary number is performed by simply changing all 1's into 0's and 0's into 1's. *i.e.*, each bit is replaced by its complement (complement of 1 is 0 and complement of 0 is 1).

So, 1's complement of a number (N) = $(2^n - 2^{-m} - N)$

i.e., each bit is subtracted from binary '1'.

i.e.,

$$1 - 0 = 1 \text{ and}$$

$$1 - 1 = 0$$

which causes the bit change, from 0 to 1 and from 1 to 0.

Example 1.53. *Find the 1's complement of $(1011)_2$.*

Solution. given number = 1011, $n = 4$, $m = 0$

1's complement of the number = $(2^4 - 2^0 - 1011) = 100$

$= 16 - 1 - 1011$

$= 15 - 1011$

$= 1111 - 1011 \Rightarrow 0100$

Example 1.54. *Find the 1's complement of 1001.1.*

Solution. given number = 1001.1, $n = 4$, $m = 1$

1's complement of the number = $(2^4 - 2^{-1} - N) = 10000.1 - 1001.1$

$\Rightarrow (0110.0)_2$

Some more examples

binary number	*1's complement*
010001	101110
11110101	00001010
1010.0101	0101.1010
1100.001	0011.110
10010.000	01101.111
10010.10	01101.01
0000.11	1111.00
0000.00	1111.11

1's complement subtraction method

Subtraction between the binary numbers can be performed by using 1's complement method. Here, instead of subtracting a number, we add the 1's complement of the number.

In the 1's complement subtraction, we have two cases.

Case (*i*) *Subtraction of smaller number from larger number.*

Procedure :

(1) Determine 1's complement of the smaller number (subtrahend).

(2) Add 1's complement to the larger number (minuend) and the result is a positive number.

(3) Carry will be generated after the addition which is called "End-Around Carry' (EAC) Remove the EAC and add it to the result.

Example 1.55. *Subtract 101011 from 111101 using 1's complement method.*

Solution.

```
                          111101
1's complement            010100
of 101011 is          EAC ①010001
                               1     Add end around carry
                          010010
```

$\Rightarrow \quad (10010)_2 = (18)_{10}$

Example 1.56. *Perform the binary subtraction for the following using 1's complement method.*

(a) 28 – 8 *(b) 30 – 25* *(c) 25.5 – 12.25* *(d) 10.625 – 8.75.*

Solution. (*a*)

$28_{10} = 11100$

$8_{10} = 01000$ 1's complement of 8 = – 8 = 10111

```
∴            11100  – Minuend
             10111  – 1's complement of subtrahend
         EAC ①10011
               + 1    Add EAC
             10100
```

$= \quad (10100)_2 = (20)_{10}$

(*b*) $30_{10} = 11110$ $25_{10} = 11001$

```
                          Minuend = 11110
    1's complement of subtrahend = 00110
                              EAC ①00100
                                     + 1   Add EAC
                                   00101
```

$\Rightarrow \quad (101)_2 = (5)_{10}$

(*c*) 25.5 = 11001.1

Binary of 12.25 = 01100.011

1's complement of 12.25 = 10011.100

```
∴             11001.10
              10011.10
         EAC ①01101.00
                     1   Add EAC
               1101.01
```

$\Rightarrow \quad (1101.01)_2 = (13.25)_{10}$

(*d*)

$$\text{Minuend } 10.625 = 1010.101$$
$$\text{Subtrahend } 8.75 = 1000.110$$
$$\text{1's complement of } 8.75 = 0111.001$$

$\therefore$

$$\begin{array}{r} 1010.101 \\ 0111.001 \\ \hline \text{EAC } \textcircled{1}0001.110 \\ 1 \quad \text{Add EAC} \\ \hline 1.111 \end{array}$$

$\Rightarrow$ $(1.111)_2 = (1.875)_2$

Case (*ii*) *Subtraction of larger number from smaller number.*

Procedure:

(1) Determine the 1's complement of the larger number (subtrahend)

(2) Add 1's complement to the smaller number (minuend)

(3) After addition, no carry will be generated but answer is in 1's complement form (negative number). To get the answer in true form, take the 1's complement of it and assign negative sign to the answer.

Example 1.57. *Perform the binary subtraction operation using 1's complement method for the following.*

(*a*) 43 – 57 (*b*) 8 – 10 (*c*) 8.75 – 10.625

(*d*) $\frac{3}{4} - \frac{13}{16}$ (*e*) $\frac{5}{8} - \frac{7}{8}$ (*f*) 11.125 – 16.875

Solution.

(*a*)

$$(57)_{10} = 111001$$
$$(43)_{10} = 101011$$

$$\begin{array}{lr} \text{Minuend} & 101011 \\ \text{1's complement of subtrahend} & 000110 \\ \hline & 110001 \end{array}$$

The result has no carry, so the answer is in 1's complement form

1's complement of (110001) is –(001110)

answer in true form = $(-14)_{10}$

(*b*) $(8)_{10} = 1000$ $(10)_{10} = 1010$

$$\begin{array}{lr} \text{Minuend} & = 1000 \\ \text{1's complement of subtrahend} & = 0101 \\ \hline & 1101 \end{array}$$

No carry, so the answer is in 1's complement form.

$\therefore$ True answer $= -(0010)_2$

$= (-2)_{10}$

(*c*) $(8.75)_{10} = 1000.110$

$(10.625)_{10} = 1010.101$

$$
\begin{array}{rr}
\text{Minuend} = & 1000.110 \\
\text{1's complement of subtrahend} = & \underline{0101.010} \\
 & 1110.000
\end{array}
$$

There is no carry, so the answer is in the 1's complement form

$\therefore$ True answer $= -(1.111)_2$

$= -(1.875)_{10}$

(*d*) $\frac{3}{4} = 0.1100$ Minuend $= 0.1100$

$\frac{13}{16} = 0.1101$ 1's complement of subtrahend $= \underline{0.0010}$

0.1110

There is no carry, so the answer is in 1's complement form.

$\therefore$ True answer $= -(0.0001)2$

$= \left(-\frac{1}{16}\right)_{10}$

(*e*) $\frac{5}{8} = 0.101$ Minuend $= 0.101$

$\frac{7}{8} = 0.111$ 1's complement of $\frac{7}{8}$ subtrahend $= \underline{0.000}$

0.101

There is no carry, so the answer is in 1's complement form.

$\therefore$ True answer $= -(0.010)_2 = (-2/8)_{10}$

(*f*) 16.875 – 11.125

16.875 = 10000.111 Minuend = 01011.001

11.125 = 01011.001 1's complement of subtrahend = $\underline{01111.000}$

01010.001

There is no carry, so the answer is in 1's complement form.

$\therefore$ True answer $= -(0101.110)_2$

$= -(5.75)_{10}$

Advantages of 1's complement subtraction

1. The 1's complement subtraction can be accomplished with a binary adder. Therefore, this method is useful in arithmetic logic circuits.

2. It is very easy to find the 1's complement of a number *i.e.*, to get the (–N), where N is a number.

Disadvantage :

1. In 1's complement method, hardware implementation is difficult and it gives the concept of negative zero.
2. It also gives one value less to represent negative values.

2's complement system (representation)

The 2's complement of any binary number is determined by adding 1 to 1's complement of that number. If N is a binary word and $\overline{N}$ is its 1's complement. Then the 2's complement of the binary word N is $\overline{N}+1$.

★ 2's complement of number = 1's complement of a number +1.

★ 2's complement form is used to represent –ve numbers.

★ 2's complement of one (1) is 1.

★ 2's complement of zero is 10.

Example 1.58. *Find the 2's complement of the following binary numbers.*

(a) $(1010)_2$ *(b)* 1111 *(c)* 11.01

Solution. *(a)*

Given binary number = 1010

1's complement of 1010 = 0101

By adding 1 to 1's complement = +1

0110

2's complement of 1010 = $(0110)_2$

(b)

Given binary number = 1111

1's of 1111 is = 0000

By adding 1 to 1's complement = 1

0001

2's complement of 1111 = $(0001)_2$

(c)

Given binary number = 11.01

1's complement of 11.01 = 00.10

By adding 1 to 1's complement = 1

00.11

2's complement of 11.01 is = $(00.11)_2$

Note : From the above examples, we can conclude that the 2's complement of a number can be obtained by leaving all least significant 0's and the first one (1) unchanged, and replacing all 1's with 0's and 0's with 1's in all other higher significant bits.

Example 1.59. *Convert the following decimal numbers into binary and find their 2's complements.*

(a) 40 *(b)* 18 *(c)* 16.5

Solution. (*a*) The binary equivalent of 40 is $(101000)_2$

1's complement of 40 = $(010111)_2$

By adding 1 to its 1's complement = +1

011000

2's complement of 40 = $(011000)_2$

(*b*) The binary equivalent of 18 = $(10010)_2$

2's complement of 18 = $(01110)_2$

(*c*) The binary equivalent of 16.5 = 10000.1

2's complement of 16.5 = $(01111.1)_2$

2's complement subtraction method

Like 1's complement subtraction in 2's complement subtraction also, the subtraction is accomplished by addition only. For subtraction of two numbers we have two cases.

Case (*i*) : *Subtraction of a smaller number from larger number.*

Procedure :

1. Determine the 2's complement of the small number (subtrahend)
2. Add 2's complement to the larger number (minuend)
3. Discard the carry generated.

Note. Always a carry is generated in this case.

Example 1.60. *Perform the following subtraction operations using 2's complement method*

(*a*) $(111001)_2 - (101011)_2$ (*b*) $(1111)_2 - (1010)_2$ (*c*) $(112)_{10} - (65)_{10}$

(*d*) 22–7 (*e*) $(100.5)_{10} - (50.75)_{10}$

Solution. (*a*) Minuend 111001 = 57

2's complement of subtrahend 010101 43

discard carry ①001110 = 14

$\therefore$ 111001 – 101011 $\Rightarrow (1110)_2 = (14)_{10}$

(*b*) Minuend = 1111

2's complement of subtrahend = 0110

discard carry ①0101

$\therefore$ 1111 – 1010 $\Rightarrow (101)_2 = (5)_{10}$

(*c*) $(112)_{10} - (65)_{10}$

binary equivalent of 112 = 1110000

binary equivalent of 65 = 1000001

Minuend = 1110000

2's complement of subtrahend 0111111

discard carry = ①0101111

112 – 65 $\Rightarrow (101111)_2 = (47)_{10}$

(*d*) 22 = 10110 and 7 = 00111

Minuend = 10110
2's complement of subtrahend = 11001
discard carry ①01111

∴ $22 - 7 \Rightarrow (1111)_2 = (15)_{10}$

(*e*) $(100.5)_{10} = (1100100.10)_2$
$(50.75)_{10} = (0110010.11)_2$

Minuend = 1100100.10
2's complement of subtrahend = 1001101.01
discard carry ①0110001.11

∴ $(100.5)_{10} - (50.75)_{10} = (110001.11)_2 = (49.75)_{10}$

Case (*ii*) : *Subtraction of a larger number from smaller number.*

Procedure :

1. Determine the 2's complement of the larger number (subtrahend)
2. Add the 2's complement to small number (minuend)
3. Answer is in 2's complement form. To get the answer in the true form, take the 2's complement and assign negative sign to the answer.

Note : Here, no carry is generated.

Example 1.61. *Perform the following subtraction operations using 2's complement method.*

(*a*) 7 − 22 (*b*) 13 − 27 (*c*) 10 − 28 (*d*) 16.5 − 24.75

Solution. (*a*) $(7)_{10}$ = 00111 Minuend = 00111
$(22)_{10}$ = 10110 2's complement of subtrahend = 01010

No carry 10001

∴ True answer = The 2's complement of (1001) is $-(1111)_2 = (-15)_{10}$

(*b*) $(13)_{10}$ = 01101 Minuend = 01101
$(27)_{10}$ = 11011 2's complement of subtrahend = 00101

No carry 10010

∴ True answer = The 2's complement of (10010) is $-(1110)_2 = -(14)_{10}$

(*c*) $(10)_{10}$ = 01010 Minuend = 01010
$(28)_{10}$ = 11100 2's complement of subtrahend = 00100

No carry 1110

∴ True answer = The 2's complement of (01110) = $(-10010)_2 = -(18)_{10}$

(*d*) 16.5 = 10000.10 Minuend = 10000.10
24.75 = 11000.11 2's complement of subtrahend = 00111.01

10111.11

True answer = 2's complement of 10111.11 = $-(01000.01)_2 = -(8.25)_{10}$

Advantage : Reduction in hardware.

Note : In 2's complement subtraction, carry indicates whether the answer is positive or negative. When the carry is 1, the answer is positive. When there is no carry, the answer is negative and is the 2's complement of the actual magnitude.

The comparison of the three signed-number systems is shown in Table 1.7.

Table 1.7. Comparison of three signed-number systems

Number system → *parameters ↓*	*Signed magnitude*	*Signed 1's complement*	*Signed 2's complement*
1. Representation of a positive number	0 (sign) magnitude	0 (sign) magnitude	0 (sign) magnitude
2. Representation of negative number	1 (sign) magnitude	$2^m - N - 1$	$2^m - N$
3. Form of zero	Plus zero = 00...0 Minus zero = 10...0	Plus zero = 00...0 Minus zero = 11...1	Plus zero = 00..00 No minus zero
4. Negative numbers in complement form	No	Yes	Yes
5. Sign bit involved in addition or subtraction	No	Yes	Yes
6. Use end-around carry	Yes	Yes	No
7. Method for detecting over flow in adding two positive numbers	Carry from the mag. bits	Negative sum	Negative sum
8. Method for detecting over flow in adding two negative numbers	Carry from the mag. bits	Positive sum	Positive sum

1.8.3 Binary Multiplication

The process of binary multiplication is almost similar to the process of decimal multiplication. *Binary multiplication* is just a matter of shifting and adding. The rules for binary multiplication are given in Table 1.8.

Table 1.8. Rules for binary multiplication

Multiplicand *A*	*Multiplier* *B*	*Product* *P*
0	0	0
0	1	0
1	0	0
1	1	1

Note : The product of two binary numbers is 1, only if both the bits are 1, otherwise it is 0.

To handle large numbers, number with several bits: Binary multiplication involves the partial products, shifting each successive partial product left to one place (bit), and then adding all the partial products.

Example 1.62. *Perform the binary multiplication for the following*

(a) 7×5 (b) 22×6 (c) 27×21

(d) $\frac{3}{8} \times \frac{1}{4}$ (e) 4.75×3.625

Solution. (a) $(7)_{10} = 111$ Multiplicand $= 111$

$(5)_{10} = 101$ Multiplier $= 101$

$$\begin{array}{r l} 111 & \\ \times\ 101 & \\ \hline 111 & \text{Partial product 1} \\ 000 & \text{Partial product 2} \\ 111 & \text{Partial product 3} \\ \hline 100011 & \text{Final product} \end{array}$$

$\Rightarrow$ $(100011)_2 = (35)_{10}$

(b) $(22)_{10} = 10110$

$(6)_{10} = \times\ 110$

$$\begin{array}{r} 132 \end{array} \quad \begin{array}{r} 00000 \\ 10110 \\ 10110 \\ \hline 10000100 \end{array} \Rightarrow (10000100)_2 = (132)_{10}$$

10000100 = Final product $\therefore$ $(10000100)_2 = (132)_{10}$

(c) $27 = 11011$

$21 = \times\ 10101$

$$\begin{array}{r} 11011 \\ 00000 \\ 11011 \\ 00000 \\ 11011 \\ \hline 1000110111 \end{array}$$

1000110111 Final product $\Rightarrow (1000110111)_2 = (567)_{10}$

(d) $\frac{3}{8} = 0.011$

$\frac{1}{4} = 0.01$

$$\begin{array}{r} 0.011 \\ \times\ 0.01 \\ \hline 0011 \\ 0000 \\ \hline 0.00011 \end{array}$$

$\Rightarrow$ $(0.00011)_2 \Rightarrow \left(\frac{3}{32}\right)_{10}$

(e)

```
           4.75  =   100.110
          3.625  = × 11.101
                  ----------
                      100110
                     000000
                    100110
                   100110
                  100110
                  ------------
                  10001.001110
```

$$= (10001.00111)_2 \Rightarrow (17.21875)_{10} = \left(17\frac{7}{32}\right)_{10}$$

1.8.4 Binary Division

Binary division is carried out in a similar manner to division using decimal numbers *i.e.*, comparison and subtraction till the final remainder is obtained.

Suppose the divisor has two bits, then, if the first three digits of the dividend are equal to, or larger than the divisor, put 1 in the quotient and subtract the divisor from the first two bits. Next, bring down the next bit of the dividend to the LSB of the remainder. If the remainder is larger than the divisor, put another 1 into the quotient. If it is smaller, insert 0. If 1 is inserted, subtract the divisor from the remainder and bring down the next bit of the dividend and repeat the procedure.

The rules for division by binary bits is as follows.

$$0 \div 1 = 0$$

$$1 \div 1 = 1$$

Note : Division by zero is not permitted.

Let us consider the different examples for binary division.

Example 1.63. *(a) Divide 50 by 5 using binary division.*

(b) Divide 11001 by 100.

(c) Divide 10.5 by 5.25 by using binary division.

(d) Divide 1100.10 by 010.

Solution. (*a*) Divisor = 5 = 101

Dividend = 50 = 110010

The quotient = 101 = $(5)_{10}$

```
          101  ← quotient
     101 )110010
          101
          -----
          00101
            101
          -----
            000  ← remainder
```

(*b*)

```
        110.01
100 ) 11001
      100
      0100
       100
       00100
         100
         000
```

Hence, result = 110.01 = $(6.25)_{10}$

(*c*) Dividend = 10.5 = 1010.10

Divisor = 5.25 = 101.01

```
               10
101.01 ) 1010.10
         101
         0000.00
          000.00
          0000
```

∴ quotient = $(10)_2$ = $(2)_{10}$

(*d*)

```
       110.01
10 ) 1100.00
     10
      10
      10
      00
      00
       01
       00
        10
        10
        00
```

result = $(110.01)_2$ = $(6.25)_{10}$

1.9. THE 9'S AND 10'S COMPLEMENTS

In any number system, the complement of a digit *a*, is denoted by *a*' or $\overline{a}$ is defined as

$$\boxed{\overline{a} = (b - 1) - a}$$

i.e., the complement $\overline{a}$ is the difference between the largest digit in base *b* and the digit *a*.

9's complement : The $(r - 1)$'s complement of a decimal system (r = 10), is called 9's complement. The 9's complement of a number N = $\overline{\text{N}} = \left[(10^n - 10^{-m}) - \text{N}\right]$

n = number of digits in integer part of the number N

m = number of digits in fractional part of the number N

So, the 9's complement of a number can be obtained by subtracting each digit from 9.

Example 1.64. *Find the 9's complement for the following*

(a) 25 (b) 2389 (c) 6291 (d) 374.68 (e) 0.3267 (f) 25.639

Solution. (*a*)

$$\begin{array}{r} 99 \\ -25 \\ \hline 74 \end{array}$$

∴ 74 is the 9's complement of 25 or $n = 2$

∴ 9's complement $= 10^2 - 1 - 25 \Rightarrow (74)_{10}$

(*b*) 9999

$$\begin{array}{r} 2389 \\ \hline (7610)_{10} \end{array}$$

Here $n = 4$ ∴ 9's complement $= 10^4 - 1 - 2389 = (7610)_{10}$

(*c*) 9's complement of (6291) Here $n = 4$ ∴ $10^4 - 1 - 6291 = (3708)_{10}$

(*d*) $(374.68)_{10}$

Here $n = 3$ ∴ 9's complement of (374.68)

$m = 2$

$= 10^3 - 10^{-2} - 374.68$

$= 999.99 - 374.68$

$= (625.31)_{10}$

(*e*) 0.3267

∴ Here $n = 0$ ∴ 9's complement of 0.3267

$m = 4$

$= 10^0 - 10^{-4} - 0.3267$

$= 1 - 10^{-4} - 0.3267$

$= 0.9999 - 0.3267$

$= (0.6732)_{10}$

(*f*) $n = 2$, $m = 3$ ∴ 9's complement of (25.639) is $10^2 - 10^{-3} - 25.639$

$= 100 - 0.001 - 25.639$

$= 99.999 - 25.639 = (74.360)_{10}$

Note : If two numbers are said to be complement to each other, then one of these numbers is positive and other is negative.

10's complement : The 10's complement of a number can be formed by leaving all the least significant 0's unchanged, subtracting the first non-zero least significant digit from 10, and subtracting all higher significant digits from 9.

∴ 10's complement of a number is obtained by adding 1 to its' 9's complement.

Note : *r*'s complement is obtained by adding 1 to (*r* – 1)'s complement since

$$\underbrace{r^n - N}_{r\text{'s complement}} = \underbrace{\left[(r^n - 1) - N\right] + 1}_{(r-1)\text{'s complement}}$$

Example 1.65. *Find the 10's complement for the following*

(a) 6250 *(b) 0123* *(c) 0.3267* *(d) 25.639*

Solution. (*a*) $n = 4$

$\therefore$ 10's complement of 6250 = $10^4 - 6250$

$= (3750)_{10}$

(*b*) $n = 3$

$\therefore$ 10's complement of $(0123)_{10} = 10^3 - 123 = (877)_{10}$

(*c*) $n = 0$

$\therefore$ 10's complement of (0.3267) = $10^0 - 0.3267$

$= (0.6733)_{10}$

(*d*) $n = 2$

$\therefore$ 10's complement of (25.639) = $10^2 - 25.639$

$= (74.361)_{10}$

We can use the 9's and 10's complements to subtract decimal numbers *i.e.*, negative numbers are represented by the complements.

Example 1.66. *Perform the following subtraction operations using 9's and 10's complements.*

(a) 83 – 25 *(b) 37 – 69* *(c) 56.25 – 17.12*

Solution. (*a*) *Direct method:*

$$\begin{array}{r} 83 \\ -25 \\ \hline 58 \end{array}$$

9's complement method:

9's complement of 25 = $10^2 - 1 - 25 = 74$

$$\therefore \quad \begin{array}{r} 83 \\ +74 \\ \hline ①57 \\ 1 \\ \hline 58 \\ \hline \end{array}$$

Here carry is generated. Add it to the result (end–around carry)

(*b*) *Direct method:*

Minuend = 37

Subtrahend = 69

Result difference = –32

9's complement method:

$$\begin{aligned} \text{Minuend} &= 37 \\ \text{9's complement of subtrahend} &= \underline{30} \\ &\ 67 \end{aligned}$$

Here no carry is generated, hence the result is in 9's complement form

$\therefore$ True answer = 9's complement of 67

$= (-32)_{10}$

(c) *Direct method:*

$$\begin{aligned} \text{Minuend} &= 56.25 \\ \text{Subtrahend} &= \underline{17.12} \\ \text{Result} &= \underline{(39.13)_{10}} \end{aligned}$$

9's complement method:

$$\begin{aligned} \text{Minuend} &= 56.25 \\ \text{9's complement of subtrahend} &= \underline{82.87} \\ &\ 139.12 \end{aligned}$$

Here carry is generated, discard it and Add

$$\begin{aligned} &①39.12 \\ &\underline{1} \text{ Add EAC} \\ &(39.13)_{10} \end{aligned}$$

10's complement method:

(a) 10's complement of 25= $10^2 - 25 = 75$

$$\begin{aligned} \text{Minuend} &= 83 \\ \text{10's complement of (25)} &= \underline{75} \\ &\ ①58 \end{aligned}$$

Here carry is generated, then discard it. The remaining is the true answer and is a +ve number.

$\therefore$ Answer = 58 = $(58)_{10}$

(b) 10's complement of 69 = $10^2 - 69 = 31$

$\therefore$

$$\begin{aligned} \text{Minuend} &= 37 \\ \text{10's complement of subtrahend} &= \underline{31} \\ &\ 68 \end{aligned}$$

Here there is no carry, so the result is a –ve number in 10's complement form.

True answer = 10's complement of 68 = $(32)_{10}$

(c) 10's complement of (17.12) = $10^2 - 17.12$

= 82.88

$\therefore$

$$\begin{aligned} \text{Minuend} &= 56.25 \\ \text{10's complement of subtrahend} &= \underline{82.88} \\ &\ ①39.13 \end{aligned}$$

Here carry is generated then, discard it, the remainder is the +ve number and true answer $\Rightarrow (39.13)_{10}$.

Note 1. The use of 9's and 10's complement is similar to the use of 1's and 2's complements respectively.

Note 2. In signed magnitude convention, the number consists of a magnitude and symbol (+ or –) or a bit (0 or 1) indicating the sign. This is the representation of signed numbers used in ordinary arithmetic. When arithmetic operations are implemented in a computer, it is more-convenient to use signed-complement system.

As an example, consider the number 6 represented in binary with eight bits. + 6 is represented with a sign bit of 0 in the left most position followed by the binary equivalent of 6 to give 00000110. To represent – 6, there are three different ways.

In signed–magnitude representation : 10000110

In signed–1's complement representation : 11111001

In signed–2's complement representation : 11111010

1.10 OCTAL ARITHMETIC

In octal number system, we can perform 4 types of arithmetic operations similar to that of decimal number system such as addition, subtraction, multiplication and division.

1.10.1 Addition

The addition of two octal digits is the same as their equivalent sum, provided the sum is less than 8. If the sum is greater than or equal to 8, then, 8 must be subtracted from the sum to obtain the sum and carry of 1 is produced.

Example 1.67. *Perform the addition of the following using octal system.*

(a) $(432)_8$ *and* $(746)_8$ *(b)* $(533.44)_8$ *and* $(471.62)_8$

(c) $(2096)_{10}$ *and* (110110101101110) *(d)* $(B6)_{16}$ *and* $(32)_{10}$

(e) $(3211)_4$ *and* $(432)_6$

Solution. *(a)* Addend 432

Augend = 746 Sum = $(1500)_8$

1500

(b) Addend = 533.44

Augend = 471.62 Sum = $(1225.26)_8$

1225.26

(c) $(2096)_{10} = (4060)_8$

$(\underline{110}\ \underline{110}\ \underline{101}\ \underline{101}\ \underline{110})_2 = (66556)_8$

∴ Addend = $(4060)_8$

Augend = 66556 = ∴ Sum = $(72636)_8$

72636

(d) $(B6)_{16} = (\underline{10}\ \underline{110}\ \underline{110})_2 = (266)_8$

$(32)_{10} = (40)_8$

Addend = 266

Augend = 40 = ∴ Sum = $(326)_8$

326

(e) $(3211)_4 = (\underline{11}\ \underline{100}\ \underline{101})_2 = (345)_8$

$(432)_6 = (164)_{10} = (244)_8$

$$\begin{array}{rl} \text{Addend} = & 345 \\ \text{Augend} = & \underline{244} = (611)_8 \\ & 611 \end{array}$$

1.10.2 Subtraction

The subtraction of one octal number from another can be done in two ways (1) direct method (2) complement method.

Example 1.68. *Perform the subtraction operation for the following using*

(i) direct octal subtraction *(ii) 8's complement*

iii) 7's complement

(a) $(73)_8 - (64)_8$ *(b)* $56.23 - 34.13$

(c) $(79)_{10} - (64)_{10}$ *(d)* $(32)_4 - (64)_8$

(e) $(46)_{10} - (A2)_{16}$ *(f)* $(1100)_2 - (100)_2$

Solution. (i) *Direct octal subtraction*

(*a*) $(73)_8 - (64)_8 = (7)_8$

(*b*) $(56.23)_8 - (34.13)_8 = (22.10)_8$

(*c*) $(79)_{10} - (64)_{10} = (117)_8 - (100)_8 = (17)_8$

(*d*) $(32)_4 - (64)_8 = (16)_8 - (64)_8 = -(46)_8$

(*e*) $(46)_{10} - (A2)_{16} = (56)_8 - (242)_8 = -(164)_8$

(*f*) $(1100)_2 - (100)_2 = (14)_8 - (4)_8 = (10)_8$

(*ii*) *8's complement subtraction*

(*a*) $(73)_8 - (64)_8$

The 8's complement of –64 is +14

$73 + 14 = ①07$

Here carry is generated, discard it

$\therefore$ Result $= (07)_8$

(*b*) $(56.23)_8 - (34.13)_8$

The 8's complement of $(-34.13)_8$ is $= 43.65$

$56.23 + 43.65 = ①22.10$

Carry generated, discard it

$= (22.10)_8$

(*c*) $(79)_{10} - (64)_{10} = (117)_8 - (100)_8$

8's complement of $(-100)_8$ is $= (700)_8$

$(117)_8 + (700)_8 = ①017$

Carry is generated and discard it

$= (17)_8$

(*d*) $(32)_4 - (64)_8 = (16)_8 - (64)_8$

The 8's complement of $(-64)_8$ is $(14)_8$

$$(16)_8 + (14)_8 = (32)_8$$

Here no carry has generated hence the result is a negative number *i.e.*, in 8's complement form. The true answer can be obtained by taking the 8's complement of the result and put minus (–) sign.

∴ 10's complement of (32) is $(46)_8$

∴ Result $= -(46)_8$

(*e*) $(46)_{10} - (A2)_{16} = (56)_8 - (242)_8$

The 8's complement of (–242) is 536

∴ 56 + 536 = 614

Here no carry has generated hence the result is a negative number *i.e.*, in 8's complement form. The true answer can be obtained by taking the 8's complement of the result and put minus (–) sign.

∴ 8's complement of 604 is $= (164)_8$

∴ Result $= -(164)_8$

(*f*) $(1100)_2 - (100)_2 = (14)_8 - (04)_8$

8's complement of $(-04)_8$ is $(74)_8$

∴ 14 + 74 = ①10

Carry is generated and discard it

∴ Result $= (10)_8$

(*iii*) *7's complement method*

(*a*) $(73)_8 - (64)_8$

The 7's complement of $(-64)_8 = (13)_8$

∴ $(73)_8 + (13)_8$ = ①06

Carry is generated add it = 06

1 – ECA

Result $= (07)_8$

(*b*) $(56.23)_8 - (34.13)_8$

The 7's complement of $(-34.13)_8$ is $= (43.64)_8$

∴ $(56.23)_8 + (43.64)_8$ = ①22.07

= 22.07

1

Result $= (22.10)_8$

(*c*) $(79)_{10} - (64)_{10} = (117)_8 - (100)_8$

The 7's of $(-100)_8$ is $= (677)_8$

= 117 + 677

Carry is generated and add it = ①016

$$= \frac{016}{1} \quad \text{EAC}$$

Result $= (17)_8$

(*d*) $(32)_4 - (64)_8 = (16)_8 - (64)_8$

Carry is generated and add it $= (16)_8 + (13)_8$

$= (31)_8$

(Here there is no carry, so the result is in 7's complement form, to obtain the true answer, get the 7's complement of it and put minus (–) sign.)

∴ 7's complement $(31)_8 = -(46)_8$

(*e*) $(46)_{10} - (A2)_{16} = (56)_8 - (242)_8$

The 7's complement of $(-242)_8$ is $= (535)_8$

$= (56)_8 + (535)_8 = (613)_8$

(Here there is no carry, so the result is in 7's complement form, to obtain the true answer, get the 7's complement of it and put minus (–) sign.)

There is no carry, so the result is in the 7's complement form.

∴ 7's complement of (613) is $= (-164)_8$

(*f*) $(1100)_2 - (100)_2 = (14)_8 - (04)_8$

The 7's complement of $(-04)_8 = (73)_8 = 14 + 73$

Carry is generated and add it = ①07

$$= \frac{07}{1}$$

$= (10)_8$

1.11 HEXADECIMAL ARITHMETIC

Arithmetic operations can be performed in hexadecimal system same as decimal system except that here base is 16.

1.11.1 Addition

The sum of two hexadecimal digits is the same as their equivalent decimal sum provided the sum is less than 16. If the sum is greater than or equal to 16, then 16 must be subtracted from the sum to obtain the hexadecimal sum, and carry of 1 is produced.

Example 1.69. *Perform the following addition operations in hexadecimal system*

(*a*) $(12A7)_{16}$ *to* $(3479)_{16}$ (*b*) $(456.7B)_{16}$ *to* $(24.A6)_{16}$

(*c*) $(786)_{10}$ *to* $(562)_8$ (*d*) *Add* $(2096)_{10}$ *to* $(110110110001)_2$

Solution. (*a*) Addend = 12A7

Augend = 3479

$(4720)_{16}$

(*b*) 456.7B

24.A6

$(47B.21)_{16}$

(*c*) $(312)_{16}$

$+ (172)_{16}$

$= (484)_{16}$

(*d*) 830H

+ CB1

$= (14E1)_{16}$

1.11.2 Subtraction

The subtraction of one hexadecimal number from another is similar to the decimal subtraction except that whenever a 1 is borrowed from a column to the left, its value is 16 and not 10.

Example 1.70.

(*a*) 34H – 1CH (*b*) D6 – 3DH (*c*) 376.4AH – 76.B3H

(*d*) $(782)_{10} - (341)_8$ (*e*) $-(123)_4 + (110110011)_2$

(*f*) FFH – EAH (*g*) 40DH – BBH

Solution. (*a*) Minuend = 34 H

Subtrahend = 1C H

18 H

(*b*) D6 Minuend

3D Subtrahend

99H Result

(*c*) 376.4A

76.B3

$(1FF.97)_H$

(*d*) $(782)_{10} = (30E)_{16}$

$(341)_8 = E1$

$\therefore (30E)_{16} - (E1) = (22D)_{16}$

(*e*) $-(123)_4 + (110110011)_2$

$(123)_4 = (1B)_{16}$

$(110110011)_2 = (1B3)_{16}$

$\therefore (1B3) - (1B) = (198)_{16}$

(*f*) FFH – EAH $= (15)_H$

(*g*) 40DH – BBH $= (352)_H$

1.12 BINARY CODES

The electronic digital systems like computers, microprocessors, ...etc., are required to handle data (represent and manipulate) which may be consisting of numbers, alphabets or special characters. But the digital systems use signals that have two distinct values and circuit elements that have two stable states. Hence the numerals, alphabets, special characters and control functions are to be converted into binary format.

The process of conversion into binary format is known as '*binary coding*'. The combination of binary bits that represent numbers, (may be decimal numbers, octal, hexadecimal number etc.), alphabets, symbols or control functions are called '*binary codes* (or) *digital codes*'. When decimal number is represented by its equivalent binary, we call it "straight binary coding". The codes must be in binary because computers can only process 1's and 0's. Representation of a group of 2^n distinct elements in a binary code requires a minimum of n bits. This is because it is possible to arrange n bits in 2^n distinct ways.

For example, a group of four distinct quantities can be represented by a two bit code, with each quantity assigned one of the following bit combinations : 00, 01, 10, 11. A group of eight elements requires a three-bit code, with each element assigned to one and only one of the following: 000, 001, 010, 011, 100, 101, 110, 111.

The examples show that the distinct bit combinations of an n-bit code can be found by counting in binary from 0 to $(2^n - 1)$. The minimum number of bits required to code 2^n distinct quantities is n, there is no maximum number of bits that may be used for a binary code.

Basically, binary codes are classified as numeric or alphanumeric. *Numeric codes* are used to represent numbers. On the other hand, *alphanumeric codes* are used to represent characters: alphabetic letters and numerals.

In these codes, a numeral is treated simply as another symbol rather than of a number or numeric value.

1.12.1 Classification of Binary Codes

The binary codes are classified in different ways as (1) Weighted Codes (2) Non-weighted Codes (3) Reflective Codes (4) Sequential Codes (5) Alphanumeric Codes (6) Error Detecting and Correcting Codes.

(1) *Weighted Codes.* The main characteristic of a *weighted code* is, each binary bit is assigned by a "*weight*" and values depend on the position of the binary bit.

The sum of the weights of these binary bits, whose value is 1 is equal to the decimal digit which they represent.

In other words, if w_1, w_2, w_3, and w_4 are the weights of the binary digits, and x_1, x_2, x_3 and x_4 are the corresponding bit values, then the decimal digit

$N = w_4\,x_4 + w_3\,x_3 + w_2\,x_2 + w_1\,x_1$ is represented by the binary sequence $x_4x_3x_2x_1$. A sequence of binary bits which represents a decimal digit is called a "*code word*". Thus $x_4x_3x_2x_1$ is a code word of N. Examples for these codes are: BCD, 8421, 6421, 4221, 5211, 3321, $84\bar{2}\bar{1}$ etc.

(2) *Non-weighted Codes.* Non-weighted codes or un-weighted codes are those codes in which the digit value does not depend upon their position *i.e.*, each digit position within the

number is not assigned fixed value. Most commonly used non-weighted codes are : unweighted BCD code, Excess–3 code and gray code.

Comparison between weighted and non-weighted codes is shown in Table 1.9.

Table 1.9. Comparison between weighted and non-weighted codes

Sl. No.	*Particulars*	*Weighted codes*	*Non-weighted codes*
1.	Weight	In this code, each bit position is assigned a specific weight.	In this code, no specific weights are assigned to the bit positions.
2.	Value	Each bit position represents a fixed value.	Each position within the binary number is not assigned any fixed value.
3.	Examples	BCD, 4221, 5211, 8421, 6421, $84\bar{2}\bar{1}$.	Examples. Ex-3 code, Gray code.
4.	Applications	These codes are used in (*a*) Data manipulation during arithmetic operations. (*b*) For input/output operations in digital circuits. (*c*) To represent the decimal digits in calculators, volt meters etc.	These codes are used (*a*) To perform certain arithmetic operations. (*b*) Shift position encodes. (*c*) Used for error detecting purpose.

(3) *Reflective Codes (or) Self Complimenting Codes.* A code is said to be a reflective code, if the code word of the 9's complement of N *i.e.,* 9 – N can be obtained from the code word of N by interchanging all the 1's as 0's and 0's as 1's. *i.e.,* the code word for 9 is the complement for the code 0, 8 for 1, 7 for 2, 6 for 3 and 5 for 4. Note that 4221, 5211 6, 4, 2, –3 and Ex–3 codes are reflective, whereas 8421 code is not.

Example. In $642\bar{3}$, the decimal 5 is represented by = 1011

9 – N = 9 – 5 = 4 decimal 4 is represented by = 0100

i.e., 9's complement of 5 is obtained just by complementing each bit in 5.

Necessary condition. A weighted code is said to be self complementing code if the sum of all weights is equal to '9'.

Totally there are 13 possible self complementing weighted codes which include positive and negative weights.

Example. 5211, 4221, 3321, $84\bar{2}1$ etc.

Note: Ex-3 code is a non-weighted self complementing code.

Reflective codes are used to perform 9's complement subtraction.

(4) *Sequential Codes.* In sequential codes, each succeeding code is one binary number greater than its preceeding code. The examples for this code are 8421, and Ex-3 codes, whereas 4221 and 5211 are not sequential codes. These codes are greatly used in mathematic manipulation of data.

(5) *Alphanumeric Codes.* Many applications of digital computers require the handling of data consisting of not only numbers, but also of letters. To represent the letters (characters), it is necessary to formulate a binary code. An alphanumeric character set is a set of elements that includes the 10 decimal digits, the 26 letters of alphabets, and a number of special characters. Such a set contains between 36 to 64 elements if only capital letters are included, or between 64 to 128 elements if both uppercase and lowercase letters are included. In the first case, we need a binary code of six bits and in the second, we need a binary code of seven bits.

The codes which represent the *alphanumeric* character set are called '*alphanumeric codes*'.

Most of these codes, however, also represent symbols and various instructions necessary for conveying intelligible information.

The most commonly used alphanumeric codes are :

ASCII (American Standard Code for Information Interchange)

EBCDIC (Extended Binary Coded Decimal Interchange Code) and

Hollerith code. All these codes are non-weighted codes.

(6) *Error Detecting and Correcting Codes.* When the binary information is transmitted from one circuit or system to another circuit or system, there is a chance of error occurance. The error may occur at the transmitter, receiver or in the channel. This means a 0 may be changed as 1 and vice-versa due to the presence of different types of noise. To maintain the data integrity between transmitter and receiver, extra bit(s) are added to the data. These extra bits allow the detection and some times correction of error(s) in the data.

The data along with extra bit(s) form the code. The codes which allow only error detection are called "error detecting codes" and codes which allow error detection and correction are called '*error detecting and correcting codes*'. Example for error detecting code is parity and error detection and correction or error correcting code is Hamming code.

Fig. 1.8 represents the classification of various digital codes (or) binary codes.

Binary code

Weighted	*Non-weighted*	*Reflective*	*Sequential*	*Alphanumeric*	*Error detecting and correcting*
(*a*) Binary–8421	(*a*) Ex-3	(*a*) 2421	8421	ASCII	(*a*) Parity
(*b*) BCD–4221, 3321, 5211 5421, 6311, 7421, $74\bar{2}\bar{1}$, $84\bar{2}\bar{1}$	(*b*) Gray	(*b*) 5211	Ex-3	EBCDIC	(*b*) Hamming code
	(*c*) Five-bit BCD codes	(*c*) Ex-3		Hollerith	

Fig. 1.8 Classification of binary codes.

1.12.2 BCD Codes or 8421 Code

BCD is an abbreviation for Binary-Coded-Decimal in which decimal digits 0 through 9 are represented by their binary equivalents using four bits *i.e.*, BCD is a numeric code in which each decimal digit is represented by a separate binary code of four bits. The most common BCD code is 8-4-2-1. In this code, the binary weights associated with four bits are 8 4 2 1 (*i.e.*, 2^3, 2^2, 2^1, and 2^0) from left to right. This means that bit 3 has weight 8, bit 2 has weight 4, bit 1 has weight 2 and bit 0 has weight 1.

The 8421 code is a *mixed-base code*; it is binary within each group of 4 bits, but it is decimal from group to group.

Note: The four bit BCD codes use only 10 of 16 possible states (because $2^4 = 16$). There are $\frac{16!}{16! \times 10!}$ possible codes, approximately 8008 codes.

Table 1.10 shows the 4 bit 8421 BCD-code used to represent a single decimal digit.

Table 1.10. 8421 BCD code

Decimal digit	*BCD code* 8	4	2	1
0	0	0	0	0
1	0	0	0	1
2	0	0	1	0
3	0	0	1	1
4	0	1	0	0
5	0	1	0	1
6	0	1	1	0
7	0	1	1	1
8	1	0	0	0
9	1	0	0	1

With four bits, 2^4 *i.e.*, sixteen numbers can be represented. But in this code, we are using only ten. The result of six codes combinations are not used *i.e.*, 1010 to 1111 are invalid in this BCD code.

In multi-digit coding, each decimal digit is individually coded with 8421 BCD code. For example, 78 in decimal can be encoded in 8421 BCD as 0111 1000.

Comparison between BCD and binary

In multi-digit coding of 8421 BCD numbers, we require 4-bits per decimal digit. Therefore, total 8-bits are required to encode 78 in 8421 BCD. When we represent the same number 78 in binary : 1001110, we require only 7 bits. The straight binary code takes the complete decimal number and represents it in binary: the BCD code converts each decimal digit to binary individually. This means that representing numbers 8421 BCD is less efficient than pure binary number system. The advantage of a BCD code is that it is easy for conversion to decimal and vice-versa. The disadvantage of BCD is, in this arithmetic operations are more complex than in pure binary.

Example 1.71. *Convert each of the following decimal numbers into BCD.*

(a) $(97)_{10}$ *(b)* $(63.4)_{10}$ *(c)* 78.216

Solution.

(*a*) $(97)_{10}$ → 1001 0111

(*b*) $(63.4)_{10}$ → 0110 0011.0100

(*c*) 78.216 → 0111 1000.001000010110

Note : If a number in any number system other than decimal is to be converted into BCD-code, first it must be converted into decimal and, from decimal to BCD. Similarly if the conversion is required from BCD to other number system, it must always be done via decimal.

Example 1.72. *Convert given numbers into BCD*

(*a*) $(7A)_{16}$ (*b*) *1000010* (*c*) *1110110.0001* (*d*) $(36)_8$

Solution.

(*a*) $(7A)_{16} = 7 \times 16 + 10 = 112 + 10 = 122$

∴ 122 = 0001 0010 0010

(*b*) 1000010 = 64 + 2 = 66

= 0110 0110

(*c*) 1110110.0001 = 64 + 32 + 16 + 4 + 2.0625

= 118.0625

= 0001 0001 1000 . 0000 0110 . 0010 0101

(*d*) $(36)_8 = 24 + 6 = 30 =$ 0110 0000

Example 1.73. *Convert the given BCD numbers into*

(*a*) *decimal* (*b*) *binary* (*c*) *octal and* (*d*) *hexadecimal*

(*i*) *10001101* (*ii*) *10100.001* (*iii*) *100110.100110*

Solution.

(*i*) (*a*) $\underset{8}{\underline{1000}}\ \underset{9}{\underline{1001}} = (89)_{10}$ (*b*) 1011001

(*c*) (131)8 (*d*) $(59)_{16}$

(*ii*) 10100.0010

(*a*) $(14.2)_{10}$ (*c*) $(16.146)_8$

(*b*) 1110.0011 (*d*) $(E.33)_{16}$

(*iii*) 100110.100110

(*a*) $(26.98)_{10}$ (*c*) $(32.7655)_8$

(*b*) 11110.1111 (*d*) $(1A.FA)_{16}$

1.12.2.1 BCD addition

We can perform BCD addition provided that in each case, the sum in any four-bit column does not exceed 9. Then only the results are valid BCD numbers.

Examples 1.74. *Add the following BCD numbers.*

(i) 1000 0100 and 0001 0011

(ii) 0101 0010 and 0011 0110

(iii) 0100 0110.010 and 01000010.001

(iv) 1001 1000 and 0010 1000

Solution.

	Binary	Decimal
(i)	10000100	84
+	00010101	15
	10011001	99
(ii)	01010010	52
+	00110110	36
	10001000	88
(iii)	0100 0110.0100	46.4
+	0100 0010.0010	42.2
	1000 1000.0110	88.6
(iv)	1001 1000	98
+	0010 1000	28
	1001.0000	126

Note: In the above examples *(i)*, *(ii)* and *(iii)* are valid BCD additions, because in all cases, the four-bit sum is equal to or less than 9.

But example *(iv)* is an invalid BCD addition. Because here four bit sum is greater than 9 and also carry out of group is generated. Hence BCD addition can't be performed.

1.12.3 Other 4-bit BCD Codes

There are various other weighted 4-bit BCD codes, each developed to have certain properties useful for special applications. Table 1.11 shows these codes, identified by the weights assigned to their bit positions. Among those 2421, 3321 and 4221 are the self complementing codes. In the 2421 BCD code, the weights are 2-4-2-1, meaning that bit 1 and bit 3 have the same weight two (2). It is sometimes referred to as 2*-4-2-1 code, where the asterisk simply distinguishes one position with weight 2 from the other. In the above three codes, two positions have same weight, there are two possible bit patterns that could be used to represent same decimal digit, but only one of these patterns is actually assigned.

The $74\bar{2}\bar{1}$ and $84\bar{2}\bar{1}$ codes are some what different than other codes. Here $\bar{2}\bar{1}$ indicates that they are negative weights. When 1 occurs in either of the two right most positions, it means that the weight of that position is subtracted, rather than added to determine the decimal value. In the above two codes, $84\bar{2}\bar{1}$ is a self complementing code.

Table 1.11. Weighted 4-bit BCD codes

Decimal	*2421*	*3321*	*4221*	*5211*	*5311*	*5421*	*6311*	*7421*	$74\bar{2}\bar{1}$	$84\bar{2}\bar{1}$
0	0000	0000	0000	0000	0000	0000	0000	0000	0000	0000
1	0001	0001	0001	0001	0001	0001	0001	0001	0111	0111
2	0010	0010	0010	0011	0011	0010	0011	0010	0110	0110
3	0011	0011	0011	0101	0100	0011	0100	0011	0101	0101

4	0100	0101	1000	0111	0101	0100	0101	0100	0100	0100
5	1011	1010	0111	1000	1000	1000	0111	0101	1010	1011
6	1100	1100	1100	1001	1001	1001	1000	0110	1001	1010
7	1101	1101	1101	1011	1011	1010	1001	1000	1000	1001
8	1110	1110	1110	1101	1100	1011	1011	1001	1111	1000
9	1111	1111	1111	1111	1101	1100	1100	1010	1110	1111

1.12.4 Non-Weighted Codes

1.12.4.1 Excess-3 code

Excess-3 code is a modified form of BCD code. The Excess-3 code can be obtained from the natural BCD code by adding 3 to each coded number *i.e.*, the digital code which is derived by adding 3 to each decimal digit and then converting the result into four-bit binary is known as *'Excess-3 code'* (XS3) (or) EX-3 code.

Here no definite weights are assigned to the four bit positions and hence this code is an unweighted code.

In this code also, we have $16(2^4)$ possible code combinations, out of these, only ten code combinations are used as shown in Table 1.12. The remaining six code combinations are invalid. These are 0000, 0001, 00010,1101, 1110 and 1111.

Table 1.12. The excess-3 code

Decimal	*BCD*	*Excess-3*
0	0000	0011
1	0001	0100
2	0010	0101
3	0011	0110
4	0100	0111
5	0101	1000
6	0110	1001
7	0111	1010
8	1000	1011
9	1001	1100

Example 1.75. *Find the Ex-3 code for the following decimal numbers.*

(a) 7 *(b)* 36 *(c)* 72.9 *(d)* 5481

Solution.

(a) The excess-3 code for the decimal 7 is

$$7 + 3 = 10 \rightarrow 1010$$

∴ The excess-3 code for the decimal 7 = 1010.

(b)
```
  3  6
 +3 +3     = 0110 1001
 ------
  6  9
```

(c)
```
  7  2 .  9     (1010 0101.1100)
 +3 +3   +3
 ----------
 10  5 . 12
```

(d)
```
 5 4 8 1     (1000 0111 1011 0100)
 3 3 3 3
 ----------
 8 7 11 4
```

Note : (1) The special feature of this unweighted code is self-complementing. That is, 1's complement of an excess-3 code number is the excess-3 code for the 9's complement of the respective decimal number.

For example: The excess-3 code for decimal 4 is 0111. Then it's one's complement is 1000. This 1000 is the excess-3 code for the decimal 5, which is the 9's complement of 4.

This self complementing property makes the excess-3 code useful in some arithmetic operations.

(2) For a digit in any number system to be converted into Ex-3 code, first it should be converted into BCD and then into Ex-3 code and vice versa *i.e.* conversions between any number system and Ex-3 can be done via BCD code (decimal).

Example 1.76. *Convert each of the following numbers of various number systems into Ex-3 code.*

(a) $(1011)_2$ *(b)* $(436)_8$ *(c)* $(3A)_{16}$ *(d)* $(1100.011)_2$ *(e) (10.8)*

Solution.

(a) 1011 = $(11)_{10}$

```
Ex-3 code =  1  1 (0100 0100)
            +3 +3
            -----
             4  4
```

(b) $(436)_8 = 64 \times 4 + 3 \times 8 + 6$

$= 256 + 24 + 6 = (286)_{10}$

```
Ex-3 code = 2 8 6 (0101 1011 1001)
            3 3 3
            -------
            5 11 9
```

(c) (3A016 = 16 × 3 + 10

= $(58)_{10}$

```
Ex-3 code =  5    8
          = +3  + 3
            -------
             8   11
          = 1000 1011
```

(d) $(1100.011)_2$ = 12.375

```
Ex-3 code = 33 333
            ---------
            45 6 10 8
          = 01000101 0110.1010 1000
```

(e) (1C.8)16 $= 16 \times 1 + 12 + \frac{8}{16}$

$= 16 + 12 + 0.5 = (28.5)_{10}$

```
Ex-3 code = 2 8 . 5      = 0101 1011 . 1000
            3 3   3
            -------
            5 11 8
```

1.12.4.2 Excess-3 addition

This code has an interesting property when used in addition. The addition operation may be explained in two cases.

Procedure : (1) Add the given binary numbers using rules for binary addition.

(2) If any group produces a decimal carry, add 0011 to that group (digit)

(3) If any group does not produce a decimal carry, subtract 0011 from that group.

Case (i) : In Ex-3 code, whenever we add decimal digits whose sum is 9 or less, an Ex-6 number results. To return to excess-3 form or true form, we must subtract 3.

Example 1.77. *Perform the following addition operations in Ex-3 code.*

(a) 2 + 5

Solution.

```
(a) 2 in Ex-3 code is    0 1 0 1
    5 in Ex-3 code is  + 1 0 0 0
    7 in Ex-6 code       1 1 0 1
    Subtract 3          – 0 0 1 1
    7 in Ex-3 code      = 1 0 1 0
```

Note : When we add two Ex-3 numbers and assume that there is no carry from each group, (*i.e.*, the sum of individual group is 9 or less), then the result is in Ex-6 form and we must subtract 3 from each group to return to excess 3 code.

Example 1.78. *Perform the following addition operations in Ex-3 code.*

(a) 46 + 33   *(b) 72.6 + 15.2*

Solution.

```
(a) 46 in Ex-3 code       =     0 1 1 1 1 0 0 1
    33 in Ex-3 code           + 0 1 1 0 0 1 1 0
    Ex-6 code 6 for 79          1 1 0 1 1 1 1 1
    Subtract 3 from           – 0 0 1 1 0 0 1 1
    each digit
    79 in Ex-3 code       =     1 0 1 0 1 1 0 0
(b) 72.6 in Ex-3 code     =     1 0 1 0 0 1 0 1 . 1 0 0 1
    15.2 in Ex-3 code     =   + 0 1 0 0 1 0 0 0 . 0 1 0 1
    Ex-6 code for 87.8          1 1 1 0 1 1 0 1 . 1 1 1 0
    Substract 3 from          – 0 0 1 1 0 0 1 1 . 0 0 1 1
    each digit
    Ex-3 code for 87.8    =     1 0 1 1 1 0 1 0 . 1 0 1 1
```

Case (ii) **:** Whenever the sum of decimal digits exceeds 9, there will be a carry from one group to the next. When this happens, the group that produced the carry will revert to 8421 form. This occurs because of the Ex-6 and six unused four-bit groups. To restore the answer to Ex-3 code, we must add 3 to the group that produced the carry.

Example 1.79. *Add the following in Ex-3 code*

(a) 28 + 36 *(b) 9 + 8* *(c) 87.6 + 12.6*

Solution.

(*a*)	2 8	= 0 1 0 1	1 0 1 1	Excess –3 code for 28
	3 6	0 1 1 0	1 0 0 1	Excess –3 code for 36
		1 1 0 0	0 1 0 0	First result
		– 0 0 1 1	+ 0 0 1 1	Subtract and add 3
		1 0 0 1	0 1 1 1	Ex-3 for 64

(*b*)	Ex-3 code for 9	= 1 1 0 0
	Ex-3 code for 8	= 1 0 1 1
	First result	1 0 1 1 1
	Add 3	0 0 1 1
		1 1 0 1 0 = Ex-3 code for 17.

(*c*)	Ex-3 code for 87.6	= 1000 0111 . 0110
		0011 0011 0011
		1011 1010 . 1001
	Ex-3 code for 12.6	0001 0010 . 0110
		0011 0011 0011
		0100 0101 . 1001

Add Ex-3 of 87.6 and 12.6

	= 1011 1010 . 1001
	0100 0101 . 1001
	100000000 . 0010
Add 0011	0011
	100000000 . 0101
	= 100.5

Advantages : (1) In the Ex-3 code, addition operation use the ordinary binary addition.

(2) The 1's and 2's complements can be used to subtract Ex-3 numbers.

1.12.4.3 Gray code

The code which exhibits only a single-bit change from one number to the next is known as '*gray code*' *i.e.*, in this code between any two successive code words, there will be change in only one position. This code is also called as '*unit distance code*' or '*cyclic code*'. As there is no specific weight assigned to each bit position, this code is an unweighted code. Gray code is also known as '*reflected code*', because n-bit gray code can be generated by reflecting the $(n-1)^{th}$ bit code.

The Table 1.13 shows the bit pattern assigned for gray code from decimal 0 to decimal 15.

Table 1.13. Gray code representation

Decimal code		*Gray code*
0	–	0000
1	–	0001
2	–	0011
3	–	0010
4	–	0110
5	–	0111
6	–	0101
7	–	0100
8	–	1100
9	–	1101
10	–	1111
11	–	1110
12	–	1010
13	–	1011
14	–	1001
15	–	1000

Gray code belongs to a class of codes called 'minimum' changes codes', in which only one bit in the code group changes when going from one step to the next. Because of this, this code is not suitable for arithmetic operations.

Another property of gray code is that the gray-coded number corresponding to the decimal number 2^n-1, for any *n*; differs from gray coded O (0000) in one bit position only (*i.e.*, at n value position). For example, for $n = 2$, 3 and 4, we see that $2^2 - 1 = 3$, 10 = 0 0 1 0 in gray code ∵ change at position 1.

$2^3 - 1 = 7 =$ 0100 in gray code ∵ change at position 3.

$2^4 - 1 = 15 =$ 1000 in gray code ∵ change at position 4.

We can observe the reflection property of the gray code in Table 1.14(*a*), (*b*) and (*c*).

Table 1.14. 2 bit gray code, 3 bit gray code, and 4 bit gray code

2–bit gray code (*a*)	*3–bit gray code* (*b*)	*4–bit gray code* (*c*)
g_1 g_0	g_2 g_1 g_0	g_3 g_2 g_1 g_0
0 0	0 0 0	0 0 0 0
0 1	0 0 1	0 0 0 1

1 1	0 1 1	0 0 1 1
1 0	0 1 0	0 0 1 0
	1 1 0	0 1 1 0
	1 1 1	0 1 1 1
	1 0 1	0 1 0 1
	1 0 0	0 1 0 0
		1 1 0 0
		1 1 0 1
		1 1 1 1
		1 1 1 0
		1 0 1 0
		1 0 1 1
		1 0 0 1
		1 0 0 0

Advantages

(1) This code is useful for Input/Ouput devices, A to D convertes, and other peripheral equipment.

(2) The Gray code exhibits a single bit change from one code number to the next. This property is useful in shaft position encoders, where error susceptibility increases with the number of bit changes between adjacent numbers in sequence.

(3) This code is useful when other codes such as binary etc. produce undefined or ambiguous results during the transition from one number to next. For example, to change from 7 to 8 *i.e.,* 0111 to 1000, four changes are needed.

(4) Very easy for conversion between gray and binary.

(5) Gray code is used to represent the analog data by continous change of a shaft postion. The shaft is partitioned into segments, and each segment is assigned a number. If the adjacent segments are made to correspond with the gray code sequence, ambiguity is eliminated when detection is sensed in the line that separtes any two segments shown in Fig. 1.9.

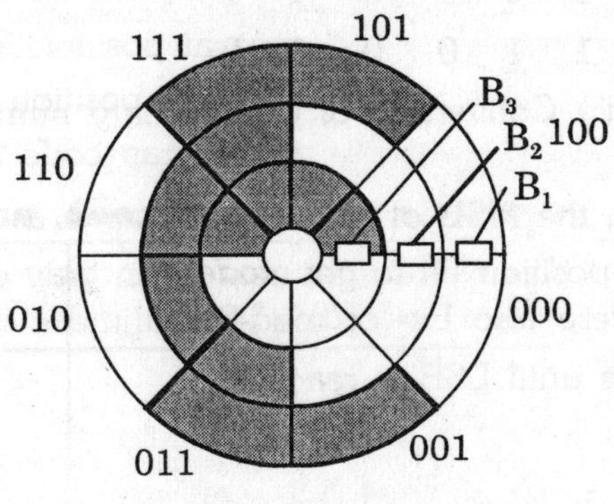

Fig. 1.9 Rotating disk.

Gray to binary conversion : The Gray to binary code conversion can be achieved using the following steps :

(1) The MSB of the binary number is same as the MSB of the gray code number. So write it down.

(2) To obtain the next binary bit, perform exclusive OR mod-2 addition or addition with neglecting carry between the just written down binary bit and the next gray code bit. Write down the result.

(3) Repeat the step 2 until all gray code bits have been completed *i.e.*, until LSB of Gray code number is reached.

Note : (1) The number of bits in a gray code = number of bits in the equivalent binary number and vice-versa.

(2) The conversion process starts from MSB and moves towards LSB.

Let g_n-------$g_1 g_0$ designate a code word in the $(n + 1)^{th}$ bit gray code. Let $b_n, ...b_1, b_0$ designate the corresponding binary number, where,

suffix 0 indicates LSB and n indicates MSB. Then, i^{th} bit of g_i can be otabined as

$$g_i = b_i \oplus b_{i+1} \qquad 0 \le i \le n-1$$

where, b_i represents the ith bit in the binary code.

$$g_n = b_n.$$

for gray-binary

$$b_n = g_n$$
$$b_i = b_{i+1} \times g_i$$

Example 1.80. *Convert the following gray code numbers into binary*

(a) 101101 *(b)* 10101111 *(c)* 101011

Solution.

(a) Gray code

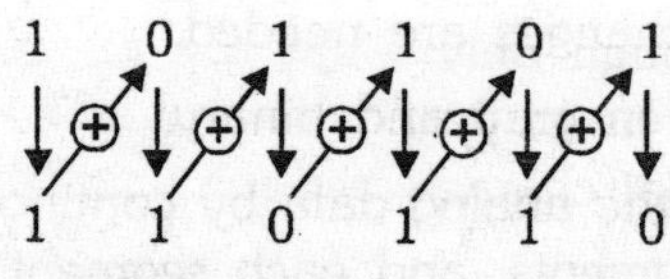

Binary code

(b) Gray code	1	0	1	0	1	1	1	1
Binary code	1	1	0	0	1	0	1	0
(c) Gray code	1	0	1	0	1	1		
Binary code	1	1	0	0	1	0		

Binary to gray conversion : Conversion of given binary number into gray code involves the following steps :

(1) Record the MSB, that is, the MSB of gray code is same as MSB of the binary number.

(2) Add this bit to the next position bit to get next bit in gray code, nelgecting carry if any and record the sum. (Here also Ex- or mod-2 arithmetic is used for addition).

(3) Continue recording sums until LSB is reached.

Example 1.81. *Convert the following binary number into gray code :*

(a) 110110 *(b)* 11001010 *(c)* 110010.

Solution.

(a)

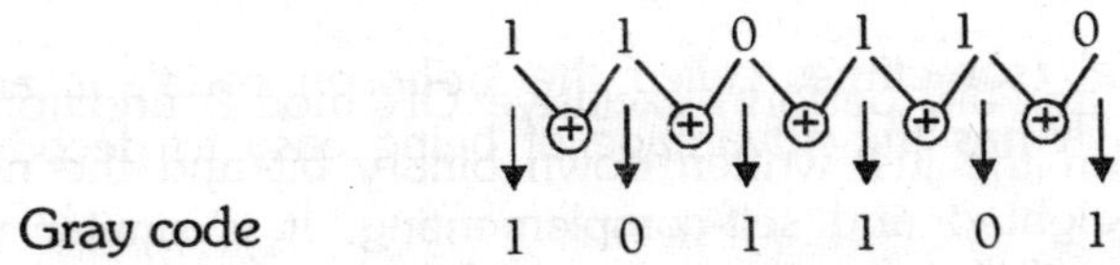

(b)

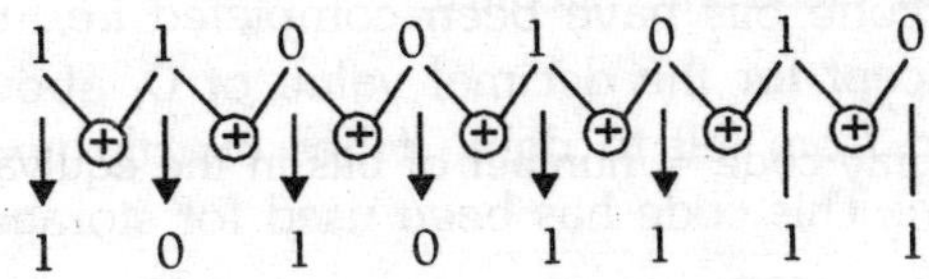

(c)

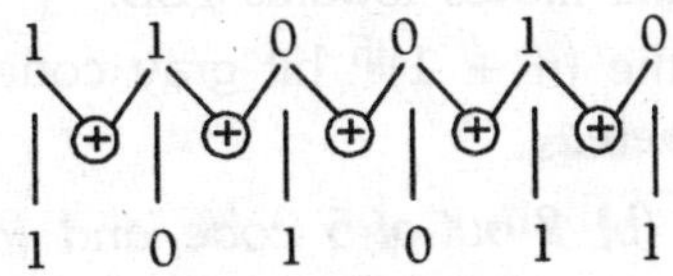

1.12.4.4 Unweighted BCD code

We know that BCD code is 8421 code. Using this code each decimal digit is represented as weighted 8421 code. If we take decimal number in which each digit is represented using 8421 code hence it becomes unweighted code. The decimal numbers after 9, that is 10, 11, 12 etc. are represented separately for each digit.

i.e., 10 in unweighted BCD is 00010000

11 in unweighted BCD is 00010001 etc.

1.13 5-BIT CODES

Five bit codes also exist. Although only 4-bits are needed to encode any decimal digit from 0 to 9, an extra-bit will allow us to decode the number more easily. Table 1.15 shows some 5-bit BCD codes having special characteristics. These special characteristics of the code are useful for error detection.

Table 1.15. Five-bit BCD codes

Decimal	*51111*	*Shift counter*	*86421*	*63210*	*(2-out of-5)*
0	00000	00000	00000	00110	00011
1	00001	00001	00001	00011	00101
2	00011	00011	00010	00101	00110
3	00111	00111	00011	01001	01001
4	01111	01111	00100	01010	01010
5	10000	11111	00101	01100	01100
6	11000	11110	01000	10001	10001
7	11100	11100	01001	10010	10010
8	11110	11000	10000	10100	10100
9	11111	10000	10001	11000	11000

The '2-out-of-5 code' is an unweighted code that has been used in telephone and communication work. Not only does it have even points, it also has exactly two 1's in each code group. Because of this, errors can be detected more reliably than with just an even-parity check.

The '*shift–counter code*' (sometimes called the '*Johnson code*') is an unweighted code used in electronic counters. It has the advantage of being easy to decode electronically.

The 51111 code is weighted and self-complementing. It is similar to the shift-counter code, so that it is easily decoded with electronic circuitry.

The 63210 code is weighted, except for the decimal value of 0, above decimal number 0, the weights are 6, 3, 2, 1, 0, read from left to right. It has exactly two 1's in each code group, allowing reliable error detection. This code has been used for storage of digital data on magnetic drums.

The 86421 is a weighted code.

Example 1.82. *Encode the following numbers*

(i) 684, (ii) 571 in (a) shift-counter code (b) 2-out-of-5 code and (c) 63210 code.

Solution.

(*i*)	684 in shift-counter code	11110	11000	01111
	in 2-out of-5 code	10001	10100	01010
	in 63210 code	10001	10100	01010
(*ii*)	571 in shift-counter code	11111	11100	00001
	in 2-out-of-5 code	01100	10010	00101
	in 63210 code	01100	10010	00011

1.14 CODES WITH MORE THAN 5-BITS

We can decode and detect errors more easily by adding more bits. Table 1.16 shows some weighted BCD codes that use more than 5-bits.

Table 1.16. More than 5-bit codes

Decimal	*5043210*	*543210*	*9876543210*
0	0100001	000001	0000000001
1	0100010	000010	0000000010
2	0100100	000100	0000000100
3	0101000	001000	0000001000
4	0110000	010000	0000010000
5	1000001	100001	0000100000
6	1000010	100010	0001000000
7	1000100	100100	0010000000
8	1001000	101000	0100000000
9	1010000	110000	1000000000

The 9876543210 code, sometimes called the '*counter code*' uses only a single 1 in each code group; this makes it easy to decode and to detect errors. The code has the disadvantage of requiring more electronic circuitry than the simpler 4-and 5-bit codes.

The 5043210 code, also called the '*biquinary code*' is occasionally used in electronic counters (note that biquinary means two-five). Each member of this code contains a group of 2 bits and a group of 5 bits. The group of two bits indicates whether the number is more or less than 5. The group of five bits denotes the count. Reliable error detection is possible because a single 1 is in the group of 2 bits and a single 1 is in the group of 5 bits.

The 543210 code is like biquinary code, except that 6 bits are used instead of 7. The most significant digit (for left) indicates whether the number is less than 5 or not. This code is sometimes used in electronic counters.

1.15 ALPHA NUMERIC CODES

In addition to numerical data, a computer must be able to handle non-numerical information. In other words, a computer should recognize codes that represent letters of the alphabet, punctuation marks and special characters as well as numbers. We can say that an alpha numeric code represents all of the various characters and functions that are found on a standard typewriter (or computer) keyboard.

The standard binary code for the alphanumeric characters is ASCII. It uses 7 bits to code ($2^7 = 128$) characters, as shown in Table 1.17. The ASCII (Pronounced "ask-ee") code contains 94 graphic characters that can be printed and 34 non printing characters used for various control functions. The graphic characters consist of the 26 upper case letters. (A through Z), 26 lower case letters (a through z), 10 numbers (0 to 9), and 32 special printable characters such as %, *, $, &, + and – so on. The 34 control characters are designed in the ASCII table with abbreviated names. They are lighted in the table with their full functional names. The control characters are used for routing data and arranging the printed text into a prescribed format. There are three types of control characters : format effectors, information separators, and communication-control characters.

Table 1.17. ASCII codes

ASCII Code	*Character*	*ASCII Code*	*Character*	*ASCII Code*	*Character*	*ASCII Code*	*Character*
00	NUL	21	!	42	B	63	c
01	SOH	22	"	43	C	64	d
02	STX	23	#	44	D	65	e
03	ETX	24	$	45	E	66	f
04	EOT	25	%	46	F	67	g
05	ENQ	26	&	47	G	68	h
06	ACK	27	'	48	H	69	i
07	BEL	28	(	49	I	6A	j
08	BS	29	)	4A	J	6B	k
09	HT	2A	*	4B	K	6C	l
0A	LF	2B	+	4C	L	6D	m

0B	VT	2C	,	4D	M	6E	*n*
0C	FF	2D	-	4E	N	6F	*o*
0D	CR	2E	.	4F	O	70	*p*
0E	SO	2F	/	50	P	71	*q*
0F	SI	30	0	51	Q	72	*r*
10	DLE	31	1	52	R	73	*s*
11	DC1 (X-on)	32	2	53	S	74	*t*
12	DC2 (Tape)	33	3	54	T	75	*u*
13	DC3 (X-off)	34	4	55	U	76	*v*
14	DC4	35	5	56	V	77	*w*
15	NAK	36	6	57	W	78	*x*
16	SYN	37	7	58	X	79	*y*
17	ETB	38	8	59	Y	7A	*z*
18	CAN	39	9	5A	Z	7B	{
19	EM	3A	:	5B	[	7C	|
1A	SUB	3B	;	5C	\	7D	} (ALT mode)
1B	ESC	3C	<	5D	]	7E	~
1C	FS	3D	=	5E	^ (↑)	7F	DEL (RUB OUT)
1D	GS	3E	>	5F	– (←)		
1E	RS	3F	?	60	‘		
1F	US	40	@	61	*a*		
20	SP	41	A	62	*b*		

Format effectors are characters that control the layout of printing. They include the familiar typewriter controls such as back space (BS), horizontal tabulation (HT), and carriage return (CR).

Information separators are used to separate the data into divisions such as paragraphs and pages. They include characters such as record separator (RS) and file separator (FS).

The *communication-control characters* are useful during the transmission of text between remote terminals. They are STX (start-of-text) and ETX (end of text) which are used to frame a text message when transmitted through telephone wires.

The seven bit code format is $b_6b_5b_4b_3b_2b_1b_0$. The first 3-bits (rightmost 3 bits) are taken from the column in which symbol appears and the last 4 bits correspond to the row. For instance, the letter 5 appears as follows : 101 0011.

ASCII is a 7-bit code, but most computers manipulate an 8-bit quantity as a single unit called a 'byte'. So the ASCII characters most often are stored one per byte., The extra bit is used for other purposes, depending on the application. For example, simple printers recognise 8-bit ASCII characters with the MSB set to 0. When used in data communication, the eighth bit may be used to indicate parity of the character.

The ASCII code is used for the transfer of alphanumeric information between a computer and input/output devices such as video terminals or printers. A computer also uses it internally to store the information that an operator types in at the computer's keyboard.

There are three common ways to display alphanumerics. They are

(a) *Discrete method*– Here, single light source produces each symbol ex : nixie tube.

(b) *Bar-matrix method* – In this method, one or more light sources may be involved in the display of a particular symbol.

(c) *Dot-matrix method* – Here, many individual sources are shaped like dots. A typical example is the 5 × 7 LED matrix.

EBCDIC codes: EBCDIC stands for Extended Binary Coded Decimal Interchange Code. It is pronounced as "ebb-see-dick". It is a standard code for large computers. It is an 8-bit-code. It provides more extensive character set than ASCII. But it does not provide for parity with an 8-bit byte. It is longer than ASCII. Table 1.18 shows EBCDIC code.

Table 1.18. EBCDIC codes

Characters or Control characters	*Hex codes for EBCDIC*	*Characters or Control characters*	*Hex codes for EBCDIC*
NUL	00	SUB	3F
SOH	01	SP	40
STH	02	BLANK	E0
ETS	03	¢	4A
HT	05	.	4B
DEL	07	,	6B
VT	0B	?	6F
FF	0C	:	7A
CR	0D	;	5E
SO	0E	!	5A
S1	0F	′	7D
DLE	10	″	7F
DC1	11	+	4E
DC2	12	–	60
DC3	13	—	6D
RES	14	*	5C
NL	15	/	61
BS	16	=	7E
CAN	18	<	4C
EM	19	>	6E
FLS	1C	(	4D
GS	1D	)	5D
RDS	1E	{	8B
US	1F	}	9B
BYP	24	[	AD

LF	25	]	DD
EOB	26	|	4F
ENQ	2D	&	50
ACK	2E	$	5B
BEL	2F	¬	5F
SYN	32	%	6C
DC4	35	#	7B
EOT	37	@	7C
NAK	3D	F	C6
a	81	G	C7
b	82	H	C8
c	83	I	C9
d	84	J	D1
e	85	K	D2
f	86	L	D3
g	87	M	D4
h	88	N	D5
i	89	O	D6
j	91	P	D7
k	92	Q	D8
l	93	R	D9
m	94	S	E2
n	95	T	E3
o	96	U	E4
p	97	V	E5
q	98	W	E6
r	99	X	E7
s	A2	Y	E8
t	A3	Z	E9
u	A4	0	F0
v	A5	1	F1
w	A6	2	F2
x	A7	3	F3
y	A8	4	F4
z	A9	5	F5
A	C1	6	F6
B	C2	7	F7
C	C3	8	F8
D	C4	9	F9
E	C5		

1.16 ERROR CORRECTING AND DETECTING CODES

The digital information in the form of binary is transmitted from one device to another or circuit (or) system to another. In this, there is a chance for occurring of error. This means a signal corresponding to 0 may change to 1 or vice-versa due to the presence of different types of noise. A reliable system must have a mechanism for detecting and correcting such errors. For this purpose, extra bit (or) more than one bit are added in the data. These extra bits allow the detection and some times correction of errors in the data.

1.16.1 Types of Errors

When ever an electromagnetic signal flows from one point to another, it is subjected to unpredictable interference from heat, magnetism and other forms of electricity. This interference can change the shape or timing of the signal. If the signal is carrying binary data, such changes can alter the meaning of the data changing 0 to 1 or 1 to 0. Bits can be changed singly or in clumps. There are three types of possible errors that may occur: single bit, multiple-bit and burst errors.

Single-bit Error : The term single bit error means that only one bit of a given data unit is changed from 1 to 0 or 0 to 1. The data unit may be a byte, character or packet etc. Example: The sender transmitting a data is 00100100, if the received message is 00000100. Here the 6th position of data is changed from 1 to 0.

Multiple-bit Error : The term multipie-bit error means that two or more non consecutive bits in data unit have changed from 1 to 0 or from 0 to 1.

The Fig. 1.10(*a*) shows an example of multiple-bit error.

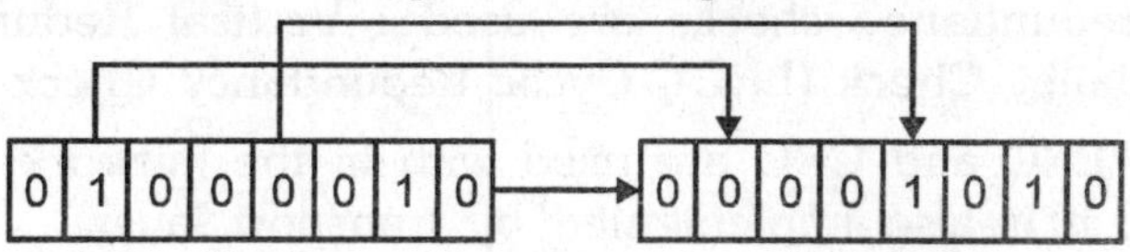

Fig. 1.10(*a*) : Multiple-bit error.

Burst Error : The term burst error means that two or more consecutive bits in the data unit have changed from 1 to 0 or 0 to 1 see Fig. 1.10(*b*).

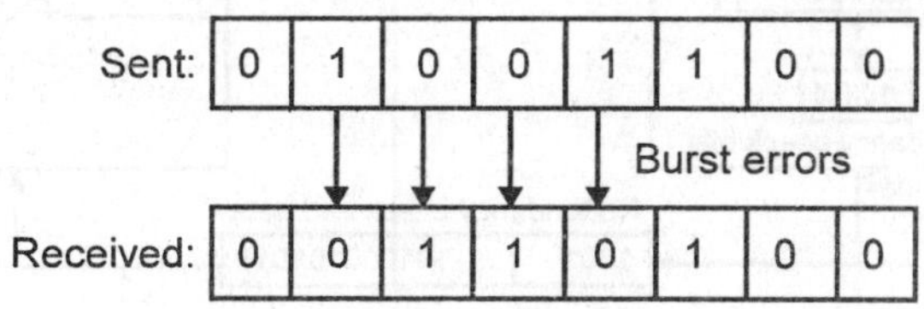

Fig. 1.10(*b*) : Burst error.

1.16.2 Error Detection

Even if we know, what type of errors can occur, we can recognize those, if we have a copy of the intended transmission for comparison. But if we don't have a copy of the original we will have no way of knowing that we have received an error. There are several mechanisms, they are exact count encoding and redundancy. These codes are used only to detect errors, not to prevent the occurrence of error or not to correct the occurred errror.

Exact-count Encoding : With exact count encoding, the no. of 1's in each data unit is same. An example of exact count encoding scheme is the ARQ code. In ARQ code each data unit (character) has three 1's in it, and therefore a simple count of the number of 1's received can determine whether a transmission error has occurred or not.

Redundancy : The redundancy involves transmitting of each data unit twice. If same data unit is not received twice in succession, a transmission error has occurred. The same concept can be used for messages. This system is accurate but it would be in supportably slow. Not only would the transmission time double, but the time it takes to compare every unit by bit must be added.

The major problem with this is, if the error has occurred at the same position of both data units, it is not possible to recognize it even though there is an error, because in comparison all bits of both data units are same.

So instead of transmitting every data unit twice, a shorter group of bits may be appended to the end of each data unit. This technique is called redundancy, because the extra bits are redundant to the information; and they are discarded as soon as the accuracy of transmission has been determined.

As soon as data unit is generated, it passes through a device called generating function, that analyzes it and adds an appropriately coded redundancy check bits. This expanded data unit is traveled over the physical channel to the receiver. The receiver puts the entire stream through a checking function, if there is no error, the data position of data unit is accepted and the redundant bits are discarded (See Fig. 1.11).

Four types of redundancy checks are used : Vertical Redundancy Check (VRC), Longitudinal Redundancy Check (LRC), Cyclic Redundancy Check (CRC) and checksum.

Generally VRC, LRC and CRC are used with in the network and checksum is used in the inter-network, it is also implemented by transport layer.

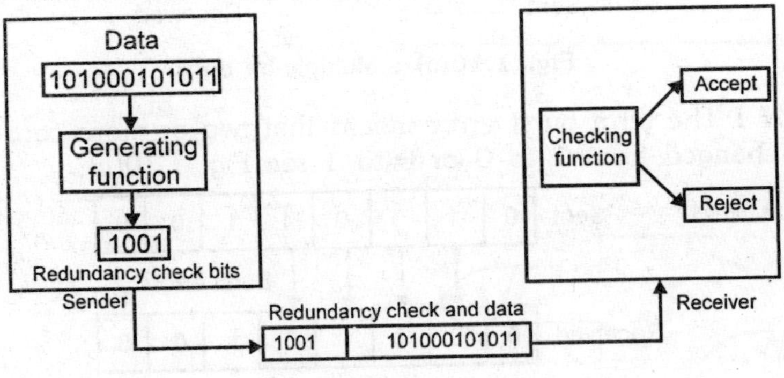

Fig. 1.11 Redundancy.

1.16.2.1 Vertical redundancy check (VRC)

It is most common and least expensive mechanism for error detection. This VRC is also called a parity check. The definition of parity is equivalence of equality. In this technique one bit is appended to each data unit, this bit is called a parity bit or redundant bit, with this bit the total number of 1's in the data unit becomes either even (even parity) or (odd parity). Both sending and receiving systems must be of same parity. If an even-parity data unit is transmitted and an odd parity is received, the receiver knows that the data unit has an error.

For example, if we consider each data unit as a ASCII character, the length of each character is seven bits, including parity, total length of data unit is eight bits.

Suppose we want to transmit a character '*C*', the binary value is 1000011 (43 hex). Before transmitting we pass it through a generating function (See Fig. 1.12). Here the generating function is a even-parity generator, it counts the number of 1's and append the appropriate parity bit *i.e.*, 1 with this parity bit the data unit is 11000011, the total number of 1's is a even number. The system now transmits this data unit to the receiver. At the receiver, the checking function counts number of 1's, if the number of 1's is an even number, then the data unit passes, otherwise it will be rejected.

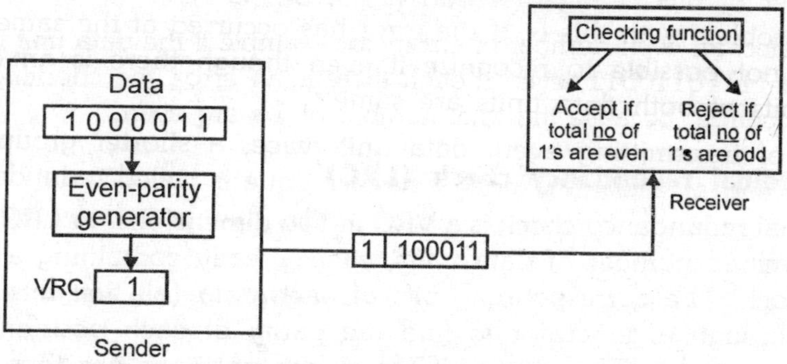

Fig. 1.12 Even parity VRC.

Two types of parity generator circuits are used as a generating function at the transmitter and checking function at the receiver: serial and parallel see Fig. 1.13.

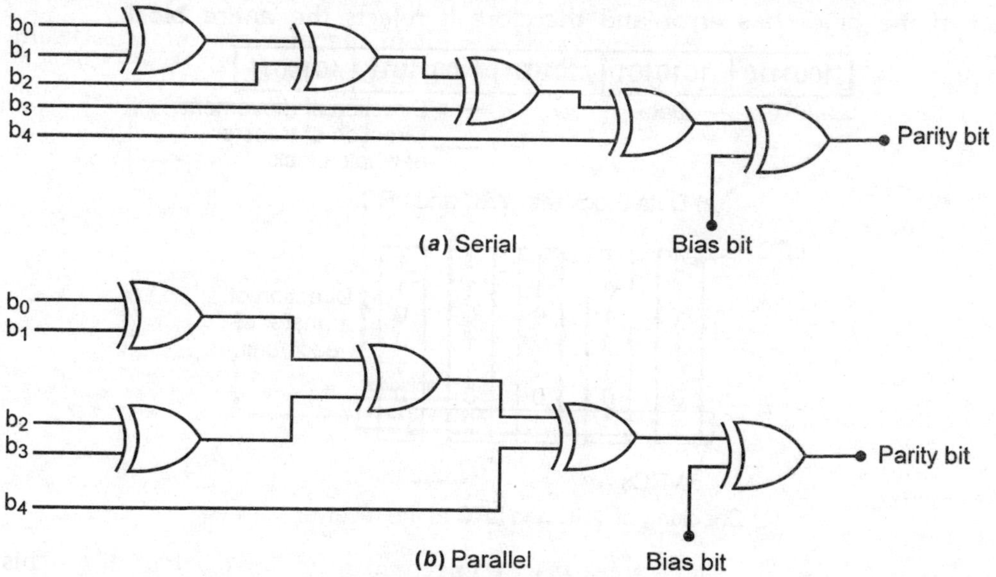

Fig. 1.13 Parity generators.

In the sequential parity generator b_0 is XORed with b_1, the result is XORed with b_2 and so on. The result of last XOR operation is compared with bias bit. For even parity, this bit made logic 0, and for odd parity it is logic '1'. The output of the circuit is a parity bit, which is appended to the data unit same as in parallel or combinational generator.

Some circuits can be used as parity checkers in the receiver. A parity checker uses the same procedure as a parity generator except that logic condition of the final comparison is used to determine if a parity violation has occurred (for odd parity a 1 indicates an error and '0' indicates no error, for even parity a 1 indicates an error and 0 indicates no error).

Reliability

1. VRC can detect all single errors. It can also detect multiple-bit or burst errors as long as the number of bits changed is odd (1, 3, 5... etc.).

2. It can not detect an even number of errors, for example if the data unit is 1100011001 1, if it has received as 11110110011, even though there is an error, the checking function at the receiver can not recognize, because the total number of 1's are even.

1.16.2.2 Longitudinal redundancy check (LRC)

The longitudinal redundancy check is a VRC in two dimensions. In LRC error detection, groups a predetermined number of data units, each already containing a VRC parity bit, together into a block. The corresponding bits of each data (all first bits, all second bits etc.) are passed through a generator to find the parity of each position. Each position then gets its own parity bit. The parity bits of all the positions are then assembled into a new data unit which is added at the end of the data block (message) see Fig. 1.14(*a*). As the data block reaches the receiver, it passes through a LRC checker. The total parity including VRC and LRC must match the parity expected. If so the receiver discards the VRC and LRC, accepts the data units. If any values are wrong, the receiver knows that some part of the block has error and therefore it rejects the entire block.

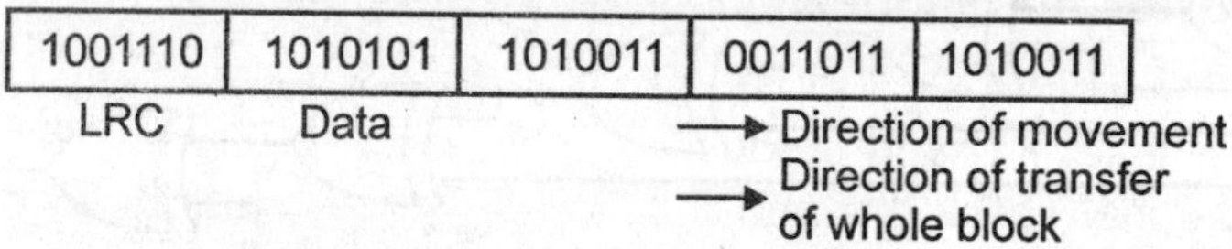

(*a*) Data Block with VRC and LRC.

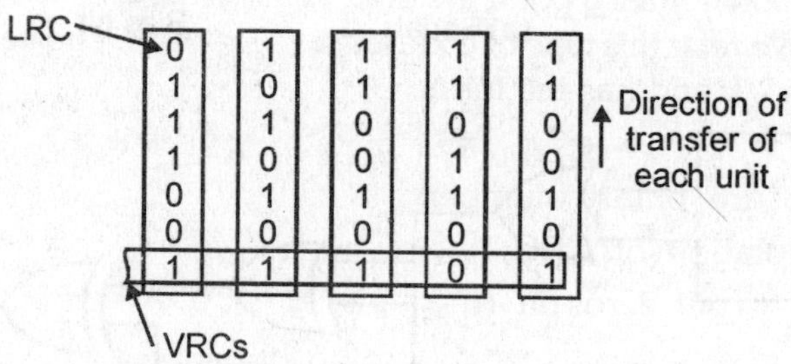

(*b*) Checking of VRC and LRC at the receiver.

Fig. 1.14

Reliability

1. It determines all single-bit, multiple-bit and burst errors except one pattern of errors.

2. If two bits in one data unit are damaged, and two bits in exactly the same positions in another data unit are also damaged, the LRC checker will not detect an error.

1.16.2.3 Cyclic redundancy check (CRC)

It is the most powerful redundancy checking technique, based on the binary division. Polynomial codes are used for generating check bits in the form of cyclic redundancy check.

The CRC bits are appended at the end of each data unit so that the resulting data unit becomes exactly divisible by a second, predetermined number. At the receiver, the incoming data unit is divided by the same number. If there is no remainder, the data unit is assumed to be intact and is therefore accepted. A remainder indicates that data unit is damaged (See Fig. 1.15(*a*)).

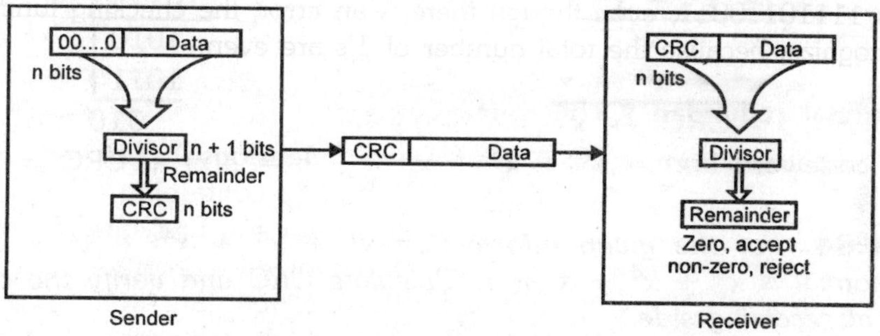

Fig. 1.15(a) CRC generator and checker.

In polynomial codes the information symbols, the code words and the error vectors are represented by polynomial with binary coefficients. The K information bits (i_{k-1}, i_{k-2},....., i_1, i_0) are used for the information polynomial of degree $k - 1$.

$$i(x) = i_{k-1}x^{k-1} + i_{k-2}\, x^{k-2} + ... + i_1\, x + i_0$$

The encoding process takes $i(x)$ and produces a code word polynomial '$b\,(x)$' that contains information bits and additional check bits and that satisfy certain pattern. To detect errors, the receiver checks for this pattern-whether it is same or not. The polynomial code specified by the generator is called generator polynomial $g\,(x)$. If the data unit has n bits in which k are information bits and $n - k$ are check bits. We refer this type of code as an (n, k) code. The generating polynomial for such a code has degree $n - k$ and has the form

$$g(x) = x^{n-k} + g_{n-k-1}\, x^{n-k-1} + ... + g_1\, x + 1.$$

where g_{n-k-1}, g_{n-k-2},..., g_1 are binary numbers.

Steps involved in calculation of CRC:

1. Multiply $i(x)$ by x^{n-k} (put zeros in $(n-k)$ low order positions)
2. Divide $x^{n-k}\, i(x)$ by $g(x)$

$$x^{n-k}\, i(x) = g(x)\; q(x) + r(x) \rightarrow \text{remainder}$$

$\qquad\qquad\qquad\qquad \llcorner\!\!\rightarrow$ quotient (for $q(x)$)

3. The remainder polynomial $r(x)$ provides the CRCs, add remainder $r(x)$ to $x^{n-1}\, i(x)$ (*i.e.*, put check bits in the $n-k$ low order position)

$$b(x) = x^{n-k}\, i(x) + r(x)$$

where $b(x)$ is a transmitted codeword/data unit.

Example 1.83. *The given information is $x^3 + x^2$ and the generating polynomial is $x^3 + x + 1$, find the CRC and transmitted code word.*

Solution. Generating polynomial $g(x) = x^3 + x + 1$.

Information: $(1, 1, 0, 0) \rightarrow i(x) = x^3 + x^2$.

Encoding $x^3\ i(x) = x^6 + x^5 = 1100000$.

$$\begin{array}{r} x^3 + x^2 + x \\ x^2 + x + 1 \overline{)\, x^6 + x^5 \qquad\qquad} \\ \underline{x^6 + x^4 + x^3 \qquad} \\ x^5 + x^4 + x^3 \qquad \\ \underline{x^5 + x^3 + x^2 \quad} \\ x^4 + x^2 \quad \\ \underline{x^4 + x^2 + x} \\ \text{CRC} \rightarrow x \end{array}$$

```
        1110
1011) 1100000
      1011
       1110
       1011
        1010
        1011
         010
```

Transmitted codeword $b(x) = x^6 + x^5 + x$ $\qquad b = b(x) + \text{CRC} = \underbrace{1100}_{\text{data}}\underbrace{010}_{\text{CRC}}$

Example 1.84. *For the given information $x^7 + x^5 + x^4 + x^2 + x + 1$ and generating polynomial is $x^5 + x^4 + x + 1$. Calculate CRC and verify the data is error received or not at receiving side.*

Solution. $g(x) = x^5 + x^4 + x^1 + x^0$ or 110011

$i(x) = x^7 + x^5 + x^4 + x^2 + x^1 + x^0$ or 10110111

Encoding $= i(x)\ x^5 = x^{12} + x^{10}\ x^9 + x^7 + x^6 + x^5$

$= 1011011100000$

```
          11010111
110011) 1011011100000
        110011
         111101
         110011
           111010
           110011
             100100
             110011
              101110
              110011
               111010
               110011
               01001 → CRC
```

Transmitted data $= b(x) = i(x) + \text{CRC}$

$$= \underbrace{10110111}_{\text{data}}\ \underbrace{01001}_{\text{CRC}}$$

At the receiver, the transmitted data is again divided by $g(x)$.

```
             11010111
110011) 1011011101001
        110011
         111101
         110011
           111010
           110011
             100110
             110011
              101010
              110011
               110011
               110011
               00000 → remainder = 0
```

no error has occurred.

Reliability

1. With CRC, approximately 99.95% of all transmission errors are detected.

2. It cannot detect those errors, where the change in bit value of a block of code is exactly the value of the divisor or its multiple values. The standard polynomial code or CRC codes are shown in Table 1.19.

Table 1.19. Standard generator polynomials

Name	*Polynomic*	*Used in*
CRC-8	$x^8 + x^2 + x + 1$	ATM header error check
CRC-10	$x^{10} + x^9 + x^5 + x^4 + x + 1$	ATM AALCRC
CRC-12	$x^{12} + x^{11} + x^3 + x^2 + x + 1$	Bisync
CRC-16	$x^{16} + x^{15} + x^2 + x + 1$	Bisync
CCITT-16	$x^{16} + x^{12} + x^5 + 1$	HDLC, XMODEM, V.41
CCITT-32	$x^{32} + x^{26} + x^{23} + x^{22} + x^{16} + x^{12} + x^{11} + x^{10} + x^8 + x^7 + x^5 + x^4 + x^2 + x + 1$	IEEE802, DOD, V.42AALs

To generate the CRC check, we can use the shift register circuit. Let us see a circuit for generating CRC bits of example 2.1 (see Fig. 1.15(*b*)) after 7 clock pulses the contents of registers are $r(x) = 010$

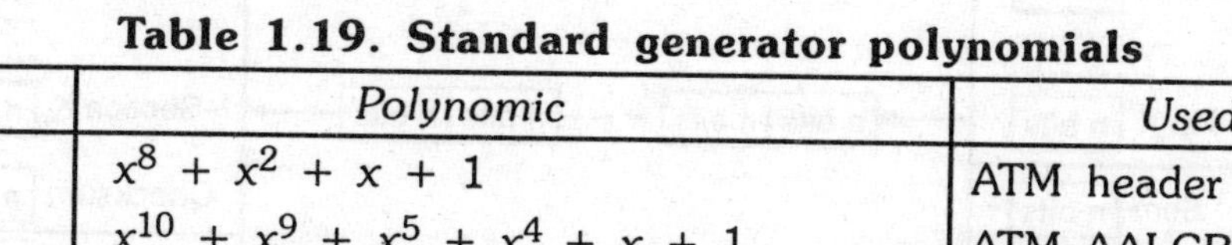

Fig. 1.15(*b*) Shift register circuit for generated polynomial.

1.16.2.4 Check sum

The error detection method used in the network higher layer protocol is called check sum. The check sum is based on the concept of redundancy. Several internet protocols (IP, TCP, UDP) use check bits to detect errors.

Check sum generator

At the sender, the checksum, is generated by the following steps:

1. The data is divided into k sections, each of n bits (usually-16)
2. Sections 1 and 2 are added together using 1's complement arithmetic in such a way that the total is also n bits long.
3. In this way all the k sections are added.
4. The final result (sum) is then complemented and appended to the end of original data unit as redundancy bits, called checksum field.
5. The extended data unit is transmitted across the network, so the sum of data segment is T, the checksum is $-T$ (see Fig. 1.16).

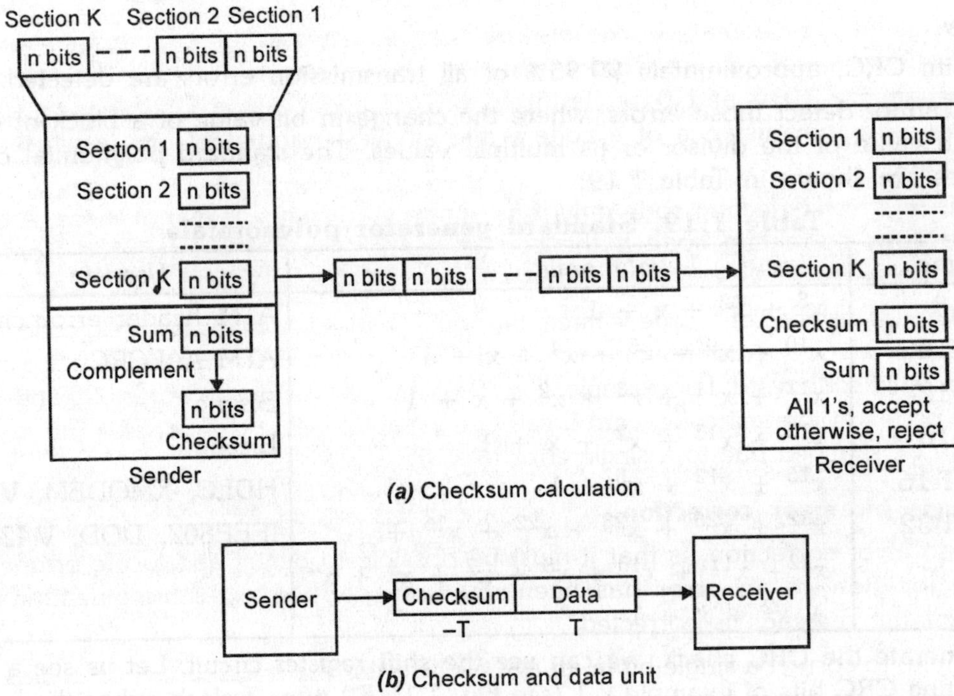

(a) Checksum calculation

(b) Checksum and data unit

Fig. 1.16

Checksum checker

The receiver subdivides the data unit and adds all the segments together. If the extended data unit is intact, the total value found by adding the data segments and the checksum field should be zero (T plus $-T$ is 0). If the result is not zero, there is an error in the data block. The actual result of the addition will be n 1's if the data is unchanged *i.e.*, represents the value -0.

Reliability

1. It detects all errors involving odd number of bits, as well as most errors involving even number of bits.

2. However, if one or more bits of a segment are errors and the corresponding bit or bits of opposite value in a second segment are also damaged, the sum of these columns will not change and receiver will not detect a problem.

1.16.3 Error Correction

The mechanisms that we have covered upto this point detect error but cannot correct errors. The error correction can be handled in two ways

1. When an error is discovered, the receiver can have the sender re-transmit the entire data unit.

2. A receiver can use an error correcting code, which automatically corrects certain errors.

The number of bits required to correct a multiple-bit or burst error is so high that in most cases it is inefficient to do so. For this reason in most cases error correction is limited to one, two or three bit errors. For a code to be error detecting, its minimum distance must be further increased. The distance between the code words is the number of digits that must change in one word so that the other word results. For example, the distance between the 1010 and 0100 is three, since the code words differ in three bit positions. The minimum distance of a code is the smallest number of bits in which any code word differ.

A code is an error detecting code only if its minimum distance is two or more. A code is said to be error correcting code if the correct code word can be deduced from the erroneous word.

If the minimum distance of a code is three, then any single error changes a valid code word into an invalid one, which is a distance one away from the original code word a distance two away from any other valid code word. For example, if we consider two valid code words 000 and 111. A single error occurs in the first code word it can be changed to 001, 010 or 100. The second code word can be changed due to a single error to 110, 101 or 011.

1.16.3.1 Single bit error correction

The key to error correction, is that it must be possible to detect and locate erroneous digits. If the location of an error has been determined, then by complementing the erroneous digit the message is corrected.

For example, to correct a single bit in an ASCII character the error correction code must determine which of the seven bits has changed. In this case we have to distinguish between eight different sates: no error, error in position 1, error in position 2 and so on up to error in position 7. To do so requires enough redundancy bits to show all eight states.

At first glance, it looks a three bit redundancy code should be adequate because three bits can show eight different states (000 to 111) and can therefore indicate the locations of eight different possibilities. But what if an error occurs in the redundancy bits themselves 7 bits of data plus 3-bits of redundancy equals to 10 bits. Three bits, however can identify only eight possibilities. Additional bits are necessary to cover all possible error locations.

1.16.3.2 Redundancy bits

To calculate the number of redundancy bits (r) required to correct a given number of data bits (m), we must find the relation between m and r. The total length of resulting code is $m + r$ bits.

If the total transmitting unit is '$m + r$' length then 'r' must be able to indicate at least $m + r + 1$ different states, of these one state means no error and $m + r$ state indicates the location of an error in each of the $m + r$ positions.

So $m + r + 1$ states must be discoverable by r–bits and r–bits indicate 2^r different states.

$$\therefore \quad 2^r \geq m + r + 1 \qquad \text{...(1.1)}$$

The relation between r and m is shown in Table 1.20.

Table 1.20. Relation between data and redundancy bits

Number of data bits (m)	*Number of redundancy bits (r)*	*Total bits (m + r)*
1	2	3
2	3	5
3	3	6
4	3	7
5	4	9
6	4	10
7	4	11

1.16.3.3 Hamming code

So far, we have examined the number of bits required to cover all of the the possible single bit error states in a transmission. But how to manipulate those bits to discover which bit (digit) has error ? A technique was developed by R.W. Hamming.

Positions of the redundancy bits

The Hamming code can be applied to data units of any length by using the equation (1.1).

For example, a seven bit ASCII code requires four redundancy bits, that can be added to the end of the data unit or interpersed with original data bits. Generally, these bits are placed in a positions 1, 2, 4 and 8 (powers of 2). These bits are referred as r_1, r_2, r_4 and r_8 (see Fig. 1.17).

11	10	9	8	7	6	5	4	3	2	1
d_7	d_6	d_5	r_8	d_4	d_3	d_2	r_4	d_1	r_2	r_1

Fig. 1.17 Positions of redundancy bits in Hamming code.

In the Hamming code, each r bit is the vertical redundancy check bit for one combination of data bits.

r_1 is selected so as to establish even parity in bit positions 1, 3, 5, 7, 9, 11

r_2 is selected to establish even parity in bit positions 2, 3, 6, 7, 10, 11

r_4 is selected to establish even parity in bit positions 4, 5, 6, 7

r_8 is selected to establish even parity in bit positions 8, 9, 10, 11

To see the pattern behind this strategies, the r_1 bit is calculated using all bit positions whose binary representation includes 1 in the right most position (at weight equal to 1) and r_2 is calculated using all bit positions with 1 in second position (at weight equal to 2) and so on (see Table 1.21).

Table 1.21. Redundancy bits calculation

Position number	*Binary rule*			
	r_8	r_4	r_2	r_1
1	0	0	0	1
2	0	0	1	0
3	0	0	1	1
4	0	1	0	0
5	0	1	0	1
6	0	1	1	0
7	0	1	1	1
8	1	0	0	0
9	1	0	0	1
10	1	0	1	0
11	1	0	1	1

See Table 1.21, under column r_4 the bit 1 contains by the positions 4, 5, 6 and 7 only as explained earlier.

Error detection and Error Correction

Now, as soon as the data unit is received, the receiver starts calculating new vertical redundancy checks using some set of bits used by the sender plus relevant parity (r) bit for each set. If in any combination, the number of 1's are even number, assign the corresponding 'c' value as '0' else '1'. Like this, after knowing the values of c_8, c_4, c_2 and c_1 the error bit location can be identified by c's position ($c_8c_4c_2c_1$). All the values of c's is zero, indicates no error has occurred.

Example 1.85. *Cosider the message is 1001101 is transmitted through the channel, obtain the redundancy bits and transmitting unit needed. Assume bit number 8 has been changed. How to locate it.*

Solution.

Data bits = 1001101

Total number of data bits = m = 7

The number of redundancy bits required r = 4 (from Table 1.20).

Total number of bits in a transmitted data = $m + r = 7 + 4 = 11$

Positions of redundancy bits is 1, 2, 4 and 8. (r_1, r_2, r_4 and r_8).

Calculation of *r*'s:

r_1 is calculated so that the bit positions 1, 3, 5, 7, 9, 11 contains an even number of 1's.

Positions	11	10	9	8	7	6	5	4	3	2	1
	m_7	m_6	m_5	r_8	m_4	m_3	m_2	r_4	m_1	r_2	r_1
Original data	1	0	0		1	1	0		1		
r_1 (to give even-parity at positions 1, 3, 5, 7, 9 and 11)	1		0		1		0		1		1
r_2 (to give even-parity at positions 2, 3, 6, 7, 10, 11)	1	0			1	1			1	0	
r_4 (to give even-parity at positions 4, 5, 6, 7)					1	1	0	0			
r_8 (to give even-parity at positions 8, 9, 10, 11)	1	0	0	1							

coded message = 1001 11 00101

As per the problem, the 8th bit is changed from 1 to 0. *i.e.*, the message is received as 10001 00101. To calculate the position of error occurred at the receiver, recalculate all the values of c_1, c_2, c_4, c_8 with the received data.

Positions	11	10	9	8	7	6	5	4	3	2	1	
Message received	1	0	0	0	1	1	0	0	1	0	1	
1, 3, 5, 7, 9, 11 bits even parity check	1		0		1		0		1		1	c_1 = 0 since even parity
2, 3, 6, 7, 10, 11 bits even parity check	1	0			1	1			1	0		c_2 = 0 since even parity
4, 5, 6, 7 bits even parity check					1	1	0	0				c_4 = 0 since even parity
8, 9, 10, 11 bits even parity check	1	0	0	1								c_8 = 1 since odd parity

Ther error location = $c_8\ c_4\ c_2\ c_1$

1 0 0 0

which means that the location of error is in position 8. To correct the error, the digit in position 8 is complemented and the correct message 10011100101 is obtained.

Example 1.86. *The Hamming code 101101101 is received. Correct it if any errors. There are four parity bits and odd parity is used.*

Solution.

Bit positions	9	8	7	6	5	4	3	2	1	
Bit designation	m_5	r_8	m_4	m_3	m_2	r_4	m_1	r_2	r_1	Error position
Received data	1	0	1	1	0	1	1	0	1	Bit value
1, 3, 5, 7 and 9 bits for odd parity check	1		1		0		1		1	$c_1 = 1$ since odd parity
2, 3, 6 and 7 bit positions checking for odd parity			1	1			1	0		Number of 1s is odd: $c_2 = 0$
Bit positions 4, 5, 6 and 7 for checking odd parity			1	1	0	1				Number of 1s is odd: $c_4 = 0$
Bit positions 8 and 9 for checking odd parity	1	0								Number of 1s is odd: $c_8 = 0$

∴ The resultant word, that is, error position $= c = c_8\, c_4\, c_2\, c_1 = 0001$. This says that the bit in the number 1 location is in error. First bit position value should be 0 instead of 1. Hence the correct code is 101101100 and the message is 11101.

Example 1.87. *Determine which bit, if any, is in error in the even parity. Hamming coded character is 1100111. Decode the message.*

Solution.

Bit positions	7	6	5	4	3	2	1	Error position
Bit designation	m_4	m_3	m_2	r_4	m_1	r_2	r_1	Bit value
Received data	1	1	0	0	1	1	1	
Bit positions 1, 3, 5, and 7 for even parity check	1		0		1		1	$c_1 = 1$
Bit positions 2, 3, 6 and 7 for even parity check	1	1			1	1		$c_2 = 0$
Bit positions 4, 5, 6 and 7 for even parity check	1	1	0	0				$c_4 = 0$

The resultant word, that is, error position

$c = c_4\ c_2\ c_1 = 001.$

This says that the bit in the number 1 location is in error. First bit positional value should be 0 instead of 1. Hence the correct code is 1100110 and the message is 1101.

Example 1.88. *Message has been coded in Hamming code for BCD and transmitted through a noisy channel. Decode the message assuming that at most a single error has occurred in each codeword. 1001001011100111101100011011.*

Solution. The given bit stream has four Hamming coded BCD codes of 7 bits each. Therefore, the received words are

1. 1001001   2. 0111001
3. 1110110   4. 0011011

For finding the error positions in each word, each word can be checked as in Ex. 1.87 and then get the correct data.

Example 1.89. *Encode the information character 01101110101 according to the 15-bit hamming code.*

Solution. The number of bits in the message is 11, hence number of redundant bits needed is 4. The location of the parity bits are (1, 2, 4, and 8). As we know that parity bits are assigned by checking message bits having 1 in the same location as parity bit in their binary location numbers.

Bit designation	m_{11}	m_{10}	m_9	m_8	m_7	m_6	m_5	r_8	m_4	m_3	m_2	r_4	m_1	r_2	r_1
	Hamming coded message														
Bit location	15	14	13	12	11	10	9	8	7	6	5	4	3	2	1
	= 011011110101110														
Data bits	0	1	1	0	1	1	1		0	1	0		1		
r_1 (to give even parity at positions 1, 3, 5, 7, 9, 11, 13, 15)	0		1		1		1		0		0		1		0
r_2 (to give even parity at positions 2, 3, 6, 7, 10, 11, 14, and 15)	0	1			1	1			0	1			1	1	
r_4 (to give even parity at positions 4, 5, 6, 7, 12, 13, 14 and 15)	0	1	1	0					0	1	0	1			

r_8 (to give even parity at positions 8, 9, 10, 11, 12, 13, 14 and 15)	0	1	1	0	1	1	1	1							

1.17 SOLVED PROBLEMS

Problem 1. *List the first 16 numbers in base 12. Use the letters A and B to represent the last two digits.*

Solution. 0, 1, 2, 3, 4, 5, 6, 7, 8, 9, A, B, 10, 11, 12, 13.

Problem 2. *What is the largest binary number that can be obtained with 16 bits ? What is its decimal equivalent ?*

Solution. The largest 16-bit binary number is 1111 1111 1111 1111 and its decimal equivalent = 65,535.

Problem 3. *Convert the following numbers.*

(a) *101110, 1110101.11, 110110100 and 11001010.0101 to base 10*

(b) *$(.121\ 21)_3$, $(4310)_5$, $(50)_7$, $(198)_{12}$, $(1431)_8$ to base 10*

(c) *(1231), (673.23), 104, 2004, $(3.1415...)_{10}$ and 175.175 to base 2.*

(d) *Convert the following decimal numbers to the indicated bases*

(i) 7562.45 to octal *(ii) 1938.257 to hexadecimal*

(iii) (1776) to base 6 *(iv) 1984 to base 8.*

(v) 3.1415... to base 8. *(vi) 0.875 into binary trinary and quinary.*

(e) *Convert the following binary numbers into octal, hexadecimal and base-4 number system.*

(i) 11011 *(ii) 100110.11001* *(iii) 1111000.0001*

(f) *Convert the following octal number into binary, hexadecimal and decimal*

(i) 347 *(ii) 444* *(iii) 576.402* *(iv) 1431*

(g) *Convert the following hexadecimal numbers into binary, octal and decimal.*

(i) AC *(ii) BD* *(iii) 2002.A3*

Solution.

(a)

$$101110 = 32 + 8 + 4 + 2 = 46$$

$$111010.11 = 32 + 16 + 8 + 2 + 0.5 + 0.25$$

$$= 58.75$$

$$110110100 = 256 + 128 + 32 + 16 + 4$$

$$= 336$$

$$11001010.0101 = 128 + 64 + 8 + 2 + 0.25 + 0.0625$$
$$= 202.3125$$

(*b*) $$(12121)_3 = 1 \times 3^0 + 2 \times 3^1 + 1 \times 3^2 + 2 \times 3^3 + 1 \times 3^4$$
$$= 1 + 6 + 9 + 54 + 81 = 151$$
$$(4310)_5 = 4 \times 5^3 + 3 \times 5^2 + 1 \times 5^1 + 0 \times 5^0$$
$$= 500 + 75 + 5 = 580$$
$$(50)_7 = 7^1 \times 5 + 0 \times 7^0 = 35$$
$$(198)_{12} = 1 \times 12^2 + 9 \times 12^1 + 8 \times 12^0$$
$$= 144 + 108 + 8 = 260$$
$$(1431)_8 = 1 \times 8^3 + 4 \times 8^2 + 3 \times 8^1 + 1 \times 8^0$$
$$= 512 + 256 + 24 + 1 = 793$$

(*c*) $$(1231)_{10} = 10011001111$$
$$(673.23)_{10} = 1010100001.001110$$
$$10^4 = (10000)_{10} = 10011101110100$$
$$2004 = 11111010100$$
$$(3.1415...)_{10} = 0011.0010001$$
$$(175.175)_{10} = 10101111.00101$$

(*d*) (*i*)

8	7562	9
	7560	−2
8	945	−1
8	118	6
8	14	6
	1	

$0.45 \times 8 = 3.60 \rightarrow$ Carry 3
$0.60 \times 8 = 4.8 \rightarrow$ Carry 4
$0.8 \times 8 = 6.4 \rightarrow$ Carry 6
$0.48 \times 8 = 3.2 \rightarrow$ Carry 3

$= 16612.3463$

(*ii*)

16	1938	
16	121	−2
	7	−10

$= 7A2$

$0.257 \times 16 = 4.112 \rightarrow$ Carry of 4
$0.112 \times 16 = 1.792 \rightarrow$ Carry of 1
$0.792 \times 16 = 12.672 \rightarrow$ Carry of C
$0.672 \times 16 = 10.752 \rightarrow$ Carry of 4
$= 0.41CA$

$\Rightarrow (7A2.41CA)_{16}$

(*iii*)

$$\begin{array}{r|l l} 6 & 1776 & \\ \hline 6 & 296 & -0 \\ \hline 6 & 49 & -1 \\ \hline & 8 & -1 \end{array} \qquad \therefore (1776)_{10} = (8110)_6$$

(*iv*)

$$\begin{array}{r|l l} 8 & 1984 & \\ \hline 8 & 248 & -0 \\ \hline 8 & 31 & -0 \\ \hline & 3 & -7 \end{array} \qquad \therefore (1984)_{10} = (3700)_8$$

(*v*)

Fot integer : $(3)_{10} = (3)_8$

For fraction : $0.4115 \times 8 = 1.1320 \rightarrow \text{carry} = 1$

$0.1320 \times 8 = 1.056 \rightarrow \text{carry} = 1$

$\therefore \quad (3.1415...)_{10} = (3.11)_8$

(*vi*)

Into binary

$0.875 \times 2 = 1.75 \rightarrow \text{carry} = 1$

$0.75 \times 2 = 1.50 \rightarrow \text{carry} = 1$

$0.5 \times 2 = 1.0 \rightarrow \text{carry} = 1$

$\therefore \quad (0.875)_{10} = (0.111)_2$

Into trinary

$0.875 \times 3 = 2.625 \rightarrow \text{carry of } 2$

$0.625 \times 3 = 1.875 \rightarrow \text{carry of } 1$

$0.875 \times 3 = 2.625 \rightarrow \text{carry of } 2$

$\therefore \quad (0.875)_{10} = (0.212)_8$

Into quinary

$0.875 \times 4 = 3.5 \rightarrow \text{carry of } 3$

$0.5 \times 4 = 2.0 \rightarrow \text{carry of } 2$

$\therefore \quad (0.875)_{10} = (0.32)_4$

(*e*) (*i*) 11011 into octal $= (33)_8$

11011 into hexadecimal $= (1B)_{16}$

11011 into base-4 $= (123)_4$

(*ii*) 100110.11001 into octal $= (46.62)_8$

into hexadecimal $= (26.C4)_{16}$

in base-4 $= (212.302)_4$

(*iii*) 1111000.0001 into octal = (170.04)

into hexadecimal = 78.1

into quinary = 1320.01

(*f*) (*i*) $(347)_8$ in binary = $(011\ 100\ 111)_2$

in hexadecimal = $(E7)_{16}$

in decimal = $(231)_{10}$

(*ii*) 444 in binary = 100 100 100

in hexadecimal = 124H

in decimal = $(292)_{10}$

(*iii*) $(576.402)_8$ in binary = $(101\ 111\ 110.100\ 000\ 010)_2$

in hexadecimal = 17E.81H

in decimal = $(382.315)_{10}$

(*iv*) $(1431)_8$ in binary = 001 100 011 001

in hexadecimal = 319H

in decimal = $(793)_{10}$

(*g*) (*i*) $(AC)_{16}$ in binary = 1010 1100

in octal = $(254)_8$

in decimal = $(172)_{10}$

(*ii*) $(BD)_{16}$ in binary = 1011 1101

in octal = $(275)_8$

in decimal = 189

(*iii*) (2002.A3) in binary = 0010000000000010.10100011

in octal = $(20002.506)_8$

in decimal = $(8194.81)_{10}$

Problem 4. *Determine the value of b (base) if*

(*a*) $(16)_{10} = (100)_b$ (*b*) $(292)_{10} = (1204)_b$

(*c*) $(211)_b = (152)_8$ (*d*) $(33)_{10} = (201)_b$

(*e*) 23 + 44 + 14 + 32 = 223

Solution.

(*a*)
$$(16)_{10} = 10\times1 + 6 = 16$$
$$(100)_b = b^2 + 0 + 0 = b^2$$
$$\therefore b^2 = 16 \qquad \boxed{\therefore b = 4}$$

(*b*)
$$(1204)_b = b^3 + 2b^2 + 4 = (292)_{10}$$
$$b^3 + 2b^2 = 288$$
$$b^2(b+2) = 288 \qquad \boxed{\therefore b = 6}$$

(*c*)
$$2b^2 + b + 1 = 8^2 + 8 \times 5 + 2$$
$$2b^2 + b + 1 = 64 + 40 + 2 = 106$$

$$2b^2 + b - 105 = 0$$

$$b = \frac{-1 \pm \sqrt{1+4\times2\times105}}{4} = \frac{-1\pm 29}{4}$$

$$= \frac{28}{4} \text{ or } \frac{-30}{4}$$

$$= 7 \text{ or } -7.5 \quad \boxed{\therefore\ b = 7} \text{ since base must be positive}$$

(*d*)

$$(33)_{10} = (201)_b$$

$$2b^2 + 1 = 33$$

$$2b^2 = 32$$

$$b^2 = 16$$

$$\therefore\ b = 4$$

(*e*)

$$23 + 44 + 14 + 32 = 223$$

$$2b + 3 + 4b + 4 + b + 4 + 3b + 2 = 2b^2 + 2b + 3$$

$$10b + 12 = 2b^2 + 2b + 3$$

$$2b^2 - 8b - 10 = 0$$

$$2b^2 - 10b + 2b - 10 = 0$$

$$2b\ (b - 5) + 2\ (b - 5) = 0$$

$$(b - 5)\ (2b + 2) = 0$$

$$b = 5 \quad \text{or} \quad b = 1$$

$$\therefore \quad \boxed{b = 5} \text{ since base must be positive.}$$

Problem 5. *Find the value of x in the following :*

(*a*) $(4085.025)_{10} = (x)_{15}$

(*b*) $(789)_{16} + (473)_8 = (x)_4$

(*c*) $(1256)_8 = (x)_2$

(*d*) $(AC.2)_{16} = (x)_8$

(*e*) $(19.125)_{10} = (x)_8$

(*f*) $(10011.11)_2 = (x)_{16}$

(*g*) $(432)_5 = (x)_7$

Solution.

(*a*) $(4085.025)_{10} = (1325.059)_{15}$

$\therefore \quad x = 1325.059$

(*b*) $(2066)_{10} + (315)_{10} = (2381)_{10}$

$= (211031)_4$

$\therefore \quad x = 211031$

(c) $(1256)_8 = (001\ 010\ 101\ 110)_2$

$\therefore$ $x = 1010101110$

(d) $(AC.2)_{16} = 101011000010$

$= (254.1)_8$

$\therefore$ $x = 254.1$

(e) $(19.125)_{10} = (23.1)_8$

$\therefore$ $x = 23.1$

(f) $(10011.11)_2 = (13.C)_{16}$

$\therefore$ $x = 13.C$

(g) $(432)_5 = (117)_{10}$

$= (225)_7$

$\therefore$ $x = 225$

Problem 6. *Each of the following arithmetic operation is correct in at least one number system. Determine the possible bases of the numbers in each operation.*

(a) $1234 + 5432 = 6666$

(b) $\sqrt{41} = 5$

(c) $\dfrac{302}{20} = 12.1$

Solution. (a) $1234 + 5432 = 6666$

This arithmetic operation is decimal arithmetic operation. Hence the base for this number system = 10. The possible bases are b = 7 to any value

(b)

$$\sqrt{41} = 5$$

$$\sqrt{4b+1} = 5$$

$$4b + 1 = 25$$

$$4b = 24$$

$\therefore$ $b = 6$ since b must be greater than 1.

(c)

$$\frac{302}{20} = 12.1$$

$$\frac{3b^2+2}{2b+0} = b + 2 + \frac{1}{b}$$

$$\frac{3b^2+2}{2b} = \frac{b^2+2b+1}{b}$$

$$3b^2 + 2 = 2b^2 + 4b + 2$$

$$b^2 - 4b = 0$$

$$b\,(b - 4) = 0$$

$\therefore$ $b = 0$ or $b = 4$ since b must be greater than 1.

Problem 7. *(a) Given the binary numbers a = 10101.1, b = 101.01, c = 1001.1. Perform the following binary operations :*

(i) a + c *(ii) a – b* *(iii) a.c* *(iv) a/b*

(b) Perform the following division in binary : 11111111/101

(c) Calculate x – y for the following cases

(i) x = 11.01 and y = 100.1

(ii) x = 100001 and y = 110

(iii) x = 1001.01 and y = 11.1

Solution.

(a) (i) a + c = 1010.1

$$\frac{\begin{array}{r}1010.1\\1001.1\end{array}}{10100.0} = 10100$$

(ii) a – b = 1010.10

$$\frac{\begin{array}{r}1010.10\\101.01\end{array}}{101.01} = 101.01$$

(iii)

$$a.c = \frac{1010.1 \times 1001.1}{\begin{array}{r}10101\\10101\\00000\\00000\\10101\end{array}}$$

$$\overline{1100011.11} = 1100011.11 = (99.75)_{10}$$

(iv) 1010.1/101.01

$$\begin{array}{r} 10\\ 10101\overline{)101010}\\ \underline{10101}\\ 0000\\ \underline{0000}\\ 0000 \end{array} = (10)_2$$

(b)

```
101)11111111 = 110011 = 51
    101
    ---
     101
     101
     -----
     00111
       101
       ---
       101
       101
       ---
       000
```

(c) (i) $x = 111.01$ $\quad \therefore x - y = 111.01$

$y = 100.10$ $\qquad\qquad\qquad\quad 100.10$

$\qquad\qquad\qquad\qquad\qquad 010.11$

$= (10.11)_2 = 2.75$

(ii) $x = 100001$ $\quad \therefore x - y = 100001$

$y = 110$ $\qquad\qquad\qquad\quad 110$

$\qquad\qquad\qquad\qquad\qquad 011011$

$= 11011 = (27)_{10}$

(iii) $x = 1001.01$ $\quad \therefore x - y = 1001.01$

$y = 11.1$ $\qquad\qquad\qquad\quad 11.10$

$\qquad\qquad\qquad\qquad\qquad 0101.11$

$= 101.11 = (5.75)_{10}$

Problem 8. *Noting that $3^2 = 9$, formulate a simple procedure for converting base–3 numbers directly to base–9. Use the procedure to convert $(2110201102220112)_3$ to base 9*

Solution. If we want to convert a number from base–3 to base–9, we have to group two bits in base–3 and write its equivalent base–9 value.

$(21\ 10\ 20\ 11\ 02\ 22\ 01\ 12)_3$

$= 6{+}1\ \ 3{+}0\ \ 6{+}0\ \ 3{+}1\ \ 0{+}2\ \ 6{+}2\ \ 0{+}1\ \ 3{+}2$

$= (73642815)_9$

Problem 9. *Perform the following subtraction operations using*

(a) direct binary subtraction	*(i) 101101 – 111*
(b) one's complement arithmetic	*(ii) 101– 110111*
(c) two's complement arithmetic	*(iii) 1001 – 11*

Solution.

(*a*) By direct substraction

(*i*) minuend = 101101

subtrahend = 111

result = 100110 = $(38)_{10}$

(*ii*) minuend = 000101

subtrahend = 110111

result = –110010 = $-(50)_{10}$

(*iii*) minuend = 1001

subtrahend = 11

result = 0110 = $(6)_{10}$

(*b*) 1's complement subtraction

(*i*) minuend = 101101

1's complement of subtrahend = 111000

= ①100101

At the EAC is generated

add it to the result = 100101

1

100110 = $(38)_{10}$

(*ii*) minuend = 000101

1's complement of subtrahend = 001000

001101

There is no carry, so the result is a negative number, the true answer can be obtained by taking the 1's complement of the result.

1's complement of (001101) = –110010 = $-(50)_{10}$

(*iii*) minuend = 1001

1's complement of subtrahend = 1100

①0101

At the carry is generated

add it to the result = 0101

1

0110 = $+(6)_{10}$

(*c*) 2's complement subtraction

(*i*) minuend = 101101

2's complement of the subtrahend = 111001

(000111) = ①100110

Here carry is generated, discard it

$\therefore$ result = 100110 = $+(38)_{10}$

(*ii*) minuend = 000101

2's complement of the subtrahend = 001001

001110

No carry is generated, hence the result is a negative number, which is in 2's complement form. The true answer can be obtained by finding the 2's complement of the result and placing a minus sign

= –110010 = $-(50)_{10}$

(*iii*) minuend = 1001

2's complement of the subtrahend = 1101

①0110

Here EAC is generated, discard it

result = 0110 = $+(6)_{10}$

Problem 10. *Perform the following subtraction operation using*

(i) 9's complement (ii) 10's complement

(a) 120 – 118 (b) 127 – 128 (c) 1– 3

Solution. (*a*) (*i*) 9's complement of 118 = 881

minuend = 120

subtrahend = 881

add EAC ①001

1

002

$\therefore$ $(2)_{10}$

(*ii*) 10's complement of 118 = 882

minuend = 120

subtrahend = 882 = $(2)_{10}$

discard carry ①002

(*b*)(*i*) 9's complement of 128 = 871

minuend = 127 – A

subtrahend = 871

998

No carry, $\therefore$ True result = 9's complement of 998 = 001 = $(-1)_{10}$

(*ii*) 10's complement of 128 = 872

minuend = 127

subtrahend = 872

999

No carry, ∴ True result = 10's complement of 999 = $(-1)_{10}$

(*c*) (*i*) 9's complement of 3 = 6

minuend = 1

subtrahend = 6

7

No carry, ∴ The true answer = 9's complement of 7 = 9 – 7 = $(-2)_{10}$

(*ii*) 10's complement of 3 = 7

minuend = 1

subtrahend = 7

8

No carry, ∴ The true answer = 10's complement of 8

= $(-2)_{10}$

Problem 11. *Perform the following arithmetic operations using 2's complement presentation for –ve numbers.*

(a) $58_{10} - 33_{10} + 14_{10} + (-45)_{10}$

(b) $(+42)_{10} + (-13)_{10} - (60)_{10} + 36_{10}$

Solution. (*a*)

Binary value of 58 = 111010

2's complement of 33 = 011111

Binary value of 14 = 001110

2's complement of 45 = 010011

①111010

To get the true find the answer 2's complement of (111010) and put (–) sign

= 000110 = $(-6)_{10}$

(*b*) Binary value of 42 = 101010

2's complement of 13 = 110011

2's complement of 60 = 000100

100100

Binary value of 36 =

Discard carry ①000101

∴ Answer = 101 = 5_{10}

Problem 12. *In the following series, same integer is expressed in different number system. Determine the missing number of the series.*

(a) 10000, 121, 100, ?, 24, 22, 20....

(b) 1110, 112, 32, 24, ?, 20, 16....

(c) 101001, 210, 111, 41, 33, ?, 25, 23....

Solution. (*a*)

The given number in binary (base–2) = 10000 = $(16)_{10}$

The given number in binary base–3 = 121 = $(16)_{10}$

The given number in binary base–4 = 100 = $(16)_{10}$

The given number in binary base–5 = $\frac{16}{5}$ = (31) = $(16)_{10}$

The given number in binary base–6 = 24 = 16_{10}

∴ missing number = $(31)_5$

(*b*) The first number in the series is $(1110)_2$

Its equivalent decimal = 14

Similarly, the second number is in base-3

∴ $(112)_3 = (14)_{10}$

The third number is in base-4

∴ $(32)_4 = (14)_{10}$

The fourth number is in base-5

∴ $(24)_5 = 14$

∴ The missing number will be in base-6

hence $(14)_{10} = (?)_6$

∴ missing number = $(22)_6$

(*c*) Here, the first, second and so on numbers are in base-2, base-3 ....

The decimal equivalent of any number is,

ex: $(210)_3 = (21)_{10}$

∴ The missing number is in base-7

hence $(21)_{10} = (x)_7$

∴ $x = 30$

∴ missing number = $(30)_7$

Problem 13. *Encode each of the 10 decimal digits 0, 1, ..., 9 by means of the following weighted binary codes.*

7 3 2 –1, 5 4 –2 –1, 8 7 –4 –2, 6 3 1 –1

Solution.

Decimal number	7 3 2 –1	5 4 –2 –1	8 7 –4 –2	6 3 1 –1
0	0 0 0 0	0 0 0 0	0 0 0 0	0 0 0 0
1	0 0 1 1	0 1 1 1	0 1 1 1	0 0 1 0
2	0 0 1 0	0 1 1 0	1 0 1 1	0 1 0 1
3	0 1 0 0	0 1 0 1	0 1 1 0	0 1 0 0
4	0 1 1 1	0 1 0 0	1 0 1 0	0 1 1 0
5	0 1 1 0	1 0 0 0	0 1 0 1	1 0 0 1
6	1 0 0 1	1 1 1 1	1 0 0 1	1 0 0 0
7	1 0 0 0	1 1 1 0	0 1 0 0	1 0 1 0
8	1 0 1 1	1 1 0 1	1 0 0 0	1 1 0 1
9	1 1 0 1	1 1 0 0	1 1 1 1	1 1 0 0

Among the above codes, '87 – 4 –2' is the self complementing code.

Problem 14. *(a) Prove that in every positively weighted code, one of the weight must be 1, second weight must be either 1 or 2, and the sum of weights must be equal to or greater than 9.*

(b) Show by listing all such codes that there are only 17 positively weighted codes, of which only 4 are, self-complementing.

Solution. (*a*) In a positively weighted code, the only way to represent the decimal integer 1 is to have atleast one weight with value 1 and the way to represent decimal 2 is by having one weight 2 (or) by having one more weight (second weight) 1.

Clearly, to represent decimal 9, the sum of the weights must be equal to or greater than 9.

(*b*) The possible weighted codes to represent decimal numbers are

(*a*) 5211	(*b*) 4311	(*c*) 5311	(*d*) 6311	(*e*) 4221
(*f*) 5221	(*g*) 6221	(*h*) 3321	(*i*) 4321	(*j*) 5321
(*k*) 6321	(*l*) 7321	(*m*) 4421	(*n*) 5421	(*o*) 6421
(*p*) 7421	(*q*) 8421			

Among the above codes, code (*a*), (*b*), (*e*) and (*h*) are sefl-complementing codes. Their sum of weights is 9.

Problem 15. *(a) Prove that in a self-complementing code, the sum of the weights must be 9.*

(b) Show the weights of three different four-bit self-complementing codes whose only negative weight is – 4.

Solution. (*a*) Assume that weights are $w_3\ w_2\ w_1$ and w_0.

∴ If sum of weights is greater than 9

i.e., $w_3 + w_2 + w_1 + w_0 > 9$, then the complement of zero is greater than 9, which indicates the violation of self-complementing code. If the sum is less than 9 *i.e.*, $w_3 + w_2 + w_1 + w_0 < 9$, which indicates that the complement of zero is less than 9, then that type of code is also not a valid one and the sum of weights must be 9 (at least) to represent all the decimal digits.

(*b*) 7 5 1–4, 8 3 2 –4 and 6 5 2 –4 are three self-complementing weighted codes with negative weight.

Problem 16. *Represent the following decimal numbers in BCD*

(i) 13597 *(ii) 93286* *and* *(iii) 99880*

Solution. (*i*) 13597 in BCD is = 00010011010110010111

(*ii*) 93286 in BCD is = 10010011001010000110

(*iii*) 99880 in BCD is = 10011001100010000000

Problem 17. *Determine the binary code for each of the ten decimal digits using a weighted code with weights 7, 4, 2 and 1.*

Solution.

Decimal digit	*7421 code*
0	0000
1	0001
2	0010
3	0011
4	0100
5	0101
6	0110
7	1000 (or) 0111
8	1001
9	1010

Problem 18. (*a*) *Obtain the (r – 1)'s complement of (base = 6)* $(452310)_6$

(b) Design a 3-bit code to represent each of the six digits of base–6 number system. Make the binary code self-complementing so that the 5's complement which is obtained by changing 1's to 0's and 0's to 1's in all the bits of the coded number.

Solution. (*a*) The 5's complement of (452310) is obtained by subtracting each digit from 5. = (103245)

(*b*) The 3–bit code of the base–6 number system is 000, 001, 010, 101, 110, 111 for the digits 0, 1, 2, 3, 4 and 5 with the weights of '221'.

Problem 19. *The following were suggested as the first code words in a cyclic code. In each case, either complete the code or show that it cannot be completed. Each code sequence must contain the set of all possible code words and the last code word must be unit distance from the first.*

(a) 000, 001, 0011, 111

(b) 000, 010, 011, 111, 101

(c) 000, 010, 110, 111

(d) 0000, 0100, 0101, 1101, 1111, 1011, 1010.

Solution. To find whether the given codes are completed or not completed, map tech is used and assumed that the three variables are ABC.

(*a*)

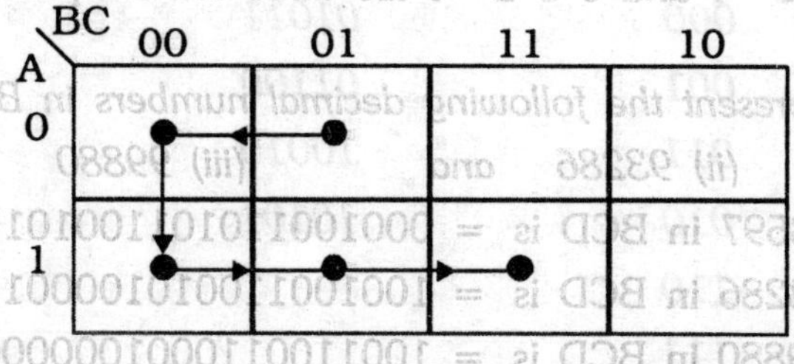

From map, we can say that the codes belong to completed (cyclic) by adding the remaining codes 101, 100, 110, 010.

(*b*)

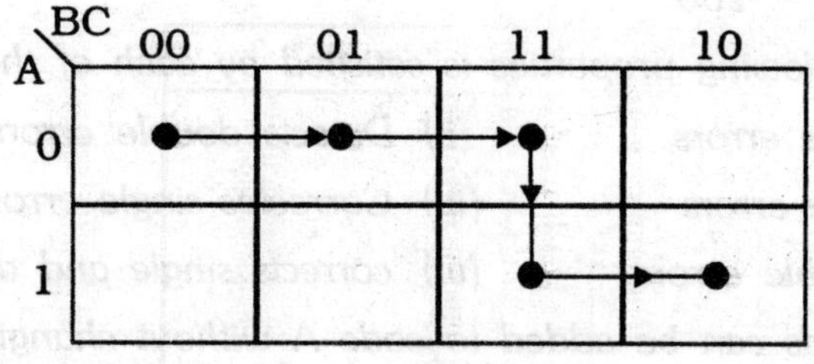

This code cannot be completed, since 010 is not adjacent to other 100 (or) 101.

(*c*)

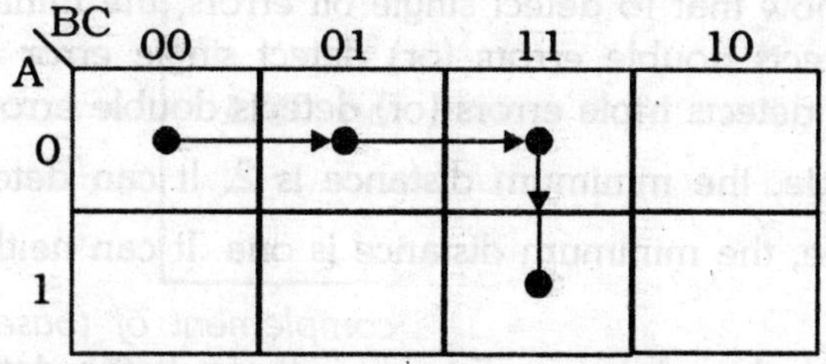

This code can be completed by adding the remaining codes. 010, 110, 100 and 101 since each code is having atleast one adjacent.

(*d*)

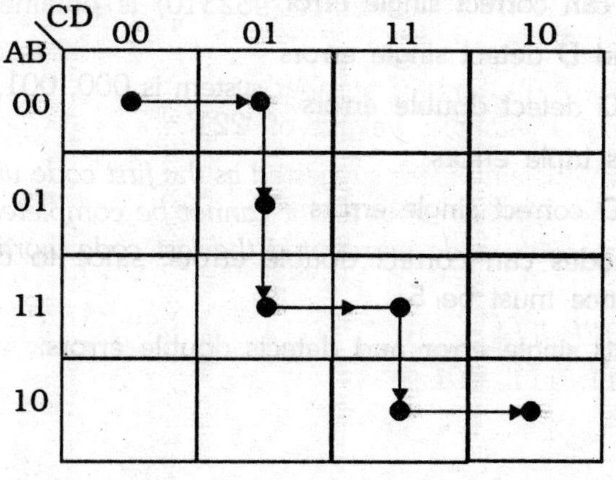

This code can be completed by adding the remaining codes 0011, 0010, 0100, 0110, 0111, 1000, 1001, 1100, 1110, since each remaining code is having atleast one adjacent.

Problem 20. *Consider the following four codes.*

Code A	*Code B*	*Code C*	*Code D*
0001	*000*	*01011*	*000000*
0010	*001*	*01100*	*001111*
0100	*011*	*10010*	*110011*
1000	*010*	*10101*	
	110		
	111		
	101		
	100		

(a) Which of the following properties is satisfied by each of the above codes.

(i) Detects single errors *(ii) Detects double errors*

(iii) Detects triple errors *(iv) Correctes single errors*

(v) corrects double errors *(vi) corrects single and detects double errors.*

(b) How many words can be added to code A without changing it's error-detection and correction capabilities ? Give a possible set of such words. Is this set unique ?

Solution. (*a*) We know that to detect single bit errors, the minimum distance = 2. If min-distance is 3, then it detects double errors (or) detect single error and correct single error. If minimum distance is 4, it detects triple errors (or) detects double errors and corrects single error.

Code A : In this code, the minimum distance is 2. It can detect single bit error.

Code B : In this code, the minimum distance is one. It can neither detect nor correct error bits.

Code C : In this code, the minimum distance is three. It can detect double errors, or detect one single error and correct single error.

Code D : In this code, the minimum distance is four. It can detect triple errors or detect double errors and can correct single error.

(*i*) Codes A, C and D detect single errors

(*ii*) Codes C and D detect double errors

(*iii*) Code D detects triple errors

(*iv*) Codes C and D correct single errors

(*v*) None of the codes can correct double errors, since to connect double errors, the minimum distance must be 5.

(*vi*) Code D corrects single error and detects double errors.

(*b*) There are four possible code words, that can be added to code A, without changing its error-detection and correction capability. Those are 1101, 0111, 1011 and 1110. This set is unique.

Problem 21. *Construct addition and multiplication tables for base–2 and base–3.*

Solution. Additionable in base–2

+	0	1
0	0	1
1	1	0

and carry = 1

Multiplication in base–2

×	0	1
0	0	0
1	0	1

Addition table in base–3 : Here, assume the two numbers are A and B, and their values may be 0, 1 or 2. After adding, two outputs are produced sum (*s*) and carry (*c*),.They are also in base–3.

A	*B*	*S*	*C*
0	0	0	0
0	1	1	0
0	2	2	0
1	0	1	0
1	1	2	0
1	2	0	1
2	0	2	0
2	1	0	1
2	2	1	1

Multiplication table for base–3

Here the output (*y*) is the product of inputs (A and B) and which is in base–3.

A	*B*	*Y* = *AB*
0	0	0
0	1	0
0	2	0
1	0	0
1	1	1
1	2	2
2	0	0
2	1	2
2	2	11

Problem 22. (*a*) *A person on SATURN possessing 18 fingers has a property worth* $(1{,}00{,}000)_{18}$. *He has 3 daughters and two sons. He wants to distribute half of the money equally to his sons and remaining half to his daughters equally. How much his each son and each daughter will get in Indian currency ?*

(*b*) *An Indian started on an expedition to SATURN with Rs. 1,00,000. The expenditure on SATURN will be in the ratio of 1 : 2 : 7 for food, clothing and travelling. How much he will be spending on each item in the currency of SATURN ?*

Solution. (*a*) Given property is in base 18.

But Indian currency is in base 10.

$$\Rightarrow (1{,}00{,}000)_{18} = 1 \times 18^5 + 0 \times 18^4 + 0 \times 18^3 + 0 \times 18^2 + 0 \times 18^1 + 0 \times 18^0$$
$$= 18{,}89{,}568$$

$\Rightarrow$ His property in Indian currency = Rs. 18,89,568

Of this money, one half is distributed equally between his *two* sons

$$\Rightarrow \quad \text{each son will get money} = \text{Rs.} \frac{18{,}89{,}568}{2} \times \frac{1}{2}$$
$$= \text{Rs. } 4{,}72{,}392$$

Other half is distributed equally between his *three* daughters

$$= \text{Rs.} \frac{18{,}89{,}568}{2} \times \frac{1}{3}$$
$$= \text{Rs. } 3{,}14{,}928$$

∴ Easch son will get Rs. 4,72,392 and each daughter will get Rs. 3,14,928.

(*b*) Total number of parts the money has to be divided into = 1 + 2 + 7 = 10.

Of this 1 part has to be spent for food.

$$\Rightarrow \quad \text{amount required for food} = \frac{1}{10} \times 1{,}00{,}000$$
$$= \text{Rs. } 10{,}000$$

$$\text{amount spent for clothing} = \frac{2}{10} \times 1{,}00{,}000$$
$$= \text{Rs. } 20{,}000$$

$$\text{amount spent for travelling} = \frac{7}{10} \times 1{,}00{,}000$$
$$= \text{Rs. } 70{,}000$$

But the currency of saturn is in base 18.

⇒	0	1	2	3	4	5	6	7	8	9
	10	11	12	13	14	15	16	17		
	↓	↓	↓	↓	↓	↓	↓	↓		
	A	B	C	D	E	F	G	H		

⇒ amount spent for food in currency of SATURN =

18	10,000	
18	555	–10
18	30	–15
	1	–12

$= (\text{ICFA})_{18}$

amount spent for clothing in currency of SATURN =

18	20,000	
18	1,111	–2
18	61	–13
	3	–7

$= (\text{37D2})_{18}$

amount spent for travelling in currency of SATURN=

18	70,000	
18	3,888	–16
18	216	–0
	12	–0

$= (\text{COOG})_{18}$

Problem 23. *The solution of the quadratic equation $x^2 - 11x + 22 = 0$ is $x = 3$ and $x = 6$. What is the base of the numbers ?*

Solution. Given quadratic equation is $x^2 - 11x + 22 = 0$

and also given, it has roots $x = 3$ and $x = 6$

in some base say 'a'.

$$\Rightarrow \quad (x - 3)(x - 6) = x^2 - 11x + 22$$

but 3 and 6 are same in any base, whose value is more than 6.

i.e.,

$$(3)_a = (3)_{10}$$

$$(6)_a = (6)_{10}$$

$$\Rightarrow \quad (x^2 - 9x + 18)_{10} \equiv (x^2 - 11x + 22)_a$$

$$\Rightarrow \quad (9)_{10} = (11)_a$$

$$\Rightarrow \quad a + 1 = 9$$

$$\Rightarrow \quad a = 8$$

(or)

$$(18)_{10} = (22)_a$$
$$18 = 2a + 2$$
$$\Rightarrow \quad 2a = 16$$
$$a = 8$$

1.18 REVIEW QUESTIONS

1. Convert the following numbers to the base indicated :

(i) $(1431)_8$ to base 10
(ii) $(0.4375)_{10}$ to binary
(iii) $(53.1575)_{10}$ to base 2
(iv) $(FACE)_{16}$ to binary
(v) $(11011.11)_2$ to decimal
(vi) $(0.00625)_{10}$ to base–2

2. Convert the following numbers to the base indicated :

(i) $(43.32)_{10}$ to binary
(ii) $(1984)_{10}$ to base 8
(iii) 11001010.0101 to base 10
(iv) $(3\ A\ 6)_{16}$ to decimal.

3. Solve for x

(i) $(1256)_8 = (x)_2$
(ii) $(19.125)_{10} = (x)_8$
(iii) $(AC.2)_{16} = x_8$
(iv) $(10011.11)_2 = x_{16}$
(v) $(432)_5 = (x)_7$
(vi) $(1431)_8 = (x)_{10}$
(vii) $(A98B)_{12}$ to $(x)_3$

4. Find the 9's and 10's complements of the following numbers :

(a) 6521
(b) 433
(c) $(935)_{11}$
(d) AC
(e) $(765)_8$

5. Perform the following binary arithmetic operations using 1's complement and 2's complements.

(a) 1101.1101 – 1011.10
(b) 101101 – 100001
(c) 01100 – 00011
(d) BD – AC
(e) $(642)_8 - (530)_8$

6. What are weighted and non-weighted codes ? Explain with an example.

7. Represent the decimal numbers 1234, 678.7 and 150.1 in

(i) binary
(ii) BCD
(iii) Excess-3 code
(iv) Gray Code

8. Explain error detecting and correcting codes with an example.

9. Write short notes on cyclic codes.

10. What is parity checking, explain its necessity and implementation.

11. Construct a seven-bit error-correcting code to represent the decimal digits by augmenting the excess-3 code and by using odd-1 parity check.

12. Determine which bit, if any, is in error in the Harming coded character 1100111. Decode the message.

13. The message below has been coded in the Harming code and transmitted through a noisy channel. Decode the message assuming that at most a single error has occurred in each code word.

1001001, 0111001, 1110110, 0011011

14. The message below has been coded in Harming code for BCD and transmitted through a noisy channel. Decode the message assuming that at most a single error has occurred in each code word.

1001001011100111101100011011.

15. Generate the even parity Harming codes for the following binary data

(*a*) 1101 (*b*) 1001

16. A seven bit Harming code is received as 1111101. What is the correct code ?

17. Give the classification of binary codes and explain.

18. Explain about self-complementing codes.

19. Solve for *x*

(*a*) $(48)_{10} = (120)_x$ (*b*) $(32)_{16} + (47)_8 + (14)_5 = (x)_6$

20. Each of the following arithmetic operators is correct in at least one number system. Determine the possible bases of the numbers in each operation.

(*i*) 11 + 23 + 47 = 103 (*ii*) 23 + 20 + 31 = 200

(*iii*) $\sqrt{24} = 4$ (*iv*) $\sqrt{34} = 5$

(*v*) $\sqrt{12} = 3$ (*vi*) $\frac{25}{15} = 1.222$

1.19 SHORT ANSWER QUESTIONS

1. Write a 4-bit binary number whose 2's complement is also same_______.

[**Ans.** 1000]

2. Is there any binary number whose 1's complement is same as original number

[**Ans.** No]

3. The base of a number system is _________

[**Ans.** It is an integer number and greater than 1, and indicates the number of symbols used in that number system.]

4. The value of a digit depends on its position in ________ number system.

[**Ans.** weighted]

5. Negative number can be represented in digital systems by using _______ methods.

[**Ans.** (*a*) sign-magnitude representation (*b*) 1's complement representation, (*c*) 2's complement representation]

6. In sign magnitude representation, _______ bit is used to represent sign.

[Ans. most significant bit]

7. In sign magnitude representation, positive sign is represented by _______ and negative sign is represented by ______

[Ans. positive sign is represented by '0' and negative sign by 1.]

8. 1's complement of a negative number is represented by _______

[Ans. $(2^n - 1 - n)$, where n = no. of bits in the number.]

9. 2's complement of a number is obtained by adding 1 to _______

[Ans. 1's complement of the number.]

10. The base of octal number system is _______

[Ans. 8]

11. The number of symbols used to represent a number in a *n*-base number system is ___

[Ans. (*n*)]

12. The maximum value of a symbol in a *n*-base number system is _______

[Ans. $n - 1$]

13. The 1's complement of a binary number system is obtained by changing

(*a*) each 0 as 1 (*b*) each 1 as 0 (*c*) both *a* and *b* (*d*) none

[Ans. both (*a*) and (*b*)]

14. _______ number system uses numerals and alphabets.

[Ans. Hexadecimal]

15. The decimal number 5 in excess-3 code can be written as _______

[Ans. 1000]

16. The number of bits required to represent a decimal number 16 in the binary form is ______

[Ans. 5]

17. The necessary condition for a weighted code to be self-complement code is ______

[Ans. the sum of it's weights must be equal to 9]

18. In the gray code, each successive code differs from it's preceding code by _______

[Ans. only one bit]

19. BCD code is _______ code

[Ans. weighted code]

20. Which of the following is not a valid octal number ?

(*a*) 732 (*b*) 64.15 (*c*) 82.3 (*d*) 11.11

[Ans. (*c*)]

21. The valid Hex number is _______

(*a*) 7H.32 (*b*) 8A.9B (*c*) 6G4.51 (*d*) 60I

[Ans. (*b*)]

22. The number system with radix 4 is called ______

[Ans. quinary]

23. ASCII code is used to represent ______

[Ans. digits, alphabets and special characters.]

24. The maximum negative number that can be represented in 2's complement form of *n*-bit number is ______ **[Ans.** -2^{n-1}]

25. The maximum positive number that can be represented by using 2's complement form (for *n*–bit number) is ______

[Ans. $(2^{n-1} - 1)$]

26. 101 + .101 = ______

[Ans. 1010]

27. The largest number that can be represented by using 8-bits is ______

[Ans. $2^8 - 1 = 255$]

28. The decimal equivalent of $(1101010)_2$ is ______

[Ans. 106]

29. Conversion from decimal to ______ requires repeated division by 8

[Ans. Octal]

30. Conversion from decimal to hex requires repeated division by ______

[Ans. 16]

31. In BCD code, each digit is converted into ______ no. bits in binary equivalent

[Ans. four]

32. $(614)_8$ in decimal is ______

[Ans. $(396)_{10}$]

33. $(72)_8$ in binary is ______

[Ans. 111010]

34. The number following (627) in octal is ______

[Ans. 630]

35. 1101001 in octal = ______

[Ans. $(151)_8$]

36. (2C) H in decimal = ______

[Ans. 44]

37. (37B) H in binary is ______

[Ans. 0011 0111 1011]

38. The next number of A9F in hexadecimal is ______

[Ans. AAO]

39. The binary number 110110110 = (______ $)_{16}$

[**Ans.** 1B6]

40. The main drawback of BCD representation is ______

[**Ans.** it requires more bits]

41. Advantage of BCD representation is ______

[**Ans.** conversion is easy]

42. Advantage of gray code is ______

[**Ans.** no intermediate states during transitions]

43. Application of gray code is ______

[**Ans.** in A to D conversions.]

44. Convert $(1010.101)_2$ to decimal

[**Ans.** 10.625]

45. Convert $(0.521)_{10}$ to octal =

[**Ans.** $(0.4126)_8$]

46. Convert $(10110001001001.\ 0010011100)_2$ to octal and hexadecimal.

[**Ans.** $(26111.116)_8$, $(2C49.27)_{16}$]

47. *r*'s complement of N = ______

[**Ans.** $r^n - N$]

48. 10's complement of 5.3276 = ______

[**Ans.** 4.6724]

49. 2's complement of (110011)2 = ______

[**Ans.** $(001101)_2$]

50. The complement of a complemented number is ______

[**Ans.** same number (or) original number]

51. Mention any one advantage of 2's complement over 1's complement.

[**Ans.** The magnitude of a negative number = $2^m - N$ which one value greater than 1's complement representation.]

52. State duality principle.

[**Ans.** Duality theorem/principle states that The dual of any Boolean function can be obtained

(*a*) By changing each or sign to an AND sign and vice-versa.

(*b*) Complementing any 0 or 1 appearing in the expression.]

53. Write De-morgan's theorems.

[**Ans.** (1) Complement of a product is equal to the sum of the individual variable complements.

$\overline{A.B} = \overline{A} + \overline{B}$]

(2) Complement of sum is equal to the product of the individual variable complements.

$\Rightarrow \overline{A+B} = \overline{A}\,.\,\overline{B}$]

54. Define absorption theorem.

[Ans. $A+\overline{A}B = A+B$]

55. What do you mean by canonical form ?

[Ans. The representation of a Boolean function in SOP or POS from, that is, each term in the expression must contain all the literals of the function either in normal form or complement form.]

56. Define a max term.

[Ans. It is a sum term which contains all literals of the function, either in complement form or normal form.]

57. Define a min term

[Ans. (A) It is a product term which contains all literals of the function either in normal form or complement form.]

58. Give the Truth table of equivalence function

Ans.

x	y	f
0	0	1
0	1	0
1	0	0
1	1	1

59. Convert $(110111)_4$ to base 8.

[Ans. $(67)_8$]

60. $(21.125)_{10}$ in base 4 is ______

[Ans. $(11.02)_4$]

61. Convert $(110111)_4$ to base '8' ______

[Ans. 67]

62. Write the condition for a weighted code to be self complementing code ______

[Ans. the sum of weights = 9]

63. What feature or characteristic makes the gray code, a cyclic code ______

[Ans. there is only one bit change between two successive code words]

64. BCD code is ______

(*a*) Error detecting code (*b*) error correcting code

(*c*) both (*d*) none

[Ans. (*d*)]

65. A code, which consists of 000, 111 only as valid code words, is able to

(*a*) detect single error (*b*) detect double error

(*c*) correct single error (*d*) all

[Ans. (*d*)**]**

66. The minimum distance in a code, for that code to be (single) error detecting code is ______

[Ans. 2 (two)**]**

67. The minimum distance of 2–out–5 code is ______

[Ans. two**]**

68. In the Hamming code 100011011110101, the message is ______

[Ans. 10001101111**]**

69. r's complement of $(25.639)_{10}$ is ______

[Ans. 74.361**]**

70. r's complement of $(r^n - N)$ is ______

[Ans. N**]**

71. What is meant by end–around–carry method ______

[Ans. It is the carry generated in binary subtraction with 1's or 2's complement method. In 1's complement the EAC is added and in 2's complement the EAC is discarded.**]**

72. For the operation $\sqrt{121} = 11$, give at least two possible bases.

[Ans. b = 10 and above**]**

73. 1234 + 6453 = 7687, for this operation the possible bases are

(*a*) 10 (*b*) 11 (*c*) 12 (*d*) all

[Ans. (*d*)**]**

74. Gray equivalent of 0101 is ______.

[Ans. 0111, position = 5**]**

75. Excess-3 code of 0011 is ______.

[Ans. 0110**]**

76. If $(26)_7 = (24)_b$, The value of b is ______.

[Ans. 8**]**

77. Two's complement of 011001 is ______.

[Ans. 100111**]**

78. 1101.1 + 1001.1 = ______.

[Ans. 10111**]**

79. $(432)_5 = (X)_7$, then the value of X is ______.

[Ans. 225**]**

80. $(1431)_8 = (X)_{10}$, then the value of X is ______.

[Ans. 793]

81. $(1010110101111)_2 = (?)_{16}$

[Ans. 15AF]

82. Write one example for Non-weighted code ?

[Ans. Gray code, Ex-3 code]

83. 10's complement of 635 is ______.

[Ans. 365]

84. The self-complementing code among the following is ( ).

(*a*) 2422 (*b*) 3322 (*c*) 8421 (*d*) 5211

[Ans. (*d*)]

85. The distance between the two code words 0010, 0001 is ______.

[Ans. two]

86. How many bits are needed to count the number 255 ?

[Ans. Seven bits]

87. $(0.100)_2 = (X)_{10}$, then X = ______.

[Ans. 0.5]

88. The received erroneous code is 1011100. Using Hamming procedure find, which bit position is in error ?

[Ans. Bit]

89. The given gray number is 1011, its equivalent binary number is ______.

(*a*) 1101 (*b*) 1000 (*c*) 1111 (*d*) 1011

[Ans. (*a*)]

90. The equivalent gray number for the binary number 1101 is ______.

(*a*) 1111 (*b*) 1011 (*c*) 1000 (*d*) 1100

[Ans. (*b*)]

91. 9's complement of 234 is ______.

(*a*) 234 (*b*) 765 (*c*) 766 (*d*) 432

[Ans. (*b*)]

92. The minimum distance between the code words is 3, then that code is used for single bit error correction and detection. (True/False).

[Ans. True]

93. The necessary condition for self-complementing code is, the sum of weights is 9. (True/False)

[Ans. True]

1.20 MULTIPLE CHOICE QUESTIONS

1. A binary number with *n*-digits all of which are unity has the value ______
(*a*) $n^2 - 1$ (*b*) 2^n (*c*) 2^{n-1} (*d*) $2^n - 1$ **[Ans.** (*d*)]

2. The 1's complement of a 4-bit number 1001 is______
(*a*) 1001 (*b*) 0110 (*c*) 1010 (*d*) 1000 **[Ans.** (*b*)]

3. The 2's complement of the decimal –4 is______
(*a*) 0100 (*b*) 1100 (*c*) 1011 (*d*) 1010 **[Ans.** (*b*)]

4. The 2's complement of a "–N" is 1011, the value of + N = ______
(*a*) 0100 (*b*) 0110 (*c*) 0101 (*d*) 0011 **[Ans.** (*c*)]

5. BCD number is obtained by ______
(*a*) Converting decimal number to binary.
(*b*) Each decimal number is represented by a 4-bit binary.
(*c*) Converting decimal to octal.
(*d*) Converting binary to decimal. **[Ans.** (*b*)]

6. The 2's complement of a 2's complement binary number is ______
(*a*) 1's complement of the binary number.
(*b*) 2's complement of the binary number.
(*c*) Original number.
(*d*) None. **[Ans.** (*c*)]

7. Given binary number is 00000111, its 2's complement in hexadecimal form is ______
(*a*) F8 (*b*) F9 (*c*) 07 (*d*) F0 **[Ans.** (*b*)]

8. ______ is the equivalent number in decimal system for the number 452.4 in the octal system
(*a*) 298.50 (*b*) 298.25 (*c*) 246.50 (*d*) 246.25 **[Ans.** (*a*)]

9. 1111.01 + 0111.10 = ______
(*a*) 11110.11 (*b*) 10110.11 (*c*) 1011.11 (*d*) 1011.10 **[Ans.** (*b*)]

10. ______ is the code word for the decimal number 7 in the weighted code of weights 642-3
(*a*) 1101 (*b*) 1111 (*c*) 0101 (*d*) 1100 **[Ans.** (*c*)]

11. Pick out the code from the following which is not a self complementing code
(*a*) 2421 code (*b*) 8421 code (*c*) 642–3 code (*d*) 3321 code **[Ans.**

12. A code with minimum distance 3 is able to ______
(*a*) Correct single error (*b*) Detect double error
(*c*) *a* and *b* (*d*) None **[Ans.** (*c*)]

13. Gray code is ______
(*a*) Non weighted code (*b*) Cyclic code (*c*) Reflected code (*d*) all the above **[Ans.** (*d*)]

14. In the Hamming code for each group of 'm' information digits, K parity checking digits are added with _______ condition.

(*a*) $2^K = m$ (*b*) $2^K = m + 1$ (*c*) $2^K \geq m+K+1$ (*d*) $2^K \leq m+K+1$

[Ans. (*c*)]

15. In the Hamming code 1001101, error has occurred _______ position.

(*a*) 4 (*b*) 5 (*c*) 1 (*d*) 7 **[Ans. (*a*)]**

16. $\sqrt{100} = 10$ is true in _______ base

(*a*) Any (*b*) 5 (*c*) 6 (*d*) 10 **[Ans. (*a*)]**

17. For a number N with n digits in the integer part and m digits in the fractional part then the number $r^n - r^{-m} - N$ is _______ **[Ans. (*a*)]**

2

Boolean Algebra

2.1 INTRODUCTION

Irish logician and mathematician Georgeboole, (an investigation into law of thought) developed a special language of digital circuits for expressing logic circuit functions as well as analysing and designing logic circuits. This algebra also provides an economical way of describing the circuitry used in digital computers. It is the mathematical foundation of switching theory and logic design.

'*Boolean algebra*', is a set of rules, laws and theorems by which logical operations can be mathematically expressed.

Or

Boolean algebra is an algebra that may be defined with a set of elements, a set of operators, and a number of unproved axioms or postulates. A *set* of elements is any collection of objects having a common property. If B is a set which consists at least two elements x and y, then $x \in$ B denotes that x is a member of the set B, and $y \notin$ B denotes that y is not an element of B. The set of three operators AND (Boolean product) operation $\odot$, the OR (Boolean sum) operation +, and the NOT (complement) operation ' are defined in the set.

2.2 BOOLEAN POSTULATES

The '*Boolean algebra*' is an algebra (B ; ., +, ' ; 0, 1) consisting of a set B (which consists at least two elements 0 and 1) together with three operations, the AND (Boolean product) operation . , the OR (Boolean sum) operation +, and the NOT (complement) operation ' defined on the set, such that for any objects/elements x, y and z of B, $x \in$ B denotes that x is an element of the set B, and $y \notin$ B denotes that y is not an element of B.

$x.y$ (the product of x and y)

$x + y$ (the sum of x and y)

and x' (the complement of x) are in B.

Here + and . are binary operators and ' (not) is a unary operator.

To satisfy the Boolean postulates, there are at least two elements $x, y \in$B such that $x \neq y$.

The postulates of a mathematical system form the assumptions from which it is possible to deduce the rules, theorems and properties of the system. For Boolean algebra, the following postulates/axioms hold.

Postulate 1 : 'Closure' : A set B is closed with respect to a binary operator, if for every pair of elements of B, the binary operator specifies a rule for obtaining a unique element of B.

For example : The set of natural numbers N = {1, 2, 3, 4, ...} is closed with respect to the binary operator plus (+) by the rules of arithmetic addition, since for any $x, y \in N$ we obtain a unique $z \in N$ by the operation, $x + y = z$.

The set of natural numbers is not closed w.r.t. the binary operator minus (–) by the rules of arithmetic subtraction because $2 - 3 = -1$ and $2, 3, \in N$ while $(-1) \notin N$.

Postulate 2 : 'Commutative Law' : The binary operators + and . on a set B are said to be commutative whenever (*a*) $x.y = y.x$ (*b*) $x + y = y + x$ for all $x, y \in B$.

Postulate 3 : 'Associative Law' : A binary operators . and + on a set B are said to be associative whenever

(*a*) $(x.y).z = x.(y.z)$

(*b*) $(x + y) + z = x + (y + z)$ for all $x, y, z \in B$

Postulate 4 : 'Distributive Law' : A binary operator . in a set B, is said to be distributive over + whenever.

(*a*) $x. (y + z) = x.y + x.z$ for all $x, y, z \in B$ and + is distributive over . when

(*b*) $x + (y.z) = (x + y) . (x + z)$

Postulate 5 : 'Absorptive Law' : The binary operators . and + on a set B, are said to be absorptive whenever

(*a*) $x. (x + y) = x$

(*b*) $x + (x.y) = x$

Postulate 6 : Inverse : A set B having the identity element e with respect to a binary operator . is said to have an inverse whenever for every $x \in B$, there exists an element $y \in B$ such that, $x.y = e$

Postulate 7 : Union Law : Zero (null, smallest) and one (universal, largest) elements :

There exists a '*unique*' element (the one element) $1 \in B$ such that for each element $x \in B$ $x.1 = 1.x = x$

There exists a unique element (the zero) $0 \in B$ such that for each element $x \in B$ $x + 0 = 0 + x = x$.

Postulate 8 : Complementary : For each $x \in B$ there exists a unique element $x' \in B$, called the *complement* of x, such that $x.x' = 0$ $x + x' = 1$.

2.3 HUNTINGTON POSTULATES

E.V. Huntington has formulated the following postulates in 1904 for Boolean algebra which is a set B of elements a, b, c and has two operators + and .

1. (*a*) Closure with respect to the operator +

   (*b*) Closure with respect to the operator .

2. (*a*) An identity element with respect to +, designated by 0 :

$$x + 0 = 0 + x = x$$

(*b*) An identity element with respect to ., designated by 1 :

$$x.1 = 1.x = x$$

3. (*a*) Commulative w.r.t. + : $x + y = y + x$

(*b*) Commulative w.r.t. . : $x.y = y.x$

4. (*a*) • is distributive over + : $x . (y + z) = (x.y) + (x.z)$

(*b*) + is distributive over . : $x + (y.z) = (x + y). (x + z)$

5. For every element $x \in$ B, there exists an element $x' \in$ B (called the complement of *x*) such that

(*a*) $x + x' = 1$ and

(*b*) $x.x' = 0$

6. There exist at least two elements $x, y \in$ B such that $x \neq y$.

Comparison between Boolean algebra and ordinary algebra

1. Huntington postulates do not include the associative law. However, this law holds good for Boolean algebra.
2. The distributive law of + over . , *i.e.*, $x + (y.z) = (x + y).(x + z)$, is valid for Boolean algebra, but not for ordinary algebra.
3. Boolean algebra does not have additive, multiplicative and inverses; therefore, there are no subtraction or division operations.
4. Complement of a number is not available in ordinary algebra but it is available in Boolean algebra.
5. Ordinary algebra deals with real numbers, which constitute an infinite set of elements. Boolean algebra deals with the as yet undefined set of elements B, but in the two-valued Boolean algebra defined with only two elements, 0 and 1.

2.4 BOOLEAN ALGEBRA SYMBOLS

To define the Boolean algebra, the following symbols are used.

1. Symbol of constants : The Boolean algebra contains only two constants 0 and 1.
2. Symbol of variables : Any characters in lowercase or uppercase *i.e.*, A, B, C, ... or *a, b, c,* ..... are used as a variables.
3. Symbol of equality : '=' which is used to indicate the equality relation between variables and/or constants . A = B means the two variables A and B are logically equal.
4. Dot sign (.) : A dot sign (.), which is also same as 'X' indicates a logical product of two variables and/or constants connected to it. If the two variables are A and B, then their logical product is A×B = A.B. Dot operation is also called as logical AND operation.
5. Plus sign (+) : It indicates the logical sum of terms so connected, if two terms are A and B and can be written as A + B or A OR B. Plus operation is also called as logical OR operation.

6. Over line (prime) : (¯) or ('). This sign of operation indicates the terms so overlined are to be complemented, if the term is A then $\overline{A}$ is a prime A (or) NOT A (or) A bar.

2.5 TWO-VALUED FOUR-ELEMENT BOOLEAN ALGEBRA

A two-valued Boolean algebra is defined on a set of two elements, B = {0, 1} with rules for the two binary operators + and . as shown in Table 2.1

Table 2.1 (*a*)

x	y	$x.y$
0	0	0
0	1	0
1	0	0
1	1	1

Table 2.1(*b*)

x	y	$x + y$
0	0	0
0	1	1
1	0	1
1	1	1

Table 2.1(*c*)

x	x'
0	1
1	0

Validation of Huntington postulates for two-valued set B = {0, 1} are

1. Closure is obvious from the tables, since the result of each operation is either 1 or 0 and 1, 0 ∈ B.
2. (*a*) 0 + 0 = 0; 0 + 1 = 1 + 0 = 1

   (*b*) 1.1 = 1; 1.0 = 0.1 = 0

which establish the two identity elements, 0 for + and 1 for . operations.

3. The commulative and distributive laws are proved in the Table 2.2.

Table 2.2

x	y	z	$x + y$	$x + z$	$x + (y.z)$	$(x + y)(x + z)$	$x.y$	$x.z$	$x(y + z)$	$(x.y) + (x.z)$
0	0	0	0	0	0	0	0	0	0	0
0	0	1	0	1	0	0	0	0	0	0
0	1	0	1	0	0	0	0	0	0	0
0	1	1	1	1	1	1	0	0	0	0
1	0	0	1	1	1	1	0	0	0	0
1	0	1	1	1	1	1	0	1	1	1
1	1	0	1	1	1	1	1	0	1	1
1	1	1	1	1	1	1	1	1	1	1

4. The complement operation is

   (*a*) $x + x' = 1$ since $0 + 0' = 0 + 1 = 1$ and

   $1 + 1' = 1 + 0 = 1$

   (*b*) $x.x' = 0$ since $0.0' = 0.1 = 0$

   $1.1' = 1.0 = 0$

Consider the four-element Boolean algebra B = {[0, *a*, *b*, 1] ; . , +, ', ; 0, 1}. The three operations are described by the following Table 2.3.

Table 2.3 (a)

.	0	a	b	1
0	0	0	0	0
a	0	a	0	a
b	0	0	b	b
1	0	a	b	1

Table 2.3 (b)

+	0	a	b	1
0	0	a	b	1
a	a	a	1	1
b	b	1	b	1
1	1	1	1	1

Table 2.3 (c)

'	
0	1
a	b
b	a
1	0

2.6 BASIC THEOREMS AND PROOFS

The Boolean theorems are proved from the postulates.

Theorem 1 : (a) $x + x = x$

(b) $x.x = x$

Proof : (a) $x + x = x$

$x + x = (x + x).1$ by postulate $x.1 = x$

$= (x + x).(x + x')$ by postulate $x + x' = 1$

$= x + x.x'$

$= x + 0$ by postulate $x + 0 = x$

$= x$

(b) $x.x = x.x + 0$

$= x.x + x.x'$

$= x.(x + x')$

$= x.1$ by postulate $x.1 = x$

$= x$

Theorem 2 : (a) $x + 1 = 1$ (b) $x.0 = 0$

Proof : (a) $x + 1 = 1.\ (x + 1)$

$= (x + x')\ .\ (x + 1)$

$= x + x'.1$

$= x + x' = 1$

(b) $x.0 = x.0 + 0$

$= x.0 + x.x'$

$= x.(x' + 0)$

$= x.x'$

$= 0$

Theorem 3 : $(x')' = x$

if $x = 0,\ x' = 1$ then $(x')' = (1)' = 0 = x$

if $x = 1,\ x' = 0$ then $(x')' = (0)' = 1 = x.$

Theorem 4 : (*a*) $x + (y+z) = (x + y) + z$

(*b*) $x \,.\, (yz) = (xy) \,.\, z$

Proof : The above theorems can be proved by '*perfect induction*' as shown in Table 2.4 (*a*) and 2.4 (*b*). Perfect induction is a method of proof where a theorem is verified for every possible combination of values that the variable may assume. All these possible combination values are tabulated for convenience in a table called *truth table* (or) table of combinations. Truth table is a table that shows all input and output possibilities of a logic circuit. In this table, every variable may assume one of the two values, 0 or 1. Truth table of 2 variables contains $2^2 = 4$ combinations :

Table 2.4 (*a*)

x	y	z	$(x + y)$	$(x + y) + z$	$y + z$	$x + (y + z)$
0	0	0	0	0	0	0
0	0	1	0	1	1	1
0	1	0	1	1	1	1
0	1	1	1	1	1	1
1	0	0	1	1	0	1
1	0	1	1	1	1	1
1	1	0	1	1	1	1
1	1	1	1	1	1	1

Table 2.4 (*b*)

x	y	z	$y\ z$	$x\ y$	$x.(y.z)$	$(x.y).z$
0	0	0	0	0	0	0
0	0	1	0	0	0	0
0	1	0	0	0	0	0
0	1	1	1	0	0	0
1	0	0	0	0	0	0
1	0	1	0	0	0	0
1	1	0	0	1	0	0
1	1	1	1	1	1	1

Theorem 5 : (*a*) $(x + y)' = x' \,.\, y'$

(*b*) $(x.y)' = x' + y'$

The proof of this theorem is shown in Table 2.5.

Proof :

Table 2.5

x	y	$x + y$	$(x + y)'$	x'	y'	$x'.y'$	$(x.y)$	$x'+y'$	$(x.y)'$
0	0	0	1	1	1	1	0	1	1
0	1	1	0	1	0	0	0	1	1
1	0	1	0	0	1	0	0	1	1
1	1	1	0	0	0	0	1	0	0

Theorem 6 : (*a*) $x + x.y = x$

(*b*) $x.(x + y) = x$

Proof : (*a*)

$$
\begin{aligned}
x + x.y &= x.1 + x.y \\
&= x\ (1 + y) \\
&= x.1 \\
&= x
\end{aligned}
$$

(*b*) $x.(x + y) = x$

$$
\begin{aligned}
x.(x+y) &= x.x + x.y \\
&= x + x.y \\
&= x.1 + x.y \\
&= x\ (1 + y) \\
&= x
\end{aligned}
$$

2.7 SWITCHING ALGEBRA

Among all the Boolean algebras, the two-element Boolean algebra B_2 ({0, 1}; ., +, ' ; 0, 1) is known as "*switching algebra*". Switching algebra is used for analysis and design of switching circuits that make up digital systems. It contains two extreme elements, the largest number represented by 1 and the smallest number represented by 0. Boolean functions defined on the switching algebra are called '*switching functions*'.

The basic postulate of switching algebra is the existance of a two-valued switching variable which can take any of the two distinct values 0 and 1. Precisely stated, if x is a switching variable, then $x \neq 0$ if and only if $x = 1$

$x \neq 1$ if and only if $x = 0$

These values are often referred to as the '*truth values*' of x.

The combination of finite number of switching variables (x, y etc.) and constants (0, 1) by means of the switching operations (+, ., and ') is called "*switching expression*". More precisely, any switching constant or variable is a switching expression, and if f_1 and f_2 are switching expressions, then f_1', f_2', $f_1 + f_2$ and $f_1 f_2$ are also switching expressions. No other combinations of variables and constants are switching functions.

The basic properties of switching algebra can be classed into three groups, according to the number of variables involved, as shown in Table 2.6.

$$n = 1$$

1.1	$x_1 + x_1 = x_1$	1.1^1	$x_1.x_1 = x_1$
1.2	$x_1 + x_1' = 1$	1.2^1	$x_1.x_1' = 0$
1.3	$1 + x_1 = x_1 + 1 = 1$	1.3^1	$1.x_1 = x_1.1 = x_1$
1.4	$0 + x_1 = x_1 + 0 = x_1$	1.4^1	$0.x_1 = x_1.0 = 0$

$$n = 2$$

2.1	$x_1 + x_2 = x_2 + x_1$	2.1^1	$x_1.x_2 = x_2.x_1$
2.2	$x_1 + x_1.x_2 = x_1$	2.2^1	$x_1.(x_1 + x_2) = x_1$
2.3	$(x_1 + x_2').x_2 = x_1.x_2$	2.3^1	$x_1.x_2' + x_2 = x_1 + x_2$

$$n = 3$$

3.1	$x_1+x_2+x_3 = (x_1+x_2) + x_3 = x_1 + (x_2 + x_3)$	3.1^1	$x_1.x_2.x_3 = (x_1.x_2).x_3 = x_1.(x_2.x_3)$
3.2	$x_1.x_2 + x_1.x_3 = x_1.(x_2+x_3)$	3.2^1	$(x_1 + x_2).(x_1 + x_3) = x_1 + x_2.x_3$
3.3	$(x_1+x_2).(x_2+x_3)\ (x_3+x_1') = (x_1+x_2)\ .\ (x_3+x_1')$	3.3^1	$x_1.x_2 + x_2.x_3 + x_3.x_1' = x_1x_2 + x_3.x_1'$

where n = number of binary variables (switching variables).

Table 2.6 properties of switching algebra.

Now, we shall present several general theorems of switching functions.

2.7.1 Generalized De Morgan's Theorem

(*a*) $(x_1 + x_2 + \ldots\ldots + x_n)' = x_1'.x_2'.\ \ldots\ldots\ .x_n'$

(*b*) $(x_1.x_2\ \ldots\ldots\ldots\ x_n)' = x_1' + x_2' + \ \ldots\ldots\ + x_n'$

De Morgan's theorem does not cover completely the relation between complementary functions. Shannon has suggested a generalization of the theorem.

2.7.2. Shannon's Theorem

$(f\ (x_1, x_2, \ldots\ldots\ x_n, +\ ,\ .))' = f\ (x_1', x_2', \ldots\ldots, x_n', .,+)$

This theorem says that the complement of any function is obtained by replacing each variable by its complement and, at the same time, by interchanging the . (Dot operation) and + (OR operation).

Proof : Since $(f\ (x_1, x_2, \ldots\ldots\ x_n, +\ ,\ .))'$ may be expressed by

$$(f\ (x_1, \ldots\ldots\ x_n, +\ ,\ .))' = \{(f_1\ (x_1, \ldots\ldots, x_n, +\ ,\ .) + f_2\ (x_1, \ldots\ldots, x_n +\ ,\ .)\}'$$
$$= (f_1\ (x_1, \ldots., x_n +\ ,\ .))'\ .\ ((f_2\ (x_1, \ldots., x_n +\ ,\ .))'$$

and/or

$$(f\ (x_1, \ldots\ldots\ x_n, +\ ,\ .))' = \{f_1\ (x_1, \ldots\ldots, x_n, +, .).f_2\ (x_1, \ldots\ldots, x_n +\ ,\ .)\}'$$
$$= (f_1\ (x_1, \ldots., x_n +\ ,\ .))' + (f_2\ (x_1, \ldots., x_n, +\ ,\ .))'$$

where f_1 and f_2 represent two partial functions of f.

2.7.3 Expansion Theorem

(*a*) $f(x_1, x_2, ..., x_n) = x_1 . f(1, x_2.....x_n) + x_1'.f(0, x_2....x_n)$

(*b*) $f(x_1 x_2, ..., x_n) = [x_1 + f(0, x_2....x_n)]\ [x_1' + f(1, x_2,.....,x_n)]$

The function of this theorem is said to be expanded with respect to x_1.

Proof : If we first substitute $x_1 = 1$ and $x_1' = 0$ and substitute $x_1 = 0$ and $x_1' = 1$ in each equation, they reduce to identities.

Any switching function in *n* variables may be developed into a series by means of the expansion theorem.

2.7.4 Theorem

(*a*) (1) $x_1.f(x_1,x_2,...x_n) = x_1.f(1, x_2.......x_n)$

(2) $x_1 + f(x_1,x_2,...x_n) = x_1 + f(0, x_2.......x_n)$

(*b*) (1) $x_1'.f(x_1,x_2,...x_n) = x_1'.f(0, x_2.......x_n)$

(2) $x_1' + f(x_1,x_2,...x_n) = x_1' + f(1, x_2.......x_n)$

2.8 DE MORGAN'S THEOREMS

De Morgan was a great logician and mathematician. He has suggested two theorems that form an important part of Boolean algebra. In the equation form, they are :

First Law : $\overline{A+B} = \overline{A}.\overline{B}$

The above theorem can be stated as 'complement of a sum is equal to the product of the individual complements'.

It means that the complement of two or more variables ORed is the same as the AND of the complements of each variable *i.e.,* For '*n*' variables.

$$\overline{A+B+C.....N} = \overline{A}.\overline{B}.\overline{C}.......\overline{N}$$

Proof : This law can be verified by using a table which is known as truth table (shown in Table 2.6).

Table 2.6

A	B	$\overline{A}$	$\overline{B}$	$A+B$	$\overline{A+B}$	$\overline{A}.\overline{B}$
0	0	1	1	0	1	1
0	1	1	0	1	0	0
1	0	0	1	1	0	0
1	1	0	0	1	0	0

From the above table, we can observe that the last two columns are same. Hence, $\overline{A+B} = \overline{A}.\overline{B}$.

Second Law : According to the De Morgan's second law, '*the complement of a product is equal to sum of individual complements.*'

If we consider two variables, A and B, we have

$$\overline{A.B} = \overline{A} + \overline{B}$$

It means that the complement of two or more variables ANDed is same as the OR of the complements of each variable.

Proof : The second law can be verified using truth table shown in Table 2.7.

Table 2.7

A	B	$\overline{A}$	$\overline{B}$	AB	$\overline{A}.\overline{B}$	$\overline{A+B}$
0	0	1	1	0	1	1
0	1	1	0	0	1	1
1	0	0	1	0	1	1
1	1	0	0	1	0	0

From the Table 2.7, it can be observed that the last two columns are equal. Hence $\overline{A.B} = \overline{A} + \overline{B}$. If there are 'N' variables, then the De Morgan's second law gives

$$\overline{A.B.C......N} = \overline{A} + \overline{B} + \overline{C} + ... + \overline{N}$$

Applications

(1) The De Morgan theorems change plus sign to '.' (product) sign and vice-versa *i.e.,* AND operation to OR operation and vice-versa.

(2) These theorems are used to convert Boolean expression of SOP form to POS form and vice-versa.

2.9 DUALITY THEOREM

The '*duality theorem*' says that starting with a Boolean relation, you can derive another Boolean relation by

1. Changing OR (operation) *i.e.,* + (plus) sign to an AND sign (operation) *i.e.,* . (dot) and vice-versa
2. Complement any 0 or 1 appearing in the expression *i.e.,* replacing constants 0 and 1 by 1 and 0, respectively.

Any statement or theorem which has this property is called, *dual,* and this quality of duality which characterizes switching algebra is known as the 'principle of duality.'

For example (1) we know that

$A + 0 = A$ and its dual relation is

$A.1 = A$ which is true.

(2) The Distributive law states that

$A. (B+C) = A.B + A.C$. By changing each OR operation (+) and AND operation (.), we get the dual relation, which is

$$A + B.C = (A+B) . (A+C)$$

The proof can be obtained by the perfect induction as shown in Table 2.8.

Table 2.8

A B C 1	A.B 2	A.C 3	BC 4	A+B 5	A+C 6	B+C 7	(A.B+A.C) 8	A(B+C) 9	(A+B).(A+C) 10	A+BC 11
0 0 0	0	0	0	0	0	0	0	0	0	0
0 0 1	0	0	0	0	1	1	0	0	0	0
0 1 0	0	0	0	1	0	1	0	0	0	0
0 1 1	0	0	1	1	1	1	0	0	1	1
1 0 0	0	0	0	1	1	0	0	0	1	1
1 0 1	0	1	0	1	1	1	1	1	1	1
1 1 0	1	0	0	1	1	1	1	1	1	1
1 1 1	1	1	1	1	1	1	1	1	1	1

In the above table, columns 8 and 9 are same and columns 10 and 11 are same, hence the above two statements are proved.

Note : (1) The duality theorem is useful to produce a new Boolean relation.

(2) The dual of SOP form is POS form and vice-versa.

2.9.1 Self Dual Function

If a Boolean function is equal to its dual function, then the given function is called self-dual function.

Example.

$$F_1 = ab + bc + ca$$

$$\text{Dual of } F_1 = F_1^D = (a+b)\ (b+c)\ (c+a)$$

$$= (ac + b)\ (c + a)$$

$$= ac + bc + ac + ab$$

$$= ab + bc + ca \qquad \text{since } F_1 = F_1^D,\ F_1 \text{ is a self-dual}$$

Example.

$$F_2 = \Sigma m\ (0, 3, 5, 6)$$

$$\text{Dual of } F_2 = F_2^D = m_0^D\ m_3^D\ m_5^D\ m_6^D$$

$$= M_{7-0}\ M_{7-3}\ M_{7-5}\ M_{7-6}$$

$$= M_7\ M_4\ M_2\ M_1$$

$$= m_0 + m_3 + m_5 + m_6$$

$$= \Sigma m\ (0, 3, 5, 6) = F_2$$

Hence F_2 is a self dual function.

2.10 SWITCHING FUNCTIONS

Let $f\ (x_1, x_2, ..., x_n)$ be a switching expression with '*n*' variables. Since each of the variables, $x_1, x_2, ..., x_n$ can independently assume one of the two values 0 or 1, there are 2^n combinations to be considered in determining the value of *f*.

In order to determine the value of an expression either '0 or 1' for all combinations, it is only necessary to substitute the values for the variables in the expression.

For example, if $f\ (a, b, c) = ab + \bar{a}c + b\bar{c}$, then, for the combination $a = 1, b = 0, c = 1$, the value of the expression is $f\ (a, b, c) = 1.0 + 1'.1 + 0.1'$

$$= 1.0 + 0.1 + 0.0 = 0$$

In a similar manner, the value of f can be computed for every combination as shown in Table 2.9.

Table 2.9. Truth table for $f(a, b, c) = ab + \bar{a}c + b\bar{c}$

a	b	c	f
0	0	0	0
0	0	1	1
0	1	0	1
0	1	1	1
1	0	0	0
1	0	1	0
1	1	0	1
1	1	1	1

For every possible combination of variables, the value of expression f is identical.

Any switching function can be represented by a truth table. The number of rows in the truth table is 2^n, where n is the number of binary variables in the function. For each row of the table, there is a value for the function equal to either 1 or 0. Thus, a *switching function* $f(x_1, x_2,...., x_n)$ is a correspondence which associates an element of the algebra with each of the 2^n combinations of the variables $x_1, x_2, ..., x_n$.

The complement $f'(x_1, x_2, ..., x_n)$ is a function whose value is 0 whenever the value of $f(x_1, x_2, ..., x_n)$ is 1, and 1 whenever the value of $f = 0$.

If a function is specified by the truth table, then it's complement can be obtained by complementing each entry in the column f.

One more simple procedure for deriving the complement of a function is to take the dual of the function and complement each variable.

Example 2.1. *Determine the complements of the following functions :*

(i) $f_1 = \overline{A}\,\overline{B}C + A\overline{B}\,\overline{C}$

(ii) $f_2 = \overline{B}(AC + \overline{AC})$

Solution. (*i*) $f_1 = \overline{A}\,\overline{B}C + A\overline{B}\,\overline{C}$

The dual of f_1 is f_D $= (\overline{A} + \overline{B} + C).(A + \overline{B} + \overline{C})$

The complement of $f_1 = f_D' = (A + B + \overline{C}).(\overline{A} + B + C)$

for verification $f_1' = \overline{(\overline{A}\,\overline{B}C + A\overline{B}\,\overline{C})}$

$$= \overline{(\overline{A}\,\overline{B}C)}.\overline{(A\overline{B}\,\overline{C})}$$

$$= (\overline{\overline{A}} + \overline{\overline{B}} + \overline{C}).(\overline{A} + \overline{\overline{B}} + \overline{\overline{C}})$$

$$= (A + B + \overline{C}).(\overline{A} + B + C) \text{ Hence verified}$$

(*ii*) $f_2 = \overline{\text{B}}.(\text{AC} + \overline{\text{A}}\,\overline{\text{C}})$

$$\text{Dual of } f_2 = \overline{\text{B}} + \left[(\text{A}+\text{C}).(\overline{\text{A}}+\overline{\text{C}})\right]$$

$$\text{Complement of } f_2 = \text{B} + \left[(\overline{\text{A}}+\overline{\text{C}}).(\text{A}+\text{C})\right]$$

2.10.1 Representation of Switching Functions

In a switching function, a binary variable may appear in it's normal form/unprimed forn (x) or in it's complement form/primed from (x' or $\overline{x}$). Now, consider the two binary variables x and y combined with an AND operation. Since each variable may appear in either form, there are four possible combinations : $\overline{x}\,\overline{y}, \overline{x}y, x\overline{y}, xy$. Each of these four terms is called a '*minterm*' or a '*standard product*' i.e., a minterm is a product term which contains all the variables either in normal or complement form.

In a similar manner, *n*-variables can be combined to form 2^n minterms.

The 2^n different minterms may be determined by a method similar to the one shown in Table 2.10 for three variables. The binary numbers from 0 to 2^n–1 are listed under the *n* variables. Each minterm is obtained from an AND term of the n-variables, with each variable being primed if the corresponding bit of the binary number is a 0 and unprimed if *a* 1. Each minterm is represented by a symbol m_J. Where J denotes the decimal equivalent of the binary number of the designated minterm.

In a similar way, *n* variables, forming an OR term, with each variable being normal or complement, provide 2^n possible combinations called 'maxterms' or standard sums *i.e.*, a maxterm is a sum term, which contains all the literals or variables in normal form or complement form. The 2^n distinct maxterms for *n* variables can be obtained in a similar way from the Table 2.10 for 3 variables. Each maxterm is obtained from an OR term of the *n* variables, with each variable being unpaired if the corresponding bit is *a* 0 and primed if *a* 1.

Note : Each maxterm is the complement of it's corresponding minterm and vice-versa.

Table 2.10. Three variable minterms and maxterms

A	B	C	*Minterms* representing term	*Minterms* designation	*Maxterms* representing term	*Maxterms* designation
0	0	0	$\overline{\text{A}}\,\overline{\text{B}}\,\overline{\text{C}}$	m_0	A + B + C	M_0
0	0	1	$\overline{\text{A}}\,\overline{\text{B}}\text{C}$	m_1	$\text{A}+\text{B}+\overline{\text{C}}$	M_1
0	1	0	$\overline{\text{A}}\text{B}\overline{\text{C}}$	m_2	$\text{A}+\overline{\text{B}}+\text{C}$	M_2
0	1	1	$\overline{\text{A}}\text{BC}$	m_3	$\text{A}+\overline{\text{B}}+\overline{\text{C}}$	M_3
1	0	0	$\text{A}\overline{\text{B}}\,\overline{\text{C}}$	m_4	$\overline{\text{A}}+\text{B}+\text{C}$	M_4
1	0	1	$\text{A}\overline{\text{B}}\text{C}$	m_5	$\overline{\text{A}}+\text{B}+\overline{\text{C}}$	M_5
1	1	0	$\text{AB}\overline{\text{C}}$	m_6	$\overline{\text{A}}+\overline{\text{B}}+\text{C}$	M_6
1	1	1	ABC	m_7	$\overline{\text{A}}+\overline{\text{B}}+\overline{\text{C}}$	M_7

There are two important forms that are used for representation of Boolean expressions. They are

(*i*) SOP (Sum-Of-Products)

(*ii*) POS (Product-Of-Sums).

2.11 SUM-OF-PRODUCTS (SOP) FORM

The SOP expression contains two or more AND functions ORed together. Each AND term (product) consists of one or more variables appearing in either complemented or uncomplemented form. All these product terms are connected with OR operation. This kind of logic expression is called as 'SOP' Sum-Of-Products.

Example. $AB + \overline{A}BC$; $x\overline{y}z + x\overline{z}$

$$xy + \overline{y}z + \overline{z}\,\overline{x} + z\overline{y}$$

2.12 PRODUCT-OF-SUMS (POS) FORM

The POS expression contains two or more OR functions ANDed together. Each OR term contains one or more variables in complemented or uncomplemented form. These sum terms are connected by AND operation. This kind of logic expression is called as 'POS' Product-Of-Sums.

The following are the examples of POS expressions

$$(A+B).(\overline{A}+B+C);(x+\overline{y}).(x+\overline{z})$$

$$(x+\overline{y}).(\overline{x}+y).(\overline{x}+\overline{z})$$

Note : (1) In SOP expression, one inversion sign cannot cover more than one variable in a product term *i.e.*, we cannot have $\overline{ABC}$ or $\overline{AB}\,C$ etc. in SOP expression

example : $\overline{ABC} + A\overline{B}C + \overline{A}B\overline{C}$ is not a SOP form.

(2) In POS expression, one inversion sign cannot cover more than one variable in a sum term *i.e.*, we cannot have $\overline{A+B+C}$ or $\overline{A+B}+C$ etc. in SOP expression

example : $\overline{(A+B+C)}.\overline{(A+B)}$ is not a POS form.

2.13 CANONICAL FORMS

A Boolean function in a canonical form may have two forms

(1) Canonical SOP (2) Canonical POS

2.13.1 Canonical SOP

It is an SOP form in which each product term contains all the variables either in complemented or uncomplemented form *i.e.*, for a 3 variable function, each product term must contain 3 variables. So, here each product term is called as minterm. This type of representation is called "sum of minterms" (or) canonical SOP (or) disjunctive normal expression.

The examples of canonical SOP for a 3 variable function are

(1) $A\overline{B}C + A\overline{BC} + \overline{A}B\overline{C}$ (2) $xyz + \overline{x}\,\overline{y}z + x\overline{y}\,\overline{z}$

2.13.2 Canonical POS

It is a POS form in which each sumterm contains all the variables either in complemented or uncomplemented form *i.e.*, for a 3-variable Boolean function, each sum term must contain 3-variables. Here, the sum is also called as "maxterm", and this type of expression is called canonical POS, or product of maxterms or conjective normal expression.

The following are the examples of canonical POS

(1) $(x+y+z).(x+\overline{y}+\overline{z}).(\overline{x}+\overline{y}+z)$

(2) $(A+B).(\overline{A}+B).(A+\overline{B})$

2.13.3 Converting SOP Expressions into Canonical SOP Form

SOP form can be converted into canonical SOP by ANDing the terms in the expression with the terms formed by ORing the literals and it's complements which are not present in that term. For example, for a three literal expression with literals A, B and C, if there is a term AC, where B is missing, then we form $(B+\overline{B})$ and AND it with AC. Therefore, we get

$$AC(B+\overline{B}) = ABC+A\overline{B}C.$$

Example 2.2. *Convert the given expression into Canonical SOP form.*

$$f = AB + BC + CA$$

Solution. $AB(C+\overline{C})+BC(A+\overline{A})+CA(B+\overline{B})$

$$= ABC + AB\overline{C} + ABC + \overline{A}BC + ABC + A\overline{B}C$$
$$= \overline{A}BC + A\overline{B}C + AB\overline{C} + ABC$$
$$= m_3 + m_5 + m_6 + m_7 = \Sigma m\ (3, 5, 6, 7)$$

Example 2.3. *Convert the given expression into Canonical SOP form.*

$$f = A + AB + ABC$$

Solution. $A(B+\overline{B})(C+\overline{C})+AB(C+\overline{C})+ABC$

$$= (AB + A\overline{B})\ (C+\overline{C}) + ABC + AB\overline{C} + ABC$$
$$= ABC + AB\overline{C} + A\overline{B}C + A\overline{B}\overline{C} + ABC + AB\overline{C} + ABC$$
$$= ABC + AB\overline{C} + A\overline{B}C + A\overline{B}\overline{C}$$
$$= m_7 + m_6 + m_5 + m_4 = \Sigma m\ (4, 5, 6, 7)$$

where Σ denotes SOP.

2.13.4 Canonical SOP Expression from Truth Table

Canonical SOP expression can be obtained from the truth table.

To obtain the Boolean expression in CSOP form, the following steps must be followed.

1. When a variable value is 1, write it in uncomplemented form in the equation.
2. When a variable value is 0, write it in complemented form in the equation.
3. Write each term of the individual input combinations that give F = 1 as a product.
4. OR (sum) all the terms which specify an input combination to obtain the SOP equation.

The SOP expression gives the active-high equation of a logic function and it is the sum of all input variable combinations that give an output of F = 1. It does not give the complement equivalent POS expression.

Example 2.4. *Convert the given expression into CSOP form*

$$f = A + AB + ABC.$$

Solution. From the table take sum of all the product terms for which $f = 1$, shown in Table 2.11.

Table 2.11

A	B	C	f
0	0	0	0
0	0	1	0
0	1	0	0
0	1	1	0
1	0	0	1
1	0	1	1
1	1	0	1
1	1	1	1

$$\therefore \quad f = A\overline{B}\,\overline{C} + A\overline{B}C + AB\overline{C} + ABC$$

Example 2.5. *Convert the given function*

$$f = AB+\overline{B}\,\overline{C}+\overline{A}C \text{ into CSOP}$$

Solution.

Table 2.12

A	B	C	AB	$\overline{B}\,\overline{C}$	$\overline{A}C$	f
0	0	0	0	1	0	1
0	0	1	0	0	1	1
0	1	0	0	0	0	0
0	1	1	0	0	1	1
1	0	0	0	1	0	1
1	0	1	0	0	0	0
1	1	0	1	0	0	1
1	1	1	1	0	0	1

$$\therefore \quad f = \overline{A}\,\overline{B}\,\overline{C} + \overline{A}\,\overline{B}C + \overline{A}BC + A\overline{B}\,\overline{C} + AB\overline{C} + ABC$$

From the Table 2.12, take the sum of all the product terms for which $f = 1$

2.13.5 Conversion of POS Expression into Canonical POS Form

The POS form can be converted into canonical POS by summing the terms in the expression with terms formed by ANDing (producting) the variable and its complement which are not present in that term. For example, for a three literal expression with literals A, B and C, if there is a term B + C where literal A is missing, then form term $A.\bar{A}$ and add it with B + C. Therefore, we get

$$B+C+A.\bar{A} = (B+C+A).(B+C+\bar{A})$$

Example 2.6. *Convert the given expressions into CPOS form.*

(a) $f = (A + B)(B + C)(C + A)$

(b) $f = A(A + B)(A + B + C)$

Solution. (a)

$$\begin{aligned} f &= (A + B)(B + C)(C + A) \\ &= (A + B + C\bar{C})(B + C + A\bar{A})(C + A + B\bar{B}) \\ &= (A+B+C)(A+B+\bar{C})(A+B+C) \\ &\quad (\bar{A}+B+C)(A+B+C)(A+\bar{B}+C) \\ &= (A+B+C)(A+B+\bar{C})(A+\bar{B}+C)(\bar{A}+B+C) \\ &= M_0 + M_1 + M_2 + M_4 \\ &= \Pi M(0, 1, 2, 4) \end{aligned}$$

(b) $(A+B.\bar{B}+C.\bar{C})(A+B+C.\bar{C})(A+B+C)$

$$\begin{aligned} &= \left((A+B)(A+\bar{B}) + C.\bar{C}\right)(A+B+C)(A+B+\bar{C})(A+B+C) \\ &= (A+B+C\bar{C})(A+\bar{B}+C\bar{C})(A+B+C)(A+B+\bar{C}) \\ &= (A+B+C)(A+B+\bar{C})(A+\bar{B}+C) \\ &\quad (A+\bar{B}+\bar{C})(A+B+C)(A+B+\bar{C}) \\ &= (A+B+C)(A+B+\bar{C})(A+\bar{B}+C)(A+\bar{B}+\bar{C}) \\ &= M_0 + M_1 + M_2 + M_3 \\ &= \Pi M(0, 1, 2, 3) \end{aligned}$$

2.13.6 CPOS Expression from a Truth Table

To obtain a Boolean expression in its product-of-sums form from a truth table, the following steps are to be followed :

1. When a variable is 1, write it down in the equation of its complement form.
2. When a variable is 0, write it down in it's uncomplemented form.
3. Write each term of the individual input combinations that give F = 0 as a sum.
4. AND all terms which specify an input combination to obtain the POS equation.

The POS expression gives the active-low equation of a logic function and it is the product of all input variable combinations that give an output of the F = 0. It does not give the complement of the equivalent SOP equation.

Consider the POS equation $F = (A+\overline{B})(\overline{A}+B) = AB + \overline{A}\,\overline{B}$

The complement of this is $\overline{F} = \overline{AB + \overline{A}\,\overline{B}} = (\overline{AB}).(\overline{\overline{A}\,\overline{B}}) = (\overline{A}+\overline{B})(A+B)$

$$= \overline{A}B + A\overline{B}$$

Example 2.7. *Find the CPOS equation of the given equation from the truth table (Table 2.13).*

$$F = A + AB + ABC$$

Table 2.13

A	B	C	f
0	0	0	0
0	0	1	0
0	1	0	0
0	1	1	0
1	0	0	1
1	0	1	1
1	1	0	1
1	1	1	1

Solution. From the table, take product of all sum terms for which F = 0

$\therefore$ $$F = (A+B+C)(A+B+\overline{C})(A+\overline{B}+C)(A+\overline{B}+\overline{C})$$

Example 2.8. *Obtain the (a) SOP (b) POS expressions for the logic function whose truth table is given in Table 2.14.*

Table 2.14

A	B	C	f
0	0	0	1
0	0	1	1
0	1	0	0
0	1	1	1
1	0	0	0
1	0	1	0
1	1	0	0
1	1	1	1

Solution. (*a*) $f = \overline{A}\,\overline{B}\,\overline{C} + \overline{A}\,\overline{B}C + \overline{A}BC + ABC$

(*b*) $f = (A+\overline{B}+C)(\overline{A}+B+C)(\overline{A}+B+\overline{C})(\overline{A}+\overline{B}+C)$

2.13.7 Truth Table from SOP Expression

Sometimes it may be convenient to obtain the truth table of an SOP expression. The truth table should have a number of combinations of input variables equal to 2^n, where n is the number of variables. Consider the expression, $f = \overline{A}\,\overline{B}\,\overline{C} + AB\overline{C} + \overline{A}B\overline{C} + \overline{A}\,\overline{B}C$. It's truth table is given in Table 2.15.

Table 2.15

A	B	C	$\overline{A}\,\overline{B}\,\overline{C}$	$AB\overline{C}$	$\overline{A}B\overline{C}$	$\overline{A}\,\overline{B}C$	f
0	0	0	1	0	0	0	1
0	0	1	0	0	0	1	1
0	1	0	0	0	1	0	1
0	1	1	0	0	0	0	0
1	0	0	0	0	0	0	0
1	0	1	0	0	0	0	0
1	1	0	0	1	0	0	1
1	1	1	0	0	0	0	0

$$f = \Sigma m\ (0,\ 1,\ 2,\ 6)$$

The input combinations for this expression are 000, 001, 010 and 110. The output for each combination is entered as 1 in the f column of the table, while all the remaining combinations are entered as 0 in the f column.

2.13.8 Truth Table from a POS Expression

Consider the POS expression, $f = (\overline{A}+B)(A+\overline{B})$. There are two variables, so the truth table must contain four combinations. Each ORed term in the equation gives *a* 0 output. The input combinations represented by the equations are 10 and 01. These two output states are entered in the f column of the truth table as 0. All the other combinations are represented by 1. The truth table is shown in Table 2.16.

Table 2.16

	A	B	$\overline{A}+B$	$A+\overline{B}$	f
0	0	0	1	1	1
1	0	1	1	0	0
2	1	0	0	1	0
3	1	1	1	1	1

$$\therefore \quad f = \Pi M\ (1,\ 2)$$

2.13.9 Conversion between Canonical Forms

The conversion of an SOP equation into the corresponding POS form is more difficult than the reverse process. The steps involved are as follows :

★ Simplify the SOP equation if possible

★ Invert the SOP equation

★ Multiply out and simplify the inverse equation

★ Invert the result.

Example 2.9. *Convert the SOP equation,* $F = A\overline{B}C + \overline{A}B\overline{C}$ *into it's POS form.*

Solution. Step 1 : $\overline{F} = \overline{A\overline{B}C + \overline{A}B\overline{C}} = (\overline{A\overline{B}C})\,(\overline{\overline{A}B\overline{C}})$

$$= (\overline{A} + \overline{\overline{B}} + \overline{C})\,(\overline{\overline{A}} + \overline{B} + \overline{\overline{C}})$$

$$= (\overline{A} + B + \overline{C})\,(A + \overline{B} + C)$$

Step 2 : $\overline{A}A + \overline{A}\,\overline{B} + \overline{A}C + AB + B\overline{B} + BC + A\overline{C} + \overline{B}\,\overline{C} + C\overline{C}$

$$= \overline{A}\,\overline{B} + \overline{A}C + AB + BC + A\overline{C} + \overline{B}\,\overline{C}$$

$$\because A\overline{A} = B\overline{B} = C\overline{C} = 0$$

$$= \overline{A}\,\overline{B} + \overline{A}C + AB + A\overline{C}$$

Step 3 : $\overline{\overline{A}\,\overline{B} + \overline{A}C + AB + A\overline{C}}$

$$= (\overline{\overline{A}\,\overline{B}})\,(\overline{\overline{A}C})\,(\overline{AB})\,(\overline{A\overline{C}})$$

$$= (A + B)\,(A + \overline{C})\,(\overline{A} + \overline{B})\,(\overline{A} + C)$$

Example 2.10. *Convert* $f = \overline{A}\,\overline{B}\,\overline{C} + \overline{A}\,\overline{B}C + \overline{A}B\overline{C} + A\overline{B}\,\overline{C} + AB\overline{C} + ABC$ *into it's POS form.*

Solution. Step 1: Simplification $= \overline{A}\,\overline{C}\,(\overline{B} + B) + A\overline{C}\,(\overline{B} + B) + \overline{A}\,\overline{B}C + ABC$

$$= \overline{A}\,\overline{C} + A\,\overline{C} + C\,(\overline{A}\,\overline{B} + AB)$$

$$= \overline{C}(A + \overline{A}) + C(\overline{A}\,\overline{B} + AB)$$

$$= \overline{C} + C\,(\overline{A}\,\overline{B} + AB)$$

$$= \overline{C} + \overline{A}\,\overline{B} + AB$$

Step 2: $\overline{\overline{C} + \overline{A}\,\overline{B} + AB}$

$$= (\overline{\overline{C}})\,(\overline{\overline{A}} + \overline{\overline{B}})\,(\overline{A} + \overline{B})$$

$$= C\,(A + B)\,(\overline{A} + \overline{B})$$

$$= C\,(A\overline{B} + B\overline{A})$$

$$= A\overline{B}C + \overline{A}BC$$

Step 3: $\overline{A\overline{B}C + \overline{A}BC}$

$$= (\overline{A} + B + \overline{C})\,(A + \overline{B} + \overline{C})$$

The other technique used for conversion when the given equation is a direct canonical form is as follows :

The complement of a function expressed as the sum of minterms equals the sum of minterms missing from the original function. This is because the original function is expressed by those minterms that make the function equal to 1, whereas it's complement is a 1 for those minterms for which the function is *a* 0.

For example, $f = (A, B, C) = \Sigma m\ (0, 2, 4, 6)$. The complement of this function can be expressed as

$$f'\ (A, B, C) = \Sigma m\ (1, 3, 5) = m_1 + m_3 + m_5$$

Now, if we make the complement of f' by De Morgan's theorem, we obtain f in a different form.

$$f = (f')' = (m_1 + m_3 + m_5)' = m_1'.m_3'.m_5'$$
$$= M_1.M_3.M_5 = \Pi\ (1, 3, 5)$$

∴ We now can conclude that the following relation is true

$$m_i' = M_i$$

i.e., the maxterm with subscript *i* is a complement of the minterm with the same subscript, *i*, and vice-versa.

In a similar way, the conversion between POS to SOP can be obtained by simply multiplying all the terms and simplifying them.

For example if $f = (A+B)\ (A+\overline{B})\ (\overline{A}+B)$ is in POS, to get into SOP form, multiply the three terms

$$\therefore \quad f = (A+A\overline{B}+BA+B\overline{B})\ (\overline{A}+B) = (A+A\overline{B}+AB)\ (\overline{A}+B) = A\overline{A}+A\overline{B}\ \overline{A}$$

$$= AB + AB\overline{A} + AB + A\overline{B}B + A\overline{A} + A\overline{B}\ \overline{A}$$

$$\therefore \quad f = AB$$

The general conversion procedure is to convert from one canonical form to another, interchange the symbols Σ and Π and lift those numbers missing from the original form. The number of missing terms is the total number of minterms or maxterms 2^n minus no. of terms given in the function where *n* is number of binary variables in the given function minus the terms in the given function.

Example 2.11. *Convert the given function,* $f = xy + \overline{x}z$ *into other canonical form.*

Solution. *Case (i)* First derive the truth table of the function, as shown in Table 2.17.

Table 2.17

x	y	z	f
0	0	0	0
0	0	1	1
0	1	0	0
0	1	1	1
1	0	0	0
1	0	1	0
1	1	0	1
1	1	1	1

$\therefore \quad f(x, y, z) = \Sigma\,(1, 3, 6, 7)$

$\therefore \quad f = \Pi\,(0, 2, 4, 5)$

Case (ii) $\quad f = xy + \bar{x}z$

Step 1 : $\quad f' = \overline{xy + \bar{x}z}$

$$= (\overline{xy})\,(\overline{\bar{x}z})$$

Step 2 : $\quad = (\bar{x} + \bar{y})\,(x + \bar{z})$

$$= \bar{x}\bar{z} + x\bar{y} + \bar{y}\bar{z}$$

Step 3 : $\quad \overline{\bar{x}\bar{z} + x\bar{y} + \bar{y}\bar{z}} = (x + z)\,(\bar{x} + y)\,(y + z)$

To convert into canonical POS form,

$(x + z + y\bar{y})\,(\bar{x} + y + z\bar{z})\,(x\bar{x} + y + z)$

$$= (x + y + z)\,(x + z + \bar{y})\,(\bar{x} + y + z)$$
$$(\bar{x} + y + \bar{z})\,(x + y + z)\,(\bar{x} + y + z)$$
$$= (x + y + z)\,(x + z + \bar{y})\,(\bar{x} + y + z)\,(\bar{x} + y + \bar{z})$$
$$= \Pi\,(0, 2, 4, 5)$$

Example 2.12. *Convert the given expressions into SOP form.*

(a) $(x + z)\,(\bar{x} + y)\,(y + z)$

(b) $(\bar{A} + B + \bar{C})\,(A + \bar{B} + C)$

Solution. Method 1 : *(a)* To convert the given POS expression into SOP form, just multiply the terms and simplify them by applying appropriate rules.

$(x + z)\,(\bar{x} + y)\,(y + z) \quad = (x\bar{x} + xy + \bar{x}z + yz)\,(y + z)$

$$= xy + \bar{x}yz + yz + xyz + \bar{x}z + yz$$
$$= xy + \bar{x}yz + yz + xyz + \bar{x}z$$
$$= xy + yz(\bar{x} + x + 1) + \bar{x}z \;=\; xy + yz + \bar{x}z$$
$$= (xy + y)\,(xy + z) + \bar{x}\bar{z} \;=\; xy + \bar{x}z$$

(b) $(\bar{A} + B + \bar{C})\,(A + \bar{B} + C)$

$$= A\bar{A} + \bar{A}\bar{B} + \bar{A}C + AB + B\bar{B} + BC + \bar{C}A + \bar{C}\bar{B} + C\bar{C}$$
$$= \bar{A}\bar{B} + \bar{A}C + AB + BC + A\bar{C} + \bar{B}\bar{C}$$
$$= \bar{B}(\bar{A} + A) + \bar{A}C + AB + BC + A\bar{C}$$
$$= \bar{B} + \bar{A}C + AB + A\bar{C} + BC$$
$$= B + A + \bar{A}C + A\bar{C} + BC$$
$$= B + A + C + A\bar{C} + BC$$
$$= A + B + C + BC$$
$$= A + B + C$$

Example 2.13. *Convert the following expression into the other canonical form*

$$f = \Sigma m\ (0, 3, 4, 6)$$

Solution. The given Boolean function,

$$f = \Sigma m\ (0, 3, 4, 6) = m_0 + m_3 + m_4 + m_6$$

The complement of the given function,

$$\bar{f} = \Sigma m\ (1, 2, 5, 7)$$
$$= (m_1 + m_2 + m_5 + m_7)$$

By applying De Morgan's theorem, the complement of $\bar{f}$ is

$$(\bar{\bar{f}}) = f = \overline{(m_1 + m_2 + m_5 + m_7)}$$
$$= \overline{m_1} . \overline{m_2} . \overline{m_5} . \overline{m_7} = M_1.M_2.M_5.M_7$$
$$= \Pi M\ (1,, 2, 5, 7)$$

Example 2.14. *Convert the following expression to the other canonical form*

$$f\ (A, B, C, D) = \Pi M\ (0, 1, 2, 5, 6, 7, 11)$$

Solution. The given function $f = \Pi M\ (0, 1, 2, 5, 6, 7, 11)$

$= M_0.M_1.M_2.M_5.M_6.M_7.M_{11}$ and the complement of the given function is

$$\bar{f} = \Pi M\ (3, 4, 8, 9, 10, 12, 13, 14, 15)$$
$$= M_3.M_4.M_8.M_9.M_{10}.M_{12}.M_{13}.M_{14}.M_{15}$$

By applying De Morgan's theorem, the complement of $\bar{f}$ is

$$(\bar{\bar{f}}) = f = \overline{M_3.M_4.M_8.M_9.M_{10}.M_{12}.M_{13}.M_{14}.M_{15}}$$
$$= \overline{M_3} + \overline{M_4} + \overline{M_8} + \overline{M_9} + \overline{M_{10}} + \overline{M_{12}} + \overline{M_{13}} + \overline{M_{14}} + \overline{M_{15}}$$
$$= m_3 + m_4 + m_8 + m_9 + m_{10} + m_{12} + m_{13} + m_{14} + m_{15}$$
$$f = \Sigma m\ (3, 4, 8, 9, 10, 12, 13, 14, 15)$$

Example 2.15. *Obtain the truth table for the POS expression*

$$F = (A + B + C)\ \ (\bar{A} + B + \bar{C}).$$

Solution. As we know that the maxterm in the given function is represented by '0' in the truth table, we have (see Table 2.18)

Table 2.18

A	B	C	f
0	0	0	0
0	0	1	1
0	1	0	0
0	1	1	1
1	0	0	1
1	0	1	1
1	1	0	1
1	1	1	1

Example 2.16. *Find the complement of the function f = (A + BC) (B + CA) in its POS form.*

Solution. Given function,

$$f = (A + BC)(B + CA)$$
$$= (A + B)(A + C)(B + C)(B + A) = (A + AB + AC + BC)(B + C)$$
$$= AB + AB + BC + AC + ABC + BC$$
$$= AB + BC + CA$$

Complement of the function $\bar{f} = \overline{AB + BC + CA}$

$$= (\bar{A} + \bar{B})(\bar{B} + \bar{C})(\bar{C} + \bar{A})$$

2.14 STANDARD FORMS

The two canonical forms of Boolean algebra SOP and POS are basic forms and those can be obtained from the function truth table.

These two forms are very *seldom the ones with the least number of literals,* because each minterm or maxterm must contain all the variables either uncomplemented or complemented.

Another way of expression of Boolean functions is in "standard form". In this configuration, the terms in the function may contain one, two, or any number of literals. There are two types of standard forms. The sum of products and product of sums.

The *sum* of products is a Boolean function containing product (AND) terms of one or more literals each. The sum denotes ORing of these terms.

Example. $f_1 = \bar{x} + yz + \bar{x}y\bar{z}$

The above SOP expression has 3 product terms of one, two and three literals each, respectively.

A product of sum is a Boolean expression containing sum (OR) terms. Each term may contain any number of literals. The *product* denotes the ANDing of these sum terms.

Example. $f_2 = \bar{x}(y + \bar{z})(\bar{x} + y + \bar{z})$. This expression has 3 sum terms with one, two and three literals, respectively.

2.14.1 Non Standard Form

A Boolean function may be expressed in *non standard form i.e.,* neither in SOP nor in POS.

example $f = (AC + BD)(\bar{A}\bar{C} + \bar{B}\bar{D})$

This can be changed into standard form by using distributive law to remove the parentheses

$f_3 = \bar{A}\bar{C}BD + \bar{B}\bar{D}AC$. The result is in SOP form.

Note : 1. The canonical SOP form or POS-form of a switching function is unique (up to commutation).

2. Two switching functions are equivalent if and only if their canonical sum-of-products forms or product-of-sums forms are identical.

2.15 SIMPLIFICATION OF BOOLEAN EXPRESSIONS

The Boolean theorems, algebra and rules studied in the previous section can be used to simplify the expressions of a logic circuit. The steps for simplification of Boolean expressions using Boolean algebra can be generalized as follows :

1. The original expression is put into the SOP form by repeated application of De Morgan's theorems and multiplication of terms.

2. Once it is in the SOP form, the product terms are checked for common factors and factoring is performed whenever possible.

Boolean rules are used simultaneously which results in the elimination of one or more terms.

Now, we will consider some examples to illustrate the above mentioned technique.

Example 2.17. *Simplify the expression*

(a) $AB+A\bar{B}.(\overline{\bar{A}\bar{C}})$ (b) $AB+A(B+C)+B(B+C)$

(c) $f = \Sigma m\ (1, 3, 5, 7)$ (d) $f = \Pi M\ (3, 5, 7)$ (e) $f = \Sigma m\ (0, 1, 2, 4, 6)$

Solution. (*a*)

$$f = AB+A\bar{B}(\overline{\bar{A}.\bar{C}})$$
$$= AB+A\bar{B}\ (A+C) = AB+A\bar{B}A+A\bar{B}C$$
$$= AB+A\bar{B}+A\bar{B}C = AB+A\bar{B}\ (1+C)$$
$$= AB+A\bar{B} = A\ (B+\bar{B})$$
$$= A$$

(*b*)

$$f = AB+A(B+C)+B(B+C)$$
$$= AB+AB+AC+B+BC$$
$$= AB+AC+BC+B$$
$$= B\ [A+C+1]\ +\ AC$$
$$= B\ [1]\ +\ AC$$
$$= B\ +\ AC$$

(c)

$$f = \Sigma m\ (1, 3, 5, 7)$$
$$f = \bar{A}\bar{B}C+\bar{A}BC+A\bar{B}C+ABC$$
$$= \bar{A}C\ (\bar{B}+B)+AC\ (\bar{B}+B)$$
$$= \bar{A}C+AC\ = C\ (A+\bar{A})$$
$$= C$$

(*d*)

$$f = \Pi M\ (3, 5, 7)$$
$$f = (A+\bar{B}+\bar{C})\ (\bar{A}+B+\bar{C})\ (\bar{A}+\bar{B}+\bar{C})$$
$$= (A\bar{A}+\bar{A}\bar{B}+\bar{C}\bar{A}+BA+B\bar{B}+B\bar{C}+A\bar{C}+\bar{B}\bar{C}+\bar{C})\ (\bar{A}+\bar{B}+\bar{C})$$
$$= (\bar{A}\bar{B}+\bar{C}\bar{A}+AB+B\bar{C}+A\bar{C}+\bar{B}\bar{C}+\bar{C})\ (\bar{A}+\bar{B}+\bar{C})$$

$$= \left(\overline{C}(1+\overline{B}+A+B+\overline{A})+\overline{A}\,\overline{B}+AB\right)(\overline{A}+\overline{B}+\overline{C})$$

$$= (\overline{C}+\overline{A}\,\overline{B}+AB)(\overline{A}+\overline{B}+\overline{C})$$

$$= \overline{A}\,\overline{C}+\overline{A}\,\overline{B}\,\overline{A}+\overline{A}AB+\overline{B}\,\overline{C}+\overline{A}\,\overline{B}\,\overline{B}+AB\overline{B}+\overline{C}+\overline{A}\,\overline{B}\,\overline{C}+AB\overline{C}$$

$$= \overline{A}\,\overline{C}+\overline{A}\,\overline{B}+0+\overline{B}\,\overline{C}+\overline{A}\,\overline{B}+0+\overline{C}+\overline{A}\,\overline{B}\,\overline{C}+AB\overline{C}$$

$$= \overline{C}(A+\overline{B}+\overline{C}+\overline{A}\,\overline{B}+AB+1)+\overline{A}\,\overline{B}$$

$$= \overline{A}\,\overline{B}+\overline{C}$$

(e) $f = \Sigma m\ (0, 1, 2, 4, 6)$

$$f = \overline{A}\,\overline{B}\,\overline{C}+\overline{A}\,\overline{B}C+\overline{A}B\overline{C}+A\overline{B}\,\overline{C}+AB\overline{C}$$

$$= \overline{B}\,\overline{C}\,(\overline{A}+A)+B\overline{C}\,(\overline{A}+A)+\overline{A}\,\overline{B}C$$

$$= \overline{B}\,\overline{C}+B\overline{C}+\overline{A}\,\overline{B}C = \overline{C}\,(B+\overline{B})+\overline{A}\,\overline{B}C$$

$$= \overline{C}+\overline{A}\,\overline{B}C = \overline{A}\,\overline{B}+\overline{C}$$

Example 2.18. *Simplify the following logic expressions using Boolean algebra.*

(a) $(A+B)(A+\overline{B})$ (b) $ABC+A\overline{B}C+AB\overline{C}$ (c) $(A+B)(A+\overline{B})(\overline{A}+C)$

(d) $A+AB+ABC+\ldots$ (e) $AB+A(B+C)+B(B+C)$

Solution. (a) $(A+B)(A+\overline{B}) = A.A+A\overline{B}+BA+B\overline{B}$

$$= A+A\overline{B}+BA \qquad \text{Since } A\text{–}A = A \text{ and } B.\overline{B} = 0$$

$$= A\,[1+\overline{B}+B] = A\,[1+1] = A.1$$

$$= A$$

(b) $ABC+A\overline{B}C+AB\overline{C} = AB\,(C+\overline{C})+A\overline{B}C$

$$= AB+A\overline{B}C = A\,[B+\overline{B}C]$$

$$= A\,[B+C]$$

(c) $(A+B)(A+\overline{B})(\overline{A}+C) = (A+A\overline{B}+BA+B\overline{B})(\overline{A}+C)$

$$= A\overline{A}+\overline{A}AB+B\overline{A}.A+\overline{A}B\overline{B}+AC+A\overline{B}C+ABC+CB\overline{B}$$

$$= 0+0+0+0+AC+A\overline{B}C+ABC+0$$

$$= AC\,[1+\overline{B}+B] = [AC\,(1+1)] = AC.1 = AC$$

(d) $A + AB + ABC + \ldots\ldots = A\,[1+B+BC+\ldots\ldots]$

$$= A\,[1] = A$$

(e) $AB + A\,(B+C) + B\,(B+C) = AB + AB + AC + B + BC$

$$= AB + AC + BC + B = AB + AC + B\,(C + 1)$$

$$= AB + B + AC = B + AC$$

Example 2.19. *Simplify the following expressions*

(a) $(\overline{A}+B)\,(A+B+D)\,\overline{D}$ (b) $ABC + A\overline{B}.(\overline{\overline{A}\,\overline{C}})$

(c) $\overline{A}C\,(\overline{\overline{A}BD})+\overline{A}B\overline{C}\,\overline{D}+A\overline{B}C$ (d) $(\overline{A}+B)\,(A+\overline{B})$

(e) $A\overline{B}C+\overline{A}BD+\overline{C}\,\overline{D}$ (f) $B\,(A+\overline{B})\,(B+C)$

(g) $\overline{AB}\,(B+C)$ (h) $B\,(A+C)\,+C$

(i) $(A+B)\,BC\,+\,A$ (j) $\overline{A\overline{\overline{B}}\,(A+C)}+\overline{\overline{A}B}\,\overline{(A+\overline{\overline{B}}+\overline{C})}$

Solution. (*a*)

$$f = (\overline{A}+B)\,(A+B+D)\,\overline{D}$$
$$= (\overline{A}A+\overline{A}B+\overline{A}D+BA+B+DB)\,\overline{D} \;=\; (\overline{A}B+\overline{A}D+BA+B+DB)\,\overline{D}$$
$$= \left[\overline{A}B+\overline{A}D+B(A+1)+DB\right]\overline{D} = \left[\overline{A}B+\overline{A}D+B(1+D)\right]\overline{D}$$
$$= (\overline{A}B+\overline{A}D+B)\,\overline{D} \;=\; \left[B\,(1+\overline{A})+\overline{A}D\right]\overline{D} \;=\; (B+\overline{A}D)\,\overline{D}$$
$$= B\overline{D}+\overline{A}D\overline{D} \;=\; B\overline{D}$$

(*b*)

$$f = ABC \;+\; A\overline{B}(\overline{\overline{A}\,\overline{C}})$$
$$= ABC+A\overline{B}\,(\overline{\overline{A}}+\overline{\overline{C}}) \;=\; ABC+A\overline{B}\,(A+C)$$
$$= ABC+A\overline{B}\;+AC\overline{B} = AC\,(B+\overline{B})\;+\;A\overline{B}$$
$$= A\left[\overline{B}+C\right]$$

(*c*)

$$f = \overline{A}C\,(\overline{\overline{A}BD})+\overline{A}B\overline{C}\,\overline{D}+A\overline{B}C$$
$$= \overline{A}C\,(A+\overline{B}+\overline{D})+\overline{A}B\overline{C}\,\overline{D}+A\overline{B}C$$
$$= \overline{A}AC+\overline{A}\,\overline{B}C+\overline{A}C\overline{D}+\overline{A}B\overline{C}\,\overline{D}+A\overline{B}C$$
$$= \overline{B}C\,(\overline{A}+A)+\overline{A}\,\overline{D}\,(C+B\overline{C})$$
$$= \overline{B}C+\overline{A}\,\overline{D}\,(C+B)$$

(*d*)

$$f = (\overline{A}+B)\,(A+\overline{B})$$
$$= \overline{A}.A+\overline{A}\,\overline{B}+AB+B.\overline{B}$$
$$= \overline{A}\,\overline{B}+AB$$

(*e*) $A\overline{B}C+\overline{A}BD+\overline{C}D$

This function can not be simplified further.

(*f*)

$$f = B\left[(A+\overline{B})\,(B+C)\right]$$
$$= B\left[AB+AC+B\overline{B}+\overline{B}C\right]$$

$$= B\left[AB+AC+\overline{B}C\right] = AB+ABC+\overline{B}C\,B$$

$$= AB\,(1+C)+0 = AB$$

(g) $$f = \overline{AB}\,(B+C)$$

$$= (\overline{A}+\overline{B})\,(B+C) = \overline{A}B+\overline{A}C+\overline{B}C$$

(h) $$f = B\,(A+C) + C$$

$$= AB + BC + C = AB + C\,(B + 1)$$

$$= AB + C = (C+B)\,(A+C)$$

(i) $$f = (A+B)\,BC + A$$

$$= ABC + BC + A = BC\,[1 + A] + A$$

$$= A + BC = (A+B)\,(A+C)$$

(j) $$f = \overline{A\overline{B}\,(A+C)}+\overline{A}B\,\overline{(A+\overline{B}+\overline{C})}$$

$$= \overline{A\overline{B}+A\overline{B}C}+\overline{A}B\,\overline{(A+\overline{B}+\overline{C})}$$

$$= (\overline{A}+B)\,(\overline{A}+B+\overline{C})+\overline{A}B.\overline{A}BC$$

$$= \overline{A}+\overline{A}B+\overline{A}\,\overline{C}+\overline{A}B+B+B\overline{C}+\overline{A}BC$$

$$= \overline{A}(1+B+B+BC+\overline{C})+B\;(1+\overline{C}) = \overline{A}+B$$

Example 2.20. *Simplify the following Boolean expressions*

(a) $xy+xyz+xy\overline{z}+\overline{x}yz$ (b) $\overline{A}\,\overline{B}\,\overline{C}+\overline{A}B\overline{C}+\overline{A}BC$

(c) $ABC+A\overline{B}C+AB\overline{C}$ (d) $AC+C\,(A+\overline{A}B)$

(e) $\overline{A}B\overline{C}D+\overline{A}BCD+ABD$ (f) $AB+\overline{A}\overline{C}+A\overline{B}C\,(AB+C)$ (g) $(A+B)\,(\overline{A}+C)\,(\overline{B}+\overline{C})$

Solution. (a) $$f = xy+xyz+xy\overline{z}+\overline{x}yz$$

$$= xy\,(1+z+\overline{z})+\overline{x}yz = xy+\overline{x}yz$$

$$= y\,(x+\overline{x}z) = y\,(x+z)$$

(b) $$f = \overline{A}\,\overline{B}\,\overline{C}+\overline{A}B\overline{C}+\overline{A}BC$$

$$= \overline{A}\,\overline{C}\,(B+\overline{B})+\overline{A}BC = \overline{A}\,\overline{C}+\overline{A}BC$$

$$= \overline{A}\,(\overline{C}+BC) = \overline{A}\,(\overline{C}+B)$$

(c) $$f = ABC+A\overline{B}C+AB\overline{C}$$

$$= AB\,(C+\overline{C})+A\overline{B}C = AB+A\overline{B}C$$

$$= A\,(B+\overline{B}C) = A\,(B+C)$$

(d) $$f = AC+C\,(A+\overline{A}B)$$

$$= AC+C\,(A+B) = AC+AC+BC$$

$$= C\,(A+B)$$

(e) $f = \overline{A}B\overline{C}D + \overline{A}BCD + ABD$

$= \overline{A}BD\,(C + \overline{C}) + ABD \;=\; \overline{A}BD + ABD$

$= BD\,(\overline{A} + A) = BD$

(f) $f = AB + \overline{AC} + A\overline{B}C\,(AB + C)$

$= AB + \overline{AC} + A\overline{B}C.AB + A\overline{B}C.C$

$= AB + \overline{AC} + A\overline{B}C \;=\; AB + \overline{A} + \overline{C} + A\overline{B}C$

$= \overline{A} + B + \overline{C} + A\overline{B} \;=\; \overline{A} + B + A + \overline{C}$

$= 1 + B + \overline{C} \;=\; 1 + \overline{C} = 1$

(g) $f = (A + B)\,(\overline{A} + C)\,(\overline{B} + \overline{C})$

$= (A\overline{A} + AC + \overline{A}B + BC)\,(\overline{B} + \overline{C})$

$= A\overline{B}C + \overline{A}\,\overline{B}B + B\overline{B}C + AC\overline{C} + \overline{A}B\overline{C} + BC\overline{C}$

$= A\overline{B}C + \overline{A}B\overline{C}$

Example 2.21. *Simplify the following Boolean expressions.*

(a) $\overline{(A + \overline{B})\,(\overline{A} + B)}$ (b) $\overline{\overline{\overline{AB}\,C}\,D}$

(c) $a + a\overline{b} + a\overline{b}\,\overline{c} + a\overline{b}\,\overline{c}\,\overline{d} + \ldots\ldots$ (d) $\overline{x}\,\overline{y}z + yz + xz$

(e) $(x + y)\,(\overline{\overline{x}\,(\overline{y} + \overline{z})}) + \overline{x}\,\overline{y} + \overline{x}\,\overline{z}$ (f) $(AB + \overline{A}C + BC)\,(A + \overline{B} + A\overline{B})$

(g) $ABC + \overline{A}B + AB\overline{C}$ (h) $(A\overline{B}C) + (\overline{B} + \overline{C})\,(\overline{B} + \overline{D}) + \overline{A + C + D}$

Solution. (a) $f = \overline{(A + \overline{B})\,(\overline{A} + B)}$

$= \overline{A + \overline{B}} + \overline{\overline{A} + B} \;=\; \overline{A}B + A\overline{B}$

(b) $f = \overline{\overline{\overline{AB}\,C}\,D} \;=\; \overline{\overline{\overline{AB}\,C}} + \overline{D}$

$= \overline{AB}\,C + \overline{D} \;=\; (\overline{A} + \overline{B})\,C + \overline{D}$

(c) $f = a + a\overline{b} + a\overline{b}\,\overline{c} + a\overline{b}\,\overline{c}\,\overline{d} + \ldots\ldots$

$= a\,(1 + \overline{b} + \overline{b}\,\overline{c} + \overline{b}\,\overline{c}\,\overline{d}) \;=\; a.1 \;=\; a$

(d) $f = \overline{x}\,\overline{y}z + yz + xz \;=\; z\,(x + y + \overline{x}\,\overline{y})$

$= z\,(x + y + \overline{y}) \;=\; z\,(x + 1) \;=\; z$

(e) $(x + y)\left[\overline{\overline{x}\,(\overline{y} + \overline{z})}\right] + \overline{x}\,\overline{y} + \overline{x}\,\overline{z} \;=\; (x + y)\,(\overline{\overline{x}\,\overline{y} + \overline{x}\,\overline{z}}) + \overline{x}\,\overline{y} + \overline{x}\,\overline{z}$

$= (x + y)\,(x + yz) + \overline{x}\,\overline{y} + \overline{x}\,\overline{z}$

$= x.x + xyz + x.y + y.yz + \overline{x}\,\overline{y} + \overline{x}\,\overline{z}$

$= x + xyz + xy + yz + \overline{x}\,\overline{y} + \overline{x}\,\overline{z} \;= x(1 + yz) + xy + yz + \overline{x}\,\overline{y} + \overline{x}\,\overline{z}$

$$= x(1+y)+yz+\bar{x}\bar{y}+\bar{x}\bar{z} \;=\; x+yz+\bar{x}\bar{y}+\bar{x}\bar{z}$$

$$= x+yz+\bar{x}(\bar{y}+\bar{z}) \;=\; [x+\bar{x}(\bar{y}+\bar{z})]+\; yz$$

$$= [x+(\bar{y}+\bar{z})]+yz \;=\; x+\bar{y}+\bar{z}+yz$$

$$= x+\bar{y}+\bar{z}+z \;=\; x+\bar{y}+1 \;=\; x+1 \;=\; 1$$

(f) $f = (AB+\bar{A}C+BC)(A+\bar{B}+A\bar{B})$

$$= ABA+A\bar{A}C+ABC+AB\bar{B}+\bar{A}BC+B\bar{B}C+ABA\bar{B}+\bar{A}CA\bar{B}+BCA\bar{B}$$

$$= AB+ABC+\bar{A}BC \;=\; AB[1+C]+\bar{A}BC \;=\; AB+\bar{A}BC$$

(g) $f = ABC+\bar{A}B+AB\bar{C} \;=\; AB(C+\bar{C})+\bar{A}B$

$$= AB+\bar{A}B \;=\; B(A+\bar{A}) \;=\; B.1 \;=\; B$$

(h) $f = A\bar{B}C+(\bar{B}+\bar{C})(\bar{B}+\bar{D})+\overline{A+C+D}$

$$= A\bar{B}C+\bar{B}\bar{B}+\bar{B}\bar{D}+\bar{C}\bar{B}+\bar{C}\bar{D}+\bar{A}\bar{C}\bar{D}$$

$$= A\bar{B}C+\bar{B}+\bar{B}\bar{D}+\bar{B}\bar{C}+\bar{C}\bar{D}[1+\bar{A}] \;=\; A\bar{B}C+\bar{B}[1+\bar{D}]+\bar{B}\bar{C}+\bar{C}\bar{D}$$

$$= A\bar{B}C+\bar{B}+\bar{B}\bar{C}+\bar{C}\bar{D} \;=\; \bar{B}[AC+1]+\bar{B}\bar{C}+\bar{C}\bar{D}$$

$$= \bar{B}+\bar{B}\bar{C}+\bar{C}\bar{D} \;=\; \bar{B}[1+C]+\bar{C}\bar{D}$$

$$= \bar{B}+\bar{C}\bar{D}$$

Example 2.22. *Prove that*

(a) $A+\bar{A}B = A+B$

(b) $AB+BC+\bar{B}C = AB+C$

(c) $AB+\bar{B}C+AC = AB+\bar{B}C$

(d) $(x+\bar{y}+xy)(x+\bar{y})\bar{x}y = 0$

(e) $(AB+C+D)(\bar{C}+D)(\bar{C}+D+E) = AB\bar{C}+D$

(f) $A+\bar{A}B+A\bar{B} = A+B$

(g) $(a+b)(\bar{a}+c)(b+c) = (a+b)(\bar{a}+c)$

(h) $xy+\bar{x}z+yz = \bar{x}z+xy$

(i) $\bar{x}\bar{y}z+yz+xz = z$

Solution. (a) $A+\bar{A}B = (A+\bar{A})(A+B)$

$= 1.(A+B) = A + B \quad \therefore \quad$ L.H.S. = R.H.S.

(b) $AB+BC+\bar{B}C = AB+C(B+\bar{B})$

$= AB + C.1 = AB + C$ Hence proved.

(c) Take R.H.S. $= AB+\bar{B}C = AB(1+C)+\bar{B}C(1+A)$

$$= AB+ABC+\overline{B}C+\overline{B}CA = AB+\overline{B}C+AC\,(B+\overline{B})$$

$$= AB+\overline{B}C+AC.1 = AB+\overline{B}C+AC$$

(*d*) $(\overline{x}+\overline{y}+xy)\,(x+\overline{y})\,\overline{x}y = (\overline{x}+\overline{y}+xy)\,(x+\overline{y})\,(\overline{x}y)$

$$= (\overline{x}+\overline{y}+y)\,(\overline{x}yx+\overline{x}y\overline{y}) = (\overline{x}+1)\,(0+0)$$

$$= 1.0 = 0 \quad \text{Hence proved.}$$

(*e*) $f = (AB+C+D)\,(\overline{C}+D)\,(\overline{C}+D+E)$

$$= (AB\overline{C}+C\overline{C}+\overline{C}D+ABD+CD+D)\,(\overline{C}+D+E)$$

$$= \left[AB\overline{C}+\overline{C}D+ABD+D\,(1+C)\right](\overline{C}+D+E)$$

$$= (AB\overline{C}+\overline{C}D+D+ABD)\,(\overline{C}+D+E)$$

$$= \left[AB\overline{C}+D\,(1+\overline{C}+AB)\right](\overline{C}+D+E)$$

$$= (AB\overline{C}+D)\,(\overline{C}+D+E)$$

$$= AB\overline{C}+D\overline{C}+ABD\overline{C}+D+AB\overline{C}E+DE$$

$$= AB\overline{C}+D\,(1+\overline{C}+ABD+E)+AB\overline{C}E$$

$$= AB\overline{C}\,(1+E)+D$$

$$= AB\overline{C}+D \quad \text{Hence proved.}$$

(*f*) $f = A+\overline{A}B+A\overline{B}$

$$= A\,(1+\overline{B})+\overline{A}B = A+\overline{A}B$$

$$= (A+\overline{A})\,(A+B) = 1.A+B$$

$$= A+B \quad \text{Hence proved.}$$

(*g*) $(a+b)\,(\overline{a}+c)\,(b+c) = (a\overline{a}+ac+b\overline{a}+bc)\,(b+c) = (ac+\overline{a}b+bc)\,(b+c)$

$$= abc+\overline{a}b+bc+ac+\overline{a}bc+bc = ac\,(b+1)+bc+\overline{a}b\,(1+c)$$

$$= \overline{a}b+bc+ac$$

R.H.S. $= (a+b)\,(\overline{a}+c)$

$$= a\overline{a}+ac+\overline{a}b+bc = \overline{a}b+bc+ac$$

$\therefore$ L.H.S. = R.H.S. Hence proved.

(*h*) $f = xy+\overline{x}z+yz = xy+\overline{x}z+yz\,(x+\overline{x})$

$$= xy+\overline{x}z+xyz+\overline{x}yz$$

$$= xy\,(1+z)+\overline{x}z\,(1+y) = xy+\overline{x}z.$$

$\therefore$ L.H.S. = R.H.S.

(*i*) $f\,(x, y, z) = \overline{x}\,\overline{y}z+yz+xz$

$$= z\,(\overline{x}\,\overline{y}+y+x) = z\,(x+y+\overline{y})$$

$$= z\,(x+1) = z.1 = z.$$

Example 2.23. *Simplify the following to least number of literals by manipulation of Boolean algebra*

(a) $A\bar{B}\bar{C}D+\bar{A}\bar{B}D+BC\bar{D}+\bar{A}B+B\bar{C}$

(b) $x+xyz+\bar{x}yz+xw+x\bar{w}+\bar{x}y$

(c) $(AB+\bar{A}C+BC)(A+\bar{B}+A\bar{B})$

(d) $(A\bar{B}C)+(\bar{B}+\bar{C})(\bar{B}+\bar{D})+\overline{(A+C+D)}$

(e) $ABC+\bar{A}B+AB\bar{C}$

(f) $\overline{ABC}+\bar{A}BC+ABC+A\bar{B}\bar{C}$

(g) $(B+\bar{C})(\bar{B}+C)+\bar{A}+B+\bar{C}$

(h) $(x+y+z)(x+y)x$

(i) $\bar{a}\bar{b}\bar{c}\bar{d}+\bar{a}\bar{b}\bar{c}d+a\bar{b}\bar{c}\bar{d}+a\bar{b}\bar{c}d$

(j) $(x+\bar{y}\bar{x})[xz+x\bar{z}(y+\bar{y})]$

Solution. (a)

$$
\begin{aligned}
f &= A\bar{B}\bar{C}D+\bar{A}\bar{B}D+BC\bar{D}+\bar{A}B+B\bar{C}\\
&= \bar{B}D(\bar{A}+A\bar{C})+BC\bar{D}+\bar{A}B+B\bar{C} = \bar{B}D(\bar{A}+\bar{C})+BC\bar{D}+\bar{A}B+B\bar{C}\\
&= \bar{B}D\bar{A}+\bar{B}\bar{C}D+BC\bar{D}+\bar{A}B+B\bar{C} = \bar{B}D(\bar{A}+\bar{C})+B(\bar{A}+\bar{C})+BC\bar{D}\\
&= (\bar{A}+\bar{C})(\bar{B}D+B)+BC\bar{D} = (\bar{A}+\bar{C})(B+D)+BC\bar{D}\\
&= \bar{A}B+B\bar{C}+\bar{A}D+\bar{C}D+BC\bar{D} = \bar{A}B+\bar{A}D+\bar{C}D+B(\bar{C}+C\bar{D})\\
&= \bar{A}B+\bar{A}D+\bar{C}D+B\bar{C}+B\bar{D}
\end{aligned}
$$

(b)

$$
\begin{aligned}
f &= x+xyz+\bar{x}yz+xw+x\bar{w}+\bar{x}y\\
f &= x(1+yz+w+\bar{w})+\bar{x}yz+\bar{x}y = x+\bar{x}y(z+1)\\
&= x+\bar{x}y = x+y
\end{aligned}
$$

(c)

$$
\begin{aligned}
&= (AB+\bar{A}C+BC)(A+\bar{B}+A\bar{B})\\
&= AAB+A\bar{A}C+ABC+AB\bar{B}+\bar{A}\bar{B}C+B\bar{B}C+A\bar{B}AB+\bar{A}CA\bar{B}+BCA\bar{B}\\
&= AB+0+ABC+0+\bar{A}\bar{B}C+0+0+0+0\\
&= AB(1+C)+\bar{A}\bar{B}C = AB+\bar{A}\bar{B}C
\end{aligned}
$$

(d) $(A\bar{B}C)+(\bar{B}+\bar{C})(\bar{B}+\bar{D})+\overline{(A+C+D)}$

$$
\begin{aligned}
f(A, B, C, D) &= A\bar{B}C+\bar{B}+\bar{B}\bar{C}+\bar{B}\bar{D}+\bar{C}\bar{D}+\bar{A}\bar{C}\bar{D}\\
&= \bar{B}(AC+1+\bar{C}+\bar{D})+\bar{C}\bar{D}(1+\bar{A})\\
&= \bar{C}\bar{D}+\bar{B}
\end{aligned}
$$

(e)

$$
\begin{aligned}
ABC+\bar{A}B+AB\bar{C} &= AB(C+\bar{C})+\bar{A}B = AB+\bar{A}B\\
&= B(A+\bar{A}) = B
\end{aligned}
$$

(f) $$f = \overline{ABC} + \overline{A}BC + ABC + A\overline{B}\,\overline{C}$$
$$= BC\,(\overline{A}+A) + \overline{A}+\overline{B}+\overline{C} + A\overline{B}\,\overline{C}$$
$$= BC + \overline{B}\left[1+A\overline{C}\right] + \overline{A}+\overline{C}$$
$$= \overline{B}+C+\overline{A}+\overline{C} = \overline{B}+\overline{A}+1$$
$$= \overline{B} + 1 = 1$$

(g) $(B+\overline{C})\,(\overline{B}+C)+\overline{A}+B+\overline{C}$
$$= B\overline{B}+\overline{B}\,\overline{C}+BC+C\overline{C}+\overline{A}+B+\overline{C}$$
$$= \overline{B}\,\overline{C}+BC+\overline{A}+B+\overline{C}$$
$$= \overline{C}\,(1+\overline{B}) + B\,(1+C) + \overline{A}$$
$$= \overline{A}+B+\overline{C}$$

(h) $(x+y+z)\,(x+y)\,(x)$
$$f\,(x, y, z) = (x + y + z)\,(x + xy)$$
$$= (x + y + z)\,[x\,(1+y)]$$
$$= x\,(x + y + z) = x.x + xy + xz$$
$$= x\,(1 + y + z) = x$$

(i) $$f = \overline{a}\,\overline{b}\,\overline{c}\,\overline{d} + \overline{a}\,\overline{b}\,\overline{c}d + a\overline{b}\,\overline{c}\,\overline{d} + a\overline{b}\,\overline{c}d$$
$$= \overline{a}\,\overline{b}\,\overline{c}\,(d+\overline{d}) + a\overline{b}\,\overline{c}\,(d+\overline{d})$$
$$= \overline{a}\,\overline{b}\,\overline{c} + a\overline{b}\,\overline{c} = \overline{b}\,\overline{c}\,(a+\overline{a}) = \overline{b}\,\overline{c}$$

(j) $(x+\overline{x}\,\overline{y})\,[(xz + x\overline{z}\,(y+\overline{y})]$
$$= (x+\overline{y})\left[xz + x\overline{z}\,(y+\overline{y})\right]$$
$$= (x+\overline{y})\,(xz + x\overline{z}) = (x+\overline{y})\left[x(z+\overline{z})\right]$$
$$= (x+\overline{y})x = x + x\overline{y}$$
$$= x\,(1+\overline{y}) = x$$

Example 2.24. *Simplify the following three variable expression using Boolean algebra.*
$$f = \Sigma m\ (0, 2, 4, 6)$$

Solution. Step 1: From the above minterms we can write the expression in SOP form as follows :
$$f = \overline{A}\,\overline{B}\,\overline{C} + \overline{A}B\overline{C} + A\overline{B}\,\overline{C} + AB\overline{C}$$

Step 2: Search for common terms for factorization and apply Boolean rules.
$$f = \overline{A}\,\overline{C}\,(\overline{B}+B) + A\overline{C}\,(B+\overline{B})$$
$$= \overline{A}\,\overline{C} + A\overline{C} = \overline{C}\,(A+\overline{A})$$
$$= \overline{C}$$

Example 2.25. *Simplify the following three variable expression using Boolean algebra*
$$f = \Pi M\ (1, 3, 5, 7).$$

Solution. Step 1 : From the maxterms, write expression in POS form and convert it into SOP form.

$$
\begin{aligned}
f &= (A+B+\bar{C})\,(A+\bar{B}+\bar{C})\,(\bar{A}+B+\bar{C})\,(\bar{A}+\bar{B}+\bar{C})\\
&= (A+B+\bar{C})\,(A\bar{A}+AB+A\bar{C}+\bar{A}\bar{B}+B\bar{B}+\bar{B}\bar{C}+\bar{C}\bar{A}+\\
&\quad B\bar{C}+\bar{C}\bar{C})\,(\bar{A}+\bar{B}+\bar{C})\\
&= (A+B+\bar{C})\,(AB+A\bar{C}+\bar{A}\bar{B}+\bar{B}\bar{C}+\bar{A}\bar{C}+B\bar{C})\,(\bar{A}+\bar{B}+\bar{C})\\
&= (A+B+\bar{C})\,(A\bar{A}B+AB\bar{B}+AB\bar{C}+A\bar{A}\bar{C}+A\bar{B}\bar{C}+A\bar{C}C\\
&\quad +\bar{A}\bar{B}\bar{C}+\bar{A}B\bar{C}+B\bar{B}\bar{C}+B\bar{C}\bar{C}+\bar{A}\bar{A}\bar{C}+\bar{A}\bar{B}\bar{C}+\bar{A}\bar{C}\bar{C}+\bar{A}\bar{C}+\bar{B}\bar{C}+\bar{C}\bar{C})\\
&= (A+B+\bar{C})\,(AB\bar{C}+A\bar{B}\bar{C}+\bar{A}\bar{B}+\bar{A}\bar{B}\bar{C}+\bar{A}B\bar{C}+B\bar{C}+\bar{A}\bar{C}+\bar{B}\bar{C}+\bar{C})\\
&= (A+B+\bar{C})\,\left[A\bar{C}\,(B+\bar{B})+\bar{A}\bar{B}\,(1+\bar{C})+B\bar{C}\,(\bar{A}+1)+\bar{A}\bar{C}+\bar{C}\,(\bar{B}+1)\right]\\
&= (A+B+\bar{C})\,(A\bar{C}+\bar{A}\bar{B}+B\bar{C}+\bar{A}\bar{C}+\bar{C})\\
&= (A+B+\bar{C})\left[\bar{C}\,(A+\bar{A})+\bar{A}\bar{B}+\bar{C}\,(B+1)\right]\\
&= (A+B+\bar{C})\,(\bar{C}+\bar{A}\bar{B}+\bar{C}) = (A+B+\bar{C})\,(\bar{A}\bar{B}+\bar{C})\\
&= A\bar{A}\bar{B}+A\bar{C}+\bar{A}\bar{B}B+B\bar{C}+\bar{A}\bar{B}\bar{C}+\bar{C}\\
&= A\bar{C}+B\bar{C}+\bar{A}\bar{B}\bar{C}+\bar{C} = \bar{C}\left[1+A+B+\bar{A}\bar{B}\right] = \bar{C}
\end{aligned}
$$

2.16 KARNAUGH MAP SIMPLIFICATION

While simplifying a switching function $f\,(x_1, x_2, \ldots, x_n)$, our aim is to find an expression $g\,(x_1, x_2, \ldots, x_n)$ which is equivalent to f and which minimizes some cost criteria. There are three criteria to determine minimal cost. They are :

1. Minimum number of appearances of literals.
2. Minimum number of literals in SOP/POS expression.
3. Minimum number of terms in an SOP expression, provided there is no other such expression with the less number of terms and with fewer literals.

Whenever we want to simplify the Boolean expression by Boolean algebra, we need better understanding of Boolean laws, rules and theorems. During the process of simplification, we have to predict each successive step. For this reason, it is difficult to simplify the expressions and sometimes we may not get unique simplified expression.

Let us consider an example for minimization of the function

$$f\,(A, B, C) = \bar{A}B\bar{C}+\bar{A}\bar{B}\bar{C}+A\bar{B}\bar{C}+\bar{A}BC+ABC+A\bar{B}C$$

While simplifying, combine first and second product terms, which yields

$$\bar{A}\bar{C}\,(B+\bar{B}) = \bar{A}\bar{C}$$

Similarly, the combinations of the second and third, fourth and fifth, fifth and sixth terms yields

$$\bar{B}\bar{C}\,(A+\bar{A}) = \bar{B}\bar{C}$$

$$BC(A+\overline{A}) = BC$$

$$AC(B+\overline{B}) = AC.$$

$\therefore$ The simplified expression is $f = \overline{A}\,\overline{C} + \overline{B}\,\overline{C} + BC + AC$, which is said to be in an irredundant form, since any attempt to reduce it, either by deleting any of the four terms or by removing a literal, will give an expression which is not equivalent to f.

An SOP expression from which no term or literal can be deleted without altering its logical value is called an irredundant or irreducible expression.

The above reduction procedure is not unique and a different combination of terms may give different reduced expressions. If we combine the 1st and 2nd, 3rd and 6th and 4th and 5th terms, we obtain an expression

$$f = \overline{A}\,\overline{C} + A\overline{B} + BC$$

In a similar manner, by combining the 1st and 4th, 2nd and 3rd, the 5th and 6th terms, we obtain third irredundant expression.

$$f(A, B, C) = \overline{A}B + \overline{B}\,\overline{C} + AC$$

While all three expressions are irredundant, only the later two are minimal. Consequently, *an irredundant expression is not necessarily minimal, nor the minimal* always *unique*. Therefore it is necessary to develop procedures for generating the set of all minimal expressions, so that the appropriate one may be selected according to other criteria.

The *map method* gives us a systematic approach for simplifying a Boolean expression *i.e.*, it provides a simple straightforward procedure for minimizing Boolean functions. This method may be regarded either as a pictorial form of a truth table or as an extension of venn diagram. This map method was first proposed by Veitch and modified by Karnaugh. Hence it is also known as "*Veitch diagram*" or the "*Karnaugh map*".

2.16.1 The Karnaugh Map

Karnaugh map is a graphical device used to simplify a logical equation or to convert a truth table to it's corresponding logic circuit in a simple, orderly process. The K–map, like a truth table, is a means for showing the relationship between logic inputs and the designed output. It contains boxes/squares called "cells". Each square represents one minterm, of 2^n possible minterms formed from 'n' variables.

Therefore, a 2-variable K–map contains $2^2 = 4$ cells, a 3-variable K–map contains $2^3 = 8$ cells, a 4-variable K–map contains $2^4 = 16$ cells, and so forth.

Fig. 2.1 shows the outlines of 2, 3 and 4 variable maps.

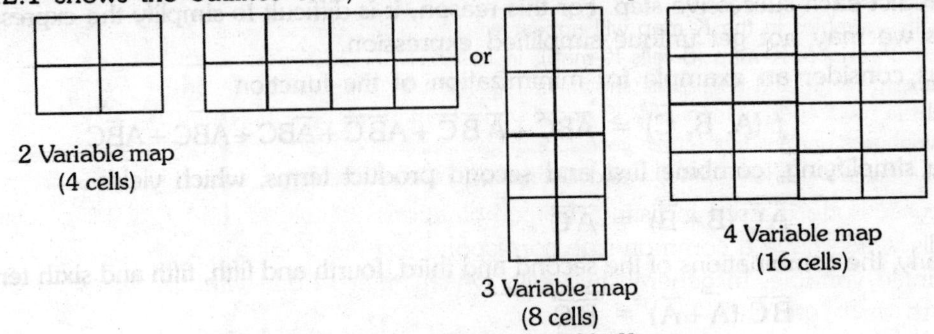

Fig. 2.1 The 2, 3 and 4 variable K–maps.

Since, any Boolean function can be expressed as a sum of minterms, it follows that a Boolean function is recognized graphically in the map from the area enclosed by those cells whose minterms are included in the function.

The minterms are assigned to the cells of a K–map by labelling each row and column of the map with a variable, with it's complement, or with a combination of variables and complements. The minterm corresponding to a given cell is the product of all variables of the corresponding row and column. Fig 2.2 shows the way to label the rows and columns of a 2, 3 and 4 variable map and the minterms corresponding to each cell.

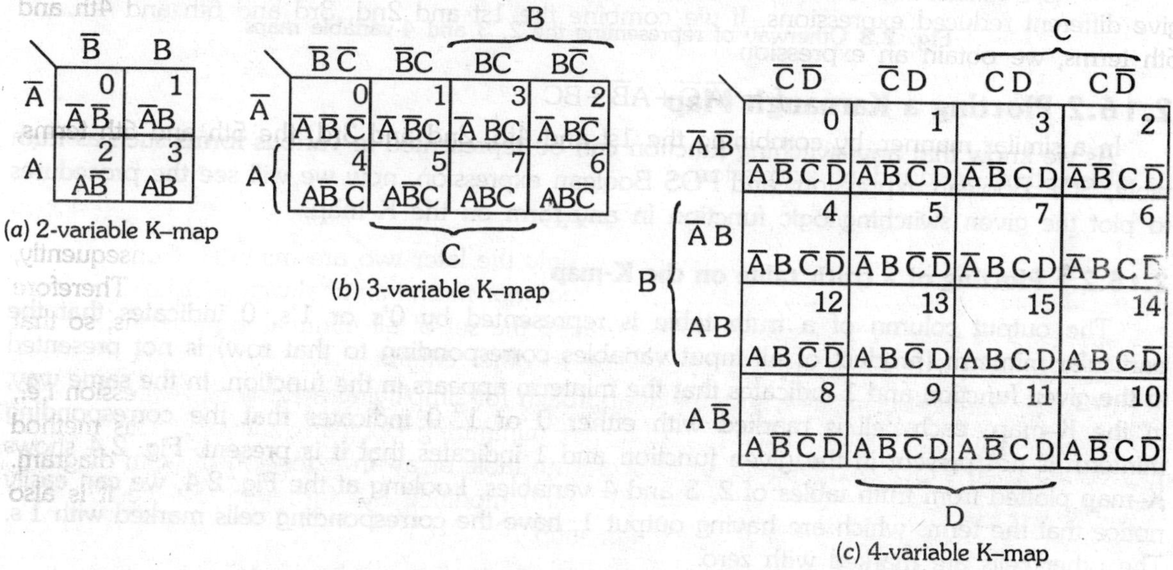

Fig. 2.2 K–maps with minterms.

The important characterstic of the K–map is, when we move from one cell to the next along any row or column, one and only one variable in the minterm changes (from complement to uncomplement form or vice-versa). If we observe in Fig. 2.2 (*b*), when moving from first row-first column, to second row-first column, *i.e.,* from $\overline{A}\,\overline{B}\,\overline{C}$ to $A\overline{B}\,\overline{C}$, there is only one change from $\overline{A}$ to A. Similarly, only one change occurs in moving from first column to second column *i.e.,* from $\overline{A}\,\overline{B}\,\overline{C}$ to $\overline{A}\,\overline{B}C$, the change here is from $\overline{C}$ to C.

Note : 1. Irrespective of the number of variables, the labels along each row and column must confirm to the single-change rule.

2. All the variables in the K–map will appear equally in complement and uncomplement forms w.r.t. all cells *i.e.,* in a 3 variable K–map (8-cells K–map), in four cells, A will appear in normal form and in four cells as a complement form, and so on (as shown in Fig. 2.2 with brackets on the map) Fig. 2.3 shows another way to label the rows and columns of 2, 3 and 4 variable maps and the corresponding minterms of each cell. Instead of writing actual product terms, corresponding shorthand minterm notations are written in the cell, the row and columns are marked with 0's and 1's instead of variables.

The cyclic code is used in listing the combinations as column and row. As a result of this coding, cells which have a common side correspond to the combinations that differ by the value of just a single variable. In general, two cells which differ in one variable value are said to be *'adjacent'* to each other.

A \ B	0	1
0	m_0	m_1
1	m_2	m_3

(*a*) 2-variable map

A \ BC	00	01	11	10
0	m_0	m_1	m_3	m_2
1	m_4	m_5	m_7	m_6

(*b*)

AB \ CD	00	01	10	11
00	m_0	m_1	m_3	m_2
01	m_4	m_5	m_7	m_6
11	m_{12}	m_{13}	m_{15}	m_{14}
10	m_8	m_9	m_{11}	m_{10}

(*c*)

Fig. 2.3 Otherway of representing the 2, 3 and 4-variable maps.

2.16.2 Plotting a Karnaugh Map

As we know that any switching function can be represented in various forms such as truth table, SOP Boolean expression, and POS Boolean expression, now we will see the procedures to plot the given switching/logic function in any form on the K–map.

2.16.2.1 Plotting of a truth table on the K-map

The output column of a truth table is represented by 0's or 1's, 0 indicates that the particular minterm (product of all input variables corresponding to that row) is not presented in the given function and 1 indicates that the minterm appears in the function. In the same way, in the K–map, each cell is marked with either 0 or 1. 0 indicates that the corresponding minterm is not present in the given function and 1 indicates that it is present. Fig. 2.4 shows K–map plotted from truth tables of 2, 3 and 4 variables. Looking at the Fig. 2.4, we can easily notice that the terms which are having output 1, have the corresponding cells marked with 1's. The other cells are marked with zero.

A	B	f
0	0	1
0	1	0
1	0	1
1	1	0

A \ B	0	1
0	1	0
1	0	1

or

A \ B	$\overline{B}$	B
$\overline{A}$	1	0
A	0	1

Fig. 2.4 (*a*) Representation of 2-variable truth table on 2-variable K–map.

A	B	C	f
0	0	0	1
0	0	1	0
0	1	0	1
0	1	1	0
1	0	0	0
1	0	1	1
1	1	0	0
1	1	1	0

AB \ C	0	1
00	1	0
01	1	0
11	0	0
10	0	1

or

A \ BC	00	01	11	10
0	1	0	0	1
1	0	1	0	0

or

A \ BC	$\overline{B}\,\overline{C}$	$\overline{B}C$	BC	$B\overline{C}$
$\overline{A}$	1	0	0	1
A	0	1	0	0

Fig. 2.4 (*b*) Representation of 3-variable truth table on 3-variable K-maps.

A	B	C	D	f
0	0	0	0	0
0	0	0	1	1
0	0	1	0	0
0	0	1	1	1
0	1	0	0	1
0	1	0	1	0
0	1	1	0	1
0	1	1	1	0
1	0	0	0	1
1	0	0	1	1
1	0	1	0	1
1	0	1	1	1
1	1	0	0	0
1	1	0	1	0
1	1	1	0	0
1	1	1	1	0

AB \ CD	00	01	11	10
00	0	1	1	0
01	1	0	0	1
11	0	0	0	0
10	1	1	1	1

or

AB \ CD	$\bar{C}\bar{D}$	$\bar{C}D$	CD	$C\bar{D}$
$\bar{A}\bar{B}$	0	1	1	0
$\bar{A}B$	1	0	0	1
AB	0	0	0	0
$A\bar{B}$	1	1	1	1

Fig. 2.4(*c*) Representation of 4-variable truth table on 4-variable K-map.

2.16.2.2 Plotting of canonical SOP form on K–map

A switching function/Boolean function in SOP form can be plotted on a K–map by placing a 1 in each cell corresponding to the minterm in the canonical SOP expression. Remaining cells are posted with zeros *i.e.,* for the products which are not present in the given canonical SOP expression. Suppose, the given function is not in a canonical SOP form, then convert it into canonical form. The following examples illustrate the above procedure.

Example 2.26. *Plot the Boolean expression* $f = \bar{A}\bar{B}\bar{C} + \bar{A}BC + AB\bar{C}$ *on the K–map.*

Solution. The given expression '*f*' has 3–variables and hence it can be plotted using 3–variable K–map (see Fig. 2.5).

The given function $f = \bar{A}\bar{B}\bar{C} + \bar{A}BC + AB\bar{C}$

$= \Sigma m\ (0, 3, 6)$

A \ BC	00	01	11	10
0	1	0	1	0
1	0	0	0	1
	m_0		m_3	m_6

Fig. 2.5(*a*)

or

	$\bar{B}\bar{C}$	$\bar{B}C$	BC	$B\bar{C}$
$\bar{A}$	1	0	1	0
A	0	0	0	1
	$\bar{A}\bar{B}\bar{C}$		$\bar{A}BC$	$AB\bar{C}$

Fig. 2.5(*b*)

Example 2.27. *Plot the Boolean expression*

$$f = \bar{A}\bar{B}CD + \bar{A}BC\bar{D} + A\bar{B}CD + AB\bar{C}\bar{D} + ABCD$$

Solution. The expression has 4-variables and hence a 4-variable K–map is enough to plot *f* as shown below (Fig. 2.6)

$$f = \Sigma m\ (3, 6, 9, 12, 15)$$

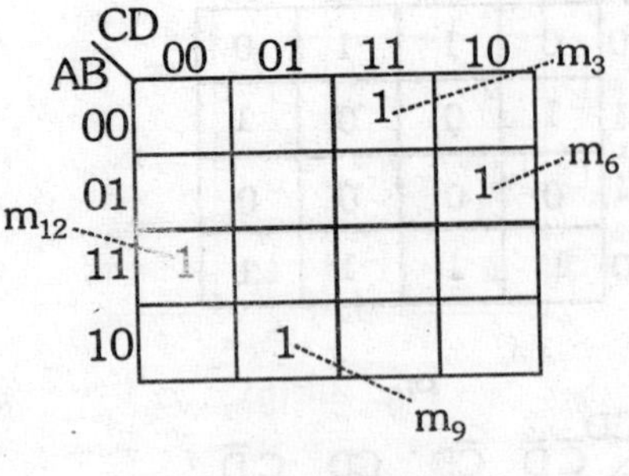

Fig. 2.6(*a*)

(or)

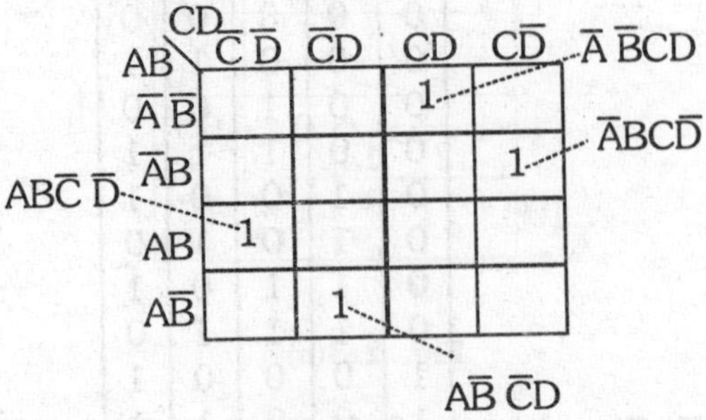

Fig. 2.6(*b*)

2.16.2.3 Plotting of canonical POS form on K–map

A Boolean function expressed in the POS form can be plotted on the K–map by placing a 0 in each cell corresponding to a maxterm in the expression. Remaining cells are filled with 1's. This procedure is illustrated in the following examples.

Example 2.28. *Plot the Boolean expression* $f = (A+B+C)\ (A+\bar{B}+C)\ (\bar{A}+B+\bar{C})\ (A+B+\bar{C})$ *on the K–map.*

Solution. The expression contains 3 variables and hence it can be plotted using a 3-variable K–map as shown below. (Fig. 2.7)

$$f = \Pi M\ (0, 1, 2, 5)$$

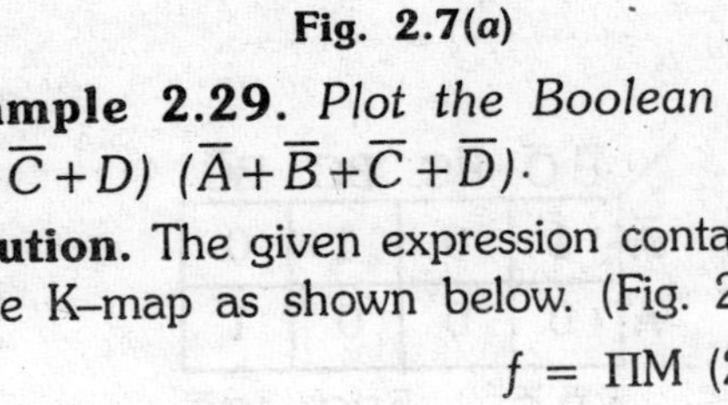

Fig. 2.7(*a*)

(or)

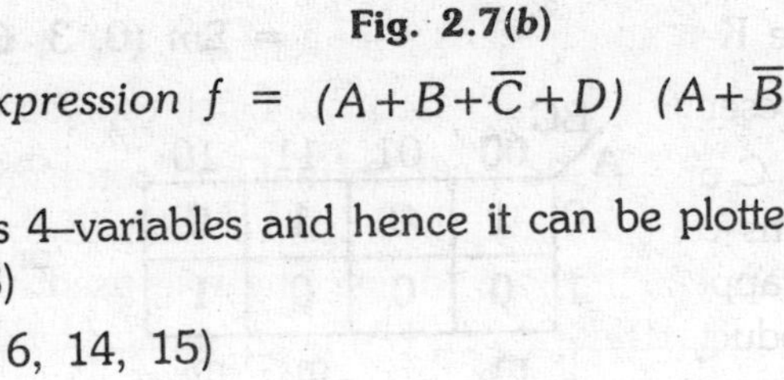

Fig. 2.7(*b*)

Example 2.29. *Plot the Boolean expression* $f = (A+B+\bar{C}+D)\ (A+\bar{B}+\bar{C}+D)\ (\bar{A}+\bar{B}+\bar{C}+D)\ (\bar{A}+\bar{B}+\bar{C}+\bar{D})$.

Solution. The given expression contains 4–variables and hence it can be plotted using a 4-variable K–map as shown below. (Fig. 2.8)

$$f = \Pi M\ (2, 6, 14, 15)$$

AB \ CD	00	01	11	10
00	1 (0)	1 (1)	1 (3)	0 (2)
01	1 (4)	1 (5)	1 (7)	0 (6)
11	1 (12)	1 (13)	0 (15)	0 (14)
10	1 (8)	1 (9)	1 (11)	1 (10)

M_2 M_6 M_{15} M_{14}

Fig. 2.8(*a*)

or

AB \ CD	C+D	C+$\overline{D}$	$\overline{C}$+$\overline{D}$	$\overline{C}$+D
A+B	1	1	1	0
A+$\overline{B}$	1	1	1	0
$\overline{A}$+$\overline{B}$	1	1	0	0
$\overline{A}$+B	1	1	1	1

Fig. 2.8(*b*)

2.16.3 Grouping of Cells for Simplification

We have seen the representation of logic function on the K–map and we also know that the minterms are marked by 1's and the maxterms are marked by 0's. Once the logic function is plotted on the K–map, we have to use grouping technique to simplify the logic function. Grouping is nothing but combining the adjacent logic cells which contain 1's or 0's. The simplification is achieved by grouping adjacent 1's and 0's in grouping of 2^x, where $x = 1, 2.....n$, where n is the number of variables in the given function. When adjacent 1's are grouped, then we get SOP form as a result and when adjacent 0's are grouped, then we get the result in POS form. Let us see the various grouping rules.

2.16.3.1 Grouping two adjacent cells (pairs)

Now, we will consider some examples for combining two adjacent cells.

(*a*) Fig. 2.9(*a*) shows the K–map of a particular 3-variable function.

A \ BC	00	01	11	10
0	0	0	1	1
1	0	0	0	0

Fig. 2.9(*a*)

The K–map contains a pair of 1's that are horizontally adjacent to each other; the first 1 cell represents $\overline{A}BC$ and second 1 cell represents $\overline{A}B\overline{C}$. Note that in these two terms, only the variable C appears in both normal and complement form ($\overline{A}$ and B remain unchanged). These two terms can be grouped/looped/combined to give a resultant that eliminates the variable C, since it appears in both forms. The rule used here is A*a* + A*a*' = A, where A denotes a single (or) product of literals and '*a*' is a single literal (variable). This can be proved as

$$f = \overline{A}BC + \overline{A}B\overline{C}$$

$$= \overline{A}B\,(C+\overline{C}) = \overline{A}B\,(1)$$

$$= \overline{A}B$$

(*b*) The same principle holds true for any pair of vertically (or) horizontally adjacent 1's. Fig. 2.9 (*b*) shows an example of two vertically adjacent 1's. The first 1 cell represents $\overline{A}\,\overline{B}C$

and second 1 cell represents $A\overline{B}C$. These two cells can be looped/combined and the variable A is eliminated, since it appears in both its uncomplemented and complemented form to give a result of

$$f = \overline{A}\,\overline{B}C + A\overline{B}C$$
$$= \overline{B}C\,(\overline{A}+A) = \overline{B}C\,(1)$$
$$= \overline{B}C$$

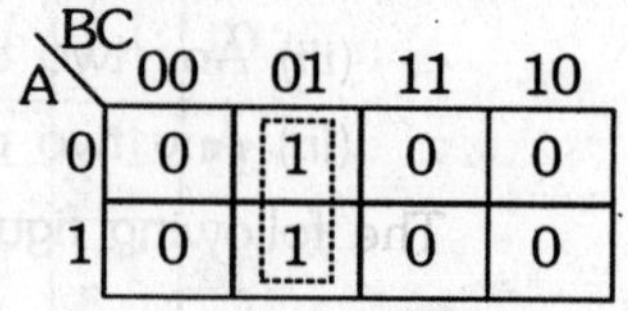

Fig. 2.9(*b*)

(*c*) In a K–map the left most column and right most column of squares are considered to be adjacent. Thus, the two 1's in these columns with a common row can be combined to eliminate one variable. This is illustrated in the Fig. 2.9 (*c*).

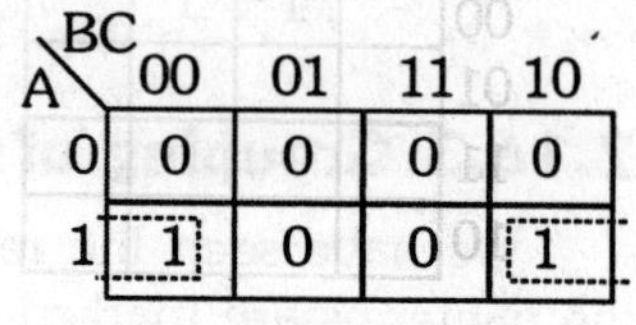

Fig. 2.9(*c*)

The left most column 1 cell represents $A\overline{B}\,\overline{C}$ and right most column 1 cell represents $AB\overline{C}$. Here the variable B has appeared in both its complemented and uncomplemented forms and hence it is eliminated as follows :

$$f = A\overline{B}\,\overline{C} + AB\overline{C}$$
$$= A\overline{C}\,(\overline{B}+B) = A\overline{C}\,(1) = A\overline{C}$$

(*d*) Let us consider another example of K–map that has two overlapping pairs of 1's. This shows that one cell can be shared between two pairs as shown in Fig. 2.9 (*d*).

$$f = \overline{A}\,\overline{B}\,\overline{C} + \overline{A}\,\overline{B}C + A\overline{B}C$$
$$= \overline{A}\,\overline{B}\,\overline{C} + \overline{A}\,\overline{B}C + \overline{A}\,\overline{B}C + A\overline{B}C$$
$$= \overline{A}\,\overline{B}\,(\overline{C}+C) + \overline{B}C\,(\overline{A}+A)$$
$$= \overline{A}\,\overline{B}\,(1) + \overline{B}C\,(1) = \overline{A}\,\overline{B} + \overline{B}C$$

Group1 Group2

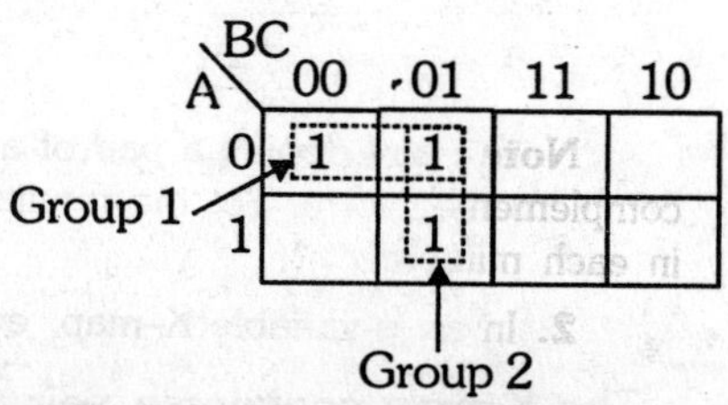

Fig. 2.9(*d*)

(*e*) Now we will consider another example in which three groups of pairs can be formed. But two pairs are enough to include all 1's present in the K–map. In this case, the third pair is not required as shown in Fig. 2.9 (*e*).

$$f = \overline{A}\,\overline{B}C + \overline{A}BC + ABC + AB\overline{C}$$
$$= \overline{A}C\,(\overline{B}+B) + AB\,(C+\overline{C})$$
$$= \overline{A}C\,(1) + AB\,(1) = \overline{A}C + AB$$

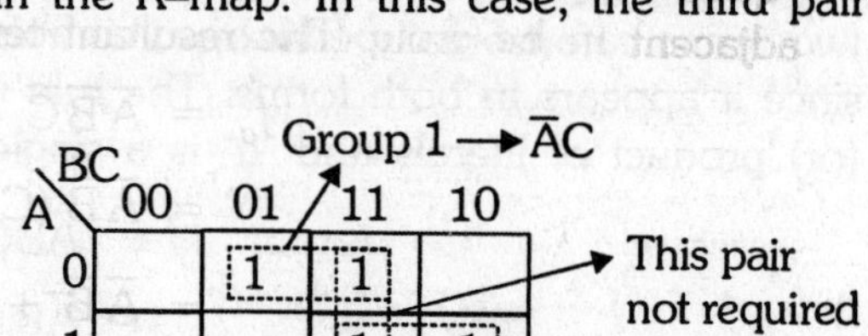

Fig. 2.9(*e*)

(*f*) Now we will see the different possible grouping of pairs in a 4-variable K–map:

(*i*) Two horizontal adjacent cells

(*ii*) Two vertical adjacent cells

(*iii*) Any two cells of first and last rows and belong to same column

(*iv*) Any two cells of first and last columns and belong to same row.

The following figure shows the examples for each case. [see Fig. 2.9 (*f*)]

AB \ CD	00	01	11	10
00		1	1	
01				
11				
10				

Fig 2.9.(*f*) (*i*)

AB \ CD	00	01	11	10
00				
01			1	
11			1	
10				

Fig. 2.9(*f*) (*ii*)

AB \ CD	00	01	11	10
00				1
01				
11				
10				1

Fig. 2.9(*f*) (*iii*)

AB \ CD	00	01	11	10
00				
01				
11	1			1
10				

Fig. 2.9(*f*) (*iv*)

$$f_1 = \bar{A}\bar{B}\bar{C}D + \bar{A}\bar{B}CD$$
$$= \bar{A}\bar{B}D\,(C+\bar{C}) = \bar{A}\bar{B}D$$
$$f_2 = \bar{A}BCD + ABCD$$
$$= BCD\,(\bar{A}+A) = BCD$$
$$f_3 = \bar{A}\bar{B}C\bar{D} + A\bar{B}C\bar{D}$$
$$= \bar{B}C\bar{D}\,(A+\bar{A}) = \bar{B}C\bar{D}$$
$$f_4 = AB\bar{C}\bar{D} + ABC\bar{D}$$
$$= AB\bar{D}\,(C+\bar{C}) = AB\bar{D}\,(1) = AB\bar{D}$$

Note : **1.** Looping a pair of adjacent 1's in a K–map eliminates the variable that appears in normal and complemented form. So, the resultant product term contains (*n*–1) variables where *n* = number of variables in each minterm.

2. In an *n*-variable K–map, every minterm contains '*n*' adjacents.

2.16.3.2 Grouping 4 adjacent cells (quads)

In K-map we can group four 1's that are adjacent to each other. This group is called "quad". Fig. 2.10 shows several examples of quads. Fig. 2.10 (*a*) shows the four 1's which are adjacent horizontally. The resultant term is

$$f_a = \bar{A}\bar{B}\bar{C} + \bar{A}\bar{B}C + \bar{A}BC + \bar{A}B\bar{C}$$
$$= \bar{A}\bar{B}\,(C+C) + \bar{A}B\,(C+\bar{C})$$
$$= \bar{A}\bar{B} + \bar{A}B = \bar{A}\,(\bar{B}+B) = \bar{A}$$

A \ BC	00	01	11	10
0	1	1	1	1
1				

Fig. 2.10(*a*)

Here the resultant term contains only one variable. Fig. 2.10 (*b*) shows another quad in a 3 variable K–map.

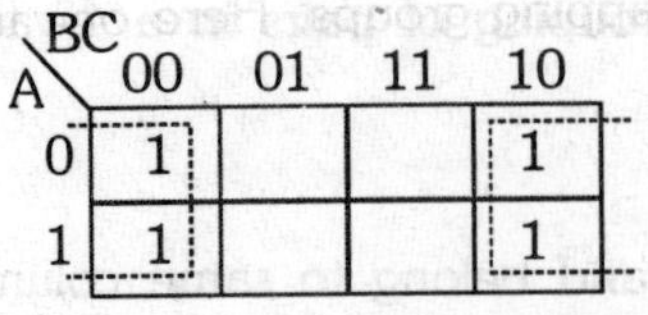

Fig. 2.10(*b*)

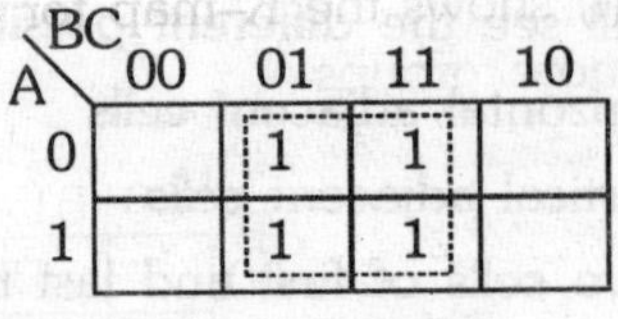

Fig. 2.10(*c*)

The resultant of Fig. 2.10 (*b*) is $\overline{A}\,\overline{B}\,\overline{C} + A\overline{B}\,\overline{C} + \overline{A}B\overline{C} + AB\overline{C}$

$$f_b = \overline{B}\,\overline{C}\,(\overline{A}+A) + B\overline{C}\,(\overline{A}+A)$$
$$= \overline{B}\,\overline{C}\,(1) + B\overline{C}\,(1) = \overline{B}\,\overline{C} + B\overline{C}$$
$$= \overline{C}\,(B+\overline{B}) = C$$

The K-map in Fig. 2.10 (*c*) contains four 1's in a square, and they are considered adjacent to each other. The result is

$$f_c = \overline{A}\,\overline{B}C + A\overline{B}C + \overline{A}BC + ABC$$
$$= \overline{B}C\,(A+\overline{A}) + BC\,(\overline{A}+A)$$
$$= \overline{B}C + BC = C\,(B+\overline{B}) = C$$

Now we will consider the different quad combinations in a 4-variable K-map.

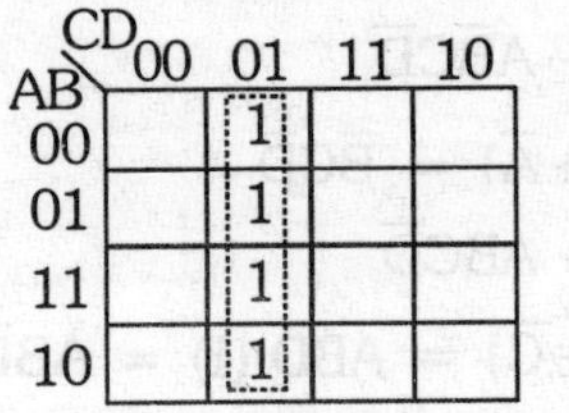

Fig. 2.10(*d*)

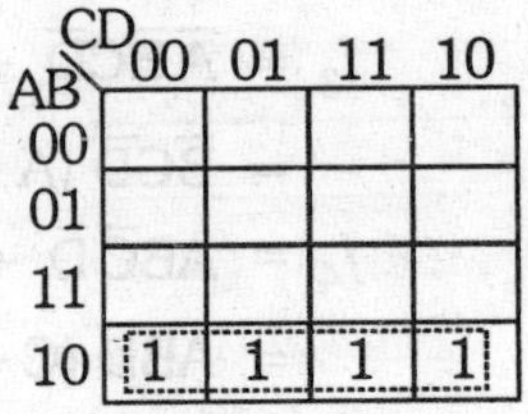

Fig. 2.10(*e*)

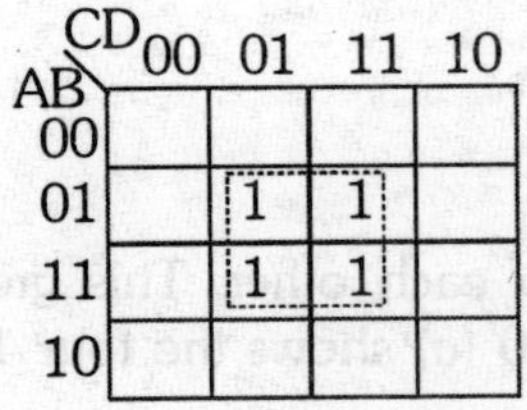

Fig. 2.10(*f*)

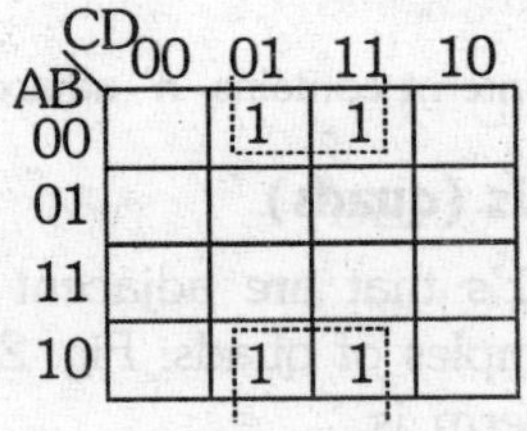

Fig. 2.10(*g*)

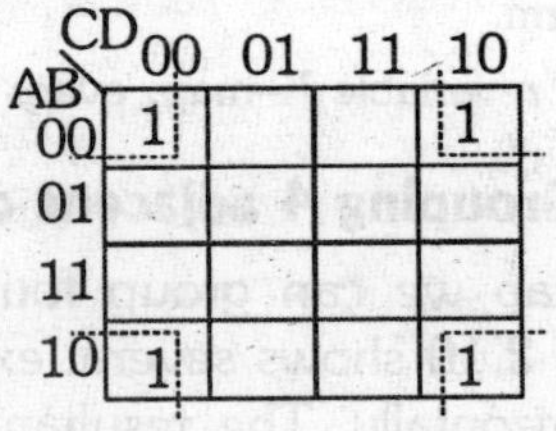

Fig. 2.10(*h*)

Fig. 2.10 (*d*) contains four 1's vertically, Fig. 2.10 (*e*) contains four 1's horizontally, Fig. 2.10 (*f*) contains four 1's in a square, Fig. 2.10 (*g*) contains four 1's which are adjacent and Fig. 2.10 (*h*) also contains four 1's at each corner of the K-map are also adjacent to each other since left most and right most columns and top and bottom rows are considered to be adjacent to each other.

$f_d = \overline{C}D \qquad f_e = A\overline{B} \qquad f_h = BD, \qquad f_g = \overline{B}D, \qquad f_h = \overline{B}\,\overline{D}$

The Fig. 2.10 (*i*) shows the K–map for overlapping groups. Here one term can be shared between two (or) more groups.

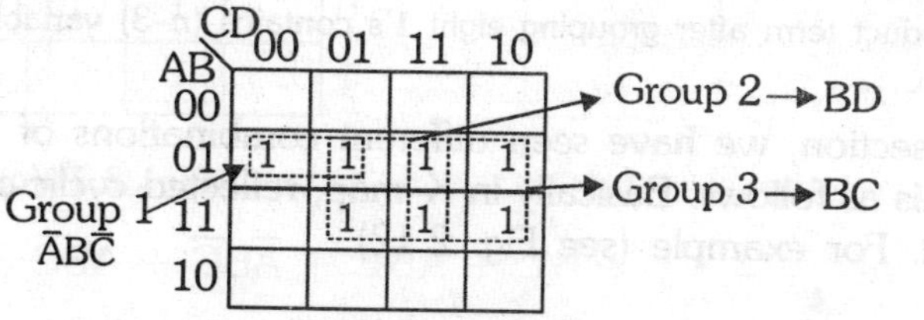

Fig. 2.10(*i*)

Note : 1. Looping a quad of 1's eliminates the two variables that appear in both normal and complemented form.

2. After grouping a quad, the resultant term contains only (*n*–2) variables, where '*n*' is the number of variables of K–map.

2.16.3.3 Grouping eight adjacent cells (octets)

A group of eight 1's that are adjacent to one another is called an 'octet'. Fig. 2.11 shows several examples of octets. When an octet is looped/grouped in a four variable map, three of the four variables are eliminated because only one variable remains unchanged. Fig. 2.11 (*a*) shows the eight 1's that are horizontally adjacent and Fig. 2.11 (*b*) shows the eight 1's that are vertically adjacent.

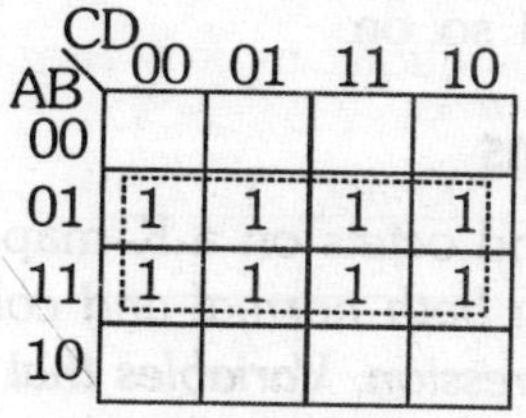

Fig. 2.11(*a*)

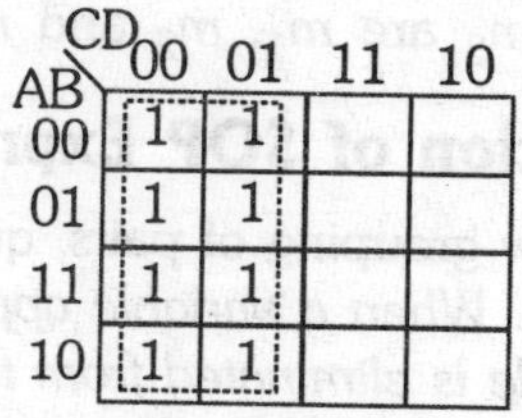

Fig. 2.11(*b*)

The K–map in Fig. 2.11 (*a*) has the following terms.

$$f_a = \bar{A}B\bar{C}\bar{D} + \bar{A}B\bar{C}D + \bar{A}BCD + \bar{A}BC\bar{D} + AB\bar{C}D + AB\bar{C}D + ABCD + ABC\bar{D}$$

$$= \bar{A}B\bar{C}(\bar{D}+D) + \bar{A}BC(D+\bar{D}) + AB\bar{C}(D+\bar{D}) + ABC(D+\bar{D})$$

$$= \bar{A}B\bar{C} + \bar{A}BC + AB\bar{C} + ABC$$

$$= \bar{A}B(\bar{C}+C) + AB(\bar{C}+C) = \bar{A}B + AB = B(A+\bar{A}) = B$$

Similarly

$$f_b = D$$

The other possible octets in a four-variable K–map are shown in Fig. 2.11 (*c*) and 2.11 (*d*).

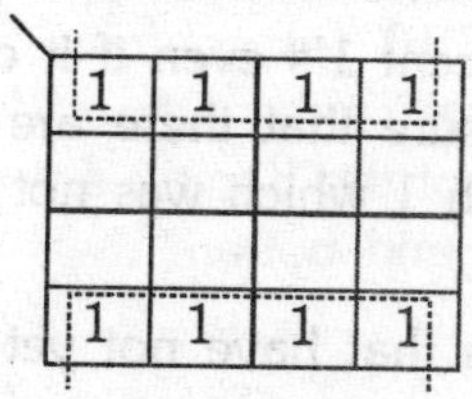

Fig. 2.11(*c*) $f_c = \bar{B}$

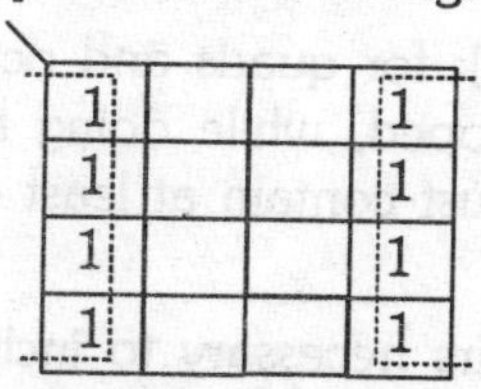

Fig. 2.11(*d*) $f_d = \bar{D}$

Note : 1. Looping an octet of 1's eliminates the three variables that appear in both normal and complemented form.

2. The resultant product term after grouping eight 1's contains (*n*–3) variables, where *n* is the number of variables in the K–map.

In the above subsection, we have seen different combinations of adjacents. One way we can find the adjacents is as follows. Basically in K–map, reflected cyclic code is used to represent the value of each cell. For example (see Fig. 2.12)

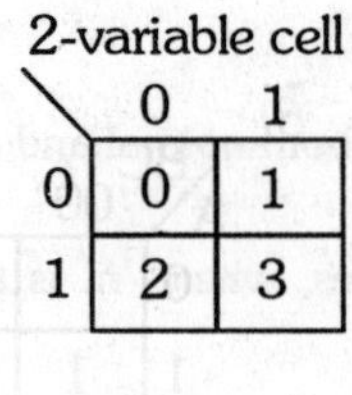

Fig. 2.12(*a*)

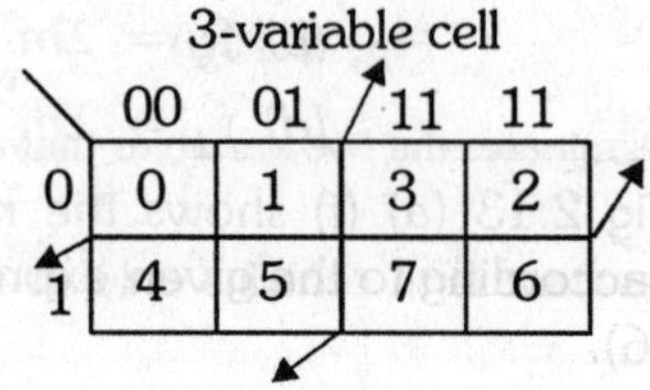

Fig. 2.12(*b*)

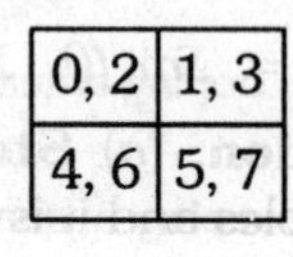

Fig. 2.12(*c*)

If we fold the K–map at the centre vertically and horizontally, the cells which coincide with each other become adjacents. After making the vertical folds to the 3-variable K–map, the resultant contains 4 cells and all the minterms written in them are adjacent. Each minterm contains 3 adjacent minterms. For example adjacents of minterm m_0 are m_1, m_2 and m_4 and adjacents of minterm m_3 are m_1, m_2 and m_7 and so on.

2.16.4 Simplification of SOP Expressions

We have seen how grouping of pairs, quads and octets on a K–map can be used to obtain a simplified expression. *When a variable appears in both normal and complemented form with in a group, that variable is eliminated from the expression. Variables that are same for all cubes of the loop must appear in the final expression.*

For example, a pair of 1's eliminates one variable, a quad of 1's eliminates two variables and an octet of 1's eliminates three variables.

From the above discussion, it is clear that a larger loop of 1's eliminates more variables. The procedure will first be outlined and then applied to several examples. The steps below are followed in using K–map method for simplifying a Boolean expression :

1. Construct the K–map and place 1's in those cells corresponding to the 1's in the truth table. Place 0's in the other cells.

2. Examine the K–map for adjacent 1's and loop those 1's which are not adjacent to any other 1's. These are called "isolated" 1's.

3. Check for those 1's which are adjacent to only one other 1 and loop such pairs.

4. Similarly check for quads and octets of adjacent 1's even if it contains some 1's that have already been looped, while doing this, make sure that there are minimum number of loops and the loop must contain at least one cell with 1 which was not covered by any other loop.

5. Loop any pairs necessary to include any 1's that have not yet been looped, making sure that there are minimum number of loops.

6. Form the simplified expression by summing all the resultant product terms of all the loops.

These steps will be followed exactly and referred to in the following examples. In each case, the resulting logic expression will be in it's simplest SOP form.

Note : Always try to form as large as possible group and it may not be the subgroup *i.e.*, which was already covered.

Example 2.30. *Find the minimal expression using K-map for the following examples :*

(a) $f_a = \Sigma m\ (1, 2, 4, 6)$ *(b)* $f_b = \Sigma m\ (0, 5, 7)$

(c) $f_c = \Sigma m\ (0, 1, 4, 5)$ *(d)* $f_d = \Sigma m\ (0, 2, 4, 6)$

Solution. (*a*) **Step 1.** Fig 2.13 (*a*) (*i*) shows the K-map for 3-variables and it is plotted according to the given expression f_a (A, B, C) = Σm (1, 2, 4, 6).

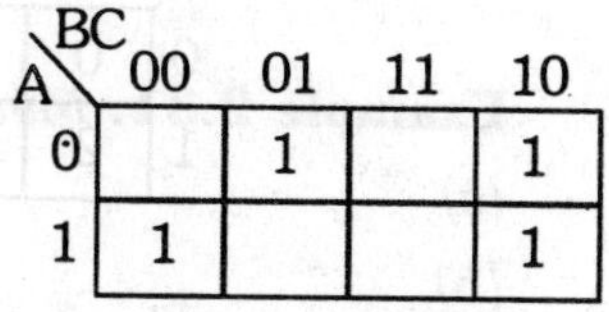

Fig. 2.13(*a*) (*i*)

Step 2. There is an isolated 1 in cell 1 and it does not have any adjacents. It is separately referred as group 1 $= \overline{A}\,\overline{B}C$.

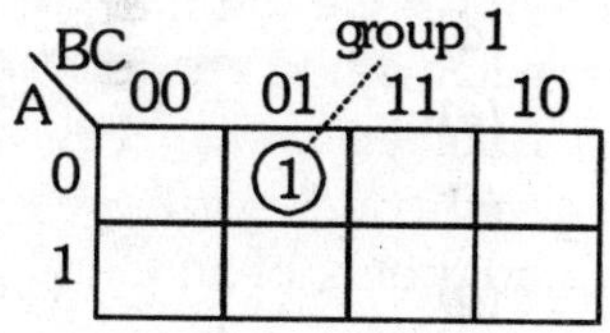

Fig. 2.13(*a*) (*ii*)

Step 3. Cell 2 and cell 6 are having 1 and they are adjacents and referred as group 2 = $B\overline{C}$. Cell 4 and cell 6 are having 1 and they are adjacents and referred by group 3 = $A\overline{C}$.

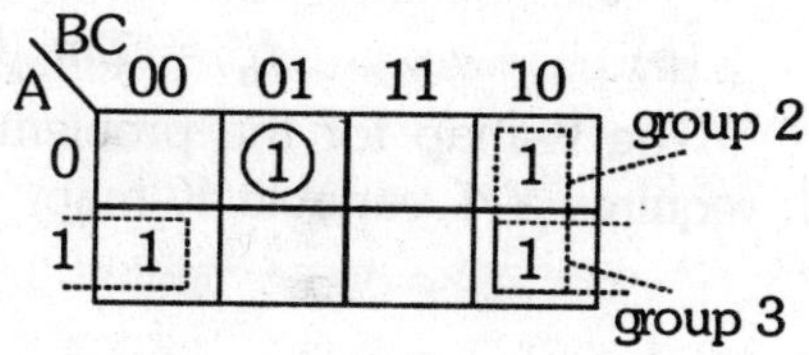

Fig. 2.13(*a*) (*iii*)

Step 4. There is no quad and octet.

Step 5. All 1's are covered.

Step 6. Each group generates a sum term to get a simplified expression for f_a.

Simplified expression from group 1 : $\overline{A}\,\overline{B}C$

group 2 : $B\overline{C}$

group 3 : $A\overline{C}$

$$\therefore \quad f_a = B\overline{C} + A\overline{C} + \overline{A}\,\overline{B}C = \overline{C}\,(B+A) + \overline{A}\,\overline{B}C$$

(*b*) $$f_b = \Sigma m\ (0, 5, 7)$$

$$f_a = \overline{A}\,\overline{B}\,\overline{C} + AB$$

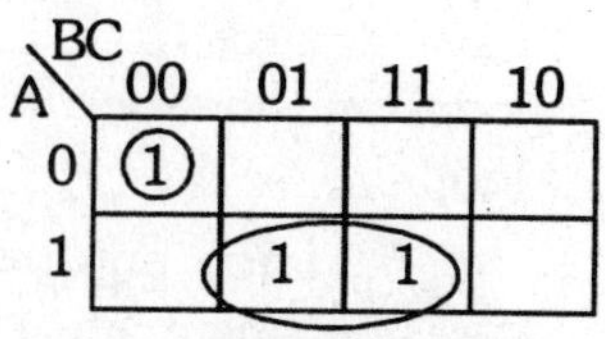

Fig. 2.13(*b*)

(c) $f_c = \Sigma m\ (0, 1, 4, 5)$

$= \overline{B}$

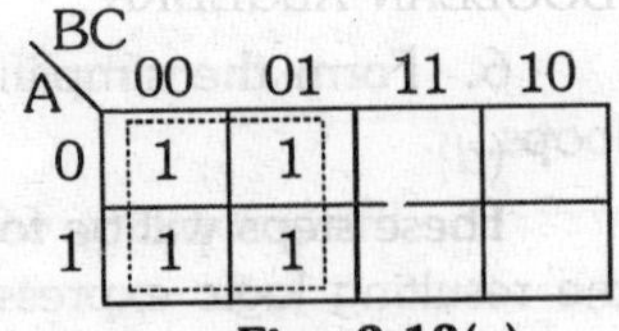

Fig. 2.13(c)

(d) $f_d = \Sigma m\ (0, 2, 4, 6)$

$f_d = \overline{C}$

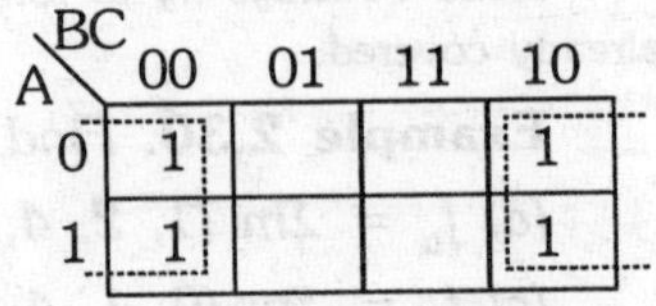

Fig. 2.13(d)

Example 2.31. *Minimize the following expressions using 4-variable K–map (see Fig. 2.14).*

(a) $f_a = \Sigma m\ (2, 5, 6, 9, 12, 13)$

(b) $f_b = \Sigma m\ (0, 1, 2, 3, 8, 9, 10, 11)$

(c) $f_c = \Sigma m\ (4, 5, 6, 7, 12, 13, 14, 15)$

(d) $f_d = \Sigma m\ (2, 6, 8, 9, 10, 11, 14)$

(e) $f_e = \Sigma m\ (0, 1, 4, 5, 8, 9, 10, 11, 14, 15)$

(f) $f_f = \Sigma m\ (0, 2, 5, 7, 8, 10, 13, 15)$

(g) $f_g = \Sigma m\ (1, 3, 4, 6, 9, 11, 12, 14)$

Solution. (a) $f_a = \Sigma m\ (2, 5, 6, 9, 12, 13)$ The K–map is shown in Fig. 2.14.

$\therefore \quad f_a = \overline{A}\,\overline{B}C\overline{D} + B\overline{C} + A\overline{C}D$

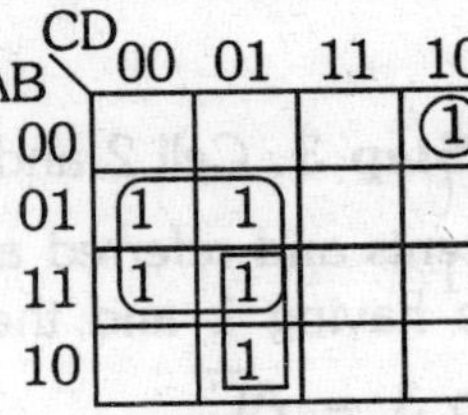

Fig. 2.14

(b) $f_b = \Sigma m\ (0, 1, 2, 3, 8, 9, 10, 11)$

The K–map for the problem is shown in Fig. 2.15.

It requires a 4–variable K–map.

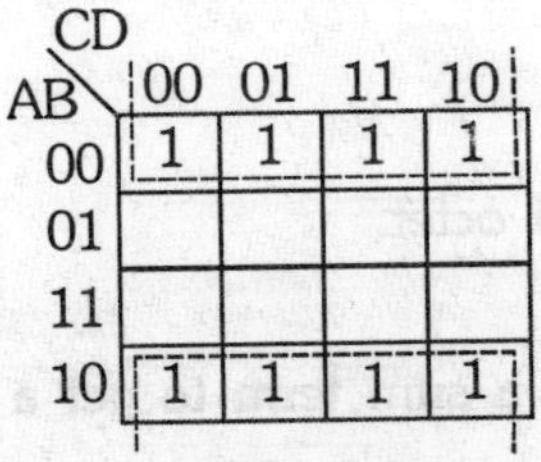

Fig. 2.15

$f_b = \overline{B}$

(c) $f_c = \Sigma m\ (4, 5, 6, 7, 12, 13, 14, 15)$

The above problem requires a 4-variable K–map as shown in Fig. 2.16.

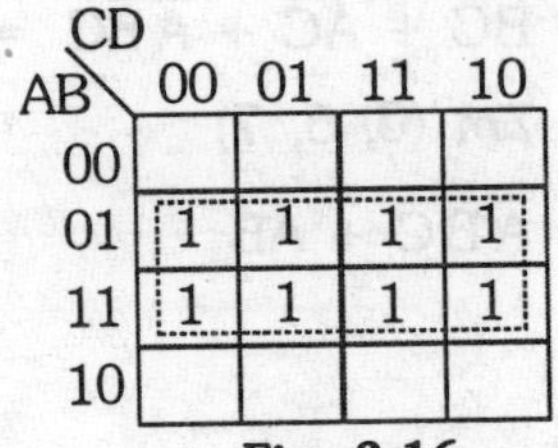

Fig. 2.16

$f_c = B$

(*d*) $f_d = \Sigma m\ (2, 6, 8, 9, 10, 11, 14)$

For this problem, we need a 4-variable K–map as shown in Fig. 2.17.

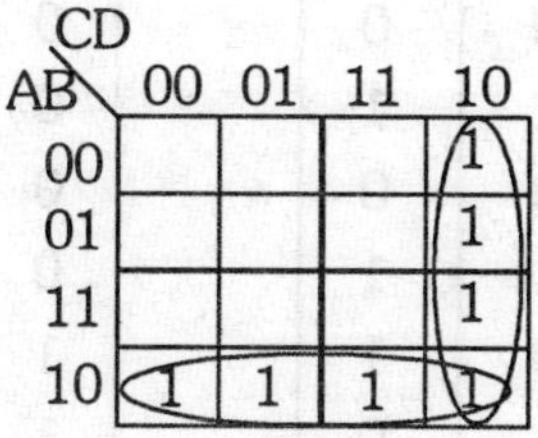

Fig. 2.17

$$f_d = C\overline{D} + A\overline{B}$$

(*e*) $f_e = \Sigma m\ (0, 1, 4, 5, 8, 9, 10, 11, 14, 15)$

For the above problem, we need a 4-variable K–map as shown in Fig. 2.18 (*a*).

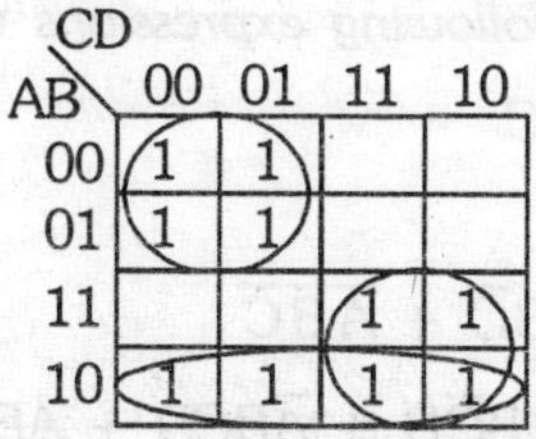

Fig. 2.18(*a*)

$$f_e = \overline{A}\,\overline{C} + A\overline{B} + AC$$

(*f*) $f_f = \Sigma m\ (0, 2, 5, 7, 8, 10, 13, 15)$

For the given problem, we need a 4-variable K–map as shown in Fig. 2.18 (*b*).

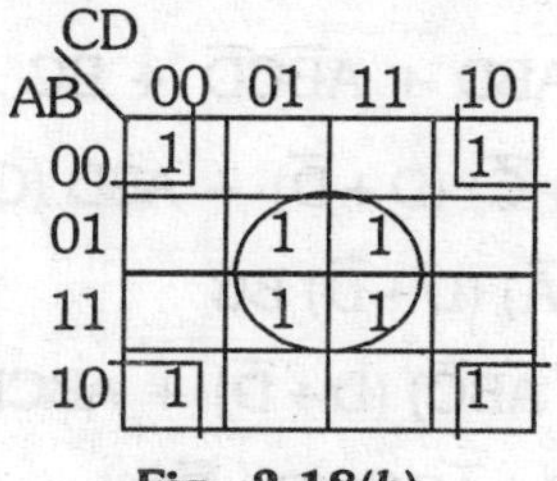

Fig. 2.18(*b*)

$$f_f = BD + \overline{B}\,\overline{D}$$

(*g*) $f_g = \Sigma m\ (1, 3, 4, 6, 9, 11, 12, 14)$

The K–map is shown in Fig. 2.18 (*c*).

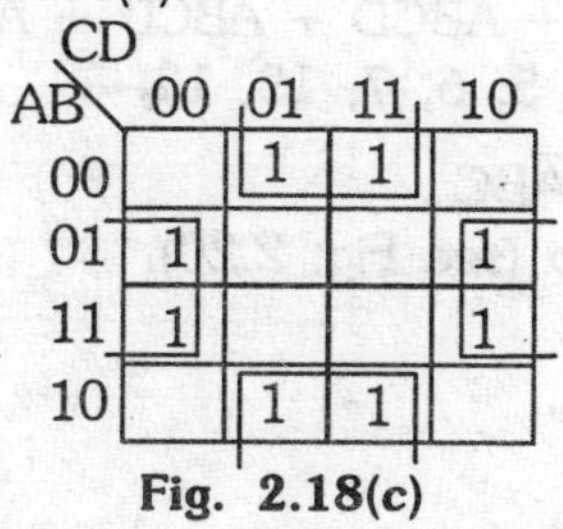

Fig. 2.18(c)

$$f_g = \overline{B}D + B\overline{D}$$

Example 2.32. *Simplify the truth table shown in Fig. 2.19 (a) and 2.19 (b) using K–map method.*

Solution.

A	B	C	f
0	0	0	0
0	0	1	1
0	1	0	0
0	1	1	1
1	0	0	1
1	0	1	0
1	1	0	1
1	1	1	0

Fig. 2.19(*a*)

A	B	C	f
0	0	0	1
0	0	1	0
0	1	0	0
0	1	1	1
1	0	0	1
1	0	1	1
1	1	0	1
1	1	1	0

Fig. 2.19(*b*)

Example 2.33. *Simplify the following expressions using K–map.*

(a) $\bar{A}B + ABD + A\bar{B}C\bar{D} + BC$

(b) $BC + A\bar{C} + AB + BCD$

(c) $ABC + \bar{A}BC + \bar{A}BC + AB\bar{C} + \bar{A}\,\bar{B}\,\bar{C}$

(d) $f = A\bar{B}\bar{C}D + \bar{A}\,\bar{B}\,\bar{C}\,\bar{D} + AB\bar{C}D + \bar{A}BCD + ABD + \bar{B}C\bar{D} + \bar{A}B\bar{C}D$

(e) $f = \bar{A}\bar{C} + A\bar{B}\bar{C} + ABC + A\bar{B}C$

(f) $f = AB + A\bar{C} + AD + A\bar{B}C + ABC$

(g) *Find the complement of the function* $(B\bar{C} + \bar{A}D)\,(A\bar{B} + C\bar{D})$ *and reduce it using K–map.*

Solution. *(a)*

$$
\begin{aligned}
f &= \bar{A}B + ABD + A\bar{B}C\bar{D} + BC \\
&= \bar{A}B\,(C+\bar{C})\,(D+\bar{D}) + ABD\,(C+\bar{C}) + A\bar{B}C\bar{D} \\
&\quad + (A+\bar{A})\,(D+\bar{D})\,BC \\
&= (\bar{A}BC + \bar{A}B\bar{C})\,(D+\bar{D}) + ABCD + AB\bar{C}D + A\bar{B}C\bar{D} \\
&\quad + (ABC + \bar{A}BC)\,(D+\bar{D}) \\
&= \bar{A}BCD + \bar{A}B\bar{C}D + \bar{A}BC\bar{D} + \bar{A}B\bar{C}\,\bar{D} + A\bar{B}C\bar{D} + ABCD + \bar{A}BC\bar{D} \\
&\qquad + ABCD + AB\bar{C}D \\
&= \bar{A}BCD + \bar{A}B\bar{C}D + ABCD + AB\bar{C}D + \bar{A}BC\bar{D} + \bar{A}B\bar{C}D \\
&= \Sigma m\,(4, 5, 6, 7, 15, 13) \\
&= BD + \bar{A}BC
\end{aligned}
$$

For simplification using K–map (see Fig. 2.20).

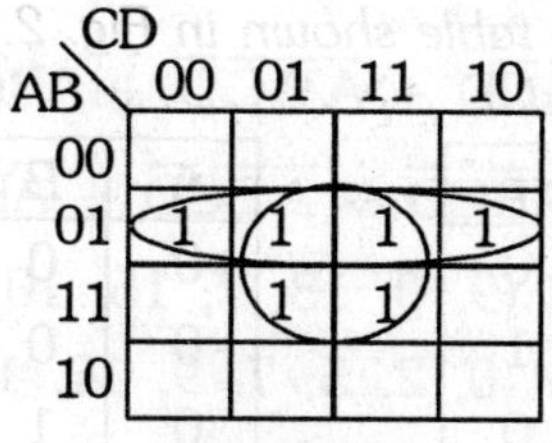

Fig. 2.20

$$f = \overline{A}B + BD$$

(*b*) $f = BC + A\overline{C} + AB + BCD$

$= (A+\overline{A})\,BC\,(D+\overline{D}) + A\,(B+\overline{B})\,\overline{C}\,(D+\overline{D})$

$+ AB\,(C+\overline{C})\,(D+\overline{D}) + (A+\overline{A})\,BCD$

$= ABCD + \overline{A}BCD + ABC\overline{D} + \overline{A}BC\overline{D} + AB\overline{C}D + AB\overline{C}\,\overline{D} + A\overline{B}\,\overline{C}D$

$+ A\overline{B}\,\overline{C}\,\overline{D} + ABCD + ABC\overline{D} + AB\overline{C}D + AB\overline{C}\,\overline{D} + ABCD + \overline{A}BCD$

$= ABCD + \overline{A}BCD + ABC\overline{D} + \overline{A}BC\overline{D} + AB\overline{C}D$

$+ AB\overline{C}\,\overline{D} + A\overline{B}\,\overline{C}D + A\overline{B}\,\overline{C}\,\overline{D}$

$= \Sigma m\,(15, 7, 14, 6, 13, 12, 9, 8)$

$= \Sigma m\,(6, 7, 8, 9, 12, 13, 14, 15)$

$= A\overline{C} + BC$

For simplification using K–map see Fig. 2.21.

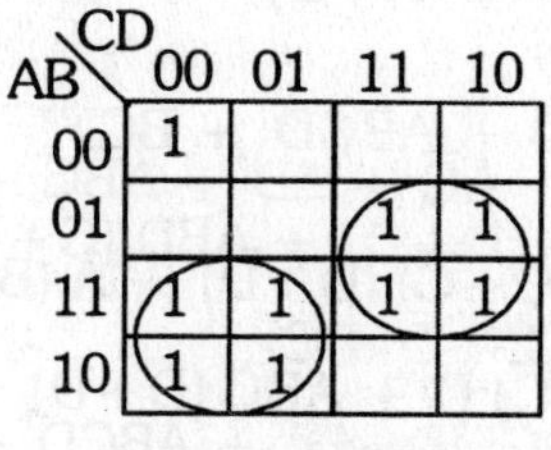

Fig. 2.21

(*c*) $f = ABC + \overline{A}\,\overline{B}C + \overline{A}BC + AB\overline{C} + \overline{A}\,\overline{B}\,\overline{C}$

$f = \Sigma\,(7, 1, 3, 6, 0)$

For simplification see Fig. 2.22.

$= \overline{A}\,\overline{B} + BC + AB$

$= B\,(A+C) + \overline{A}\,\overline{B}$

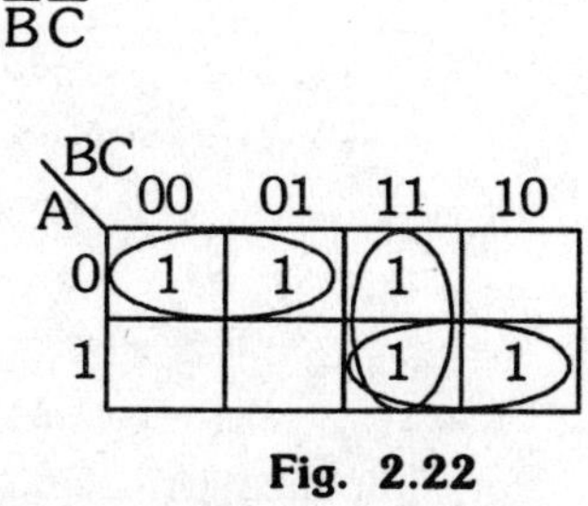

Fig. 2.22

(*d*) $f = A\overline{B}\,\overline{C}D + \overline{A}\,\overline{B}\,\overline{C}\,\overline{D} + AB\overline{C}D + \overline{A}BCD + ABD\,(C+\overline{C})$

$(A+\overline{A})\,\overline{B}C\overline{D} + \overline{A}BC\overline{D}$

$$= A\overline{B}\,\overline{C}D + \overline{A}\,\overline{B}\,\overline{C}\,\overline{D} + AB\overline{C}D + \overline{A}BCD + ABCD + AB\overline{C}D$$

$$+ A\overline{B}C\overline{D} + \overline{A}\,\overline{B}C\overline{D} + \overline{A}B\overline{C}D$$

$$= \Sigma m\ (9, 0, 13, 7, 15, 10, 2, 5)$$

$$= \Sigma m\ (0, 2, 5, 7, 9, 10, 13, 15)$$

For simplification (see Fig. 2.23)

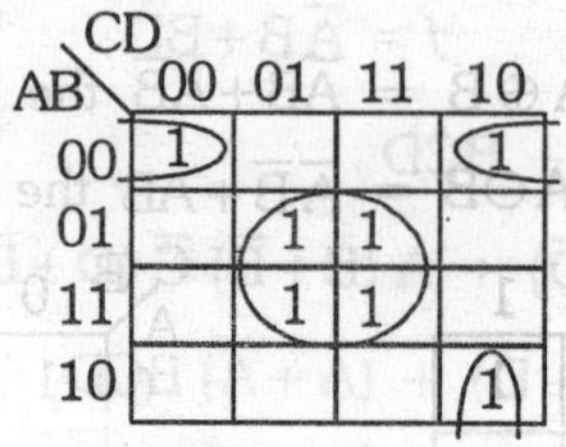

Fig. 2.23

(e) $$f = \overline{A}\,\overline{C} + A\overline{B}\,\overline{C} + ABC + A\overline{B}C$$

$$= \overline{A}\,\overline{C}\,(B+\overline{B}) + A\overline{B}\,\overline{C} + ABC + A\overline{B}C$$

$$= \overline{A}B\overline{C} + \overline{A}\,\overline{B}\,\overline{C} + A\overline{B}\,\overline{C} + ABC + A\overline{B}C$$

$$= \Sigma f\ (2, 0, 4, 7, 5)$$

$$= \Sigma f\ (0, 2, 4, 5, 7) \text{ (see Fig. 2.24)}$$

$$= \overline{B}\,\overline{C} + AC + \overline{A}\,\overline{C}$$

$$= AC + \overline{C}\,(A+\overline{B})$$

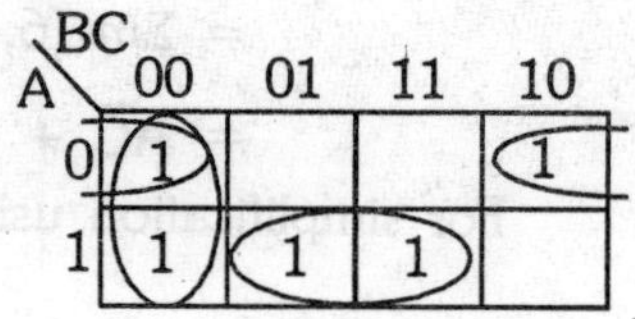

Fig. 2.24

(f) $$f = AB + A\overline{C} + AD + A\overline{B}C + ABC$$

$$= AB\,(C+\overline{C})\,(D+\overline{D}) + A\,(B+\overline{B})\,\overline{C}\,(D+\overline{D}) + A\,(B+\overline{B})$$

$$(C+\overline{C})\,D + A\overline{B}C\,(D+\overline{D}) + ABC\,(D+\overline{D})$$

$$= ABCD + ABC\overline{D} + AB\overline{C}D + AB\overline{C}\,\overline{D} + AB\overline{C}D + A\overline{B}\,\overline{C}D$$

$$+ A\overline{B}\,\overline{C}D + A\overline{B}\,\overline{C}\,\overline{D} + ABCD + A\overline{B}CD + AB\overline{C}D + A\overline{B}\,\overline{C}D$$

$$+ A\overline{B}CD + A\overline{B}C\overline{D} + ABCD + ABC\overline{D}$$

$$= ABCD + ABC\overline{D} + AB\overline{C}D + AB\overline{C}\,\overline{D}$$

$$+ A\overline{B}\,\overline{C}D + A\overline{B}\,\overline{C}\,\overline{D} + A\overline{B}CD + A\overline{B}C\overline{D}$$

$$= \Sigma m\ (15, 14, 13, 12, 9, 8, 11, 10)$$

$$= \Sigma\ 8, 9, 10, 11, 12, 13, 14, 15$$

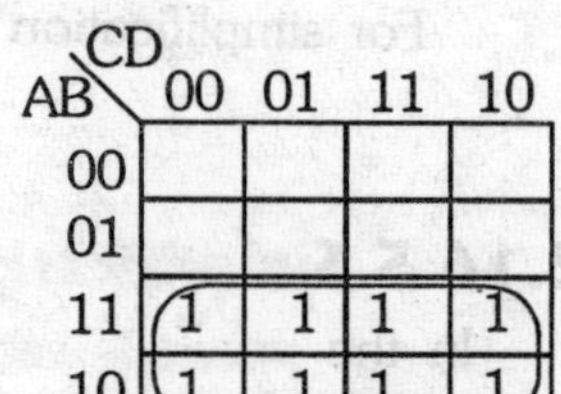

Fig. 2.25

For simplification using K-map (see Fig. 2.25)

$$f = A$$

(g) $$f = (B\overline{C} + \overline{A}D)\,(A\overline{B} + C\overline{D})$$

$$= B\overline{C}A\overline{B} + \overline{A}DA\overline{B} + B\overline{C}C\overline{D} + \overline{A}DC\overline{D} = 0$$

$$\therefore \quad \overline{f} = (\overline{B} + C)\,(A + \overline{D}) + (\overline{A} + B)\,(\overline{C} + D)$$

$$= 1$$

Example 2.34. *Plot the K–map for EX– OR and EX– NOR functions of 2, 3 and 4 variables.*

Solution. *For 2–variables*

f = EX–OR of 2 variables = $A \oplus B = \overline{A}B + A\overline{B}$ the K–map is shown in Fig. 2.26 (*a*).

EX–NOR of 2 variables = $A \odot B = \overline{A}\,\overline{B} + AB$ the K–map is shown in Fig. 2.26 (*b*).

A \ B	0	1
0		1
1	1	

Fig. 2.26(*a*)

A \ B	0	1
0	1	
1		1

Fig. 2.26(*b*)

For 3–variables

EX–OR of 3 variables = $A \oplus B \oplus C = \overline{A}\,\overline{B}C + \overline{A}B\overline{C} + A\overline{B}\,\overline{C} + ABC$ See Fig. 2.27 (*a*)

EX–NOR 3 variables = $A \odot B \odot C = \overline{A}\,\overline{B}\,\overline{C} + \overline{A}BC + A\overline{B}C + AB\overline{C}$ See Fig. 2.27 (*b*)

A \ BC	00	01	11	10
0		1		1
1	1		1	

Fig. 2.27(*a*)

A \ BC	00	01	11	10
0	1		1	
1		1		1

Fig. 2.27(*b*)

For 4–variables

EX–OR for 4 variables is $A \oplus B \oplus C \oplus D$ See Fig. 2.28 (*a*)

EX–NOR for 4 variables is $A \odot B \odot C \odot D$ See Fig. 2.28 (*b*)

AB \ CD	00	01	11	10
00		1		1
01	1		1	
11		1		1
10	1		1	

EX-OR

Fig. 2.28(*a*)

AB \ CD	00	01	11	10
00	1		1	
01		1		1
11	1		1	
10		1		1

EX-NOR

Fig. 2.28(*b*)

2.16.5 Simplification of POS Expressions

In the previous section, we have considered the Boolean expressions in SOP form and grouped 2, 4, and 8 adjacent ones to get the simplified expression. Sometimes, it is more simpler to find the expression in POS form rather than SOP and also it can be expressed with less number of literals. The procedure to find the simplified expression in POS is same as SOP except plotting of 0's instead of 1's. The steps involved are given below :

1. Construct the K–map and place 0's in cells corresponding to the 0's in the truth table or maxterms in the POS expression.

2. Check the K–map for adjacent 0's and encircle those 0's which are not adjacent to any-other 0's. Those are called isolated 0's.

3. Check for those 0's which are adjacent to only one other 0 and group such pairs.

4. Check for quads and octets of adjacent 0's even if it contains some 0's that have already been grouped. While doing this, make sure that there are minimum number of groups.

5. Combine any pair necessary to include any 0's that have not yet been grouped.

Form the simplified expression f in SOP form by summing the resultant product terms from each group. Now, apply the De Morgan's theorem on f, to produce the simplified expression in POS form.

Example 2.35. *Minimize the following expression in POS form*

(*a*) $f = \Pi M\ (0, 1, 3, 4, 7)$

(*b*) $f = \Pi M\ (0, 4, 6, 7, 8, 12, 13, 14, 15)$

Solution. (*a*) **Step 1:** Fig. 2.29 shows the 3 variable K–map and it is plotted according to the given expression.

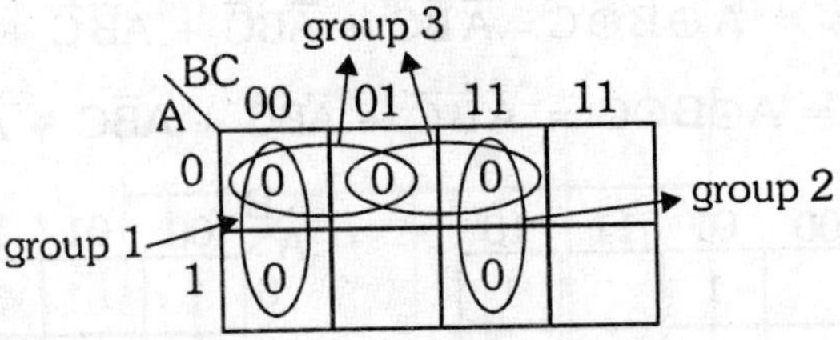

Fig. 2.29

Step 2: There are no isolated 0's.

Step 3: The cell 4 is adjacent to cell 0, which give a group 1. Cell 3 and cell 7 are adjacents and are combined and give a group 2.

Step 4: There are no quads or octets.

Step 5: The 0 in the cell 1 can be combined with 0 in the cell 3 or cell 0 to form a pair. This pair is referred to as group 3.

Step 6: All 0's are covered.

$$f \text{ in SOP} = \text{SOP of group 1} + \text{group 2} + \text{group 3}$$

$$= \overline{B}\,\overline{C} + \overline{A}\,\overline{B} + BC$$

$$f \text{ in POS} = \overline{\overline{B}\,\overline{C} + \overline{A}\,\overline{B} + BC}$$

$$= (\overline{B}+\overline{C})\,(B+C)\,(A+B)$$

(*b*) See Fig. 2.30 $\quad f = \overline{C}\,\overline{D} + AB + BC$

$$f \text{ in POS} = \overline{\overline{C}\,\overline{D} + AB + BC}$$

$$= (C+D)\,(\overline{A}+\overline{B})(\overline{B}+\overline{C})$$

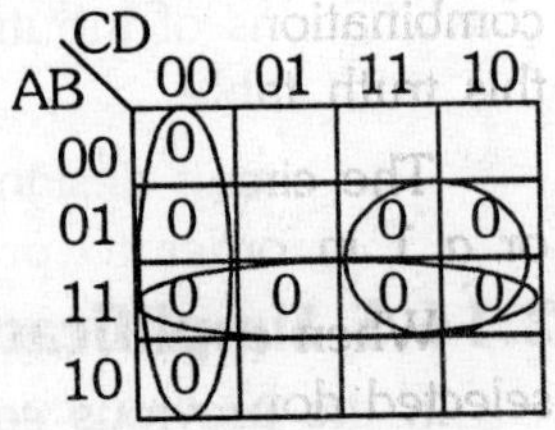

Fig. 2.30

2.16.6 Don't Care Conditions/Cannot Happen States

So far the functions considered have been completely specified for every input combination. In those functions, for every combination, the function value may be 1 or 0.

Some logic circuits can be designed so that there are certain input conditions for which there are no specified output levels, usually because these input conditions will never occur. In other words, there will be certain combinations of input levels where we "don't care". In such cases, the output level is not defined, it can be either HIGH or LOW. These output levels are indicated by 'X' or '*d*' or 'ϕ' in the truth tables and are called *"don't care outputs"* and such combinations are called *don't care conditions.*" Such situation may occur when the variables are not mutually independent; that is, dependency among the variables may preclude the occurrence of certain combinations, for which, consequently, the value of the function will not be specified.

Particularly, it may happen in "coding circuits". When $x_1, x_2, ..., x_n$ are variable designations inputs to a switching circuit and when, $f(x_1, x_2, ..., x_n)$ designate it's outputs, it often occurs that for certain input combinations the value of the output is unspecified. Such don't care combinations may be specified in either of two ways, *i.e.*, 0 or 1. An incompletely specified function containing K don't care combinations actually corresponds to a class of 2^K distinct functions.

Let us consider an example of truth table shown in Table 2.19.

Table 2.19

A	*B*	*C*	*f*
0	0	0	0
0	0	1	1
0	1	0	0
0	1	1	0
1	0	0	1
1	0	1	×
1	1	0	×
1	1	1	×

Here, outputs are defined for input conditions from 000 to 100 and for the remaining combinations of inputs, output is not defined, hence these are called don't care conditions for this truth table.

The circuit designer is free to make the output for any 'don't care' conditions either *a* 0 or *a* 1 in order to produce simplest output expression.

When employing the map of an incompletely specified function, we assign the value 1 to selected don't care combinations and the value 0 to others, in such a way to increase the size of the selected subcube/group/loop whenever possible.

Note: No subcube/loop can be found containing only don't care cells, because it is not required that the function equal to 1 for those combinations.

For example, the K map for above truth table is shown in Fig. 2.31 (*a*) with 'x' placed in cells 5, 6 and 7.

A \ BC				
	1	1	0	0
	1	X	X	X

Fig. 2.31(*a*)

A \ BC	00	01		
0	1	1	0	0
1	1	1	0	0

Fig. 2.31(*b*)

Here, the don't care in cell 5 is taken as 1, so that we can form a quad, (see Fig. 2.31 (*b*)) and the remaining two don't care cells are treated as 0.

Using these don't cares the simplified Boolean expression is

$f = \overline{B}$, otherwise the expression is

$$f = \overline{B}\overline{C} + \overline{A}\overline{B} = \overline{B}\,(\overline{A}+\overline{C})$$

Example 2.36. *Find the reduced SOP for the following functions.*

(*a*) $f(A, B, C) = \Sigma m\ (0, 1, 5) + \Sigma d\ (4, 7)$

(*b*) $f(A, B, C, D) = \Sigma m\ (0, 1, 3, 7, 15) + \Sigma d\ (2, 11, 12)$

(*c*) $f(A, B, CD) = \Sigma m\ (4, 6, 7, 13, 14) + \Sigma d\ (5, 10, 12, 15)$

Solution. (*a*) To form a quad, the don't care in cell 4 is treated as 1 and the remaining cell 7 is treated as 0. The simplified expression is $f = \overline{B}$ (see Fig. 2.32)

A \ BC	00	01	11	10
0	1	1		
1	X	1	X	

Fig. 2.32

(*b*) (1) To form a horizontal quad with cells 0, 1, 2 and 3, the don't care value in cell 2 is treated as 1 (see Fig. 2.33).

(2) To form a vertical quad with cells 3, 7, 11 and 14, the don't care value in cell 11 is treated as 1.

The remaining cell 12 value is treated as '0'.

The simplified expression is $f = \overline{A}\overline{B} + CD$.

AB \ CD	00	01	11	10
00	1	1	1	X
01			1	
11	X		1	
10			X	

Fig. 2.33

(*c*) Here, if we consider the don't care cells 5, 12 and 15 value as 1, then we can form an octet. The remaining don't care cell 10 value is considered as 0 (see Fig. 2.34).

The simplified expression is $f = B$.

AB \ CD	00	01	11	10
00				
01	1	×	1	1
11	×	1	×	1
10				×

Fig. 2.34

Example 2.37. *Reduce the following function using K-map technique in the POS form.*

$$f\,(A, B, C, D) = \Pi M\,(0, 2, 6, 8, 12) + \Pi d\,(3, 4, 7, 10, 14)$$

Solution. Step 1: Fig. 2.35 shows the K-map for four variables and it is plotted according to the given max terms and don't care terms.

Step 2: There are no isolated 0's.

Step 3: There is no such 0 which is adjacent to only one other 0.

Step 4: By taking don't cares as 0's, we have one octet. These are referred to as subcube 1.

Step 5: All 0's have already been covered. The simplified expression is

$$f = \overline{D}$$

AB \ CD	00	01	11	10
00	0			0
01	×			0
11	0			×
10	0			×

Subcube 1

Fig. 2.35

2.16.7 Five and Six Variable K-maps

Karnaugh maps are most convenient to simplify 2, 3 and 4 variable expressions. They can also be used to simplify 5 and 6 variable boolean expressions. For the functions of seven or more variables, the map is very large and its value as an effective tool in the minimization procedure decreases, since it is very difficult to keep the track of adjacents.

2.16.7.1 Five variable K-map

The minimization procedure described so far with respect to functions of two, three or four variables can be extended to the case of five variables.

The 5-variable K-map contains $2^5 = 32$ cells. Instead of representing a single 32–cell map, two 16–cell K–maps are generally used. If the variables are A, B, C, D and E, the two identical 16–cell maps contain B, C, D and E variables with one 16–cell map for A = 1 and other 16–cell map for A = 0 ($\overline{A}$). This is shown in Fig. 2.36.

BC \ DE	00	01	11	10
00	0	1	3	2
01	4	5	7	6
11	12	13	15	14
10	8	9	11	10

Fig. 2.36(*a*)

A = 0

BC \ DE	00	01	11	10
00	16	17	19	18
01	20	21	23	22
11	28	29	31	30
10	24	25	27	26

Fig. 2.36(*b*)

A = 1

It is noted that in order to identify the adjacent grouping in the 5–variable maps, we must imagine that the two maps superimposed on one another; not 'hinged' or 'mirror image'. Every cell in one map is adjacent to the corresponding cell in the other map, because only one variable *i.e.*, A changes between such corresponding cells. Thus, row and column on one map is adjacent to the corresponding row and column on the other map, and also some rules are used for adjacents within one 16–cell map. This is illustrated in Fig. 2.36. (*c*) and (*d*).

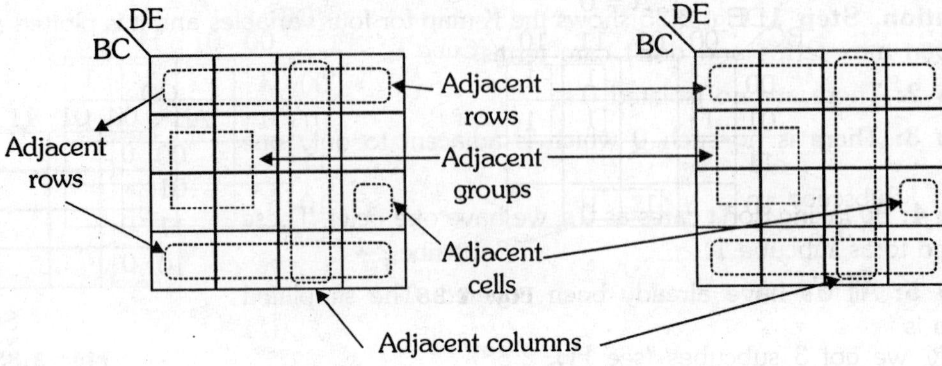

Fig. 2.36(c) **Fig. 2.36(*d*)**

A = 0 A = 1

Fig. 2.36 5-variable K-map with adjacents.

The other way to represent a five variable map is as a single 32-cell map as shown in Fig. 2.37.

AB \ CDE	000	001	011	010	110	111	101	100
00	0	1	3	2	6	7	5	4
01	8	9	11	10	14	15	13	12
11	24	25	27	26	30	31	29	28
10	16	17	19	18	22	23	21	20

Fig. 2.37

Each cell, in addition to it's being adjacent to four other cells, can be combined with a fifth cell on the other side of the centre symmetry line.

For example, (1) Cell 11 contains 5 adjacent cells, those are cell 3, cell 9, cell 10, cell 27 and cell 15.

(2) Cell 17 contains 5 adjacent cells, those are cell 16, 19, 1, 25 and cell 21.

Example 2.38. *Simplify the following Boolean functions using K–map technique.*

(*a*) $f\,(A, B, C, D, E) = \Sigma\, m\,(0, 2, 3, 4, 6, 7, 9, 11, 16, 18, 19, 20, 22, 23, 25, 27)$

(*b*) $f\,(A, B, C, D, E) = \Sigma\, m\,(0, 4, 7, 8, 9, 10, 11, 16, 24, 25, 26, 27, 29, 31)$

(*c*) $f\,(A, B, C, D, E) = \Sigma\, m\,(0, 4, 8, 12, 16, 18, 20, 22) + \Sigma d\,(24, 26, 28, 30, 31)$

Solution. (*a*)

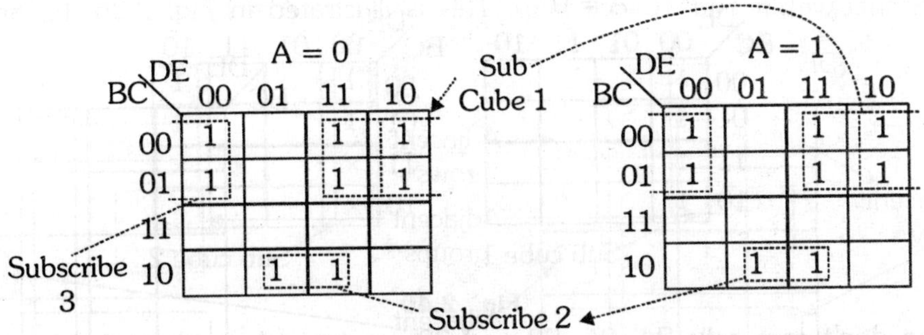

Fig. 2.38

Totally, we got 3 subcubes (see Fig. 2.38)

Subcube 1 is an octet, the resultant is = $\overline{B}D$

Subcube 2 is a quad, the resultant is = $B\overline{C}E$

Subcube 3 is a quad, the resultant is = $\overline{B}\,\overline{D}\,\overline{E}$

$\therefore \quad f = \overline{B}D + B\overline{C}E + \overline{B}\,\overline{D}\,\overline{E}$

(*b*)

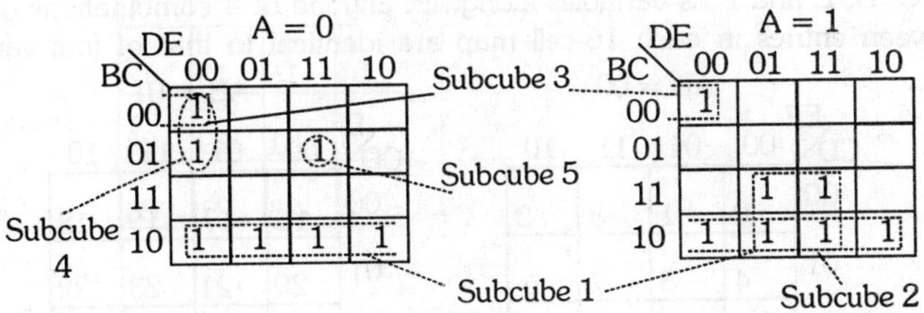

Fig. 2.39

Totally, we have 5–subcubes after grouping the adjacents (see Fig. 2.39)

Subcube 1 is an octet which gives = $B\overline{C}$

Subcube 2 is a quad which gives = ABE

Subcube 3 is a pair which gives = $\overline{B}\,\overline{C}\,\overline{D}\,\overline{E}$

Subcube 4 is a pair which gives = $\overline{A}\,\overline{B}\,\overline{D}\,\overline{E}$

Subcube 5 is single which gives = $\overline{A}\,\overline{B}CDE$

$$f = B\overline{C} + ABE + \overline{B}\,\overline{C}\,\overline{D}\,\overline{E} + \overline{A}\,\overline{B}\,\overline{D}\,\overline{E} + \overline{A}\,\overline{B}CDE$$

(c)

A = 0

BC \ DE	00	01	11	10
00	1			
01	1			
11	1			
10	1			

Sub cube 1

A = 1

BC \ DE	00	01	11	10
00	1			1
01	1			1
11	×		×	×
10	×			×

Sub cube 2

Fig. 2.40

Here, the don't care cells 24, 26, 28 and 30 are treated as 1's to form a subcube 1 of octet size, which results (see Fig. 2.40)

$$f = \overline{D}\,\overline{E}$$

The don't care cells 26 and 30 are taken as 1 to form a subcube 2 of octet size, which results = $A\overline{E}$. The remaining don't care cell 31 is treated as zero.

$$\therefore \quad f = \overline{D}\,\overline{E} + A\overline{E} = \overline{E}\,(A + \overline{D})$$

2.16.7.2 Six-variable K-map

A six variable K-map contains $2^6 = 64$ cells. These cells are divided into four identical 16-cell maps as shown in Fig 2.41. If the variables are A, B, C, D, E and F, then each 16-cell map contains C, D, E and F as variables alongwith anyone of 4 combinations of A and B. The adjacents between entries in each 16-cell map are identical to that of four-variable map.

AB = 00

CD \ EF	00	01	11	10
00	0	1	3	2
01	4	5	7	6
11	12	13	15	14
10	8	9	11	10

AB = 01

CD \ EF	00	01	11	10
00	16	17	19	18
01	20	21	23	22
11	28	29	31	30
10	24	25	27	26

CD \ EF	00	01	11	10
00	32	33	35	34
01	36	37	39	38
11	44	45	47	46
10	40	41	43	42

AB = 10

CD \ EF	00	01	11	10
00	48	49	51	50
01	52	53	55	54
11	60	61	63	62
10	56	57	59	58

AB = 11

Fig. 2.41 6–variable K–map.

Fig. 2.42. will give the different possible adjacent cells.

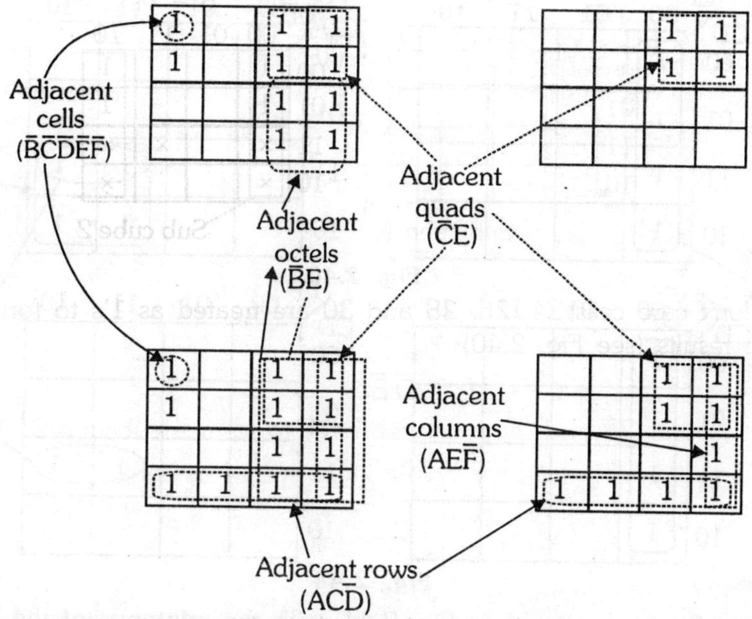

Fig. 2.42 A 6-variable K-map with different possible adjacents.

The other possible way for representing a 6-variable K-map is shown in Fig. 2.43.

ABC \ DEF	000	001	011	010	110	111	101	100
000	0	1	3	2	6	7	5	4
001	8	9	11	10	14	15	13	12
011	24	25	27	26	30	31	29	28
010	16	17	19	18	22	23	21	20
110	48	49	51	50	54	55	53	54
111	56	57	59	58	62	63	61	60
101	40	41	43	42	46	47	45	44
100	32	33	35	34	38	39	37	36

Fig. 2.43. 6-variable K-map as a single block of 64-cells.

Example 2.39. *Simplify the given Boolean functions using K-map.*

(a) $f\ (A, B, C, D, E, F) = \Sigma m\ (0, 4, 8, 12, 26, 27, 30, 31, 32, 36, 40, 44, 62, 63)$

(b) $f\ (A, B, C, D, E, F) = \Sigma m\ (9, 11, 25, 27, 41, 43, 56, 57, 59)$

$+ \Sigma d\ (8, 10, 13, 15, 29, 30, 31, 58, 61)$

Solution.

(*a*)

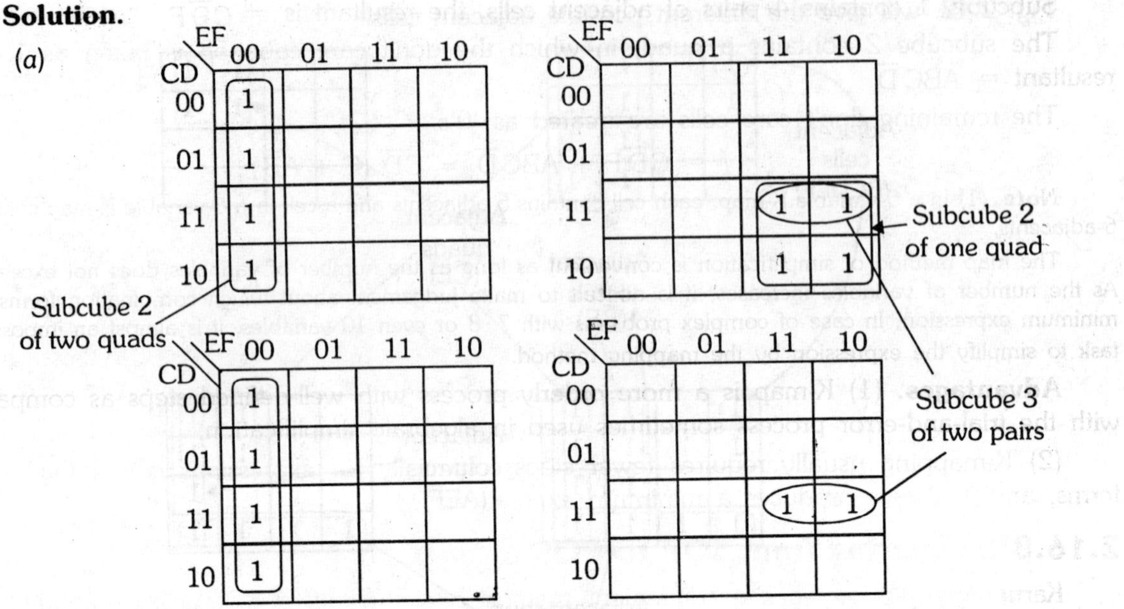

Fig. 2.44

The 6-variable K-map is shown in Fig. 2.44, with the minterms of the given expression. Finally, the resultant contains 3-subcubes.

Subcube 1 contains two quads, the resultant is = $\overline{B}\,\overline{E}\,\overline{F}$

Subcube 2 is formed to be one quad, the resultant is = $\overline{A}BCE$

Subcube 3 is formed from two pairs, the resultant expression is = BCDE

$$\therefore \quad f = \overline{B}\,\overline{E}\,\overline{F} + \overline{A}BCE + BCDE$$

$$= \overline{B}\,\overline{E}\,\overline{F} + BCE + (\overline{A} + D)$$

(*b*) The 6-variable K-map with the given function is shown in Fig. 2.45. It contains two subcubes.

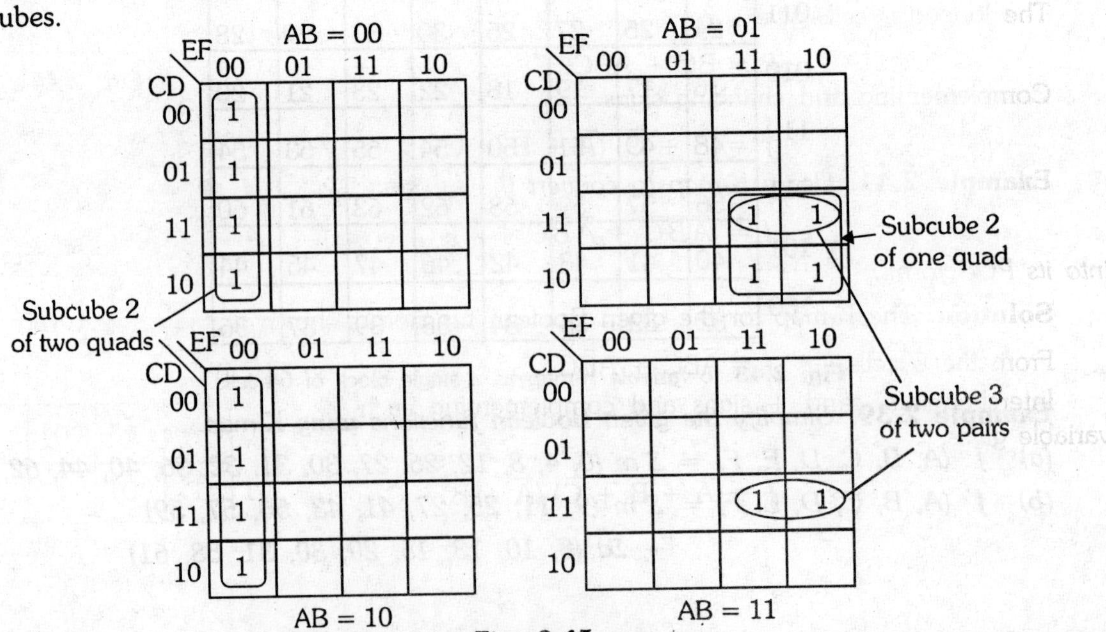

Fig. 2.45

Subcube, 1 contains 4 pairs of adjacent cells, the resultant is = $C\overline{D}F$

The subcube 2 contains a quad, in which the don't care cube 58 is taken as 1, the resultant = $ABC\overline{D}$

The remaining don't care cells are treated as '0's.

$$\therefore \quad f = C\overline{D}F + ABC\overline{D} = C\overline{D}\,(F + AB)$$

Note. (1) In a 5-variable K-map, each cell contains 5 adjacents and a cell in a 6-variable K-map contains 6-adjacents.

The map method of simplification is convenient as long as the number of variables does not exceed 5. As the number of variables increases, it is difficult to make judgement about which combination forms the minimum expression. In case of complex problems with 7, 8 or even 10 variables, it is almost an impossible task to simplify the expression by the mapping method.

Advantages. (1) K-map is a more orderly process with well-defined steps as compared with the trial-and-error process sometimes used in algebraic simplification.

(2) K-mapping usually requires fewer steps, especially for expressions containing many terms, and it always produces a minimum expression.

2.16.8 Converting from SOP to POS form Using K-map

Karnaugh map provides a convenient method of converting a SOP expression into it's corresponding POS form. The procedure is as follows :

(1) Map the SOP expression

(2) Loop the 0 cells to obtain the simplified equation

(3) For each group of 0 cells, read off the POS term, interchanging . and + signs and complementing all variables.

Example 2.40. *Find the POS form of the given function* $f = \overline{B}\,\overline{C} + \overline{A}B + ABC$

Solution. The K-map for the given Boolean expression is shown in Fig 2.46.

A \ BC	00	01	11	10
0	1	0	1	1
1	1	0	1	0

Fig. 2.46

The looped 0 cells give

$$f' = \overline{B}C + AB\overline{C}.$$

Complementing and changing signs

$$f = (B + \overline{C})\,(\overline{A} + \overline{B} + C)$$

Example 2.41. *Use a K-map to convert the function*

$$f = \overline{A}\,\overline{B}\,\overline{C} + \overline{A}\,\overline{B}C + \overline{A}B\overline{C} + A\overline{B}\,\overline{C} + AB\overline{C} + ABC$$

into its POS form.

Solution. The K-map for the given Boolean function is shown in Fig. 2.47

From the 0 cells, $f = \overline{A}BC + A\overline{B}C$.

Interchanging . and + signs and complementing each variable gives,

$$f = (A + \overline{B} + \overline{C})\,(\overline{A} + B + \overline{C})$$

A \ BC	00	01	11	10
0	1	1	0	1
1	1	0	1	1

Fig. 2.47

2.16.9 Converting from POS to SOP form Using K-map

It is also possible to employ a K-map to convert an equation from it's POS form to it's SOP form. The procedure is as follows :

(1) Map the POS equation inserting 0 into the relevant cells.

(2) Loop the cells containing 1's.

(3) Read-off the simplified SOP equation.

Example 2.42. *Use a K-map to convert the* $F = (A + \overline{B} + \overline{C})\ (\overline{A} + B + \overline{C})$ *into its SOP form.*

Solution. The K-map with the given maxterms is shown in Fig 2.48.

A \ BC	00	01	11	10
0	1	1	0	1
1	1	0	1	1

Fig. 2.48

Looping of 1 cells gives 3 subcubes, the resultant in SOP form is

$$f = \overline{C} + \overline{A}\,\overline{B} + AB$$

$$= \overline{C} + \overline{A}\,\overline{B} + AB$$

Example 2.43. *Convert the given Boolean functions into the SOP form*

(a) $f = (A + B)\ (\overline{B} + C)$

(b) $f = (A + B)\ (A + \overline{C})\ (\overline{A} + \overline{B})\ (\overline{A} + C)$

Solution. (*a*) The K-map for the given Boolean function is shown in Fig. 2.49

The looping of 1 cells gives

$$f = A\overline{B} + BC$$

A \ BC	00	01	11	10
0	0	0	1	0
1	1	1	1	0

Fig. 2.49

(*b*) The K-map for the given Boolean function is shown in Fig. 2.50

From the 1 cells,

$$f = \overline{A}B\overline{C} + A\overline{B}C$$

A \ BC	00	01	11	10
0	0	0	0	1
1	0	1	0	0

Fig. 2.50

2.17 QUINE-McCLUSKEY OR TABULATION PROCEDURE

In order to manipulate the functions of a larger number (more than 6) of variables, more systematic procedure, preferably one which can be carried out by a computing machine, is necessary. The suggested technique is a *tabulation method*. It is a specific step-by-step procedure that is guaranteed to produce a simplified standard form expression for a function. It is quite suitable for hand computation and is prone to mistakes because of it's routine, monotonous process. The tabulation method was first formulated by Quine and later improved by McCluskey. It is also known as the '*Quine*-McCluskey' method.

The tabular method of simplification consists of two parts. The first is to find an exhaustive search of all the terms that are candidates for inclusion in the simplified function. These terms are called prime implicants.

The second operation is to choose among the prime implicants, those that give an expression with the least number of literals.

2.17.1 Minimal Functions and their Properties

Some terms related to minimal function and irredundant expression are given below and neither of them is necessarily unique.

(1) **Covers.** A switching function $f(x_1, x_2, ..., x_n)$ is said to cover another function $g(x_1, x_2, ..., x_n)$, f covers g, then it has a 1 in every row of the truth table in which g has a 1.

If f covers g and at the same time, g covers f, then f and g are equivalent. It is denoted by $f \supseteq g$.

(2) **Implicant.** It denotes the set of adjacent minterms or the simplified product term obtained from the minterms of set.

Let $f(x_1, x_2, ..., x_n)$ be a switching function and $h(x_1, x_2, ..., x_n)$ be a product of literals. If f covers h, then h is said to imply f, or h is said to be an implicant of f. Implication is often denoted by $h \rightarrow f$.

Example. If $f = AB + CD$ and $h = ABC'$, then f covers h, and h implifies f.

(3) **Prime Implicant (PI).** An implicant is called a prime implicant if it is not a subset of another implicant of the function

Or

A prime implicant P, of a function f is a product term which is covered by f, such that deletion of any literal from the P results in a new product which is not covered by f.

Or

P is a prime implicant if and only if P implies f, but does not imply any product with fewer literals which in turn also implies f.

Generally the set of all prime implicants of a function f will be denoted by $P = \{P_1, P_2, ..., P_i\}$

Example. If $f = x'y + xz + \overline{y}\overline{z}$, then $x'y$ is a prime implicant,

Since it is covered by f and neither x' nor y alone implies f.

(4) **Essential Prime Implicant (EPI).** A prime implicant which includes at least 1 cell that is not included in any other prime implicant is called essential prime implicants.

Or

A prime implicant P of a function f is said to be an essential prime implicant, if it covers at least one minterm of f which is not covered by any other prime implicant.

Since every minterm of f must be covered by an expression for f, all essential prime implicants must be contained in any irredundant expression for this function.

(5) **Selective Prime Implicants (SPI).** Suppose if all the EPI have not covered the total minterms of a function, then it is necessary to add some more PI. We have to select PI among

the remaining PI so that the minimal expression must contain less number of literals. Those PI's are called selective prime implicants. The selection can be done in two ways.

To obtain the minimal expression for the given function :

(1) Find all the PI of the expression.

(2) Determine the EPI and include them in the minimal expression and remove all PI which are covered by the EPI.

(3) If the set of EPI covers all the minterms of f, then it is unique minimal expression. Otherwise, select additional prime implicants so that the function f is covered completely and the total number and size of the PI added are minimal.

Example 2.44. *Minimize the function* $f(A, B, C, D) = \Sigma(0, 1, 3, 5, 7, 8, 9, 10, 11, 12, 14)$

Solution. K-map for the function is shown in Fig. 2.51.

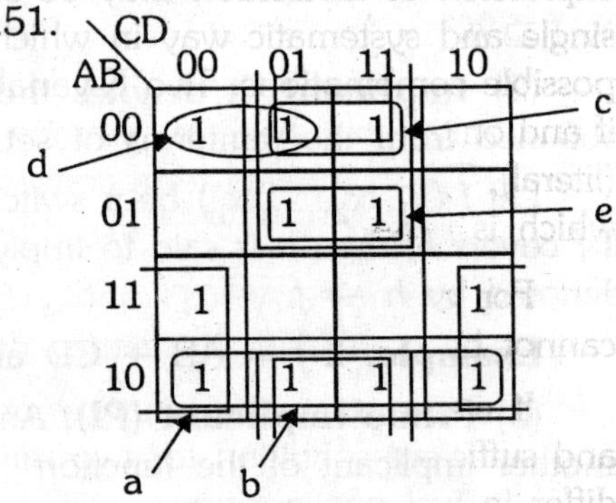

Fig. 2.51

From the K-map, it can be observed that we have 6-prime implicants.

They are $a = A\overline{D}$ $\quad d = \overline{A}\overline{B}\overline{C}$

$b = A\overline{B}$ $\quad e = \overline{A}D$

$c = \overline{B}D$

Out of 6-prime implicants, 'f' is covered totally by c and e and e is covered totally by b and c, hence f and e PI are not essential.

$\therefore$ The remaining PI a, b, c, and d are called EPI.

$$f = a + b + c + d$$
$$= A\overline{D} + A\overline{B} + \overline{A}D + \overline{A}\overline{B}\overline{C}$$

The above four EPI covered all the minterms. This minimal expression is unique.

Example 2.45. *Simplify the function using K-map.*

$$f(A, B, C, D) = \Sigma m(0, 6, 8, 13, 15) + \Sigma d(2, 5, 7, 9, 10, 12, 14).$$

Solution. The K-map for the given Boolean function is shown in the Fig. 2.52.

From the K-map, we have 6 prime implicants

$a = A\overline{C}$ $\quad c = C\overline{D}$ $\quad e = ABD$

$b = BD$ $\quad d = \overline{B}\overline{C}\overline{D}$ $\quad f = \overline{B}\overline{D}$

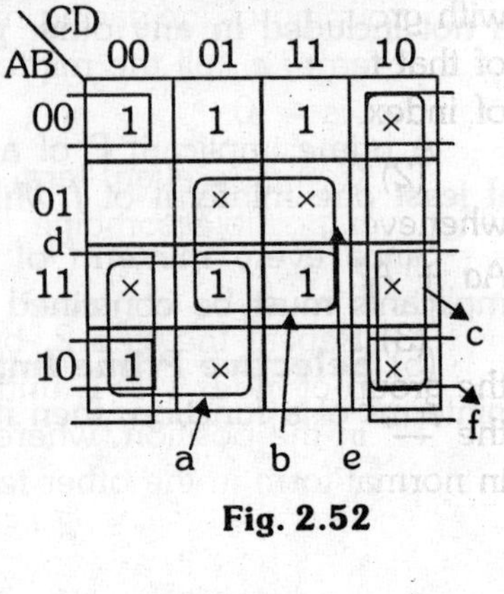

Fig. 2.52

Among all these PI, a can be covered by f and e and b can be covered by e similarly d is covered by f. So, except for these three prime implicants, the remaining PI c, e and f are called EPI and EPI's are covering all the minterms of the function. Hence the minimal expression is unique.

$$f = C\overline{D} + ABD + \overline{B}\,\overline{D}$$
$$= \overline{D}(\overline{B} + C) + ABD$$

Two procedures are used in Q-M method.

(1) The binary representation.

(2) The decimal representation.

Now we will see the details of each method.

2.17.2 The Binary Representation

The basic idea in this procedure is the use of the combining theorem $Aa + Aa' = A$ on all adjacent pairs of terms which yield the set of all prime implicants, from which a minimum expression of a function may be selected. The main task of this procedure is to determine a single and systematic way in which terms can (or cannot) be combined, and to carry out all possible combinations. Two K-variable terms can be combined into a single (K-1) variable term if and only if they have in common (K-1) identical literals and differ only in just a single variable (literal). The combined term consists of the product of the (K-1) literals, while the one variable, which is uncomplemented in one term and complemented in the other, is deleted.

For example, $\overline{A}\,\overline{B}CD$ and $\overline{A}BCD$ can be combined to $\overline{A}CD$ but $\overline{A}\,\overline{B}CD$ and $\overline{A}\,\overline{B}\,\overline{C}\,\overline{D}$ cannot be combined, since they differ by two variables (*i.e.*, C and D).

If we consider the binary representation of the minterms, we observe that the necessary and sufficient condition for two minterms to be combinable is that their binary representations differ in just one position.

For example, the representation for $\overline{A}\,\overline{B}CD$ and $\overline{A}BCD$ are 0011 and 0111 respectively. The combined term is denoted by 0–11 (or) 0211, where dash, (or) 2 indicates that the particular literal (B) has been absorbed and the combined term is $\overline{A}CD$. The terms $\overline{A}\,\overline{B}CD$ and $\overline{A}\,\overline{B}\,\overline{C}\,\overline{D}$, cannot be combined, since their binary representations 0011 and 0000 differ in two positions, *i.e.*, in the third and fourth place. For the binary representations of two minterms to be different in just one position, it is necessary that the number of their 1's differ by exactly one.

The procedure for finding the minimal expression :

(1) Arrange all minterms in groups, such that all terms in the same group have same number of 1's in their binary representation. Start with the least number of 1's and continue with grouping of increasing number of 1's. The number of 1's in each term is called the index of that term *i.e.*, all the minterms of same index are placed in a same group. The lowest value of index is = 0.

(2) Compare every term of the lowest-index group with each term in the successive group; whenever possible combine the two terms of different groups by using the combine theorem $Aa + A\overline{a} = A$.

(3) Repeat the step 2, by comparing each term in a group of index i with every term in the group of index $i + 1$, until all minterms are completed in all groups. While combining place the '—' in the position, where the same variable appears in complement form in one term and in normal form in the other term as explained earlier. A check mark (✓) is placed next to every

term which has been combined with atleast one term (Each term may be combined with several terms, but only a single check is enough).

(4) The terms generated in the above steps are now compared in the same fashion. A new term is generated by combining two terms which differ by only a single 1 and whose dashes are in the same position. This process continues until no further combinations are possible.

The remaining unchecked terms constitute the set of EPI of the function.

Example 2.46. *Simplify the following boolean function by using Quine–McCluskey method*

(a) $f(A, B, C, D) = \Sigma m\,(0, 2, 3, 6, 7, 8, 10, 12, 13)$

(b) $f(A, B, CD) = \Sigma m\,(0, 1, 2, 5, 7, 8, 9, 10, 13, 15)$

Solution. (a) (1) List all the minterms in the binary form and arrange them in the form of groups according to the number of 1's contained as shown in Table 2.20. This is done by grouping the minterms into 2 sections by using horizontal lines. First section contains the number with no 1's in it. The second section contains those numbers that have only one and so on.

Table 2.20

Minterm	Binary representation ABCD	No. of 1's	Minterms	Index	Binary representation ABCD
m_0	0000	0	m_0	0	0000 √
m_2	0010	1	m_2	1	0010 √
m_3	0011	2	m_8		1000 √
m_6	0110	2	m_3		0011 √
m_7	0111	3	m_6	2	0110 √
m_8	1000	1	m_{10}		1010 √
m_{10}	1010	2	m_{12}		1100 √
m_{12}	1100	2	m_7	3	0111 √
m_{13}	1101	3	m_{13}		1101 √

(2) Compare each binary term with every term in the adjacent next higher category. If they differ only by one position put a check mark and copy the term into the next column with '—' in the place where the variable is unmatched, which is shown in the Table 2.21.

Table 2.21

(a)

Minterms group	*Binary representation*
0, 2	00 – 0 √
0, 8	– 000 √
2, 3	001 – √
2, 6	0 – 10 √
2, 10	– 010 √
8, 10	10 – 0 √
8, 12	1 – 00 – P
3, 7	0 – 11 √
6, 7	011 – √
12, 13	110 – Q

(b)

Minterms	*Binary representation*
0, 4, 8, 10	– 0 – 0 – R
2, 3, 6, 7	0 – 1 – S

(3) Apply the same process for the resultant column and continue these cycles until no further elimination of literals. There are four uncombined terms and are named as P, Q, R and S.

These are called as PI.

(4) List all the PI in the Table 2.22 called prime implicant table.

Table 2.22 PI Table

Prime implicants	*Binary representation* *ABCD*	*Literal representation*
8, 12	1 – 00	$A\overline{C}\overline{D}$
12, 13	110 –	$AB\overline{C}$
0, 2, 8, 10	– 0 – 0	$\overline{B}\overline{D}$
2, 3, 6, 7	0 – 1 –	$\overline{A}C$

(5) Now select the minimum number of prime implicants which must cover all the minterms using prime implicant chart.

The Prime Implicant chart is a pictorial representation of all PI and their minterm functions. It consists of array of x columns and y rows, where x designates the number of minterms in the given function and y designates the number of PI. The entries in the ith row in the chart consist of ×'s placed at its intersection with the column. Corresponding minterm is covered by the ith prime implicant.

The problem now is to select a minimal subset of prime implicants, such that each column contains at least one x in the rows corresponding to the selected subset and the total number of literals in the prime implicants selected is as small as possible.

If any column contains just a single ×, the PI corresponding to the row in which this × appears is essential, and consequently must be included in any irredundant expression of f. This × is circled, and a check mark is placed next to the EPI. The row which corresponds to an EPI is referred to as an "*essential row*". Once the EPI are obtained, check whether they are covering all the minterms or not. If they are covering, the resultant expression is unique. See Table 2.23.

Table 2.23

Name	*Prime implicants*	*Literal √ representation*	*Minterms*								
			m_0	m_2	m_3	m_6	m_7	m_8	m_{10}	m_{12}	m_{13}
P	8, 12	$A\overline{C}\overline{D}$						×		×	
Q	12, 13 √	$AB\overline{C}$								×	⊗
R	0, 2, 8, 10 √	$\overline{B}\overline{D}$	⊗	×				×	⊗		
S	2, 3, 6, 7 √	$\overline{A}C$		×	⊗	⊗	⊗				

In the columns m_0, m_3, m_6, m_7, m_{10}, and m_{13} contain only one × and their corresponding PI must be included in the minimal function and they are called EPI.

Here Q, R and S are EPI.

These 3 EPI are covering all the minterms. So the minimal expression is

$$f = Q + R + S$$

$$= AB\overline{C} + \overline{B}\overline{D} + \overline{A}C$$

Example 2.47. $f = (A, B, C, D) = \Sigma m\ (0, 1, 2, 5, 7, 8, 9, 10, 13, 15)$.

Solution. Step 1: List the minterms in the binary form and arrange them according to categories of 1's (as shown in Table 2.24).

Table 2.24

Minterm	*Binary representation*	*No. of 1's*	*Minterm*	*Index*	*Binary representation ABCD*
m_0	0000	0	m_0	0	0000 √
m_1	0001	1	m_1	1	0001 √
m_2	0010	1	m_2		0010 √
m_5	0101	2	m_8		1000 √
m_7	0111	3	m_5	2	0101 √
m_8	1000	1	m_9		1001 √
m_9	1001	2	m_{10}		1010 √
m_{10}	1010	2	m_7	3	0111 √
m_{13}	1101	3	m_{13}		1101 √
m_{15}	1111	4	m_{15}	4	1111 √

Step 2: Compare each binary term with every term of the next index and if they differ by one position, put a check mark and copy the term in the next column with '–' in the position they differed (as shown in Table 2.25).

Table 2.25

Minterms	Binary representation A B C D
0, 1	000 – √
0, 2	00 –0 √
0, 8	– 000
1, 5	0 – 01 √
1, 9	– 001 √
2, 10	– 010 √
8, 9	100 – √
8, 10	10 – 0 √
5, 7	01 –0
5, 13	– 101 √
9, 13	1 – 01 √
7, 15 – 111	13, 15, 11 – 1

Table 2.25(*a*)

Minterms	Binary representation	
0, 1, 8, 9	– 00 – P	
0, 2, 8, 10	– 0 – 0	Both are same Q
0, 8, 2, 10	– 0 – 0	
1, 5, 9, 13	– – 01	Both are same R
1, 9, 5, 13	– – 01	
5, 7, 13, 15	– 1 – 1	Both are same S
5, 13, 7, 15	– 1 – 1	

Step 3: List the prime implicants as shown in Table 2.26

Table 2.26

Prime implicants	Binary representation A B C D	Literal representation
0, 1, 8, 9	– 00 –	$\overline{B}\,\overline{C}$
0, 2, 8, 10	– 0 – 0	$\overline{B}\,\overline{D}$
1, 5, 9, 13	– – 01	$\overline{C}D$
5, 7, 13, 15	– 1 – 1	BD

Step 4: Select EPI/minimum number of PI which must cover all minterm from PIC shown in Table 2.27.

Table 2.27

Sl. No.	PI	Literal	m_0	m_1	m_2	m_5	m_7	m_8	m_9	m_{10}	m_{13}	m_{15}
P	0, 1, 8, 9	$\overline{B}\,\overline{C}$	×	×				×	×			
Q	0, 2, 8, 10	$\overline{B}\,\overline{D}$	×		⊗			×		⊗		
R	1, 5, 9, 13	$\overline{C}D$		×		×			×		×	
S	5, 7, 13, 15	BD				×	⊗				×	⊗

m_2 and m_{10} is covered by only one PI – Q. So it is EPI

m_7 and m_{15} is covered by only one PI – S. So it is EPI

∴ By using Q and S the minterms covered are 0, 2, 8, 10, 5, 7, 13, 15.

The two EPI cover all the minterms except 1 and 9, those minterms can be covered by using either P or R.

∴ Minimal expression = f = P + Q + S (or) Q + R + S

∴ $$f = \overline{B}\,\overline{C} + \overline{B}\,\overline{D} + BD \text{ (or)}$$

$$f = \overline{B}\,\overline{D} + \overline{C}D + BD$$

Example 2.48. *Minimize the given expression using Q – M method*

$$f = \overline{A}BD + A\overline{B}D + ABC\overline{D}$$

Solution. To find the minimal expression using Q-M method, first it is necessary to find the canonical SOP form of f.

$$f = \overline{A}BD\,(C + \overline{C}) + A\overline{B}D\,(C + \overline{C}) + ABC\overline{D}$$

$$= \Sigma\, m\,(11, 9, 14, 7, 5)$$

$$= \Sigma\, m\,(5, 7, 9, 11, 14)$$

Step 1: (See Table 2.28). List the minterms in the binary form and arrange them according to the no. of 1's.

Table 2.28

Minterms	Binary representation	No. of 1's	Minterm	Index	Binary representation ABCD
m_5	0101	2	m_5	2	0101 √
m_7	0111	3	m_9		1001 √
m_9	1001	2	m_7	3	0111 √
m_{11}	1011	3	m_{11}		1011 √
m_{14}	1110	3	m_{14}		1110 — P

Step 2: (See Table 2.29) Compare each binary term with the every term of the next index and if they differ by one position, put check mark and copy the term in the next column with – in the position that they differ.

Table 2.29

Minterms	*Binary representation ABCD*
5, 7	01 – 1 Q
9, 14	10 – 1 R

Step 3: List all prime implicants (See Table 2.30)

Table 2.30

P. I	*Binary representation*	*Literal representation*
14	1110	$ABC\overline{D}$
5, 7	01 – 1	$\overline{A}BD$
9, 11	10 – 1	$A\overline{B}D$

Step 4: (See Table 2.31) Select minimum number of EPI's to cover all the minterms.

Table 2.31

Sl. No.	*P. I*	*Literal Representation*	m_5	m_7	m_9	m_{11}	m_{14}
P	14 √	$ABC\overline{D}$					×
Q	5, 7 √	$\overline{A}BD$	×	×			
R	9, 11 √	$A\overline{B}D$			×	×	

Here three PI are EPI

$$\therefore \quad f = ABC\overline{D} + (\overline{A}B + A\overline{B})\,D$$

Example 2.49. *Simplify the following Boolean function using Q–M method.*

$$f(A, B, C, D) = \Sigma\, m\,(1, 2, 3, 5, 9, 12, 14, 15) + \Sigma\, d\,(4, 8, 11)$$

Solution.

Note: Don't care conditions are used to find the PI, but it is not compulsory to include don't care terms in the final expression.

Step 1: (See Table 2.32)

Table 2.32

Minterms	*Binary representation*	*No. of 1's*	*Minterm*	*Index*	*Binary representation*
m_1	0001	1	m_1		0001 √
m_2	0010	1	m_2	1	0010 √
m_3	0011	2	md_4		0100 √
m_5	0101	2	md_8		1000 √
m_9	1001	2	m_3		0011 √

m_{12}		1100	2	m_5	2	0101 √
m_{14}		1110	3	m_9		1001 √
m_{15}		1111	4	m_{12}		1100 √
m_4		0100	1	m_{14}	3	1110 √
m_8		1000	1	m_{11}		1011 √
md_{11}		1011	3	md_{15}	4	1111 √

Step 2: See Table 2.33 (*a*)

Table 2.33(*a*)

Minterms	*Binary representation* *ABCD*
1, 3	00 – 1 √
1, 5	0 – 01 – P
1, 9	– 001 √
2, 3	001 – Q
4, 5	010 – R
4, 12	– 100 – S
8, 9	100 – T
8, 12	1– 00 – U
3, 11	– 011 √
9, 11	10 – 1 √
12, 14	11 – 0 – V
11, 15	1 – 11 – W
14, 15	111 – X

Table 2.33(*b*)

Minterms	*Binary representation*
1, 3, 9, 11	– 0 – 1 –Y
1, 9, 3, 11	– 0 – 1

Step 3: See Table 2.34

Table 2.34

Sl. No.	*P. I*	*B. R*
P	1, 5	0 – 01
Q	2, 3	001 –
R	4, 5	010 –
S	4, 12	– 100
T	8, 9	100 –

U	8, 12	1 – 00
V	12, 14	11 – 0
W	11, 15	1 – 11
X	14, 15	111 –
Y	1, 3, 9, 11	– 0 – 1

Step 4: See Table 2.35

Table 2.35

Sl.No.	*P. I.*	*Literal representation*	m_1	m_2	m_3	dm_4	m_5	dm_8	m_9	dm_{11}	12	14	15
P	1, 5	$\overline{A}\,\overline{C}D$	×				×						
√Q	2, 3	$\overline{A}\,\overline{B}C$		⊗	×								
R	4, 5	$\overline{A}B\overline{C}$				×	×						
S	4, 12	$B\overline{C}\,\overline{D}$				×					×		
T	8, 9	$A\overline{B}\,\overline{C}$						×	×				
U	8, 12	$A\overline{C}\,\overline{D}$						×			×		
V	12, 14	$AB\overline{D}$									×	×	
W	11, 15	ACD											×
√X	14, 15	ABC										×	×
Y	1, 3, 9, 11	$\overline{B}D$	×		×				×	×			

Here only Q is a EPI which covers 2 and 3

To cover the minterm 1, 5, 9, 12, 14, 15

We may select PI Y → to cover 1 and 9

X → to cover 14, 15

P → to cover 1, 5

U → to cover 8, 12

f = P + Q + U + X + Y (or)

f = Q + Y + R + S + X (or)

f = Q + Y + V + X + R (or)

f = Q + Y + P + V + X

f = Q + Y + S + R + Y and so on.

2.17.3 The Decimal Representation

Tabulation method can be adopted by using decimal code for the minterms rather than their binary representation. With this, the tedious manipulation that one must undergo when using the tabulation method by binary representation is reduced. This method that uses subtraction of decimal numbers instead of the comparing and matching of binary numbers.

We note that each 1 in a binary number represents the coefficient multiplied by a power of 2. When two minterms are the same in every position except only one position, the minterm with the extra 1 must be larger than the number of the other minterm by a power of 2. Therefore, two minterms can be combined only if they differ by a power of 2, that is, only if the difference between their decimal codes is 2^i. The combined term consists of the same literals as the minterms, with the exception of the variable whose weight is 2^i, which is deleted.

For example, the combination of minterms 0 and 8 is written as 0, 8 (8) and 1 and 5 is written as 1, 5(4).

In the earlier case, the literal appears at 2^3 = 8th place and in the later case, the literal belongs to 2^2 = 4th place, will be deleted.

(1) The condition that the decimal codes of two combinable terms must differ by a power of 2 is necessary but not sufficient. Two minterms whose code differ by a power of 2, but which have the same index cannot be combined, since they differ by more than one variable.

(2) Similarly, if a minterm with smaller index has a higher decimal value than another minterm whose index is higher, then the two minterms can not be combined, although they may differ by a power of 2.

For example, the minterms 8 and 7 and their indexes are 1 and 3 respectively, can not be combined, since they differ in four variables.

Except these two conditions, the tabulation procedure using decimal representation is completely analogous to that using the binary representation.

Example 2.50. *Find the PI using Tabulation method in the given function*

f = *(A, B, C, D, E)* = *Σ m (1, 2, 12, 13, 15, 17, 18, 19, 20, 21, 23, 24, 25, 27, 29, 31)*

Solution. Step 1: Arrange all the minterms according to the number of 1's and group the minterms of same index (shown in Table 2.36)

Table 2.36

Minterm	*Binary representation*	*No. of 1's*	*Minterms*	*Index*
1	00001	1	√ 1	1
2	00010	1	√ 2	
12	01100	2	√ 12	
13	01101	3	√ 17	
15	01111	4	√ 18	2
17	10001	2	√ 20	
18	10010	2	√ 24	
19	10011	3	√ 13	
20	10100	2	√ 19	3
21	10101	3	√ 21	
23	10111	4	√ 25	
24	11000	2	√ 15	

25	11001	3	√ 23	4
27	11011	4	√ 27	
29	11101	4	√ 29	
31	11111	5	31	5

Step 2: Combine every possible minterm of lower index with all other minterms of the next index group and put the check mark on the combined minterms (see Table 2.37).

Table 2.37

(a)

Minterms	*Difference*
1, 17	(16) – P_1
2, 18	(16) – P_2
12, 13	(1) – P_3
17, 19	(2) √
17, 21	(4) √
17, 25	(8) √
18, 19	(1) – P_4
20, 21	(1) – P_5
24, 25	(1) – P_6
13, 15	(2) √
13, 29	(16) √
19, 23	(4) √
29, 27	(8) √
21, 23	(2) √
21, 29	(8) √
25, 27	(2) √
25, 29	(4) √
15, 31	(16) √
23, 31	(8) √
27, 31	(4) √
29, 31	(2) √

(b)

Minterms	*Difference*
(17, 19, 21, 23)	(2, 4) √
(17, 19, 25, 27)	(2, 8) √
(17, 21, 25, 29)	(4, 8) √
(13, 15, 29, 31)	(2, 16) – P_7
(19, 23, 27, 31)	(4, 8) √
(21, 23, 29, 31)	(2, 8) √
(25, 27, 29, 31)	(2, 4) √

(c)

17, 21, 25, 29, 19, 23, 27, 31	(4, 8) – P_8

Step 3: Prime Implicant chart (see Table 2.38) to find Essential Prime Implicants.

Table 2.38

	Decimal representation	Literal representation
√ P_1	1, 17 (16)	$\overline{B}\,\overline{C}\,\overline{D}E$
√ P_2	2, 18 (16)	$\overline{B}\,\overline{C}D\overline{E}$
√ P_3	12, 13 (1)	$\overline{A}BC\overline{D}$
P_4	18, 19 (1)	$A\overline{B}\,\overline{C}D$
√ P_5	20, 21 (1)	$A\overline{B}C\overline{D}$
√ P_6	24, 25 (1)	$AB\overline{C}\,\overline{D}$
√ P_7	13, 15, 29, 31 (2, 16)	BCE
√ P_8	17, 19, 21, 23, 25, 27, 29, 31	AE

Prime impli-cant	√ *1*	√ *2*	√ *12*	√ *13*	√ *15*	√ *17*	√ *18*	√ *19*	√ *20*	√ *21*	√ *23*	√ *24*	√ *25*	√ *27*	√ *29*	√ *31*
P_1	⊗					×										
P_2	⊗					×										
P_3		⊗	×													
P_4						×	×									
P_5								⊗	×							
P_6											⊗	×		×	×	
P_7			×	⊗												
P_8					×		×		×	⊗		×	⊗	×	×	

$P_1\ P_2\ P_3\ P_5\ P_6\ P_7$ and P_8 EPI.

All the minterms are covered by this. The unique minimal expression is

$$f = P_1 + P_2 + P_3 + P_5 + P_6 + P_7 + P_8$$

$$= \overline{B}\,\overline{C}\,\overline{D}E + \overline{B}\,\overline{C}D\overline{E} + \overline{A}BC\overline{D} + A\overline{B}C\overline{D} + AB\overline{C}\,\overline{D} + BCE + AE$$

Example 2.51. $f(w, x, y, z) = \Sigma\, 1, 4, 9, 10, 15 + \Sigma\, d\,(6, 7, 8, 11)$.

Solution. Step 1: (See Table 2.39) Arrange all the minterms according to the number of 1's and group the minterms of same index.

Table 2.39

Minterm	*Binary representation*	*No. of 1's*	*Minterms*	*Index*
m_1	0001	1	√1	
m_4	0100	1	√4	1
m_9	1001	2	√8	
m_{10}	1010	2	√9	
m_{15}	1111	4	√10	2
md_6	0110	2	√6	
md_7	0111	3	√7	
md_8	1000	1	√11	3
md_{11}	1011	3	√15	4

Step 2: (See Table 2.40). Combine the every possible minterm of lower index with the all minterms of the next index group and put check mark on the combined production term.

(a)

Table 2.40

Minterms	*Decimal difference*
1, 9	(8) $-P_1$
4, 6	(2) $-P_2$
8, 9	(1) √
8, 10	(2) √
9, 11	(2) √
6, 7	(1) $-P_3$
10, 11	(1) √
7, 15	(8) $-P_4$
11, 15	(4) $-P_5$

(b)

Minterms	*Decimal difference*
(8, 9, 10, 11)	(1, 2) – P_6

Step 3: (See Table 2.41). Arrange all prime implicants in the Prime Implicant Chart to find the Essential Prime Implicants.

Table 2.41

Prime implicant	Decimal representation	Binary representation	Literal	1	4	⑥	⑦	⑧	9	10	⑪	15
P_1	1, 9 (8)	– 001	$\overline{B}\,\overline{C}D$	⊗					×			
P_2	4, 6 (2)	01 – 0	$\overline{A}B\overline{D}$		⊗	×						
P_3	6, 7 (1)	011 –	$\overline{A}BD$			×	×					
P_4	7, 15 (8)	– 111	BCD				×					×
P_5	11, 15 (4)	1 – 11	ACD								×	×
P_6	8, 9, 10, 11 (1, 2)	10 –	$A\overline{B}$					⊗ ×	×	×	×	

Here P_1, P_2 and P_6 are EPI. The minterms covered by them are 1, 9, 4, 6, 8, 10, 11. The uncovered minterm is 15.

To cover 15 we may choose either P_4 or P_5 as selective P.I.

$$f = P_1 + P_2 + P_6 + P_4$$

(or)

$$= P_1 + P_2 + P_6 + P_5$$

$$= \overline{B}\,\overline{C}D + \overline{A}B\overline{D} + A\overline{B} + BCD \text{ (or)}$$

$$= \overline{B}\,\overline{C}D + \overline{A}B\overline{D} + A\overline{B} + ACD$$

Sometimes, it is difficult to find the selective PI to cover the uncovered minterms of the function by the EPI.

For this, we must use a technique called reduction of the chart, which is explained in the following examples.

Example. $f = (A, B, C, D, E) = \Sigma\, m\,(1, 3, 4, 5, 6, 7, 10, 11, 12, 13, 14, 15, 18, 19, 20, 21, 22, 23, 25, 26, 27)$

Step 1: See Table 2.42. Arrange all the minterms according to the number of 1's and group the minterms of same index.

Table 2.42

Minterms	Binary representation	No. of 1's	Minterms	Index
m_1	00001	1	√ 1	
m_3	00011	2	√ 4	1

m_4	00100	1
m_5	00101	2
m_6	00110	2
m_7	00111	3
m_{10}	01010	2
m_{11}	01011	3
m_{12}	01100	2
m_{13}	01101	3
m_{14}	01110	3
m_{15}	01111	4
m_{18}	10010	2
m_{19}	10011	3
m_{20}	10100	2
m_{21}	10101	3
m_{22}	10110	3
m_{23}	10111	4
m_{25}	11001	3
m_{26}	11010	3
m_{27}	11011	4

√ 3 √ 5 √ 6 √ 10 √ 12 √ 18 √ 20	2
√ 7 √ 11 √ 13 √ 14 √ 19 √ 21 √ 22 √ 25 √ 26	3
√ 15 √ 23 √ 27	4

Step 2(a): See Table 2.43. Combine the every possible minterm of lower index with the all minterms of the next index group and put the checkmark on the combined minterms.

Table 2.43

Minterms	*Decimal difference*
1, 3	(2) √
1, 5	(4) √
4, 5	(1) √
4, 6	(2) √
4, 12	(8) √
4, 20	(16) √
3, 7	(4) √

Minterms	*Decimal difference*
12, 14	(2) √
18, 19	(1) √
18, 22	(4) √
18, 26	(8) √
20, 21	(1) √
20, 22	(2) √
7, 15	(8) √

3, 11	(8) √
3, 19	(16) √
5, 7	(2) √
5, 13	(8) √
5, 21	(16) √
6, 7	(1) √
6, 14	(8) √
6, 22	(16) √
10, 11	(1) √
10, 14	(4) √
10, 26	(16) √
12, 13	(1) √
7, 23	(16) √
11, 15	(4) √
11, 27	(16) √
13, 15	(2) √
14, 15	(1) √
19, 23	(4) √
19, 27	(8) √
21, 23	(2) √
22, 23	(1) √
25, 27	(2) – P_1
26, 27	(1) √

Step 2 (*b*): See Table 2.44

Table 2.44

Minterms	*Decimal difference*
1, 3, 5, 7	(2, 4) – P_2
4, 5, 6, 7	(1, 2) √
4, 5, 12, 13	(1, 8) √
4, 5, 20, 21	(1, 16) √
4, 6, 12, 14	(2, 8) √
4, 6, 20, 22	(2, 16) √
4, 12, 5, 13	(8, 1) √
4, 12, 6, 14	(8, 2) √
3, 7, 11, 15	(4, 8)–P_3
3, 7, 19, 23	(4, 16)–P_4
3, 11, 19, 27	(8, 16)–P_5
5, 7, 13, 15	(2, 8) √
5, 7, 21, 23	(2, 16) √
6, 7, 14, 15	(1, 8) √
6, 7, 22, 23	(1, 16) √
10, 11, 14, 15	(1, 4) P_6
10, 26, 11, 27	(1, 16) P_7
12, 13, 14, 15	(1, 2) √
18, 19, 22, 23	(1, 4) P_8
18, 26, 19, 27	(8, 1) P_9
20, 21, 22, 23	(1, 2) √
4, 5, 6, 7, 12, 13, 14, 15	(1, 2, 8) P_{10}
4, 5, 6, 7, 20, 21, 22, 23	(1, 2, 16) P_{11}

Step 3: PIC (See Table 2.45). Totally it has 11PI and now we can find the EPI by plotting PIC as shown in Table 2.45.

Table 2.45

Product terms (Prime implicants)	Literal representation	Minterms																				
		(1)	3	4	5	6	7	10	11	(12)	(13)	14	15	18	19	(20)	(21)	22	23	(25)	26	27
P_1 – 25, 27 (2)	$AB\bar{C}E$ √																			(×)		×
P_2 – 1,3,5,7 (2,4)	$\bar{A}\bar{B}E$ √	(×)	×		×		×															
P_3 – 3,7,11,15 (4,8)	$\bar{A}DE$		×				×		×				×									
P_4 – 3,7,19,23 (4,16)	$\bar{B}DE$		×				×								×				×			
P_5 – 3,11,19,27 (8,16)	$\bar{C}DE$		×						×			×	×									
P_6 – 10,11,14,15 (1,4)	$\bar{A}BD$							×	×			×	×									
P_7 – 10,11,26,27 (1,16)	$B\bar{C}D$							×	×												×	×
P_8 – 18,19,22,23 (1,4)	$A\bar{B}D$													×	×			×	×			
P_9 – 18,19,26,27 (1,8)	$A\bar{C}D$													×	×						×	×
P_{10} – 4,5,6,7,12,13 14,15 (1,2,8)	$\bar{A}C$ √			×	×	×	×			(×)	(×)	×	×									
P_{11} – 4,5,6,7,20,21 22,23 (1,2,16)	$\bar{B}C$ √			×	×	×	×									(×)	(×)	×	×			

From the PIC, we note that there are 4 EPI P_1, P_2, P_{10} and P_{11} and they cover the minterms 1, 3, 4, 5, 6, 7, 12, 13, 14, 15, 20, 21, 22, 23, 25 and 27 and to cover the remaining minterms 10, 11, 18, 19 and 26.

To cover the above minterms, use the reduced chart, which contains uncovered minterms and remaining PI. *i.e.,* which is obtained by removing the essential rows and columns covered. [The rows (PI) and columns (minterms) are said to be essential if the particular PI is an essential PI]

Table 2.46. Reduced PI chart

	10	11	18	19	26
P_3		×			
P_4				×	
P_5		×		×	
P_6	×	×			
P_7	×	×			×
P_8			×	×	
P_9			×	×	×

Table 2.46 shows the reduced chart, here all the rows are not essential. Some of them may be removed.

For example, here, P_4 is covering the minterm 19 but P_5 is covering 11 and 19, so the removal of row P_4 can not prevent us from finding minimal expression. Here, P_5 covers the minterm covered by P_4 and an extra also and hence P_4 can be replaced by P_5. Here P_5 is called a dominate row.

A row X of a prime implicant chart is said to dominate another row Y of that chart if X covers every column covered by Y, *i.e.,* if row X dominates row Y and the PI corresponding to row X does not have more literals than the PI corresponding to row Y, then row Y can be deleted from the chart.

Thus rows P_3 and P_4 in the Table 2.46 are deleted by a dominated row P_5, similarly rows P_6 and P_8 are removed because they are dominated by rows P_7 and P_9 respectively. The final reduced chart *i.e.,* final chart is shown in Table 2.47. It contains 3 rows, P_5, P_7 and P_9 and 5 columns (indicates minterms uncovered by EPI).

Table 2.47

	√ 10	√ 11	√ 18	√ 19	√ 26
P_5		×		×	
√ P_7	⊗	×			×
√ P_9			⊗	×	×

From the final chart, minterm 10 is covered by only one PI *i.e.,* P_7 and similarly minterm 18 by P_9. So, these two PI must be included in the minimal expression. These two PI will cover all the remaining minterms.

∴ Minimal expression = EPI's + Selected Prime Implicants.

$$f = P_1 + P_2 + P_7 + P_9 + P_{10} + P_{11}$$
$$= AB\overline{C}E + \overline{A}BE + B\overline{C}D + A\overline{C}D + \overline{A}C + \overline{B}C$$
$$= AB\overline{C}E + \overline{A}BE + \overline{C}D(A + B) + C(\overline{A} + \overline{B})$$

Now, we will see another method for finding the Selected Prime Implicants from PIC *i.e.*, dominated columns.

Prime implicant charts can also be reduced by removing certain columns. Consider the same example, for column 10 and 11 in Table 2.46. Inorder to cover the column 10, either P_6 or P_7 must be selected, whereby column 11 is also automatically covered by the same rows, since it has '×' mark in these rows. Here column 11 is also covered by P_3, but it is not covering column 10.

A column *x* in a PIC is said to dominate another column *y* of that PIC, if *x* has an × mark in every row in which *y* has an ×. *i.e.*, if a column *x* dominates column *y*, then column *x* can be deleted from the chart without affecting search for a minimal expression. In fact, removal of dominated column doesn't prevent us from finding all minimal expressions.

Note : When reducing columns, the dominating one are removed, incase of rows, the dominated rows are deleted.

2.17.3.1 The branching method

This is another technique used for simplification of PIC, when the given function doesn't have any EPI, dominated rows or dominated columns.

The procedure of *branching method* can be explained with an example.

Example. $f(A, B, C, D) = \Sigma m\,(0, 1, 5, 7, 8, 10, 14, 15)$

Step 1: See Table 2.48

Table 2.48

Minterm	*Binary representation*	*No. of 1's*	*Minterm*	*Index*
0	0000	0	√ 0	0 √
1	0001	1		
5	0101	2	√ 1	1
7	0111	3	√ 8	
8	1000	1	√ 5	2
10	1010	2	√ 10	
14	1110	3	√ 7	3
			√ 14	
15	1111	4	√ 15	4

Step 2: (See Table 2.49) It has totally eight prime implicants

Table 2.49

Minterms	*Decimal difference*
(0, 1)	(1) P_1
(0, 8)	(8) P_2
1, 5	(4) P_3
8, 10	(2) P_4
5, 7	(2) P_5
10, 14	(4) P_6
7, 15	(8) P_7
14, 15	(1) P_8

Step 3: See Table 2.50 (*a*). Prime Implicant Chart to find Essential Prime Implicants.

Table 2.50(*a*)

EPI	*Decimal rotation*	*Literal difference*	*Binary representation ABCD*	√ *0*	√ *1*	*5*	*7*	*8*	*10*	*14*	*15*
P_1	0, 1	(1) ($\overline{A}\overline{B}\overline{C}$)	000 –	×	×						
P_2	0, 8	(8) ($\overline{B}\overline{C}\overline{D}$)	– 000	×				×			
P_3	1, 5	(4) ($\overline{A}C\overline{D}$)	0 – 10		×	×					
P_4	8, 10	2 ($A\overline{B}\overline{D}$)	10 – 0					×	×		
P_5	5, 7	(2) ($\overline{A}BD$)	01 – 1			×	×				
P_6	10, 14	(4) ($AC\overline{D}$)	1 – 10						×	×	
P_7	7, 15	(8) BCD	– 111				×				×
P_8	14, 15	(1) ABC	111 –							×	×

The function has no EPI, because each minterm is covered by two rows.

No dominating rows, since each row is covering only 2 minterms. No dominating columns, because each column is covered by only 2 rows.

Here, each PI is covered by 2 minterms and each minterm is covered by 2 PI. Such a chart is called "*cyclic PIC*".

In order to find the minimal expression for the given function, it is necessary to make an arbitrary selection of one row and they apply the reduction procedure. Consider, for example, column 1 in the chart (minterm 1) can be covered by either P_1 or P_3 PI. Consequently one of these must be included in the minimal expression.

If a row P_1 is arbitrarily chosen, the chart of Table 2.50 (*b*) results.

Table 2.50(*b*) Reduced chart after section of row P_1

	5	7	8	10	14	15
P_2			×			
P_3	×					
P_4			×	×		
P_5	×	×				
P_6				×	×	
P_7		×				×
P_8					×	×

By selecting P_1, it is covering minterms 0 and 1. So, those two columns are not included in the reduced chart 2.50 (*b*).

In this chart, row P_2 is dominated by row P_4 and row P_3 is dominated by row P_5. So, rows P_4 and P_5 are selected (see Table 2.50 *c*).

Table 2.50(c)

	√ 5	√ 7	√ 8	√ 10	14	15
P_4			(×)	×		
P_5	(×)	×				
P_6				×	×	
P_7		×				×
P_8					×	×

By those two rows, the minterms covered are 5, 7, 8 and 10. The remaining minterms are shown in final chart. (see Table 2.50 *d*)

Table 2.50(*d*) Final chart

	14	15
P_6	×	
P_7		×
P_8	×	×

Here, we can observe that P_6 and P_7 are dominated by P_8.

$$\therefore \quad f = P_1 + P_4 + P_5 + P_8$$

$$= \overline{A}\,\overline{B}\,\overline{C} + A\overline{B}\,\overline{D} + \overline{A}BD + ABC$$

The entire process is now repeated by selecting row P_3. The removal of row P_3 results in chart 2.50(*e*).

Table 2.50(*e*)

	0	7	8	10	14	15
P_1	×					
P_2	×		×			
P_4			×	×		
P_5		×				
P_6				×	×	
P_7		×				×
P_8					×	×

Here P_1 is dominated by row P_2 and row P_5 is dominated by row P_7. So we can select P_2 and P_7 and remove P_1 and P_5 (see Table 2.51 *f*).

Table 2.50(*f*)

	√	√	√			√
	0	7	8	10	14	15
√ P_2	⊗		×			
P_4			×	×		
P_6				×	×	
√ P_7		⊗				×
P_8					×	×

The remaining two minterms 10, 14 may be covered and for that the selection of PI was shown in Table 2.50 (*g*) *i.e.*, final chart.

Table 2.50(*g*)

	10	14
P_4	×	
P_6	×	×
P_8		×

Here, row P_6 is dominating P_4 and P_8. So P_6 must be included in the minimal expression.

$$\therefore \quad f = P_2 + P_3 + P_6 + P_7$$

$$= \overline{B}\,\overline{C}\,\overline{D} + \overline{A}C\overline{D} + AC\overline{D} + BCD$$

2.18 REDUCED-DIMENSION MAP

As we have seen, the K-map is an invaluable tool in studying switching functions. We have used map entries of 0, 1 and don't care (x), but now let us consider including variable terms as entries in our maps.

Previously all variables were used as map variables, with the dimension of the map equal to the number of variables. However, we can reduce the dimension of our maps by entering map variables as entries on a "reduced dimension map (RDM)". These entries are called "map-entered variables".

This can be illustrated with an example : Consider the function f = AD + BC and its four-variable map shown here.

The 'f' contains the following minterms and are shown in map, (See Fig. 2.53 *a*).

$$f = \Sigma\, m\ (6, 7, 9, 11, 13, 14, 15)$$

AB \ CD	00	01	11	10
00	1			
01			1	1
11		1	1	1
10		1	1	

(*a*) K-map

↔

AB \ C	0	1
00		
01		1
11	D	1
10	D	D

(*b*) RDM

Fig. 2.53 Function f.

Using D as a map-entered variable, D can be eliminated as a map variable, there by reducing the map to a three-variable map. Entries on the RDM are made after examining the locations on the original map where the map-entered variable D is true and where false, with other map variables held constant.

The location of the entry on the RDM is the map-variable combination that was held constant while examining the value of the function when the map entered variable is equal to '0' and when the map-entered variable is equal to 1.

(1) When the function is 0 in both locations, a '0' is entered on the RDM.

(2) When the function is 1 in both locations, a 1 is entered on the RDM.

(3) If the function is 1 when the map-entered variable is 1 and 0 when the map-entered variable is 0, the map-entered variable is entered on the RDM.

(4) If the function is 0 when the map-entered variable is 1 and the function is 1 when the map-entered variable is 0, the complement of the map-entered variable is placed on the RDM.

Table 2.51 lists RDM entries for the cases that might be encountered in generating RDM's.

Table 2.51. Rules for eliminating a map entered variable of a K-map or a RDM

Value of function for map-entered variable location		RDM entry
x = 0	x = 1	
0	0	0
0	1	x
1	0	$\bar{x}$
1	1	1
0	F	F.x
1	F	$\bar{x}$ + F
F	0	$\bar{x}$F
F	1	x + F
F	$\bar{F}$	x ⊕ F
$\bar{F}$	F	x ⊕ F
F	F	F
F	G	$\bar{x}$F + xG
F	F + G	F + xG
F + G	F	F+$\bar{x}$G

Value of function for map-entered variable location		RDM entry
x = 0	x = 1	
F	F.G	F ($\bar{x}$ + G)
F.G	F	F (x + G)
F + G	F + H	F + $\bar{x}$G + xH
F.G	F.H	F ($\bar{x}$G + xH)
F + G	F.H	FH + $\bar{x}$ (F + G)
F.G	F + H	FG + x (F + H)
0	–	(0, x)*
1	–	(1, $\bar{x}$)*
–	0	(0, $\bar{x}$)*
–	1	(1, x)*
–	F	(x.F, $\bar{x}$ + F)*
F	–	($\bar{x}$F, x + F)*

where " – " indicates don't care, (*a*, *b*)* indicates that the RDM may be either *a* or *b*, which ever most simplifies the RDM.

Example 2.52. *Find the RDM maps of 4-variable, 3-variable for the given 5-variable K-map. (see Fig. 2.54).*

$$f = \Sigma\, m\,(3, 8, 9, 12, 13, 21, 22, 29)$$

AB \ CD

		1	
1	1		
1	1		

E = 0

		1	1
		1	

E = 1

Fig. 2.54

Solution. First take the E as a map-entered variable and the 4-variable RDM is shown in Fig. 2.55 (*a*).

AB \ CD	00	01	11	10
00			$\bar{E}$	
01		E		E
11	$\bar{E}$	1		
10	$\bar{E}$	$\bar{E}$		

Fig. 2.55(*a*)

AB \ C	0	1
00		$D\bar{E}$
01	DE	$D\bar{E}$
11	$\bar{E}$+D	
10	$\bar{E}$	

Fig. 2.55(*b*)

While constructing the 3-variable RDM, from the 4-variable RDM, now consider here D as a map-entered variable and the resultant RDM is shown in Fig 2.55 (*b*).

Example 2.53. *Find the 3-variable RDM for the given 4-variable RDM shown in Fig 2.56 (a).*

Solution.

CD \ AB	00	01	11	10
00			1	
01	E		1	
11		1	1	
10			F	

Fig. 2.56(*a*)

C \ AB	00	01	11	10
0	DE		1	
1		D	D+F	

Fig. 2.56(*b*)

Fig 2.56 (*b*) shows the reduced *i.e.,* 3-variable RDM. From this, we see that RDM can be reduced further. Further reductions can reduce a map until only a single equation remains.

Reading of reduced-dimension maps

A RDM is read in three steps

(1) All map-entered variables are set to zero and the terms from the 1's and don't-care entries read from the map.

(2) The terms containing the map-entered variables are obtained by considering 1's as don't care entries. The terms from the encirclements are ANDed with the expression within the circle.

(3) The function is then read by combining the results of steps 1 and 2.

The following examples illustrate the map-reading process.

Example 2.54. *The function is represented in Fig. 2.57, write the function.*

C \ AB	00	01	11	10
0	1			D
1	D	1	$\bar{D}$	D

Fig. 2.57

Solution. **Step 1:** $\bar{B}D$

Step 2: $\bar{B}\bar{C}\bar{D}$

Step 3: $\bar{B}D + BC\bar{D}$.

Example 2.55. *Find the function represented in Fig. 2.58.*

Solution. **Step 1:** A

Step 2: $\overline{B}C$

Step 3: $A + \overline{B}C$.

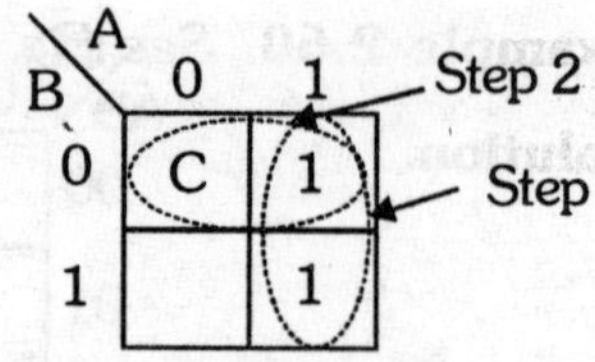

Fig. 2.58

Example 2.56. *Find the function represented in Fig. 2.59.*

Solution. **Step 1:** B

Step 2: $\overline{C}D + AC\overline{D}$

Step 3: $B + \overline{C}D + AC\overline{D}$.

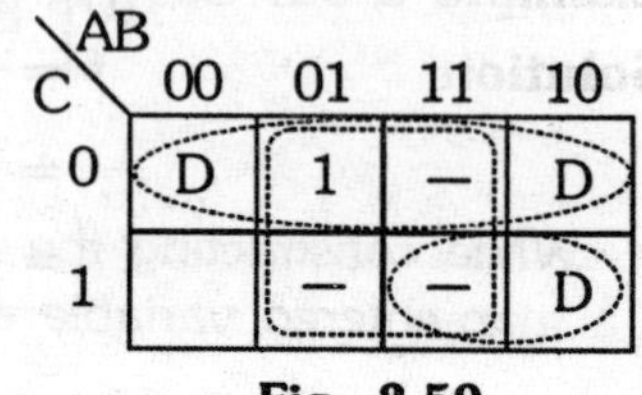

Fig. 2.59

Example 2.57. *See Fig. 2.60.*

Solution. **Step 1:** $\overline{B}D$

Step 2: $\overline{A}DE + ABC\,(E.F)$

Step 3: $\overline{B}D + \overline{A}DE + ABCEF$.

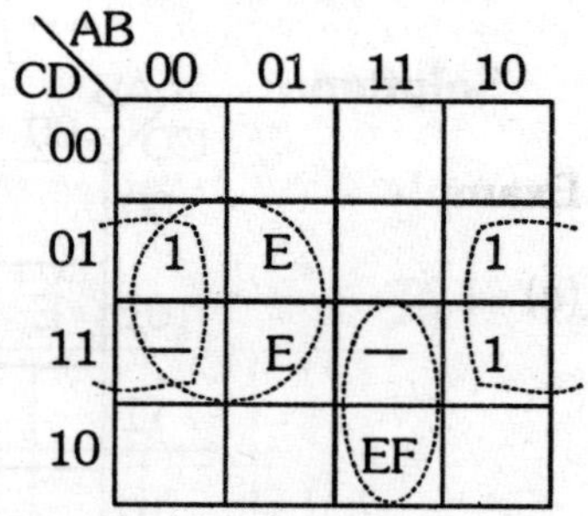

Fig. 2.60

Example 2.58. *Obtain minimized expression for the function represented on the RDM of Fig. 2.61.*

Solution.

$$f = B\overline{D}\,(E + F) + BDE$$

$$= BE\,(\overline{D} + D) + B\overline{D}F$$

$$= BE + B\overline{D}F.$$

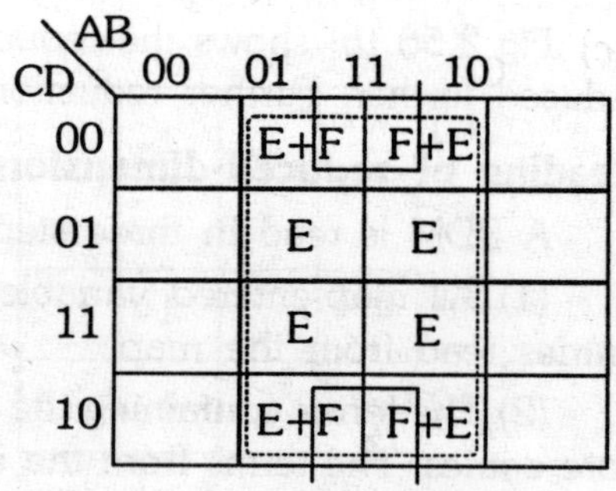

Fig. 2.61

Example 2.59. *See Fig. 2.62.*

Solution.

$$f = A\overline{B}C.D + ABC$$

$$= AC\,(\overline{B}D + B)$$

$$= AC\,(D + B).$$

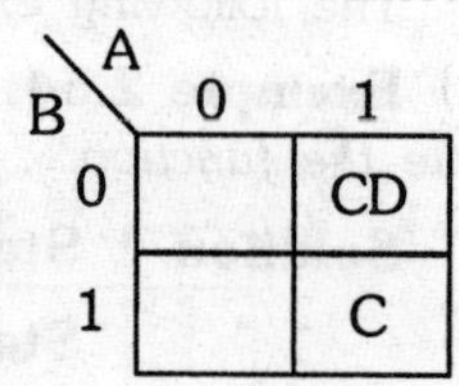

Fig. 2.62

Example 2.60. *See Fig. 2.63.*

Solution.

$$f = ADC + \overline{A}C$$
$$= C(\overline{A} + AD)$$
$$= C(A + D)$$

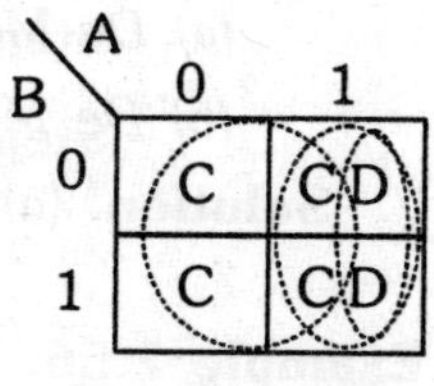

Fig. 2.63

Example 2.61. *See Fig. 2.64.*

Solution.

$$f = \overline{A}C + A\overline{B}\,(C + D) + ABCE$$
$$= \overline{A}C + A\overline{B}C + A\overline{B}D + ABCE$$
$$= C(\overline{A} + A\overline{B} + ABE) + A\overline{B}D$$
$$= C\left[\overline{A} + A(\overline{B} + BE)\right] + A\overline{B}D$$
$$= C\left[\overline{A} + A(\overline{B} + E)\right] + A\overline{B}D$$
$$= C(\overline{A} + \overline{B} + E) + A\overline{B}D.$$

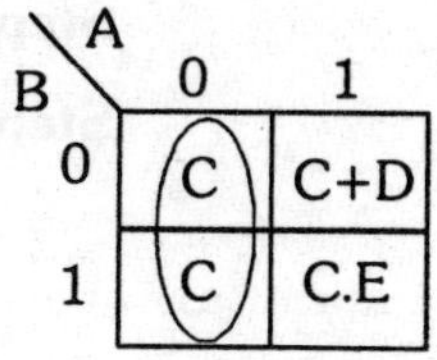

Fig. 2.64

Example 2.62. *Reduce the number of map variables to represent the functions :*

(*a*)

C \ AB	00	01	11	10
0	1			1
1	1	1		

Fig. 2.65(*a*)

(*b*)

C \ AB	00	01	11	10
0	1	1		
1			1	1

Fig. 2.65(*b*)

(*c*)

CD \ AB	00	01	11	10
00	1		1	1
01		1		1
11	1		1	
10		1	1	1

Fig. 2.65(c)

Solution.

(*a*)

C \ A	0	1
0	$\overline{B}$	$\overline{B}$
1	1	

Fig. 2.66(*a*)

(*b*)

C \ A	0	1
0	1	0
1	0	1

Fig. 2.66(*b*)

(*c*)

C \ AB	00	01	11	10
0	$\overline{D}$	D	$\overline{D}$	1
1	D	$\overline{D}$	1	$\overline{D}$

Fig. 2.66(c)

2.19 SOLVED PROBLEMS

Problem 1. *Demonstrate by means of truth tables the validity of the following identities.*

(a) De Morgan's theorem for three variables $(xyz)' = x' + y' + z'$

(b) De Morgan's theorem for second law $= (x + y + z)' = x'y'z'$

Solution. (*a*) Proof for 3 variables De Morgan's theorem by truth table (see Table 2.52).

Table 2.52

xyz	$x'\,y'z'$	xyz	$(xyz)'$	$x' + y' + z'$
000	111	0	1	1
001	110	0	1	1
010	101	0	1	1
011	100	0	1	1
100	011	0	1	1
101	010	0	1	1
110	001	0	1	1
111	000	1	0	0

From the truth table $(xyz)' = x' + y' + z'$.

(*b*) For proof of $(x + y + z)' = x'.y'.z'$ (See Table 2.53).

Table 2.53

xyz	$x'y'z'$	$x + y + z$	$(x + y + z)'$	$x'.y'.z'$
000	111	0	1	1
001	110	1	0	0
010	101	1	0	0
011	100	1	0	0
100	011	1	0	0
101	010	1	0	0
110	001	1	0	0
111	000	1	0	0

From the truth table, $(x + y + z)' = x'.y'.z'$ (Table 2.54).

Problem 2. *Simplify the following Boolean expression to a minimum number of literals:*

(*a*) $x'y' + xy + x'y$ (*b*) $(x + y)\,(x + y')$

(*c*) $x' + xy + xz' + xy'z'$ (*d*) $x'y + xy' + xy + x'y'$

(*e*) $x'yz + xz$ (*f*) $xy + x\,(wz + wz')$

(*g*) $xyz + \bar{x}y + xy\bar{z}$

Solution. (*a*)

$$\begin{aligned} f(x, y) &= x'y' + xy + x'y = x'(y + y') + xy \\ &= x'.1 + xy = x' + xy = (x' + x)(x' + y) \\ &= x' + y. \end{aligned}$$

(b) $f = (x + y)(x + y') = x + xy + xy' + 0$
$= x(1 + y) + xy' = x + xy'$
$= x(1 + y') = x.1 = x.$

(c) $f(x,y,z) = x' + xy + xz' + xy'z'$
$= (x' + x)(x' + y) + xz' + xy'z'$
$= x' + y + xz'(1 + y') = x' + y + xz'$
$= (x' + x)(x' + z') + y$
$= x' + y + z'.$

(d) $f(x,y) = x'y + xy' + xy + x'y' = x'y + xy + xy' + x'y'$
$= y(x + x') + y'(x + x') = y.1 + y'.1 = y + y'$
$= 1.$

(e) $f(x,y,z) = x'yz + xz = z(x + x'y)$
$= z\left[(x + x^1)(x + y)\right] = z(x + y).$

(f) $f = xy + x(wz + w\bar{z}) = xy + xw(z + \bar{z})$
$= xy + xw = x(y + w).$

(g) $f = xyz + x'y + xy\bar{z} = xy(z + z) + x'y$
$= xy + x'y = y(x + \bar{x}) = y.$

Problem 3. *Reduce the following Boolean expression to the indicated number of literals:*

(a) A'C' + ABC + AC' to three literals.

(b) (x' . y' + z)' + z + xy + wz to three literals.

(c) A'B (D' + C'D) + B (A + A'CD) to one literal.

(d) (A' + C) (A' + C') (A + B + C'D) to four literals.

Solution. (a) $f(A, B, C) = A'C' + ABC + AC' = A'C' + AC' + ABC$
$= C'(A + A') + ABC = C' + ABC$
$= (C' + AB)(C + C') = (AB + C').$

(b) $f = (w, x, y, z) = (x'y' + z)' + z + xy + wz$
$= (x'y' + z)' + z + xy + wz$
$= \overline{x'y'}.z' + z + xy + wz$
$= (x + y)z' + z + xy + wz$
$= xz' + yz' + z + xy + wz$
$= xz' + yz' + z(1 + w) + xy$
$= xz' + yz' + z + xy$
$= (x + z) + (z + z') + yz^1 + xy$
$= x + z + yz' + xy$
$= x(1 + y) + (z + y)(z + z')$
$= x + y + z.$

(c) $f = (A, B, C, D) = A'B\,(D' + C'D) + B\,(A + A'CD)$

$= A'B\left[(D' + C')\,(D' + D)\right] + B\left[(A + A')\,(A + CD)\right]$

$= A'B\,(D' + C') + B\,(A + CD)$

$= A'BD' + A'BC' + AB + BCD$

$= B\,(A'D' + A'C' + A + CD)$

$= B\left[\overline{A}\,\overline{D} + \overline{A}\,\overline{C} + A + CD\right]$

$= B\left[\overline{A}\,\overline{D} + \overline{C} + A + CD\right] = B\left[\overline{D} + A + \overline{C} + CD\right]$

$= B\left[\overline{D} + A + \overline{C} + D\right]$

$= B\left[1 + A + \overline{C}\right] = B\,[1]$

$= B.$

(d) $f = (A, B, C, D) = (\overline{A} + C)\,(\overline{A} + \overline{C})\,(A + B + \overline{C}D)$

$= (\overline{A} + \overline{A}\,\overline{C} + \overline{A}C + C\overline{C})\,(A + B + \overline{C}D)$

$= \overline{A}\,(1 + \overline{C} + C)\,(A + B + \overline{C}D)$

$= \overline{A}\,(A + B + \overline{C}D) = \overline{A}A + \overline{A}B + \overline{A}\,\overline{C}D$

$= \overline{A}\,(B + \overline{C}D).$

Problem 4. *Find the complement of the function $F = A + BC$; then show that $F.\overline{F} = 0$ and $F + \overline{F} = 1$*

Solution. Given that $F = A + BC$

$\overline{F} = \overline{A + BC} = \overline{A}(\overline{B} + \overline{C})$

$F.\overline{F} = (A + BC)\,\overline{A}\,(\overline{B} + \overline{C})$

$= (A\overline{A} + \overline{A}BC)\,(\overline{B} + \overline{C})$

$= \overline{A}BC\,(\overline{B} + \overline{C})$

$= \overline{A}BC\overline{B} + \overline{A}BC\overline{C}$

$= 0 + 0 = 0$ Hence proved.

$F + \overline{F} = A + BC + \overline{A}\,(\overline{B} + \overline{C})$

$= A + BC + \overline{A}\,\overline{B} + \overline{A}\,\overline{C}$

$= A + \overline{B} + BC + \overline{A}\,\overline{C}$

$= A + \overline{C} + \overline{B} + BC$

$= A + (\overline{C} + B) + \overline{B} + (\overline{C} + C)$

$$= A + \overline{C} + B + \overline{B}$$
$$= 1 + A + \overline{C}$$
$$= 1 \quad \text{Hence proved.}$$

Problem 5. *Find the complements of the following expressions :*

(a) $xy^1 + x^1y$

(b) $(AB^1 + C)\, D^1 + E$

(c) $AB\,(C^1D + CD^1) + A^1B^1\,(C^1 + D)\,(C + D^1)$

(d) $(x + y^1 + z)\,(x^1 + z^1)\,(x + y)$

Solution. *(a)* $f(x, y) = xy^1 + x^1y \rightarrow$ To find $\bar{f}$, complement each variable, and change + as . and . as +

$$\bar{f} = \overline{xy' + x'y} = (\overline{xy'})\,(\overline{x'y})$$
$$= (\bar{x} + y)\,(x + \bar{y})$$
$$= x.\bar{x} + xy + \bar{x}\,\bar{y} + y.\bar{y}$$
$$= xy + \bar{x}\,\bar{y}.$$

(b) $F = (A\overline{B} + C)\,\overline{D} + E$ To find $\bar{f}$, complement each variable, and change + as . and . as +

$$= \left[\overline{(A\overline{B} + C)} + D\right]\overline{E}$$
$$= \left[(\overline{A} + B)\,\overline{C} + D\right]\overline{E}$$
$$= (\overline{A}\,\overline{C} + B\overline{C} + D)\,\overline{E}$$
$$= \overline{A}\,\overline{C}\,\overline{E} + B\overline{C}\,\overline{E} + DE.$$

(c) $F = AB\,(C'D + CD') + A'B'\,(C' + D)\,(C + D')$

To find $\overline{F}$, complement each variable and change + as . and . as +

$$\overline{F} = \left[\overline{A} + \overline{B} + (C + \overline{D})\,(\overline{C} + D)\right]\left[(A + B) + (C\overline{D} + \overline{C}D)\right]$$
$$= (\overline{A} + \overline{B} + C\overline{C} + \overline{C}\,\overline{D} + CD + D\overline{D})\,(A + B + C\overline{D} + \overline{C}D)$$
$$= (\overline{A} + \overline{B} + \overline{C}\,\overline{D} + CD)\,(A + B + C\overline{D} + \overline{C}D)$$
$$= A\overline{A} + A\overline{B} + A\overline{C}\,\overline{D} + ACD + \overline{A}B + B\overline{B} + B\overline{C}\,\overline{D}$$
$$+ BCD + \overline{A}C\overline{D} + \overline{A}\,\overline{C}D + \overline{B}C\overline{D} + \overline{B}\overline{C}D + C\overline{D}\,\overline{C}\,\overline{D}$$
$$+ CDC\overline{D} + \overline{C}D\overline{C}\,\overline{D} + \overline{C}D.CD$$
$$= A\,(\overline{B} + \overline{A}\,\overline{D} + CD) + B\,(\overline{A} + \overline{C}\,\overline{D} + CD)$$
$$+ \overline{A}C\overline{D} + \overline{A}\,\overline{C}D + \overline{B}C\overline{D} + \overline{B}\overline{C}D$$
$$= \overline{B} + A\overline{C}\,\overline{D} + ACD + B\overline{A} + B\overline{C}\,\overline{D} + BCD$$

$$+ \overline{A}C\overline{D} + \overline{A}\,\overline{C}D + \overline{B}C\overline{D} + \overline{B}\,\overline{C}D$$

(d) $F = (x + y' + z)(\overline{x} + \overline{z})(x + y)$

$\overline{F} = \overline{x}y\overline{z} + xz + \overline{x}\,\overline{y}$

Problem 6. *For the Boolean function F given in the truth table, find the following :*

(a) List the minterms of the function

(b) List the minterms of F'

(c) Express F in SOP form

(d) Simplify the function to a minimal number of literals, shown in Table 2.54.

Table 2.54

x	*y*	*z*	*f*
0	0	0	0
0	0	1	0
0	1	0	1
0	1	1	1
1	0	0	0
1	0	1	0
1	1	0	1
1	1	1	1

Solution. From the given truth table,

(*a*) The minterms in the given function

$$F = \Sigma\, m\,(2, 3, 6, 7)$$

(*b*) Minterms in $\overline{F} = \Sigma\, m\,(0, 1, 4, 5)$

(*c*) F in canonical SOP $= \overline{x}y\overline{z} + \overline{x}yz + xy\overline{z} + xyz$

(*d*) $F = \overline{x}y\overline{z} + \overline{x}yz + xy\overline{z} + xyz$

$= \overline{x}y(z + \overline{z}) + xy(z + \overline{z})$

$= \overline{x}y + xy$

$= y(x + \overline{x}) = y$

Problem 7. *Given the following Boolean function*

$$F = B\overline{C}D + \overline{B}\,\overline{C}D + \overline{A}BC + A\overline{B}C + ABC$$

(a) Obtain the truth table of the function.

(b) Simplify the function to minimum number of literals using Boolean algebra.

(c) Obtain the truth table of the simplified function and show that is same as the one in part (a).

Solution. (*a*) Truth table for F (shown in Table 2.55)

Table 2.55

ABCD	$\overline{A}\,\overline{B}\,\overline{C}\,\overline{D}$	$B\overline{C}D$	$\overline{B}\,\overline{C}D$	$\overline{A}BC$	$A\overline{B}C$	ABC	*F*
0000	1111	0	0	0	0	0	0
0001	1110	0	1	0	0	0	1
0010	1101	0	0	0	0	0	0
0011	1100	0	0	0	0	0	0
0100	1011	0	0	0	0	0	0
0101	1010	1	0	0	0	0	1
0110	1001	0	0	1	0	0	1
0111	1000	0	0	1	0	0	1
1000	0111	0	0	0	0	0	0
1001	0110	0	1	0	0	0	1
1010	0101	0	0	0	1	0	1
1011	0100	0	0	0	1	0	1
1100	0011	0	0	0	0	0	0
1101	0010	1	0	0	0	0	1
1110	0001	0	0	0	0	1	1
1111	0000	0	0	0	0	1	1

$$F = \Sigma m\ (1, 5, 6, 7, 9, 10, 11, 13, 14, 15)$$

(*b*)

$$F = B\overline{C}D + \overline{B}\,\overline{C}D + \overline{A}BC + A\overline{B}C + ABC$$

$$= \overline{C}D\,(B + \overline{B}) + BC\,(A + \overline{A}) + AC\,(\overline{B} + B)$$

$$= \overline{C}D + BC + AC$$

$$= \overline{C}D + C\,(A + B).$$

(*c*) Truth table of simplified F (shown in Table 2.56)

Table 2.56

ABCD	$\overline{C}D$	*(A + B)*	*C (A + B)*	*F*
0000	0	0	0	0
0001	1	0	0	1
0010	0	0	0	0
0011	0	0	0	0
0100	0	1	0	0
0101	1	1	0	1
0110	0	1	1	1
0111	0	1	1	1

1000	0	1	0	0
1001	1	1	0	1
1010	0	1	1	1
1011	0	1	1	1
1100	0	1	0	0
1101	1	1	0	1
1110	0	1	1	1
1111	0	1	1	1

$$F = \Sigma\ (1, 5, 6, 7, 9, 10, 11, 13, 14, 15)$$

$\therefore$ The truth table of original function and minimised function are same.

Problem 8. *Simplify the following algebraic expressions :*

(a) $\bar{x} + \bar{y} + xy\bar{z}$

(b) $A + \bar{A}B + \bar{A}\bar{B}C + \bar{A}\bar{B}\bar{C}D + ...$

(c) $BC + \bar{C}\bar{D} + AB\bar{D}$

(d) $\bar{w}\bar{x} + \bar{x}\bar{y} + \bar{x}\bar{z} + yz$

(e) $(\bar{x} + xy\bar{z}) + (\bar{x} + xy\bar{z})(x + \bar{x}\bar{y}z)$

Solution. *(a)*

$$f = x' + y' + xyz'$$
$$= x' + (y' + y)(y' + xz') = x' + y' + xz'$$
$$= (x' + x)(x' + \bar{z}) + \bar{y}$$
$$\boxed{f = \bar{x} + \bar{y} + \bar{z}}$$

(b)

$$f = A + \bar{A}B + \bar{A}\bar{B}C + \bar{A}\bar{B}\bar{C}D + ...$$
$$= A + \bar{A}(\bar{B} + \bar{B}C + \bar{B}\bar{C}D + ...)$$
$$= A + B + \bar{B}(C + \bar{C}D + ...) = A + \bar{B} + C + \bar{C}(D + ...)$$
$$\boxed{f = A + B + C + ...}$$

(c)

$$f = BC + \bar{C}\bar{D} + AB\bar{D}$$
$$= BC + \bar{C}\bar{D} + ABC\bar{D} + AB\bar{C}\bar{D}$$
$$= BC + \bar{C}\bar{D}(1 + AB) + ABC\bar{D}$$
$$= BC + \bar{C}\bar{D} + ABC\bar{D} = BC(1 + A\bar{D}) + \bar{C}\bar{D}$$
$$\boxed{f = BC + \bar{C}\bar{D}}$$

(d)

$$f = \bar{w}\bar{x} + \bar{x}\bar{y} + \bar{x}\bar{z} + yz$$
$$= \bar{x}\bar{y} + \bar{x}\bar{z} + yz + \bar{w}\bar{x}\bar{y}\bar{z} + \bar{w}\bar{x}\bar{y}z + \bar{w}\bar{x}y\bar{z} + \bar{w}\bar{x}yz$$
$$= \bar{x}\bar{y}(1 + \bar{w}z + \bar{w}\bar{z}) + \bar{x}\bar{z}(1 + \bar{w}y) + yz(1 + \bar{w}\bar{x})$$
$$\boxed{f = \bar{x}\bar{y} + \bar{x}\bar{z} + yz}$$

(e) $$\begin{aligned} f &= (\bar{x} + xy\bar{z})(\bar{x} + xy\bar{z})(x + \bar{x}\bar{y}z) \cdot \\ &= (\bar{x} + xy\bar{z})[1 + (x + \bar{x}\bar{y}z)] \\ &= \bar{x} + xy\bar{z} \\ &= \bar{x} + y\bar{z} \\ &= (\bar{x} + y)(\bar{x} + \bar{z}) \end{aligned}$$

Problem 9. *Find the complements of the following expressions and then simplify them :*

(a) $\bar{x}(\bar{y} + \bar{z})(x + y + \bar{z})$

(b) $(A + \overline{B}\,\overline{C})(B + \overline{A}\,\overline{C})(C + \overline{A}\,\overline{B})$

(c) $\overline{A} + (\overline{B} + C + \overline{C}\,\overline{D})(B + \overline{C}D)$

Solution. (a) $$f = \bar{x}(\bar{y} + \bar{z})(x + y + \bar{z})$$
$$\begin{aligned} \bar{f} &= x + (yz) + (\bar{x}\bar{y}z) \\ &= x + yz + \bar{x}\bar{y}z = (x + \bar{y}z)(x + \bar{x}) + yz \\ &= x + \bar{y}z + yz = x + z(y + \bar{y}) \end{aligned}$$
$$\boxed{\bar{f} = x + z}$$

(b) $$f = (A + \overline{B}\,\overline{C})(B + \overline{A}\,\overline{C}) + (C + \overline{A}\,\overline{B})$$
$$\bar{f} = \overline{A}(B + C) + \overline{B}(A + C) + \overline{C}(A + B)$$

(c) $$\bar{f} = \overline{A} + (\overline{B} + C + \overline{C}\,\overline{D})(B + \overline{C}D)$$
$$\bar{f} = A\{B\overline{C}(C + D) + \overline{B}(C + \overline{D})\}$$
$$\boxed{\bar{f} = A\{B\overline{C}D + \overline{B}C + \overline{B}\,\overline{D}\}}$$

Problem 10. *Prove that :*

(a) $xy + \bar{x}\bar{y} + \bar{x}yz = xy\bar{z} + \bar{x}\bar{y} + yz$

(b) $xyz + w\bar{y}\bar{z} + wxz = xyz + w\bar{y}\bar{z} + wx\bar{y}$

Solution. (a) $xy + \bar{x}\bar{y} + \bar{x}yz = xy\bar{z} + \bar{x}\bar{y} + yz$

$$\begin{aligned} \text{L.H.S.} &= xy + \bar{x}\bar{y} + \bar{x}yz = y[(x + \bar{x}z) + \bar{x}\bar{y}] \\ &= y(x + z) + \bar{x}\bar{y} = xy + yz + \bar{x}\bar{y} \end{aligned}$$
$$\begin{aligned} \text{R.H.S.} &= xy\bar{z} + \bar{x}\bar{y} + yz = y(z + x\bar{z}) + \bar{x}\bar{y} \\ &= y(z + x) + \bar{x}\bar{y} = xy + yz + \bar{x}\bar{y} \end{aligned}$$

$\therefore$ L.H.S. = R.H.S.

(b) $xyz + w\bar{y}\bar{z} + wxz = xyz + w\bar{y}\bar{z} + wx\bar{y}$

$$\begin{aligned} \text{L.H.S.} &= xyz + w\bar{y}\bar{z} + wxz \\ &= wxyz + \bar{w}xyz + wx\bar{y}\bar{z} + w\bar{x}\bar{y}\bar{z} + wxyz + wx\bar{y}z \\ &= wxyz + \bar{w}xyz + wx\bar{y}\bar{z} + w\bar{x}\bar{y}\bar{z} + wx\bar{y}z \\ &= \Sigma m\ (15, 7, 12, 8, 13) \\ &= \Sigma m\ (7, 8, 12, 13, 15) \end{aligned}$$

$$\text{R.H.S.} = xyz + w\bar{y}\bar{z} + wx\bar{y}$$
$$= wxyz + \bar{w}xyz + wx\bar{y}\bar{z} + w\bar{x}\bar{y}\bar{z} + wx\bar{y}z + wxyz$$
$$= wxyz + \bar{w}xyz + wx\bar{y}\bar{z} + w\bar{x}\bar{y}\bar{z} + wx\bar{y}z$$
$$= \Sigma m\ (15, 7, 12, 8, 13)$$
$$= \Sigma m\ (7, 8, 12, 13, 15)$$

Hence L.H.S. = R.H.S.

Problem 11. *Given,* $x\bar{y} + \bar{x}y = z$, *show that*

$$x\bar{z} + \bar{x}z = y$$

Solution. Given $z = x\bar{y} + \bar{x}y$

$$\therefore \quad \bar{z} = \overline{x\bar{y} + \bar{x}y}$$
$$= (\bar{x} + y)\ (x + \bar{y})$$
$$x\bar{z} + \bar{x}z = x(\bar{x} + y)\ (x + \bar{y}) + \bar{x}(x\bar{y} + \bar{x}y)$$
$$= (x\bar{x} + xy)\ (x + \bar{y}) + \bar{x}xy + x\bar{y}$$
$$= xy\ (x + \bar{y}) + \bar{x}y$$
$$= xy + \bar{x}y = y(x + \bar{x})$$
$$= y \quad \text{Hence Proved.}$$

Problem 12. *Find the value of the two-valued variables A, B, C and D by solving the set of simultaneous equations*

$$A' + AB = 0 \quad \text{...(1)}$$
$$AB = AC \quad \text{...(2)}$$
$$AB + AC' + CD = C'D \quad \text{...(3)}$$

Solution. From equation (1)

$$\bar{A} + AB = 0$$
$$\Rightarrow \quad (\bar{A} + A)\ (\bar{A} + B) = 0$$
$$\Rightarrow \quad \bar{A} + B = 0$$
$$\because \quad \bar{A} = 0$$
$$\because \quad A = 1 \text{ and } B = 0$$

From equation (2)

$$AB = AC \quad 1.0 = 1.C$$
$$\therefore \quad C = 0$$

From equation (3)

$$AB + A\bar{C} + CD = \bar{C}D \qquad A = 1$$
$$1.0 + 1.1 + 0.D = 1.D \qquad B = 0$$

$$0 + 1 + 0 = D \qquad C = 0$$

$$\therefore \quad D = 1 \qquad D = 1$$

Problem 13. *Prove that if* $\overline{w}x + y\overline{z} = 0$, *then*

$$wx + \overline{y}(\overline{w} + \overline{z}) = wx + xz + \overline{x}\,\overline{z} + \overline{w}\,\overline{y}z$$

Solution. Adding $\overline{w}x + y\overline{z}$ on both sides.

$$\begin{aligned}
\text{L.H.S.} &= wx + \overline{y}\,\overline{w} + \overline{y}\,\overline{z} + \overline{w}x + y\overline{z} \\
&= x(w + \overline{w}) + \overline{y}\,\overline{w} + \overline{z}(y + \overline{y}) \\
&= x + \overline{z} + \overline{y}\,\overline{w} \\
\text{R.H.S.} &= wx + xz + \overline{x}\,\overline{z} + \overline{w}\,\overline{y}z + \overline{w}x + y\overline{z} \\
&= x(w + \overline{w}) + xz + \overline{x}\,\overline{z} + \overline{w}\,\overline{y}z + y\overline{z} \\
&= x + xz + \overline{x}\,\overline{z} + \overline{w}\,\overline{y}z + y\overline{z} \\
&= x(1 + z) + \overline{x}\,\overline{z} + \overline{w}\,\overline{y}z + y\overline{z} \\
&= x + \overline{x}\,\overline{z} + \overline{w}\,\overline{y}z + y\overline{z} = x + \overline{z} + \overline{w}\,\overline{y}z + y\overline{z} \\
&= \overline{z}(1 + y) + x + \overline{w}\,\overline{y}z = \overline{z} + x + \overline{w}\,\overline{y}z \\
&= x + \overline{z} + \overline{w}\,\overline{y}
\end{aligned}$$

$\therefore$ L.H.S. = R.H.S. Hence proved.

Problem 14. *The dual* f_d *of a function* $f(x_1, x_2, ..., x_n)$ *is obtained by interchanging the operations of logical addition and multiplication and by interchanging the constants 0 and 1 within any expression for that function.*

(a) Show that $\quad f_d = f'(x_1', x_2', ..., x_n')$

(b) Find a 3-variable function that is it's own dual i.e., self dual.

(c) Prove that for any function f and any two-valued variable A, which may or may not be a variable in f, the function g = Af + A'fd is self-dual.

Solution. (*a*) $f(x_1, x_2, x_3, ..., x_n, 0, 1, +, .)$

$$f_d = f(x_1, x_2, x_3, ..., x_n, 1, 0, ., +)$$

Complement of $\quad f = f'(x_1, x_2, x_3, ..., x_n, 0, 1, +, .)$

$$= f\left(x_1', x_2', x_3', ..., x_n', 1, 0, ..., +\right)$$

Complement of the function $\Rightarrow f'\left(x_1', x_2', x_3', ..., x_n', 0, 1, +, .\right)$

$$= f(x_1, x_2, ..., x_n, 1, 0, ., +)$$

$$= f_d(x_1, x_2, ..., x_n)$$

(*b*) There are two different self dual functions for 3-variables.

$$f_1 = ab + bc + ca$$

$$f_2 = \overline{a}\,\overline{b}\,\overline{c} + \overline{a}bc + a\overline{b}c + \overline{a}bc \text{ (or)}$$

$$abc + a\overline{b}\,\overline{c} + \overline{a}bc + \overline{a}\,\overline{b}c$$

(c) A self dual function (g) can be obtained by selecting any function f (2 or 3 variables) and a two-valued variable 'A' which is not a variable in f, and substituting $g = Af + A'f_d$.

For example, $f = a + b$ and the two-valued variable is A = C.

$$\therefore \quad f_d = a \cdot b$$

$$g = c\,(a + b) + c'\,(ab)$$

$$= ac + bc + abc'$$

$$= a\,(c + bc') + bc = a\,(b + c) + bc$$

$$= ab + bc + ca \text{ is self dual function.}$$

Problem 15. *A safe has 5 locks, v, w, x, y and z, all of which must be unlocked for the safe to open. The keys to the locks are distributed among five executives in the following manner :*

Mr. A has keys for locks v and x

Mr. B has keys for locks v and y

Mr. C has keys for locks w and y

Mr. D has keys for locks x and z

Mr. E has keys for locks v and z

(a) Determine the minimal number of executives required to open the safe.

(b) Find all the combinations of executives that can open the safe. Write an expression f(A, B, C, D, E) which specifies when the safe can be opened as a function of what executives are present.

(c) Who is the "essential executive" without whom the safe cannot be opened ?

Solution. (a) To open the door, we must open all locks. The key 'V' can be opened by executives A or B or E

$$\therefore \quad v = A + B + E$$

Similarly, $w = C\,;\ x = A + D,\ y = B + C\,;\ z = D + E$

(b) The minimal number of executives required to open the safe (all locks)

$$f = v + w + x + y + z$$

$$= (A + B + E)\ (C)\ (A + D)\ (B + C)\ (D + E)$$

$$= ACD + ACE + BCD + CDE$$

The minimum number of executives required = 3

(c) Here, each term contains executive 'C' So, C is an essential executive and without him the safe cannot be opened.

Problem 16. *You are presented with a set of requirements under which an insurance policy will be issued. The applicant must be :*

(i) A married female 25 years old or over, or

(ii) A female under 25, or

(iii) A married male under 25 who has not been involved in a car accident, or

(iv) A married male who has been involved in a car accident, or

(v) A married male 25 years or over who has not been involved in a car accident.

The variables w, x, y and z assume the truth value 1 in the following cases :

w = 1, if applicant has been involved in a case accident.

x = 1, if applicant is married.

y = 1, if applicant is a male.

z = 1, if applicant is under 25.

(a) You are asked to find an algebraic expression which assumes the value 1 whenever the policy should be issued.

(b) Simplify algebraically the above expression and suggest a simpler set of requirements.

Solution. (*a*) A policy should be issued under any one of the following conditions.

(*i*) A married female 25 years old or over $= x\bar{y}\bar{z}$

(*ii*) A female under 25 $= \bar{y}z$

(*iii*) A married male under 25 who has not been involved in a car accident

$$= xy\bar{w}z = \bar{w}xyz$$

(*iv*) A married male who has been involved in a car accident $= wxy$

(*v*) A married male of 25 years or over who has not been involved in a car accident

$$= xy\bar{z}\bar{w} = \bar{w}xy\bar{z}$$

$$\therefore \quad f = x\bar{y}\bar{z} + \bar{y}z + \bar{w}xyz + wxy + \bar{w}xy\bar{z}$$

$$f = x\bar{z}(\bar{y} + \bar{w}y) + \bar{y}z(1 + \bar{w}x) + xy(w + z)$$

$$= x\bar{z}\bar{y} + x\bar{z}\bar{w} + \bar{y}z + xyw + xyz$$

$$= \bar{y}(x\bar{z} + z) + x\bar{z}\bar{w} + xyw + xyz$$

$$= \bar{y}x + \bar{y}z + x\bar{z}\bar{w} + xyw + xyz$$

$$= \bar{y}z + x(\bar{y} + wy + yz + \bar{w}\bar{z}) = \bar{y}z + x$$

$$= \bar{y}z + x(\bar{y} + w + z + \bar{w}z) = \bar{y}z + x(\bar{y} + w + z + \bar{w})$$

$$\boxed{f = \bar{y}z + x}$$

(*b*) To issue a policy either applicant may be married (or) female under 25 years.

Problem 17. *Prove that if a + b = a + c and a' + b = a' + c, then b = c.*

Solution.

$$b = b + aa' = (b + a)(b + a')$$
$$= (a + c)(c + a')$$
$$= c + aa'$$
$$= c \quad \text{Hence proved.}$$

Problem 18. *Prove if a + b = a + c and ab = ac, then b = c.*

Solution.

$$b = b + aa' = (a + b)(a' + b) = (a + c)(a' + b)$$
$$= aa' + a'c + ba + bc$$
$$= a'c + ac + bc$$
$$= c(a' + a + b)$$
$$= c(1 + b) = c \quad \text{Hence proved.}$$

Problem 19. *Find all minimal four-variable functions which assumes the value 1 when the minterms 4, 10, 11, 13 are equal to 1, and assume the value when the minterms are 1, 3, 6, 7, 8, 9, 12, 14 are equal to 1.*

Solution. Let us assume the variables are A, B, C, D

$f = 1$ for the minterms 4, 10, 11, 13

$f = 0$ for the minterms 1, 3, 6, 7, 8, 9, 12 and 14

$f = d$ for the minterms 0, 2, 5, 15

$\therefore \quad f = \Sigma m\ (4, 10, 11, 13) + \Sigma d\ (0, 2, 5, 15)$

AB \ CD	00	01	11	10
00				×
01	1	×		
11		1	×	
10			1	1

$$f = \bar{A}B\bar{C} + ABD + A\bar{B}C$$

Problem 20. *Each of the following functions actually represents a set of four functions, corresponding to the various assignments of the don't care terms.*

$$f_1\ (a, b, c, d) = \Sigma m\ (1, 3, 4, 5, 9, 10, 11) + \Sigma d\ (6, 8)$$

$$f_2\ (a, b, c, d) = \Sigma m\ (0, 2, 4, 7, 8, 15) + \Sigma d\ (9, 12)$$

(a) Find $f_3 = f_1 \cdot f_2$

(b) Find $f_4 = f_1 + f_2$

Solution.

(a) $f_1 =$

ab \ cd	00	01	11	10
00		1	1	
01	1	1		×
11				
10	×	1	1	1

$$f_1 = \bar{a}b\bar{d} + a\bar{b} + \bar{b}d$$

$f_2 =$

ab \ cd	00	01	11	10
00	1			1
01	1		1	
11	×		1	
10	1	×		

$$f_2 = \bar{c}\bar{d} + \bar{a}\,b\,\bar{d} + bcd$$

$$f_3 = f_1 \cdot f_2 = (\bar{a}b\bar{d} + a\bar{b} + \bar{b}d)\,(\bar{c}\,\bar{d} + \bar{a}\,b\,\bar{d} + bcd)$$
$$= \bar{a}b\bar{c}\,\bar{d} + a\bar{b}\bar{c}\,\bar{d}$$

$\therefore$ $$f_3 = \Sigma m\ (4,\ 8)$$

(*b*) $$f_4 = \Sigma m\ (0, 1, 2, 3, 4, 5, 7, 8, 9, 10, 11, 15) + \Sigma d\ (6, 12)$$

Problem 21. *Let $f = \Sigma m\ (5, 6, 13)$ and $f_1 = \Sigma m\ (0, 1, 2, 3, 5, 6, 8, 9, 10, 11, 13)$. Find f_2 such that $f = f_1 . f_2'$.*

Solution.

$$f = (f_1 . f_2')$$
$$f' = f_1' + f_2$$
$$f' = \Sigma m\ (0, 1, 2, 3, 4, 7, 8, 9, 10, 11, 12, 14, 15)$$
$$f_1' = \Sigma m\ (4, 7, 12, 14, 15)$$

$\therefore$ $$f_2 = \Sigma m\ (0, 1, 2, 3, 8, 9, 10, 11)$$

Problem 22. *Find the simplest function g (A, B, C, D) that will make the function f = A'BC + (AC + B)D + g (A, B, C, D) self dual.*

Solution. To find the dual function f_d for the function f, + and **.** are interchanged, for each 1 cell in the map of f there must be a 0 at the complementary variable value in the map of f_d.

For example, if f has a 1 in cell 1010, then,

f_d has a 0 in cell 0101. The K-map for f without g is shown below.

$$f = A'BC + ACD + BD$$
$$f = \Sigma m\ (5, 6, 7, 11, 13, 15)$$

If $f = 1$ at cells 5, 6, 7, 11, 13, 15 and

$f = 0$ at cells 10, 9, 8, 4, 2, 0 for f_d.

Place '0' the cells of the map.

AB \ CD	00	01	11	10
00	0			0
01	0	1	1	1
11		1	1	
10	0	0	1	0

In the map, only form cells are vaccant, in that two cells must be filled with 1 and remains two with 0. Now our task is to select which two cells must be filled with 1 so that the g (A, B, C, D) should be minimum.

Case (i) If one place the 1 at cells 12 and 14

The minimal SOP expression for f = Ab + BD + ACD + BC

= AB + A'BC + BD + AC + ABC

$$\boxed{g = AB}$$

Case (ii) If we place the 1 at cells 1 and 3

The minimal SOP expression for $f = BD + \overline{A}D + \overline{A}BC + ABC$

$$\therefore \quad \boxed{g = \overline{A}D}$$

$$\therefore \quad \boxed{g = AB \text{ (or) } \overline{A}D}$$

Problem 23. *Prove that if x and y are switching variables, then*

(a) $x + y = x \oplus y \oplus xy$ *(b)* $x \oplus 1 = x'$

Solution. (*a*) $x \oplus y \oplus xy$

$$\begin{aligned}
&= (x'y + xy') \oplus xy \\
&= (x'y + xy')\,(xy)' + (x'y + xy')'\,xy \\
&= (x'y + xy')\,(x' + y') + (x + y')\,(x' + y)\,(xy) \\
&= x'y + xy' + (xy + x'y')\,xy \\
&= x'y + xy' + xy \\
&= y\,(x + x') + x\,(y + y') \\
&= x + y
\end{aligned}$$

(*b*)
$$\begin{aligned}
x \oplus 1 &= x.1' + 1.x' \\
&= x.0 + 1.x' \\
&= 0 + x' \\
&= x' \qquad \text{Hence proved.}
\end{aligned}$$

Problem 24. *Given the function* $f\,(x_1x_2x_3x_4) = \Sigma m\,(4, 6, 7, 15) + \Sigma d\,(2, 3, 5)$. *Realize f in the form shown below i.e., find correspondence* x_i*'s and a, b, c, d and determine the functions A, B and D*

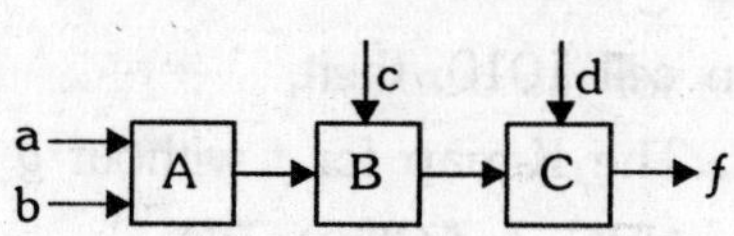

Solution. The K-map for f is shown below

$$\begin{aligned}
f &= \overline{x}_1x_2 + x_2x_3x_4 \\
&= x_2\,(\overline{x}_1 + x_3x_4) \\
&= (x_3x_4 + \overline{x}_1)x_2
\end{aligned}$$

This can be realized by

x_1x_2 \ x_3x_4	00	01	11	10
00			×	×
01	1	×	1	1
11			1	
10				

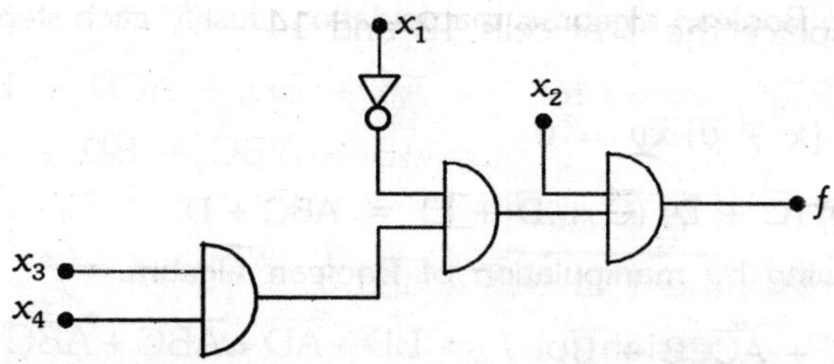

$\therefore$ A = 2 input AND gate B = 2 input OR gate

C = 2 input AND gate

$a = x_3,\ b = x_4,\ c = \overline{x}_1,\ d = x_2$

2.20 REVIEW QUESTIONS

1. Simplify the following expressions using Boolean algebra rules :

(*a*) $F = \overline{A + \overline{B} + \overline{C} + \overline{D}} + \overline{A}BC\overline{D}$

(*b*) $F = (A + B)(\overline{A} + B)(A + \overline{B})$

(*c*) $F = (A + \overline{B} + \overline{C})(\overline{A} + \overline{B} + \overline{C})$

2. Obtain the truth table for the POS expression

$$F = (A + \overline{B} + C)(\overline{A} + B + C)(\overline{A} + B + C)$$

3. (*a*) Show that $(AB + \overline{C})(\overline{A} + \overline{B})C = 1$

(*b*) Show that $\overline{B(AD + \overline{BC})}(\overline{A} + \overline{D})\overline{C}(\overline{A} + \overline{D}) = 0$

(*c*) Show that $(B + D)(D + C)(A + D) = D + ABC$

(*d*) Show that $(B + D)(A + D)(B + C)(A + C) = AB + CD$

(*e*) $\overline{A}BC\overline{D} + \overline{A}BCD + A\overline{B}\,\overline{C}D + A\overline{B}CD + AB\overline{C}D + ABCD = AD + \overline{A}BC$

4. (*a*) Express $F = AB\overline{C} + A\overline{B}C + \overline{A}\,\overline{B}\,\overline{C}$ in POS form.

(*b*) Express $F = (A + B + \overline{C})(A + \overline{B} + C)(\overline{A} + \overline{B} + \overline{C})$ in SOP form.

5. Obtain the Boolean equation for the output F_1 and F_2 of the circuit, whose truth table in given in Table 2.57.

Table 2.57

A	B	C	F_1	F_2
0	0	0	1	1
0	0	1	1	0
0	1	0	1	0
0	1	1	0	0
1	0	0	1	0
1	0	1	0	1
1	1	0	1	1
1	1	1	0	1

6. Verify the following Boolean algebraic manipulation. Justify each step with reference to a postulate or theorem.

(*i*) $(x + \bar{y} + xy)(x + \bar{y})\bar{x}y = 0$

(*ii*) $(AB + C + D)(\bar{C} + D)(\bar{C} + D + E) = AB\bar{C} + D$

7. Simplify the following by manipulation of Boolean algebra.

(*i*) $(A + \bar{B}\bar{A})(AC + A\bar{C}CB + \bar{B})$

(*ii*) $A + ABC + \bar{A}BC + AD + A\bar{D} + \bar{A}B$

8. State and prove the De Morgan's theorem.

9. Find the minimal SOP and POS expressions for the functions given below :

(*i*) $f_1\,(w, x, y, z) = \Sigma m\,(0, 1, 2, 3, 8, 9, 10, 11)$

(*ii*) $f_2\,(A, B, C, D) = \Pi m\,(1, 4, 5, 6, 11, 12, 13, 14, 15)$

(*iii*) $f_3\,(A, B, C, D, E) = \Sigma m\,(0, 1, 2, 3, 4, 6, 9, 10, 15, 16, 17, 18, 19, 20, 23, 25, 26, 31)$

10. Determine the minimal SOP and POS realisations for the incompletely specified functions, using K-map :

(*i*) $f_1 = \Sigma m\,(4, 5, 7, 12, 14, 15) + d\,(3, 8, 10)$

(*ii*) $f_2 = \Sigma m\,(1, 4, 8, 9, 13, 14, 15) + d\,(2, 3, 11, 12)$

(*iii*) $f_3 = \Pi M\,(0, 2, 4, 9, 12, 15) + \Pi d\,(1, 5, 7, 10)$

11. Simplify the following Boolean expressions using K-map :

(*i*) $F = \bar{A}\bar{B}\bar{C} + \bar{B}C\bar{D} + \bar{A}BC\bar{D} + A\bar{B}\bar{C}$

(*ii*) $F = \bar{A}B\bar{C} + \bar{A}\bar{B}\bar{D} + \bar{A}\bar{C}D + AC + BCD$

12. Determine the complement of the functions :

(*a*) $F_1 = A\bar{B}C + \bar{A}BD$

(*b*) $F_2 = A\bar{B}C + \bar{A}\bar{B}\bar{C} + \bar{A}\bar{B}C + A\bar{B}\bar{C} + ABC$

using (*i*) De Morgan's rule (*ii*) a K-map.

13. What is a K-map ? Illustrate it with an example.

14. What are SOP and POS forms ? Explain with an example.

15. Define prime implicant and EPI. Minimize the following function by tabulation/tabular method :

$$f\,(A, B, C, D, E) = \Sigma m\,(0, 1, 2, 3, 4, 6, 9, 10, 15, 16, 17, 18, 19, 20, 23, 25, 26, 31)$$

16. Use Q-M method to determine the set of PI and EPI and obtain the possible minimal expressions for the following function.

$$F\,(A, B, C, D, E) = \Sigma m\,(8, 12, 13, 18, 19, 21, 22, 24, 25, 28, 30, 31) + \Sigma d\,(1, 2, 4, 6, 7, 11, 26)$$

2.21 SHORT ANSWER QUESTIONS

1. Two value boolean algebra is called as ______

[**Ans.** switching algebra]

2. Write dual for consensus theorem.

[**Ans.** $AB + \overline{A}C + BC = AB + A\overline{C}$ is cansensus theorem]

3. $A + \overline{A}B =$ ______

[**Ans.** $A + B$ $(A + B)(\overline{A} + C)(B + C) = (AB)(A + \overline{C})$ is the dual of cansensus theorem.]

4. Boolean algebra follows huntington post lates. (TRUE/FALSE)

[**Ans.** True]

5. $f(x, y, z) = xy + xz + y\overline{z} + \overline{y}\overline{z} + x$, when $x = 0, y = 1, z = 1, f(x, y, z) =$ ______

[**Ans.** $f(x, y, z) = 0$]

6. The minimal expression of a boolean function is covered by only essential prime implicant, then it has ________ number of minimal forms.

[**Ans.** one]

7. If $a + b = 1$, $ab = 0$ then $a =$ —; $b =$ —

[**Ans.** $\overline{b}, \overline{a}$]

8. ______ law states that $a + a = a$ and $a.a = a$

[**Ans.** Idempotence]

9. $a + ab = a$ is ______ law

[**Ans.** absorption]

10. According to Demorgan's theorem, $(a + b)' =$ ________

[**Ans.** a'.b']

11. According to Demorgan's theorem, $(ab)' =$ ________

[**Ans.** $a' + b'$]

12. According to the Boolean algebra, $a + 1 = a.1 =$ ______

[**Ans.** a]

13. The sum of 2^n minterms of n-variable function is ______

[**Ans.** 1]

14. The product of any two different n-variable minterms is ______

[**Ans.** 0]

15. The sum of any two different n-variable miaxterms is ____

[**Ans.** 1]

16. In Prime Implecant Chart if each PI covers two minterm and each minterms is covered by two Prime Implicant, such a chart is called ______

[**Ans.** "Cyclic Prime Implecant Chart"]

17. The product of all 2^n max terms of n variables is ______

[**Ans.** 0]

18. A Boolean function is equal to its dual, then the function is called ______

[**Ans.** Self-Dual Function (SDF)]

19. The dual of m_3 is ______

[**Ans.** M_4]

20. The dual of m_0 is ______

[**Ans.** M_7]

21. ________ code is used to represent a max term in the k-map.

[**Ans.** Cyclic]

22. ______ law is used to combine cells in the *k*-map
[**Ans.** $Aa + A\bar{a} = A$]

23. The number of cells in a 4-variable *k*-map is ______
[**Ans.** 16]

24. The adjacent cells for a minterm 5 in a 4–variable *k*-map are ______
[**Ans.** 1, 4, 7, 13]

25. The number of rows in a truth table of *n*-variable is ______
[**Ans.** 2^n]

26. $f(A, B, C) = \Sigma m\,(1, 2, 4, 7)$. This can be expressed in the other canonical form as ______
[**Ans.** $\pi\,(0, 4, 5, 6)$]

27. The SOP form a function is $\overline{A}BC+\overline{A}\,\overline{B}C+ABC$, ______ is its equivalent POS form.
[**Ans.** $(A+\overline{B}+\overline{C})\,(A+B+\overline{C})\,(\overline{A}+\overline{B}+\overline{C})$]

28. $f(A, B, C) = \Sigma m\,(0, 1, 2, 3, 4, 5, 6, 7)$, the minimal expression is ______
[**Ans.** 1]

29. The min term $\overline{A}\,\overline{B}CD$ is the representation of ______ decimal number
[**Ans.** 3]

30. The max term $(\overline{A}+\overline{B}+C+D)$ is the representation of ______ decimal number.
[**Ans.** 12]

31. The dual of the boolean function
$A + B + C = (A + B)\,(A + C)$ is ______
[**Ans.** $A\,(B + C) = AB + AC$]

32. $\overline{\overline{A}.\overline{B}.C} =$ ______
[**Ans.** $(A + B + \overline{C})$]

33. $(\overline{A + \overline{B + C}}) = A$ ______
[**Ans.** $A\,(B + C)$]

34. $(A+ B)\,(\overline{A}+\overline{B}) =$ ______
[**Ans.** $A + B$]

35. $1 + A + AB + ABC + \ldots\ldots =$ ______
[**Ans.** 1]

36. In a K-map, the group of 4 adjacent cells reduces to ______ number of literals in the resultant product.
[**Ans.** two]

37. The max term corresponding to decimal number 8 is ______
[**Ans.** $\overline{A}+B+C+D$.]

38. $AB+\overline{A}C+BC =$ ______
[**Ans.** $AB + \overline{A}C$]

2.21.1 Multiple Choise Questions

1. Using the distributive law, $x + yz =$ ______

(*a*) $xy + z$ (*b*) $x + z$ (*c*) $(x + y)$ (*d*) $xy + xz$
[**Ans.** (*c*)]

2. ______ is the dual for $x + xy = x$

(*a*) $x\,(xy) = x$ (*b*) $x\,(x + y) = \overline{X}$ (*c*) $x\,(xy) = \overline{n}$ (*d*) $x\,(x + y) = x$
[**Ans.** (*d*)]

3. _______ is known as consensus theorem

(*a*) $AB+\bar{A}C+BC=AB+\bar{A}C$ (*b*) $AB+\bar{A}C=(A+C)(\bar{A}+B)$

(*c*) $(A+B)(\bar{A}+C)=AC+\bar{A}B$ (*d*) $A+\bar{A}B=A+B$

[Ans. (*a*)**]**

4. Using Demorgan's theorem $\overline{(x+y)}$ = _______

(*a*) $\bar{x}.\bar{y}$ (*b*) $\overline{xy}$ (*c*) $x+y$ (*d*) $x+\bar{y}$

[Ans. (*a*)**]**

5. The value of $F(x, y, z) = xy\ \bar{z}+x\bar{y}z+xy+y\bar{z}+z+\bar{z}x$ for the combinations in which $z = 1$, is _______

(*a*) 0 (*b*) 1 (*c*) Cannot be said (*d*) None

[Ans. (*b*)**]**

6. Minterm is a _______

(*a*) A sum term which contains each variable as factors in either complemented or uncomplemented form

(*b*) A product term which contains each variable as factors in either complemented or uncomplemented form

(*c*) A product term which contains each variable in only complemented form

(*d*) A product term which contains each variable in uncomplemented form.

[Ans. (*b*)**]**

7. If $f(x, y, z) = \Sigma\ (0, 2, 3, 6, 7)$ is the compact disjunctive form of a function then the compact conjunctive form of the same function is

(*a*) $f(x, y, z) = \Sigma\ (1, 4, 5)$ (*b*) $f(x, y, z) = \pi\ (0, 2, 3, 6, 7)$

(*d*) $f(x, y, z) = \pi\ (1, 4, 5)$ (*d*) None

[Ans. (*c*)**]**

8. The main theorem that is being applied in the Karnaugh map simplification is

(*a*) Demorgan's theorem (*b*) Consensus theorem

(*c*) Combining theorem, $Aa + A\bar{a} = A$ (*d*) None.

[Ans. (*c*)**]**

9. In Karnaugh map, _______ code is used for fixing the value of a minterm

(*a*) Excess–3 (*b*) 2421 (*c*) 642–3 (*d*) Cyclic

[Ans. (*d*)**]**

10. In the minimization using K–map, sometimes one cell will be covered by more than one subcube, this can be justified by ______ law theoretically.

(*a*) Associate law (*b*) Distributive law

(*c*) Commutative (*d*) Idempotent law

[Ans. (*d*)**]**

11. A + B' = A'.B' states ______ theorem

(*a*) De Morgan's theorem (*b*) Consensus (*c*) Distributive (*d*) Associative

[Ans. (*a*)**]**

12. Simplify the expression : A + AB + ABC + ABCD + ABCDE + ______ =

(*a*) A (*b*) B (*c*) C (*d*) D

[Ans. (*a*)**]**

13. The logic expression for Y (A, B, C) = Σm (0, 2, 3, 6) is equivalent to ______

(*a*) Σm (1, 4, 5, 7) (*b*) ΠM (0, 2, 3, 6) (*c*) ΠM (1, 4, 5, 7) (*d*) none

[Ans. (*c*)**]**

14. Write the product of sum expression of (A, B, C) = Σm (1, 2 4)

(*a*) Σm (0, 3, 5, 6) (*b*) Σm (1, 2, 4) (*c*) ΠM (0, 3, 5, 6, 7) (*d*) none

[Ans. (*c*)**]**

15. Write the sum of product expression for f = ΠM (2, 3, 4, 5) is

(*a*) Σm (2, 3, 4, 5) (*b*) Σm (0, 1, 6, 7) (*c*) ΠM (0, 1, 6, 7) (*d*) none

[Ans. (*b*)**]**

16. An example of canonical SOP is ______

(*a*) ABC + BC + AB (*b*) AB (*c*) ABC + AB (*d*) AB'C + AB'C

[Ans. (*d*)**]**

17. An example of canonical POS is

(*a*) AB + BC (*b*) (A + B' + C') (B + A')

(*c*) (A + B + C) (A + B + C) (B + C) (*d*) (A + B + C) (A + B + C)

[Ans. (*d*)**]**

18. The dual of the function $f = xy + yz$ is

(*a*) $(xy)\ (y + z)$ (*b*) $(x + y)\ (yz)$ (*c*) $x + y + z$ (*d*) $(x + y).(y + z)$

[Ans. (*d*)**]**

19. Which fo the following statement is not correct

(*a*) $x + xy = x$ (*b*) $x\ (x + y) = x$

(*c*) $xy + xy = x$ (*d*) $zx + zxy = zx + zy$

[Ans. (*d*)**]**

20. '1' is an identical element for ______ logic operation

(*a*) Dot (*b*) Plus (*c*) Complement (*d*) None

[Ans. (*a*)**]**

21. According to distributive law, $x + yz$ = ______

(*a*) $xy + z$ (*b*) $x\ (y + z)$ (*c*) $(x + y)\ .\ (x + z)$ (*d*) xyz

[Ans. (*c*)**]**

22. The number of calls in a 4-variable K-map is ______

(*a*) 4 (*b*) 16 (*c*) 8 (*d*) 3

[Ans. (*b*)**]**

23. For each cell, number of adjacent cells in a 4-variable K-map is ______

(*a*) 2 (*b*) 4 (*c*) 3 (*d*) 8

[Ans. (*b*)**]**

24. Which of the following code is used in K-map for representing the minterms ?

(*a*) 8421 (*b*) Gray code (*c*) Excess-3 code (*d*) BCD

[Ans. (*b*)**]**

25. Simplify $xy + xz + yz$ = ______

(*a*) $xy + yz$ (*b*) $xz + yz$ (*c*) x (*d*) $xy + xz$

[Ans. (*d*)**]**

26. Simplify the function F (A, B, C) = ABC + ABC + ABC

(a) AB + BC (b) AC + CB (c) ABC (d) A (B + C)

[**Ans.** (c)]

27. The minterm corresponding to decimal number 13 is

(a) ABCD (b) A + B + C' + D (c) A' + B + C + D (d) ABC'D

[**Ans.** (d)]

28. In a digital control system, there are three inputs *a, b, c* the output should be present when at least two inputs are present. The Boolean expression for output is

(a) a'bc' + ab'c' + a'b'c (b) a'b'c' + a'bc' + a'bc + a'b'c

(c) ab + bc + ac (d) a'b + b'c + ac'

[**Ans.** (c)]

29. Six variable function contain ______ number of cells in the K-map

(a) 8 (b) 16 (c) 32 (d) 64

[**Ans.** (d)]

30. What is the decimal position of the minterm A'B'C'D' in the K-map

(a) 2 (b) 0 (c) 15 (d) 14

[**Ans.** (b)]

31. If *f* (*w, x, y, z*) = Σ*m* (1, 3, 5, 7) the simplified equation is

(a) w'z (b) wz' (c) wz (d) w'z'

[**Ans.** (a)]

32. If f (A, B, C, D) = 1 then the K-map contains ______ number of logic 1's is

(a) 4 (b) 8 (c) 16 (d) 32

[**Ans.** (c)]

33. In Quine-Mccluskey method, two terms from adjacent groups are combinable if their binary representation differs by a ______ digit in the same position.

(a) single (b) two (c) four (d) none

[**Ans.** (a)]

34. Which of the following is a combining theorem?

(a) Aa + Aa' = A (b) A'a + A'a' = A' (c) A'a' + a'a = A (d) none

[**Ans.** (a)]

35. The complement of the function is F = xy'z + x'yz

(a) (x' + y + z') (x + y' + z') (b) x'y (x' + y' + z)

(c) (x' + y' + z) . (x + y' + z') (d) (x' + y + z') (x' + y' + z')

[**Ans.** (a)]

36. In a 4-variable K-map if we covered 8-adjacent cells as a group than the resultant term contains ______ number of variables.

(a) 4 (b) 3 (c) 2 (d) 1

[**Ans.** (d)]

37. The following two equations are true as per Boolean arithmetic operations A'B + B = 1 and BC = 0, then the value of A is ______

(a) 0 (b) 1 (c) 2 (d) don't care.

[**Ans.** (d)]

38. The value of C is ______

(*a*) 0 (*b*) 1 (*c*) 2 (*d*) don't care.

[Ans. (*a*)]

39. In a tabular method, the minimal expression of a function contains ______

(*a*) Only Prime implicants

(*b*) Prime Implicants and Essential Prime Implicants

(*c*) Only Essential Prime Implicants (*d*) All variables.

[Ans. (*b*)]

40. In a 4-variable K-map, the function contains all minterms then the minimal expression is ______

(*a*) A (*b*) 1 (*c*) 0 (*d*) don't care.

[Ans. (*b*)]

41. A function f (A, B, C) contains minterms 1, 2, 3, 5, 6 and 7, its complement contains ______

(*a*) Σm (0, 4) (*b*) Σm (1, 2, 3, 5, 6, 7)

(*c*) ΠM (1, 2, 3, 5, 6, 7) (*d*) ΠM (0, 4)

[Ans. (*a*)]

42. The Boolean expression (A' + B) (A + C') (B' + C') simplified to ______

(*a*) (A + B) C' (*b*) (A + B') C' (*c*) (A' + B) C' (*d*) none

[Ans. (*c*)]

43. Boolean expression (A + C') (B' + C') simplifies to ______

(*a*) C' + A'B (*b*) C' (A' + B) (*c*) BC' + A'B (*d*) none

[Ans. (*a*)]

44. The Boolean expression A'B + A'BC simplifies to ______

(*a*) A'C (*b*) A'B (*c*) C (*d*) none

[Ans. (*b*)]

45. In a K-map, each given minterm is represented by ______

(*a*) 1 (*b*) 0 (*c*) don't care (*d*) none

[Ans. (*a*)]

46. In a K-map, each given maxterm is represented by ______

(*a*) 1 (*b*) 0 (*c*) don't care (*d*) none

[Ans. (*b*)]

47. The given maxterm is A + B + C, its equivalent binary represented is

(*a*) 101 (*b*) 010 (*c*) 111 (*d*) 000

[Ans. (*d*)]

48. The given maxterm is ABC, its equivalent binary represented is

(*a*) 101 (*b*) 010 (*c*) 111 (*d*) 000

[Ans. (*c*)]

49. In the K-map, if two cells are said to be adjacent, they are differentiated by ______ number of variables.

(*a*) 1 (*b*) 2 (*c*) *n* (*d*) *n*–1

[Ans. (*a*)]

50. Looping a pair of adjacent 1's in a K-map eliminates ______ number of variables.

(*a*) 1 (*b*) 2 (*c*) *n* (*d*) *n*–1

[Ans. (*a*)]

51. In the *n*-variable K-map, after looping a quad of adjacent 1's, the resultant term contains _____ number of variables.

(*a*) 1 (*b*) 2 (*c*) *n*–1 (*d*) *n*–2

[**Ans.** (*d*)]

52. The don't care in K-map covered in a looping, them its value is treated as _____

(*a*) 1 (*b*) 0 (*c*) 2 (*d*) none

[**Ans.** (*a*)]

53. The number of cells in a 5-variable K-map is _____

(*a*) 6 (*b*) 64 (*c*) 32 (*d*) 63

[**Ans.** (*c*)]

54. A.B. = B.A. is _____ Law.

(*a*) Commutative (*b*) Associative (*c*) Distributive (*d*) Absorption.

[**Ans.** (*a*)]

55. Union law states that _____

(*a*) $x.1 = 1.x = x$ (*b*) $x.(x + y) = x$ (*c*) $a.(b.c) = (a.b).c$ (*d*) x + xbar + 1

[**Ans.** (*a*)]

56. An identity element with respect to +, is designated by _____

(*a*) 0 (*b*) 1 (*c*) *a* (*d*) none

[**Ans.** (*a*)]

57. In two valued Boolean algebra, if $x = 1$, then $x + x =$ _____

(*a*) 1 (*b*) 0 (*c*) 2 (*d*) undefined.

[**Ans.** (*a*)]

58. In two valued Boolean algebra, if $x = 1$, then $x.x =$ _____

(*a*) 0 (*b*) 1 (*c*) 2 (*d*) undefined.

[**Ans.** (*a*)]

59. $(x + y)' =$ _____

(*a*) $x'.y'$ (*b*) $x.y$ (*c*) $x + y$ (*d*) $(x.y)'$

[**Ans.** (*a*)]

60. $(a.b)' = a' + b'$ is _____ theorem.

(*a*) De Morgan's (*b*) Shannon's (*c*) Expansion (*d*) Boolean

[**Ans.** (*a*)]

61. The duel of the given function can be obtained by _____

(*a*) Changing + and . operations (*b*) complementing 0 and 1

(*c*) both (*d*) none.

[**Ans.** (*c*)]

62. If F = A (B + C), then its duel is _____.

(*a*) A + BC (*b*) AB + BC (*c*) ABC (*d*) AB.

[**Ans.** (*a*)]

63. The function and its duel are same then the function is called as _____

(*a*) Self duel (*b*) complement (*c*) same (*d*) inverse.

[**Ans.** (*a*)]

64. The complement of "minterm 4" _____

(*a*) Maxterm 4 (*b*) minterm 0 (*c*) minterm 13 (*d*) maxterm 13.

[**Ans.** (*a*)]

65. Which of the following is the non standard form.

(*a*) AB' + (B + A') (*b*) (A + B') (A' + B) (*c*) (A + B) (A' + B') (*d*) AB' + BA'

[Ans. (*a*)]

66. The resultant of the function cannot be further simplified is called ______

(*a*) minimal (*b*) irredundant (*c*) either *a* or *b* (*d*) none.

[Ans. (*c*)]

2.21.2 True or False

1. A row X in a Prime Implicant Chart is said to dominate another row Y of the chart if X covers every column covered by Y.

[Ans. True]

2. A column X in a Prime Implicant Chart is said to be dominate another column of that Prime Implicant Chart, if X has covers every row of Y and some other rows.

[Ans. True]

3. A minterm is product term, which contains all the variables either in complemented or uncomplemented form.

[Ans. True]

4. A maxterm is sum term, which contains all the variables either in complemented or uncomplemented form.

[Ans. True]

5. In the minimal expression it is necessary include in the resultant of a loop which covers only don't care terms.

[Ans. False]

6. An Implicant is called a Prime Implicant, if it is not a subset of any other Implicant of the functions.

[Ans. True]

7. A Prime Implicant, which includes at least a 1 cell that is not covered by any other Prime Implicant, is called Essential Prime Implicant.

[Ans. True]

8. The Essential Prime Implicant of the function covers all the minterms then the resultant expression is a unique expression.

[Ans. True]

9. $(x + y) + z = x + (b + z)$ is associative law.

[Ans. True]

10. Huntington postulates do not include the associative law.

[Ans. True]

11. Boolean algebra does not have additive, multiplicative and inverses.

[Ans. True]

12. Any character of uppercase of lowercase can be used as variables in the Boolean algebra.

[Ans. True]

13. According to the two-valued Boolean algebra $1 + 1 = 1$.

[Ans. True]

14. According to the two-valued Boolean algebra $x + x' = 1$.

[Ans. True]

15. According to the two-valued Boolean algebra $x + x = x$.

[Ans. True]

16. According to the two-valued Boolean algebra $x + 1 = 1$.

[**Ans.** True]

17. According to the two-valued Boolean algebra $(x')' = x$.

[**Ans.** True]

18. Truth table gives the relation between all possible combinations inputs and output.

[**Ans.** True]

19. The two-element Boolean algebra is called as switching algebra.

[**Ans.** True]

20. The complement of a sum is equal to the product of the individual complements is De Morgan's theorem.

[**Ans.** True]

2.21.3 Fill in the Blanks

1. In a *n*-variable K-map, each cell contains ______ number of adjacent.

[**Ans.** *n*]

2. A cell with 1 in a K-map is not having any adjacent cells, then that cell is called as ______.

[**Ans.** isolated cell]

3. The output is not defined for a particular input combination is called ______.

[**Ans.** don't care combination]

4. The efficient procedure used for simplification of a more than 6 variable Boolean function is called ______.

[**Ans.** tabulation method]

5. The ncessary condition to combine the two minterms is ______.

[**Ans.** both minterm values must be differentiated by power of 2]

6. $x = xy =$ ______

[**Ans.** X]

7. AB + A + 1 ______

[**Ans.** 1]

8. A Prime Implicant Chart in which each minterm is covered by two Prime Implicants and each Prime Implicant is covered only two minterms is called ______.

[**Ans.** Cyclic Prime Implicant Chart]

9. The map entered variables in the K-map are ______.

[**Ans.** 0, 1 and don't care]

10. In ______ map variables can be entered in the RD map.

[**Ans.** reduced dimension map]

11. The complement of the function $F(x, y, z) = xy + yz$ is ______.

[**Ans.** $(x' + y')(y' + z')$]

12. The duel of the function $F(x, y, z) = x'y + y'z$ is ______.

[**Ans.** $(x + y)(x + z)$]

13. A function and its duel are same, then the function is called as ______.

[**Ans.** self duel function]

14. ______ are the adjacent cells for the minterm 6 in the 4-variable K-map.

[**Ans.** Minterms 2, 4, 7 and 14]

15. SOP expression for the given function $F = (A + B + C)(A' + B')$ is ______.

[**Ans.** $AB' + A'B + A'C + B'C$]

16. The minterm presented in the given function can be represented in the K-map by ______.
[Ans. 1]

17. The expression for the two variable function, whose output is 1 for even number of input 1's is ______.
[Ans. AB + A'B']

18. The expression for the two variable function, whose output is 0 for even number of input 1's is ______.
[Ans. (x + y') + (y + z')]

19. The complement of any function is obtained by replacing each variable by its complement and, at the same time, by interchanging the dot operation and plus operation, this is ______ theorem.
[Ans. Shannon's theorem]

20. What are the map entered variables in the RDM.
[Ans. 0, 1 and all the variables of the function]

21. Any Boolean algebra theorem can be proved by ______ method.
[Ans. perfect induction]

22. Canonical SOP is also called as ______.
[Ans. sum of minterms or disjunctive normal expression]

23. Canonical POS is also called as ______.
[Ans. sum of maxterms or conjunctive normal expression]

3

Design of Combinational Circuits

The basic example of a combinational circuit is a 'logic gate'. The logic gate is an electronic circuit whose inputs and output are logically related. These logic gates can be used to implement any Boolean function *i.e.,* they are basic building blocks of digital system. According to the type of logical operation between inputs and output, basically there are 8 standard gates, which are used in digital circuit design. They are : AND, OR, INVERT (or) COMPLEMENT, TRANSFER (or) BUFFER, NAND, NOR, EXCLUSIVE-OR (XOR (or) EX–OR) and EXCLUSIVE NOR (Equivalence).

Before going to study the details of these logic gates, let us first see the different logical operators.

3.1 LOGICAL OPERATORS

The logical operators are used to represent and solve logical expressions. There are three basic logical operators : NOT/INVERT, AND and OR.

3.1.1 Logical Operator NOT/INVERT

The inversion (complementing or negation) operator is written as a bar (or) prime over it's argument (literal or variable). Sometimes, it is written as "NOT". Thus, inverse of a argument A is $\bar{A}$ or NOT A or A'. The NOT operation can be realized by using a switching circuit for better understanding. For this, Fig. 3.1 is considered and the analogy of NOT is shown in Table 3.1

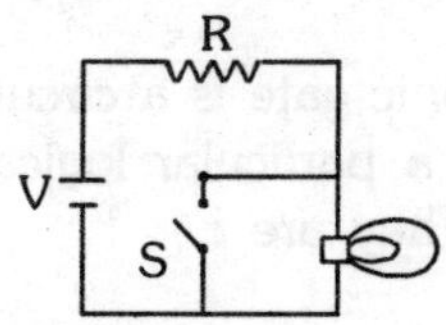

Fig. 3.1 Switching circuit.

Table 3.1

Switch condition (Input)	*Lamp condition (Output)*
Switch open (low)	Lamp ON (high)
Switch close (high)	Lamp OFF (low)

When switch 'S' is open, lamp is 'ON' and when switch 'S' is close, lamp is OFF.

3.1.2 Logical Operator AND

It is denoted by × (or). Sometimes, these signs may be omitted, for example A.B, A×B or AB has same meaning. A.B may be read as A AND B (or) A Times B. The logic operator AND can be represented by a switching-circuit as shown in Fig. 3.2 and it's analogy is shown in Table 3.2.

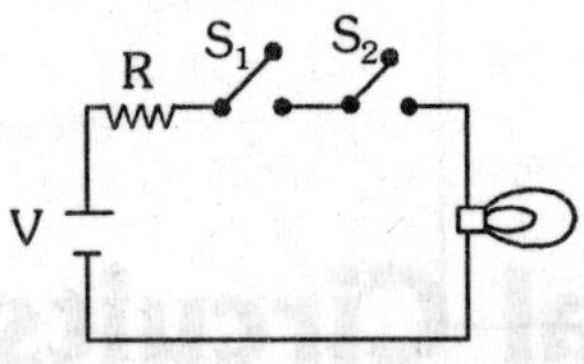

Fig. 3.2

Table 3.2

Inputs		*Outputs*
S_1	S_2	Lamp
Open (L)	Open (L)	OFF (low)
Open (L)	Close (H)	OFF (low)
Close (H)	Open (L)	OFF (low)
Close (H)	Close (H)	ON (high)

It is clear that the lamp will be ON only when both switches are closed.

3.1.3 Logical Operator OR

It is represented by '+'. 'A+B' can be read as A or B. The switching circuit realization of OR function is shown in Fig. 3.3 and it's analogy in Table 3.3.

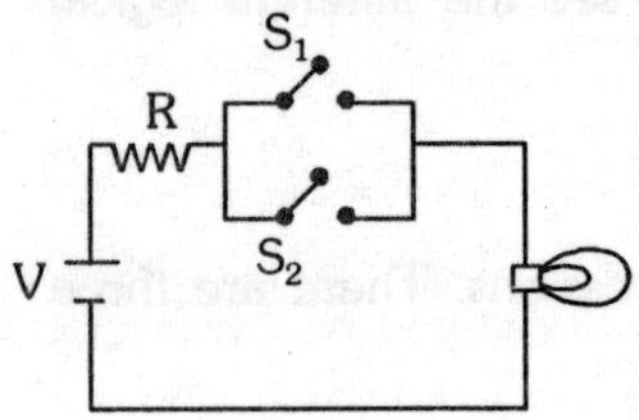

Fig. 3.3

Table 3.3

Inputs		*Output*
S_1	S_2	Lamp
Open (L)	Open (L)	OFF (low)
Open (L)	Close (H)	ON (high)
Close (H)	Open (L)	ON (high)
Close (H)	Close (H)	ON (high)

The lamp will be OFF only when both S_1 and S_2 are open and in remaining all other conditions, the lamp is ON.

3.2 LOGIC GATES

These are the basic elements that makeup a digital system. The electronic gate is a circuit that is able to operate on a number of binary inputs in order to perform a particular logical function. Basically, two types of logics are used in the digital electronics. They are :

(1) Positive logic

(2) Negative logic

The Boolean 0 and 1 do not represent actual number but instead represent the state of a voltage variable, or what is called it's logic level. A voltage in a digital circuit is said to be at the logic level 0 or logic level 1, depending on it's actual numerical value. Sometimes, logic 0 and logic 1 may be called as shown in Table 3.4.

Table 3.4

Logic 0	*Logic 1*
False	True
Off	ON
Low	High
No	Yes
Open switch	Close switch

Positive Logic. Positive logic is that in which out of two voltages, one which is more positive (or) greater is taken as logic 1 and other voltage is taken as logic 0.

Negative Logic. Negative logic is that in which out of two voltages, one which is more negative or less is taken as logic 1 and other as logic 0.

Unless or otherwise specified, all the signals are dealt with positive logic.

The gate is a digital circuit with one or more inputs, but only one output. The gate can be constructed from diodes, transistors and resistors. Now, let us consider the symbols and the truth tables of gates and assume that each gate may have one (x) or two (x, y) binary inputs and one binary output (F). Any logic gate operation can be easily understood with the help of "truth table". A truth table is a table that shows all the input–output possibilities of logic circuit.

3.2.1 AND Gate

If two inputs x and y are combined using the AND operation, the result F, can be represented as $F = x.y$. In this expression, the . (dot) sign stands for Boolean AND operation. The logic symbol for two input and one output AND gate is shown in Fig. 3.4 (*a*) and four input AND gate is shown in Fig. 3.4 (*b*).

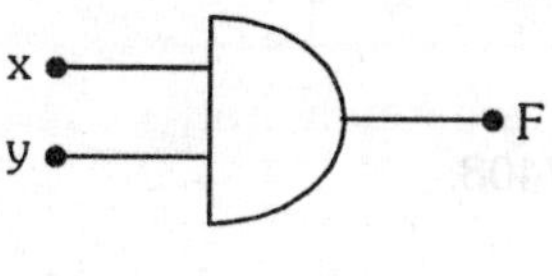

Fig. 3.4(*a*) Two-input AND gate.

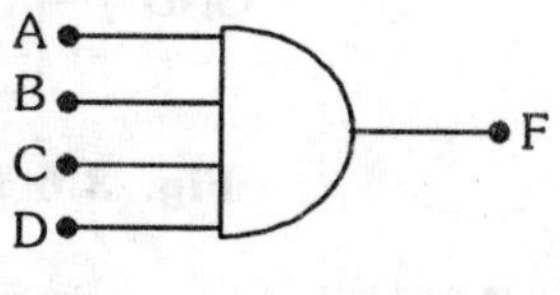

Fig. 3.4(*b*) Four-input AND gate.

The AND gate can have any number of inputs greater than one. The operation of the AND gate is such that the output is HIGH only when all of the inputs are HIGH. When any of the input is LOW, the output is LOW. The Fig. 3.5 shows the two input AND gate with all four possibilities of input combinations and the resulting output for each.

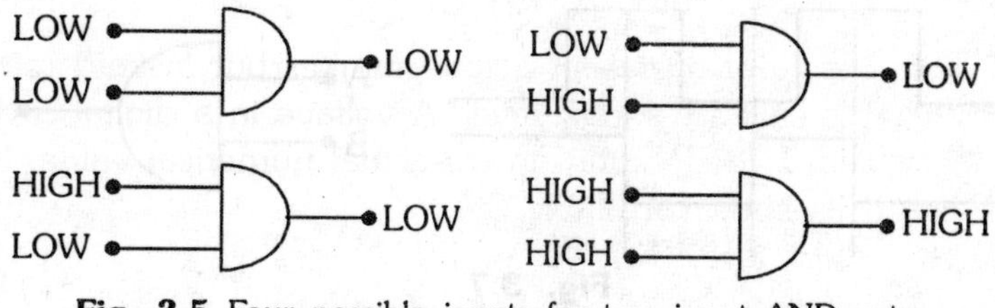

Fig. 3.5 Four possible inputs for two-input AND gate.

The truth table for a two–input AND gate is shown in Table 3.5.

Table 3.5. Truth table for 2–input AND gate

Inputs		*Output*
x	*y*	*F*
0	0	0
0	1	0
1	0	0
1	1	1

Conclusion :

1. The AND operation is performed like ordinary multiplication of 1's and 0's.
2. An output equal to 1 occurs only for the single case where all inputs are 1.
3. The output is 0 for any case where one or more inputs are 0.
4. For an *n*-input AND gate, the O/P F = *x.y.z.......*
5. With AND operation 1.1 = 1, 1.1.1. so on = 1.

Standard Package : Fig. 3.6 shows the pin diagram of a IC7408, a TTL quad–2 input AND gate. This IC contains four 2–input AND gates and it is a 14–pin dual-line package.

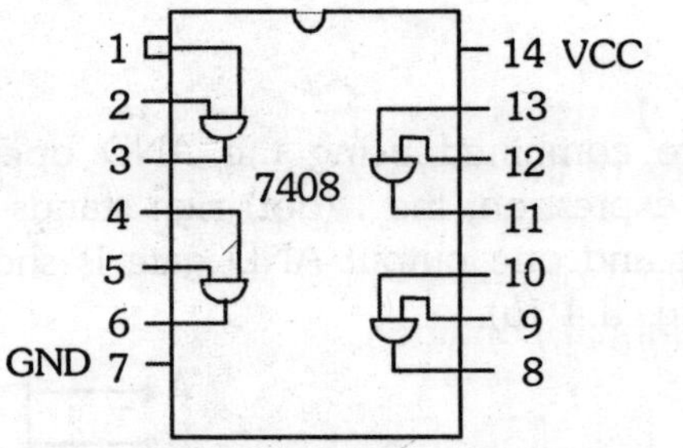

Fig. 3.6 Pin diagram of IC7408.

Pulsed Operation : In many applications, the inputs to a gate may be voltages that change with time between the two logic levels and are called as pulsed waveforms. In studying the pulsed operation of the AND gate, we consider the inputs with respect to each other in order to determine the output level at any given time. Following examples illustrate this operation.

Example 3.1. *Determine the output F from the AND gate for the given input waveforms shown in Fig. 3.7.*

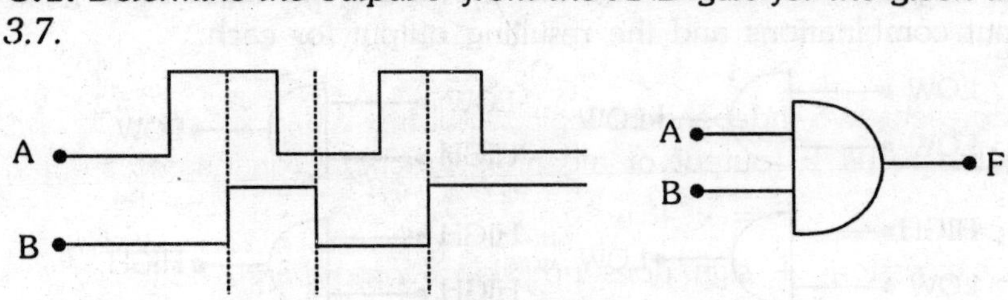

Fig. 3.7

Solution. The output of an AND gate is determined by realizing that it will be high only when both inputs are high at the same time. For the inputs, the output is high during t_2 and t_6 periods only. In remaining times, the output is 0 as shown in Fig. 3.8.

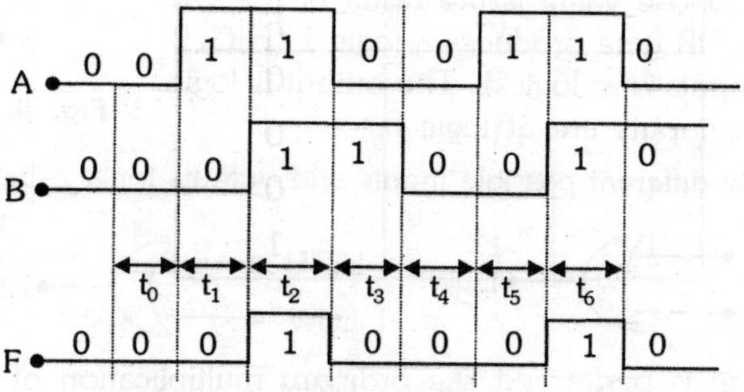

Fig. 3.8 Output waveform.

Example 3.2. *Determine the output waveform for the AND gate shown in Fig. 3.9.*

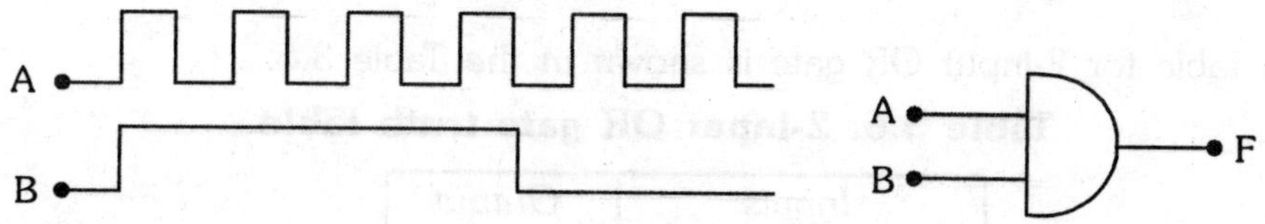

Fig. 3.9

Solution. The output F = 1, when both A and B are one at the sametime. Waveform F is 0 whenever B is 0, regardless of the signal at A. When B = 1, F waveform is same as A. Thus, we can use the input B as a control input whose logic level determines whether the waveform A gets through to the F output or not. Here, the AND gate is used as an "inhibit circuit". We can say that B = 0 is the inhibit condition which produces a 0 output. (see Fig. 3.10)

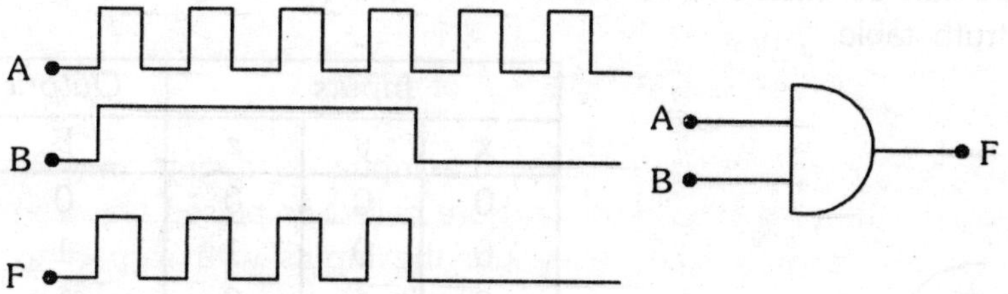

Fig. 3.10

3.2.2 OR Gate

Let x and y represent two independent logic variables. When x and y are combined using the OR operation, the result F (output or algebraic function) can be represented by

$$F = x + y$$

In the above expression, + sign does not stand for ordinary addition, it is for logical addition.

An OR gate is a digital circuit that has two or more inputs and one output and output is equal to the OR sum of the inputs. Fig. 3.11 shows the logic symbol for 2–input OR gate.

The inputs x and y are logic voltage levels and the output F is a logic voltage level whose value is the result of the OR operation on x and y. An OR gate produces a logic 1 (HIGH) output when any of the inputs is at logic 1. The output is logic 0 (low) only when all the inputs are at logic 0.

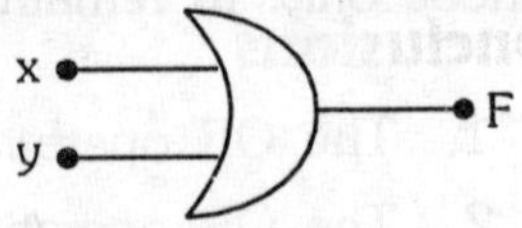

Fig. 3.11 Two-input OR gate.

Fig. 3.12. Illustrate the different possible inputs and outputs for a 2–input OR gate.

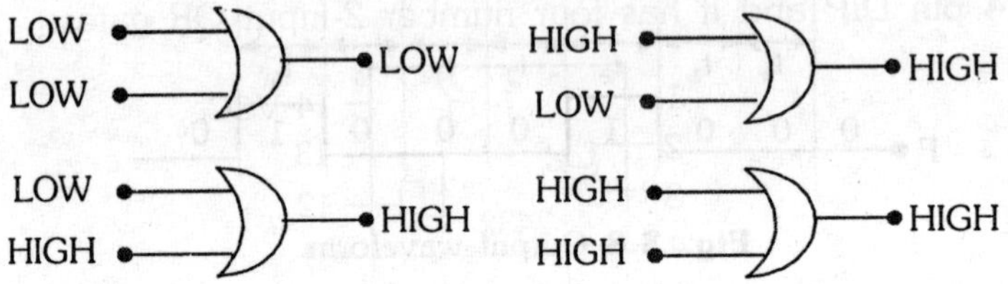

Fig. 3.12 Four possible inputs to 2-input OR gate.

The truth table for 2-input OR gate is shown in the Table 3.6.

Table 3.6. 2-input OR gate truth table

Inputs		*Output*
x	*y*	*F*
0	0	0
0	1	1
1	0	1
1	1	1

This same idea can be extended to more than two inputs. Fig. 3.13 shows a three input OR gate and it's truth table.

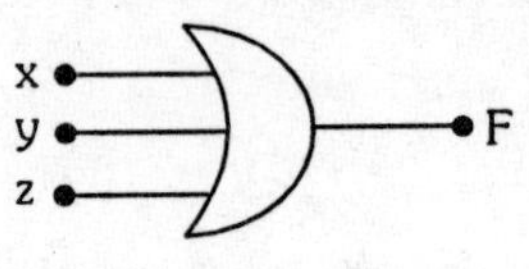

(*a*) 3–input OR gate

Inputs			*Output*
x	y	z	F
0	0	0	0
0	0	1	1
0	1	0	1
0	1	1	1
1	0	0	1
1	0	1	1
1	1	0	1
1	1	1	1

$F = x + y + z.$

(*b*) Truth table for 3-input OR gate

Fig. 3.13

For an *n*-input OR gate, the output 'F' can be expressed as $F = x + y + z + \ldots\ldots$ So the output of any OR gate can be expressed as the OR of it's various inputs.

Conclusions

1. The OR operation produces a result of 1 when any of the input variables is 1.
2. The OR operation produces a result of 0 only when all the input variables are 0.
3. With OR operation 1+1 = 1, 1+1+1+ ... = 1.

Standard Package : Fig. 3.14 shows the pin diagram of IC 7432, a TTL quad 2-input OR gate. It is a 14 pin DIP and it has four number 2-input OR gates.

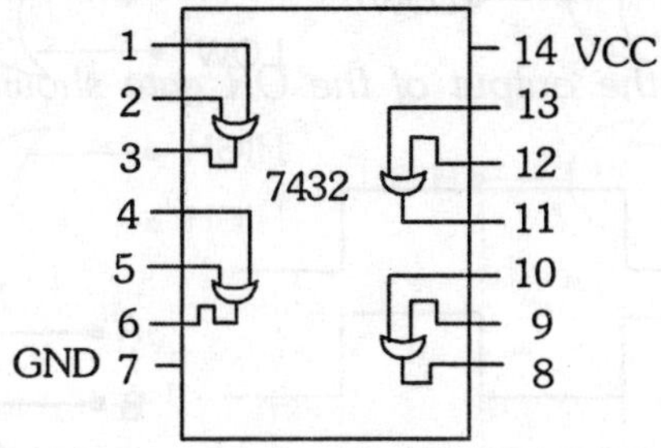

Fig. 3.14 Pin configuration of IC 7432.

Pulsed Operation : In many applications, the inputs of OR gate may not be single bit but sometimes, they may be pulses. The operation of OR gate under pulsed inputs is illustrated in the following examples.

Example 3.3. *Determine the OR gate output for the Fig. 3.15.*

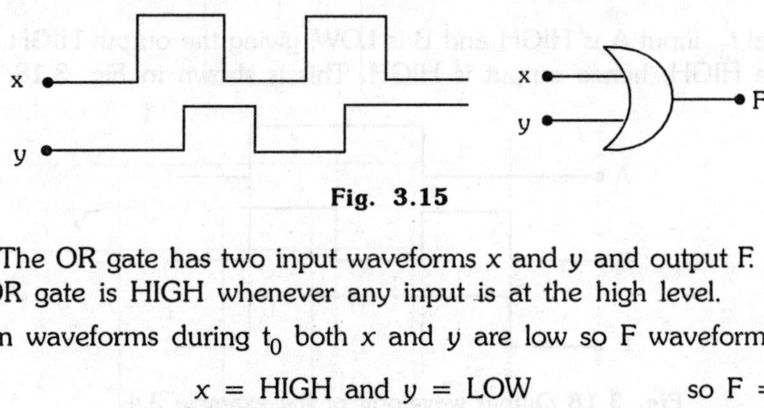

Fig. 3.15

Solution. The OR gate has two input waveforms *x* and *y* and output F. As we know that the output of OR gate is HIGH whenever any input is at the high level.

In the given waveforms during t_0 both *x* and *y* are low so F waveform is Low.

During t_1	x = HIGH and y = LOW	so F = HIGH
During t_2	Both are HIGH	so F = HIGH
During t_3	x = LOW and y = HIGH	so F = HIGH
During t_4	Both are LOW	so F = LOW
During t_5	x = HIGH and y = LOW	so F = HIGH and
During t_6	x and y are HIGH	so F = HIGH.

The waveform for F is shown in Fig. 3.16.

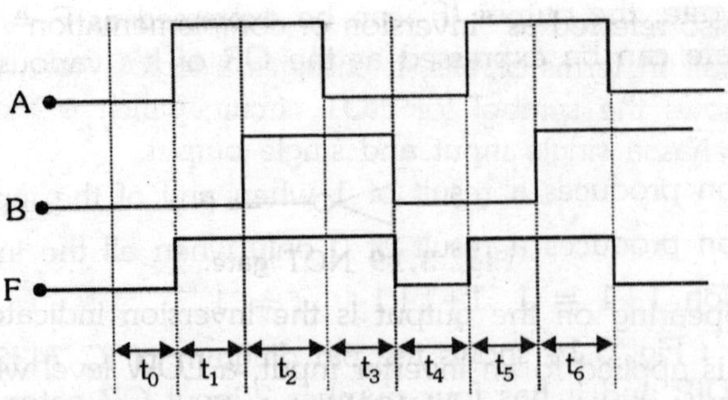

Fig. 3.16 Waveforms for example 3.3.

Example 3.4. *Determine the output of the OR gate shown in Fig. 3.17.*

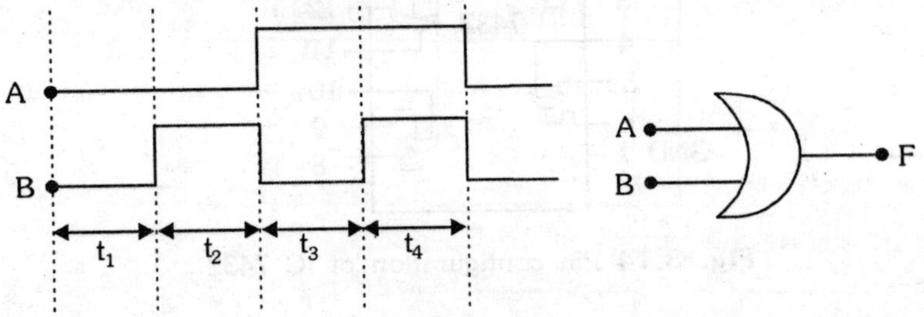

Fig. 3.17

Solution. During the time interval t_1, both inputs A and B are LOW (0), making output LOW (0). During the time interval t_2, input A is LOW but input B is HIGH, hence output is HIGH.

During interval t_3, input A is HIGH and B is LOW, giving the output HIGH. At the interval t_4, both inputs are HIGH, hence output is HIGH. This is shown in Fig. 3.18.

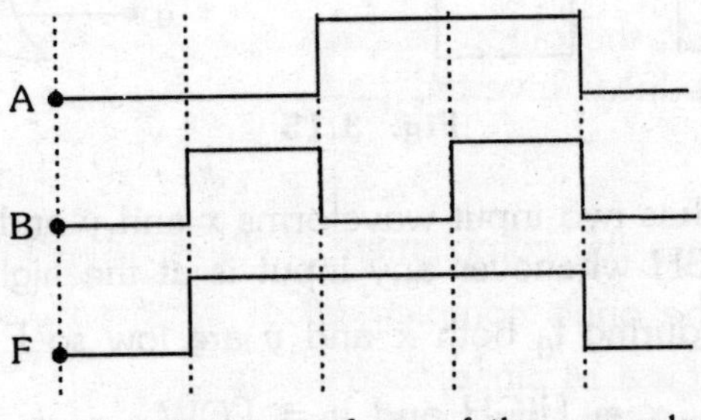

Fig. 3.18 Output waveform of the example 3.4.

3.2.3 NOT Gate

NOT operation is performed on a single input variable. If x is an input variable to the NOT gate, the output (F) can be expressed as

$$F = \overline{x} \text{ (or) } x'$$

This expression can be read as : "F equals NOT x". or "F equals the inverse of x". or "F equals the complement of x".

NOT operation is also referred as "inversion or complementation". It changes the one logic level to it's opposite level. In terms of bits, it changes a logic 1 to a logic 0 and a logic 0 to a logic 1. Fig. 3.19 shows the symbol for NOT circuit, which is more commonly called as "INVERTER". It always has a single input and single output.

Fig. 3.19 NOT gate.

The bubble [o] appearing on the output is the inversion indicator.

When HIGH level is applied to an inverter input, a LOW level will appear on it's output. When LOW level is applied to it's input, a HIGH level will appear on it's output. The truth table for the NOT gate is shown in Table 3.7.

Table 3.7. Truth table for NOT gate

Input	*Output*
x	F
0	1
1	0

The Fig. 3.20 shows the output of an inverter for a pulse input.

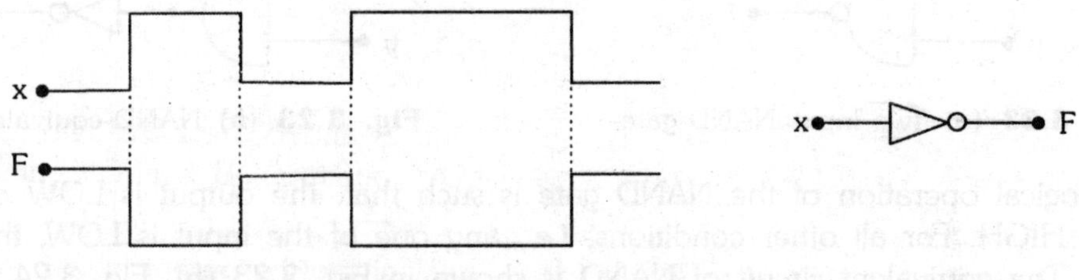

Fig. 3.20 Output of a NOT gate for pulsed input.

When the input is LOW, the output is HIGH, and when the input is HIGH, the output is LOW, thus the inverter produces inverted pulses.

Conclusions

1. The output is always the complement of the input variable.
2. It is used to get the one's complement of the given binary number.

Standard Package : It is a 14 pin, dual in line package. It contains, TTL hex inverters. The Fig. 3.21 shows the pin diagram of IC7404.

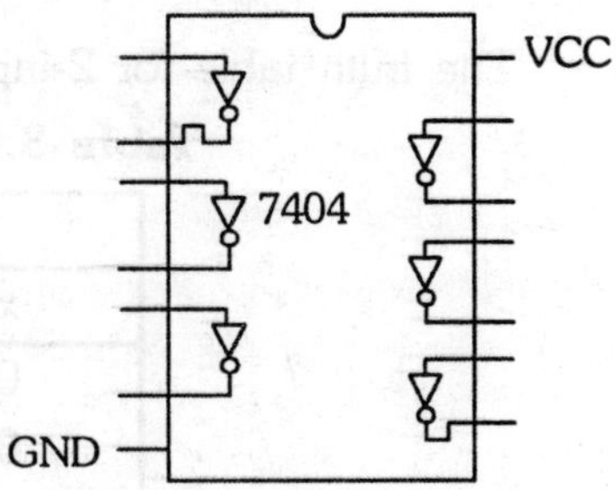

Fig. 3.21 Pin diagram of 7404.

3.2.4 Buffer (OR) Transfer Gate

It is a single input single output gate. If 'x' is an input variable to the BUFFER gates, an output of the gate (F) can be expressed as F = x.

The logic symbol of the BUFFER gate is shown in Fig. 3.22. When input is HIGH, output is HIGH and when input is LOW, output is also LOW. The truth table for it is shown in Table 3.8.

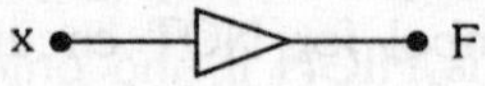

Fig. 3.22 Buffer gate.

Table 3.8

Input	*Output*
x	F
0	0
1	1

From the truth table, the output of BUFFER gate is always same as input *i.e.*, it just transfers the input as output.

Conclusion

1. The output is same as input.

3.2.5 NAND Gate

The term NAND is a contraction of NOT-AND. It is nothing but an AND function with a complemented output. The standard logic symbol for a two-input NAND gate is shown in Fig. 3.23 (*a*). It is same as AND gate symbol except the small circle (bubble) on it's output.

If x and y are the inputs of the NAND gate, then output F is represented by

$$F = \overline{xy}$$

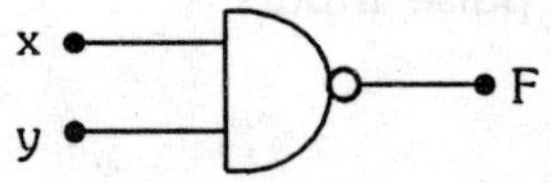

Fig. 3.23 (*a*) Two-input NAND-gate

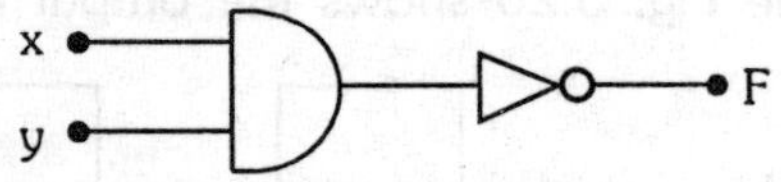

Fig. 3.23 (*b*) NAND-equivalent.

The logical operation of the NAND gate is such that, the output is LOW only when all inputs are HIGH. For all other conditions, *i.e.*, any one of the input is LOW, the output will be HIGH. The equivalent circuit of NAND is shown in Fig. 3.23 (*b*). Fig. 3.24 illustrates the logical operations of a two-input NAND gate for all the four input combinations.

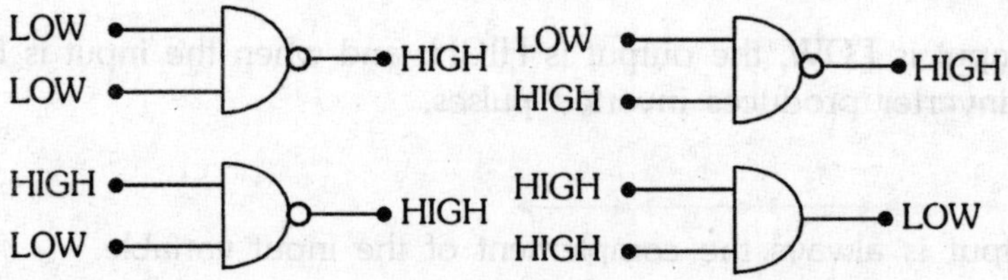

Fig. 3.24 Four possible combinations for a two-input NAND gate.

The truth table for 2-input NAND gate is shown in Table 3.9.

Table 3.9. Truth table for 2-input NAND gate

Inputs		*Output*
x	*y*	F
0	0	1
0	1	1
1	0	1
1	1	0

Conclusions

1. The output of a NAND gate is LOW only when all inputs are HIGH.
2. The output of a NAND gate is HIGH if and only if at least one of the inputs is Low.
3. It is cascading of AND and NOT gates, *i.e.*, NOT of AND = NAND.
4. It is called a universal gate because any gate or Boolean function can be constructed with this gate.
5. For multiple inputs, the output F = $\overline{x.y.z....}$

Standard Package : Fig. 3.25 shows the pin diagram of a IC7400, a TTL, quad 2-input NAND gate. It contains four 2-input NAND gates inside a 14-pin DIP.

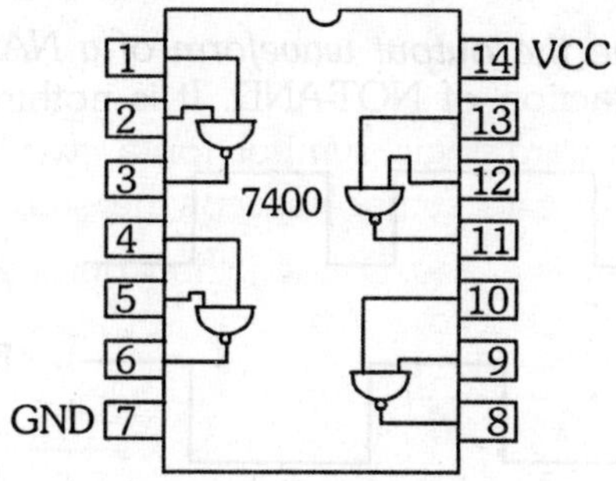

Fig. 3.25 2-input NAND gate.

The pulsed operation of the NAND gate is explained with the following examples.

Example 3.5. *Draw the output waveforms of a 2-input NAND gate for the given input waveforms shown in Fig. 3.26.*

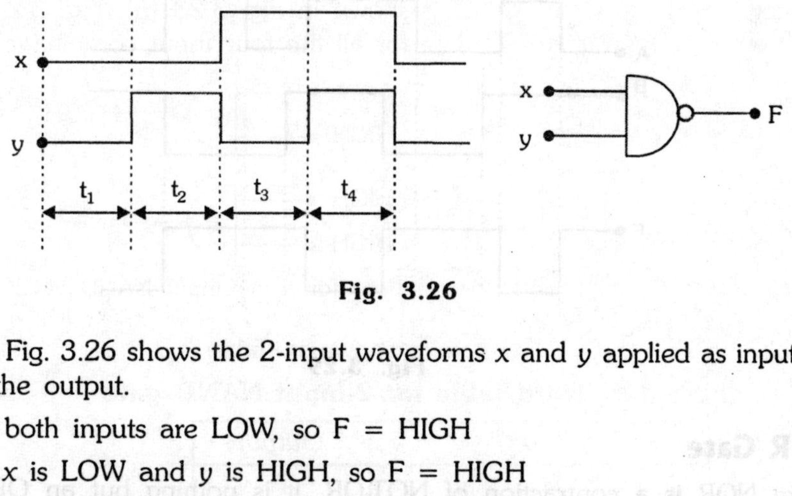

Fig. 3.26

Solution. Fig. 3.26 shows the 2-input waveforms *x* and *y* applied as inputs to the NAND gate and F is the output.

During t_1, both inputs are LOW, so F = HIGH

During t_2, *x* is LOW and *y* is HIGH, so F = HIGH

During t_3, *x* is HIGH and *y* is LOW, so F = HIGH

During t_4, both inputs are HIGH, so F = LOW

This is shown in Fig. 3.27.

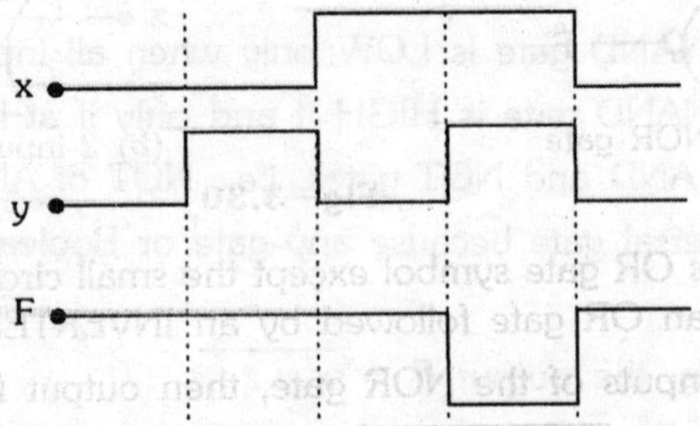

Fig. 3.27 Output waveform for the example 3.5.

Example 3.6. *Determine the output waveform of a NAND gate having the inputs shown in Fig. 3.28.*

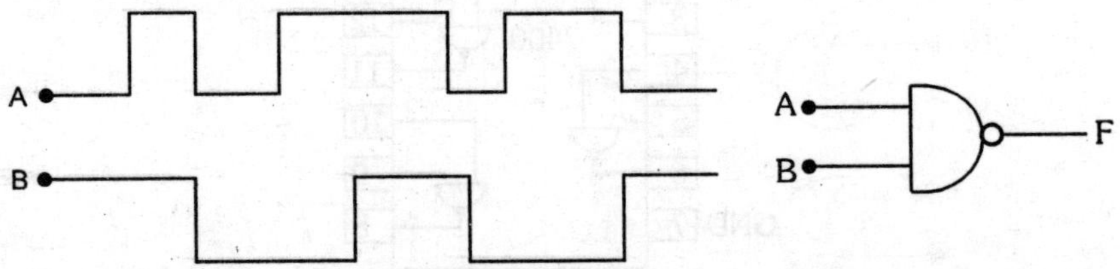

Fig. 3.28

Solution. The output can be determined by several ways. One way is to first draw the output for an AND gate, and then invert it. Another way is to use NAND truth table. The output waveform for the example 3.6 is shown in Fig. 3.29.

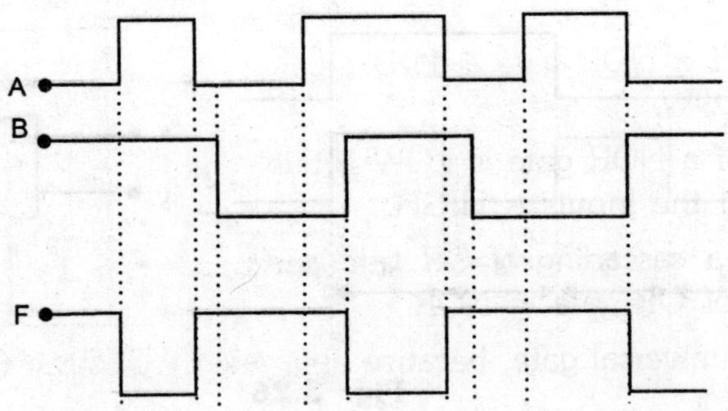

Fig. 3.29

3.2.6 NOR Gate

The term NOR is a contraction of NOT-OR. It is nothing but an OR function with a complemented output. The standard logic symbol for a two-input NOR gate is shown in Fig. 3.30 (*a*) and it's equivalent is shown in Fig. 3.30 (*b*).

(*a*) 2-input NOR gate (*b*) 2-input NOR gate equivalent

Fig. 3.30

NOR gate is same as OR gate symbol except the small circle (bubble) on it's output. Thus, NOR gate operates like an OR gate followed by an INVERTER.

If x and y are the inputs of the NOR gate, then output F is represented by

$$F = \overline{x + y}$$

The logic operation of NOR is such that the output is HIGH only when all inputs are LOW. For remaining all other conditions, the output is LOW. Fig. 3.31 illustrates the logical operations of a 2-input NOR gate for all four input combinations.

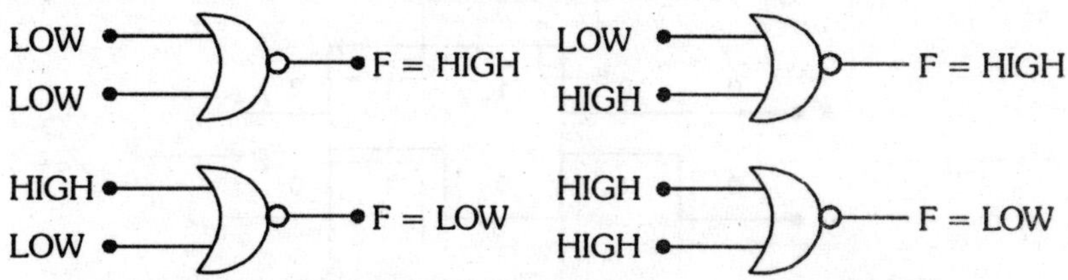

Fig. 3.31 Four possible combinations for a 2-input NOR gate.

The truth table for 2-input NOR gate is shown in Table 3.10.

Table 3.10. Truth table of a 2-input NOR gate

Inputs		Output
x	y	F
0	0	1
0	1	0
1	0	0
1	1	0

Conclusions

1. The output of a NOR gate is HIGH only when all inputs are LOW.

2. The output of a NOR gate is LOW if and only if at least one of the inputs is HIGH.

3. NOR gate is a cascading of OR gate and NOT gate *i.e.*, NOT of OR gate = NOR.

4. It is called a universal gate, because any gate or Boolean function can be constructed with this gate.

5. For multiple inputs, the output $F = \overline{x + y + z + ...}$

Standard Package : Fig. 3.32 shows the pin diagram of a 7402, a TTL quad 2-input NOR gate. This IC contains four 2-input NOR gates inside a 14-pin DIP.

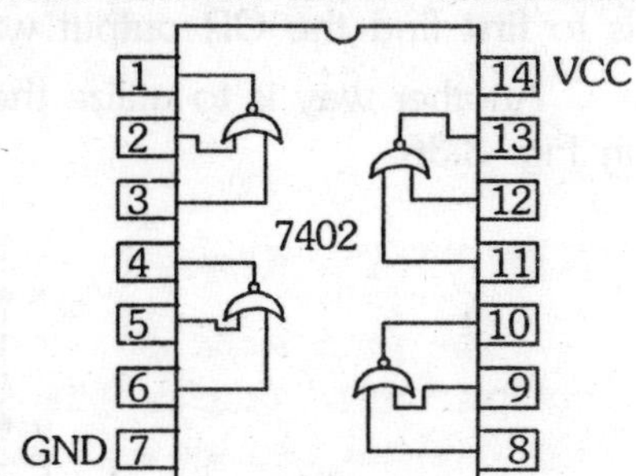

Fig. 3.32 Pin diagram of a IC 7402.

The pulsed operation of the NOR gate is explained with the following examples.

Example 3.7. *Draw the output waveform for the 2-input NOR gate shown in Fig. 3.33.*

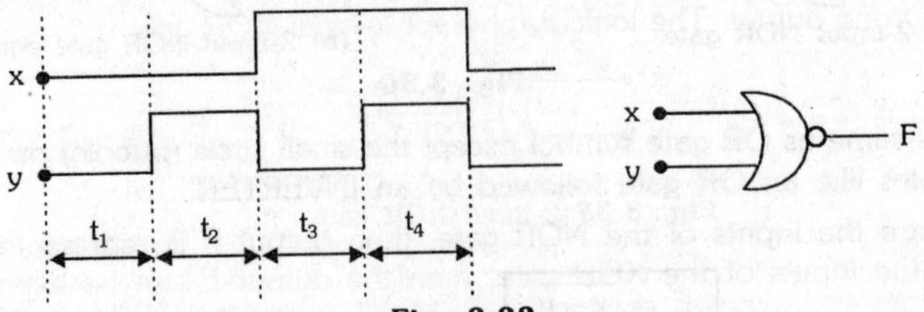

Fig. 3.33

Solution. Fig. 3.33 shows the waveforms of two inputs, x and y to a NOR gate. In the duration t_1, both inputs are LOW, so output (F) is HIGH. During t_2, t_3 and t_4, the output is LOW as shown in Fig. 3.34.

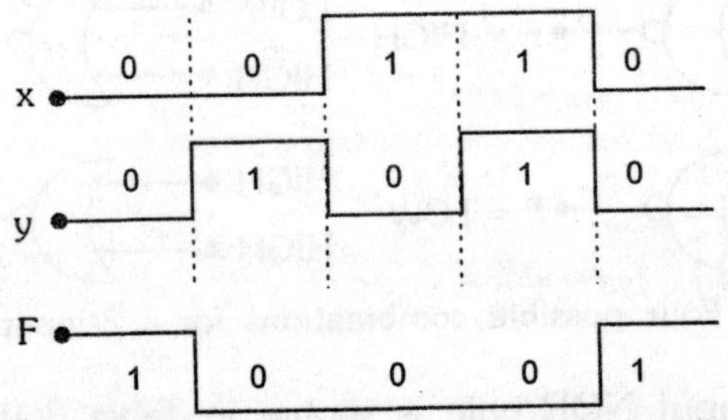

Fig. 3.34 Output waveform for the example 3.7.

Example 3.8. *Determine the waveform at the output of a NOR gate for the input waveforms shown in Fig. 3.35.*

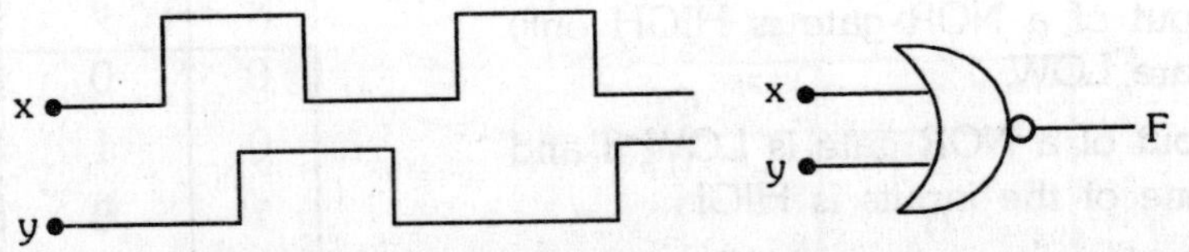

Fig. 3.35

Solution. There are several ways to determine the NOR gate output waveform. One way is to first find the OR output waveform and then invert it.

Another way is to utilize the truth table of the NOR gate. The output waveform is shown in Fig. 3.36.

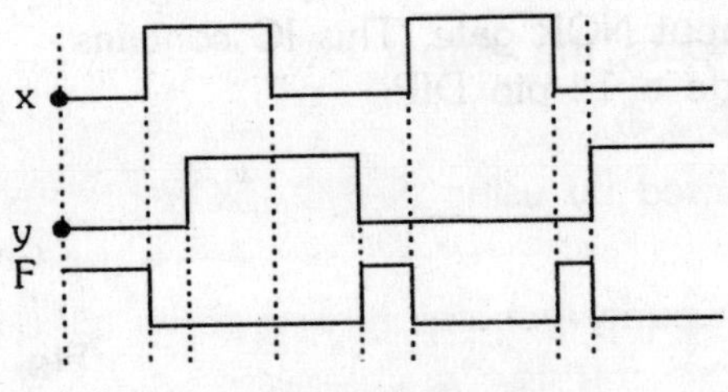

Fig. 3.36

3.2.7 Exclusive-OR gate

The Exclusive-OR gate is an expansion of EX-OR gate. The EX-OR (XOR) gate has two or more inputs and only one output. The logic symbol for 2-input XOR gate is shown in Fig. 3.37.

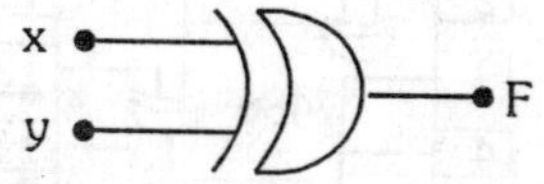

Fig. 3.37 2-input XOR gate.

If x and y are the inputs of the XOR gate, then the output F can be expressed as

$$F = x'y + xy' = x \oplus y$$

The output F is HIGH, when odd number of inputs are 1's and for all other conditions, the output is LOW. Fig. 3.38 shows the logic operation of 2-input EX-OR gate for all the four possible input combinations.

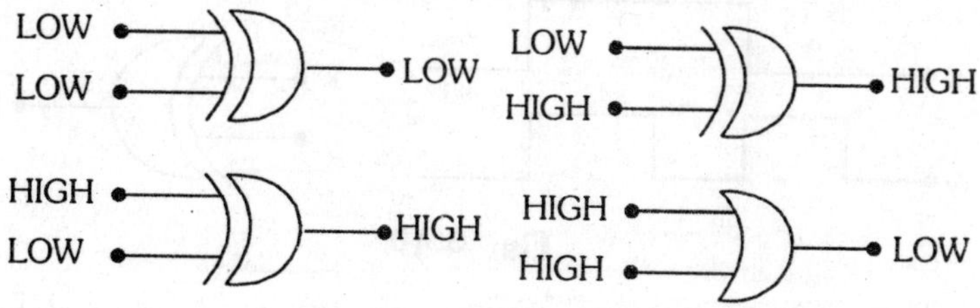

Fig. 3.38 Four possible inputs for two input EX-OR gate.

The truth table for a 2-input EX-OR gate is shown in Table 3.11.

Table 3.11. Truth table for 2-input XOR gate

Inputs		*Output*
x	y	F
0	0	0
0	1	1
1	0	1
1	1	0

Conclusions

1. EX-OR gate is used to find the number of one's in the input. If number of 1's are even, the output is '0' else the output is 1.
2. For many number of inputs, the output

$$F = x \oplus y \oplus z \oplus ......$$

3. EX-OR gate can be realised by using NAND, NOR or (AND and NOT) or (OR and NOT) gates.
4. These gates are used as parity generators and parity checkers.

Standard Package : The IC 74LS86 is a quad 2-input EX-OR gate. It contains four

2-input EX-OR gates in a 14 pin, DIP as shown in Fig. 3.39.

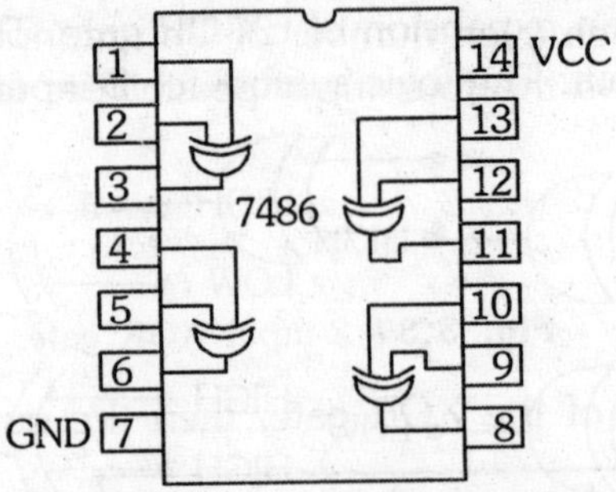

Fig. 3.39 Pin diagram of 7486.

The pulsed operation of EX-OR gate is illustrated in the following example.

Example 3.9. *Draw the output wave for the 2-input EX-OR gate shown in Fig. 3.40.*

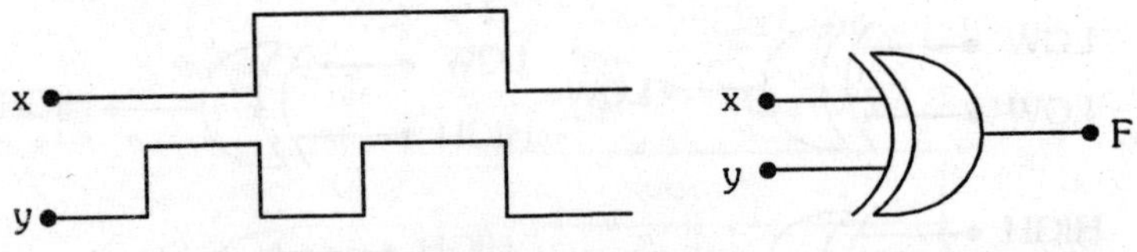

Fig. 3.40

Solution. During the time interval t_1, both inputs x and y are LOW, *i.e.*, even number of inputs are LOW, so the output is LOW (0). During the interval t_2, x is LOW and y is HIGH so the output F = 1. During the interval t_3, x is HIGH and y is LOW, so the output F = 1 and finally during the interval t_4, both inputs are HIGH, so the output is LOW (0). The output waveform is shown in Fig. 3.41.

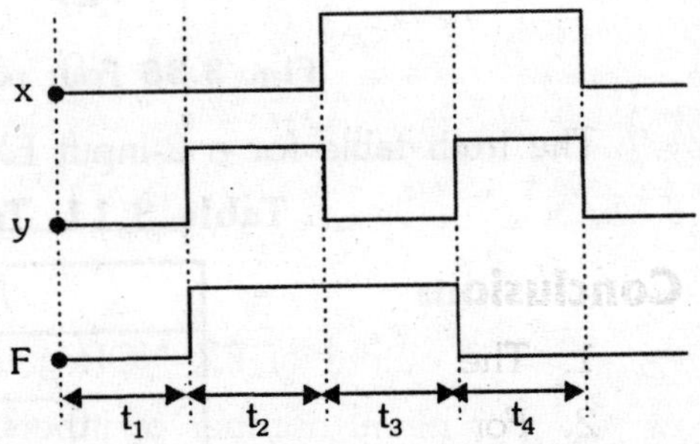

Fig. 3.41 Output (F) waveform.

3.2.8 Exclusive-NOR Gate

The term EX-NOR is a contraction of NOT–XOR, NOT exclusive OR gate. It is logically equivalent to the EX–OR gate with an inverter at the output. It is also called coincidence. It has two input terminals and one output terminal. The logic symbol for 2-input EX–NOR gate is shown in Fig. 3.42.

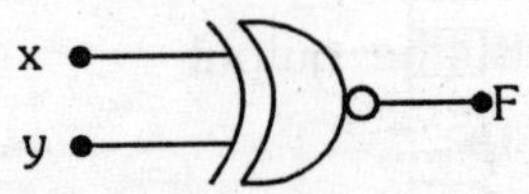

Fig. 3.42 2-input EX–NOR gate.

If x and y are the two inputs of the XNOR gate, then the output F can be expressed as

$$F = xy + x'y' = x \odot y$$

The output F is HIGH, when even number of 1's are input, and remaining all other cases the output is 0. Fig. 3.43 shows the four input possibilities for the EX–NOR gate.

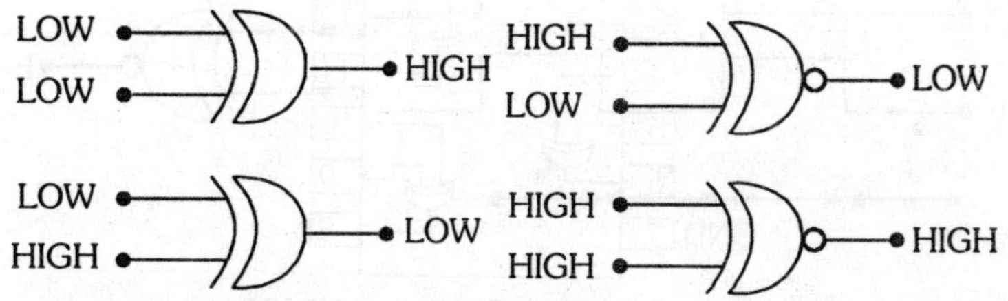

Fig. 3.43 Four possible inputs for two input EX–NOR gate.

The truth table for 2–input EX–NOR gate is shown in Table 3.12. This table can be extended for any number of inputs. Regardless of the number of inputs, the output is HIGH when even number of inputs are HIGH or when all inputs are LOW.

Table 3.12. Truth table for 2–input EX–NOR gate

Inputs		*Output*
x	*y*	F
0	0	1
0	1	0
1	0	0
1	1	1

Conclusions

1. The output of EX–NOR gate is HIGH only when even number of inputs are high.
2. For many number of inputs, the output $F = x \odot y \odot z \odot$ ..... and so on.
3. EX–NOR gate can be realized by NAND/NOR or AND, OR and NOT gates.

Standard Package : Fig. 3.44 shows the pin diagram of IC74266, a TTL quad–2 input EX–NOR gate. It contains four 2–input EX–NOR gates in a 14 pin–DIP.

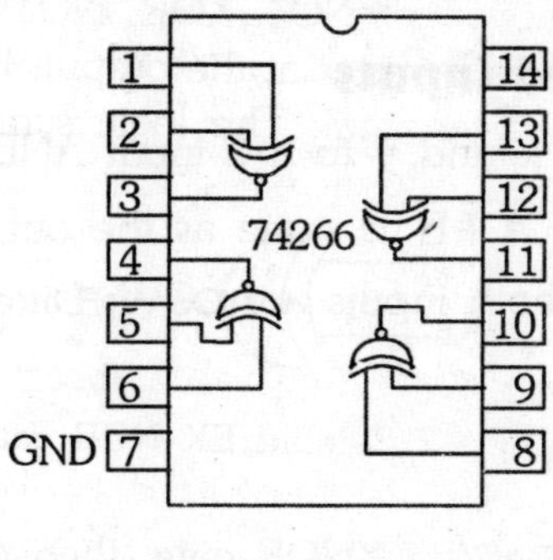

Fig. 3.44 Pin diagram of 74266.

The pulsed operation of an EX–NOR gate is illustrated in the following example.

Example 3.10. *Draw the output waveform of the 2-input EX–NOR gate shown in Fig. 3.45.*

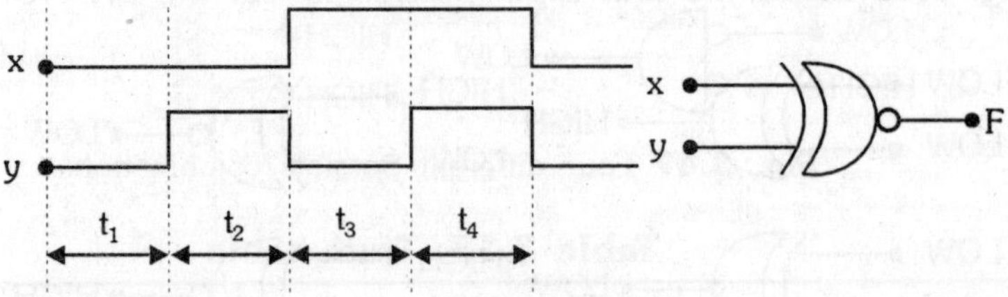

Fig. 3.45

Solution. During the time interval t_1, both inputs are LOW, so output is 1. (HIGH)
During the time interval t_2, input x is LOW and y is HIGH, so the output F = LOW (0).
During the time interval t_3, x is HIGH and y is LOW so F = 0 (LOW).
During the time interval t_4, both inputs are HIGH so F = HIGH (1) shown in Fig. 3.46.

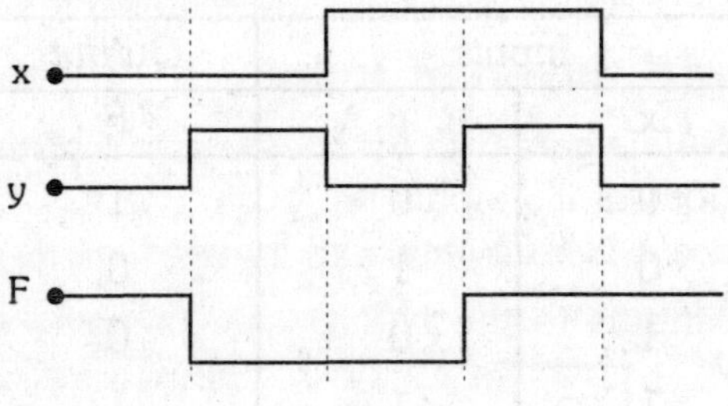

Fig. 3.46 Output waveform for example 3.10.

3.3 GATES WITH INVERTED INPUTS

Each of the various logic functions can always be performed by two gates whose function is different from the function required. Two of these combinations have already been met : the AND gate followed by a NOT gate gives the NAND logical function, and OR gate followed by a NOT gate gives the NOR function. Various other gate combinations are also possible and any of them may sometimes be convenient to employ.

3.3.1 AND Gate with Inverted Inputs

Table 3.13 shows that if inputs x and y to a 2-input AND gate are inverted, the output of the circuit is $\boxed{F = \overline{A}.\overline{B}}$ and $F = \overline{A + B}$ is same as the output of a NOR gate.

The output, F = 1 only when both inputs are LOW. Different possible inputs are shown in Fig. 3.47.

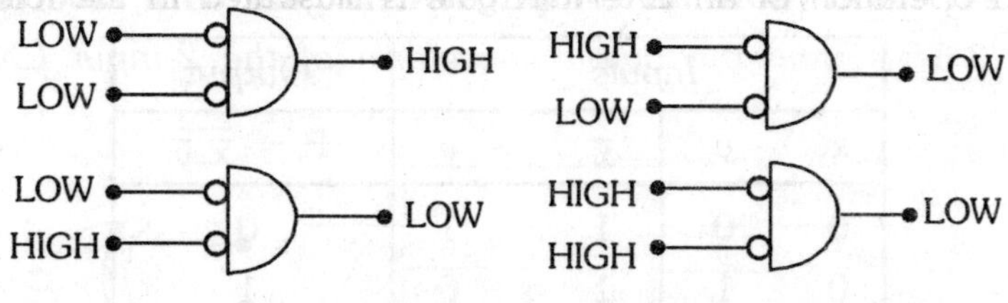

Fig. 3.47 Four different possible combinations.

Table 3.13. Truth table

Inputs		*Inverted Inputs*		*Output*
x	y	$\overline{x}$	$\overline{y}$	F
0	0	1	1	1
0	1	1	0	0
1	0	0	1	0
1	1	0	0	0

3.3.2 NAND Gate with Inverted Inputs

The truth table of a NAND gate with inverted inputs is given in Table 3.14. The output of the gate is LOW only when both of it's inputs are LOW. If either, or both input (s) is/are HIGH, the output is also HIGH. Hence a NAND gate with inverted inputs performs the logical OR function.

If $\overline{x}$ and $\overline{y}$ are the inputs to the NAND gate then the output (F) of the NAND gate is

$$F = \overline{\overline{x}.\overline{y}} = x + y$$

The logic symbol is shown in Fig. 3.48.

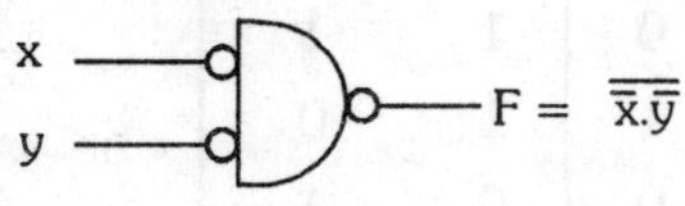

Fig. 3.48

The four possible combinations of inputs are shown in Fig. 3.49.

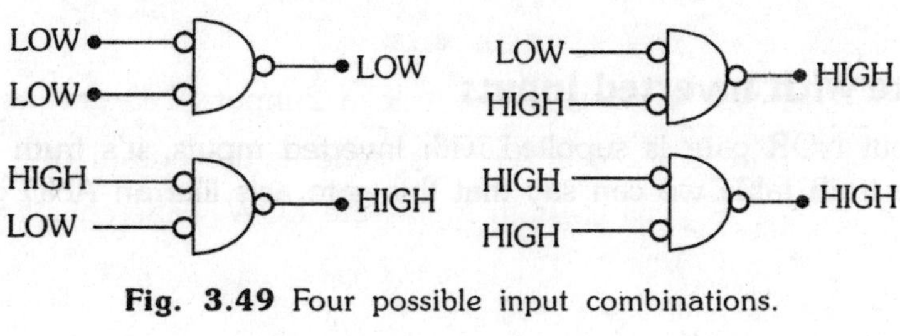

Fig. 3.49 Four possible input combinations.

Table 3.14. Truth table for 2-input NAND gate with inverted input

Inputs				*Output*
x	y	$\overline{x}$	$\overline{y}$	$F = \overline{\overline{x}.\overline{y}}$
0	0	1	1	0
0	1	1	0	1
1	0	0	1	1
1	1	0	0	1

3.3.3 OR gate with Inverted Inputs

The truth table for OR gate with inverted inputs is shown in Table 3.15.

The logic diagram is shown in Fig. 3.50.

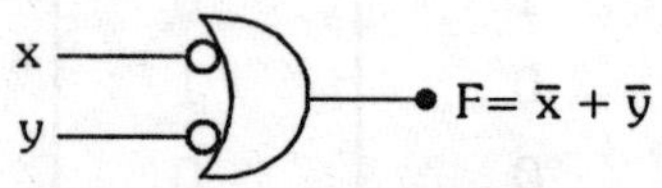

Fig. 3.50

The two inputs are $\overline{x}$ and $\overline{y}$ and the output F of the OR gate is

$$F = \overline{x} + \overline{y} \quad \text{i.e.,} \quad F = \overline{x.y}$$

Table 3.15. Truth table for 2-input OR gate with inverted input

Inputs				*Output*
x	y	$\overline{x}$	$\overline{y}$	$F = \overline{x} + \overline{y}$
0	0	1	1	1
0	1	1	0	1
1	0	0	1	1
1	1	0	0	0

From the truth table, it is clear that the OR gate with the inverted inputs will function as a NAND gate.

3.3.4 NOR gate with Inverted Inputs

When a 2-input NOR gate is supplied with inverted inputs, it's truth table is shown in Table 3.16. By the truth table we can say that this gate acts like an AND gate.

Table 3.16. Truth table

Inputs				*Outputs*
x	y	$\bar{x}$	$\bar{y}$	$F = \overline{\bar{x} + \bar{y}}$
0	0	1	1	0
0	1	1	0	0
1	0	0	1	0
1	1	0	0	1

When the input $\bar{x}$ and $\bar{y}$ are the inputs of the NOR gate, then the output (F) of it is

$$F = \overline{\bar{x} + \bar{y}} = xy$$

This is shown in Fig. 3.51.

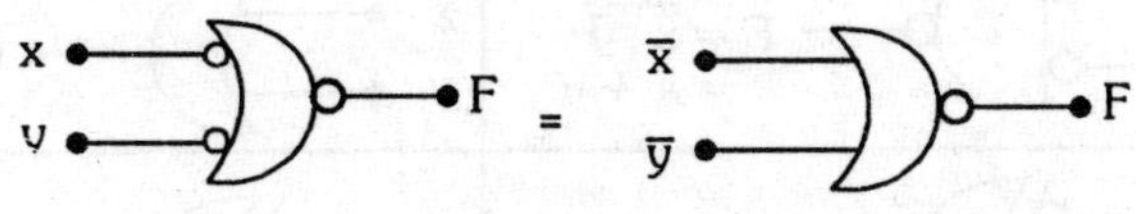

Fig. 3.51 NOR gate with inverted inputs.

The four possible input combinations are shown in Fig. 3.52.

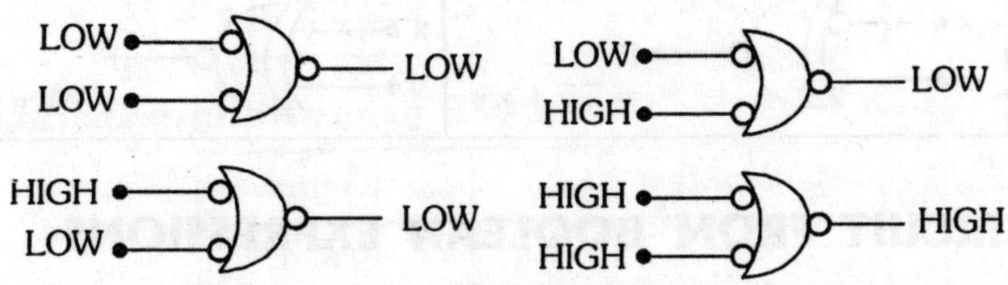

Fig. 3.52 Four possible input combinations.

The output F =1, when both inputs are 1 and is '0' for all other conditions.

Note : (1) The EX-OR gate and EX–NOR gate with inverted inputs will function as same.

(2) The AND, OR and NOT gates are called basic gates and NAND and NOR gates are called universal gates.

(3) Table 3.17 shows the equivalence relation between the gates with the inverted inputs (alternate symbols).

(4) The equivalences of standard and alternate symbols are valid for gates with any number of inputs.

(5) The standard and alternate symbols for each gate represent the same physical circuit.

Table 3.17. Equivalence relation between gates

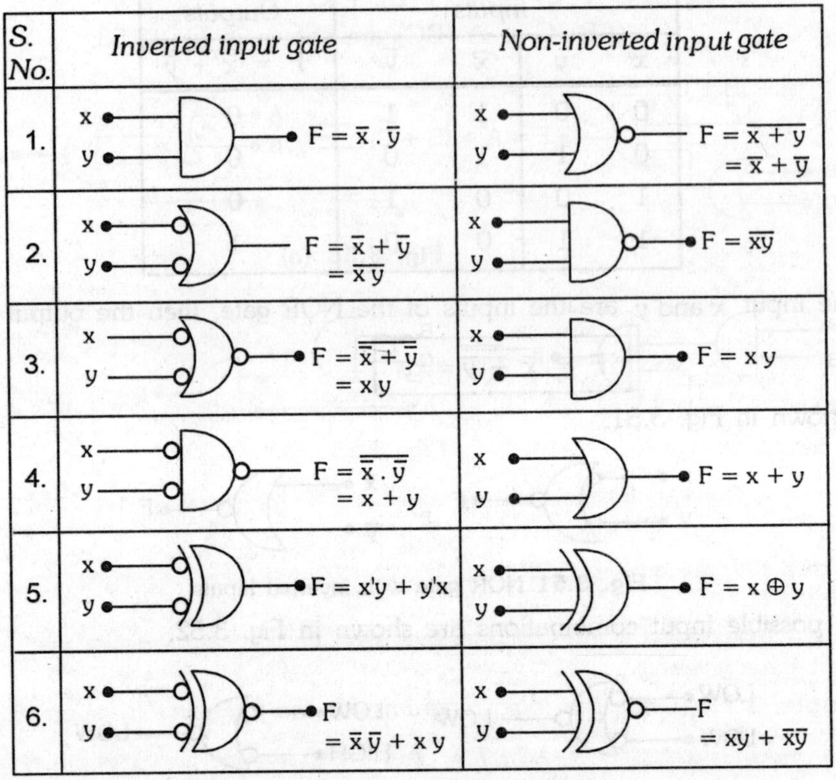

S. No.	Inverted input gate	Non-inverted input gate
1.	$F = \overline{x} . \overline{y}$	$F = \overline{x + y}$ $= \overline{x} + \overline{y}$
2.	$F = \overline{x} + \overline{y}$ $= \overline{x\,y}$	$F = \overline{xy}$
3.	$F = \overline{\overline{x} + \overline{y}}$ $= x\,y$	$F = x\,y$
4.	$F = \overline{\overline{x} . \overline{y}}$ $= x + y$	$F = x + y$
5.	$F = x'y + y'x$	$F = x \oplus y$
6.	$F = \overline{x}\,\overline{y} + x\,y$	$F = xy + \overline{x}\overline{y}$

3.4 LOGIC CIRCUIT FROM BOOLEAN EXPRESSIONS

Combinational logic circuits are made by connecting the basic logic gates together to form a new, more complex, logic function. When a logic circuit is designed, a logic expression is first obtained, simplified by using Boolean rules or K-map if possible and then a logic diagram that will implement the wanted logic operation is drawn.

The basic process is simple, draw one gate for each logic operation in the Boolean equation and ensure that each gate is given the correct input (*s*). For example, if we need a circuit that is defined by x = A.B.C, we would immediately know that all that was needed was a three-input AND gate.

If we needed a circuit that is defined by $x = A + \overline{B}$, we would use a two-input OR gate with a NOT gate on one of the inputs (B).

Suppose if we want to construct a circuit whose output is $x = AB + \overline{B}C + A B\overline{C}$. This expression contains three terms which are ORed together. This tells us that a 3-input OR gate is needed and to represent each term, we need 3 AND gates in which one is a 3-input gate.

Example 3.11. *Draw the logic diagrams for the following Boolean laws*

Associative Laws

Solution. Associative Laws.

1. $A + (B + C) = (A + B) + C$
2. $(AB)\,C = A\,(BC)$ are shown in Fig. 3.53.

1. A + (B + C)

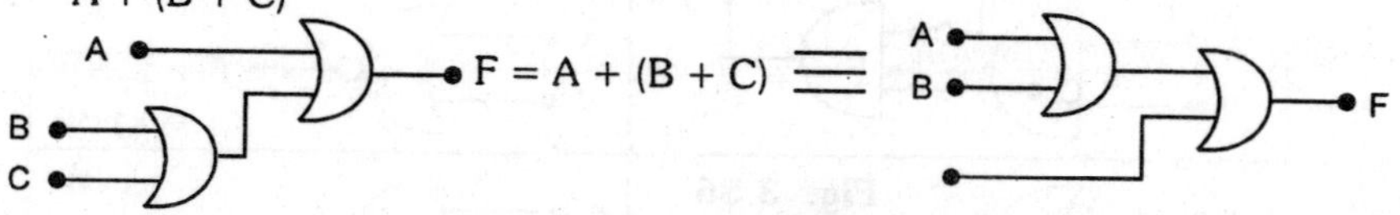

Fig. 3.53 (*a*)

2.

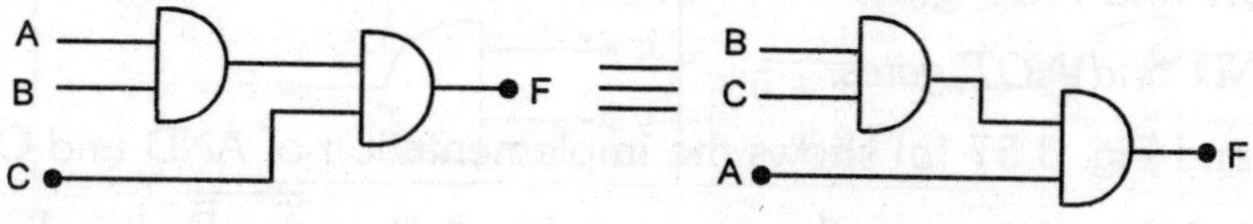

Fig. 3.53 (*b*)

Note : Any Boolean function can be expressed in two forms

1. SOP form 2. POS form

If the given expression is in SOP form, then that function can be realized by using AND-OR-NOT logic (two/three levels).

If the given expression is in POS form, then that function can be realized by using OR-AND-NOT logic (two or three levels).

Example 3.12. *Draw the logic diagrams for the following Boolean expressions*

(a) $f = A.B + C$ *(b)* $f = AC + B\overline{C} + \overline{A}BC$ *(c)* $f = AB + \overline{B}C$

Solution. *(a)* $f = A.B + C$, it's realization is shown in Fig. 3.54.

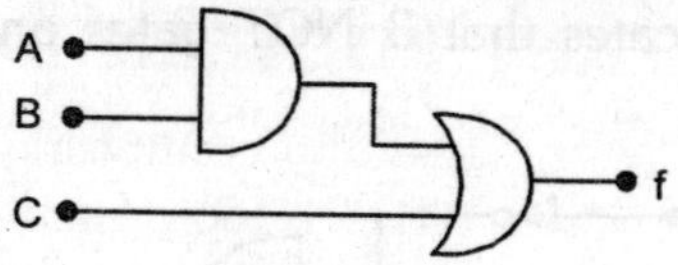

Fig. 3.54

(b) The realization of the function, $f = AC + B\overline{C} + \overline{A}BC$ is shown in Fig. 3.55. It contains 3 AND gates and one OR gate.

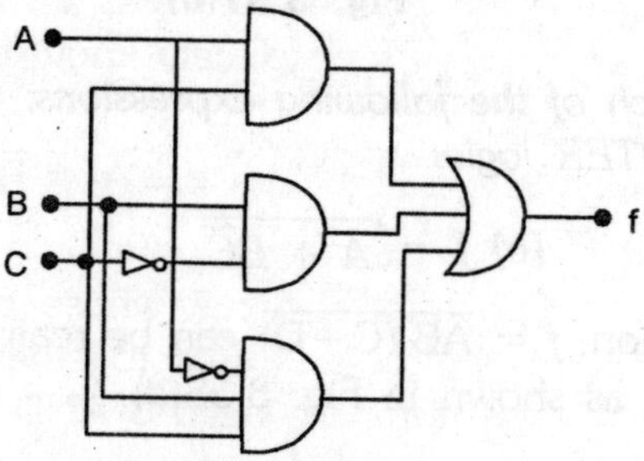

Fig. 3.55

(c) $f = AB + \overline{B}C$. The realization of this function is shown in Fig. 3.56. It has 2 AND gates, one NOT gate and one OR gate.

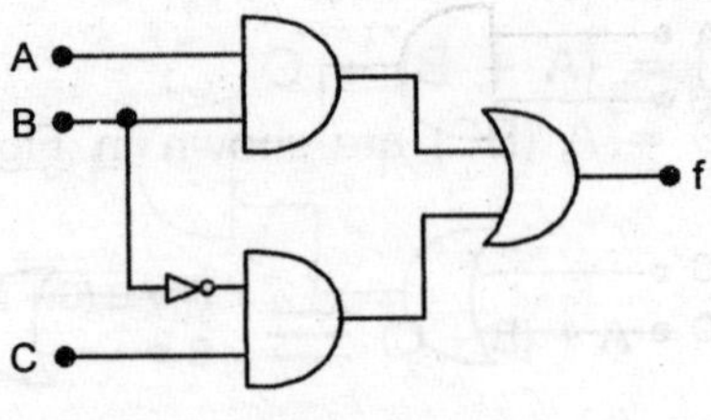

Fig. 3.56

Example 3.13. *Show how to implement*

(i) AND gate with OR and NOT gates.

(ii) OR gate with AND and NOT gates.

Solution. Fig. 3.57 (*a*) and Fig. 3.57 (*b*) shows the implementation of AND and OR gates.

(*i*) If A and B are inputs of the AND gate, then output, f = A.B = $\overline{\bar{A}+\bar{B}}$. It indicates that 3 NOT gates and one OR gate are needed.

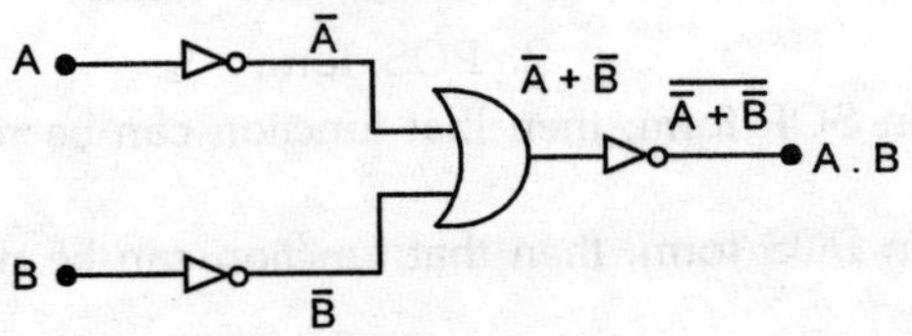

Fig. 3.57(*a*)

(*ii*) If A and B are inputs of the OR gate, then output, f = A + B. This can be represented as f = A+B = $\overline{\bar{A}.\bar{B}}$, which indicates that 3 NOT gates and one AND gate are needed to implement the function.

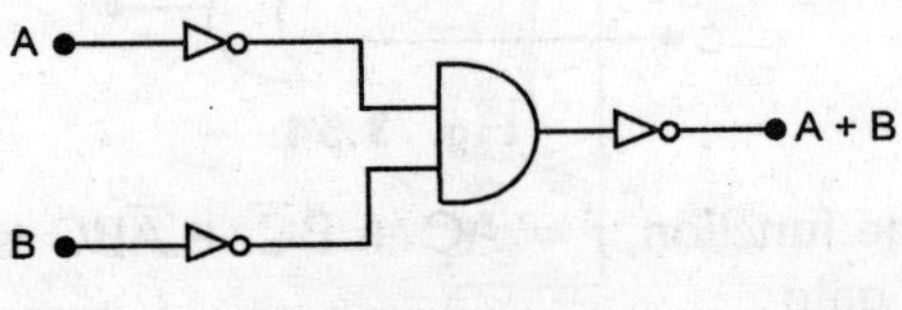

Fig. 3.57(*b*)

Example 3.14. *For each of the following expressions, construct the corresponding logic circuit, using AND/OR/INVERTER logic.*

(*a*) $f = \overline{AB\,(C+D)}$ (*b*) $f = \overline{A + B\bar{C}}$

Solution. (*a*) The function, $f = \overline{AB\,(C+D)}$ can be realized by using two AND gates, one OR gate and one NOT gate as shown in Fig. 3.58(*a*).

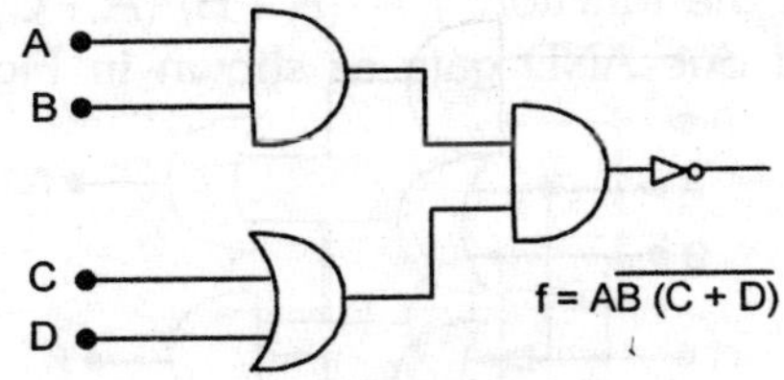

Fig. 3.58(*a*)

(*b*) The function, $f = \overline{A + B\overline{C}}$ can be realized by using one AND gate, one OR gate and two NOT gates as shown in Fig. 3.58(*b*).

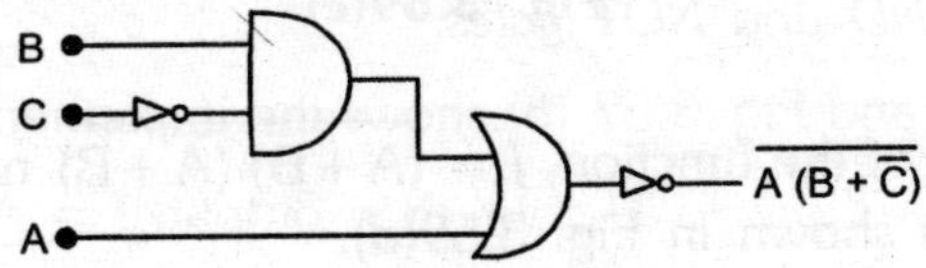

Fig. 3.58(*b*)

Example 3.15. *Construct the logic diagram for the given Boolean expressions.*

(*a*) $f = (A+B)\,(\overline{A}+B)\,(A+\overline{B}+C)$ (*b*) $\overline{(A+\overline{B})\,(\overline{A}+B)}$

(*c*) $(A+B)\,(\overline{A}+C)\,(\overline{B}+\overline{C})$ (*d*) $\overline{(A+\overline{B})\,(\overline{A}+B)}$ (*e*) $(A+B)\,(\overline{A}+C)$

Solution. (*a*) To implement the function, $f = (A+B)\,(\overline{A}+B)\,(A+\overline{B}+C)$, we need 3 OR gates, one AND gate and two NOT gates as shown in Fig. 3.59(*a*).

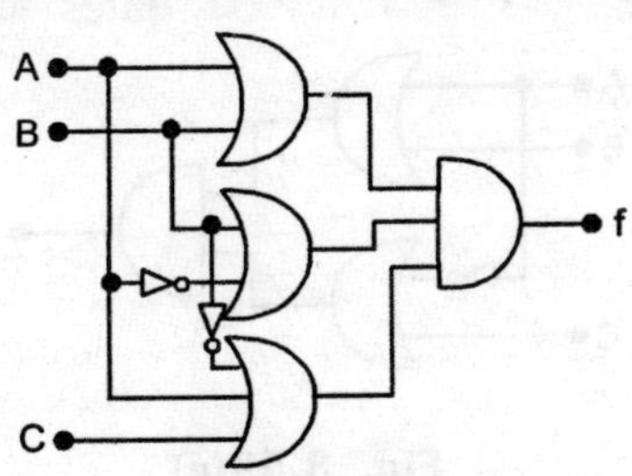

Fig. 3.59(*a*)

(*b*) To implement the function, $f = \overline{(A+\overline{B})}\,(\overline{A}+B)$, we need three NOT gates, 2 OR gates and one AND gate shown in Fig. 3.59(*b*).

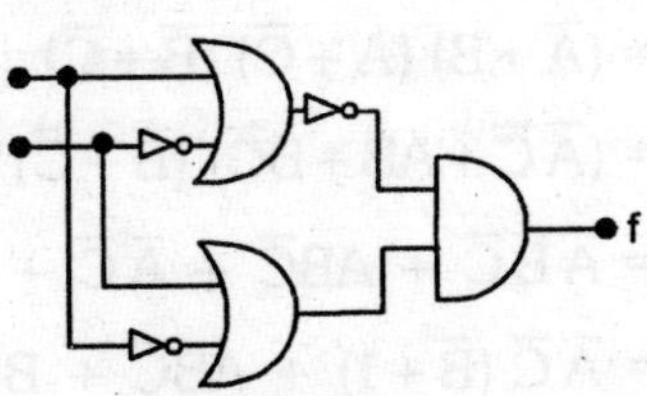

Fig. 3.59(*b*)

(*c*) The implementation of the function, $f = (A+B)\,(\overline{A}+C)\,(\overline{B}+\overline{C})$ can be done by using 3 OR gates, 3 NOT gates and one AND gate as shown in Fig. 3.59(*c*).

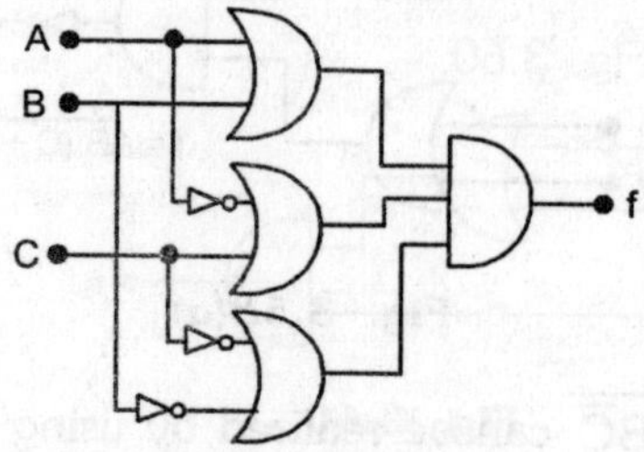

Fig. 3.59(*c*)

(*d*) The implementation of the function, $f = \overline{(A+\overline{B})\,(\overline{A}+B)}$ needs two OR gates, two NOT gates and one AND gate as shown in Fig. 3.59(*d*).

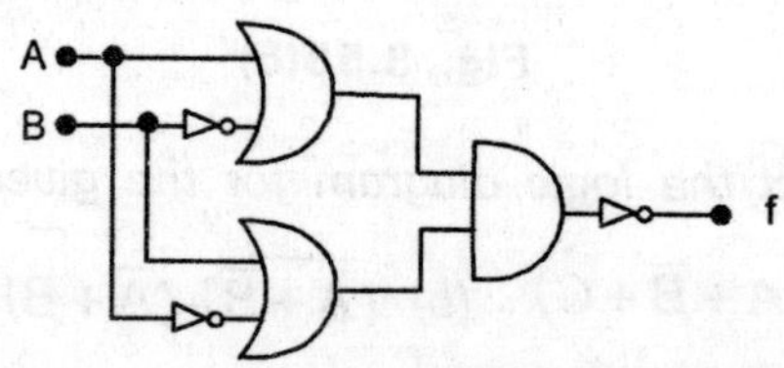

Fig. 3.59(*d*)

(*e*) The function, $f = (A+B)\,(\overline{A}+C)$ can be implemented using two OR gates, one AND gate and one NOT gate as shown in Fig. 3.59(*e*).

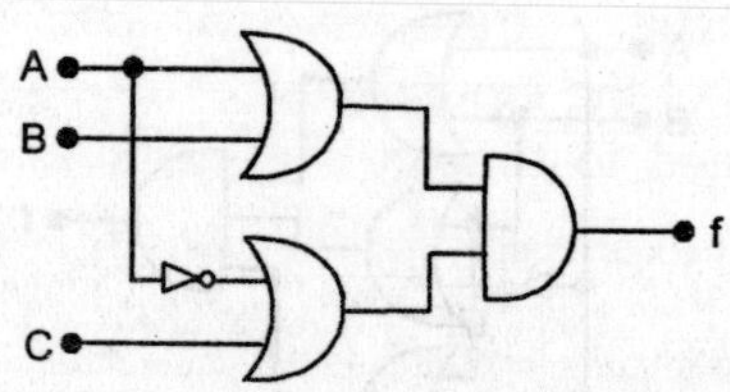

Fig. 3.59(*e*)

Example 3.16. *Simplify the following Boolean equation and realize the same using combinational gates.*

$$f = (\overline{A}+B)\,(A+\overline{C})\,(\overline{B}+\overline{C})$$

Solution.

$$f = (\overline{A}+B)\,(A+\overline{C})\,(\overline{B}+\overline{C})$$

$$= (\overline{A}\,\overline{C}+AB+B\overline{C})\,(\overline{B}+\overline{C}) \qquad \text{since } A.\overline{A}=0$$

$$= \overline{A}\,\overline{B}\,\overline{C} + AB\overline{C} + \overline{A}\,\overline{C} + B\overline{C} \qquad \text{since } B.\overline{B}=0 \text{ and } \overline{C}.\overline{C}=\overline{C}$$

$$= \overline{A}\,\overline{C}\,(\overline{B}+1) + AB\overline{C} + B\overline{C}$$

$$= \overline{A}\,\overline{C} + B\overline{C}\,(A+1)$$

$$= \overline{C}\,(\overline{A} + B)$$

The logic diagram is shown in Fig. 3.60.

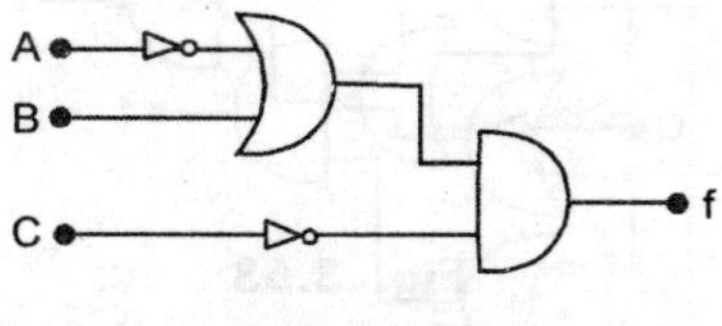

Fig. 3.60

Example 3.17. *Draw the output wave for the circuit shown in Fig. 3.61.*

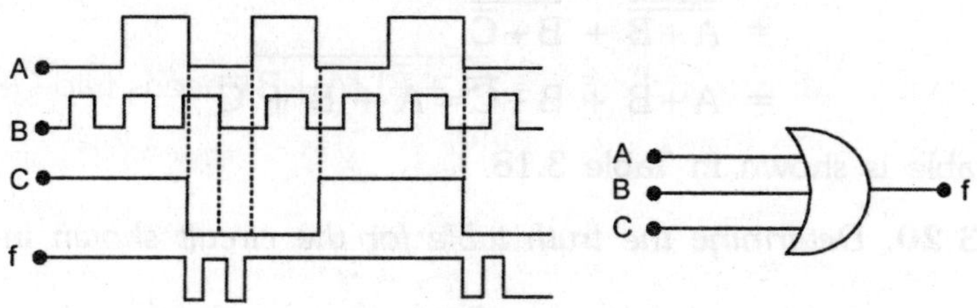

Fig. 3.61

Solution. The output is shown in the Fig. 3.61.

3.5 ANALYSIS OF CIRCUITS

A combinational circuit is analysed by proceeding from it's input terminals through to it's output terminal. The Boolean equation of each gate is written down and combined to obtain an equation for the output of the circuit. A truth table will then give the output of the circuit for each of the possible combinations of the input signals. The logic function at the intermediate points in the circuit can be determined to simplify the determination of the output function.

Example 3.18. *Write a Boolean expression for the output, f in the Fig. 3.62 and determine the truth table.*

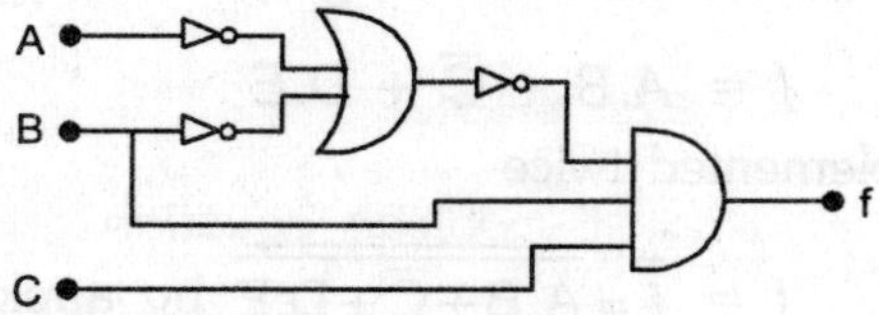

Fig. 3.62

Solution. The Boolean expression for the output, *f* is

$$f = \overline{(\overline{A}+\overline{B})}\,BC$$

$$= \overline{\overline{A}}\,\overline{\overline{B}}.BC = AB.BC = ABC$$

Truth Table

A	B	C	F
0	0	0	0
0	0	1	0
0	1	0	0
0	1	1	0
1	0	0	0
1	0	1	0
1	1	0	0
1	1	1	1

Example 3.19. *Determine the truth table for the circuit shown in Fig. 3.63.*

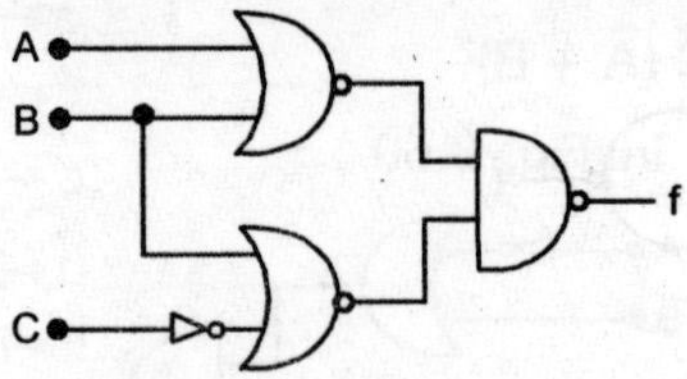

Fig. 3.63

Solution. $f = \overline{\overline{A+B} \,.\, \overline{B+\overline{C}}}$

$= \overline{\overline{A+B}} + \overline{\overline{B+\overline{C}}}$

$= A+B + B+\overline{C} = A + B + \overline{C}$

The truth table is shown in Table 3.18.

Table 3.18

A	B	C	*f*
0	0	0	1
0	0	1	0
0	1	0	1
0	1	1	1
1	0	0	1
1	0	1	1
1	1	0	1
1	1	1	1

Example 3.20. *Determine the truth table for the circuit shown in Fig. 3.64.*

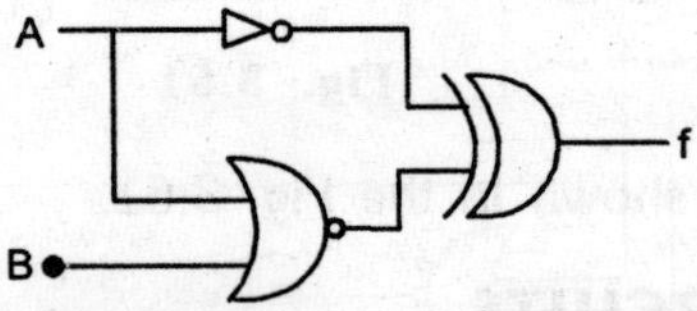

Fig. 3.64

Solution. The $f = \overline{A} \oplus \overline{A+B}$

$= A\,(\overline{A+B}) + (A+B).A'$

$= A\,(\overline{A}.\overline{B}) + A'B = A'B$

The truth table is shown in Table 3.19.

Table 3.19

A	B	*f*
0	0	0
0	1	1
1	0	0
1	1	0

Example 3.21. *Implement the function using (a) NAND gates and (b) NOR gates only.*

$$f = A.B + \overline{C} + D.E$$

Solution. (*a*) *f* is complemented twice

$f = \overline{\overline{f}} = \overline{\overline{A.B+\overline{C}+D.E}}$ by applying De Morgan's theorem

$$f = \overline{\left(\overline{(A.B)}.\overline{\overline{C}}.\overline{D.E}\right)}$$

$$= \overline{\overline{A.B}.C.\overline{DE}}$$

This can be realized as shown in Fig. 3.65(*a*)

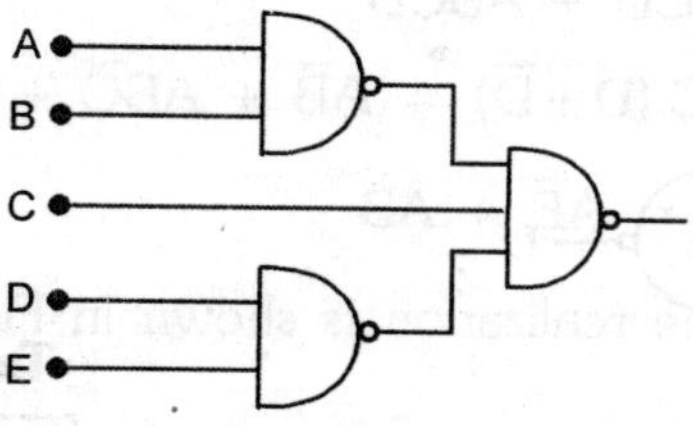

Fig. 3.65(a)

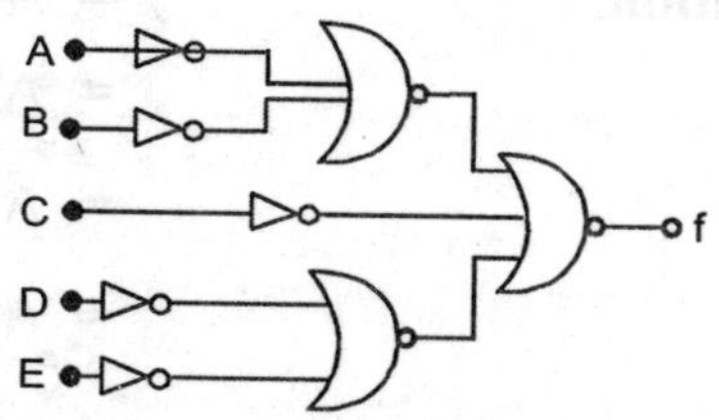

Fig. 3.65(b)

(*b*)
$$f = \overline{\overline{A.B}} + \overline{C} + \overline{\overline{D.E}}$$
$$= \overline{\overline{A}+\overline{B}} + \overline{C} + \overline{\overline{D}+\overline{E}}$$

This can be realized as shown in Fig. 3.65(*b*).

Example 3.22. *For the combinational logic circuit shown in Fig. 3.66, obtain the simplified expressions for the input variables V, W, X, Y and Z in terms of A, B and C.*

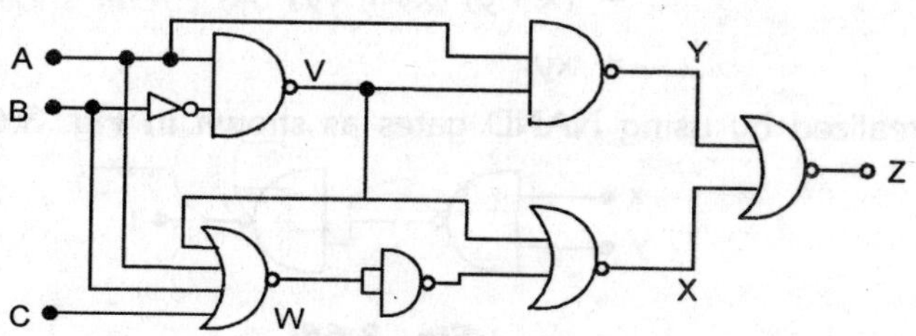

Fig. 3.66

Solution.
$$V = \overline{A\overline{B}} = \overline{A} + \overline{\overline{B}} = \overline{A} + B$$
$$W = \overline{(\overline{A}+B) + A + B + C}$$
$$= \overline{\overline{A}+B} \,.\, \overline{A}.\overline{B}.\overline{C}$$
$$= (A.\overline{B})\, \overline{A}.\overline{B}.C = 0$$
$$X = \overline{V+\overline{W}} = \overline{V}\,.\,W = 0$$
$$Y = \overline{V.A} = \overline{A.(\overline{A}+B)} = \overline{A} + A\overline{B}$$
$$= \overline{A} + \overline{B}$$
$$Z = \overline{X+Y} = \overline{0+\overline{A}+\overline{B}}$$
$$= A.B.1$$
$$= AB.$$

Example 3.23. *Simplify the following logic expression and realize it using NAND gates.*
$$f = A\overline{B} + AB\overline{C} + ABCD + ABC\overline{D}$$

Solution.

$$f = A\overline{B} + AB\overline{C} + ABCD + ABC\overline{D}$$
$$= A\overline{B} + AB\overline{C} + ABC\,(D+\overline{D}) = A\overline{B} + AB\overline{C} + ABC$$
$$= A\overline{B} + AB\,(C+\overline{C}) = A\overline{B} + AB$$
$$= A\,(B+\overline{B}) = A$$

The realization is shown in Fig. 3.67.

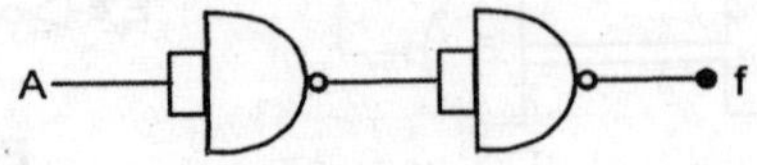

Fig. 3.67

Example 3.24. *Simplify the Boolean expression* $f = (x+y)\,(x+\overline{y})\,(\overline{x}+y)$ *and realize using NAND gates only.*

Solution.

$$f = (x+y)\,(x+\overline{y})\,(\overline{x}+y)$$
$$= (x+y)\,(x\overline{x} + \overline{x}\,\overline{y} + xy + y\overline{y})$$
$$= (x+y)\,(xy + \overline{x}\,\overline{y})$$
$$= xy$$

This can be realized by using NAND gates as shown in Fig. 3.68.

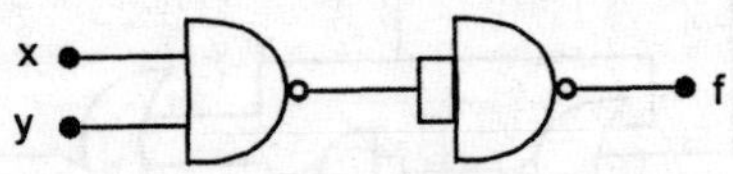

Fig. 3.68

Example 3.25. *Find the Boolean function for* f_1 *and* f_2 *in terms of A and B as shown in Fig. 3.69 and identify the circuit.*

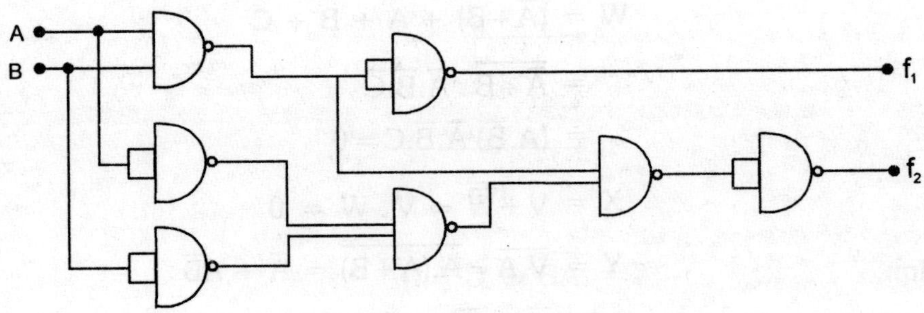

Fig. 3.69

Solution. First, we will give numbers to the gates in the circuit as shown in Fig. 3.70.

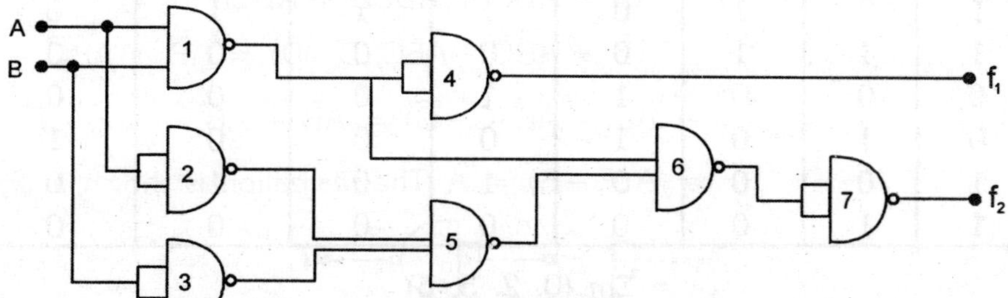

Fig. 3.70

The output of $NAND_1$ is $\overline{AB}$ and output of $NAND_4$ is $\overline{\overline{AB}} = AB$. $\therefore$ $\boxed{f_1 = AB}$

The output of $NAND_2$ and $NAND_3$ are $\overline{A}$ and $\overline{B}$ respectively.

$$\text{Output of } NAND_5 = \overline{\overline{A}.\overline{B}} = A + B$$

$$\text{Output of } NAND_6 = \overline{(A+B).(\overline{A}+\overline{B})}$$

$$= \overline{(A+B)} + \overline{(\overline{A}+\overline{B})}$$

$$= \overline{A}\,\overline{B} + AB$$

$$\text{Output of } NAND_7 = f_2 = \overline{\overline{A}\,\overline{B} + AB} = \overline{A}B + A\overline{B}$$

$\therefore$ The circuit functions like a half adder,

where f_2 = SUM and f_1 = CARRY.

Example 3.26. *Determine the logic expression of the output f for the circuit shown in Fig. 3.71 and also draw the truth table.*

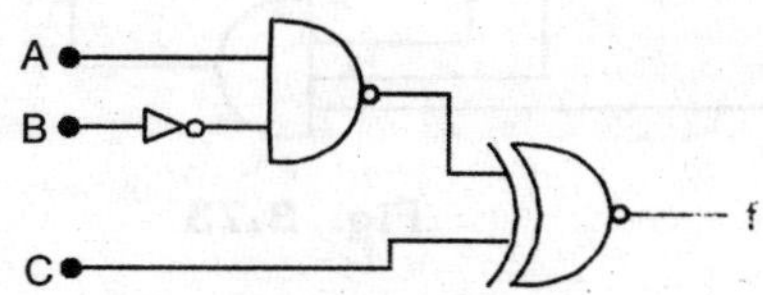

Fig. 3.71

Solution.

$$f = \overline{A\overline{B}} \oplus C = (\overline{A}+B) \oplus C$$

$$= \overline{C}\,(\overline{A}+B) + \overline{(\overline{A}+B)}.C$$

$$= \overline{A}\,\overline{C} + B\overline{C} + A\overline{B}.C$$ The truth table is shown in Table 3.20.

Table 3.20

A	B	C	$\overline{A}$	$\overline{B}$	$\overline{C}$	$\overline{A}\,\overline{C}$	$B\overline{C}$	$A\overline{B}C$	f
0	0	0	1	1	1	1	0	0	1
0	0	1	1	1	0	0	0	0	0

0	1	0	1	0	1	1	1	0	1
0	1	1	1	0	0	0	0	0	0
1	0	0	0	1	1	0	0	0	0
1	0	1	0	1	0	0	0	1	1
1	1	0	0	0	1	0	1	0	1
1	1	1	0	0	0	0	0	0	0

$$\therefore\ f = \Sigma m\ (0, 2, 5, 6)$$

Example 3.27. *Determine the output expression for the circuit shown in Fig. 3.72.*

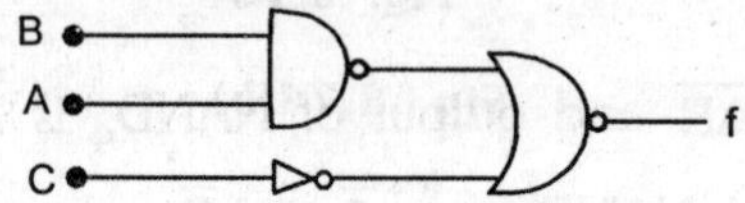

Fig. 3.72

Solution. The output of NAND gate is $\overline{AB}$

The output of NOR gate is $\overline{\overline{AB}+C} = f$

$$\therefore\ f = \overline{\overline{AB}}.\overline{C} = AB\overline{C}.$$

Example 3.28. *Convert the given circuit shown in Fig. 3.73 using only NOR gates.*

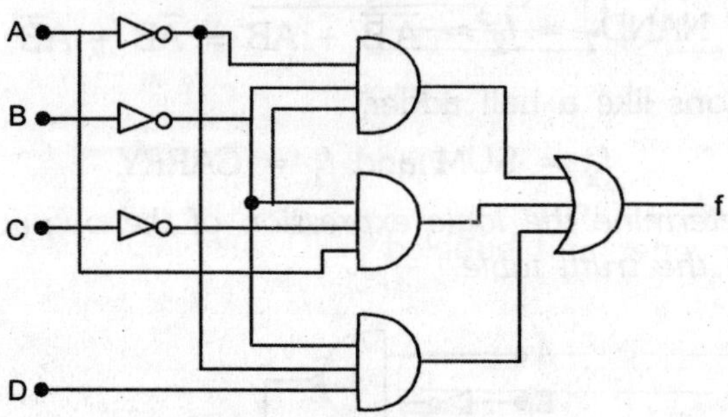

Fig. 3.73

Solution. (*i*) Adding the bubbles to the output of OR gate and the inputs of AND gates, the circuit is shown in Fig. 3.74(*a*).

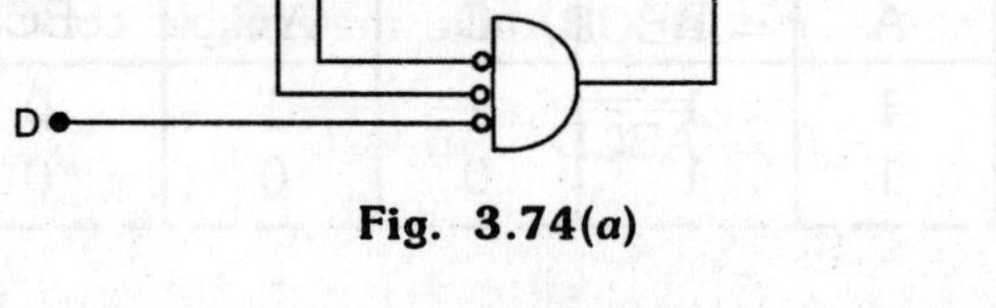

Fig. 3.74(*a*)

(*ii*) Adding an inverter on each line that has received bubble (see Fig. 3.74(*b*)).

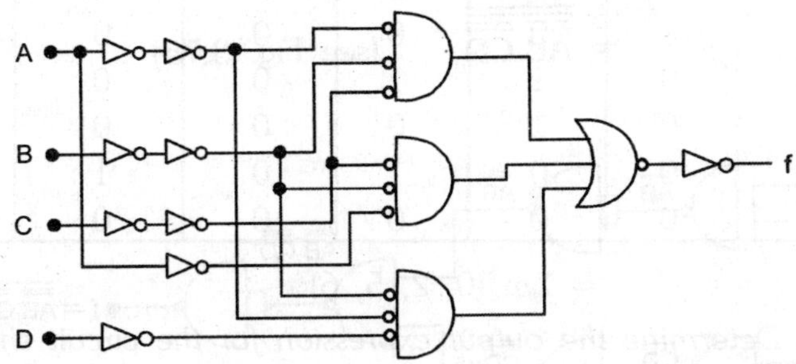

Fig. 3.74(*b*)

(*iii*) Eliminating the double inversion, we have (see Fig. 3.74(*c*)).

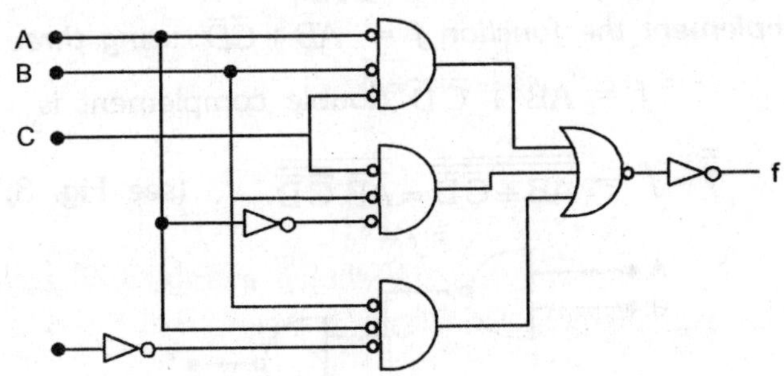

Fig. 3.74(*c*)

(*iv*) Replacing the inverter and bubbled AND with NOR (see Fig. 3.74(*d*)).

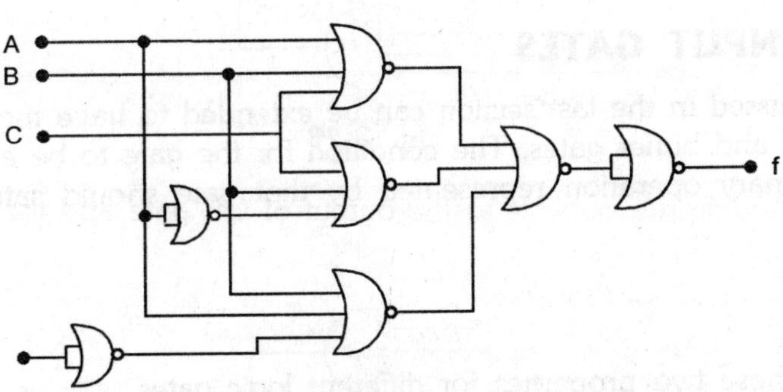

Fig. 3.74(*d*)

Example 3.29. *Implement f = ABCD using two-input NAND gates.*

Solution. f = ABCD, take the double complement

$$\bar{\bar{f}} = f = \overline{\overline{ABCD}} = \overline{\overline{AB} + \overline{CD}}$$

$$= \overline{\overline{AB}}.\overline{\overline{CD}}$$

$$= \overline{\overline{\overline{AB}.\overline{CD}}} \quad \text{(see Fig. 3.75)}$$

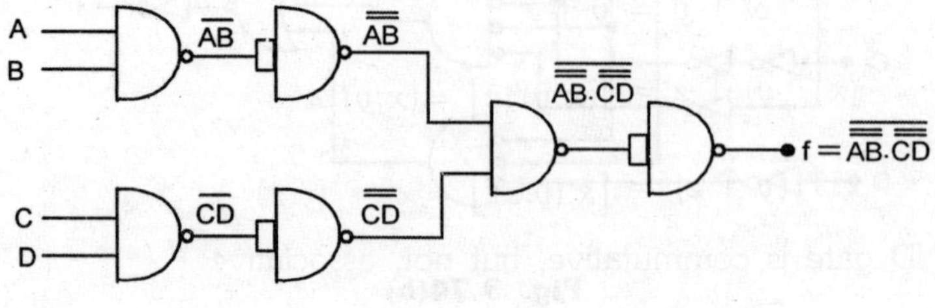

Fig. 3.75

Example 3.30. *Implement the function* $f = AB + \overline{C}\overline{D}$ *using three NAND gates only.*

Solution. $f = AB + \overline{C}\,\overline{D}$ double complement is

$$\overline{\overline{f}} = f = \overline{\overline{AB + \overline{C}\,\overline{D}}} = \overline{\overline{AB}.\overline{\overline{C}\,\overline{D}}} \quad \text{(see Fig. 3.76)}$$

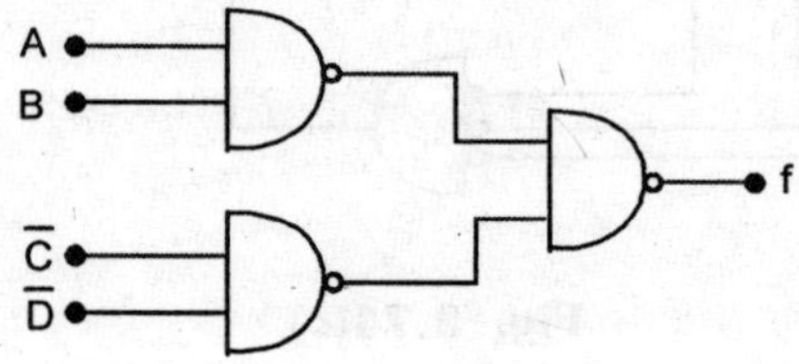

Fig. 3.76

3.6 MULTIPLE INPUT GATES

All the gates discussed in the last section can be extended to have more than two inputs, except for the inverter and buffer gates. The condition for the gate to be extended to multiple inputs, is that, the binary operation represented by that gate should satisfy two properties, which are

1. Commutative
2. Associative

Let us examine these two properties for different logic gates.

(*a*) **For the OR gate,**

$$x + y = y + x \text{ [Commutative]}$$

and $$(x + y) + z = x + (y + z) = x + y + z. \text{ [Associative]}$$

Hence, the OR gate satisfies these 2 properties.

(*b*) **For AND gate,**

$$x \cdot y = y \cdot x$$
$$(x \cdot y) \cdot z = x \cdot (y \cdot z) = x \cdot y \cdot z$$

Hence, the above 2 gates can be extended to 3 or more variables.

(*c*) **For NAND gate,**

$$x \uparrow y = y \uparrow x$$

and
$$(x \uparrow y) \uparrow z = \left[(x.y)'.z\right]' = (x.y) + z'$$
$$x \uparrow (y \uparrow z) = \left[x.(y.z)'\right]' = x' + (y.z)$$

$\therefore$ The NAND gate is commutative, but not, associative.

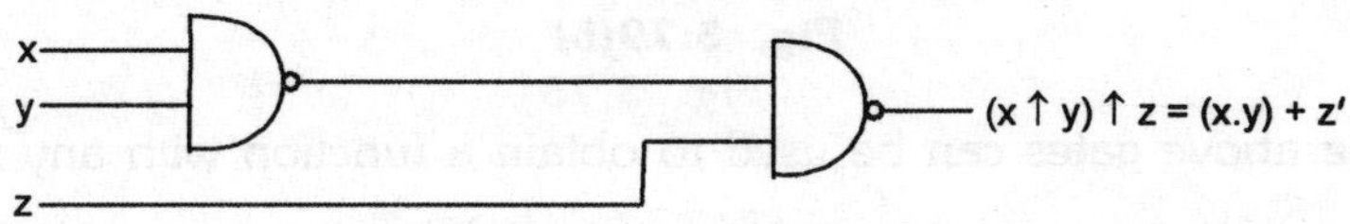

Fig. 3.77(*a*)

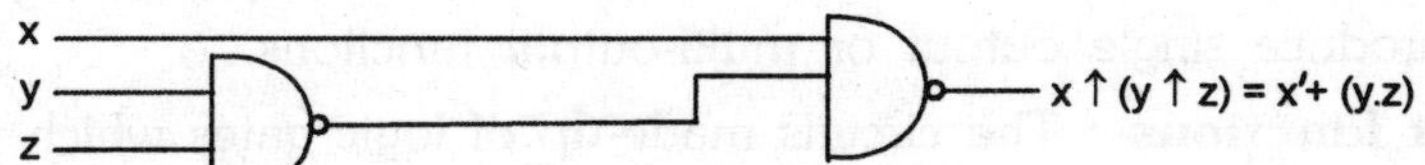

Fig. 3.77(*b*)

(*d*) **For NOR gate,**

$$x \downarrow y = y \downarrow x$$

and,
$$(x \downarrow y) \downarrow z = \left[(x+y)'+z\right]' = (x+y).z' = x.z' + y.z'$$
$$x \downarrow (y \downarrow z) = \left[x+(y+z)'\right]' = x'.(y+z) = x'.y + x'.z$$

$\therefore$ The NOR gate is commutative, but not associative.

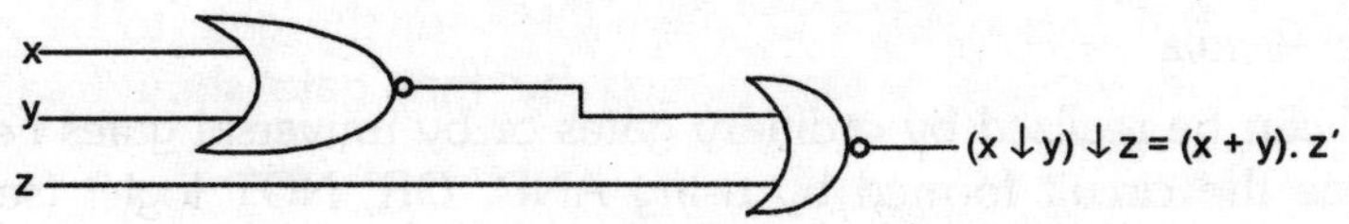

Fig. 3.78(*a*)

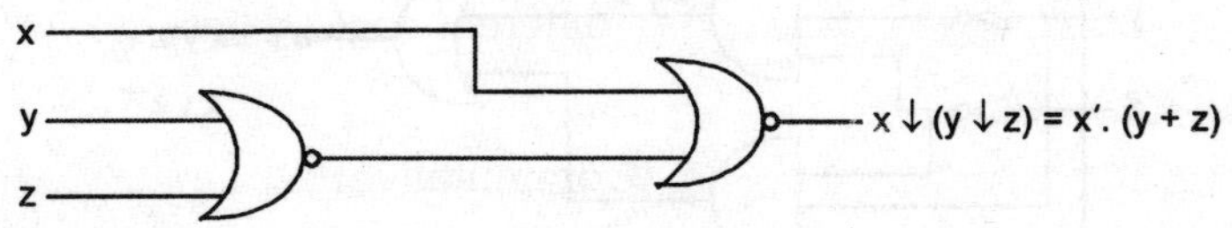

Fig. 3.78(*b*)

To overcome this difficulty, the multiple input NOR gate is defined as complemented OR gate, and the NAND gate is defined as complemented AND gate. Hence, we have,

$x \downarrow y \downarrow z = (x + y + z)'$ (see Fig. 3.79 (*a*)

and, $x \uparrow y \uparrow z = (x \cdot y \cdot z)'$ (see Fig. 3.79 (*b*)

Three input NOR gate :

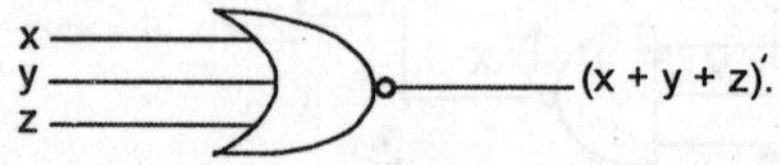

Fig. 3.79(*a*)

Three input NAND gate :

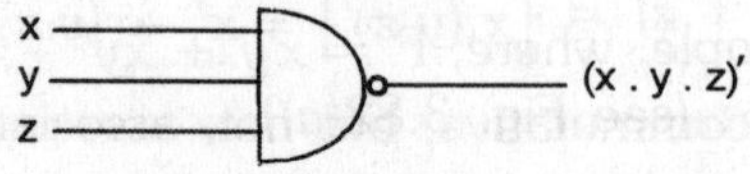

Fig. 3.79(*b*)

Similarly, all the above gates can be used to obtain a function with any number of inputs.

3.7 SINGLE OUTPUT AND MULTI-OUTPUT FUNCTIONS

The logic gates can be used to obtain the functions in Boolean algebra. The circuits obtained so may produce single output or multi-output functions.

Single output functions : The circuits made-up of logic gates which produce only one function at the output, with any number of input variables is called as a single output function circuit. It can be represented diagramatically as below (see Fig. 3.80).

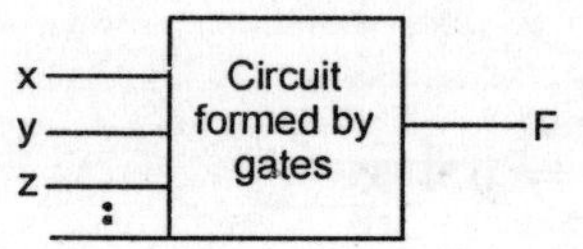

Fig. 3.80. Block diagram of a single input function.

Here, *x*, *y*, *z* and so on represent the inputs to the circuit which is formed by using logic gates. F is the single function which can be obtained at the output of the circuit.

Now, let us consider a few examples of the single output functions.

1. $F = x'\ y'\ z' + xyz'$

This function F can be realized by ordinary gates or by universal gates *i.e.*, NAND or NOR gates. Let us first see the circuit formed by using AND, OR, NOT logic. (see Fig. 3.81 (*a*)).

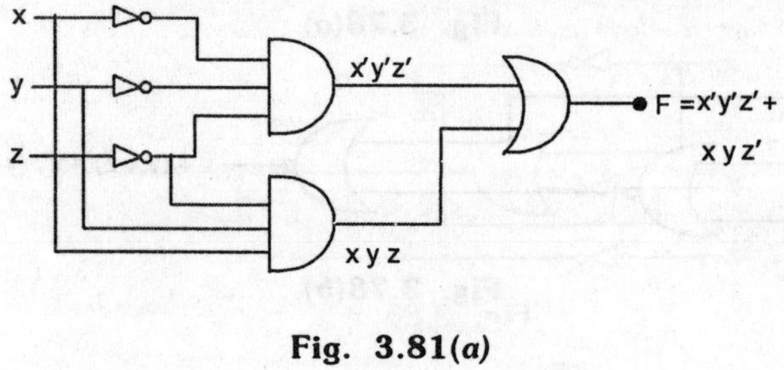

Fig. 3.81(*a*)

This can be realized by using NAND gates as below (see Fig. 3.81(*b*)).

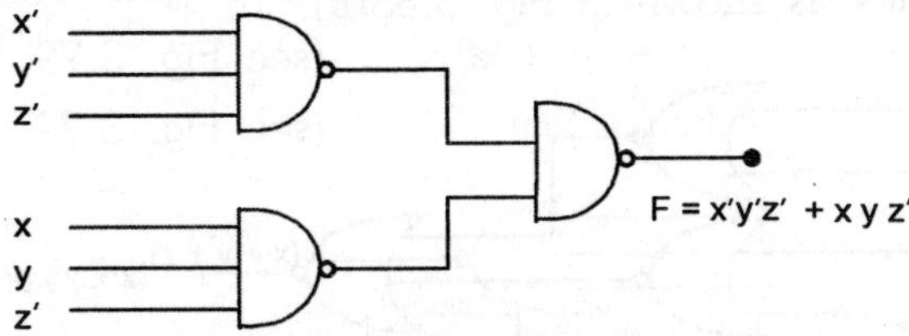

Fig. 3.81(*b*)

2. Consider another example, where, F = $x'y + xy' + z$. This function, F can be realized by AND, OR logic as follows : (see Fig. 3.82(*a*)).

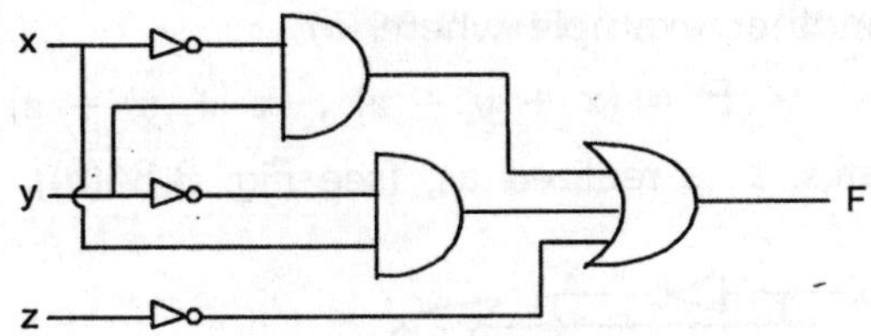

Fig. 3.82(*a*)

The realization of F by using NAND is as shown (Fig. 3.82(*b*)).

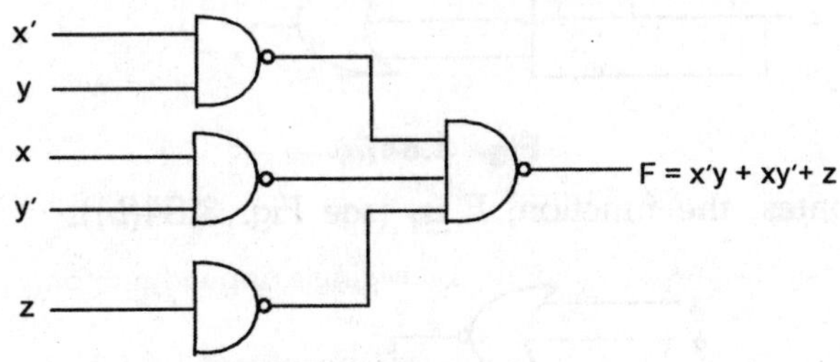

Fig. 3.82(*b*)

The above 2 examples are SOP forms. Even the POS forms are single output functions, which can be realized either by using OR-AND logic or NOR gates.

Let us consider a few examples of POS forms (see Fig. 3.83(*a*)).

1. F = $(x + y').(x' + y)\ z'$

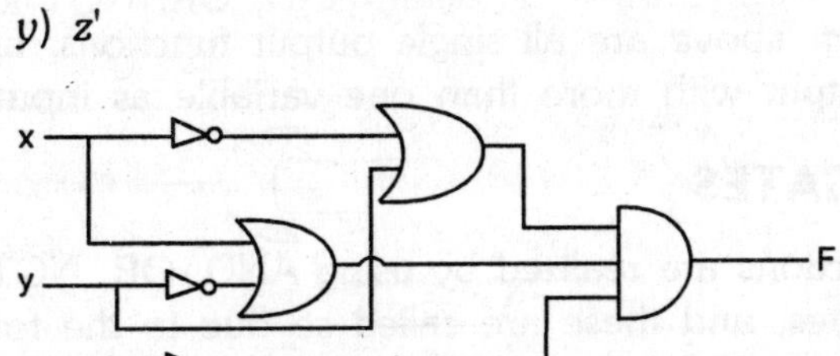

Fig. 3.83(*a*)

The above is the realization using AND, OR, NOT logic. The same function, F, can be realized by using NOR gates as shown in Fig. 3.83(*b*).

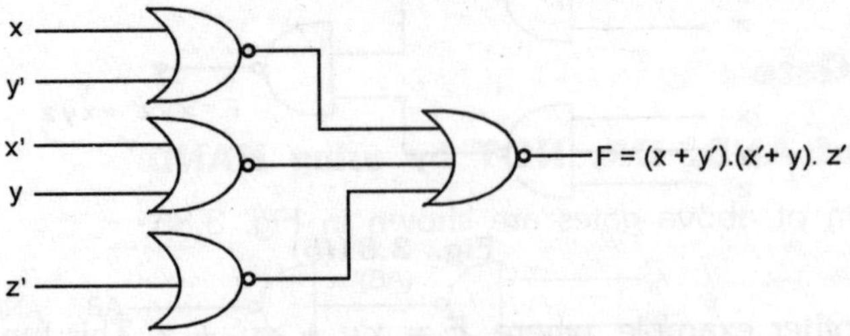

Fig. 3.83(*b*)

2. Let us consider another example where,

$$F = (x + y + z) \cdot (x' + y' + z)$$

By using ordinary gates, F is realized as, (see Fig. 3.84(*a*)).

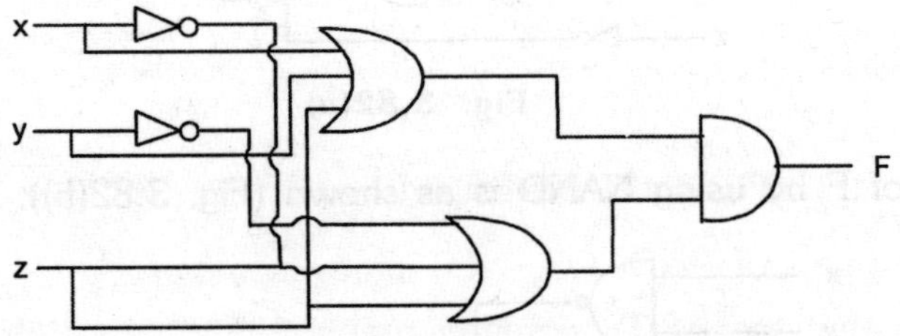

Fig. 3.84(*a*)

By using universal-gates, the function, F is, (see Fig. 3.84(*b*)).

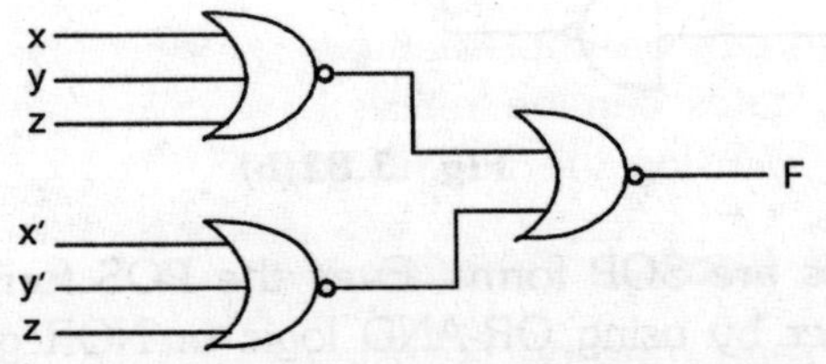

Fig. 3.84(*b*)

The examples shown above are all single output functions, as all the circuits produce a single function as an output with more than one variable as input.

3.8 UNIVERSAL GATES

All the Boolean functions are realized by using AND, OR, NOT logic gates. But, there are gates called universal gates, and these are called so due to the following reasons :

(*i*) Any logic gate, either it be AND, OR, NOT, can be realized by using the universal gate.

(*ii*) Any Boolean function can be realized by using these universal gates and (*iii*) Any digital circuit can be realized by using these gates.

NAND, NOR gates are called as universal gates. Now, let us examine how these gates are universal.

3.8.1 NAND Gate

(*i*) Realization of AND, OR, NOT by using NAND

The realization of above gates are shown in Fig. 3.85.

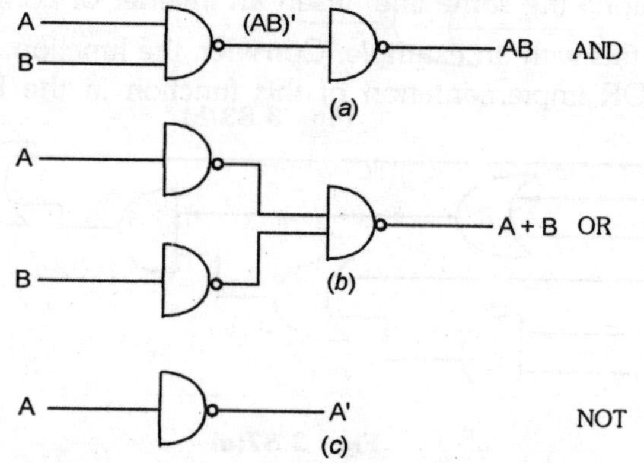

Fig. 3.85. Realization of AND, OR, NOT gates.

Thus, the realization of AND, OR, NOT gates is shown.

It can be seen that the NOT operation requires only one NAND gate.

The AND gate requires 2 NAND gates. The first produces the inverted AND, and the second acts as an inverter to produce the normal output.

The OR gate is realized by a NAND gate, with additional inverters in each input.

Any Boolean function can be implemented by using AND, OR, NOT logic gates. But, we have seen that all these gates can be realized by using NAND gates. Hence, any Boolean function can be realized by using the NAND gates.

(*ii*) Implementation of Boolean function

To implement a Boolean function by using NAND logic, it is first necessary to obtain a simplified Boolean function in terms of Boolean operators, and then convert the function to the NAND logic. It is convenient to use 2 alternate graphic symbols to facilitate the conversion to NAND logic as below. The *AND-INVERT* graphic symbol consists of an AND graphic-symbol followed by a small-circle. The *INVERT-OR* graphic-symbol (see Fig. 3.86*a*) consists of an OR graphic symbol that is preceeded by small circles in all inputs. Any of these symbols can be used to represent a NAND-gate. (See Fig. 3.86(*b*))

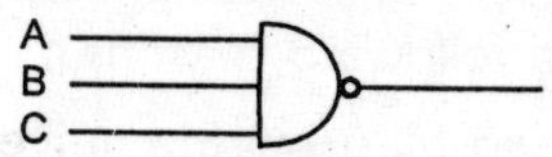

(*i*) *AND-INVERT*

Fig. 3.86(*a*)

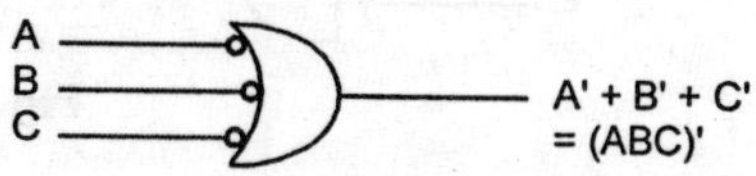

(*ii*) *INVERT-OR*

Fig. 3.86(*b*)

The steps to be followed to obtain a multilevel NAND-diagram from a Boolean-expression are as follows :

(*i*) From the given Boolean expression, draw the logic diagram with AND, OR and inverter gates. Assume that both the normal and the complement inputs are available.

(*ii*) Convert all AND gates to NAND gates with AND–invert symbols.

(*iii*) Convert all OR gates to NAND gates with invert–OR symbols.

(*iv*) Check all small circles in the diagram. For every small circle that is not compensated by another small circle along the same line, insert an inverter or complement the input variable.

Let us demonstrate this with an example. Consider, the function, F = A + (B'+C).(D'+BE'). First, we see the AND-OR implementation of this function in the Fig. 3.87.

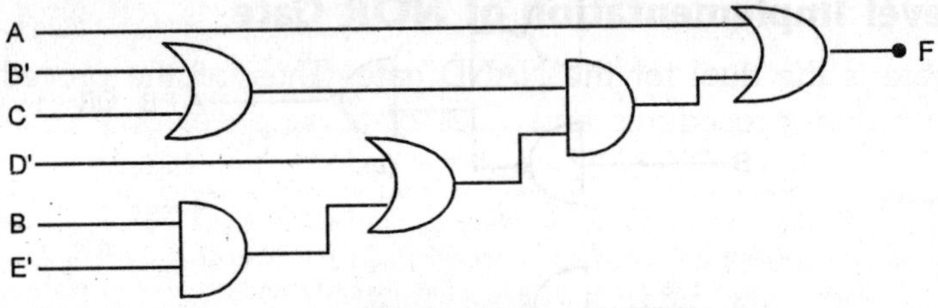

Fig. 3.87(*a*)

This circuit is a multilevel implementation. The above logic diagram can be converted into a NAND circuit. The procedure requires that we change every AND gate to an AND-invert graphic symbol and every OR gate to an invert-OR graphic-symbol. The NAND circuit performs the same logic as the AND–OR circuit as long as the complementing circles do not change the value of the function. Any connection between an output of a gate that has a complementing circle and the input of another gate that also has a complementing circle, represents double complementation and does not change the logic of the circuit. However, the small circles associated with inputs, A, B', C and D' cause extra complementations that are not compensated with other small circles along the same line. We can insert inverters after each of these inputs, or, complement the literals to obtain A', B, C' and D. Thus, the NAND implementation using AND-invert and invert-OR is as below (see Fig. 3.87(*b*)).

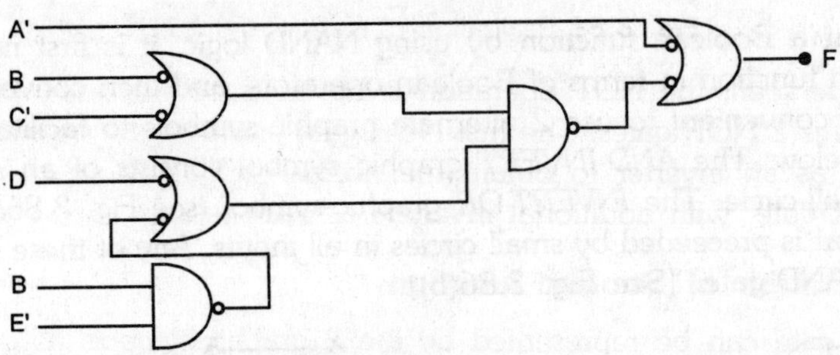

Fig. 3.87(*b*)

Replacing the AND-invert and invert-OR graphic symbols with their equivalent NAND gates, we have, F as (see Fig. 3.87(c)).

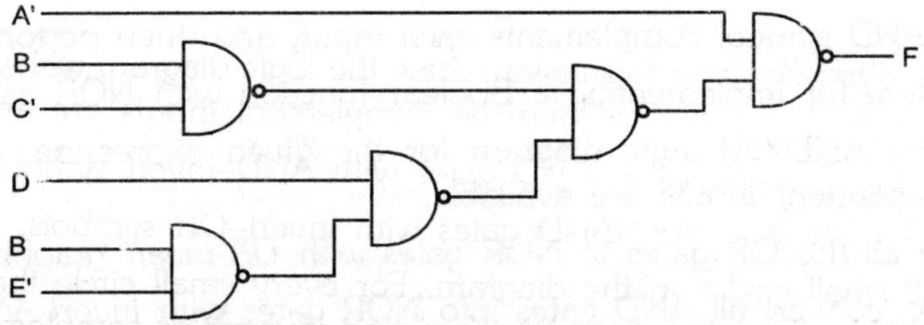

Fig. 3.87(*c*)

Hence, the NAND gate has satisfied all the properties to be called as a *universal gate.*

3.8.2 Multilevel Implementation of NOR Gate

The NOR gate is the dual for the NAND gate. Thus, all the procedures and the rules for NOR logic are the dual procedures and rules of those developed for NAND logic.

(*i*) Universal gate

The NOR gate is universal because any Boolean function can be implemented with it. The implementation of AND, OR, NOT gates by using, NOR logic is as shown in Fig. 3.88.

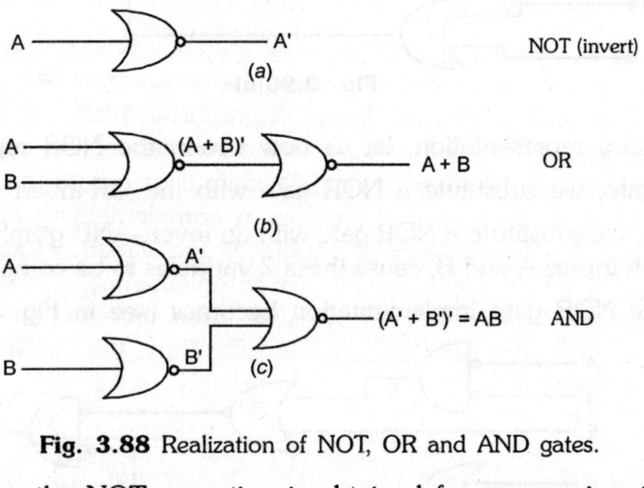

Fig. 3.88 Realization of NOT, OR and AND gates.

As it can be seen, the NOT operation is obtained from a one-input NOR gate. The OR operation requires 2 NOR gates. The first NOR gate produces the inverted-OR, and the second NOR gate acts as an inverter to obtain the normal output. The AND operation is obtained through a NOR gate, with additional inverters at each input.

(*ii*) Implementation of Boolean-functions

The NOR gate can be represented by the 2 graphic-symbols as below: (see Fig. 3.89)

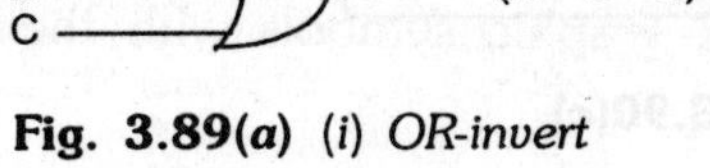

Fig. 3.89(*a*) (*i*) *OR-invert*

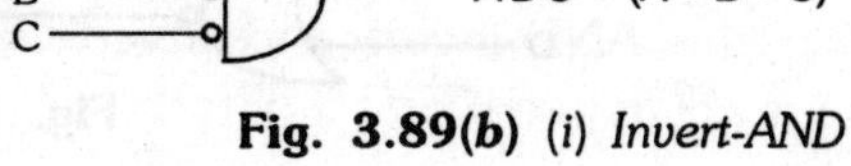

Fig. 3.89(*b*) (*i*) *Invert-AND*

The OR-invert defines the NOR operation as an OR followed by a complement.

The invert-AND symbol complements each input, and, then performs an AND operation.

The procedure for implementing a Boolean-function with NOR gates is as follows :

(*i*) Draw the AND-OR logic diagram for the given expression. Assume that both the normal and complement inputs are available.

(*ii*) Convert all the OR gates to NOR gates with *OR-invert* graphic symbols.

(*iii*) Similarly, convert all AND gates into NOR gates with *invert-AND* graphic symbols.

(*iv*) Any small circle that is not compensated by another small circle along the same line needs an inverter or the complementation of the input variable.

Consider, the Boolean-function given by,

$$F = (AB + E) . (C + D)$$

Let us now draw the AND-OR implementation of the above function is shown in Fig. 3.90(*a*).

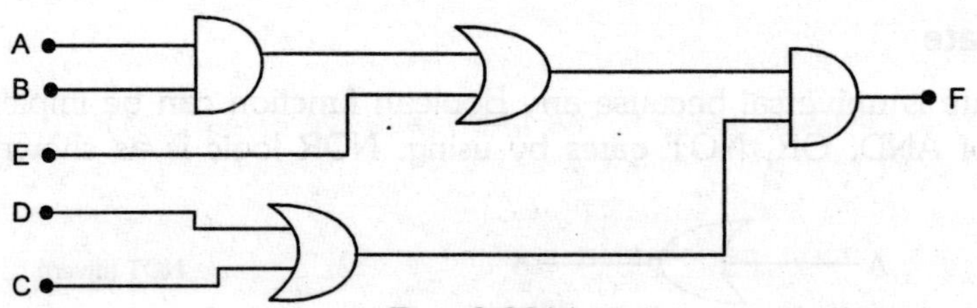

Fig. 3.90(*a*)

From this ordinary representation, let us now obtain the NOR representation.

For each OR gate, we substitute a NOR gate with the OR-invert graphic-symbol.

Each AND gate, we substitute a NOR gate with an invert-AND graphic symbol. The 2 small circles associated with inputs A and B, cause these 2 variables to be complemented to A' and B'.

Hence, now the NOR gate implementation becomes (see in Fig. 3.90(*b*)) :

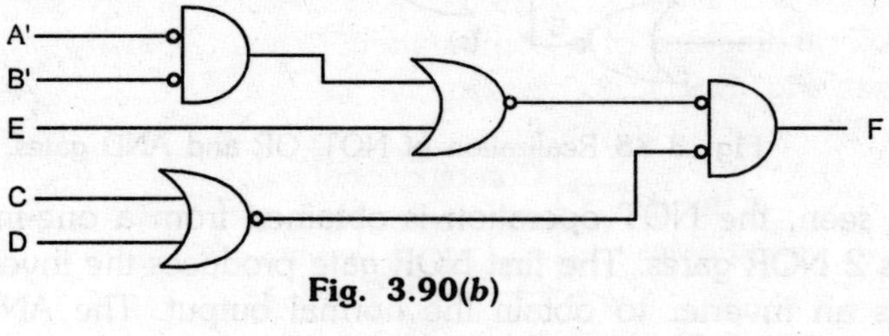

Fig. 3.90(*b*)

Making use of only the NOR gates, the circuit can be redrawn as shown in Fig. 3.90(*c*).

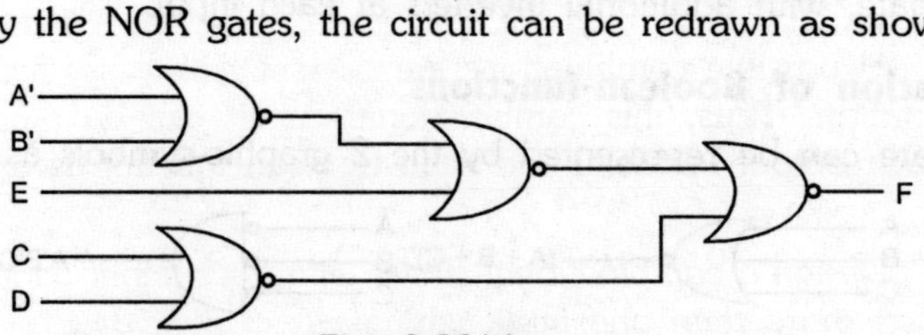

Fig. 3.90(*c*)

3.9 DESIGN OF COMBINATIONAL CIRCUITS

The design of combinational-circuits ends in a set of Boolean functions from which the logic diagram can be obtained. The design involves the following steps :

1. The first step is the determination of the number of available input variables and required output variables.
2. The input and output variables are assigned letter symbols.
3. The truth table is formed which defines relationships between the input and output variables.
4. The simplified Boolean function for each output is obtained.
5. The logic diagram is drawn.

The Boolean functions thus obtained can be simplified by any of the methods like algebraic manipulation, map method or tabulation procedure. However, the finally obtained design has to satisfy the criteria like : minimum number of gates, minimum number of inputs to a gate, minimum propagation time of the signal through the circuit, minimum number of interconnections and limitations of driving capabilities of each gate.

The digital systems perform a number of information processing tasks. The basic arithmetic operations are also implemented by using these combinational circuits. Now, we proceed with the implementation of these basic functions at that instant. It is independent of the previous inputs. A combinational circuit performs a specific information processing operation which is specified by Boolean functions.

Whereas the circuits whose outputs depend not only on present inputs, but also on past inputs are known as sequential circuits. Here, the circuit behaviour must be specified by a time sequence of inputs and internal-states. Hence, here the memory elements are needed inorder to store the information of the previous states.

A combinational circuit, thus, consists of input variables, logic gates and the output variables. The logic gates accept the inputs and generate the signals which are given to outputs. Both the input and output signals are represented by the binary signals, *i.e.*, either a logic 0 or a logic 1.

Therefore, the block diagram of a combinational circuit with *n* inputs and *m* outputs can be represented as (see Fig. 3.91)

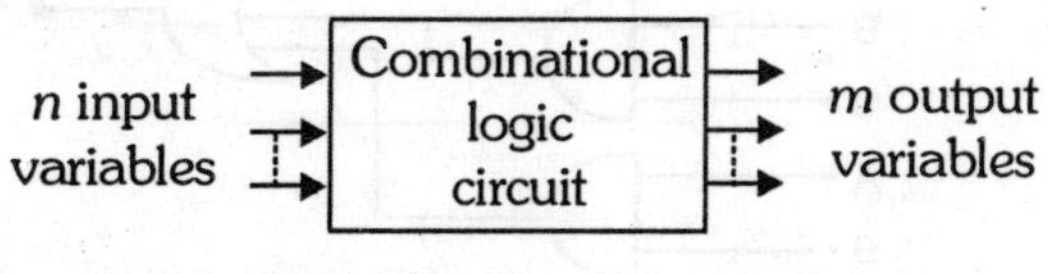

Fig. 3.91 Block diagram of a combinational circuit.

For *n* variables, 2^n possible combinations are present. For every input combination, there is an output combination. Thus, this combinational circuit can be described by *m* Boolean functions, one for each output variable, and each Boolean function is expressed in terms of *n* input variables.

Example 3.31. *Design a combinational circuit with three input variables that will produce a logic 1 output when more than one input variable is at logic 1.*

Solution. As per the given problem, there are three input variables named as A, B and C and one output variable *f*. The relationship between input variables and output variable can be tabulated as shown in truth Table 3.21. For the minterms 3, 5, 6, and 7, the output = 1 (since for these minterms the more than one input variable at logic 1).

Table 3.21

A	*B*	*C*	*f*
0	0	0	0
0	0	1	0
0	1	0	0
0	1	1	1
1	0	0	0
1	0	1	1
1	1	0	1
1	1	1	1

The simplification of the truth table for output '*f*' using K-map is shown in Fig. 3.92.

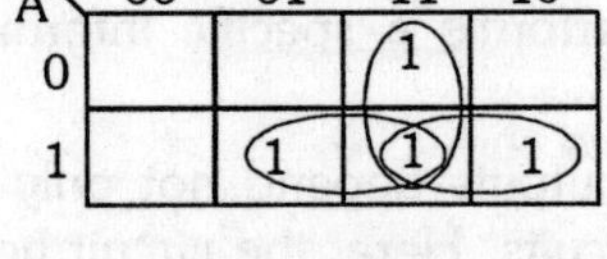

Fig. 3.92

f = AC + BC + AB

The logic diagram is shown in Fig. 3.93.

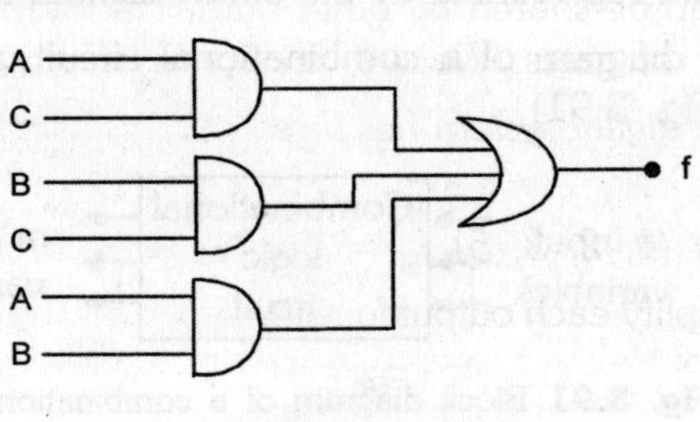

Fig. 3.93

Example 3.32. *Design a combinational circuit with three input variables whose output will produce 1 when the input value is even number.*

Solution. The given problem must have 3 input variables A, B and C and one output variable '*f*'. *f* = 1 for even input values. So, the function *f* = Σ (0, 2, 4, 6). The truth table is shown in Table 3.22.

Table 3.22

A	B	C	f
0	0	0	1
0	0	1	0
0	1	0	1
0	1	1	0
1	0	0	1
1	0	1	0
1	1	0	1
1	1	1	0

The simplification of the Table 3.22 is shown in Fig. 3.94 by using K-map.

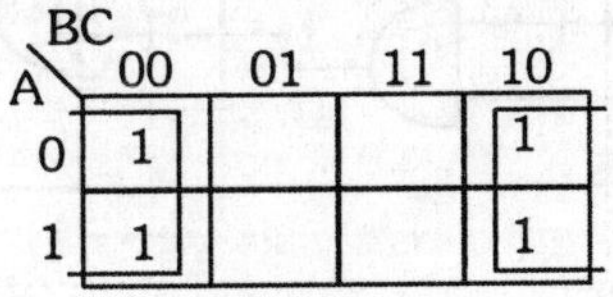

Fig. 3.94

$$\boxed{f = \overline{C}}$$

The realization of the circuit is shown in Fig. 3.95.

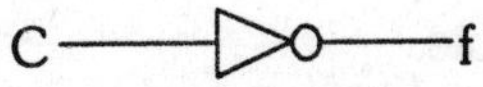

Fig. 3.95

3.9.1 Multiple Output Functions

In many digital applications, many times multiple outputs are derived from the same input variables. In this case, we can simplify and draw the logic diagram of each function separately. Sometimes, the simplified output functions may have common terms. The common term used by one output function can be shared by other output functions. This sharing of common terms reduces the total number of gates. We will see some of the examples for this.

Example 3.33. *The digital system has three outputs. Simplify and draw the logic diagrams for the output functions.*

$$f_1 = \Sigma m\ (1, 2, 3, 5)\ f_2 = \Sigma m\ (2, 3, 4, 5) \text{ and } f_3 = \Sigma\ (1, 3, 5, 7)$$

Solution. Let us simplify each output function separately by using K-map as shown in Fig. 3.96.

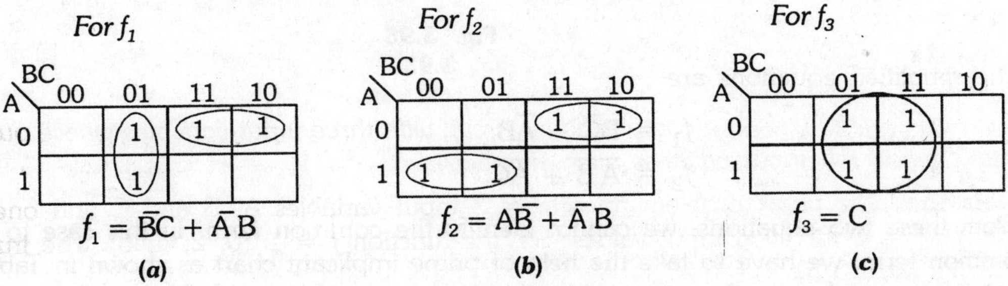

Fig. 3.96

The simplified equations are $f_1 = \overline{B}C + \overline{A}B$

$$f_2 = A\overline{B} + \overline{A}B$$

$$f_3 = C$$

In these equations, $\overline{A}B$ term is common in both f_1 and f_2. So, we can reduce one gate, by sharing the same product term by both functions f_1 and f_2. The logic diagram is shown in Fig. 3.97.

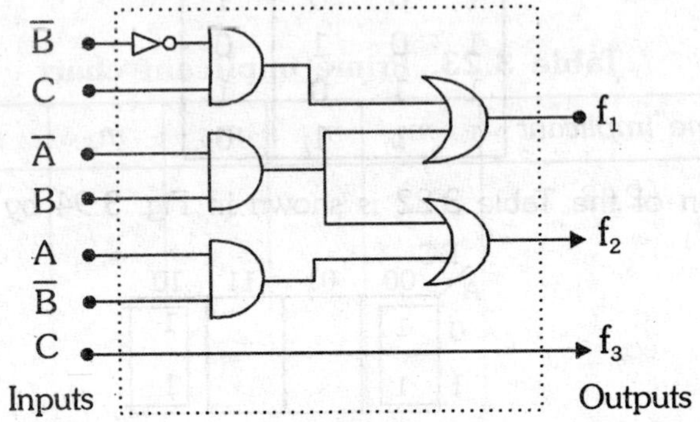

Fig. 3.97

In this example, we have identified easily the common term, but sometimes it is required to use special techniques to identify the common terms. This is explained in the following example.

Example 3.34. *Simplify the following functions and draw the logic diagram for the same.*

$$f_1 = f(A, B, C) = \Sigma(2, 3, 6)$$

$$f_2 = f(A, B, C) = \Sigma m(0, 1, 3)$$

Solution. Let us simplify each output function by the use of K-maps as shown in Fig. 3.98.

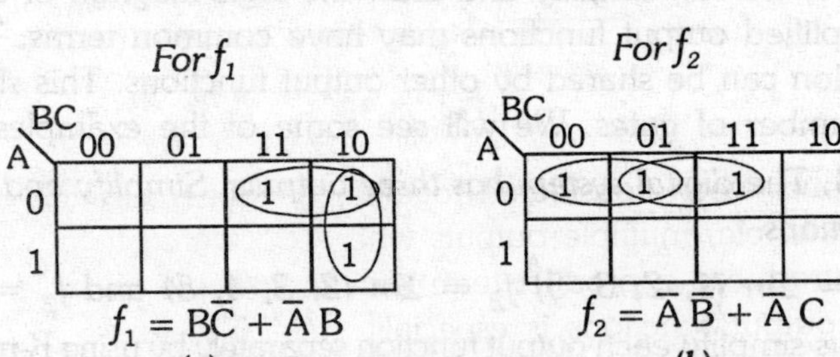

(a) (b)

Fig. 3.98

The simplified equations are

$$f_1 = B\overline{C} + \overline{A}B$$

$$f_2 = \overline{A}\,\overline{B} + \overline{A}C$$

From these two equations, we cannot identify the common term. In this case to identify the common term, we have to take the help of prime implicant chart as shown in Table 3.23.

It has totally four prime implicants *i.e.*, f_1 has two prime implicants (2, 3) and (3, 7) and f_2 has two prime implicants (0, 1) and (1, 3). From the table, we can see that minterm m_3 ($\overline{A}BC$) is common to both outputs and can therefore be shared. In the K-maps, if the pair corresponds to minterm m_3 is not formed, then we get the simplified functions as

$$f_1 = B\overline{C} + \overline{A}BC \text{ and}$$
$$f_2 = \overline{A}\,\overline{B} + \overline{A}BC$$

Table 3.23. Prime implicant chart

Prime implicant		m_0	m_1	m_2	m_3	m_6
f_1	2, 3			•	•	
	3, 7				•	•
f_2	0, 1	•	•			
	1, 3		•		•	

The Fig. 3.99 shows the logic diagram for the multiple outputs.

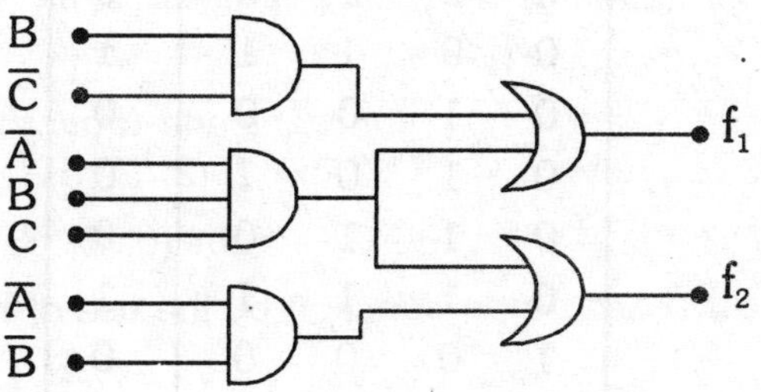

Fig. 3.99

The multiple outputs when considered together do not guarantee a minimum number of ICs. In the previous example, multiple outputs when considered together reduced one product term, however they introduced a product term with three input variables. The reduction of the product terms in the output functions is especially helpful when we realize the digital circuits using PALs or PLDs and so on.

3.9.2 Which Gate Representation to Use

In some logic-circuits, designers use only the standard logic-gate symbols in their circuit schematics. While this practice is not incorrect, it does nothing to make the circuit operation easier to follow. Proper use of the alternate gate symbols in the circuit diagram can make the circuit operation much clearer. This is illustrated by considering an example shown in Fig. 3.100.

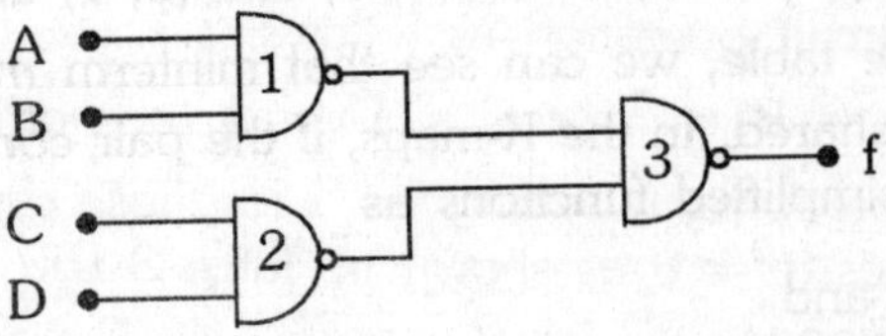

Fig. 3.100(a) Original circuit using standard NAND gate only.

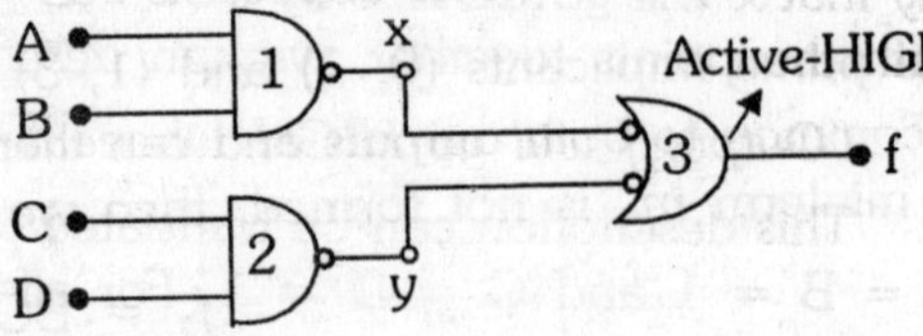

Fig. 3.100(b) Equivalent representation where output is active-HIGH.

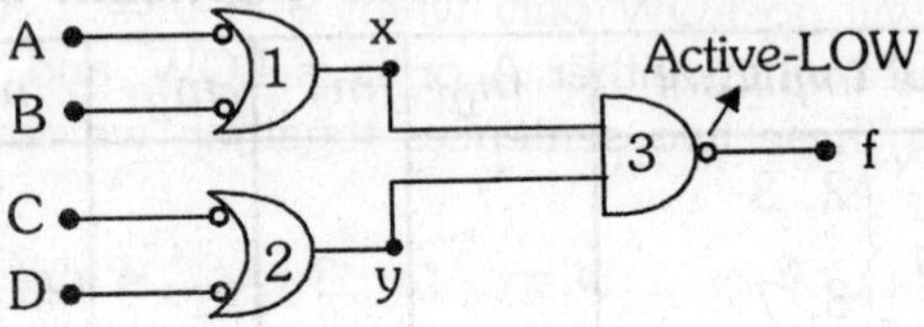

Fig. 3.100(c) Equivalent representation where output *f* is active-LOW.

A	B	C	D	*f*
0	0	0	0	0
0	0	0	1	0
0	0	1	0	0
0	0	1	1	1
0	1	0	0	0
0	1	0	1	0
0	1	1	0	0
0	1	1	1	1
1	0	0	0	0
1	0	0	1	0
1	0	1	0	0
1	0	1	1	1
1	1	0	0	1
1	1	0	1	1
1	1	1	0	1
1	1	1	1	1

Fig. 3.100(d) Truth table.

Fig. 3.100(*a*) contains 3 standard NAND gates connected to produce an output *f* that depends on inputs A, B, C and D. While this diagram is logically correct, it does not facilitate any understanding of how the circuit functions.

Case (*i*) : The Fig. 3.100 (*b*) was obtained from original circuit diagram by replacing NAND gate 3 with it's alternate symbol. In this diagram, (*i*) output *f* is taken from a NAND gate symbol that has an active-HIGH output. Thus, we can say that *f* will go HIGH when either *x* or y is LOW. (*ii*) x and y are the outputs of NAND gates having active LOW outputs. We can say that x will go LOW only if A = B = 1 and y will go LOW only if C = D = 1. Combining both the points together, we can describe the circuit operation as follows :

"Output f will go HIGH whenever either A = B = 1 or C = D = 1 (or both)"

This description can be translated to truth-table form by setting f = 1 for those cases where A = B = 1 and C = D = 1. For all other cases, f = 0 as shown in Table 3.100 (*d*).

Case (*ii*) : The representation of Fig. 3.100 (c) is obtained from the original circuit diagram by replacing NAND gates 1 and 2 by their alternate symbols. In this equivalent representation, the f output is taken from a NAND gate that has an active-LOW output. Thus, (*i*) we can say that f will go LOW only when x = y = 1, (*ii*) since x and y are active-HIGH outputs, x will be HIGH when either A or B is LOW, and y will be HIGH when either C or D is LOW. Combining these two sentences together, we can describe the circuit operation as follows:

"Output f will go LOW only when A or B is LOW and C or D is LOW"

This description can be translated to truth table form by making f = 0 for all cases where at least one of the A or B inputs is LOW and at the same time at least one of the C or D inputs is LOW. For all other cases, f = 1. This resultant truth table is same as that obtained for the circuit diagram of Fig. 3.100 (*b*).

3.9.3 Which Circuit Diagram Should be Used

The answer depends on the particular function being performed by the circuit output.

If the circuit is being used to cause some action when output f goes to the 1 state, then we can say that f is to be active-HIGH and the circuit diagram of Fig. 3.100 (*b*) type is used.

On the other hand, if the circuit is being used to cause some action when f goes to the '0' state, then f is to be active–LOW and the circuit diagram of Fig. 3.100 (*c*) should be used. There will be some situations where both output states are used to produce different actions and either one can be considered to be the active state. For these cases, either circuit representation can be used.

Bubble Placement : In Fig. 3.100 (*b*), the symbols for NAND gates 1 and 2 were choosen to have active-LOW outputs to match the active-LOW inputs of NAND gate 3.

In Fig. 3.100 (*c*), the NAND gates 1 and 2 were chosen to have active-HIGH outputs to match the active-HIGH inputs of NAND gate 3. This leads to the following general rule for preparing logic-circuit schematics:

"Whenever possible, choose gate symbols so that bubble outputs are connected to bubble inputs, and not bubble outputs to not bubble inputs". This is illustrated in the following examples.

Example 3.35. *The logic circuit in Fig. 3.101 (a) is being used to activate a fan when its output f goes HIGH. Modify the circuit diagram so that it more effectively represents the circuit operation.*

Solution. If f = 1 will activate the fan, f is to be active-HIGH. Thus, the AND gate 2 symbol does not have to be changed. The NOR gate symbol to be changed to with the alternate symbol with a non-bubble (active-HIGH) output to match the non-bubble input of AND gate 2 as shown in Fig. 3.101 (*b*). Note that the circuit now has non-bubble outputs connected to the non-bubble inputs of gate 2.

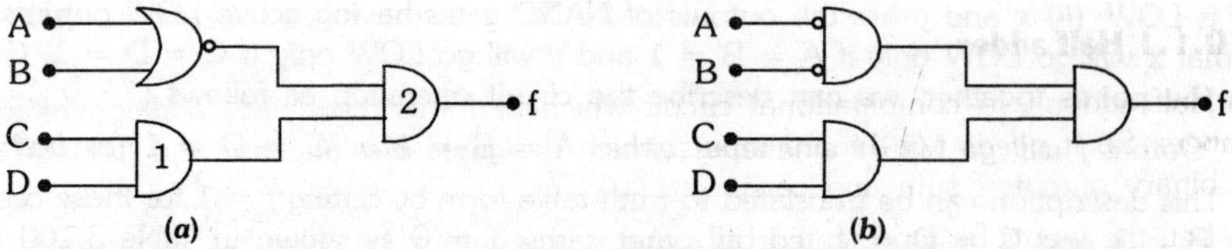

Fig. 3.101

Example 3.36. *When the output of the logic circuit in Fig. 3.102 (a) goes LOW, it activates another logic circuit. Modify the circuit diagram to more effectively represent the circuit operation.*

Solution.

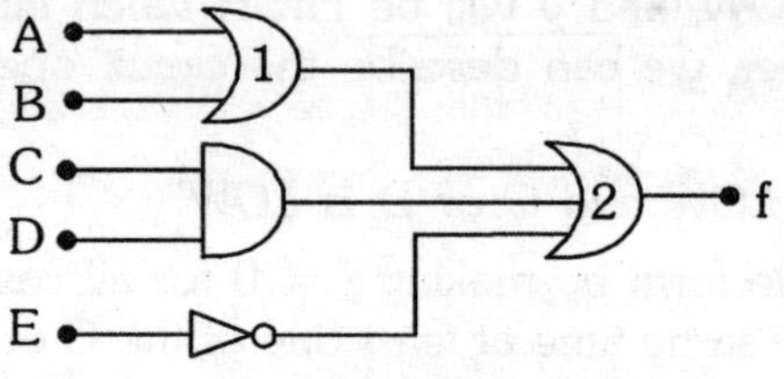

Fig. 3.102(*a*)

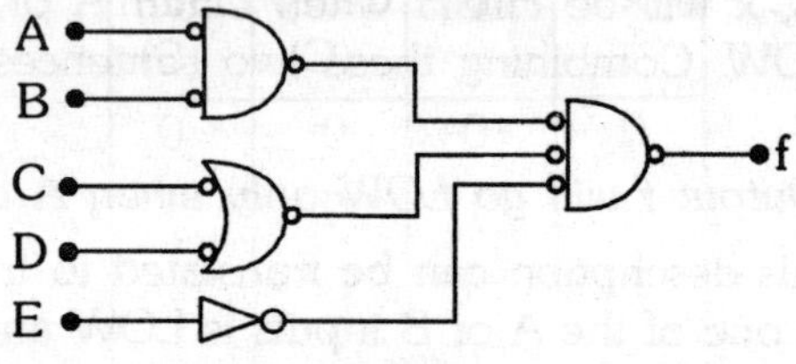

Fig. 3.102(*b*)

Since *f* is to be active-LOW, the symbol for OR gate 2 has to be changed to its alternate symbols as shown in Fig. 3.102 (*b*). The new OR gate 2 symbol has bubble inputs, and AND gate and OR gate 1 symbols have to be changed to bubbled outputs. The NOT gate has already a bubble output. Now the circuit has all bubble outputs connected to bubble inputs of gate 2.

3.10 ARITHMETIC CIRCUITS

The logic circuits which are used for performing the digital arithmetic operations such as addition, subtraction, multiplication and division are called "arithmetic circuits". These circuits are common in many digital systems. Note that the regular logic gates will be wired together to form 'adders' and 'subtractors'.

3.10.1 Adders

Digital computers perform various arithmetic operations. The most basic operation is addition of two binary numbers. The simple addition consists of four possible operations, they are :

$$0 + 0 = 0$$
$$0 + 1 = 1$$
$$1 + 0 = 1$$
$$1 + 0 = 10$$

The first three operations produce the result whose length is one bit. This result is called as "sum". But, when, both the bits are 1 (as shown in last operation), the result is 2 bits. The higher significant bit is called "carry" and the lower significant bit is called "sum".

There are different types of adders that are used in digital systems in various applications. They are :

1. Half adder, 2. Full adder, 3. Parallel binary adder, 4. Look-ahead carry adder, 5. Serial adder, 6. BCD adder, 7. Excess-3 adder.

Now, we will have the detailed explanation of each adder.

3.10.1.1 Half adder

Half adder is a combinational circuit which performs the binary addition of two binary numbers. So it needs two binary inputs which designate the 'augend' and 'addend' bits; and two binary outputs : sum and carry.

Let, the inputs be represented by x and y, the outputs be given by S (sum) and C (carry). The block schematic diagram is shown in Fig. 3.103 and the truth table is shown in Table 3.24.

Table 3.24. Truth table for half adder

Inputs		*Outputs*	
x	*y*	Carry (C)	Sum (S)
0	0	0	0
0	1	0	1
1	0	0	1
1	1	1	0

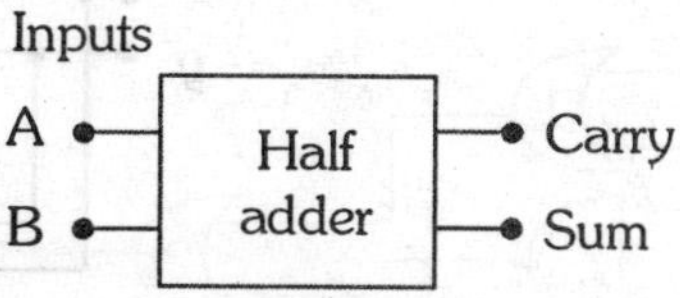

Fig. 3.103 Block schematic of half adder.

It can be seen that the carry = 1, only when both inputs x and y are 1. The simplified expressions for sum and carry are obtained from K-map as shown in Fig. 3.104.

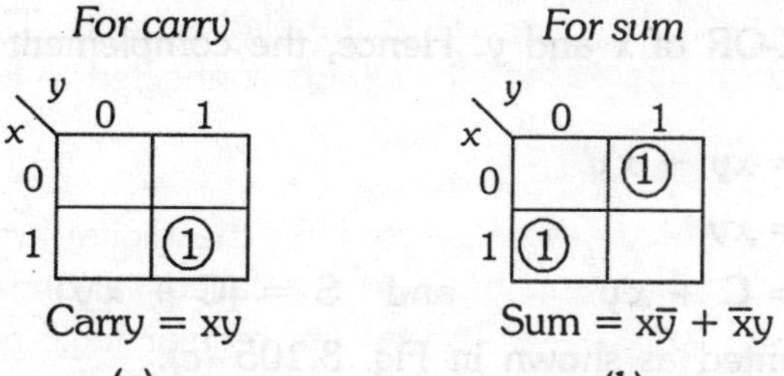

Fig. 3.104 K-map simplifications for half adder outputs.

(*i*) The output expressions are

$$S = x'y + xy'$$
$$C = xy$$

The logic diagram implementation is shown in Fig. 3.105 (*a*), as are four other implementations for a half-adder. They all achieve the same result as far as their input-output behaviour is concerned.

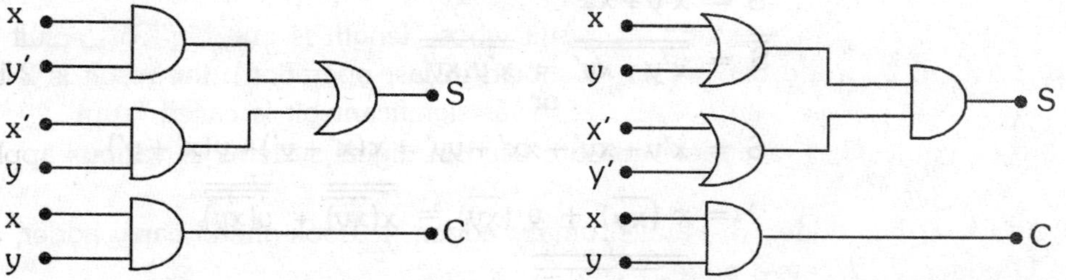

(*a*) $S = xy' + x'y$
$C = xy$

(*b*) $S = (x + y)(x' + y')$
$C = xy$

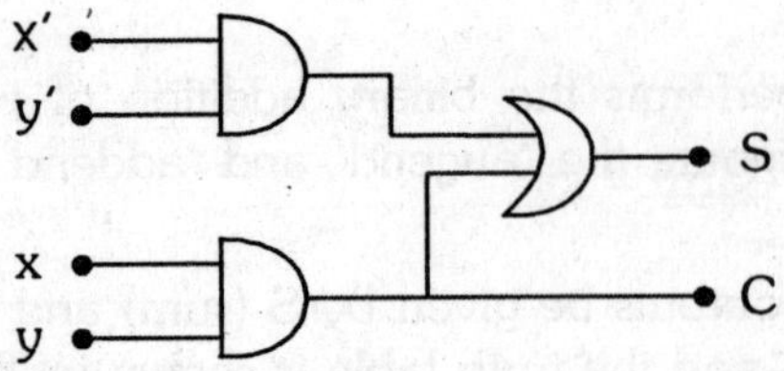

(c) S = (C + x'y')'
C = xy

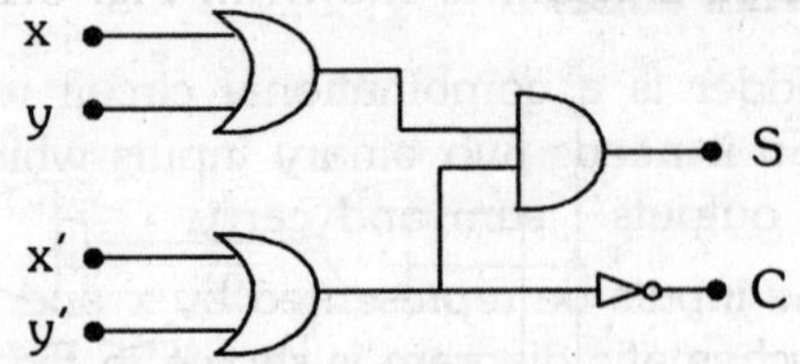

(d) S = (x + y) (x' + y')
= (x' + y')'

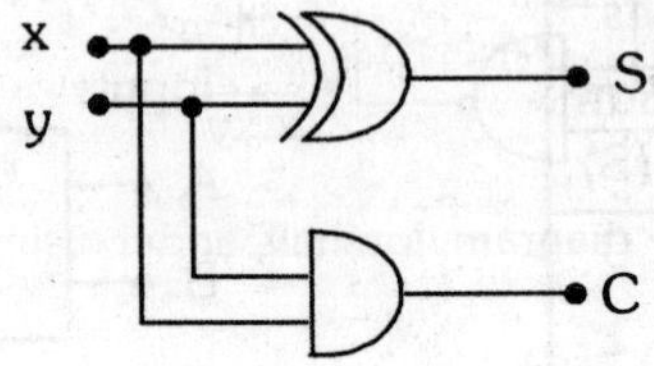

(e) S = $x \oplus y$
C = xy

Fig. 3.105 Various implementation of a half-adder.

(*ii*) If we express the S in POS form and C in SOP form, we have S = (*x* + *y*) (*x*' + *y*'), C = *xy* can be implemented as shown in Fig. 3.105 (*b*).

(*iii*) Also, S is the EX-OR of *x* and *y*. Hence, the complement of S is the equivalence of *x* and *y*.

$\therefore$ S' = *xy* + *x*'*y*'

C = *xy*

$\therefore$ S' = C + *x*'*y*' and S = (C + *x*'*y*')'

This can be implemented as shown in Fig. 3.105 (*c*).

(*iv*) We can represent both S and C in POS form as

S = (*x* + *y*) (*x*' + *y*')

C = (*x*' + *y*'). The logic diagram for these is shown in Fig. 3.105 (*d*).

(*v*) Half adder can be implemented using an EX-OR and an AND gates as shown in Fig. 3.105 (*e*).

Half adder using only NAND gates

$$S = x'y + xy'$$

$$\overline{\overline{S}} = \overline{\overline{x'y + xy'}} = \overline{\overline{x'y}\cdot\overline{xy'}}$$

or

$$S = x'y + xy' + xx' + yy' = x(x' + y') + y(x' + y')$$

$$= x\,\overline{(xy)} + y\,\overline{(xy)} = \overline{\overline{x\overline{(xy)}}} + \overline{\overline{y\overline{(xy)}}}$$

$$= \overline{\overline{x\overline{(xy)}}\cdot\overline{y\overline{(xy)}}}$$

$$C = \overline{\overline{xy}}$$

The logic diagram is shown in Fig. 3.106.

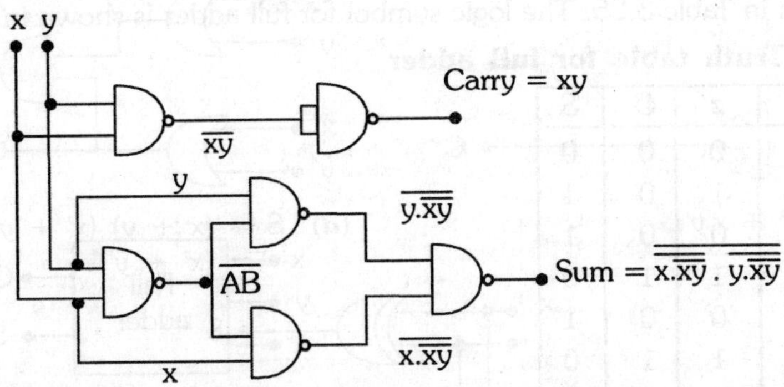

Fig. 3.106 Logic diagram for half adder using NAND gates only.

Half adder using only NOR gates

To draw the logic diagram using NOR gates, the S and C must be represented in POS form. Thus :

$$S = \bar{x}y + x\bar{y}$$

$$= \bar{x}y + x\bar{y} + x\bar{x} + y\bar{y} = \bar{x}(x+y) + \bar{y}(x+y)$$

$$= \bar{x}\,\overline{\overline{(x+y)}} + \bar{y}\,\overline{\overline{(x+y)}} = \overline{x+\overline{(x+y)}} + \overline{y+\overline{(x+y)}}$$

$$= \overline{\overline{\overline{x + \overline{(x+y)}} + \overline{y + \overline{(x+y)}}}}$$

$$C = AB = \overline{\bar{A}+\bar{B}}$$

The logic diagram is shown in Fig. 3.107.

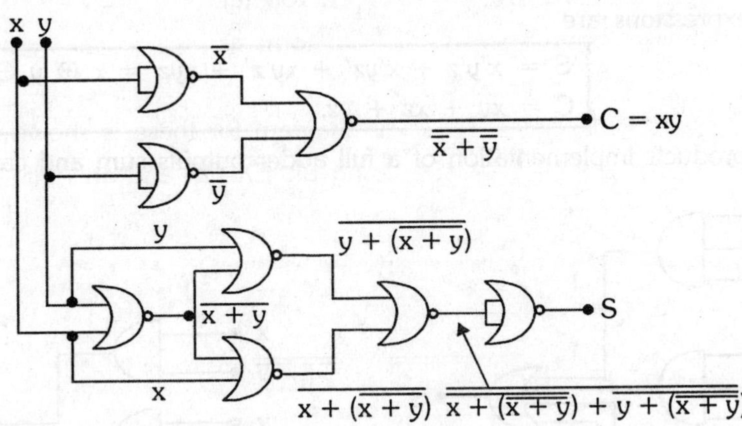

Fig. 3.107 Logic diagram for half adder using NOR gates only.

3.10.1.2 Full-adder

Full-adder is a combinational circuit that performs the addition of 3 binary digits. The input variables are represented by x, y and z; where x and y represent the 2 input variables to be added, and z represents the carry from the previous lower significant position. The arithmetic sum of the

3 binary digits varies from 0 to 3. Hence, 2 bits are required to represent the output, C and S. The truth table is shown in Table 3.25. The logic symbol for full adder is shown in Fig. 3.108.

Table 3.25. Truth table for full adder

x	y	z	C	S
0	0	0	0	0
0	0	1	0	1
0	1	0	0	1
0	1	1	1	0
1	0	0	0	1
1	0	1	1	0
1	1	0	1	0
1	1	1	1	1

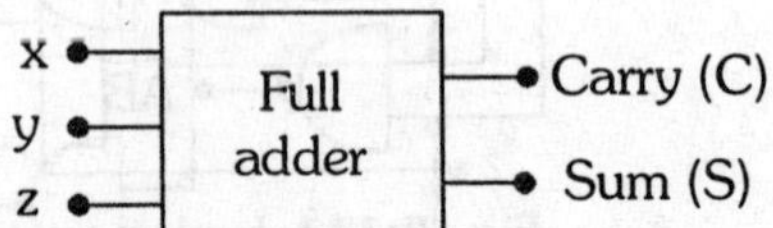

Fig. 3.108 Logic symbol for full adder.

All the possible input combinations of the 3 input-variables are shown here. The output S represents the sum of the 3 input variables, and C represents the carry. It can be seen that S =1, only when one input is 1 or when all the 3 inputs are equal to 1. The C output is 1, when 2 or 3 inputs are equal to 1. Two Boolean functions can be obtained, each to represent, either S or C. By using the truth table above, we can simplify the functions by using Karnaugh maps as shown in Fig. 3.109.

Fig. 3.109 K-maps for sum and carry.

The output expressions are

$$S = x'y'z + x'yz' + xy'z' + xyz = x \oplus y \oplus z$$

and

$$C = xy + xz + yz.$$

The sum of products implementation of a full adder outputs sum and carry are shown in Fig. 3.110.

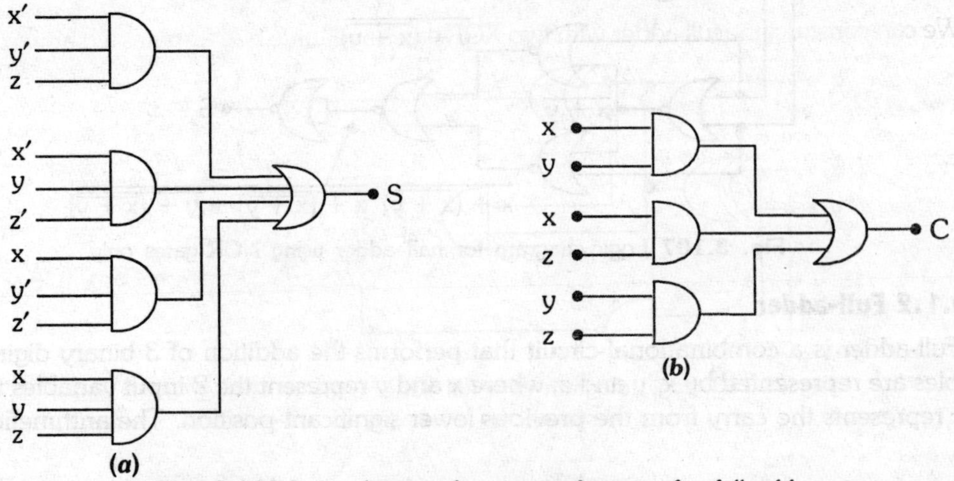

Fig. 3.110 Logic diagram for sum and carry of a full adder.

As the name suggests, the full-adder can be implemented by using 2 half-adders and one OR gate. The implementation of a full-adder by using 2 half-adders is as shown in Fig. 3.111

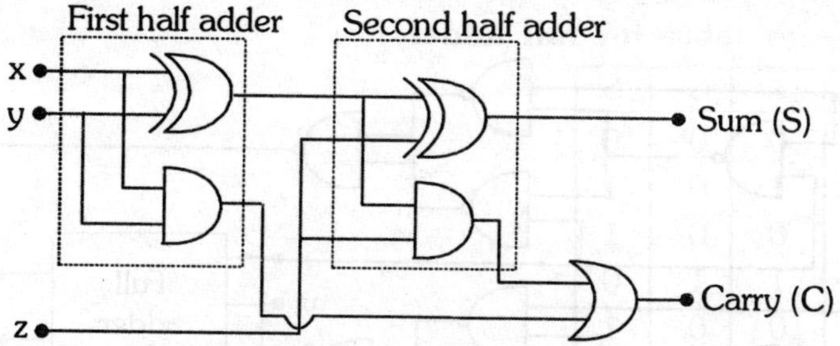

Fig. 3.111 Implementation of full adder with two half adders.

Here,
$$S = z \oplus (x \oplus y)$$
$$= z'(xy' + x'y) + z(xy' + x'y)'$$
$$= z'(xy' + x'y) + z(xy + x'y')$$
$$= xy'z' + z'yz' + xyz + x'y'z$$

The carry-output is,

$$C = z(xy' + x'y) + xy = xy'z + x'yz + xy = xy + yz + zx$$

The construction of full adder using two half adders and OR gate in the block schematic is shown in Fig. 3.112.

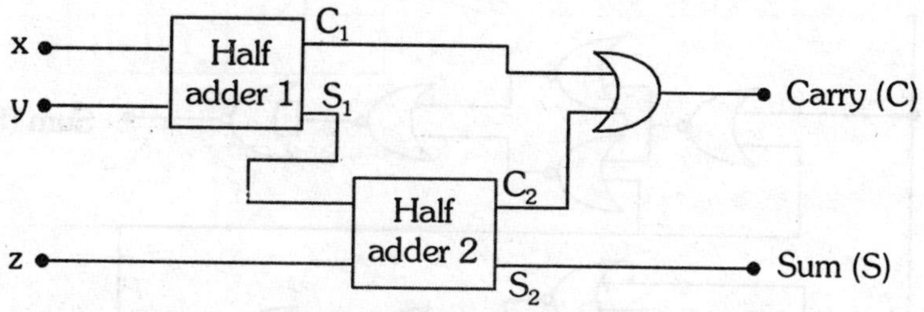

Fig. 3.112 Full adder construction with two half adders (block schematic).

(*i*) Full adder with XOR and NAND gates

We can construct the full adder with two XOR gates and three NAND gates as shown in Fig. 3.113.

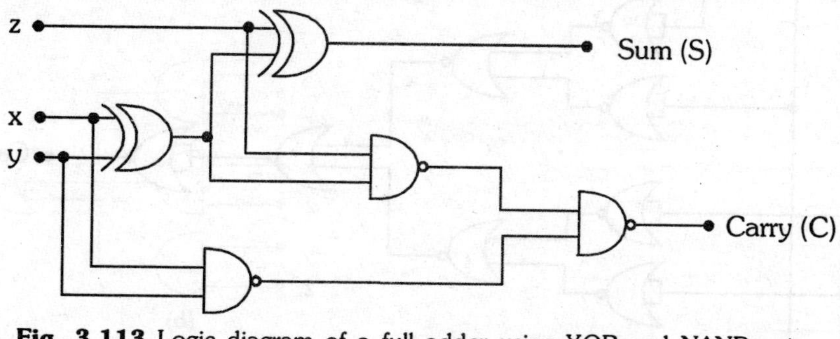

Fig. 3.113 Logic diagram of a full adder using XOR and NAND gates.

(*ii*) Construction of full adder with only NAND gates

The full adder can be constructed by using 11 NAND gates as shown in Fig. 3.114.

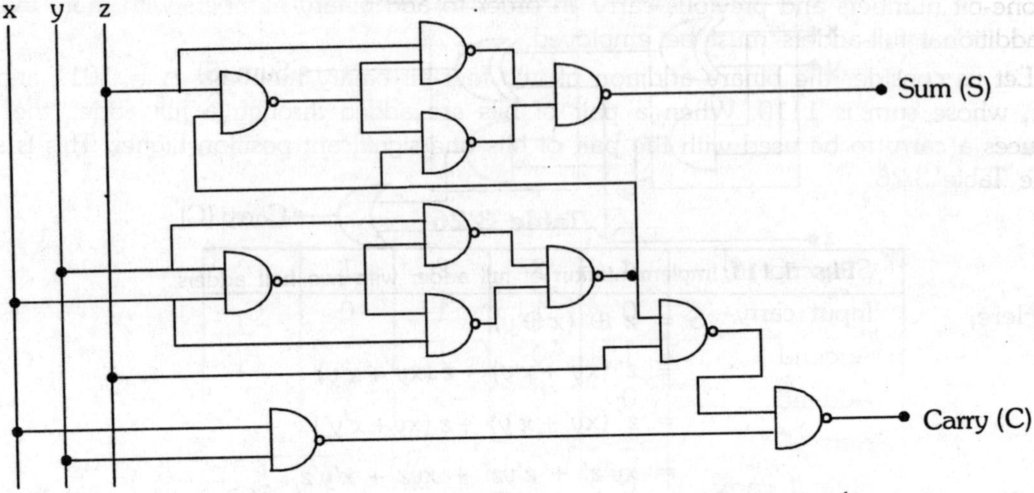

Fig. 3.114 Construction of full adder using NAND gates only.

(*iii*) Construction of full adders using only NOR gates

The construction of full adder using NOR gates only is shown in Fig. 3.115.

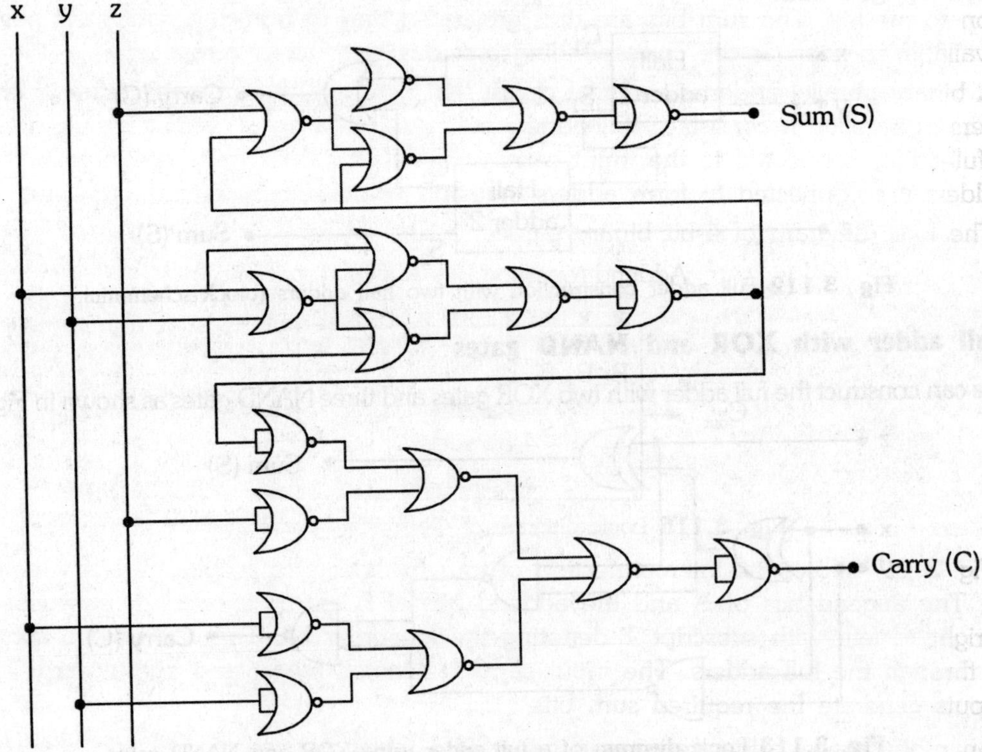

Fig. 3.115 Full adder using NOR gates only.

3.10.1.3 Parallel binary adder

As we know that a single full-adder is capable of adding three one bit binary numbers (or) two one-bit numbers and previous carry, in order to add binary numbers with more than one bit, additional full-adders must be employed.

Let us consider the binary addition of two four-bit binary numbers, A = 1011 and B = 0011, whose sum is 1110. When a pair of bits are added through a full adder, the circuit produces a carry to be used with the pair of bits one significant position higher. This is shown in the Table 3.26.

Table 3.26

Subscript (i)	*4*	*3*	*2*	*1*	
Input carry	0	1	1	0	C_i
Augend	1	0	1	1	A_i
Addend	0	0	1	1	B_i
Sum	1	1	1	0	S_i
Output carry	0	0	1	1	C_{i+1}

The bits are added with full-adders, starting from the least significant position (subscript 1), to form a sum bit and carry bit. The input carry C_1 in the least significant position must be '0'. The value of C_{i+1} in a given significant position is the output carry of the full-adder. This value is transferred into the input carry of the full-adder that adds the bits one higher-significant position to the left. The sum bits are thus generated starting from the right most position and are available as soon as the corresponding previous carry bit is generated.

A binary parallel adder is a digital circuit that produces the arithmetic sum of two binary numbers in parallel. It consists of full-adders connected in a chain, with the output carry from each full-adder connected to the input carry of the next full-adder in the chain. Thus several full adders are connected to form adders that add several binary bits at one time.

The logic diagram of 4-bit binary parallel adder is shown in Fig. 3.116.

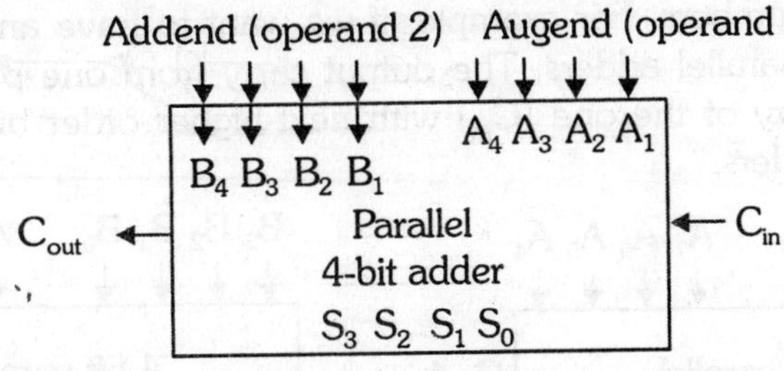

Fig. 3.116 Logic diagram of 4-bit binary parallel adders.

Fig. 3.117 shows the interconnection of four full-adders to provide a 4-bit binary parallel adder. The augend bits of A and the addend bits of B are designated by subscript numbers from right to left, with subscript 1 denoting the low-order bit. The carriers are connected in chain through the full adders. The input carry is denoted by C_1 and output carry is C_5. The S outputs generate the required sum bits.

An *n*-bit parallel adder requires *n* full-adders, which is used to add two binary numbers of length '*n*'. The four bit full adder is an example of an MSI function.

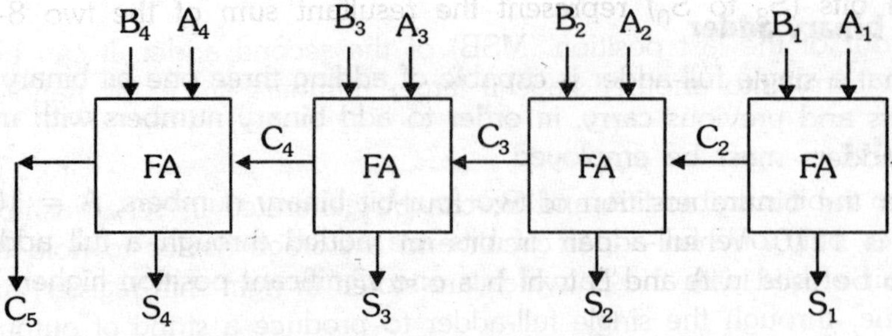

Fig. 3.117 4-bit binary parallel adder.

Note : Design of this circuit by classical method would require a truth table with 2^9 = 512 entries, since there are nine inputs to the circuit.

By using an iterative method of cascading an already known function, we are able to obtain the simple and well organized implementation.

The IC7483 and 74LS283 are a TTT medium scale integrated circuit which is a 4-bit binary parallel adder shown in Fig. 3.117. The inputs of this IC are two 4-bit numbers A_3, A_2, A_1, A_0 and B_3, B_2, B_1, B_0 and carry C_{in} 0 into the LSB position. The outputs are the sum bits S_3, S_2, S_1, S_0 and the carry C_{out} 3, output of the MSB position.

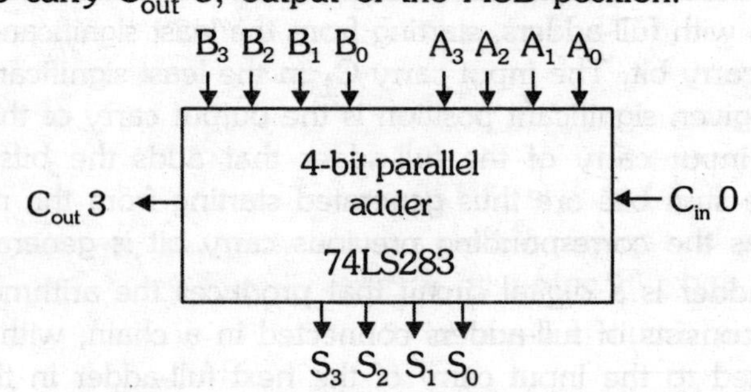

Fig. 3.118 Functional symbol of the IC 74LS283.

Two or more parallel-adder blocks can be connected in cascading to accommodate the addition of larger binary numbers. For example, if we want to have an 8-bit adder, we can get it by cascading two 4-bit parallel adders. The output carry from one package must (C_{out} 3) be connected to the input carry of the one (C_{in}) with next-higher order bits. Fig. 3.119 shows the cascading of two 4-bit adders.

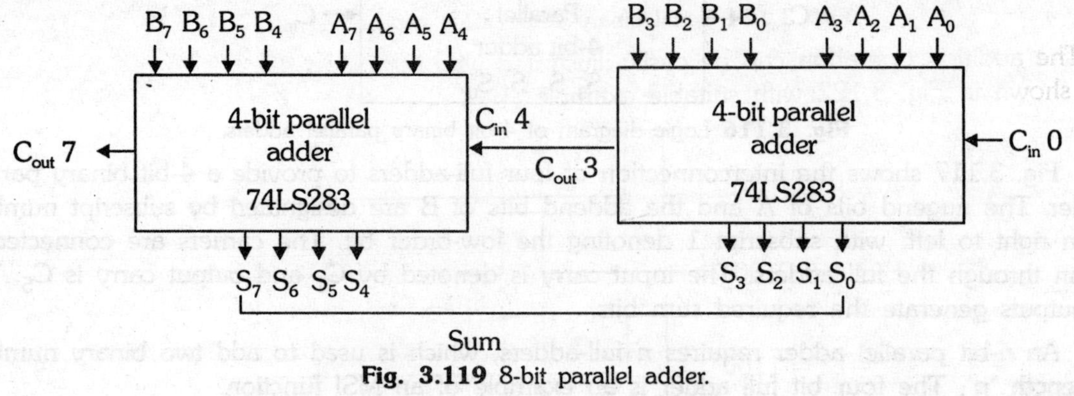

Fig. 3.119 8-bit parallel adder.

The eight sum bits (S_8 to S_0) represent the resultant sum of the two 8-bit numbers. C_{out} 7 is the carry out of the last position (MSB) of the second adder. It can be used as an overflow bit or as a carry into another stage if larger numbers are to be handled.

3.10.1.4 Serial adder

The sum of two *n*-bit numbers, A and B, can be generated in serial fashion. The serial addition method uses only one full-adder circuit and a storage device to hold the generated output carry. The pair of bits in A and B (which are stored in right shift register) are transferred serially, one at a time, through the single full-adder to produce a string of output bits for the sum. The stored output carry from one pair of bits is used as an input carry for the next pair of bits as shown in Fig. 3.120.

The addition of two *n*-bit numbers starts from the least significant bits and progresses step by step to the most significant bits in the different time sequences. Initially the 'D' flip-flop is cleared.

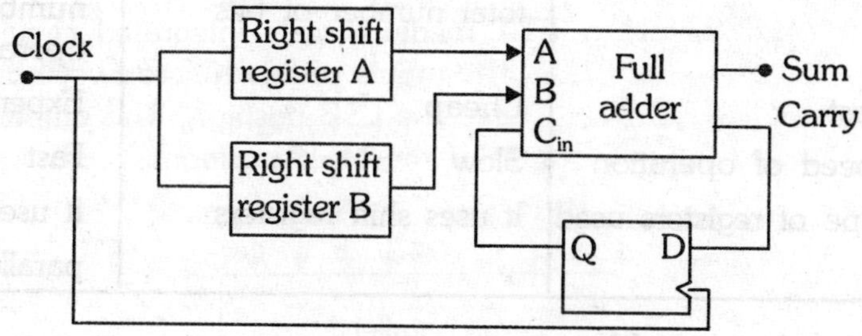

Fig. 3.120 Serial adder.

The serial adder can be represented by another way also as shown in Fig. 3.121, which contains two half-adders and OR gate with a delay circuit.

The time delay of the circuit is equal to the time T between the input pulses.

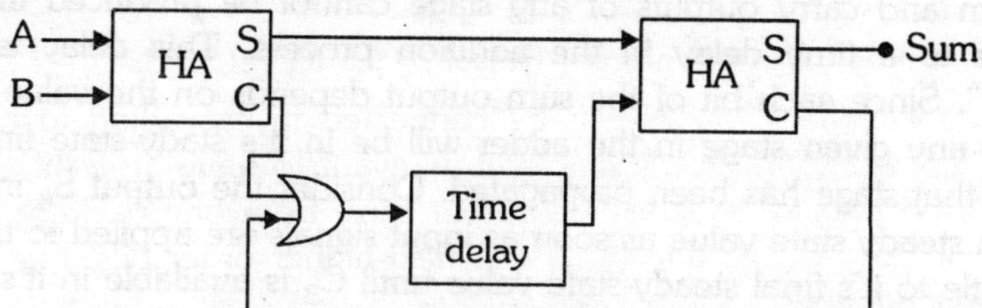

Fig. 3.121 Serial binary adder using two half-adders.

The addition operation can be easily illustrated with respect to the pulse trains for A and B as shown in Fig. 3.122 with suitable example.

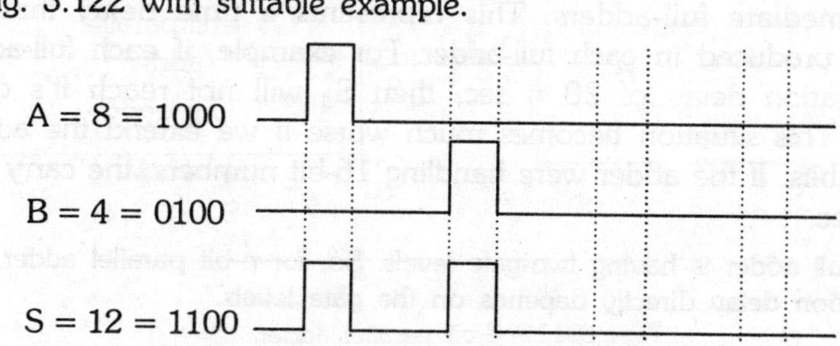

Fig. 3.122 Example for serial addition.

3.10.1.5 Comparison between serial and parallel binary adders

The comparison between serial adder and parallel adder is shown in Table 3.27.

Table 3.27. Comparison between serial and parallel binary adders

Sl. No.	*Function*	*Serial adder*	*Parallel adder*
1.	Process of addition	Only one bit at a time, as one after another	All bits are added at a time
2.	Num of full adders used	One	Equal to number bits in the binary number i.e., one for each bit addition
3.	Total time	The time required for addition depends on the total number of bits	Time required for addition does not depends on the number of bits in each operand
4.	Cost	Cheap	Expensive
5.	Speed of operation	Slow	Fast
6.	Type of registers used	It uses shift registers	It uses registers with parallel load capability.

3.10.1.6 Look-ahead carry adder

The addition of two binary numbers in parallel implies that all the bits of the augend and the addend are available for computation at the same time. In a parallel adder, the carry output of each stage is connected to the carry input of the next higher-order stage (ripple carry). Therefore, the sum and carry outputs of any stage cannot be produced until the input carry occurs. This leads to a time delay in the addition process. This delay is known as "*carry propagation delay*". Since each bit of the sum output depends on the value of the input carry, the value of S_i in any given stage in the adder will be in it's stady-state final value only after the input carry to that stage has been propagated. Consider the output S_4 in Fig. 3.117. Inputs A_4 and B_4 reach a steady state value as soon as input signals are applied to the adder. But input carry does not settle to it's final steady-state value until C_3 is available in it's steady-state value. Similarly, C_3 has to wait for C_2, and so on down to C_1. Thus, only after the carry propagates through all stages will the last output S_4 and carry C_5 settle to their final steady-state value.

This means that adder will not produce correct result until LSB carry has propagated through the intermediate full-adders. This represents a time delay that depends on the propagation delay produced in each full-adder. For example, if each full-adder is considered to have a propagation delay of 20 *n* sec, then S_4 will not reach it's correct value until 80 *n* sec (20×4). This situation becomes much worse if we extend the adder circuit to add greater number of bits. If the adder were handling 16-bit numbers, the carry propagation delay could be 320 *n* sec.

Note : Each full adder is having two-gate levels. So, for *n*-bit parallel adder, there are '2*n*' gate levels. The propagation delay directly depends on the gate levels.

The carry propagation time is the limiting factor on the speed with which two numbers are added in parallel. Although a parallel adder will always have same value at it's output terminals, the outputs will not be correct unless the signals are given enough time to propagate through the gates connected from the inputs to the outputs.

1. One method for reducing the carry propagation delay time is to employ faster gates with reduced delays. But, physical circuits have a limit to their capability.
2. Another solution is to increase the equipment complexity in such a way that the carry delay time is reduced. There are several techniques for reducing the carry propagation time in a parallel adder. The most widely used technique employs the principle of "*look ahead carry*". This method utilizes logic gates to look at the lower-order bits of the augend and addend to see if a higher-order carry is to be generated. It uses two functions carry generate and carry propagate.

Consider the circuit of the full adder shown in Fig. 3.123. If we define two new binary variables:

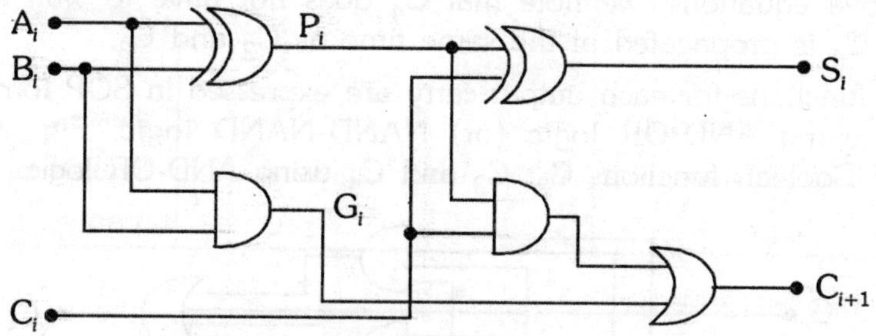

Fig. 3.123 Full-adder circuit.

Table 3.28

A_i	B_i	C_i	C_{i+1}	*Condition*
0	0	0	0	No carry
0	0	1	0	generate
0	1	0	0	
0	1	1	1	Carry
1	0	0	0	propagate
1	0	1	1	
1	1	0	1	Carry
1	1	1	1	generate

$$P_i = A_i \oplus B$$

$$G_i = A_i B_i$$

The output sum and carry can be expressed as

$$S_i = P_i \oplus C_i$$

$$C_{i+1} = G_i + P_i C_i$$

where G_i is called "carry generate" and it produces an output carry when both A_i and B_i are one, regardless of the input carry.

P_i is called a "carry propagate" because it is the term associated with the propagation of the carry from C_i to C_{i+1}. The truth table for this purpose is shown in Table 3.27. Now, we will write the Boolean function for the carry output of each stage and substitute for each C_i, it's value from the previous equations.

$$C_2 = G_1 + P_1C_1$$

$$\begin{aligned} C_3 &= G_2 + P_2C_2 \\ &= G_2 + P_2(G_1 + P_1C_1) \\ &= G_2 + G_1P_2 + P_2P_1C_1 \end{aligned}$$

$$\begin{aligned} C_4 &= G_3 + P_3C_3 = G_3 + P_3(G_2 + G_1P_2 + P_2P_1C_1) \\ &= G_3 + P_3G_2 + P_3P_2G_1 + P_3P_2P_1C_1 \end{aligned}$$

From the above equations, we note that C_4 does not have to wait for C_3 and C_2 to propagate; infact, C_4 is propagated at the same time as C_2 and C_3.

The Boolean functions for each output carry are expressed in SOP form. Thus, they can be implemented using AND-OR logic (or) NAND-NAND logic. Fig. 3.124 shows the implementation of Boolean functions C_2, C_3 and C_4 using AND-OR logic.

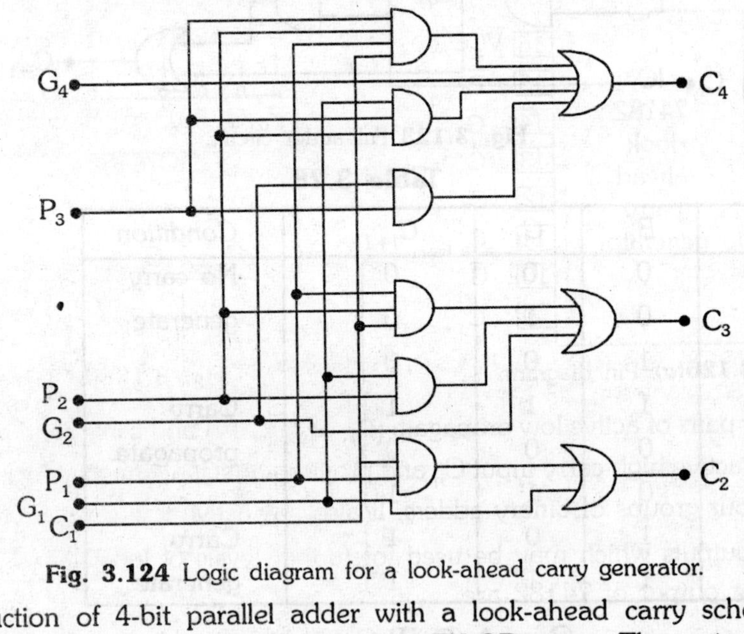

Fig. 3.124 Logic diagram for a look-ahead carry generator.

The construction of 4-bit parallel adder with a look-ahead carry scheme is shown in Fig. 3.124. Here, each sum output requires two EX-OR gates. The output of first EX-OR generates P_i, and the AND gate generates G_i. The carries are generated using look-ahead carry generator and applied as inputs to the second EX-OR gate. Other input to the EX-OR gate is P_i. Thus second EX-OR gate generates sum outputs. Each output is generated after a delay of two levels of gate. Thus outputs S_2 through S_4 have equal propagation delay times.

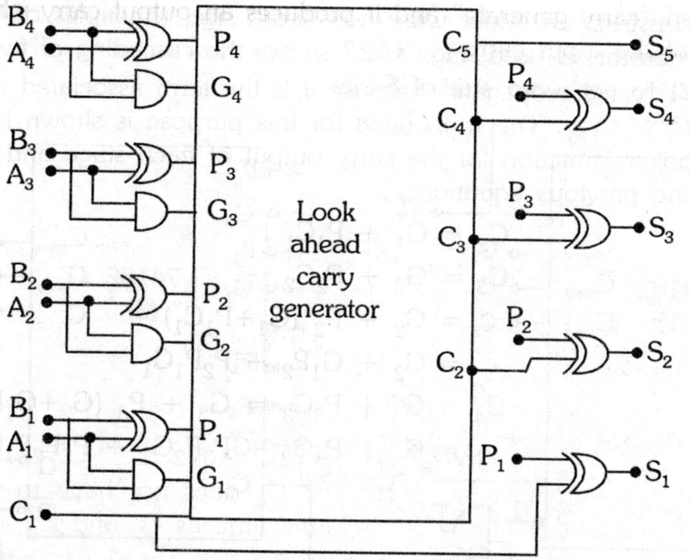

Fig. 3.125 4-bit parallel adder with look ahead carry generator.

Look ahead carry generator is an MSI circuit. An example for it is IC74182. The pin diagram is shown in Fig. 3.126 (*a*) and logic symbol in Fig. 3.126 (*b*).

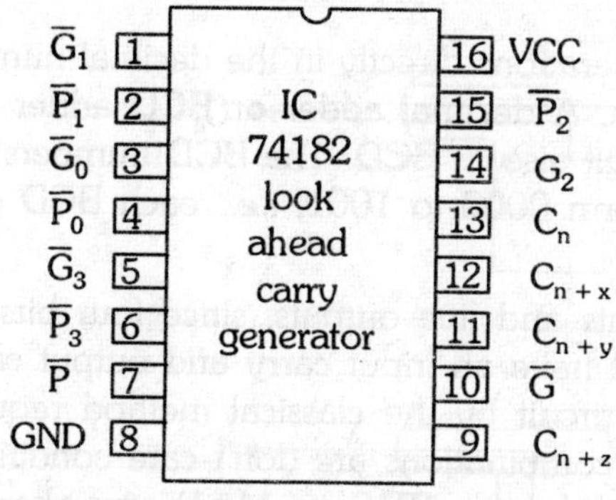

Fig. 3.126(*a*) Pin diagram.

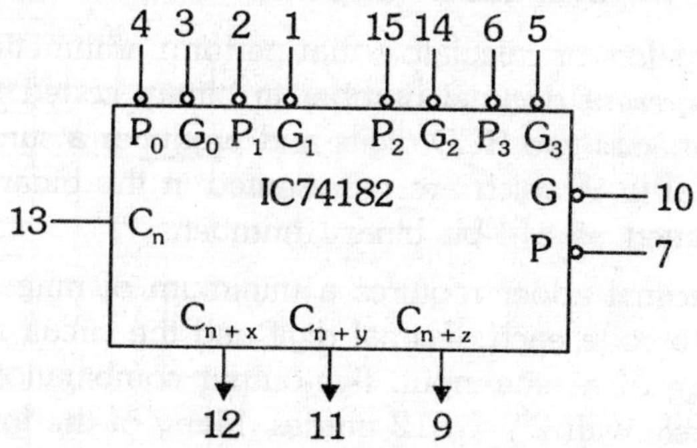

Fig. 3.126(*b*) Logic symbol.

It has four pairs of active low propagate ($\overline{P}_0$, $\overline{P}_1$, $\overline{P}_2$, $\overline{P}_3$) and carry generate ($\overline{G}_0$, $\overline{G}_1$, $\overline{G}_2$, $\overline{G}_3$) signals and an active high carry input C_n and provides anticipated active high carries (C_{n+x}, C_{n+y}, C_{n+z}) across four groups of binary adders. It also has active low carry propagate ($\overline{P}$) and carry generate ($\overline{G}$) outputs which may be used for further levels of look ahead. The logic equations provided at the output of 74182 are

$$C_{n+x} = G_0 + P_0 C_n$$

$$C_{n+y} = G_1 + P_1 G_0 + P_1 P_2 C_n$$

$$C_{n+z} = G_2 + P_2 G_1 + P_2 P_1 G_0 + P_2 P_1 P_0 C_n$$

$$\overline{G} = \overline{G_3 + P_2 G_2 + P_3 P_2 G_1 + P_3 P_2 P_1 G_0}$$

$$\overline{P} = \overline{P_3 P_2 P_1 P_0}$$

If we want to construct an adder with more word size, (multiples of 4) the cascading of look ahead carry generators is used. Fig. 3.127 shows the cascading of two look ahead carry generators (IC74182) to get word size of 8-bits.

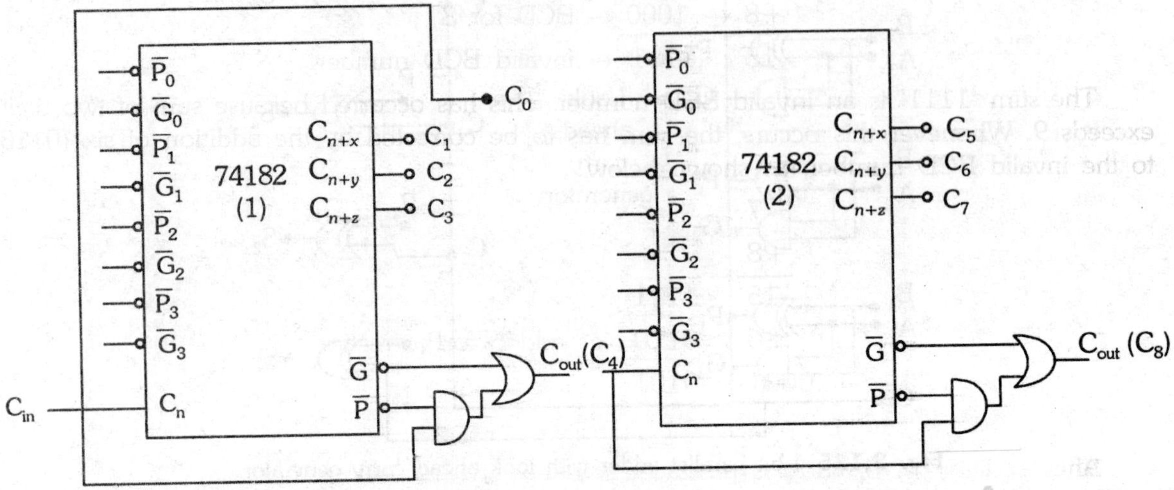

Fig. 3.127 Cascading of two look ahead carry generators.

3.10.1.7 Decimal adder (or) BCD adder

Computers or calculators that perform arithmetic operations directly in the decimal number system represent decimal number in binary coded form. A decimal adder or BCD adder is a circuit that adds two BCD digits and produces a sum digit also in BCD. The BCD numbers use 10 digits; 0 to 9 which are represented in the binary form 0000 to 1001, *i.e.*, each BCD digit is represented as a 4-bit binary number.

A decimal adder requires a minimum of nine inputs and five outputs, since four bits are required to code each decimal digit and the circuit must have an input carry and output carry. The design of a nine-input, five-output combinational circuit by the classical method requires a truth table with $2^9 = 512$ entries. Many of the input combinations are don't care conditions, since each binary code input has 6 combinations invalid (*i.e.*, 1010 to 1111). An alternate procedure is to add the numbers with full-adders, taking into consideration the fact that six combinations in each 4-bit input are not used. The output must be modified so that only those binary combinations that are valid combinations of the decimal code are generated.

When we want to write BCD number say 732, it can be represented as: 0111 0011 0010. Here, we should note that BCD number cannot be greater than 9. The addition of BCD numbers can be best understood by considering the three cases that occur when two BCD numbers are added.

Case (*i*) : Sum equal to 9 or less with carry 0

Let us consider additions of 3 and 6 in BCD

$$\begin{array}{rrl} 6 & -0110 & \leftarrow \text{BCD for } 6 \\ +3 & -0011 & \leftarrow \text{BCD for } 3 \\ \hline 9 & 1001 & \leftarrow \text{BCD for } 9 \end{array}$$

The addition is carried out as in normal binary addition.

Case (*ii*) : Sum greater than 9 with carry 0

Let us consider addition of 7 and 8 in BCD.

```
 7 ← 0111 ← BCD for 7
+8 ← 1000 ← BCD for 8
-------------------------
15   1111 ← invalid BCD number.
```

The sum '1111' is an invalid BCD number. This has occurred because sum of two digits exceeds 9. Whenever this occurs, the sum has to be corrected by the addition of six (0110) to the invalid BCD number, as shown below.

```
  7
 +8
----
 15        1111
 +6        0110  ← add 6 for correction
      ---------
      0001 0101
        1    5   ← BCD for 15.
```

After addition of 6, carry is produced into the second decimal position.

Case (*iii*) : Sum is less than or equal to 9 with carry 1

Let us consider addition of 8 and 9 in BCD.

```
 8 1000       ← BCD for 8
+9 1001       ← BCD for 9
---------
17
0001 0001     ← incorrect BCD
```

In this case, the result (0001 0001) is valid BCD number, but it is incorrect. To set the correct BCD result, correction factor of 6 has to be added to the least significant digit sum, as shown below.

```
 8 1000          ← BCD for 8
+9 1001          ← BCD for 9
-----------
17 0001 0001     ← incorrect BCD result
   0000 0110     ← add 6 for correction
   ---------
   0001 0111     ← BCD for 17
```

The Summary of the BCD addition

1. Add two BCD numbers using ordinary binary addition.
2. If four-bit sum is equal to or less than 9, no correction is needed. The sum is a BCD number.
3. If four-bit sum is greater than or if carry is generated from the four bit sum, the sum is invalid. To correct it, add 0110 to the four bit sum. If a carry is resulted from this addition, add it to the next higher-order BCD digit.

Thus to implement the BCD adder we require:

1. 4-bit binary adder for initial addition.
2. Logic circuit to detect sum greater than 9 and
3. One more 4-bit adder to add 0110 to the sum if sum is greater than 9 or carry is one (1).

The binary adder will form the sum in binary and produce a result that may range from 0 to 19. (9+9+1). These binary numbers are labelled by symbols K, Z_8, Z_4, Z_2 and Z_1 and are listed in Table 3.29, where K is the carry (output). The subscripts under the letter Z represent the weights 8, 4, 2 and 1 that can be assigned to the four bits in the BCD code. The first column in the table lists the binary sums as they appear in the outputs of a 4-bit binary adder. The output sum of two decimal digits must be represented in BCD and should appear in the form listed in the second column of the table.

The logic circuit that detects the necessary condition can be determined from the Table 3.29 and the simplified expression can be obtained by using K-maps as shown in Fig. 3.128.

1. It is obvious that a correction is needed when the binary sum has an output carry, K = 1. The other six combinations from 1010 to 1111 need a correction.

Table 3.29. BCD adder truth table

Binary sum					*BCD sum*					
K	Z_8	Z_4	Z_2	Z_1	C	S_8	S_4	S_2	S_1	Decimal
0	0	0	0	0	0	0	0	0	0	0
0	0	0	0	1	0	0	0	0	1	1
0	0	0	1	0	0	0	0	1	0	2
0	0	0	1	1	0	0	0	1	1	3
0	0	1	0	0	0	0	1	0	0	4
0	0	1	0	1	0	0	1	0	1	5
0	0	1	1	0	0	0	1	1	0	6
0	0	1	1	1	0	0	1	1	1	7
0	1	0	0	0	0	1	0	0	0	8
0	1	0	0	1	0	1	0	0	1	9
0	1	0	1	0	1	0	0	0	0	10
0	1	0	1	1	1	0	0	0	1	11
0	1	1	0	0	1	0	0	1	0	12
0	1	1	0	1	1	0	0	1	1	13
0	1	1	1	0	1	0	1	0	0	14
0	1	1	1	1	1	0	1	0	1	15
1	0	0	0	0	1	0	1	1	0	16
1	0	0	0	1	1	0	1	1	1	17
1	0	0	1	0	1	1	0	0	0	18
1	0	0	1	1	1	1	0	0	1	19

$$C = Z_8 Z_4 + Z_8 Z_2$$

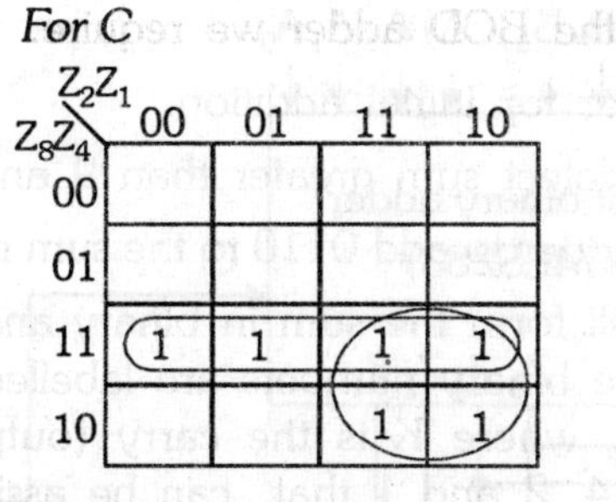

Fig. 3.128 Simplification for C.

The condition for a correction and an output carry can be expressed by the Boolean function

$$C = K + Z_8Z_4 + Z_8Z_2$$

The Fig. 3.129 shows the block diagram of a BCD adder. First, two numbers together with input carry, are added in the top 4-bit binary adder (1) to produce binary sum. When the output carry equals to zero (*i.e.*, when sum is ≤ 9 and $K = 0$), nothing (0000) is added to the binary sum. When it is equal to one (*i.e.*, when sum > 9, or $K = 1$), binary 0110 is added to the binary sum in the bottom 4-bit binary adder (2). The output carry generated by the bottom binary adder can be ignored, since it supplies information already available at the output-carry terminal.

Note : (1) To achieve shorter propagation delays, BCD adder includes the necessary circuits for look-ahead carries.

(2) The decimal parallel adder that adds '*n*' decimal digits needs *n* BCD adder stages. The output carry from one stage must be connected to the input carry of the next higher order stage.

The Fig. 3.130 shows the 8-bit BCD adder (cascading of two 4-bit BCD adders) by using the 4-bit binary adders (74283) and correcting circuit.

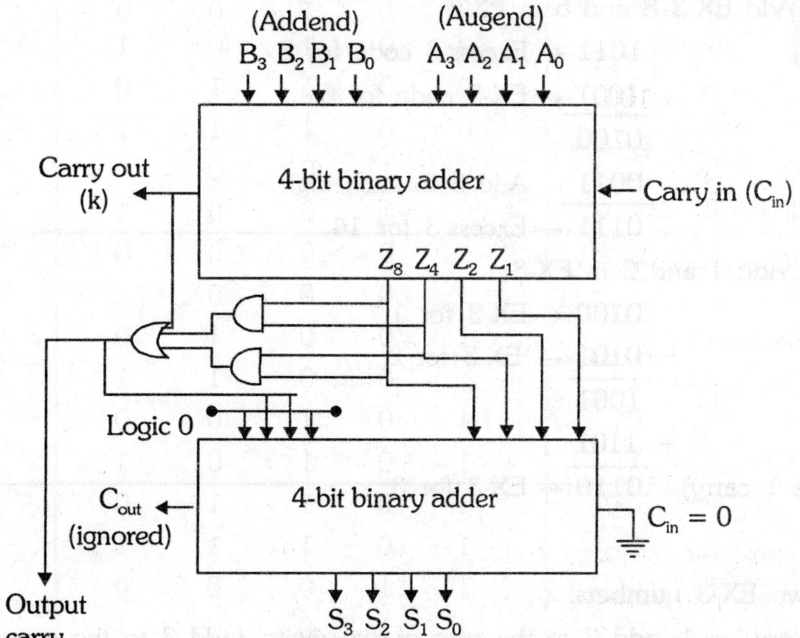

Fig. 3.129 Block diagram of BCD adder.

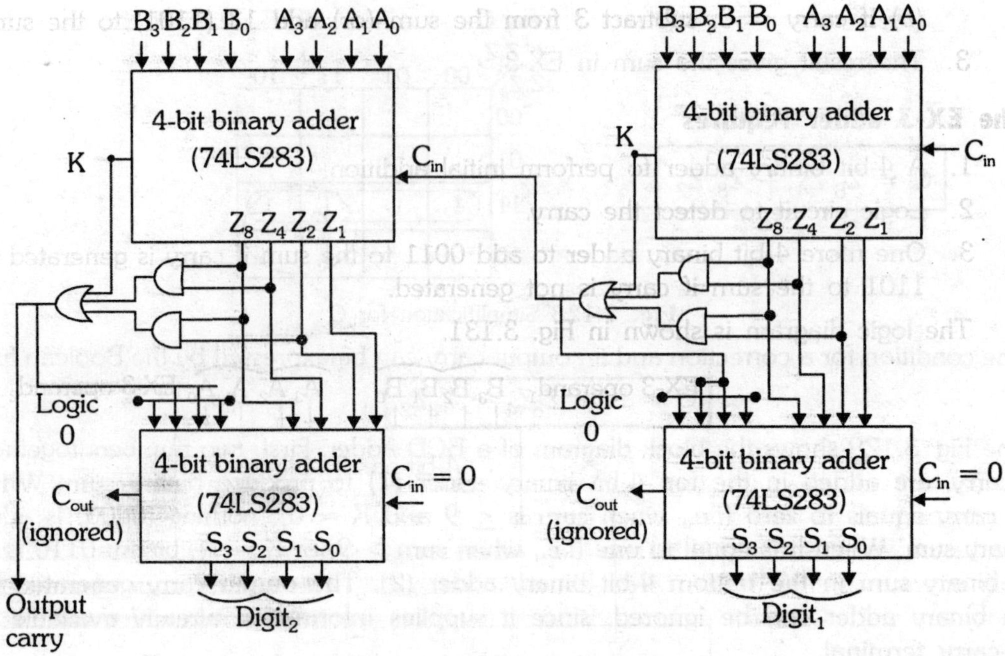

Fig. 3.130 8-bit BCD adder.

3.10.1.8 Excess-3 adder

The Excess-3 adder is an adder which adds two EX-3 numbers and produces the result also in EX-3 numbers only. Before going to see what are the corrections to be done to the binary adder, let us see some examples of EX-3 addition.

Case (*i*). Add EX-3 8 and 6 in EX-3.

```
              1011 ← Excess-3 code for 8
            + 1001 ← EX-3 code for 6
              ----
  0001        0100
  0011        0011   Add 3 to each digit
  ----        ----
  0100        0111 ← Excess-3 for 14.
```

Case (*ii*). Add 1 and 2 in EX-3

```
              0100 ← EX-3 for 1
              0101 ← EX-3 for 2
              ----
              1001
            + 1101
              ----
(ignore 1 carry) 0110 ← EX-3 for 3
```

Summary

1. Add two EX-3 numbers.
2. (*a*) If carry = 1, add 3 to the sum of two digits. (add 3 to the carry also)

(*b*) If carry = 0, subtract 3 from the sum (or) add 13 (1101) to the sum.

3. The result gives the sum in EX-3.

The EX-3 adder requires

1. A 4-bit binary adder to perform initial addition.
2. Logic circuit to detect the carry.
3. One more 4-bit binary adder to add 0011 to the sum if carry is generated (or) to add 1101 to the sum if carry is not generated.

The logic diagram is shown in Fig. 3.131.

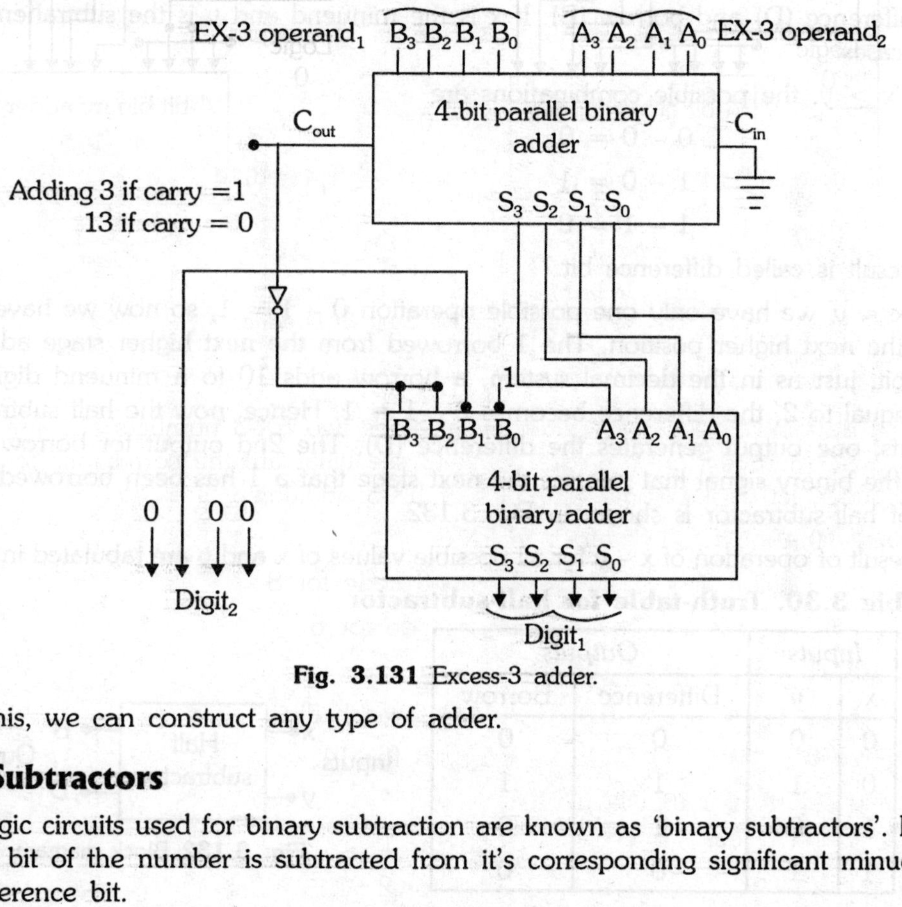

Fig. 3.131 Excess-3 adder.

Like this, we can construct any type of adder.

3.10.2 Subtractors

The logic circuits used for binary subtraction are known as 'binary subtractors'. Here each subtrahend bit of the number is subtracted from it's corresponding significant minuend bit to form a difference bit.

If the minuend bit is smaller than the subtrahend bit, a 1 is borrowed from the next significant position. The fact that a 1 has been borrowed must be conveyed to the next higher pair of bits by means of a binary signal coming out (output) of a given stage and going into (input) the next higher stage.

The subtraction consists of four elementary operations, which are :

$$0 - 0 = 0$$

$$0 - 1 = 1 \text{ with 1 borrow}$$

$$1 - 0 = 1$$
$$1 - 1 = 0$$

Like adder circuits, there are different subtractors :

1. Half subtractor
2. Full subtractor
3. Parallel subtractor
4. BCD subtractor
5. Excess-3 subtractor
6. Binary adder-subtractor

3.10.2.1 Half subtractor

A half subtractor is a combinational circuit that subtracts two binary bits and produces two outputs, difference (D) and borrow (B). If x is the minuend and y is the subtrahend, then we have two cases:

1. If $x \geq y$, the possible combinations are

$$0 - 0 = 0$$
$$1 - 0 = 1$$
$$1 - 1 = 0$$

The result is called difference bit.

2. If $x < y$, we have only one possible operation $0 - 1 = 1$, so now we have to borrow a 1 from the next higher position. The 1 borrowed from the next higher stage adds 2 to the minuend bit, just as in the decimal system, a borrow adds 10 to a minuend digit. With the minuend equal to 2, the difference becomes $2 - 1 = 1$. Hence, now the half subtractor needs two outputs, one output generates the difference (D). The 2nd output for borrow (B), which generates the binary signal that informs the next stage that a 1 has been borrowed. The block diagram of half-subtractor is shown in Fig. 3.132.

The result of operation of $x - y$ for all possible values of x and y are tabulated in Table 3.30.

Table 3.30. Truth-table for half-subtractor

Inputs		*Outputs*	
x	y	Difference	Borrow
0	0	0	0
0	1	1	1
1	0	1	0
1	1	0	0

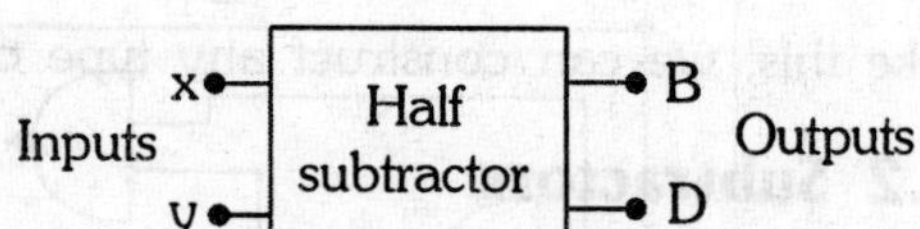

Fig. 3.132 Block diagram.

1. The output $B = 0$ when $x \geq y$ and $B = 1$ for $x < y$ *i.e.*, $x = 0$ and $y = 1$

2. The output D of the half-subtractor is '0' when both inputs are same *i.e.*, $x = y = 0$ (or) $x = y = 1$ and $D = 1$ if both inputs are opposite *i.e.*, $x = 0, y = 1$ (or) $x = 1, y = 0$.

The Boolean expression for the outputs of half subtractor can be determined as follows by using K-maps shown in Fig. 3.133.

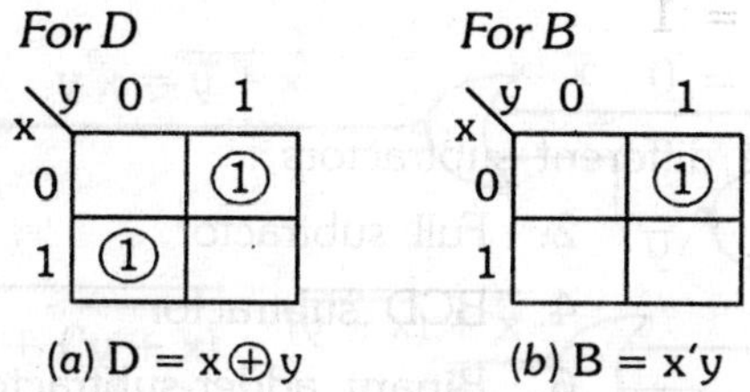

Fig. 3.133 K-maps for difference and borrow.

The Boolean functions for two outputs of the half-subtractor are :

$$D = x'y + xy' = x \oplus y$$

$$B = x'y$$

The 'D' output is the result of arithmetic operation '$2B + x - y$'.

Note : 1. The logic expression D is same as sum (S) expression of the half-adder.

The logic diagram for half-subtractor is shown in Fig. 3.134.

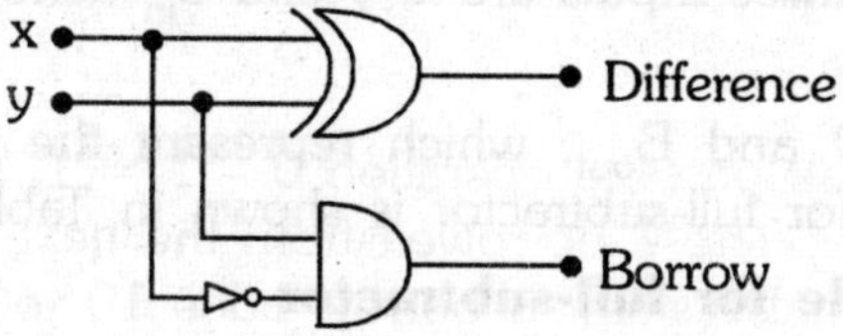

Fig. 3.134 Logic diagram for half subtractor.

(a) Half-subtractor using NAND gates only

Fig. 3.135 represents the half-subtractor using NAND gates. It requires totally seven NAND gates :

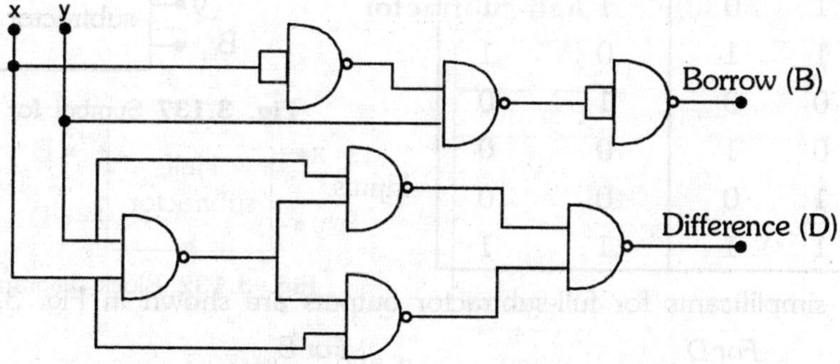

Fig. 3.135 Half-subtractor using NAND gates.

(b) Half subtractor using NOR gates

Fig. 3.135 shows the half-subtractor using NOR gates only. It requires seven NOR gates.

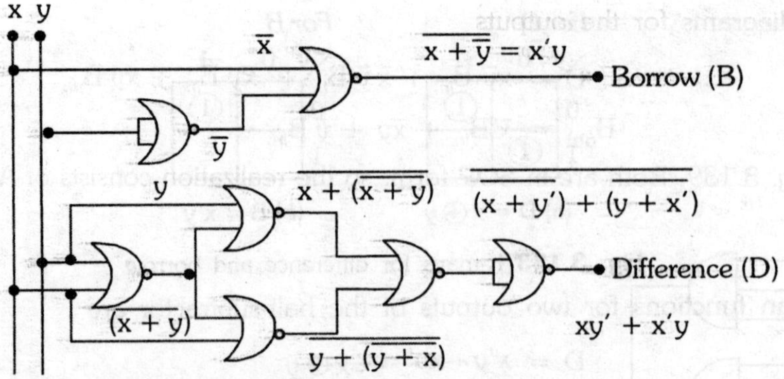

Fig. 3.136 Half-subtractor using NOR gates only.

3.10.2.2 Full-subtractor

A full subtractor is a combinational circuit that performs a subtraction of two bits, taking into account borrow of the lower significant stage. This circuit has three inputs and two outputs as shown in Fig. 3.137. The three inputs are *x*, *y* and B_{in} denoting minuend, subtrahend, and previous borrow, respectively.

The two outputs are, D and B_{out}, which represent the difference and output borrow respectively. The truth table for full-subtractor is shown in Table 3.31.

Table 3.31. Truth table for full-subtractor

Inputs			*Outputs*	
x	*y*	B_{in}	*D*	B_{out}
0	0	0	0	0
0	0	1	1	1
0	1	0	1	1
0	1	1	0	1
1	0	0	1	0
1	0	1	0	0
1	1	0	0	0
1	1	1	1	1

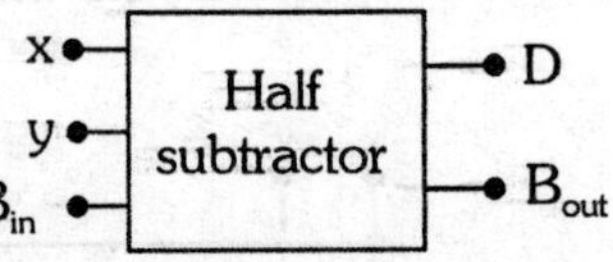

Fig. 3.137 Symbol for full-subtractor.

The K-map simplificants for full-subtractor outputs are shown in Fig. 3.138.

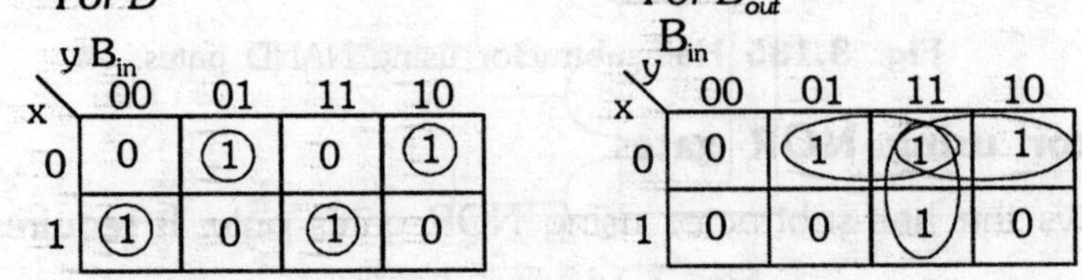

$D = x\bar{y}\bar{B}_{in} + \bar{x}\bar{y}B_{in} + xyB_{in} + \bar{x}y\bar{B}_{in}$

$B_{out} = \bar{x}B_{in} + \bar{x}y + yB_{in}$

Fig. 3.138 K-map simplifications for D and B_{out}.

The logic diagrams for the outputs

$$D = x\bar{y}\,\bar{B}_{in} + \bar{x}\,\bar{y}\,B_{in} + \bar{x}y\,\bar{B}_{in} + xy\,B_{in}$$

and

$$B_{out} = \bar{x}\,B_{in} + \bar{x}y + y\,B_{in}$$

are shown in Fig. 3.139. Both are in SOP form, so the realization consists of AND-OR (*i.e.*, two level gate) logic.

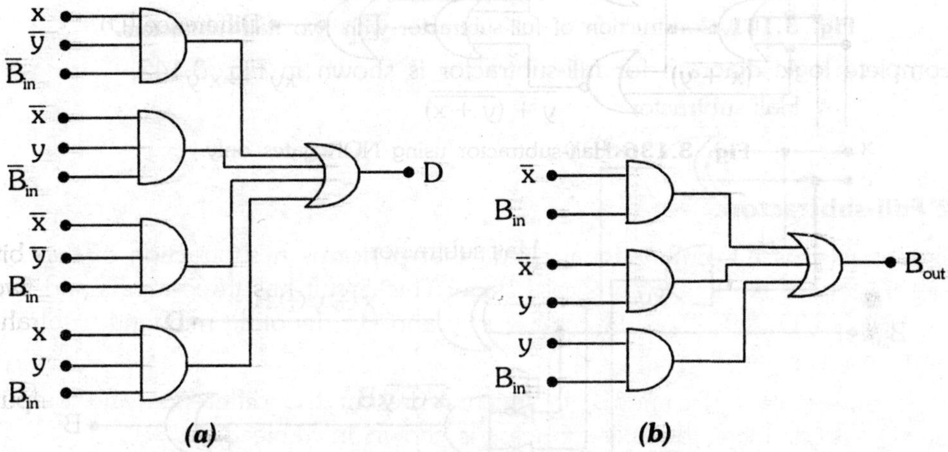

Fig. 3.139 Logic diagrams for difference and borrow of full subtractor.

The Boolean function 'D' can be simplified as

$$\begin{aligned} D &= \bar{x}\,\bar{y}\,B_{in} + \bar{x}y\,\bar{B}_{in} + x\bar{y}\,\bar{B}_{in} + xy\,B_{in} \\ &= B_{in}\,(\bar{x}\,\bar{y} + xy) + \bar{B}_{in}\,(\bar{x}y + xy') \\ &= B_{in}\,(x \odot y) + \bar{B}_{in}\,(x \oplus y) \\ &= B_{in}\,(\overline{x \oplus y}) + \bar{B}_{in}\,(x \oplus y) \\ &= x \oplus y \oplus B_{in} \end{aligned}$$

with this, the logic diagram is shown in Fig. 3.140.

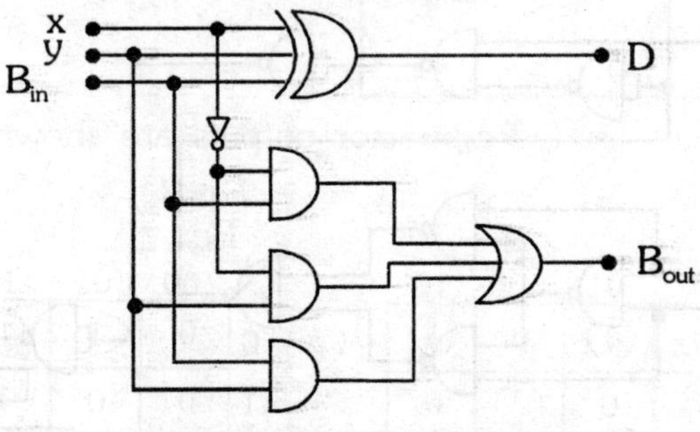

Fig. 3.140 Full-subtractor.

A full-subtractor can also be implemented with two half-subtractors and one OR gate, as shown in Fig. 3.141.

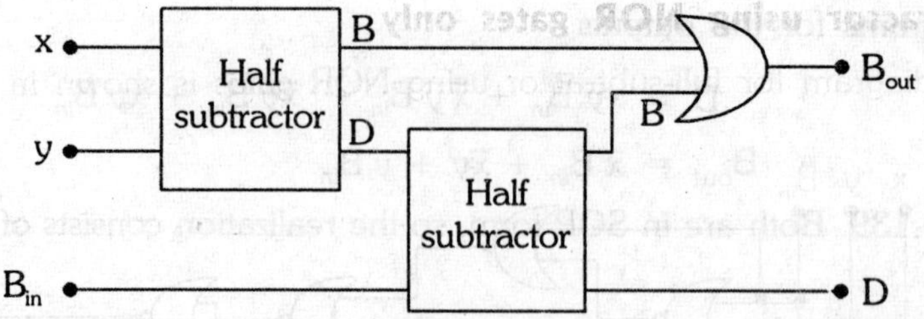

Fig. 3.141 Construction of full-subtractor with two half-subtractors.

The complete logic diagram for full-subtractor is shown in Fig. 3.142.

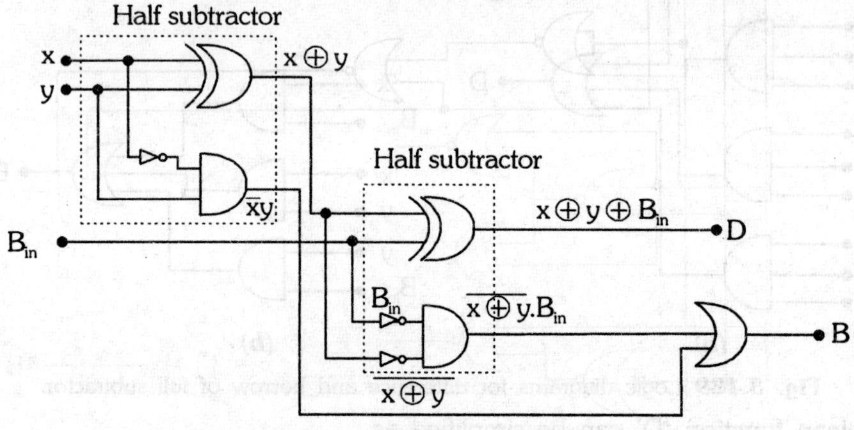

Fig. 3.142 Logic diagram for full-subtractor.

(*a*) Full subtractor using NAND gates only

The logic diagram for full subtractor using NAND gates is shown in Fig. 3.143 and it contains 17 NAND gates.

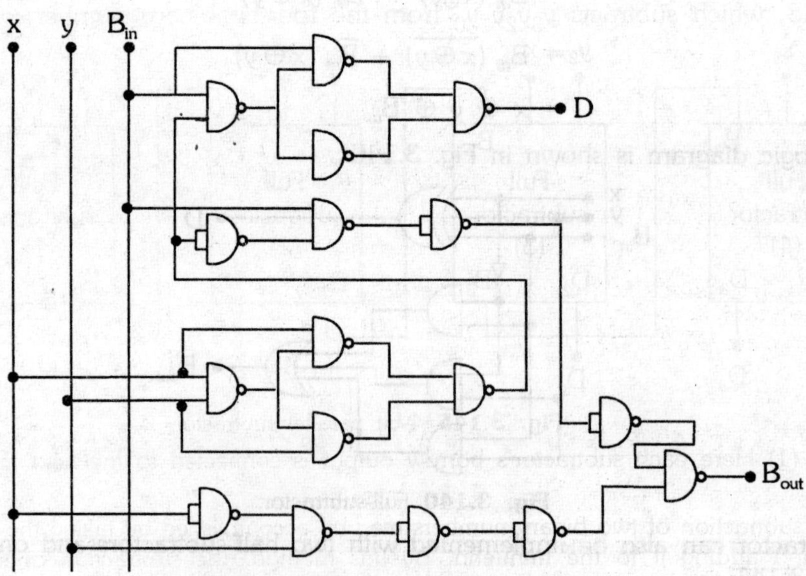

Fig. 3.143 Full subtractor using NAND gates.

(*b*) Full-subtractor using NOR gates only

The logic diagram for full-subtractor using NOR gates is shown in Fig. 3.144. It requires 16 NOR gates.

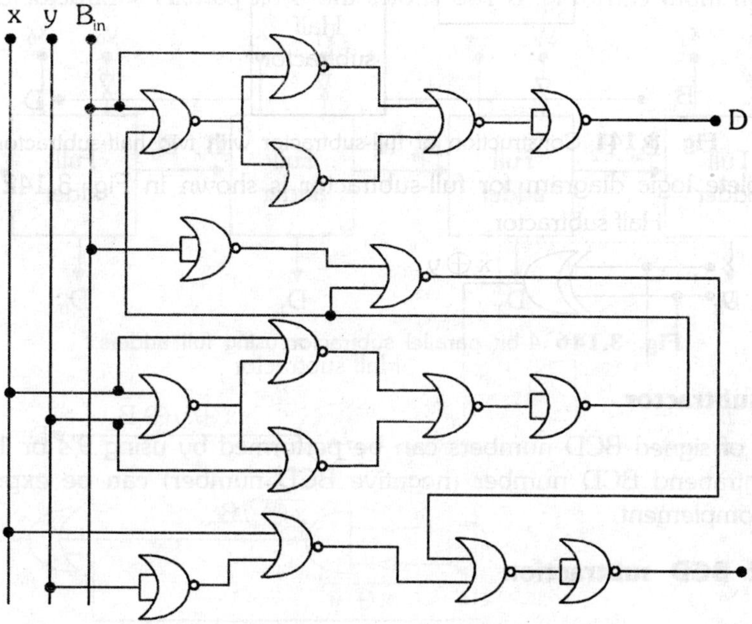

Fig. 3.144 Full-subtractor using NOR gates.

3.10.2.3 Parallel subtractor

The full subtractors are wired together to form a "parallel binary subtractor". The number of full subtractors required depends on the number of bits. 4-bit parallel subtractor is shown in Fig. 3.145, which subtracts $y_3y_2y_1y_0$ from the four-bit binary number $x_3x_2x_1x_0$.

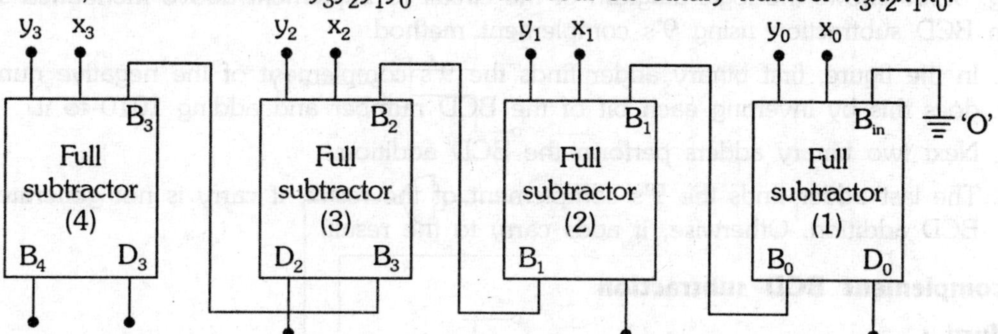

Fig. 3.145 4-bit parallel subtractor.

Note : (1) Here each subtractor's borrow output is connected to the next more significant bit's borrow input.

(2) The subtraction of two binary numbers may be accomplished by taking the complement of the subtrahend and adding it to the minuend. By this method, the subtraction operation becomes an addition operation which requires full-adders.

The subtraction $x - y$ can be done by taking 2's complement of y and adding it to x. The 2's complement can be obtained by taking 1's complement and adding one to the least significant pairbits. The 1's complement can be obtained with inverters and a one can be added to the sum through input carry Fig. 3.146 shows the 4-bit parallel subtractor using full adders.

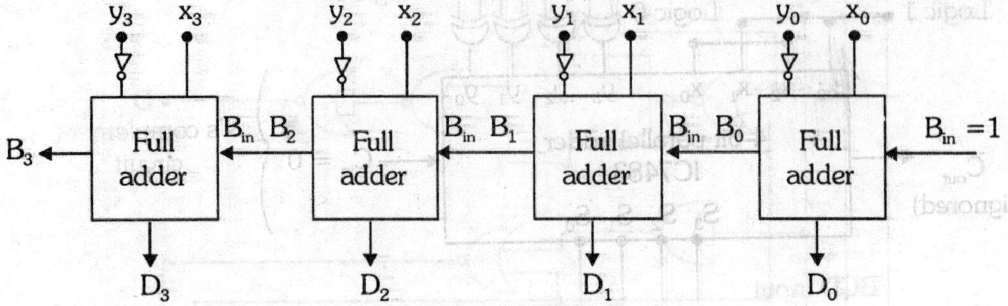

Fig. 3.146 4-bit parallel subtractor using full adders.

3.10.2.4 BCD subtractor

The addition of signed BCD numbers can be performed by using 9's or 10's complement methods. The subtrahend BCD number (negative BCD number) can be expressed by taking the 9's or 10's complement.

9's complement BCD subtraction

Procedure :

1. Take the 9's complement of the negative number.
2. Add two numbers using BCD addition.
3. If a carry is generated, add carry to the result. Otherwise, find the 9's complement of the result.

Fig. 3.147 shows the logic diagram of the circuit to implement above mentioned steps to perform BCD subtraction using 9's complement method.

1. In the figure, first binary adder finds the 9's complement of the negative number. It does this by inverting each bit of the BCD number and adding 1010 to it.
2. Next two binary adders perform the BCD addition.
3. The last adder finds the 9's complement of the result, if carry is not generated after BCD addition. Otherwise, it adds carry to the result.

10's complement BCD subtraction

Procedure :

1. Find the 10's complement of a negative number.
2. Add two numbers using BCD addition.
3. If a carry is not generated, find the 10's complement of the result, otherwise ignore the carry.

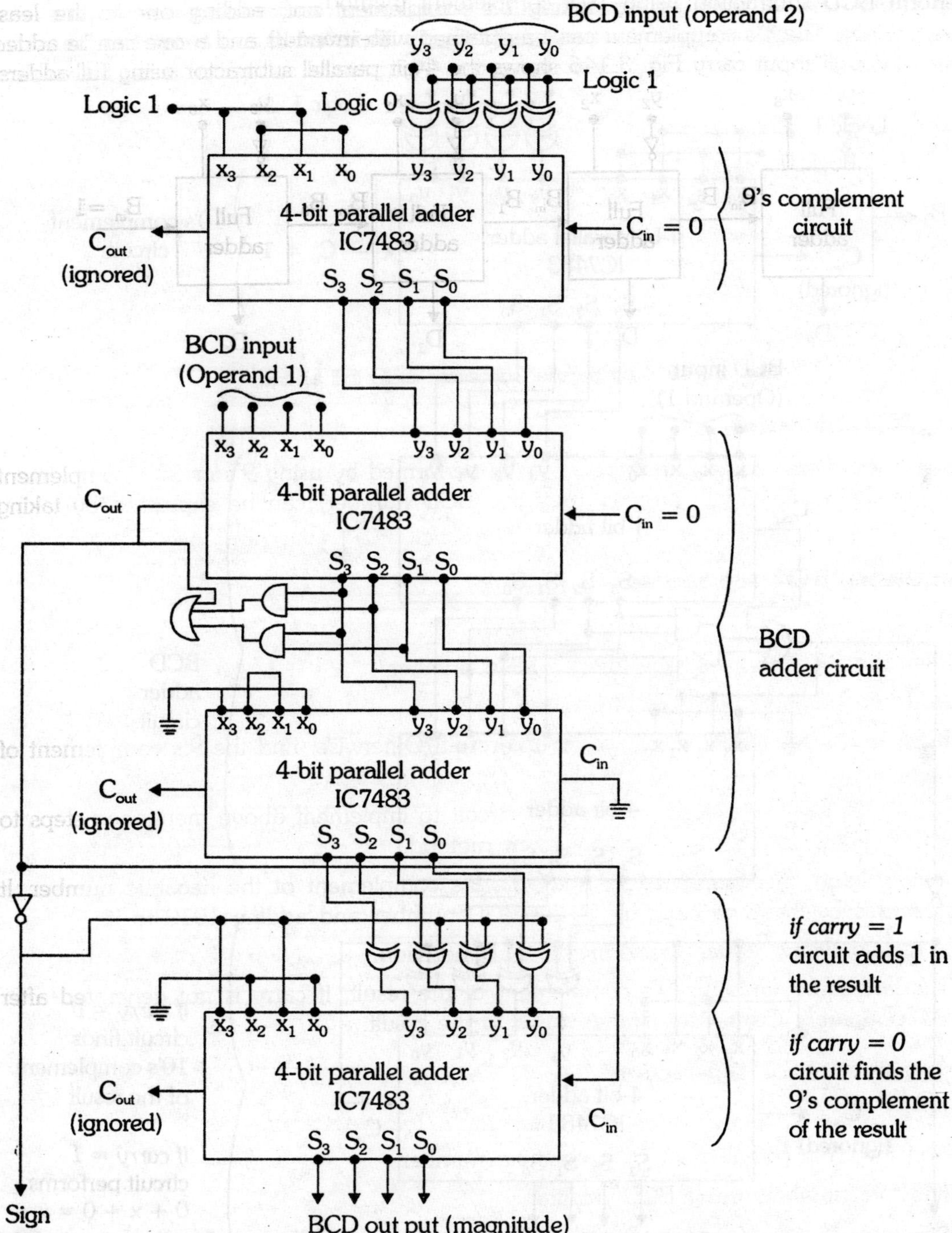

Fig. 3.147 4-bit BCD subtractor using 9's complement method.

Fig. 3.148 shows the logic diagram of the circuit to implement the above mentioned steps to perform BCD subtraction using 10's complement method.

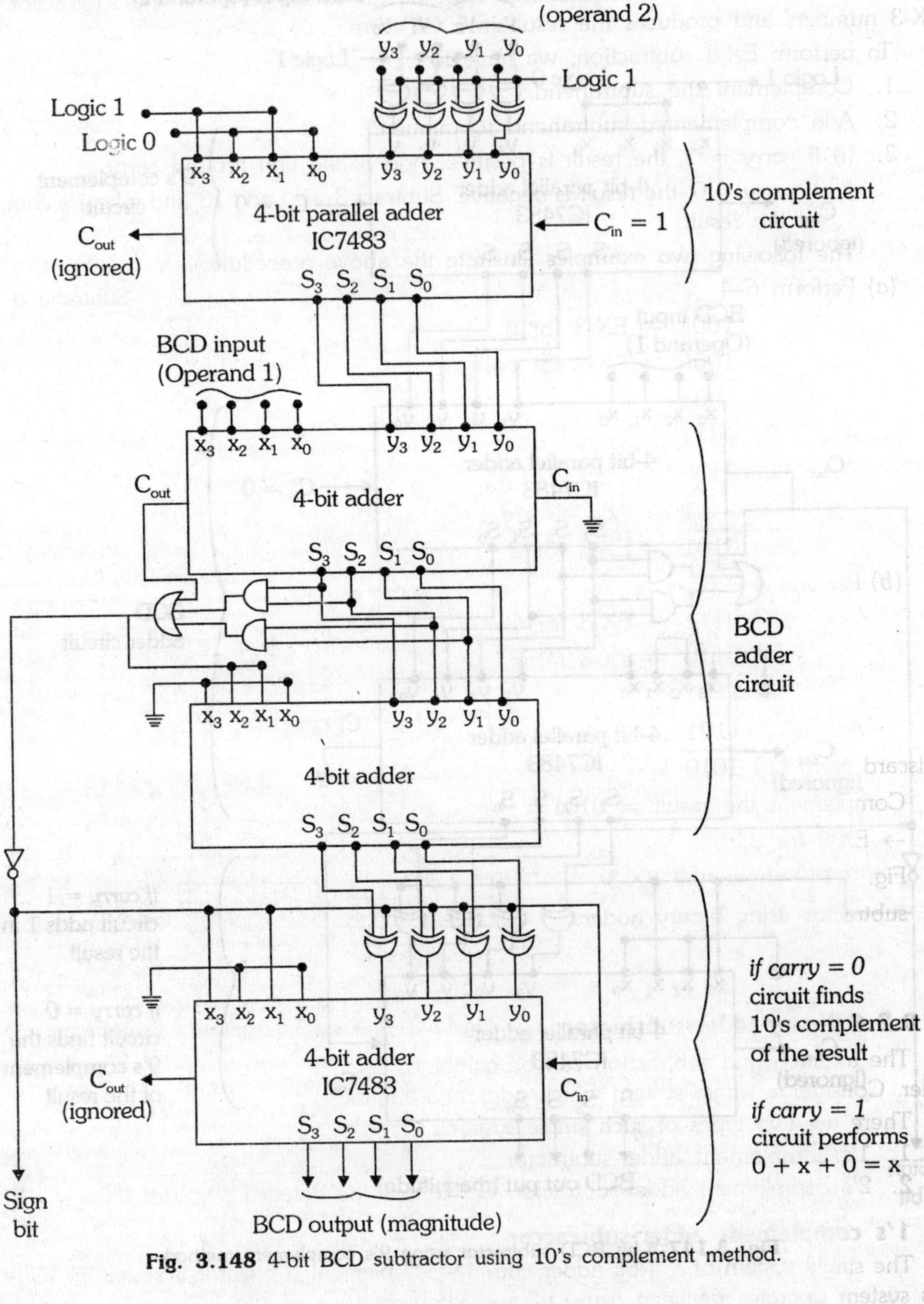

Fig. 3.148 4-bit BCD subtractor using 10's complement method.

3.10.2.5 Excess-3 subtractor

EX-3 subtractor is a combinational logic circuit which performs the subtraction of two EX-3 numbers and produces the result in EX-3 form.

To perform EX-3 subtraction, we have to

1. Complement the subtrahend.
2. Add complemented subtrahend to minuend.
3. (*i*) If carry = 1, the result is positive. Add 3 and end around carry.
   (*ii*) If carry = 0, the result is negative. Subtract 3, *i.e.*, add 13 and take 1's complement of the result.

The following two examples illustrate the above procedure.

(*a*) Perform 6–4

1001 → EX-3 for 6
1000 → EX-3 for (–4)
(1) 0001
Add 3 0011
0100
1 Add discard carry
0101 → EX-3 for 2

(*b*) Perform 4 – 6

0111 → EX-3 for 4
0110 → EX-3 from (–6)
1101
Add 13 1101
Discard carry (1) 1010
Complement the result = 0101
→ EX-3 for 2.

Fig. 3.149 shows the EX-3 subtractor using binary adders.

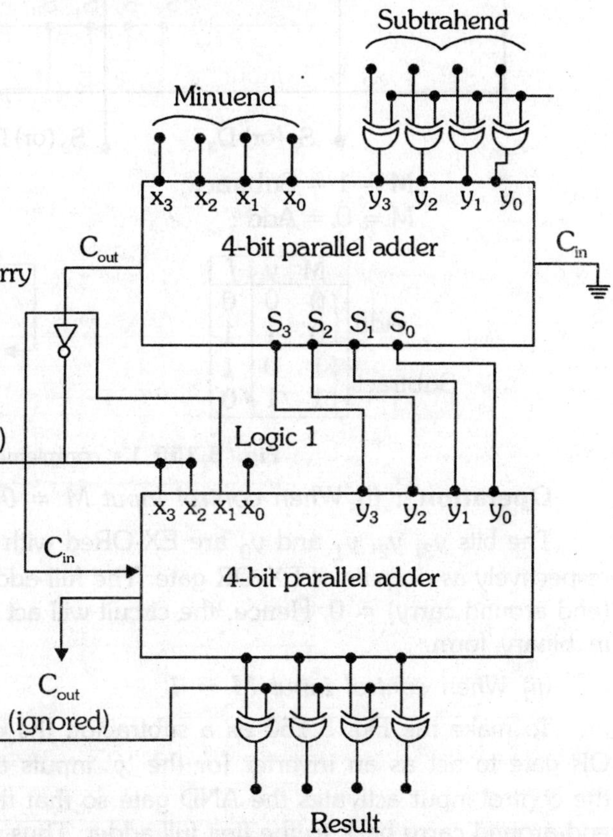

Fig. 3.149 Excess-3 subtractor using 4-bit binary adders.

3.10.2.6 Binary adder-subtractor

The addition and subtraction can be combined into one circuit with one common binary adder. Consider a single system which adds and subtracts.

There are two types of such single systems

1. 1's complement adder-subtractor.
2. 2's complement adder-subtractor. Let us see the detailed study of each.

(*i*) 1's complement adder-subtractor

The single system of a 4-bit adder cum 1's complement subtractor is shown in Fig. 3.150. This system operates between $y_3y_2y_1y_0$ and $x_3x_2x_1x_0$.

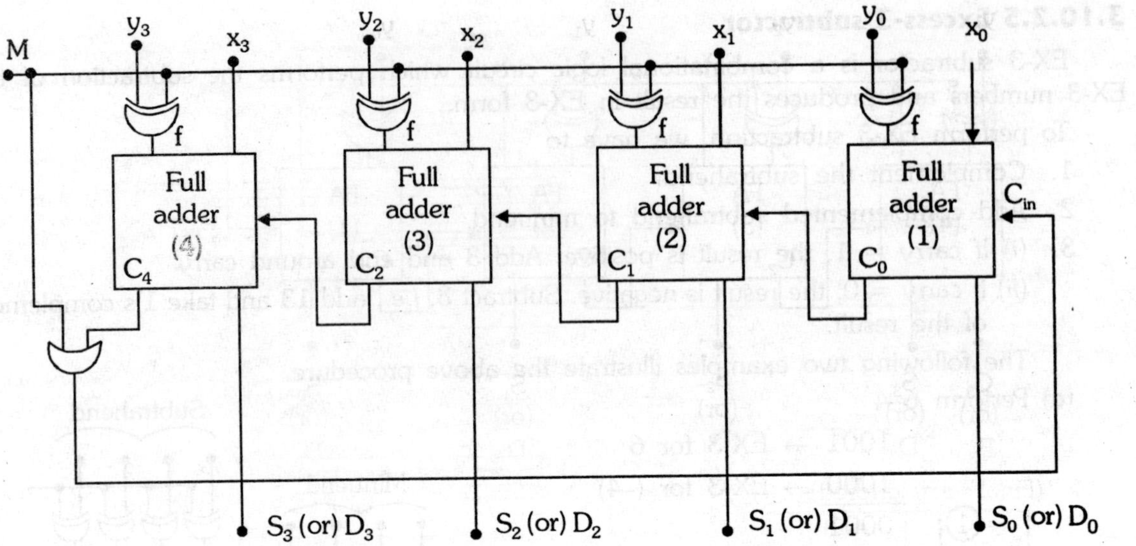

M = 1 = Subtract ;
M = 0 = Add ;

	M	y	f
Add	0	0	0
	0	1	1
Subtract	1	0	1
	1	1	0

$$\begin{array}{r} +x_3x_2x_1x_0 \\ -y_3y_2y_1y_0 \\ \hline S_3S_2S_1S_0 \\ \text{or} \\ D_3D_2D_1D_0 \end{array}$$

Fig. 3.150 1's complement adder subtractor.

Operation : (*i*) *When control input M = 0*

The bits y_3, y_2, y_1 and y_0 are EX-ORed with '0' and produces $(y_3 \oplus 0 = y_3)$ $y_3y_2y_1$ and y_0 respectively as outputs of EX-OR gate. The full-adders receive the value of y and the input carry (end around carry) = 0. Hence, the circuit will act as a "adder". The result appears as $S_3S_2S_1S_0$ in binary form.

(*ii*) *When control input M = 1*

To make the Fig. 3.150 as a subtractor, the control input must be 1. This causes the EX-OR gate to act as an inverter for the 'y' inputs to the full adders. (*i.e.*, $y \oplus 1 = \bar{y}$). The 1 on the control input activates the AND gate so that the carry output of the full adder can take the end-around carry back to the first full adder. Thus, subtractor subtracts the binary input $y_3y_2y_1y_0$ from $x_3x_2x_1x_0$. The difference appears as $D_3D_2D_1D_0$ in binary form.

(*ii*) 2's complement adder-subtractor

The single system of a 4-bit adder cum 2's complement subtractor is shown in Fig. 3.151. It operates between two 4-bit numbers: $x_3x_2x_1x_0$ and $y_3y_2y_1y_0$. In case of subtraction : $y_3y_2y_1y_0$ is subtracted from $x_3x_2x_1x_0$.

Operation : (*i*) When control M = 0, the circuit will act as a 4-bit binary adder (since $y \oplus 0 = y$). Then, bits $y_3y_2y_1y_0$ pass through the XOR gates without inversion. Therefore, the unit adds the number $x_3x_2x_1x_0$ to $y_3y_2y_1y_0$. The result appears as $S_3S_2S_1S_0$ in the binary form. If M = 0, input carry = 0, and final carry output is neglected.

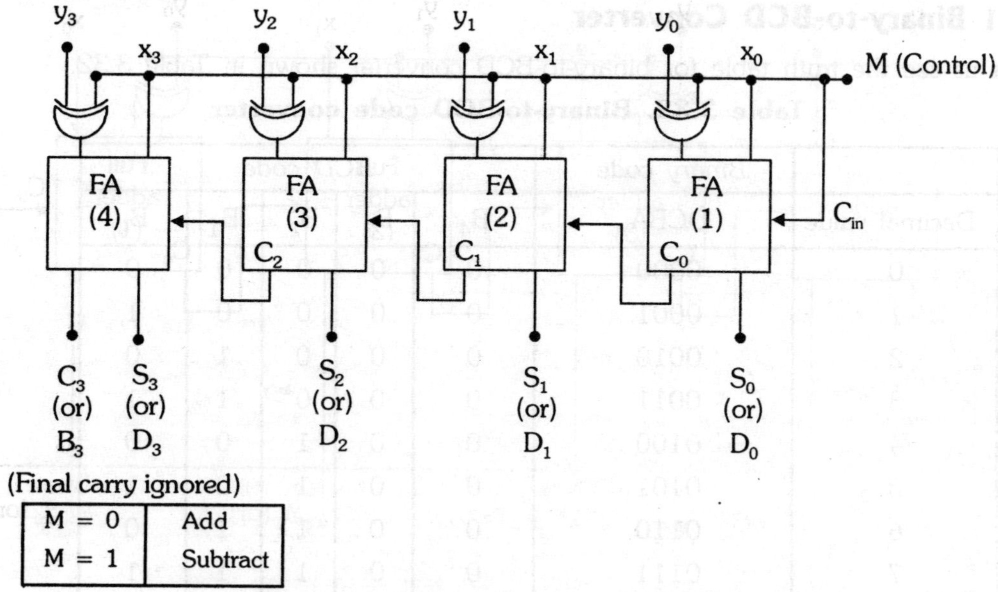

Fig. 3.151 2's complement adder subtractor.

(*ii*) When M = 1, this causes the XOR gate to act as an inverter for the *y* inputs to the full adders $(y \oplus 1 = \overline{y})$. The input to each full adder is *x* – *y*. If M = 1, the input carry = 1. The high M adds a 1 to the first full adder. This addition of 1 to the 1's complement of *y* bits forms 2's complement of *y*. Then, the circuit performs the subtraction operation as per 2's complement. The result appears as $D_3D_2D_1D_0$ in binary form. The final carry output of the last full adder is not used or ignored.

3.11 CODE CONVERTERS

The availability of a large different types of codes for the same discrete elements of information result in the use of different codes by different digital systems. Sometimes, it may happen that the use of output of one number system as the input to another number system. Hence it is necessary to design a conversion circuit and insert it between the two systems.

Thus a code converter is a circuit, which accepts the input information in one binary code, converts it and produces an output into another binary code *i.e.*, which makes the two systems compatible, even though each uses a different binary code.

To convert from binary code *x* to binary code *y*, the input lines must supply the bit of combination of elements as specified by code *x* and the output lines must generate the corresponding bit combination of code *y*. A combinational circuit performs this transformation by means of logic gates. The general block diagram of a code converter is shown in Fig. 3.152.

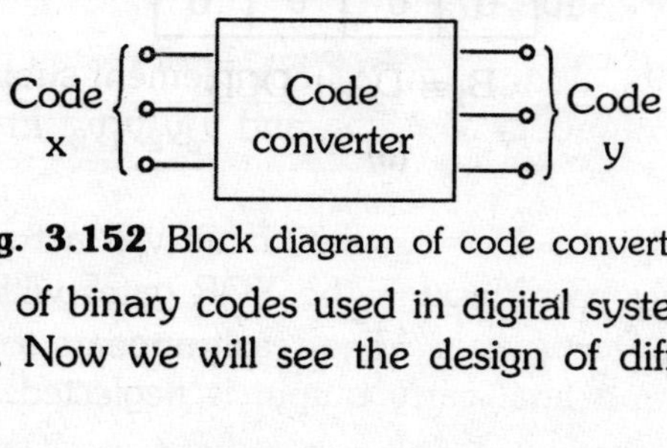

Fig. 3.152 Block diagram of code converter.

There are a wide variety of binary codes used in digital systems. Some of these codes are BCD, EX-3, gray and so on. Now we will see the design of different code converters.

3.11.1 Binary-to-BCD Converter

Let us see the truth table for binary-to-BCD converter shown in Table 3.32.

Table 3.32. Binary-to-BCD code converter

	Binary code	*BCD code*				
Decimal value	DCBA	B_4	B_3	B_2	B_1	B_0
0	0000	0	0	0	0	0
1	0001	0	0	0	0	1
2	0010	0	0	0	1	0
3	0011	0	0	0	1	1
4	0100	0	0	1	0	0
5	0101	0	0	1	0	1
6	0110	0	0	1	1	0
7	0111	0	0	1	1	1
8	1000	0	1	0	0	0
9	1001	0	1	0	0	1
10	1010	1	0	0	0	0
11	1011	1	0	0	0	1
12	1100	1	0	0	1	0
13	1101	1	0	0	1	1
14	1110	1	0	1	0	0
15	1111	1	0	1	0	1

To draw the logic diagram for code converter, each of the output functions are simplified by using K-maps as shown in Fig. 3.153.

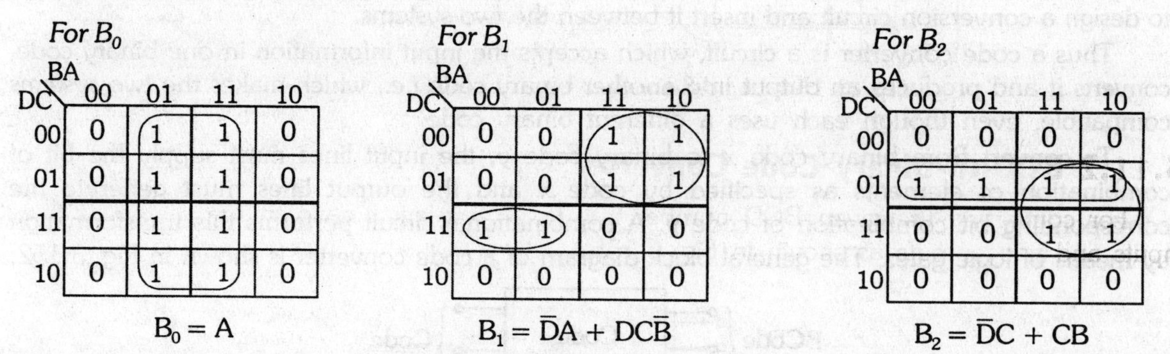

(*a*) (b) (c)

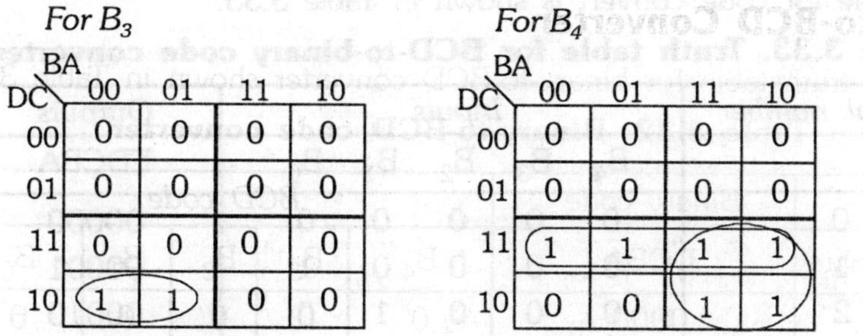

$B_3 = \bar{D}C\bar{B}$ (*d*) $B_4 = DC + DB = D(C + B)$ (*e*)

Fig. 3.153 K-map simplifications for binary to-BCD code converter outputs.

The logic diagram for the code converter is shown in Fig. 3.154.

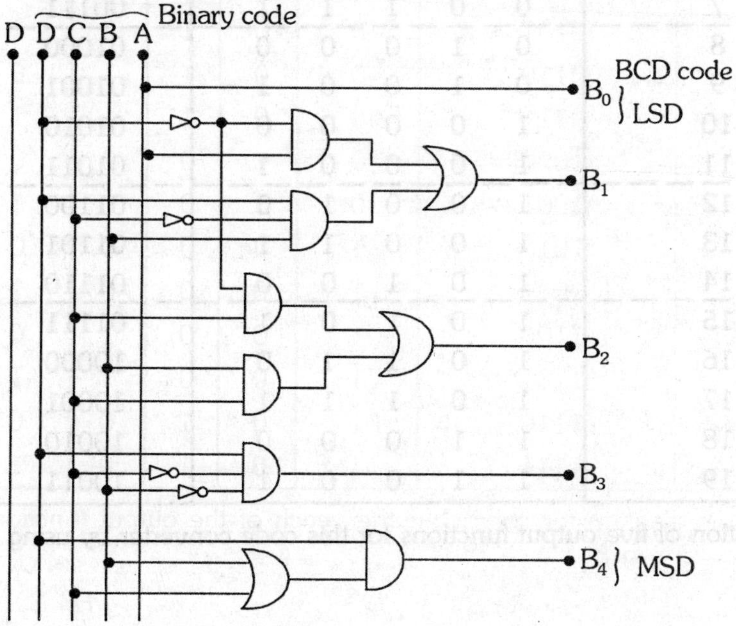

Fig. 3.154 Logic diagram : binary-to-BCD converter.

3.11.2 BCD-to-Binary Code Converter

For converting the given BCD number into it's equivalent binary, the circuit needs five inputs and five outputs as shown in Fig. 3.155.

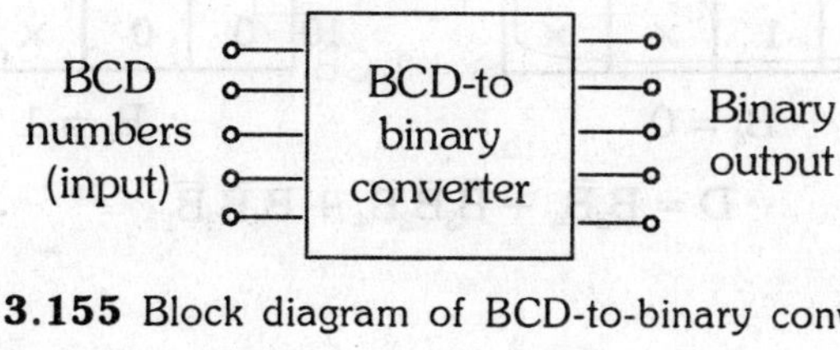

Fig. 3.155 Block diagram of BCD-to-binary converter.

The truth table for code convert is shown in Table 3.33.

Table 3.33. Truth table for BCD-to-binary code converter

Decimal number	*Inputs*					*Outputs*
	B_4	B_3	B_2	B_1	B_0	EDCBA
0	0	0	0	0	0	00000
1	0	0	0	0	1	00001
2	0	0	0	1	0	00010
3	0	0	0	1	1	00011
4	0	0	1	0	0	00100
5	0	0	1	0	1	00101
6	0	0	1	1	0	00110
7	0	0	1	1	1	00111
8	0	1	0	0	0	01000
9	0	1	0	0	1	01001
10	1	0	0	0	0	01010
11	1	0	0	0	1	01011
12	1	0	0	1	0	01100
13	1	0	0	1	1	01101
14	1	0	1	0	0	01110
15	1	0	1	0	1	01111
16	1	0	1	1	0	10000
17	1	0	1	1	1	10001
18	1	1	0	0	0	10010
19	1	1	0	0	1	10011

The simplification of five output functions for this code converter by using K-maps is shown in Fig. 3.156.

For E

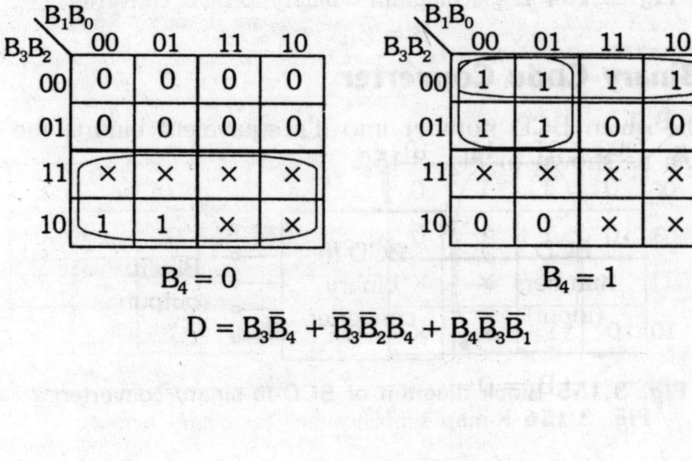

$$D = B_3\overline{B}_4 + \overline{B}_3\overline{B}_2B_4 + B_4\overline{B}_3\overline{B}_1$$

For D

B_3B_2 \ B_1B_0	00	01	11	10
00	0	0	0	0
01	0	0	0	0
11	×	×	×	×
10	0	0	×	×

$B_4 = 0$

B_3B_2 \ B_1B_0	00	01	11	10
00				
01			1	1
11	×	×	×	×
10	1	1	×	×

$B_4 = 1$

$$E = B_4B_3 + B_4B_2B_0$$

For C

B_3B_2 \ B_1B_0	00	01	11	10
00	0	0	0	0
01	1	1	1	1
11	×	×	×	×
10	0	0	×	×

$B_4 = 0$

B_3B_2 \ B_1B_0	00	01	11	10
00	0	0	1	1
01	1	1	0	0
11	×	×	×	×
10	0	0	×	×

$B_4 = 1$

$$C = \bar{B}_4B_2 + B_2\bar{B}_1 + B_4\bar{B}_2B_1$$

For B

B_3B_2 \ B_1B_0	00	01	11	10
00	0	0	1	1
01	0	0	1	1
11	×	×	×	×
10	0	0	×	×

$B_4 = 0$

B_3B_2 \ B_1B_0	00	01	11	10
00	1	1	0	0
01	1	1	0	0
11	×	×	×	×
10	1	1	×	×

$B_4 = 1$

$$B = \bar{B}_4B_1 + B_4\bar{B}_1$$

For A

B_3B_2 \ B_1B_0	00	01	11	10
00	0	1	1	0
01	0	1	1	0
11	×	×	×	×
10	0	1	×	×

$B_4 = 0$

B_3B_2 \ B_1B_0	00	01	11	10
00	0	1	1	0
01	0	1	1	0
11	×	×	×	×
10	0	1	×	×

$B_4 = 1$

Fig. 3.156 K-map simplifications for binary outputs.

$$\boxed{A = B_0}$$

The Boolean expression output functions are given by :

$$A = B_0$$

$$B = B_1 \oplus B_4$$

$$C = B_4'B_2 + B_2B_1' + B_4B_2'B_1$$

$$D = B_3\overline{B}_4 + \overline{B}_3\overline{B}_2B_4 + B_4\overline{B}_3B_1$$

$$E = B_4B_3 + B_4B_2B_0$$

The combined logic diagram for the multiple output code converter is shown in Fig. 3.157.

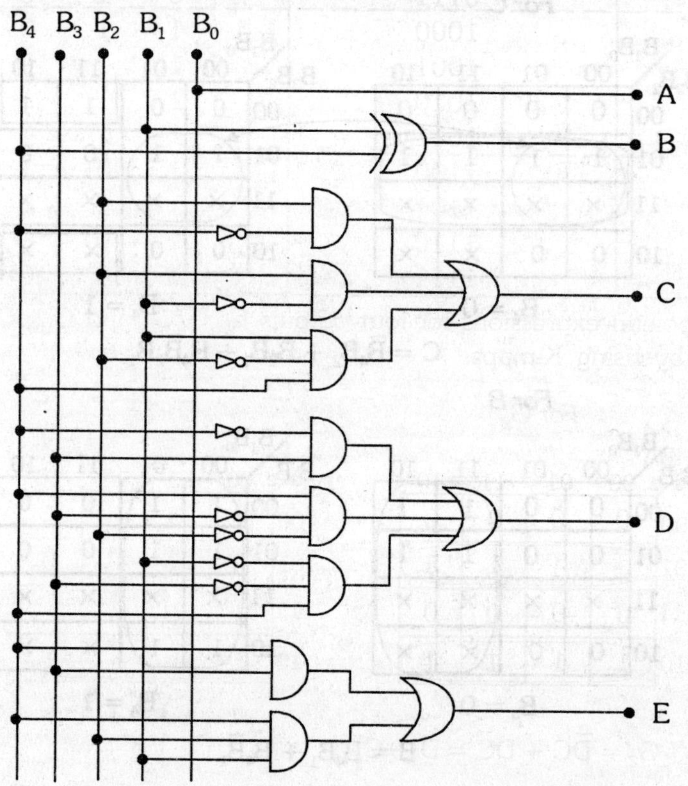

Fig. 3.157 Logic diagram for binary-to-BCD code converter.

3.11.3 Binary-to-Gray Code Converter

The main advantage of gray code is that, only one bit in the numerical representation changes between the successive numbers. The Table 3.34 shows the truth table for Binary-to-Gray code conversion. The conversion circuit must contain four-inputs and four-outputs as shown in Fig. 3.158.

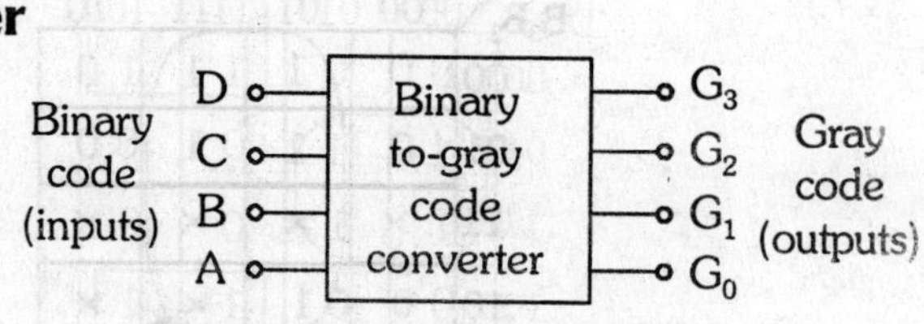

Fig. 3.158 Block diagram for binary-to-gray code converter.

Table 3.34. Truth-table for binary-to-gray code converter

Decimal	*Binary code*	*Gray code*			
	DCBA	G_3	G_2	G_1	G_0
0	0000	0	0	0	0
1	0001	0	0	0	1
2	0010	0	0	1	1
3	0011	0	0	1	0
4	0100	0	1	1	0
5	0101	0	1	1	1
6	0110	0	1	0	1
7	0111	0	1	0	0
8	1000	1	1	0	0
9	1001	1	1	0	1
10	1010	1	1	1	1
11	1011	1	1	1	0
12	1100	1	0	1	0
13	1101	1	0	1	1
14	1110	1	0	0	1
15	1111	1	0	0	0

The simplified Boolean expressions for four outputs $G_3, G_2 G_1$ and G_0 of the converter are shown in Fig. 3.159 by using K-maps.

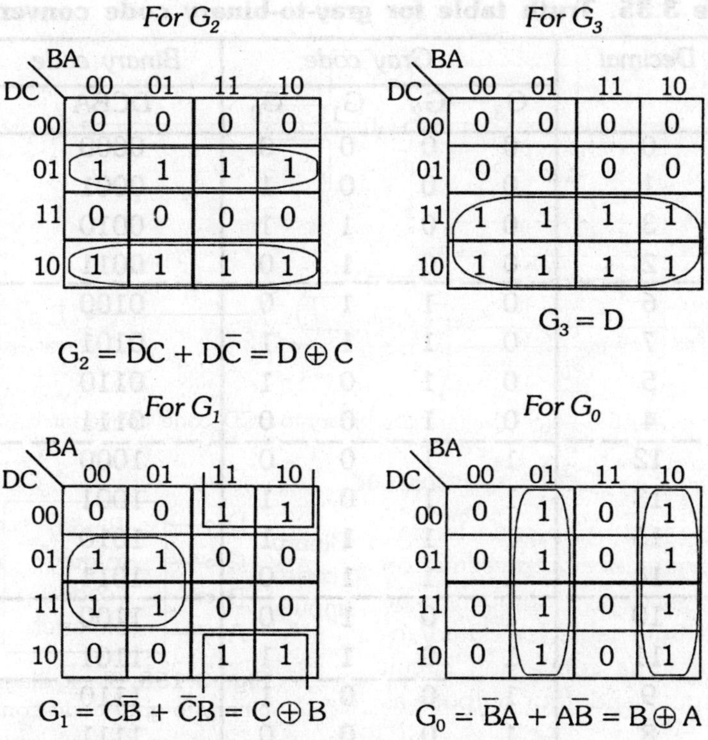

Fig. 3.159 K-maps for binary-to-gray code converter outputs.

The logic diagrams for the binary-to-gray code converter outputs.

$$G_0 = A \oplus B \qquad G_2 = C \oplus D$$
$$G_1 = B \oplus C \qquad G_3 = D$$

are shown in Fig. 3.160.

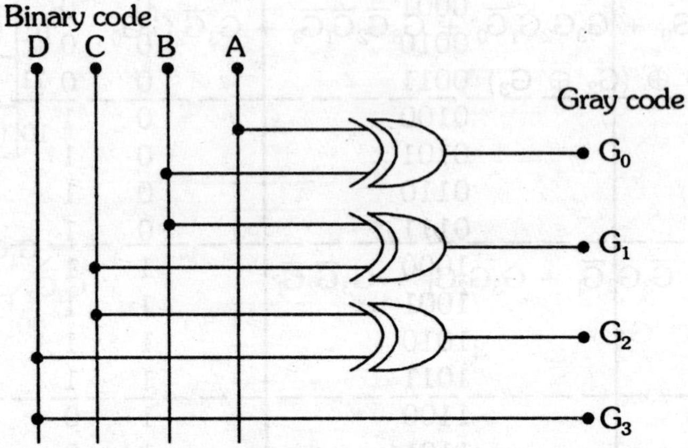

Fig. 3.160 Logic diagram for binary-to-gray code converter.

3.11.4 Gray Code-to-Binary Code Converter

The gray-to-binary code converter contains four inputs G_3, G_2, G_1 and G_0 and four outputs D, C, B and A. The truth table for gray-to-binary code converter is shown in Table 3.35.

Table 3.35. Truth table for gray-to-binary code converter

Decimal	*Gray code*				*Binary code*
	G_3	G_2	G_1	G_0	DCBA
0	0	0	0	0	0000
1	0	0	0	1	0001
3	0	0	1	1	0010
2	0	0	1	0	0011
6	0	1	1	0	0100
7	0	1	1	1	0101
5	0	1	0	1	0110
4	0	1	0	0	0111
12	1	1	0	0	1000
13	1	1	0	1	1001
15	1	1	1	1	1010
14	1	1	1	0	1011
10	1	0	1	0	1100
11	1	0	1	1	1101
9	1	0	0	1	1110
8	1	0	0	0	1111

The simplified expressions for four binary code out of the gray-to-binary code converter are as shown in Fig. 3.161.

$$A = \overline{G_3}\,\overline{G_2}\,\overline{G_1}G_0 + \overline{G_3}\,\overline{G_2}G_1\overline{G_0} + \overline{G_3}G_2\overline{G_1}\,\overline{G_0} + \overline{G_3}G_2G_1G_0$$
$$+ G_3G_2\overline{G_1}G_0 + G_3G_2G_1\overline{G_0} + G_3\overline{G_2}\,\overline{G_1}\,\overline{G_0} + G_3\overline{G_2}G_1G_0$$
$$A = (G_0 \oplus G_1) \oplus (G_2 \oplus G_3)$$

For A

G_3G_2 \ G_1G_0	00	01	11	10
00	0	1	0	1
01	1	0	1	0
11	0	1	0	1
10	1	0	1	0

$$B = \overline{G_3}\,\overline{G_2}G_1 + \overline{G_3}G_2\overline{G_1} + G_3G_2G_1 + G_3\overline{G_2}\,\overline{G_1}$$
$$= G_1 \oplus G_2 \oplus G_3$$

For B

G_3G_2 \ G_1G_0	00	01	11	10
00	0	0	1	1
01	1	1	0	0
11	0	0	1	1
10	1	1	0	0

For C

G_3G_2 \ G_1G_0	00	01	11	10
00	0	0	0	0
01	1	1	1	1
11	0	0	0	0
10	1	1	1	1

$$C = \overline{G_3}G_2 + G_3\overline{G_2}$$
$$C = G_2 \oplus G_3$$

For D

G_3G_2 \ G_1G_0	00	01	11	10
00	0	0	0	0
01	0	0	0	0
11	1	1	1	1
10	1	1	1	1

$$D = G_3$$

Fig. 3.161 K-map simplifications for gray-to-binary code converter.

The four outputs of gray-to-binary code converter are

$A = (G_0 \oplus G_1) \oplus (G_2 \oplus G_3)$

$B = G_1 \oplus G_2 \oplus G_3$

$C = G_2 \oplus G_3$

$D = G_3$

The logic diagram for the above four outputs is shown in Fig. 3.162.

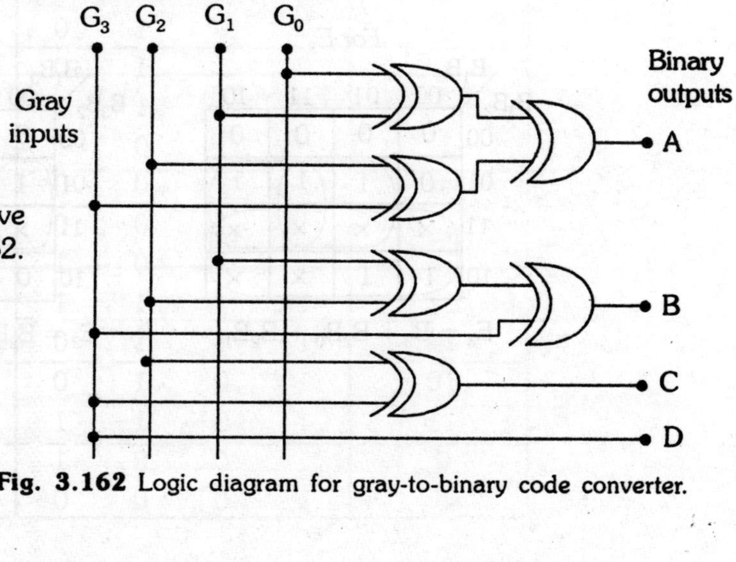

Fig. 3.162 Logic diagram for gray-to-binary code converter.

3.11.5 BCD-to-Excess-3 Code Converter

The Excess-3 code can be obtained from the BCD code by adding 3 to each coded number. The code converter contains four inputs and four outputs as shown in Fig. 3.163.

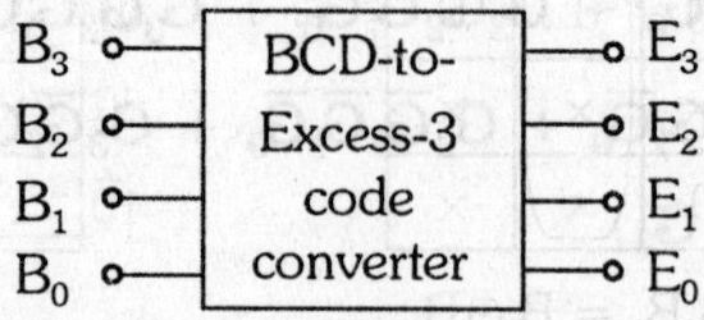

Fig. 3.163 Block diagram of BCD-to-EX-3 code converter.

The truth table for BCD-to-Excess 3 code is shown in Table 3.36.

Table 3.36. Truth table for BCD-to-Excess 3 code converter

Decimal number	*BCD number code*				*Excess-3 code*			
	B_3	B_2	B_1	B_0	E_3	E_2	E_1	E_0
0	0	0	0	0	0	0	1	1
1	0	0	0	1	0	1	0	0
2	0	0	1	0	0	1	0	1
3	0	0	1	1	0	1	1	0
4	0	1	0	0	0	1	1	1
5	0	1	0	1	1	0	0	0
6	0	1	1	0	1	0	0	1
7	0	1	1	1	1	0	1	0
8	1	0	0	0	1	0	1	1
9	1	0	0	1	1	1	0	0

For E_3

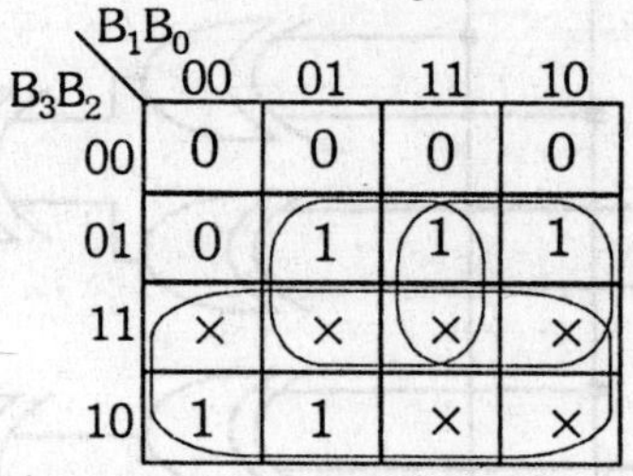

$E_3 = B_3 + B_2B_0 + B_2B_1$

For E_2

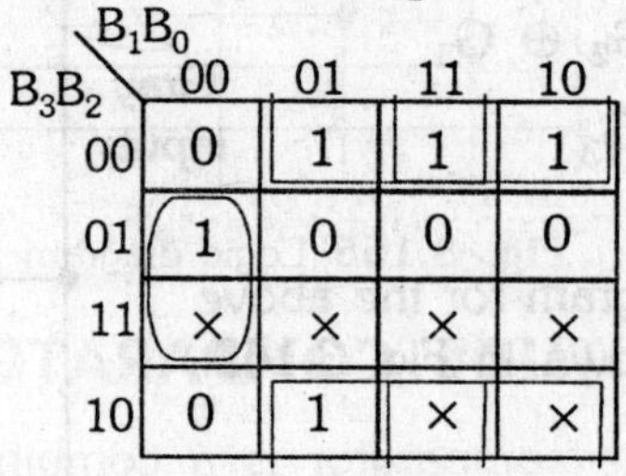

$E_2 = \bar{B}_2B_0 + \bar{B}_2B_1 + B_2\bar{B}_1\bar{B}_0$

For E_1

B_3B_2 \ B_1B_0	00	01	11	10
00	1	0	1	0
01	1	0	1	0
11	×	×	×	×
10	1	0	×	×

$E_1 = \overline{B}_1\overline{B}_0 + B_0B_1 = B_0 \odot B_1$

For E_0

B_3B_2 \ B_1B_0	00	01	11	10
00	1	0	0	1
01	1	0	0	1
11	×	×	×	×
10	1	0	×	×

$E_0 = \overline{B}_0$

Fig. 3.164 K-maps for BCD-to-Excess 3 code converter outputs.

The logic diagram for the outputs

$$E_3 = B_3 + B_2\,(B_0 + B_1)$$
$$E_2 = B_2\overline{B}_1\overline{B}_0 + \overline{B}_2\,(B_0 + B_1)$$
$$E_1 = B_0 \odot B_1$$
$$E_0 = \overline{B}_0$$

is shown in Fig. 3.165.

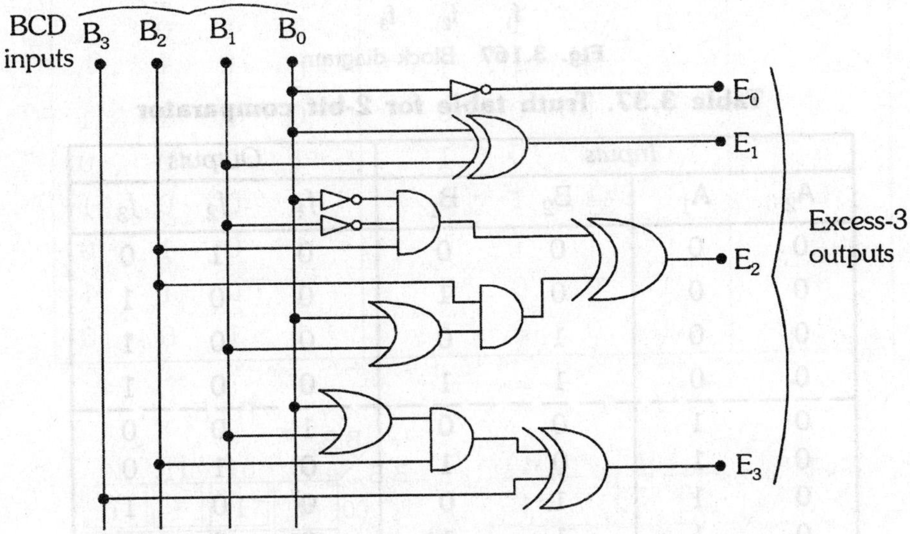

Fig. 3.165 Logic diagram for BCD-to-Excess 3 code converter.

3.12 MAGNITUDE COMPARATOR

A magnitude comparator is a combinational circuit designed primarily to compare the relative magnitude of two binary numbers A and B. The result of this comparison is specified by three binary variables that indicate, whether A > B, A = B or A < B. Fig. 3.166 shows the block diagram of an *n*-bit comparator. It receives two *n*-bit numbers as input and produces 3 outputs. Depending on the magnitudes of the two numbers, one of the outputs will be high.

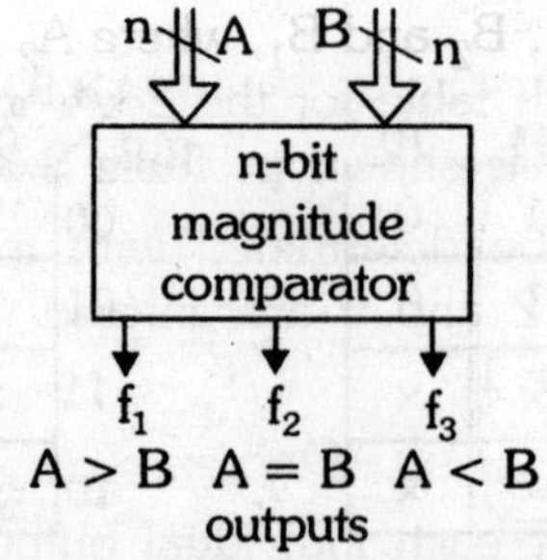

Fig. 3.166 Block diagram of *n*-bit comparator.

The output
$$f_1 = 1 \text{ if } A > B \text{ and}$$
$$f_2 = 1 \text{ if } A = B \text{ and}$$
$$f_3 = 1 \text{ if } A < B$$

As an example, consider an elementary two-bit comparator, as shown in Fig. 3.167.

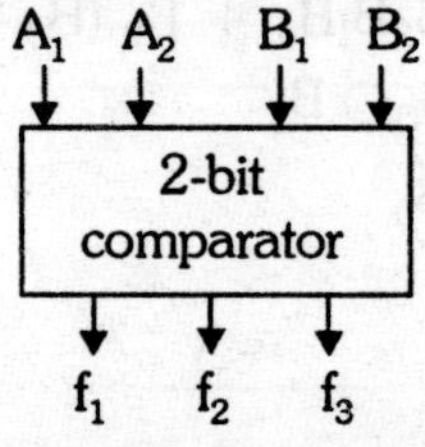

Fig. 3.167 Block diagram.

Table 3.37. Truth table for 2-bit comparator

Inputs				*Outputs*		
A_2	A_1	B_2	B_1	f_1	f_2	f_3
0	0	0	0	0	1	0
0	0	0	1	0	0	1
0	0	1	0	0	0	1
0	0	1	1	0	0	1
0	1	0	0	1	0	0
0	1	0	1	0	1	0
0	1	1	0	0	0	1
0	1	1	1	0	0	1
1	0	0	0	1	0	0
1	0	0	1	1	0	0
1	0	1	0	0	1	0
1	0	1	1	0	0	1
1	1	0	0	1	0	0
1	1	0	1	1	0	0
1	1	1	0	1	0	0
1	1	1	1	0	1	0

The circuit has four inputs A_2, A_1, B_2 and B_1, where A_2 and B_2 denote the most significant bits of A and B respectively. The truth table for the 2-bit comparator (here 2-bit means, each number to be compared is of 2-bits) is shown in Table 3.37.

(*i*) The logical equations may be determined with the use of the maps as shown in Fig. 3.168 (*a*). Where the values 1, 2 and 3 are entered in the appropriate cells to denote, respectively $f_1 = 1$, $f_2 = 1$ and $f_3 = 1$.

Or

(*ii*) The K-map simplifications for each individual output are shown in Fig. 3.168 (*b*).

A_2A_1 \ B_2B_1	00	01	11	10
00	2	3	3	3
01	1	2	3	3
11	1	1	2	1
10	1	1	3	2

Fig. 3.168(*a*) Map for f_1, f_2, f_3.

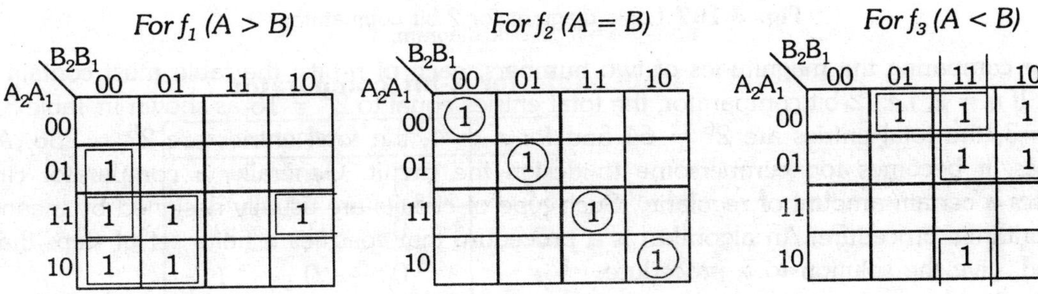

$f_1 = A_2\overline{B}_2 + A_1\overline{B}_2B_1 + A_2A_1\overline{B}_1$

$f_2 = \overline{A}_2\overline{A}_1\overline{B}_2\overline{B}_1 + \overline{A}_2A_1\overline{B}_2B_1 + A_2A_1B_2B_1 + A_2\overline{A}_1B_2\overline{B}_1 = (A_1 \odot B_1)\ (A2 \odot B2)$

$f_3 = \overline{A}_2B_2 + \overline{A}_2\overline{A}_1B_1 + \overline{A}_1B_2B_1$

Fig. 3.168 (*b*) K-maps for simplification of comparator outputs.

The logic diagram for the three output functions is shown in Fig. 3.169.

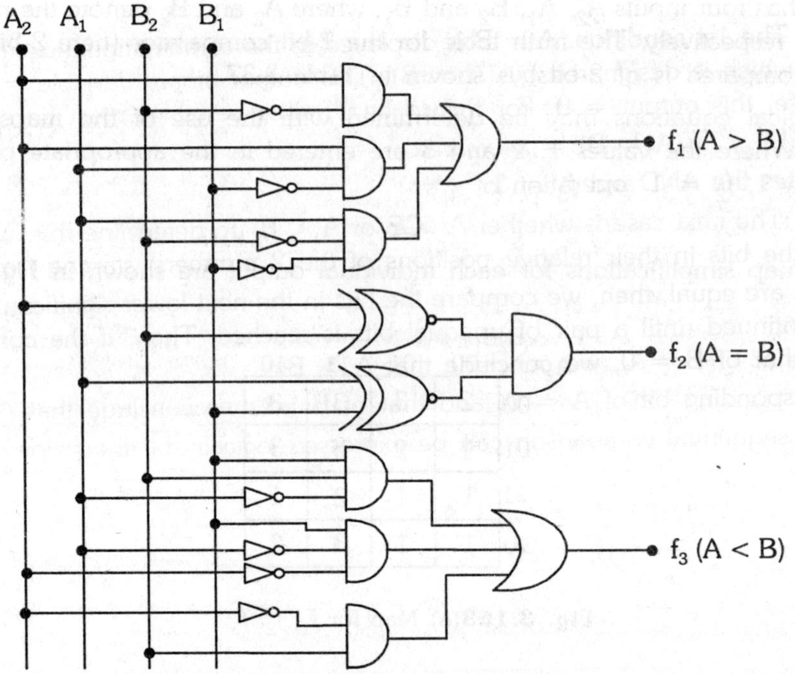

Fig. 3.169 Logic diagram for 2-bit comparator.

For comparing the magnitudes of two numbers, each of *n*-bits, the table must contain 2^{2n} entries. If $n = 2$, *i.e.*, 2-bit comparator, the total entries equal to $2^4 = 16$ as shown in Table 3.37. For $n = 3$, the total entries are $2^6 = 64$ and for $n = 4$, the total entries are $2^8 = 256$. As *n* increases, it becomes too cumbersome to design the circuit. Generally, a comparator circuit possesses a certain amount of regularity. Such type of circuits are usually designed by means of an algorithmic procedure. An algorithm is a procedure that specifies a finite set of steps that, if followed, give the solution to a procedure.

We illustrate this procedure for designing a 4-bit comparator. Consider the two 4-bit numbers A and B, where,

$$A = A_3\ A_2\ A_1\ A_0$$

and

$$B = B_3\ B_2\ B_1\ B_0\ ;$$

such that A_0 and B_0 are the least significant bits (LSB) and A_3 and B_3 are MSB. As can be seen, the 2 numbers are equal, only if all the pairs of the significant digits are equal, *i.e.*, if $A_3 = B_3$ and $A_2 = B_2$ and $A_1 = B_1$ and $A_0 = B_0$.

When the numbers are binary, then, the equality relation of each pair of bits can be expressed by the equivalence function, as

$x_i = A_i.B_i + A_i'.B_i'$; where, $i = 0, 1, 2, 3$ and $x_i = 1$, only when the bits in the position *i* are equal, *i.e.*, if both are 1's or 0's.

So, now we have to design a combinational circuit for 3 conditions :

(1) A = B

(2) A > B

(3) A < B.

Case (*i*) : The 1st condition, A = B is displayed in the combinational-circuit by an output binary-variable which is designated by (A = B). This output binary variable = 1, if A and B are equal or else, this output = 0. For this to be 1, all x_i variables must be equal to 1. *i.e.*;

$$(A = B) = x_3.x_2.x_1.x_0$$

This indicates the AND operation of x_i's.

Case (*ii*) : The next case is whether A > B or A < B. To determine this, we examine the magnitudes of the bits in their relative positions of the 2 numbers, starting from the MSB.

If the 2 bits are equal, then, we compare the bits in the next lower significant position. This procedure is continued until a pair of unequal bits is reached. Then, if the corresponding bit of A = 1 and that of B = 0, we conclude that A > B.

If the corresponding bit of A = 0, and that of B =1, we conclude that A < B.

Hence, the sequential comparison can be expressed logically by the following 2 Boolean-functions :

$$(A > B) = A_3B_3' + x_3A_2B_2' + x_3x_2A_1B_1' + x_3x_2x_1A_0B_0'$$

and

$$(A < B) = A_3'B_3 + x_3A_2'B_2 + x_3x_2A_1'B_1 + x_3x_2x_1A_0'B_0$$

∴ Now the other 2 binary-output variables are (A > B) and (A < B). Whichever variable = 1, then, that particular condition is valid.

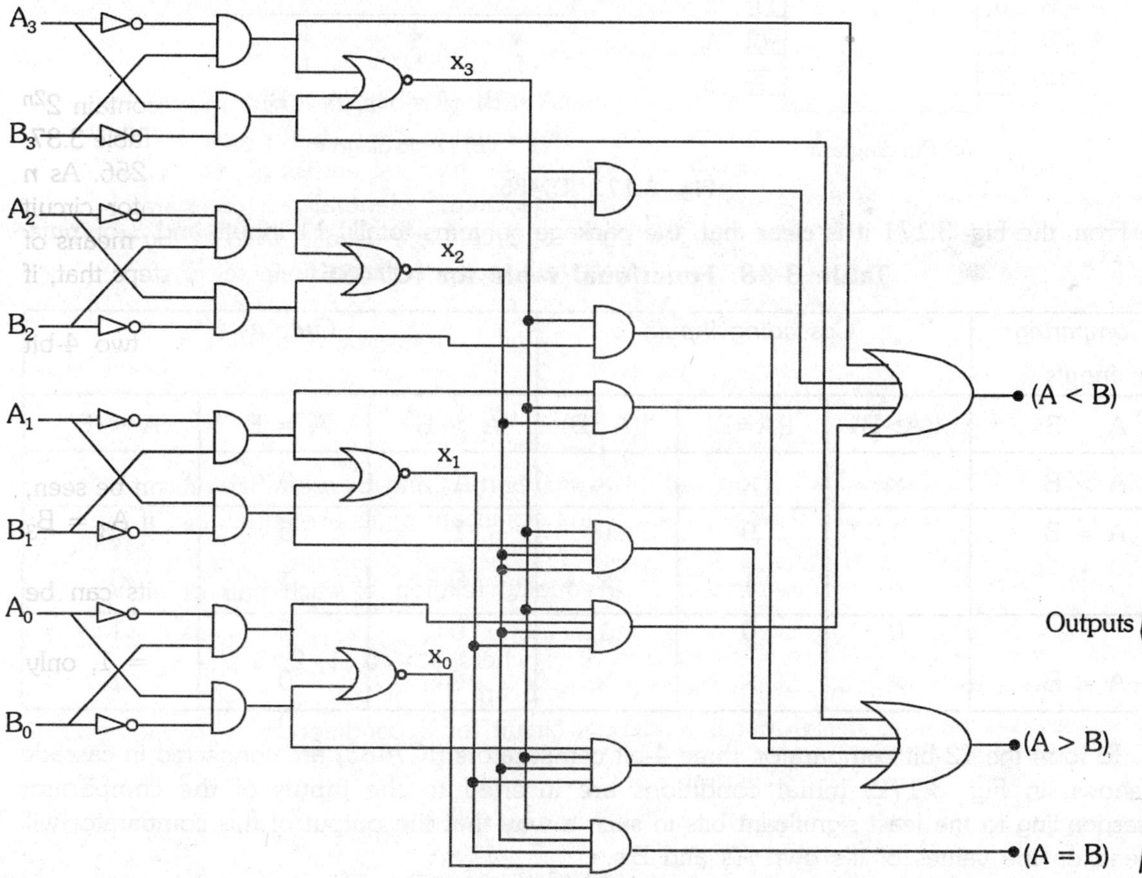

Fig. 3.170 A 4-bit magnitude comparator.

As can be seen from the combinational circuit shown in Fig. 3.170, first the 4 x-outputs *i.e.*, x_4, x_3, x_2, x_1 are generated by equivalence operations and then they are applied to the AND gate to generate the output binary variable (A = B).

The other 2 output variables, (A > B), (A < B) are obtained by generating the Boolean-functions as listed above. The design procedure has a regular pattern. Hence, similarly, this procedure can be extended for the comparison of *n*-bit numbers.

IC7485 : The IC7485 is a MSI package and is a 4-bit compator. It can be used to compare two 4-bit binary words. These ICs can be cascaded to compare words of almost any length. It's 4-bit inputs are weighted ($A_3 - A_0$) and ($B_3 - B_0$), where A_3 and B_3 are MSBs.

Fig. 3.171 shows the logic symbol and pin diagram of IC7485. The operation is described in Table 3.38.

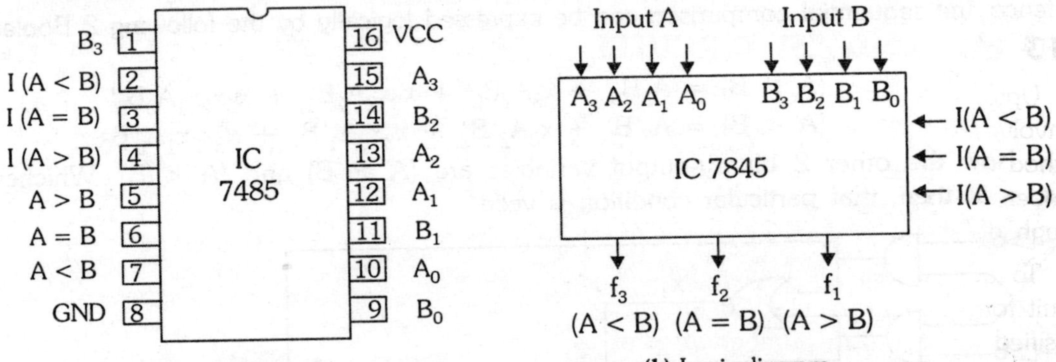

(*a*) Pin diagram (*b*) Logic diagram

Fig. 3.171 IC7485.

From the Fig. 3.171 it is clear that, the package contains totally 11 inputs and 3 outputs.

Table 3.38. Functional table for IC7485

Comparing inputs	*Cascading inputs*			*Outputs*		
A B	I(A>B)	I(A=B)	I(A<B)	A > B	A = B	A < B
A > B	×	×	×	1	0	0
A = B	1	0	0	1	0	0
	×	1	×	0	1	0
	0	0	1	0	0	1
A < B	×	×	×	0	0	1

To form the 12-bit comparator, three 4-bit comparators (IC7485) are connected in cascade as shown in Fig. 3.172. Initial conditions are inserted to the inputs of the comparator corresponding to the least significant bits in such a way that the output of this comparator will depend on the values of it's own A's and B's.

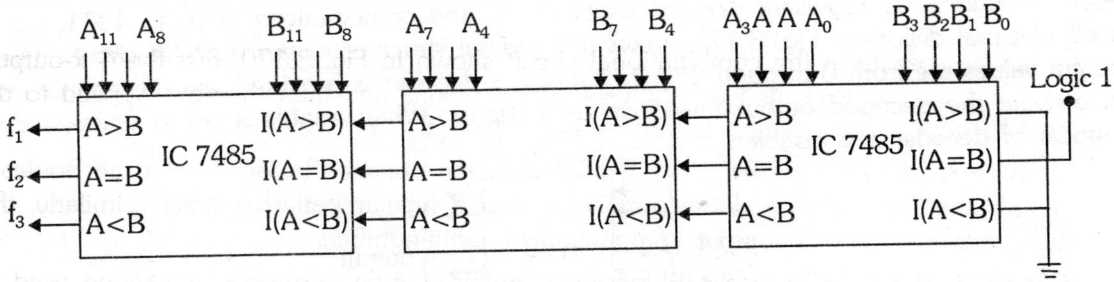

Fig. 3.172 12-bit comparator.

The two 12-bit numbers A and B can be represented as

$A = A_{11}..... A_0$ and

$B = B_{11}.... B_0$ where A_{11} and B_{11} are MSB's and A_0 and B_0 are LSB's.

3.13 MSI AND LSI CIRCUITS

Upto now (last sections), we have seen traditional method of combinational circuit design. It involves simplification and realization using logic gates with criteria of low-cost circuit. This method of designing is effective for smaller circuits. When the circuit complexity is more, more number of gates are required with more number of interconnections between them, even though designing of such circuits may be time consuming and less reliable.

To avoid these problems, several combinational circuits are available in the integrated circuit form which are used in the design of many digital systems. The integrated circuits are classified according to number of gates as :

1. Small Scale Integration (SSI)
2. Medium Scale Integration (MSI)
3. Large Scale Integration (LSI)
4. Very large Scale Integration (VLSI)

The number of gates in an IC limits the number of pins, typically SSI circuits contain 14 or 16.

Most of the combinational circuits available in MSI form are used to perform the specific digital functions such as : multiplexing, demultiplexing, addition, subtraction, decoding, comparison, parity encoding etc. We have already seen some of them in the previous sections, now we will see some other in this section.

3.13.1 Decoders

We know that with *n*-bits, we can represent a maximum of 2^n distinct combinations in the binary format. A decoder is a multiple-input and multiple-output combinational logic circuit which converts coded inputs into coded outputs, where the input and output codes are different. The input code generally has fewer bits than the output code. Each input code word produces a different output code word, *i.e.*, there is a one-to-one mapping from input code words into output code words. For every combination of input values, only one output will be equal to 1 at the same time. Generally, if the decoder has *n*-input lines, then the maximum number of unique output lines are 2^n. If the *n*-bit decoded information has unused or don't care combinations, the decoder output will have fewer than 2^n outputs. For example, in BCD code, 4-bit combinations

0000 through 1001 represent decimal digits (0–9), and combinations 1010 to 1111 are not used (decimal decoder). Fig. 3.173 shows the general structure of the decoder circuit. The 2^n output values are from 0 through $2^n - 1$. Usually, a decoder is provided with an enable input, to activate the decoded output based on data inputs. When enable input is unasserted, all outputs of decoder are disabled.

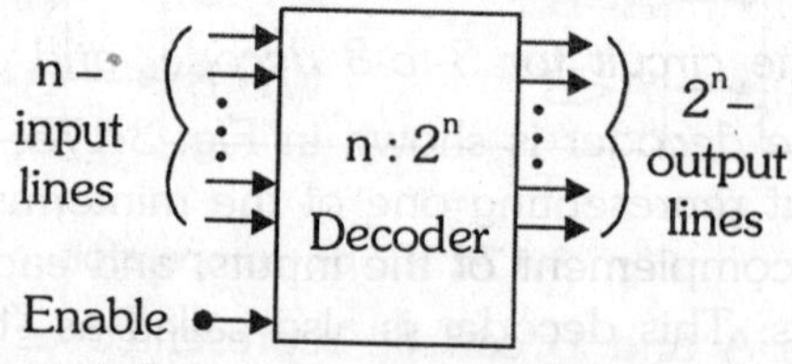

Fig. 3.173 General structure of decoder.

Generally decoder is represented by "*n-to-m lines*" or "*n×m lines*" or *n : m lines* where n = number of inputs, m = number of output lines, maximum of 2^n.

As an example, consider the 2-to-4 line decoder circuit of Fig. 3.174. Here, 2 inputs are decoded into four outputs, each output representing one of the minterms of the 2 input variables. The two inputs are represented by A and B and outputs are represented by Y_0, Y_1, Y_2 and Y_3.

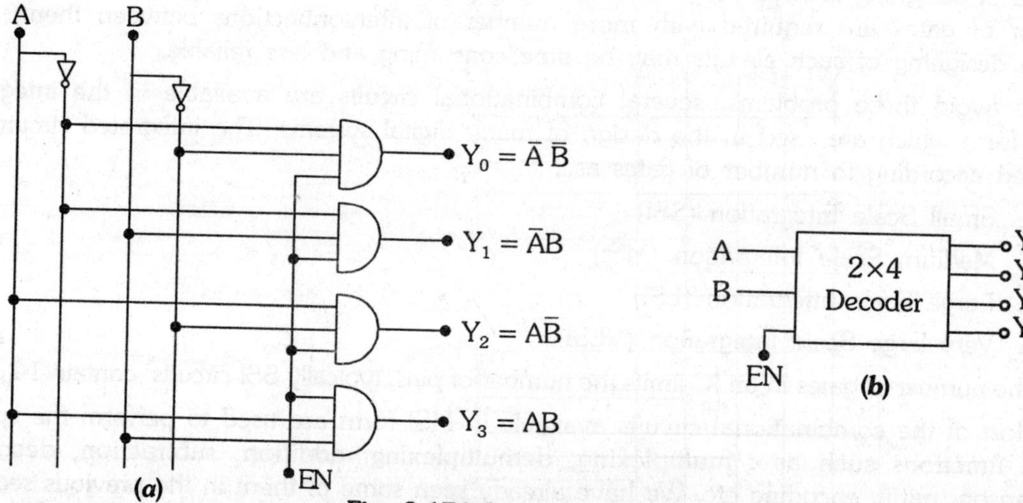

Fig. 3.174 2-to-4 line decoder.

Table 3.39 shows the truth table for 2-to-4 decoder.

If enable input is 1 (EN = 1), one of the outputs Y_0 to Y_3 is active for a given input. The output Y_0 is active *i.e.*, $Y_0 = 1$, when inputs A = B = 0, the output Y_1 is active when A = 0 and B = 1 and so on. If EN = 0, then all the outputs are 0.

Table 3.39. Truth table for 2-to-4 decoder

Inputs			*Outputs*			
EN	A	B	Y_3	Y_2	Y_1	Y_0
0	×	×	0	0	0	0
1	0	0	0	0	0	1

1	0	1	0	0	1	0
1	1	0	0	1	0	0
1	1	1	1	0	0	0

Example 3.37. *Draw the circuit for 3-to-8 decoder and explain.*

Solution. The 3-to-8 line decoder is shown in Fig. 3.175. The three inputs are decoded into eight outputs, each output representing one of the minterms of the 3-input variables. The three NOT gates provide the complement of the inputs, and each one of the eight AND gates generates one of the minterms. This decoder is also called as "binary-to-octal converter". The three input variables may represent a binary number, and outputs will be then represented by the eight digits in the octal number system.

The input-output relationship of the 3-to-8 decoder is shown Table 3.39. Observe that the output variables are mutually exclusive, because only one output can be equal to 1 at any one time. The output line whose value is equal to 1 represents the minterm equivalent of the binary number presently available in the input lines. The three input variables are A, B and C and the output variables are Y_0, Y_1, Y_2, Y_3, Y_4, Y_5, Y_6 and Y_7.

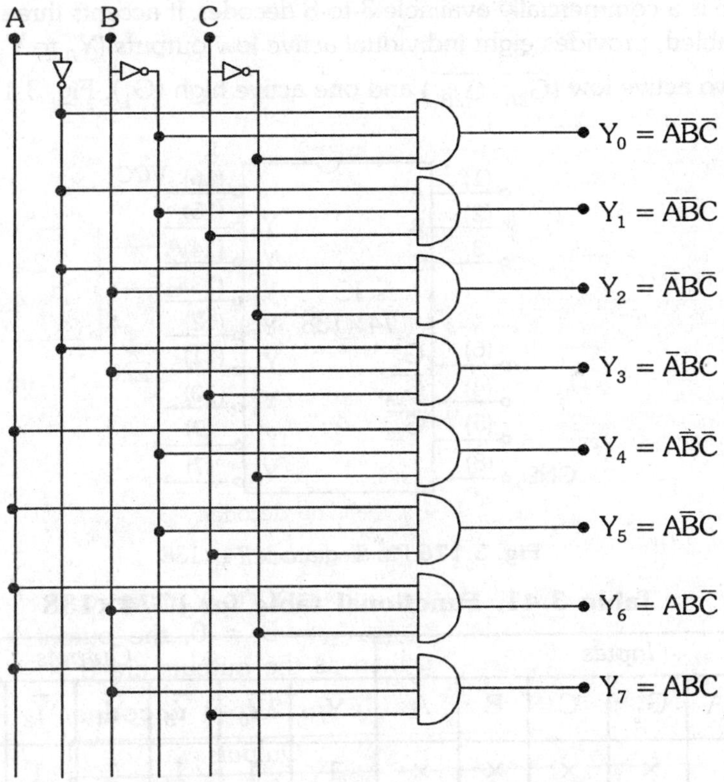

Fig. 3.175 Logic diagram of 3-to-8 line decoder.

Table 3.40. Truth table of a 3-to-8 line decoder

Inputs			*Outputs*							
A	B	C	Y_0	Y_1	Y_2	Y_3	Y_4	Y_5	Y_6	Y_7
0	0	0	1	0	0	0	0	0	0	0
0	0	1	0	1	0	0	0	0	0	0
0	1	0	0	0	1	0	0	0	0	0
0	1	1	0	0	0	1	0	0	0	0
1	0	0	0	0	0	0	1	0	0	0
1	0	1	0	0	0	0	0	1	0	0
1	1	0	0	0	0	0	0	0	1	0
1	1	1	0	0	0	0	0	0	0	1

3.13.1.1 The 74×138 3-to-8 decoder

The 74×138 is a commercially available 3-to-8 decoder. It accepts three binary inputs (A, B, C) and when enabled, provides eight individual active low outputs (Y_0 to Y_7). The device has 3 enable inputs : two active low ($\overline{G_{2A}}$, $\overline{G_{2B}}$) and one active high (G_1). Fig. 3.176 and Table 3.41.

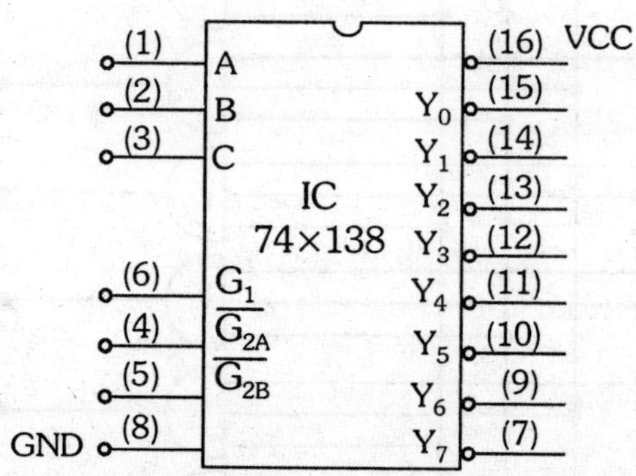

Fig. 3.176 Pin diagram of 74×138.

Table 3.41. Functional table for IC74×138

Inputs						*Outputs*							
G_2B	G_2A	G_1	C	B	A	$\overline{Y_7}$	$\overline{Y_6}$	$\overline{Y_5}$	$\overline{Y_4}$	$\overline{Y_3}$	$\overline{Y_2}$	$\overline{Y_1}$	$\overline{Y_0}$
1	×	×	×	×	×	1	1	1	1	1	1	1	1
×	1	×	×	×	×	1	1	1	1	1	1	1	1
×	×	0	×	×	×	1	1	1	1	1	1	1	1
0	0	1	0	0	0	1	1	1	1	1	1	1	0

0	0	1	0	0	1	1	1	1	1	1	1	0	1
0	0	1	0	1	0	1	1	1	1	1	0	1	1
0	0	1	0	1	1	1	1	1	1	0		1	1
0	0	1	1	0	0	1	1	1	0	1	1	1	1
0	0	1	1	0	1	1	1	0	1	1	1	1	1
0	0	1	1	1	0	1	0	1	1	1	1	1	1
0	0	1	1	1	1	0	1	1	1	1	1	1	1

3.13.1.2 Cascading of decoders

Binary decoders can be cascaded to get the higher order decoders. For example,

(*i*) 4-to-16 decoder can be obtained

(*a*) By using two 3-to-8 decoders.

(*b*) By using two 2-to-4 decoders and 16 AND gates.

(*c*) By using five 2-to-4 decoders.

(*ii*) The 5-to-32 decoder can be obtained

(*a*) By using one 2-to-4 and four 3-to-8 decoders.

(*b*) By using two 4-to-16 decoders and so on.

5-to-32 decoder using one 2-to-4 and four 3-to-8 decoders

The block diagram for 5-to-32 decoder is shown in Fig. 3. 177 as commercially 5-to-32 ICs are not available. So by using cascading of 3-to-8 decoders and 2-to-4 decoder, we can get the 5-to-32 decoder. The five inputs are A, B, C, D and E where E is the MSB. The 32 outputs are represented by Y_0 and Y_{31}. The inputs of 2-to-4 decoder are D and E, and the four outputs connected enable inputs of four 3-to-8 decoders. All the three inputs (A, B, C) of four decoders are connected as a common as shown in Fig. 3.178. When DE = 00, the decoder (1) is selected and according to the value of ABC, the corresponding output line is selected. If DECBA = 00101, the $Y_5 = 1$,

and if DE = 01 decoder 2 is selected.

if DE = 10 decoder 3 is selected.

if DE = 11 decoder 4 is selected.

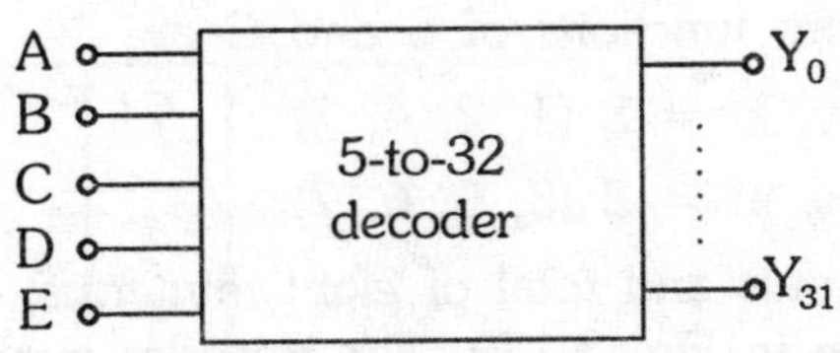

Fig. 3.177 Block diagram of 5-to-32 decoder.

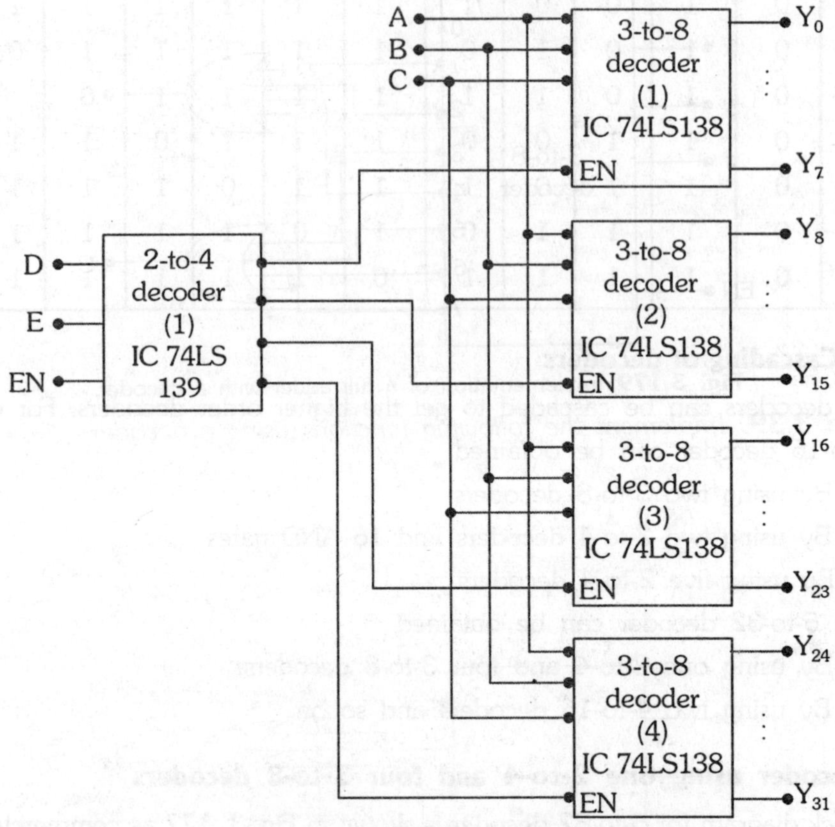

Fig. 3.178 5-to-32 decoder using 74LS138 and 74LS139.

3.13.1.3 Realization of multiple output functions using decoder

The combination of decoder and external logic gates can be used to implement single or multiple output functions. The decoder provides the 2^n minterms of n-input variables. Since any Boolean function can be expressed in cannonical SOP form, one can use a decoder to generate the minterms and an external OR gate to form the sum. In this way, any combinational circuit with n-inputs and m outputs can be implemented with n-to-2^n line decoder and m OR gates. This is illustrated in the following example.

Example 3.38. *Implement the full-adder circuit using a decoder and two OR gates.*

Solution. The full adder contains two outputs sum (S) and carry (C). From the truth tables of full adder, the output functions of S and C are :

$$S(x, y, z) = \Sigma(1, 2, 4, 7)$$
$$C(x, y, z) = \Sigma(3, 5, 6, 7)$$

Since there are three inputs and total of eight minterms, we need a 3-to-8 line decoder. The implementation is shown in Fig. 3.179. The decoder generates the eight minterms for x, y, z. The OR gate for output S forms the sum of minterms 1, 2, 4 and 7. The OR gate for output C forms the sum of minterms 3, 5, 6 and 7.

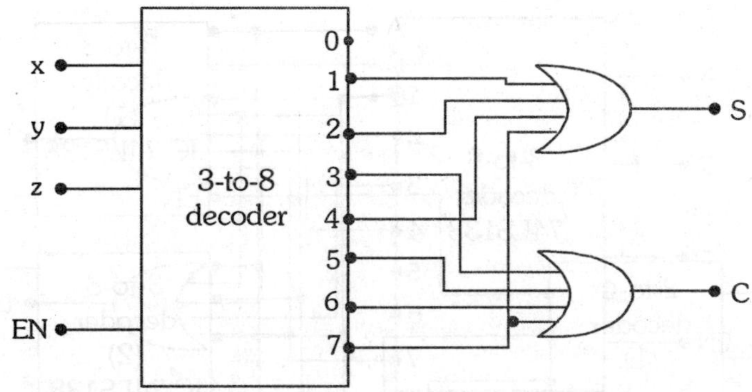

Fig. 3.179 Implementation of a full adder with a decoder.

Example 3.39. *Implement the following functions using a decoder*

$$F_1(x, y, z) = \Sigma m\ (0, 1, 3, 7)$$

$$F_2(x, y, z) = \Sigma m\ (2, 3, 7)$$

Solution. Fig. 3.180 shows the implementation of the desired multiple output function using the decoder (3-to-8).

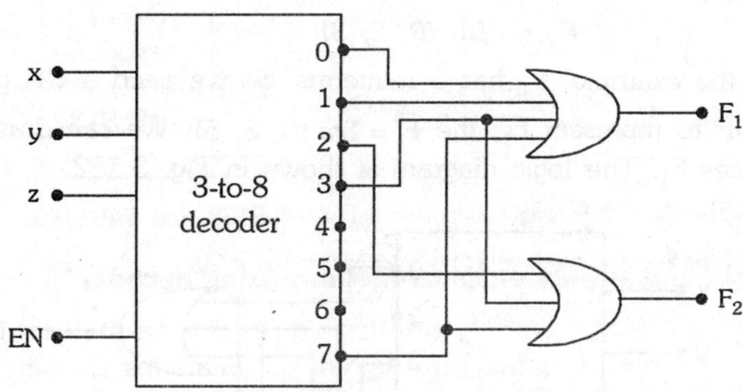

Fig. 3.180

Example 3.40. *Implement the full subtractor using a decoder and OR gates.*

Solution. From the truth table of the full-subtractor, we have the two functions borrow and difference. These two functions can be represented in the SOP form as :

$$D(x, y, z) = \Sigma m\ (1, 2, 4, 7)$$

$$B(x, y, z) = \Sigma m\ (1, 2, 3, 7).$$

The logic diagram for the full-subtractor is shown in Fig. 3.181.

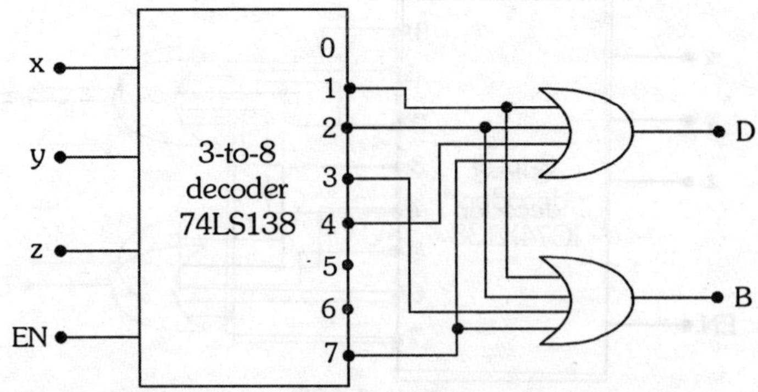

Fig. 3.181 Implementation of full subtractor using decoder.

Note : A function with a long list of minterms requires an OR gate with a large number of inputs. A function F having a list K-minterms can be expressed in its complemented form F', with 2^n–K minterms. If the number of minterms in a function is greater than $2^n/2$, F' can be expressed with fewer minterms than required for F. In such a case, it is advantageous to use a NOR gate to sum the minterms of F'. The output of NOR gate will generate the normal output of F.

Example 3.41. *Implement the following functions using a decoder, OR gate and NOR gate*

$$F_1 = \Sigma m\ (1, 2, 4, 6, 7)$$

$$F_2 = \Sigma m\ (0, 2, 3)$$

Solution. In the example, F_1 has 5 minterms. So we need a OR gate with five inputs. Instead of using OR to represent F_1, the $\overline{F_1} = \Sigma m\ (0, 3, 5)$. We can draw the $\overline{F_1}$ using NOR gate, which produces F_1. The logic diagram is shown in Fig. 3.182.

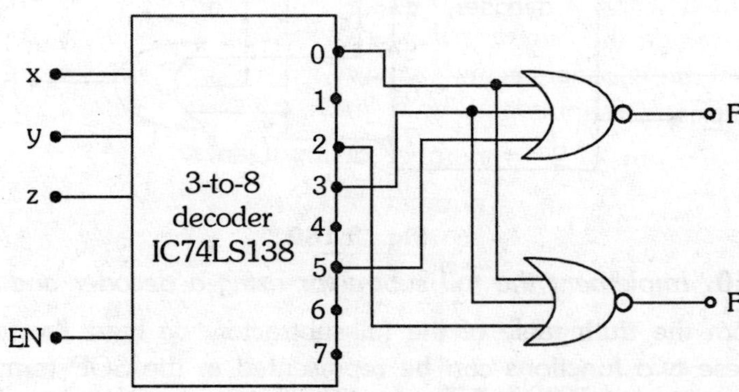

Fig. 3.182

Example 3.42. *Implement the following functions using decoder and AND gates :*

$$f_1\ (x, y, z) = \Pi\ (0, 2, 3, 7)$$

$$f_2\ (x, y, z) = \Pi\ (2, 4, 6)$$

Solution. To implement the above two functions, we need a decoder and two AND gates, each AND output represents one function in POS form as shown in Fig. 3.183.

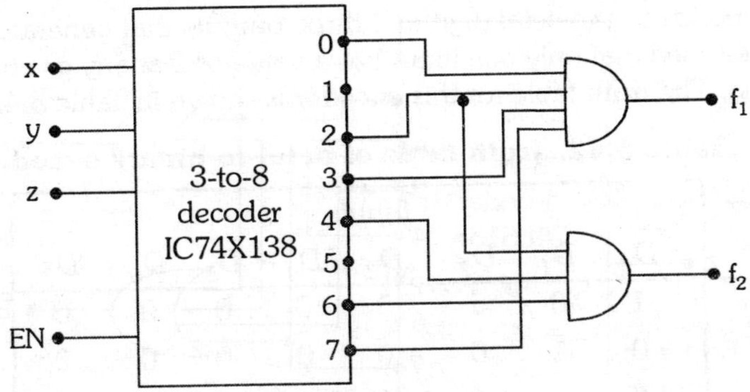

Fig. 3.183

3.13.1.4 Applications of decoder

Decoders have a wide variety of applications in digital technology.

1. They may be used to route input data to a specified output line ex : addressing core memory, where input data are to be stored in (or read from) a specified memory location.
2. They can be used as code converters, ex : binary-to-octal, BCD-to-7 segment display decoder.
3. These are used for data distribution, *i.e.*, demultiplexing.
4. Decoders are also used as building blocks in implementing arbitrary switching functions.

3.13.2 Encoders

An 'encoder' is a multiple-input, multiple-output combinational logic circuit. It performs the inverse operation of a decoder. An encoder has 2^n (or fewer) input lines and n-output lines. The output lines generate the binary code corresponding to the input value. Fig. 3.184 shows the block diagram of the encoder-circuit. Generally, an encoder can be represented (or) specified by number of inputs and outputs as 2^n : *n lines (or)* 2^n*-to-n (or)* $2^n \times n$ *lines or m : n lines.*

where $m \leq 2^n$ = number of input lines

n = number of output lines.

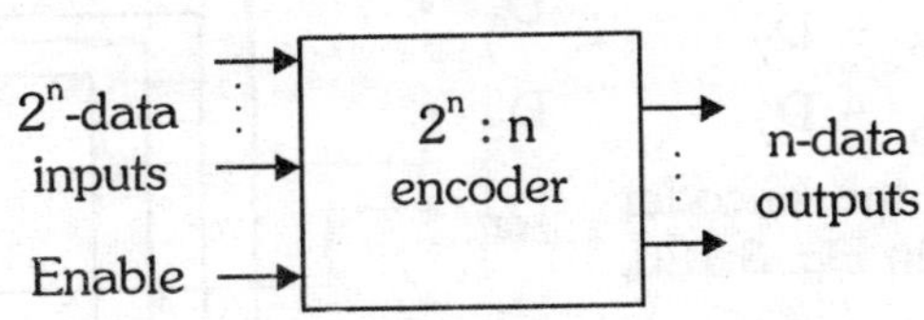

Fig. 3.184 Block diagram of encoder.

In the encoder, the number of output lines (n) are always less than the number of input lines (m or 2^n).

8-to-3 line encoder

This is a simple and basic encoder. It has eight inputs and three outputs. The logic symbol of this type of encoder is shown in Fig. 3.185. It is also called as "*octal-to-binary*" encoder. Each input of the encoder corresponds to the octal digit and three outputs that generate the corresponding binary number. It is assumed that only one input has a value of 1 at any given time; otherwise the circuit has no meaning. The truth table for this encoder is shown in Table 3.42.

Table 3.42. Truth table of octal-to-binary encoder

Octal number	*Inputs*								*Outputs*		
	D_0	D_1	D_2	D_3	D_4	D_5	D_6	D_7	A	B	C
0	1	0	0	0	0	0	0	0	0	0	0
1	0	1	0	0	0	0	0	0	0	0	1
2	0	0	1	0	0	0	0	0	0	1	0
3	0	0	0	1	0	0	0	0	0	1	1
4	0	0	0	0	1	0	0	0	1	0	0
5	0	0	0	0	0	1	0	0	1	0	1
6	0	0	0	0	0	0	1	0	1	1	0
7	0	0	0	0	0	0	0	1	1	1	1

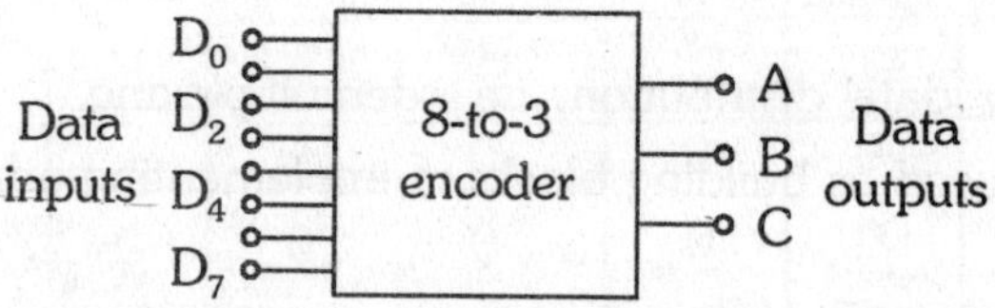

Fig. 3.185 Logic symbol of 8-to-3 encoder.

The encoder can be implemented with OR gates whose inputs are determined directly from the truth table. The output C is equal to 1 when the input octal digit is 1 or 3 or 5 or 7, the output B = 1, for the input octal digits 2, 3, 6 or 7 and the output A = 1 for the input octal digits 4, 5, 6 or 7.

These conditions can be expressed by the Boolean equations as :

$$A = D_4 + D_5 + D_6 + D_7$$
$$B = D_2 + D_3 + D_6 + D_7$$
$$C = D_1 + D_3 + D_5 + D_7$$

The implementation of the encoder with three OR gates is shown in Fig. 3.186.

Note : (1) The encoder has one limitation that only one input can be active at any given time. If two inputs are active 1 simultaneously, the output produces an undefined combination. For example, if D_2 and D_5 are 1 simultaneously, the output of the encoder will be 111 because all three outputs are equal to 1. This does not represent binary 2 or binary 5. To resolve this ambiguity, encoder circuit must establish a priority to ensure that only one input is encoded.

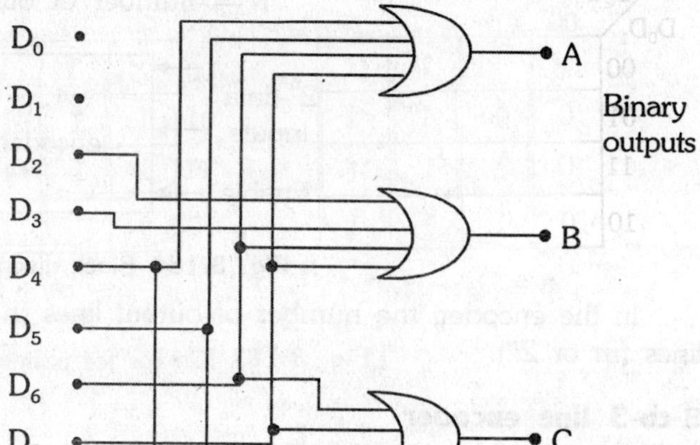

Fig. 3.186 Logic diagram for octal-to-binary encoder.

(2) Another ambiguity in the octal-to-binary encoder is that an output with all 0's is generated when all inputs are '0'. The problem is, the output with all 0's is also generated when $D_0 = 1$. This can be resolved by providing an additional output that specifies the condition that none of the inputs is active.

3.13.2.1 Priority encoder

A priority encoder is an encoder circuit that includes priority function. If two or more inputs are equal to 1 at the same time, the input having the highest priority will take precedence. Let us consider an example of a 4-bit priority encoder. It contains 4 inputs and three outputs. The four inputs are D_0, D_1, D_2 and D_3 where D_3 is the input with highest priority and D_0 is the lowest priority. The two outputs A, B indicate the binary value and one more output (V) represents the validity of output. It is called as "*valid-output indicator*".

V = 1 only when one or more inputs are equal to 1.

V = 0 if all inputs are zero.

The truth table for the 4-bit priority encoder is shown in Table 3.43.

Table 3.43. Truth table of 4-bit priority encoder

Inputs				*Outputs*		
D_0	D_1	D_2	D_3	A	B	V
0	0	0	0	×	×	0
1	0	0	0	0	0	1
ϕ	1	0	0	0	1	1
ϕ	ϕ	1	0	1	0	1
ϕ	ϕ	ϕ	1	1	1	1

The simplified expressions for the three outputs (A, B and V) are shown in Fig. 3.187.

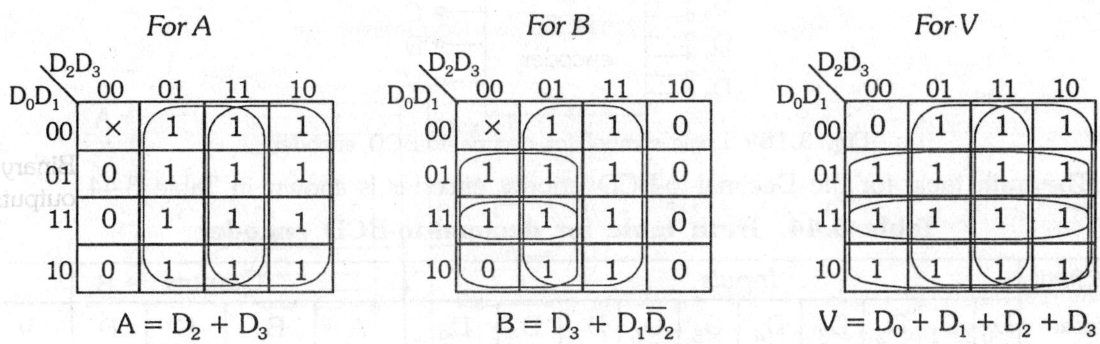

Fig. 3.187 K-maps for priority encoder outputs.

The output functions of the priority encoder can be represented as :

$$A = D_2 + D_3$$
$$B = D_3 + D_1\overline{D}_2$$
$$V = D_0 + D_1 + D_2 + D_3$$

The logic diagram for the above equations is shown in the Fig. 3.188.

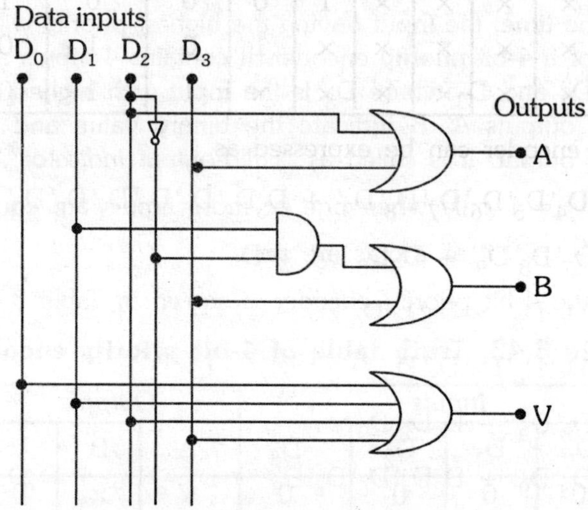

Fig. 3.188 Logic diagram of the priority encoder.

Example 3.43. *Design a decimal-to-BCD encoder.*

Solution. It is a priority encoder, which contains ten-inputs D_0 to D_9 and four outputs A, B, C and D and one more valid output. The logic symbol for it is shown in Fig. 3.189.

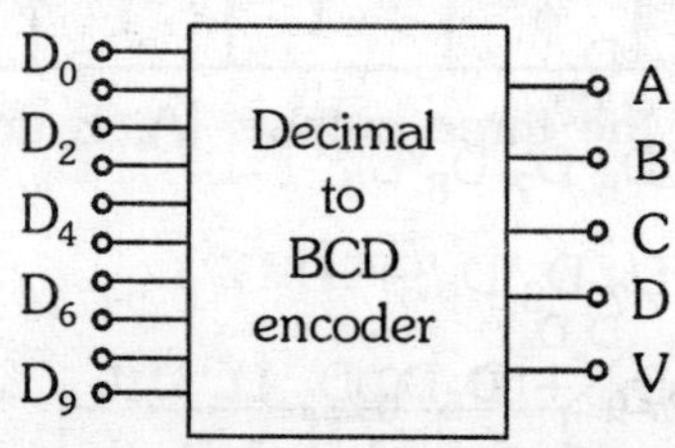

Fig. 3.189 Logic symbol for decimal-to-BCD encoder.

The truth table for the Decimal-to-BCD priority encoder is shown in Table 3.44.

Table 3.44. Truth table for decimal-to-BCD encoder

Decimal	*Inputs*										*Outputs*				
value	D_0	D_1	D_2	D_3	D_4	D_5	D_6	D_7	D_8	D_9	A	B	C	D	V
	0	0	0	0	0	0	0	0	0	0	0	0	0	0	0
0	1	0	0	0	0	0	0	0	0	0	0	0	0	1	1
1	×	1	0	0	0	0	0	0	0	0	0	0	1	0	1
2	×	×	1	0	0	0	0	0	0	0	0	0	1	1	1

3	×	×	×	1	0	0	0	0	0	0	0	0	1	1	1
4	×	×	×	×	1	0	0	0	0	0	0	1	0	0	1
5	×	×	×	×	×	1	0	0	0	0	0	1	0	1	1
6	×	×	×	×	×	×	1	0	0	0	0	1	1	0	1
7	×	×	×	×	×	×	×	1	0	0	0	1	1	1	1
8	×	×	×	×	×	×	×	×	1	0	1	0	0	0	1
9	×	×	×	×	×	×	×	×	×	1	1	0	0	1	1

The outputs of the encoder can be expressed as :

$$D = D_1D_2'D_3'D_4'D_5'D_6'D_7'D_8'D_9' + D_3D_4'D_5'D_6'D_7'D_8'D_9'$$

$$+ D_5D_6'D_7'D_8'D_9' + D_7D_8'D_9' + D_9$$

$$D = D_1D_2'D_4'D_6'D_8' + D_3D_4'D_6'D_8' + D_5D_6'D_8' + D_7D_8' + D_9$$

$$C = D_2D_3'D_4'D_5'D_6'D_7'D_8'D_9' + D_3D_4'D_5'D_6'D_7'D_8'D_9'$$

$$+ D_6D_7'D_8'D_9' + D_7D_8'D_9'$$

$$= D_2D_4'D_5'D_8'D_9' + D_3D_4'D_5'D_8'D_9' + D_6D_8'D_9' + D_7D_8'D_9'$$

$$B = D_4D_5'D_6'D_7'D_8'D_9' + D_5D_6'D_7'D_8'D_9' + D_6D_7'D_8'D_9' + D_7D_8'D_9'$$

$$= D_4D_8'D_9' + D_5D_8'D_9' + D_6D_8'D_9' + D_7D_8'D_9'$$

$$= D_8'D_9'(D_4 + D_5 + D_6 + D_7)$$

$$A = D_8D_9^1 + D_9 = D_8 + D_9$$

$$V = D_0D_1'D_2'D_3'D_4'D_5'D_6'D_7'D_8'D_9' + D_1D_2'D_3'D_4'D_5'D_6'D_7'D_8'D_9'$$

$$+ D_2D_3'D_4'D_5'D_6'D_7'D_8'D_9' + D_3D_4'D_5'D_6'D_7'D_8'D_9'$$

$$+ D_4D_5'D_6'D_7'D_8'D_9' + D_5D_6'D_7'D_8'D_9' + D_6D_7'D_8'D_9' + D_7D_8'D_9'$$

$$+ D_8D_9' + D_9$$

$$= D_0 + D_1 + D_2 + D_3 + D_4 + D_5 + D_6 + D_7 + D_8 + D_9.$$

The epressions for outputs of the parity encoder are :

$$A = D_8 + D_9$$

$$B = D_8'D_9'(D_4 + D_5 + D_6 + D_7)$$

$$C = D_4'D_5'D_8'D_9'(D_2 + D_3) + D_8'D_9'(D_6 + D_7)$$

$$D = D_4'D_6'D_8'(D_1D_2' + D_3) + D_8'(D_5D_6' + D_7) + D_9.$$

The logic diagram is shown in Fig. 3.190.

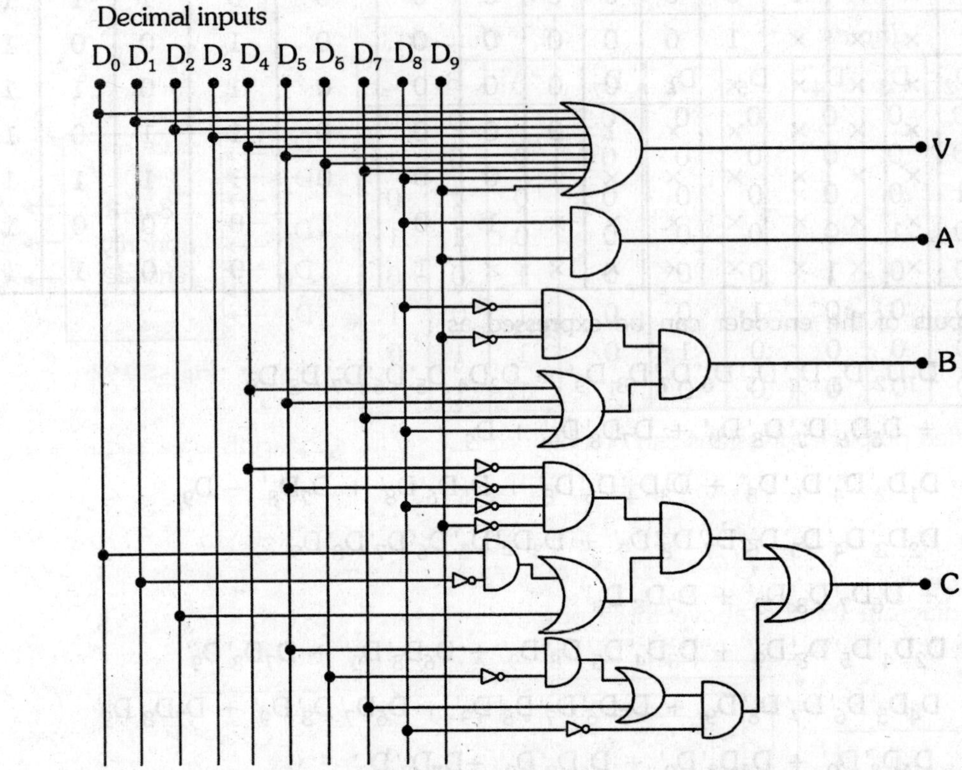

Fig. 3.190 Logic diagram for BCD encoder.

It is very difficult to find the simplified expressions for outputs A, B, C and D, since each output is a function of 10-input variables. For this, we have an MSI package of decimal to BCD encoder. The logic symbol of encoder IC74LS147 is shown in Fig. 3.191.

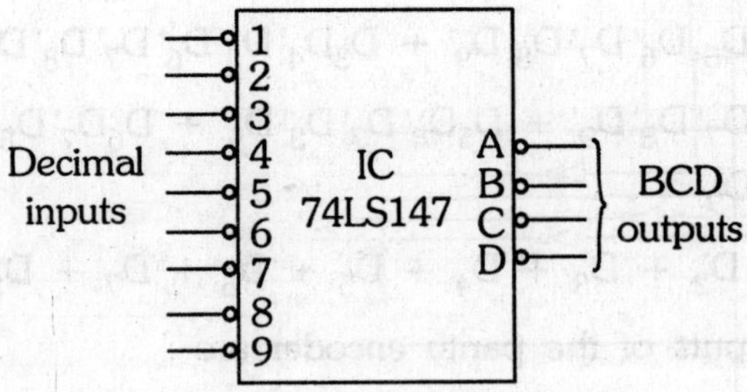

Fig. 3.191 Logic symbol for 74LS147.

The inputs and outputs are in inverted logic (active low signals).

Example 3.44. *Design an octal-to-binary priority encoder.*

Solution. The octal-to-binary priority encoder contains eight inputs and three outputs. The block diagram is shown in Fig. 3.192. It is also called as 8-to-3 priority encoder. The truth table for it is shown in Table 3.45. The example for this is IC74LS148.

Table 3.45. Truth table

Inputs								*Outputs*		
D_0	D_1	D_2	D_3	D_4	D_5	D_6	D_7	A	B	C
1	0	0	0	0	0	0	0	0	0	0
0	1	0	0	0	0	0	0	0	0	1
0	0	1	0	0	0	0	0	0	1	0
0	0	0	1	0	0	0	0	0	1	1
0	0	0	0	1	0	0	0	1	0	0
0	0	0	0	0	1	0	0	1	0	1
0	0	0	0	0	0	1	0	1	1	0
0	0	0	0	0	0	0	1	1	1	1

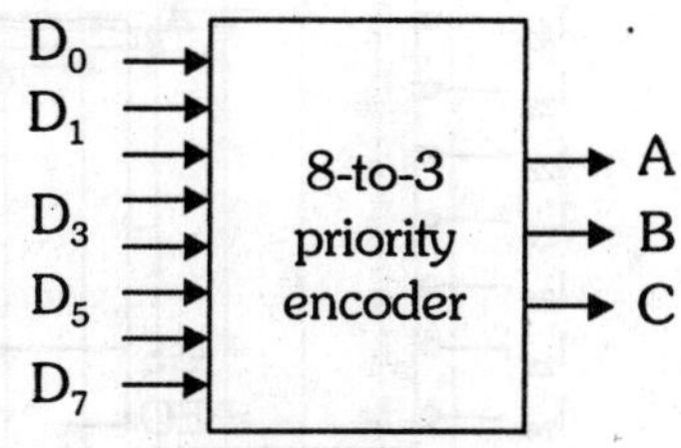

Fig. 3.192

The simplified output expressions can be obtained directly from the truth table and they are :

$$A = D_4 + D_5 + D_6 + D_7$$

$$B = D_2D_4'D_5' + D_3D_4'D_5' + D_6 + D_7$$

$$C = D_1D_2'D_4'D_6' + D_3D_4'D_6' + D_5D_6' + D_7 \cdot$$

The logic diagram for the above three outputs are shown in Fig. 3.193.

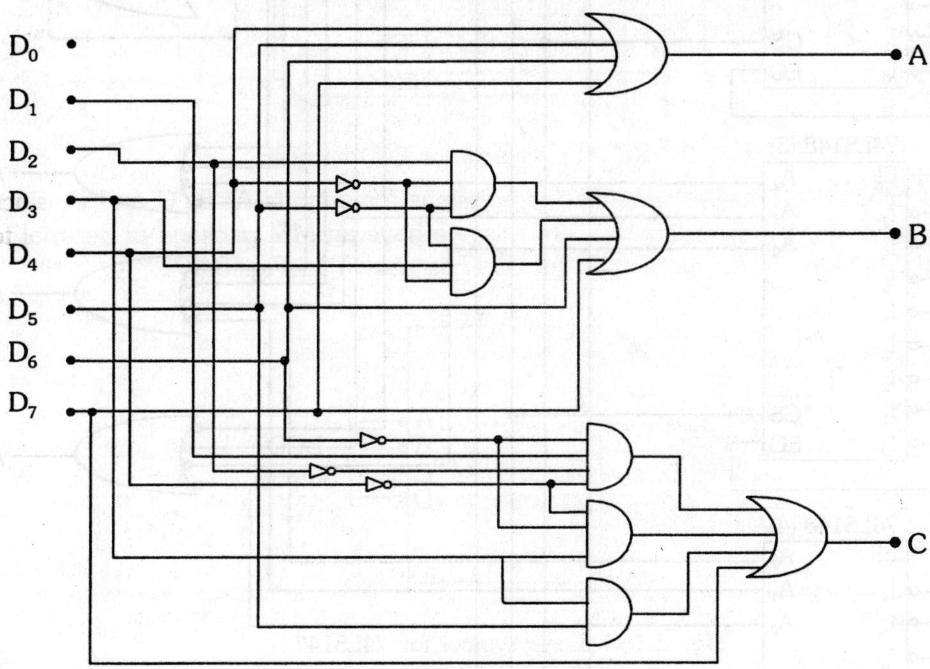

Fig. 3.193 Logic diagram.

3.13.2.2 Cascading of encoders

Encoders can be cascaded to form a larger one from smaller encoders. This can be explained with an example of implementation of the 32-to-5 encoder using four 8-to-3 encoders. The logic diagram is shown in Fig. 3.194. Here we have taken 4 numbers of 74LS148

(8-to-3 encoder). The inputs are named as I_0 to I_{31} and outputs are named as A_0–A_4.

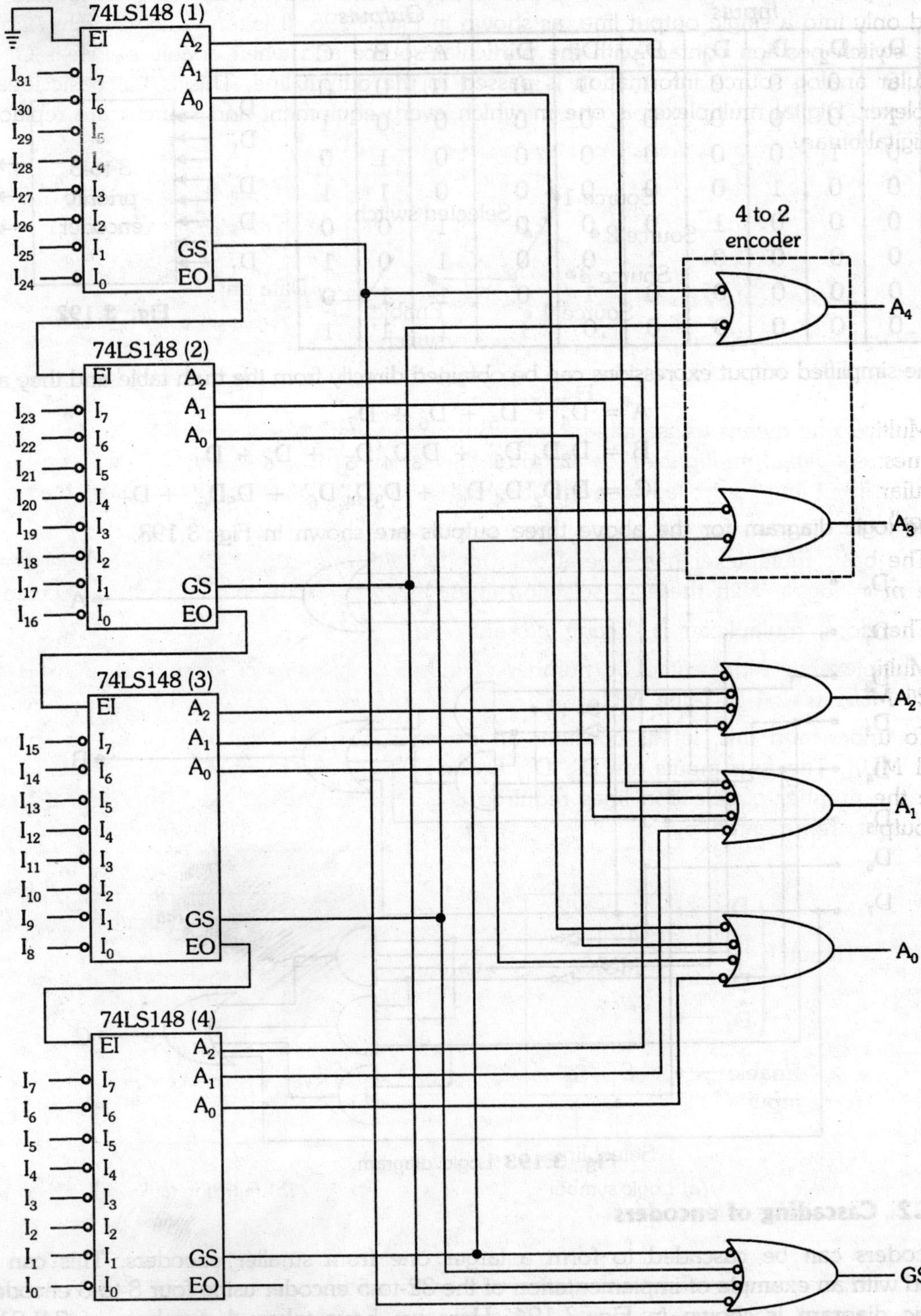

Fig. 3.194 32-input to 5 output priority encoder using 4 numbers 8-to-3 encoders.

3.13.3 Multiplexers

Multiplexer is a digital switch. It allows digital information from several sources to be routed only into a single output line, as shown in Fig. 3.195. It is an analog switch. According to the switch position contact with the particular source and when enable switch is ON, that particular analog source information is passed to the output line. This is the basic idea of a multiplexer. Digital multiplexer is one in which every equipment and sources are replaced by the digital/binary.

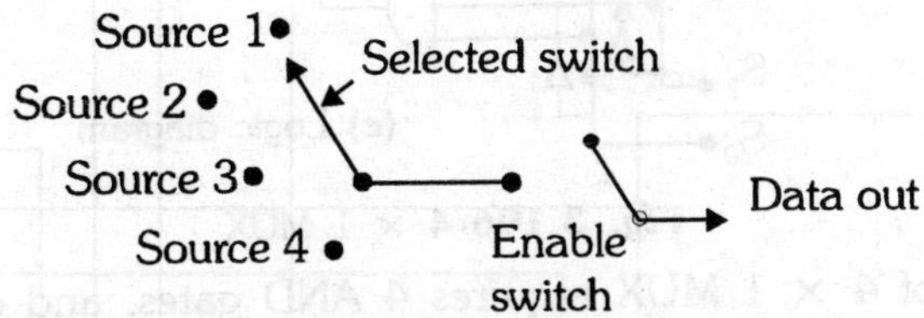

Fig. 3.195 Selector switch.

Multiplexing means transmitting a large number of units over a smaller number of channels (or) lines. A "*digital multiplexer*" is a combinational circuit that selects binary information of a particular input line controlled by a set of selection lines. A multiplexer is also called as a "*data selector*".

The basic multiplexer has several data inputs (2^n), one output line and m selection lines where $m = \log_2{}^n$. With these m selection lines, any one input combination is selected.

Therefore, multiplexer is "*many-into-one*".

Multiplexer is represented by number of inputs × number of outputs as $2^n \times 1$ MUX, *1-of-2^n MUX* (or) 4-to-1-line MUX.

To understand this, let us consider an example of 4 × 1 MUX. Fig. 3.196 shows the 4 × 1 MUX. The four inputs are D_0, D_1, D_2 and D_3. As the number of inputs are four (2^2), hence the number of selection lines required is 2, and are named as S_1 and S_0. It has only one output, represented by Y.

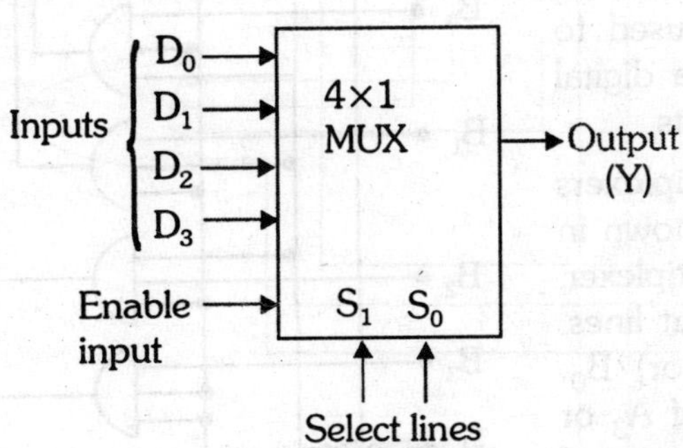

(a) Logic symbol

Select lines		Output
S_1	S_0	Y
0	0	D_0
0	1	D_1
1	0	D_2
1	1	D_3

(b) Function table

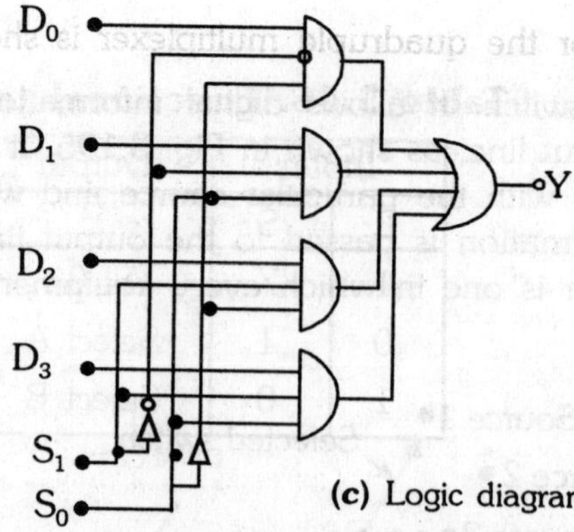

(c) Logic diagram

Fig. 3.196 4 × 1 MUX

The logic diagram of 4 × 1 MUX requires 4 AND gates, and each AND gate is having one of the inputs. The selection lines are decoded to select one of the AND gate. The logic diagram is shown in Fig. 3.196 (*a*) and its functional table is shown in Fig. 3.196 (*b*).

The functional table lists the input to output path for each possible bit combination of selection lines.

Case (*i*) : When $S_1 S_0 = 00$, the AND gate associated with input D_0 is selected, the output of other three AND gates is 0, hence $Y = D_0$

Case (*ii*) : When $S_1 S_0 = 10$, the AND gate associated with input D_2 is selected, and the output of remaining other three AND gates is 0, hence $Y = D_2$ and so on.

If the select input lines are driven by output lines of a sequential counter, then multiplexers can be used as a "*sequential data selector*".

Multiplexer IC may have an '*enable*' input to control the operation of the unit. When enable input is active, the circuit functions normally, otherwise, the outputs are disabled. The enable input (sometimes called strobe) can be used to expand two or more multiplexer IC's to a digital multiplexer with a larger number of inputs.

In some cases, two (or) more multiplexers are enclosed with in one IC package, as shown in Fig. 3.197. It is a quadruple 2-to-1 line multiplexer, each capable of selecting one or two input lines. Output, Y_0 can be equal to either A_0 (or) B_0. Similarly output Y_3 may have the value of A_3 or B_3 and so on. The selection line S selects one of the two lines in all four multiplexers. The enable line E enables the multiplexers in the '0' state and disable all the multiplexers in the 1 state.

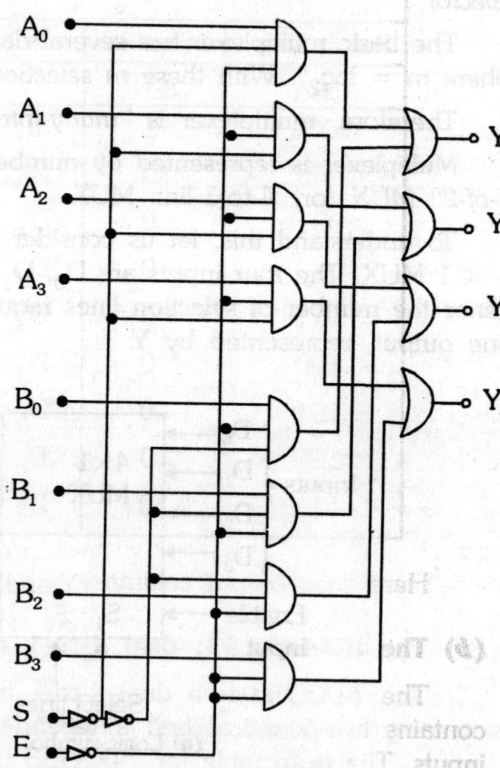

Fig. 3.197 Quadruple 2-to-1 line multiplexer.

The functional table for the quadruple multiplexer is shown in Table 3.46.

Table 3.46. Functional table

Inputs		*Output*
E	S	Y
1	X	All 0's
0	1	Select A
1	0	Select B

3.13.3.1 IC multiplexers

(*a*) The 74 XX 151; 8-to-1 multiplexer

It is a 8-to-1 multiplexer. It has eight inputs and two outputs, one is active high and other is active low. The Fig. 3.198 shows the logic symbol for 74XX151. It has three select inputs, A, B and C which select one of the eight inputs.

The truth table for 74XX151 is shown in Table 3.47.

Table 3.47. Truth table

Input		*Outputs*	
Select lines	Enable	X	$\overline{X}$
CBA	$\overline{EN}$		
XXX	1	0	1
000	0	D_0	$\overline{D_0}$
001	0	D_1	$\overline{D_1}$
010	0	D_2	$\overline{D_2}$
011	0	D_3	$\overline{D_3}$
100	0	D_4	$\overline{D_4}$
101	0	D_5	$\overline{D_5}$
110	0	D_6	$\overline{D_6}$
111	0	D_7	$\overline{D_7}$

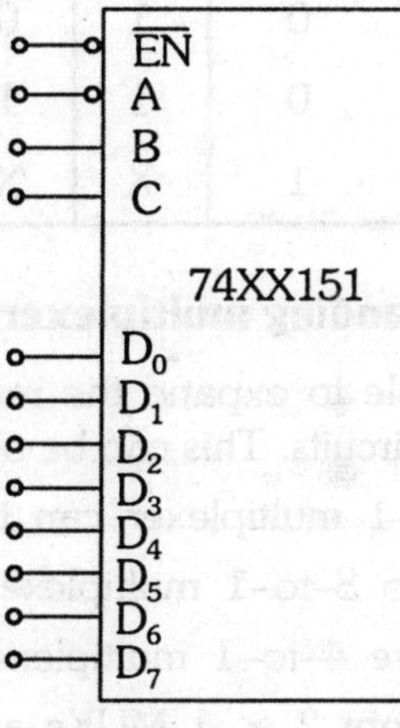

Fig. 3.198. Logic symbol for 74XX151.

Here each output column (Y and $\overline{Y}$) is represented by D_n and $\overline{D}_n$.

(*b*) The IC74XX153: dual 4-to-1 multiplexer

The 74XX153 is a dual 4-to-1 multiplexer. Its logic symbol is shown in Fig. 3.199. It contains two identical and independent 4-to-1 multiplexers. Each multiplexer has separate inputs. The truth table for 74XX153 is shown in Table 3.48.

Table 3.48. Truth table

Inputs				*Outputs*	
1 EN	2 EN	B	A	1Y	2Y
0	0	0	0	$1D_0$	$2D_0$
0	0	0	1	$1D_1$	$2D_1$
0	0	1	0	$1D_2$	$2D_2$
0	0	1	1	$1D_3$	$2D_3$
0	1	0	0	$1D_0$	0
0	1	0	1	$1D_1$	0
0	1	1	0	$1D_2$	0
0	1	1	1	$1D_3$	0
1	0	0	0	0	$2D_0$
1	0	0	1	0	$2D_1$
1	0	1	0	0	$2D_2$
1	0	1	1	0	$2D_3$
1	1	X	X	0	0

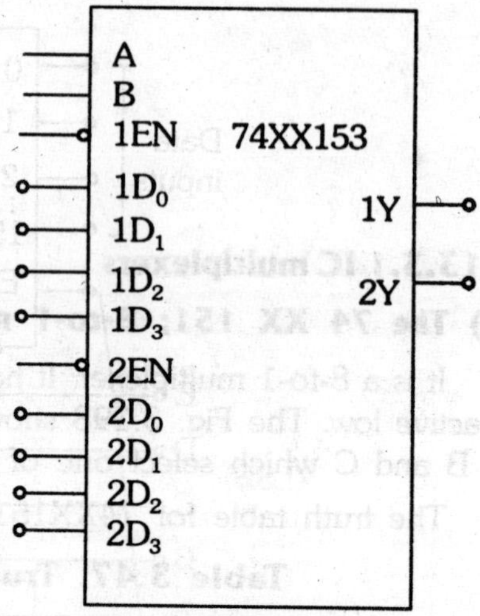

Fig. 3.199 Logic symbol for IC74XX153.

3.13.3.2 Expanding multiplexers

It is possible to expand the range of inputs for multiplexer beyond the available range in the integrated circuits. This can be obtained by interconnecting several multiplexers. For example,

(*a*) 16–to–1 multiplexer can be obtained by using

(*i*) Two 8–to–1 multiplexer and one OR gate (or) one 2X1MUX

(*ii*) Five 4–to–1 multiplexers.

(*iii*) Eight 2 × 1 MUXs and one 8 × 1MUX

(*b*) 32–to–1 multiplexer can be obtained by cascading of

(*i*) two 16 × 1 MUXs and one 2 × 1 MUX (or) OR gate

(*ii*) Four 8 × 1 MUXs and one 4 × 1 MUX and so on.

To illustrate this, we will consider one example.

Example 3.45. *Design a 32 : 1 multiplexer using two 16 : 1 multiplexers.*

Solution. A 32 : 1 multiplexer can be designed by using two 16 : 1 multiplexers. A 32 : 1 multiplexer will have five selection lines A, B, C, D and E, where A is the MSB. If A is connected to the enable input of one of the 16 : 1 multiplexers while the enable input of the other multiplexer is connected to $\overline{A}$. Thus when A = 0, MUX_1 is selected and when A = 1, MUX_2

is selected (single EN are active low inputs). The first 16 inputs (D_0–D_{15}) are connected to MUX_1 and second 16 inputs (from D_{16}–D_{31}) are connected to MUX_2. Now the outputs of two multiplexers are given as input to OR gate. Thus, the system functions as a 32 : 1 MUX_1 as shown in Fig. 3.200.

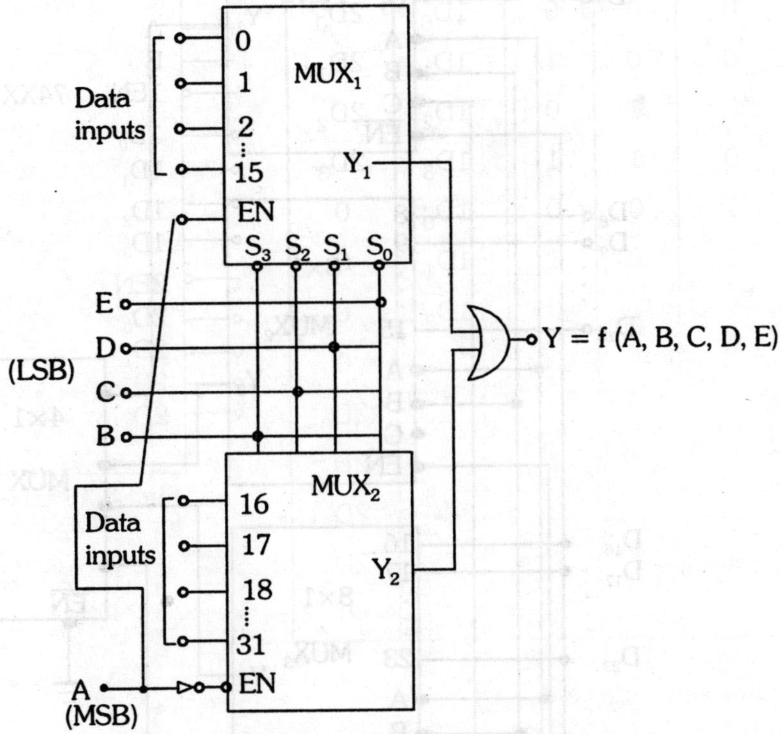

Fig. 3.200 32×1 MUX using 16×1 MUXs.

Example 3.46. *Design a 32 × 1 MUX using four 8 × 1 MUXs, and one 4 × 1 MUX.*

Solution. The 32 × 1MUX can be obtained by using four 8 × 1 MUX and one 4 × 1 MUX. The 32 × 1 MUX has five selection lines A, B, C, D and E. A 8 × 1 MUX has three selection lines A, B and C. The most significant D and E bits are used to select one of the four multiplexers by using 2 × 4 decoder. The four 8 × 1 multiplexers produce four outputs. These four outputs are connected to 4 × 1 multiplexer. The logic diagram is shown in Fig. 3.201.

The inputs of MUX_1 are $D_0 - D_7$, MUX_2 are $D_8 - D_{15}$, MUX_4 are $D_{16} - D_{23}$ and MUX_4 are $D_{24} - D_{31}$.

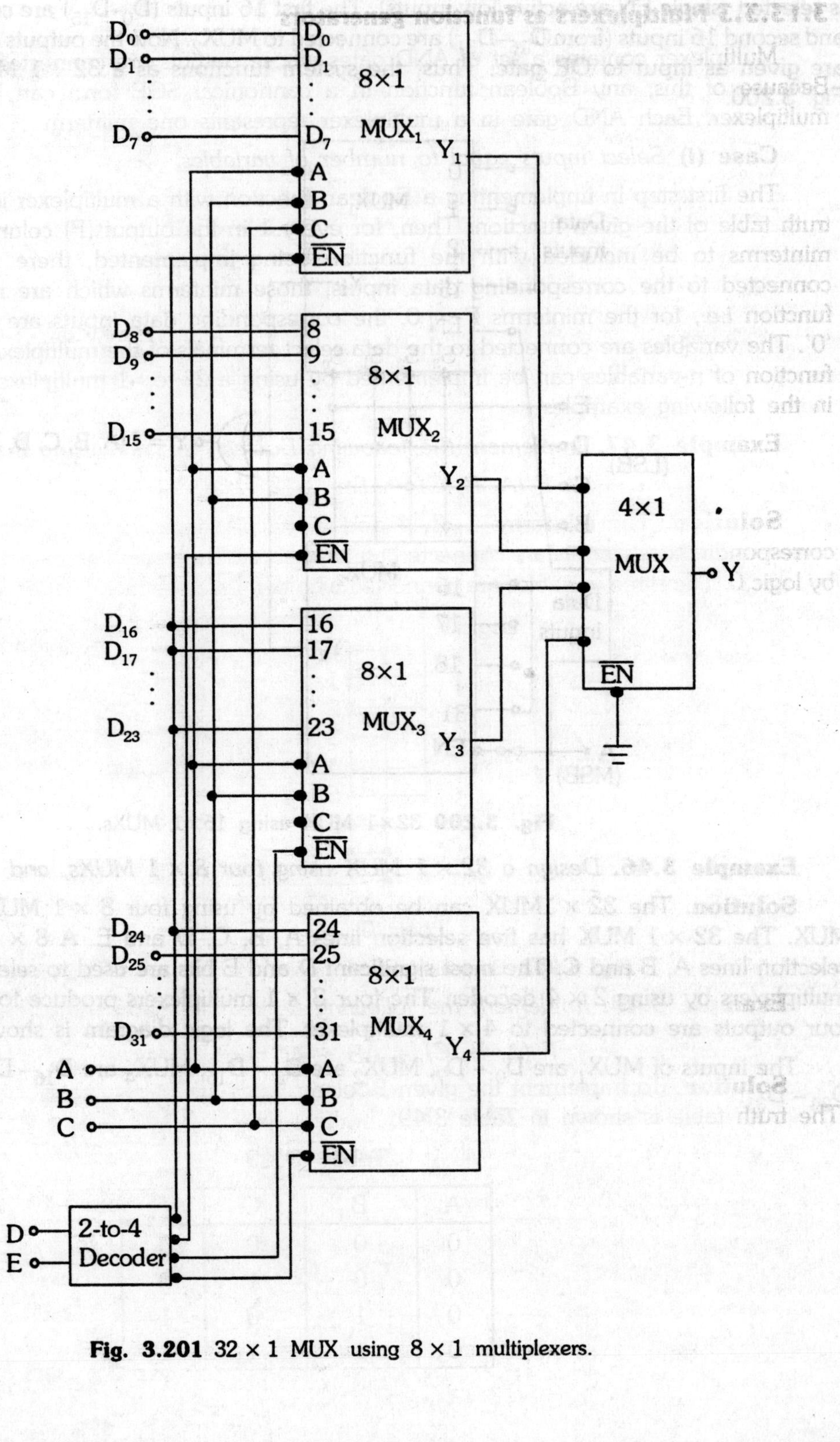

Fig. 3.201 32 × 1 MUX using 8 × 1 multiplexers.

3.13.3.3 Multiplexers as function generators

Multiplexer contains a set of AND gates whose outputs are connected to single OR gate. Because of this, any Boolean function in a cannonical SOP form can be realized using a multiplexer. Each AND gate in a multiplexer represents one minterm.

Case (i) *Select inputs equal to number of variables.*

The first step in implementing a Boolean function with a multiplexer is to write down the truth table of the given function. Then, for every 1 in the output (F) column of the table, *i.e*, minterms to be included with the function being implemented, there must be a logic 1 connected to the corresponding data inputs; those minterms which are not included in the function *i.e.*, for the minterms F = 0, the corresponding data inputs are connected by logic '0'. The variables are connected to the data select terminals of the multiplexer. Thus, a Boolean function of *n*-variables can be implemented by using a 2^n to –1 multiplexer. This is illustrated in the following examples.

Example 3.47. *Implement the following Boolean function using 8 × 1 MUX*

$$F\ (A,\ B,\ C) = \Sigma m\ (1,\ 2,\ 4,\ 5).$$

Solution. The minterms 1, 2, 4 and 5 are presented in the given function, so their corresponding data inputs are connected by logic 1, and the remaining data inputs are connected by logic 0. The three variables are connected with the three selection lines as shown in Fig. 3.202.

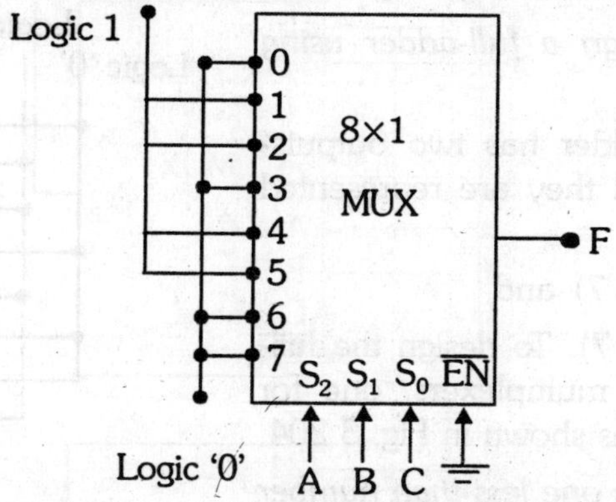

Fig. 3.202 Boolean function implementation with a multiplexer.

Example 3.48. *Implement the following Boolean function using 8 × 1 multiplexer.*

$$F\ (A,\ B,\ C) = \bar{A}B + AC + B\bar{C}$$

Solution. To implement the given Boolean function, first we have to find its truth table. The truth table is shown in Table 3.49.

Table 3.49

A	B	C	F
0	0	0	0
0	0	1	0
0	1	0	1
0	1	1	1

1	0	0	0
1	0	1	1
1	1	0	1
1	1	1	1

$\therefore$ $F = \Sigma m\ (2, 3, 5, 6, 7)$.

The Fig. 3.203 shows the logic circuit implementation.

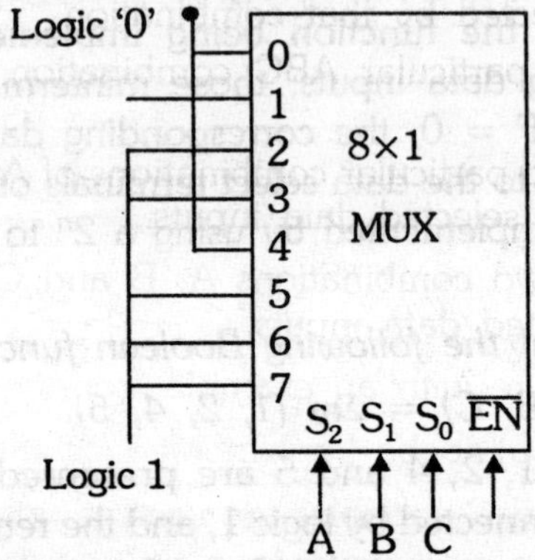

Fig. 3.203 Logic implementation of Boolean function.

Example 3.49. *Design a full-adder using 8 × 1 MUX.*

Solution. The full adder has two outputs-sum (S) and carry (C) and they are represented by

$S = \Sigma m\ (1, 2, 4, 7)$ and

$C = \Sigma m\ (3, 5, 6, 7)$. To design the full-adder we need two 8×1 multiplexers, one for sum and another for carry as shown in Fig. 3.204.

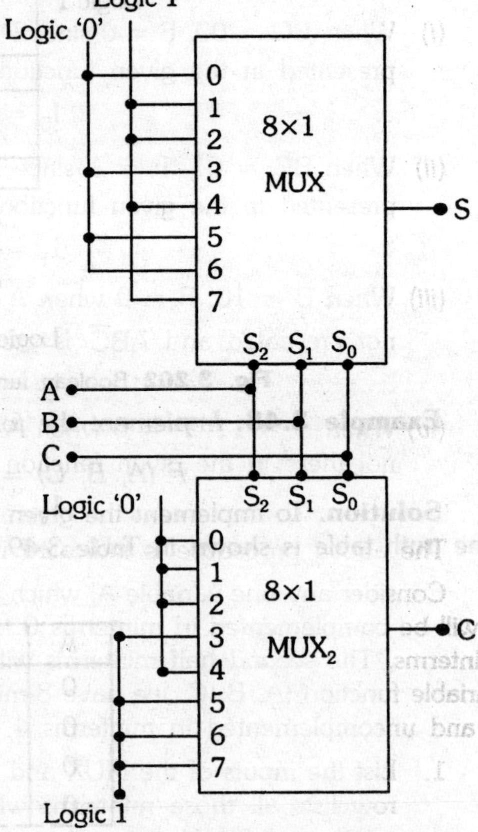

Fig. 3.204 Full adder using 8 × 1 multiplexers.

Case (*ii*) *Select inputs one less than number of variables.*

A multiplexer with *n*- select inputs can also be employed to implement a Boolean function that has $n + 1$ variables.

The *n*-variables are connected to the selection lines of the multiplexer and the remaining single variable of the function is used for the inputs of the multiplexer.

If A is this variable, the inputs of the multiplexer are choosen as either A, $\overline{A}$, 0 or 1 (earlier we have only two values to the inputs of the multiplexer *i.e.*, 0 or 1). By use of these four values for the inputs and by connecting the other variables to the selection lines, we can implement the Boolean function with a multiplexer. In this way, it is possible to generate any function of $n + 1$ variables with a 2^n-to-1 multiplexer.

Suppose that the logic function F (A, B, C, D) is to be implemented, where D is the MSB used to connect as a input to the multiplexer, the design procedure is as follows :

1. Write down the truth table of the logic function. Each combination of A, B and C must occur for both D = 0 and for D = 1.
2. F = 0, both times an ABC combination occurs, then the logic 0 must be connected to the data inputs selected by that combination.
3. F = 1, both times a particular ABC combination occurs, connect logic 1 to the selected data inputs.
4. F is different for the two particular combinations of A, B and C and F = D each time, then connect D to the selected data inputs.
5. F is different for the two combinations A, B and C and F = $\overline{D}$, then $\overline{D}$ must be connected to the selected data inputs.

To demonstrate this procedure with an example, consider the function of three variables f (A, B, C) = Σ (1, 3, 5, 6) to be implemented by using 4 × 1 MUX.

Let us assume that the B and C variables are used to connect the selection lines *i.e.*, B to S_1 and C to S_0. The inputs of the multiplexer may be 0, 1, A and $\overline{A}$. The implementation of this function is shown in Fig. 3.205 (*a*) and its truth table in 3.205 (*b*).

(*i*) When BC = 00, F = 0 since D_0 and D_4 = 0 (*i.e.*, minterms $\overline{A}\,\overline{B}C$ and $\overline{A}BC$ are not presented in the given function.)

$$\therefore \qquad I_0 = 0 \text{ (which is an input line of } 4 \times 1 \text{ MUX)}$$

(*ii*) When BC = 01, F = 1 since D_1 and D_5 = 1 (*i.e.*, minterms $\overline{A}\,\overline{B}C$ and $A\overline{B}C$ are presented in the given function)

$$\therefore \qquad I_1 = 1$$

(*iii*) When B = 10, F = 0 when A = 0 and F = 1 when A = 1 (*i.e.*, minterms $\overline{A}B\overline{C}$ is not presented and $AB\overline{C}$ is presented)

$$\therefore \qquad I_2 = A$$

(*iv*) When B = 11, F = 0 when A = 1 and F = 1 when A = 0 (*i.e.*, minterms ABC is not there in the given function but $\overline{A}BC$ is there in the function).

$$\therefore \qquad I_3 = \overline{A}$$

The above procedure is indicated in the implementation table.

Consider any one variable A, which is a highest order position in the sequence of variables, it will be complemented in minterms 0 to $(2^{n/2} - 1)$, which comprise the first half in the list of minterms. The second half minterms will have their A variable uncomplemented. For a three variable function, A, B, C, we have 8-minterms. Variable A is complemented in minterms 0 to 3 and uncomplemented in minterms 4 to 7.

1. List the inputs of the MUX and under them list all the minterms in two rows. The first row lists all those minterms where A is uncomplemented, and second row all the minterms with A uncomplemented as shown in Fig. 3.205 (*c*).

2. Circle all the minterms of the function and inspect each column separately.
   (*a*) If two minterms in a column are not circled, apply '0' to corresponding MUX input
   (*b*) If two minterms are circled, apply 1 to corresponding MUX input.
   (*c*) If the bottom minterm (A) is circled and the top ($\overline{A}$) is not circled, apply A to the corresponding MUX input.
   (*d*) If the top minterm ($\overline{A}$) is circled and the bottom (A) is not circled, apply $\overline{A}$ to the corresponding MUX input. The resultant table is called an "*implementation table*".

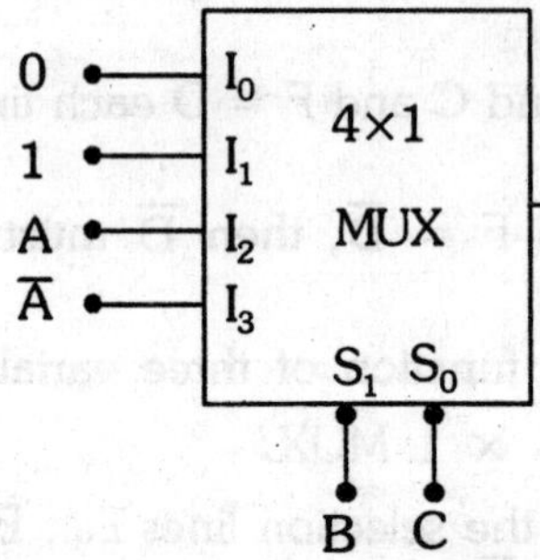

(*a*) MUX implementation

Minterm	ABC	F
0	000	0
1	001	1
2	010	0
3	011	1
4	100	0
5	101	1
6	110	1
7	111	0

(*b*) Truth table

	(00) I_0	(01) I_1	(10) I_2	(11) I_3
(0) A'	0	①	2	③
(1) A	4	⑤	⑥	7
Inputs to MUX →	0	1	A	$\overline{A}$

(*c*) Implementation table

Fig. 3.205 Example of Σm (1, 3, 5, 6).

Example 3.50. *Implement the following function*

$$f\ (A, B, C, D) = \Sigma m\ (0, 1, 3, 4, 7, 10, 14)$$

using

(1) 16 : 1 MUX

(2) 8 : 1 MUX

(3) 4 : 1 MUX

(4) 2 × 1 MUX

Note: (1) Any one variable of the function can be taken as input to the MUX and remaining variables are connected to the selection lines.

(2) More than one variable can also be taken as a input to the MUX. For example, a 4–variable function with 2–variables as input can be implemented by a 4 × 1 MUX (or) by taking 3–variables as input.

We can implement the same function by using a 2×1 MUX. But here we need some combinational gates at the input of MUX.

Solution. Given function

$$f\ (A, B, C, D) = \Sigma m\ (0, 1, 3, 4, 7, 10, 12, 14)$$

(*a*) Implementations with 16 × 1 MUX

The minterms which are given in the function are connected by logic 1 and remaining minterms are connected by logic '0' shown in Fig. 3.206 (*a*).

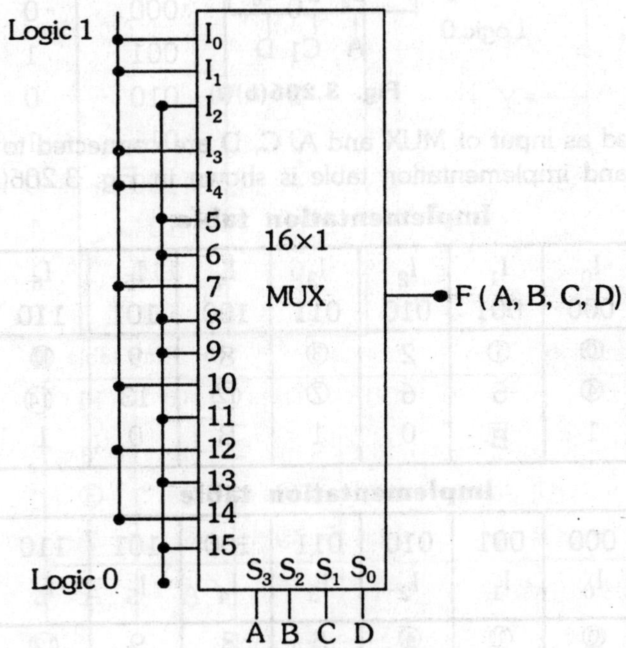

Fig. 3.206(*a*)

Implementation table

	(000) I_0	(001) I_1	(010) I_2	(011) I_3	(100) I_4	(101) I_5	(110) I_6	(111) I_7
$\bar{A}$	⓪	①	2	③	④	5	6	⑦
A	8	9	⑩	11	⑫	13	⑭	15
	$\bar{A}$	$\bar{A}$	A	$\bar{A}$	1	0	A	$\bar{A}$

(*b*) Implementation with 8 × 1 MUX

(*i*) Assume 'A' is connected as MUX input and B, C, D are select lines, the implementation table and logic diagram is shown in Fig. 3.206 (*b*) (*i*).

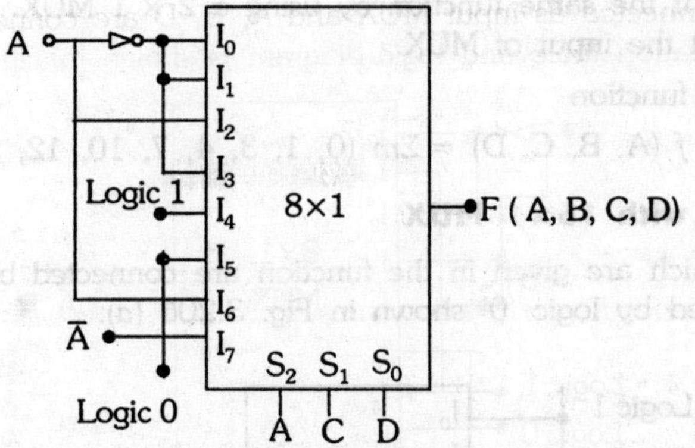

Fig. 3.206(*b*)(*i*)

(*ii*) If B is connected as input of MUX and A, C, D are connected to the select lines. The logic diagram and implementation table is shown in Fig. 3.206(*b*)(*ii*).

Implementation table

	I_0 000	I_1 001	I_2 010	I_3 011	I_4 100	I_5 101	I_6 110	I_7 111
$\overline{B}$(0)	⓪	①	2	③	8	9	⑩	11
B(1)	④	5	6	⑦	⑫	13	⑭	15
MUX Inputs	1	$\overline{B}$	0	1	B	0	1	0

Implementation table

	000 I_0	001 I_1	010 I_2	011 I_3	100 I_4	101 I_5	110 I_6	111 I_7
$\overline{C}$	⓪	①	④	5	8	9	⑫	13
C	2	③	6	⑦	⑩	11	⑭	15
MUX Inputs	$\overline{C}$	1	$\overline{C}$	C	C	0	1	$\overline{A}$

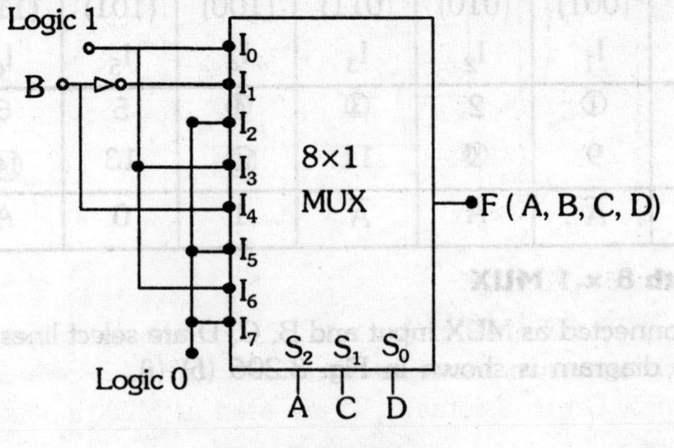

Fig. 3.206(*b*)(*ii*)

(*iii*) If C is connected as input MUX and A, B, D are connected to the select lines. The implementation table and logic diagram is shown in Fig. 3.206 (*b*) (*iii*).

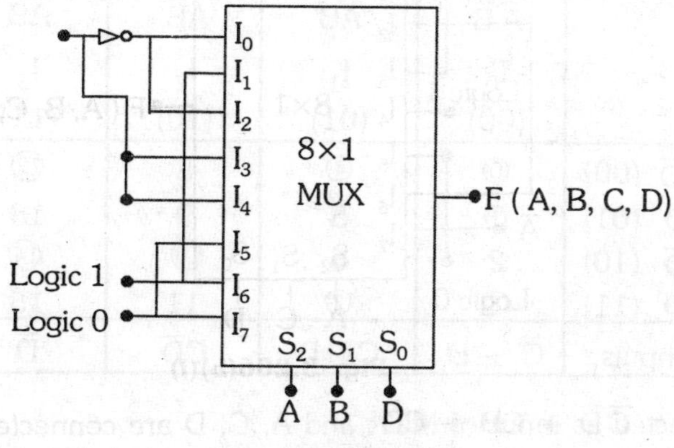

Fig. 3.206(*b*)(*iii*)

(*iv*) If D is connected as input of MUX and A, B, C are connected to the select lines. The implementation table and logic diagram is shown in Fig. 3.206(*b*)(*iv*).

Implementation table

	000 I_0	001 I_1	010 I_2	011 I_3	100 I_4	101 I_5	110 I_6	111 I_7
$\overline{D}$	⓪	2	④	6	8	⑩	⑫	⑭
D	①	③	5	⑦	9	11	13	15
MUX Inputs	1	D	$\overline{D}$	D	0	$\overline{D}$	$\overline{D}$	$\overline{D}$

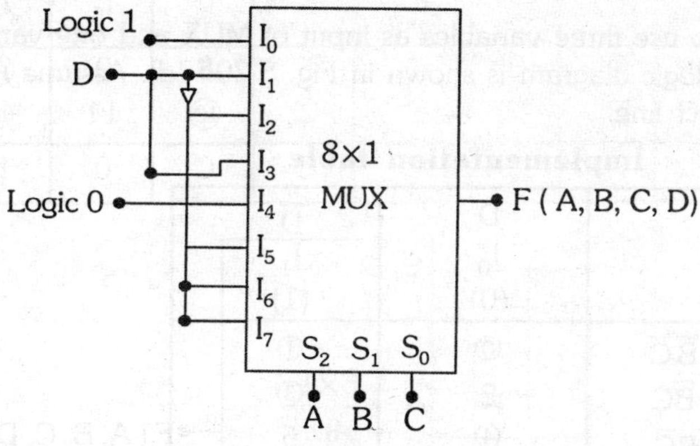

Fig. 3.206(*b*)(*iv*)

(c) Implementation with 4×1 MUX

Here we have to take two variables as MUX input and remaining two variables are connected to select lines. Let us assume C, D are used as MUX inputs and variables A, B are connected to select lines.

The implementation table and logic diagram is shown in Fig. 3.206(c).

Implementation table

	$\bar{A}\bar{B}$ I_0 (00)	$\bar{A}B$ I_1 (01)	$A\bar{B}$ I_2 (10)	AB I_3 (11)
$\bar{C}\bar{D}$ (00)	⓪	④	8	⑫
$\bar{C}D$ (01)	①	5	9	13
$C\bar{D}$ (10)	2	6	⑩	⑭
CD (11)	③	⑦	11	15
MUX Inputs	$\bar{C} + D$	$C \odot D$	$C\bar{D}$	$\bar{D}$

First row $(I_0) = \bar{C}\bar{D} + \bar{C}D + CD$

$= \bar{C}(\bar{D} + D) + CD$

$= \bar{C} + CD$

$= \bar{C} + D$

Second row $(I_1) = \bar{C}\bar{D} + CD$

$= C \odot D$

Third row $(I_2) = C\bar{D}$

Fourth row $(I_3) = \bar{C}\bar{D} + C\bar{D}$

$= \bar{D}$

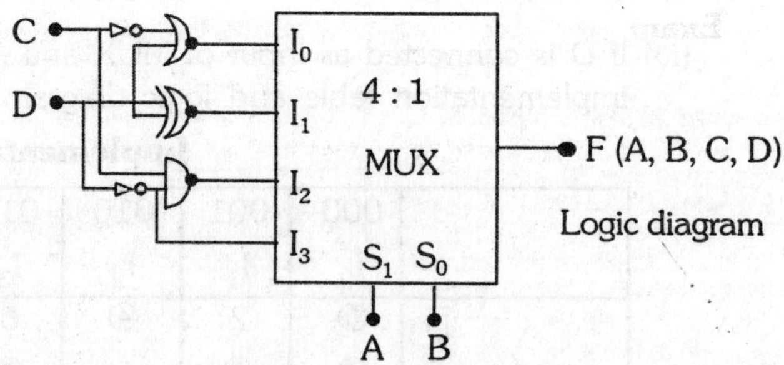

Fig. 3.206(c)

(d) Implementation using 2 × 1 MUX

Here we have to use three variables as input of MUX and one variable as select line. The implementation and logic diagram is shown in Fig. 3.205 (d). Assume A, B and C are used as inputs and D as select line.

Implementation table

	D I_0 (0)	$\bar{D}$ I_1 (1)
$\bar{A}\bar{B}\bar{C}$	⓪	①
$\bar{A}\bar{B}C$	2	③
$\bar{A}B\bar{C}$	④	5
$\bar{A}BC$	6	⑦
$A\bar{B}\bar{C}$	8	9
$A\bar{B}C$	⑩	11
$AB\bar{C}$	⑫	13
ABC	⑭	15
MUX Inputs	$A \odot C + BC$	$\bar{A}(\bar{B} + C)$

From first row, $I_0 = \overline{A}\,\overline{B}\,\overline{C} + \overline{A}B\overline{C} + A\overline{B}C + AB\overline{C} + ABC$

$= \overline{A}\,\overline{C} + AC + BC$

From second row, $I_1 = \overline{A}(\overline{B} + C)$

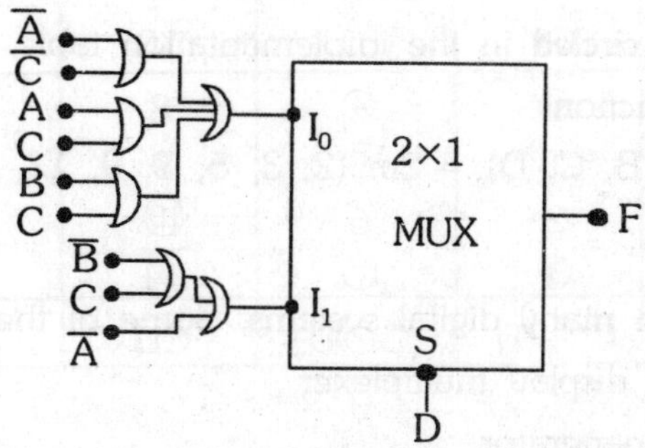

Fig. 3.206(*d*)

Example 3.51. *Realize the Boolean function for the given Fig. 3.207.*

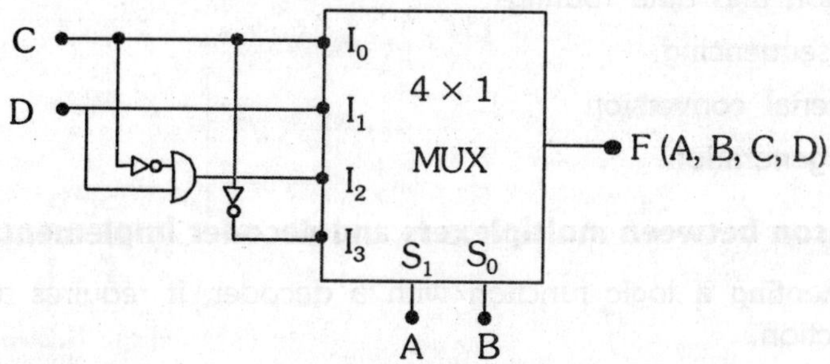

Fig. 3.207

Solution. The implementation table for the above figure is shown in Table 3.50.

Table 3.50. Implementation table

	I_0	I_1	I_2	I_3
$\overline{C}\,\overline{D}$	0	4	8	⑫
$\overline{C}D$	1	⑤	⑨	⑬
$C\overline{D}$	②	6	10	14
CD	③	⑦	11	15
MUX Inputs	C	D	$\overline{CD}$	$\overline{C}$

From the figure

$I_0 = C$ $\qquad I_2 = \overline{CD}$

$I_1 = D$ $\qquad I_3 = \overline{C}$

(*a*) To get $I_0 = C$, the minterms 2 and 3 must be present in the function.

(*b*) To get $I_1 = D$, the minterms 5 and 7 must be present in the function.

(c) To get $I_2 = \overline{C}D$, the minterm '9' must be present in the function and

(d) To get $I_3 = \overline{C}$, the minterms 12 and 13 must be present in the given function.

All these minterms are circled in the implementation table.

From the table, the function

$$F(A, B, C, D) = \Sigma m\ (2, 3, 5, 7, 9, 12, 13)$$

3.13.3.4 Applications

Multiplexers are used in many digital systems. Some of them are

1. The seven-segment display multiplexer
2. The logic function generator
3. Digital counter with multiplexed displays
4. Data selection and data routing
5. Operation sequencing
6. Parallel-to-serial conversion
7. Waveform generation.

3.13.3.5 Comparison between multiplexers and decoder implementation

1. For implementing a logic function with a decoder, it requires an OR gate for each output function.
2. Single decoder is enough to implement many functions, since decoder generates all the possible minterms.
3. In the MUX method, it uses smaller-size units but requires one multiplexer for each output function.

Conclusion

1. The combinational circuits with small number of outputs should be implemented with multiplexers. Combinational circuits with many output functions should be implemented by the decoder.
2. Generally, decoders are mostly used for decoding binary information and multiplexers are mostly used to form a selected path between multiple sources and a single destination.

3.13.4 Demultiplexer

Demultiplexer is a combinational circuit that performs a reverse multiplexer function. The demultiplexer is a logic circuit that receives information on a single line and transmits this information on one of 2^n possible output lines. The selection of a specific output line is controlled by the bit values of '*n*' selection lines. A decoder with an enable input can function as a demultiplexer. The idea of a demultiplexer can best be understood by the Fig. 3.207. Clearly, any one of the four outputs can be connected to the input.

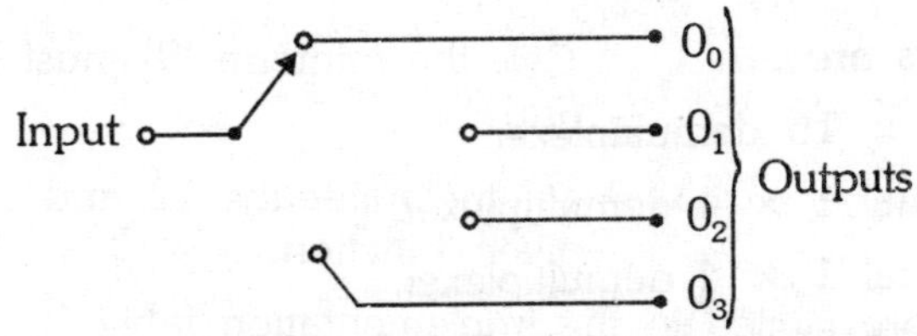

Fig. 3.208 Demultiplexer principle.

The demultiplexer is represented by 1 : 4 DEMUX (or) 1×4 DEMUX or 1–to – 4 DEMUX. General representation is 1 : 2^n Demux.

To understand this, let us take an example of 1×4 DEMUX. Demultiplexer is also called as "DATA DISTRIBUTORS".

The block diagram of 1×4 DEMUX is shown in Fig. 3.209 (*a*) and its logic diagram is shown in Fig. 3.209(*b*).

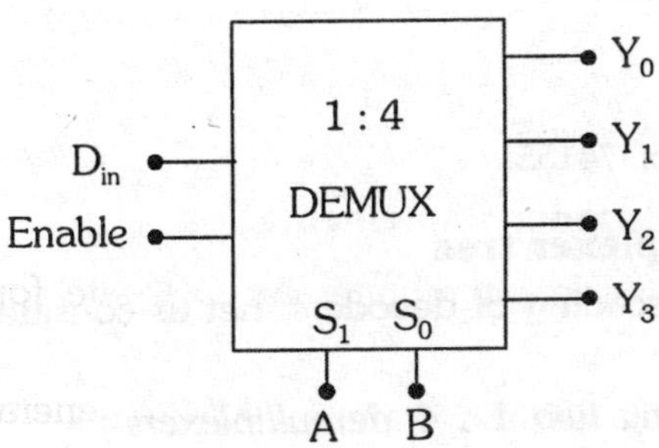

Fig. 3.209(*a*) Block diagram.

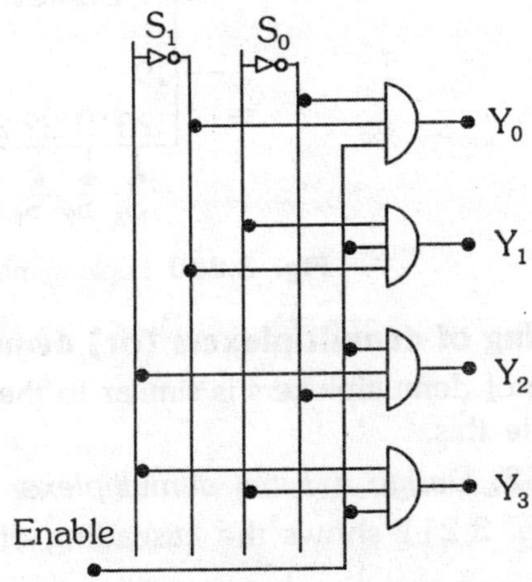

Fig. 3.209(*b*) Logic diagram.

The truth table for 1 : 4 Demux is shown in Table 3.51.

Table 3.51. Truth table for 1 : 4 demultiplexer

Enable	*Selection lines*		*Input*	*Outputs*			
EN	S_1	S_0	D_{in}	Y_0	Y_1	Y_2	Y_3
0	×	×	×	0	0	0	0
1	0	0	0	0	0	0	0
1	0	0	1	1	0	0	0
1	0	1	0	0	0	0	0
1	0	1	1	0	1	0	0
1	1	0	0	0	0	0	0
1	1	0	1	0	0	1	0
1	1	1	0	0	0	0	0
1	1	1	1	0	0	0	1

3.13.4.1 Some important demultiplexer ICs :

The IC demultiplexers are :

(*a*) 74154 → 1 × 16 demultiplexer.

(*b*) 74155 → Dual 1 × 4 demultiplexer.

(*c*) 74156 → Dual 1 × 4 demultiplexer.

The logic symbol of 74155 is shown in Fig. 3.210.

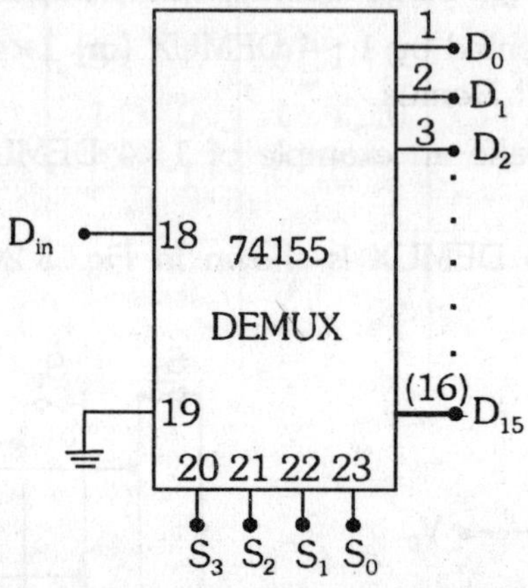

Fig. 3.210 Logic symbol of 74155.

3.13.4.2 Cascading of demultiplexers (or) demultiplexer tree

The cascading of demultiplexers is similar to the cascading of decoders. Let us consider an example to illustrate this.

Example 3.52. *Design a 1 : 8 demultiplexer using two 1 : 4 demultiplexers.*

Solution. Fig. 3.211 shows the cascading of two, 1 : 4 demultiplexers to form 1 : 8 demultiplexer.

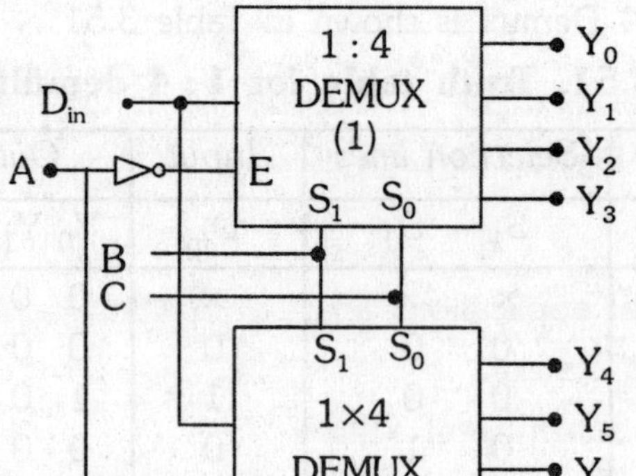

Fig. 3.211 Cascading of demultiplexers.

Example 3.53. *Implement a full adder using demultiplexer.*

Solution. Let us see the truth table of the full adder as shown in Table 3.52.

Table 3.52

Decimal value	*A*	*B*	C_{in}	*S*	C_{out}
0	0	0	0	0	0
1	0	0	1	1	0
2	0	1	0	1	0
3	0	1	1	0	1
4	1	0	0	1	0
5	1	0	1	0	1
6	1	1	0	0	1
7	1	1	1	1	1

$\therefore$ $$S = \Sigma m\ (1, 2, 4, 7)$$

and $$C_{out} = \Sigma m\ (3, 5, 6, 7)$$

The logic diagram is shown in Fig. 3.212, which requires a 1 : 8 Demux and 2-OR gates.

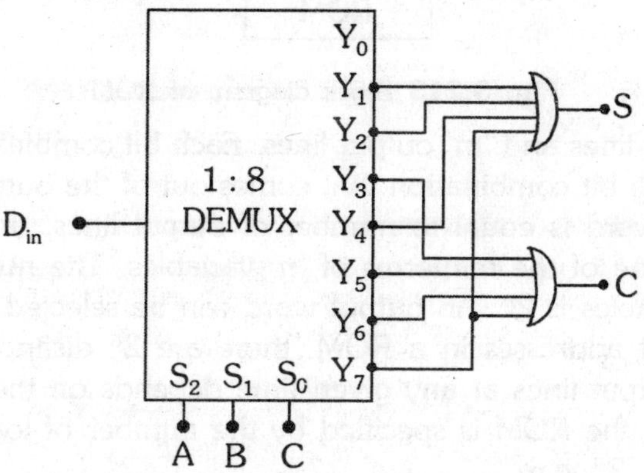

Fig. 3.212 Logic diagram of full adder.

3.13.4.3 Applications

Demultiplexer has several applications apart from data distribution. They are

1. in security monitoring system.
2. in synchronous data transmission system.

3.14 LOGIC IMPLEMENTATION USING PLD'S

Till now, we have studied several MSI chips, but in many applications, the designer needs a special type of IC. It is very difficult to manufacture different IC's to satisfy all the designers, since, the operation (or) function of an any MSI device is fixed. There are many situations, however, for which there are no LSI or VLSI chips to achieve the necessary logic functions.

These situations have typically required designers to interconnect several LSI and VLSI chips to achieve the necessary function. This situation led to the development of programmable logic devices (PLD's), which can be configured by the individual user for his specific application. PLD is an IC that contains large number of gates, flip-flops and registers that are interconnected on the chip. This IC is said to be programmable because the specific function IC is determined by interconnecting required contacts. Basically, there are three types of programmable devices, which are :

1. Read Only Memory (ROM)
2. Programmable Logic Array (PLA)
3. Programmable Array Logic (PAL)

3.14.1 Read Only Memory (ROM)

Read only memory (ROM) is a device that includes both the decoder and OR gates within a single IC package. The connection between outputs of the decoder and the inputs of the OR gates can be specified for each particular configuration. Fig. 3.213 shows the block diagram of ROM.

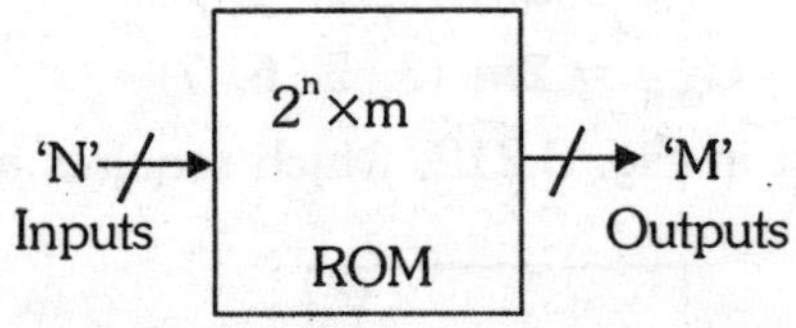

Fig. 3.213 Block diagram of ROM.

It consists of '*n*' input lines and '*m*' output lines. Each bit combination of the input variables is called on '*address*'. Each bit combination that comes out of the output lines is called a '*word*'. The number of bits per word is equal to number of output lines, *m*. The address specified in binary number denotes one of the minterms of '*n*' variables. The number of distinct addresses possible with *n*-input variables is 2^n. An output word can be selected by a unique address, and since there are 2^n distinct addresses in a ROM, there are 2^n distinct words in the ROM. The word available on the output lines at any given time depends on the address value applied to the input lines. Generally, the ROM is specified by the number of locations (words) × number of output lines (bits) *i.e.*, $2^n \times m$.

Let us consider a 32 × 4 ROM. This ROM consists of 32 words of 4 bit each. The total capacity of ROM is 128 bits *i.e.*, number of bits stored in the unit. This means that there are four output lines and five input ($2^5 = 32$) lines. There are only five-input variables required and with the five-variables, we can specify 32 addresses or minterms. For each address input, there is a unique selected word. If the input address is "00000", word number '0' is selected and applied to the output lines. If the input address is "11111", word number 31 is selected and applied to the output lines.

Internally, the ROM is a combinational circuit with AND gates connected as a decoder and a number of OR gates equal to the number of outputs in the unit. Fig. 3.214 shows the internal logic construction of a 32 × 4 ROM. The five input variables are decoded in 32 lines by means of 32 AND gates and 5 inverters. Each output of the decoder represents one of the minterms of a function of five variables. The 32 outputs of the decoder are connected through fuses to each OR gate *i.e.,* each OR gate is connected by 32 outputs of the decoder through the fuses that can be blown as desired.

The ROM is a two-level implementation in sum of minterms form. It does not have to be an AND-OR implementation, but it can be any other possible two-level minterm implementation. Let us see the AND-OR and AND-OR-INVERTER implementation of ROM in Fig. 3.215, which shows the 4×2 ROM.

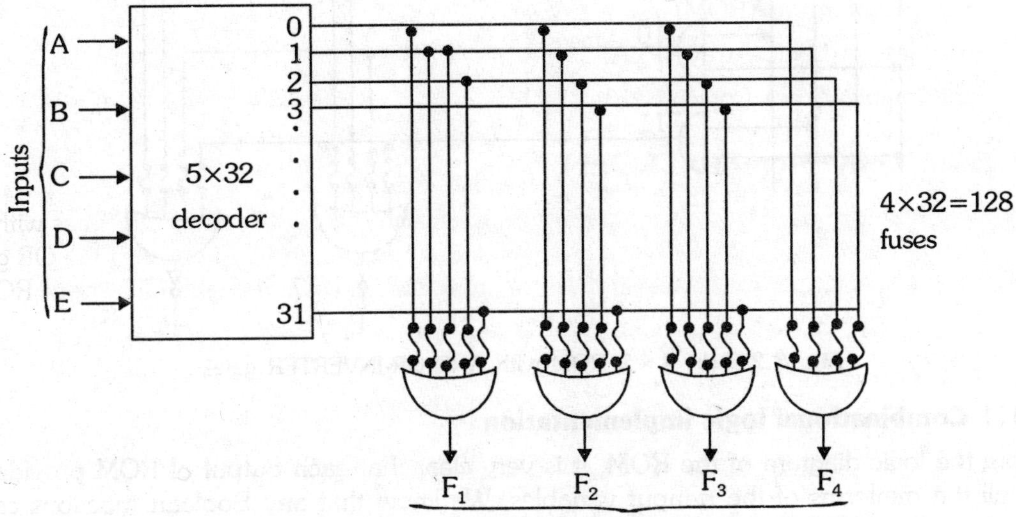

Fig. 3.214 Logic construction of 32 × 4 ROM.

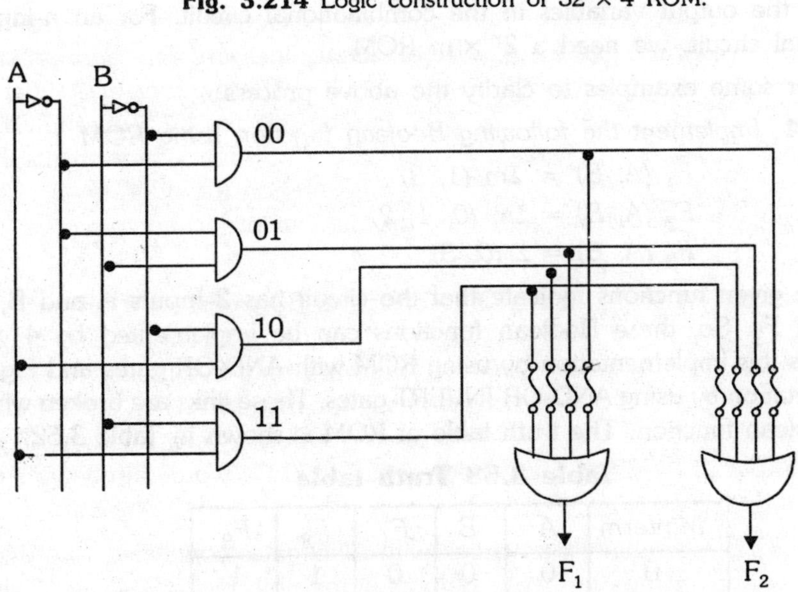

Fig. 3.215(*a*) 4 × 2 ROM with AND-OR gates.

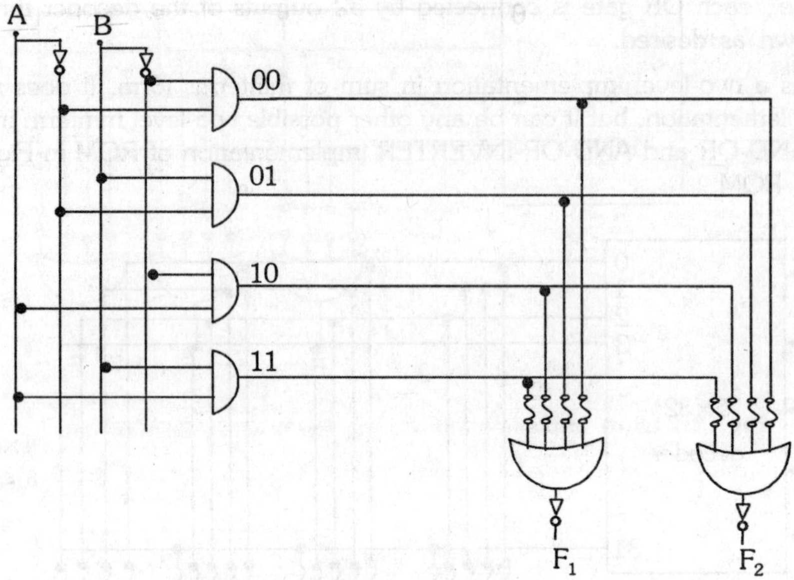

Fig. 3.215(*b*) 4 × 2 ROM with AND-OR-INVERTER gates.

3.14.1.1 Combinational logic implementation

From the logic diagram of the ROM, it is very clear that each output of ROM provides the sum of all the minterms of the *n*-input variables. We know that any Boolean functions can be expressed in sum of minterms form. If any minterm is not present in the given function, that particular link will be broken and each ROM output can be made to represent the Boolean function of one of the output variables in the combinational circuit. For an *n*-input and *m*-output combinational circuit, we need a $2^n \times m$ ROM.

Let us consider some examples to clarify the above process.

Example 3.54. *Implement the following Boolean function using ROM*

$$F_1\ (A, B) = \Sigma m\ (1, 3)$$
$$F_2\ (A, B) = \Sigma m\ (0, 1, 2)$$
$$F_3\ (A, B) = \Sigma\ (0, 3)$$

Solution. The given functions indicate that the circuit has 2 inputs A and B, and three outputs F_1, F_2 and F_3. So, these Boolean functions can be implemented by 4 × 3 ROM. Fig. 3.215 (*a*) shows this implementation by using ROM with AND-OR gates and Fig. 3.215 (*b*) shows the implementation by using AND-OR-INVERT gates. Those links are broken which are not included in the Boolean function. The truth table of ROM is shown in Table 3.52.

Table 3.53 Truth table

Minterm	*A*	*B*	F_1	F_2	F_3
0	0	0	0	1	1
1	0	1	1	1	0
2	1	0	0	1	0
3	1	1	1	0	1

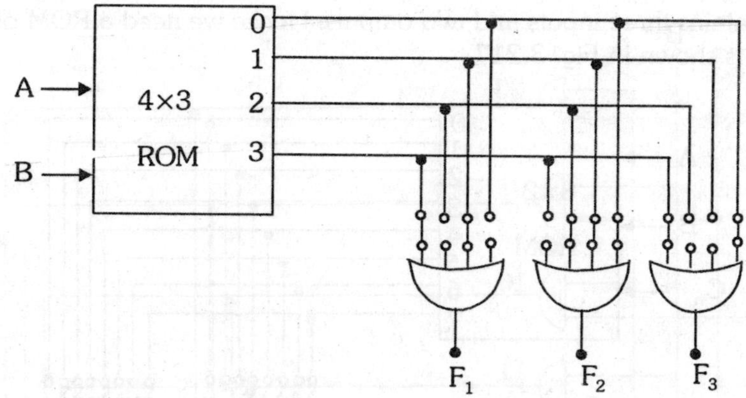

Fig. 3.216(*a*) Implementation using ROM with AND-OR gates.

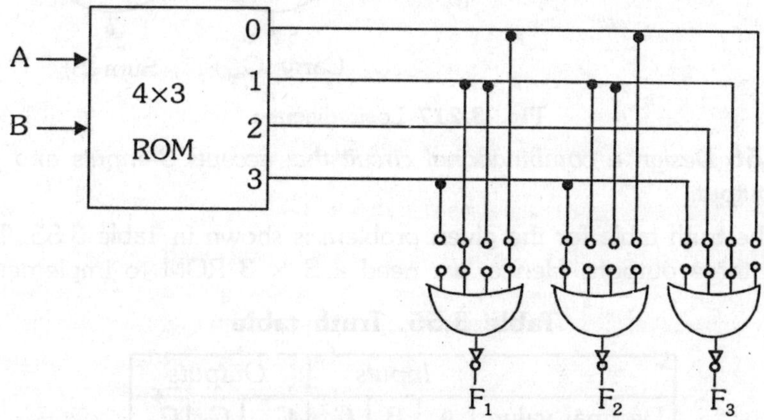

Fig. 3.216(*b*) Implementation using ROM with AND-OR-INVERT gates.

Example 3.55. *Design a full-adder using a ROM.*

Solution. The truth table for full adder is shown in Table 3.54.

Table 3.54. Truth table for full adder

A	B	C_{in}	S	C_{out}
0	0	0	0	0
0	0	1	1	0
0	1	0	1	0
0	1	1	0	1
1	0	0	1	0
1	0	1	0	1
1	1	0	0	1
1	1	1	1	1

$$S = \Sigma m\ (1, 2, 4, 7)$$

$$C_{out} = \Sigma m\ (3, 5, 6, 7)$$

The circuit contains three inputs and two outputs. Hence we need a ROM of 8×2 ($2^3 \times 2$). The logic diagram is shown in Fig. 3.217.

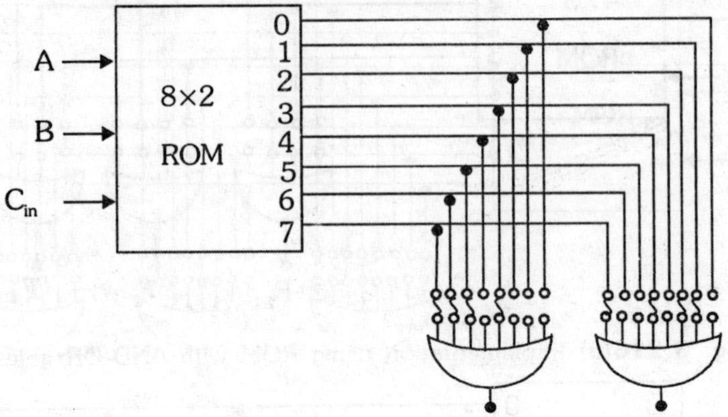

Fig. 3.217 Logic diagram.

Example 3.56 *Design a combinational circuit that accepts 3–inputs and produces its 1's complement as output.*

Solution. The truth table for the given problem is shown in Table 3.55. The given circuit has 3 inputs and three outputs. Hence, we need a 8×3 ROM to implement it.

Table 3.55. Truth table

	Inputs			*Outputs*		
Decimal value	A	B	C	C_1	C_2	C_3
0	0	0	0	1	1	1
1	0	0	1	1	1	0
2	0	1	0	1	0	1
3	0	1	1	1	0	0
4	1	0	0	0	1	1
5	1	0	1	0	1	0
6	1	1	0	0	0	1
7	1	1	1	0	0	0

The output functions

$$C_1 = \Sigma m\ (0, 1, 2, 3)$$
$$C_2 = \Sigma m\ (0, 1, 4, 5)$$
$$C_3 = \Sigma m\ (0, 2, 4, 6)$$

The logic diagram is shown in Fig. 3.218.

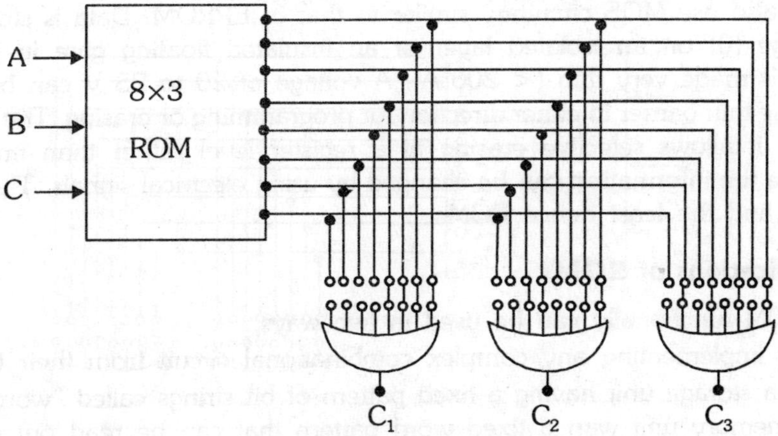

Fig. 3.218 Logic diagram.

3.14.1.2 Types of ROMs

There are four types of ROMs, masked ROM, PROM, EPROM and EEPROM.

(*a*) Masked ROMs

All the interconnections are made by the designer in the manufacturing process. For this, the customer must submit his requirements in the form of truth table to the designer. The manufacturer makes the corresponding mask for the paths to produce the 1's and 0's according to the customer's truth table. This process is costly because the vendor charges the customer a special fee for custom masking a ROM. This programming is economical if large quantities of the same ROM configuration are to be manufactured.

(*b*) Programmable Read Only Memory (PROM)

It is more economical for small quantities. PROM allows the user to store data or program. PROMs use the fuses with material like nichrome and polycrystalline. The user can blow these fuses by passing around 20 to 50 mA of current for the period of 5 to 20 μ sec. The blowing of fuses according to the truth table is called "programming of ROM". This allows the user to program the unit in the laboratory by using PROM programmers to achieve the desired relationship between input address and stored words. The PROM programmer selectively burns the fuses according to the bit pattern to be stored. This process is also known as "burning of PROM". The PROMs are one time programmable. Once programmed, the information or logic stored is permanent.

(*c*) Erasable Programmable Read Only Memory (EPROM)

Erasable PROMs use the MOS technique. They store 0 or 1 as a charge in the buried layer of the IC chip. EPROMs can be programmed by the user with a special EPROM programmer. The main advantage of this type is, we can erase the stored data in the EPROM by exposing the chip to ultraviolet light through its quartz window for 15 to 20 minutes.

In EPROM, it is not possible to erase selective information; when erased, the entire information is lost. The chip can be reprogrammed. This memory is ideally suitable for experimental projects, product developments and college laboratories, since these chips can be reused many times.

(*d*) Electrically Erasable Programmable Read Only Memory (EEPROM)

EEPROMs also use MOS circuitary similar to that of EPROM. Data is stored as a charge (1) or no charge (0) on an isolated layer or an insulated floating gate in the device. The insulating layer is made very thin (< 200°A). A voltage of 20 to 25 V can be used to move charges across the thin barrier in either direction for programming or erasing. The main advantage of EEPROM is, it allows selective erasing at a register level rather than erasing the entire information since the information can be changed by using electrical signals. The EEPROMs are most expensive and the least dence ROMs.

3.14.1.3 Applications of ROMs

1. The ROM functionally can be used in two ways,
   (*a*) For implementing any complex combinational circuit from their truth tables.
   (*b*) As a storage unit having a fixed pattern of bit strings called "words" *i.e.*, ROM is a memory unit with a fixed word pattern that can be read out only.
2. These are useful for converting one binary code to another (such as ASCII to EBCDIC and vice-versa).
3. Used to perform arithmetic functions such as multipliers, (for display of characters in a CRT)
4. Used in the design of control units of digital systems. A control unit that utilizes a ROM to store binary control information is called "microprogrammed control unit".

Demerits of ROM

The combinational circuits do not use all the minterms every time. Occasionally, they have don't care conditions. Don't care condition when implemented with a ROM becomes an address input that will never occur. The words as the don't care addresses need not be programmed and may be left in their original state (all 0's or all 1's). The result is that not all the bit patterns available in the ROM are used, which may be considered a waste of available equipment. For example, consider a combinational circuit that converts 12-bit code to 8-bit ASCII code. The input consists of 12 lines designated by 0, 1, 2, ...11. The site of ROM for implementing the code converter must be 4096 × 8, since there are 12 inputs and 8-outputs. There are only 128 valid entries for the code converter, and all other input combinations are don't care conditions. Thus, only 128 words of 4096 available are used. The remaining 3968 words of ROM are not used and are thus wasted.

3.14.2 Programmable Logic Array (PLA)

For cases where the number of don't care conditions are more or expensive, it is more convenient to use a second type of PLD called a "*Programmable Logic Array* (PLA)".

The PLA is similar to a ROM in concept ; however it does not provide full decoding of the variables and does not generate all the minterms as in the ROM. The PLA replaces the decoder by group of AND gates, each of which can be programmed to generate a product term of the input variables.

In PLA, both AND gate and OR gate arrays are programmable, since both gates have fuses at the inputs. Generally PLA size can be represented or specified by $n \times k \times m$ where

n = number of inputs

k = number of product terms and

m = number of outputs.

The Fig. 3.219 shows the block diagram of PLA. It consists of n–inputs, m–outputs (or) sum terms and k–product terms. The product terms constitute a group of K-AND gates and the sum terms constitute a group of m–OR gates. Fuses inserted are provided

(a) Between all n–inputs and their complements and inputs of AND gates.

(b) Between outputs of AND gates and the inputs of the OR gates.

(c) In the output inverters to allow the output function to be generated in the AND-OR form or in the AND-OR-INVERT form. When inverter is bypassed by link we get AND–OR form, else AND-OR-INVERTER implementation.

The total number of programmable fuses in a $n \times k \times m$ PLA are :

(a) Between inputs and AND gates $= 2\,n \times k$

(b) Between AND gates and OR gates $= k \times m$

(c) At the output $m = m$

$\therefore$ Total fuses $= 2n \times k + k \times m + m$, whereas in case of ROM $2^n \times m$.

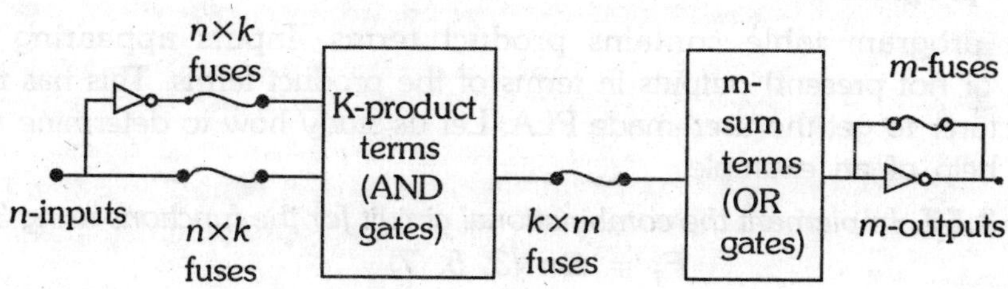

Fig. 3.219 Block diagram of PLA.

The Fig. 3.220 shows the internal construction of PLA with 3 inputs, 3 product terms and two outputs.

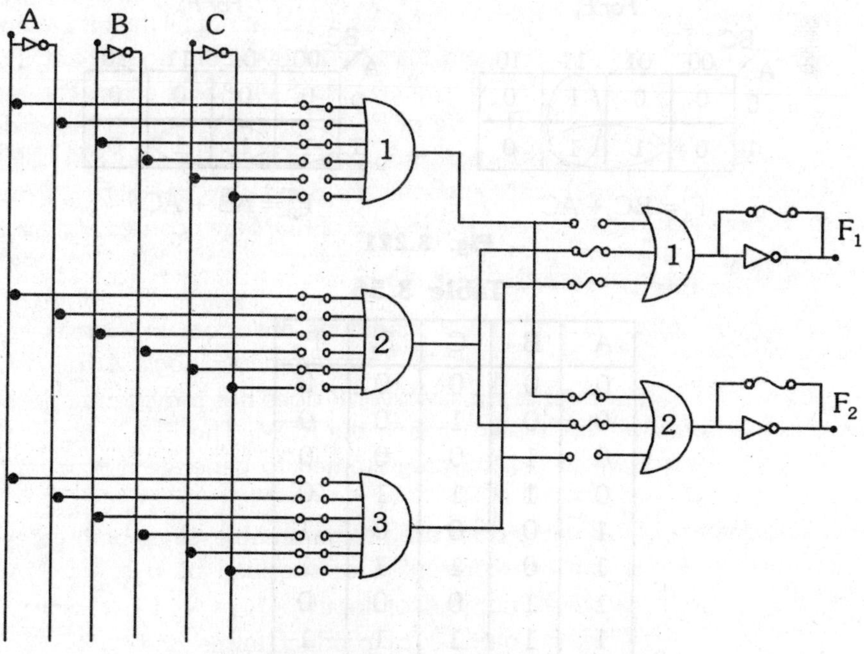

Fig. 3.220 Internal construction of PLA.

3.14.2.1 Types of PLAs

Like ROM, there are two types of PLAs according to their programming :

1. Mask–programmable PLA.
2. Field–programmble PLA.

(*a*) Mask–programmable PLA

In the mask–programmable PLA, the customer must submit a PLA program table to the manufacturer. This table is used by the vendor to produce a user–made PLA that has the required internal paths between inputs and outputs.

(*b*) Field–programmable PLA

The FPLA can be programmed by the user by means of certain procedures. Here every individual user can program their application in the FPLA independently.

3.14.2.2 PLA program table

The PLA program table contains product terms. Inputs appearing (complement, uncomplement or not present) outputs in terms of the product terms. This has to be submitted to the manufacturer to get the user–made PLA. Let us study how to determine a PLA program table with the help of an example.

Example 3.57. *Implement the combinational circuit for the functions using* $3 \times 3 \times 2$ *PLA.*

$$F_1 = \Sigma m\ (3,\ 5,\ 7)$$
$$F_2 = \Sigma m\ (4,\ 5,\ 7)$$

Solution. The truth table for the given Boolean functions is shown in Table 3.56. The K–map simplifications are shown in Fig. 3.221.

For F_1 *For F_2*

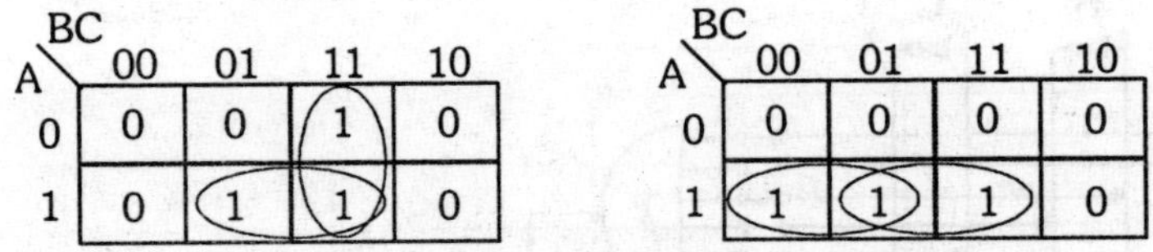

$F_1 = BC + AC$ $F_2 = A\bar{B} + AC$

Fig. 3.221

Table 3.56

A	B	C	F_1	F_2
0	0	0	0	0
0	0	1	0	0
0	1	0	0	0
0	1	1	1	0
1	0	0	0	1
1	0	1	1	1
1	1	0	0	0
1	1	1	1	1

The output equations are

$$F_1 = AC + BC$$

$$F_2 = A\overline{B} + AC$$

There are three distinct product terms in this combinational circuit, they are $A\overline{B}$, AC and BC. The circuit has 3 inputs, and 3 product terms and two outputs.

The PLA program table is shown in Table 3.57. It consists of 3 columns. First column lists the product terms, second column specifies the required paths between inputs and AND gates, third column specifies the paths between AND gates and the OR gates. Under each output variable, we write a T (for true) if output inverter is by passed, and C (for complement) if the function is to be complemented with the output inverter. The Boolean terms listed at left of the table are not a part of the PLA table, they are included for reference only.

For each product term, the input is marked with 1, 0 or – (dash). If a variable in the product term appears in the normal form, the corresponding input variable is marked with a 1. If it appears in it's complemented form, the corresponding input variable is marked with a 0. If the variable is absent in the product term, it is marked with a dash. Each product term is associated with an AND gate. The paths between the inputs and the AND gates are specified under the column heading inputs

(*a*) A '1' in the input column specifies a path from the corresponding input to the input of the AND gate that forms a product term.

(*b*) A '0' in the input column specifies a path from the corresponding complemented input to the input of the AND gate.

(*c*) A dash (–) specifies no connection.

The logic diagram is shown in Fig. 3.220.

Table 3.57. PLA program table

	Product term	*Inputs* A B C	*Outputs* F_1	F_2	
AB′	1	1 0 –	–	1	
AC	2	1 – 1	1	1	
BC	3	– 1 1	1	–	
			T	T	T/C

The paths between the AND and OR gates are specified under the column heading outputs. The output variables are marked with 1's for all those product terms that formulate the function. F_1 is marked with 1's for the product terms 2 and 3 and with a dash (–) for product term 1. Each product term that has a 1 in the output column requires a path from the corresponding AND gate to the output OR gate. Finally, T, indicates that the fuse across the output inverter remains intact.

Example 3.58. *Implement the combinational circuit defined by the functions* F_1 *(A, B, C)* $= \Sigma\ (3, 5, 6, 7)$, $F_2 = \Sigma\ (0, 2, 4, 7)$ *by using a* $3 \times 4 \times 2$ *PLA.*

Solution. The outputs can be simplified by using K-maps as shown in Fig. 3.222.

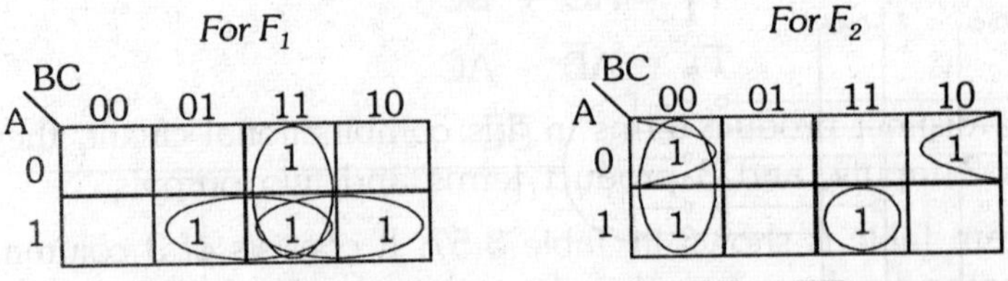

$F_1 = AB + AC + BC$ $\quad\quad$ $F_2 = \overline{B}\,\overline{C} + \overline{A}\,\overline{C} + ABC$

Fig. 3.222

The two output functions contain totally 6-product terms, but the given PLA has only 4 product terms. So, it is not possible to implement directly. To reduce the total number of product terms, let us see the complement of the functions as shown in Fig. 3.223.

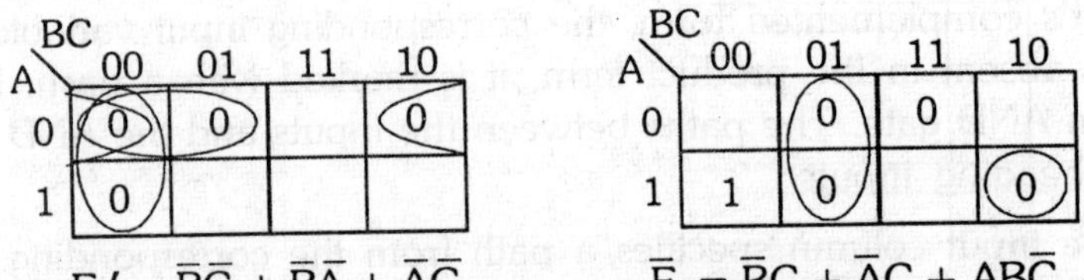

$F_1' = \overline{B}\,\overline{C} + \overline{B}\,\overline{A} + \overline{A}\,\overline{C}$ $\quad\quad$ $\overline{F}_2 = \overline{B}C + \overline{A}C + AB\overline{C}$

Fig. 3.223

By taking F_1' and F_2, totally we get only 4 product terms, they are : $\overline{B}\,\overline{C}$, $\overline{A}\,\overline{B}$, $\overline{A}\,\overline{C}$ and ABC. The PLA program table is shown in Table 3.58.

Table 3.58. PLA program table

	Product term	Inputs *A B C*	Outputs F_1	F_2
$\overline{B}\,\overline{C}$	1	– 0 0	1	1
$\overline{A}\,\overline{C}$	2	0 – 0	1	1
$\overline{A}\,\overline{B}$	3	0 0 –	1	–
ABC	4	1 1 1	0	1
			C	T

The logic diagram for it is shown in Fig. 3.224.

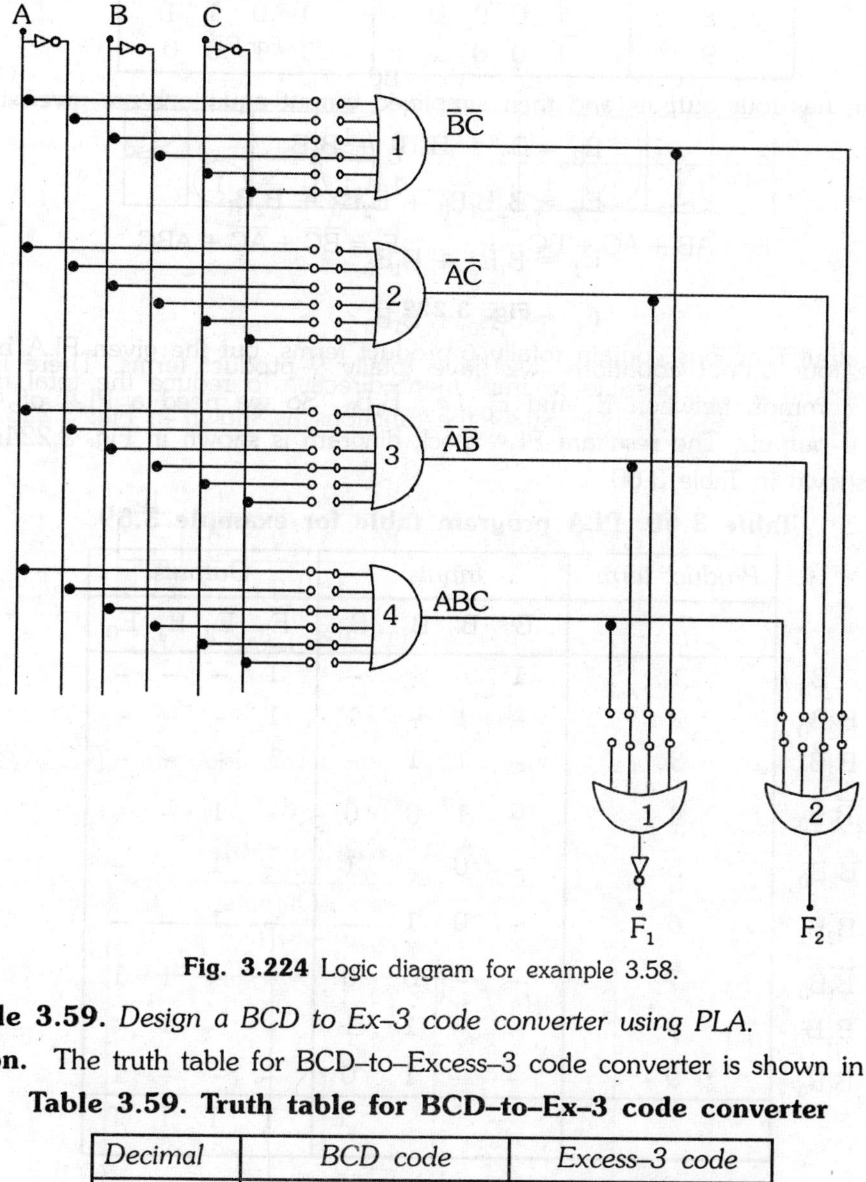

Fig. 3.224 Logic diagram for example 3.58.

Example 3.59. *Design a BCD to Ex–3 code converter using PLA.*

Solution. The truth table for BCD–to–Excess–3 code converter is shown in Table 3.59.

Table 3.59. Truth table for BCD–to–Ex–3 code converter

Decimal	*BCD code*	*Excess–3 code*
	B_3 B_2 B_1 B_0	E_3 E_2 E_1 E_0
0	0 0 0 0	0 0 1 1
1	0 0 0 1	0 1 0 0
2	0 0 1 0	0 1 0 1
3	0 0 1 1	0 1 1 0
4	0 1 0 0	0 1 1 1
5	0 1 0 1	1 0 0 0
6	0 1 1 0	1 0 0 1

7	0 1 1 1	1 0 1 0
8	1 0 0 0	1 0 1 1
9	1 0 0 1	1 1 0 0

The circuit has four outputs and their simplified output equations are given below.

$$E_3 = B_3 + B_2B_0 + B_2B_1$$

$$E_2 = B_2\overline{B}_1\overline{B}_0 + \overline{B}_2B_0 + \overline{B}_2B_1$$

$$E_1 = \overline{B}_1\overline{B}_0 + B_1B_0$$

$$E_0 = \overline{B}_1\overline{B}_0 + B_1\overline{B}_0$$

From the four output equations, we have totally 9–product terms. There is only one product term common between E_1 and E_0 *i.e.*, $\overline{B}_1\overline{B}_0$. So we need a PLA of 4 inputs, 9 products and 4 outputs. The resultant PLA block diagram is shown in Fig. 3.224(*a*) and the PLA table is shown in Table 3.60.

Table 3.60. PLA program table for example 3.59

	Product term	*Inputs*				*Outputs*			
		B_3	B_2	B_1	B_0	E_3	E_2	E_1	E_0
B_3	1	1	–	–	–	1	–	–	–
B_2B_0	2	–	1	–	1	1	–	–	–
B_2B_1	3	–	1	1	–	1	–	–	–
$B_2\overline{B}_1\overline{B}_0$	4	–	1	0	0	–	1	–	–
$\overline{B}_2B_0$	5	–	0	–	1	–	1	–	–
$\overline{B}_2B_1$	6	–	0	1	–	–	1	–	–
$\overline{B}_1\overline{B}_0$	7	–	–	0	0	–	–	1	1
B_1B_0	8	–	–	1	1	–	–	1	–
$B_1\overline{B}_0$	9	–	–	1	0	–	–	–	1
						T	T	T	T

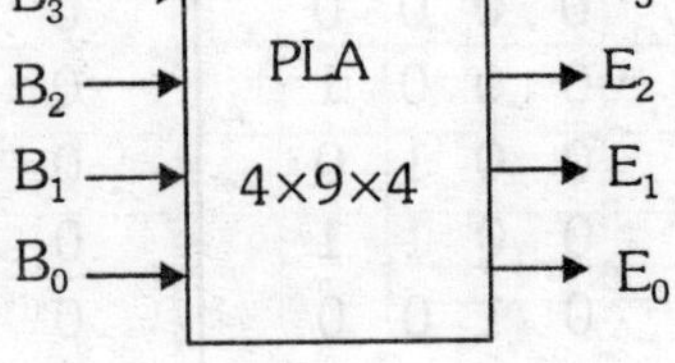

Fig. 3.224(*a*) Block diagram of BCD–to–Ex–3 code converter using PLA.

3.14.3 Programmable Array Logic (PAL)

Programmable logic devices have many gates interconnected through many of electronic fuses. It is sometimes convenient to draw the internal logic of such devices in a compact form referred to as "array logic". Fig. 3.225 shows the conventional and array logic symbols for a multiple-input AND gate.

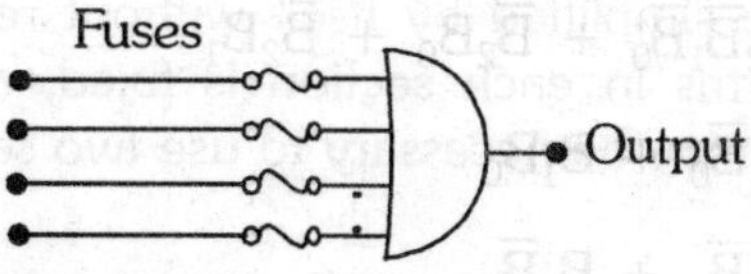

Fig. 3.225 Conventional symbol.

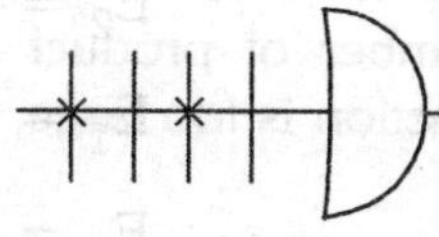

Fig. 3.226 Array logic symbol.

The array logic symbol shown in Fig. 3.226 uses a single horizontal line connected to the gate input and multiple vertical lines to indicate the individual inputs. Each intersection between horizontal line and vertical line indicates the fuse connection.

The PLA is a programmable AND array and programmable OR array. However, PLA is a programmable AND gate and fixed OR array. Because only AND gates are programmable, the PAL is easier to program, but is not as flexible as the PLA Fig. 3.227 shows the array logic of a typical PAL. It has four inputs and four outputs. Each input has a buffer and an inverter gate.

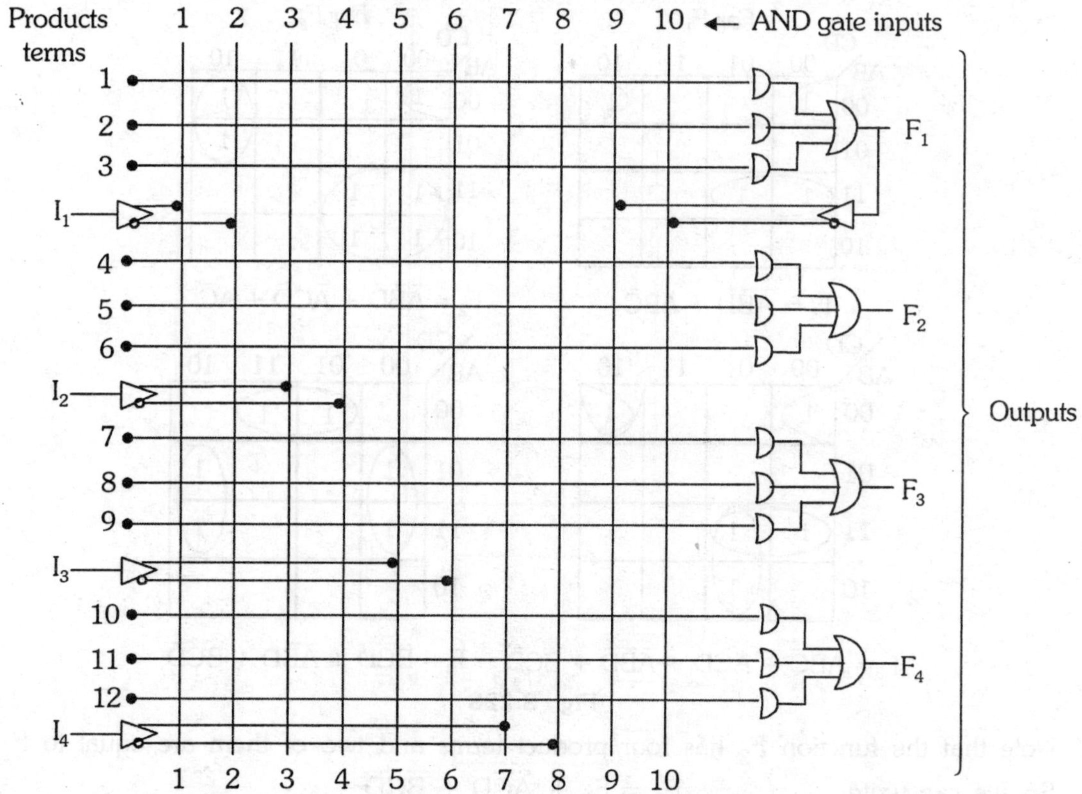

Fig. 3.227 PAL with 4-inputs, 4-outputs and 3-AND-OR gate sections.

It is to be noted that two gates are shown with one composite graphic symbol with normal and complement outputs. There are four sections. Each section has three programmable AND gates and one fixed OR gate. The output of section 1 is connected to a buffer-inverter gate and then fed back into the inputs of the AND gates, through fuses. This allows the logic designer to feed an output function back as an input variable to create a new function. Such PALs are called as "*programmable I/O PALs*'. In the PAL, product terms cannot be shared among two or more OR gates. Therefore, each function can be simplified by itself without regard to the common product terms. The number of product terms in each section is fixed, and if the number of product terms in the function is too large, it may be necessary to use two sections two implement one Boolean function.

Let us see the implementation of a combinational circuit using PAL with the help of examples.

Example 3.60. *Implement the following Boolean functions using PAL.*

$$F_1\ (A, B, C, D) = \Sigma m\ (0, 2, 12, 13)$$
$$F_2\ (A, B, C, D) = \Sigma m\ (0, 2, 6, 8, 9, 12, 13)$$
$$F_3\ (A, B, C, D) = \Sigma m\ (0, 2, 9, 10, 12, 13)$$
$$F_4\ (A, B, C, D) = \Sigma m\ (1, 3, 4, 6, 12, 14)$$

Solution. The simplification of four output functions is shown in Fig. 3.228.

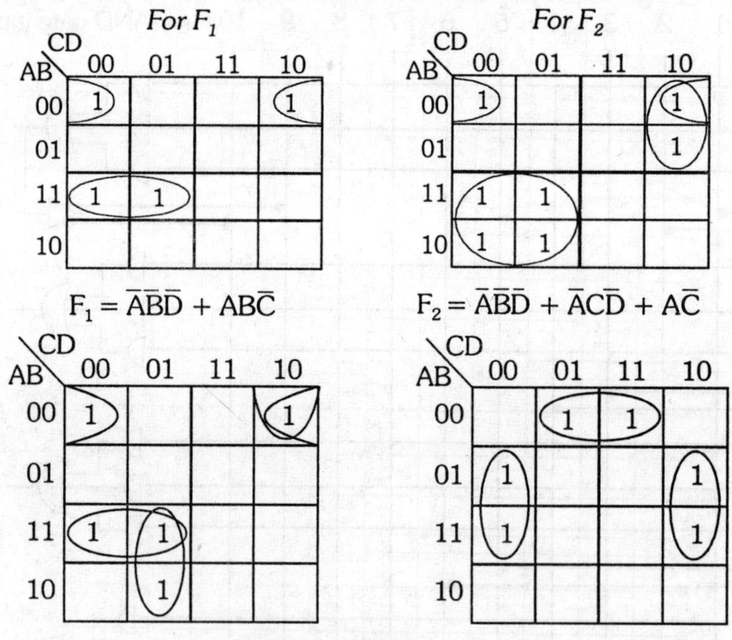

$F_1 = \bar{A}\bar{B}\bar{D} + AB\bar{C}$ $\quad$ $F_2 = \bar{A}\bar{B}\bar{D} + \bar{A}C\bar{D} + A\bar{C}$

$F_3 = AB\bar{C} + A\bar{C}D + \bar{A}\bar{B}\bar{D} + \bar{B}C\bar{D}$ $\quad$ $F_4 = B\bar{C}\bar{D} + \bar{A}\bar{B}D + BC\bar{D}$

Fig. 3.228

Note that the function F_3 has four product terms and two of them are equal to F_1.

So, we can write $\quad F_3 = F_1 + A\bar{C}D + \bar{B}C\bar{D}$

The program table for PAL is similar to the PLA table. The Table 3.61 shows the PAL program table with product terms, AND inputs and outputs.

Table 3.61. PLA program table

Product term	A	B	C	D	F_1	Outputs
1	0	0	–	0	–	
2	1	1	0	–	–	$F_1 = \bar{A}\bar{B}\bar{D} + AB\bar{C}$
3	–	–	–	–	–	
4	0	0	–	1	–	
5	0	–	1	0	–	$F_2 = \bar{A}\bar{B}D + \bar{A}C\bar{D} + A\bar{C}$
6	1	–	0	–	–	
7	1	–	0	1	–	
8	–	0	1	0	–	$F_3 = F_1 + A\bar{C}D + \bar{B}C\bar{D}$
9	–	–	–	–	1	
10	–	1	–	0	–	
11	0	0	–	1	–	$F_4 = \bar{A}\bar{B}D + \bar{B}\bar{D}$

The logic diagram for this is shown in Fig. 3.229.

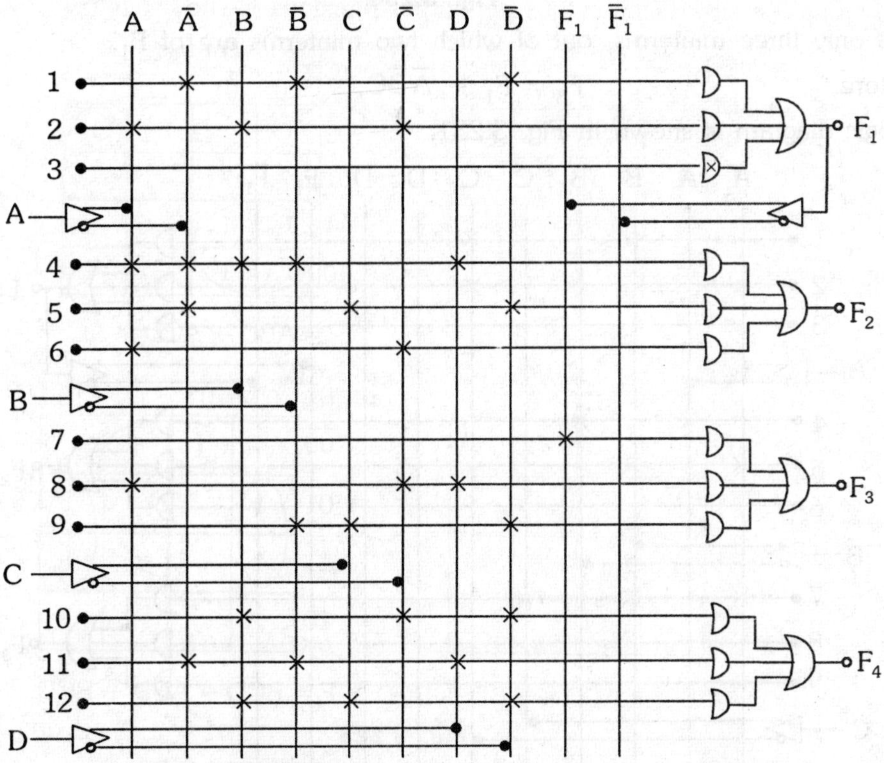

Fig. 3.229 PLA logic diagram for example 3.58.

Example 3.61. *Tabulate the PAL programmable table for the four Boolean functions listed below. Also draw the logic diagram.*

$$F_1(A, B, C) = \Sigma m(2, 4, 6)$$
$$F_2(A, B, C) = \Sigma m(0, 1, 6, 7)$$
$$F_3(A, B, C) = \Sigma m(1, 2, 4, 6)$$
$$F_4(A, B, C) = \Sigma m(1, 2, 5, 7)$$

Solution. The simplified equations by using K-map are shown in Fig. 3.230.

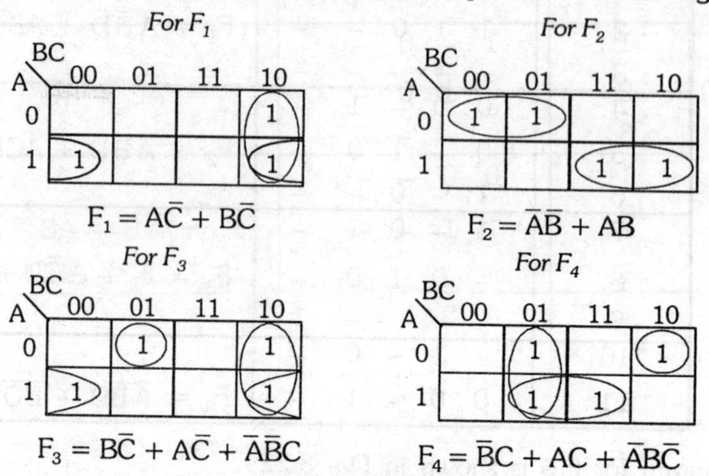

Fig. 3.230

F_3 has only three minterms, out of which two minterms are of F_1.

Therefore, $$F_3 = F_1 + \bar{A}\bar{B}C.$$

The logic diagram is shown in Fig. 3.231.

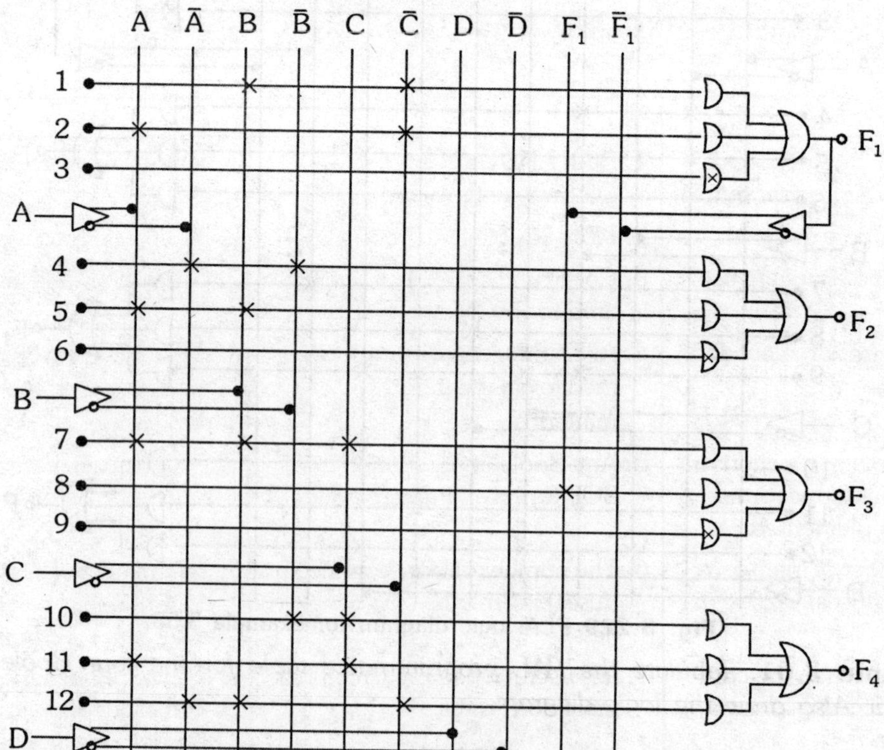

Fig. 3.231 PLA logic diagram for example 3.61.

The PLA program table is shown in Table 3.62.

Table 3.62. PLA program table for example 3.61

Product terms	*Inputs* A	B	C	F_1	*Outputs*
1	1	–	0	–	
2	–	1	0	–	$F_1 = A\overline{C} + B\overline{C}$
3	–	–	–	–	
1	0	0	–	–	
2	1	1	–	–	$F_2 = \overline{A}\,\overline{B} + AB$
3	–	–	–	–	
1	0	0	1	–	
2	–	–	–	1	$F_3 = F_1 + \overline{A}\,\overline{B}C$
3	–	–	–	–	
1	–	0	1	–	
2	1	–	1	–	$F_4 = \overline{B}C + AC + \overline{A}BC$
3	0	1	0	–	

3.14.4 Comparison between PROM, PLA and PAL

The comparison is shown in Table 3.63.

Table 3.63

PROM	*PLA*	*PAL*
1. AND array is fixed and OR array is programmable.	1. Both AND array and OR array are programmable.	1. AND array is programmable and OR array is fixed.
2. Flexible to the designer.	2. More flexible to the designer.	2. Flexible to the designer.
3. Cheap and simple to use.	3. Costly and complex as compared to PROM and PAL.	3. Cheaper and simple.
4. All minterms are decoded.	4. AND array can be programmed to get the desired minterms.	4. AND array can be programmed to get the desired minterms.
5. Only Boolean functions in sum of minterms forms can be implemented.	5. Any Boolean function in the SOP form can implemented.	5. Any Boolean function in the SOP form can be implemented.
6. The number of inter-connections (fuses) is $2^n \times m$ (more).	6. The number of inter-connections (fuses) is $2n \times K + K \times m + K$ (less).	6. The number of inter-connectiotions (fuses) for a typical PLA is $24 \times n + 12$ (medium).
7. It is useful when the combination circuit uses more minterms.	7. It is useful when the circuit has more don't cares.	7. It is useful when the circuit has 4 functions and atleast any one function is part of the other.

3.15 DISADVANTAGES IN ASYNCHRONOUS CIRCUITS

Generally, three types of problems may occur in asynchronous sequential circuits, they are: cycles, Races and Hazards.

3.15.1 CYCLES

If an input change induces a feedback transition through more than one unstable state, then such situation is called a "cycle". Here circuit moves via a unique sequence of unstable states and never comeback to stable state. The following example gives the explanation about cycles.

Table 3.64 PS and NS table.

Present state A_n B_n	*Next state* (A_{n+1}, B_{n+1})			
	Input (I_1I_2)			
	00	*01*	*11*	*10*
00	00①			
01	00			
11	10			
10	11	②		

In the above table, Let I_1 and I_2 are the input variables A_n and B_n are present state variables and A_{n+1} and B_{n+1} are the next state variables. The circled entries represent stable states and uncircle entries represent unstable states.

Let assume the circuit belongs to the above PS/NS table is initially in state 2 (stable state) *i.e.*, the corresponding input I_1I_2 = 01 and present state A_nB_n = 10.

If I_2 changes from 1 to 0 *i.e.*, I_1I_2 = 00 and A_nB_n = 10, the next state becomes 11.

→ Now, if the present state A_nB_n = 11 and I_1I_2 = 00, then the next state is '01'.

→ If the present state is 01, and I_1I_2 = 00, then the next state is 00.

→ If I_1I_2 = 00 and A_nB_n = 00 then the next state A_{n+1}, B_{n+1} = 00 is a stable state.

Thus a circuit goes via a sequence of unstable states when there is a change in the input variables values, such situation is called "cycle". If an asynchronous sequential circuit entries a cycle, the core must be taken to entire that the circuit terminates in a stable states, otherwise, the circuit goes from one unstable state to another state until a new change in the input occurs.

3.15.2 RACES

When two or more internal variables (feedback variables) change value of a result of single input changing state *i.e.* in response to a change in an input variable, then a 'Race' condition exist in an asynchronous sequential circuit.

There are two types of races, namely (1) critical races (2) non-critical races.

Critical races

Consider an example of PS and NS stable of a sequential circuit to demonstrate the problem of critical races.

Table 3.65 Present and Next state tables.

Present state A_n B_n	*Next state* (A_{n+1}, B_{n+1}) *Input* (I_1I_2) 00	01	11	10
00	11(1)	00(5)	11(9)	11(13)
01	01(2)	11(6)	11(10)	00(14)
11	10(3)	11(7)	11(11)	10(15)
10	10(4)	00(8)	11(12)	10(16)

Assume that the sequential circuit is in state (5) *i.e.*, stable state (whose I_1I_2 = 01 and A_nB_n = 00 and corresponding next state A_{n+1} B_{n+1} = 00), if now I_2 is changes from 1 to 0, the next state variables A_{n+1} B_{n+1} changes from 00 to 11. But, due to unequal propagation delay, one of the next state variables 1 and other variables does not change, hence the NS value A_{n+1} B_{n+1} = 01 or 10.

Case 1. If A_{n+1} B_{n+1} = 01, then the circuit goes to stable state 2.

Case 2. If A_{n+1} B_{n+1} = 10, then the circuit goes to stable state 4.

Therefore, for an input change *i.e.*, I_1I_2 from 01 to 00, depending on the relative switching speed of the next state variables. If A_{n+1} and B_{n+1}, the final state may be either state 2 or state 4. But both final states are wrong for the given change in the input variables. This situation called a "critical race". The critical races must be eliminated (or) avoided in an asynchronous circuits.

Note: (1) When you design a feed-back sequential circuit, you must ensure that its PS and NS table does not contain any critical races. Otherwise the sequential circuit may operate unpredictable, with the next state for racy transitions depending on factory temperature and voltage etc.

(2) For an input change, a particular sequential circuit having critical race problem may always go to the some wrong state. Also, identical circuits with different propagation delays may go to different wrong states.

Non-critical races

For the above table, assume that the circuit is in stable state 5 (A_{n+1} B_{n+1} = 00), than the input change from I_1I_2 = 00 to 11, then the next states of the circuit are switch from 00 to 11. Again due to unequal propagation delay, A_{n+1} B_{n+1} becomes either 01 or 10. Hence, the circuit will go to state 10 or state 12, and then finally to state 11, which is a stable state. Here, incepective of the NS variables *i.e.*, A_{n+1} and B_{n+1} going to 1, the some correct final state 11 is reached. Here the corect final state is reached after transition through unstable states is called "Non-Critical race". This types of non-critical races can be tolerated in an asynchronous circuits.

3.15.3 HAZARDS

In analysis of combinational logic circuits, we ignore the circuit delay and predict only the steady state behaviour. That is, the circuit output is a function of its inputs under assumption that the inputs have been stable for a long time, relative to the delays of the elements of the circuit.

With the effect of the circuit delays, the transient behaviour of the combinational circuit may vary from the predicted steady state analysis. With this, the circuit may produce a short pulse of a output, also called of "glitch", at that time, when the steady state analysis predicts that there is no change in the output.

A "Hazard" is a unwanted transient *i.e.*, spike or glich, that occurs due to unequal path or unequal propagation delays through a combinational circuit. There are two types of hazards, namely (1) Static hazard (2) Dynamic hazard. These two hazards occur in a combinational circuits, similarly "essential hazard" a third type, occur in the sequential circuits.

The Fig. 3.232 shows the Hazards.

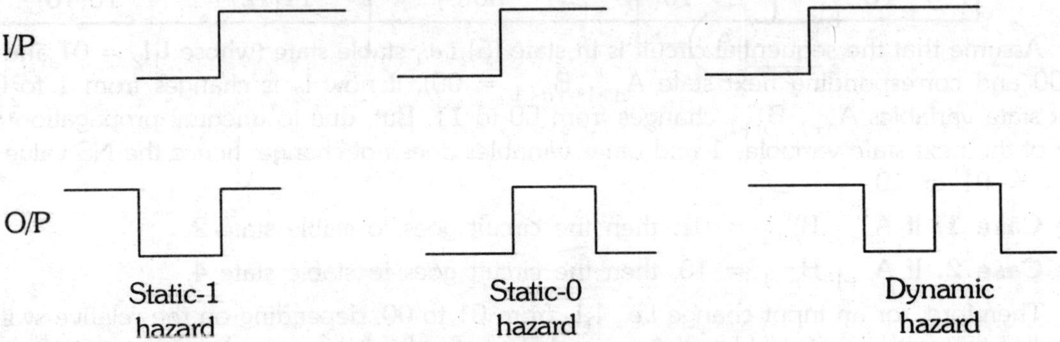

Fig. 3.232 types of hazards.

(*a*) **Static hazard.** Static hazard is a condition which results in a single momentary incorrect output due to change in a single input variable when the output is expected to remain in the same state. There are two types, static-1 hazard and static-0 hazard.

Static-1 hazard. If the output momentarily goes to state '0' when the output is expected to remain in state '1', as per the steady state analysis.

Static-0 hazard. If the output momentarily goes to state '1' when the output is expected to remain in state '0' as per the steady state analysis.

To understand the static hazard, consider the K-Map of the Boolean function $f(a, b, c) = \Sigma(0, 1, 2, 6)$ the corresponding circuit is shown in Fig. 3.233.

From the below diagram, when abc = 000, the output f = 1 due to 1 output from the AND Gate 1(A–). Now, if the input b is changes from 0 to 1, causes abc = 000 to 010, the output f should be in 1 due to 1 at the output of Lower AND gate (A_2). Due to change in the value of input variable, the output f is supported to remain in HIGH state. But, due to unequal propagation delay, if the A_1 gate output changes to 0 shortly before the A_2 gate output becomes '1', then during this short period, the output f = 0 momentarity. This situation is called static hazard.

A static hazard can be removed by covering the adjacent cells with a redundant grouping that overloops both grouping. For the above circuit, static hazard can be eliminated by covering adjacent cells (000, 010 shown in dotted lines). This gives a term $\bar{a}\,\bar{c}$, which overloops both $\bar{a}\,b$ and $b\,\bar{c}$ groupings. The modified circuit with $\bar{a}\,\bar{c}$ is shown in Fig. 2.233(*c*).

Now, if the input (abc) changes 000 to 010, the output will remain at '1' state, because the output of new AND gate is high for both input combinations. Here, though, there is a change in propagation delay, since the lower AND gate output is 1, the output of the circuit is 1 only, without any hazard.

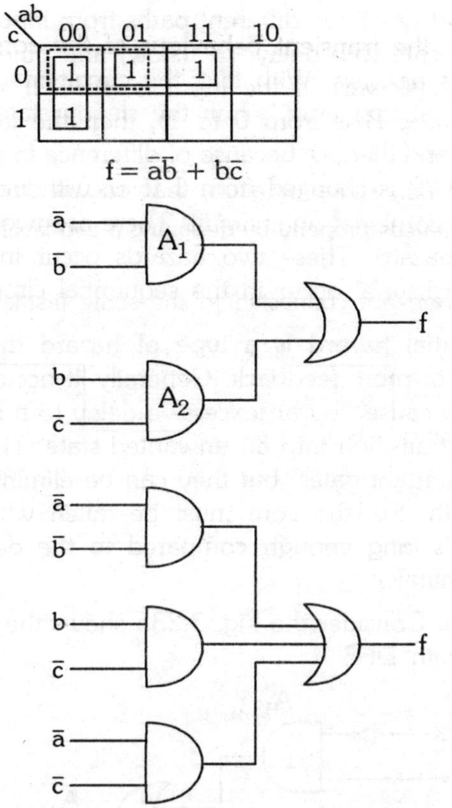

(a) K-map for $f = \Sigma(0, 1, 2, 6)$

(b) Logic diagram with static hazard

(c) Logic diagram without static hazard

Fig. 3.233

Note: While designing a static hazard free circuit, all adjacent input combinations, having some output occur with some subcube of the corresponding function *i.e.*, every pair of adjacent 1 cells and every pair of adjacent '0' cells in the K-map of a switching function should be covered by at least one subcube.

(*b*) **Dynamic hazard.** When the output is supported to change from 0 to 1 or from 1 to 0, the circuit may go three or more transients and produce more than one spike or glitch. Such multiple glitch situation is called of "dynamic hazards". Dynamic hazards can be eliminated by some method of used in static hazards. Here multiple output transitions can occur if there are multiple paths with different delays from the changing inputs to the changing output.

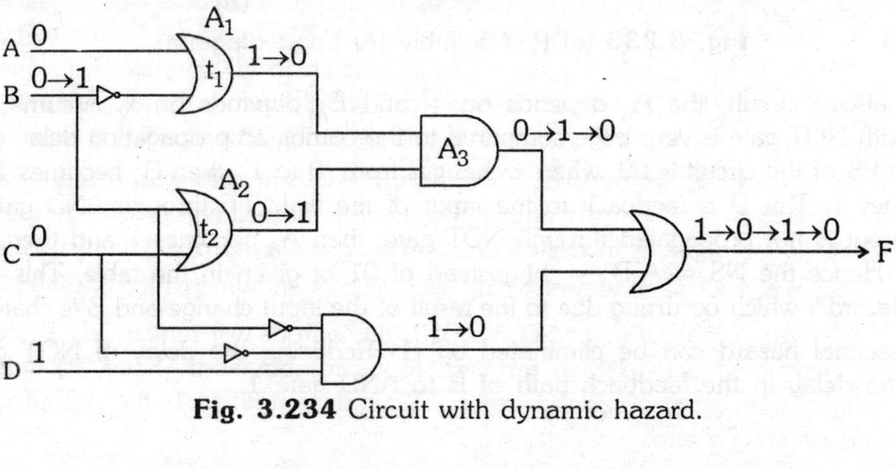

Fig. 3.234 Circuit with dynamic hazard.

The Fig. 3.234 shown the circuits, that has three different paths from input A to the output F. One of the path goes through OR (A_1) gate with delay 't_1'' (slow) and other path is through OR gate (A_2) with propagation delay 't_2' (slower). If the input is (ABCD) = 0001 then the output is 1. If any change in the input (*i.e.*, B is from 0 to 1), then the input is 0101, the O/P AND (A_3) gate is changes from 0 to 1 and then, 0, because of difference in propagation delay of two OR gate (A_1 and A_2). Finally the O/P is changed from 1 to 0, with more glitch change.

Note: (1) Dynamic hazards can be eliminateds properly by designing a two-level AND-OR (or) OR-AND logic circuits.

(2) The variables and its complements are not connected to the some first-level gate.

(*c*) **Essential Hazards.** The Essential hazard is a type of hazard that exists only in asynchronous sequential circuits with two or more feedback. Generally, it occur in toggling type circuits. It is an operational error generally caused by an excessive delay to a feedback variable in response to an input change, leadge a transition into an unwanted state. This type of hazard can not be eliminated by adding any redundant gates, but they can be eliminated by adjusting the amount of delay in the affected path. So, the core must be taken while designing the feedback path delay, so that the delay is long enough compared to the delay of the other signals that originate from the input terminals.

Elimination of Essential Hazards. Consider the Fig. 3.235 shows the PS and NS table and its corresponding sequential circuit with DFF.

PS	NS(A_nB_n)	
AB	x = 0	x = 1
00	00	01
01	11	01
11	11	11

(*a*)

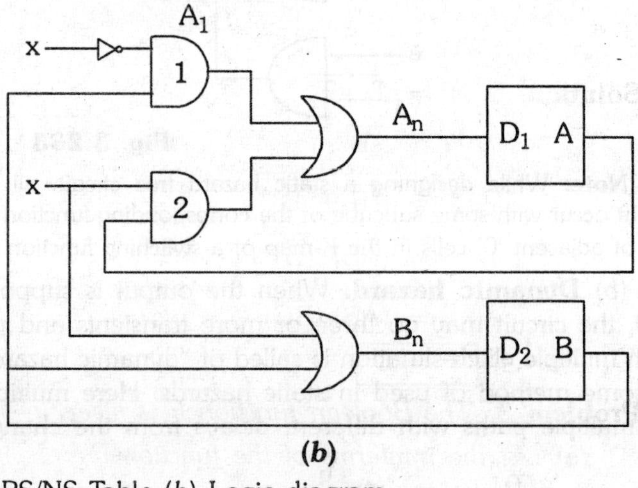

(*b*)

Fig. 3.235 (*a*) PS/NS Table (*b*) Logic diagram.

In the above circuit, the A_n depends on $\bar{x}$ and B_n depends on x. Assume, the delay associated with NOT gate is very large compared to the combined propagation delay of FFS and Gates. If the PS of the circuit is 00, when x changes from '0 to 1', then B_n becomes 1 and there by A becomes 1. This B is feedback to the input of the flight FF through AND gate 1. If the change in input is not propagated through NOT gate, then A_n becomes 1 and there by A also becomes 1. Hence the NS = A_nB_n = 11 instead of 01 of given in the table. This is called of "Essential Hazard", which occurring due to the result of the input change and B is changing faster.

This essential hazard can be eliminated by (1) Reducing the delay of NOT gate or (2) Increasing the delay in the feedback path of B to AND gate 1.

3.16 SOLVED PROBLEMS

Problem 1. *Draw the output waveform of the circuit.*

Solution.

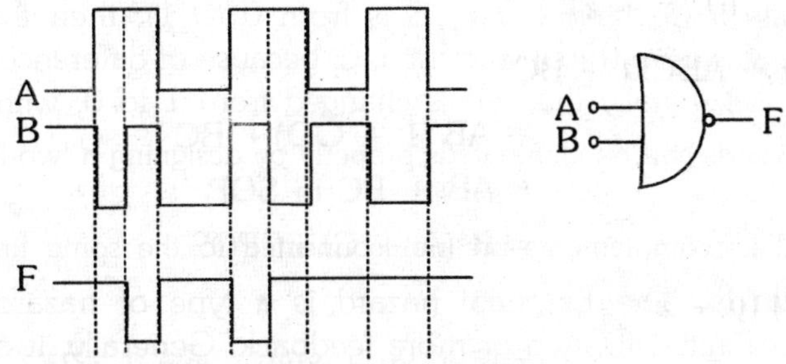

Problem 2. *Determine the truth table for the circuit.*

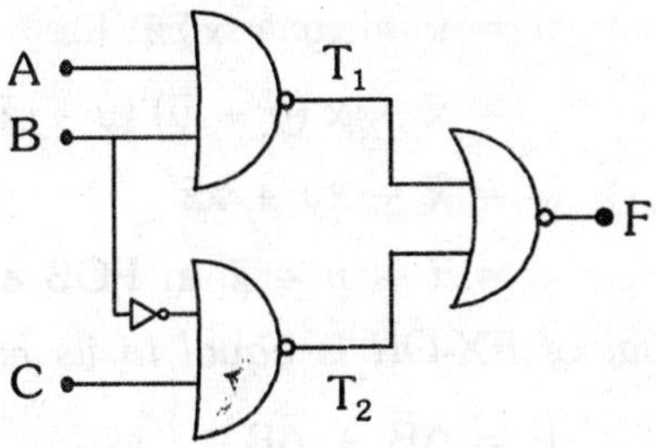

Solution. $F = \overline{T_1 + T_2}$

$T_1 = \overline{AB}$ $\qquad T_2 = \overline{\overline{B}C}$

$\therefore$ $F = \overline{\overline{AB} + \overline{\overline{B}C}}$

$= (\overline{\overline{AB}}) . (\overline{\overline{\overline{B}C}})$

$= (AB)\,(\overline{B}C)$

$\boxed{F = 0}$

A	B	C	F
0	0	0	0
0	0	1	0
0	1	0	0
0	1	1	0
1	0	0	0
1	0	1	0
1	1	0	0
1	1	1	0

Problem 3. *The Boolean function F is given in the truth table, find the following :*

(a) List the minterms of the function F.

(b) List the minterms of F'.

(c) Express F in sum of minterm form.

(d) Simplify the function to an expression with minimum number of literals.

A	B	C	F
0	0	0	1
0	0	1	1
0	1	0	0
0	1	1	1
1	0	0	1
1	0	1	0
1	1	0	0
1	1	1	0

Solution. (*a*) $F = \Sigma m\ (0, 1, 3, 4)$

(*b*) $F' = \Sigma m\ (2, 5, 6, 7)$

(*c*) $F = \overline{A}\,\overline{B}\,\overline{C} + \overline{A}\,\overline{B}C + \overline{A}BC + A\overline{B}\,\overline{C}$

(*d*) $F = \overline{A}\,\overline{B} + \overline{A}BC + A\overline{B}\,\overline{C}$

$= \overline{A}\,\overline{B} + \overline{A}C + \overline{B}\,\overline{C}$

Problem 4. *Convert the following expressions into SOP and POS forms.*

(a) $(AB + C)(B + \overline{C}D)$

(b) $\bar{x} + x(x + \bar{y})(y + \bar{z})$

Solution. *(a)* $AB + AB\overline{C}D + BC$

$$= AB(1 + \overline{C}\overline{D}) + BC$$

$$= AB + BC \text{ in SOP}$$

$$= B(A + C) \text{ in POS.}$$

(b) $\bar{x} + (xy + x\bar{y})(y + \bar{z})$

$$= \bar{x} + xy + x\bar{y}y + xy\bar{z} + x\bar{y}\bar{z}$$

$$= \bar{x} + xy + xy\bar{z} + x\bar{y}\bar{z}$$

$$= \bar{x} + xy + x\bar{y}\bar{z}$$

$$= \bar{x} + x(y + \bar{y})(y + \bar{z})$$

$$= \bar{x} + xy + x\bar{z}$$

$$= \bar{x} + y + \bar{z} \text{ in POS and SOP forms.}$$

Problem 5. *Show that dual of EX-OR is equal to its complement.*

Solution. $\therefore F = A\bar{B} + \bar{A}B$

The dual of $F = F_D = (A + \bar{B})(\bar{A} + B) = \bar{A}\bar{B} + AB$

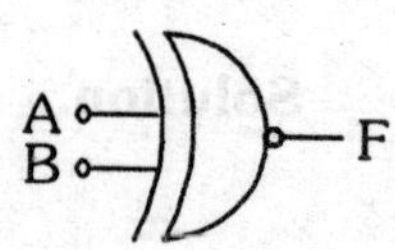

The complement of $F = \bar{F} = (A\bar{B} + \bar{A}B)' = (\overline{A\bar{B}})(\overline{\bar{A}B})$

$$= (\bar{A} + B)(A + \bar{B}) + \bar{A}\bar{B} + AB$$

$$\boxed{\therefore \bar{F} = F_D}$$

Problem 6. *In a TTL SSI 14 pins package, 2 pins are reserved for power and ground, other 12–pins are used for input and output terminals. Determine the number of gates that can be enclosed in one package if it contains the following type :*

(a) Two-input EX-OR gates (b) Three-input OR gates (c) Four-input NOR gates (d) Five-input NAND gates (e) Eight-input AND gates.

Solution.

(a) For a two input and one out EX-OR gate, totally we need 3 pins, hence 4 gates can be enclosed.

(b) Three input OR gate requires 4-pins, and hence 3 gates can be enclosed.

(c) A four-input NOR gate requires totally 5–pins, so only two–NOR gates can be enclosed, remaining two pins are not connected.

(d) A five–input NAND gate requires totally 6–pins, so only two–NAND gates can be enclosed.

(e) An eight–input AND gate requires 9–pins, so only one gate can be enclosed and remaining 3–pins are not connected.

Problem 7. *Show that a positive logic NAND gate is a negative logic NOR gate and vice-versa.*

Solution. We can prove this with the help of truth tables.

(*a*) Truth table for positive logic NAND

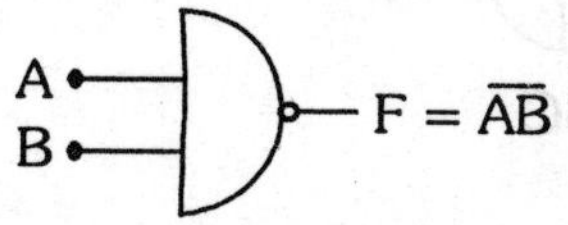

A	B	F
0	0	1
0	1	1
1	0	1
1	1	0

Truth table for negative logic NOR

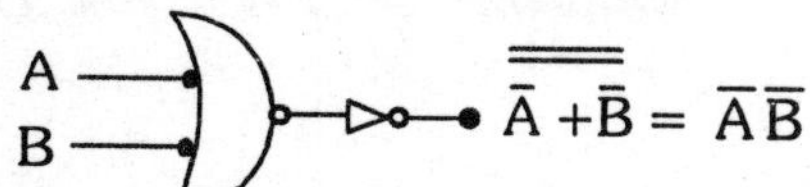

A	B	F
0	0	1
0	1	1
1	0	1
1	1	0

(*b*) Truth table for negative logic NAND

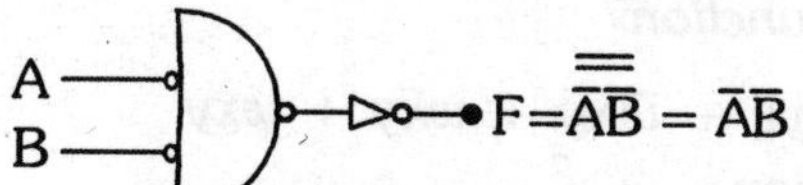

A	B	F
0	0	1
0	1	0
1	0	0
1	1	0

Truth table for positive logic NOR

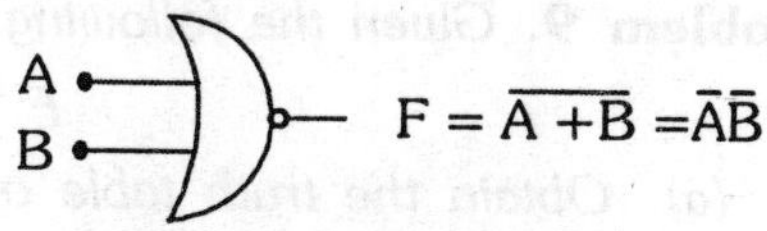

A	B	F
0	0	1
0	1	0
1	0	0
1	1	0

Hence proved.

Problem 8. *Draw the logic diagram corresponding to Boolean expressions without simplifying them*

(*a*) $B\bar{C} + AB + ACD$ (*b*) $(A + B)(C + D)(\bar{A} + B + D)$

(*c*) $(AB + \bar{A}\bar{B})(C\bar{D} + \bar{C}\bar{D})$

Solution. (*a*) The function requires 3 AND gates. Among them, two are 2–input AND gates and one 3–input AND gate and one OR gate. The logic diagram is

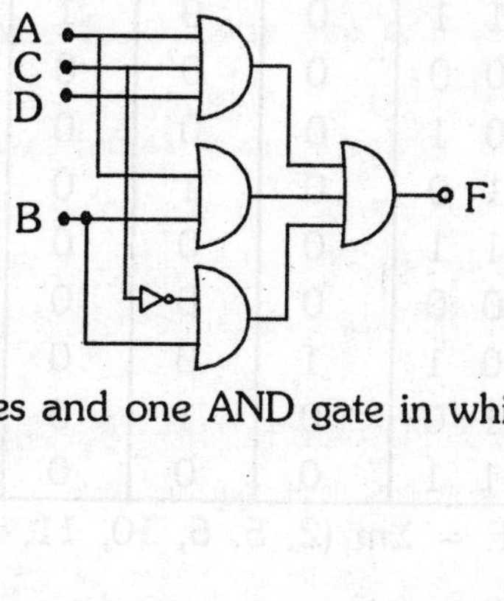

(*b*) The function requires 3 OR gates and one AND gate in which one OR-gate is 3-input one.

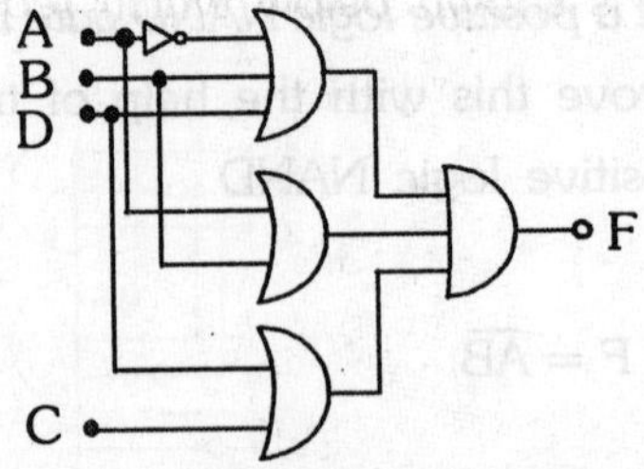

(c) The function requires three gates, one EX–OR, one EX–NOR and one AND gate.

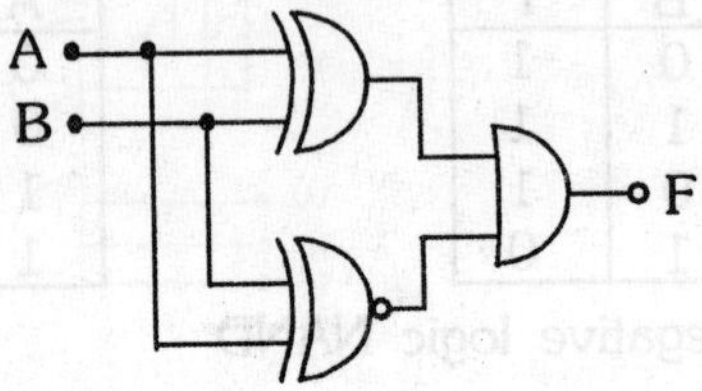

Problem 9. *Given the following Boolean function*

$$F = x\bar{y}z + \bar{x}y\bar{z} + \bar{w}xy + w\bar{x}y + wxy$$

(a) Obtain the truth table of the function.
(b) Draw the logic diagram using original Boolean expression.
(c) Simplify the function to minimal number of literals and draw it's logic diagram.
(d) Obtain the truth table for the minimal expression and show that it is same as the one in part (a).

Solution. (a) The truth table for the function $F = x\bar{y}z + \bar{x}y\bar{z} + \bar{w}xy + w\bar{x}y + wxy$ is shown below.

w x y z	$x\bar{y}z$	$\bar{x}y\bar{z}$	$\bar{w}xy$	$w\bar{x}y$	wxy	F
0 0 0 0	0	0	0	0	0	0
0 0 0 1	0	0	0	0	0	0
0 0 1 0	0	1	0	0	0	1
0 0 1 1	0	0	0	0	0	0
0 1 0 0	0	0	0	0	0	0
0 1 0 1	1	0	0	0	0	1
0 1 1 0	0	0	1	0	0	1
0 1 1 1	0	0	1	0	0	0
1 0 0 0	0	0	0	0	0	0
1 0 0 1	0	0	0	0	0	0
1 0 1 0	0	1	0	1	0	1
1 0 1 1	0	0	0	1	0	1
1 1 0 0	0	0	0	0	0	0
1 1 0 1	1	0	0	0	0	1
1 1 1 0	0	1	0	0	1	1
1 1 1 1	0	0	0	0	1	1

$\therefore$ $F = \Sigma m\ (2, 5, 6, 10, 11, 13, 14, 15)$

(*b*) The logic diagram is shown below which requires 4 AND gates, 4 inverters and one OR gate.

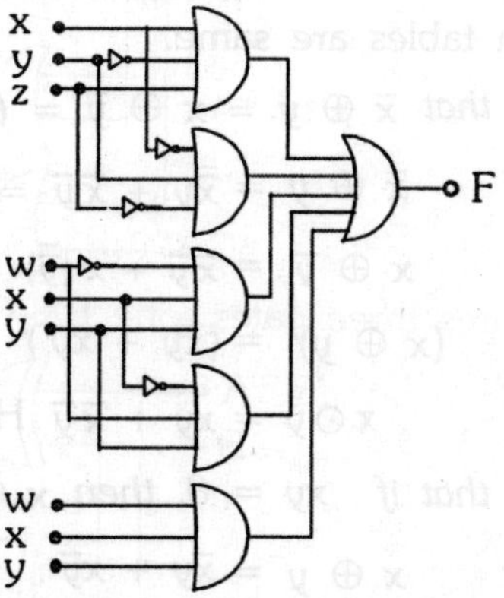

(*c*) The simplified expression can be obtained by using K-map as shown below.

AB \ CD	00	01	11	10
00				1
01		1		1
11		1	1	1
10			1	1

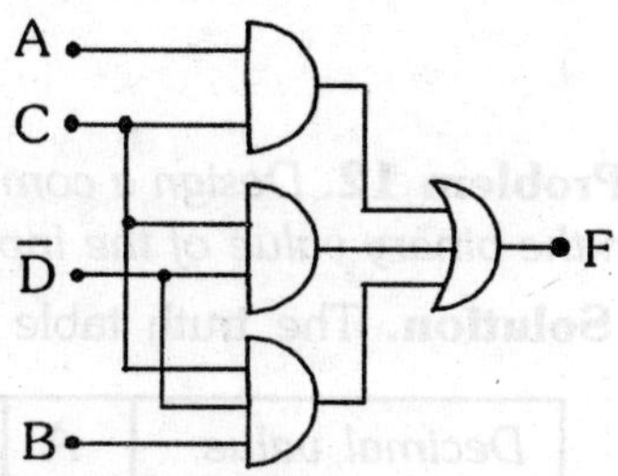

$$F = AC + C\bar{D} + B\bar{C}D$$

The logic diagram is shown above.

(*d*) The truth table for the simplified expression is shown below.

A B C D	AC	$C\bar{D}$	$B\bar{C}D$	F
0 0 0 0	0	0	0	0
0 0 0 1	0	0	0	0
0 0 1 0	0	1	0	1
0 0 1 1	0	0	0	0
0 1 0 0	0	0	0	0
0 1 0 1	0	0	1	1
0 1 1 0	0	1	0	1
0 1 1 1	0	0	0	0
1 0 0 0	0	0	0	0
1 0 0 1	0	0	0	0
1 0 1 0	1	0	0	1
1 0 1 1	1	0	0	1
1 1 0 0	0	0	0	0
1 1 0 1	1	0	0	1
1 1 1 0	1	1	1	1
1 1 1 1	1	0	0	1

From the table

$$F = \Sigma m\ (2, 5, 6, 10, 11, 13, 14, 15)$$

Hence both truth tables are same.

Problem 10. *Prove that* $\bar{x} \oplus y = x \oplus \bar{y} = (x \oplus y)' = x\,.\,y.$

Solution.

$$\bar{x} \oplus y = \bar{\bar{x}}y + \bar{x}\bar{y} = xy + \bar{x}\bar{y}$$

$$x \oplus \bar{y} = \bar{x}\bar{y} + x(\bar{\bar{y}}) = \bar{x}\bar{y} + xy$$

$$(x \oplus y)' = (\bar{x}y + xy')' = \bar{x}\bar{y} + xy$$

$$x \odot y = xy + \bar{x}\bar{y} \text{ Hence proved.}$$

Problem 11. *Show that if* $xy = 0$, *then* $x \oplus y = x + y.$

Solution.

$$x \oplus y = \bar{x}y + x\bar{y}$$

$$= \bar{x}y + x\bar{y} + xy + xy$$

$$= y(x + \bar{x}) + x(y + \bar{y})$$

$$= x + y \text{ Hence proved.}$$

Problem 12. *Design a combinational circuit that has 3–inputs and one output. The output = 1 when the binary value of the input is less than 3. The output is logic – 0 otherwise.*

Solution. The truth table for the combinational circuit is shown below.

Decimal value	A	B	C	F
0	0	0	0	1
1	0	0	1	1
2	0	1	0	1
3	0	1	1	0
4	1	0	0	0
5	1	0	1	0
6	1	1	0	0
7	1	1	1	0

The logic diagram is

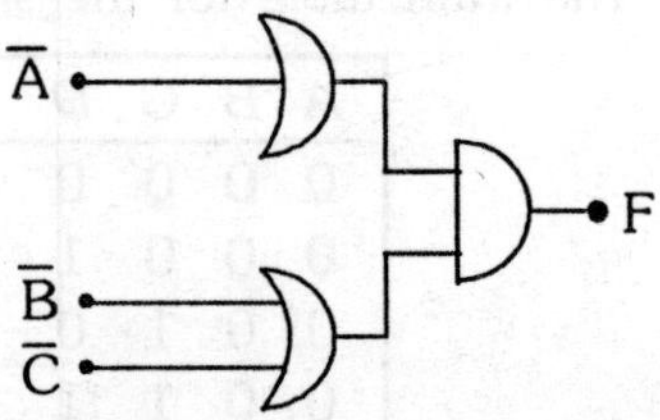

$$\therefore \quad F = \bar{A}\bar{B} + \bar{A}\bar{C}$$

$$= \bar{A}(\bar{B} + \bar{C})$$

Problem 13. *Design a combinational circuit with three inputs A, B and C and three outputs x, y and z. When the binary inputs are 0, 1, 2 and 3, the binary output is one greater than the input. When the binary inputs are 4, 5, 6 and 7, the binary output is one less than the input.*

Solution. The truth table for the required combinational circuit is shown below.

Decimal number	*Inputs* A	B	C	*Outputs* x	y	z
0	0	0	0	0	0	1
1	0	0	1	0	1	0
2	0	1	0	0	1	1
3	0	1	1	1	0	0
4	1	0	0	0	1	1
5	1	0	1	1	0	0
6	1	1	0	1	0	1
7	1	1	1	1	1	0

The simplified expressions for outputs *x*, *y* and *z* are given below from the K–maps.

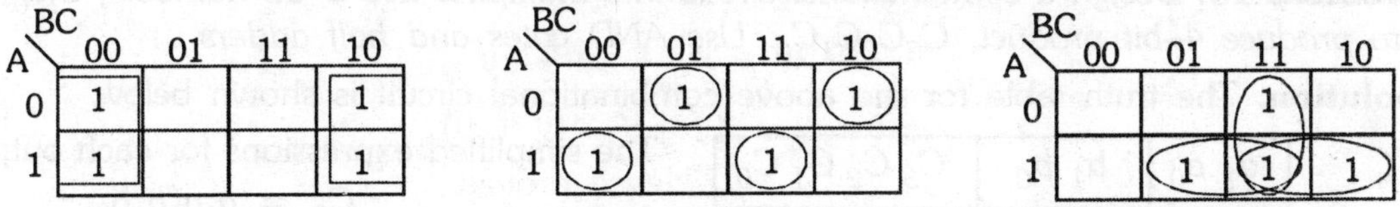

$Z = \overline{C}$ $\qquad$ $y = A \oplus B \oplus C$ $\qquad$ $x = AC + BC + AB$

The logic diagram for the outputs is shown below.

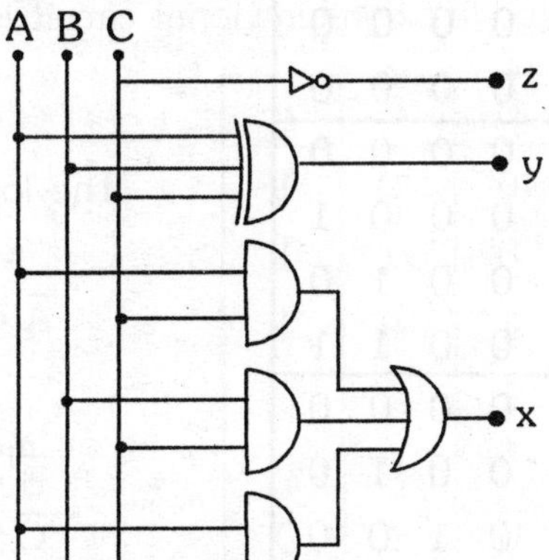

Problem 14. *Design a 3-input combinational circuit whose output is equal to 1 if the input variables have more 1's than 0's. The output is '0' otherwise.*

Solution. The required combination, circuit requires three inputs and one output and its truth table is shown below.

A	*B*	*C*	*F*
0	0	0	0
0	0	1	0
0	1	0	0
0	1	1	1
1	0	0	0
1	0	1	1
1	1	0	1
1	1	1	1

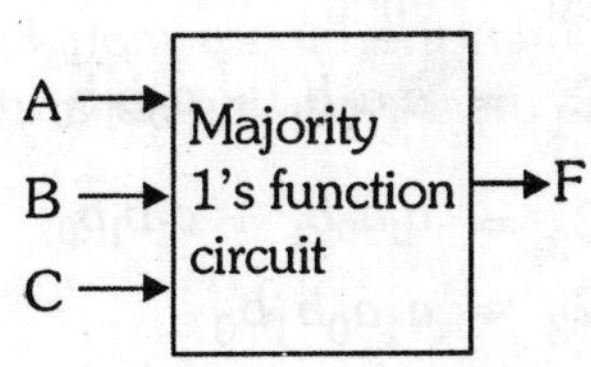

The simplified expression for K is

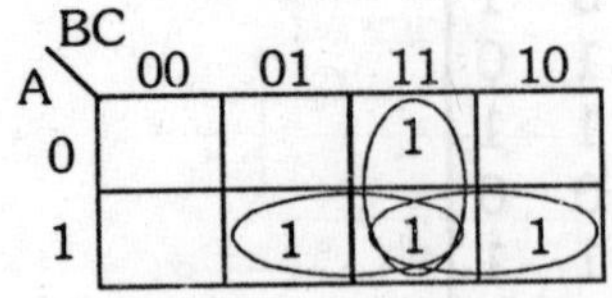

$F = AB + BC + AC$

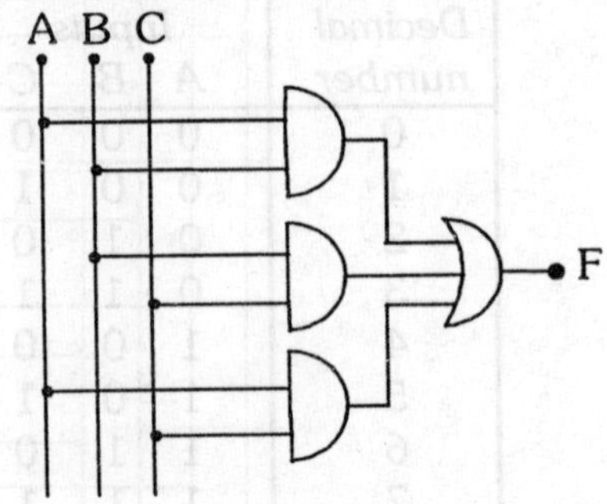

The logic diagram is shown above.

Problem 15. *Design a combinational circuit that multiplies two 2–bit numbers, a_1a_0, and b_1b_0, to produce 4–bit product, $C_3C_2C_1C_0$. Use AND gates and half adders.*

Solution. The truth table for the above combinational circuit is shown below.

a_1 a_0	b_1 b_0	C_3 C_2 C_1 C_0
0 0	0 0	0 0 0 0
0 0	0 1	0 0 0 0
0 0	1 0	0 0 0 0
0 0	1 1	0 0 0 0
0 1	0 0	0 0 0 0
0 1	0 1	0 0 0 1
0 1	1 0	0 0 1 0
0 1	1 1	0 0 1 1
1 0	0 0	0 0 0 0
1 0	0 1	0 0 1 0
1 0	1 0	0 1 0 0
1 0	1 1	0 1 1 0
1 1	0 0	0 0 0 0
1 1	0 1	0 0 1 1
1 1	1 0	0 1 1 0
1 1	1 1	1 0 0 1

The simplified expressions for each output is

$$C_3 = a_1a_0b_1b_0$$

For C_0

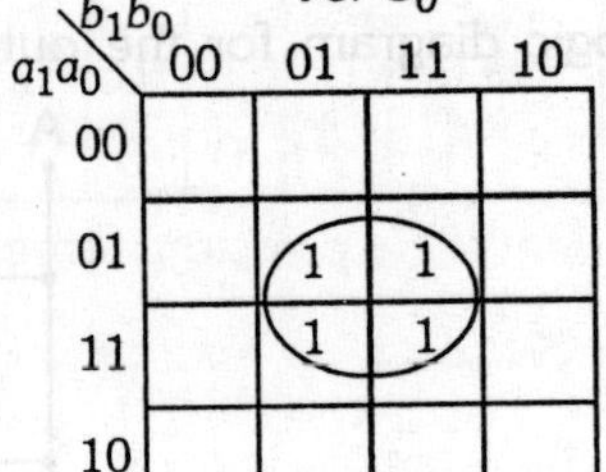

$C_0 = a_0b_0$

For C_1

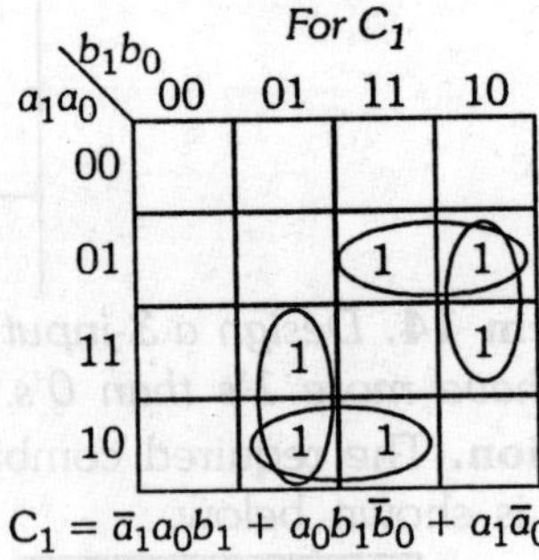

$C_1 = \bar{a}_1a_0b_1 + a_0b_1\bar{b}_0 + a_1\bar{a}_0b_0 + a_1\bar{b}_1b_0$

For C_2

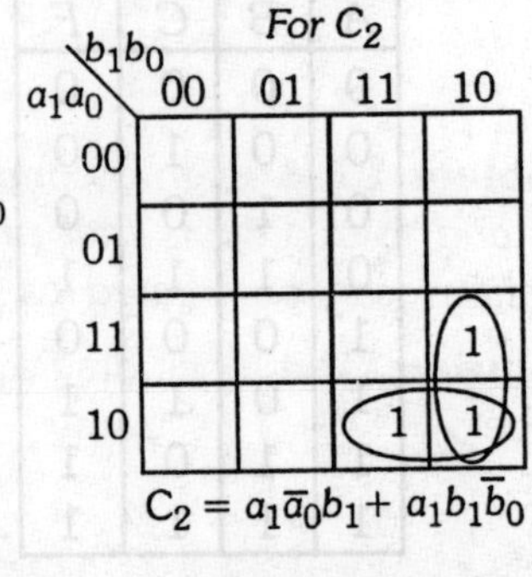

$C_2 = a_1\bar{a}_0b_1 + a_1b_1\bar{b}_0$

$$\therefore \quad C_0 = a_0b_0$$

$$C_1 = \bar{a}_1a_0b_1 + a_0b_1\bar{b}_0 + a_1\bar{a}_0b_0 + a_1\bar{b}_1b_0$$

$$C_2 = a_1\bar{a}_0b_1 + a_1b_1\bar{b}_0$$

$$C_3 = a_1a_0b_1b_0$$

The logic diagram is shown below.

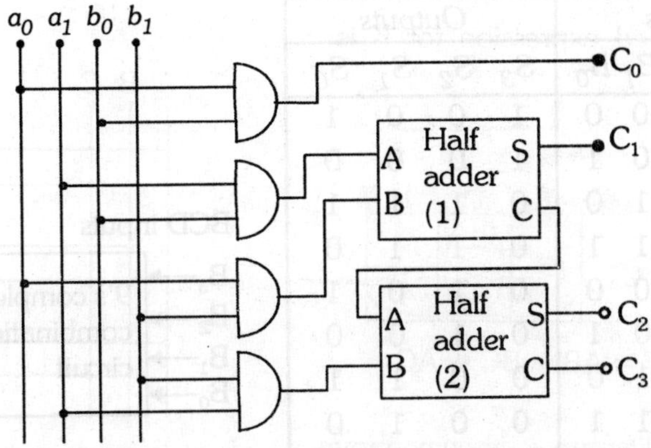

Problem 16. *Design a combinational circuit with two inputs and four outputs. The output binary number should be the square of the input binary number.*

Solution. The block diagram and truth table for the combinational circuit are shown below.

A	B	S_3	S_2	S_1	S_0
0	0	0	0	0	0
0	1	0	0	0	1
1	0	0	1	0	0
1	1	1	0	0	1

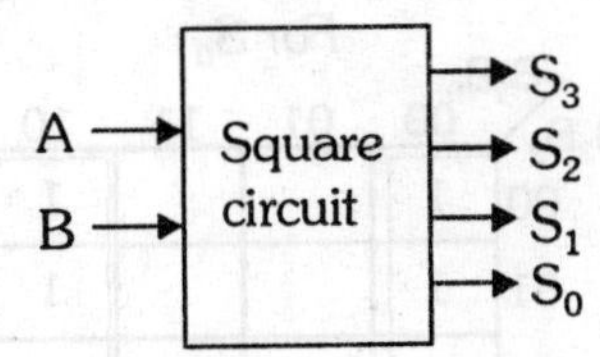

The output expression for each output is

$$S_0 = B$$

$$S_1 = 0$$

$$S_2 = A\overline{B}$$

and
$$S_3 = AB.$$

The logic diagram is shown below.

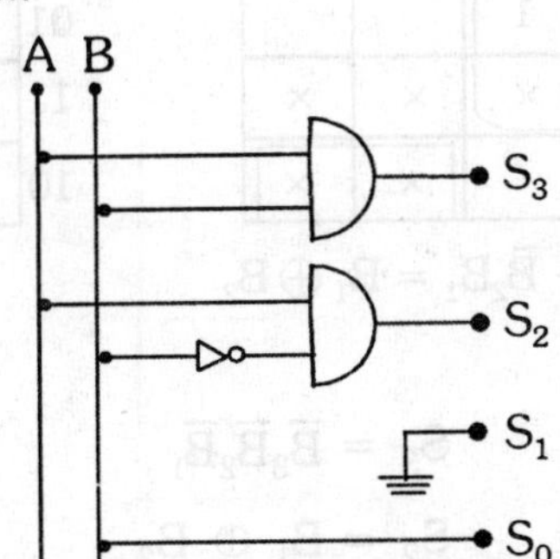

Problem 17. *Design a combinational circuit with four-inputs that represent a decimal digit in BCD and four outputs that produce the 9's complement of the input digit (The six unused combinations can be treated as don't care conditions).*

Solution. The block diagram and the truth table for the required combinational circuit are shown below.

Inputs				Outputs			
B_3	B_2	B_1	B_0	S_3	S_2	S_1	S_0
0	0	0	0	1	0	0	1
0	0	0	1	1	0	0	0
0	0	1	0	0	1	1	1
0	0	1	1	0	1	1	0
0	1	0	0	0	1	0	1
0	1	0	1	0	1	0	0
0	1	1	0	0	0	1	1
0	1	1	1	0	0	1	0
1	0	0	0	0	0	0	1
1	0	0	1	0	0	0	0

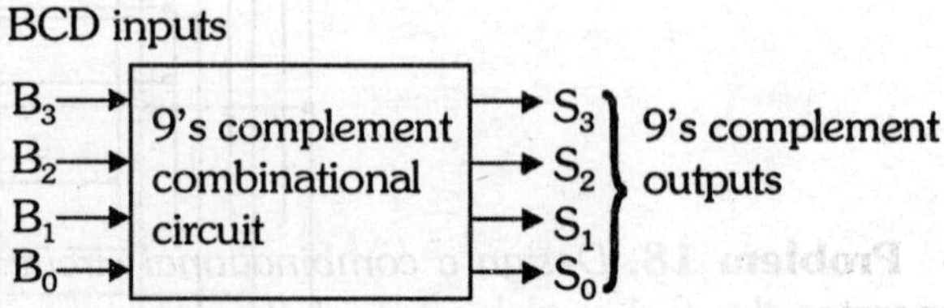

The simplified expressions for the outputs can be obtained from the K-maps.

For S_0

B_3B_2 \ B_1B_0	00	01	11	10
00	1			1
01	1			1
11	×	×	×	×
10	1		×	×

$$S_0 = \bar{B}_0$$

For S_1

B_3B_2 \ B_1B_0	00	01	11	10
00			1	1
01			1	1
11	×	×	×	×
10			×	×

$$S_1 = B_1$$

For S_2

B_3B_2 \ B_1B_0	00	01	11	10
00			1	1
01	1	1		
11	×	×	×	×
10			×	×

$$S_2 = B_2\bar{B}_1 + \bar{B}_2B_1 = B_1 \oplus B_2$$

For S_3

B_3B_2 \ B_1B_0	00	01	11	10
00	1	1		
01				
11	×	×	×	×
10			×	×

$$S_3 = \bar{B}_3\bar{B}_2\bar{B}_1$$

The four outputs are

$$S_3 = \bar{B}_3\bar{B}_2\bar{B}_1$$

$$S_2 = B_1 \oplus B_2$$

$$S_1 = B_1$$

$$S_0 = \bar{B}_0$$

The logic diagram for this combinational circuit shown below.

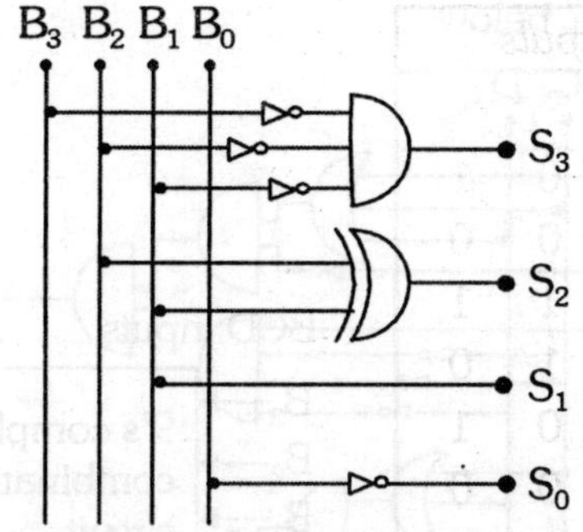

Problem 18. *Design a combinational circuit with four inputs and four outputs. The output generates the 2's complement of the input binary.*

Solution. The block diagram and the truth table for the above circuit are shown below.

Decimal digit	Input A	B	C	D	Output C_3	C_2	C_1	C_0
0	0	0	0	0	0	0	0	0
1	0	0	0	1	1	1	1	1
2	0	0	1	0	1	1	1	0
3	0	0	1	1	1	1	0	1
4	0	1	0	0	1	1	0	0
5	0	1	0	1	1	0	1	1
6	0	1	1	0	1	0	1	0
7	0	1	1	1	1	0	0	1
8	1	0	0	0	1	0	0	0
9	1	0	0	1	0	1	1	1
10	1	0	1	0	0	1	1	0
11	1	0	1	1	0	1	0	1
12	1	1	0	0	0	1	0	0
13	1	1	0	1	0	0	1	1
14	1	1	1	0	0	0	1	0
15	1	1	1	1	0	0	0	1

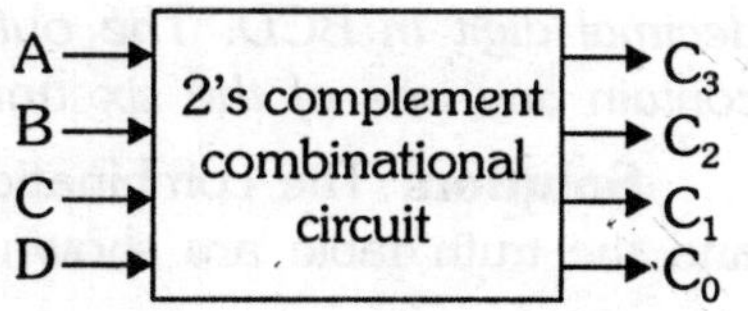

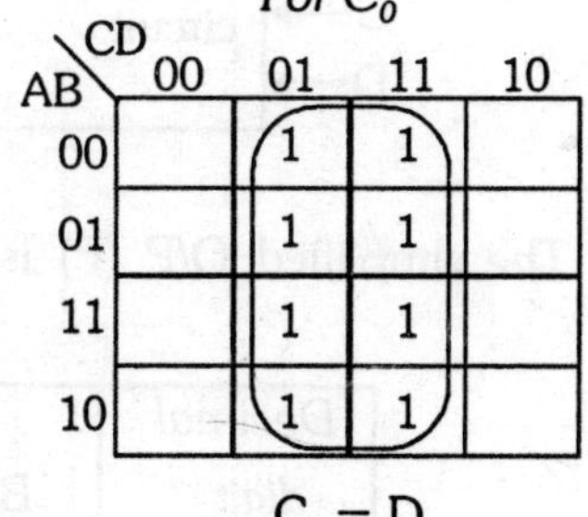

$C_0 = D$

The simplified four outputs can be obtained by using K-maps as shown above.

Similarly,

$$C_3 = \overline{A}B + \overline{A}C + \overline{A}D + A\overline{B}\,\overline{C}\,\overline{D} = \overline{A}(B + C + D) + A\overline{B}\,\overline{C}\,\overline{D}$$

$$C_2 = \overline{B}C + \overline{B}D + B\overline{C}\,\overline{D} = \overline{B}(C + D) + B\overline{C}\,\overline{D}$$

$$C_1 = C\overline{D} + \overline{C}D = C \oplus D$$

The logic diagram is shown below.

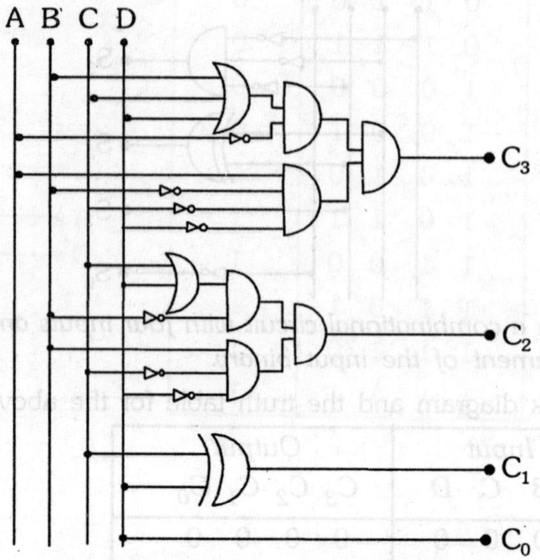

Problem 19. *Design a combinational circuit that detects an error in the representation of decimal digit in BCD. The output of the circuit must be equal to logic – 1 when the inputs contain any one of the six unused bit combinations in the BCD code.*

Solution. The combinational circuit has four inputs and one output. The block diagram and the truth table are shown below.

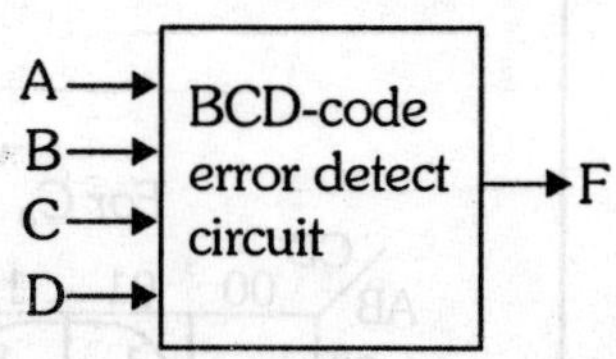

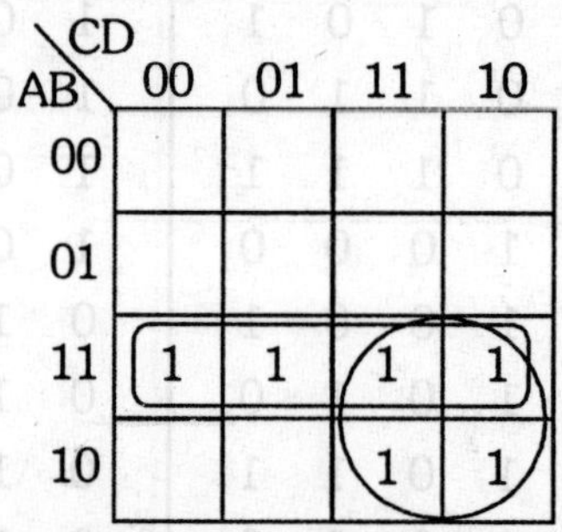

The simplified O/P (F) is shown above.

$$F = AB + AC = A\,(B + C)$$

Decimal digit	Input $B_3\ B_2\ B_1\ B_0$	Output F
0	0 0 0 0	0
1	0 0 0 1	0
2	0 0 1 0	0
3	0 0 1 1	0
4	0 1 0 0	0
5	0 1 0 1	0

6	0 1 1 0	0
7	0 1 1 1	0
8	1 0 0 0	0
9	1 0 0 1	0
10	1 0 1 0	1
11	1 0 1 1	1
12	1 1 0 0	1
13	1 1 0 1	1
14	1 1 1 0	1
15	1 1 1 1	1

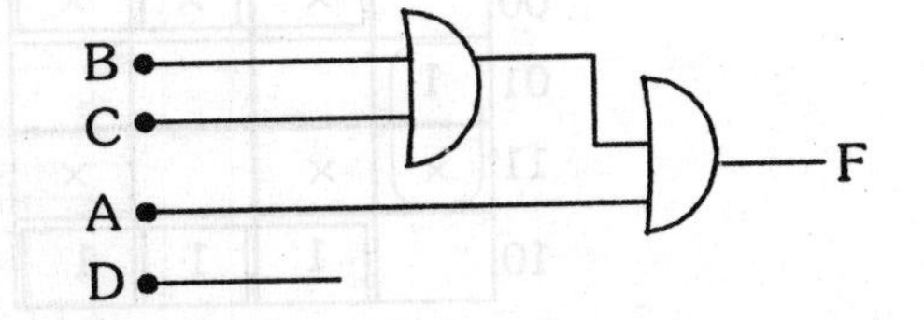

The logic diagram is shown above.

Problem 20. *Design a combinational circuit that converts a decimal digit from the 8 4 – 2 – 1 code to BCD code.*

Solution. From the problem we can realize that circuit must contain 4–inputs and 4–outputs. The block diagram and the truth table are shown below.

Decimal digit	*Inputs* A 8	B 4	C $\bar{2}$	D $\bar{1}$	*Outputs* B_3	B_2	B_1	B_0
0	0	0	0	0	0	0	0	0
7	0	1	1	1	0	0	0	1
6	0	1	1	0	0	0	1	0
5	0	1	0	1	0	0	1	1
4	0	1	0	0	0	1	0	0
11	1	0	1	1	0	1	0	1
10	1	0	1	0	0	1	1	0
9	1	0	0	1	0	1	1	1
8	1	0	0	0	1	0	0	0
15	1	1	1	1	1	0	0	1

The output equations can be simplified by using K-maps as shown below. The don't cares are

$$d = \Sigma m\ (1,\ 2,\ 3,\ 12,\ 13,\ 14)$$

For B_0

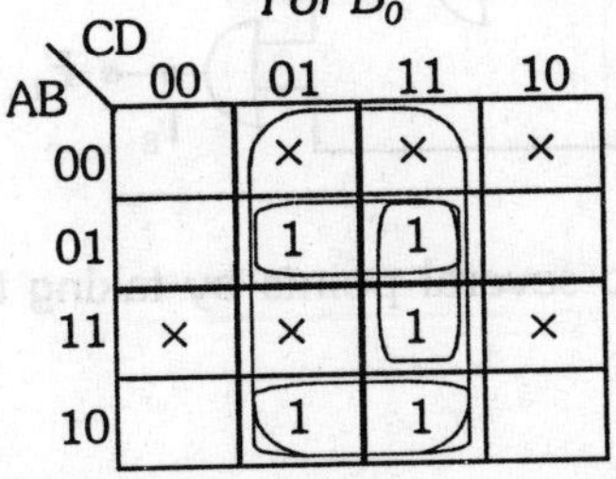

$B_0 = D$

For B_1

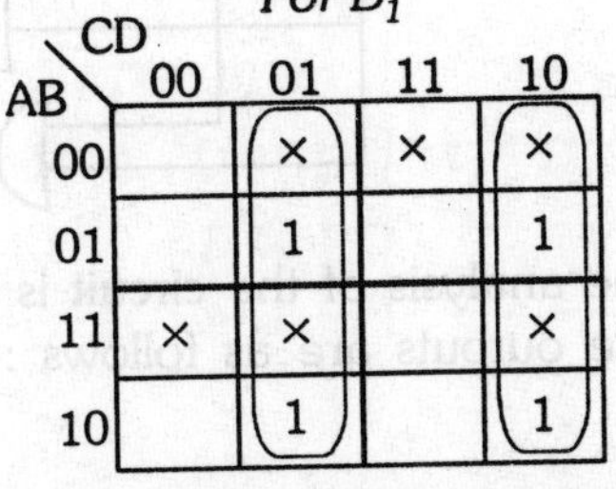

$B_1 = \bar{C}D + C\bar{D} = C \oplus D$

For B_2

AB \ CD	00	01	11	10
00		×	×	×
01	1			
11	×	×		×
10		1	1	1

$$B_2 = \overline{B}C + \overline{B}D + B\overline{C}\overline{D}$$

For B_3

AB \ CD	00	01	11	10
00		×	×	×
01				
11	×	×	1	×
10	1			

$$B_3 = A\overline{C}\overline{D} + AB$$

The logic diagram is shown in the below figure.

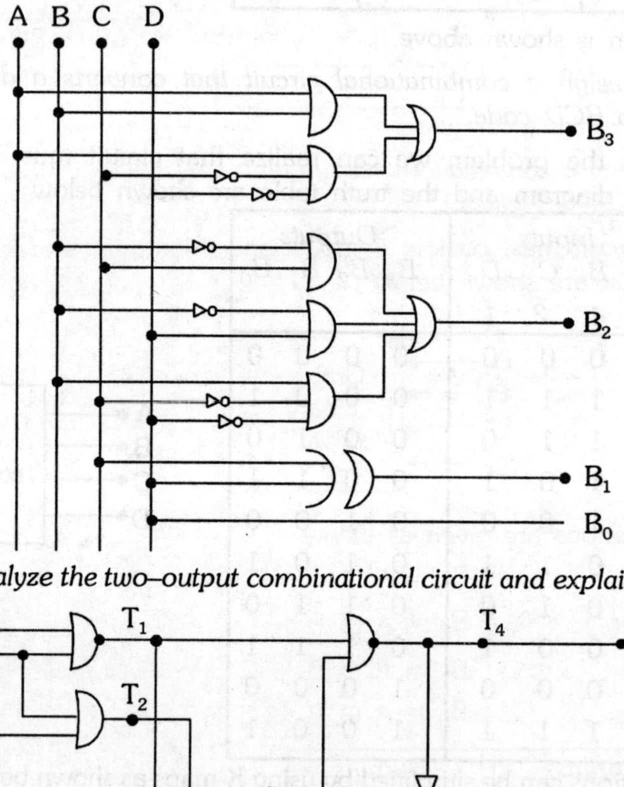

Problem 21. *Analyze the two–output combinational circuit and explain the circuit operation.*

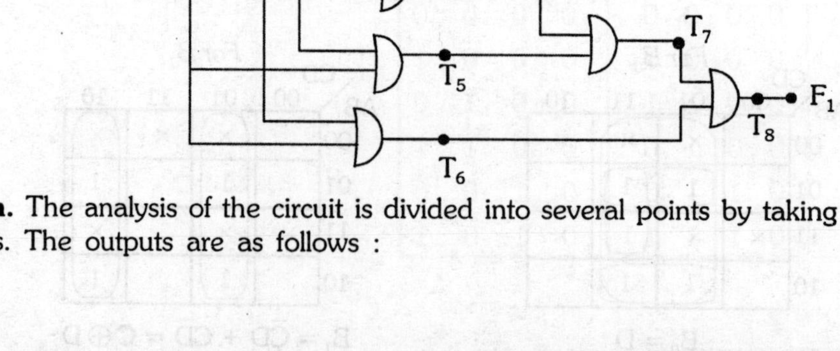

Solution. The analysis of the circuit is divided into several points by taking the output at different gates. The outputs are as follows :

$$T_1 = AB$$
$$T_2 = A + B$$
$$T_3 = C\,(A + B)$$
$$T_4 = T_1 + T_3$$
$$= AB + C\,(A + B) = AB + BC + CA$$
$$\boxed{\therefore\ F_2 = AB + BC + CA}$$
$$T_5 = T_2 + C = A + B + C$$
$$T_6 = T_1 \,.\, C = (A\,.\,B)\,C = ABC$$
$$T_7 = \overline{T}_4 \,.\, T_5$$
$$= \overline{(AB + BC + CA)}\ (A + B + C)$$
$$T_8 = T_6 + T_7 = ABC + \overline{(AB + BC + CA)}\ (A + B + C)$$
$$\boxed{F_1 = A \oplus B \oplus C}$$

∴ The circuit acts like a full adder.

Problem 22. *Design a combinational circuit that converts 4–bit gray code to 4–bit straight binary number using EX–OR gates.*

Solution. The circuit must contain 4–inputs and 4–outputs. The block diagram and the truth table of the circuit are shown below :

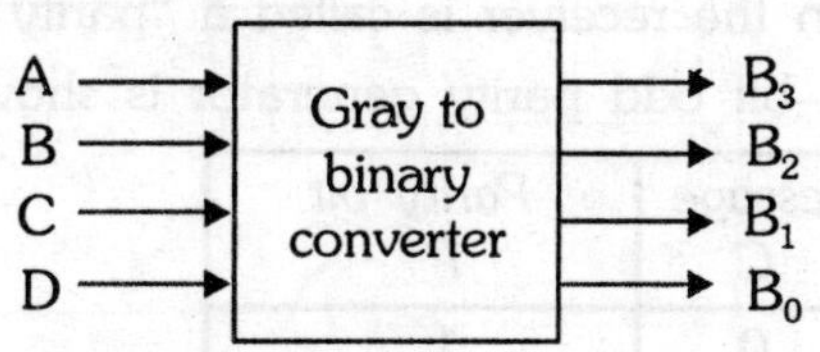

The output expressions are given as below :

$$B_3 = A$$
$$B_2 = A \oplus B$$
$$B_1 = A \oplus B \oplus C$$
$$B_0 = A \oplus B \oplus C \oplus D$$

Decimal digit	*Inputs* A	B	C	D	*Outputs* B_3	B_2	B_1	B_0
0	0	0	0	0	0	0	0	0
1	0	0	0	1	0	0	0	1
2	0	0	1	1	0	0	1	0
3	0	0	1	0	0	0	1	1
4	0	1	1	0	0	1	0	0
5	0	1	1	1	0	1	0	1

6	0 1 0 1	0 1 1 0
7	0 1 0 0	0 1 1 1
8	1 1 0 0	1 0 0 0
9	1 1 0 1	1 0 0 1
10	1 1 1 1	1 0 1 0
11	1 1 1 0	1 0 1 1
12	1 0 1 0	1 1 0 0
13	1 0 1 1	1 1 0 1
14	1 0 0 1	1 1 1 0
15	1 0 0 0	1 1 1 1

The logic diagram is shown below :

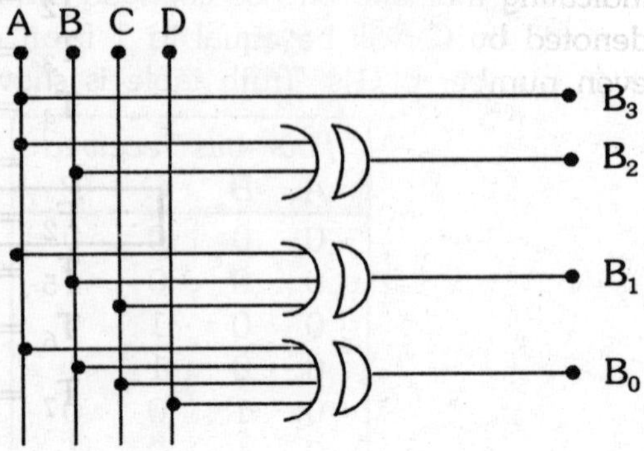

Problem 23. *Design the circuit of 3–bit parity generator and the circuit of 4–bit parity checker using odd parity bit.*

Solution. A parity bit is used for the purpose of detecting errors during transmission of binary data. Parity bit is an extra bit included with a binary message to make the number of 1's either even or odd.

The message, including parity bit, is transmitted and then checked at the receiver end for errors. An error is detected if the checked parity does not correspond with the one transmitted.

The circuit that generates the parity bit in the transmitter is called "parity generator". The circuit that checks the parity in the receiver is called a "parity checker."

The truth table for the 3–bit odd parity generator is shown below.

Three-bit message A	B	C	*Parity bit* P
0	0	0	1
0	0	1	0
0	1	0	0
0	1	1	1
1	0	0	0
1	0	1	1
1	1	0	1
1	1	1	0

The output expression P can be simplified by using K–map as :

A \ BC	00	01	11	10
0	(1)		(1)	
1		(1)		(1)

$= A \odot B \odot C$

The logic diagram is shown below.

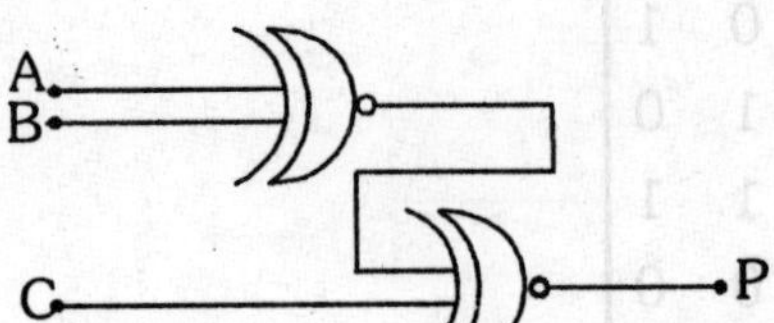

Any errors during the transmission *i.e.*, if four bits received have an even number of 1's, indicating that one bit has changed during the transmission. The output of the parity checker, denoted by C, will be equal to 1 if an error occurs, that is, if the four bits received have an even number of 1's. Truth table is shown below.

Four–bits received				*Parity checker*
A	*B*	*C*	*P*	*C*
0	0	0	0	1
0	0	0	1	0
0	0	1	0	0
0	0	1	1	1
0	1	0	0	0
0	1	0	1	1
0	1	1	0	1
0	1	1	1	0
1	0	0	0	0
1	0	0	1	1
1	0	1	0	1
1	0	1	1	0
1	1	0	0	1
1	1	0	1	0
1	1	1	0	0
1	1	1	1	1

The simplified expression for the parity checker, C, can be obtained by using K–map as shown below.

AB \ CP	00	01	11	10
00	1		1	
01		1		1
11	1		1	
10		1		1

$C = A \odot B \odot C \odot P$

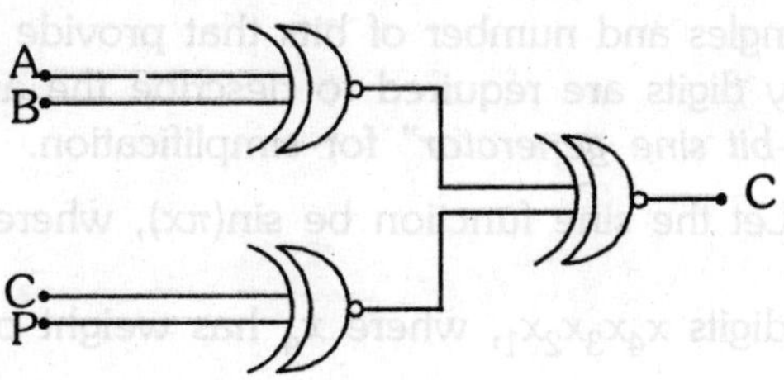

The logic diagram is shown above.

Note: Even parity generator and checker circuits are designed/implemented by exclusive–OR gates.

Problem 24. *Draw logic diagram of EX-OR gate with (a) NAND gates only (b) NOR gates only.*

Solution. (*a*) The EX–OR gate using only NAND gates is shown below.

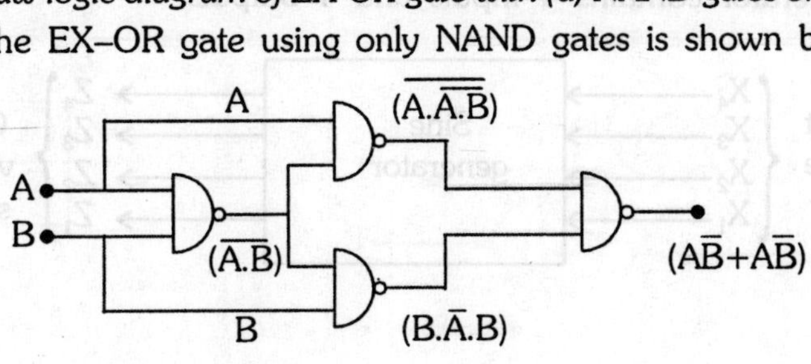

(*b*) The EX-OR gate using NOR gates is shown below.

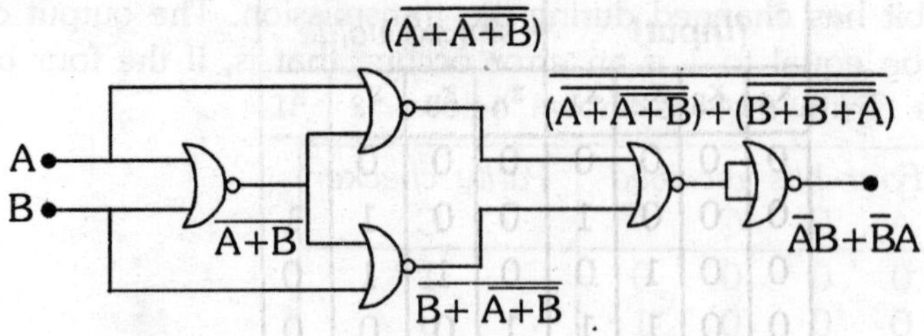

$$f = \bar{A}B + \bar{B}A = \bar{A}B + \bar{B}A + A\bar{A} + B\bar{B}$$

$$= \bar{A}(B + A) + \bar{B}(A + B)$$

$$= \bar{A}\,\overline{\overline{(A + B)}} + \bar{B}\,\overline{\overline{(A + B)}}$$

$$= \overline{A + \overline{(A + B)}} + \overline{B + \overline{(A + B)}}$$

$$= \overline{\overline{\overline{\left(A + \overline{(A + B)}\right)} + \overline{\left(B + \overline{(A + B)}\right)}}}.$$

Problem 25. *Design a circuit which generates a positive half sign wave.*

Solution. Trigonometric functions can be generated by using combinational circuits. Combinational sine wave generators are used whenever the sine function must be evaluated fast and repeatedly.

A combinational sine generator receives as its inputs the angle, and in turn, produces the sine of that angle. The angle given in radians converted to binary and the sine value is produced in binary. The accuracy of the calculation is a function of number of bits that describe the angles and number of bits that provide the sine values. Practical applications at least eight binary digits are required to describe the angle or sine values. Here we are considering only "*four-bit sine generator*" for simplification.

Let the sine function be sin(πx), where $0 \le x \le 1$. The angle 'x' will be described by the four digits $x_4x_3x_2x_1$, where x_4 has weight of $\frac{1}{2}$, x_3 weight of $\frac{1}{4}$, $x_2 = \frac{1}{8}$ and $x_1 = \frac{1}{16}$.

For example to specify the angle of 45°, *x* must be (π/4) *i.e.*, $\pi/4$ = 45°. The *x* must be 0100. To represent an angle 30°, *x* must be $\pi/6$. However it is impossible to represent this value precisely with 4–bits and the closest possible value is 3/16 = 0011. Like this way for '15' different angles, may be 15 different outputs values are tabulated in the truth table. The block diagram of sine generator contains 4–inputs and 4–outputs.

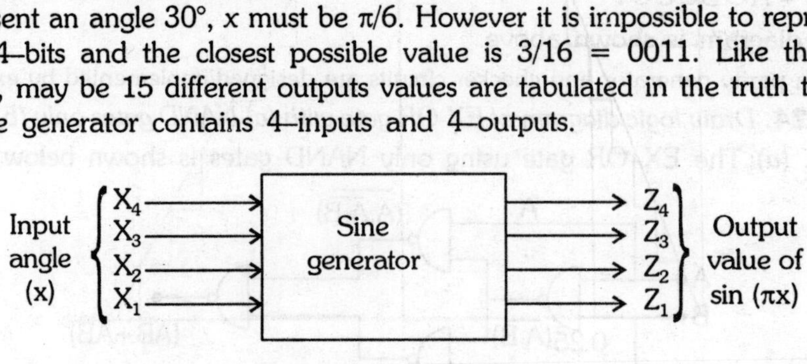

Angle (Input)				*Outputs sin(πx) value*			
x_4	x_3	x_2	x_1	z_4	z_3	z_2	z_1
0	0	0	0	0	0	0	
0	0	0	1	0	0	1	1
0	0	1	0	0	1	1	0
0	0	1	1	1	0	0	0
0	1	0	0	1	0	1	1
0	1	0	1	1	1	0	1
0	1	1	0	1	1	1	0
0	1	1	1	1	1	1	1
1	0	0	0	1	1	1	1
1	0	0	1	1	1	1	1
1	0	1	0	1	1	1	0
1	0	1	1	1	1	0	1
1	1	0	0	1	0	1	1
1	1	0	1	1	0	0	0
1	1	1	0	0	1	1	0
1	1	1	1	0	0	1	1

The output value of sin (πx) is represented by $z = z_4z_3z_2z_1$, where $0 \leq z \leq 1$, z_4 has weight $\frac{1}{2}$, z_3 has weight $\frac{1}{4}$, $z_2 = \frac{1}{8}$ and $z_1 = \frac{1}{16}$ and so on. For an angle of 30° *i.e.*, $x = 0011$, the $z = 1000 = 0.5$. The waveform is shown below.

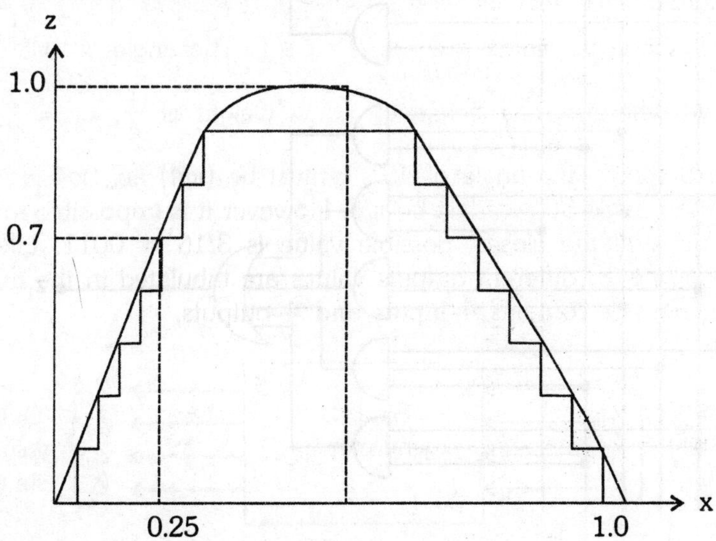

The logic diagram is shown in the figure.

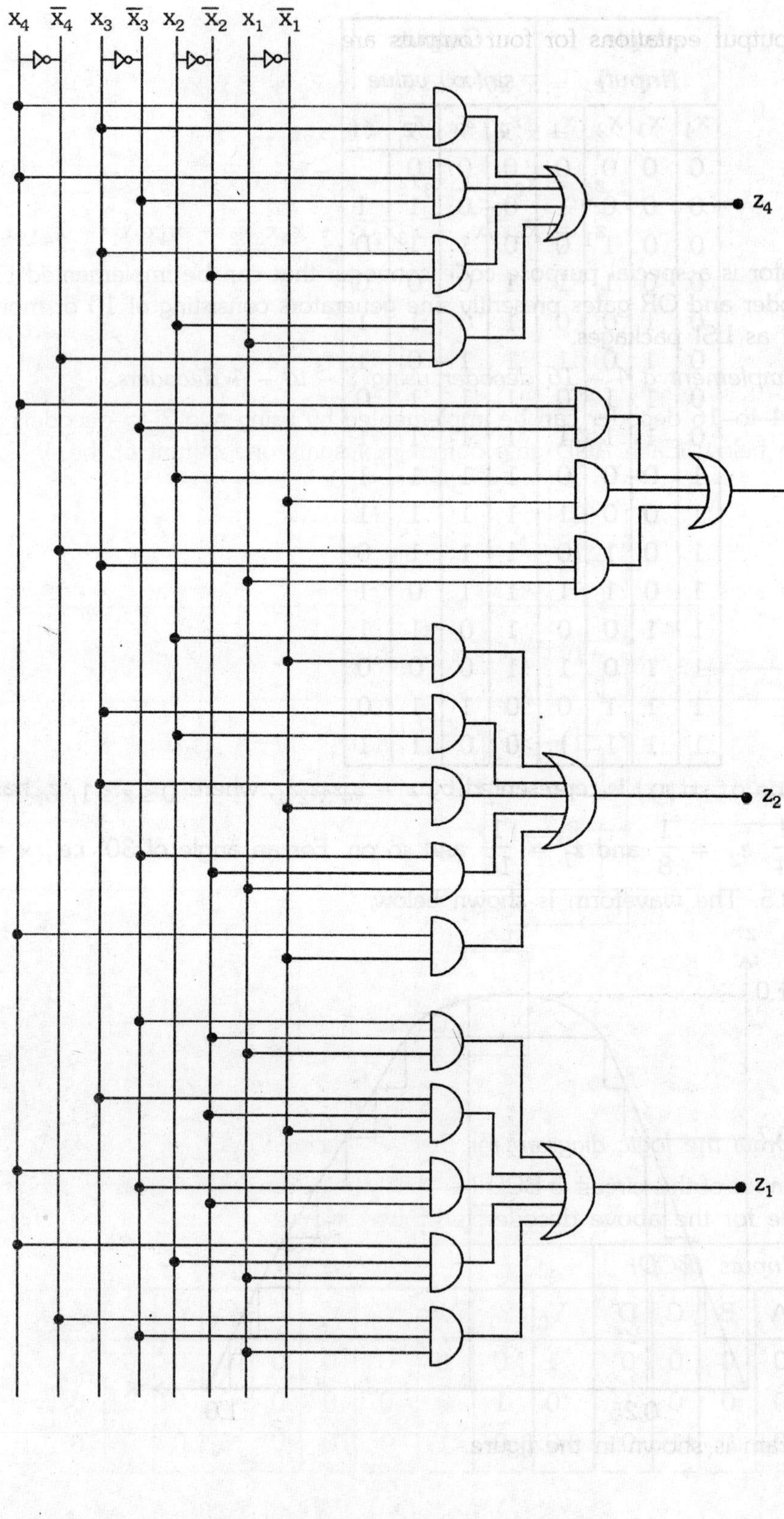
x_4 $\bar{x}_4$ x_3 $\bar{x}_3$ x_2 $\bar{x}_2$ x_1 $\bar{x}_1$
z_4
z_3
z_2
z_1

The simplified output equations for four outputs are

$$z_4 = \bar{x}_4 x_3 + x_4 \bar{x}_3 + x_3 \bar{x}_2 + \bar{x}_4 x_2 x_1$$

$$z_3 = x_4 \bar{x}_3 + x_2 \bar{x}_1 + \bar{x}_4 x_3 x_1$$

$$z_2 = x_2 \bar{x}_1 + x_3 x_2 + x_3 \bar{x}_1 + \bar{x}_3 \bar{x}_2 x_1 + x_4 \bar{x}_1$$

$$z_1 = \bar{x}_3 \bar{x}_2 x_1 + x_3 \bar{x}_2 \bar{x}_1 + x_4 \bar{x}_3 \bar{x}_2 + x_4 x_2 x_1 + \bar{x}_4 x_3 x_1$$

The sine generator is a special purpose code converter that can be implemented by using the 4 – to – 16 decoder and OR gates presently sine generators consisting of 10 or more input bits can be obtained as LSI packages.

Problem 26. *Implement a 4 – 16 decoder using 2 – to – 4 decoders.*

Solution. The 4–to–16 decoder can be implemented by using two 2–to–decoders and 16 AND gates as shown below. Each AND gate output represents one output of the 4 – to – 16 decoder.

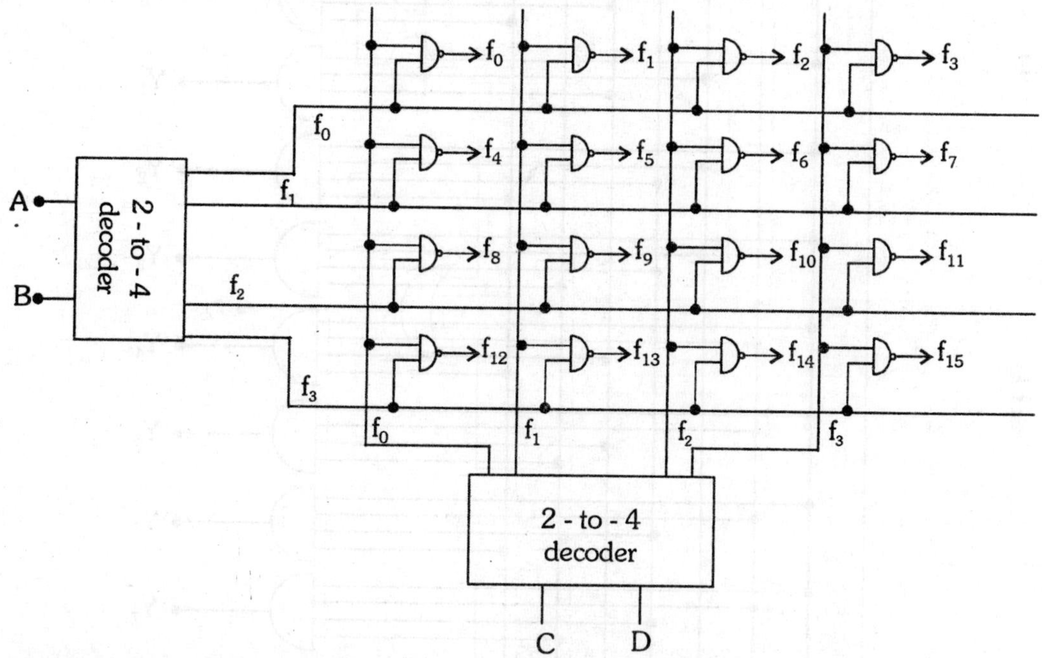

Problem 27. *Draw the logic diagram for BCD to decimal decoder.*

Solution. The input of this circuit is BCD (4 – bits) and the output is its equivalent decimal value. The truth table for the above decoder is shown below. •

Inputs (BCD)				*Outputs (decimal)*									
A	B	C	D	Y_0	Y_1	Y_2	Y_3	Y_4	Y_5	Y_6	Y_7	Y_8	Y_9
0	0	0	0	1	0	0	0	0	0	0	0	0	0
0	0	0	1	0	1	0	0	0	0	0	0	0	0
0	0	1	0	0	0	1	0	0	0	0	0	0	0

0	0	1	1	0	0	0	1	0	0	0	0	0	0
0	1	0	0	0	0	0	0	1	0	0	0	0	0
0	1	0	1	0	0	0	0	0	1	0	0	0	0
0	1	1	0	0	0	0	0	0	0	1	0	0	0
0	1	1	1	0	0	0	0	0	0	0	1	0	0
1	0	0	0	0	0	0	0	0	0	0	0	1	0
1	0	0	1	0	0	0	0	0	0	0	0	0	1

The logic diagram is shown below, which uses 10 AND gates.

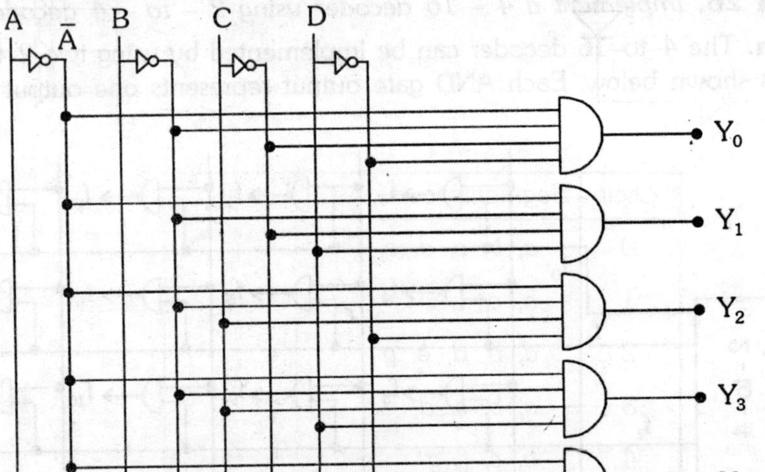

Problem 28. *Design a BCD to 7 segment display decoder.*

Solution. In most practical applications, seven segment displays are used to give a visual indication of the output states of digital systems, such as decade counter, latches etc. These outputs are usually in four bit BCD form, and are thus not suitable for directly driving seven segment displays. Let us tabulate the segments activated during each digit display in common cathode.

The basic seven segment display format is using LED's shown below

The common cathode is shown below.

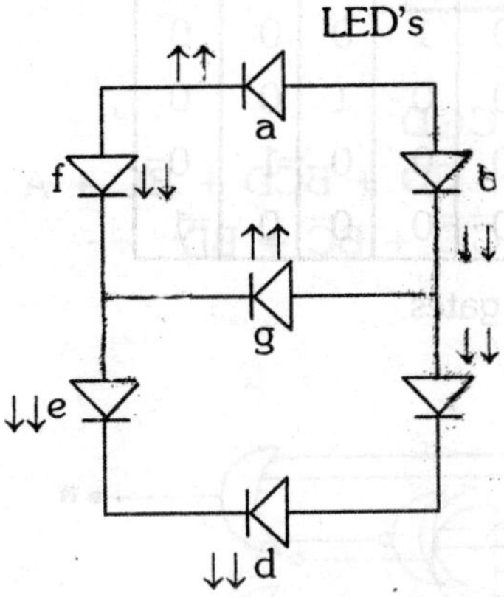

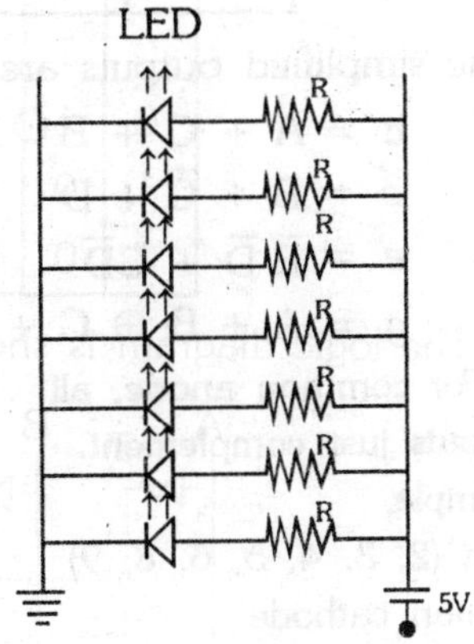

Digit	Segments activated	Display
0	a, b, c, d, e, f	0
1	b, c	1
2	a, b, d, e, g	2
3	a, b, c, d, g	3
4	b, c, f, g	4
5	a, c, d, f, g	5
6	a, c, d, e, f, g	6
7	a, b, c	7
8	a, b, c, d, e, f, g	8
9	a, b, c, d, f, g	9

From the above table we can determine the truth table for BCD–to–7 segment decoder.

Digit	A	B	C	D	a	b	c	d	e	f	g
0	0	0	0	0	1	1	1	1	1	1	0
1	0	0	0	1	0	1	1	0	0	0	0
2	0	0	1	0	1	1	0	1	1	0	1
3	0	0	1	1	1	1	1	1	0	0	1
4	0	1	0	0	0	1	1	0	0	1	1
5	0	1	0	1	1	0	1	1	0	1	1
6	0	1	1	0	1	0	1	1	1	1	1

7	0	1	1	1	1	1	1	0	0	0	0
8	1	0	0	0	1	1	1	1	1	1	1
9	1	0	0	1	1	1	1	1	0	1	1

The simplified outputs are given below.

$a = A + C + B \odot D$

$b = \bar{B} + C \odot D$

$c = B + \bar{C} + D$

$d = \bar{B}\bar{D} + C\bar{D} + B\bar{C}D + \bar{B}C + A$

$e = \bar{B}\bar{D} + C\bar{D}$

$f = A + \bar{C}\bar{D} + B\bar{C} + B\bar{D}$

$g = A + B \oplus C + C\bar{D}$

Note: For common anode, all the outputs just complement. For example,

$g = \Sigma m\ (2, 3, 4, 5, 6, 8, 9)$

in common cathode

and $g = \Sigma m\ (0, 1, 7)$

in common anode and so on for all other outputs.

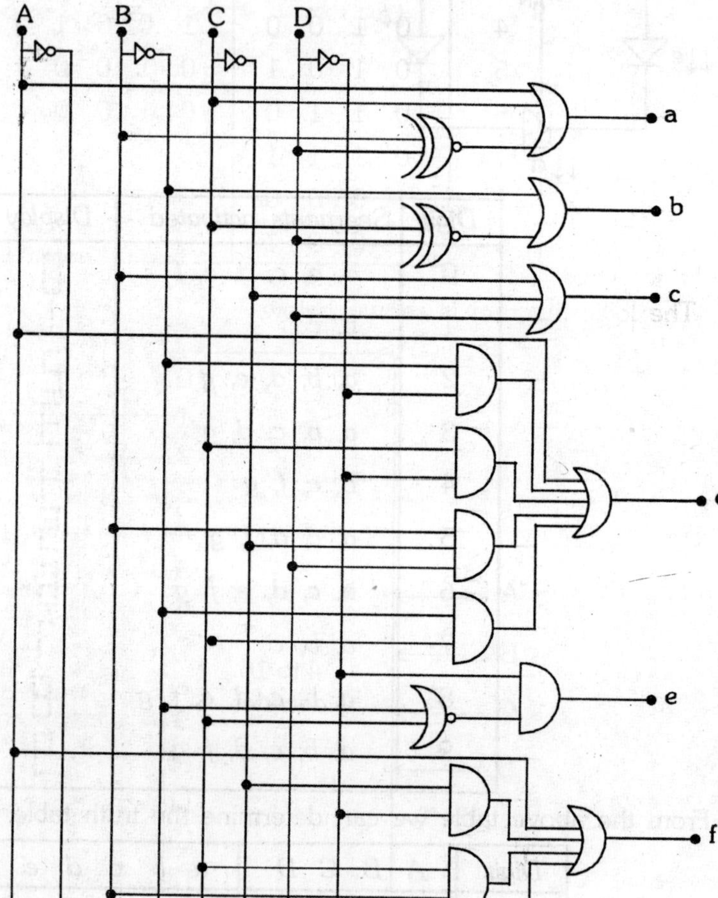

The logic diagram is shown above.

Problem 29. *Implement BCD–to–7 segment decoder for common anode using 4–to–16 decoder.*

Solution. The truth table for BCD – to – 7 segment decoder for common anode is shown below.

Digit	A B C D	a b c d e f g
0	0 0 0 0	0 0 0 0 0 0 1
1	0 0 0 1	1 0 0 1 1 1 1
2	0 0 1 0	0 0 1 0 0 1 0
3	0 0 1 1	0 0 0 0 1 1 0
4	0 1 0 0	1 0 0 1 1 0 0
5	0 1 0 1	0 1 0 0 1 0 0
6	0 1 1 0	0 1 0 0 0 0 0
7	0 1 1 1	0 0 0 1 1 1 1
8	1 0 0 0	0 0 0 0 0 0 0
9	1 0 0 1	0 0 0 0 1 0 0

The logic diagram is shown below.

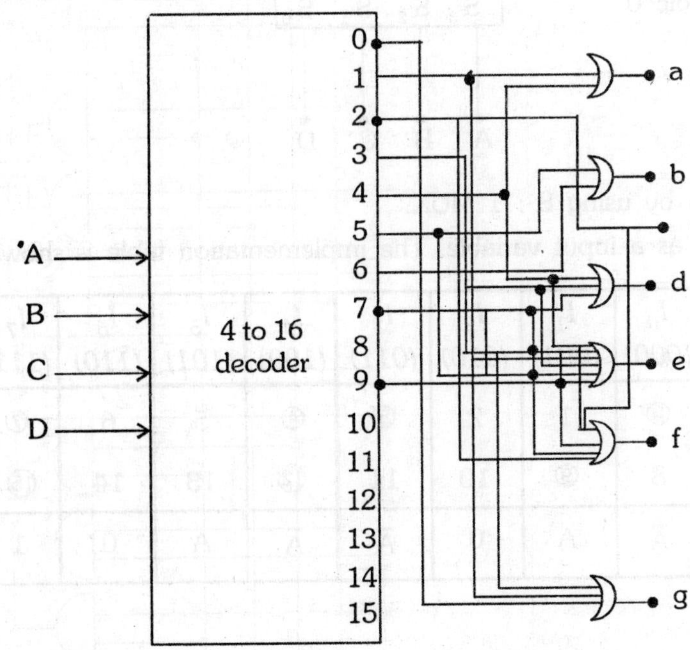

Problem 30. *Realize the function using (a) 16 : 1 MUX (b) 8 : 1 MUX (c) 4 : 1 MUX for the function f (A, B, C, D) = Σm (0, 3, 4, 7, 9, 13, 15)*

Solution. (*a*) The function has 4–input variables and maximum of 16 minterms we can generate. The implementation by 16 : 1 MUX is done directly.

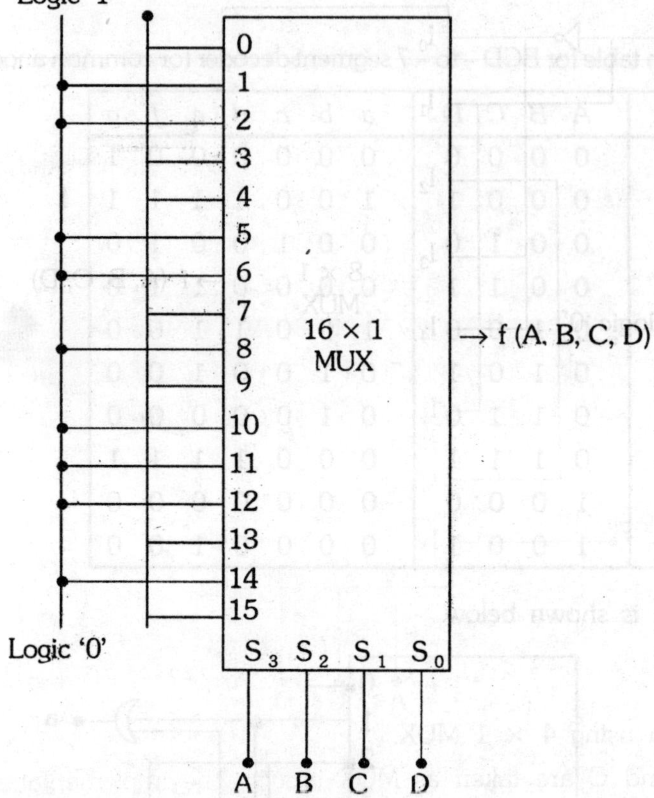

(*b*) Implementation by using 8 : 1 MUX.

Let us consider 'A' as a input variable. The implementation table is shown below.

	I_0 (000)	I_1 (001)	I_2 (010)	I_3 (011)	I_4 (100)	I_5 (101)	I_6 (110)	I_7 (111)
$\bar{A}$	⓪	1	2	③	④	5	6	⑦
A	8	⑨	10	11	⑫	13	14	⑮
Data MUX inputs	$\bar{A}$	A	0	$\bar{A}$	$\bar{A}$	A	0	1

The logic diagram is shown below.

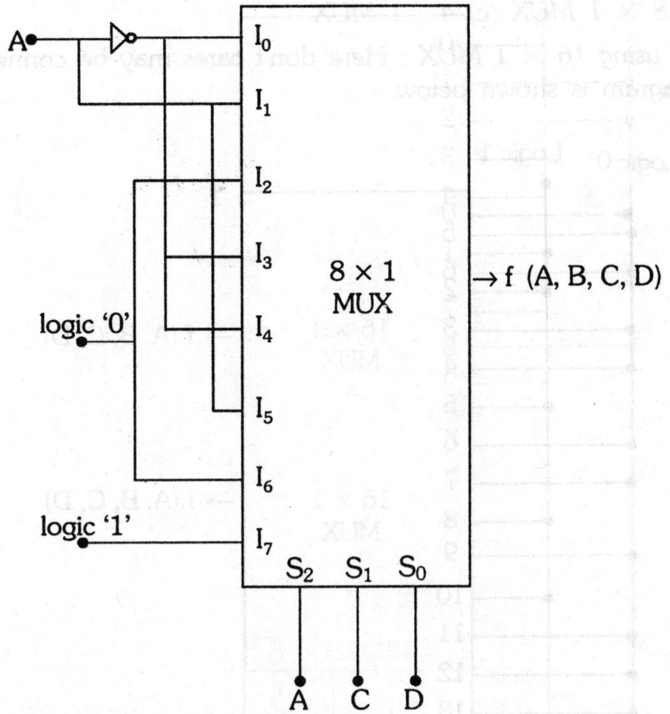

(*c*) Implementation using 4 × 1 MUX.

The variables D and C are taken as MUX inputs. The implementation table and logic diagram are shown below.

	00 I_0	01 I_1	10 I_2	11 I_3
$\overline{C}\,\overline{D}$	⓪	④	8	12
$\overline{C}D$	1	5	⑨	⑬
$C\overline{D}$	2	6	10	14
CD	③	⑦	11	⑮
MUX Inputs	$C \odot D$	$C \odot D$	$\overline{C}D$	D

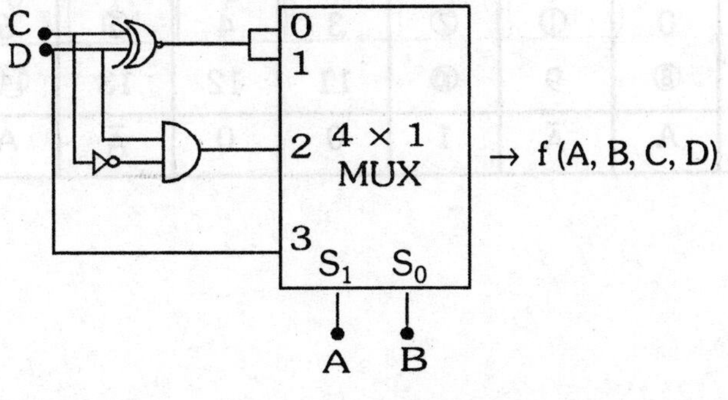

Problem 31. *Realize the function* $f = \Sigma m\ (1, 2, 5, 8, 10, 14) + d\ (6, 7, 15)$ *using (a) 16 × 1 MUX (b) 8 × 1 MUX (c) 4 : 1 MUX.*

Solution. *(a) By using 16 × 1 MUX* : Here don't cares may be connected to logic 1 or logic '0'. The logic diagram is shown below.

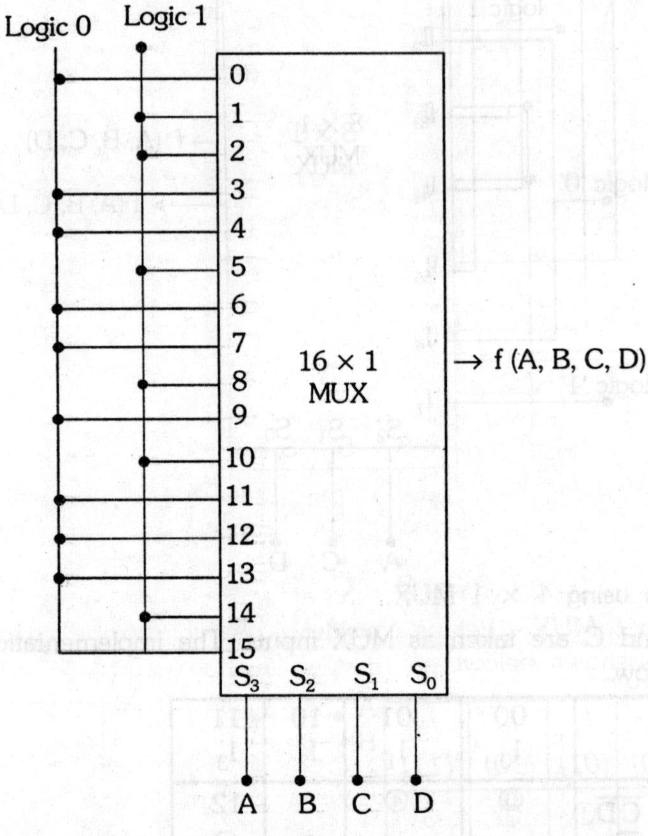

(b) By using 8 × 1 MUX: Let us consider 'A' as a input variable. The implementation table is shown below.

	I_0 (000)	I_1 (001)	I_2 (010)	I_3 (011)	I_4 (100)	I_5 (101)	I_6 (110)	I_7 (111)
$\bar{A}$	0	①	②	3	4	⑤	6	7
A	⑧	9	⑩	11	12	13	⑭	15
MUX inputs	A	$\bar{A}$	1	0	0	$\bar{A}$	A	0

The logic diagram is shown below.

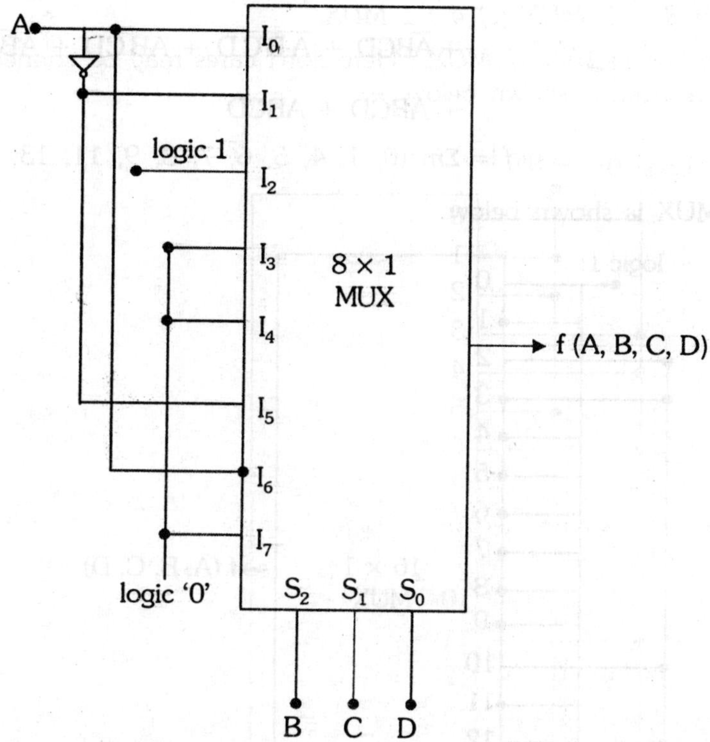

(c) *By using 4 × 1 MUX* : Let us consider A and B are the input variables and the implementation table is shown below.

	I_0 (00)	I_1 (01)	I_2 (10)	I_3 (11)
$\bar{A}\bar{B}$	0	①	②	3
$\bar{A}B$	4	⑤	⑥	7
$A\bar{B}$	⑧	9	⑩	11
AB	12	13	⑭	15
MUX inputs	$A\bar{B}$	$\bar{A}$	1	0

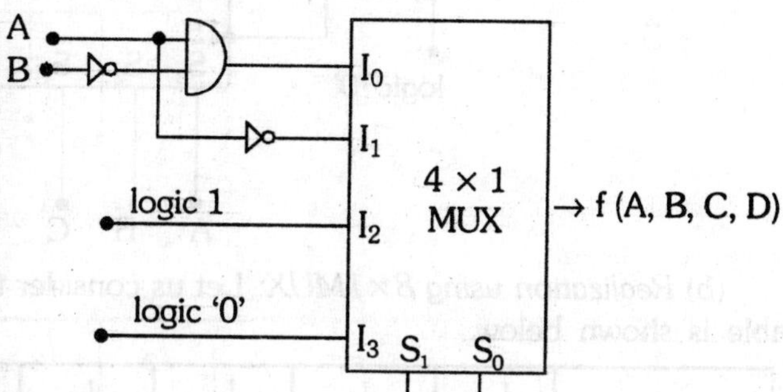

The logic diagram is shown above.

Problem 32. *Realize the function f (A, B, C, D) = $\bar{A}B + \bar{B}\bar{C} + AD$ using (a) 16 × 1 MUX (b) 8 × 1 MUX (c) 4 × 1 MUX.*

Solution. $f(A, B, C, D) = \bar{A}B + \bar{B}\bar{C} + AD$

$$= \bar{A}B(C + \bar{C}) + \bar{B}\bar{C}(A + \bar{A})(D + \bar{D})$$

$$+ AD(B + \bar{B})(C + \bar{C})(D \quad \bar{D})$$

$$= \overline{A}\,\overline{B}\,\overline{C}\,\overline{D} + \overline{A}\,\overline{B}\,\overline{C}D + \overline{A}B\overline{C}\,\overline{D} + \overline{A}B\overline{C}D + \overline{A}BC\overline{D}$$

$$+ \overline{A}BCD + \overline{A}\,\overline{B}\,\overline{C}\,\overline{D} + A\overline{B}\,\overline{C}D + A\overline{B}CD$$

$$+ AB\overline{C}D + ABCD$$

$$f = \Sigma m\ (0, 1, 4, 5, 6, 7, 8, 9, 11, 13, 15)$$

(*a*) The 16×1 MUX is shown below.

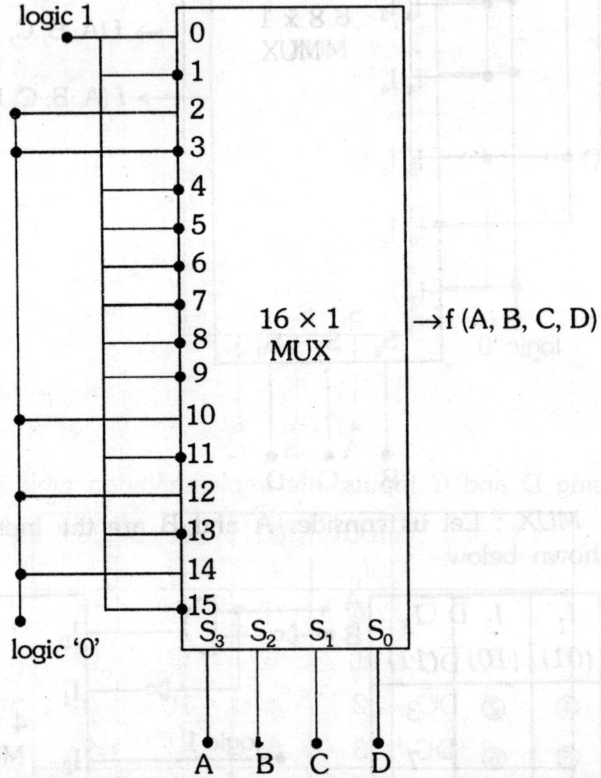

(*b*) *Realization using 8×1MUX*: Let us consider the variable D as input, the implementation table is shown below.

	I_0 (000)	I_1 (001)	I_2 (010)	I_3 (011)	I_4 (100)	I_5 (101)	I_6 (110)	I_7 (111)
$\overline{D}$	⓪	2	④	⑥	⑧	10	12	14
D	①	3	⑤	⑦	⑨	⑪	⑬	⑮
MUX inputs	1	0	1	1	1	D	D	D

The realization is shown below.

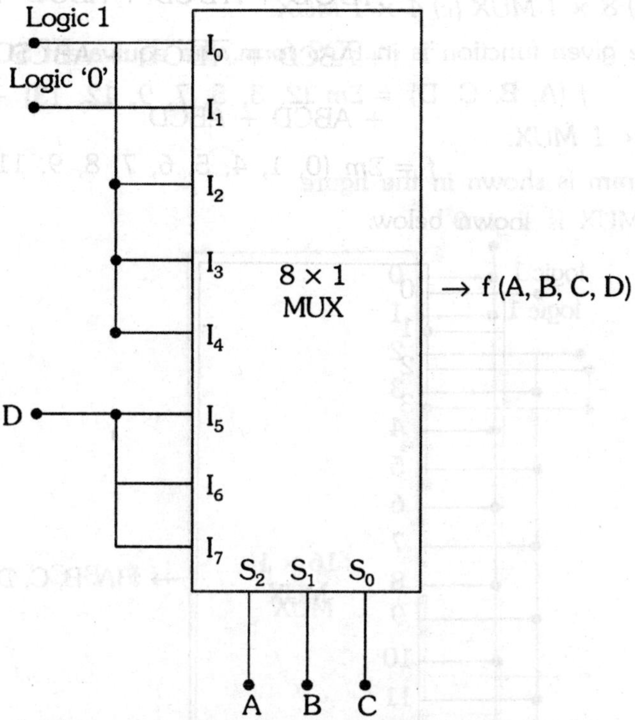

(*c*) Realization using D and C inputs, the implementation table is shown below.

	(00) I_0	(01) I_1	(10) I_2	(11) I_3
$\overline{D}\,\overline{C}$	⓪	④	⑧	12
$\overline{D}C$	①	⑤	⑨	⑬
$D\overline{C}$	2	⑥	10	14
DC	3	⑦	⑪	⑮
MUX inputs	$\overline{D}$	1	$\overline{D} + C$	C

The logic diagram is shown below.

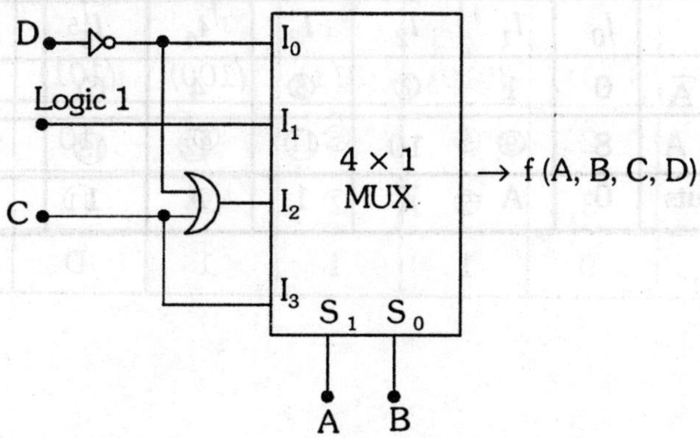

Problem 33. *Realize the function* $f(A, B, C, D) = \Pi(1, 4, 6, 10, 14) + d(0, 8, 11, 15)$ *using (a)* 16×1 *MUX (b)* 8×1 *MUX (c)* 4×1 *MUX.*

Solution. The given function is in POS form, the equivalent SOP is

$$f(A, B, C, D) = \Sigma m\ (2, 3, 5, 7, 9, 12, 13) + d\ (0, 8, 11, 15).$$

(a) Using 16×1 *MUX:*

The logic diagram is shown in the figure.

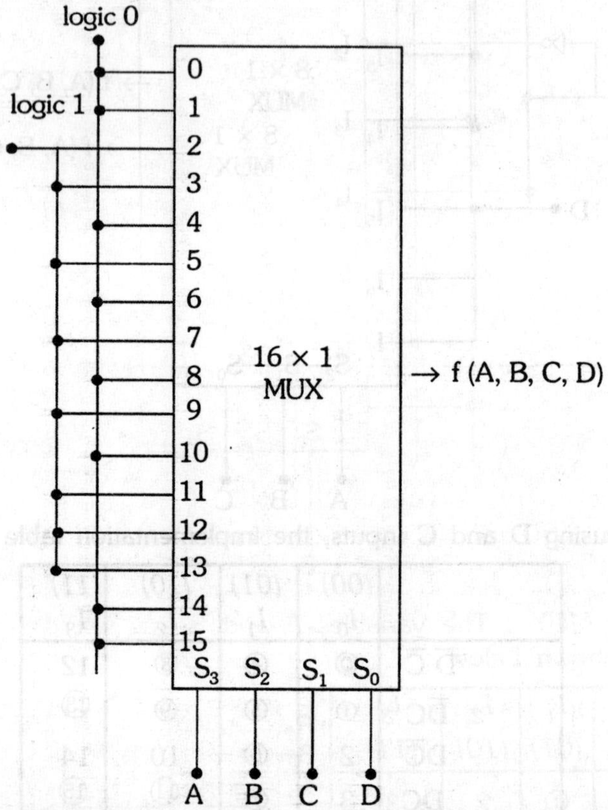

(b) Using 8×1 *MUX:*

The implementation table is shown below.

	I_0	I_1	I_2	I_3	I_4	I_5	I_6	I_7
$\bar{A}$	0	1	②	③	4	⑤	6	⑦
A	8	⑨	10	⑪	⑫	⑬	14	⑮
MUX inputs	0	A	$\bar{A}$	1	A	1	0	1

The logic diagram is shown below.

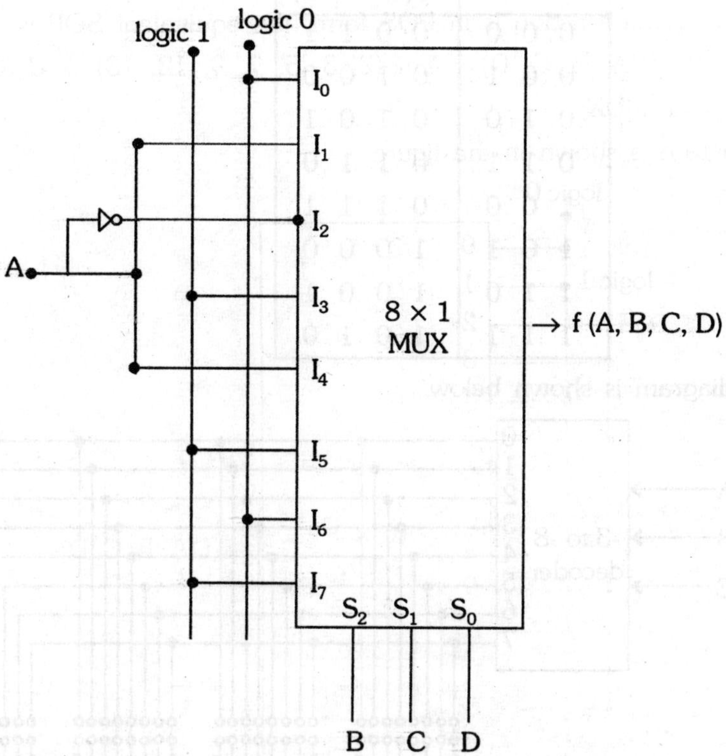

(c) *Using 4 × 1 MUX* : The variables A and B are considered as MUX inputs, the implementation table shown below.

	I_0 (00)	I_1 (01)	I_2 (10)	I_3 (11)
$\bar{A}\bar{B}$	⓪	①	2	3
$\bar{A}B$	④	⑤	⑥	⑦
$A\bar{B}$	⑧	⑨	10	⑪
AB	12	⑬	14	⑮
MUX Inputs	$\bar{A} + \bar{B}$	1	$\bar{A}B$	A + B

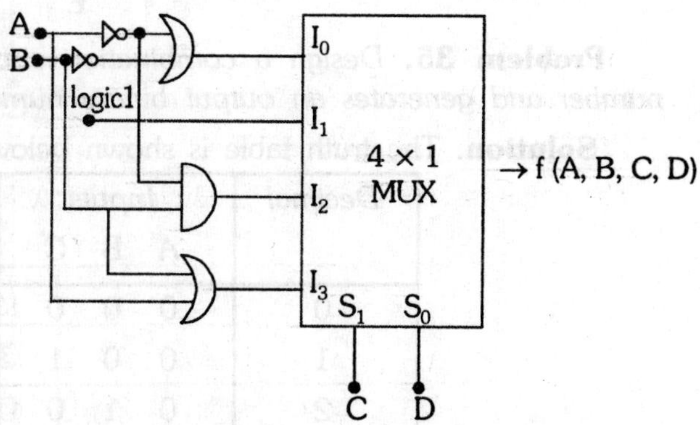

The logic diagram is shown above.

Problem 34. *Design a combinational circuit using a ROM. The circuit accepts 3–bit binary number and generates its equivalent excess–3 code.*

Solution. The truth table of the given combinational circuit is shown below. The ROM has 3–inputs and 4–outputs. The size of the ROM required is $2^3 \times 4$ bits.

Inputs	*Outputs*
A B C	E_3 E_2 E_1 E_0
0 0 0	0 0 1 1
0 0 1	0 1 0 0
0 1 0	0 1 0 1
0 1 1	0 1 1 0
1 0 0	0 1 1 1
1 0 1	1 0 0 0
1 1 0	1 0 0 1
1 1 1	1 0 1 0

The logic diagram is shown below.

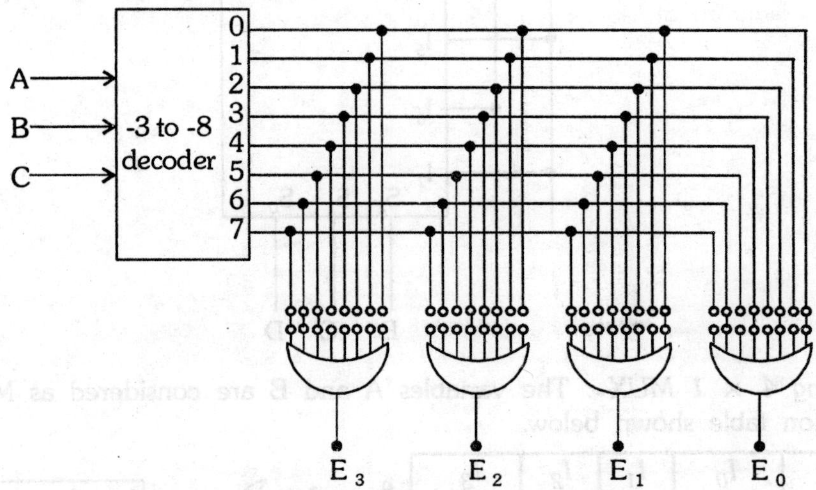

Problem 35. *Design a combination circuit using a ROM. The circuit accepts a 3–bit number and generates an output binary number equal to the square of the input number.*

Solution. The truth table is shown below.

Decimal	*Inputs*			*Outputs*						*Decimal*
	A	*B*	*C*	B_5	B_4	B_3	B_2	B_1	B_0	
0	0	0	0	0	0	0	0	0	0	0
1	0	0	1	0	0	0	0	0	1	1
2	0	1	0	0	0	0	1	0	0	4
3	0	1	1	0	0	1	0	0	1	9
4	1	0	0	0	1	0	0	0	0	16
5	1	0	1	0	1	1	0	0	1	25
6	1	1	0	1	0	0	1	0	0	36
7	1	1	1	1	1	0	0	0	1	49

$$B_0 = C$$
$$B_1 = 0$$
$$B_2 = \overline{A}B\overline{C} + AB\overline{C} = B\overline{C}$$
$$B_3 = \overline{A}BC + A\overline{B}C = C\,(A \oplus B)$$
$$B_4 = A\overline{B}\,\overline{C} + A\overline{B}C + ABC = A\,(\overline{B}\,\overline{C} + \overline{B}C + BC)$$
$$= \overline{A}\,(\overline{B}\,\overline{C} + C) = \overline{A}\,(\overline{B} + C) \quad B_5 = AB$$

$\therefore$

$$B_0 = \Sigma m\ (1, 3, 5, 7) \qquad B_3 = \Sigma m\ (3, 5)$$
$$B_1 = 0 \qquad B_4 = \Sigma m\ (4, 5, 7)$$
$$B_2 = \Sigma m\ (2, 6) \qquad B_5 = \Sigma m\ (6, 7)$$

To implement the above circuit we need a ROM of capacity $2^3 \times 6$. The below figure shows the block diagram of logic circuit. But two outputs are derived directly, so the ROM of $2^3 \times 4$ is enough.

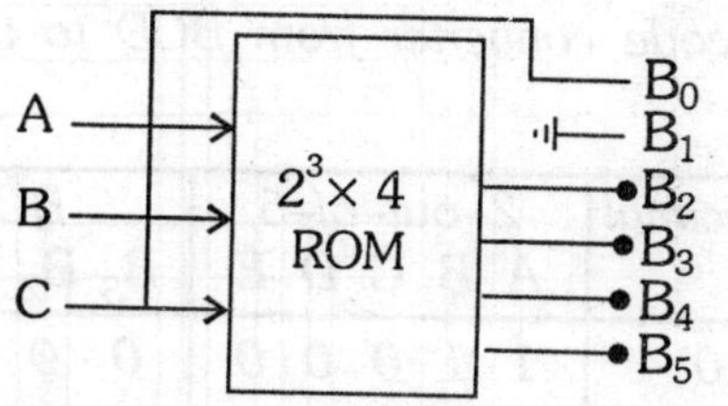

Problem 36. *Express F_1 and F_2 as functions of A, B, C and D.*

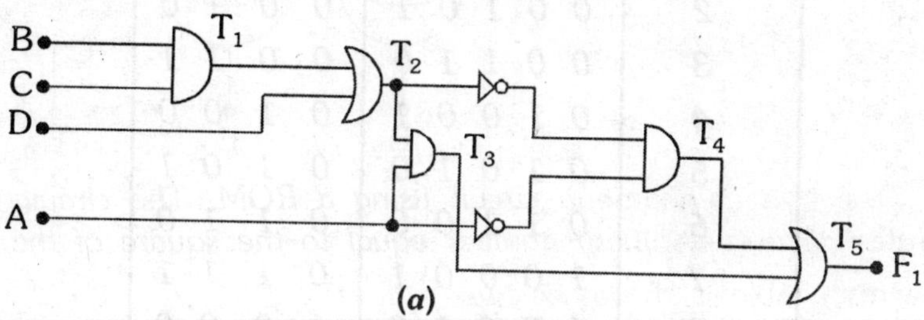

(a)

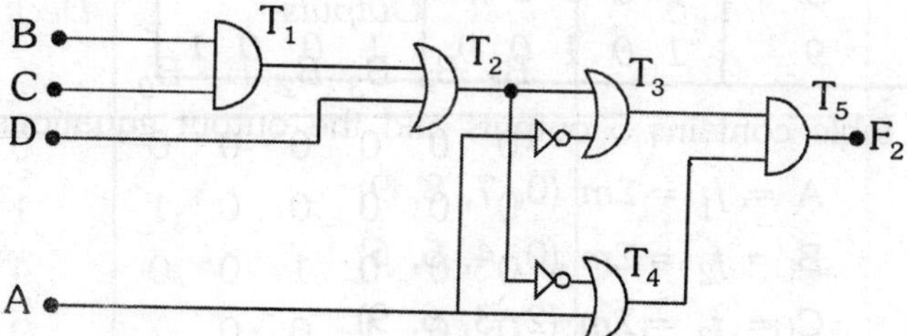

(b)

Solution. The output figure *(a)* = T_1 shown below.

$$T_1 = BC$$
$$T_2 = D + BC$$
$$T_3 = A\,(D + BC) = T_2 \,.\, A$$

$$T_4 = \overline{(D + BC)}\,.\,\overline{A} = \overline{T_2}\overline{A}$$

$$T_5 = T_3 + T_4 = A\,(D + BC) + \overline{A}\,.\,\overline{(D + BC)}$$

$$= A \odot (D + BC)$$

(*b*)

$$T_1 = BC$$

$$T_2 = D + BC$$

$$T_3 = T_2 + \overline{A}$$

$$T_4 = \overline{T_2} + A$$

$$T_5 = T_3 \,.\, T_4 = (T_2 + \overline{A})\,(\overline{T_2} + A)$$

$$= \overline{T_2}\overline{A} + T_2 A$$

$$= T_2 \odot A$$

$$= A \odot (D + BC)$$

Problem 37. *Draw the code converter from BCD to the 2–out–of–5 code. (Note. The table is shown below).*

Decimal	*2–out–of–5* $A\ B\ C\ D\ E$	*BCD* $B_3\ B_2\ B_1\ B_0$
0	*1 1 0 0 0*	*0 0 0 0*
1	*0 0 0 1 1*	*0 0 0 1*
2	*0 0 1 0 1*	*0 0 1 0*
3	*0 0 1 1 0*	*0 0 1 1*
4	*0 1 0 0 1*	*0 1 0 0*
5	*0 1 0 1 0*	*0 1 0 1*
6	*0 1 1 0 0*	*0 1 1 0*
7	*1 0 0 0 1*	*0 1 1 1*
8	*1 0 0 1 0*	*1 0 0 0*
9	*1 0 1 0 0*	*1 0 0 1*

Solution. The truth table contains 5 outputs and the output equations are

$$A = f_1 = \Sigma m\ (0, 7, 8, 9)$$

$$B = f_2 = \Sigma m\ (0, 4, 5, 6)$$

$$C = f_3 = \Sigma m\ (2, 3, 6, 9)$$

$$D = f_4 = \Sigma m\ (1, 3, 5, 8)$$

$$E = f_5 = \Sigma m\ (1, 2, 4, 7)$$

This circuit can be implemented using a 4 – to – 16 decoder.

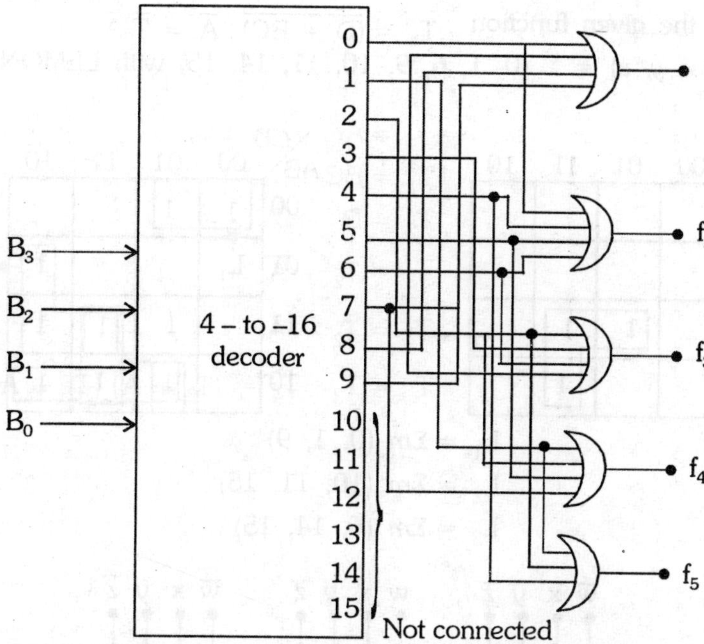

Problem 38. *Construct addition and multiplication tables for base 3.*

Solution. Assume that A and B are the two number to be add or multiply. The value of each digit may be 0, 1 or 2. The tables for addition and multiplication are given below.

A	B	Sum	Carry
0	0	0	0
0	1	1	0
0	2	2	0
1	0	1	0
1	1	2	0
1	2	0	1
2	0	2	0
2	1	0	1
2	2	1	1

Addition table

A	B	F = AB
0	0	0
0	1	0
0	2	0
1	0	0
1	1	1
1	2	2
2	0	0
2	1	2
2	2	11

Multiplication table

Problem 39. *A certain four–input gate, called a "LEMON" gate, realize the switching function LEMON (A, B, C, D) = BC (A + D) assuming that the input variables are available in both primed and unprimed form :*

Show a realization of the function

f (w, x, y, z) = Σ (0, 1, 6, 9, 10, 11, 14, 15) only with three LEMON gates and one OR gate.

Solution. Each LEMON gate covers the minterms in the L–shape.

So, the K–map for LEMON gate is shown below.

To implement the given function

$f(w, x, y, z) = \Sigma(0, 1, 6, 9, 10, 11, 14, 15)$ with LEMON gates, the K–map is shown below.

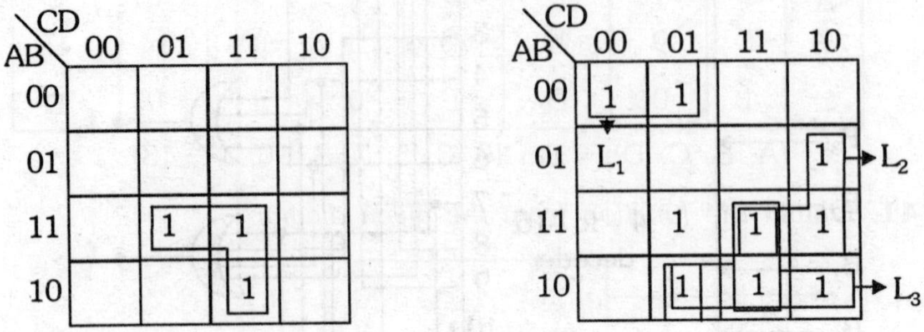

$$L_1 = \Sigma m(0, 1, 9)$$
$$L_2 = \Sigma m(10, 11, 15)$$
$$L_3 = \Sigma m(6, 14, 15)$$

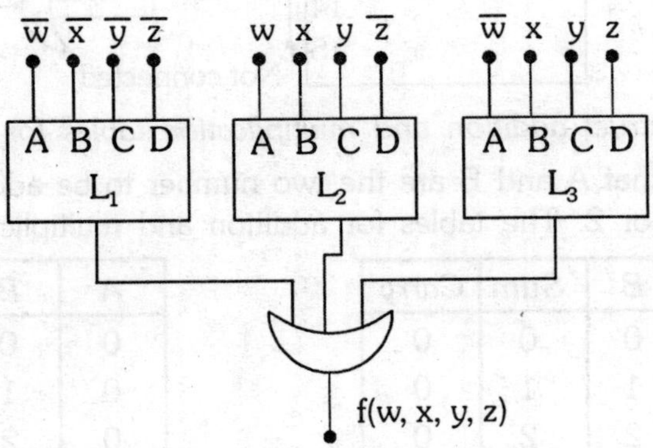

Problem 40. *Realize the function for the given circuit.*

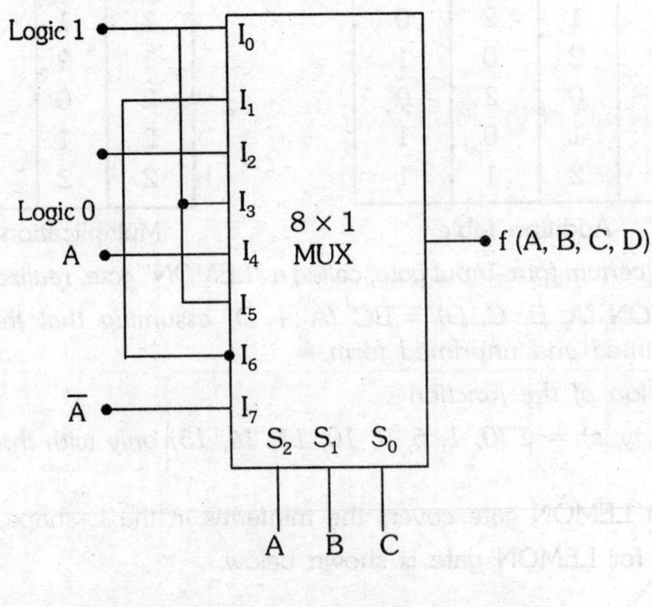

Solution. From the above figure we can draw the implementation table as below.

	I_0	I_1	I_2	I_3	I_4	I_5	I_6	I_7
$\bar{A}$	⓪	1	2	③	4	⑤	6	⑦
A	⑧	9	⑩	⑪	12	⑬	14	15
	1	0	D	1	0	1	0	$\bar{D}$

$\therefore$ $f(A, B, C, D) = \Sigma m\ (0, 3, 5, 7, 8, 10, 11, 13)$

Problem 41. *Obtain the function from the given diagram.*

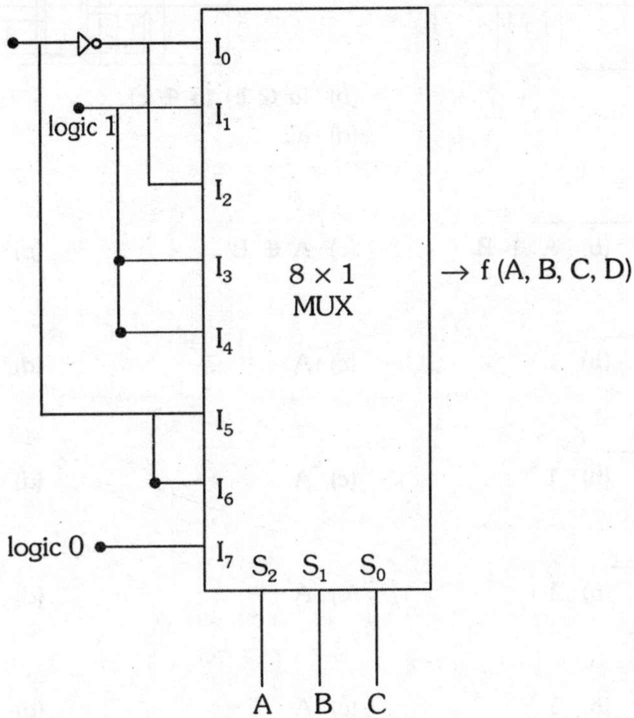

Solution. The inputs of the MUX are

$$I_0 = I_2 = \bar{D} \qquad I_5, I_6 = D$$

$$I_1 = I_3 = I_4 = 1 \qquad I_7 = 0$$

The implementation table is given below.

	I_0	I_1	I_2	I_3	I_4	I_5	I_6	I_7
$\bar{D}$	⓪	②	④	⑥	⑧	10	12	14
D	1	③	5	⑦	⑨	⑪	⑬	15
	$\bar{D}$	1	$\bar{D}$	1	1	D	D	0

$\therefore$ $f = \Sigma m\ (0, 2, 4, 6, 8, 3, 7, 9, 11, 13)$

3.17 SHORT ANSWER QUESTIONS

1. $a \oplus b =$ ________

(a) $ab + \overline{a}\overline{b}$ (b) $a'b$
(c) $\overline{a}\overline{b} + ab$ (d) $(a' + b')(a + b)$

[**Ans.** (d)]

2. $(a \oplus b)' =$ ________

(a) $\overline{a}\overline{b} + ab$ (b) $(a' + b)(a + b')$
(c) $a' \oplus b$ (d) all

[**Ans.** (d)]

3. $a(b \oplus c) =$ ________

(a) $ab \oplus ac$ (b) $(a \oplus b)(a \oplus c)$
(c) $a \oplus b \oplus c$ (d) all

[**Ans.** (a)]

4. $A \oplus B \oplus AB =$ ________

(a) $A \oplus B$ (b) $A' \oplus B$ (c) $A \oplus B'$ (d) $A + B$

[**Ans.** (d)]

5. $A \oplus A =$ ________

(a) 0 (b) 1 (c) A (d) $\overline{A}$

[**Ans.** (a)]

6. $A \oplus \overline{A} =$ ________

(a) 0 (b) 1 (c) A (d) $\overline{A}$

[**Ans.** (b)]

7. $A \oplus 0 =$ ________

(a) 0 (b) 1 (c) A (d) $\overline{A}$

[**Ans.** (c)]

8. $A \oplus 1 =$ ________

(a) 0 (b) 1 (c) A (d) $\overline{A}$

[**Ans.** (d)]

9. $A \oplus A \oplus \ldots \oplus A =$ ________ for even number of A's

(a) 0 (b) 1 (c) A (d) $\overline{A}$

[**Ans.** (a)]

10. $A \oplus A \oplus \ldots \oplus A =$ ________ for odd number of A's

(a) 0 (b) 1 (c) A (d) $\overline{A}$

[**Ans.** (b)]

11. If an input A is given to an inverter, the output will be

(a) A (b) $\overline{A}$ (c) 1 (d) 0

[**Ans.** (b)]

12. Which gate is suitable for bit comparison ?
(*a*) EX-OR gate (*b*) AND (*c*) EX-NOR (*d*) OR
[Ans. (*c*)]

13. The difference bit output of a half-subtractor is same as :
(*a*) difference bit of a full-subtractor (*b*) sum bit of a half-adder
(*c*) sum bit of a full-adder (*d*) carry bit of a half-adder
[Ans. (*b*)]

14. A full-adder can be realized by
(*a*) one half-adder, two OR gates (*b*) two-half adders, one OR gate
(*c*) two half adders, two OR gates (*d*) one half adders, one OR gate
[Ans. (*b*)]

15. The sum expression in a half adder is
(*a*) $x \oplus y$ (*b*) $(x + y)(x' + y')$ (*c*) $(c + x'y')'$ (*d*) all
[Ans. (*d*)]

16. The carry expression for a full adder is
(*a*) $x \oplus y + z$ (*b*) $xy + yz + xz$ (*c*) $xyz + \bar{x}\bar{y}z$ (*d*) none
[Ans. (*b*)]

17. The number of full adders in a 4-bit parallel adder is
(*a*) one (*b*) two (*c*) four (*d*) 16
[Ans. (*c*)]

18. The number of full adders required to add two 4-bit datas in a serial adder
(*a*) one (*b*) two (*c*) four (*d*) sixteen
[Ans. (*a*)]

19. In a look ahead carry adder, the carry of each stage (except initial stage) is a function of
(*a*) carry propagate (*b*) carry generate (*c*) both (*d*) none
[Ans. (*c*)]

20. BCD-adder requires — number of full adders.
(*a*) one (*b*) two (*c*) three (*d*) four
[Ans. (*b*)]

21. EX–3 adder requires — number of full adders
(*a*) one (*b*) four (*c*) three (*d*) two
[Ans. (*d*)]

22. The — number of NAND gates are required to realize a half subtractor
(*a*) five (*b*) six (*c*) seven (*d*) eight
[Ans. (*c*)]

23. Full subtractor can be realized by using
(*a*) one half-subtractor and one OR gate (*b*) two half-subtractors and one OR gate
(*c*) one half-subtractor and one AND gate (*d*) two half-subtractors and one AND gate
[Ans. (*b*)]

24. BCD subtractor requires

(*a*) one parallel adder (*b*) two parallel adders
(*c*) three parallel adders (*d*) four parallel adders

[Ans. (*d*)**]**

25. EX-3 subtractor can be implemented by using — number of parallel adders

(*a*) one (*b*) two (*c*) three (*d*) four

[Ans. (*b*)**]**

26. Magnitude comparator is a circuit that compares

(*a*) magnitude of two numbers (*b*) magnitude of three numbers
(*c*) magnitude of four numbers (*d*) sign of the two numbers

[Ans. (*a*)**]**

27. A decoder with '*n*'-inputs produces maximum of — number of minterms

(*a*) 2^n (*b*) $2^n–1$ (*c*) $2^{n–1}$ (*d*) $2n$

[Ans. (*a*)**]**

28. Any Boolean function can be implemented if it can be represented be — form

(*a*) SOP (*b*) POS (*c*) cannonical SOP (*d*) cannonical POS

[Ans. (*c*)**]**

29. For economical realization, decoder is used to realize a function which contains ________

(*a*) less number of don't cares (*b*) more number of don't cares
(*c*) for any case (*d*) none

[Ans. (*a*)**]**

30. A 6–to–64 decoder can be obtained by cascading of ________

(*a*) four numbers 4–to–16 decoders and one 2 : 4 decoder
(*b*) five numbers 4–to–16 decoders
(*c*) three numbers 4–to–16 decoders and two numbers 2 : 4 decoder
(*d*) can not possible

[Ans. (*a*)**]**

31. The general representation of an encoder is ________ (where $m \neq n$ and $2^n > m$)

(*a*) $2^n : m$ (*b*) $m : 2^n$ (*c*) $n : 2^n$ (*d*) $2^n : n$

[Ans. (*d*)**]**

32. MUX is ________ implementation

(*a*) AND–OR (*b*) OR–AND (*c*) NAND–OR (*d*) NOR–AND

[Ans. (*a*)**]**

33. Multiplexer is represented by________

(*a*) $2^n \times 1$ (*b*) $2^n \times n$ (*c*) $n \times 2^n$ (*d*) 1×2^n

[Ans. (*a*)**]**

34. Demultiplexer is represented by________

(*a*) $2^n \times 1$ (*b*) $2^n \times n$ (*c*) $n \times 2^n$ (*d*) 1×2^n

[Ans. (*d*)**]**

35. ROM is ________ implementation

(*a*) AND–OR (*b*) AND–OR–INVERT

(*c*) either *a* or *b* (*d*) none

[Ans. (*c*)**]**

36. PLA can be represented by________

Ans. Number of inputs, outputs and product terms

37. ROM contains ________

(*a*) programmable AND array and programmable OR array

(*b*) programmable AND array and non-programmable OR array

(*c*) non-programmable AND array and programmable OR array

(*d*) non-programmable AND array and non-programmable OR array

[Ans. (*c*)**]**

38. PLA contains ________

(*a*) programmable AND array and programmable OR array

(*b*) non-programmable AND array and programmable OR array

(*c*) non-programmable AND array and non-programmable OR array

(*d*) programmable AND array and non-programmable OR array

[Ans. (*a*)**]**

39. PAL contains ________

(*a*) programmable AND array and non-programmable OR array

(*b*) programmable AND array and programmable OR array

(*c*) non-programmable AND array and programmable OR array

(*d*) non-programmable AND array and non-programmable OR array

[Ans. (*a*)**]**

40. What is the number of inputs, outputs of a decoder that accepts 64 different input combinations?

(*a*) 5 (*b*) 6 (*c*) 64 (*d*) 8

[Ans. (*b*)**]**

41. Which of the following statements refers to decoder ?

(*a*) Only one output can be activated at any time

(*b*) Has more outputs than inputs

(*c*) Can be used as a BCD to 7 segment display interface

(*d*) All the above

[Ans. (*d*)**]**

42. Which of the following statements are belongs to an encoder ?

(*a*) Has more inputs than outputs

(*b*) We can get the priority of the inputs

(*c*) Only one input is considered at any time

(*d*) All

[Ans. (*d*)**]**

43. Which of the following statements belongs to a MUX ?

(*a*) Has more inputs than outputs

(*b*) Uses SELECT inputs

(*c*) Used in parallel to serial conversion

(*d*) All

[Ans. (*d*)**]**

44. Which of the following statements refers to a DEMUX ?

(*a*) Uses SELECT inputs

(*b*) Only one of its output can be active at any time

(*c*) Can be used to route an input signal to one of several possible outputs

(*d*) All the above

[Ans. (*d*)**]**

45. Can more than one decoder output be activated at one time ?

Ans. No

46. What is a code converter ?

Ans. Code converter takes input data represented in one type of binary code and converts it to another type of binary code.

47. How many binary outputs would a three-digit BCD–to–binary converter have ?

(*a*) 3 (*b*) 4 (*c*) 10 (*d*) 12

[Ans. (*c*)**]**

48. If in the switching circuit, switches A, B have value 0 for OFF, 1 for ON and the output Y has 0 for 0 volts, 1 for 5 V then the expression for Y is________

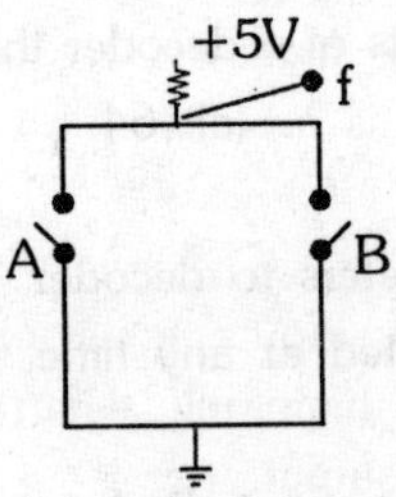

(*a*) AB (*b*) A + B (*c*) $\overline{A+B}$ (*d*) none

[Ans. (*c*)**]**

49. For the circuit, taking open as 0 and closed as 1, the expression for the circuit, Y is given by

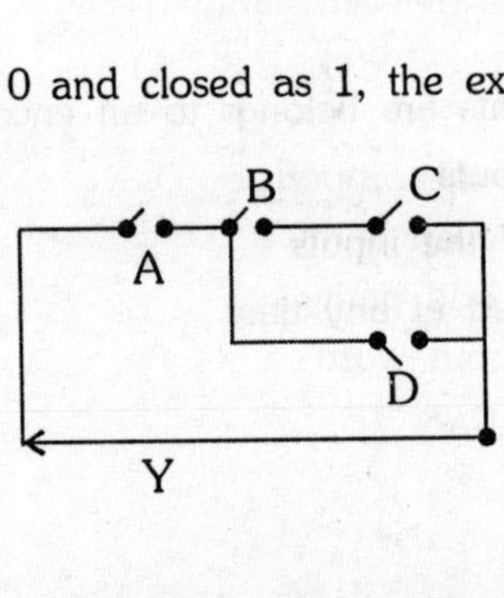

(a) A + (B+C) D (b) A + BC + D (c) A (BC + D) (d) none

[Ans. (c)**]**

50. In regard to NAND gate, the following four statements are made.

(i) It is equivalent to an AND followed by a NOT gate.
(ii) If all inputs to it are low, the output is low.
(iii) If all inputs to it are high, the output is low.
(iv) NAND operation is equivalent to OR operation of these, the only true statements are

(a) i, ii (b) i, iii (c) i, iv (d) ii, iv

[Ans. (b)**]**

51. In regard to the NOT gate, the following statements are made.

(i) It has one or more inputs.
(ii) It has only one output.
(iii) The output value is always opposite to the input.
(iv) It may have many outputs of these four statements, the two statements are only.

(a) 1, 2 (b) 1, 3 (c) 2, 3 (d) 1, 4

[Ans. (c)**]**

52. In regard to OR gate the following statements are made :

(i) It has only one input.
(ii) The output is low only when all inputs are low.
(iii) The output is high, if any one input is high.
(iv) It has one or more outputs

Of these four statements, the true statements are only.

(a) 1, 4 (b) 1, 2 (c) 2, 4 (d) 2, 3

[Ans. (d)**]**

53. The EX–OR gate for two inputs has the properties

(i) if one input is low and the other high, the output is high.
(ii) if both inputs are high, the output is high.
(iii) if both inputs are low, the output is high.
(iv) the output is 1 only when the inputs are different.

(a) 1, 4 (b) 2, 3 (c) 2, 4 (d) 3, 4

[Ans. (a)**]**

54. In regard to the EX-OR gate with many inputs, consider the following statements.

(i) If the inputs are two, the output is high only when the inputs are different.
(ii) Even parity input produces a low output.
(iii) Odd parity inputs words produce a low output.
(iv) The gate recognised only words of even parity.

(a) 1, 2 (b) 3, 4 (c) 1, 3 (d) 2, 4

[Ans. (a)**]**

55. EX-OR followed by an inverter has the property

(a) output high for input word of even parity
(b) output high for input word of odd parity
(c) output is 1's complement of input word
(d) output is 2's complement input word.

[Ans. (a)**]**

56. A half adder has
(*a*) one input and one output (*b*) two inputs and two outputs
(*c*) two inputs and one output (*d*) three inputs and two outputs
[Ans. (*b*)**]**

57. In the full adder
(*a*) the SUM is 1 when there are an odd number of input 1's
(*b*) the SUM is 1 when there are an even number of input 1's
(*c*) the carry is 1 when there are an odd number of input 1's
(*d*) the carry is 1 when there are an even number of input 1's.
[Ans. (*a*)**]**

58. In the half adder, SUM and carry are represented by ____, ____ respectively
(*a*) $x \oplus y, xy$ (*b*) $xy, x \oplus y$ (*c*) $x + y, xy$ (*d*) $x + y, x \oplus y$
[Ans. (*a*)**]**

59. In sign magnitude numbers,
(*a*) the MS bit stands for sign (*b*) the LS bit is a sign
(*c*) the MS bit is a '0' (*d*) the MS-bit is a '1'
[Ans. (*a*)**]**

60. The value of 2^5 in octal system is
(*a*) 40 (*b*) 20 (*c*) 400 (*d*) 200
[Ans. (*a*)**]**

61. The expression A (A + B) = ________
(*a*) A + B (*b*) A (*c*) B (*d*) AB
[Ans. (*b*)**]**

62. If binary multiplication is to be implemented by using a two input gate, then it should be________
(*a*) EX-OR gate (*b*) NOR gate (*c*) AND gate (*d*) OR gate
[Ans. (*c*)**]**

63. The number of Boolean functions that can be generated by *n*-variables is equal to____
(*a*) 2^n (*b*) 2^{2^n} (*c*) 2^{n-1} (*d*) -2^n
[Ans. (*b*)**]**

64. The Boolean expression $A \oplus B \oplus A$ is equal to _____
(*a*) $AB + \overline{A}B$ (*b*) $\overline{A}B + A\overline{B}$ (*c*) B (*d*) $\overline{A}$
[Ans. (*c*)**]**

65. The Boolean expression $\overline{A}B + A\overline{B} + AB$ = ________
(*a*) A + B (*b*) $\overline{AB}$ (*c*) $\overline{A+B}$ (*d*) AB
[Ans. (*a*)**]**

66. 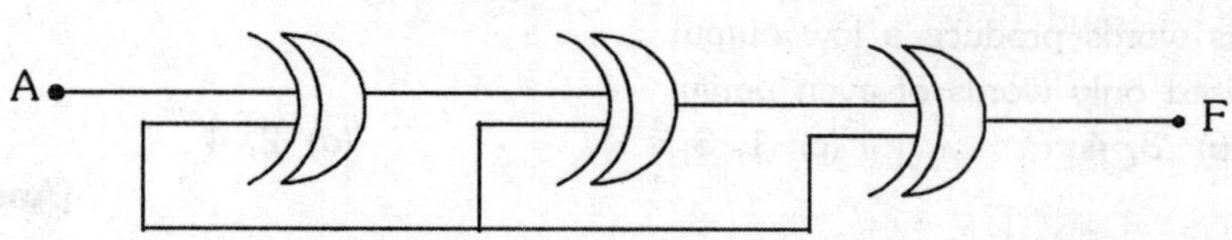

For this circuit the output F = ________
(*a*) 0 (*b*) 1 (*c*) A (*d*) $\overline{A}$
[Ans. (*a*)**]**

67. The minimum number of NAND gates required to implement the function $F = (\overline{x} + \overline{y})(z + w)$
(*a*) 3 (*b*) 4 (*c*) 5 (*d*) 6
[Ans. (*b*)**]**

68. The operation which is commutative but not associative is ______

(*a*) AND (*b*) OR (*c*) EX-OR (*d*) NAND

[Ans. (*d*)**]**

69. The dual function of A $(\overline{B}C + BC + B\overline{C})$ is ________

(*a*) $A + (\overline{B} + C)(B + C)(B + \overline{C})$ (*b*) $\overline{A}(\overline{B}C + \overline{B}\,\overline{C} + B\overline{C})$

(*c*) $A\overline{B}\,\overline{C} + ABC + AB\overline{C}$ (*d*) $A + \overline{B}\,\overline{C} + BC + B\overline{C}$

[Ans. (*a*)**]**

70. An XOR gate with 6 terms contains _____ number of minterms

(*a*) 6 (*b*) 12 (*c*) 64 (*d*) 32

[Ans. (*d*)**]**

3.18 REVIEW QUESTIONS

1. Briefly explain the operation of a carry look ahead adder.

2. Construct a full adder using half adders. Differentiate between serial adder and parallel adder.

3. Write short note on magnitude comparator.

4. Design a combinational circuit that accepts a three-bit binary number and generates on output binary equal to the cube of the input number.

5. Realize the full adder using minimum number of NAND gates.

6. Implement the logic function using (*a*) 16 × 1 MUX (*b*) 8 × 1 MUX (*c*) 4 × 1 MUX and

$$f(A, B, C, D) = \Sigma m\,(1, 2, 4, 5, 14, 15)$$

7. Define a multiplexer and show how a MUX may be used as a sequence data selector.

8. Implement the following functions using 4-to-16 line decoder

(*a*) $F(A, B, C, D) = \Sigma m\,(0, 1, 2, 5, 7, 9, 11, 14)$

(*b*) $F(A, B, C, D) = \Sigma m\,(2, 3, 4, 5, 10, 11, 12)$

9. What is a decoder and how is it different from the demultiplexer ?

10. What are encoders and explain 1 × 4 encoder with logic diagram *i* functional table ?

11. Explain the working of MUX and DEMUX.

12. Realize the functions using a 6 × 10 × 4 PLA

(*a*) $F_1(A, B, C, D, E, F) = \Sigma m\,(0, 1, 7, 8, 9, 10, 11, 15, 19, 23, 27, 31, 32, 33, 35, 39, 40, 41, 47, 63)$

(*b*) $F_2 = \Sigma m\,(8, 9, 10, 11, 12, 14, 21, 25, 40, 41, 42, 43, 44, 46, 57, 59)$

13. Write short notes on ROM.

14. Write short notes on PAL and PLA.

15. Realize the given functions using PAL

(*a*) $f_1 = \overline{A}B\overline{D}F + C\overline{D}E\overline{F} + AC\overline{E}F + A\overline{B}CD\overline{F}$

(*b*) $f_2 = A\overline{B}CD\overline{E} + ABC\overline{C}\,\overline{D}E$

16. Write short notes on ROM, PROM and EPROMS.

17. Design a switching circuit that converts the 4-bit binary code into a 4-bit gray code using ROM.

18. Assume that the exclusive OR gate has a propagation delay of 20 *n* sec and that the AND (or) OR gates have a propagation delay of 10 *n* sec. What is the total propagation delay time in the 4-bit adder ?

19. Design a combination circuit that generates 10's complement of a BCD digit.

20. Draw the 2-to-4-line decoder with only NOR gates.

21. Specify the size of a ROM needed for the following combinational circuits.

(*a*) A binary multiplexer that multiplies two 4-bit numbers.

(*b*) A 4 bit adder-subtractor

(*c*) ABCD-to-seven segment decoder.

22. Implement following four Boolean functions using ROM, PAL and PLA

(*a*) f_1 (A, B, C) = Σ (1, 2, 4, 6)

(*b*) f_2 (A, B, C) = Σ (0, 1, 6, 7)

(*c*) f_3 (A, B, C) = Σ (2, 6)

(*d*) f_4 (A, B, C) = Σ (1, 2, 3, 5, 7)

23. Design a combinational circuit that converts a decimal digit from the 2421 code to the $8\ 4\ \overline{2}\ \overline{1}$ code.

24. Manipulate the following Boolean expression in such a way so that it can be implemented using EX-OR and AND gates only.

$$f = A\overline{B}C\overline{D} + \overline{A}BC\overline{D} + A\overline{B}\,\overline{C}D + \overline{A}B\overline{C}D$$

25. Design a combinational circuit which is defined by the following three Boolean functions with decoder and external gates.

$$F_1 = \overline{A}\,\overline{B}\,\overline{C} + AC$$

$$F_2 = A\overline{B}\,\overline{C} + \overline{A}B$$

$$F_3 = \overline{A}\,\overline{B}C + AB$$

26. Construct a 5 × 32 decoder with four 3 × 8 decoders and one 2 × 4 decoder.

27. Design a combinational circuit that compares two 4-bit numbers A and B to check if they are equal. The circuit has one output *f*, so that $f = 1$, if A = B and $f = 0$ if A ≠ B.

28. Draw the logic diagram of an octal-to-binary priority encoder.

29. An 8 × 1 MUX has inputs A, B, and C connected to the selection inputs. The data inputs I_0 to I_7 are as follows : $I_1 = I_2 = I_7 = 0$, $I_3 = I_5 = 1$; $I_0 = I_4 = D$; and $I_6 = \overline{D}$. Determine the Boolean function that the MUX implements.

30. An 8 × 1 MUX has inputs A, B and C connected to the selection inputs S_2, S_1 and S_0 respectively. The data inputs I_0 through I_7 are as follows : $I_1 = I_2 = I_7 = 0$, $I_3 = I_5 = 1$; $I_0 = I_4 = D$; and $I_6 = \overline{D}$. Determine the Boolean expression that the MUX implements.

31. How does an encoder differ from a decoder ?

32. Explain the operation of a nibble multiplexer.

33. Design a parallel binary multiplexer that multiplies a 4-bit number B = $B_3B_2B_1B_0$ by a 3-bit number A = $A_2A_1A_0$ to form a product C = $C_6C_5C_4C_3C_2C_1C_0$.

34. How does a priority encoder differ from an ordinary encoder ?

35. Design a parity generator and checker circuit for an 5-bit data with even priority.

4

Introduction to Sequential Circuits

4.1 INTRODUCTION

So far, we have studied the analysis and design of combinational circuits, whose outputs at any instant of time are entirely dependent upon the inputs present at that time. It constitutes only a part of digital systems. There are many applications in which digital outputs are required to be generated in accordance with the present inputs and previous state of the circuit. This requirement cannot be satisfied using a combinational logic system. These applications require outputs to be generated that are not only dependent on the present input conditions, but they also depend upon the past history of these inputs. The past history is provided by feedback from the outputs back to the input. These circuits include the memory elements and are called as "sequential circuits".

The block diagram of a sequential circuit is shown in Fig. 4.1. It consists of a combinational circuit to which memory elements are connected to form a feedback path.

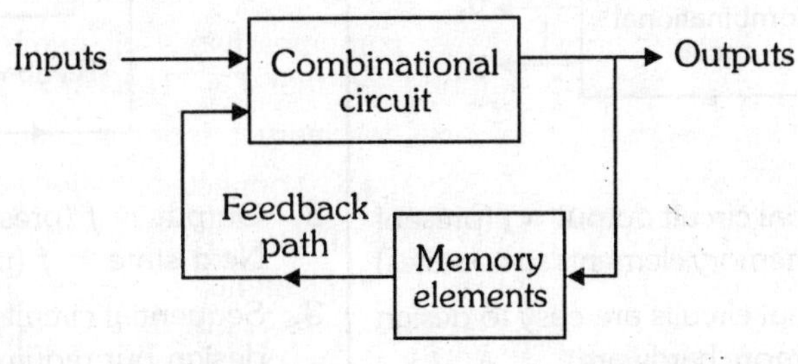

Fig. 4.1 Block diagram of a sequential circuit.

The memory elements are devices capable of storing binary information within them. The binary information stored in the memory elements at any given time defines the "state" of the sequential circuit. The sequential circuit receives the binary information from external inputs. These inputs and the present state of memory elements determine the binary value of the output of the circuit. They also determine the condition for changing the state in the memory elements. Thus, the next state of the memory elements is also a function of the external inputs,. and the present state.

A sequential circuit is specified by a time sequence of inputs, outputs and internal states. The sequential circuits can be classified in two ways depending on the timing of their signals; They are :

1. Synchronous sequential circuits
2. Asynchronous sequential circuits.

A synchronous sequential circuit is a system whose behaviour can be defined from the knowledge of its signals at the discrete instants of time, *i.e.*, signals can affect the memory elements only at discrete instants of time. These are also called as "clocked sequential circuits."

In asynchronous sequential circuits, change in input signals can affect memory element at any instant of time; *i.e.*, the behaviour of an asynchronous sequential circuit depends upon the order in which its input signals change and can be affected at any instant of time. These are also called as unclocked sequential circuits. The memory elements used in both cases are flip-flops which are capable of storing 1-bit binary information. A flip-flop circuit has two outputs, one for the normal value and other for the complement value of the bit stored in it. Before going to study the operation and types of flip-flops it is better to introduce the comparison between combinational and sequential circuits (see Table 4.1) and comparison between synchronous and asynchronous sequential circuits (see Table 4.2).

Table 4.1

Sl. No.	*Combinational circuits*	*Sequential circuits*
1.	A circuit whose output is dependent only on the inputs at that instant (or) The circuit that does not contain any memory elements.	1. A circuit whose output depends not only on the present inputs, but also on the past history of the inputs (or) A circuit that contains at least one memory element.
	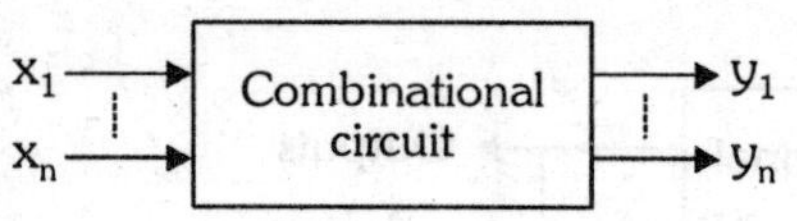	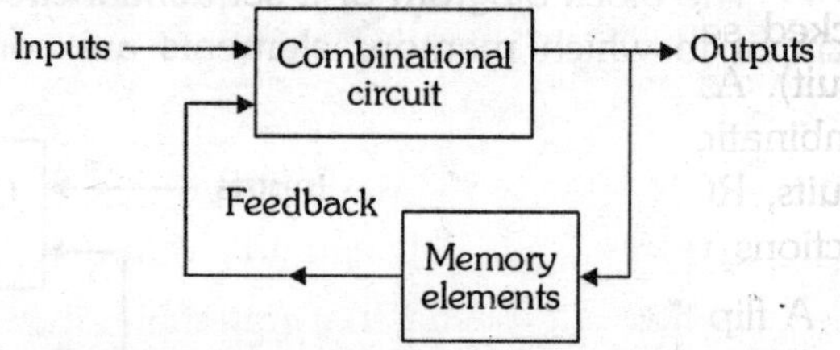
2.	Combinational circuit output = *f* (present inputs) (No memory elements so no states)	2. Output = *f* (present input, present state) Next state = *f* (present input, present state)
3.	Combinational circuits are easy to design but require more hardware.	3. Sequential circuits are comparatively harder to design but require less hardware.
4.	Combinational circuits are faster in speed, because, delay between input and output is due to propagation delay of gates.	4. Sequential circuits are slower than the combinational circuits because of memory elements (or) delay elements.
5.	More expensive circuit.	5. Cheaper circuit.
6.	Designer has less flexibility since the output depends only on the present inputs.	6. Designer has more flexibility because, the output depends on both present input and past history of input.
7.	Example : Parallel adder.	7. Example : Serial adder.
8.	The behaviour is defined by the set of output functions only.	8. The behaviour is defined by the set of output functions and next state functions.

Table 4.2. Comparison between asynchronous and synchronous sequential circuit

Synchronous sequential circuits	*Asynchronous sequential circuits*
1. A synchronous sequential circuit is a system whose behaviour is defined from the knowledge of its signals at discrete instants of time, *i.e.*, the input signals can affect the memory elements upon the activation of the clock pulse.	1. The behaviour of asynchronous sequential circuit depends upon the order in which the inputs change, and the state of the circuits can be affected at any instant of time, *i.e.*, the change in the input signals can affect memory element at any instant of time.
2. Here the memory elements are clocked flip-flops.	2. The memory elements are either unclocked flip-flops (or) time delay elements.
3. Synchronization is employed by the help of clock pulses.	3. Here no synchronisation, hence it is a combinational circuit with feedback.
4. The maximum operating speed of clock depends on time delays involved.	4. Because of absence of clock, asynchronous circuits can operate faster than synchronous circuits.
5. Easier to design.	5. More difficult to design.

4.2 FLIP-FLOPS

We know that a sequential circuit needs at least one memory element to store the present state of the circuit. This information must be stored at least one clock signal period (in case of clocked sequential circuit) or until the input signals change (in case of unclocked sequential circuit). As soon as we said memory device what we will get into mind is ROM. But ROM is combinational, according to the above discussion of storage element needed in sequential circuits, ROM has no memory. (The memory of a ROM refers the fact that it "memorizes" the functions relationship between the output variables and the input variables).

A flip-flop is a circuit that maintains the binary state indefinitely until directed by an input signal to switch states. Therefore, the storage device must retain the stored information until we are ready to use it. The basic unit for this storage is a flip-flop (abbreviated as FF). It is a device with two stable states *i.e.*, output is either to 0V (logic 0) or + 5 V d.c (logic 1). A flip-flop maintains its output state (either 0 or 1) until directed by an input signal to change its state. It always stores 1-bit information. So it is also called 1-bit memory unit.

4.2.1 Basic Building Block

Now we will see a storage device capable of storing 1-bit. The basic digital memory circuit is obtained by cross-coupling of two NOT gates N_1 and N_2 as shown in Fig. 4.2.

The circuit may be in $y = 0$ ($\bar{y} = 1$) state called '0' state or it may be in $y = 1$ ($\bar{y} = 0$), called '1' state. These states are stable and consistant with the interconnections.

When $y = 0$ (assume) then A = 1, because through NOT gate y becomes 0.

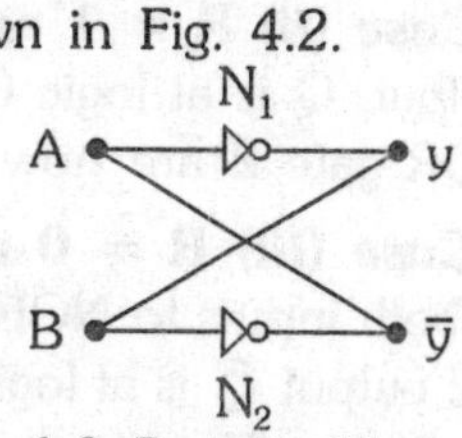

Fig. 4.2 Basic building block.

$\overline{y} = 1$, then B = 0, as it passes through NOT gate $\overline{y} = 1$.

These outputs are ture for long time. Since this information is locked or latched, hence this block is called as a latch or 1-bit memory element or 1-bit storage cell. The output of each gate is connected to the input of the other, and this feedback combination is called a cross-coupled connection. Since the flip-flop has two states, it is also called a binary (or) bistable multivibrator. This circuit (latch) does not provide any mechanism to write data into it. Hence there is a necessity to modify the basic building block.

There are many types of flip-flops. The major differences among various types of flip-flops are in the number of inputs they process and the manner in which the inputs affect their output state. The types of flip-flops are :

(1) RS-flip-flop (2) JK-flip-flop (3) D-flip-flop (4) T-flip-flop.

4.2.2 RS-Flip-flop

The simplest type of flip-flop is the Reset-Set flip-flop (RS). It can be constructed from either two NAND gates or two NOR gates.

RS-flip-flop using NOR gates. The Fig. 4.3 shows the RS flip-flop using two NOR gates. The two NOR gates are cross coupled so that output of NOR gate 1 is connected to one of the inputs of NOR gate 2 and vice-versa. The flip-flop has two outputs Q and $\overline{Q}$, and two inputs, set and reset.

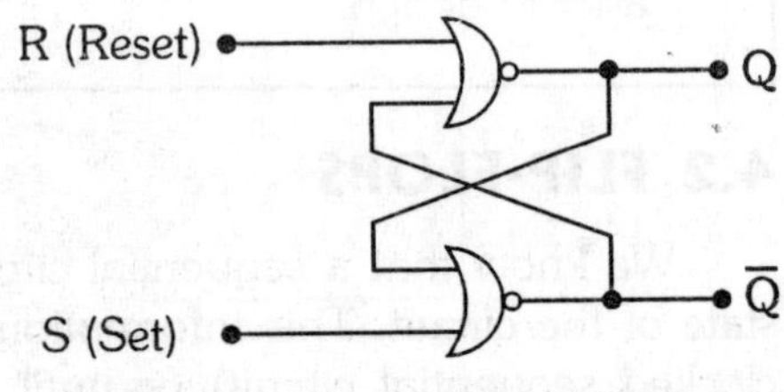

Fig. 4.3 RSFF using NOR gates.

As we know that, logic 1 at any input of a NOR gate forces its output to a logic 0. Let us see the operation of this circuit for various input possibilities.

***Case (i)* R = 0 and S = 1.** In this case, S input of NOR gate 2 is at logic 1, hence its output, $\overline{Q}$ is at logic 0 which indicates that both inputs of NOR gate 1 are at logic 0. So that its output, Q is at logic 1, as shown in Fig. 4.3(*a*).

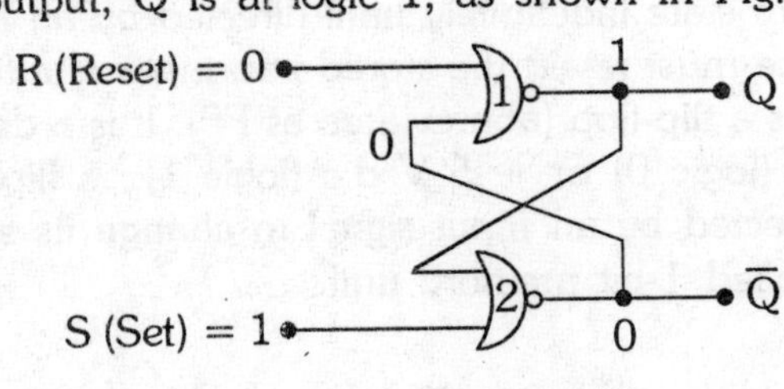

Fig. 4.3(*a*)

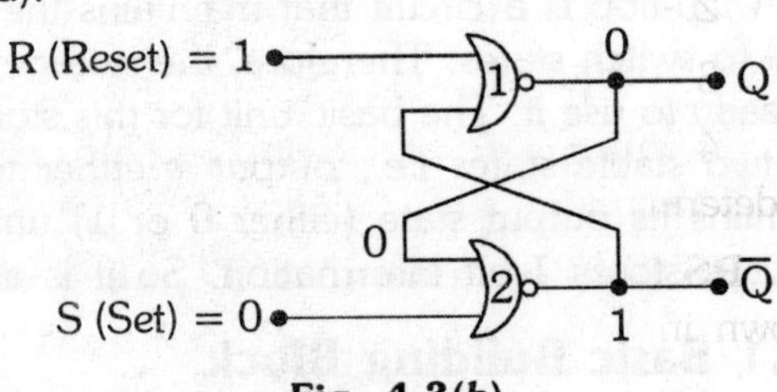

Fig. 4.3(*b*)

***Case (ii)* R = 1 and S = 0.** In this case, R input of the NOR gate 1 is at logic 1, hence its output, Q is at logic 0. This indicates (because of cross couple connection) that both inputs of NOR gate 2 are now at logic 0, so its output, $\overline{Q}$ is at logic 1, as shown in Fig. 4.3 (*b*).

***Case (iii)* R = 0 and S = 0.** Let us assume that initially Q = 1 and $\overline{Q}$ = 0. With $\overline{Q}$ = 0, both inputs to NOR gate 1 are at logic 0. So its output, Q is at logic 0. With Q = 1, NOR gate 2 output $\overline{Q}$ is at logic 0. Thus the output is same as initial value. This is same for Q = 0 and $\overline{Q}$ = 1 also. The diagram shown in Fig. 4.3 (*c*).

***Case (iv)* R = 1 and S = 1.** When both inputs are at logic 1, they force the outputs of both NAND gates to logic '0' *i.e.*, Q = 0 and $\overline{Q}$ = 0. (Since we know that Q is a logic binary variable and $\overline{Q}$ is its complement, then Q ≠ $\overline{Q}$ for any condition).

So we call this condition (or) state as an "indeterminate state or prohibited state". Thus in normal operation, this condition must be avoided. This is shown in Fig. 4.3(*d*).

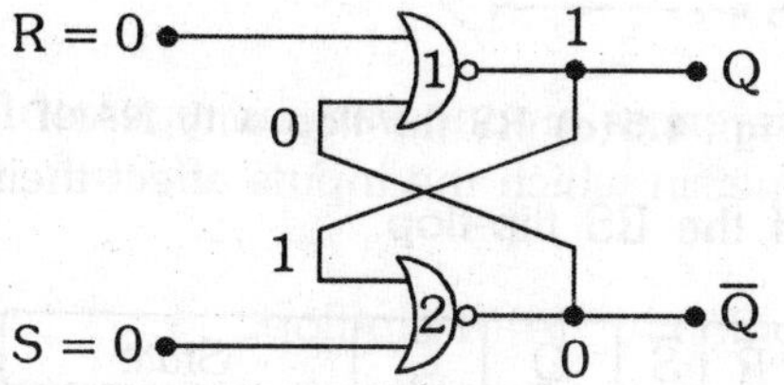

Fig. 4.3(*c*)

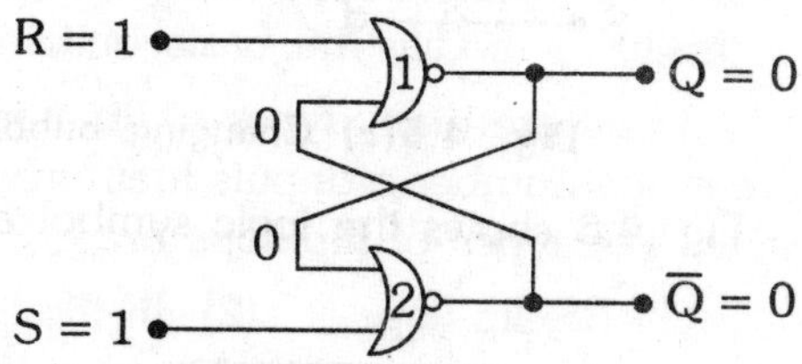

Fig. 4.3(*d*)

The symbol of RS flip-flop is shown in Fig. 4.4 (*a*) and its truth table is shown in Fig. 4.4(*b*).

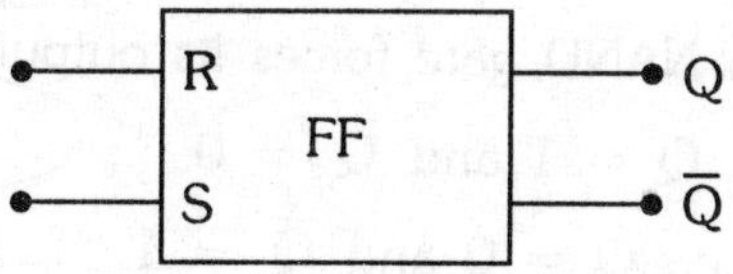

Fig. 4.4(*a*) Symbol.

S R	Q $\overline{Q}$	State
0 0	NC NC	No change
0 1	0 1	Set
1 0	1 0	Reset
1 1	φ φ	Indeterminate

Fig. 4.4(*b*) Truth table.

NC–No change

Summary. 1. When both inputs are low, the output does not change and flip-flop remains latched in its last state. This state is also called as "no change or inactive state".

2. When R = 0, S = 1, The Q output of flip-flop is at logic 1 (Set state)

3. When R = 1, S = 0, the $\overline{Q}$ output of flip-flop is at logic 1 (Reset state)

4. When R and S inputs both are high, output is unpredictable. This state is called "indeterminate state".

RS flip-flop using NAND gates. We can construct the RS flip-flop using NAND gates as shown in Fig. 4.5.

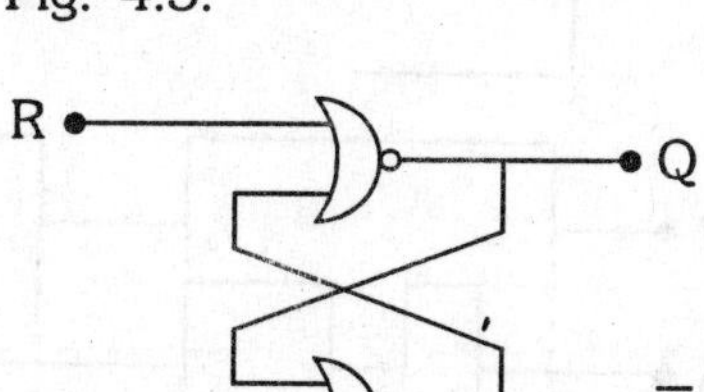

Fig. 4.5(*a*) Basic circuit.

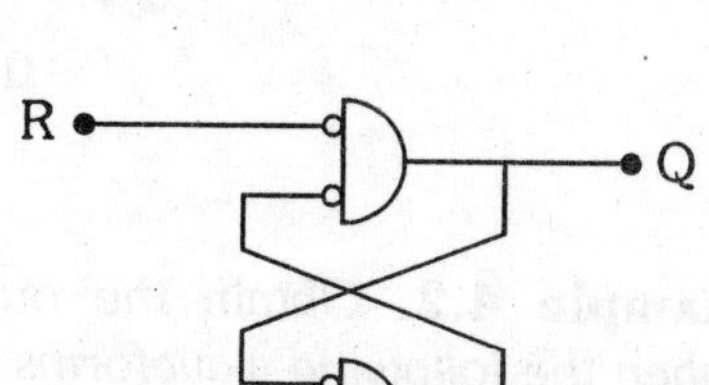

Fig. 4.5(*b*) With bubbled AND gate.

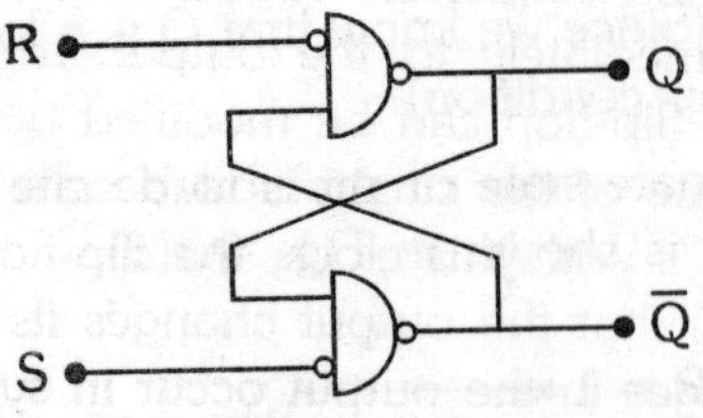

Fig. 4.5(c) Changing bubble.

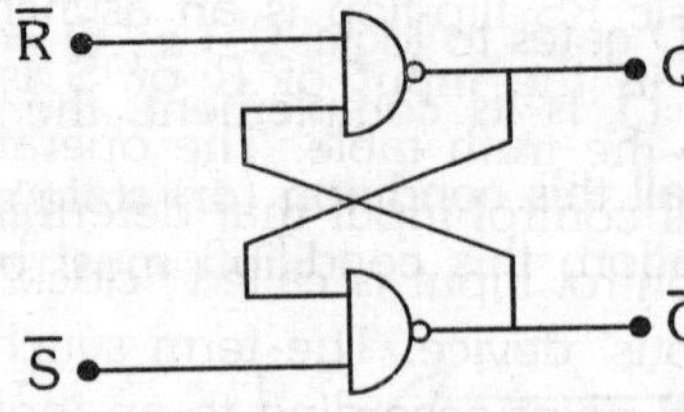

Fig. 4.5(d) RS flip-flop with NAND gates.

Fig. 4.6 shows the logic symbol and truth table of the RS flip-flop.

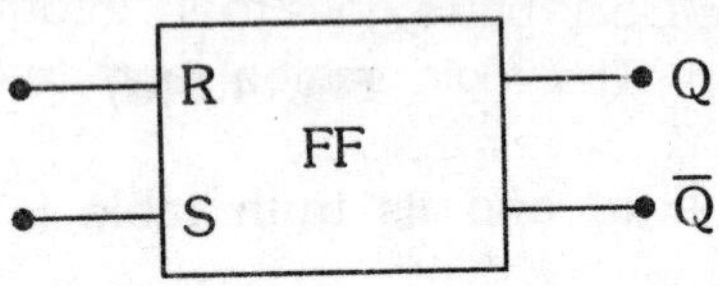

Fig. 4.6(a) Logic symbol.

$\overline{R}$	$\overline{S}$	Q	$\overline{Q}$	*State*
1	1	NC	NC	No change
1	0	1	0	Set
0	1	0	1	Reset
0	0	1	1	Undeterminate

Fig. 4.6(b) Truth table.

Operation : We know that a low on any input to a NAND gate forces its output high.

(*i*) If $\overline{S}$ = 1, $\overline{R}$ = 0, it will set the $\overline{R}\,\overline{S}$ flip-flop *i.e.*. Q = 1 and $\overline{Q}$ = 0.

(*ii*) If $\overline{S}$ = 0, $\overline{R}$ = 1, it will reset the $\overline{R}\,\overline{S}$ flip-flop *i.e.*, Q = 0 and $\overline{Q}$ = 1.

(*iii*) If $\overline{S}$ = $\overline{R}$ = 1, the flip-flop will remain in its previous state.

(*iv*) If $\overline{S}$ = $\overline{R}$ = 0, the flip-flop output is unpredictable.

Example 4.1. *The following waveforms are applied to the inputs of RS flip-flop. Determine output waveform. Assume initially Q = 1 (see Fig. 4.7).*

Solution. See Fig. 4.7

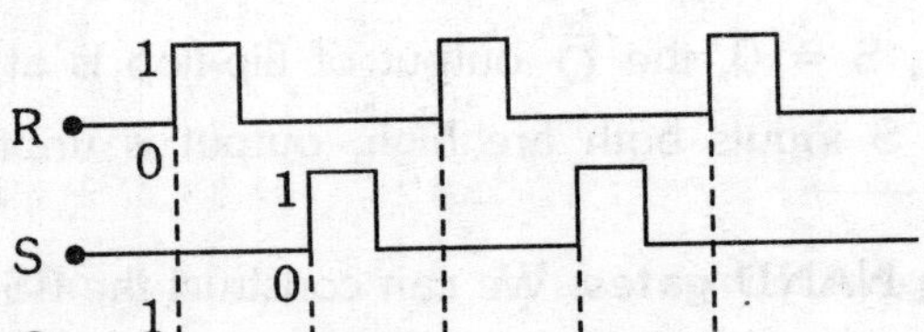

Fig. 4.7

Example 4.2. *Obtain the output waveform, when the following waveforms are applied to the RS flip-flop. (Assume initially Q = 1) (see Fig. 4.7 (a))*

Solution. See Fig. 4.7(a)

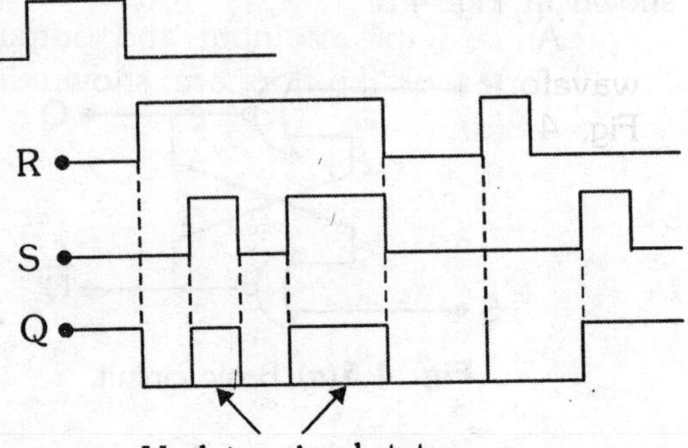

Fig. 4.7(a)

4.2.3 Clocked RS-Flip-Flop

The basic RS flip-flop is an asynchronous transparent sequential circuit. This means that any change in the input of R or S is transmitted immediately to the output at Q and $\overline{Q}$ according to the truth table. The operation of the basic flip-flop can be modified by providing an additional control input that determines when the state of the circuit is to be changed. This additional control input is called "clock" or "clock pulse". With this clock, the flip-flop is called a 'synchronous' device. The term synchronous indicates that the output changes its state only at a specified point according to an input, *i.e.*, the changes in the output occur in synchronous with the clock. By adding gates to the input of the basic circuit, the flip-flop can be made to respond to input levels during the occurrence of a clock pulse.

An RS-flip-flop with a clock pulse is shown in Fig. 4.8(*a*). It consists of basic NOR flip-flop circuit and two AND gates at the input. The AND gates remain at '0' as long as the clock pulse (CP) is '0', regardless of the R and S inputs. When the clock pulse goes to 1, information from R and S inputs is allowed to reach the basic flip-flop. The logic symbol and truth table for clocked RS-flip-flop are shown in Fig. 4.8 (*b*) and 4.8 (*c*) respectively.

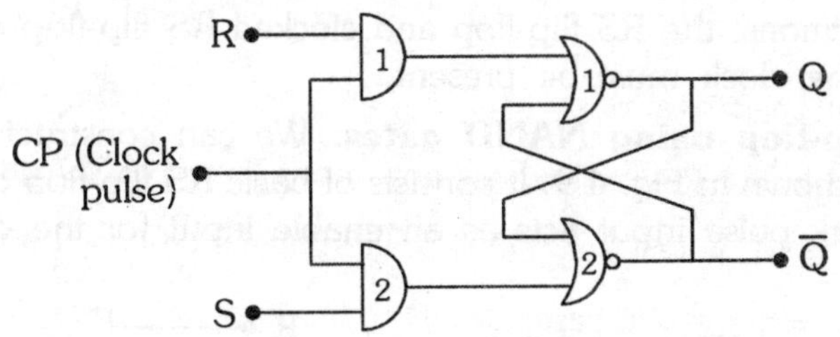

Fig. 4.8(*a*)

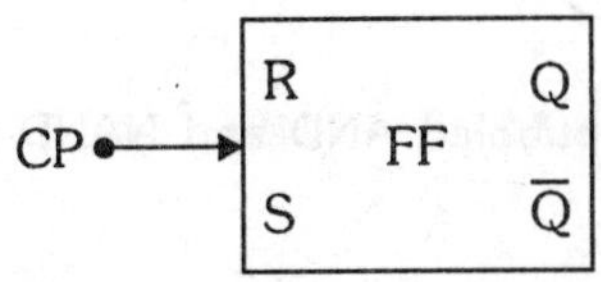

Fig. 4.8(*b*) Logic symbol.

CP	*R*	*S*	*Q*	$\overline{Q}$	*State*
0	0	0	NC	NC	No change
0	0	1	NC	NC	No change
0	1	0	NC	NC	No change
0	1	1	NC	NC	No change
1	0	0	NC	NC	No change
1	0	1	1	0	Set
1	1	0	0	1	Reset
1	1	1	×	×	Indeterminate

Fig. 4.8(*c*) Truth table.

An example of input and output waveforms of flip-flop are shown in Fig. 4.8(*d*).

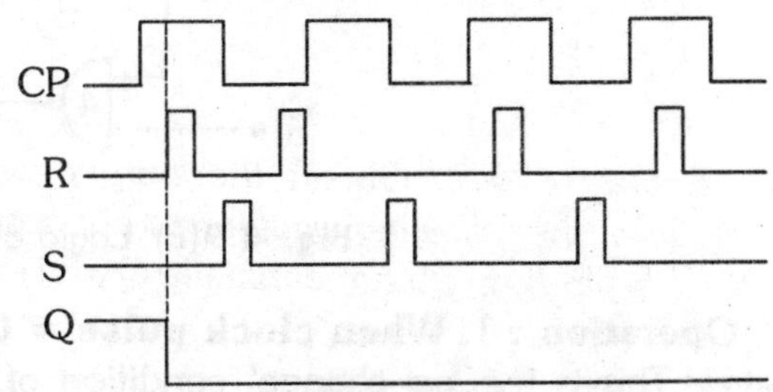

Fig. 4.8(*d*)

Operation : 1. When clock is absent (or) CLK = 0, the circuit will work as an RS-flip-flop.

2. In the presence of clock *i.e.*, CLK = 1.

(*a*) **If R = 0 and S = 0 and assume that Q = 1.** The output of both AND gates are '0' which are given as one of inputs to the NOR gate. If Q = 1 and $\overline{Q}$ = 0, the output of NOR gate 1 = 1 and NOR gate 2 = 0, *i.e.*, same as the previous values. This state is called 'no change' state.

(*b*) **If R = 1, S = 0.** The output of AND gate 1 = 1 and AND gate 2 is '0'. With these inputs, the NOR gate 1 output = 0 and NOR gate 2 output is 1. This state is called "reset" state.

(*c*) **If R = 0, S = 1.** The output of AND gate 1 is '0' and AND gate 2 is 1. With these inputs the NOR gate 1 output = 1 and NOR gate 2 output = 0. This state is called "set state".

(*d*) **If R = 1, S = 1.** The output of both AND gates is 1. With these inputs, irrespective of other input, the output of both NOR gates is '0', *i.e.*, Q = 0 and $\overline{Q}$ = 0, which is an invalid output. This state is called 'indeterminate state'.

For all the 4-conditions, the RS flip-flop and clocked RS flip-flop outputs are same, but in clocked RS-flip-flop, the clock must be present.

Clocked RS flip-flop using NAND gates. We can construct the clocked RS-flip-flop using NAND gates as shown in Fig. 4.9. It consists of basic RS flip-flop circuit and two additional NAND gates. The clock pulse input acts as an enable input for the other two inputs.

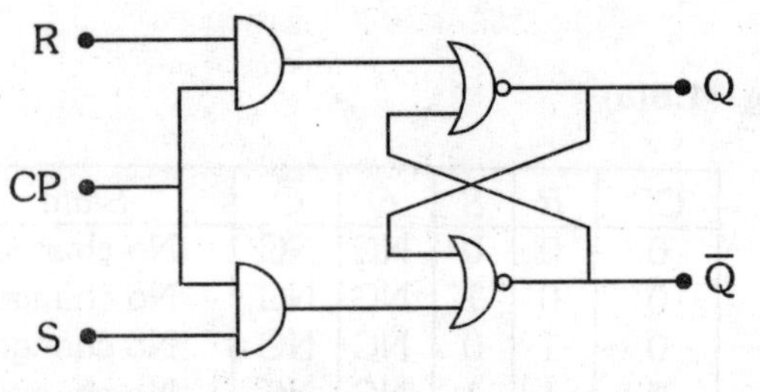

Fig. 4.9(*a*) Basic circuit.

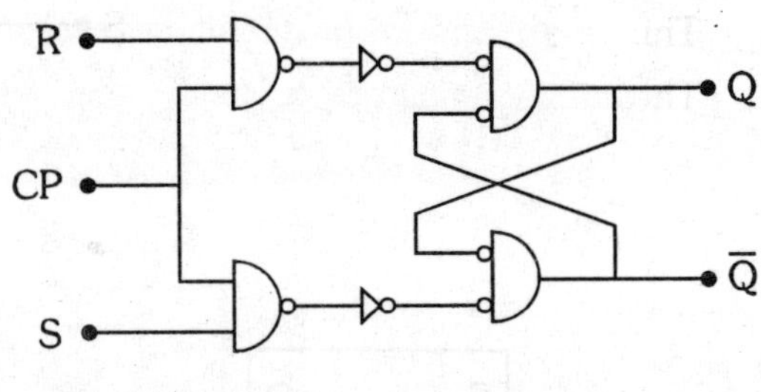

Fig. 4.9(*b*) With bubbled AND and NAND gates.

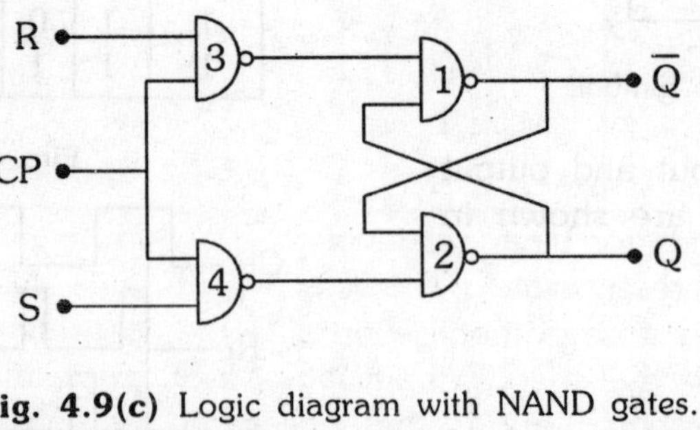

Fig. 4.9(*c*) Logic diagram with NAND gates.

Operation : 1. **When clock pulse = 0.** The output of NAND gates 3 and 4 stay at logic 1 level. This is the 'no change' condition of the basic flip-flop.

2. **When clock pulse (CP) = 1.** In this condition, the S or R input is allowed to reach the output.

(*a*) **When R = 0 and S = 0 (Assume Q = 1 and $\overline{Q}$ = 0).** In this condition, the output of NAND gates 3 and 4 are at logic 1, with Q = 1. The both inputs of NAND gate 1 are at logic 1, then its output *i.e.*, $\overline{Q}$ = 0. The NAND gate 2 output *i.e.*, Q = 1. Hence the output remains same. This state is called "no change" state.

(*b*) **R = 1 and S = 0.** In this condition, the output of NAND gate 3 is at logic 0 and output of NAND gate 4 is at logic 1. With these inputs, the output of NAND gate 1 is at logic 1 *i.e.*, $\overline{Q}$ = 1 and the output of the NAND gate 2 is at logic 0 *i.e.*, Q = 0. This state is called "reset" state.

(*c*) **If R = 0 and S = 1.** In this condition, the output of NAND gate 4 is at logic 0 and output of NAND gate 3 is at logic 1. With these inputs, the output of NAND gate 1 and NAND gate 2 are logic 0 and logic 1 respectively. This state is called "set" state.

(*d*) **If R = 1 and S = 1.** In this condition, the output of both NAND gate 3 and NAND gate 4 are at logic O. With these inputs, the output of both NAND gate 1 and NAND gate 2 is at logic 1. *i.e.*, Q = 1 and $\overline{Q}$ = 1, which is invalid. This state is called 'indeterminate state'. When CP input goes back to 0 (while S and R are maintained at logic 1), it is not possible to determine the next state, as it depends on whether the output of gate 3 or gate 4 goes to 1 first. This indeterminate state makes the circuit difficult to manage and it is seldom used in practice.

This is an important circuit because all other flip-flops are constructed from it.

The characteristic table of the flip-flop is shown in Table 4.3.

Table 4.3. Truth table for clocked RS flip-flop

Q(t)	*R*	*S*	*Q (t + 1)*
0	0	0	0
0	0	1	1
0	1	0	0
0	1	1	Indeterminate
1	0	0	1
1	0	1	1
1	1	0	0
1	1	1	Indeterminate

In the table, Q(*t*) stands for binary state of the flip-flop before the application of a clock pulse, referred to as the "present state". The R and S columns give the possible values of the inputs, and Q (*t* + 1) is the state of the flip-flop after the application of a single pulse, referred to as the "next state". Note that the CP input is not included in the characteristic table. If the present state is Q and the inputs S and R, the application of a single clock pulse causes the flip-flop to go to the next state, Q (*t* + 1).

The characteristic equation is derived from the k-map shown in Fig. 4.10 (*a*). This equation specifies the value of the next state as a function of the present state and the inputs.

$$Q(t+1) = S + \overline{R}Q \text{ and}$$
$$SR = 0$$

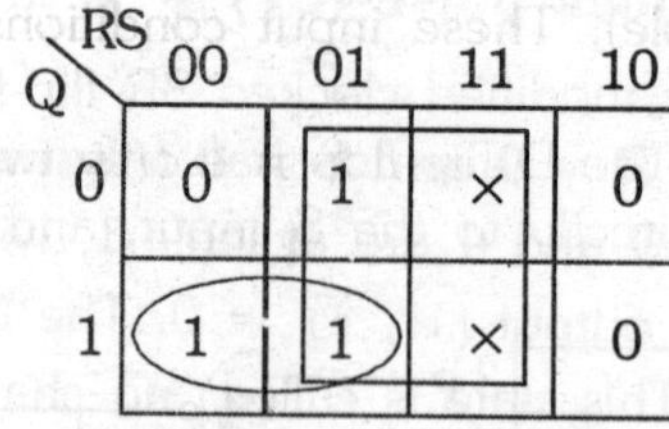

Fig. 4.10(*a*) Characteristic equation.

The two indeterminate states are marked with don't care conditions in the map. The relation RS = 0, must be included as a part of the characteristic equation to specify that both R and S cannot be equal to 1 simultaneously. The graphic/logic symbol of the RS-flip flop is shown in Fig. 4.10 (*b*). It is a rectangular shape block with three inputs S, R and CP and two outputs Q and $\overline{Q}$.

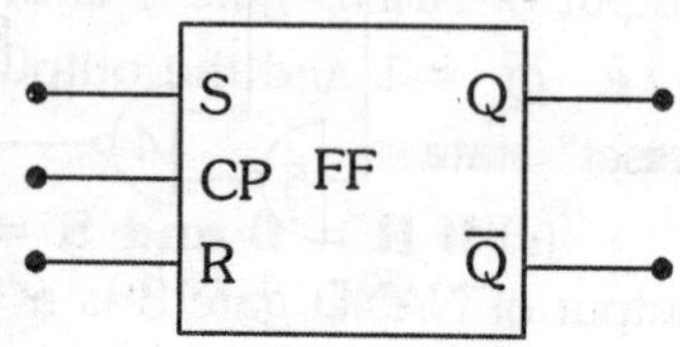

Fig. 4.10(*b*) Graphic symbol.

4.2.4 SR Flip-Flop

The equivalent NAND gate RS flip-flop can be modified to get SR flip-flop, as shown in Fig. 4.11.

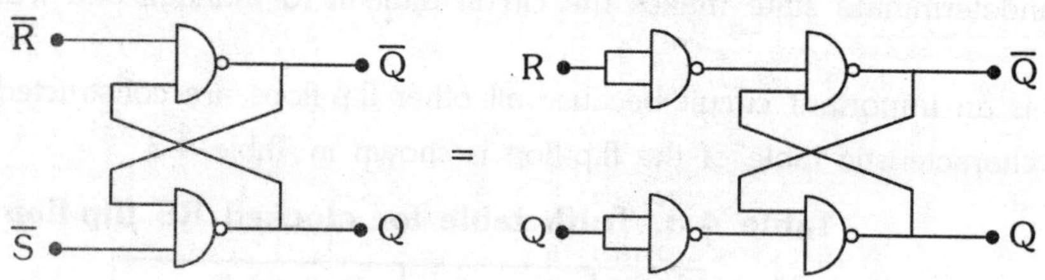

Fig. 4.11 SR flip-flop.

The truth table for SR flip-flop is same as the truth table for RS flip-flop. Fig. 4.12 (*a*) shows the logic symbol and Fig. 4.12 (*b*) shows the truth table for SR flip-flop. The state transition diagram is shown in Fig. 4.12 (*c*).

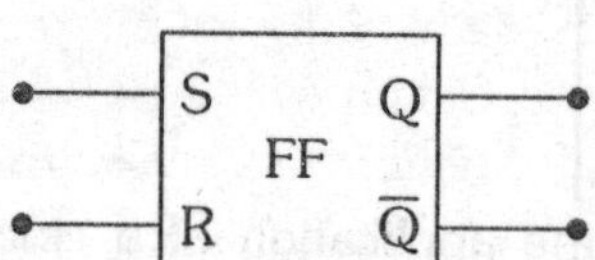

Fig. 4.12(*a*) Logic symbol.

S	R	Q	$\overline{Q}$	State
0	0	NC	NC	No change
0	1	0	1	Set
1	0	1	0	Reset
1	1	ϕ	ϕ	Indeterminate

Fig. 4.12(*b*) Truth table.

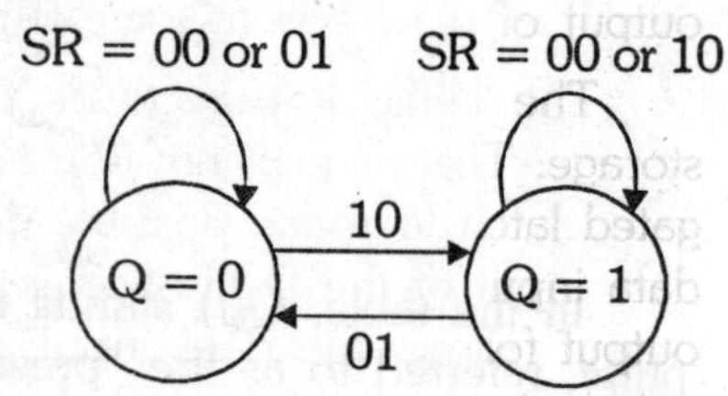

Fig. 4.12(*c*) State transition diagram.

4.2.5 D Flip-Flop

By seeing the truth table of RS flip-flop, we can realize that when both inputs are same, the output either does not change or it is unpredictable (If inputs are 00, no change and inputs

11, unpredictable). These input conditions can be avoided by making them complement of each other. The modified clocked SR flip-flop is known as D flip-flop. The Fig. 4.13(*a*) shows the D flip-flop. The D flip-flop has only two inputs D and CP and two outputs Q and $\overline{Q}$. The D input goes directly to the S input, and its complement is applied to the R-input.

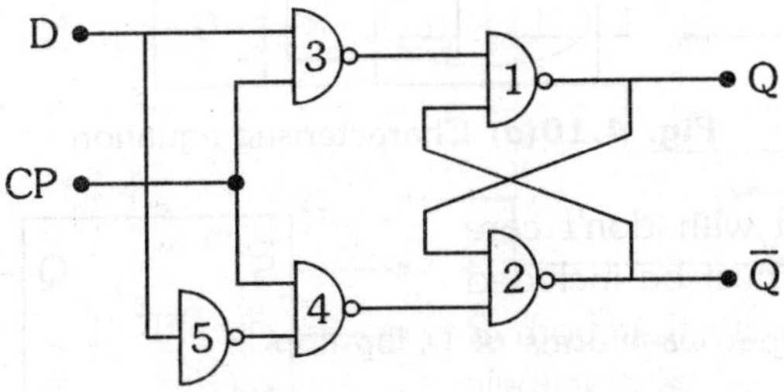

Fig. 4.13(*a*) Logic diagram.

Q	D	Q(t + 1)
0	0	0
0	1	1
1	0	0
1	1	1

Fig. 4.13(*b*) Characteristic table.

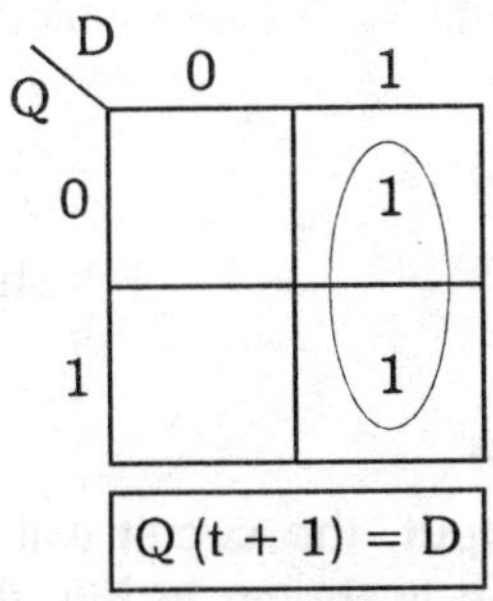

Fig. 4.13(c) Characteristic equation.

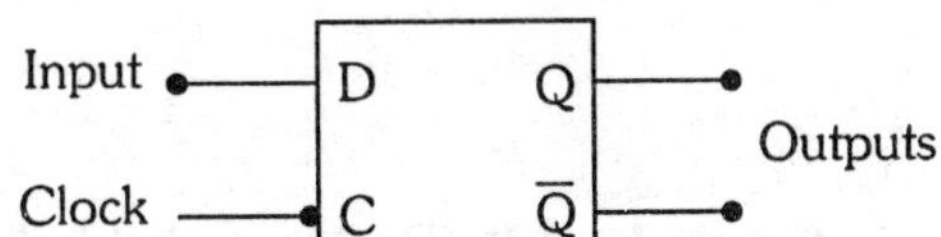

Fig. 4.13(*d*) Logic symbol.

Operation : 1. When clock pulse is 0. The outputs of gate 3 and gate 4 are at logic 1 level and the circuit cannot change state regardless of the value of D.

2. **When CP = 1.**

(*a*) **If D = 1.** The output of gate 3 is at logic 0 and gate 4 is at logic 1. With this, the output of gate 1 is at logic 1 and gate 2 is at logic 0, *i.e.*, Q = 1.

(*b*) **If D = 0.** The output of gate 3 is at logic 1 and gate 4 is at logic 0. With this, the output of gate 1 is at logic 0 and gate 2 is at logic 1, *i.e.*, Q = 0.

The D-flip flop receives the destination from its ability to hold "data" into its internal storage. This type of flip-flop is sometimes called a "gated D-latch". The CP input enables the gated latch to make possible data entry into the circuit. The binary information present at the data input of the D flip-flop is transferred to the Q output when the CP input is enabled. The output follows the data input as long as the clock pulse is in 1 state with just delay equal to gates propagation time. Because of this, this flip-flop is called "delay flip-flop".

When the clock pulse goes to 0, the binary information that was present at the data input at the time the pulse transition occurred is retained at the Q output until the pulse input is enabled again. The characteristic table for the D flip-flop is shown in Fig. 4.13(*b*). From the table, we conclude that the next state of the flip-flop is independent of the present state and is equal to the input D.

The characteristic equation is obtained from Fig. 4.13 (*c*) and the graphic symbol (edge triggered) is shown in Fig. 4.13 (*d*).

The input and output waveforms relation is shown in Fig. 4.13 (*e*).

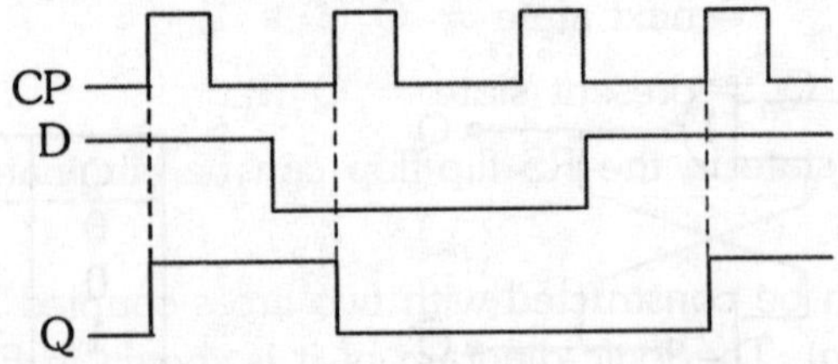

Fig. 4.13(*e*) Input and output waveforms of D flip-flop

The state diagram of the D flip-flop is shown in Fig. 4.13 (*f*).

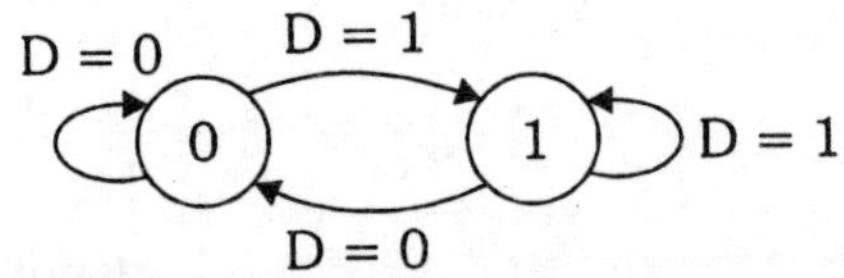

Fig. 4.13(*f*)

Example 4.3. *Design a divide-by – 2 circuit using D flip-flops.*

Solution. If D flip-flop has its $\overline{Q}$ output connected to its D input, the circuit will devide by two the signal applied to its clock input. The necessary connection is shown in Fig. 4.14(*a*).

(*a*) Suppose initially Q = 1 and $\overline{Q}$ = 0 = D.

The first clock pulse (=1) will cause the circuit to switch to Q = 0, $\overline{Q}$ = D = 1.

The second clock pulse (=1) will cause the circuit to switch to Q = 1; $\overline{Q}$ = D = 0 and so on.

The circuit waveforms are given in Fig. 4.14 (*b*) from which it is clear that the output pulse waveform has a frequency of one half that of the input waveform.

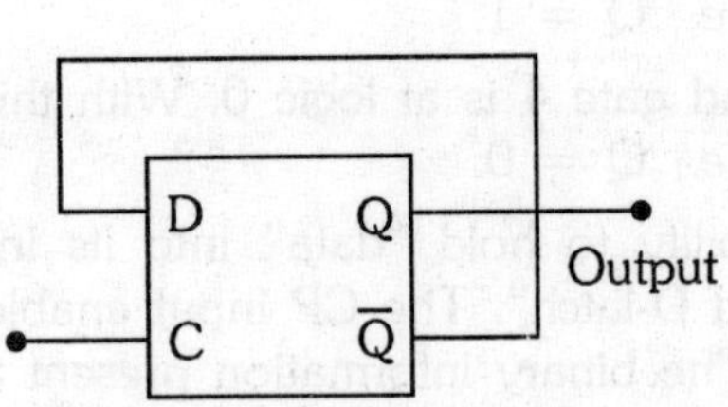

Fig. 4.14(*a*)

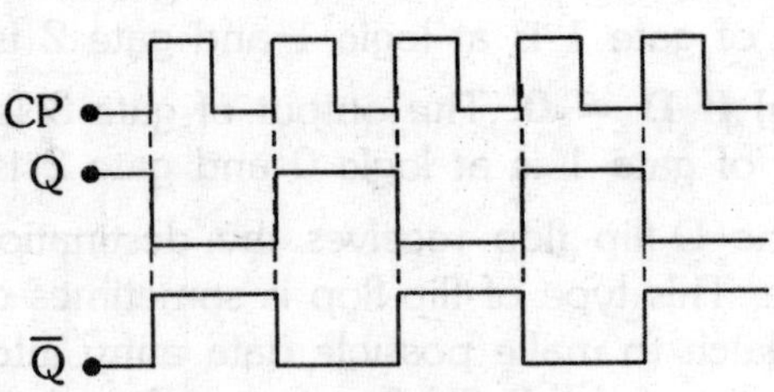

Fig. 4.14(*b*)

(*b*) Same as for initially Q = 0 and $\overline{Q}$ = D = 1.

4.2.6 J K Flip-Flop

The JK flip-flop is a refinement of the RS flip-flop. In JK flip-flop, the indeterminate state in the RS flip-flop is defined. The inputs J and K behave like inputs S and R to set and reset the flip-flop, respectively. When both inputs J and K are equal to 1, the flip-flop, switches to its complement state, that is, if Q = 0, it switches to Q= 1, and vice-versa. This condition can be written as

$$\boxed{Q_{n+1} = \overline{Q}_n \text{ (or) } Q\,(t + 1) = \overline{Q}\,(t)}$$

where Q_{n+1} = next state = Q (t + 1)

Q_n = present state = Q (t).

Here, the indeterminate state in the RS flip-flop can be eliminated by using the feedback technique.

A clocked JK flip-flop can be constructed with two cross-coupled NOR gates and two AND gates as shown in Fig. 4.15 (*a*). The logic diagram of it is shown in Fig. 4.15 (*b*). Fig. 4.15 (*c*) shows the logic symbol for clocked JK flip-flop and its truth table is shown in Fig. 4.15 (*d*).

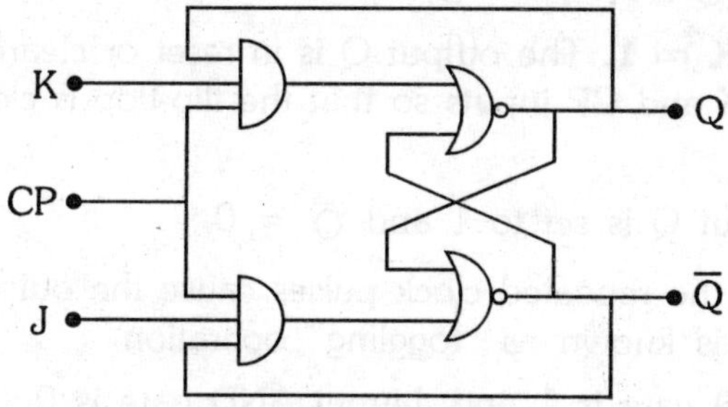

Fig. 4.15(*a*) Clocked JK flip-flop.

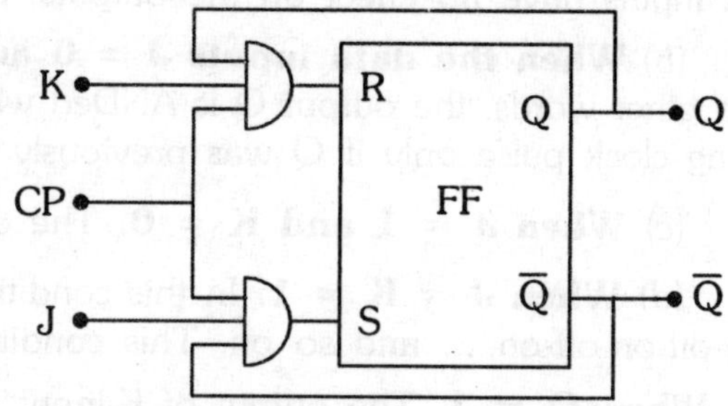

Fig. 4.15(*b*) Logic diagram.

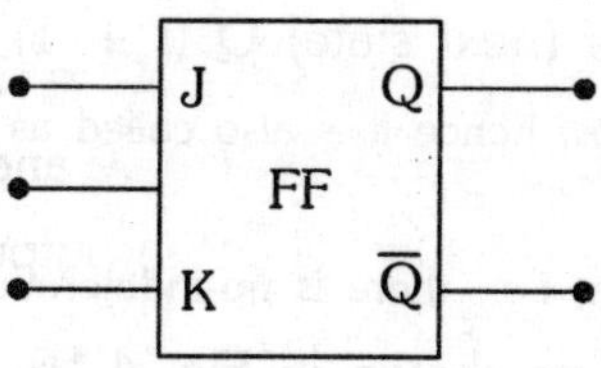

Fig. 4.15(*c*) Logic symbol.

Q	*J*	*K*	*Q (t + 1)*
0	0	0	0
0	0	1	0
0	1	0	1
0	1	1	1
1	0	0	1
1	0	1	0
1	1	0	1
1	1	1	0

≡

J	*K*	Q_{n+1}
0	0	Q_n
0	1	0
1	1	1
1	0	Q_n

Fig. 4.15(*d*) Characteristic table/truth table of JK flip-flop.

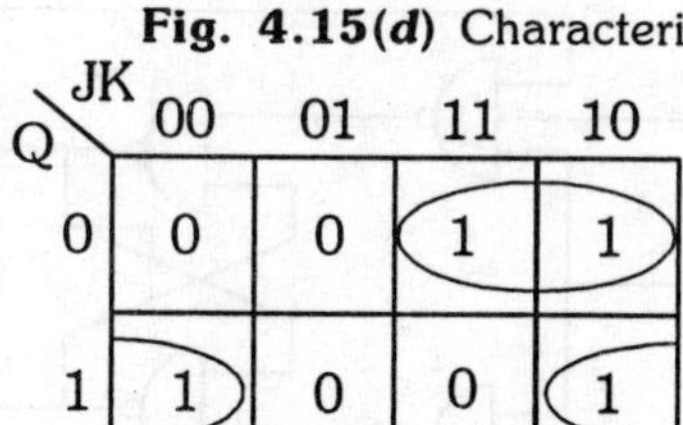

$$Q\,(t + l) = J\overline{Q} + Q\overline{K}$$

$$\boxed{Q\,(t + 1) = J\overline{Q} + Q\overline{K}}$$

Fig. 4.15(*e*) Characteristic equation.

* The output Q is ANDed with K and CP inputs so that the flip-flop is cleared during a clock pulse only if Q was previously 1.

* Similarly the output $\overline{Q}$ is ANDed with J and CP inputs so that the flip-flop is set with a clock pulse only when $\overline{Q}$ was previously 1.

* When both J and K are 1, the input pulse is transmitted through one AND gate only; the one whose input is connected to the flip-flop. Output that is previously equal to 1.

Operation : 1. As long as the CP input is at LOW level, the data inputs have no effect on the outputs. Hence, the output holds the previous data. This state is called "no change state".

2. When CP is HIGH level.

(*a*) **If data inputs J = K = 0.** The flip remains in "no change or hold" state, in which data inputs have no effect on the outputs, *i.e.*, $Q(t+1) = Q$.

(*b*) **When the data inputs J = 0 and K = 1.** The output Q is in reset or cleared to 0. In other words, the output Q is ANDed with K and CP inputs so that the flip-flop is cleared during clock pulse only if Q was previously 1.

(*c*) **When J = 1 and K = 0.** The output Q is set to 1 and $\overline{Q} = 0$.

(*d*) **When J = K = 1.** In this condition, the repeated clock pulses cause the output to turn-off-on-off-on..... and so on. This condition is known as "toggling" operation.

When Q = 1. The output of K-input AND gate is 1 and J-input AND gate is 0. Thus, $Q = 0$ and $\overline{Q} = 1$.

When Q = 0. The output of K-input AND gate is 0 and J input AND gate is 1. Thus, $Q = 1$ and $\overline{Q} = 0$.

Thus, when J = K = 1 in either case, the output state (next state) $Q(t+1) = \overline{Q}(t)$.

Note : (1) The JK flip-flop has the features of all other flip-flops, hence it is also called as "universal" flip-flop.

(2) It produces an output for each and every input combination, *i.e.*, there is no-indeterminate state.

The JK flip-flop can be constructed using NAND gates as shown in Fig. 4.16.

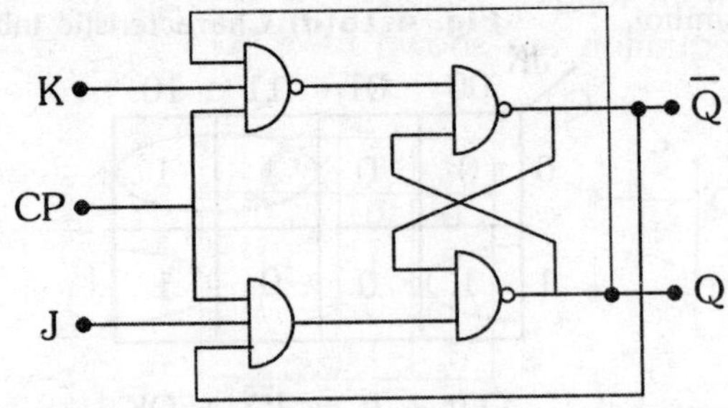

Fig. 4.16 JK flip-flop using NAND gates only.

The relation between input and output waveforms for clocked JK flip-flop is shown in Fig. 4.17 (*a*).

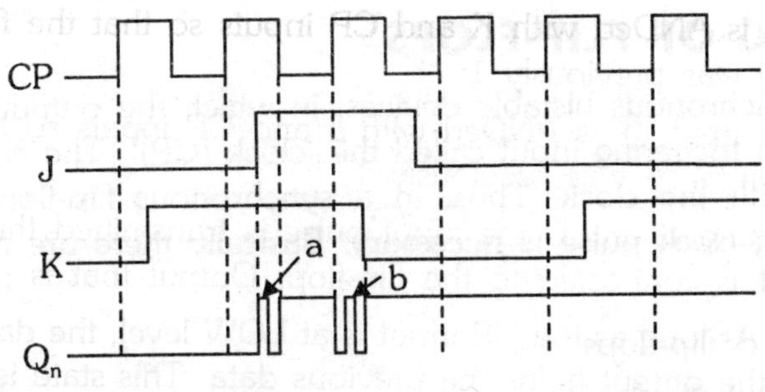

Note : *a* and *b* are toggle condition output.

Fig. 4.17(*a*) Input and output waveforms of clocked JK flip-flop.

The state diagram of the JK flip-flop is shown in Fig. 4.17(*b*).

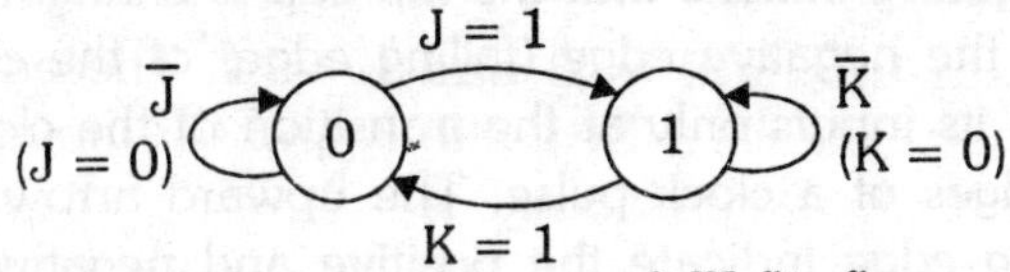

Fig. 4.17(*b*) State diagram of JK flip-flop.

4.2.7 T Flip-Flop

The T flip-flop is also known as "toggle flip-flop". It is a modification of JK flip-flop. The T-flip-flop is made from a JK flip-flop merely by connecting its J and K terminals together. This means that at all times J = K. The clocked T flip-flop is shown in Fig. 4.18.

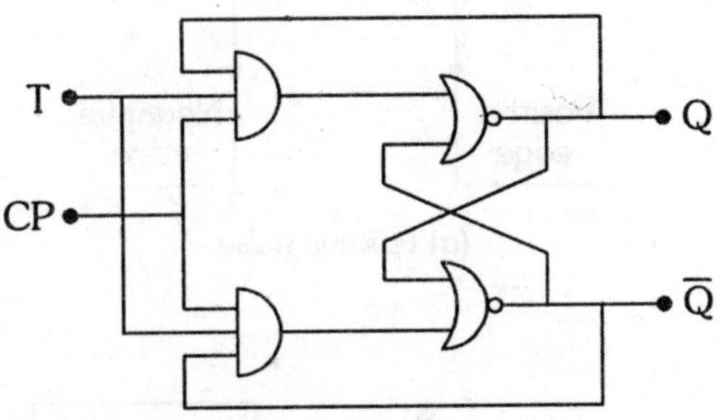

Fig. 4.18 Clocked T flip-flop.

1. **When CP = 0.** The outputs of both AND gates are 0 irrespective of the value of T. The output of the circuit is same as the previous output. Therefore Q (*t* + 1) = Q (*t*).

2. **When CP = 1.** (*a*) **If T = 0.** Both AND gates are disabled and hence there is no change in output, *i.e.*, Q (*t* + *l*) = Q (*t*)

(*b*) **If T = 1.** (*i.e.*, J = K = 1) The outputs of both AND gates depend on the previous output and finally the output toggles, *i.e.*, Q (*t* + 1) = $\overline{Q}$ (*t*). The logic symbol, characteristic table and characteristic equation are shown the Fig. 4.19.

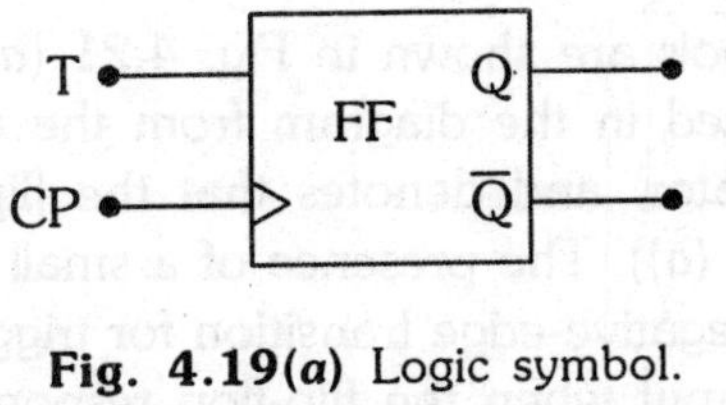

Fig. 4.19(*a*) Logic symbol.

Q_n	T	Q_{n+1}
0	0	0
0	1	1
1	0	1
1	1	0

≡

T	Q_{n+1}
0	Q_n
1	$\overline{Q}_n$

Q_n \ T	0	1
0		1
1	1	0

$$Q_{n+1} = T\overline{Q}_n + \overline{T}Q_n$$

Fig. 4.19(*b*) Truth table.

The state diagram of T flip-flop is shown in Fig. 4.19(*c*).

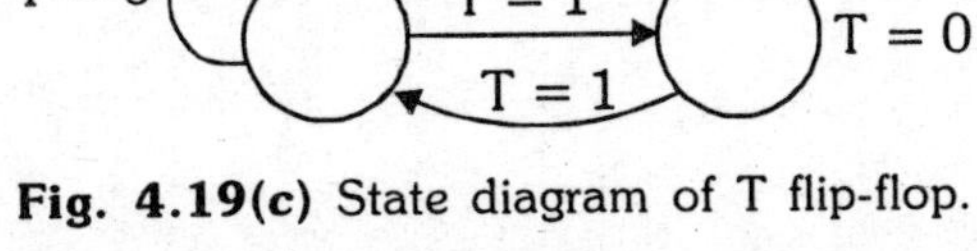

Fig. 4.19(*c*) State diagram of T flip-flop.

4.3 TRIGGERING OF FLIP-FLOPS

Flip-flops are synchronous bistable devices, in which the output changes the state only at a specified point on a triggering input called the 'clock (CP)'. The changes in the output occur in synchronisation with the clock. Thus, in a synchronous flip-flop, for change of state, an external synchronising clock pulse is necessary. Basically there are two types of triggered flip-flops, they are :

(*a*) Edge-triggered flip-flops

(*b*) Level triggered flip-flops.

4.3.1 Edge-Triggered Flip-Flops

The term 'edge-triggering' means that the flip-flop is changing the state during the positive edge (rising edge) or at the negative edge (falling edge) of the clock pulse. This indicates that the output is sensitive to its inputs only at the transition of the clock pulse. Fig. 4.20 shows the positive and negative edges of a clock pulse. The upward arrow at the rising edge and down ward arrow at the falling edge indicate the positive and negative edge triggering respectively.

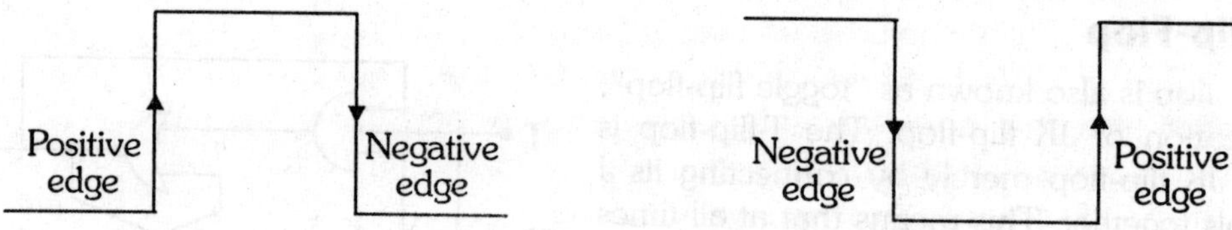

(*a*) Positive pulse (*b*) Negative pulse

Fig. 4.20

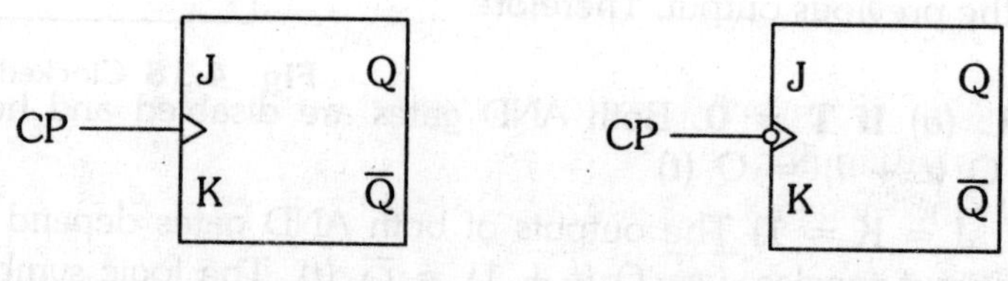

(*a*) Positive edge tugged flip-flop (*b*) Negative edge tugged flip-flop

Fig. 4.21 Edge triggered flip-flops.

The positive and negative edge triggered flip-flop symbols are shown in Fig. 4.21 (*a*) and Fig. 4.21 (*b*) respectively. The clock pulse input is recognized in the diagram from the arrow head-shape symbol. This is a symbol of a 'dynamic indicator' and denotes that the flip-flop responds to a positive edge transition of the clock (Fig. 4.21 (*a*)). The presence of a small circle outside the block along the dynamic indicator designates a negative-edge transition for triggering the flip-flop. The letter symbol 'CP' is used for the clock input when the flip-flop responds to a pulse transition.

There are 2 ways to make the flip-flop respond only to a pulse transition (positive or negative edges). They are (*a*) Use of capacitive coupling.

(*b*) Pulse transition detector.

(*a*) **Capacitive coupling.** In this configuration, an RC (Registor capacitor) circuit is inserted in the clock input of the flip-flop, as shown in Fig. 4.22. The RC values are selected such that the circuit has a short time constant, thus converting rectangular clock pulses into narrow spikes. A positive edge emerges from such a circuit in a positive spike, and negative edge emerges in a negative spike. Because of AND gates, the circuit does not respond to the negative edges and it responds only to positive edges and hence it is a positive edge triggered circuit.

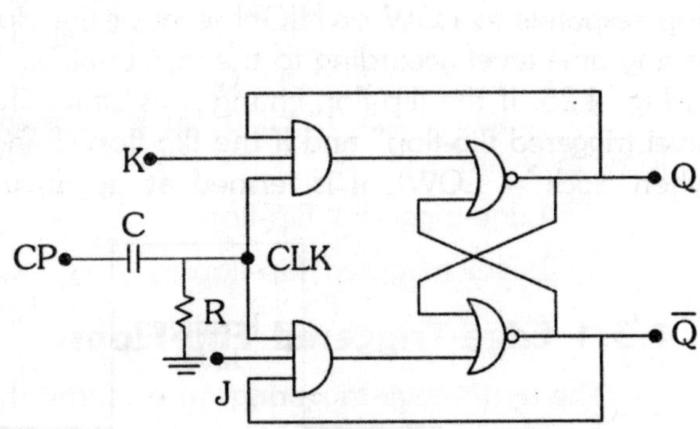

Fig. 4.22 Positive edge triggered flip-flop.

(*b*) **Pulse transition detector.** The purpose of this circuit is to produce a very short duration spike on the positive-going transition of the clock pulse. In Fig. 4.23, there is a small delay on one input of NAND gate, so that the inverted clock pulse arrives at the gate input a few nanoseconds after the true clock pulse. This produces an output spike with a duration of only a few nanoseconds. To obtain the negative going transition of the clock pulse, we have to either invert the clock pulse (or) remove the inverter at the output of NAND gate.

Fig. 4.23 Logic circuit of a pulse transition detector.

Thus the edge triggering can be given to the flip-flop and depending on the state of the inputs, the operation of the flip-flop can be determined. Fig. 4.24 shows the input and output waveforms for positive and negative edge triggered JK flip-flop.

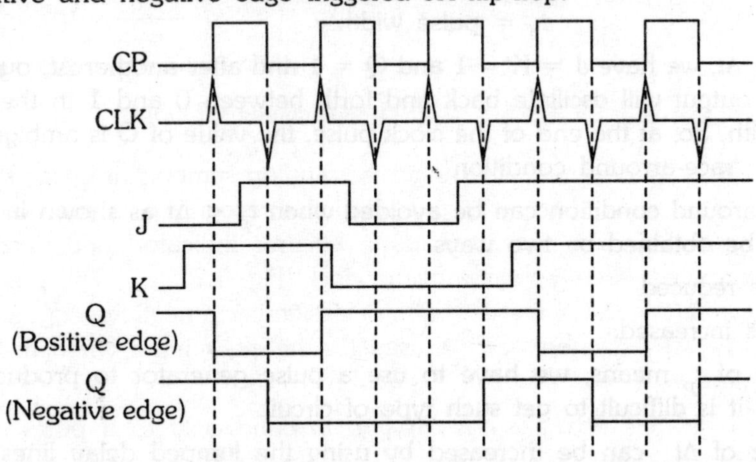

Fig. 4.24 Input and output waveforms for positive and negative edge triggered JK flip-flop.

4.3.2 Level Triggered Flip-Flop

The positive and negative clocking are generally known as 'level triggering'. Then the flip-flop responds to LOW or HIGH level of the clock. The output of the flip-flop changes the state at any time level according to the input values. The positive level and negative levels are shown in Fig. 4.25. If the flip-flop changes its state when the clock is positive, it is termed as "positive level triggered flip-flop" and if the flip-flop changes its state only when the clock is negative (*i.e.*, when clock = LOW), it is termed as negative level triggered flip-flop.

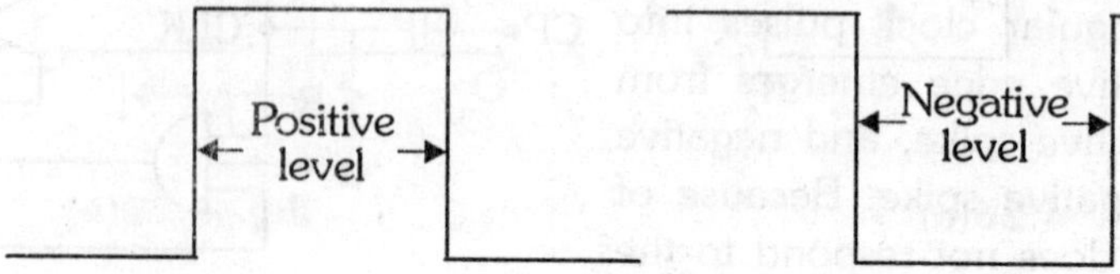

Fig. 4.25

The main drawback of level triggering is that as long as the clock is positive or negative, the flip-flop changes its state more than once or many times for the change in inputs. If the inputs are made stable for the entire clock duration, then the output changes only once. On the other hand, if the frequency of input change is higher than the input clock frequency, the output of the flip-flop undergoes multiple changes, when the clock is positive or negative. When the clock becomes unasserted, the output of the flip-flop remains in the last state.

Note : The level triggering can not be used for T and JK flip-flops.

4.3.3 Race Around Condition

It is important to note that in JK flip-flop, the output is fedback to the input, and therefore change in the output results change in the input, if the inputs J = K = 1 and Q = 0.

When a clock pulse with a width 't_p' as shown in Fig. 4.26 (*a*) is applied, the output will change from 0 to 1 after the time interval Δt,

where Δt = propagation delay of two level NAND gates

t_p = pulse width.

Now, after Δt, we have J = K = 1 and Q = 1 and after another Δt, output Q will become 0. Hence, the output will oscillate back and forth between 0 and 1 in the duration t_p of the clock pulse width. So, at the end of the clock pulse, the value of Q is ambiguous. This situation is known as a "race-arround condition".

The race around condition can be avoided when $t_p < \Delta t$ as shown in Fig. 4.26 (*b*). This condition can be obtained by two ways

(1) If t_p is reduced

(2) If Δt is increased.

Reduction of t_p means, we have to use a pulse generator to produce less pulse width waveform, but it is difficult to get such type of circuit.

The value of Δt can be increased by using the lumped delay lines in series with the feedback connection, which is worthless.

There are two solutions for reducing race around condition :

(1) use of edge triggering, which we have already studied.

(2) master-slave flip-flop configuration.

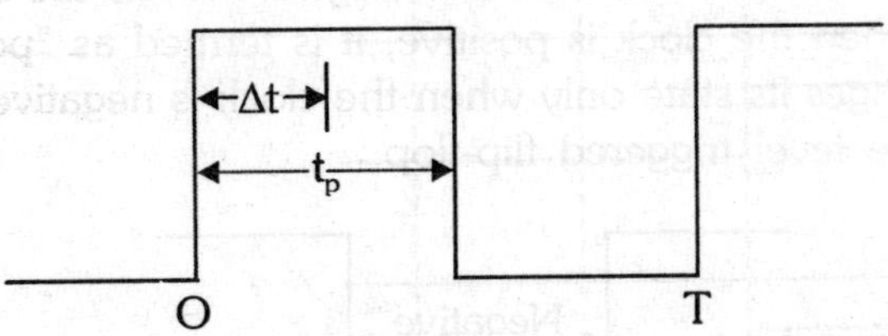

Fig. 4.26(*a*)

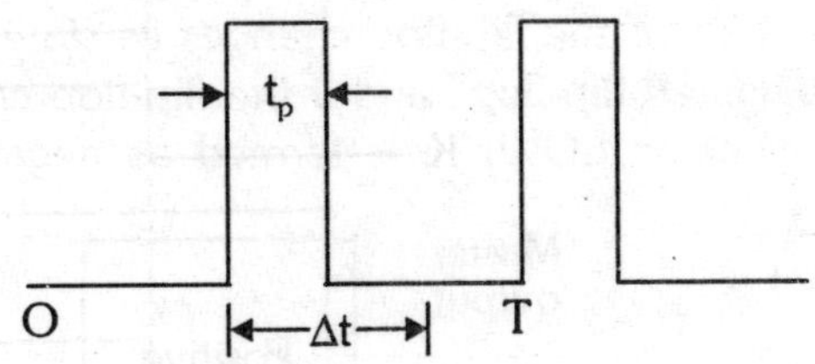

Fig. 4.26(*b*)

4.3.4 Master-Slave Flip-Flop

A master-slave flip-flop can be constructed using two JK flip-flops connected serially. The first flip-flop serves as the "master" and second as a "slave", and the overall circuit is referred to as a "master-slave flip-flop". The logic diagram of a JK master-slave flip-flop is shown in Fig. 4.27 (*a*). It consists of a master flip-flop, a slave flip-flop, and an inverter. The first flip-flop *i.e.*, master, is driven by the positive edge of the clock pulse; the second flip-flop, called slave, is driven by the negative edge of the clock pulse.

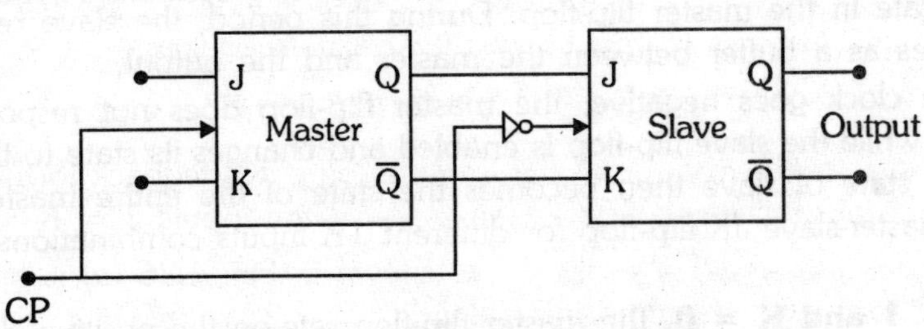

Fig. 4.27(*a*) Master slave JK flip-flop.

(1) When CP = 1, *i.e.*, clock input of master is 1, the master acts according to its JK inputs, but the slave does not respond, since it requires a negative edge at the clock input.

(2) When CP = 0, *i.e.*, clock input of slave is 1, the slave flip-flop copies the master outputs, but the master does not respond, since it requires a positive edge at its clock input. Thus, master-slave flip-flop does not have race around problem.

The output state of the master flip-flop is determined by the J and K inputs at the positive clock pulse. The output state of the master is then transferred as an input to the slave flip-flop. The slave flip-flop uses these inputs at the negative clock pulse to determine the output state. Fig. 4.27 (*b*) illustrates the operation of the master-slave flip-flop.

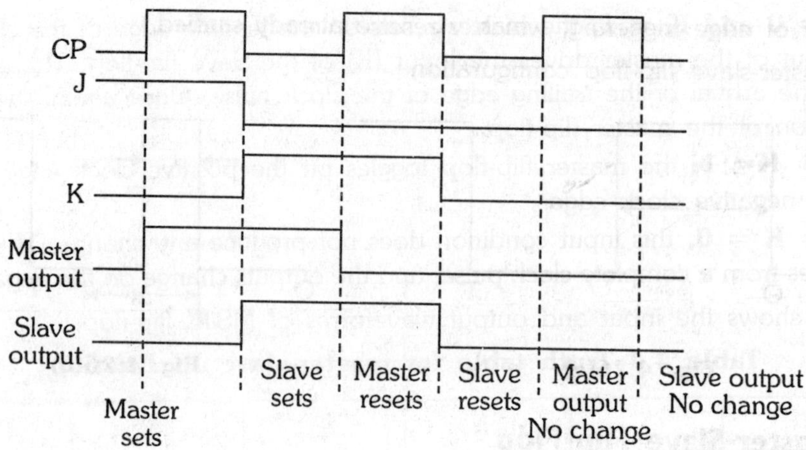

Fig. 4.27(*b*) Input and output waveforms for master-slave flip-flop.

The master-slave combination can be constructed for any type of flip-flop. The master-slave JK flip-flop constructed using NAND gates is shown in Fig. 4.28. It consists of two flip-flops connected in series. NAND gates-1 through 4 form the master flip-flop and NAND gates-5 through 8 form the slave flip-flop. When clock is positive, a change in J and K inputs causes a change of state in the master flip-flop. During this period, the slave remains in its previous state and serves as a buffer between the master and the output.

When the clock goes negative, the master flip-flop does not respond, *i.e.*, it maintains previous state, while the slave flip-flop is enabled and changes its state to that of the master flip-flop. The new state of slave then becomes the state of the entire master-slave flip-flop. The operation of master-slave JK flip-flop for different J-K inputs combinations can be explained as follows :

(*a*) **If J = 1 and K = 0,** the master flip-flop sets on the positive clock edge. The HIGH Q (1) output of the master drives the input (J) of the slave. So, when the negative clock edge hits, the slave also sets. The slave flip-flop copies the action of the master flip-flop.

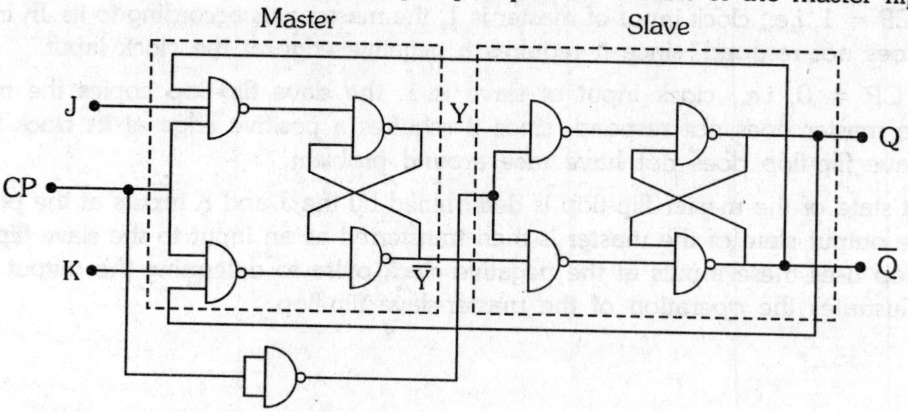

Fig. 4.28 Clocked master-slave JK flip-flop using NAND gates.

(b) **If J = 0 and K = 1,** the master resets on the leading edge of the clock pulse. The HIGH $\overline{Q}$ output of the master drives the input (K) of the slave flip-flop. Then, the slave flip-flop resets at the arrival of the trailing edge of the clock pulse. Once again, the slave flip-flop copies the action of the master flip-flop.

(c) **If J = K= 1,** the master flip-flop toggles on the positive clock edge and the slave toggles on the negative clock edge.

(d) **If J = K = 0,** this input condition does not produce any change. Thus master-slave flip-flop operates from a complete clock pulse, and the outputs change on the negative transition.

Table 4.4 shows the input and output waveforms of MSJK flip-flop.

Table 4.4 Truth table for master-slave JK flip-flop

CP	Q_n	*J*	*K*	*Y*	Q_{n+1}
↑	0	0	0	0	NC
↓	0	0	0	NC	0
↑	0	0	1	0	NC
↓	0	0	1	NC	0
↑	0	1	0	1	NC
↓	0	1	0	NC	1
↑	0	1	1	1	NC
↓	0	1	1	NC	1
↑	1	0	0	1	NC
↓	1	0	0	NC	1
↑	1	0	1	0	NC
↓	1	0	1	NC	0
↑	1	1	0	1	NC
↓	1	1	0	NC	1
↑	1	1	1	0	NC
↓	1	1	1	NC	0

4.4. ASYNCHRONOUS OR DIRECT INPUTS

For the flip-flops discussed so far, the S R, J K, D and T inputs called "control inputs", are also called "synchronous inputs". For such synchronous inputs, the flip-flop changes its state in synchronization with the clock input only. So, the synchronous control inputs must be used along with a clock to trigger a change in the flip-flop.

In addition to the above synchronous inputs, most of the clocked flip-flops have one or more asynchronous inputs, whose effect on the flip-flop output is independent of the synchronous inputs and clock input. Asynchronous inputs can be used to set the flip-flop to the 1 state, or clear the flip-flop to the 0-state at any time, regardless of the condition at the other inputs. These inputs are also called as "direct-inputs". These inputs are connected directly into the latch portion of any flip-flop so that they override the effect of synchronous inputs and the clock.

Fig. 4.29 shows a clocked J K flip-flop with $\overline{\text{PRESET}}$ ($\overline{\text{PR}}$) and $\overline{\text{CLEAR}}$ ($\overline{\text{CLR}}$) or "direct set" and "direct reset" inputs. These are active LOW inputs. An active LOW level at the preset input will SET the flip-flop, and an active LOW level on the clear input will RESET it. These asynchronous inputs are activated by a 0 level, as indicated by the small bubble on the flip-flop symbol. Both the preset and clear inputs must be kept HIGH for synchronous operation.

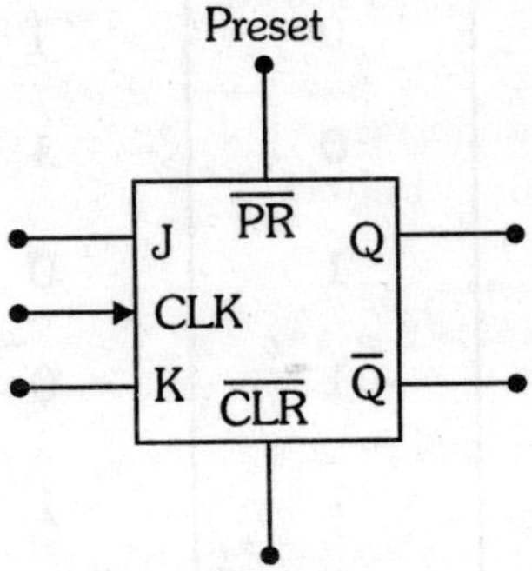

Fig. 4.29 Logic symbol for JK flip-flop with active LOW preset and clear inputs.

Table 4.5 shows how these inputs operate.

Note : LOW levels on preset and clear should not be applied simultaneously, since it leads to an ambiguous condition.

Table 4.5. Truth table

Inputs		*Outputs*		*Action*
Preset	Clear	Q_{n+1}	$\overline{Q_{n+1}}$	
0	0	×	×	Forbidden
0	1	1	0	Set
1	0	0	1	Reset/Clear
1	1	Q_n	$\overline{Q_n}$	No change

The logic diagram for an edge triggered JK flip-flop with preset $\overline{PR}$ and clear $\overline{CLR}$ inputs is shown in Fig. 4.30.

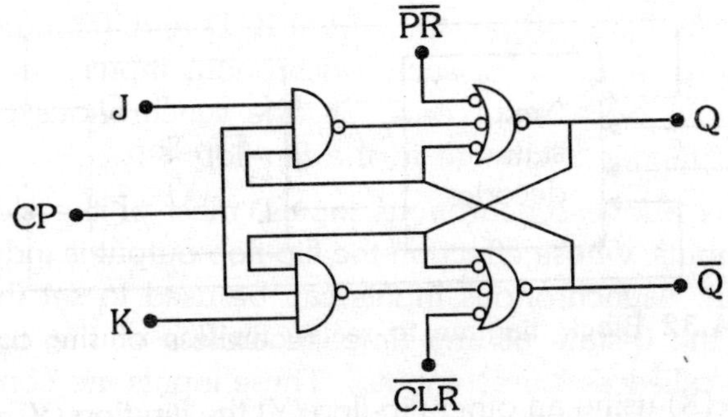

Fig. 4.30 Logic diagram for a basic JK flip-flop with active LOW preset and clear inputs.

From Fig. 4.30, it is clear that if a logic 0 is applied to the $\overline{PR}$ (preset) input, the flip-flop output (Q) will become 1 regardless of the status of the other inputs. Similarly, a logic LOW (0) at the $\overline{CLR}$ (clear) input holds the $\overline{Q}$ output of the flip-flop in the HIGH state, *i.e.*, Q = 0, Thus, asynchronous inputs can be used to hold the flip-flop in a particular state for any desired interval of time.

Some flip-flops have asynchronous inputs that are activated by logic 1 (active HIGH) rather than by logic 0's. The small bubble on the preset and clear input is not shown for these flip-flops.

Input and output waveforms shown in Fig. 4.31 illustrate the operation of preset and clear inputs. In this Fig. 4.31, J and K inputs are assumed at logic 1. Thus flip-flop's output toggles for each positive edge of the clock.

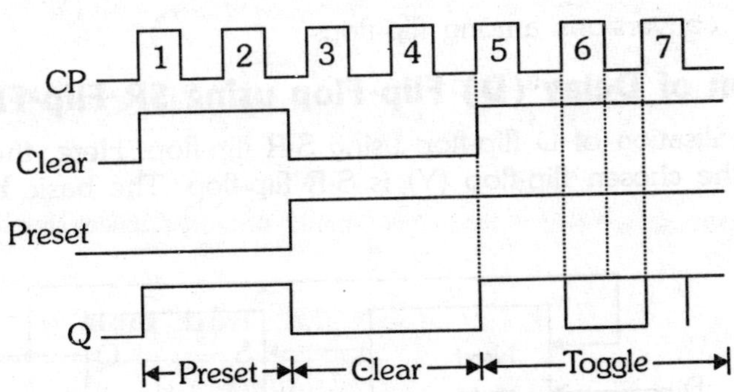

Fig. 4.31 Operation of $\overline{\text{preset}}$ and $\overline{\text{clear}}$ inputs.

During the clock pulse 1 and 2, clear = 1 and preset = 0, keeping the flip-flop set regardless of the synchronous inputs. For clock pulses 3 and 4, clear = 0 and preset = 1 and the flip-flop output is reset. During the clock pulses 5, 6 and 7, toggle operation occurs because J and K both are high with preset and clear inactive.

4.5 FLIP-FLOP CONVERSIONS

It is possible to implement one flip-flop circuit using any other flip-flop with some additional gates. The general block diagram connection for such realisation is shown in Fig. 4.32.

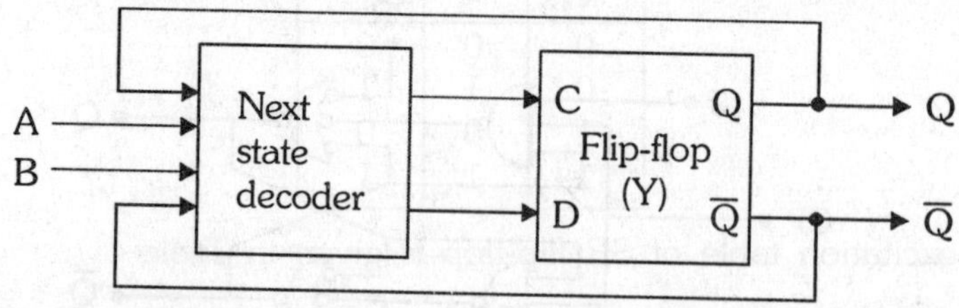

Fig. 4.32 Block diagram to realise flip-flop-x using flip-flop *y*.

To realise a flip-flop (X) using an other flip-flop (Y) the flip-flop (Y) along with a combinational circuit, called 'NEXT STATE DECODER', is used, which functions like a flip-flop (X). For this to be realised, for each set of inputs of flip-flop (X), *i.e.*, A and B and present state (Q_n) we have to find the inputs to the flip-flop (Y), *i.e.*, C and D, that will cause the flip-flop (Y) to make transition into the proper next state (Q_{n+1}) of the flip-flop (X). These inputs to the flip-flop (Y) are called the "next state code". The design procedure for such realisation can be summarised in the following steps :

(1) Obtain the clear statement description of the desired flip-flop (X).

(2) Obtain a present state–next state (PS–NS) table for the desired flip-flop (X).

(3) Using the excitation table (or) application table of the chosen flip-flop (Y), append the next state code or the excitation input values to the above present state-next-state table.

(4) Using K-maps, simplify the logic expressions for excitation inputs of flip-flop (Y) and design the next state decoder logic.

(5) Draw a circuit for the desired flip-flop (X) using next state decoder logic and the chosen flip-flop (Y) as shown in the Fig 4.32.

Let us see few conversions among flip-flops.

4.5.1 Realisation of Delay (D) Flip-Flop using SR Flip-Flop

Consider the realisation of D flip-flop using S-R flip-flop. Here, the desired flip-flop (X) is delay flip-flop and the chosen flip-flop (Y) is S-R flip-flop. The basic block diagram is shown in Fig. 4.33.

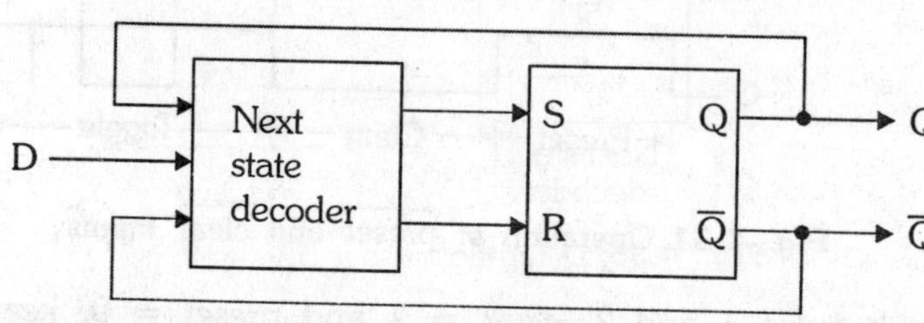

Fig. 4.33 Block diagram of D flip-flop using S-R flip-flop.

Step 1. The present state and next state table for D flip-flop is shown in Table 4.6.

Table 4.6. PS-NS table for D-flip-flop

Q_n	D	Q_{n+1}
0	0	0
0	1	1
1	0	0
1	1	1

Step 2. The excitation table of SR flip-flop is given in Table 4.7.

Table 4.7

Q_n	T	Q_{n+1}	*Excitation inputs*	
			S	R
0	0	0	0	×
0	1	1	1	0
1	0	0	0	1
1	1	1	×	0

Step 3. To design the next state decoder, the simplified expressions can be obtained by using K-maps from the excitation table as shown in Fig. 4.34

Fig. 4.34 K-maps for S and R inputs.

Step 4. The logic diagram for the next state decoder must produce an output in such a way that S = D and R = $\overline{D}$. The conversion diagram is shown in Fig. 4.35.

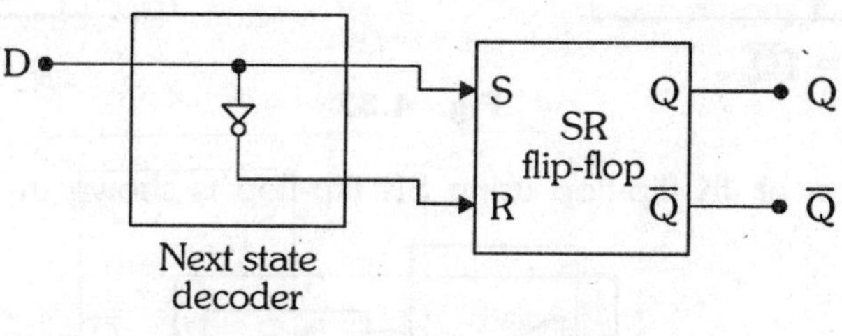

Fig. 4.35 D flip-flop using SR flip-flop.

4.5.2 Realization of JK Flip-Flop using SR Flip-Flop

Here, the desired flip-flop (×) is a JK flip-flop and the chosen flip-flop (Y) is an SR flip-flop. The block diagram of JK flip-flop using SR flip-flop is shown in Fig. 4.36.

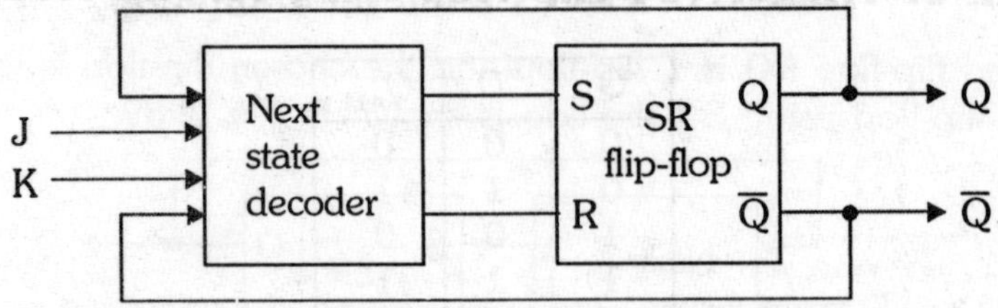

Fig. 4.36 Block diagram of JK flip-flop using SR flip-flop.

Step 1. The present state – next state table for JK flip-flop is shown in Table 4.8 (*a*).

Step 2. The excitation table for JK flip-flop using SR flip-flop is shown in Table 4.8 (*b*).

Table 4.8(*a*) Truth table

Qn	*J*	*K*	Q_{n+1}
0	0	0	0
0	0	1	0
0	1	0	1
0	1	1	1
1	0	0	1
1	0	1	0
1	1	0	1
1	1	1	0

Table 4.8(*b*)

Qn	*J*	*K*	Q_{n+1}	*Excitation inputs*	
				S	*R*
0	0	0	0	0	×
0	0	1	0	0	×
0	1	0	1	1	0
0	1	1	1	1	0
1	0	0	1	×	0
1	0	1	0	0	1
1	1	0	1	×	0
1	1	1	0	0	1

Step 3. The simplified expressions for S and R are shown in Fig. 4.37 (K-maps).

For S

Q_n \ JK	00	01	11	10
0	0	0	1	1
1	×	0	0	×

$S = T\overline{Q}_n$

For R

Q_n \ JK	00	01	11	10
0	×	×	0	0
1	0	1	1	0

$R = Q_nK$

Fig. 4.37

Step 4. Realization of JK flip-flop using SR flip-flop is shown in Fig. 4.38.

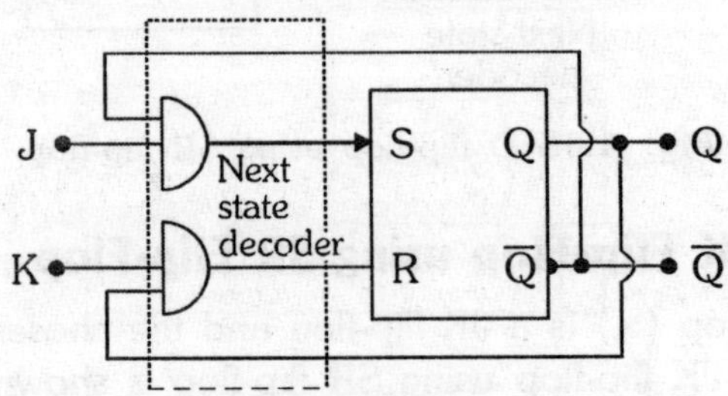

Fig. 4.38

4.5.3 Realization of Trigger (T) Flip-Flop using SR Flip-Flop

Here, the desired flip-flop (X) is T flip-flop and the chosen flip-flop is (Y) SR-flip-flop. The block diagram of T flip-flop using SR flip-flop is shown in Fig. 4.39.

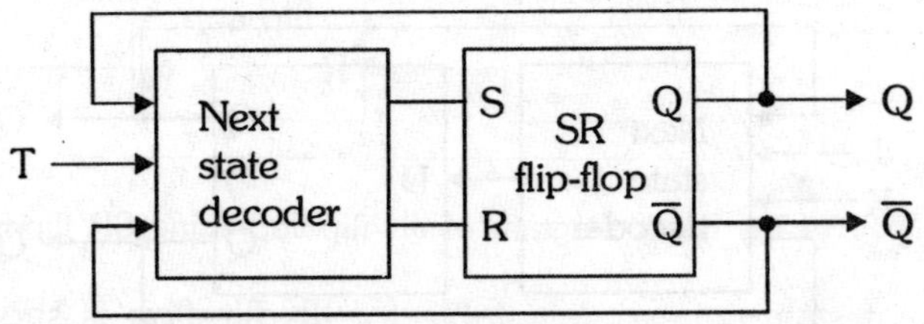

Fig. 4.39 Block diagram of T flip-flop using SR flip-flop.

Step 1. The present state-next state table for T flip-flop (truth table) is shown in Table 4.9(*a*).

Step 2. The SR values for the above truth table are shown in Table 4.9 (*b*).

Table 4.9(*a*) Truth table

Q_n	D	Q_{n+1}
0	0	0
0	1	1
1	0	1
1	1	0

Table 4.9(*b*) Excitation table

Q_n	T	Q_{n+1}	Excitation inputs	
			S	R
0	0	0	0	×
0	1	1	1	0
1	0	1	×	0
1	1	0	0	1

Step 3. The simplified outputs (*i.e.*, inputs to SR flip-flop) from the next state decoder can be derived using K-maps, as shown in Fig. 4.40

Fig. 4.40 K-maps.

Step 4. The realization of T flip-flop using SR flip-flop is shown in Fig. 4.41.

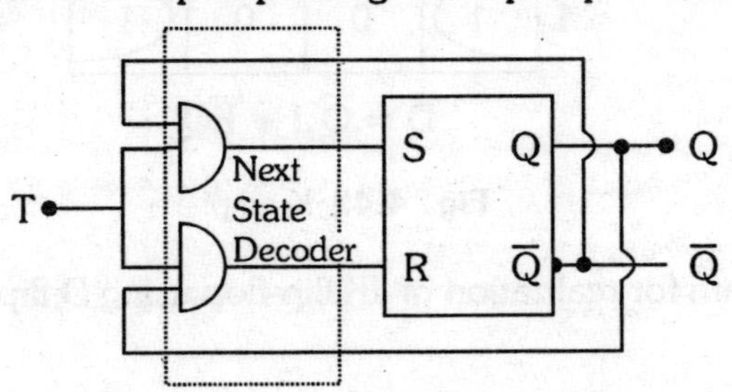

Fig. 4.41 T flip-flop using SR flip-flop.

4.5.4 Realization of JK Flip-Flop using D Flip-Flop

Here the desired flip-flop (X) is a JK flip-flop and the chosen flip-flop (Y) is a D flip-flop. The block diagram of this realisation is shown in Fig. 4.42.

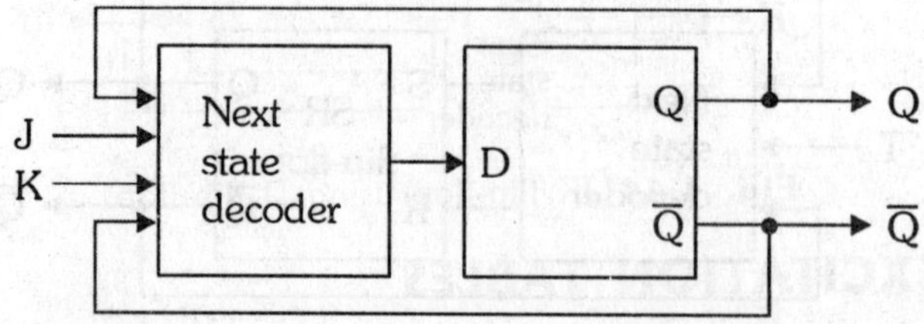

Fig. 4.42 Block diagram of JK flip-flop using D flip-flop.

Step 1. The present state-next state (truth table) for JK flip-flop is shown in Table 4.10(*a*).

Step 2. The D input values for the Table 4.10(*a*) are augmented in Table 4.10(*b*).

Table 4.10(*a*) JK flip-flop truth table

Q_n	J	K	Q_{n+1}
0	0	0	0
0	0	1	0
0	1	0	1
0	1	1	1
1	0	0	1
1	0	1	0
1	1	0	1
1	1	1	0

Table 4.10(*b*) Excitation table

Q_n	J	K	Q_{n+1}	*Excitation input (D)*
0	0	0	0	0
0	0	1	0	0
0	1	0	1	1
0	1	1	1	1
1	0	0	1	1
1	0	1	0	0
1	1	0	1	1
1	1	1	0	0

Step 3. The simplified output (D) from the next state decoder can be obtained by using K-map as shown in Fig. 4.43.

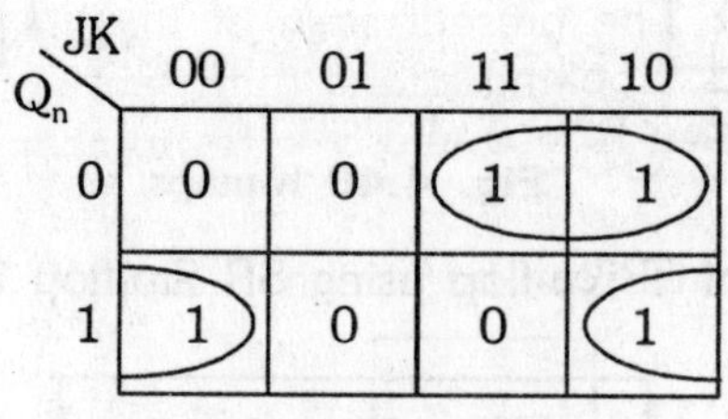

$$D = \overline{Q}_n J + \overline{K} Q_n$$

Fig. 4.43 K-map.

Step 4. The logic diagram for realization of JK flip-flop using D flip-flop is shown in Fig. 4.44.

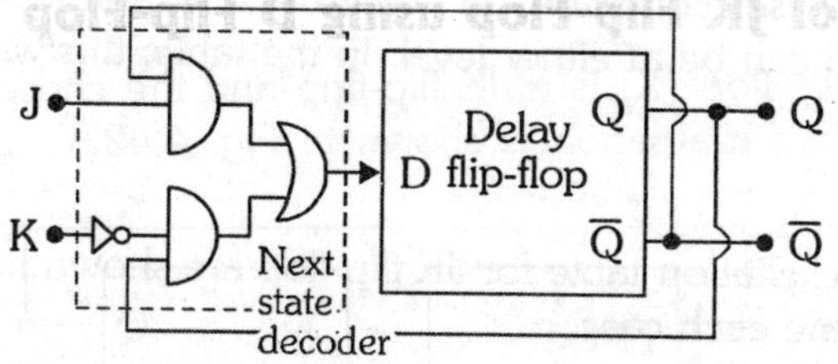

Fig. 4.44 JK flip-flop using D flip-flop.

4.6 FLIP-FLOP EXCITATION TABLES

Sometimes we know the present state and next state, there we need a table that lists the required inputs for a given state change. Such a table is known as an excitation table of the flip-flop. We can derive the excitation tables for flip-flops from their truth tables. The excitation table consists of two columns Q_n and Q_{n+1}, and a column for each input to show how the required transition can be achieved. Let us see the truth tables and excitation tables for RS, JK, D and T flip-flops.

4.6.1 RS Flip-Flop

Tables 4.11(*a*) and (*b*) show the truth table and excitation table for RS flip-flop, respectively. As shown in the table, there are four possible transitions from the present state (Q_n) to the next state (Q_{n+1}). For each transition, the required inputs can be derived from the information available in the truth table.

Table 4.11(*a*) RS truth table

R	S	Q_{n+1}
0	0	Q_n
0	1	1
1	0	0
1	1	Indeterminate

Table 4.11(*b*) RS excitation table

Q_n	Q_{n+1}	R	S
0	0	×	0
0	1	0	1
1	0	1	0
1	1	0	×

(*a*) **0 → 0 Transition** : The present state of the flip-flop is 0 and is to remain in '0' when a clock pulse is applied. Looking at the truth table of RS flip-flop, we can understand that this can happen either when R = 0 and S = 0 (no change condition) or when R = 1 and S = 0.

Thus, S has to be at 0, but R can be at either level (X). The table indicates this with a '0' under S and an 'X' under R.

(*b*) **0 → 1 Transition** : The present state is 0 and is to change to 1. This can happen when S = 1 and R = 0 (set condition). Therefore, S has to be 1 and R has to be 0 for this transition to occur.

(*c*) **1 → 0 Transition** : The present state is 1 and is to change to a 0. This can happen only when S = 0 and R = 1. (Reset condition). Therefore S has to be 0 and R has to be 1 for this transition to occur.

(*d*) **1 → 1 Transition** : The present state is 1 and is to remain 1. This can happen either when S = 1 and R = 0 (set condition) or when S = 0 and R = 0 (no change condition). Thus R has to be 0, but S can be at either level. In the table, this was indicated with a 'X' under S and '0' under R.

4.6.2 JK Flip-Flop

The truth table and excitation table for JK flip-flop are shown in Table 4.12 (*a*) and 4.12 (*b*) respectively. Let us examine each case.

Table 4.12(*a*) Truth table

J	K	Q_{n+1}
0	0	Q_n
0	1	0
1	0	1
1	1	$\overline{Q}_n$

Table 4.12(*b*) Excitation table

Q_n	Q_{n+1}	J	K
0	0	0	×
0	1	1	×
1	0	×	1
1	1	×	0

(*a*) **0 → 0 Transition** : When both present state and next state are 0, the J input must be 0 and the K-input can be either 0 or 1, which is represented by X in the table.

(*b*) **0 → 1 Transition** : Here the present state is 0 and is to change to 1. This can happen when J = 1 and K = 0 (set) (or) when J = 1 and K = 1 (Toggle condition). Thus J has to be 1, but K can be either 0 or 1.

(*c*) **1 → 0 Transition** : The present state is 1 and is to change to a 0. This can happen when J = 0 and K = 1 (or) when J = K = 1. Thus K has to be 1 and J may be either '0' or 1.

(*d*) **1 → 1 Transition** : When both present state and next state are 1, the K input must be 0 and J input can be 0 or 1.

Note : From the excitation table of the JK flip-flop, we can see that it has more don't care terms. Usually more don't care terms simplify the function.

4.6.3 D Flip-Flop

Tables 4.13 (*a*) and (*b*) show the truth table and excitation table for D flip-flop, respectively. In D-flip-flop, the next state is always equal to the D input and it is independent of the present state. Therefore D must be 0 if Q_{n+1} has to be 0, and 1 if Q_{n+1} has to be 1, regardless of the value of Q_n.

Table 4.13(*a*) Truth table

D	Q_{n+1}
0	0
1	1

Table 4.13(*b*) Excitation table

Q_n	Q_{n+1}	D
0	0	0
0	1	1
1	0	0
1	1	1

4.6.4. T Flip-Flop

The Tables 4.14 (*a*) and (*b*) show the truth table and excitation table for T flip-flop.

(*a*) For 0 → 0 and 1 → 1 transitions T = 0 and

(*b*) For 0 → 1 and 1 → 0 transitions T must be 1.

Table 4.14(*a*) Truth table

T	Q_{n+1}
0	Q_n
1	$\overline{Q}_n$

Table 4.14(*b*) Excitation table

Q_n	Q_{n+1}	T
0	0	0
0	1	1
1	0	1
1	1	0

4.7 APPLICATIONS OF FLIP-FLOPS

Flip-flops find wide applications in counter circuits, frequency dividers, shift and storage registers. A wide variety of serial decoding, comparison and timing functions can be accomplished using flip-flops. Using flip-flops, we can generate a variety of one-shots.

An operation that occurs very frequently in digital system is transfer of information from one flip-flop or a group of flip-flops to another flip-flop or another group of flip-flops.

4.8 ANALYSIS OF SYNCHRONOUS SEQUENTIAL CIRCUITS

The behaviour of a sequential circuit can be determined from the inputs, the outputs and the states of its flip-flop. The outputs and next state are both a function of its inputs and the present state. The analysis of a sequential circuit consists of obtaining a table or diagram for the time sequence of inputs, outputs and internal states. The success of analysis or design of sequential circuit depends on the aids and systematic techniques such as state tables, state diagrams and state equations used in these processes.

4.8.1. State Table

Fig. 4.45 shows a clocked sequential circuit. It has one input variable X, output variable 'Y', and two clocked JK flip-flops. The flip-flops are labelled as A and B, and their outputs are labelled as A and $\overline{A}$, and B and $\overline{B}$, respectively. The output Y can be determined by Boolean expression $\boxed{Y = XA\overline{B}}$ which is derived using three input AND gate.

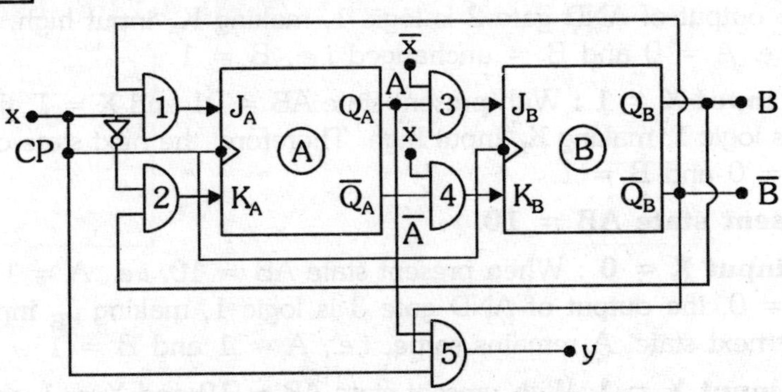

Fig. 4.45 An example of sequential circuit.

For understanding the given sequential circuit, now we start analysing the circuit.

Table 4.15 shows the state table for the given sequential circuit. It represents the relationship between input, output and flip-flop states. It consists of three columns : present state, next state and output.

(*a*) The "present state" specifies the state of the flip-flops before occurance of a clock pulse.

(*b*) The "next state" is the state of flip-flop after the application of a clock pulse.

(*c*) The output section gives the value of the output variables during the present state. Both next state and output sections have two columns representing two possible input conditions X = 0 and X = 1.

Table 4.15

Present state	*Next state*		*Output*	
	AB	*AB*	*(Y)*	
AB	X = 0	X = 1	X = 0	X = 1
00	00	10	0	0
01	01	00	0	0
10	11	10	0	1
11	01	11	0	0

We can derive the state table as follows :

Step 1. Find the next state for all possible present states and input states.

(*a*) **If present state AB = 00.**

(*i*) **If Input X = 0 :** When present state is 00, *i.e.*, A = 0 and B = 0 and X input = 0, outputs of all AND gates are logic 0. Therefore, J and K inputs for both flip-flops are 0, giving the next state same as present state.

(*ii*) **If Input X = 1 :** When present state is 00, *i.e.*, A = 0 and B = 0 and X = 1, AND gates 1 and 4 produce a logic 1 signal at the J_A input and K_B input. Therefore, in the next state A is set and B is cleared, *i.e.*, AB = 10.

(*b*) **If present state AB = 01.**

(*i*) **If input X = 0 :** When present state is 01, *i.e.*, A = 0 and B = 1 and *x* = 0, the output of AND gate 2 is logic 1, making K_A input high. Therefore, in the next state, A = 0 and B = unchanged *i.e.*, B = 1

(*ii*) **If input X = 1 :** With present state AB = 01 and X = 1, the output of AND gate 4 is logic 1, making K_B input high. Therefore, the next state of A is unchanged, *i.e.*, A = 0 and B = 1.

(*c*) **If present state AB = 10.**

(*i*) **If input X = 0 :** When present state AB = 10, *i.e.*, A = 1 and B = 0 and input X = 0, the output of AND gate 3 is logic 1, making J_B input high. Therefore, in the next state, A remains same, *i.e.*, A = 1 and B = 1.

(*ii*) **If input X = 1.** With present state AB = 10 and X = 1, the output of AND gate

1 is logic 1, making the J_A input high. With this, the next state of A = 1 and B remains unchanged *i.e.*, B = 0.

(*d*) **If present state AB = 11.**

(*i*) **If input X = 0 :** With AB = 11 and X = 0, the outputs of AND gate 2 and AND gate 3 are logic 1, making K_A and J_B inputs high. Therefore, in the next state A = 0 and B = 1.

(*ii*) **If input X = 1 :** With AB = 11 and X = 1, the outputs of all AND gates are logic 0. Therefore, the next states of A and B remain unchanged.

Step 2. Determine the entries in the output section. For this, we have to examine the AND gate 5 for all possible present states and inputs.

(*i*) If X = 0, regardless of present state, output Y = 0

(*ii*) If X = 1, output Y = 1, only when A = 1 and B = 0.

Thus, the state table of any sequential circuit can be obtained by the same procedure used in the above example. This example contains 2 flip-flops and 1 input variable, and 1 output, producing 4 rows, 2 columns in the next state and output sections. In general, a sequential circuit with *m*-flip flops and *n*-input variables produces 2^m rows, and one for each state and 2^n columns, one for each input combination in the next state and output sections of the state table.

4.8.2 State Diagram

"State diagram" is a graphical representation of a state table. Fig. 4.46 shows the state diagram for sequential circuit shown in Fig. 4.45. Here each state is represented by a circle, and transition between states is indicated by directed lines connecting the circles. A directed line connecting a circle with itself indicates that next state is same as present state. The binary number inside each circle identifies the state represented by the circle. The directed lines are labelled with two binary numbers separated by a symbol '/' (slash). The input value that causes the state transition is labelled first and the output value during the present state is labelled after the symbol '/'.

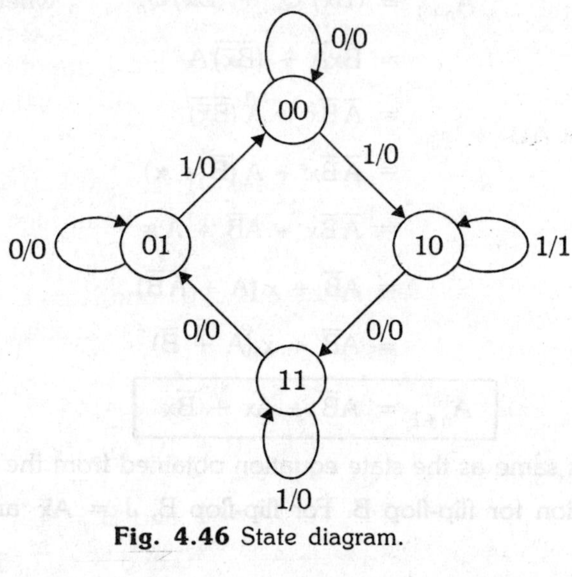

Fig. 4.46 State diagram.

4.8.3 State Equations

A state equation is also known as "application equation". It is an algebraic expression that specifies the conditions for a flip-flop state transition. The left side of the equation represents the next state of the flip-flop and the right side, a Boolean expression that specifies the present state conditions that makes the next state equal to 1. The state equation can be derived from the state table or logic diagram. Determination of state equations is shown in Fig. 4.47.

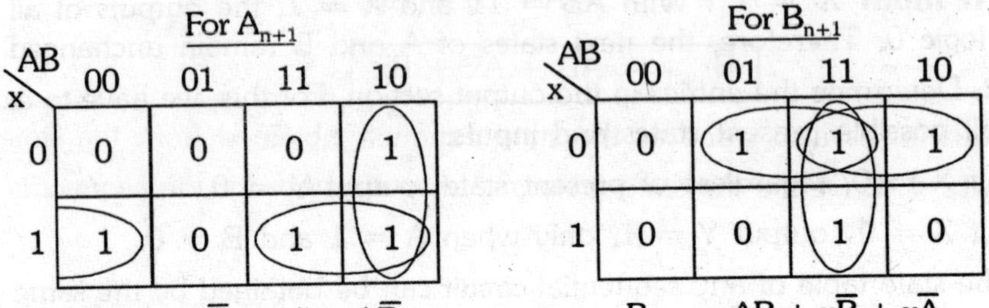

$A_{N+1} = x\bar{B} + xA + A\bar{B}$ $\quad$ $B_{N+1} = AB + x\bar{B} + x\bar{A}$

Fig. 4.47

In the K-map, 1 is marked to represent the input and present state conditions that cause the flip-flop to go to a next state of 1. For example, present states 01, 10 and 11 with X = 0 and present state 11 with X = 1 cause the flip-flop B to go to a next state of 1, therefore these entries are marked '1' in the K-map for the flip-flop.

Derivation of state equation from logic diagram. The state equation can be derived directly from the logic diagram. Looking at Fig. 4.45, we can see that the signal for J input of the flip-flop A is generated by the function $\bar{B}x$ and the signal for input K by the function $B\bar{x}$. Substituting $J = \bar{B}x$ and $K = B\bar{x}$ into a JK flip-flop, characteristic equation given by

$$Q_{n+1} = J\bar{Q}_n + \bar{K}Q_n \text{ we obtain,}$$

$$A_{n+1} = (\bar{B}x)\,\bar{Q}_n + (\overline{B\bar{x}})\,Q_n \qquad \text{where } Q_n = A$$

$$= \bar{B}x\bar{A} + (\overline{B\bar{x}})\,A$$

$$= \bar{A}\bar{B}x + A\,(\overline{B\bar{x}})$$

$$= \bar{A}\bar{B}x + A\,(\bar{B} + x)$$

$$= \bar{A}\bar{B}x + A\bar{B} + Ax$$

$$= A\bar{B} + x\,(A + \bar{A}\bar{B})$$

$$= A\bar{B} + x\,(A + \bar{B})$$

$$\boxed{A_{n+1} = A\bar{B} + Ax + \bar{B}x}$$

This equation is same as the state equation obtained from the state table. Similarly, we can find the state equation for flip-flop B. For flip-flop B, $J = A\bar{x}$ and $K = \bar{A}x$.

Therefore the state equation of flip-flop B is given as

$$B_{n+1} = A\bar{x}\,\bar{B} + (\overline{\bar{A}x})\,B$$

$$= A\bar{x}\,\bar{B} + (A + \bar{x})\,B$$

$$= A\bar{x}\,\bar{B} + AB + B\bar{x}$$

$$= \bar{x}(A\bar{B} + B) + AB$$

$$= \bar{x}(A + B) + AB$$

$$\boxed{B_{n+1} = AB + A\bar{x} + B\bar{x}}$$

This equation is also same as the state equation obtained from the state table.

Example 4.4. *Derive the state table and state diagram for the sequential circuit shown in Fig. 4.48.*

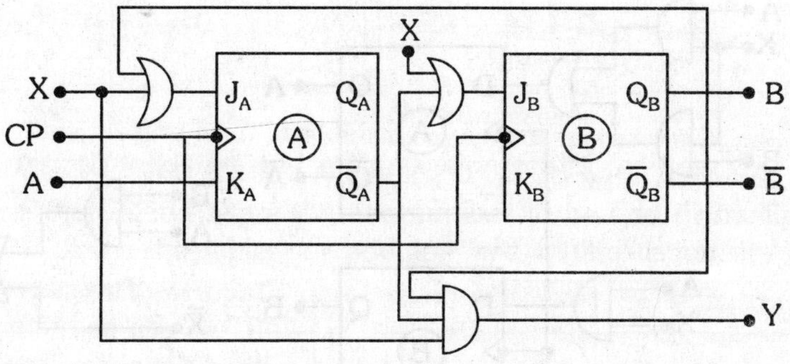

Fig. 4.48

Solution. The given circuit has two flip-flops, A and B, one input X and one output 'Y'. So, it produces a state table with 4 rows as shown in Table 4.16.

Table 4.16

Present state		*Next state*				*Output*	
		X = 0		X = 1		X = 0	X = 1
A	B	A	B	A	B	*y*	*y*
0	0	0	1	1	1	0	0
0	1	1	0	1	0	0	1
1	0	0	1	0	0	0	0
1	1	0	0	0	0	0	0

The state diagram for the sequential circuit is shown in Fig. 4.49.

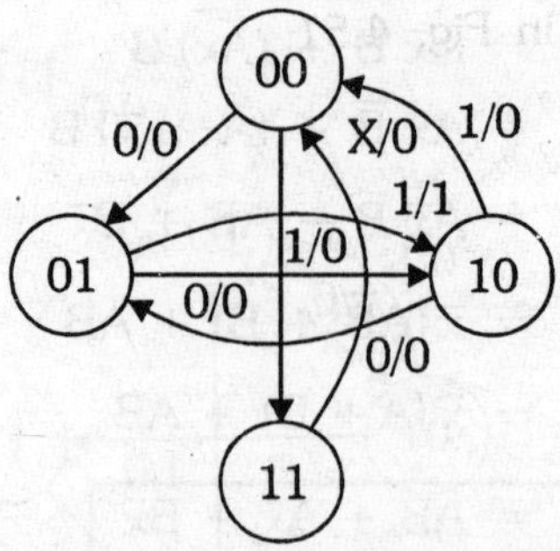

Fig. 4.49 State diagram.

Note : The directed line between states 11 to 00 is labelled with x/0, where x = 0 or x = 1.

Example 4.5. *Derive the state table and state diagram for the sequential circuit shown in Fig. 4.50.*

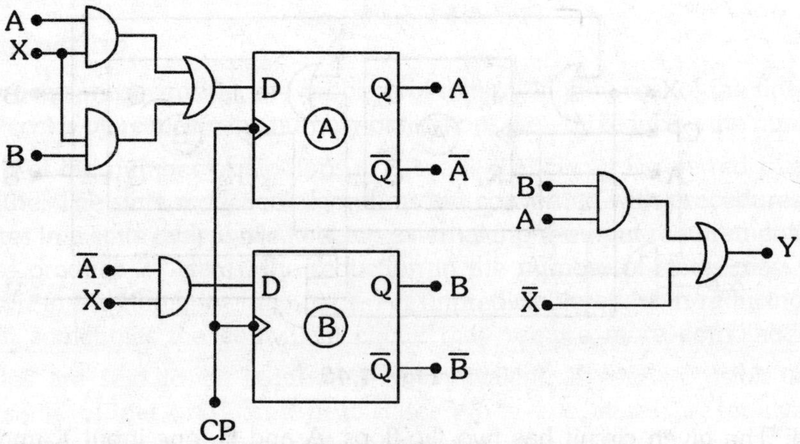

Fig. 4.50

Solution. The state table can be derived using characteristic equations of the flip-flops.

$$A_{n+1} = AX + BX$$
$$B_{n+1} = \overline{A}X$$
$$y = (A + B)\,\overline{X}.$$

The state table is shown in Table 4.17, which contains four rows and 3 columns, the next state and output has two subcolumns each.

Table 4.17 State table

Present state AB	*Next state AB*		*Output (Y)*	
	X=0	X=1	X=0	X=1
00	00	01	0	0
01	00	11	1	0
10	00	10	1	0
11	00	10	1	0

The state diagram is shown in Fig. 4.51.

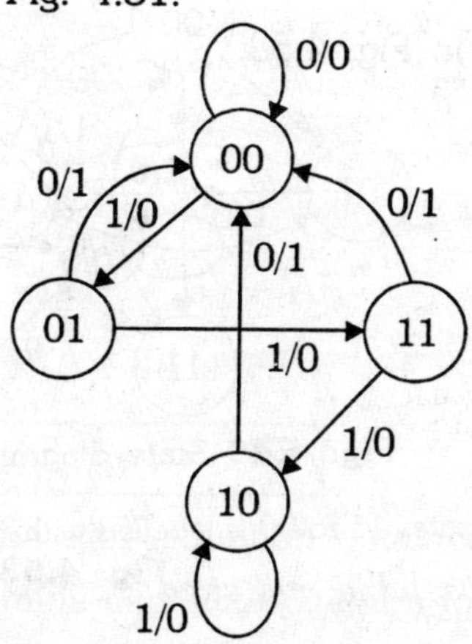

Fig. 4.51 State diagram.

Example 4.6. *Derive the state table and state diagram for sequential circuit shown in Fig. 4.52.*

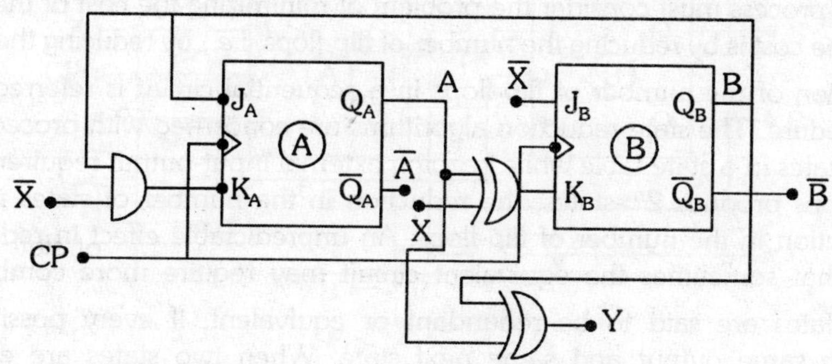

Fig. 4.52 Synchronous sequential circuit.

Solution. The flip-flop input functions for the sequential circuit are

$J_A = B \qquad J_B = \bar{X} \qquad y = A \oplus B$

$K_A = B\bar{X} \qquad K_B = A \oplus X$

The state table is shown in Table 4.18.

Table 4.18. State table

Present state	*Next state*		*Output*
	X = 0	*X = 1*	*(y)*
A B	A B	A B	
0 0	0 1	0 0	0
0 1	1 1	1 0	1
1 0	1 1	1 0	1
1 1	0 0	1 1	0

The state diagram is shown in Fig. 4.53.

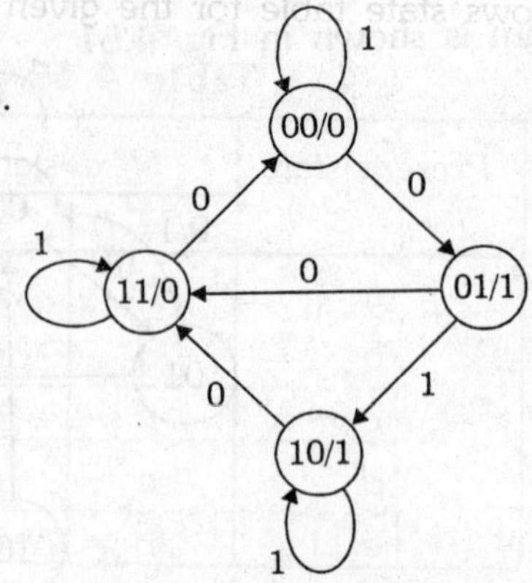

Fig. 4.53 State diagram.

Note : Here each circle is coded with state binary number/output.

4.8.4 State Reduction

Any design process must consider the problem of minimizing the cost of the final circuit. One way to reduce the cost is by reducing the number of flip-flops, *i.e.*, by reducing the number of states.

The reduction of the number of flip-flops in a sequential circuit is referred to as the "state-reduction" procedure. The state reduction algorithms are concerned with procedures for reducing the number of states in a state table while keeping external input-output requirements unchanged. Since '*m*' flip-flops produce 2^m-states, the reduction in the number of states may (or may not) result in a reduction in the number of flip-flops. An unpredictable effect in reducing the number of flip-flops is that sometimes the equivalent circuit may require more combinational gates.

The two states are said to be redundant or equivalent, if every possible set of inputs, generate exactly same output and same next state. When two states are equivalent, one of them can be removed without altering the input-output relationship.

Let us illustrate the state reduction technique with an example. We start with a sequential circuit whose specification is given in the state diagram of Fig. 4.54. Here each state is denoted by a letter symbol instead of it's binary value, because in state reduction technique internal states are not important, but input-output sequences are important. The procedure contains two steps.

Step 1. Finding the state table for the given state diagram.

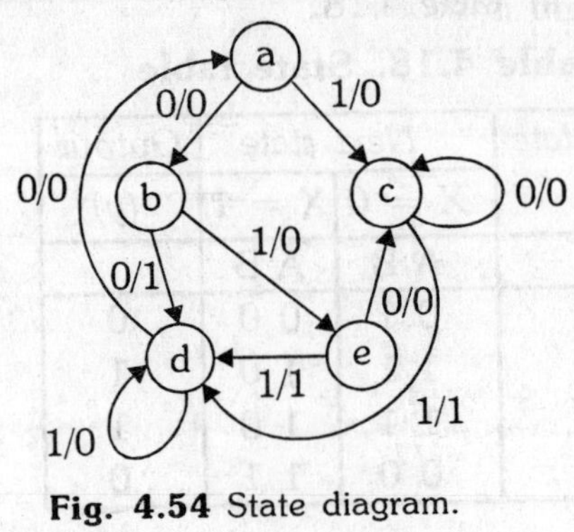

Fig. 4.54 State diagram.

Table 4.19 shows state table for the given state diagram.

Table 4.19 State table

Present state	Next state		Output	
	x = 0	x = 1	x = 0	x = 1
a	b	c	0	0
b	d	e	1	0
c	c	d	0	1
d	a	d	0	0
e	c	d	0	1

Step 2. Finding equivalent states.

Equivalent states are the states, which have same next states and outputs for all possible input combinations. In the given circuit, there are two input combinations : $x = 0$ and $x = 1$. From the state table, we can easily identify that states '*c*' and '*e*' are equivalent. This is because *c* and *e* both states go to states *c* and *d* and have outputs of 0 and 1 for $x = 0$ and $x = 1$, respectively. Therefore, state '*e*' can be removed and replaced by state *c*. The final reduced table is shown in Table 4.20 and the state diagram for the reduced table consists of only 4-states and shown in Fig. 4.55.

Table 4.20 Reduced state table

Present state	Next state		Output	
	x = 0	x = 1	x = 0	x = 1
a	b	c	0	0
b	d	c	1	0
c	c	d	0	1
d	a	d	0	0

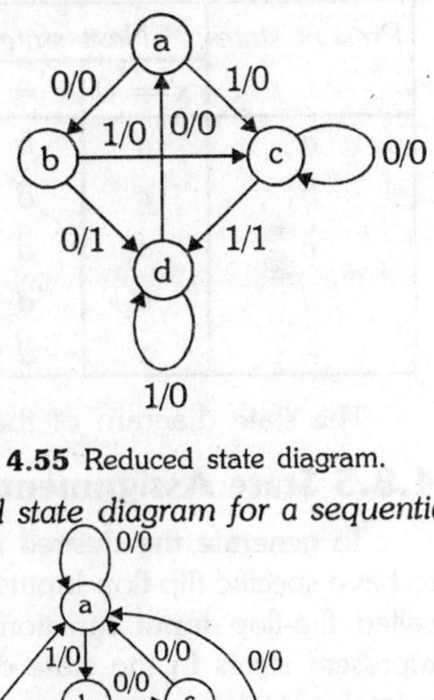

Fig. 4.55 Reduced state diagram.

Example 4.7. *Obtain the reduced state table and reduced state diagram for a sequential circuit whose state diagram is shown in Fig. 4.56.*

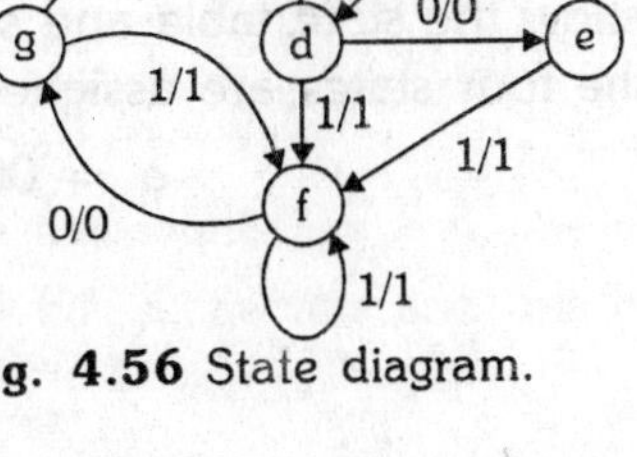

Fig. 4.56 State diagram.

Solution. The given state diagram contains seven states, one input and one output. The state table is shown in Table 4.21 (*a*).

Table 4.21(*a*) State table

Present state	*Next state*		*Output*	
	x = 0	*x = 1*	*x = 0*	*x = 1*
a	*a*	*b*	0	0
b	*c*	*d*	0	0
c	*a*	*d*	0	0
d	*e*	*f*	0	1
e	*a*	*f*	0	1
f	*g*	*f*	0	1
g	*a*	*f*	0	1

Table 4.21(*b*) Reduced state table

Present state	*Next state*		*Output*	
	x = 0	*x = 1*	*x = 0*	*x = 1*
a	*a*	*b*	0	0
b	*c*	*d*	0	0
c	*a*	*d*	0	0
d	*e*	*f*	0	1
e	*a*	*f*	0	1
f	*e*	*f*	0	1

From the state table, it is clear that states *e* and *g* are equivalent. So, state *g* is removed and replaced by state *e*. The reduced state table is shown in Table 4.21 (*b*). From the reduced state table, the states *d* and *f* are equivalent, hence '*f*' can be replaced by *d* and it can be removed. Then finally reduced state table is shown in Table 4.21 (*c*).

Table 4.21(c) Reduced state table

Present state	*Next state*		*Output*	
	x = 0	*x = 1*	*x = 0*	*x = 1*
a	*a*	*b*	0	0
b	*c*	*d*	0	0
c	*a*	*d*	0	0
d	*e*	*d*	0	1
e	*a*	*d*	0	1

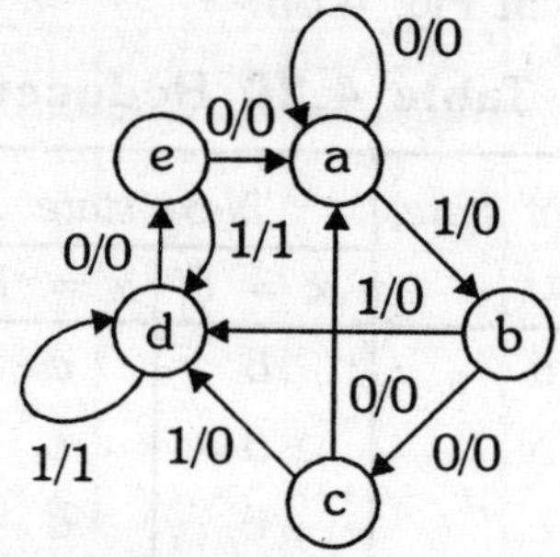

Fig. 4.57 Reduced state diagram.

The state diagram of the reduced state table is shown in Fig. 4.57.

4.8.5 State Assignment

To generate the desired next state for a particular present state and inputs, it is necessary to have specific flip-flop inputs. These flip-flop inputs are described by a set of Boolean functions called flip-flop input functions. To determine the flip-flop input functions, it is necessary to represent states in the state diagram using binary values instead of alphabets. This procedure is known as "state assignment". We must assign binary values to the states in such a way that it is possible to implement flip-flop input functions using minimum logic gates. Table 4.22 and Fig. 4.58 show the state table and state diagram of Fig. 4.55 after state assignments. One way in which the four states are assigned is as follows :

$$a = 00,\ b = 01,\ c = 10 \text{ and } d = 11.$$

Table 4.22 State table after assignment

Present state	*Next state*		*Output*	
	x = 0	x = 1	x = 0	x = 1
a (00)	01	10	0	0
b (01)	11	10	1	0
c (10)	10	11	0	1
d (11)	00	11	0	0

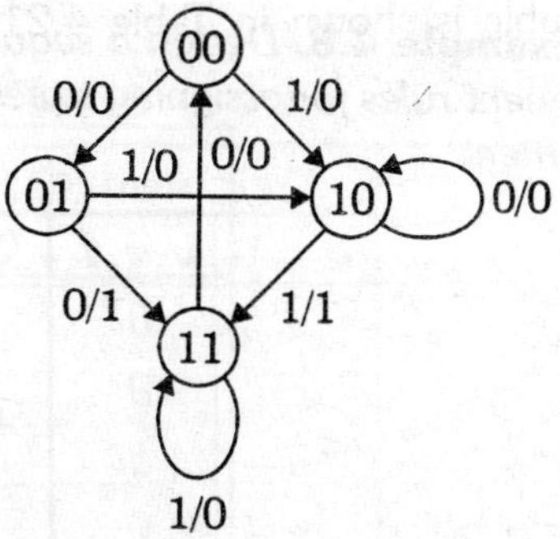

Fig. 4.58 State diagram after state assignment.

4.8.5.1 Rules for state assignments

We can assign any state with any binary value, but this state assignment is the most important parameter in reducing the cost of the final circuit. It is better to follow some rules for state assignment. Basically there are two rules for making state assignments.

Rule 1. States having the same NEXT STATES for a given input condition should have assignment in such a way so that they can be grouped into logically adjacent cells in a K-map.

Fig. 4.59 shows an example for the rule 1. Here, there are four states, whose next state is same (000). If we assign the binary values for those four states as 100, 110, 111 and 101, then we can combine them logically in a K-map.

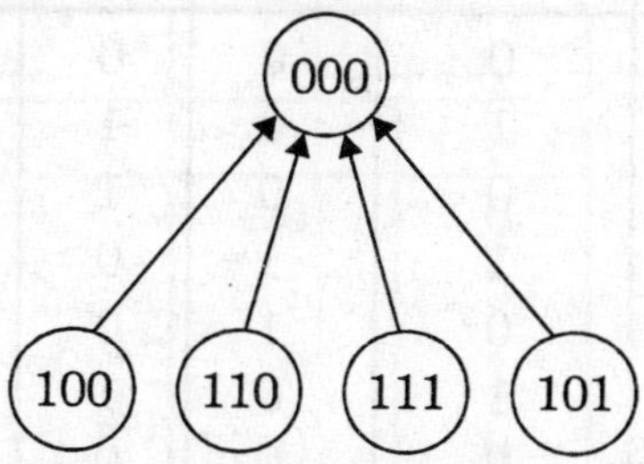

Fig. 4.59 Example for rule 1.

Rule 2. States that are the next states of a single state should be assigned in such a way that they can be grouped into logically adjacent cells in a K-map.

Fig. 4.60 shows the example for rule 2. Here there are four states originating from the same state (000). If we assign the 4-states as 100, 110, 111 and 101, then we can combine them logically in a K-map.

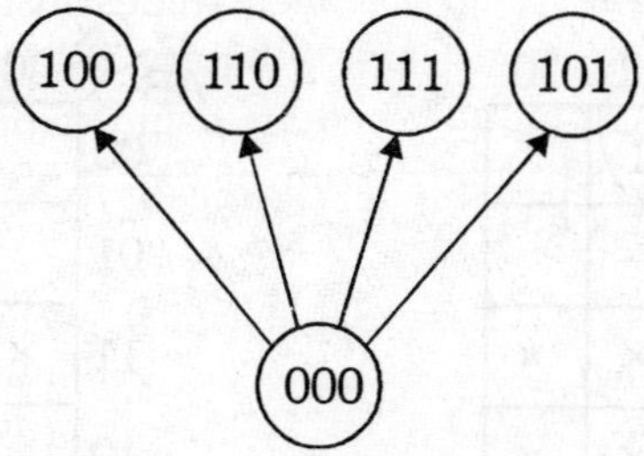

Fig. 4.60 Example for rule 2.

Example 4.8. *Design a sequential circuit for a state diagram shown in Fig. 4.61. Use state assignment rules for assigning states and compare the required combinational circuit with random assignment.*

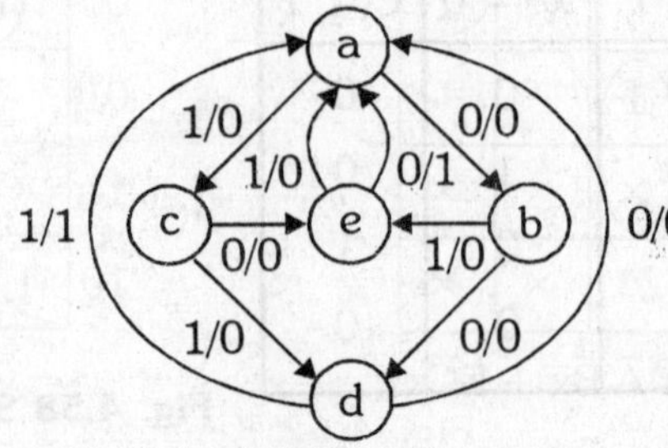

Fig. 4.61

Solution. The given state table has 5 states, *a*, *b*, *c*, *d* and *e*, and one input and one output. By applying the random state assignment, we assign as :

a = 000 b = 001 c = 010 d = 011 e = 100. The excitation table with this assumption is given in Table 4.23.

Table 4.23 Excitation table

Present state			*Input*	*Next state*			*Output*
A_n	B_n	C_n	x	A_{n+1}	B_{n+1}	C_{n+1}	y
0	0	0	0	0	0	1	0
0	0	0	1	0	1	0	0
0	0	1	0	0	1	1	0
0	0	1	1	1	0	0	0
0	1	0	0	1	0	1	0
0	1	0	1	0	1	1	0
0	1	1	0	0	0	0	0
0	1	1	1	0	0	0	1
1	0	0	0	0	0	0	1
1	0	0	1	0	0	0	0

The simplified expressions for outputs are shown in Fig. 4.62 using K-maps.

For A_{n+1}

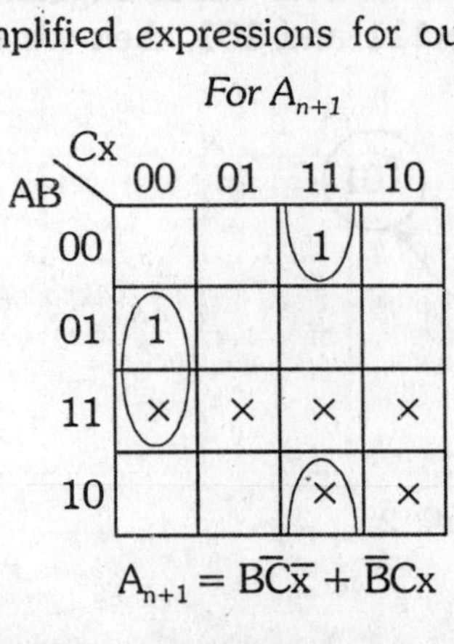

$A_{n+1} = B\overline{C}\overline{x} + \overline{B}Cx$

For B_{n+1}

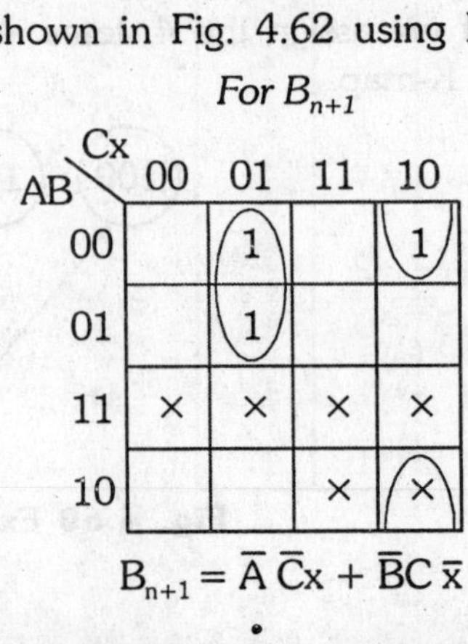

$B_{n+1} = \overline{A}\,\overline{C}x + \overline{B}C\,\overline{x}$

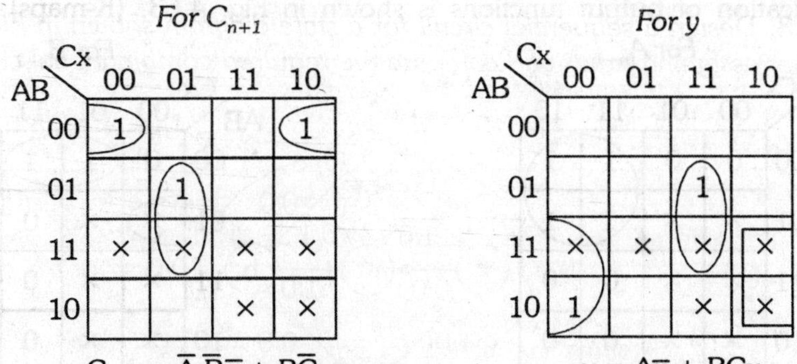

$C_{n+1} = \overline{A}\,\overline{B}\overline{x} + B\overline{C}x$ $\qquad$ $y = A\overline{x} + BCx$

Fig. 4.62 K-maps.

To implement the above four output functions we need the following combinational circuits extra rather than three flip-flops.

Three input AND gates – 7
Two input AND gates – 1
Two input OR gates – 4

Totally 12 gates with 31 inputs

Now we apply the state assignment rules and compare the results.

According to rule 1 : 'state *e*' and 'state *d*' must be adjacent.

According to rule 2 : 'state *b*' and 'state *c*' must be adjacent. With this, the new state assignments are

$$a = 000 \quad b = 001 \quad c = 011 \quad d = 101 \quad e = 111.$$

The new excitation table is shown in Table 4.24.

Table 4.24 Excitation table

Present state			*Input*	*Next state*			*Output*
A_n	B_n	C_n	x	A_{n+1}	B_{n+1}	C_{n+1}	y
0	0	0	0	0	0	1	0
0	0	0	1	0	1	1	0
0	0	1	0	1	0	1	0
0	0	1	1	1	1	1	0
0	1	1	0	1	1	1	0
0	1	1	1	1	0	1	0
1	0	1	0	0	0	0	0
1	0	1	1	0	0	0	1
1	1	1	0	0	0	0	1
1	1	1	1	0	0	0	0

The simplification of output functions is shown in Fig. 4.63. (K-maps)

For A_{n+1}

AB \ Cx	00	01	11	10
00	0	0	1	1
01	×	×	1	1
11	×	×	0	0
10	×	×	0	0

$A_{n+1} = \bar{A}C$

For B_{n+1}

AB \ Cx	00	01	11	10
00	0	1	1	0
01	×	×	0	1
11	×	×	0	0
10	×	×	0	0

$B_{n+1} = \bar{A}\,\bar{B}x + \bar{A}B\,\bar{x}$

For C_{n+1}

AB \ Cx	00	01	11	10
00	1	1	1	1
01	×	×	1	1
11	×	×	0	0
10	×	×	0	0

$C_{n+1} = \bar{A}$

For y

AB \ Cx	00	01	11	10
00	0	0	0	0
01	×	×	0	0
11	×	×	0	1
10	×	×	1	0

$y = AB\bar{C} + A\bar{B}x$

Fig. 4.63 K-maps.

The logic circuit for these outputs need 3 flip-flops and the following logic gates :

Three input AND gates – 4
Two input AND gates – 1
Two input OR gates – 2

} Totally the circuit needs 7 gates and 18 inputs.

Thus, simply by applying the rules for state assignment, the number of combinational gates required is reduced.

4.9 DESIGN PROCEDURE

In this section, we will see the design procedure for sequential circuits. The first stage of design procedure is "stating the word description" and the last stage is "drawing/obtaining the logic diagram." The procedure for designing clocked sequential circuits is summarised by a list of consecutive recommended steps :

1. Define the word description of the problem to which we have to design the logic diagram. This may be informed by a state diagram, a timing diagram or other pertinent information.
2. Obtain the state table of the circuit from the given information.
3. The number of states may be reduced by using state reduction methods if the given

sequential circuit can be characterized by input–output relationships independent of the number of states.

4. Assign the binary values to each state in the state table using state assignment rules (if possible).
5. Determine the number of flip-flops needed and assign a letter symbol to each flip-flop.
6. Choose the type of flip-flop to be used according to the application.
7. From the state table, obtain the circuit excitation and output tables.
8. Using the K-map, RDM or any other simplified method, derive the circuit output functions and the flip-flop input functions.
9. Draw the logic diagram.

Now we will see the detailed description of each step.

The word specification of the circuit behaviour is given by assuming that the logic circuit designer is familiar with digital logic terminology. It is necessary that the designer use intuition and experience to arrive at the correct interpretation of the circuit specifications, because word description may be incomplete in exact. From this we obtain the state diagram and from it we obtain the state table.

The reduction of number of states and binary value assignment to each state gives the reduction in combinational circuit requirement.

The number of flip-flops required to design any sequential circuit depends on the number of states. For example, n-flip flops are required if the circuit is having maximum 2^n states. The unused states are taken as don't care conditions during the design of combinational circuit.

The type of the flip-flop used depends on the application of the designing circuit.

* Use of D flip-flop is advisable for applications such as transfer of data *i.e.*, in shift registers.

* T-type flip-flops are used in the applications involving complementation (toggling) *i.e.*, binary counters.

* JK type flip-flops are used for general applications.

The external output information is specified in the output section of the state table. From it, we can derive the circuit output functions. By using the excitation table, we can obtain the flip-flop input equations. To get more familiar with entire procedure, let us see some examples.

Example 4.9. *A sequential circuit has one input and one output and its state diagram is shown in Fig. 4.64. Design the sequential circuit using (i) T flip-flops (ii) D flip-flops and (iii) JK flip-flops.*

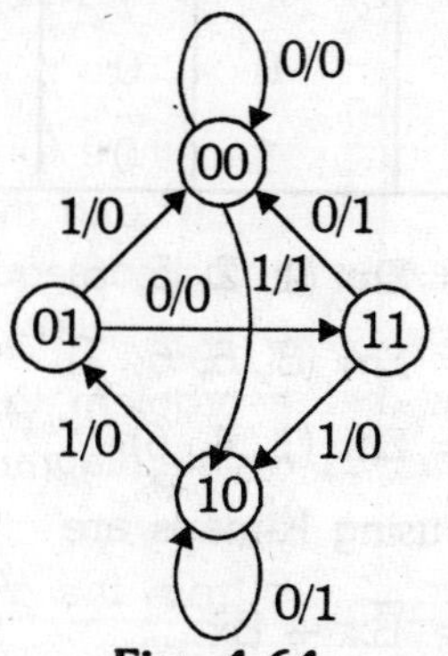

Fig. 4.64

Solution. The given state diagram consists of four states. It has one input (x) and one output z. The state table for this circuit is shown in Table 4.25. From the table , it is clear that there are no equivalent states. Therefore, there is no reduction in the state diagram. As the state diagram contains 4-states, it requires 2 flip-flops and are named as A and B.

Table 4.25 State table

Present state		Next state				Output	
		$x = 0$		$x = 1$		$x = 0$	$x = 1$
A	B	A	B	A	B	y	y
0	0	0	0	1	0	0	1
0	1	1	1	0	0	0	0
1	0	1	0	0	1	1	0
1	1	0	0	1	0	1	0

(*i*) **Design using T flip-flop.** For the design of the circuit using T-flip flop we need the excitation table first and it is shown in Table 4.26, from which we can find the flip-flop input equations and circuit output function.

Table 4.26 Excitation table

Present state		Input	Next state		Flip inputs		Output
A	B	x	A	B	T_A	T_B	y
0	0	0	0	0	0	0	0
0	0	1	1	0	1	0	1
0	1	0	1	1	1	0	0
0	1	1	0	0	0	1	0
1	0	0	1	0	0	0	1
1	0	1	0	1	1	1	0
1	1	0	0	0	1	1	1
1	1	1	1	0	0	1	0

From the state table, $T_A = \Sigma m\ (1, 2, 5, 6)$

$$T_B = \Sigma m\ (3, 5, 6, 7)$$

$$y = \Sigma m\ (1, 4, 6)$$

The simplified expressions by using K-maps are

$$T_A = \overline{B}x + B\overline{x}$$

$$T_B = AB + Bx + Ax$$

$$y = A\bar{x} + \bar{A}\bar{B}x$$

The logic diagram is shown in Fig. 4.65.

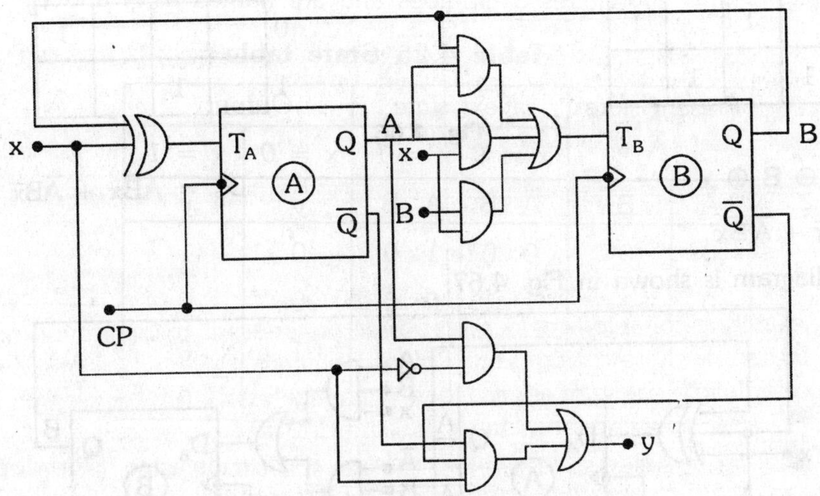

Fig. 4.65 Logic diagram using T flip-flop.

(*ii*) **Design using D-flip flop.** The excitation table is shown in Table 4.27.

Table 4.27 Excitation table

Present state		*Input*	*Next state*		*Flip flop inputs*		*Output*
A	B	x	A	B	D_A	D_B	y
0	0	0	0	0	0	0	0
0	0	1	1	0	1	0	1
0	1	0	1	1	1	1	0
0	1	1	0	0	0	0	0
1	0	0	1	0	1	0	1
1	0	1	0	1	0	1	0
1	1	0	0	0	0	0	1
1	1	1	1	0	1	0	0

From the table,

$$D_A = \Sigma m\ (1, 2, 4, 7)$$

$$D_B = \Sigma m\ (2, 5)$$

$$y = \Sigma m\ (1, 4, 6)$$

The flip-flop inputs and output expressions can be obtained from K-maps as shown in Fig. 4.66.

For D_A

A \ Bx	00	01	11	10
0		1		1
1	1		1	

For D_B

A \ Bx	00	01	11	10
0				1
1		1		

Fig. 4.66

$D_A = A \oplus B \oplus x$

$D_B = A\overline{B}x + \overline{A}B\overline{x}$

$y = A\overline{x} + \overline{A}\,\overline{B}x$

The logic diagram is shown in Fig. 4.67.

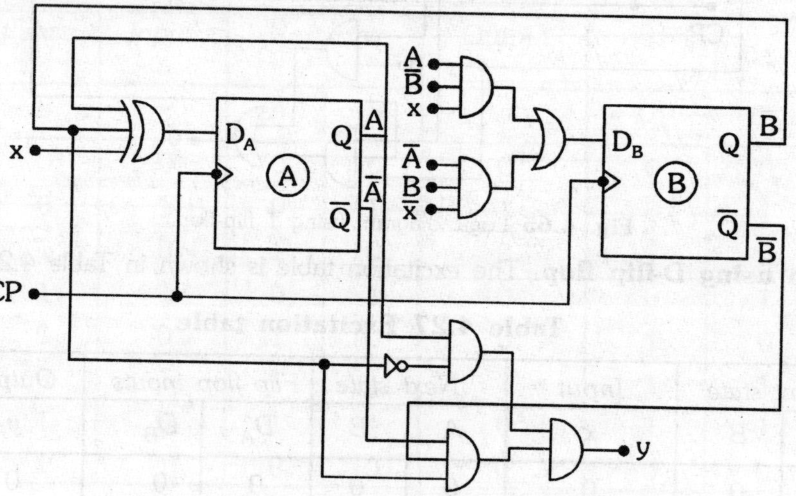

Fig. 4.67 Circuit design using D flip-flops.

(*iii*) **Design using JK flip-flop.** The excitation table for JK flip-flop is shown in Table 4.28.

Table 4.28

Present state	*Input*	*Next state*	*Flip-flop inputs*		*Output*
A B	*x*	*A B*	J_A K_A	J_B K_B	*y*
0 0	0	0 0	0 ×	0 ×	0
0 0	1	1 0	1 ×	0 ×	1
0 1	0	1 1	1 ×	× 0	0
0 1	1	0 0	0 ×	× 1	0
1 0	0	1 0	× 0	0 ×	1
1 0	1	0 1	× 1	1 ×	0
1 1	0	0 0	× 1	× 1	1
1 1	1	1 0	× 0	× 1	0

$$J_A = \Sigma m\ (1, 2) + \Sigma d\ (4, 5, 6, 7)$$
$$K_A = \Sigma m\ (5, 6) + \Sigma d\ (0, 1, 2, 3)$$
$$J_B = \Sigma m\ (5) + \Sigma d\ (2, 3, 6, 7)$$
$$K_B = \Sigma m\ (3, 6, 7) + \Sigma d\ (0, 1, 4, 5)$$
$$y = \Sigma m\ (1, 4, 6)$$

The flip-flop input equations and circuit output equations are given below

$$J_A = B \oplus x \qquad K_A = B \oplus x$$
$$J_B = Ax \qquad K_B = A + x$$
$$y = A\bar{x} + \bar{A}\bar{B}x$$

The logic diagram is shown in Fig. 4.68.

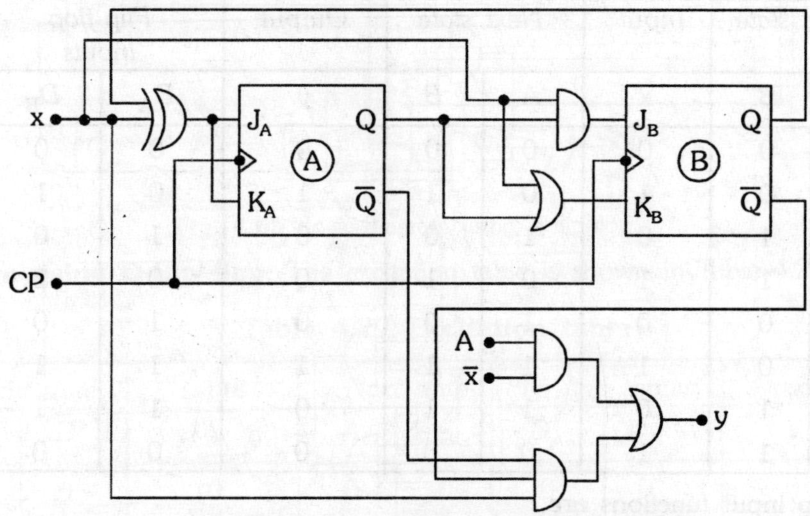

Fig. 4.68 Logic diagram using JK flip-flop.

Example 4.10. *Design the clocked sequential circuit for the given state diagram shown in Fig. 4.69.*

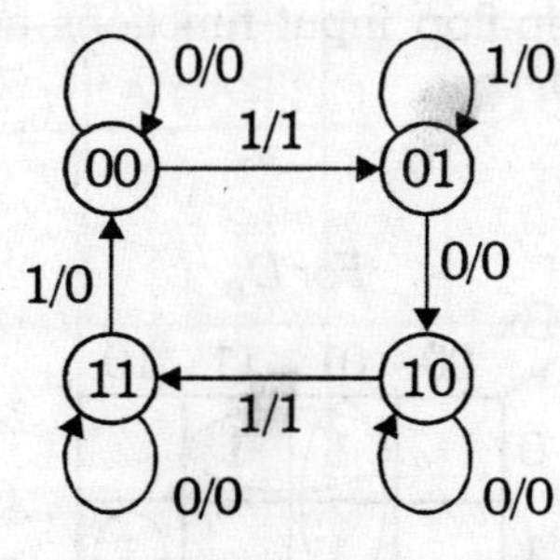

Fig. 4.69

Solution. The given state diagram contains 4 states and its equivalent state table is shown in Table 4.29.

Table 4.29. State table

Present state		*Next state*		*Output*	
A	*B*	x = 0	x = 1	x = 0	x = 1
0	0	0 0	0 1	0	1
0	1	1 0	0 1	0	0
1	0	1 0	1 1	0	1
1	1	1 1	0 0	0	0

Design with D flip-flops. The excitation table for the logic diagram is shown in Table 4.30.

Table 4.30. Excitation table

Present state		*Input*	*Next state*		*Output*	*Flip-flop inputs*	
A	*B*	*x*	*A*	*B*	*y*	D_A	D_B
0	0	0	0	0	0	0	0
0	0	1	0	1	1	0	1
0	1	0	1	0	0	1	0
0	1	1	0	1	0	0	1
1	0	0	1	0	0	1	0
1	0	1	1	1	1	1	1
1	1	0	1	1	0	1	1
1	1	1	0	0	0	0	0

The flip-flop input functions are

$$D_A = \Sigma m\ (2, 4, 5, 6)$$

$$D_B = \Sigma m\ (1, 3, 5, 6)$$

$$y = \Sigma m\ (1, 5)$$

The simplified expressions for flip-flop input functions and circuit output are obtained by using K-maps as shown in Fig. 4.70.

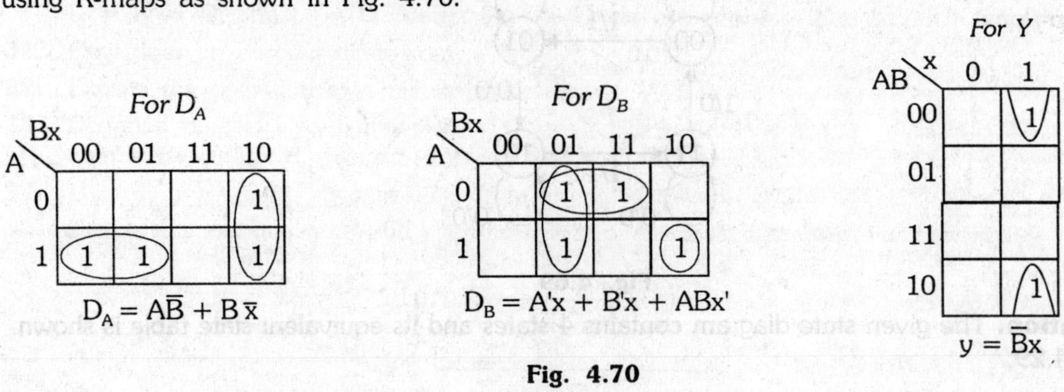

Fig. 4.70

The logic diagram is shown in Fig. 4.71.

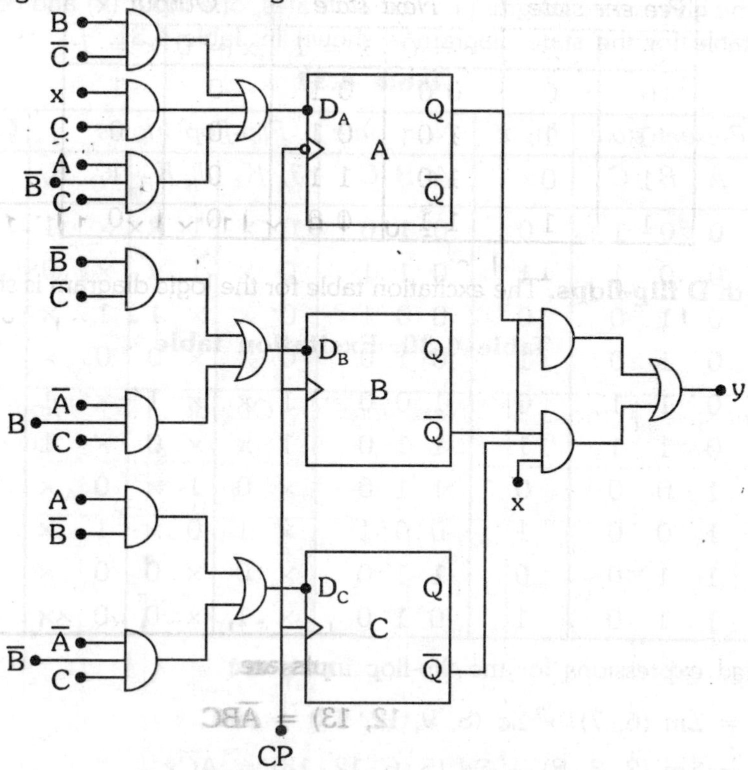

Fig. 4.71 Logic diagram using D flip-flops.

4.9.1 Design with Unused States

A clocked circuit with 'm flip-flops' will have 2^m states. There are occasions when a sequential circuit may use less than this maximum number of states. States that are not used in specifying the sequential circuit are not listed in the state table. When simplifying the input functions to flip-flops and circuit output function, the unused states are treated as don't care conditions.

Let us consider one example. First, we will design the given sequential circuit without using unused states and then we will design the given sequential circuit using these states.

Example 4.11. *Design a sequential circuit for the state diagram shown in Fig. 4.72. Using JK flip-flops.*

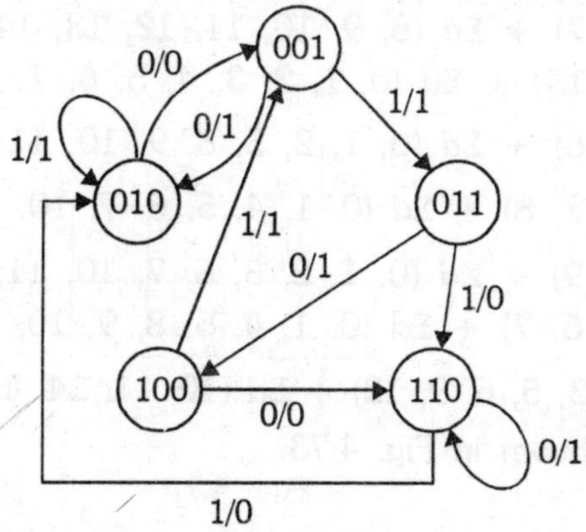

Fig. 4.72

Solution. The given state diagram contains 5 states, one input (x) and one output (y). The equivalent state table for the state diagram is shown in Table 4.31.

Table 4.31

Decimal value	*Present state* A B C	*Input* x	*Next state* A B C	*Flip-flop inputs* J_A	K_A	J_B	K_B	J_C	K_C	*Output* (y)
2	0 0 1	0	0 1 0	0	×	1	×	×	1	1
3	0 0 1	1	0 1 1	0	×	1	×	×	0	1
4	0 1 0	0	0 0 1	0	×	×	1	1	×	0
5	0 1 0	1	0 1 0	0	×	×	0	0	×	1
6	0 1 1	0	1 0 0	1	×	×	1	×	1	1
7	0 1 1	1	1 1 0	1	×	×	0	×	1	0
8	1 0 0	0	1 1 0	×	0	1	×	0	×	0
9	1 0 0	1	0 0 1	×	1	0	×	1	×	1
12	1 1 0	0	1 1 0	×	0	×	0	0	×	1
13	1 1 0	1	0 1 0	×	1	×	0	0	×	0

The simplified expressions for the flip-flop inputs are

$$J_A = \Sigma m\ (6, 7) + \Sigma d\ (8, 9, 12, 13) = \overline{A}BC$$

$$J_B = \Sigma m\ (2, 3, 8) + \Sigma d\ (5, 6, 12, 13) = A\overline{C}x$$

$$J_C = \Sigma m\ (4, 9) + \Sigma d\ (2, 3, 6, 7) = \overline{A}B\overline{C}\,\overline{x} + A\overline{B}\,\overline{C}x$$

$$K_A = \Sigma m\ (9, 13) + \Sigma d\ (2, 3, 4, 5, 6, 7) = A\overline{C}\,\overline{x} + \overline{A}C$$

$$K_B = \Sigma m\ (4, 6) + \Sigma d\ (2, 3, 8, 9) = \overline{A}B\overline{C}\,\overline{x} + \overline{A}C\,\overline{x}$$

$$K_C = \Sigma m\ (2, 6, 7) + \Sigma d\ (4, 5, 8, 9, 12, 13) = \overline{A}\,\overline{B} + \overline{A}Cx$$

$$y = \Sigma m\ (2, 3, 5, 6, 9, 12) = AB\overline{C}\,\overline{x} + \overline{A}B\overline{C} + A\overline{B}\,\overline{C}x + \overline{A}\,\overline{B}C + \overline{A}C\,\overline{x}$$

Let us see the circuit design by taking unused states as don't care conditions. The simplified flip-flop input functions are

$$J_A = \Sigma m\ (6, 7) + \Sigma d\ (8, 9, 10, 11, 12, 13, 14, 15) = BC$$

$$J_B = \Sigma m\ (9, 13) + \Sigma d\ (0, 1, 2, 3, 4, 5, 6, 7, 10, 11, 14, 15) = x$$

$$K_B = \Sigma m\ (4, 6) + \Sigma d\ (0, 1, 2, 3, 8, 9, 10, 11, 14, 15) = \overline{A}\,\overline{x}$$

$$K_A = \Sigma m\ (2, 3, 8) + \Sigma d\ (0, 1, 4, 5, 6, 7, 10, 11, 12, 13, 14, 15) = \overline{x} + C$$

$$J_C = \Sigma m\ (4, 9) + \Sigma d\ (0, 1, 2, 3, 6, 7, 10, 11, 14, 15) = \overline{A}\,\overline{x} + \overline{B}x$$

$$K_C = \Sigma m\ (2, 6, 7) + \Sigma d\ (0, 1, 4, 5, 8, 9, 10, 11, 12, 13, 14, 15) = B + x$$

$$y = \Sigma m\ (2, 3, 5, 6, 9, 12) + \Sigma d\ (10, 11, 14, 15) = AB\overline{x} + A\overline{C}x + \overline{B}x + C\overline{x}$$

The logic diagram is shown in Fig. 4.73.

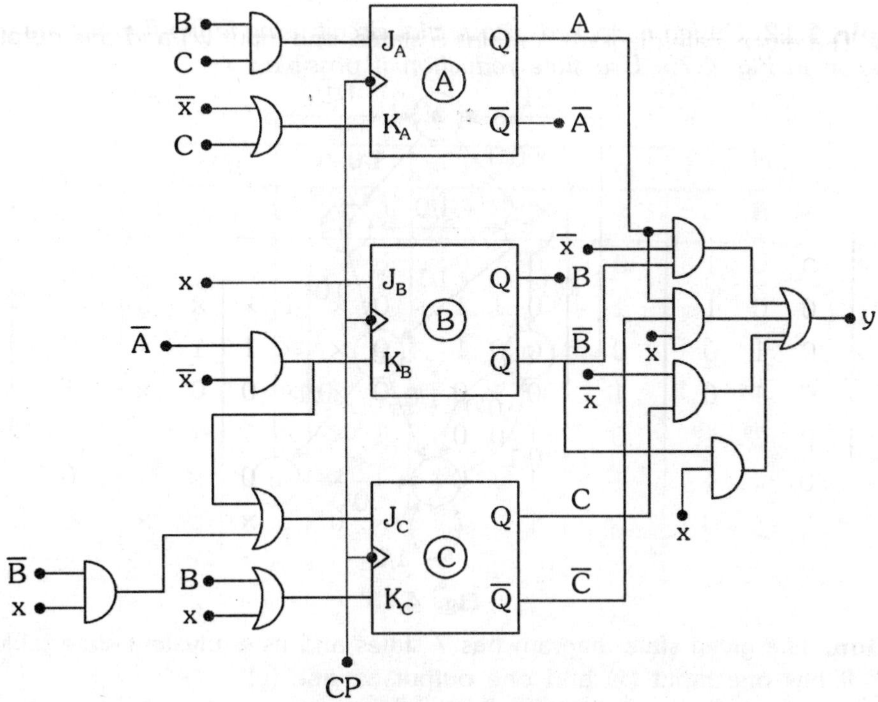

Fig. 4.73 Logic diagram.

From this problem, it is clear that the design of logic circuit using unused states reduces the number of gates required, while simplifying the flip-flop input functions and circuit output function.

Let us analyse the example to see whether the next state from the unused states is a valid state or not. The unused states are 000, 101 and 111. These states are included in the state diagram and (Fig. 4.74) transitions shown using input functions derived from the K-maps.

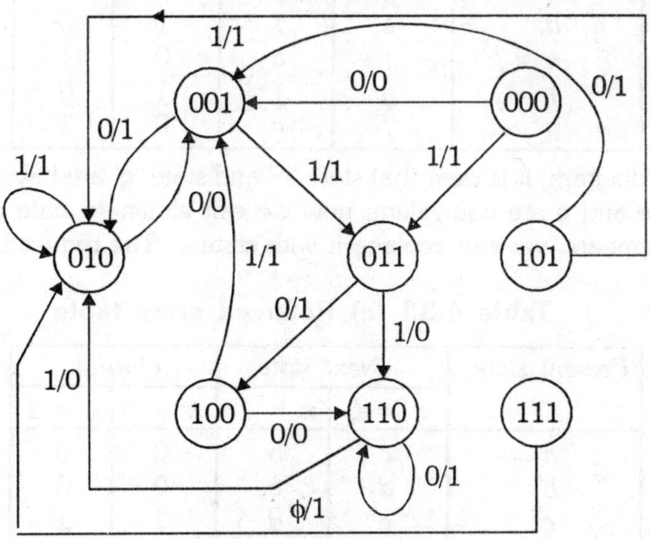

Fig. 4.74 State diagram.

Example 4.12. *Design a clocked sequential circuit using D flip-flop for the following state diagram shown in Fig. 4.75. Use state reduction if possible.*

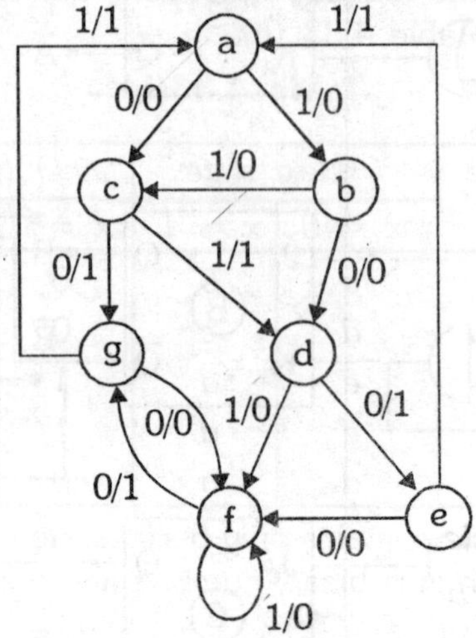

Fig. 4.75

Solution. The given state diagram has 7 states and its equivalent state table is shown in Table 4.32. It has one input (*x*) and one output assume (*y*).

Table 4.32. State table

Present state	*Next state*		*Output*	
	x = 0	x = 1	x = 0	x = 1
a	*c*	*b*	0	0
b	*d*	*c*	0	0
c	*g*	*d*	1	1
d	*e*	*f*	1	0
e	*f*	*a*	0	1
f	*g*	*f*	1	0
g	*f*	*a*	0	1

From the state diagram, it is clear that state '*e*' and state '*g*' are having same next states and outputs. So, states *e* and *g* are equivalent, now we can eliminate state *g* in the state table and wherever state *g* appears, we can replace it with state *e*. The reduced state table is shown in Table 4.33 (*a*).

Table 4.33 (*a*) Reduced state table

Present state	*Next state*		*Output*	
	x = 0	x = 1	x = 0	x = 1
a	*c*	*b*	0	0
b	*d*	*c*	0	0
c	*e*	*d*	1	1
d	*e*	*f*	1	0
e	*f*	*a*	0	1
f	*e*	*f*	1	0

From this reduced state table, states *d* and *f* are equivalent. So state *f* can be eliminated and wherever state *f* appears in the reduced state table, that is replaced by state *d*. The finally reduced state table is shown in Table 4.33 (*b*).

Table 4.33(*b*)

Present state	Next state		Output	
	x = 0	x = 1	x = 0	x = 1
a	*c*	*b*	0	0
b	*d*	*c*	0	0
c	*e*	*d*	1	1
d	*e*	*d*	1	0
e	*d*	*a*	0	1

Finally, the reduced state table contains only 5 states. So we need 3 flip-flops. Each state is assigned by 3-bit binary number and the binary numbers are assigned to the states as shown below.

a = 000 b = 001 c = 010 d = 100 e = 011.

The excitation table with D flip-flop is shown in Table 4.34.

Table 4.34 Excitation table

Decimal value	Present state $Q_A\ Q_B\ Q_C$	Input x	Next state and flip-flop inputs Q_{A+1}	Q_{B+1}	Q_{C+1}	Output (y)
0	0 0 0	0	0	1	0	0
1	0 0 0	1	0	0	1	0
2	0 0 1	0	1	0	0	0
3	0 0 1	1	0	1	0	0
4	0 1 0	0	0	1	1	1
5	0 1 0	1	1	0	0	1
8	1 0 0	0	0	1	1	1
9	1 0 0	1	1	0	0	0
6	0 1 1	0	1	0	0	0
7	0 1 1	1	0	0	0	1

The flip-flop input equations of circuit output function are given below.

$$D_A = \Sigma m\ (2, 5, 6, 9) = B\overline{C} + xC + A\overline{B}C$$

$$D_B = \Sigma m\ (0, 3, 4, 8) = \overline{B}\,\overline{C} + \overline{A}BC$$

$$D_C = \Sigma m\ (1, 4) = A\overline{B} + \overline{A}\,\overline{B}C$$

$$y = \Sigma m\ (4, 5, 8) = A\overline{B} + x\overline{B}\,\overline{C}$$

The logic diagram is shown in Fig. 4.76.

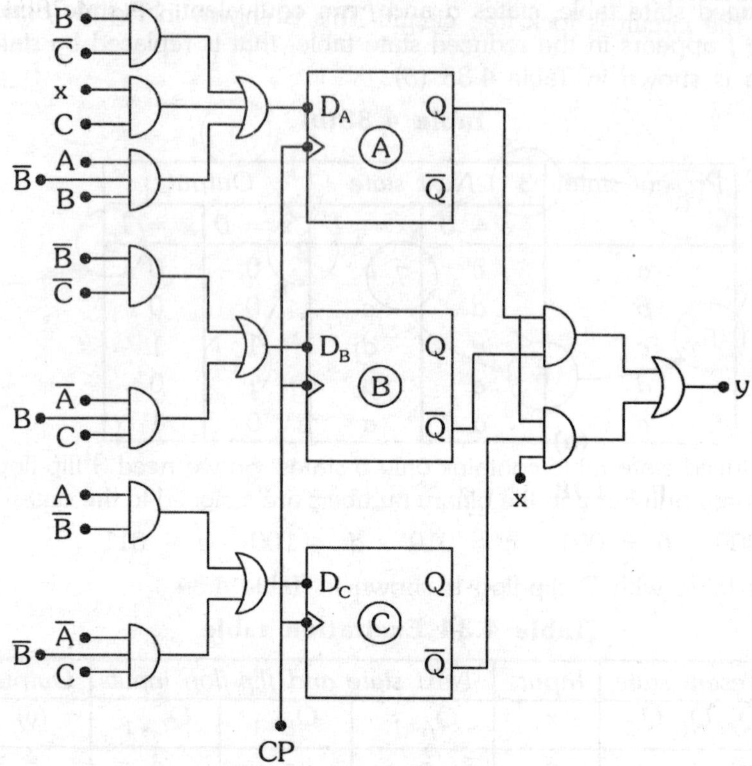

Fig. 4.76 Logic diagram.

4.9.2 Lockout Condition

In a counter, if the next state of some unused state is again an unused state and if by chance the counter happens to find itself in the unused states and never arrived at a used state, then the counter is said to be in the lockout condition. This is illustrated in Fig. 4.77.

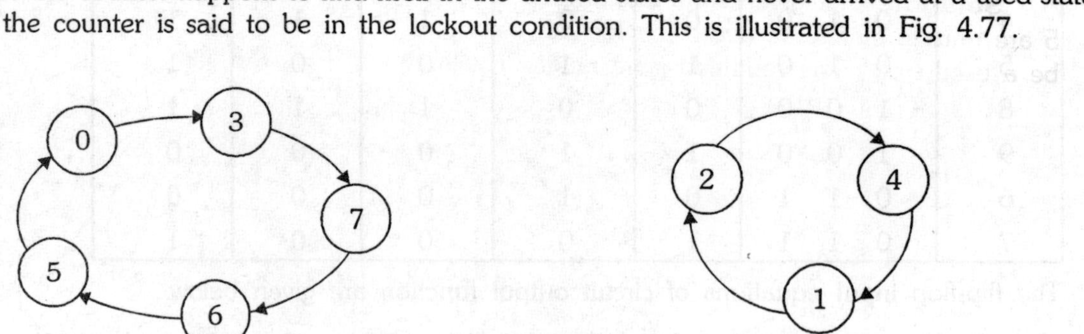

(*a*) Desired sequence

(*b*) Unused state forming lockout.

Fig. 4.77

The circuit that goes in lockout condition is called "bushless circuit". To make sure that the counter will come to the initial state from any unused state, the additional logic circuit is necessary. To ensure that the lockout does not occur, the counter should be designed by forcing the next to be the initial state from the unused states as shown in Fig. 4.78(*a*).

It is not necessary to force all unused states into initial state. Forcing any one state is sufficient. Thus the circuit is lock free circuit. This is shown in Fig. 4.78(*b*).

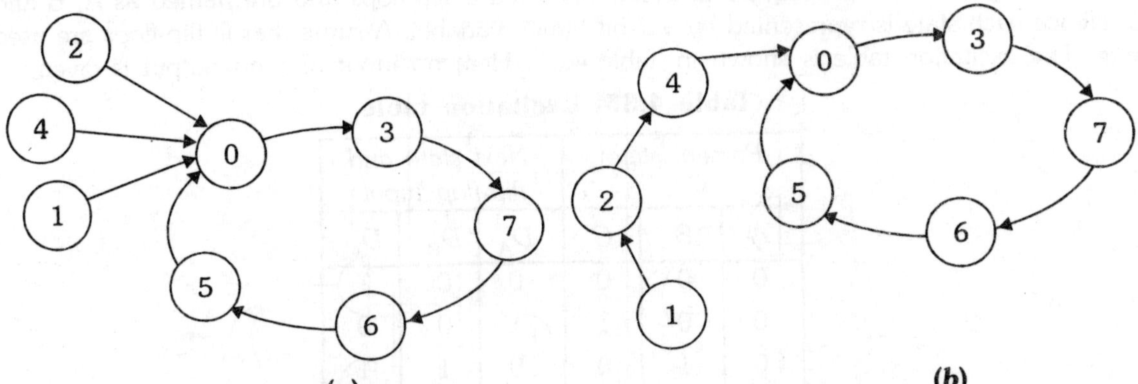

Fig. 4.78 Modified state diagram without lockout condition.

Example 4.13. *Design a clocked sequential circuit for the given state diagram in Fig. 4.79. Avoid the lockout condition.*

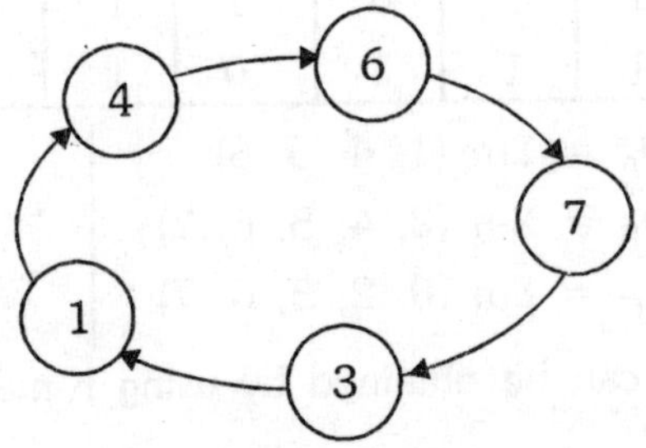

Fig. 4.79

Solution. The given state diagram has 5 states and the remaining three states 0, 2 and 5 are unused states. To avoid the lockout condition, the next states of the unused state must be a used state. The modified state diagram is shown in Fig. 4.80.

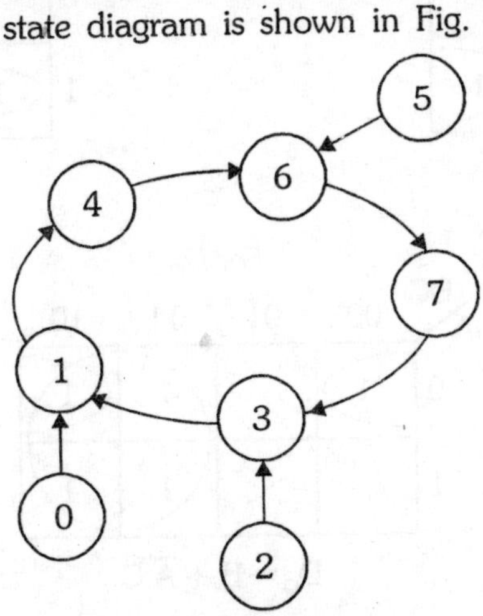

Fig. 4.80

The next state of '0' is state 1 and for 2 is state 3 and for state 5 is state 6.

To design the clocked sequential circuit, we need 3 flip-flops and are named as A, B and C. Hence each state is represented by a 3-bit binary variable. Assume that D flip-flops are used here. The excitation table is shown in Table 4.35. Here no input and no output is given.

Table 4.35. Excitation table

Present state			*Next state and flip-flop inputs*		
A	B	C	D_A	D_B	D_C
0	0	0	0	0	1
0	0	1	1	0	0
0	1	0	0	1	1
0	1	1	0	0	1
1	0	0	1	1	0
1	0	1	1	1	0
1	1	0	1	1	1
1	1	1	0	1	1

$$\therefore \quad \left.\begin{aligned} D_A &= \Sigma m\ (1, 4, 5, 6) \\ D_B &= \Sigma m\ (2, 4, 5, 6, 7) \\ D_C &= \Sigma m\ (0, 2, 3, 6, 7) \end{aligned}\right]$$

The simplified expressions can be obtained by using K-maps as shown in Fig. 4.81.

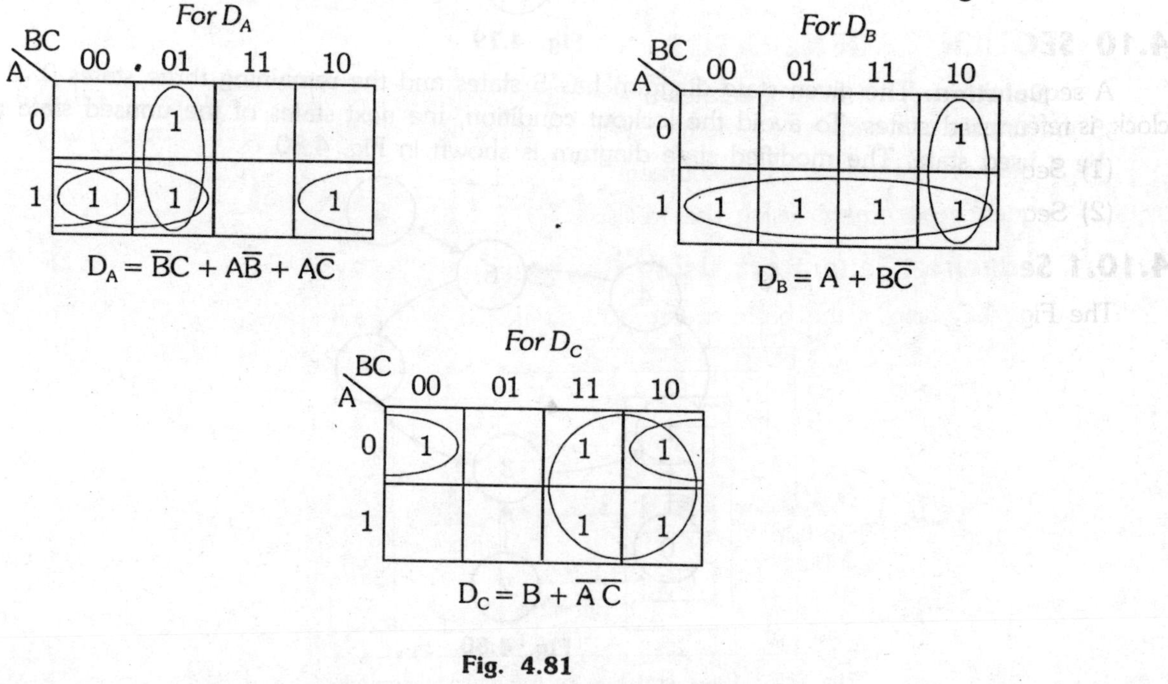

$D_A = \overline{B}C + A\overline{B} + A\overline{C}$

$D_B = A + B\overline{C}$

$D_C = B + \overline{A}\,\overline{C}$

Fig. 4.81

The logic diagram is shown in Fig 4.82.

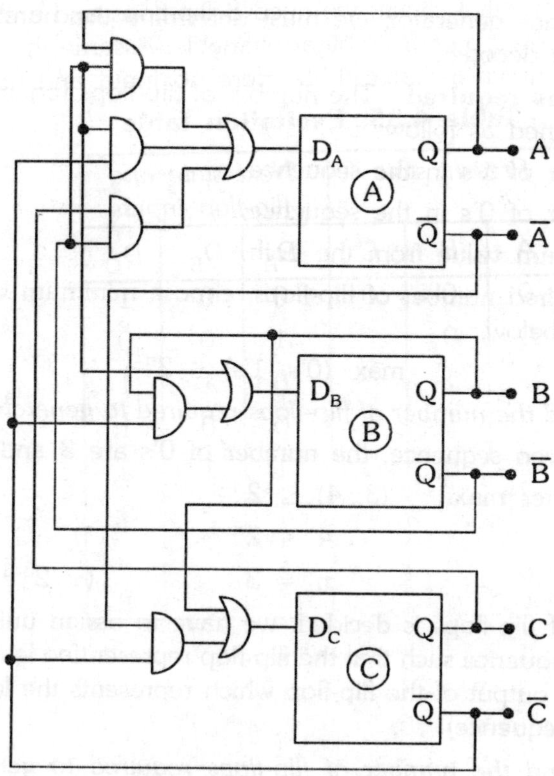

Fig. 4.82 Logic diagram.

4.10 SEQUENCE GENERATOR

A sequential circuit which generates a prescribed sequence of bits, synchronous with the clock, is referred to as a sequence generator. We can construct sequence generators by two ways:

(1) Sequence generators using counters

(2) Sequence generators using shift registers.

4.10.1 Sequence Generators Using Counters

The Fig. 4.83 shows the basic structure of sequence generator using counters.

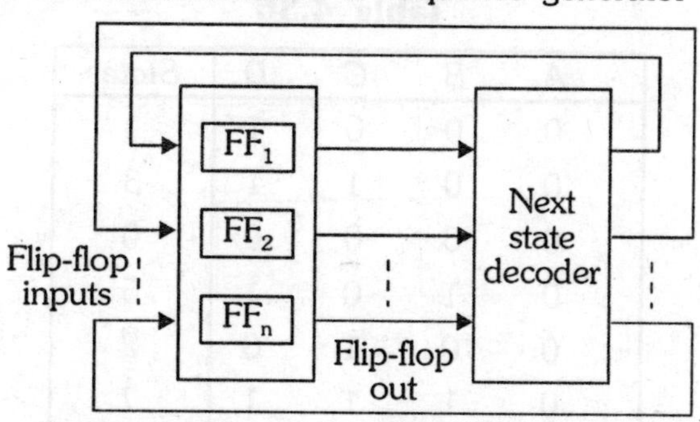

Fig. 4.83 Basic structure of sequence generator.

It contains two stages : (1) Flip-flop and (2) Next state decoder.

For design of sequence generator, we must determine the number of flip-flops and the logic circuit for next state decoder.

Number of flip-flops required : The number of flip-flops required to generate particular sequence can be determined as follows :

1. Find the number of 1's in the sequence.
2. Find the number of 0's in the sequence.
3. Take the maximum value from the both.
4. If 'n' is the required number of flip-flops, choose minimum value of 'n' to satisfy the equation given below,

$$\text{max. } (0\text{'s}, 1\text{'s}) \leq 2^{n-1}$$

Example 4.14. *Find the number of flip-flops required to generate the sequence 1101010.*

Solution. In the given sequence, the number of 0's are 3 and number of 1's are 4.

The equation becomes max. $(3, 4) \leq 2^{n-1}$

$$4 \leq 2^{n-1}$$

$$n = 3 \qquad (\because 2^{3-1} = 2^2 = 4)$$

Once the number of flip-flops is decided, we have to assign unique states corresponding to each bit in the given sequence such that the flip-flop representing least significant bit generates the given sequence (The output of the flip-flop which represents the least significant bit is used to represent the given sequence).

Example 4.15. *Find the number of flip-flops required to generate the sequence 110 1011 and design the circuit using JK flip-flops.*

Solution. The number of flip-flops required are : number of 0's are 2 and number of 1's are 5.

Hence, max. $(2, 5) \leq 2^{n-1}$

$$5 \leq 2^{n-1}$$

$$\boxed{\therefore \; n = 4}$$

We need four flip-flops and are named as A, B, C and D. The desired sequence is generated by the D flip-flop output, which is shown in Table 4.36.

Table 4.36

A	B	C	D	States
0	0	0	1	1
0	0	1	1	3
0	0	0	0	0
0	1	0	1	5
0	0	1	0	2
0	1	1	1	7
1	0	0	1	9

Assign the values to the remaining flip-flops in such a way that the total combination will produce always a new state and totally we have seven states, the maximum value of it is 9. The state diagram for it is shown in Fig. 4.84.

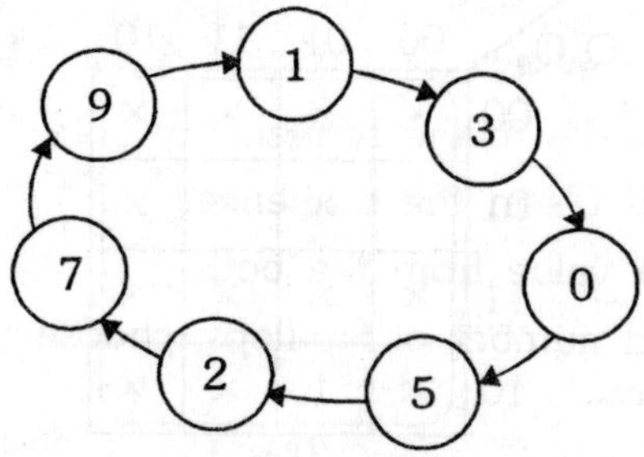

Fig. 4.84 State diagram.

The excitation table is shown in Table 4.37.

Table 4.37. Excitation table

Present state				*Next state*				*Flip-flop input conditions*			
Q_A	Q_B	Q_C	Q_D	Q_{A+1}	Q_{B+1}	Q_{C+1}	Q_{D+1}	$J_A K_A$	$J_B K_B$	$J_C K_C$	$J_D K_D$
0	0	0	0	0	1	0	1	0 ×	1 ×	× 1	1 ×
0	0	0	1	0	0	1	1	0 ×	0 ×	1 ×	× 0
0	0	1	0	0	1	1	1	0 ×	1 ×	× 0	1 ×
0	0	1	1	0	0	0	0	0 ×	0 ×	× 1	× 1
0	1	0	0	×	×	×	×	× ×	× ×	× ×	× ×
0	1	0	1	0	0	1	0	0 ×	× 1	1 ×	× 1
0	1	1	0	×	×	×	×	× ×	× ×	× ×	× ×
0	1	1	1	1	0	0	1	1 ×	× 1	× 1	× 0
1	0	0	0	×	×	×	×	× ×	× ×	× ×	× ×
1	0	0	1	0	0	0	1	× 1	0 ×	0 ×	× 0
1	0	1	0	×	×	×	×	× ×	× ×	× ×	× ×
1	0	1	1	×	×	×	×	× ×	× ×	× ×	× ×
1	1	0	0	×	×	×	×	× ×	× ×	× ×	× ×
1	1	0	1	×	×	×	×	× ×	× ×	× ×	× ×
1	1	1	0	×	×	×	×	× ×	× ×	× ×	× ×
1	1	1	1	×	×	×	×	× ×	× ×	× ×	× ×

K-map simplification. The flip-flop inputs are obtained as shown in Fig. 4.85.

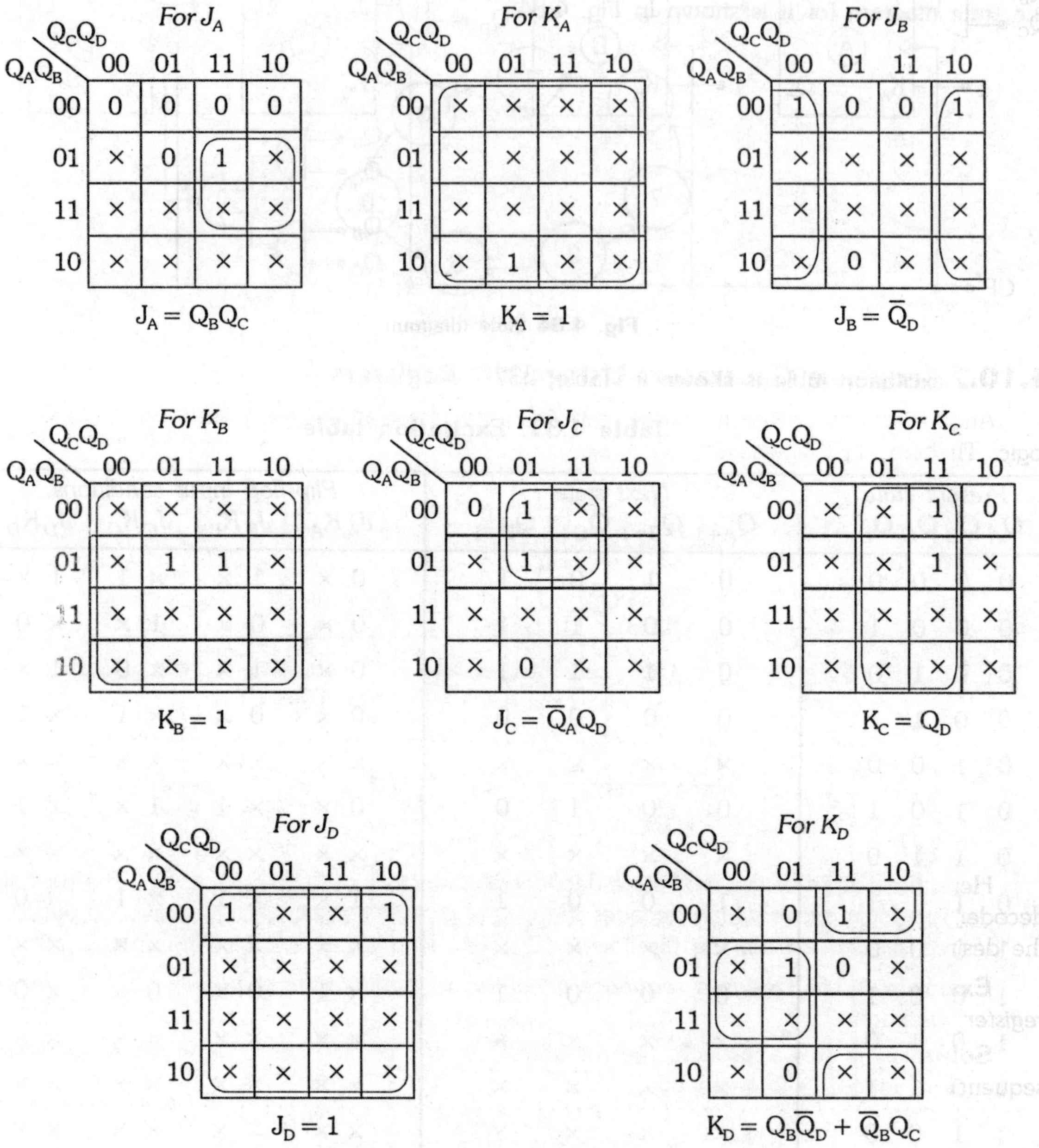

Fig. 4.85

Logic Diagram : The logic diagram is shown in Fig. 4.86.

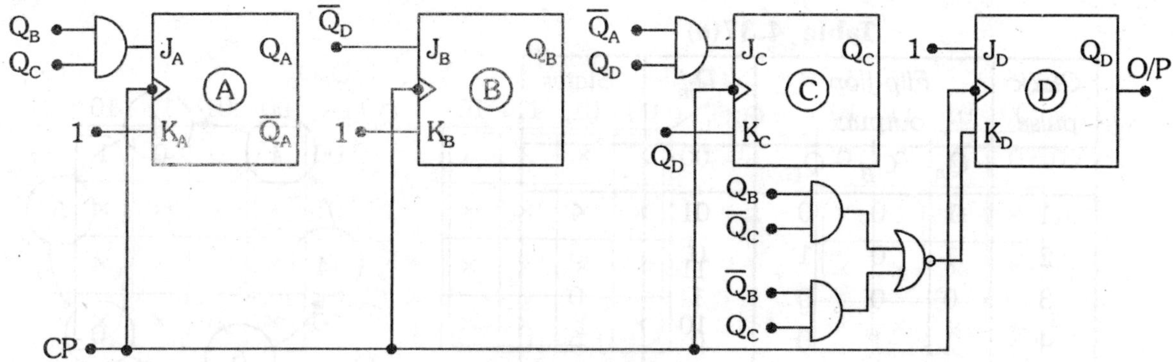

Fig. 4.86 Logic diagram.

4.10.2 Sequence Generators Using Shift-Registers

Another design procedure for sequence generator is a shift register with next state decoder logic. The Fig. 4.87 shows the block diagram of sequence generator using shift register.

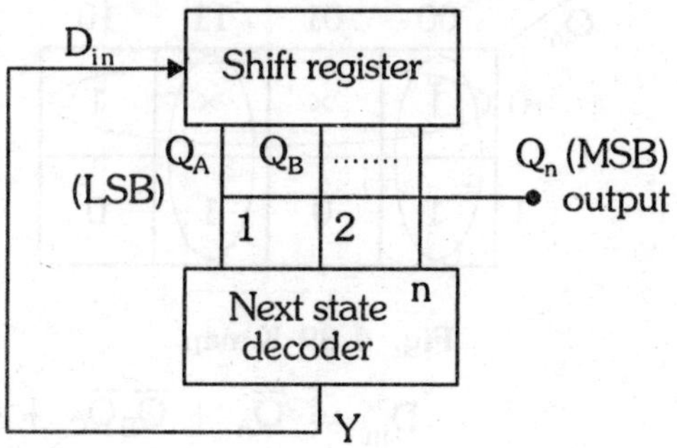

Fig. 4.87

Here, the output of the next state decoder is a function of Q_A, Q_B, ..., Q_n. The next state decoder is a logic circuit which decodes the output of shift register and generates output to get the desired sequence from the flip-flop A (Q_A) *i.e.*, least significant bit flip-flop.

Example 4.16. *Design a sequence generator to generate the sequence 1101011 by shift register method.*

Solution. In this approach, the minimum number of flip-flops *n*, required to generate a sequence of length N is given by

$$N \leq 2^n - 1$$

where N = number of bits in the sequence.

In this example N = 7, therefore, the minimum value of *n*, which may generate the sequence is :

$$7 \leq 2^n - 1$$

$$\boxed{\therefore\ n = 3}$$

With the three flip-flops, the sequence generation is shown in Table 4.37(*a*). The state diagram is shown in Fig. 4.88.

Table 4.37(*a*)

Clock pulse	*Flip-flop outputs*			D_{in}	*States*
	Q_A	Q_B	Q_C		
1	1	0	0	1	4
2	1	0	1	0	5
3	0	0	0	1	0
4	1	1	0	0	6
5	0	1	0	1	2
6	1	1	1	1	7

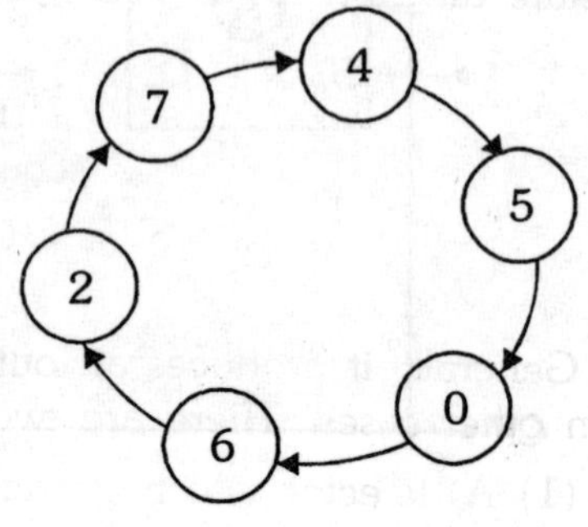

Fig. 4.88 State diagram.

The K-map simplification for D_{in} is shown in Fig. 4.89.

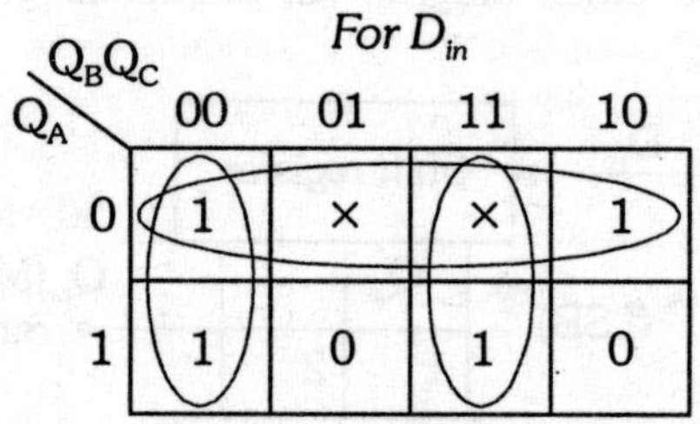

Fig. 4.89 K-map.

$$D_{in} = \overline{Q}_A + \overline{Q}_B\overline{Q}_C + Q_BQ_C$$

$$= \overline{Q}_A + Q_B \odot Q_C$$

The logic diagram is shown in Fig. 4.90. The initial state is 100. So, the input for D_A D_B and D_C are 100 respectively.

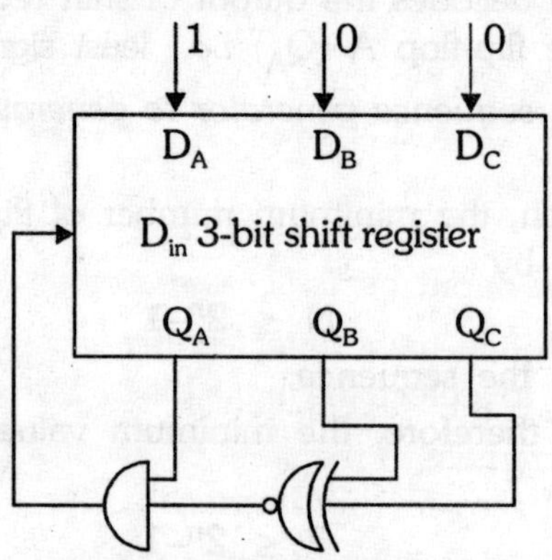

Fig. 4.90 Logic diagram.

4.11 SEQUENCE DETECTOR

The "sequence detector" is a clocked sequential circuit or machine which is used to detect the desired binary sequence. Once we know the sequence which is to be detected, we can draw the state diagram for it. From the state diagram, we can design the circuit (see Fig. 4.91)

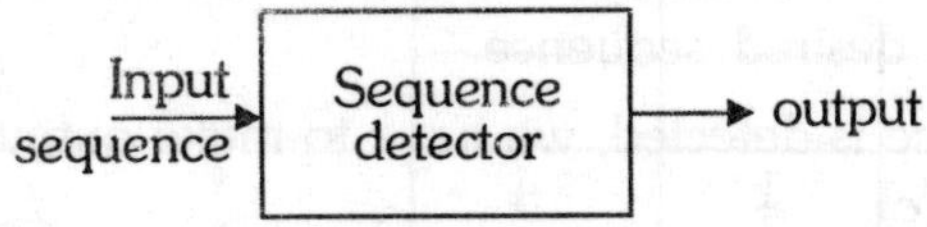

Fig. 4.91 Block diagram.

Generally it produces an output = 1, whenever it detects the desired input sequence and '0' in other cases. There are two types of detectors :

(1) A detector which detects over-lapping input sequence and

(2) A detector which detects non-overlapping input sequence.

The following example illustrates how to determine the state diagram from a given sequence.

Example 4.17. *Design a sequence detector which detects the sequence 100011.*

Solution. Generally the number of states in the state diagram is equal to the number of bits in the sequence. In this example we have six bits in the sequence, so we have six states in the state diagram.

Once the number of states is known, we have to draw the directed lines with inputs and outputs as weights between the two states. Let us start the drawing of directed lines, assuming (arbitrarily) initial state is A.

State A : In this state, the machine can receive either an input-0 or 1. For each of these, inputs, an arc is drawn originating in state A and terminating in the appropriate next state, as shown in Fig. 4.92(*a*).

When input is 1, we have detected first bit in the sequence, hence we have to go to next-state to detect the next bit in the sequence.

When input is 0, we have to remain in state A, because bit '0' is not the first bit in the sequence. In both cases the output is '0' since we have not yet detected all the bits in the sequence.

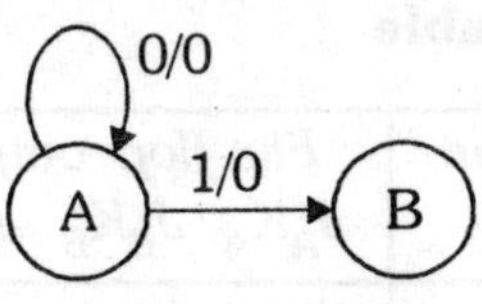

Fig. 4.92(*a*)

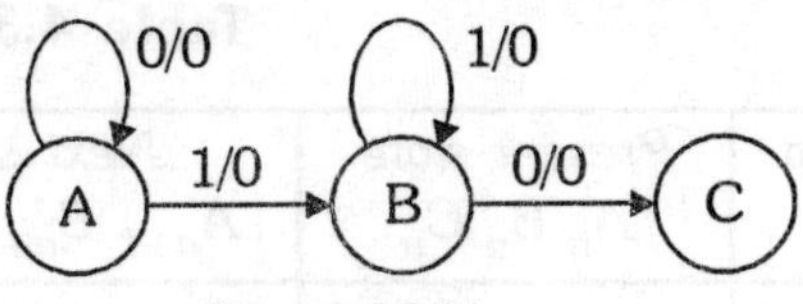

Fig. 4.92(*b*)

The arc labelled 1/0 indicates input bit is '1' and output is '0'.

State B : When input is '0', we have detected the second, bit in the sequence. Hence, we have to go to next state (C) to detect the next bit in the sequence. (see Fig 4.92(*b*))

When input is 1, we have to remain in the state B, because 1 which we have detected may start the sequence. Output is still zero for both cases.

State C : When input is 0, we have detected the third bit in the sequence, hence we have to go next state (D) to detect the next bit in the sequence.

When input is 1, we have to go to state B, because 1 which we have detected may not be the bit in the sequence to be detected but it may be the start bit of the sequence, hence we are moving to state B. The output is still zero.

State D to state F : As explained for state A, state B and state C, if the desired bit is detected, we have to go for the next state, otherwise we have to go to the previous state from where we can continue the desired sequence.

When complete sequence is detected, we have to make output 1 and go to the initial state. This is shown in Fig. 4.92 (*c*).

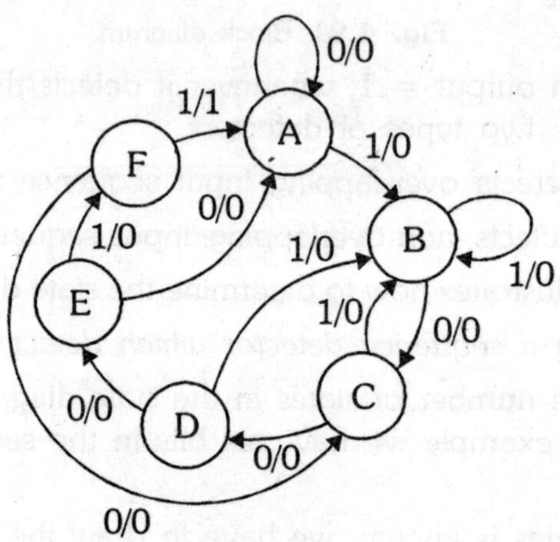

Fig. 4.92(*c*)

To determine the logic diagram, let us assume state assignments as

A = 000, B = 001 C = 010
D = 011, E = 100 and F = 101

To design the logic diagram, we have 6 states, so we need three flip-flops. Assume that flip-flops are named as A, B and C and one output *y* and input *x*. The excitation table is shown in Table 4.38. The flip-flop input equations can be obtained by using K-maps as shown in Fig. 4.93.

Table 4.38 Excitation table

Input y	Present state $A_n\ B_n\ C_n$	Next state $A_{n+1}\ B_{n+1}\ C_{n+1}$	Output z	Flip-flop outputs $J_AK_A\ J_BK_B\ J_CW_C$
0	0 0 0	0 0 0	0	0 × 0 × 0 ×
0	0 0 1	0 0 1	0	0 × 1 × × 1
0	0 1 0	0 1 1	0	0 × × 0 1 ×
0	0 1 1	1 0 0	0	1 × × 1 × 1
0	1 0 0	0 0 0	0	× 1 0 × 0 ×
0	1 0 1	0 1 0	0	× 1 1 × × 1

1	0 0 0	0 0 1	0	0× 0× × ×
1	0 0 1	0 0 1	0	0× 0× 1 0
1	0 1 0	0 0 1	0	0× ×1 1×
1	0 1 1	0 0 1	0	0× ×1 ×0
1	1 0 0	1 0 1	0	×0 0× 1×
1	1 0 1	0 0 0	1	×1 0× ×1

K-map simplification

For J_A

xA \ BC				
	0	0	1	0
	×	×	×	×
	×	×	×	×
	0	0	0	0

$J_A = \overline{x}BC$

For K_W

xA \ BC				
	×	×	×	×
	1	1	×	×
	0	1	×	×
	×	×	×	×

$K_A = \overline{x} + C$

For J_B

xA \ BC				
	0	1	×	×
		1	×	×
			×	×
			×	×

$J_B = \overline{x}C$

For K_B

xA \ BC				
	×	×	1	0
	×	×	×	×
	×	×	×	×
	×	×	1	1

$K_B = x + C$

For J_C

xA \ BC				
	0	×	×	1
	0	×	×	×
	1	×	×	×
	1	×	×	1

$J_C = x + B$

For K_C

xA \ BC				
	×	1	1	×
	×	1	×	×
	×	1	×	×
	×	0	0	×

$K_C = \overline{x} + A$

For Y

xA \ BC				
	0	0	0	0
	0	0	×	×
	0	1	×	×
	0	0	0	0

$Y = xAC$

Fig. 4.93 K-maps.

Logic diagram. The logic diagram shown in Fig 4.94.

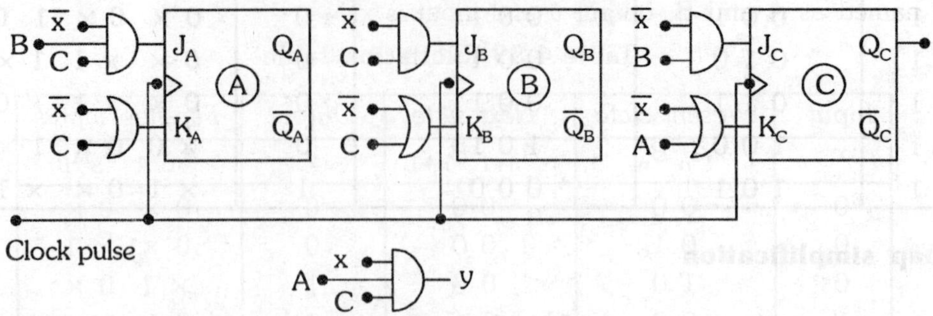

Fig. 4.94 Logic diagram.

Example 4.18. *Design a sequence detector to detect the following sequence using JK flip-flops (use Mealy machine). The sequence is '110'.*

Solution. The given sequence has 3-bits. So, we require 3 states in the state diagram.

Let us start by assuming initial state is 'A'.

State A : In this state, the machine may receive either '0' or '1'.

When input is 1, we have detected the first bit in the sequence. Hence, we have to go to the next state to detect the next bit in the sequence.

When input is 0, we have to remain in the state 'A' because bit '0' is not the first bit in the sequence. In both cases, the output is 0, since we have not yet detected all the bits in the given sequence.

State B : When input is 1, we have detected the second bit in the sequence, hence we have to go to the next state to detect the next bit in the sequence.

When input is '0', we have to go to the state 'A', to detect the first bit in the sequence *i.e.*, 1. In both cases, the output is still zero.

State C : When input is '0', we have detected the last bit in the sequence, hence we have to go to the initial state to detect the next sequence and make output high indicating that the sequence is detected.

When input is 1, we have to go to state B, 1 which we have detected may be the second 1 in the sequence (thus we are not moving to state A). The state diagrams are shown in Fig. 4.95.

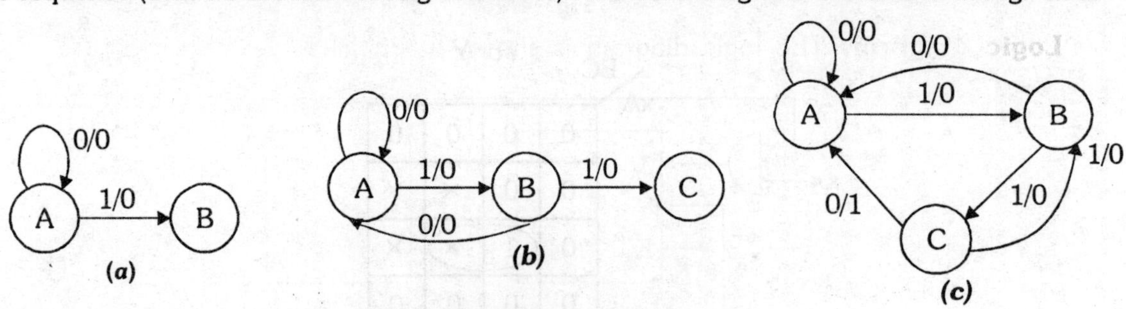

Fig. 4.95 Complete state diagram.

Before designing, first we go to state assignment.

Let us assume A = 00, B = 01 and C = 10.

The excitation table is shown in Table 4.39. Totally, we have three states. So we need B flip-flops named as A and B, output y and input is x.

Table 4.39 Excitation table

Input	*Present state*	*Next state*	*Output*	*Flip-flop inputs*
x	A_n B_n	A_{n+1} B_{n+1}	y	J_AK_A J_BK_B
0	0 0	0 0	0	0 × 0 ×
0	0 1	0 0	0	0 × × 1
0	1 0	0 0	1	× 1 0 ×
1	0 0	0 1	0	0 × 1 ×
1	0 1	1 0	0	1 × × 1
1	1 0	1 0	0	× 0 0 ×

K-map simplification. Flip-flop input functions are shown in Fig. 4.96.

For J_A

x \ AB	00	01	11	10
0	0	0	×	×
1	0	1	×	×

$J_A = xB$

For K_A

x \ AB	00	01	11	10
0	×	×	×	1
1	×	×	×	1

$K_A = \overline{X}$

For J_B

x \ AB	00	01	11	10
0	0	×	×	0
1	1	×	×	1

$J_B = X\overline{A}$

For K_B

x \ AB	00	01	11	10
0	×	1	×	×
1	×	1	×	×

$K_B = 1$

Fig. 4.96

Logic diagram. The logic diagram is shown in Fig. 4.97.

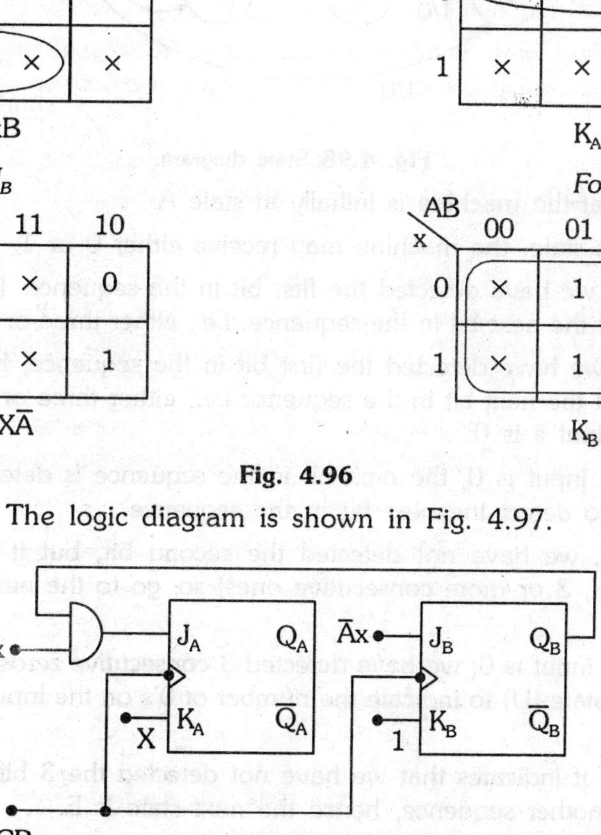

Fig. 4.97 Logic diagram.

Example 4.19. *Information bits are encoded on a single line x, so as to synchronise with a clock. Bits are encoded so that 3 or more consecutive 1's or 3 or more consecutive 0's should never appear on the line x. An error indicating sequential circuit is to be designed to indicate an error by generating 1 on the output line z. Coinciding with the third of every sequence of three zeros or three ones. For example, if three or more consecutive ones appear, the output remains one for the third and subsequent clock cycles till bit zero is appeared on the x and vice-versa also.*

(a) Draw the state diagram for the error detector.

(b) Reduce the state diagram if possible.

(c) Design with D flip-flops.

Solution. (*a*) The state table is shown in Fig. 4.98.

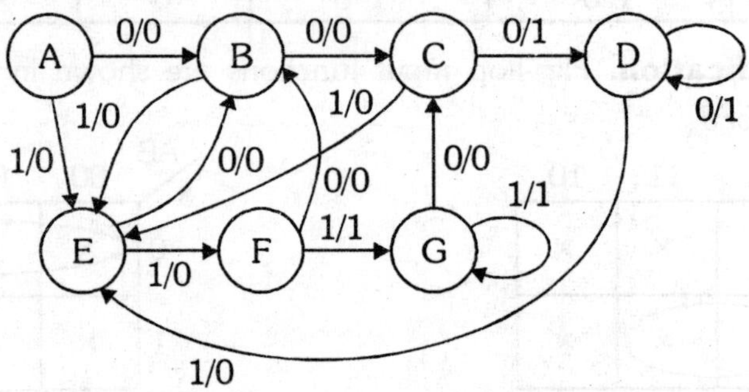

Fig. 4.98 State diagram.

Let us assume that the machine is initially at state A.

State A : In this state, the machine may receive either 0 or 1.

When input is 0, we have detected the first bit in the sequence, hence we have to go to next state (B) to detect the next bit in the sequence, *i.e.*, either three or more consecutive zeros.

When input is 1,we have detected the first bit in the sequence, hence we have to go to next state (E) to detect the next bit in the sequence *i.e.*, either three or more consecutive ones. In both cases, the output z is 0.

State B : When input is 0, the next bit in the sequence is detected, hence we have to go to next state (C) to detect the next bit in the sequence.

When input is 1, we have not detected the second bit, but it may be the first bit of another sequence. (*i.e.*, 3 or more consecutive ones) so, go to the next state E. In both cases the output z is 0.

State C : When input is 0, we have detected 3 consecutive zeros. Hence, the output z = 1, and go to the next state (D) to indicate the number of 0's on the input line by indicating 1 on the output line.

When input is 1, it indicates that we have not detected the 3 bit of the sequence, but it may the first bit of another sequence, hence the next state is E.

In this way, all the 7 states are interconnected by the directed lines. To design the circuit we need 3 flip-flops named as A, B and C. The state table is shown in Table 4.40 (*a*).

Table 4.40(a)

Present state	Next state		Output	
	x = 0	x = 1	x = 0	x = 1
A	B	E	0	0
B	C	E	0	0
C	D	E	1	0
D	D	E	1	0
E	B	F	0	0
F	B	G	0	1
G	B	G	0	1

(*b*) From the above state table, we can find that states C = D and G = F. Hence, replace D by C and G by F. The reduced state table is shown in Table 4.40 (*b*).

Table 4.40(b)

Present state	Next state		Output	
	x = 0	x = 1	x = 0	x = 1
A	B	E	0	0
B	C	E	0	0
C	C	E	1	0
E	B	F	0	0
F	B	F	0	1

(*c*) Assume the state assignments for

A = 000, B = 001, C = 010,

E = 011 and F = 100.

The excitation table is shown in Table 4.41.

Table 4.41. Excitation table

Input	Present state			Next state			Flip-flop inputs			Output
x	Q_A	Q_B	Q_C	Q_{A+1}	Q_{B+1}	Q_{C+1}	D_A	D_B	D_C	y
0	0	0	0	0	0	1	0	0	1	0
0	0	0	1	0	1	0	0	1	0	0
0	0	1	0	0	1	0	0	1	0	1
0	0	1	1	0	0	1	0	0	1	0
0	1	0	0	0	0	1	0	0	1	0
1	0	0	0	0	1	1	0	1	1	0
1	0	0	1	0	1	1	0	1	1	0
1	0	1	0	0	1	1	0	1	1	0
1	0	1	1	1	0	0	1	0	0	0
1	1	0	0	1	0	0	1	0	0	1

K-map simplification.

$$D_A = xQ_A + xQ_BQ_C = x\,(Q_A + Q_BQ_C)$$

$$D_B = \overline{Q}_RQ_C + Q_B\overline{Q}_C + x\overline{Q}_A\overline{Q}_C$$

$$\therefore\; D_B = Q_B \oplus Q_C + x\overline{Q}_A\overline{Q}_C$$

$$D_C = \overline{x}\overline{Q}_B\overline{Q}_C + \overline{x}Q_BQ_C + x\overline{Q}_A\overline{Q}_B + xQ_B\overline{Q}_C$$

$$= \overline{x}\,(Q_B \odot Q_C) + x\,(\overline{Q}_A\overline{Q}_B + Q_B\overline{Q}_C)$$

$$Z = xQ_A + \overline{x}Q_B\overline{Q}_C$$

For D_A

xQ_A \ Q_BQ_C	00	01	11	10
00	0	0	0	0
01	0	×	×	×
11	1	×	×	×
10	0	0	1	0

For D_B

xQ_A \ Q_BQ_C	00	01	11	10
00	0	1	0	1
01	0	×	×	×
11	0	×	×	×
10	1	1	0	1

For D_C

xQ_A \ Q_BQ_C	00	01	11	10
00	1	0	1	0
01	1	×	×	×
11	0	×	×	×
10	1	1	0	1

For D_Z

xQ_A \ Q_BQ_C	00	01	11	10
00	0	0	0	1
01	×	×	×	×
11	1	×	×	×
10	0	0	0	0

Fig. 4.99

Logic diagram. The logic diagram is shown in Fig. 4.100.

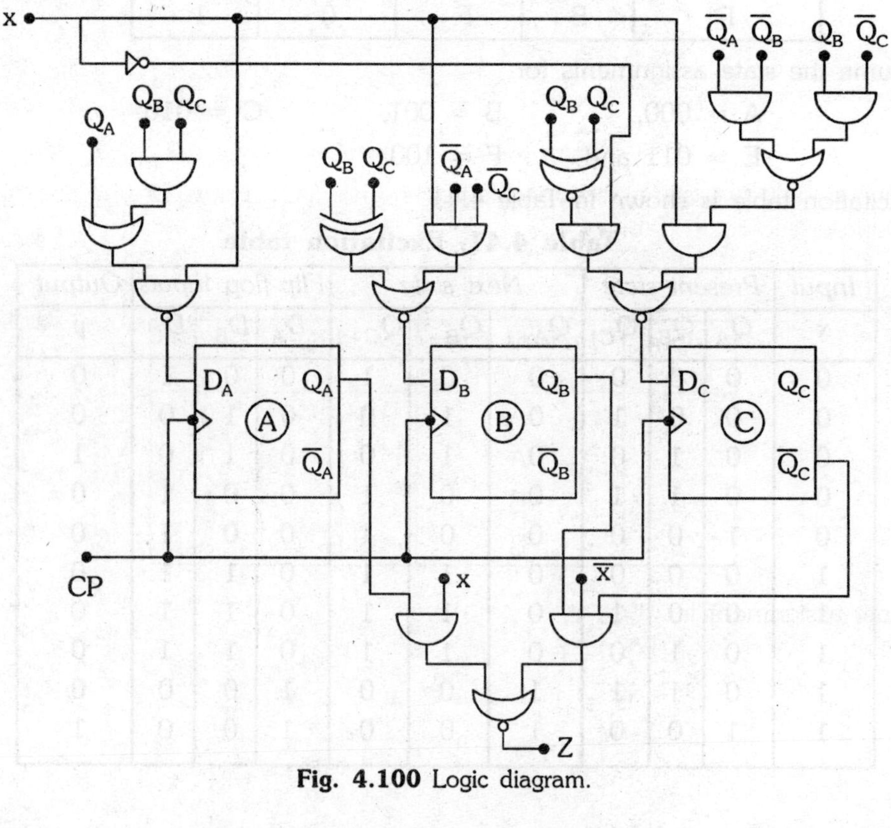

Fig. 4.100 Logic diagram.

Example 4.20. *Design and construct a Mealy circuit which has single input and single output terminals. Output is 1 if the input is a sequence of alternate 1's and 0's. Assume that input is properly synchronised with the clock.*

Solution. Suppose, if the given sequence is 10 10 10, then the output is 010101...

From the problem, the machine has to recognize the alternate 1's and 0's. For every '10' input, it must produce an output $y = 1$. So it has two states A and B. Assume initially the machine in state A.

In state A : In this state, the machine can receive either an input 0 or 1.

When input is 1, we have detected first bit (one bit) in the sequence, hence we have to go to the next state (B) to detect the next bit.

When input is '0', we have to remain in the same state (*i.e.*, state A), because bit '0' is not the first bit in the sequence. In both cases, the output is '0'.

In state B : When input is 1, it indicates that, it has not detected the second bit 1 but it may be first bit of the coming sequence. So, the machine will remain in the same state (B). When input is 0, it indicates that it has detected the second bit in the sequence, hence we have to move to next state (previous state (state A) while producing output 1 to detect the next sequence in the input.

The state diagram is shown in Fig. 4.101.

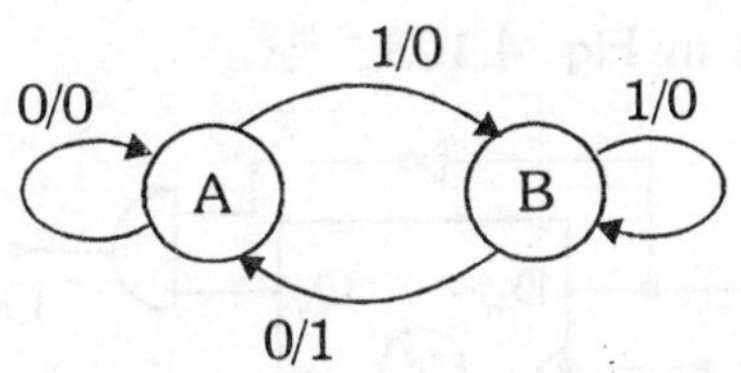

Fig. 4.101 State diagram.

Since it has only two states, so one flip-flop is enough. Assume D-type and name it as y. The state table is shown in Table 4.42.

Table 4.42

Input x	*Present state*	*Next state*	*Output* (z)
0	A	A	0
1	A	B	0
0	B	A	1
1	B	B	0

The state assignment is,

if A = 0

then B = 1.

The excitation table is shown in Table 4.43.

Table 4.43

Input x	*Present state* y	*Next state (Flip-flop y+ inputs)*	*Output* (z)
0	0	0	0
1	0	1	0
0	1	0	1
1	1	1	0

The K-map simplification is shown in Fig 4.102

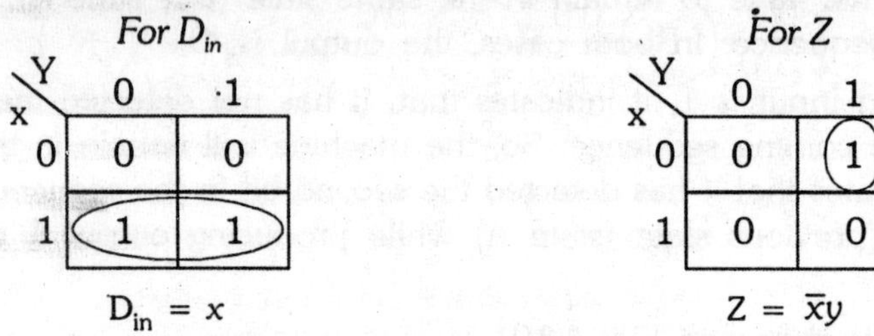

Fig. 4.102

The logic diagram is shown in Fig. 4.103.

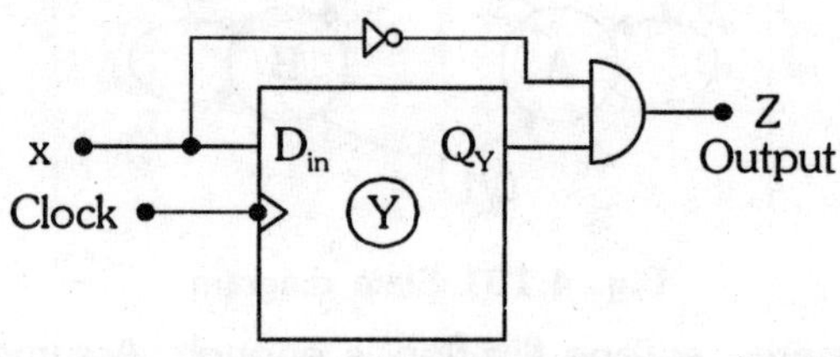

Fig. 4.103 Logic diagram.

Example 4.21. *Construct the state diagram for a two input, two output sequential machine which is to produce an output z = 1 with occurrance of an input 1 following a string of two or three consecutive input 0's. At all other times the output is to be zero. Realize using SR flip-flops.*

Solution. Let us start with the initial state A.

State A : In this state the machine may receive either 0 or 1. When input is 0, it indicates that we have detected the first bit, so go to next state (state B).

When input is 1 the machine remains in the same state. In both cases the output is (z) = 0.

State B : When input is '0', we have detected the second bit, so go to next state (state C). If input is 1, we have not detected any required sequence, so go to the initial state(A) to detect the required sequence (to start the initial operation). In both cases the output = 0.

State C : When it receives '0', it indicates that, we have detected the third bit, so go to next state (D). If input is 1, that means, we have detected the third bit as well as the required sequence. So, we have to go to the initial state (A) and output (z) = 1.

State D : If input is 0, we have detected the fourth bit (in case of 4 zeros) and go to the next state. (state E).

If input is 1, we have detected the fourth bit as well as the sequence, so we have to go the initial state to detect next sequence and output is 1.

State E : If input is 1, we have not detected the fifth bit and go to the initial state to detect next sequence and output = 0. If input is '0', we have not detected the fifth bit, but it may be the part of the sequence to be detected. (may be the 4th bit) So machine has to stay in the same state.

The state diagram is shown in Fig. 4.104.

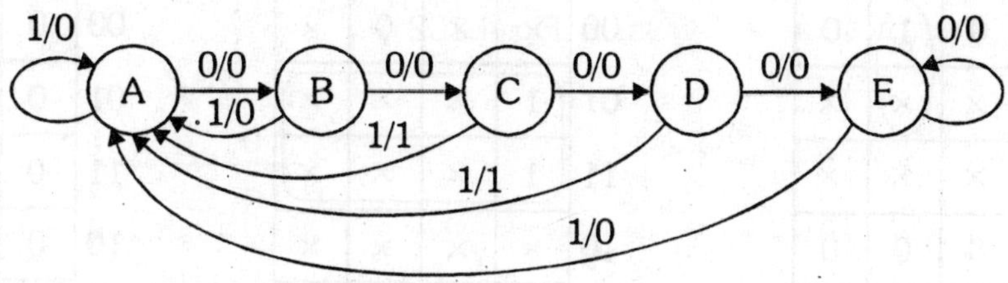

Fig. 4.104 State diagram.

The state table is shown in Table 4.44

Table 4.44 State table

Present state	Next state, Z	
	x = 0	x = 1
A	B, 0	A, 0
B	C, 0	A, 0
C	D, 0	A, 1
D	E, 0	A, 1
E	E, 0	A, 0

For state assignment, let us assume that A = 000, B = 001, C = 010, D = 011 and E = 100. We need three flip-flops named as A, B and C.

The excitation table is shown in Table 4.45.

Table 4.45 Exitation table

Input x	Present state Q_A Q_B Q_C	Next state Q_{A+1} Q_{B+1} Q_{C+1}	Out-put	Flip-flop inputs S_AR_A	S_BR_B	S_CR_C
0	0 0 0	0 0 1	0	0 ×	0 ×	1 0
0	0 0 1	0 1 0	0	0 ×	1 0	0 1
0	0 1 0	0 1 1	0	0 ×	× 0	1 0
0	0 1 1	1 0 0	0	1 0	0 1	0 1
0	1 0 0	1 0 0	0	× 1	0 ×	0 ×

1	0	0	0	0	0	0	0	0 ×	0 ×	0 ×
1	0	0	1	0	0	0	0	0 ×	0 ×	0 1
1	0	1	0	0	0	0	1	0 ×	0 1	0 ×
1	0	1	1	0	0	0	1	0 ×	0 1	0 ×
1	1	0	0	0	0	0	0	0 1	0 ×	0 ×

The K-map simplifications for flip-flop inputs are shown in Fig. 4.105.

For S_A

$xQ_A \backslash Q_BQ_C$	00	01	11	10
00	0	0	1	0
01	×	×	×	×
11	0	×	×	×
10	0	0	0	0

$S_A = \bar{x}Q_BQ_C$

For R_A

$xQ_A \backslash Q_BQ_C$	00	01	11	10
00	×	×	0	×
01	1	×	×	×
11	1	×	×	×
10	×	×	×	×

$R_A = Q_A$

For S_B

$xQ_A \backslash Q_BQ_C$	00	01	11	10
00	0	1	0	×
01	0	×	×	×
11	0	×	×	×
10	0	0	0	0

$S_B = \bar{x}\,\bar{Q}_BQ_C$

For R_B

$xQ_A \backslash Q_BQ_C$	00	01	11	10
00	×	0	1	0
01	×	×	×	×
11	×	×	×	×
10	×	×	1	1

$R_B = x + Q_BQ_C$

For S_C

$xQ_A \backslash Q_BQ_C$	00	01	11	10
00	1	0	0	1
01	0	×	×	×
11	0	×	×	×
10	0	0	0	0

$S_C = \overline{x}\overline{Q}_A\overline{Q}_C$

For R_C

$xQ_A \backslash Q_BQ_C$	00	01	11	10
00	0	1	1	0
01	×	×	×	×
11	×	×	×	×
10	×	1	×	×

$R_C = Q_C$

Fig. 4.105 K-map simplifications.

The logic diagram is shown in Fig 4.106.

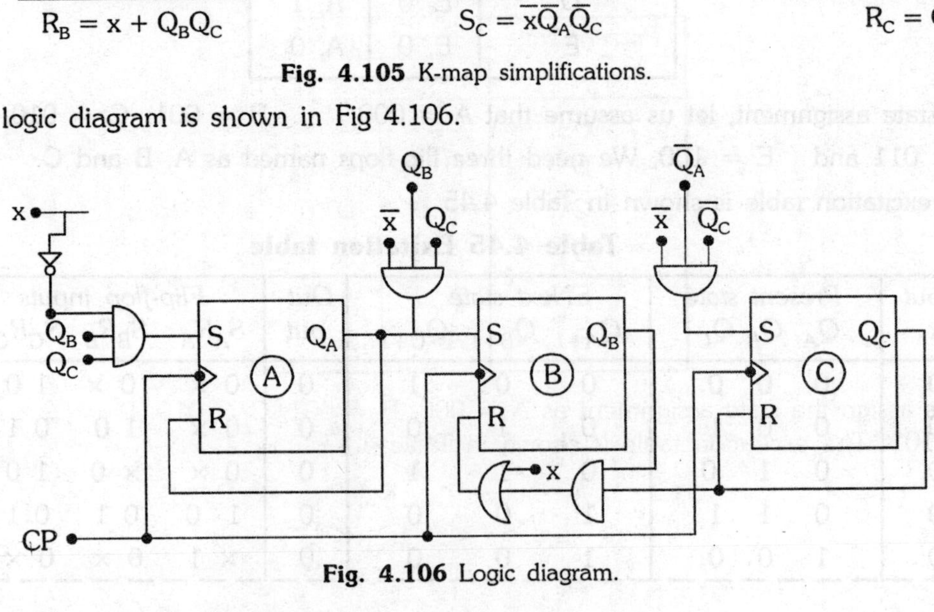

Fig. 4.106 Logic diagram.

Example 4.22. *Draw a two input, two output synchronous sequential circuit which examines the input sequence in non-overlapping strings of three inputs each and produces a one output coincident with the last input of the string if and only if the string consisted of either 2 or 3 1's. For example if the input sequence is 010101110, the required output sequence is 000001001. Use SR flip-flops in your realization.*

Solution. The input is 010 101 110, in the first three bits only one 1 is there so output is 000 upto first three bits, in the second 3-bit group, two 1's are there so output is 001. Similarly in the 3rd group, two 1's are there so the output is 001. The state diagram is shown in Fig. 4.107 which contains 6 states, so it requires three flip-flops for design, and are named as A, B and C.

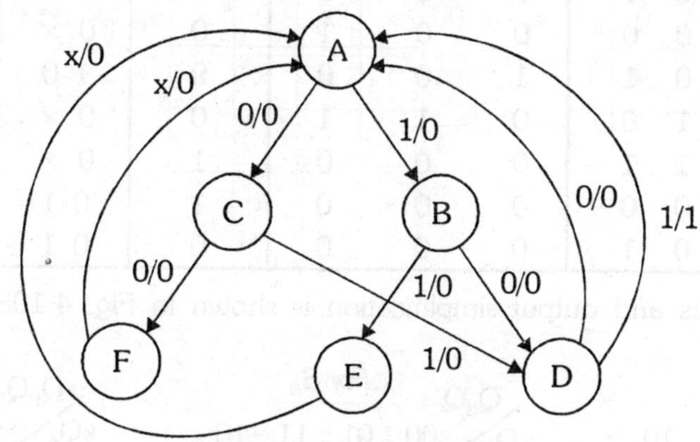

Fig. 4.107 State diagram.

Note : 'x' means (0 or 1).

The state table for the above figure is shown in Table 4.46.

Table 4.46 State table

Present state	*Next state Z*	
	x = 0	*x = 1*
A	C, 0	B, 0
B	D, 0	E, 0
C	F, 0	D, 0
D	A, 0	A, 1
E	A, 1	A, 1
F	A, 0	A, 0

Let us assign the state assignment as A = 000, B = 001, C = 010, D = 011, E = 100 and F = 101. The excitation table is shown in Table 4.47.

Table 4.47. Excitation table

Input x	*Present state* Q_A	Q_B	Q_C	*Next state* Q_{A+1}	Q_{B+1}	Q_{C+1}	*Output* z	*Flip-flop inputs* S_AR_A	S_BR_B	S_CR_C
0	0	0	0	0	1	0	0	0 ×	1 0	0 ×
0	0	0	1	0	1	1	0	0 ×	1 0	× 0
0	0	1	0	1	0	1	0	1 0	0 1	1 0
0	0	1	1	0	0	0	0	0 ×	0 1	0 1
0	1	0	0	0	0	0	1	0 1	0 ×	0 ×
0	1	0	1	0	0	0	0	0 1	0 ×	0 1
1	0	0	0	0	0	1	0	0 ×	0 ×	1 0
1	0	0	1	1	0	0	0	1 0	0 ×	0 1
1	0	1	0	0	1	1	0	0 ×	× 0	1 0
1	0	1	1	0	0	0	1	0 ×	0 1	0 1
1	1	0	0	0	0	0	1	0 1	0 ×	0 ×
1	1	0	1	0	0	0	0	0 1	0 ×	0 1

The flip-flop inputs and output simplification is shown in Fig. 4.108.

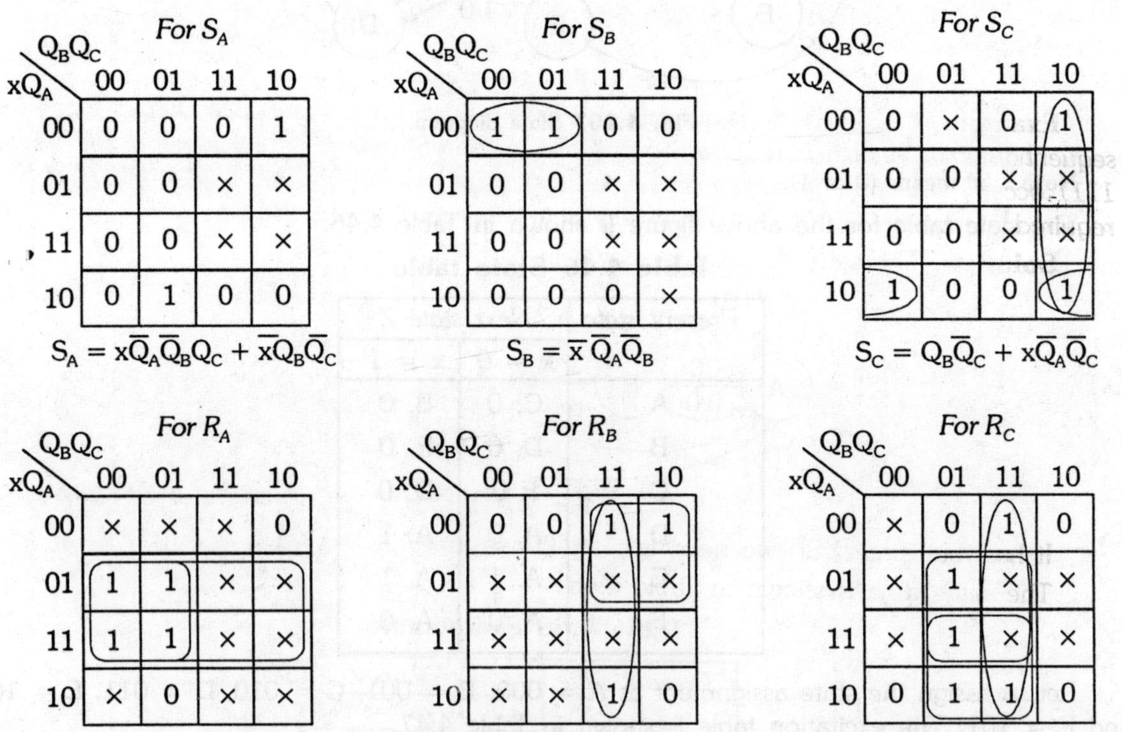

$S_A = x\overline{Q}_A\overline{Q}_BQ_C + \overline{x}Q_B\overline{Q}_C$

$S_B = \overline{x}\,\overline{Q}_A\overline{Q}_B$

$S_C = Q_B\overline{Q}_C + x\overline{Q}_A\overline{Q}_C$

$R_A = Q_A$

$R_B = \overline{x}Q_B + Q_BQ_C$

$R_C = Q_BQ_C + Q_BQ_C + xQ_C$

Fig. 4.108 K-maps.

The logic diagram is shown in Fig. 4.109.

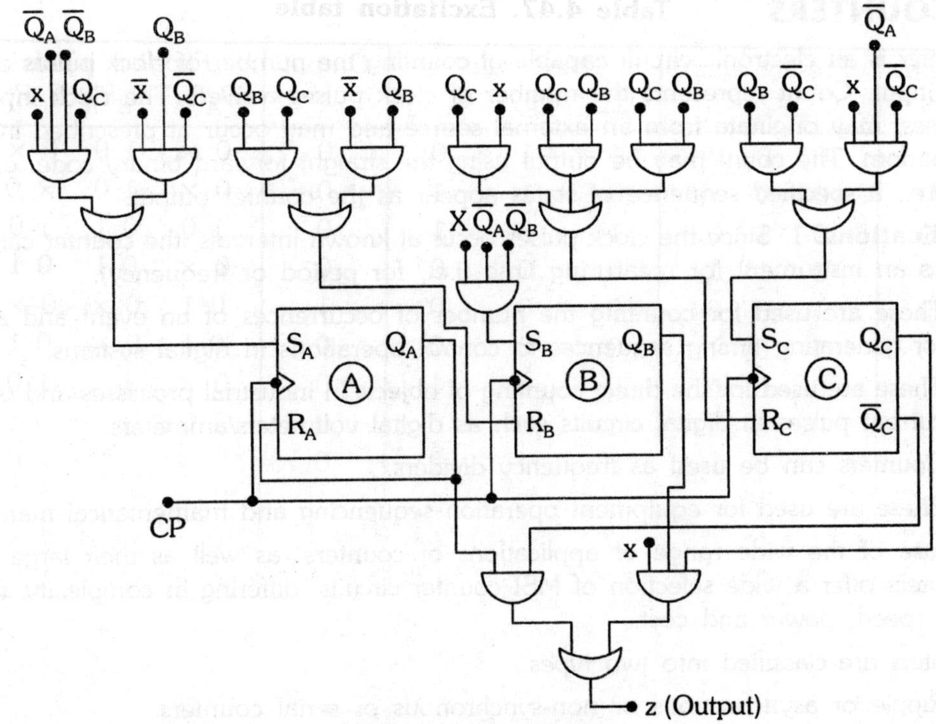

Fig. 4.109 Logic diagram.

Example 4.23. *A long sequence of pulses enters a two input, two output synchronous sequential circuit, which is required to produce an output pulse z = 1 whenever the sequence 1111 occurs. Overlapping sequences are accepted, for example, if the input is 01011111... the required output is 00000011... Draw the state diagram and state table.*

Solution. The state diagram is shown in Fig. 4.110

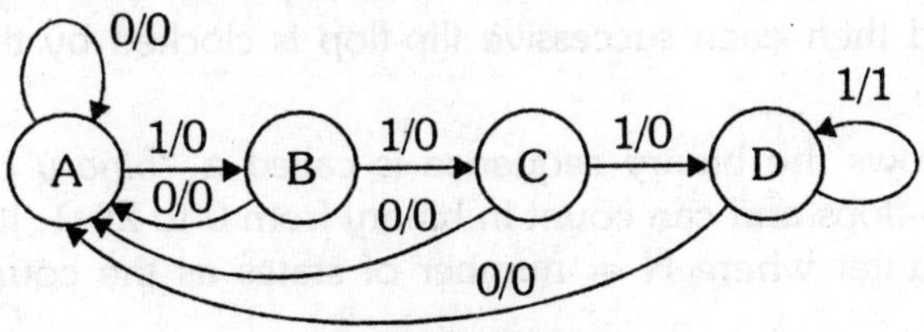

Fig. 4.110

It has four states. So we need two flip-flops named as A and B.

The state table is shown in Table 4.48.

Table 4.48 State table

Present state	*Next state, Z*	
	x = 0	*x = 1*
A	A, 0	B, 1
B	A, 0	C, 1
C	A, 0	D, 1
D	A, 0	D, 1

4.12 COUNTERS

Counter is an electronic circuit capable of counting the number of clock pulses arriving at it's clock input. Count represents the number of clock pulses arrived. The clock input, called *count pulses,* may originate from an external source and may occur at prescribed intervals of time at random. The count may be output using the straight forward binary code, or may be in BCD, *i.e.,* a specified sequence of states appear as the counter output.

Applications. 1. Since the clock pulses occur at known intervals, the counter can be used as an instrument for measuring time (*i.e.,* for period or frequency).

2. These are used for counting the number of occurrences of an event and are useful for generating timing sequences to control operations in digital systems.
3. These are used for the direct counting of objects in industrial processes and of voltage/ current pulses in digital circuits such as digital voltmeters/ammeters.
4. Counters can be used as frequency dividers.
5. These are used for equipment operation sequencing and mathematical manipulation.

Because of the wide range of applications of counters, as well as their large demand, manufacturers offer a wide selection of MSI counter circuits, differing in complexity, functional versatility, speed, power and cost.

Counters are classified into two types.

1. Ripple or asynchronous or non-synchronous or serial counters.
2. Synchronous or parallel counters.

Usually they are either JK-type or D-type.

In synchronous counter, the common clock input is connected to all the flip-flops and thus they are clocked simultaneously. In asynchronous counters, each flip-flop is triggered by the previous flip-flop, and the counter has a cumulative setting time. Hence in asynchronous counters, the flip-flops are not clocked simultaneously. The first flip-flop is clocked by the external clock pulse and then each successive flip-flop is clocked by the Q or $\overline{Q}$ output of the previous flip-flop.

A counter that follows the binary sequence is called a *"binary counter"*. An *n*-bit binary counter consists of *n*-flip-flops and can count in binary from 0 to $2^n–1$. It is also called as Modulus counter (or) Mod-N counter where N = number of states as the counter output = 2^n.

4.12.1 Asynchronous Or Ripple Counters

To design the ripple counter, the number of flip-flops required depends on the number of states. The number of the output states of the counter is called the "Modulus" (MOD) of the counter. The maximum number of states of a counter is 2^n, where *n* is the number of flip-flops in the counter. If we have two flip-flops, then the maximum possible number of output states of that counter is 2^2 *i.e.,* 4. Now we can name that counter as MOD-4 or Modulus-4 counter.

4.12.1.1 MOD-4 ripple counter

MOD-4 ripple counter is a 4–bit asynchronous counter, which consists of four states due to its two number of flip-flops. The state diagram is shown in Fig. 4.111(*a*). Fig. 4.111(*b*) shows the MOD-4 ripple counter, with the two numbers of JK flip-flops. Notice that the clock signal is connected to the clock input of only first stage flip-flop. The clock input of the second stage flip-flop is triggered by the Q_A output of the first stage. Note that the first flip-flop changes it's state only when triggered by the negative edge of each clock pulse, but the second flip-flop changes it's state only when triggered by the negative edge of each clock pulse, but the second flip-flop changes its state only when triggered by the negative-going transition of the Q_A output. Because of the inherent propagation delay time through a flip-flop, the transition of the input clock pulse and the transition of the output Q_A can never occur at exactly the same time. Therefore, the two flip-flops are never simultaneously triggered which results in asynchronous counter operation. Thus, the effect of an input clock pulse 'ripples' through the counter, taking sometime, due to propagation delays and reaches the last flip-flop. Hence the asynchronous counters are also called as "ripple counters".

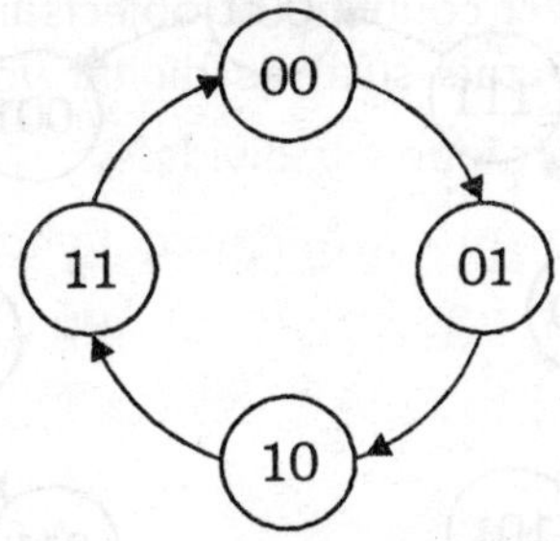

Fig. 4.111 ***(a)*** State diagram for two-bit ripple counter.

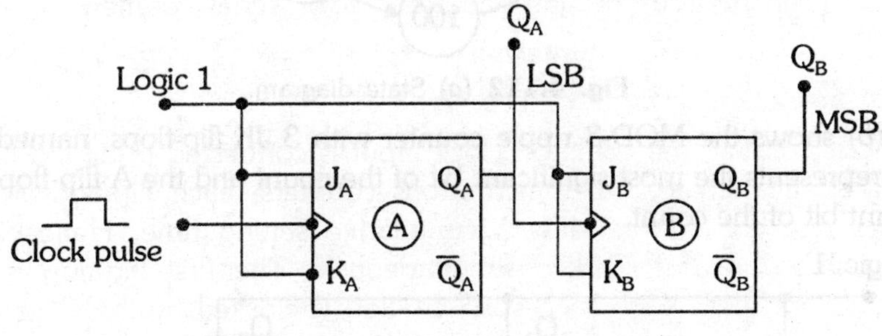

Fig. 4.111 ***(b)*** Two-bit ripple counter.

Fig 4.111 (*c*) shows the timing diagram for two bit asynchronous counter. It illustrates the changes in the state of the flip-flop outputs in response to the clock.

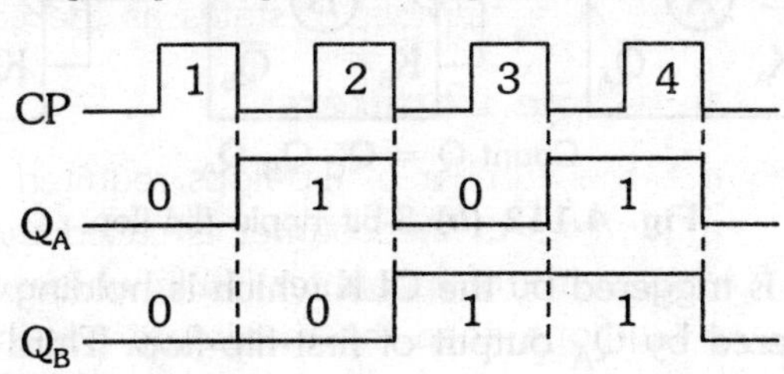

Fig. 4.111 (c) Timing diagram of two-bit ripple counter.

Note : BA is the count in each state.

4.12.1.2 Module-8 ripple counter

It is a 3-bit asynchronous counter and contains eight states from 000 to 111 and uses three flip-flops. The Fig. 4.112 (*a*) shows the state diagram of the 3-bit ripple counter. As can be seen from diagram, the binary states are indicated inside the circles and the directed lines between the circles are not marked with input-output values as in other state diagrams. The state transitions in clocked sequential circuits occur during a clock pulse; the flip flop remains in it's present state if no clock pulse occurs. For this reason the clock pulse variable CP does not appear explicitly as an input variable in a state diagram or state table. With this reason, the state diagram of a counter does not have the input-output values along the directed lines. The only input to the circuit is clock, and the outputs are directly specified by the present states of the flip-flops. The next state of the counter depends entirely on it's present state, and the state transition occurs when every clock pulse occurs.

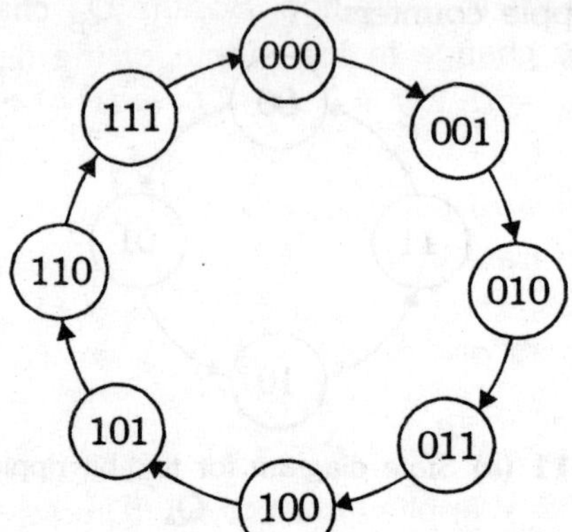

Fig. 4.112 (*a*) State diagram.

Fig. 4.112 (*b*) shows the MOD-3 ripple counter with 3 JK flip-flops, named as A, B and C, where C output represents the most significant bit of the count and the A flip-flop output indicates the least significant bit of the count.

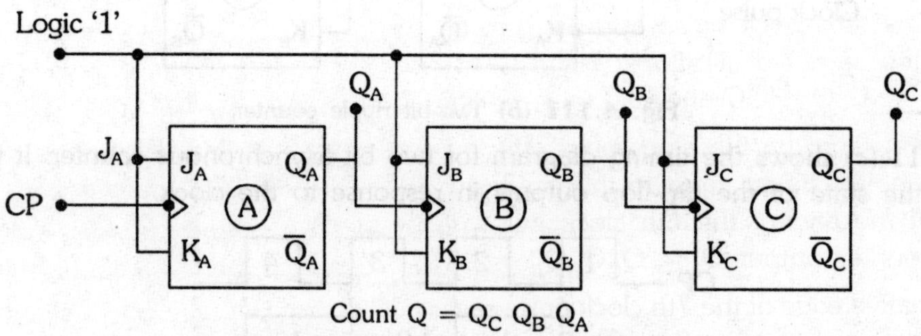

Fig. 4.112 (*b*) 3-bit ripple flip-flop.

Here, first flip-flop (A) is triggered by the CLK which is holding the least significant bit. The second flip-flop (B) is triggered by Q_A output of first flip-flop. Third flip-flop (C) is triggered by the Q_B output of the second flip-flop. So, in ripple counters, the flip-flop output transition serves as a source for triggering other flip-flops.

Operation. 1. The basic operation of the counter is studied by applying 8-clock pulses one by one to the first flip-flop (A). Observe the output of each flip-flop (ABC) *i.e.*, counter for every clock pulse. Note that the two inputs (J and K) of each flip-flop are connected in toggle condition (*i.e.*, J = K = 1) and assume that the counter is initially at RESET. At this state, the counter output is $Q = Q_CQ_BQ_A = 000$.

2. When the negative-going edge of the first clock pulse occurs, the output of the flip-flop Q_A goes to HIGH. (*i.e.*, Q_A is changing from 0 to 1). It has no effect on the second (B) and third flip-flop (C), their outputs are same as previous state because to trigger them, a negative transition pulse is required. Thus after the falling edge of the first clock pulse, $Q_A = 1$, $Q_B = 0$ and $Q_C = 0$. At this state, the counter output Q = 001.

3. For the negative going edge of the second clock pulse, the Q_A goes to LOW (*i.e.*, change in the output from 1 to 0). This indicates that the Q_A changes to negative edge triggering which is applied as a clock to lip-tlop. So, the Q_B changes from 0 to 1. But, this change in Q_B does not affect any change in the output of the flip-flop C. After the negative edge of the second clock pulse, $Q_A = 0$, $Q_B = 1$ and $Q_C = 0$. At this stage, the counter output Q = 010.

4. For the negative going edge of the third clock pulse, the Q_A changes from 0 to 1 *i.e.*, LOW to HIGH, which is a positive transition pulse which does not affect the flip-flop B. Hence the output of flip-flop B = Q_B is same as the previous state. As there is no change in the Q_B, also, there is no change in the output of flip-flo *i.e.*, Q_A. At the end of the 3rd clock pulse, the outputs of flip-flops are $Q_A = 1$, $Q_B = 1$ and $Q_C = 0$. Hence, the count $Q = Q_CQ_BQ_A = 011$.

5. When the fourth clock pulse is applied, the output Q_A goes to LOW, which is a negative transition pulse (HIGH to LOW) is applied as a clock to the flip-flop B. With the application of this pulse, the output of flip-flop B changes (Q_B) from HIGH to LOW, which is another negative transition pulse, applied as a clock to the flip-flop C. With this, the output of flip-flop C, Q_C changes from LOW to HIGH. After the fourth clock pulse, the outputs are $Q_A = 0$, $Q_B = 0$, and $Q_C = 1$. Hence, the counter output $Q = Q_CQ_BQ_A = 100$.

6. When the fifth clock pulse appears, the output Q_A goes to HIGH from LOW, which does not affect the output of flip-flops B and C. Hence, at the end of the 5th clock pulse the outputs are $Q_A = 1$, $Q_B = 0$ and $Q_C = 1$, so the counter $Q = Q_CQ_BQ_A = 101$.

7. When applying the 6th clock pulse, the output Q_A goes to LOW, which is a negative transition pulse applied as an input to the flip-flop B. With this, the Q_B changes from LOW to HIGH. Hence at the end of the 6th clock pulse, the outputs are $Q_A = 0$, $Q_B = 1$ and $Q_C = 1$, and the counter output $Q = Q_CQ_BQ_A = 110$.

8. The negative edge of the 7th clock pulse causes the Q_A to go to HIGH, which is a positive transition pulse, which does not affect the Q_B and hence Q_C also. After the 7th clock pulse, the outputs are $Q_A = 1$, $Q_B = 1$ and $Q_C = 1$, then the counter output $Q .= Q_CQ_BQ_A = 111$.

9. The negative-going edge of the 8th clockpulse causes Q_A to LOW, which is a negative transition pulse changes the output of flip-flop to LOW and which is also a negative transition pulse applied as a clock to the flip-flip C. Hence, Q_C changes from HIGH to LOW. After the 8th clock pulse, the outputs are $Q_A = 0$, $Q_B = 0$ and $Q_C = 0$. At the stage, the counter output $Q = Q_CQ_BQ_A = 000$.

Note that the counter has now recycled back to it's initial state, *i.e.*, all three flip-flops are in RESET. The operation of the 3-bit ripple counter is shown in the Table 4.49.

Table 4.49

Count sequence			Conditions for complementing
Q_C	Q_B	Q_A	Flip-flops
0	0	0	Initial condition
0	0	1	At the end of 1st clock pulse negative edge transition.
0	1	0	(Q_A changes from 0 to 1) At the end of 2nd clock pulse, negative edge transition Q_A changes from 1 to 0 and Q_B from 0 to 1.
0	1	1	At the end of 3rd clock pulse Q_A changes from 0 to 1, and no change in Q_B and Q_C.
1	0	0	At the end of 4th clock pulse, Q_A changes from 1 to 0, Q_B changes from 1 to 0 and Q_C from 0 to 1.
1	0	1	At the end of the 5th clock pulse, Q_A changes from 0 to 1 and no change in Q_B and Q_C.
1	1	0	At the end of the 6th clock pulse, Q_A changes from 1 to 0 and Q_B changes from 0 to 1 and no change in Q_C.
1	1	1	At the end of the 7th clock pulse, Q_A changes from 0 to 1 and no change in Q_B and Q_C.

The timing diagram of the 3-bit ripple counter is shown in Fig. 4.112 (*c*).

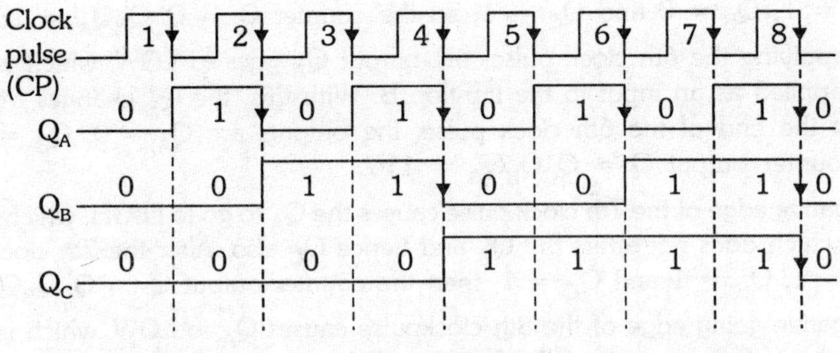

Fig. 4.112 (c) Timing diagram of the 3-bit ripple counter.

Frequency : The waveforms in Fig. 4.112 (*c*) illustrate the fact that the waveform at Q_A is one-half the frequency of the clock (CP), Q_B is one-half of the Q_A and is one-fourth of the clock frequency, or Q_C is one-eighth of the clock frequency or one-fourth of the Q_A or one-half of the Q_B.

If CP = 1000 Hz, then Q_A = 500 Hz, Q_B = 250 Hz and Q_C = 125 Hz. Hence the 3-bit ripple counter is also called as divide-by-8-counter.

Conclusion : For every negative edge of the clock there is a change in Q_A, either 0 to 1 or 1 to 0. So, the Q_A output frequency is one-half of the clock pulse. For every negative edge of the Q_A, there is a change in Q_B either 1 to 0 or 0 to 1. So Q_B output frequency is one-half of Q_A. For every negative edge of Q_B pulse, there is a change in Q_C.

In Fig. 4.112 (*c*) timing diagram, we have not considered the propagation delay of flip-flops, for simplicity. If we consider the propagation delay of the flip-flops, we get this timing diagram as shown in Fig. 4.112 (*d*).

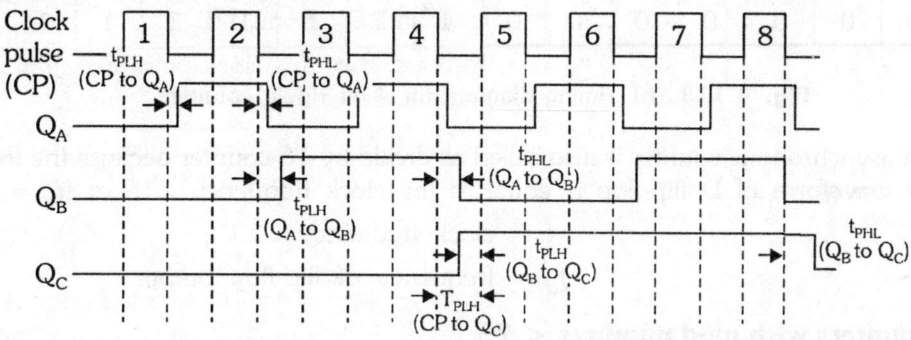

Fig. 4.112 (*d*) Timing diagram of a 3-bit ripple counter with propagation delay of each flip-flop.

From the Fig. 4.112 (*d*), we can see that the propagation delay of the first stage is added in the propagation delay of the second stage to decide the transition time for the third stage. This cumulative delay of an asynchronous counter limits the use of number of flip-flops in a single counter, which limits the maximum count of the counter.

4.12.1.3. A 4-bit asynchronous counter or four-stage or module-16 or divide-by-16 counter

A four-bit asynchronous counter is obtained by the use of 4 JK-flip-flops, named as A, B, C and D, where D flip-flop holds the most significant bit of the count and flip-flop A holds the least significant of the count. The 4-bit ripple counter and the timing diagrams are shown in the Fig. 4.113 (*a*) and 4.113 (*b*).

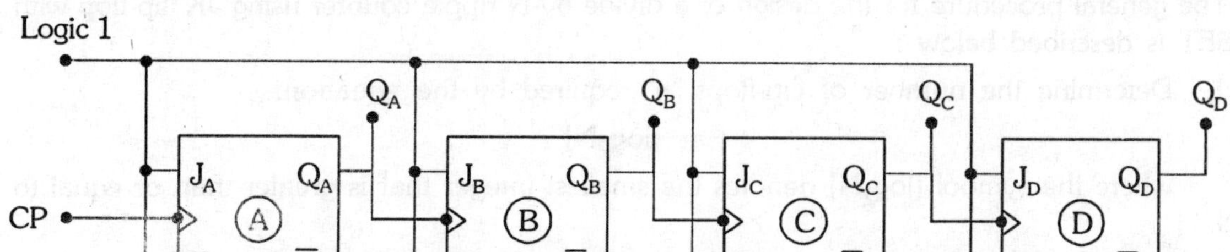

Count Q = $Q_D\ Q_C\ Q_B\ Q_A$

Fig. 4.113 (*a*) Logic diagram.

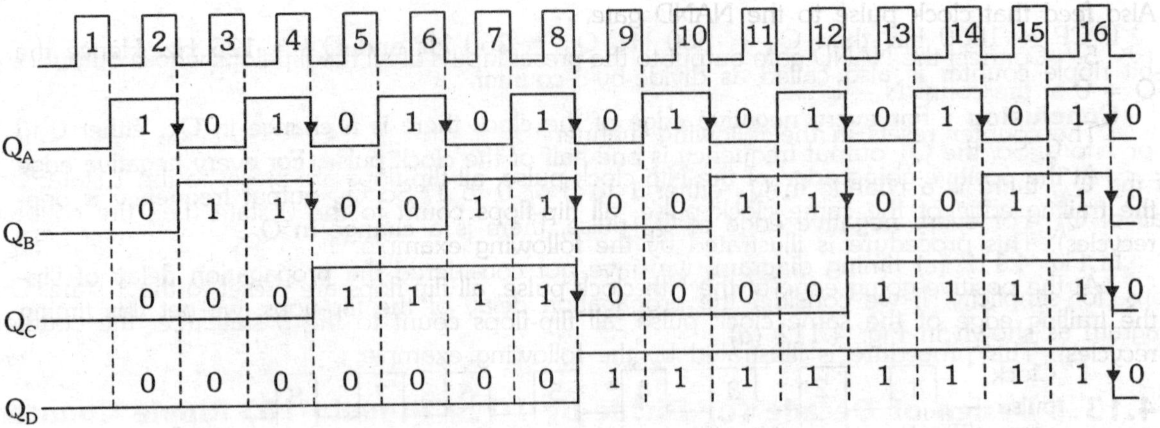

Fig. 4.113 (*b*) Timing diagram for 4-bit ripple counter.

The 4-bit asynchronous counter is also called as divide-by-16 counter because the frequency of the output waveform of D flip-flop is equal to the clock frequency ÷ 16 or ($f_d = f/16$)

where f = clock frequency

f_d = frequency of flip-flop output.

4.12.1.4 Counters with mod numbers < 2^n

Counters can also be designed to have a number of states in their sequence less than 2^n. The resulting sequence is called a "truncated sequence". To obtain the truncated sequence, it is necessary to force the counter to recycle before going through all of it's normal states.

Whenever we want to count to a base N, which is not a power of 2, then we have to consider 'divide-by' factor, which is same as the modulus. These counters are called "divide-by-N" counters. The basic counter is modified to produce the required MOD numbers less than 2^n by allowing the counter to "*skip states*" that are not part of the counting sequence. One of the most common method for doing this is making all flip-flops to reset after the count N with the help of feedback gate in the circuit. Such feedback is provided by a NAND gate, in which the output provides all clear inputs in parallel.

4.13 DESIGN OF DIVIDE-BY-N RIPPLE COUNTER

The general procedure for the design of a divide-by-N ripple counter using JK flip-flop with PRESET is described below :

1. Determine the number of flip-flops 'n' required by the equation.

$$n = [\log_2 N]$$

where the symbol $[\log_2 N]$ denotes the smallest integer that is greater than or equal to $\log_2 N$.

2. Connect the n-flip-flops as a ripple counter.
3. Find the binary representation of N – 1, where N is the maximum count you need by the counter.
4. Connect all flip-flop outputs that are 1 at the counter (N – 1) as inputs to a NAND gate.

Also feed that clock pulse to the NAND gate.

5. Connect the NAND gate output to the preset inputs of all the flip-flops whose output *i.e.*, Q = 0 at the count N – 1.

The counter resets in the following manner :

At the positive-going edge of the Nth clock pulse, all flip-flops are preset to the 1 state. On the trailing edge of the same clock pulse, all flip-flops count to the 0 state (*i.e.*, the counter recycles). This procedure is illustrated by the following example.

At the positive-going edge of the Nth clock pulse, all flip-flops are preset to the 1 state. On the trailing edge of the same clock pulse, all flip-flops count to the 0 state (*i.e.* the counter recycles). This procedure is illustrated by the following example.

4.13.1 Design of Decade (or Divide-by-10 or Modulo-10) Ripple Counter

For $$N = 10,\ n = \left[\log_2^{10}\right] = [3.322] = 4.$$

Thus, we need four negative edge triggered JK flip-flops. Connect them as a 4-bit ripple coutner.

Since N = 10, (indicates that, the counter has totally 10 states, *i.e.*, 0 to 9) where 9 is the maximum count in a MOD-10 counter then N – 1 = 9 and its binary equivalent is 1001. The outputs Q_A and Q_D and clock are given as an input to the NAND gate and its output is connected to the preset of the flip-flops B and C. The decade ripple counter is shown in Fig. 4.114.

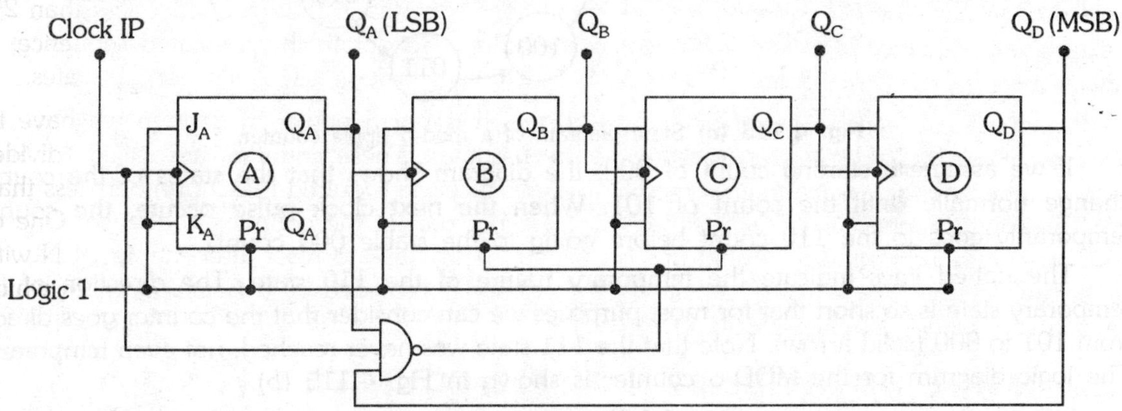

Fig. 4.114 Decade ripple counter.

One more popular method for designing the ripple counter whose count is <2^n is by using the clear input.

1. Determine the number n of flip-flop required by the equation $n = [\log_2 N]$, where N = number of states in the counter and N – 1 is the maximum count of the counter and $[\log_2 N]$ denotes the smallest integer that is greater than or equal to $\log_2 N$.

2. Connect the n-flip-flops as a ripple counter.

3. Find the binary representation of N.

4. Connect all flip-flops outputs that are 1 at the count N of the counter, as an input to the NAND gate.

5. Connect the NAND gate output to the clear inputs of all the flip-flops. The counter resets in the following manner.

As long as the NAND gate output is HIGH (the output of NAND gate is HIGH for all counts except for the count = N), it will have no effect on the counter. When it goes LOW (at the count = N), however, it will clear all the flip-flops so that the counter immediately goes to the 000 state.

This is illustrated clearly by the following example.

4.13.2 Design of MOD-6 Ripple Counter

1. Here, N = 6 and the number of flip-flops required is n = [$\log_2 6$] = 3. Thus use of 3 JK flip-flops is necessary and they are named as A, B and C. Assume that these three flip-flops are negative edge triggered flip-flops.

2. The state diagram of the MOD-6 counter is shown in Fig. 4.115(*a*).

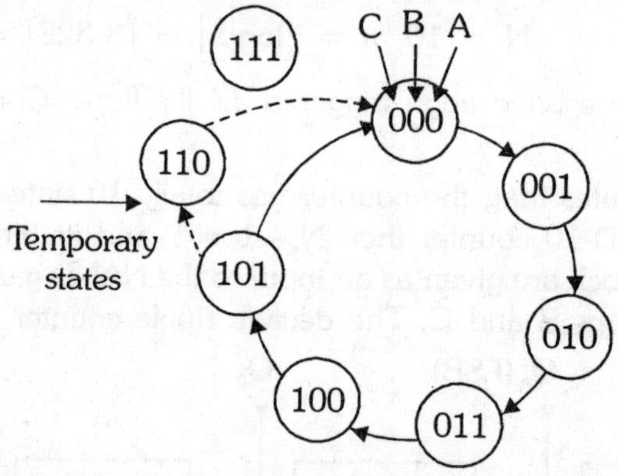

Fig. 4.115 (*a*) State diagram of a mod-6 ripple counter.

If we assume a starting count of 000, the diagram shows that the states of the counter change normally until the count of 101. When the next clock pulse occurs, the counter temporarily goes to the 110 count before going to the stable 000 count.

The dotted lines indicate the temporary nature of the 110 state. The direction of this temporary state is so short that for most purposes we can consider that the counter goes directly from 101 to 000 (solid arrow). Note that the 111 state was never reached, not even temporarily. The logic diagram for the MOD-6 counter is shown in Fig. 4.115 (*b*).

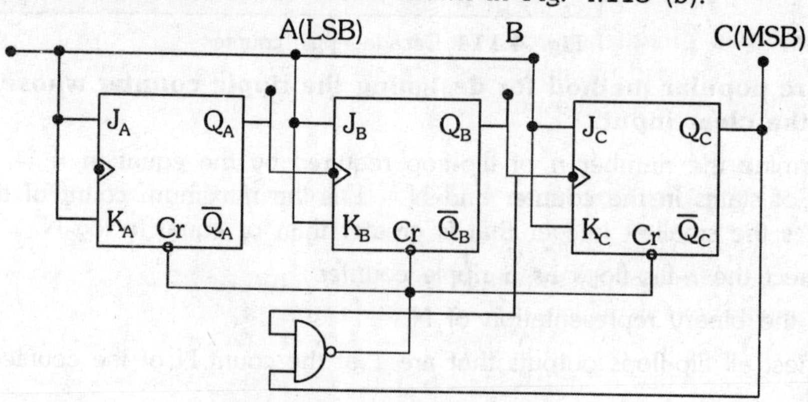

Fig. 4.115 (*b*) Logic diagram.

The inputs to the NAND gate are the outputs of the flip-flops B and C, so, the NAND gate output is LOW when $Q_B = 1$ and $Q_C = 1$, until that the NAND gate output is HIGH (which does not affect the clear input of any flip-flop). This condition will occur when the counter goes from 101 state to the 110 state (6th clock pulse). The LOW signal at the NAND gate output will immediately (generally a few nanoseconds) clear the counter to 000 state. Once the flip-flops have been cleared, the NAND output goes back to HIGH, since the $Q_B = 1$ and $Q_C = 1$ condition no longer exists.

The counting sequence is, therefore 000, 001, 010, 011, 100, 101 and (110 temporary state), 000. This type of counter is sometimes called "permuting counter", since it's natural count has been permuted (changed).

Although the counter goes to the 110 state, it remains for only few nanoseconds before it recycles to 000. Thus, we can essentially say that this counter counts from 000 to 101 and then recycles to 000. It essentially skips 110 and 111 so that it goes through only six different (000 to 101) states, thus, it is called MOD-6 counter.

The timing diagram of the MOD-6 counter is shown in Fig. 4.115(*c*).

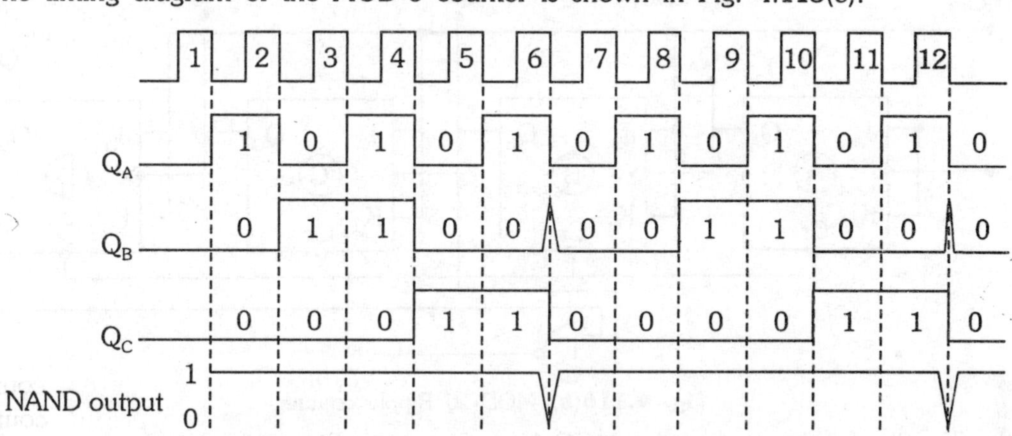

Fig. 4.115 (c) Timing diagram of MOD-6 ripple counter.

Notice that the waveform at the B output contains a "*spike or glitch*" caused by the momentary occurrence of the 110 state before clearing. This glitch is narrow and so would not produce any visible indication on numerical displays. It could, however, cause a problem if the B output is being used to drive other circuitry outside the counter. It should also be noted that the C flip-flop output has a frequency equal to the 1/6th of the input frequency; in other words, this MOD-6 counter has divided the input frequency by six. The waveform at flip-flop C output is not a symmetrical square wave, because it is HIGH for only two clock cycles and LOW for four clock cycles.

4.13.3 MOD-10 Ripple Counter / Decade Counter (or) BCD Counter

A decimal counter follows a sequence of ten states and returns to 0 after the count of 9. The counters with ten states in their sequence called are 'decade counters'. These counters are useful in display applications in which BCD is required for conversion to a decimal read out. A counter with a count sequence of 0 (0000) through 9 (1001) is called a BCD counter, because it's ten-states sequence is the BCD code. The sequence of states is shown in the state diagram of Fig. 4.116(*a*).

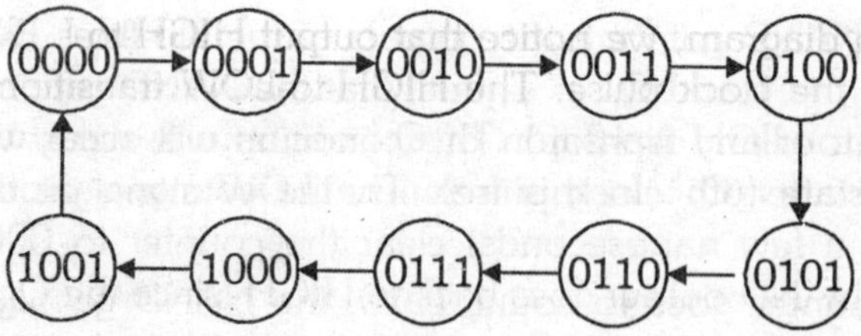

Fig. 4.116(*a*) State diagram of a decimal BCD counter.

A four-bit asynchronous decade counter with four JK flip-flops is shown in the Fig. 4.116 (*b*).

Operation. 1. The operation of the decade counter is performed by application of ten clock pulses. The clock input is connected to the first flip-flop only. The second, third and fourth flip-flops are triggered by the outputs Q_A, Q_B and Q_C respectively. The output Q of the counter is given by

$$Q = Q_D Q_C \; Q_B Q_A$$

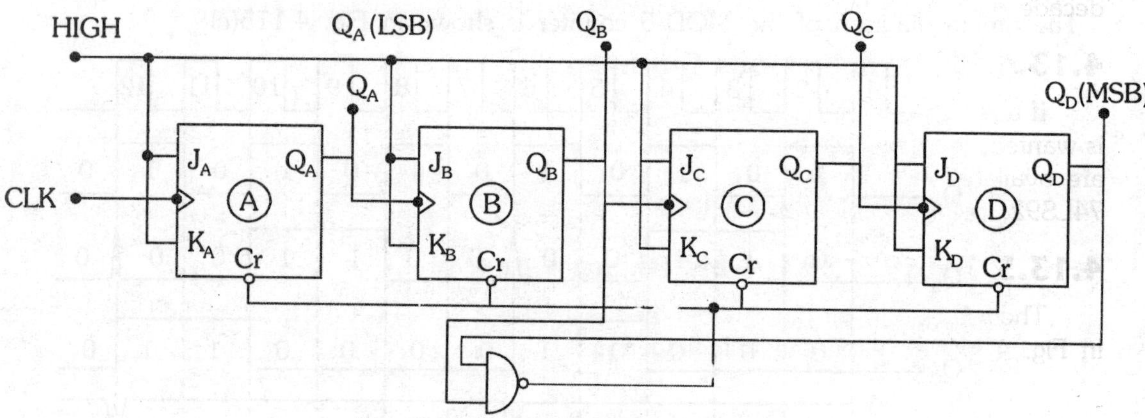

Fig. 4.116(*b*) MOD-10 Ripple counter.

2. The timing diagram of the MOD-10 is shown in Fig 4.116 (*c*).

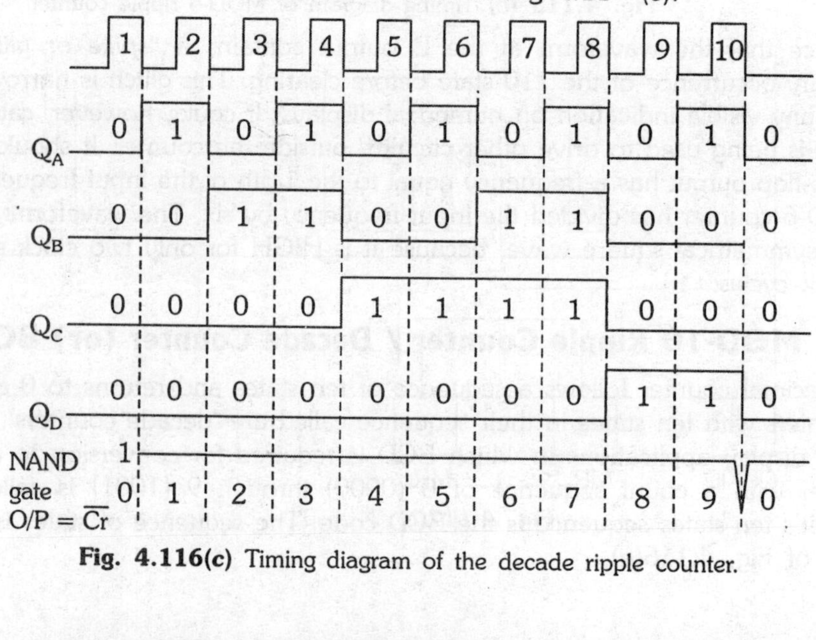

Fig. 4.116(c) Timing diagram of the decade ripple counter.

From the timing diagram, we notice that output HIGH-to-LOW transition of Q_A occurs only one delay time after the clock pulse. The HIGH-to-LOW transition of Q_B occurs one delay time after Q_A. The LOW-to-HIGH transition of Q_C occurs one delay time after Q_B. The Q_D occurs only one delay time after Q_C. In this manner, the counting process continues from the count 0000 to 1001.

3. When the counter goes to count 1010, the NAND gate output goes LOW and all flip-flops are in clear condition. Thus, when 10th clock pulse occurs the counter output Q is 0000 instead of 1010.

The BCD counter of Fig. 4.116 (*b*) is a decade counter, since it counts from 0 to 9. To count in decimal from 00 to 99, we need a two-decade counter. To count from 000 to 999, we need a three-decade counter. Multiple decade counters can be constructed by connecting BCD counters in cascade, for each decade. In the three-decade counter the inputs to the second and third decade counter come from Q_D of the previous decade. When Q_D in one decade goes from 1 to 0, it triggers the count for the next higher-order decade while it's own decade goes from 9 to 0. For instance, the count after 499 will be 500.

4.13.4 IC Asynchronous Counters

If a regular count sequence is required, an IC counter can be used. If an irregular sequence is wanted, then an IC flip-flop can be used to build it. Many four-bit non-synchronous counters are available in both the TTL and CMOS logic family. Four examples are the 74LS90, the 74LS92, the 74LS93 and 74HC93.

4.13.5 74LS93/293 4-bit Binary Counter

The 74LS93, 4-bit binary counter is shown in Fig. 4.117(*a*) The simplified symbol is shown in Fig. 4.117(*b*). Note the following points.

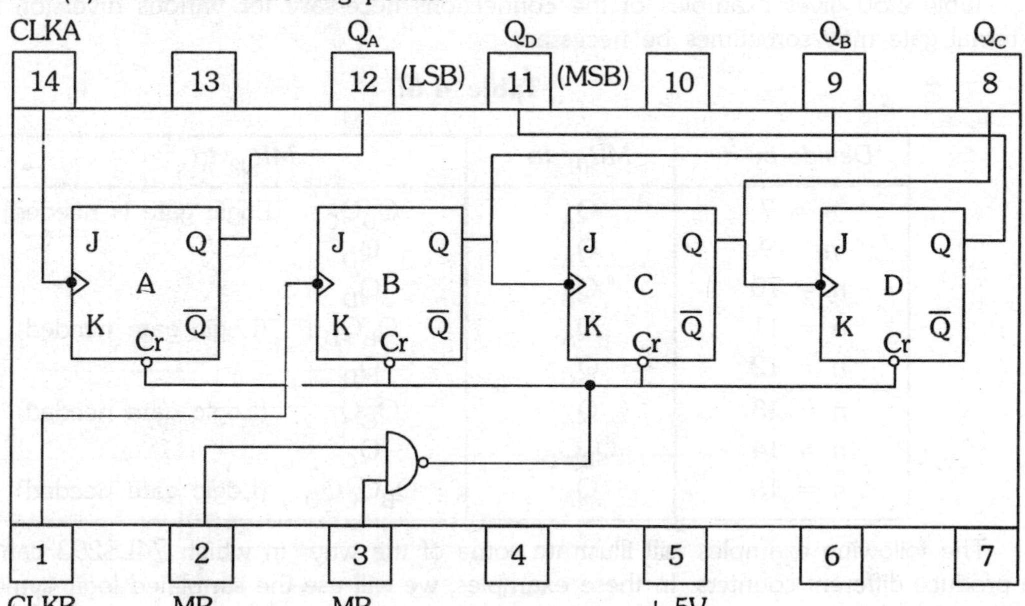

Fig. 4.117(*a*) 74LS593 4-bit binary counter.

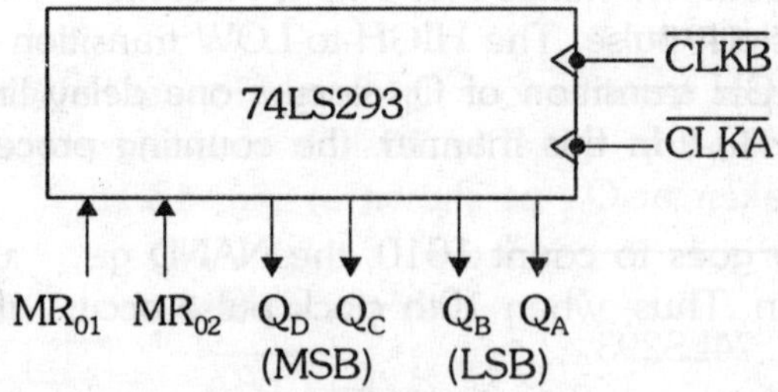

Fig. 4.117(*b*) Logic symbol.

1. The 74LS293 has four JK flip-flops with outputs Q_A, Q_B, Q_C and Q_D, where Q_A is the LSB and Q_D is the MSB. The flip flops are shown arranged with the LSB on the left.

2. Each flip-flop has a CP (clock pulse) input, the clock inputs to Q_A and Q_B, labelled CLKA and CLKB, respectively, are externally accessable and these are active-low signals.

3. Each flip-flop has an asynchronous CLEAR input, Cr. These are connected together to the output of a NAND gate, whose inputs are MR_{01} and MR_{02}, where MR stands for (Master reset). Both MR inputs must be HIGH to clear the counter to 0000.

4. Flip-flops Q_B, Q_C and Q_D are already connected as a 3-bit ripple counter. Flip-flop Q_A is not connected to anything internally. This allows the user the option of either connecting Q_A to Q_B to form a 4-bit counter (or) Q_A separately if desired.

The 74LS293 can be modified using the reset method, to give a count other than 2, 8 or 16. The circuit will reset at the required count if the Q outputs that are then at 1 are fed back to the reset inputs MR_{01} and MR_{02}.

Table 4.50 gives examples of the connections necessary for various diversion ratios. An external gate may sometimes be necessary.

Table 4.50

Devide-by-n	MR_{01} *to*	MR_{02} *to*
$n = 7$	Q_A	Q_BQ_C (Logic gate is needed)
$n = 9$	Q_A	Q_D
$n = 10$	Q_B	Q_D
$n = 11$	Q_A	Q_BQ_D (Logic gate needed)
$n = 12$	Q_C	Q_D
$n = 13$	Q_A	Q_CQ_D (Logic gate needed)
$n = 14$	Q_BQ_D	Q_C
$n = 15$	Q_A	$Q_BQ_CQ_D$ (Logic gate needed)

The following examples will illustrate some of the ways in which 74LS293 can be wired to produce different counters. In these examples, we will use the simplified logic symbol shown in Fig. 4.117 (*b*).

Example 4.24. *Show how 74LS293 should be connected to operate as a MOD-16 counter with a 10 kHz clock input.*

Solution. MOD-16 counter requires four flip-flops. So, we have to connect the Q_A output to CLKB, the clock input of flip-flop Q_B. The 10 kHz pulses are applied to CLKA, the clock input of Q_A. The output is taken at Q_D as shown in Fig. 4.118.

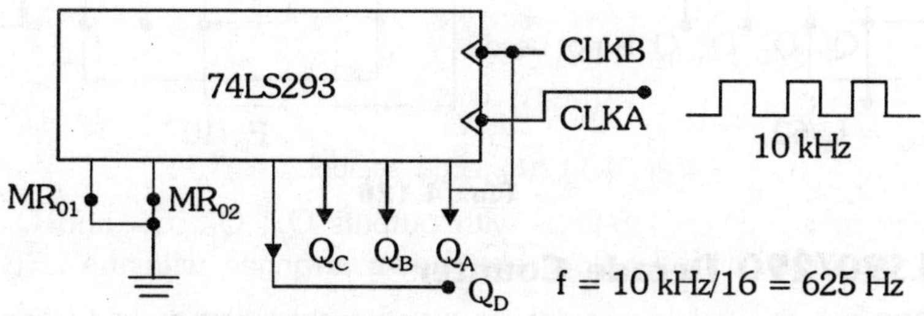

Fig. 4.118 74LS293 as a MOD-16 counter.

Example 4.25. *Show how 74LS293 works as a MOD-10 counter with 10 kHz clock input.*

Solution. MOD-10 counter requires four flip-flops, and so again we need to connect Q_A to CLKB. This time, we want the counter to recycle back to 0000 after the count 1001 *i.e.*, when it tries to go to the count of 1010 (ten). Thus Q_D and Q_B outputs have to be connected to the master reset inputs; when they both go HIGH at count 1010, the NAND output will immediately reset the counter to 0000. The circuit diagram is shown in Fig. 4.119.

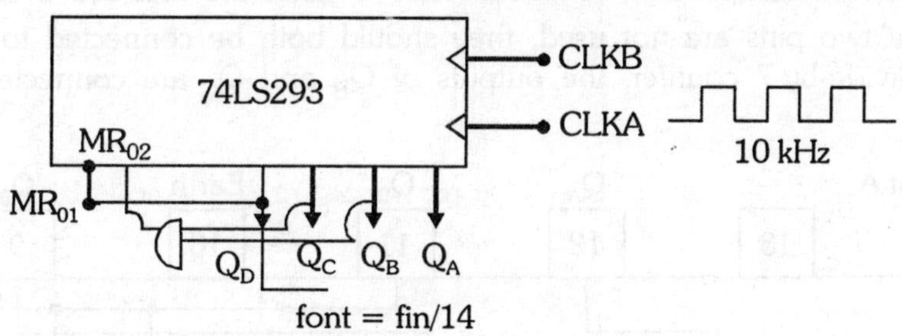

Fig. 4.119 Circuit diagram.

Example 4.26. *Show how 74LS293 works as a MOD-14 counter with 10 kHz clock input.*

Solution. It needs four flip-flops. So, Q_A is connected to CLKB. When the counter reaches the count of 1110(14), the Q_D, Q_C and Q_B outputs are high. To ensure the counter is in reset, we need an extra two-input AND gate, whose inputs are from Q_C and Q_B and it's output is connected to MR_{01} as shown in Fig. 4.119.

Example 4.27. *Show how 74LS293 can be used as a MOD-60 counter.*

Solution. The circuit must contain two 74293s to divide the input frequency by 60 *i.e.*, in two steps, one is used as a MOD-10 counter, so that the output frequency of it's Q_D is $f_{in}/10$. This signal is connected to CLKB input of the second 74293 counter, which is wired (connected) as a MOD-6 counter (hence Q_A is not being used). Thus, the Q_D output of the second counter will have a frequency, $f_{out} = \frac{f_{in}/10}{6} = f_{in}/60$ is shown in Fig. 4.120.

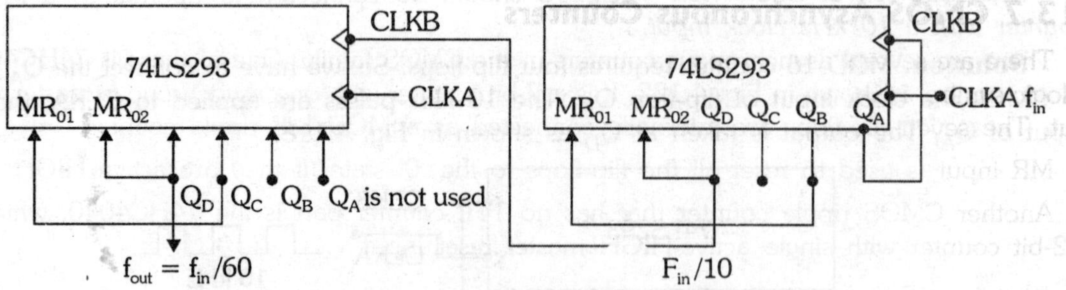

Fig. 4.120

4.13.6 74LS90/290 Decade Counter

The 74LS290 is a decade counter that consists of one JK flip-flop which acts as a divide by 2 circuit (A) and a 3-stage counter (B, C, D flip-flops) that is internally connected to give a count of 5. The pin connections are shown in Fig. 4.121. To obtain the maximum count *i.e.*, MOD-10, the Q_A and the input B/D are externally connected. The overall division ratio is $2 \times 5 = 10$.

It has two master reset terminals, MR_{01} and MR_{02}. It has two other pins labelled as R_{91} and R_{92}, which can be used to set the counter to a count of 9. This feature is useful because it ensures that the circuit will start to count from '0' after the first acti: e clock transition.

When these two pins are not used, they should both be connected to earth. To convert the IC into a divide-by-7 counter, the outputs of Q_B and Q_C are connected to R_{91} and R_{92} respectively.

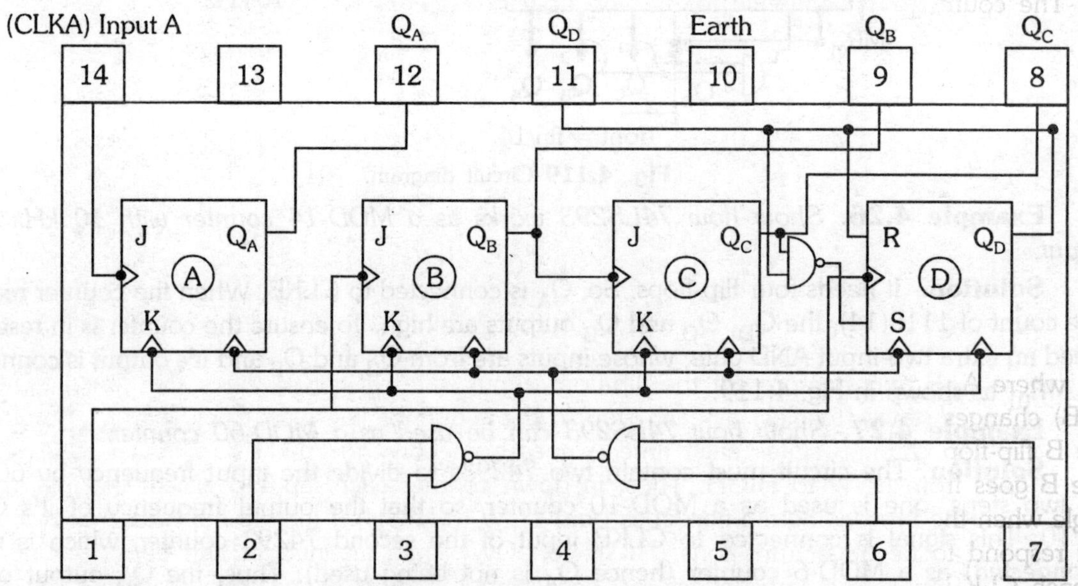

Fig. 4.121 Decade counter.

4.13.7 CMOS Asynchronous Counters

There are several asynchronous counters in the CMOS family. One of these is 74HC4024. It's logic symbol is shown in Fig. 4.122 It is a 7-bit counter with one asynchronous master reset input. The seven flip-flops are internally connected as a MOD-128 ripple counter.

MR input is used to reset all the flip-flops to the '0' state, if they are active HIGH.

Another CMOS ripple counter that has no TTL counter part is the 74HC4040, which is a 12-bit counter with single active-HIGH master reset input.

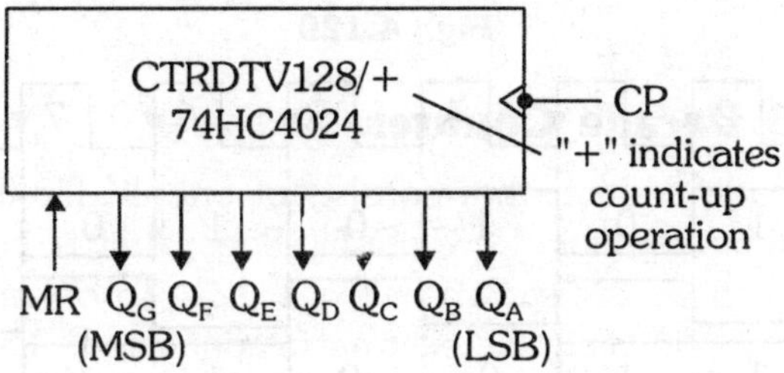

Fig. 4.122 Logic symbol for the CMOS 74HC4024 ripple counter.

4.13.8 Asynchronous Down Counter

All the counters studied so far have counted upward from zero; that is, they were 'upcounters'. Now we will see the procedure for the design of 'down counters', which will count downward from a maximum count to zero.

The count down sequence for a 3-bit down counter is :

Count	C B A
7	1 1 1
6	1 1 0
5	1 0 1
4	1 0 0
3	0 1 1
2	0 1 0
1	0 0 1
0	0 0 0

Recycles (from 000 back to 111)

where A, B and C represent the flip-flop output states. It can be seen that the A flip-flop (LSB) changes states (toggles) at each step in the sequence just as it does in the up counter. The B flip-flop changes states each time A goes from LOW to HIGH; C changes states each time B goes from LOW to HIGH. Thus, in down counter each flip-flop except the first, must toggle when the preceeding flip-flop goes from LOW to HIGH. If the flip-flops have CLK inputs that respond to negative transitions (HIGH to LOW), then an inverter can be placed infront of each CLK input. The same effect can be accomplished by diving each flip-flop CLK input from the inverted output of the preceeding flip-flop. This is shown in Fig. 4.123(*a*) for a MOD-8 down counter.

The transition diagrams of each flip-flop are shown in Fig. 4.123(*b*) and their corresponding state diagrams are shown in Fig. 4.123(*c*).

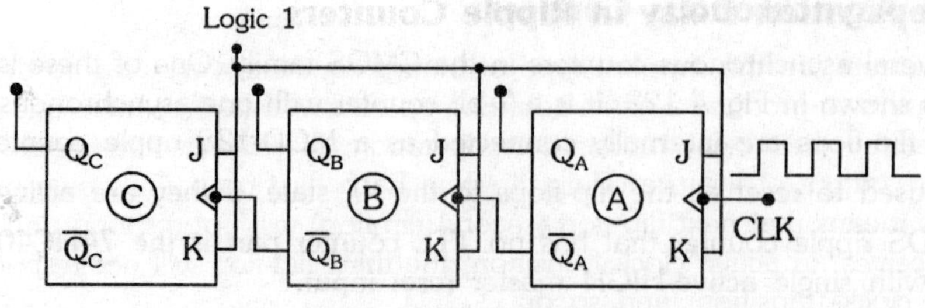

Fig. 4.123 ***(a)*** 3-bit down counter.

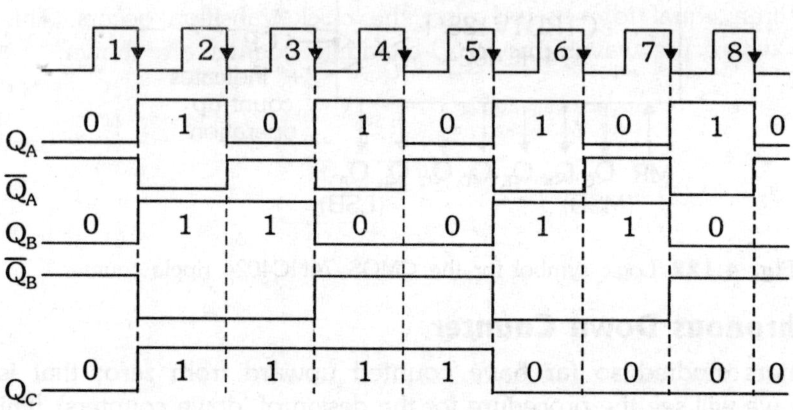

Fig. 4.123 **(*b*)** Transition diagram for 3-bit down counter.

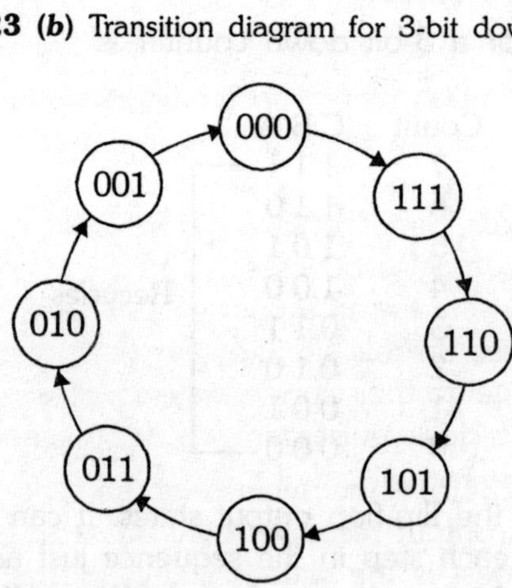

Fig. 4.123 **(*c*)** State diagram for a 3-bit asynchronous ripple counter.

The major application of down counter is in situations where it must be known when a desired number of input pulses has occurred. In these situations, the down counter is preset to the desired count and then allowed to count down as the pulses are applied. When the counter reaches zero, it is detected by a logic gate whose output then indicates that the preset number of pulses has occurred.

4.13.9 Propagation Delay in Ripple Counters

Ripple counters are the simplest type of binary counters, very easy to design and require fewest components to produce a given counting operation. But, they have one major drawback, which is caused by their basic principle of operation. Here, each flip-flop is triggered by the transition at the output of the preceding flip-flop. Each flip-flop has it's own propagation delay time (*tpd*), this means that next flip-flop (second flip-flop) will not respond until a time *tpd* after the flip-flop receives an active clock transition; the third flip-flop will not respond until the time *2tpd* after the clock transition; and so on.

The propagation delays of the flip-flops accumulate, so that *n*th flip-flop can not change states until a time equal to $n \times tpd$ after the clock transition occurs. This is illustrated in Fig. 4.124 (*a*) where the waveforms for 3-bit ripple counter are shown.

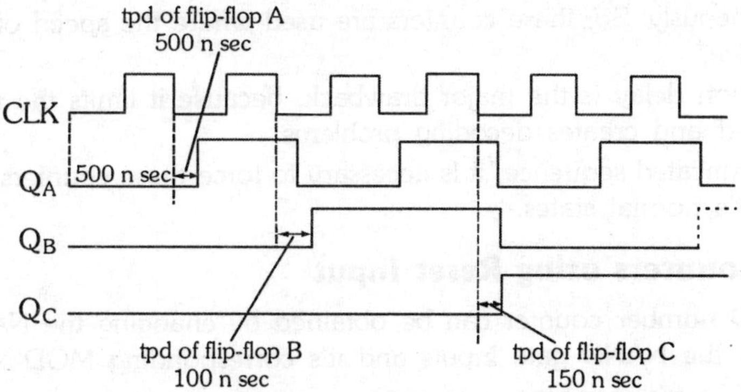

Fig. 4.124 (*a*) Waveforms of 3-bit ripple counter with Pd.

Fig. 4.124 (*a*) shows a situation where an input pulse occurs every 500 n sec, *i.e.*, clock period = 500 n sec and it is assumed that each flip-flop has a propagation delay of 50 n sec *i.e.*, *tpd* = 50 n sec. Observe that the A flip-flop output toggles 50 n sec after the negative transition of each input clock pulse. Similarly, B toggles 50 n sec after A goes from 1 to 0 (negative transition of A's output), and C toggles 50 n sec after B goes from 1 to 0 (negative transition of B's output). As a result, when the fourth input negative clock pulse occurs, the C output goes HIGH after a delay of 150 n sec. In this situation, counter does operate properly in the sense that the flip-flops do eventually get to their correct states, representing the binary count. However, the situation works if the input clock pulses are applied at much higher frequencies.

The waveforms in Fig. 4.124(*b*) show what happens if the input pulses occur once every 100 n sec. Again each flip-flop output responds only 50 n sec after the 1-to-0 transition at it's CLK input. One particular interest is the situation after the falling edge of the "*fourth*" input clock pulse, where flip-flop C output does not go HIGH until 150 n sec later, which is the same time that the A output goes HIGH in response to the *fifth* clock pulse.

In other words the condition C = 1, B = 0 and A = 0 (*i.e.*, count 100) never appears because the input frequency is too high. This could cause a serious problem if this conditon were suspended to be used to control some other operation in a digital system.

Problems such as this can be avoided if the period between input pulses is made longer than the total propagation delay of the counter. That is,

$$T_{clock} \geq n \times tpd \quad \text{– 3.1}$$

where n = number of flip-flops

$$\therefore \quad f_{max} = \frac{1}{n \times tpd} \qquad \text{... 3.2}$$

where, f_{max} = maximum input clock frequency can be applied.

4.13.10 Advantages and Disadvantages of Asynchronous Counters

Advantages : 1. Simple and straight forward operation and construction and these counters are constructed by suitably interconnecting a number of JK or D flip-flops.

2. They require fewer components.

3. Very easy to design any MOD-counter.

Disadvantages : 1. The speed of operation is low, because each flip-flop in the counter is not clocked simultaneously. So, these counters are used where the speed of operation is not of particular importance.

2. The propagation delay is the major drawback, because it limits the rate at which the counter can be clocked and creates decoding problems.

3. To obtain a truncated sequence, it is necessary to force these counters to recycle before going through all of it's normal states.

4.13.11 MOD Counters using Reset Input

Any desired MOD number counter can be obtained by changing the NAND gate inputs. The Table 4.51 shows the NAND gate inputs and it's corresponding MOD-N counter.

Table 4.51. NAND gate inputs for MOD-n counters

NAND gate inputs	*Counter*
Q_A	MOD-1 counter
Q_B	MOD-2
Q_A, Q_B	MOD-3
Q_4	MOD-4
Q_A, Q_C	MOD-5
Q_B, Q_C	MOD-6
Q_A, Q_B, Q_C	MOD-7

Example 4.28. *A counter has 15 states, 0000 through 1110. If the input frequency is 60 kHz, what will be its output frequency?*

Solution. $\frac{60 \text{ kHz}}{15} = 4 \text{ kHz}.$

Example 4.29. *The tpd of each flip-flop is 50 n sec. Determine the max-operating frequency for MOD-16 ripple counter.*

Solution. We know that MOD-16 ripple counter uses 4 flip-flops, each flip-flop with *tpd* = 50 nsec,

$$\therefore \qquad f_{max} = \frac{1}{n \times tpd} = \frac{1}{4 \times 50\, n\,sec} = 5 \text{ MHz.}$$

4.14 SYNCHRONOUS COUNTERS

In a 'synchronous counter', all the flip-flops change their state simultaneously, the operation of each stage being initiated by the clock. This is done by connecting the input clock to each flip-flop of the counter; hénce all flip-flops are clocked simultaneously. These counters are classified according to (*a*) sequence of states, (*b*) number of states or (*c*) number of flip-flops (stages) used in the counter. Here, all flip-flops are triggered simultaneously. Hence, these counters are called "parallel counters".

Synchronous binary counter : In a synchronous binary counter, we must note the following:

1. Clock pulse is applied common to all flip-flops.

2. The flip-flop in the lowest-order position is complemented with every clock pulse and the JK inputs are maintained at logic 1 (to get the toggling action)

3. The flip-flop in any other position is complemented with a pulse, provided all the bits in the lower-order positions are equal to 1.

4. The binary count dictates the next bit to be complemented.

Let us illustrate the above procedure with an example of 2-bit synchronous counter.

4.14.1 2-bit Synchronous Counter

Two bit synchronous counter is shown in Fig. 4.125(*a*). It has two JK flip-flops, namely A and B. Both are negative edge triggered flip-flops.

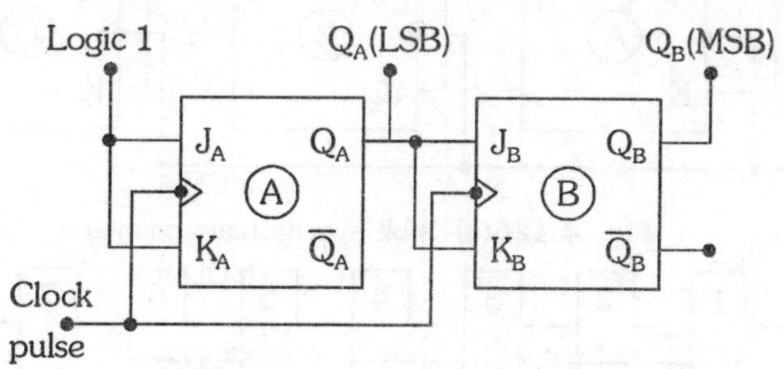

Fig. 4.125(*a*) Two-bit Synchronous counter.

Here, the clock signal is connected in parallel to clock inputs of both the flip-flops. But the Q_A output of the first stage is used to drive the J and K inputs of the second stage (flip-flop B).

Operation : Initially assume that $Q_A = Q_B = 0$. For the first clock pulse negative edge, the flip-flop A will toggle, because $J_A = K_A = 1$, whereas flip-flop B output will remain zero, because $J_B = K_B = 0$ (since upto this pulse the $Q_A = 0$) and $Q_A = 1$ only at the end of the pulse *i.e.*, negative edge of the pulse). So, after the first clock pulse, $Q_A = 1$ and $Q_B = 0$.

At the negative going edge of the second, both flip-flops will toggle because they both have a toggle condition on their J and K inputs (*i.e.*, $J_A = K_A = J_B = K_B = 1$). Thus after second clock pulse, $Q_A = 0$ and $Q_B = 1$.

At the negative going edge of the third clock pulse, flip-flop A toggles making $Q_A = 1$, but flip-flop B remains set *i.e.*, $Q_B = 1$. Hence, at the end of the third clock pulse, $Q_A = Q_B = 1$.

Finally, at the leading edge of the fourth clock pulse, both flip-flops toggle as their JK inputs are at logic 1. This results in $Q_A = Q_B = 0$ and the counter is recycled back to its original state. The timing details are shown in Fig. 4.125(*b*) and 4.125(*c*).

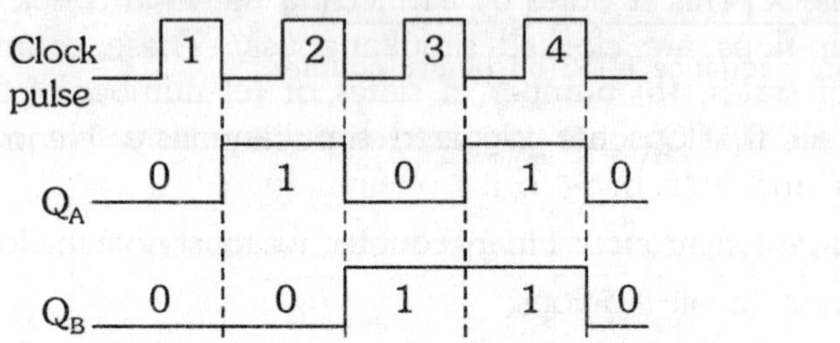

Fig. 4.125(*b*) Timing diagram of 2-bit synchronous counter.

CP	Q_B	Q_A
0, 4	0	0
1, 5	0	1
1, 6	1	0
3, 7	1	1

Fig. 4.125(*c*) State sequence of 2-bit synchronous counter.

4.14.2 A 3-bit Synchronous Binary Counter

Fig 4.126(*a*) shows 3-bit synchronous binary counter and it's timing diagram in 4.126(*b*). The state sequence for this counter is shown in 4.126(*c*).

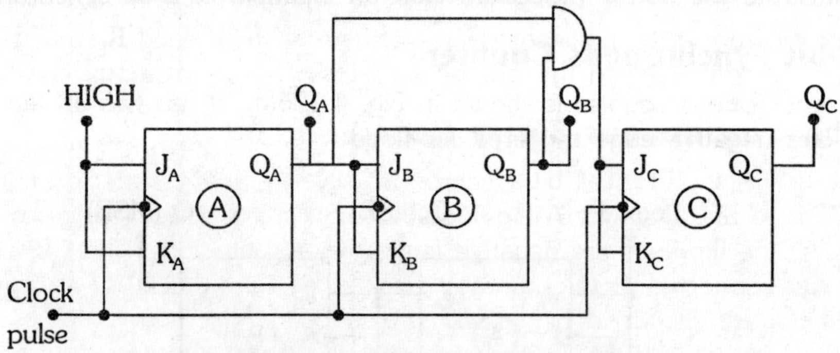

Fig. 4.126(*a*) 3-bit synchronous counter.

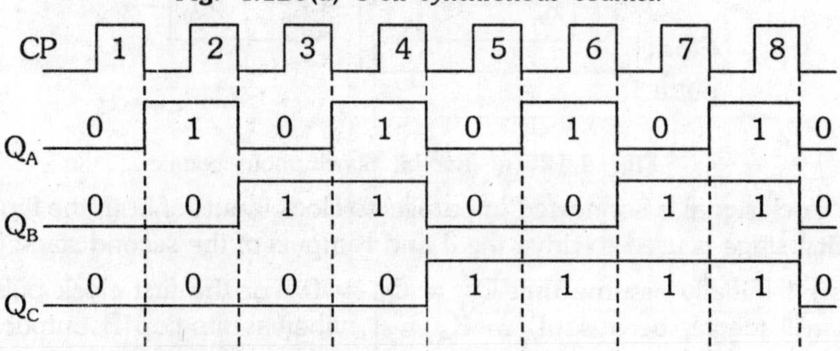

Fig. 4.126(*b*) Timing diagram for 3-bit synchronous counter.

CP	Q_C	Q_B	Q_A
0	0	0	0
1	0	0	1
2	0	1	0
3	0	1	1
4	1	0	0
5	1	0	1
6	1	1	0
7	1	1	1

Fig. 4.126(*c*) State sequence for 3-bit binary counter.

Looking at Fig. 4.126 (*b*), we can see that Q_A changes on each clock pulse as we progress from it's original state to it's final state and then back to it's original state.

To produce this operation, flip-flop A is held in the toggle made by connecting J_A and K_A inputs to HIGH. Now, we will see what flip-flop B does. Flip-flop B toggles, when Q_A is 1. When Q_A is a 0, flip-flop B is in the no-change mode and remains in it's present state.

From the Fig. 4.126 (*c*), we can notice that flip-flop C has to change it's state only when Q_B and Q_A both are at logic 1. This condition is detected by AND gate and applied to the J_C and K_C inputs of flip-flop C. Whenever both Q_A and Q_B are HIGH, the output of the AND gate makes the J and K inputs of flip-flop C HIGH, and flip-flop C toggles on the clock pulse. At other times (*i.e.*, other than $Q_A = Q_B = 1$), the J and K inputs of flip-flop C are held LOW by the AND gate output, and the flip-flop does not change it's state.

In the logic diagram of 2-bit synchronous counter, so, the J_B and K_B are connected by Q_A and J_C and K_C are connected to the output of AND gate, whose inputs are Q_A and Q_B.

4.14.3 4-bit Synchronous Binary Counter

The Fig. 4.127(*a*) shows the 4-bit synchronous counter. This counter contains four JK flip-flops, namely A, B, C and D where A flip-flop is for LSB count and D flip-flop is for MSB count of the counter. All the flip-flops are negative edge triggered flip-flops. The CLK input terminals of all flip-flops are connected to the common clock pulse. The timing diagram for this counter and state sequence are shown in Fig. 4.127 (*b*) and 4.127 (*c*) respectively.

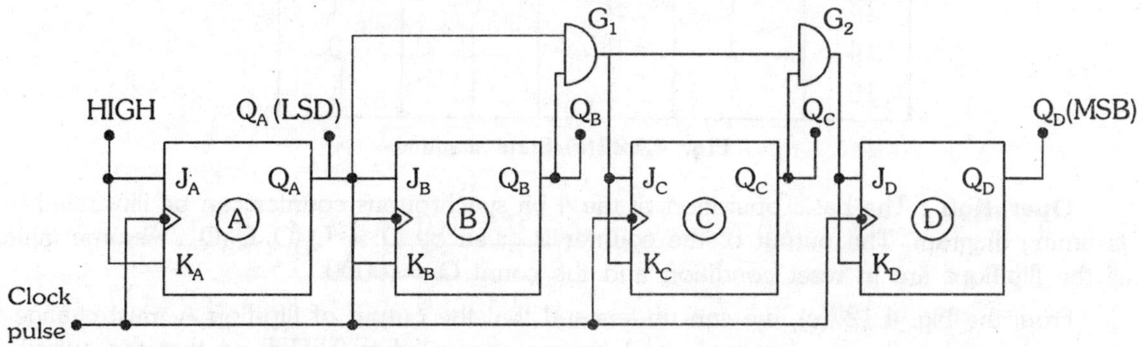

Fig. 4.127(*a*) 4-bit synchronous counter (ripple carry).

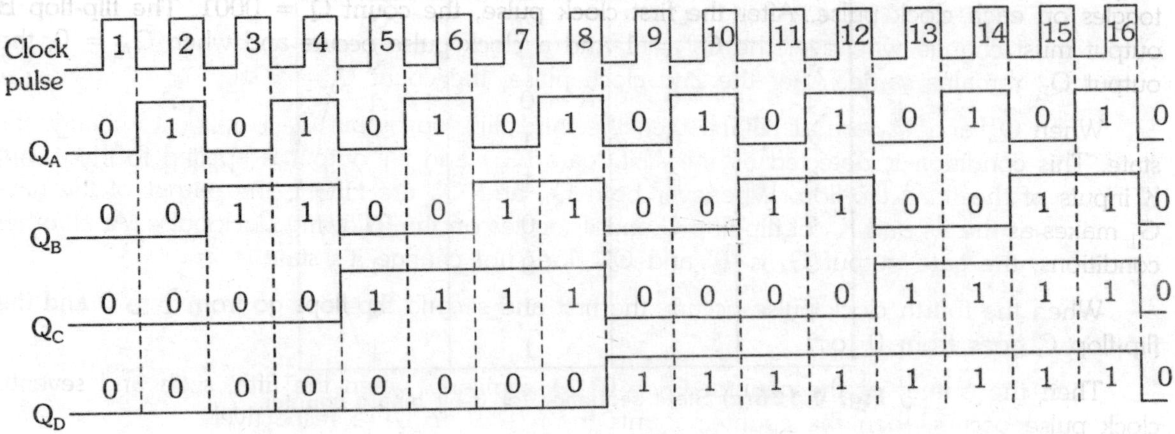

Fig. 4.127(*b*) Timing diagram of 4-bit synchronous counter.

CP	Q_D	Q_C	Q_B	Q_A
0	0	0	0	0
1	0	0	0	1
2	0	0	1	0
3	0	0	1	1
4	0	1	0	0
5	0	1	0	1
6	0	1	1	0
7	0	1	1	1
8	1	0	0	0
9	1	0	0	1
10	1	0	1	0
11	1	0	1	1
12	1	1	0	0
13	1	1	0	1
14	1	1	1	0
15	1	1	1	1

Fig. 4.127(*c*) State sequence.

Operation : The basic operation of the 4-bit synchronous counter can be illustrated from its timing diagram. The output of the counter is given by $Q = Q_DQ_CQ_BQ_A$. Assume initially, all the flip-flops are in reset condition and the count Q = 0000.

From the Fig. 4.127(*c*), we can understand that the output of flip-flop A must change for every clock pulse. So, the inputs J_A and K_A are connected to HIGH, so that the output Q_A

toggles on each clock pulse. After the first clock pulse, the count Q = 0001. The flip-flop B output must change whenever the Q_A = 1 and a clock pulse occurs and when Q_A = 0, the output Q_B remains same. After the 2nd clock pulse, the count Q = 0010.

When Q_A and Q_B are at HIGH, then the third flip-flop output Q_C is made to change it's state. This condition is detected by the AND gate (G_1) and it's output is applied to the J and K inputs of the third flip-flop. Whenever both Q_A and Q_B are HIGH, the output of the gate G_1 makes as the J_C and K_C of flip-flop C and it toggles on the following clock pulse. At all other conditions, the gate output G, is '0' and Q_C does not change it's state.

When the fourth clock pulse occurs, the first and second flip-flops go from 1 to 0 and the flip-flop C goes from 0 to 1.

Then the output of the counter Q = 0100. Similarly, when the fifth, sixth and seventh clock pulse occurs, then the counter counts from 0101 to 0111 respectively.

From the Fig. 4.127(*d*), we can observe that whenever Q_A, Q_B and Q_C are at HIGH, the fourth flip-flop D changes its output. This condition is detected by AND gate (G_2), so that when a clock pulse occurs, the D flip-flop will change the state. Note that for all other times, the J_D and K_D inputs are LOW and it is no-change condition.

Then, at the end of eigth clock pulse the counter output Q = 1000. Similarly from the ninth clock pulse to fifteenth clock pulse, the counter counts from 1001 to 1111. When the sixteenth clock pulse arrives, the four flip-flops change from 1 to 0, then the output of the counter Q = 0000.

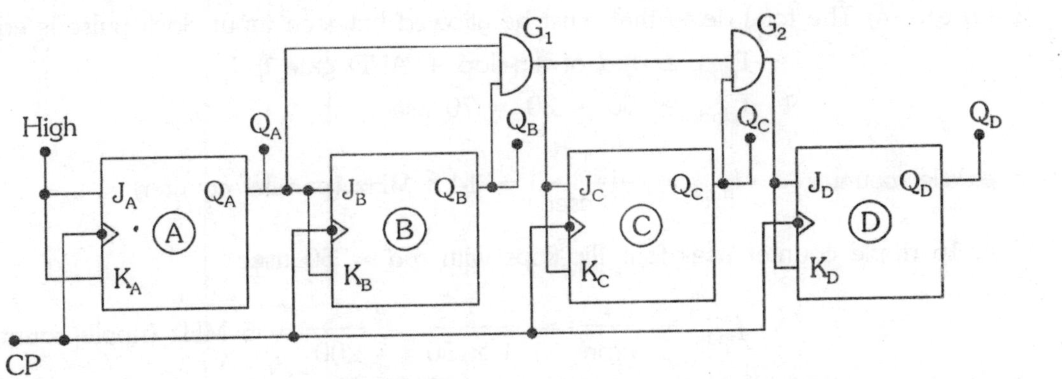

Fig. 4.127(*d*) 4-bit synchronous counter with parallel carry.

4.14.4 Maximum Frequency Calculations

Let us assume that there are four (*n* = 4) flip-flops in the counter and each flip-flop in the counter has a propagation delay (T_p) of 50 nsec. The decoding of any state (T_s) requires 100 nsec. The propagation delay (T_g) from input to output of each control gate is 50 nsec. Then the maximum clock frequencies which can be applied to ripple counter, synchronous counter are tabulated in the Table 4.52.

Table 4.52 Formulae for maximum clock frequencies

Type of the counter	*Maximum clock frequency*	*Example* $n = 4, T_p = 50$ nsec $T_s = 100$ nsec, $T_g = 50$ nsec
1. Ripple counter	$\frac{1}{f_{max}} \geq N(T_p) + T_s$	$\frac{1}{f} \geq 300$ nsec $\Rightarrow f \leq 3.67$ MHz
2. Synchronous counter with ripple carry	$\frac{1}{f_{max}} \geq T_p + (N-2)T_g$	$\frac{1}{f_{max}} \geq 150$ nsec $\Rightarrow f_{max} \leq 6.67$ MHz
3. Synchronous counter with parallel carry	$\frac{1}{f_{max}} > T_p + T_g$	$\frac{1}{f_{max}} \geq 100$ nsec $\Rightarrow f_{max} = 10$ MHz

Example 4.30. *(a) Determine f_{max} for a 4-bit synchronous counter, if tpd for each flip-flop is 50 nsec and tpd of AND gate is 20 nsec. Compare this with f_{max} for a MOD-16 ripple counter.*

(b) What has to be done to convert this counter to MOD-32?

(c) Determine f_{max} for MOD-32 parallel counter.

Solution. (*a*) The total delay that must be allowed between input clock pulse is equal to

$$T_{clock} \geq tpd \text{ of flip-flop} + \text{AND gate } tpd$$

$$\therefore \quad T_{clock} \geq 50 + 20 = 70 \text{ nsec}$$

In parallel counter,
$$f_{max} = \frac{1}{70 \text{ nsec}} = 14.3 \text{ MHz (parallel counter)}$$

MOD-16 ripple counter uses four flip-flops with *tpd* = 50 nsec

$$\therefore \quad f_{max} = \frac{1}{ntpd} = \frac{1}{4 \times 50} = \frac{1}{200} = 5 \text{ MHz (ripple counter)}$$

(*b*) To convert this 4-bit synchronous counter to MOD-32 counter, a fifth flip-flop must be added, since $2^5 = 32$. The CLK input of this flip-flop is also tied to the input clock pulses. It's J_E and K_E inputs are fed by the output of a four-input AND gate whose inputs are Q_A, Q_B, Q_C and Q_D.

(*c*) f_{max} = 14.3 MHz (since in parallel counter the f_{max} is independent on the number of flip-flops).

Comparison between synchronous and asynchronous counter

The Table 4.53 shows the comparison between synchrounous and asynchronous counters.

Table 4.53

Sl. No.	*Synchronous counters*	*Asynchronous counters*
1.	In this type, there is no connection between output of first flip-flop and clock input of the next flip-flop.	In this type, the counter flip-flops are connected in such a way that output of first flip-flop drives the clock for the next flip-flop.
2.	All flip-flops are clocked simultaneously.	All the flip-flops are not clocked simultaneously.
3.	Logic circuit is more simple even for more number of states.	Design involves complex logic circuit as the number of states increases.
4.	As clock is simultaneously given to all the flip-flops, so there is no problem of propagation delay.	Low speed, since clock must be propagated through a number of flip-flops before it reaches last flip-flop.

4.14.5 Design of Synchronous Counters

There is a situation, however, where a counter is required that follows a sequence that is not counting binary; for example 000, 010, 101, 001, 110, 000, ...

Several methods exist for designing counters that follow arbitrary sequences. Generally JK, T or D flip-flops are used in it.

Basic idea. In synchronous counters, all of the flip-flops are clocked at the same time and input of each flip-flop in the counter must be at the correct level to ensure that each flip-flop goes to the correct state. For example, assume that the present state is (CBA) 101, if next state is (CBA) 011. When the next clock pulse occurs, the inputs of the flip-flops must beat the correct levels that will cause flip-flop C to change from 1 to 0, flip-flop B from 0 to 1, and flip-flop A from 1 to 1 *i.e.*, (No change).

Design procedure. We will now go through a complete synchronous counter design procedure. Although we will do it for a specific counting sequence, the same steps can be followed for any desired sequence.

Step 1. Determine the number of flip-flops needed, *i.e.*, desired number of bits and the desired counting.

Step 2. Choose the type of flip-flop to be used, either JK, D or T etc.

Step 3. Draw the state transitions showing all possible states, including those that are not a part of the desired counting sequence.

Step 4. Use state transitions to obtain the state table that lists all present states and next states.

Step 5. From the state table, derive the circuit excitation table.

Step 6. Use K-map or any other simplification method to derive the circuit as flip-flop input functions *i.e.*, design the logic circuits to generate the levels required at each flip-flop input.

Step 7. Draw the logic diagram, *i.e.*, implement the final expression.

Note : The design procedure of a sequential circuit involves few additional steps such as state reduction and state assignment. For simplicity, these steps are wantedly not considered here.

4.14.6 Modulus-N Synchronous Counter

The counter with *n*-flip-flops has maximum mod number 2^n. For example, 3-bit binary counter is a MOD-8 counter. The basic counter can be modified to produce MOD numbers less than 2^n by allowing the counter to skip those that are normally a part of the counting sequence.

Let us discuss the design of modulus-N counter using synchronous counters.

Example 4.31. *Design a MOD-5 synchronous counter using JK flip-flop and implement it. Also, construct a timing diagram.*

Solution. Flip-flops required are

$$2^n \geq N$$

Here, $N = 5$

$\therefore$ $2^n \geq 5$

$\Rightarrow$ $n = 3$

i.e., three flip-flops are required and assume that three flip-flops are JK and are named as A, B and C. State diagram for MOD-5 counter are shown in Fig. 4.128(*a*) for the counts 101, 110 and 111 the next state is not defined and is considered as don't care.

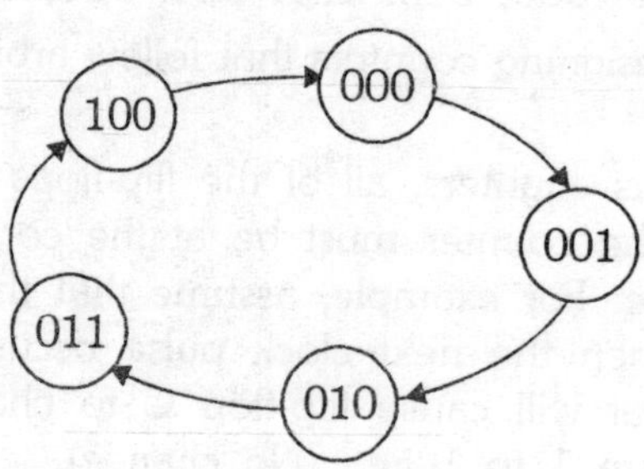

Fig. 4.128(*a*) State diagram of MOD-5 synchronous counter.

The excitation table for JK flip-flop is shown in Table 4.54(*a*).

Table 4.54(*a*) Excitation table for JK flip-flop

Q_n	Q_{n+1}	J	K
0	0	0	×
0	1	1	×
1	0	×	1
1	1	×	0

The state table for MOD-5 counter is shown in Table 4.54(*b*) which contains present state, next state and flip-flop inputs.

Table 4.54(*b*) State table

Present state			*Next state*			*Flip-flop inputs*					
Q_C	Q_B	Q_A	Q_{C+1}	Q_{B+1}	Q_{A+1}	J_C	K_C	J_B	K_B	J_A	K_A
0	0	0	0	0	1	0	×	0	×	1	×
0	0	1	0	1	0	0	×	1	×	×	1
0	1	0	0	1	1	0	×	×	0	1	×
0	1	1	1	0	0	1	×	×	1	×	1
1	0	0	0	0	0	×	1	0	×	0	×
1	0	1	×	×	×	×	×	×	×	×	×
1	1	0	×	×	×	×	×	×	×	×	×
1	1	1	×	×	×	×	×	×	×	×	×

K-map simplification. The simplification procedure is shown in Fig. 4.128(*b*).

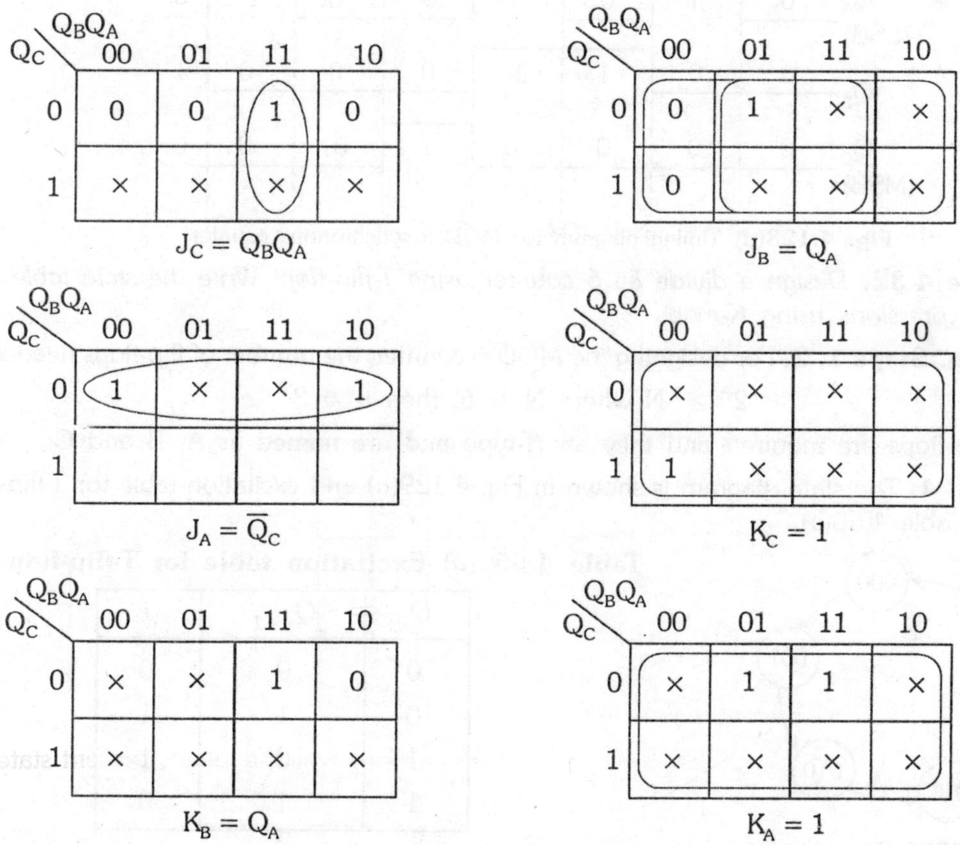

Fig. 4.128(*b*) K-maps for flip-flop inputs.

Logic diagram. The logic diagram is shown in Fig. 4.128(*c*).

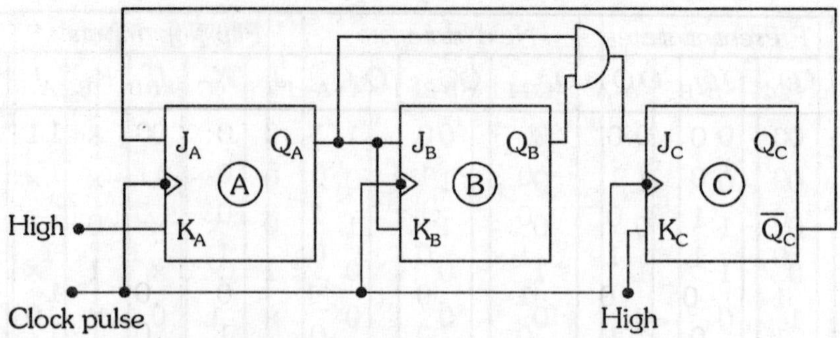

Fig. 4.128(*c*) Logic diagram of MOD-5 counter.

Timing diagram. The timing diagram for MOD-5 counter is shown in Fig. 4.128(*d*).

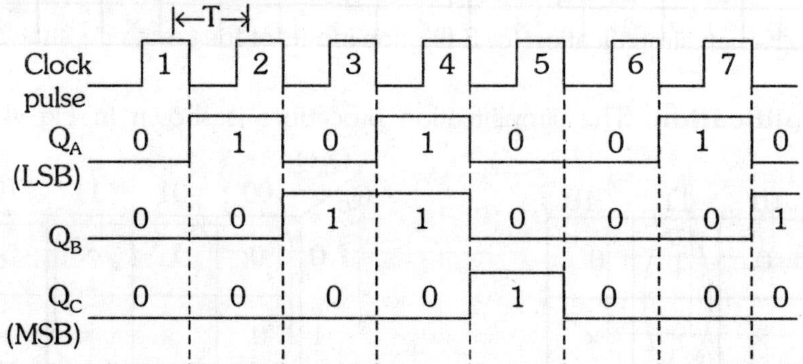

Fig. 4.128(*d*) Timing diagram for MOD-5 synchronous counter.

Example 4.32. *Design a divide by 6 counter using T-flip-flop. Write the state table and reduce the expressions using K-map.*

Solution. Steps 1, 2. For designing the MOD-6 counter, the number of flip-flops needed is,

$$2^n \geq N \text{ where } N = 6, \text{ then } n = 3$$

i.e., three flip-flops are required and they are T-type and are named as A, B and C.

Steps 3, 4. The state diagram is shown in Fig. 4.129(*a*) and excitation table for T-flip-flop is shown in Table 4.55(*a*).

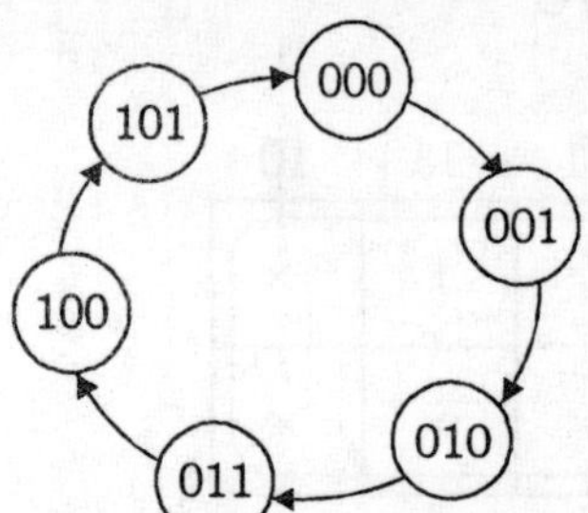

Fig. 4.129(*a*) State diagram.

Table 4.55 (*a*) Excitation table for T-flip-flop

Q_n	Q_{n+1}	T
0	0	0
0	1	1
1	0	1
1	1	0

Step 5. The state table for MOD-6 synchronous counter is shown in Table 4.55 (*b*).

Table 4.55(*b*)

Present state			*Next state*			*Flip-flop inputs*		
Q_C	Q_B	Q_A	Q_{C+1}	Q_{B+1}	Q_{A+1}	T_C	T_B	T_A
0	0	0	0	0	1	0	0	1
0	0	1	0	1	0	0	1	1
0	1	0	0	1	1	0	0	1
0	1	1	1	0	0	1	1	1
1	0	0	1	0	1	0	0	1
1	0	1	0	0	0	1	0	1
1	1	0	×	×	×	×	×	×
1	1	1	×	×	×	×	×	×

Step 6. The K-map simplification for 3 flip-flop inputs of the counter is shown in Fig. 4.129 (*b*).

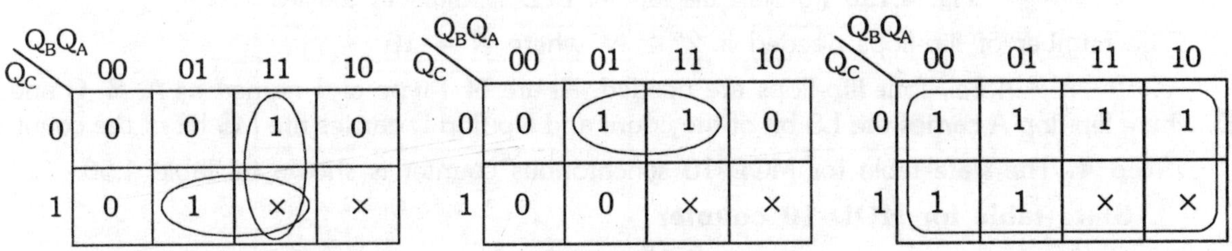

$T_C = Q_CQ_A + Q_BQ_A$ $\quad$ $T_B = \overline{Q}_CQ_A$ $\quad$ $T_A = 1$

Fig. 4.129(*b*)

Step 7. The logic diagram for MOD-6 is shown in Fig. 4.129(*c*).

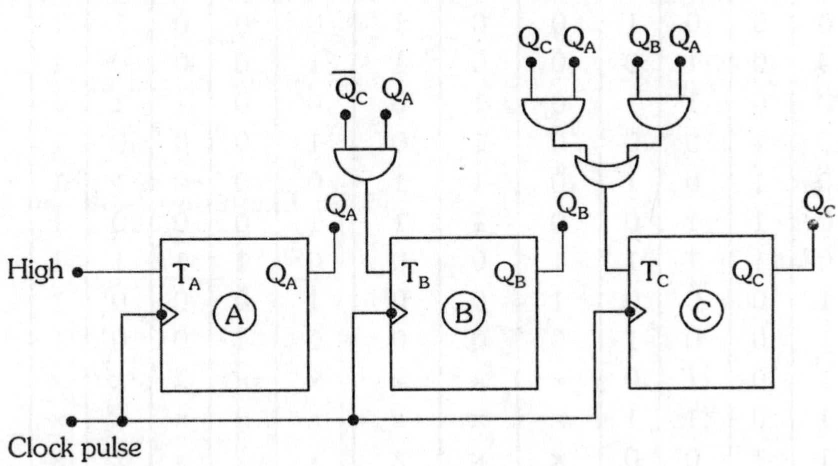

Fig. 4.129(c)

Example 4.33. *Design a synchronous BCD-counter (or) decade counter (or) MOD-10 counter using T flip-flop.*

Solution. We know that in the decade counters we have to truncate normal counter sequence after the count 1001. The counter has to reset after 1001 count instead of going for 1010 count. This counter counts in binary-coded decimal from 0000 to 1001 and back to 0000, hence it is called BCD counter. A BCD counter does not have a regular pattern as in a straight binary count. To design the circuit of a BCD counter, it is necessary to go through the design procedure. Here, we will not use clear input to reset the counter after state 1010.

Step. 1, 2, 3 : The state diagram is shown in Fig. 4.130 (*a*).

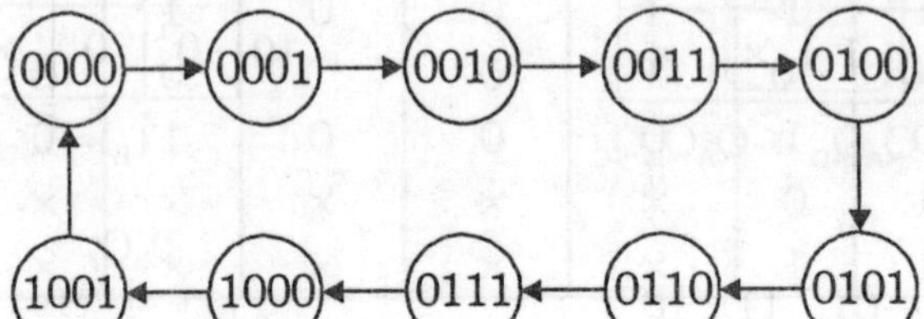

Fig. 4.130 (*a*) State diagram for BCD synchronous counter.

The number of flip-flops needed is $2^n \geq N$, where N = 10

$\therefore$ $n = 4$ *i.e.*, four flip-flops are needed, all are of T-type and named as A, B, C and D, where flip-flop A carries the LS bit of the count and flip-flop D carries the MS bit of the count.

Step 4. The state table for MOD-10 synchronous counter is shown in Table 4.56.

State table for MOD-10 counter

Table 4.56

Present state				*Next state*				*Flip-flop inputs*			
Q_D	Q_C	Q_B	Q_A	Q_{D+1}	Q_{C+1}	Q_{B+1}	Q_{A+1}	T_D	T_C	T_B	T_A
0	0	0	0	0	0	0	1	0	0	0	1
0	0	0	1	0	0	1	0	0	0	1	1
0	0	1	0	0	0	1	1	0	0	0	1
0	0	1	1	0	1	0	0	0	1	1	1
0	1	0	0	0	1	0	1	0	0	0	1
0	1	0	1	0	1	1	0	0	0	1	1
0	1	1	0	0	1	1	1	0	0	0	1
0	1	1	1	1	0	0	0	1	1	1	1
1	0	0	0	1	0	0	1	0	0	0	1
1	0	0	1	0	0	0	0	1	0	0	1
1	0	1	0	×	×	×	×	×	×	×	×
1	0	1	1	×	×	×	×	×	×	×	×
1	1	0	0	×	×	×	×	×	×	×	×
1	1	0	1	×	×	×	×	×	×	×	×
1	1	1	0	×	×	×	×	×	×	×	×
1	1	1	1	×	×	×	×	×	×	×	×

Step 5. The K-map simplification for each flip-flop input is shown in Fig 4.130(*b*).

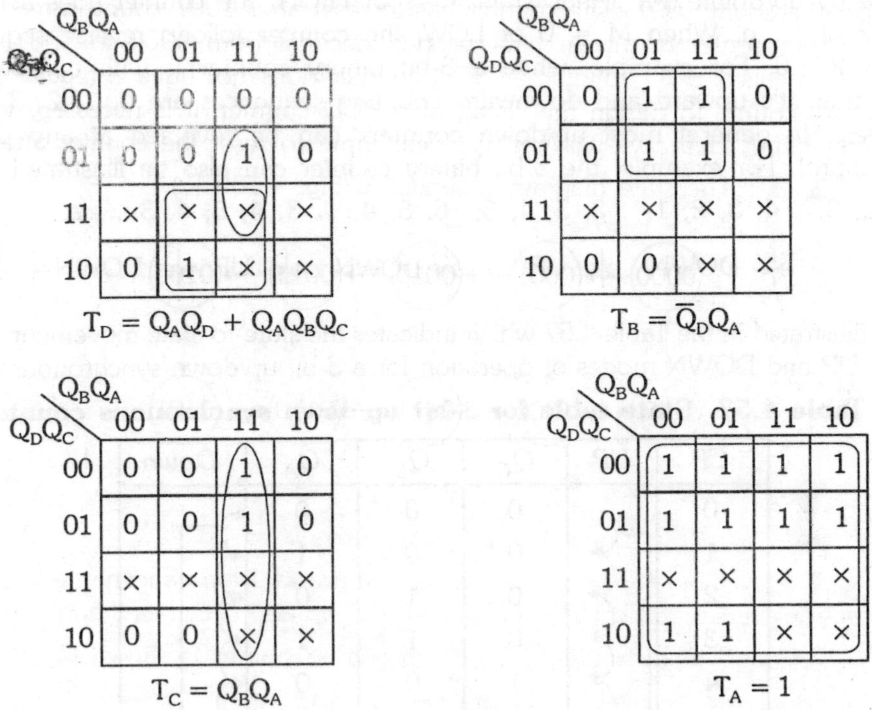

Fig. 4.130(*b*)

The logic diagram for the BCD counter is shown in Fig. 4.130(*c*).

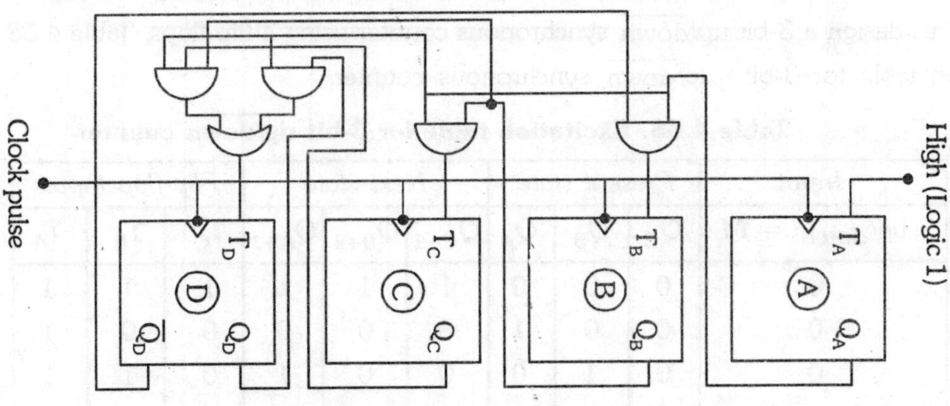

Fig. 4.130(*c*) Logic diagram for MOD-10 counter.

4.14.7 UP/$\overline{\text{Down}}$ Counter

The counter which is capable of progressing in either direction, *i.e.*, in increasing order or decreasing order through a certain counting sequence is known as "up/$\overline{\text{down}}$ counter".

This counter is also called as "bidirectional counter". Usually up/down operation of the counter is controlled by up/$\overline{\text{down}}$ (M) signal when M=1 or HIGH, the counter goes in UP sequence *i.e.*, 0, 1, 2, 3, ... *n*. When M = 0 or LOW, the counter follows reverse sequence *i.e.*, *n*, *n* – 1, *n* – 2 ...0. For example, when a 3-bit binary counter is with up/down sequential operations, then it's upward and downward counting sequences are 0, 1, 2, 3 and 3, 2, 1, 0 respectively. In general most up/down counters can be reversed at any point of their sequence (count). For example, the 3-bit binary counter can also be illustrated as follows :

0, 1, 2, 3, 4, 3, 2, 1, 2, 3, 4, 5, 6, 5, 4, 3, 4, 5, 4, 3... etc.

← UP ✱ DOWN ✱ UP ✱ DOWN ✱ UP ✱ DOWN→|

This is illustrated in the Table 4.57 which indicates the state-to-state movement of the counter for both it's UP and DOWN modes of operation for a 3-bit up/down synchronous counter.

Table 4.57. State table for 3-bit up/down synchronous counter

CP	*UP*	Q_C	Q_B	Q_A	*Down*
0		0	0	0	
1		0	0	1	
2		0	1	0	
3		0	1	1	
4		1	0	0	
5		1	0	1	
6		1	1	0	
7		1	1	1	

Let us design a 3-bit up/$\overline{\text{down}}$ synchronous counter using T-flip-flops. Table 4.58 shows the excitation table for 3-bit up/$\overline{\text{down}}$ synchronous counter.

Table 4.58. Excitation table for 3-bit up/$\overline{\text{down}}$ counter

Input	*Present state*			*Next state*			*Flip-flop inputs*		
up/$\overline{\text{down}}$ = M	Q_C	Q_B	Q_A	Q_{C+1}	Q_{B+1}	Q_{A+1}	T_C	T_B	T_A
0	0	0	0	1	1	1	1	1	1
0	0	0	1	0	0	0	0	0	1
0	0	1	0	0	0	1	0	1	1
0	0	1	1	0	1	0	0	0	1
0	1	0	0	0	1	1	1	1	1
0	1	0	1	1	0	0	0	0	1
0	1	1	0	1	0	1	0	1	1
0	1	1	1	1	1	0	1	0	0

1	0	0	0	0	0	1	0	0	1
1	0	0	1	0	1	0	0	1	1
1	0	1	0	0	1	1	0	0	1
1	0	1	1	1	0	0	1	1	1
1	1	0	0	1	0	1	0	0	1
1	1	0	1	1	1	0	0	1	1
1	1	1	0	1	1	1	0	0	1
1	1	1	1	0	0	0	1	1	1

Step 1, 2. It requires 3 flip-flops, namely A, B and C. Let us take all are negative edge triggered T-flip-flops, where flip-flop A holds the LS bit of the count and flip-flop C holds the MS bit of the count.

Step 3. The state diagram of the 3-bit up/down synchronous counter is shown in Fig. 4.131 (*a*).

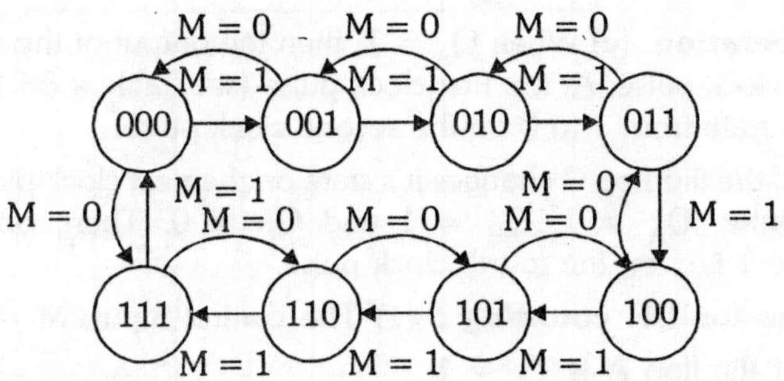

Fig. 4.131(*a*) State diagram of the 3-bit up/down counter.

Step 4. The K-map simplification for three flip-flop inputs are shown in Fig. 4.131(*b*).

MQ_C \ Q_BQ_A	00	01	11	10
00	1	0	0	0
01	1	0	0	0
11	0	0	1	0
10	0	0	1	0

$T_C = \overline{M}\overline{Q}_B\overline{Q}_A + MQ_AQ_B$

MQ_C \ Q_BQ_A	00	01	11	10
00	1	0	0	1
01	1	0	0	1
11	0	1	1	0
10	0	1	1	0

$T_B = \overline{M}\overline{Q}_A + MQ_A$

MQ_C \ Q_BQ_A	00	01	11	10
00	1	1	1	1
01	1	1	1	1
11	1	1	1	1
10	1	1	1	1

$T_A = 1$

Fig. 4.131(*b*) K-map simplifications for flip-flop inputs.

Step 6. The logic diagram for 3-bit up/down counter is shown in Fig. 4.131(*c*). It has 3 T-flip-flops and some gates to perform the flip-flop input functions.

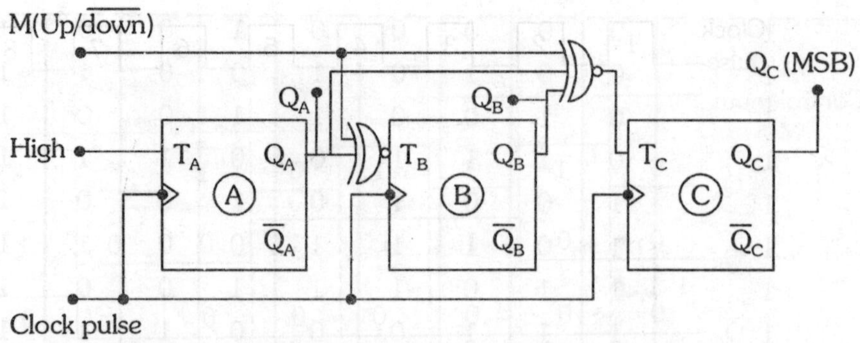

Fig. 4.131(c) 3-bit synchronous up/down counter.

Operation. From the Fig. 4.131 (*c*), notice that the up/down(M) is a control input, which determines the basic operation of the counter. When this input is HIGH, we have UP mode of operation and when it is low, we have DOWN mode of operation. The output of the counter is $Q = Q_CQ_BQ_A$.

UP mode of operation. (*a*) When $Q_A = 1$, then the output of the second flip-flop changes it's state on the next clock pulse. At the first clock pulse (see Table 4.58), $Q_A = 1$ and $Q_B = 0$. Then Q_B changes it's state from 1 to 0 on the second clock pulse.

(*b*) The output of the flip-flop C changes it's state on the next clock pulse when $Q_A = Q_B = 1$. At the third clock pulse, $Q_A = 1$, $Q_B = 1$ and $Q_C = 0$. Then, on the next clock pulse, Q_C changes from 0 to 1 *i.e.*, on the fourth clock pulse.

The Conditions for UP counting : (1) The control input, M = 1.

(2) The input of flip-flop A is $T_A = 1$

(3) The input of flip-flop B is $T_B = MQ_A$ and

(4) The input of flip-flop C is $T_C = MQ_AQ_B$

DOWN mode operation : (*a*) The output of flip-flop A changes on every clock pulse.

(*b*) The output of flip-flop B is Q_B changes its state on the next clock pulse, when $Q_A = 0$.

(*c*) The output of flip-flop C (Q_C) changes it's state on the next clock pulse when $Q_A = Q_B = 0$.

Conditions for DOWN counting

1. The control input M = 0 (LOW)
2. The input of flip-flop A (T_A) = 1
3. The input of flip-flop B (T_B) = $\overline{M}\,\overline{Q_A}$
4. The input of flip-flop C (T_C) = $\overline{M}\,\overline{Q_A}\,\overline{Q_B}$

Step 7. The timing diagram for the 3-bit synchronous up/down counter is shown in Fig. 4.131(*d*), where control input M is changed from 1 to 0 after the three clock pulses. Note that the counter starts at all 0's state and upto three clock pulses it acts as a up counter and after that it works as a down counter.

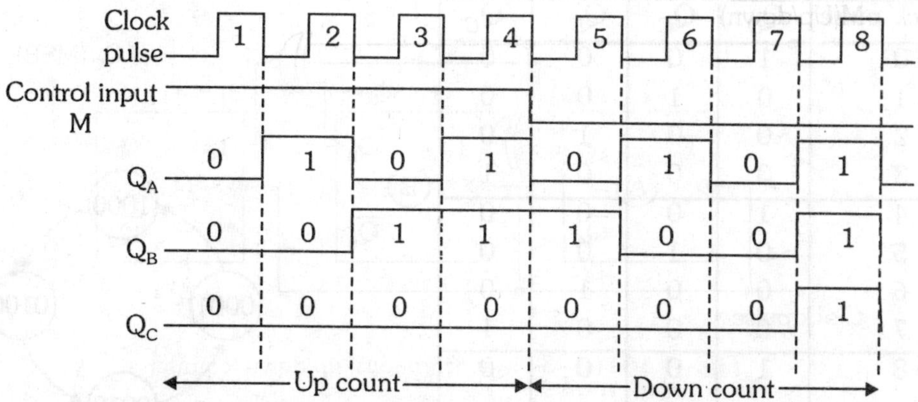

Fig. 4.131 (*d*) Timing diagram for 3-bit synchronous up/down counter.

4.15 SHIFT COUNTERS (OR) SHIFT REGISTER COUNTERS

During the construction of counters, whenever the output of last flip-flop of the counter is connected back to the control input of the first flip-flop of the counter, it is known as "*direct feedback*" (or) "*circulating shift register*", so that last flip-flop shifts its value into the first flip-flop. This type of arrangement results in "*ring counters*".

If the outputs of the last flip-flop are first or inverted output of last flip-flop is crossed over and then connected back to the control inputs of the first flip-flop. This technique is known as "inverse feedback". The counters with this property are called "shift counters."

Fig. 4.132 (*a*) shows the 4-bit ring counter using D flip-flops. The flip-flops are connected so that information shifts from left to right and back around from Q_A to Q_D. *i.e.*, MS flip-flop to LS flip-flop. In most instances, only a single 1 is in the register and it is made to circulate around the register as long as the clock pulses are applied. For this reason it is called a "ring counter".

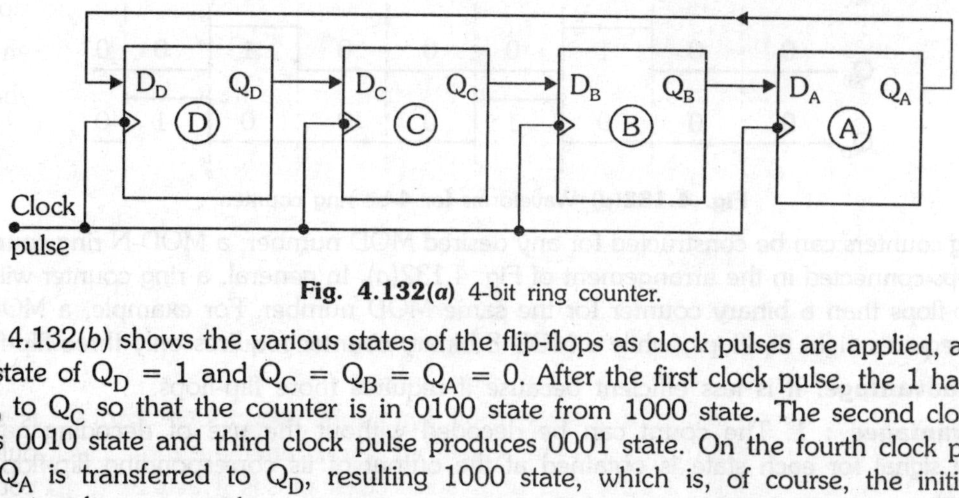

Fig. 4.132(*a*) 4-bit ring counter.

Fig. 4.132(*b*) shows the various states of the flip-flops as clock pulses are applied, assuming starting state of $Q_D = 1$ and $Q_C = Q_B = Q_A = 0$. After the first clock pulse, the 1 has shifted from Q_D to Q_C so that the counter is in 0100 state from 1000 state. The second clock pulse produces 0010 state and third clock pulse produces 0001 state. On the fourth clock pulse the 1 from Q_A is transferred to Q_D, resulting 1000 state, which is, of course, the initial state. Subsequent pulses cause the sequence to repeat. The state diagram for this is shown in Fig. 4.132(*c*).

Clock pulse	Q_D	Q_C	Q_B	Q_C
0	1	0	0	0
1	0	1	0	0
2	0	0	1	0
3	0	0	0	1
4	1	0	0	0
5	0	1	0	0
6	0	0	1	0
7	0	0	0	1
8	1	0	0	0
...	...	...	...	...
...	...	...	...	...
...	...	...	...	...

Fig. 4.132(*b*) Sequence table.

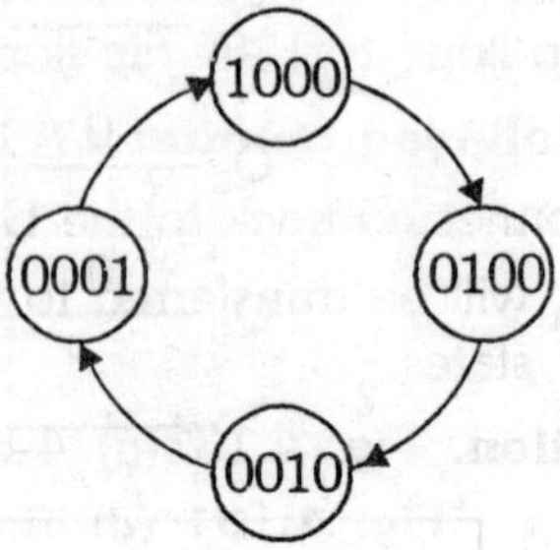

Fig. 4.132(c) State diagram.

This counter functions as MOD-4 counter since it has four distinct states before the sequence repeats. Although this circuit does not progress through the normal binary counting sequence, it is still a counter because each count corresponds to a unique set of flip-flop states. Note that each flip-flop output waveform has a frequency equal to one fifth of the clock frequency, since this is a MOD-4 ring counter as shown in Fig. 4.132(*d*).

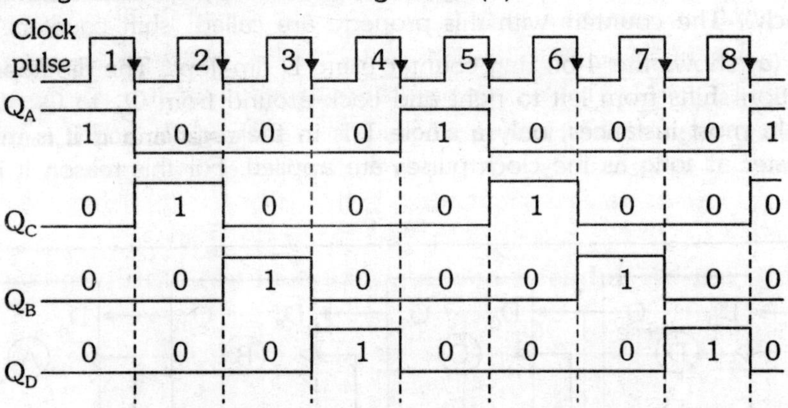

Fig. 4.132(*d*) Waveforms for 4-bit ring counter.

Ring counters can be constructed for any desired MOD number; a MOD-N ring counter uses N flip-flops connected in the arrangement of Fig. 4.132(*a*). In general, a ring counter will require more flip-flops than a binary counter for the same MOD number. For example, a MOD-8 ring counter requires eight flip-flops, while a MOD-8 binary counter requires only three flip-flops.

Disadvantage. It is less efficient because it requires more flip-flops.

Advantages : 1. The count can be decoded without the use of decoding gates. The decoding signal for each state is obtained at the output of its corresponding flip-flop.

2. This is used in applications where the counter is being used to control the sequencing of operations in a system.

4.15.1 Johnson Counter

The basic ring counter can be modified slightly to produce another type of counter, called "*shift counter*". It is having somewhat different and unique properties. The '*Johnson*' or '*twisted ring counter*' is constructed exactly like a normal ring counter except that the inverted output of the last flip-flop is connected as the input of the first flip flop. This can be constructed by using SR flip-flops (or) JK flip-flops (or) D flip-flops.

4-bit Johnson counter : A four bit Johnson counter is shown in Fig. 4.133(*a*). Note that the $\overline{Q}_A$ is connected back to the D input of flip-flop D. This means that the inverse of the level stored in Q_A will be transferred to Q_D on the clock pulse. This arrangement produces a unique sequence of states.

Operation. Fig. 4.133(*a*) 4-bit Johnson counter.

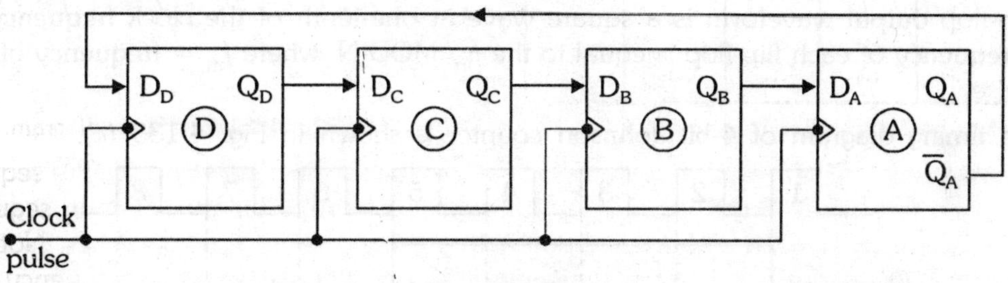

Fig. 4.133(*a*)

The Johnson-counter operation is easy to analyze. Initially, all flip-flops are cleared, so all the outputs Q_D, Q_C, Q_B and Q_A are zero. The output of the last stage Q_A is zero. Therefore it's complement output $\overline{Q}_D = 1$. This is connected back to the D input of the first stage. So D_D is one. The first falling edge of the clock produces $Q_D = 1$ and $Q_C = 0$, $Q_B = 0$ and $Q_A = 0$. Since D_A, D_B and D_C are zero, the next clock pulse produces $Q_D = 1$, $Q_C = 1$, $Q_B = 0$ and $Q_A = 0$. The sequence of states is summarized in the Table 4.59(*a*). After 8 states, the same sequence is repeated.

Table 4.59 (*a*) Four-bit Johnson counter sequence (or) state sequence table

Clock pulse	Q_D	Q_C	Q_B	Q_A
0	0	0	0	0
1	1	0	0	0
2	1	1	0	0
3	1	1	1	0
4	1	1	1	1
5	0	1	1	1
6	0	0	1	1
7	0	0	0	1

In general, we can say that an *n*-stage Johnson counter will produce a modulus of 2 × *n*, where *n* is the number of stages (*i.e.*, flip-flops) in the counter, *i.e.*, for a given MOD number, a Johnson counter requires only half the number of flip-flops that a ring counter requires.

The state diagram of a 4-bit Johnson counter is shown in Fig. 4.133(*b*).

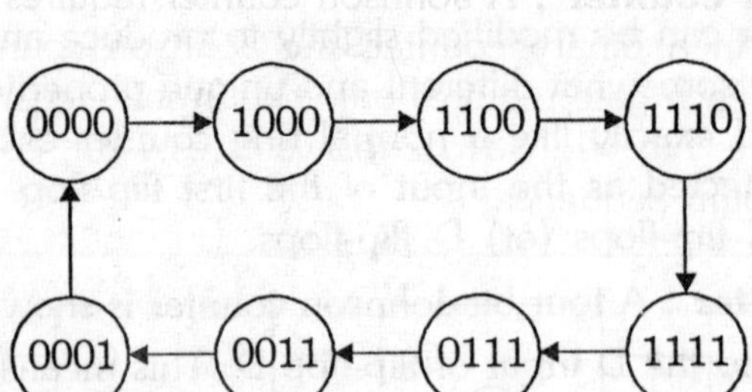

Fig. 4.133(*b*) State diagram for 4-bit Johnson counter.

For example, if we connect five flip-flops, the result is a MOD-10 Johnson counter, where each flip-flop output waveform is a square wave at one-tenth of the clock frequency. So, the output frequency of each flip-flop is equal to the f_{in}/ MOD-N, where f_{in} = frequency of the input clock pulse.

The timing diagram of 4-bit Johnson counter is shown in Fig. 4.133 (*c*).

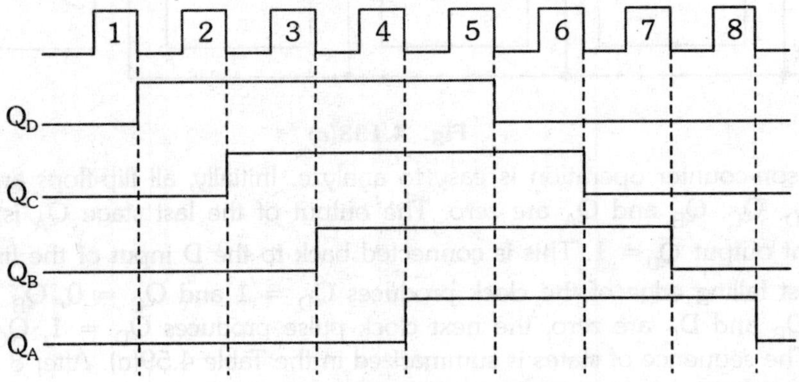

Fig. 4.133(*c*) Timing diagram.

If we design a counter of five bit sequence, it has a total of ten states as shown in Table 4.59(*b*).

Table 4.59(*b*) 5-bit Johnson counter sequence

Clock pulse	Q_E	Q_D	Q_C	Q_B	Q_A
0	0	0	0	0	0
1	1	0	0	0	0
2	1	1	0	0	0
3	1	1	1	0	0
4	1	1	1	1	0
5	1	1	1	1	1
6	0	1	1	1	1
7	0	0	1	1	1
8	0	0	0	1	1
9	0	0	0	0	1

Decoding a Johnson counter : A Johnson counter requires decoding gates, whereas a ring counter does not. As in a binary counter, the Johnson counter uses one logic gate to decade for each count, but each gate requires only two inputs, regardless of the number of flip-flops in the counter. The Fig. 4.133(*d*) shows the 4-bit counter with decader and Table 4.59(*c*) shows the count sequence.

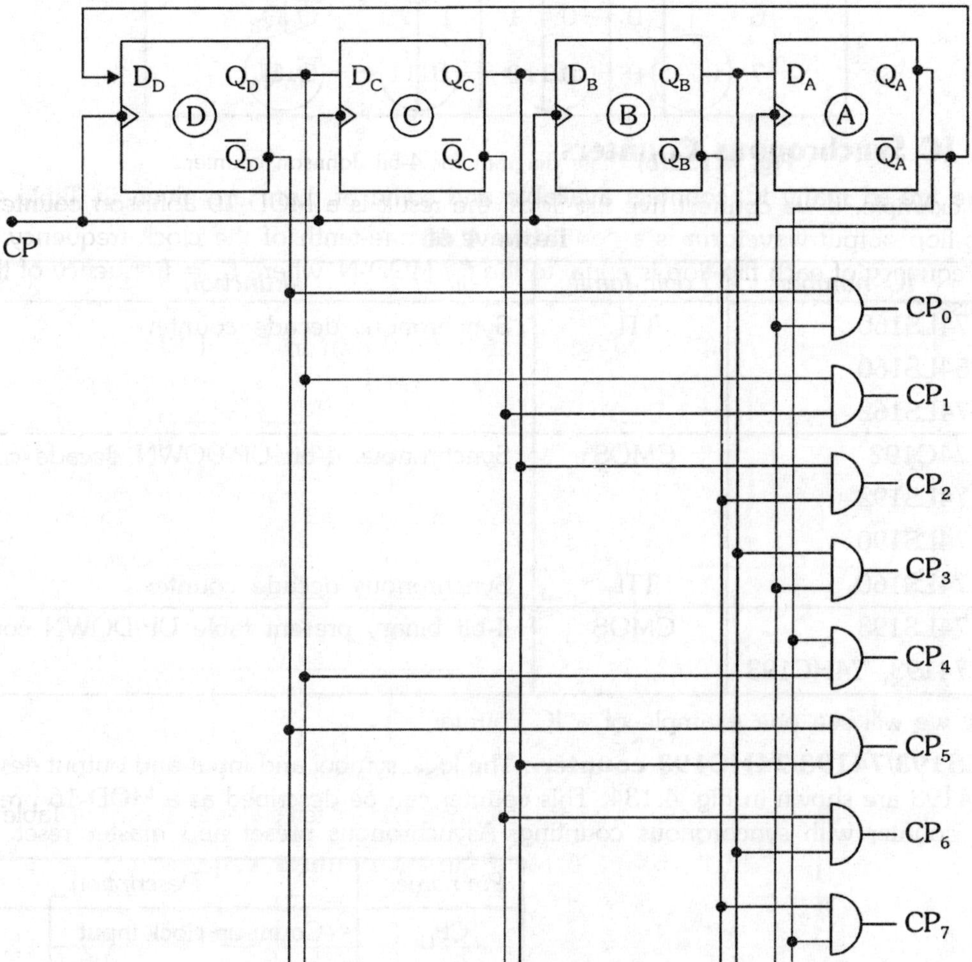

Fig. 4.133(*d*) Johnson counter with decader.

Table 4.59(c) Count sequence and required decoding

Clock pulse	Q_D	Q_C	Q_B	Q_A	*AND gate required for output*
0	0	0	0	0	$\overline{Q}_D\overline{Q}_A$
1	1	0	0	0	$Q_D\overline{Q}_C$
2	1	1	0	0	$Q_C\overline{Q}_B$

3	1	1	1	0	$Q_B\overline{Q}_A$
4	1	1	1	1	Q_DQ_A
5	0	1	1	1	$\overline{Q}_DQ_C$
6	0	0	1	1	$\overline{Q}_CQ_B$
7	0	0	0	1	$\overline{Q}_BQ_A$

4.15.2 IC Synchronous Counters

There are so many IC counters available and some of them are given in Table 4.60.

Table 4.60

IC number	*Logic family*	*Function*
74LS160	TTL	Synchronous decade counter
54LS160		
74LS162		
74C192	CMOS	Synchronous 4-bit UP/DOWN decade counter
74LS192		
74LS190		
74LS160	TTL	Synchronous decade counter
74LS193	CMOS	4-bit binary present table UP/DOWN counter
74193, 74HC193		

Now we will see one example of a IC counter.

74LS193/74193/74HC193 counter : The logic symbol and input and output description for the 74193 are shown in Fig. 4.134. This counter can be described as a MOD-16 presettable up/down counter with synchronous counting. Asynchronous preset and master reset.

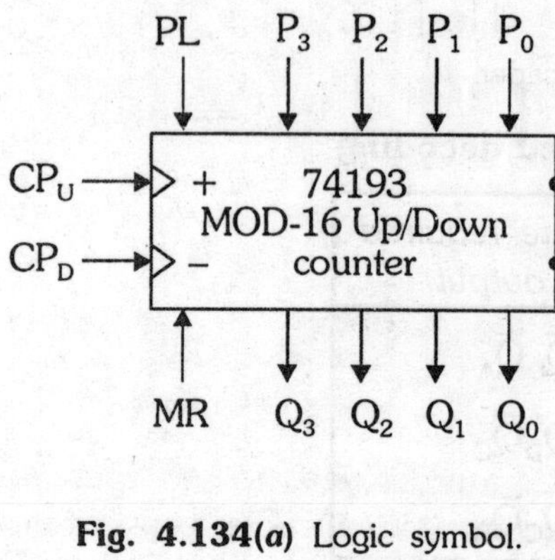

Fig. 4.134(*a*) Logic symbol.

Pin names	*Description*
CP_U	Count-up clock input
CP_D	Count-down clock input
MR	Asynchronous master reset input
$\overline{PL}$	Asynchronous parallel load input
$P_0 - P_3$	Parallel data inputs
$Q_0 - Q_3$	Flip-flop outputs
$T\overline{C}_D$	Terminal count down (borrow) output
$T\overline{C}_U$	Terminal count-up (carry) output

Fig. 4.134(*b*) Input/Output description.

MR	PL	CP_U	CP_D	Mode
H	×	×	×	Asyn C_1 reset
L	L	×	×	Asyn C_1 preset
L	H	H	H	No change
L	H	↑	H	Count-up
L	H	H	↑	Count-down

Fig. 4.134(c) Mode select table.

H = HIGH, L = LOW, × = Don't care, ↑ = PGT.

Clock inputs-CP_U and CP_D : The counter will respond to the positive going transitions at one of two clock inputs. CP_U is the count-up clock input. The counter will increment on each of the clock pulses to a maximum count of 1111; then it recycles to 0000 and starts over. CP_D is the count-down clock input. When pulses are applied to this input, the counter will decrement on each PGT to a minimum count of 0000, then it recycles to 1111 and starts over.

Master reset (MR) : The master reset is an active HIGH asynchronous input that resets the counter to the 0000 state. As long as MR = 1, the counter will hold 0000.

Preset inputs : The counter flip-flops can be preset to the logic levels present on the parallel data inputs P_3 through P_0 by momentarily pulsing the parallel load input PL from HIGH to LOW. This is an asynchronous preset that overrides the counting operation. PL will have no effect, however, if the MR input is in its active HIGH state.

Count outputs : The current count is always preset at the flip-flop outputs Q_3 through Q_0, where Q_0 is the LSB of the count and Q_3 is the MSB of the count.

Terminal count output : The terminal count outputs are used when two or more 74193s are connected in a cascade *i.e.*, to produce a larger MOD number.

In the count-up mode, the $\overline{TC}_U$ output of the lower-order counter is connected to the CP_U input of the next higher-order counter.

In the count-down mode, the $\overline{TC}_D$ output of the lower-order counter is connected to the CP_D input of the next-higher order counter. The logic for $\overline{TC}_U$ and $\overline{TC}_D$ are shown in Fig. 4.135(*a*) and 4.135(*b*) respectively.

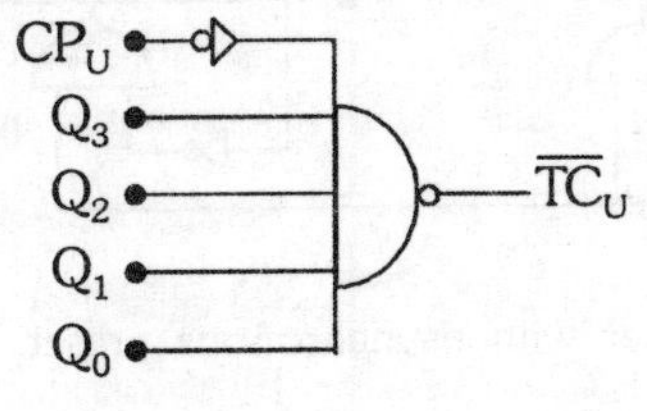

Fig. 4.135(*a*)

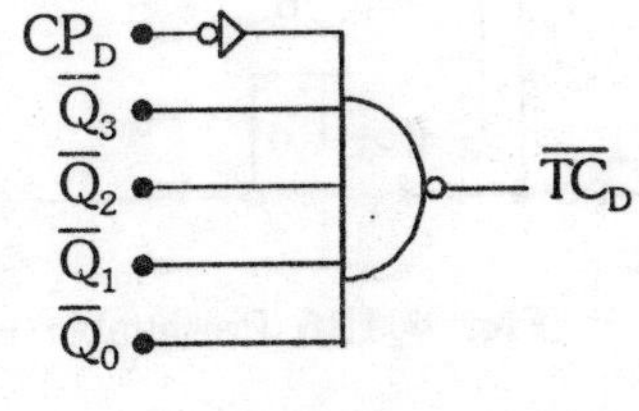

Fig. 4.135(*b*)

4.16 PRESETTABLE COUNTERS

Many parallel or synchronous counters available as ICs are designed to be "*presettable*", in other words; they can be preset to any desired starting count either asynchronously (independent of the clock signal) or synchronously (on the transition of the clock signal). This presetting operation is also referred to as "loading" the counter.

Fig 4.135 shows the logic circuit for a 3-bit presettable parallel up counter. The counter is loaded with any desired count at any time by doing the following :

(1) Apply the desired count to the parallel data inputs P_4, P_1 and P_0.

(2) Apply active LOW pulse to parallel load input, $\overline{PL}$

By this, the levels of P_2, P_1 and P_0 are transfered to Q_2, Q_1 and Q_0 respectively and asynchronously. The effect of the CLK input is disabled as long as $\overline{PL}$ is in its active-low state, since each flip-flop will have one of its asynchronous inputs activated while $\overline{PL} = 0$. Once $\overline{PL}$ returns HIGH, the flip-flops can respond to their CLK input and can resume the counting-up operation starting from the count that was loaded into the counter.

For example, if $P_2 = 1$, $P_1 = 0$, $P_0 = 1$, while $\overline{PL}$ is HIGH, these parallel data inputs have no effect. If clock pulses are present, the counter will perform the normal count-up operation. Now, if $\overline{PL}$ is pulsed to LOW at the count 010 (*i.e.*, $Q_2 = 0$, $Q_1 = 1$ and $Q_0 = 0$). This LOW at $\overline{PL}$ will produce LOWs at CLEAR input of Q_1 and at the PRESET inputs of Q_2 and Q_0 so that the counter will go to the 101 count regardless of what is occuring at the CLK input.

The count will hold at 101 until $\overline{PL}$ is deactivated (returned to HIGH); at that time the counter will resume counting up clock pulses from the count of 101. Examples of IC's which contain asynchronous presetting are 74190, 74191, 74192 and 74193 of TTL family and 74HC190, 74HC191, 74HC192 and 74HC193 from CMOS family.

Synchronous presetting counters are also available, where counter is preset on the active transition of the same clock signal that is used for counting. Examples of ICs from TTL is 74160, 74161, 74162 and 74163 and from CMOS are74HC160, 74HC161, 74HC162 and 74HC163.

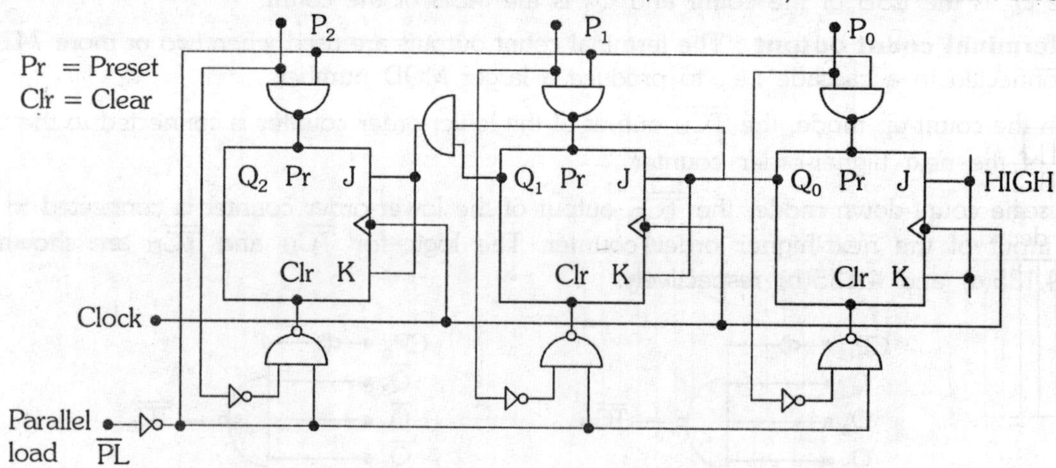

Fig. 4.136 Presettable parallel counter with asynchronous preset.

4.17 COUNTER APPLICATIONS

The digital counters are very useful and versatile devices that are found in many applications. Basically, these are used for generating timing variables to sequence and control the operations in a digital system.

1. **The Digital clock :** It is a very common example of a counter application in time keeping system. The 74LS160 decade counters (synchronous) are used for implementation of various counting operations, exclusively for hours, minutes and seconds.

2. **Frequency dividers :** The counters can be used as programmable frequency dividers. By presetting any number into the counter and counting to the maximum (UP counting) or minimum (DOWN counting) count, the division is achieved. This division operation is greatly used in various digital systems.

3. **Auto parking control :** This is a simple application of the UP/DOWN counter to solve the parking problem. It monitors and displays the available space in a parking garage.

4. **Industrial digital control system :** Counters are used in various industrial applications such as counting objects, speed measurement, frequency measurement etc.

Now we will see the two common applications in detail.

1. **Digital clock :** The Fig 4.137 shows the block diagram of a digital clock that displays seconds, minutes and hours. Here 1 Hz clock frequency is applied to the divide-by-6-counter and divide by 10 counter, which forms a divide by 60 counter. This counter counts seconds from 00–59. Ripple carry from this counter is applied to the enable input of another pair of divide by–60 (10 × 6) counter. This counter is used to count minutes from 00 to 59.

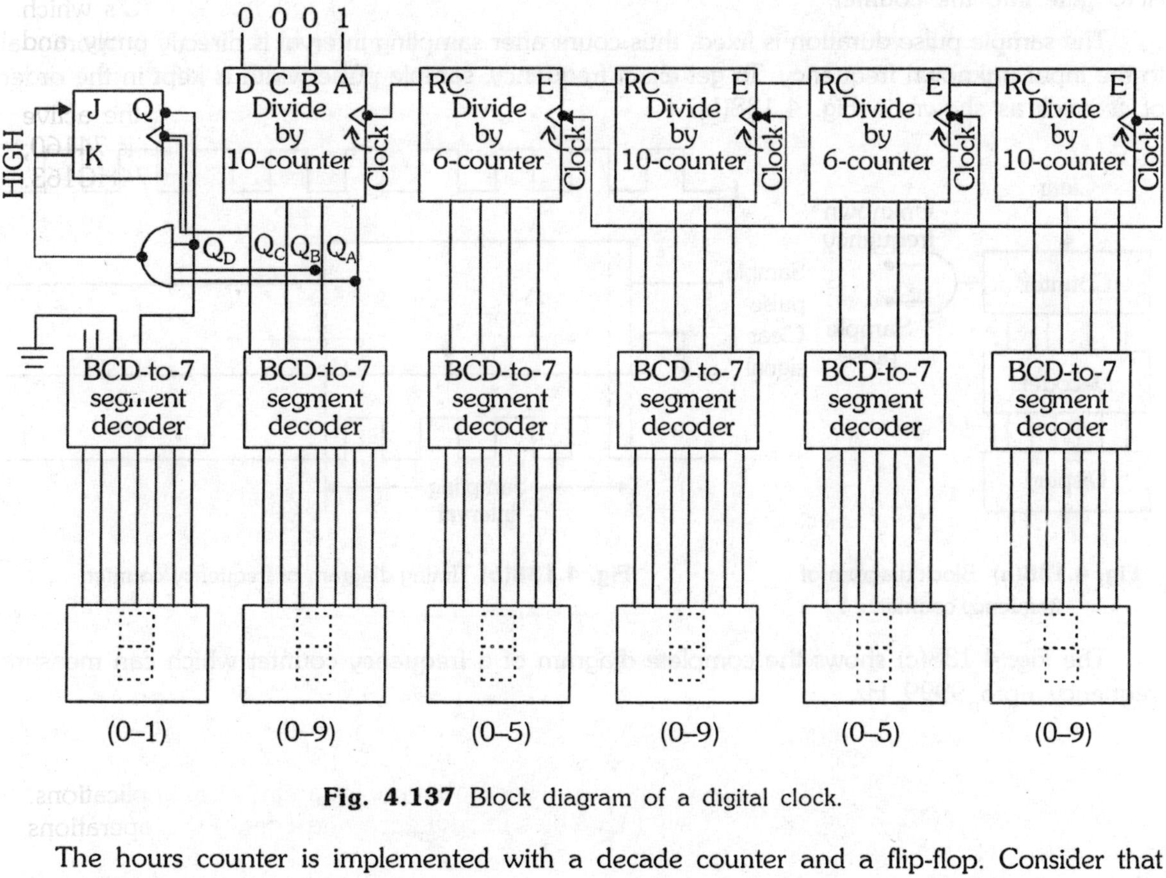

Fig. 4.137 Block diagram of a digital clock.

The hours counter is implemented with a decade counter and a flip-flop. Consider that initially both the decade counter and the flip-flops are RESET, and the NAND gate output is HIGH. The decade counter advances from 0 to 9, and as it recycles from 9 back to 0, the flip-flop is toggle to the SET state by the HIGH-to-LOW transition of Q_D.

At this moment, count is 10 decimal (decade counter in the 0 state and the flip-flop is SET). After next two clock pulses, the total count advances to 11 decimal and then to 12 decimal. In state 13, the Q_B and Q_A outputs of the decade counter are HIGH, the flip-flop is still SET, and thus output of NAND gate goes to LOW. This activates the asynchronous LOAD input decade of the decade counter, presetting the decade counter to state 1 by the data inputs (0001). The output of NAND gate is also connected to J input, thus flip-flop resets. This logic makes counter to recycle from 12 to 1 rather than 0.

The outputs of the counter are connected to the BCD-to-7 segment drivers, by which we can have the display of count.

2. **Frequency counter :** A frequency counter is a circuit that can measure and display the frequency of the signal. The diagram is shown in Fig. 4.138(*a*). It consists of a counter with it's associated decoder/display circuit and an AND gate. The input of AND gate contains the pulses with unknown frequency, f_i, and as sample (reference) pulse with a known duration which controls how long the pulses with unknown frequency are allowed to pass through the AND gate into the counter.

The sample pulse duration is fixed, thus count after sampling interval is directly proportional to the input unknown frequency. To get exact frequency, sample pulse width is kept in the order of seconds as shown in Fig. 4.138(*b*).

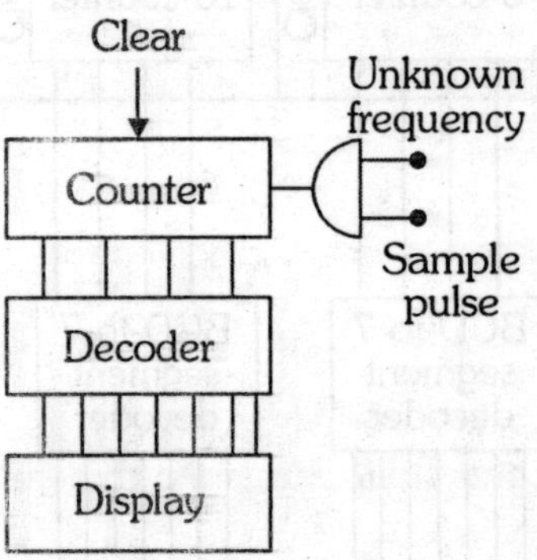

Fig. 4.138(*a*) Block diagram of frequency counter.

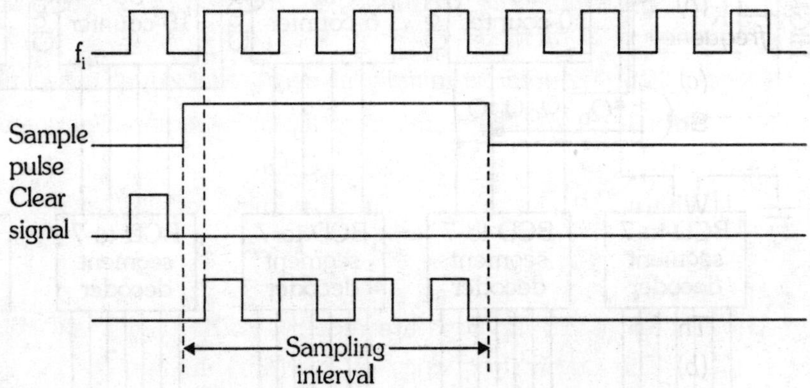

Fig. 4.138(*b*) Timing diagram of frequency counter.

The Fig. 4.138(*c*) shows the complete diagram of a frequency counter which can measure frequency upto 9999 Hz.

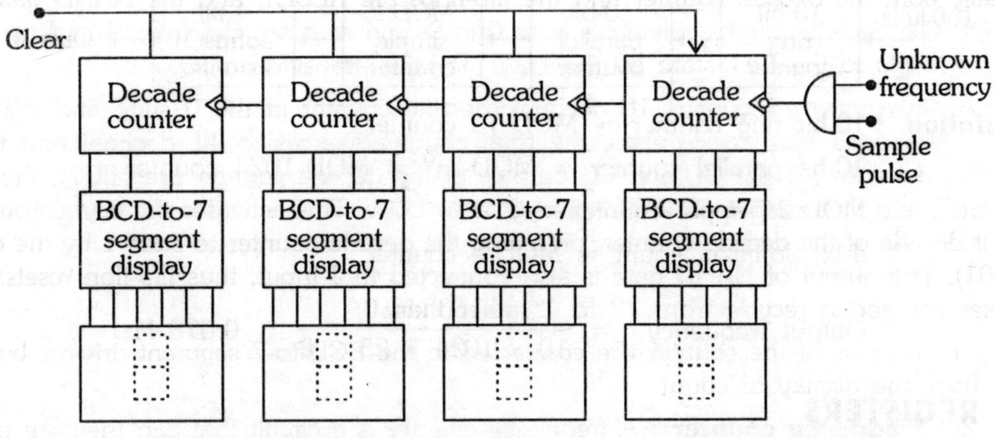

Fig. 4.138(c) Complete diagram of a frequency counter.

Example 4.34. *A 16 MHz square wave clocks a 5 bit ripple counter, what is the frequency at the output of the last flip-flop and what is the duty cycle of the same ?*

Solution. The output frequency $= \frac{16}{2^5} = \frac{16 \text{ MHz}}{32} = 500$ kHz and duty cycle = 50%.

Example 4.35. *(a) How many number of flip-flops are required to build a binary counter that counts from 0 to 1023 ?*

(b) Determine the frequency at the output of last (MSB) flip-flop for an input clock frequency of 2 MHz.

(c) If the counter is initially at zero, what count will it hold after 2060 clock pulses ?

Solution. (*a*) The number of flip-flops required is given by

$$2^n \geq \text{Count}$$

Where n = number of flip-flops.

$$\therefore \quad 2^n \geq 1023$$

$$n = 10$$

Therefore, 10 flip-flops are required to build a binary counter that counts from 0 – 1023.

(*b*) The output frequency of the MSB flip-flop is

$$f_{\text{out}} = \frac{2 \text{ MHz}}{2^{10}} = \frac{2 \text{ MHz}}{1024} = 1953.125 \text{ Hz.}$$

(*c*) The count after 2060 clock pulse is, 1024 clock pulses are needed to complete one full count from 0000 to 1023 and come back to 0000.

$$\therefore \quad \text{Count on the counter} = \frac{2060}{1024} = 12$$

Example 4.36. *Determine the output frequency of a cascaded combination of 10-bit ring counter, 10-bit parallel counter, MOD-25 ripple counter and 4-bit Johnson counter. The input frequency is 160 kHz given to 10-bit ring counter.*

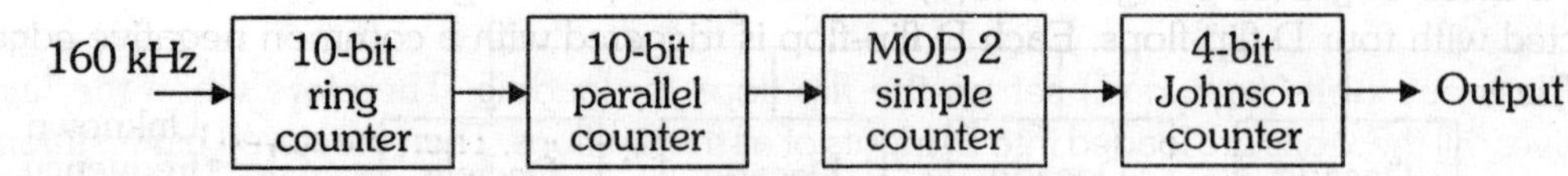

Solution. 10-bit ring counter = MOD-10 counter.

10-bit parallel counter = MOD-2^{10} = MOD-1024 counter

MOD-25 ripple counter

4-bit Johnson count = MOD-8 counter

$$\therefore \quad \text{Output frequency} = \frac{160 \text{ kHz}}{10 \times 1024 \times 25 \times 8} = 0.078 \text{ Hz.}$$

4.18 REGISTERS

As we know that flip-flop is a binary cell capable of storing one bit information. They can be connected together to perform counting operations. Such a group of flip-flops is called "counter". An array of flip-flops is used to store the binary information known as "registers". Each flip-flop stores one bit information. So an N-bit register has a group of N-flip-flops and is capable of storing any binary information number containing *n*-bits. In other words, the number of flip-flops are required as equal to the number of bits in a binary word.

(A) **Need of a register :** There are two functional characteristics for a register, they are (1) Temporary storage (2) Shifting capability.

Generally registers are used for temporary storage of data of binary information within a digital system. This storage capability makes it an important type of memory device.

Registers are also used for shifting the binary information stored in it. The shifting capability of a register permits the movement of data from stage to stage or into or out of the register upon the application of clock pulses. Then we can name the register as "shift register". These two functional characteristics make the register an important type of sequential logic circuit and widely used within the digital system. Thus a register, in general, is used solely for storing and shifting data (1's or 0's) entered into it from an external source and possesses no characteristic internal sequence of states.

(B) (*i*) **Concept of storing the data in a register :** The concept of storing the binary information either 1 or 0 is illustrated in Fig. 4.139. A binary 1 is applied as a input, and a clock pulse is applied, that stores 1 by setting the flip-flop. When the input 1 is removed, note that the flip-flop remains in the SET state, thereby storing the 1. The same procedure is applied for the storage of 0.

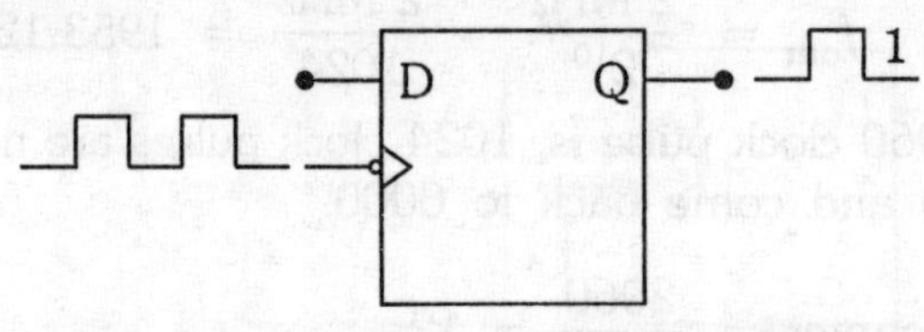

Fig. 4.139 Flip-flop as a storage element.

By wiring some of the flip-flops, we can form a register. The storage capacity of a register is the number of bits (1's and 0's) of digital data that it can retain. Each stage of a register represents one bit of storage capacity and therefore the number of stages in a register determines its total '*storage capacity*'.

(*ii*) **Buffer register** : Fig. 4.140(*a*) shows the simplest register known as buffer register constructed with four D flip-flops. Each D flip-flop is triggered with a common negative edge clock pulse. The input x-bits ($x_0x_1x_2x_3$) set up the flip-flops for loading. Therefore when the first clock pulse arrives, all the *x*-bits are loaded into D-inputs of all the flip-flops. Then the stored binary information is $Q = Q_3Q_2Q_1Q_0 = x_3x_2x_1x_0$. For example, if we want to store a 4-bit binary information 1001 in the buffer register, then these bits are applied at x-bits as $x_3 = 1$, $x_2 = 0$, $x_1 = 0$ and $x_0 = 1$. Whenever the first clock pulse is applied, then the stored binary word = Q = 1001.

In this buffer register, there is no control over input as well as output bits. For this purpose, consider a "controlled buffer register' as shown in Fig. 4.140 (*b*). This register consists of a group of flip-flops and gates that effect their transition. Here the flip-flops hold the binary information and the gates hold the control information. This control information deals when and how new information is transferred into the register.

Whenever clear signal is at HIGH, all the flip-flops are in reset and the circuit is ready for operation whenever the clear signal returns to LOW. The LOAD control input determines the entire circuit operation.

When LOAD signal is at LOW, then *x*-bits cannot reach the flip-flops. The contents of register are unchanged at this condition.

When LOAD signal is HIGH, the *x*-bits are transmitted to the data inputs and the flip-flops are ready for loading in the presence of clock pulse. The stored binary information is

$$Q = Q_3\ Q_2\ Q_1\ Q_0 = x_3x_2x_1x_0.$$

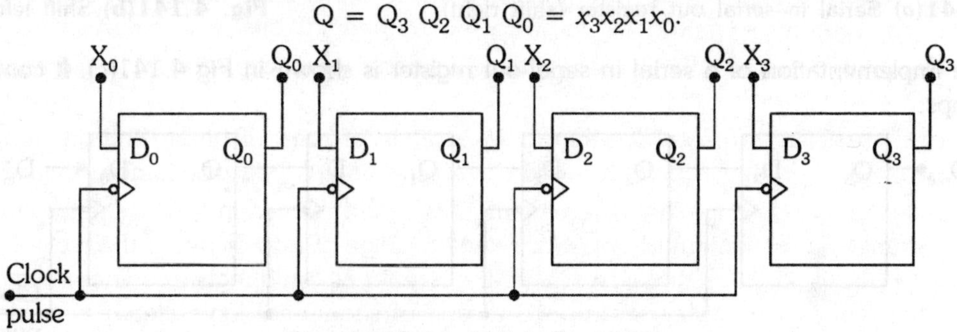

Fig. 4.140 (*a*) 4-bit buffer register.

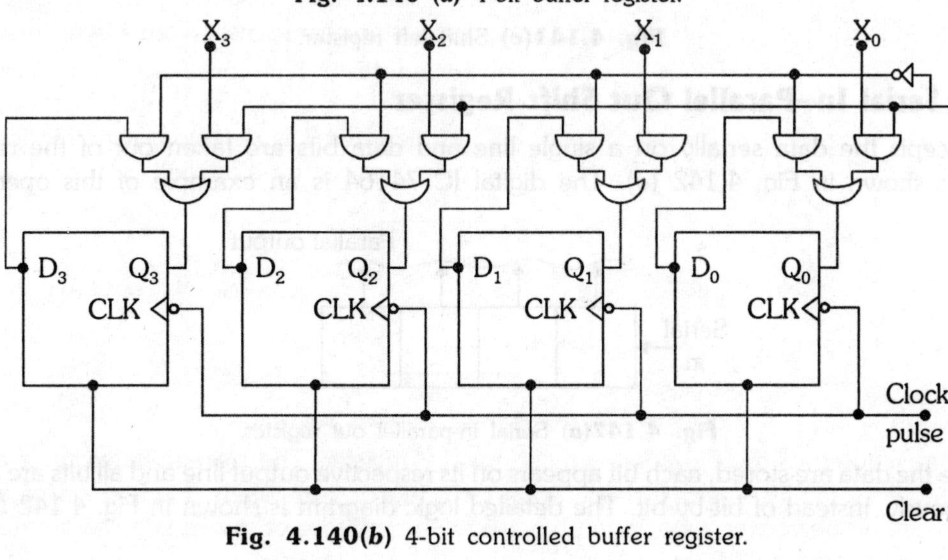

Fig. 4.140(*b*) 4-bit controlled buffer register.

(C) **Data movement in registers :** Basically data can be entered in two forms.

1. Serial (one bit at a time *i.e.*, bit-by-bit) 2. Parallel (all the bits simultaneously) and are called as serial form and parallel form respectively. The data movement in the registers is concerned with the two characteristics. They are

1. How the binary information is loaded into the register ?

2. How to read the data from the register ?

According to that, the registers are classified into four possible modes of operations. They are

(*a*) Serial in–serial out (*b*) Serial in–parallel out (*c*) Parallel in–serial out (*d*) Parallel in–parallel out.

4.18.1 Serial In–Serial Out Shift Register

It accepts the data serially *i.e.* one bit at a time on a single line as shown in Fig. 4.141(*a*) and 4.141(*b*). It produces the stored information on its output in the serial form. The IC example for it is 7491, which contains SR-flip-flops.

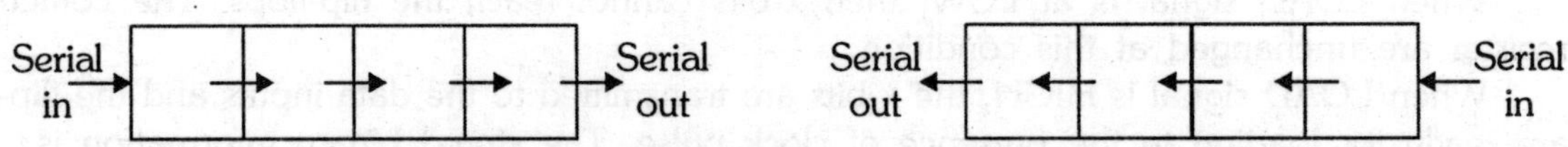

Fig. 4.141(*a*) Serial in–serial out register (shift right). **Fig. 4.141(*b*)** Shift left.

The implementation of a serial in serial-out register is shown in Fig 4.141(*c*). It contains four D-flip-flops.

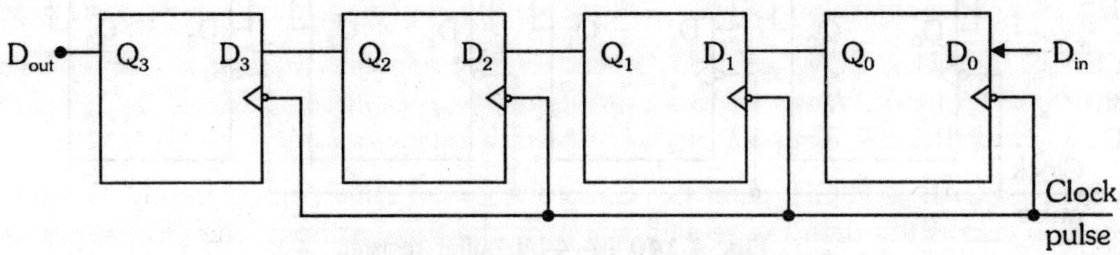

Fig. 4.141(c) Shift left register.

4.18.2 Serial In–Parallel Out Shift Register

It accepts the data serially on a single line and data bits are taken out of the register in parallel as shown in Fig. 4.142 (*a*). The digital IC 74164 is an example of this operation.

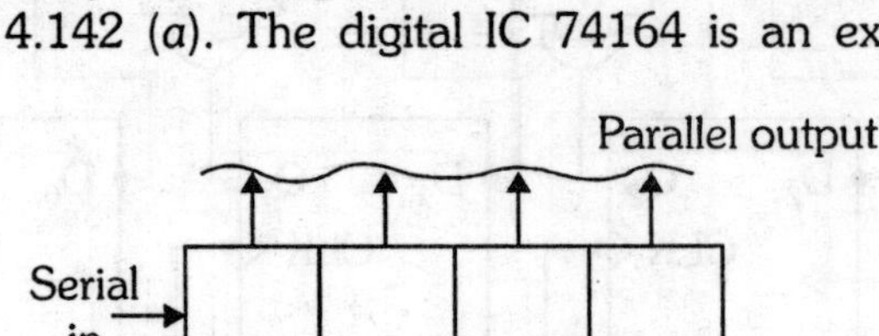

Fig. 4.142(*a*) Serial in-parallel out register.

Once the data are stored, each bit appears on its respective output line and all bits are available simultaneously, instead of bit-by-bit. The detailed logic diagram is shown in Fig. 4.142(*b*).

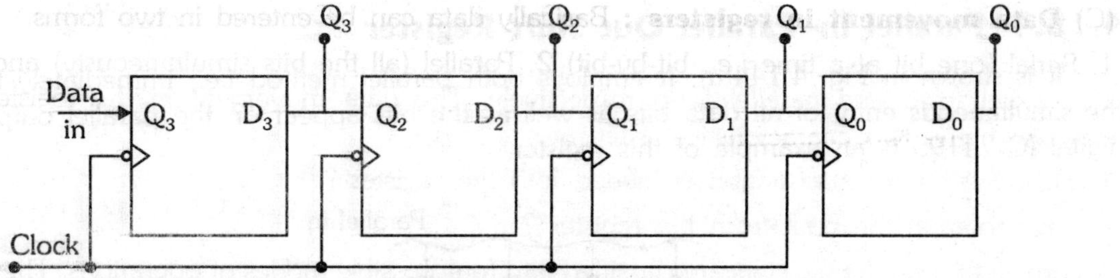

Fig. 4.142(b)

4.18.3 Parallel In–Serial Out Shift Register

It is shown in Fig. 4.143(*a*). In this type, the bits are entered in parallel *i.e.*, simultaneously into their respective stages on a parallel line. It produces the stored information on its output in serial form. The digital IC 74165 is an example of this register.

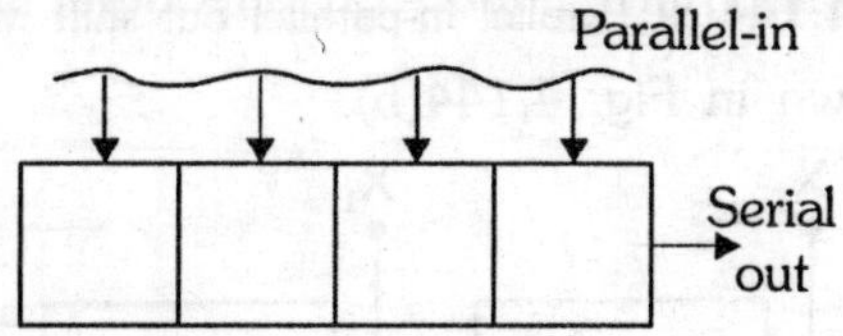

Fig. 4.143(a) Parallel in-serial out shift register.

The Fig. 4.143(*b*) illustrates a four-bit parallel in serial out register. There are four input lines x_3, x_2, x_1, and x_0 for entering data, in parallel, into the register.

SHIFT/$\overline{\text{LOAD}}$ is the control input, which allows shift or loading data operation of the register.

When SHIFT/$\overline{\text{LOAD}}$ is LOW, gates G_1G_2 and G_3 are enabled, allowing the input data to the respective filp-flop input. When a clock pulse is applied, the flip-flop with D = 1, will SET and with D = 0 will RESET. Thus all bits are stored simultaneously.

When SHIFT/$\overline{\text{LOAD}}$ is HIGH, gates G_1, G_2 and G_3 are disabled and gates G_4, G_5 and G_6, are enabled. This allows the data bits to shift left from one stage to next. The OR gates at each flip-flop inputs allows either LOAD or SHIFT operation.

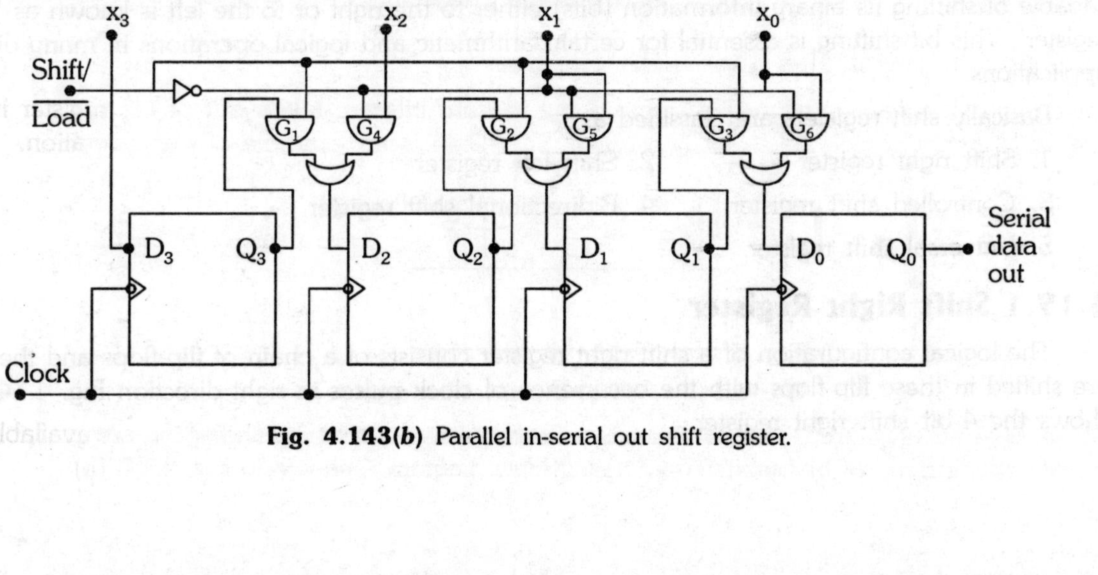

Fig. 4.143(b) Parallel in-serial out shift register.

4.18.4 Parallel In–Parallel Out Shift Register

It is shown in Fig. 4.144(*a*). It employs both parallel method *i.e.*, immediately following the simultaneous entry of all data bits as well as the bits appear on the parallel outputs. The digital IC 74195 is an example of this register.

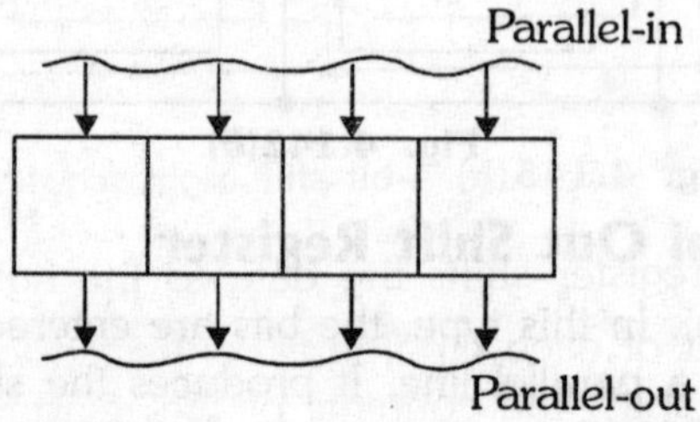

Fig. 4.144(*a*) Parallel in-parallel out shift register.

The logic diagram is shown in Fig. 4.144(*b*).

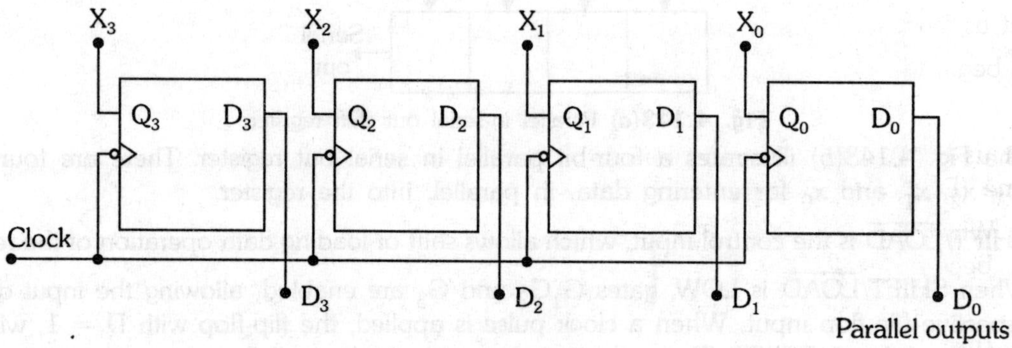

Fig. 4.144(*b*)

4.19 SHIFT REGISTERS

From the study of data movement in the register, we know that the register which is capable of shifting its binary information (bits) either to the right or to the left is known as 'shift register'. This bit shifting is essential for certain arithmetic and logical operations in many digital applications.

Basically shift registers are classified as

1. Shift right register
2. Shift left register
3. Controlled shift register
4. Bidirectional shift register
5. Universal shift register.

4.19.1 Shift Right Register

The logical configuration of a shift right register consists of a chain of flip-flops and the bits are shifted in these flip-flops with the occurrence of clock pulses in right direction Fig. 4.145(*a*) shows the 4-bit shift right register.

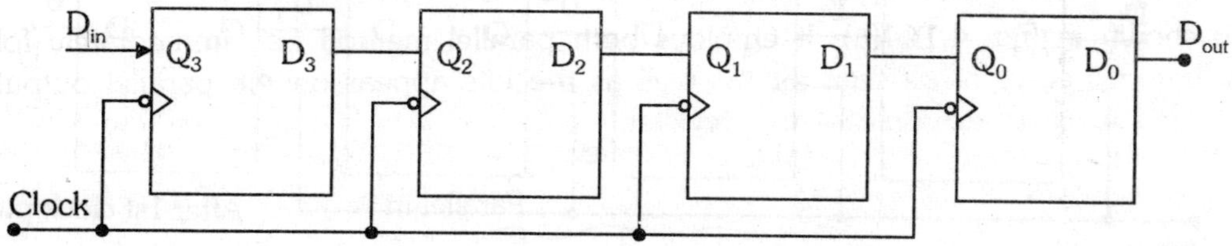

Fig. 4.145(*a*) 4-bit shift right register.

On each clock pulse, the register shifts the data to the right.

Operation : The Q output of each flip-flop is connected to the D-input of the flip-flop at its right side. Each clock pulse shifts the contents of the register one bit position to the right. The data input in serial form determines what goes into the left most flip-flop during the shift. The data output is taken in serial form from the output of the right most flip-flop prior to the application of a pulse.

Let us illustrate the operation by taking the four bit binary number 1111 into the register.

To begin the operation ensure that the register is initially cleared, by giving data output as

$$Q = Q_3Q_2Q_1Q_0 = 0000.$$

Let us begin with right most bit of the given 4-bit word; *i.e.*, this 1 is put onto the data input line (D_{in}) of flip-flop 3.

1. When the first clock pulse is applied, the left most flip-flop sets, and the stored word becomes

$$Q = Q_3Q_2Q_1Q_0 = 1000$$

2. When the second clock pulse occurs, the second bit 1 is entered into the left most flip-flop (D_{in}). The 1 in that flip-flop is shifted to next flip-flop (FF_2), since output of Q_3 is connected to D_2.

Then the stored binary word

$$Q = Q_3Q_2Q_1Q_0 = 1100$$

3. Similarly for the third and fourth clock pulses, the stored binary words are 1110 and 1111 respectively. The Fig. 4.145(*b*) illustrates the input and output condition of each flip-flop stage upon the application of each clock pulse.

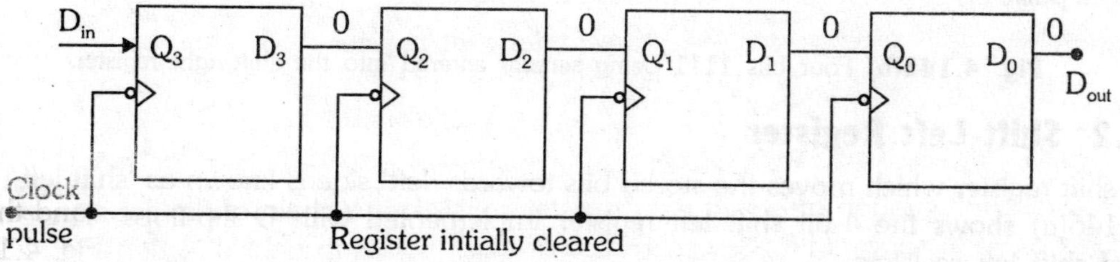

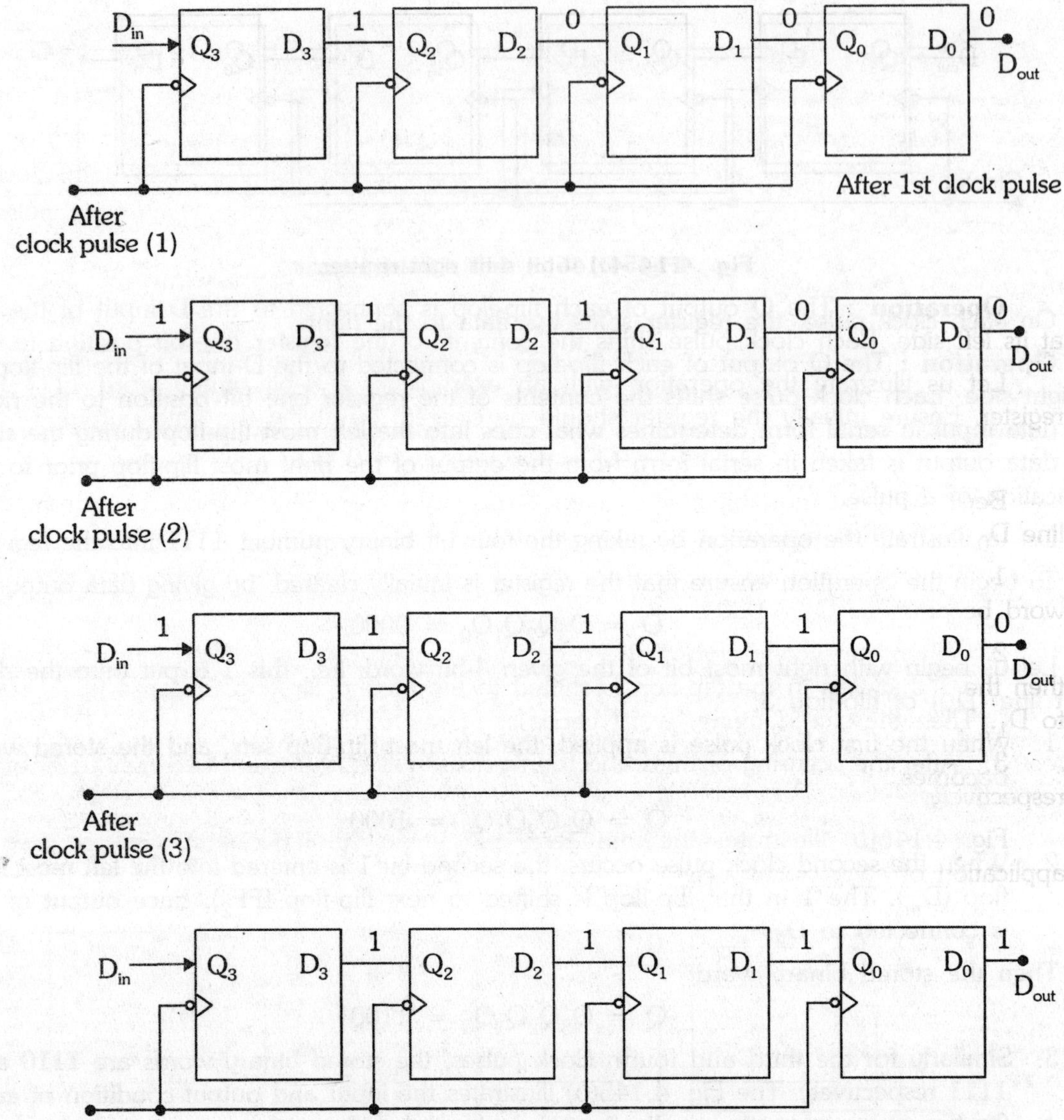

Fig. 4.145(*b*) Four bits 1111 being serially entered into the shift-right register.

4.19.2 Shift Left Register

A shift register which moves the stored bits towards 'left' side is known as 'shift left register'. Fig. 4.146(*a*) shows the 4-bit shift left register implemented with D flip-flops. There are two types of shift left registers

(*a*) Serial in-serial out (*b*) Serial in-parallel out

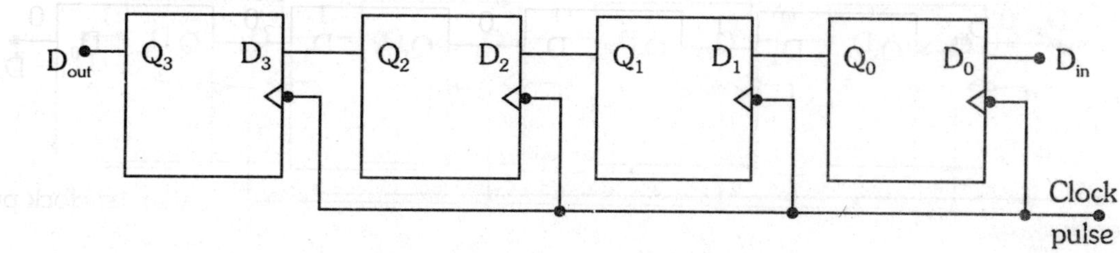

Fig. 4.146(*a*) 4-bit shift-left register.

Operation : The Q output of each flip-flop is connected to the D input of the flip-flop at its left side. Each clock pulse shifts the contents of the register one bit position to the left.

Let us illustrate the operation with an example of a 4-bit binary word 1111 into the register. Ensure initially the register should be cleared, by giving the data output as

$$Q = Q_3Q_2Q_1Q_0 = 0000.$$

Begin with left most bit of the given 4-bit word *i.e.*, 1. This 1 is put onto the data input line D_0 of the first flip-flop (right most flip-flop).

1. When the first clock is applied, the flip-flop is SET, by storing logic 1. Then the stored word becomes as Q = 0001.

2. When the second clock pulse occurs, the second bit 1 is entered into the first flip-flop, then the 1 that was in the flip-flop is shifted to the second flip-flop; because Q_0 is connected to D_1. Then the stored binary word becomes Q = 0011.

3. After the occuring of third and fourth clock pulses, the count becomes 0111 and 1111 respectively.

Fig. 4.146(*b*) illustrates the input and output condition for each flip-flop stage upon application of each clock pulse.

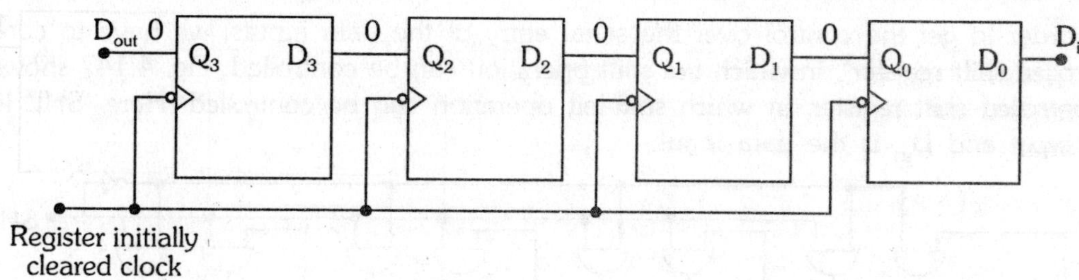

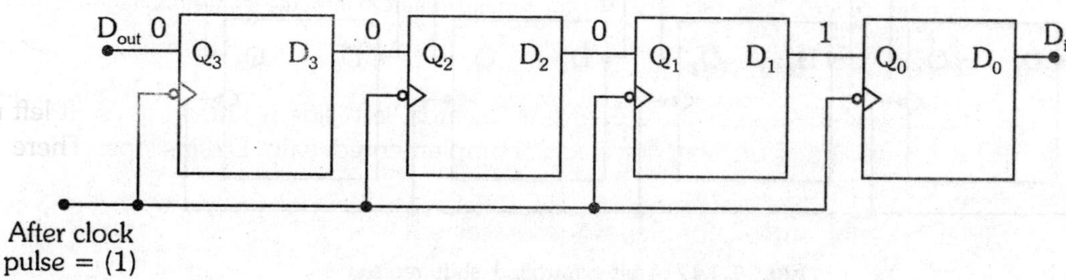

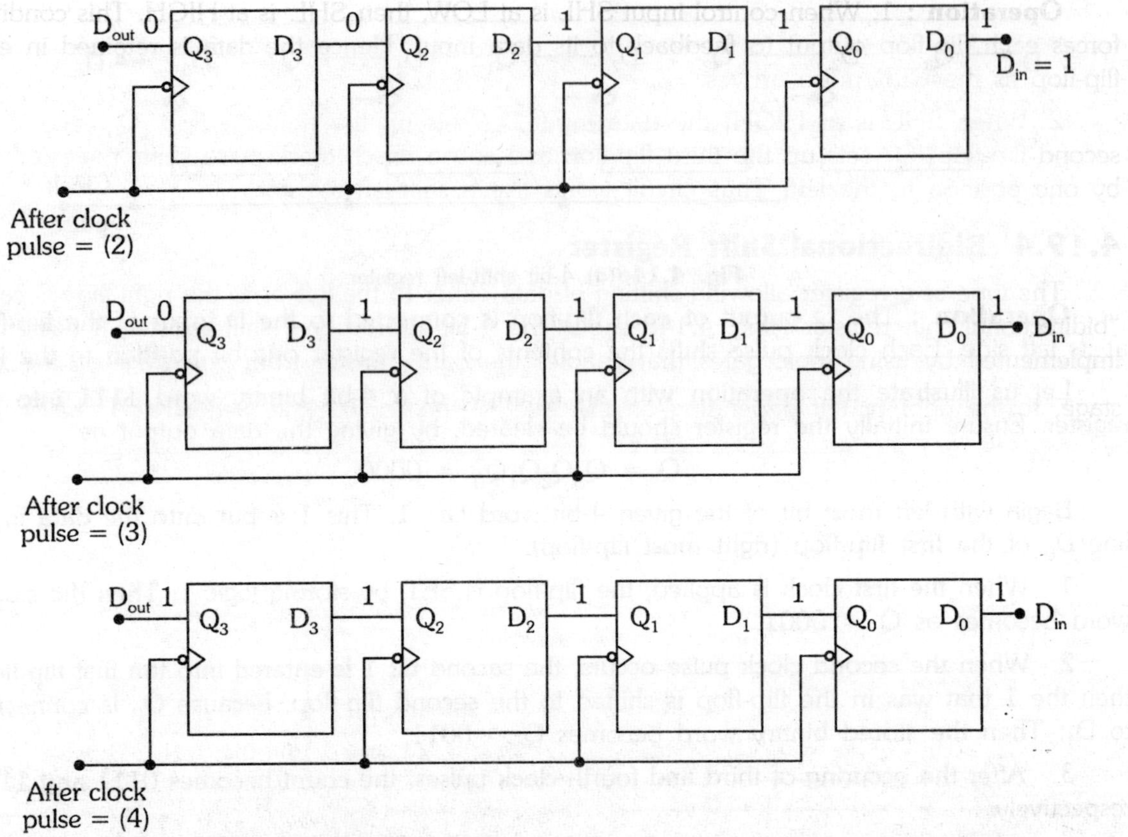

Fig. 4.146(*b*) Four bits 1111 being serially entered into the shift left register.

4.19.3 Controlled Shift Register

In order to get the control over the serial entry of the data inputs, we have to consider a 'controlled shift register', in which the shift operation can be controlled. Fig. 4.147 shows the 4-bit controlled shift register, in which shift-left operation can be controlled. Here 'SHL' is the control input and D_{in} is the data input.

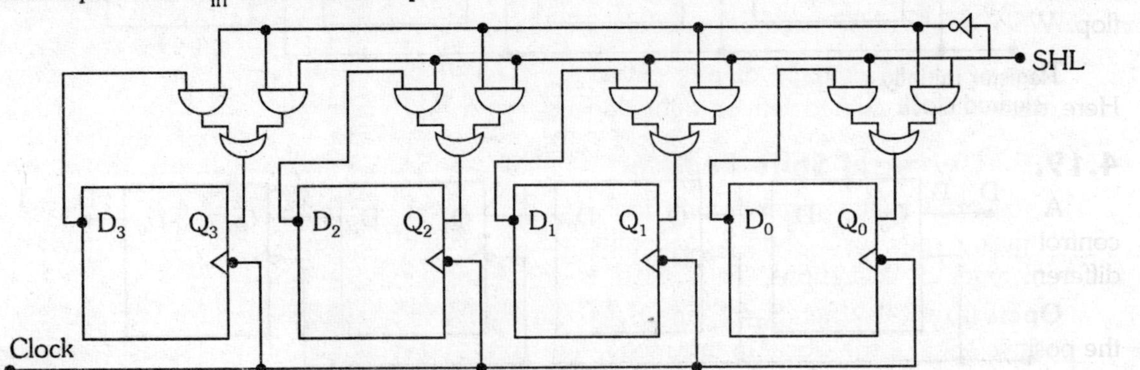

Fig. 4.147 4-bit controlled shift register.

Operation : 1. When control input SHL is at LOW, then $\overline{SHL}$ is at HIGH. This condition forces each flip-flop output to feedback to its data input. Hence the data is retained in each flip-flop as the CLK pulse arrives.

2. When SHL is at HIGH, the data input D_{in} sets up the right flip-flop, Q_0 sets up the second flip-flop, Q_1 sets up the third flip-flop and so on. Each clock pulse shifts the stored bit by one position to the left. Thus circuit works like a shift-left register.

4.19.4 Bidirectional Shift Register

The type of a register allowing shifting of data either to the left or to the right side is called "bidirectional shift register". Fig. 4.148 shows the 4-bit bidirectional shift register. It has circuit implemented by using logic gates that enables the data transfer from one stage to the next stage, to the right or left depending on the level of the control line. *i.e.*, RIGHT / $\overline{\text{LEFT}}$.

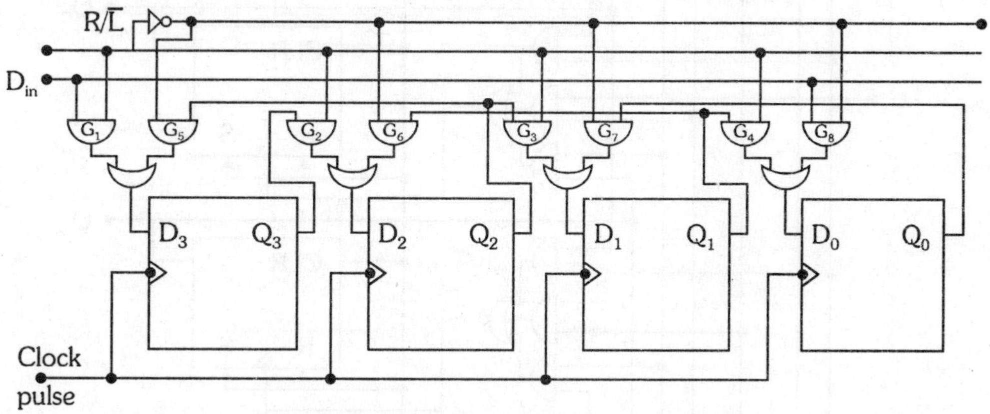

Fig. 4.148 4–bit bidirectional shift register.

Operation : 1. When control input $R/\overline{L}$ is at HIGH level, the gates G_1 to G_4 are enabled and the state of the Q output of each flip-flop is passed through to the D input of the respective flip-flop. When clock pulse occurs, then the data is shifted by one place to the right.

2. When the control input $R/\overline{L}$ is at LOW level, then the gates G_5 to G_8 are enabled and the state of the Q output of each flip-flop is passed through to the D_{input} of the preceding flip-flop. When clock pulse occurs then data is shifted by one place to the left.

Thus when $R/\overline{L}$ is high, data is shifted to right and when R/W is low, data is shifted to left. Here data can be shifted left or right depending on the control input.

4.19.5 Universal Shift Register

A universal shift register is one that has both serial and parallel inputs. It contains two control pins, named as S_0 and S_1 according to the value of S_0 and S_1, the register will perform different types of operations. The 74194 is an example for it.

Operation : 1. When $S_0=1$, and $S_1=0$, the shift right operation is done synchronously with the positive edge of the clock. In this mode, the serial data is entered at the shift-right-serial input.

2. When $S_0=0$ and $S_1=1$, then shift left operation is done in synchrony with the clock. In this mode the serial data is entered at the 'shift-left serial input'.

3. When $S_0 = S_1 = 1$, parallel loading operation is done in synchronous with a positive transition of the clock.

4. When $S_0 = S_1 = 0$, then we have inhibit clock operation. In this mode, the operation will inhibit the shifting of data in the shift register.

The block diagram, pin configuration, truth table and truth table of 74194 is shown in Fig. 4.149(*a*), 4.149(*b*) and 4.149(*c*) respectively.

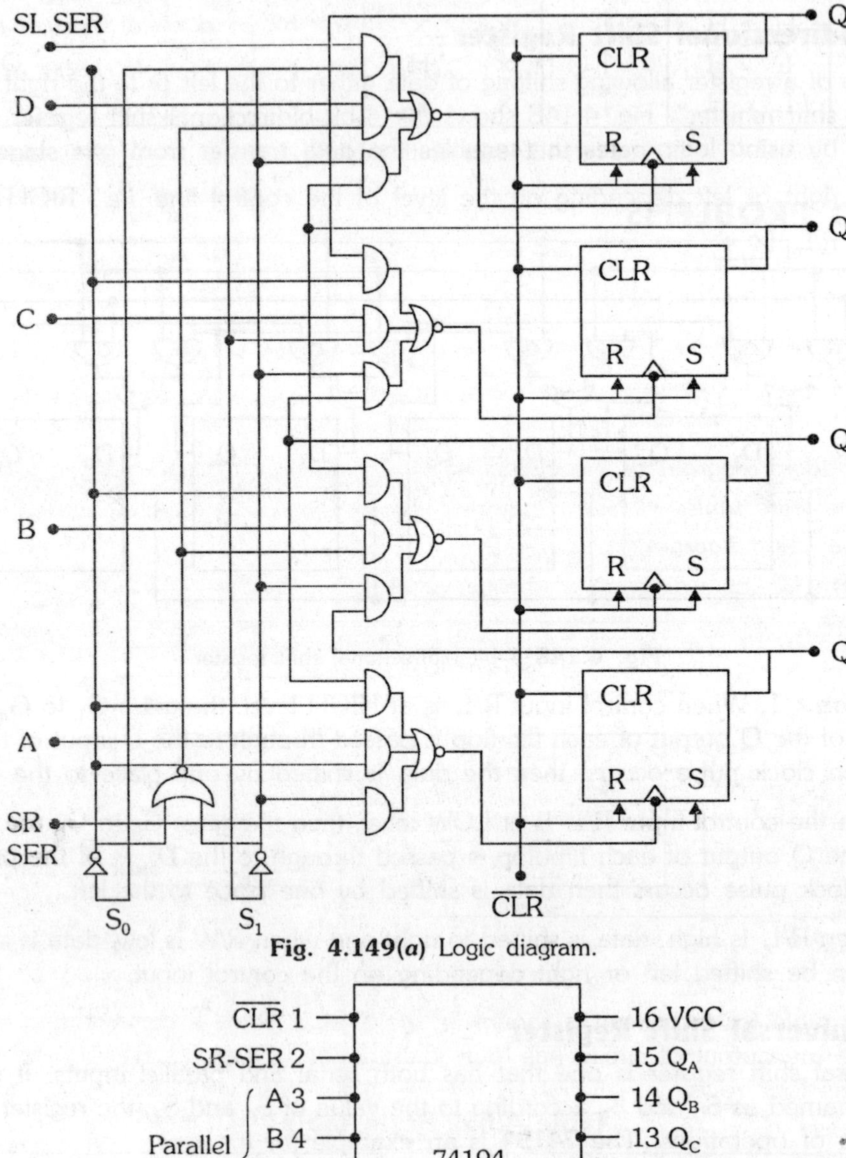

Fig. 4.149(*a*) Logic diagram.

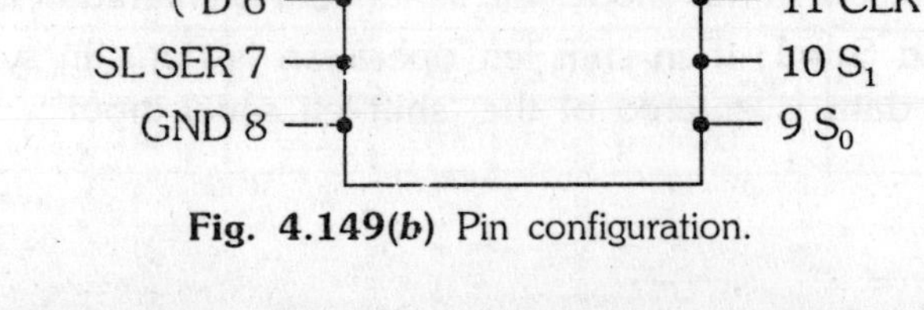

Fig. 4.149(*b*) Pin configuration.

Inputs										Outputs			
Clear	Mode		Clock	Serial		Parallel							
	S_0	S_1		Left	Right	A	B	C	D	Q_A	Q_B	Q_C	Q_D
L	×	×	×	×	×	×	×	×	×	L	L	L	L
H	×	×	L	×	×	×	×	×	×	Q_{A0}	Q_{B0}	Q_{C0}	Q_{D0}
H	H	H	↑	×	×	a	b	c	d	a	b	b	b
H	L	H	↑	×	H	×	×	×	×	H	Q_{A0}	Q_{B0}	Q_{C0}
H	L	H	↑	×	L								

Fig. 4.149(c) Truth table.

4.20 SOLVED PROBLEMS

Problem 1. *A sequential circuit with two D flip-flops, A and B; two inputs x and y; and one output z, specified by the following next state and output equations.*

$$A_{n+1} = \bar{x}y + xA_n$$

$$B_{n+1} = \bar{x}B_n + xA_n$$

$$z = B_n$$

(a) Draw the logic diagram of the circuit.

(b) Derive the state table.

(c) Derive the state diagram.

Solution. (*a*) The logic diagram of the circuit is shown in Fig. 4.150.

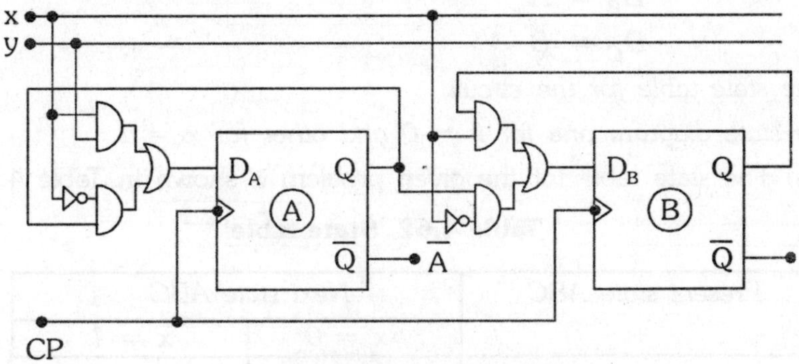

Fig. 4.150 Logic diagram.

(*b*) The state table for the circuit is shown in Table 4.61. Since it contains two inputs, so the next state and the output of circuit has four columns (*i.e.*, *xy* = 00, 01, 10 and 11). The state table contains four rows.

Table 4.61. State table

Present state	Next state (AB)				Output (z)			
AB	xy = 00	01	10	11	00	01	10	11
00	00	10	00	00	0	0	0	0

01	01	11	00	00	1	1	1	1
10	00	10	11	11	0	0	0	0
11	01	11	11	11	1	1	1	1

(c) The state diagram is shown in Fig. 4.151.

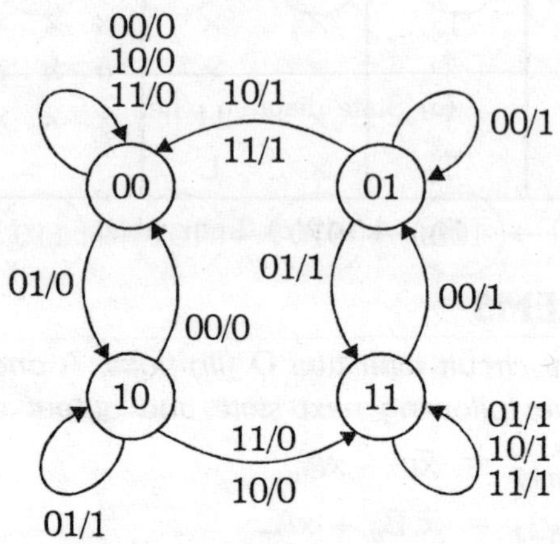

Fig. 4.151 State diagram.

Problem 2. *A sequential circuit has three D flip-flops, A, B and C, and one input, x. It is described by the following flip-flop input functions.*

$$D_A = (B\overline{C} + \overline{B}C)\,x + (BC + \overline{B}\,\overline{C})\,\overline{x}$$
$$D_B = A$$
$$D_C = B$$

(a) Derive the state table for the circuit.

(b) Draw the state diagram one for x = 0 and other for x = 1.

Solution. (*a*) The state table for the given problem is shown in Table 4.62.

Table 4.62. State table

Present state ABC	*Next state ABC*	
	x = 0	*x = 1*
000	100	000
001	000	100
010	001	101
011	101	001
100	110	010
101	010	110
110	011	111
111	111	011

(b) The state diagram is shown in Fig. 4.152.

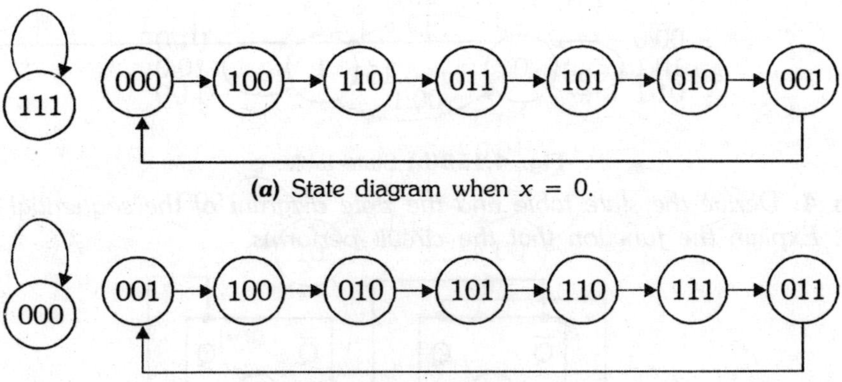

(b) State diagram when x = 1.

Fig. 4.152

Problem 3. *A sequential circuit has one flip-flop, Q ; two inputs, x and y; one output, S. It consists of a full-adder circuit connected to a D flip-flop, as shown in Fig. 4.153(a). Derive the state table and state diagram of the sequential circuit.*

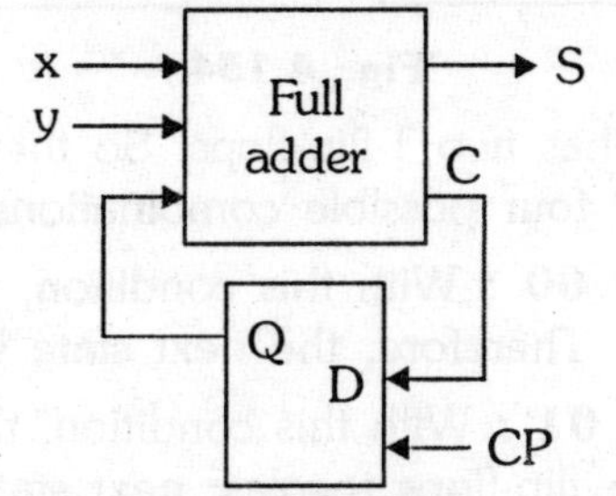

Fig. 4.153(a)

Solution. It has only one flip-flop; two inputs and one output. The state table for the above circuit is shown in Table 4.63.

Table 4.63. State table

Present state	*Inputs*		*Next state*	*Output*
Q	*x*	*y*	*D = C*	*S*
0	0	0	0	0
0	0	1	0	1
0	1	0	0	1
0	1	1	1	0
1	0	0	0	1
1	0	1	1	0
1	1	0	1	0
1	1	1	1	1

The state diagram is shown in Fig. 4.153(*b*).

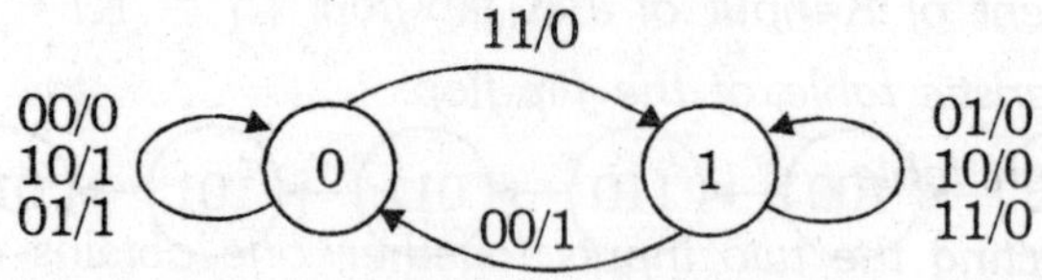

Fig. 4.153(*b*) State table.

Problem 4. *Derive the state table and the state diagram of the sequential circuit shown in Fig. 4.154. Explain the function that the circuit performs.*

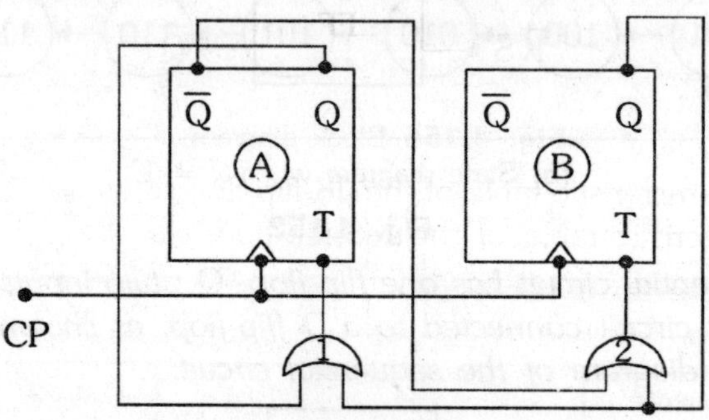

Fig. 4.154

Solution. The given circuit has two T-flip-flops. So they produce 4 rows and no inputs. Let us find the next states for all four possible combinations.

(*a*) **If present state AB = 00 :** With this condition, the input of OR gate 1 is logic 0 and OR gate 2 output is logic 1. Therefore, the next state is AB = 01.

(*b*) **If present state AB = 01 :** With this condition, the output of both OR gates 1 are at logic 1. So the output of both flip-flops toggles; next state = 10.

(*c*) **If present state AB = 10 :** With this condition, the output of OR gate 1 is logic 1 and OR gate 2 is logic 0. The next state is A = 0 B = 0.

(*d*) **If present state is AB = 11 :** With this condition, the output of both OR gates is 1, so output toggles, hence next state is AB = 00.

The state table is shown in Table 4.64. The state diagram is shown in Fig. 4.155.

Table 4.64. State table

Present state	*Next state*
AB	*AB*
00	01
01	10
10	00
11	00

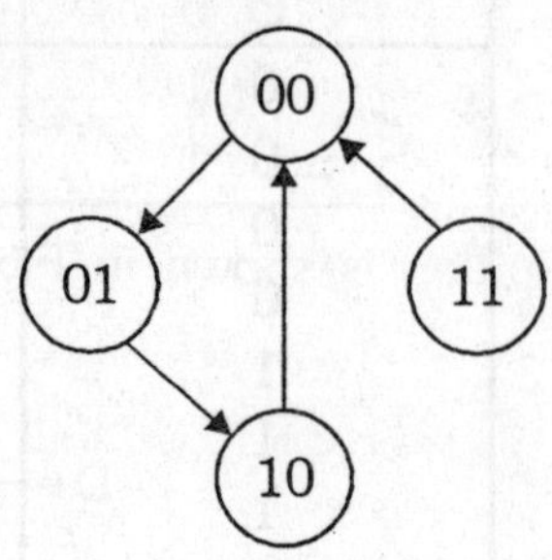

Fig. 4.155 State table.

Problem 5. *A 'MS' flip-flop has two inputs 'M and S'. Input 'M' behaves like a J and S behaves like the complement of K-input of J K flip-flop.* ($S = \overline{K}$).

(a) Tabulate the characteristic table of the flip-flop.

(b) Tabulate the excitation table.

(c) Show that by connecting the two inputs together, one obtains a D flip-flop.

Solution. (*a*) The block diagram of the MS-flip-flop is shown in Fig. 4.156.

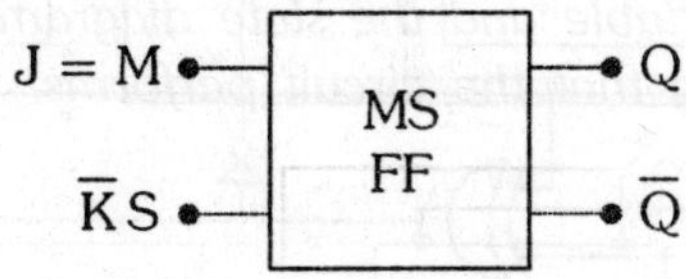

Fig. 4.156 Block diagram.

As we know that the characteristic table of the JK-flip-flop shown in Table 4.65(*a*) and from it we can derive the characteristic table of the MS-flip-flop, shown in Table 4.65(*b*).

Table 4.65(*a*) Characteristic table of JK-flip-flop

Inputs		*Next state*
J	*K*	*Q (t + 1)*
0	0	Q (*t*)
0	1	0
1	0	1
1	1	$\overline{Q}$ (*t*)

Table 4.65(*b*) Characterisitic table of MS-flip-flop

M = J	*S* = $\overline{K}$	*Q (t + 1)*
0	0	0
0	1	Q (*t*)
1	0	$\overline{Q}$ (*t*)
1	1	1

(*b*) The excitation table is shown in Table 4.66.

Table 4.66 Excitation table for MS-flip-flop

Present state	*Next state*	*Inputs values*
Q (t)	*Q (t + 1)*	*M S*
0	0	0 ×
0	1	1 ×
1	0	× 0
1	1	× 1

(*c*) The D-flip-flop is shown in Fig. 4.157.

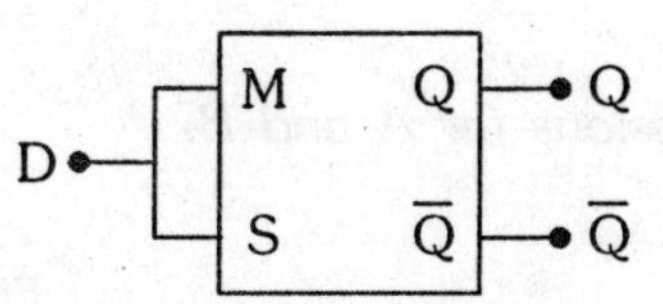

Fig. 4.157 D flip-flop.

Problem 6. *A sequential circuit has two JK flip-flops, one input x, and one output z. The logic diagram of the circuit is shown in Fig. 4.158. Derive the state table and state diagram of the circuit.*

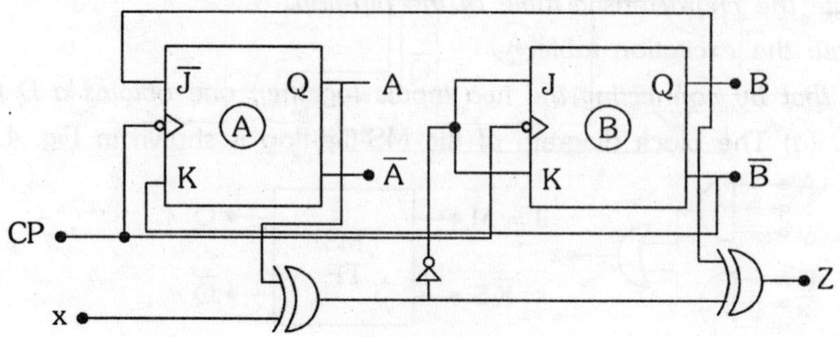

Fig. 4.158

Solution. The given circuit has two flip-flops, hence it has totally 4 states. By analysing the circuit, we can calculate the next states for each present state when $x = 0$ and $x = 1$. The equivalent state table is shown in Table 4.67. The state diagram is shown in Fig. 4.159.

Table 4.67. State table

Present state	*Input*	*Next state*	*Output*
A B	x	A B	z
0 0	0	0 1	0
0 0	1	0 0	1
0 0	0	1 0	1
0 1	1	1 1	0
1 0	0	0 0	1
1 0	1	0 1	0
1 1	0	1 1	0
1 1	1	1 0	1

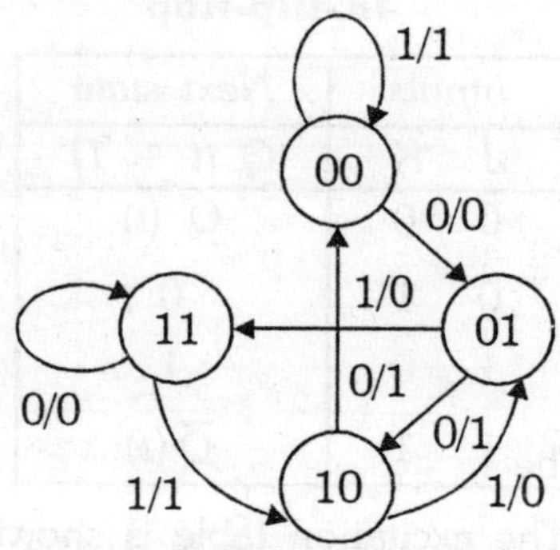

Fig. 4.159 State table.

Problem 7. *A sequential circuit has two JK-flip-flops. A and B ; two inputs, x and y ; and one output, z. The flip-flop input functions and circuit output functions are as follows :*

$$J_A = Bx + \overline{B}\overline{y} \qquad K_A = \overline{B}x\overline{y}$$

$$J_B = \overline{A}x \qquad K_B = A + x\overline{y}$$

$$z = Axy + B\overline{x}\,\overline{y}$$

(a) Draw the logic diagram of the circuit.

(b) Obtain the state table.

(c) Derive next state expressions for A and B.

Solution. (*a*) From the given flip-flop input equations, the logic diagram is shown in Fig. 4.160.

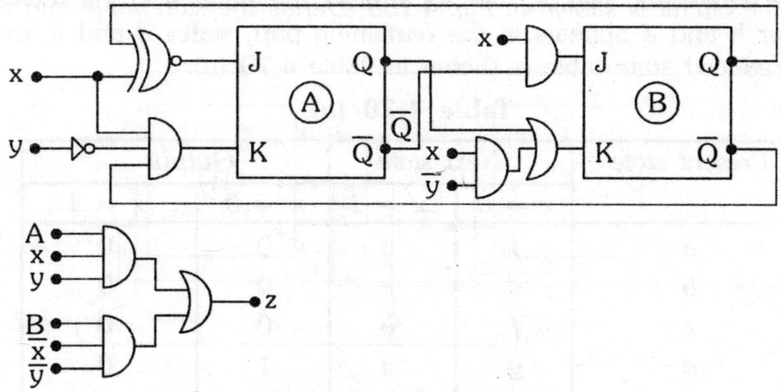

Fig. 4.160 Logic diagram.

(*b*) The state table of the circuit is shown in Table 4.68

Table 4.68

Present state	*Next states (AB)*				*Output (z)*			
AB	00	01	10	11	00	01	10	11
00	00	01	11	11	0	0	0	0
01	10	01	10	10	0	0	0	0
10	10	10	00	10	0	0	0	0
11	10	10	10	10	0	0	0	0

(*c*) The next state equations of A and B can be calculated by using K-map.

$$A\ (t + 1) = xB + \overline{x}A + yA + \overline{y}\overline{A}\,\overline{B}$$

$$B\ (t + 1) = x\overline{A}\,\overline{B} + (\overline{x} + \overline{y})\,\overline{A}B.$$

Problem 8. *Reduce the number of states in the following state table (Table 4.69) and tabulate the reduced state table.*

Table 4.69

Present state	*Next state*		*Output*	
	x = 0	*x = 1*	*x = 0*	*x = 1*
a	*f*	*b*	0	0
b	*d*	*c*	0	0
c	*f*	*e*	0	0
d	*g*	*a*	1	0
e	*d*	*c*	0	0
f	*f*	*b*	1	1
g	*g*	*h*	0	1
h	*g*	*a*	1	0

Solution. By observing the state table, it is clear that states *d* and *h* are equivalent and states *b* and *e* are equivalent. Hence, we can remove the states *h* and *e* rows from the state table and wherever *h* and *e* appears in the remaining part, states *d* and *b* are to be written respectively. The reduced state table is shown in Table 4.70 (*a*).

Table 4.70 (*a*)

Present state	*Next state*		*Output*	
	x = 0	*x = 1*	*x = 0*	*x = 1*
a	*f*	b	0	0
b	*d*	c	0	0
c	*f*	b	0	0
d	*g*	*a*	1	0
f	*f*	b	1	1
g	*g*	*d*	0	1

From the state Table 4.70(*a*), the states *a* and *c* are equivalent. Hence state '*c*' can be removed and the final reduced state table is shown in Table 4.70 (*b*).

Table 4.70 (*b*) Final reduced state table

Present state	*Next state*		*Output*	
	x = 0	*x = 1*	*x = 0*	*x = 1*
a	*f*	b	0	0
b	*d*	*a*	0	0
d	*g*	*a*	1	0
f	*f*	b	1	1
g	*g*	*d*	0	1

Problem 9. *Analyze the circuit of Fig. 4.161 and prove that it is equivalent to a T flip-flop.*

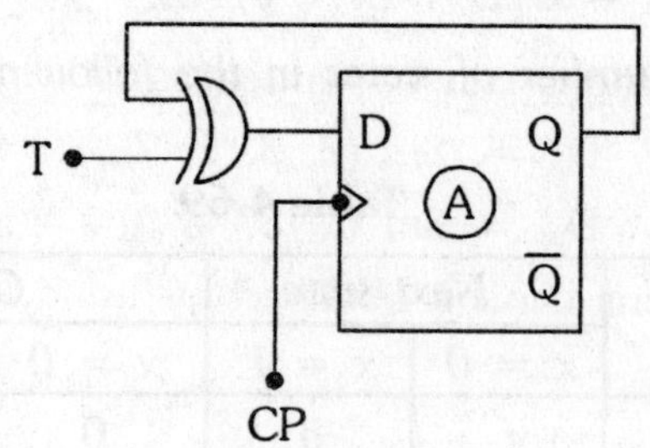

Fig. 4.161

Solution. The circuit has one input T and one D flip-flop D.

(*a*) **When Q = 0.** (*i*) If T = 0, output of EX-OR gate is 0 and next state is '0'

(*ii*) If T = 1, output of EX-OR is 1 and next state is 1.

(*b*) **When Q = 1.** (*i*) If T = 0, the output of EX-OR gate is 1 and next state is 1.

(*ii*) If T = 1, the output of EX-OR gate is 0 and next state is 0.

The state table is shown in Table 4.71.

Table 4.71

Present state	*Input*	*Next state*
A (t)	*T*	*A (t + 1)*
0	0	0
0	1	1
1	0	1
1	1	0

From the table, it is clear that it is a characteristic table for T flip-flop. Hence, the given circuit acts as a T flip-flop.

Problem 10. *Design a sequential circuit with two D flip-flops, A and B, and one input x. When x = 0, state of the circuit is same. When x = 1, the circuit goes through the state transitions from 00 to 01 to 11 to 10 back to 00, and repeats.*

Solution. From the given problem, it is clear that the circuit has 4-states and one input. The state diagram is shown in Fig. 4.162.

The equivalent state table is shown in Table 4.72.

Table 4.72

Present state	*Next state*	
	x = 0	*x = 1*
00	00	01
01	01	11
10	10	00
11	11	10

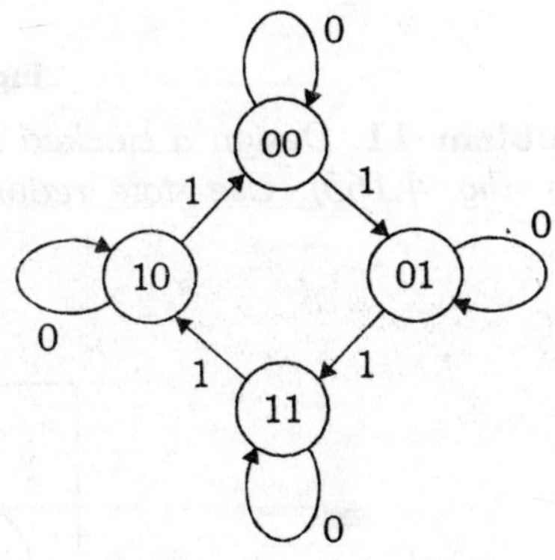

Fig. 4.162 State table.

The flip-flop input functions can be written as

$$D_A = \Sigma m\ (2, 3, 5, 7)$$

$$D_B = \Sigma m\ (1, 3, 4, 5)$$

The Boolean expressions can be obtained by using K-maps as shown in Fig. 4.163.

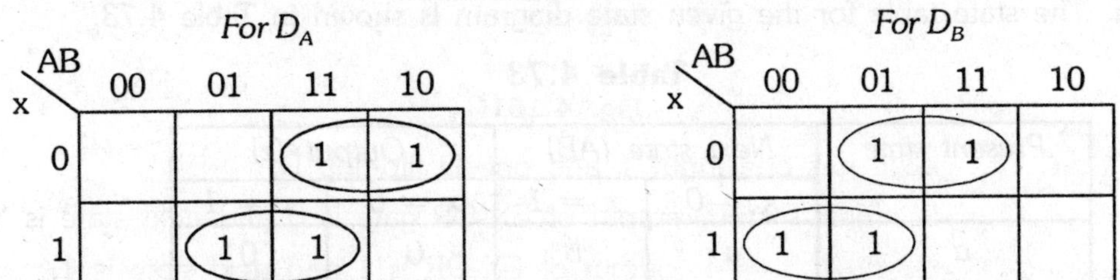

$D_A = \bar{x}A + xB$ $\qquad$ $D_B = \bar{x}A + \bar{x}B$

Fig. 4.163

The logic diagram is shown in Fig. 4.164.

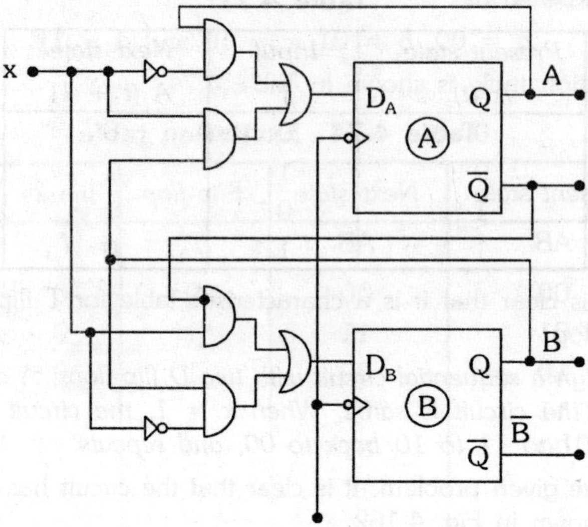

Fig. 4.164 Logic diagram.

Problem 11. *Design a clocked sequential circuit using T flip-flops for the following state diagram (Fig. 4.165). Use state reduction if possible.*

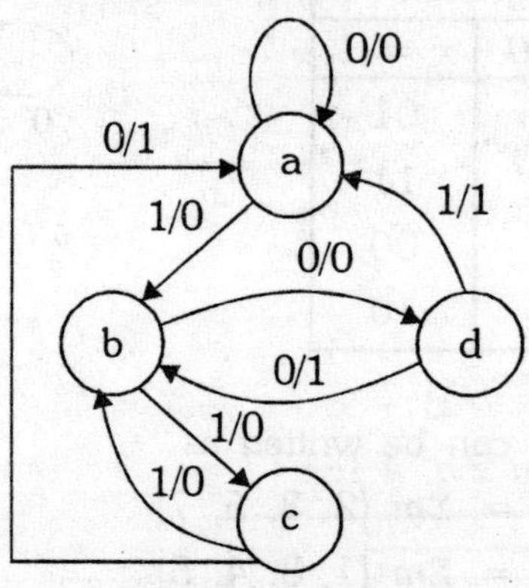

Fig. 4.165

Solution. The state table for the given state diagram is shown in Table 4.73.

Table 4.73

Present state	*Next state (AB)*		*Output (z)*	
	x = 0	*x = 1*	*x = 0*	*x = 1*
a	*a*	*b*	0	0
b	*d*	*c*	0	0
c	*a*	*b*	1	0
d	*b*	*a*	1	1

From the Table 4.73, it is clear that no two states are equivalent. The binary assignment for each state may be assigned as

$$a = 00,\ b = 01,\ c = 10 \text{ and } d = 11.$$

With this, the excitation table is shown in Table 4.74.

Table 4.74 Excitation table

Input	*Present state*	*Next state*	*Flip-flop*	*Inputs*	*Output*
x	*AB*	*AB*	T_A	T_B	z
0	00	00	0	0	0
0	01	11	1	0	0
0	10	00	1	0	1
0	11	01	1	0	1
1	00	01	0	1	0
1	01	10	1	1	0
1	10	01	1	1	0
1	11	00	1	1	1

The flip-flop inputs are

$$T_A = \Sigma m\ (1, 2, 3, 5, 6, 7)$$
$$T_B = \Sigma m\ (4, 5, 6, 7)$$

The circuit output function is

$$z = \Sigma m\ (2, 3, 7).$$

$$\therefore \quad T_A = A + B$$
$$\therefore \quad T_B = x$$
$$z = AB + \bar{x}A = A(\bar{x} + B)$$

The logic diagram is shown in Fig. 4.166.

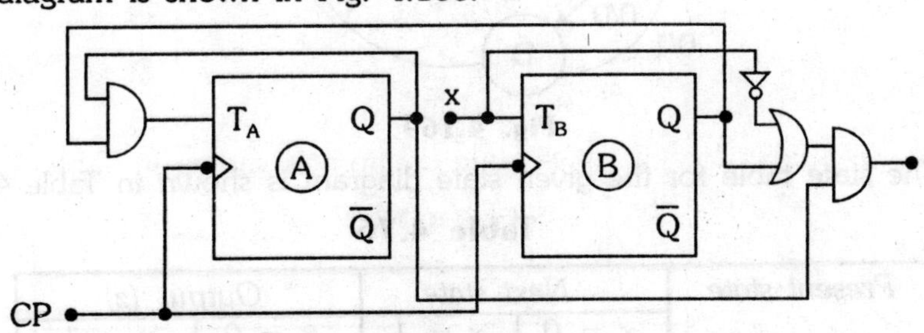

Fig. 4.166

Problem 12. *Realize the given T flip-flop in such a way that it acts as a D flip-flop.*

Solution. Here the desired flip-flop is a D-flip-flop and the chosen flip-flop is T flip-flop. The block diagram is shown in Fig. 4.167.

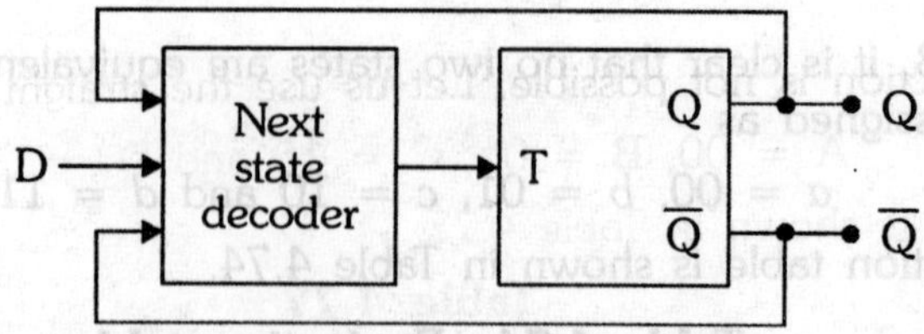

Fig. 4.167 Block diagram.

The excitation table is shown in Table 4.75. The logic diagram is shown in Fig. 4.168.

Table 4.75

Q_n	D	Q_{n+1}	*Excitation input T*
0	0	0	0
0	1	1	1
1	0	0	1
1	1	1	0

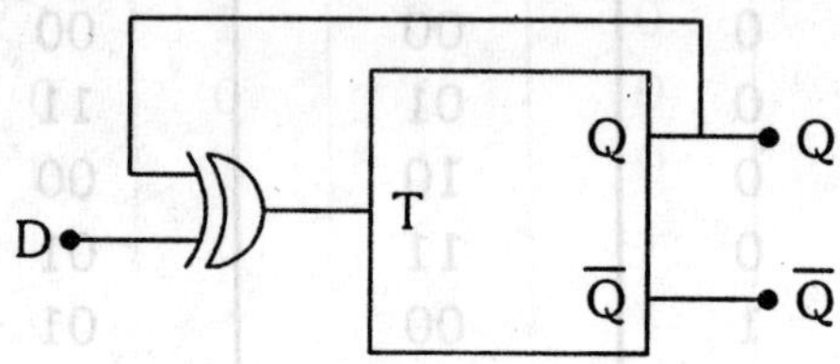

Fig. 4.168 Logic diagram.

$$T = \overline{Q}_n D + \overline{D} Q_n$$

$$T = Q_n \oplus D.$$

Problem 13. *Design a clocked sequential machine using T flip-flops for the following state diagram (Fig. 4.169). Use the state reduction procedure and use the straight binary state assignment.*

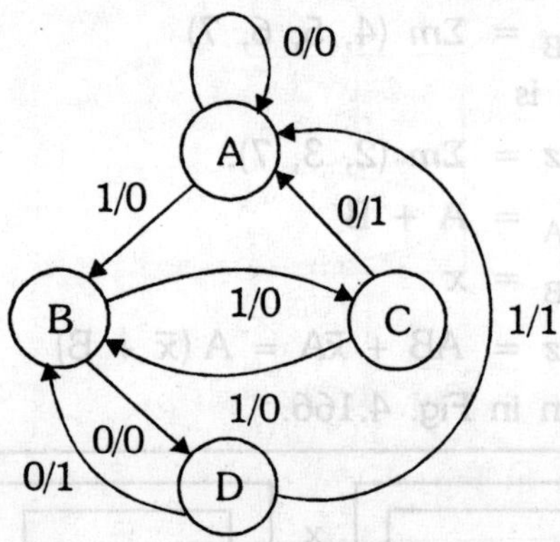

Fig. 4.169

Solution. The state table for the given state diagram is shown in Table 4.76.

Table 4.76

Present state	*Next state*		*Output (z)*	
	x = 0	*x = 1*	*x = 0*	*x = 1*
A	A	B	0	0
B	D	C	0	0
C	A	B	1	0
D	B	A	1	1

Here the state reduction is not possible. Let us use the straight binary assignment as

A = 00, B = 01, C = 10 and D = 11.

The excitation table is shown in Table 4.77.

Table 4.77

Input	*Present state*		*Next state*		*Flip-flop inputs*		*Output*
x	Q_A	Q_B	Q_{A+1}	Q_{B+1}	T_A	T_B	z
0	0	0	0	0	0	0	0
0	0	1	1	1	1	0	0
0	1	0	0	0	1	0	1
0	1	1	0	1	1	0	1
1	0	0	0	1	0	1	0
1	0	1	1	0	1	1	0
1	1	0	0	1	1	1	0
1	1	1	0	0	1	1	1

The flip-flop input functions and circuit output functions can be derived by using K-map as shown in Fig. 4.170.

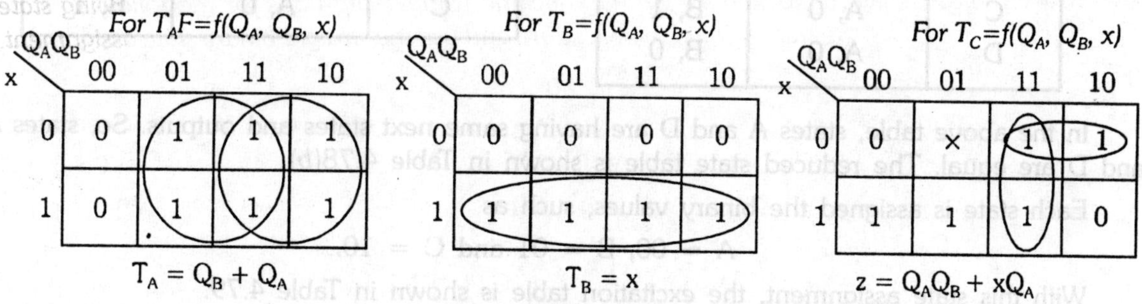

Fig. 4.170 K-maps.

The logic diagram is shown in Fig. 4.171.

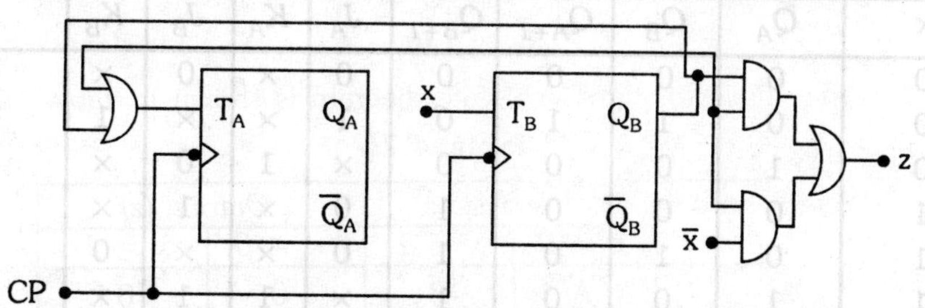

Fig. 4.171 Logic diagram.

Problem 14. *Design a clocked sequential machine using JK-flip-flop for the state diagram shown in Fig. 4.171. Use state reduction if possible.*

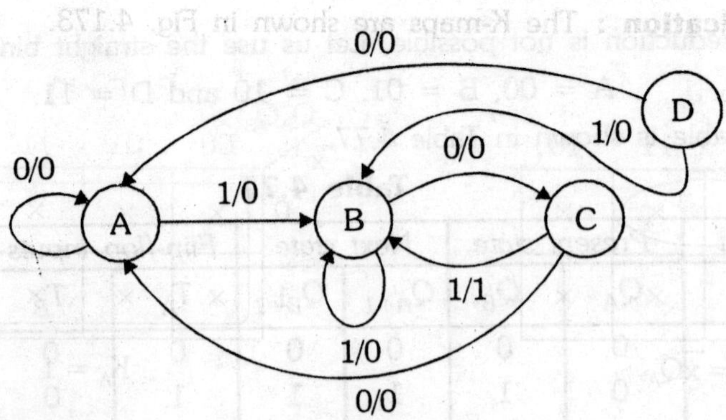

Fig. 4.172

Solution. The state table is shown in Table 4.78(*a*).

Table 4.78 (*a*)

Present state	*Next state* x = 0	*Output(z)* x = 1
A	A, 0	B, 0
B	C, 0	B, 0
C	A, 0	B, 1
D	A, 0	B, 0

Table 4.78 (*b*)

Present state	*Next state* x = 0	*Output* x = 1
A	A, 0	B, 0
B	C, 0	B, 0
C	A, 0	B, 1

In the above table, states A and D are having same next states and outputs. So, states A and D are equal. The reduced state table is shown in Table 4.78(*b*).

Each state is assigned the binary values, such as

A = 00, B = 01 and C = 10.

With this state assignment, the excitation table is shown in Table 4.79.

Table 4.79

Input	*Present state*		*Next state*		*Flip-flop inputs*				*Outputs*
x	Q_A	Q_B	Q_{A+1}	Q_{B+1}	J_A	K_A	J_B	K_B	z
0	0	0	0	0	0	×	0	×	0
0	0	1	1	0	1	×	×	1	0
0	1	0	0	0	×	1	0	×	0
1	0	0	0	1	0	×	1	×	0
1	0	1	0	1	0	×	×	0	0
1	1	0	0	1	×	1	1	×	1

K-map simplification : The K-maps are shown in Fig. 4.173.

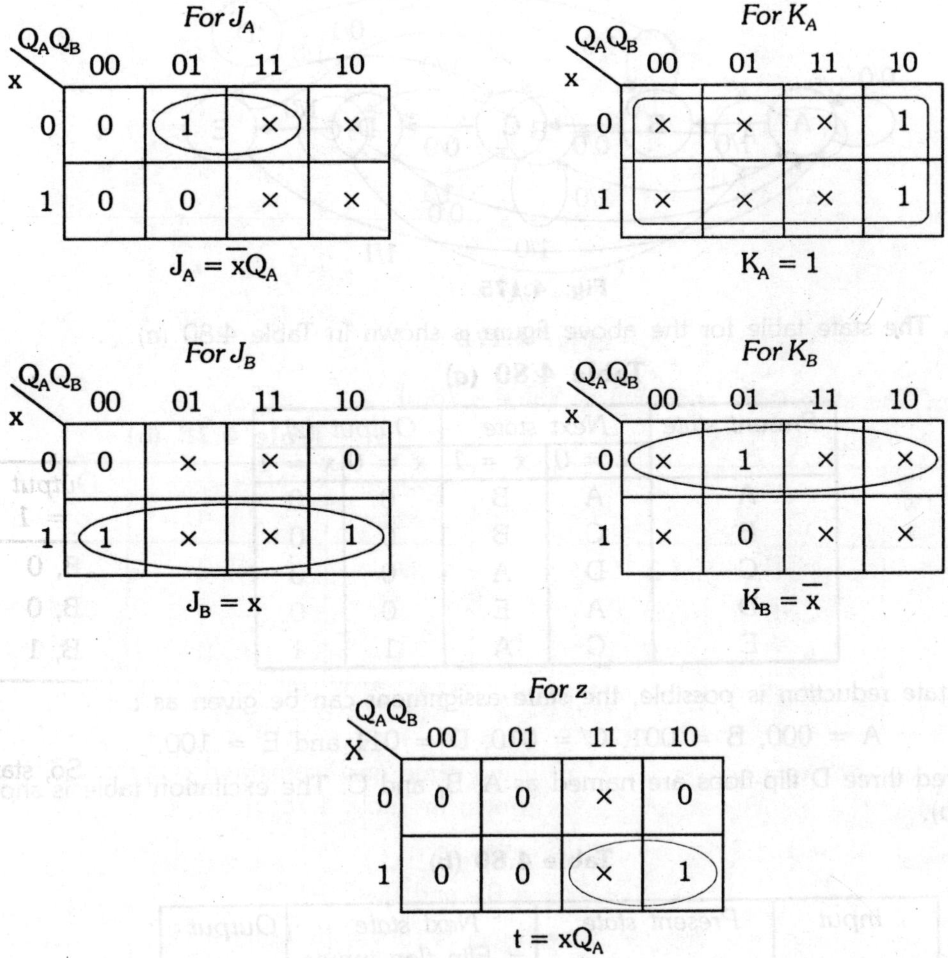

Fig. 4.173

Logic diagram : The logic diagram is shown in Fig. 4.174.

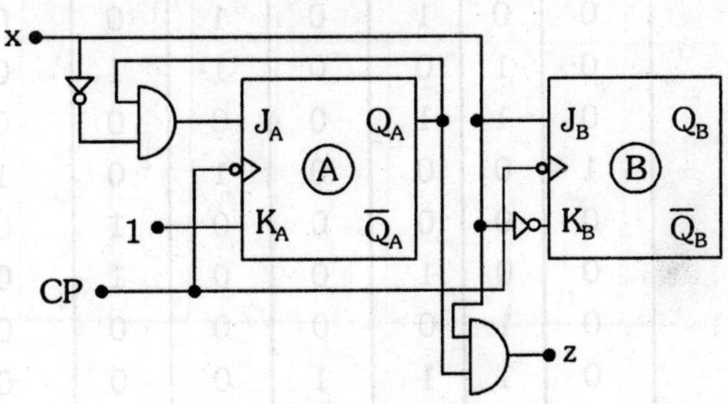

Fig. 4.174

Problem 15. *Design the sequential circuit using D flip-flops for the state diagram given in Fig. 4.175.*

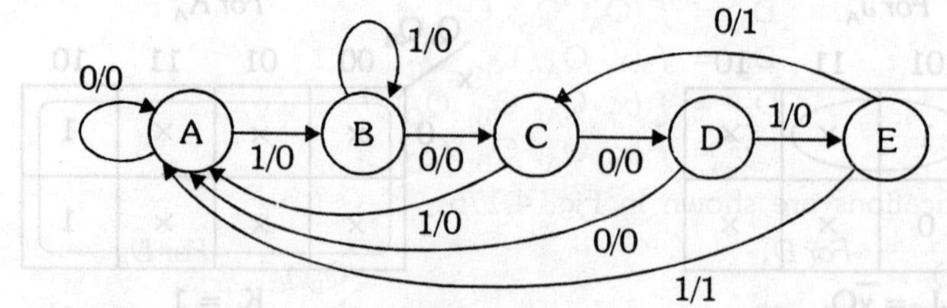

Fig. 4.175

Solution. The state table for the above figure is shown in Table 4.80 (*a*)

Table 4.80 (*a*)

Present state	*Next state*		*Output (z)*	
	x = 0	*x = 1*	*x = 0*	*x = 1*
A	A	B	0	0
B	C	B	0	0
C	D	A	0	0
D	A	E	0	0
E	C	A	1	1

Here no state reduction is possible, the state assignment can be given as :

A = 000, B = 001, C = 010, D = 011 and E = 100.

The required three D flip-flops are named as A, B, and C. The excitation table is shown in Table 4.80(*b*).

Table 4.80 (*b*)

Input	*Present state*			*Next state = Flip flop inputs*			*Output*
	Q_A	Q_B	Q_C	Q_{A+1}	Q_{B+1}	Q_{C+1}	z
0	0	0	0	0	0	0	0
0	0	0	1	0	1	0	0
0	0	1	0	0	1	1	0
0	0	1	1	0	0	0	0
0	1	0	0	0	1	0	1
1	0	0	0	0	0	1	0
1	0	0	1	0	0	1	0
1	0	1	0	0	0	0	0
1	0	1	1	1	0	0	0
1	1	0	0	0	0	0	1

The flip-flop input equations and output of the circuit are

$$D_A = f(Q_A, Q_B, Q_C, x)$$
$$D_B = f(x, Q_A, Q_B, Q_C)$$
$$D_C = f(x, Q_A, Q_B, Q_C)$$
$$y = f(x, Q_A, Q_B, Q_C)$$

K-map simplifications are shown in Fig. 4.176.

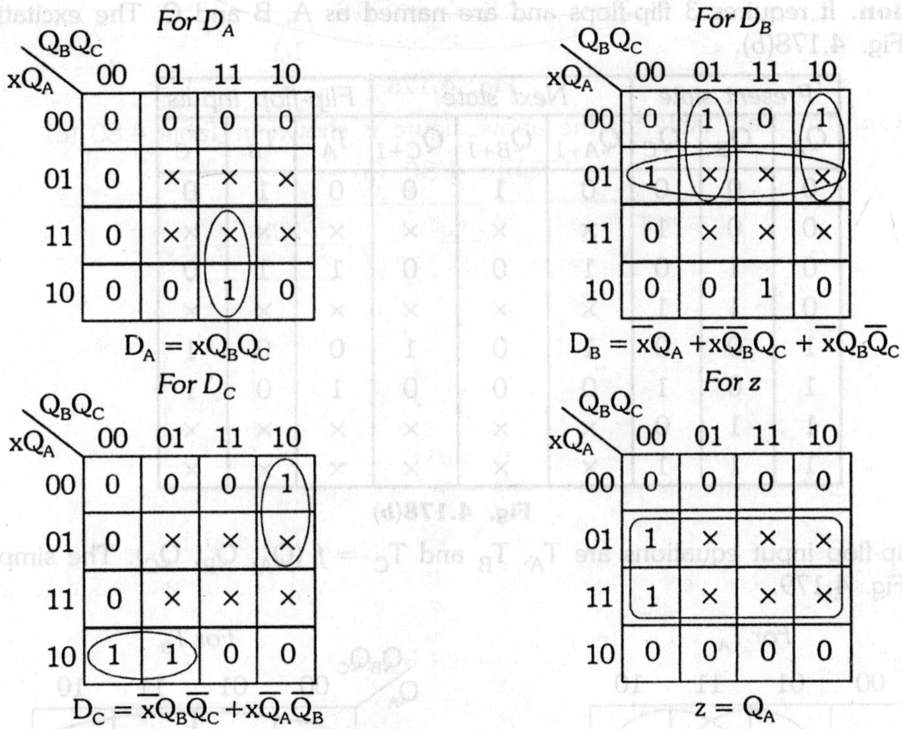

Fig. 4.176

The logic diagram is shown in Fig. 4.177.

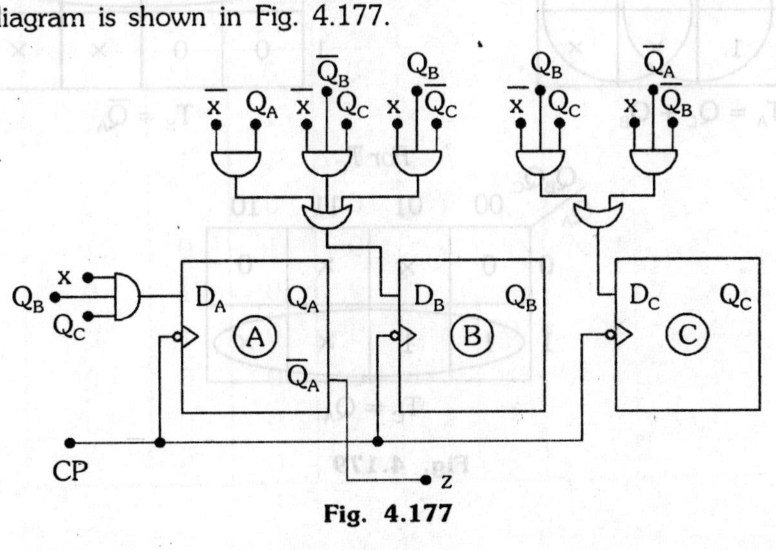

Fig. 4.177

Problem 16. *Design a sequence generator using T-flip-flops for the given sequence. Check for lock-out condition.*

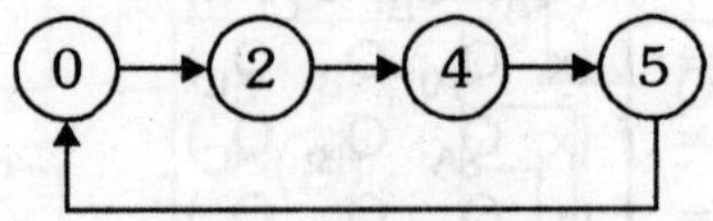

Fig. 4.178(*a*)

Solution. It requires 3 flip-flops and are named as A, B and C. The excitation table is shown in Fig. 4.178(*b*).

Present state			*Next state*			*Flip-flop inputs*		
Q_A	Q_B	Q_C	Q_{A+1}	Q_{B+1}	Q_{C+1}	T_A	T_B	T_C
0	0	0	0	1	0	0	1	0
0	0	1	×	×	×	×	×	×
0	1	0	1	0	0	1	1	0
0	1	1	×	×	×	×	×	×
1	0	0	1	0	1	0	0	1
1	0	1	0	0	0	1	0	1
1	1	0	×	×	×	×	×	×
1	1	1	×	×	×	×	×	×

Fig. 4.178(*b*)

The flip-flop input equations are T_A, T_B and $T_C = f(Q_A, Q_B, Q_C)$. The simplification is shown in Fig. 4.179.

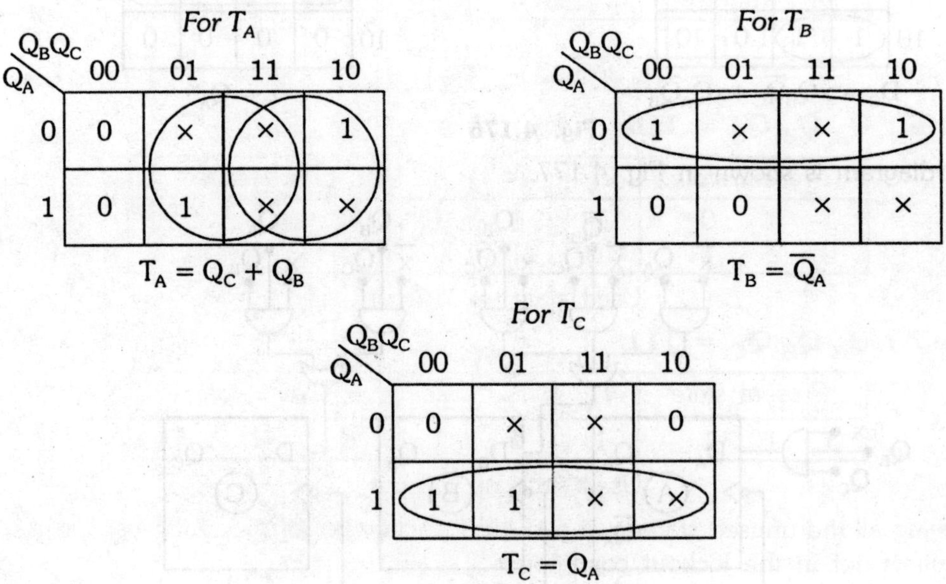

Fig. 4.179

The logic diagram is shown in Fig. 4.180.

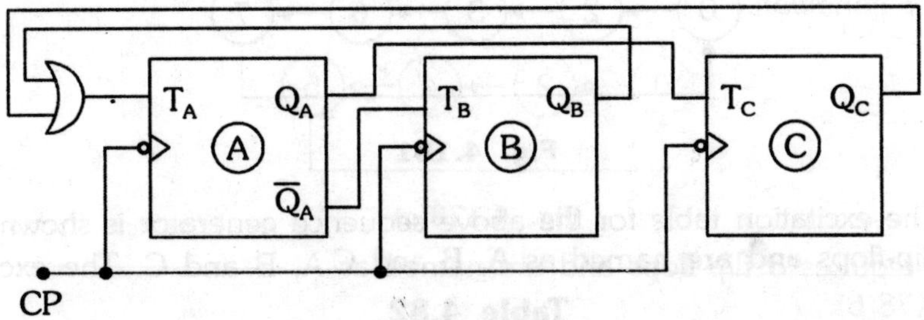

Fig. 4.180

Checking for lock out condition : The lockout conditions is shown in Table 4.81.

Table 4.81

For state 1 : $Q_A\ Q_B\ Q_C$ = 001

Present state			Next state			T_A	T_B	T_C
Q_A	Q_B	Q_C	Q_{A+1}	Q_{B+1}	Q_{C+1}	(C+B)	$\overline{A}$	A
0	0	1	1	1	1	1	1	0
1	1	1	0	1	0	1	0	1

Here 010 = state 2 (used state) so no lockout.

For state 3 : $Q_A\ Q_B\ Q_C$ = 011

Present state			T_A	T_B	T_C	Next state		
Q_A	Q_B	Q_C	(Q_C+Q_B)	$\overline{Q}_A$	Q_A	Q_{A+1}	Q_{B+1}	Q_{C+1}
0	1	1	1	1	0	1	0	1

Used state so no lockout

For state 6 : $Q_A\ Q_B\ Q_C$ = 110

Present state			T_A	T_B	T_C	Next state		
1	1	0	1	0	1	0	1	1
0	1	1	1	1	0	1	0	1

Used state no lockout

For state 7 : $Q_A\ Q_B\ Q_C$ = 111

Present state			T_A	T_B	T_C	Next state		
1	1	1	1	0	1	0	1	0

Used state no lockout

After verifying all the unused states, we can say that they go to one of the used states and hence the circuit is not in the lockout condition.

Problem 17. *Design a sequence generator to generate the following sequence using D-flip-flop and check for lockout condition.*

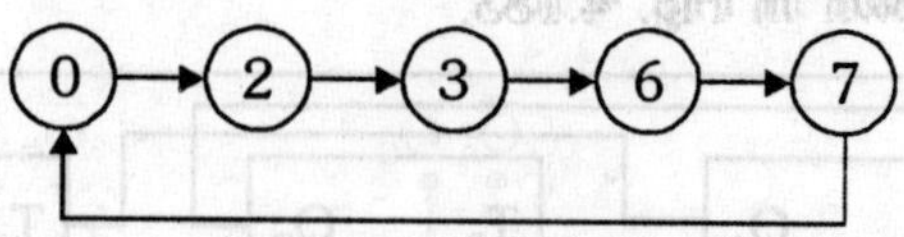

Fig. 4.181

Solution. The excitation table for the above sequence generator is shown in Table 4.82. We need three flip-flops and are named as A, B and C.

Table 4.82

Present state			*Next states flip-flop Input*		
Q_A	Q_B	Q_C	Q_{A+1}	Q_{B+1}	Q_{C+1}
0	0	0	0	1	0
0	0	1	×	×	1
0	1	0	0	1	1
0	1	1	1	1	0
1	0	0	×	×	×
1	0	1	×	×	×
1	1	0	1	1	1
1	1	1	0	0	0

The flip-flop input functions are given by using K-map simplification as shown in Fig. 4.182.

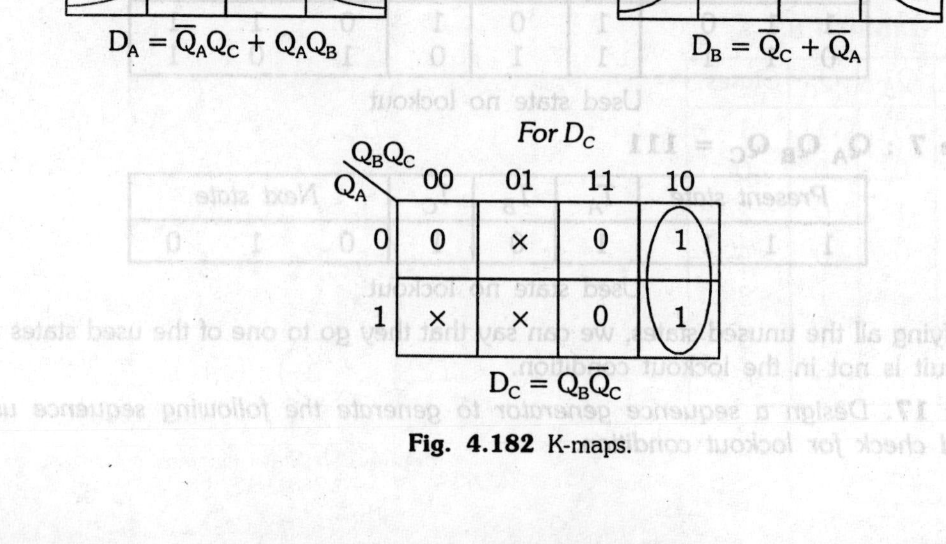

Fig. 4.182 K-maps.

The logic diagram is shown in Fig. 4.183.

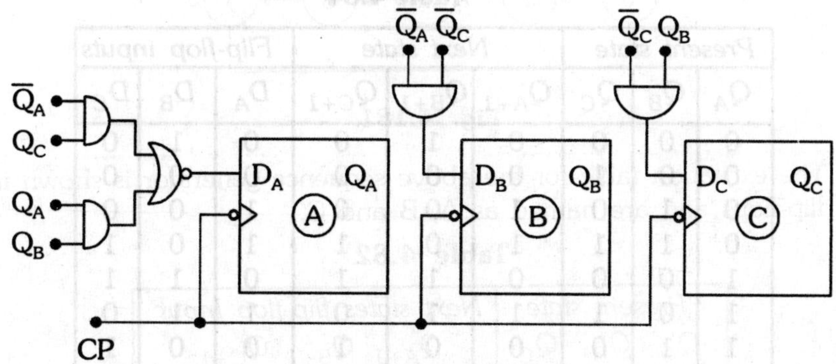

Fig. 4.183 Logic diagram.

Problem 18. *Design a pulse train generator for the following waveform.*

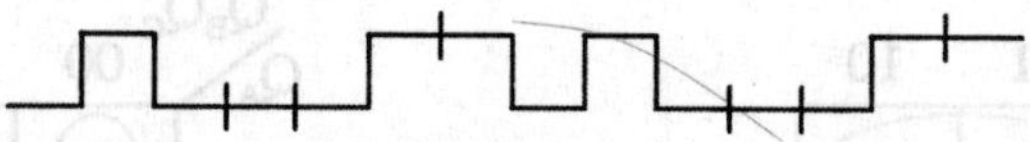

Fig. 4.184(a)

Solution. The above figure (pulse train) can be represented with binary states as shown in Fig. 4.184(*b*).

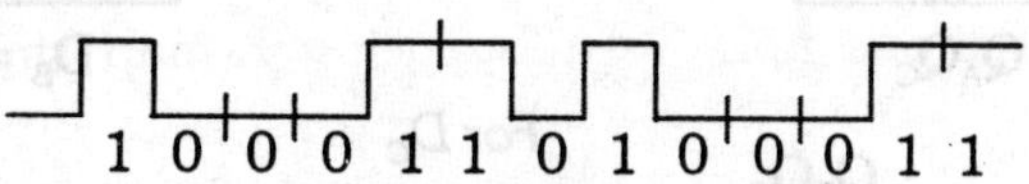

Fig. 4.184(b)

The pulse train has a sequence 1000110 and this sequence is repeated. It has 7-bits. Hence, it requires 7 states. Therefore, we need at least 3 flip-flops. We can name then as A, B and C, we will use the output of flip-flop C as a pulse train and assign unique states corresponding to each bit in the sequence as shown in Table 4.83. The state diagram is shown in Fig. 4.185.

Table 4.83

Q_A	Q_B	Q_C	*States*
0	0	1	1
0	0	0	0
0	1	0	2
1	0	0	4
0	1	1	3
1	0	1	5
1	1	0	6

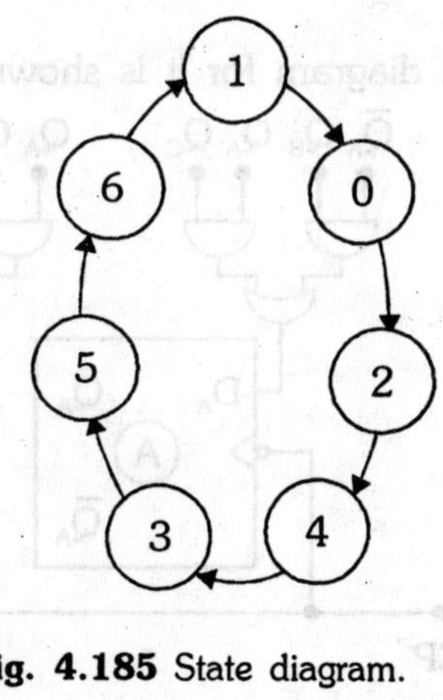

Fig. 4.185 State diagram.

Let us consider the three flip-flops are of D-type and the excitation table is shown in Table 4.84.

Table 4.84

Present state			*Next state*			*Flip-flop inputs*		
Q_A	Q_B	Q_C	Q_{A+1}	Q_{B+1}	Q_{C+1}	D_A	D_B	D_C
0	0	0	0	1	0	0	1	0
0	0	1	0	0	0	0	0	0
0	1	0	1	0	0	1	0	0
0	1	1	1	0	1	1	0	1
1	0	0	0	1	1	0	1	1
1	0	1	1	1	0	1	1	0
1	1	0	0	0	1	0	0	1
1	1	1	×	×	×	×	×	×

K-map simplification : The K-maps are shown in Fig. 4.186.

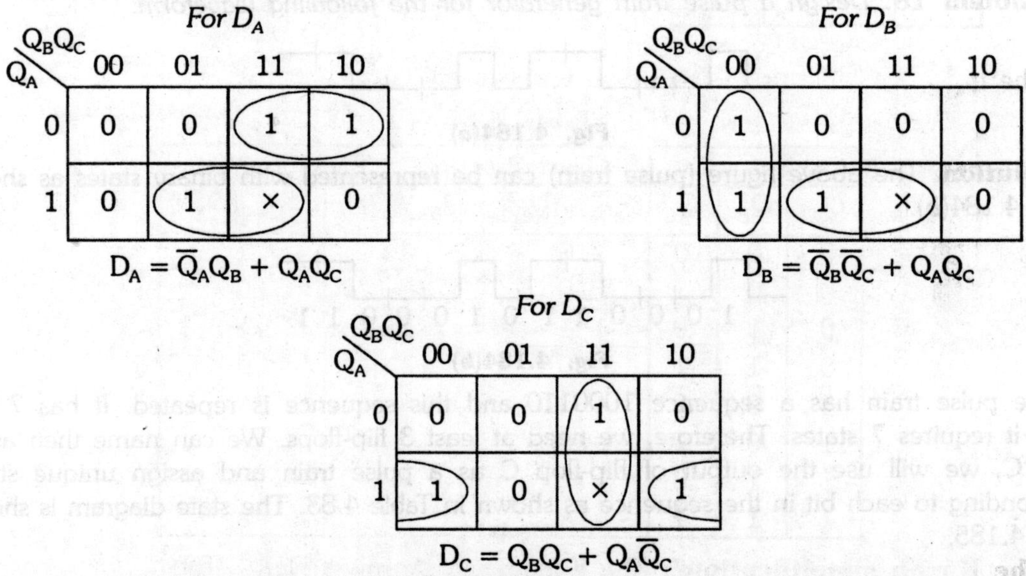

$D_A = \overline{Q}_A Q_B + Q_A Q_C$

$D_B = \overline{Q}_B \overline{Q}_C + Q_A Q_C$

$D_C = Q_B Q_C + Q_A \overline{Q}_C$

Fig. 4.186 K-maps.

The logic diagram for it is shown in Fig. 4.187.

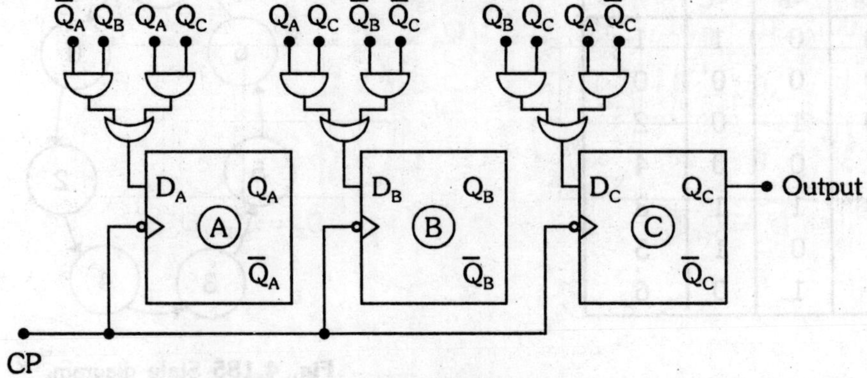

Fig. 4.187

Problem 19. *The following pulse train "10101" is to be generated. Draw the logic diagram of the counter.*

Solution. The given sequence is '10101'. It is a five bit sequence. Therefore, we need at least three flip-flops, and assume that they are named as A, B and C. We will use the output of flip-flop C as a pulse train and assign unique states corresponding to each bit in the sequence. The state assignment table is shown in Table 4.85. The state diagram is shown in Fig. 4.188.

Table 4.85

Q_A	Q_B	Q_C	*States*
0	0	1	1
0	0	0	0
0	1	1	3
0	1	0	2
1	0	1	5

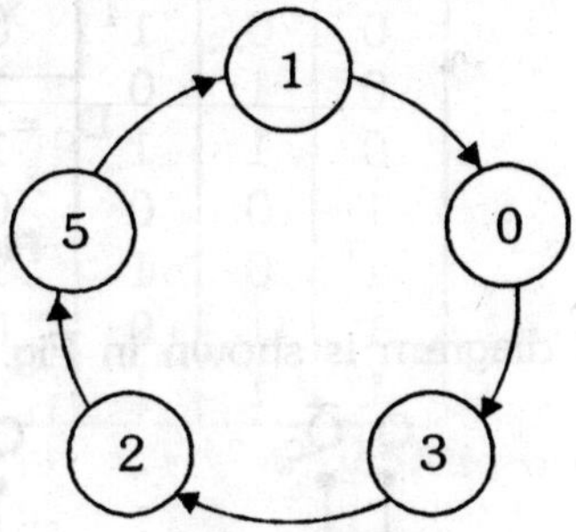

Fig. 4.188

The excitation table using D flip-flop is shown in Table 4.86.

Table 4.86

Present state			*Next state (or) Flip-flop inputs*		
Q_A	Q_B	Q_C	$Q_{A+1}(D_A)$	$Q_{B+1}(D_B)$	$Q_{C+1}(D_C)$
0	0	0	0	1	1
0	0	1	0	0	0
0	1	0	1	0	1
0	1	1	0	1	0
1	0	0	×	×	×
1	0	1	0	0	1
1	1	0	×	×	×
1	1	1	×	×	×

The K-map simplification : The K-maps are shown in Fig. 4.189.

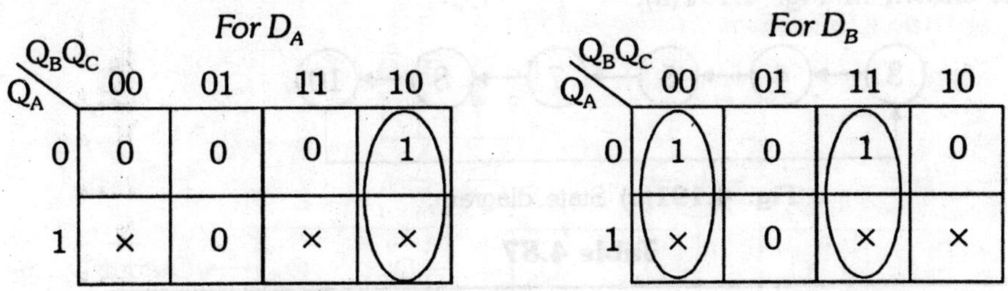

$D_A = Q_B\overline{Q}_C$ $\qquad$ $D_B = Q_B \odot Q_C$

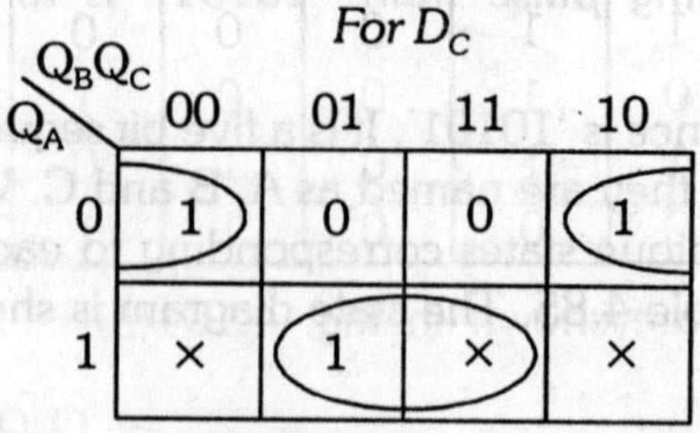

$$D_C = \overline{Q}_A\overline{Q}_C + Q_AQ_C = Q_A \odot Q_C$$

Fig. 4.189 K-maps.

The logic diagram is shown in Fig. 4.190.

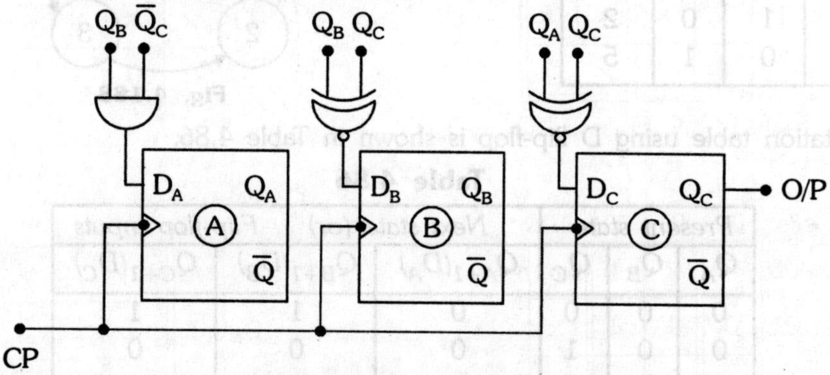

Fig. 4.190 Logic diagram.

Problem 20. *Design a non-sequential ripple counter which will go through the states 3, 4, 5, 7, 8, 9, 10, 3, 4, 5, ...*

Draw the logic diagram.

Solution. The given sequence is 3, 4, 5, 7, 8, 10 and it is repeated. The maximum count in this is '10'. Hence, we need 4 flip-flops of T type and named as A, B, C and D. The D flip-flop output is taken as a required sequence output. The excitation table is shown in Table 4.87. The state diagram is shown in Fig. 4.191(*a*).

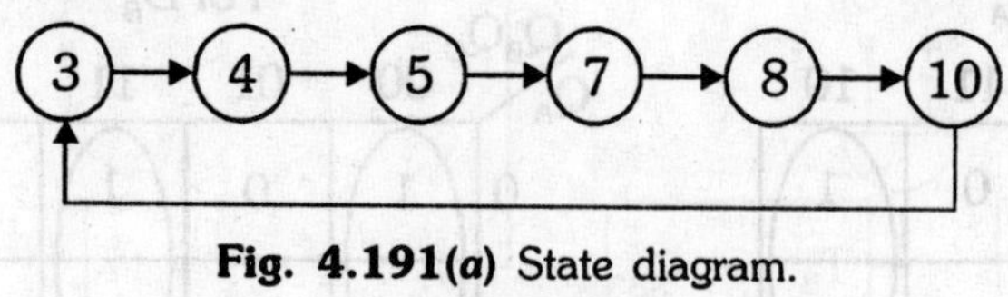

Fig. 4.191(*a*) State diagram.

Table 4.87

Present state				*Next state*				*Flip-flop inputs*			
Q_A	Q_B	Q_C	Q_D	Q_{A+1}	Q_{B+1}	Q_{C+1}	Q_{D+1}	T_A	T_B	T_C	T_D
0	0	1	1	0	1	0	0	0	1	1	1
0	1	0	0	0	1	0	1	0	0	0	1

0	1	0	1	0	1	1	1	0	0	1	0
0	1	1	1	1	0	0	0	1	1	1	1
1	0	0	0	1	0	0	1	0	0	0	1
1	0	0	1	1	0	1	0	0	0	1	1
1	0	1	0	0	0	1	1	1	0	0	1

The K-map simplifications are shown in Fig. 4.191(*b*).

For T_A

Q_AQ_B \ Q_CQ_D	00	01	11	10
00	×	×	0	×
01	0	0	1	×
11	×	×	×	×
10	0	0	×	1

$T_A = Q_BQ_C + Q_AQ_C = Q_C(Q_A + Q_B)$

For T_B

Q_AQ_B \ Q_CQ_D	00	01	11	10
00	×	×	1	×
01	0	0	1	×
11	×	×	×	×
10	0	0	×	0

$T_B = Q_CQ_D$

For T_C

Q_AQ_B \ Q_CQ_D	00	01	11	10
00	×	×	1	×
01	0	1	1	×
11	×	×	×	×
10	0	1	×	0

$T_C = Q_D$

For T_D

Q_AQ_B \ Q_CQ_D	00	01	11	10
00	×	×	1	×
01	1	0	1	×
11	×	×	×	×
10	1	1	×	1

$T_D = \overline{Q}_D + Q_C + \overline{Q}_B$

Fig. 4.191(*b*) K-maps.

The circuit diagram is shown in Fig. 4.192.

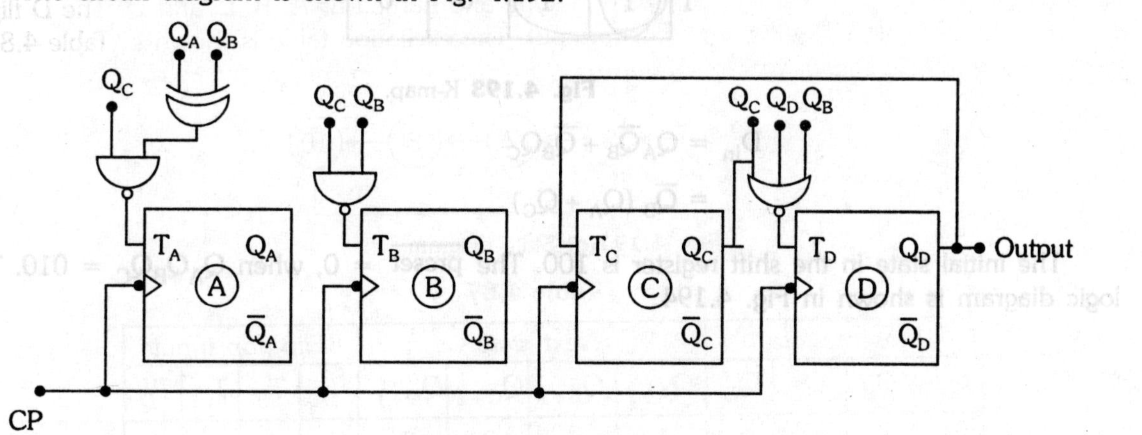

Fig. 4.192 Logic diagram.

Problem 21. *Design a pulse train generator using shift register for the following pulse train .... 111010......*

Solution. The number of flip-flops (n) required is given by

$$N \leq 2^n - 1$$

where $N = 6 \quad \therefore \quad n = 3.$

The sequence generation table is shown in Table 4.88. Here three flip-flops are A, B and C. The flip-flop A output is taken as sequence.

Table 4.88

CP	*Flip-flop outputs* Q_A	Q_B	Q_C	D_{in}	$\overline{Preset}$	*States*
1	1	0	0	1	1	4
2	1	0	1	1	1	5
3	1	1	0	0	1	6
4	0	0	0	1	1	0
5	1	1	1	0	1	7
6	0	0	1	0	1	1
1	0	1	0	×	0	2
1	1	0	0	1	1	4

The K-map for D_{in} simplification is shown in Fig. 4.193.

$Q_A \backslash Q_BQ_C$	00	01	11	10
0	1	0	×	×
1	1	1	0	0

Fig. 4.193 K-map.

$$D_{in} = Q_A\overline{Q}_B + \overline{Q}_BQ_C$$

$$= \overline{Q}_B\,(Q_A + Q_C)$$

The initial state in the shift register is 100. The $\overline{\text{preset}} = 0$, when $Q_AQ_BQ_C = 010$. The logic diagram is shown in Fig. 4.194.

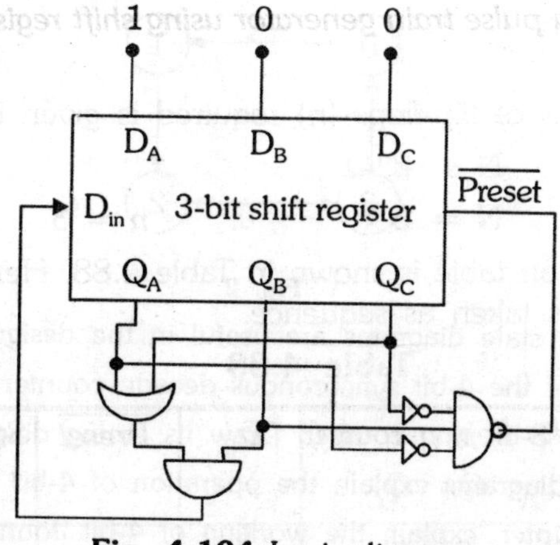

Fig. 4.194 Logic diagram.

4.21. REVIEW QUESTIONS

1. Draw and explain the working of the following flip-flops
   (*a*) Clocked T flip-flop (*b*) JK (*c*) Clocked JK.
2. Write short notes on the following :
   (*a*) Edge triggered flip-flop (*b*) Master slave flip-flop.
3. Differentiate between synchronous and asynchronous circuits.
4. Differentiate between combinational and sequential circuits.
5. Obtain the following realisations
   (*a*) T flip-flop from D flip-flop (*b*) D flip-flop from SR flip-flop.
6. What is a race around condition? How it is avoided?
7. Draw the JK master-slave flip-flop and explain how to eliminate the race around condition?
8. Write short notes on RS flip-flop using NAND gates.
9. What are the advantages of direct inputs in the flip-flops?
10. Design a sequential circuit for the following state diagram (Fig. 1).

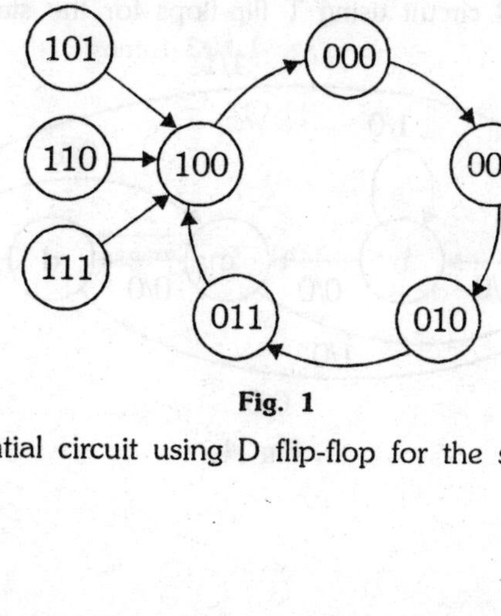

Fig. 1

11. Design a clocked sequential circuit using D flip-flop for the state diagram shown in Fig 2.

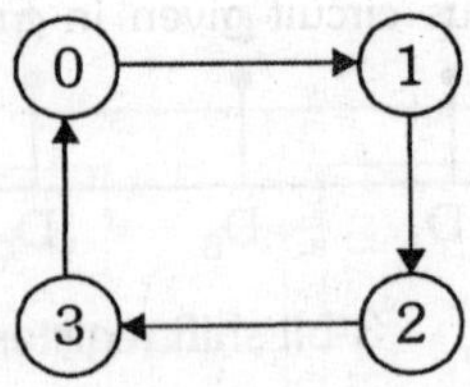

Fig. 2

12. Explain briefly how the state diagrams are useful in the design of the logic circuits.
13. Explain the operation of the 4-bit synchronous decade counter with the help of neat diagrams.
14. Explain the working of 8-bit ring counter. Draw its timing diagram.
15. With the help of neat diagrams explain the operation of 4-bit UP/DOWN synchronous counter.
16. What is a Johnson counter, explain the working of 4-bit Johnson counter?
17. Draw and explain the operation of the 4-bit parallel-in-parallel-out shift register.
18. Design a decade counter to count in the EX-3 code sequence.
19. Design a MOD-12 counter using D-flip-flop.
20. Design a counter that goes through states 1, 2, 4, 8, 3, 5, 7, 1, 2 .... using JK flip-flop.
21. Design a BCD counter with JK flip-flops.
22. Design a clocked sequential circuit using D flip-flops for the state diagram shown in Fig. 3. Use state reduction if possible. Make proper state assignment.

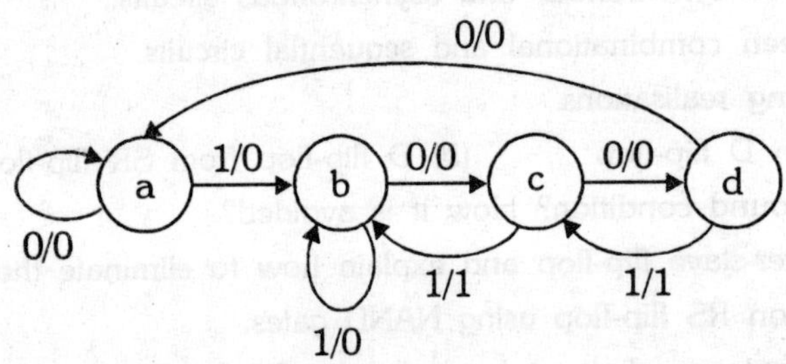

Fig. 3

23. Design a clocked sequential circuit using T flip-flops for the state diagram given in Fig. 4.

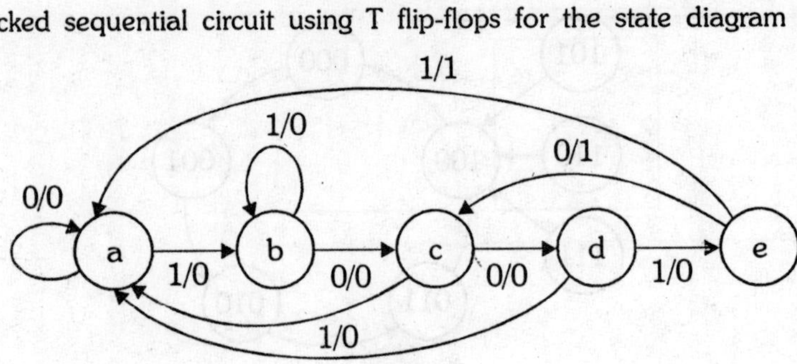

Fig. 4

24. Analyze the following synchronous circuit given in Fig. 5.

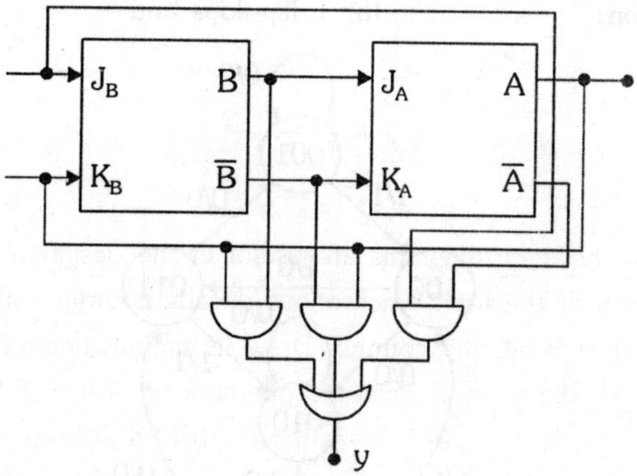

Fig. 5

25. Construct the state diagram for a two input, two output sequential circuit which is to produce an output $z = 1$ with occurrance of an input 1 following a string of two or three consecutive input 0's. At all other times the output is to be zero, realize using JK flip-flops.

26. Draw the state table and state diagram for the given clocked sequential circuit shown in Fig. 6.

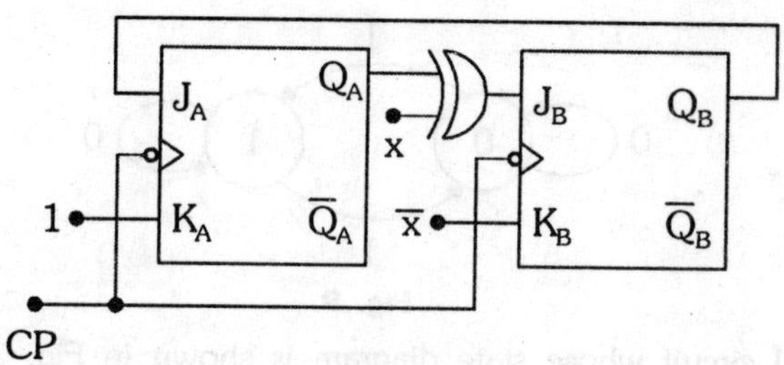

Fig. 6

27. Analyze the given circuit in Fig. 7 and draw its state table and state diagram.

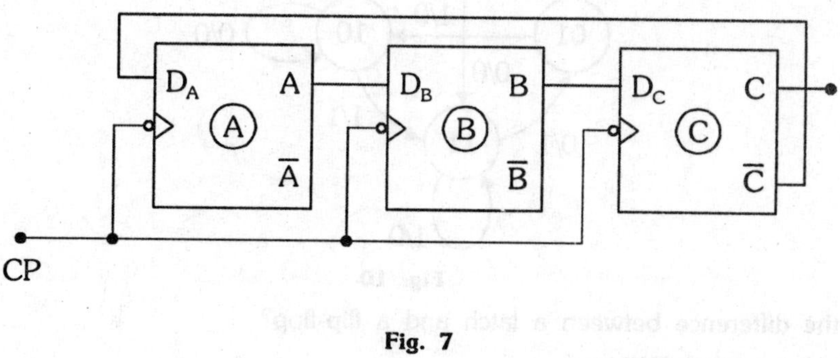

Fig. 7

28. Design the clocked sequential circuit for the given state diagram shown in Fig. 8.
Using (*a*) JK flip-flops (*b*) T flip-flops and (*c*) D flip-flops

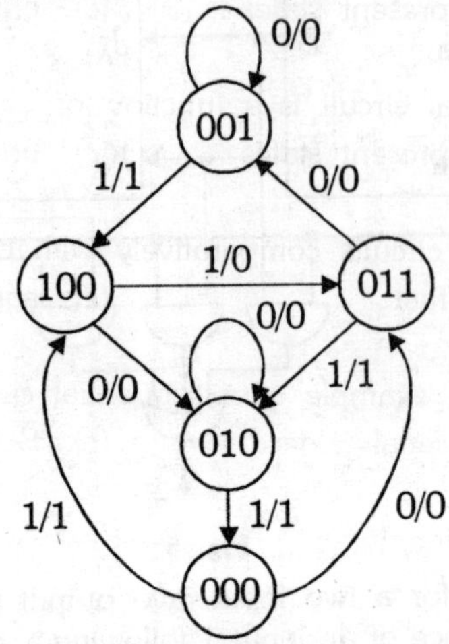

Fig. 8

29. Design a sequential circuit that checks the parity of the serial data. The state diagram is shown in Fig. 9.

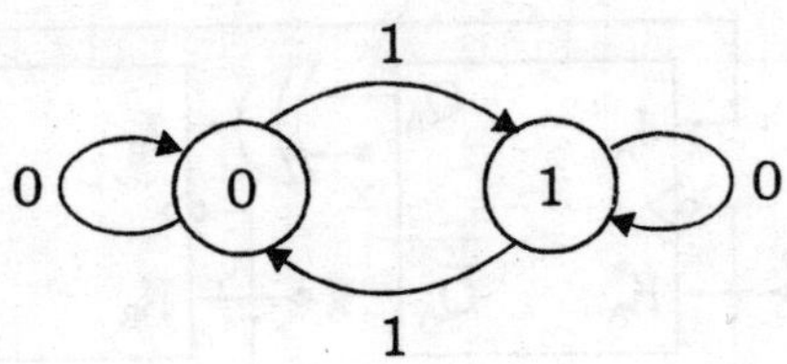

Fig. 9

30. Design a sequential circuit whose state diagram is shown in Fig. 10.

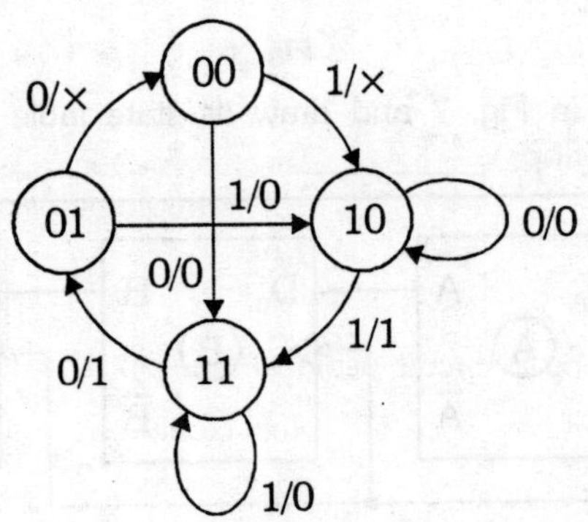

Fig. 10

31. What is the difference between a latch and a flip-flop?

32. Design an octal UP/DOWN counter.

4.22 SHORT ANSWER QUESTIONS

1. A sequential circuit is one, whose output depends on__________.

(*a*) present inputs (*b*) present states (*c*) both (*d*) none

[Ans. (*c*)**]**

2. The next state in a sequential circuit is a function of__________.

(*a*) present inputs (*b*) present states (*c*) both (*d*) none

[Ans. (*c*)**]**

3. The speed of the sequential circuits comparatively with combinational circuits is__________.

(*a*) slow (*b*) fast (*c*) equal (*d*) none

[Ans. (*a*)**]**

4. Which of the following is an example of a sequential circuit__________?

(*a*) parallel adder (*b*) serial adder (*c*) BCD adder (*d*) look ahead adder

[Ans. (*b*)**]**

5. Sequential circuits are classified in __________ways.

(*a*) three (*b*) four (*c*) five (*d*) two

[Ans. (*a*)**]**

6. Which of the following is true with respect to synchronous sequential circuits?

(*a*) It contains at least one feedback path
(*b*) Clocked flip-flops are used
(*c*) The speed of the circuit depends on the speed of the clock
(*d*) All the above

[Ans. (*d*)**]**

7. The characteristic equation of RS flip-flop is__________.

(*a*) $S + RQ$ (*b*) $\overline{S} + RQ$ (*c*) $S + \overline{R}Q$ (*d*) $SQ + \overline{R}$

[Ans. (*c*)**]**

8. The characteristic equation of D-flip-flop is__________.

(*a*) D (*b*) $\overline{D}$ (*c*) DQ (*d*) $D\overline{Q}$

[Ans. (*a*)**]**

9. The next state of D-flip flop is__________.

(*a*) same as input (*b*) complement of input
(*c*) independent of input (*d*) none

[Ans. (*a*)**]**

10. While designing a divide by-2 circuit using D-flip-flop, the D-input is connected by__________.

(*a*) Q (*b*) $\overline{Q}$ (*c*) clock (*d*) none

[Ans. (*b*)**]**

11. The characteristic equation of JK-flip-flop is__________.

(*a*) $JQ+K\overline{Q}$ (*b*) $J\overline{Q}+\overline{K}Q$ (*c*) $\overline{J}Q+\overline{KQ}$ (*d*) $\overline{J}Q+K\overline{Q}$

[Ans. (*b*)**]**

12. When both inputs of JK-flip-flop are "0" then the next state is__________.

(*a*) 0 (*b*) 1 (*c*) $Q(t)$ (*d*) $\overline{Q}(t)$

[Ans. (*c*)**]**

13. When both inputs of JK flip-flop are 1, then the next state Q $(t + 1)$ is________.

(*a*) 0 (*b*) 1 (*c*) Q(t) (*d*) $\overline{Q}$

[Ans. (*d*)**]**

14. The characteristic equation of T flip-flop is________.

(*a*) $T\overline{Q}+\overline{T}Q$ (*b*) $TQ+\overline{T}\overline{Q}$ (*c*) TQ (*d*) $T\overline{Q}$

[Ans. (*a*)**]**

15. When is the next state of the T flip-flop complement of the present state (*i.e.*, T value) ?

(*a*) 0 (*b*) 1 (*c*) both (*d*) none

[Ans. (*b*)**]**

16. In which flip-flops race around condition occurs ?

(*a*) JK (*b*) T (*c*) Both (*d*) None

[Ans. (*c*)**]**

17. How many number of flip-flops are used in master-slave flip-flop ?

(*a*) Two (*b*) Three (*c*) One (*d*) Four

[Ans. (*a*)**]**

18. What is the advantage of MS flip-flop ?

(*a*) Race around problem is reduced (*b*) More number of flip-flops are required

(*c*) Less propagation delay (*d*) none. **[Ans.** (*a*)**]**

19. What is the condition for race around problem ?

(*a*) J = K = 1 (*b*) $\Delta t < tp$ (*c*) Both (*d*) None

[Ans. (*c*)**]**

20. Which of the following are asynchronous inputs ?

(*a*) Preset (*b*) Clear (*c*) Both (*d*) None

[Ans. (*d*)**]**

21. In active low direct inputs, which of the conditions is not used ?

(*a*) Preset = clear = 0 (*b*) Preset = 0, clear 1

(*c*) Preset = 1, clear = 0 (*d*) Preset = clear = 1

[Ans. (*a*)**]**

22. While converting the SR flip-flop into D flip-flop, the SR inputs are________.

(*a*) S = D, R = $\overline{D}$ (*b*) S = $\overline{D}$, R = D (*c*) S = R = D (*d*) S = R = $\overline{D}$

[Ans. (*a*)**]**

23. While converting the SR flip-flop into JK flip-flop, the SR inputs are________.

(*a*) S = R = J (*b*) S = R = K (*c*) S = $J\overline{Q}$, R = QK (*d*) S = QK, R = $J\overline{Q}$

[Ans. (*c*)**]**

24. If the present state = 0, next state = 0, then the values of R and S are________.

(*a*) 0,0 (*b*) 1,0 (*c*) 0,1 (*d*) ×,0

[Ans. (*d*)**]**

25. If the values of R and S are 0, 1 respectively, then the present state and next states are ________.

(*a*) 1, 1 (*b*) 0, 0 (*c*) 0, 1 (*d*) 1, 0

[Ans. (*c*)**]**

26. The table which contains present state, next state and flip-flop inputs is called ________.
(*a*) truth table (*b*) characteristic table
(*c*) excitation table (*d*) realization table
[Ans. (*c*)]

27. In JK-flip, under what condition does 0 to 1 transition occur ________ ?
(*a*) J = K = 1 (*b*) J = K = 0 (*c*) J = × K = 0 (*d*) J = 1, K = ×
[Ans. (*d*)]

28. If J = × and K = 1, what are the present state and next states ?
(*a*) 0, 1 (*b*) 1, 0 (*c*) 0, 0 (*d*) 1, 1
[Ans. (*b*)]

29. If the next state is 1, then the D input is ________.
(*a*) 0 (*b*) 1 (*c*) ϕ (*d*) none
[Ans. (*b*)]

30. In T-flip-flop, both states are same, then the T value is________.
(*a*) 0 (*b*) 1 (*c*) ϕ (*d*) none.
[Ans. (*a*)]

31. Shifting a binary data to right by one bit position using shift right register, results________.
(*a*) addition by 2 (*b*) subtraction of 2 (*c*) multiplication by 2 (*d*) division by 2
[Ans. (*d*)]

32. A counter with 6 flip-flop can have the maximum number of states are________.
(*a*) 6 (*b*) 256 (*c*) 64 (*d*) 8
[Ans. (*c*)]

33. The maximum number of flip-flops required for a MOD-10 counter________.
(*a*) 10 (*b*) 4 (*c*) 3 (*d*) 5
[Ans. (*b*)]

34. The maximum count value that can be obtained by a counter which is having 5-flip-flops is ________.
(*a*) 32 (*b*) 31 (*c*) 5 (*d*) none
[Ans. (*b*)]

35. The maximum modulo number that can be obtained by a ripple counter using 4 flip-flops is ________.
(*a*) 16 (*b*) 4 (*c*) 32 (*d*) 15
[Ans. (*a*)]

36. A MOD-6 synchronous counter is designed by using JK flip-flop, the number of counts skipped by it is ________.
(*a*) 6 (*b*) 5 (*c*) 3 (*d*) 2
[Ans. (*d*)]

37. The number of flip-flops required for a MOD-16 ring counter are________.
(*a*) 2 (*b*) 4 (*c*) 15 (*d*) 16
[Ans. (*d*)]

38. A BCD counter has________different states.
(*a*) 3 (*b*) 4 (*c*) 10 (*d*) 9
[Ans. (*c*)]

39. The output frequency of a decade counter, when it is clocked by 100 kHz signal is__________.

(*a*) 10 kHz (*b*) 50 kHz (*c*) 25 kHz (*d*) 1 kHz

[Ans. (*a*)]

40. The output frequency of a MOD-16 counter, clocked by a 10 kHz clock signal is__________.

(*a*) 2.5 kHz (*b*) 6.25 kHz (*c*) 625 kHz (*d*) 10 kHz

[Ans. (*c*)]

41. Which type of counter requires more number of flip-flops for a given MOD number is__________?

(*a*) ripple counter (*b*) ring counter (*c*) Johnson counter (*d*) BCD counter

[Ans. (*b*)]

42. In__________type of register, we have access only to left most and right most flip-flops.

(*a*) serial in-serial out (*b*) serial in-parallel out

(*c*) parallel in-serial out (*d*) parallel in-serial out

[Ans. (*a*)]

43. The maximum frequency to which a MOD-16 ripple counter using 4 JK flip flops with propagation delay of 50nsec is __________.

(*a*) 4 MHz (*b*) 3.2 MHz (*c*) 50 MHz (*d*) 5 MHz

[Ans. (*d*)]

5

Capabilities and Minimisation of Sequential Machines

So far we have considered combinational switching circuits in which the outputs are functions of only the present circuit inputs. In most digital systems, however additional circuits are necessary, capable of storing information and data, and also of performing some logical (or) mathematical operations upon these data. The outputs of these circuits at any given time are functions of the external inputs, as well as the stored information at that time. Such circuits are called sequential circuits.

Conventionally, the term *sequential machine* refers to the abstract model which represents the sequential circuit. In many cases, however these terms are used interchangeably.

5.1 FINITE STATE MACHINE

A finite state machine (or finite automation) is an abstract model describing the synchronous sequential machine. It is the basis for understanding and development of the various computation structures.

5.1.1 The Finite State Model-Basic Definitions

The behaviour of a finite state machine is described as a sequence of events that occur at discrete instants, designated $t = 1, 2, 3$, etc. Suppose that a machine 'M' has been receiving input signals and has been responding by producing output signals. If now, at time t we were to apply an input signal $x\ (t)$ to M, its response $z(t)$ would depend on $x(t)$, as well as on past inputs to M.

In practice, for a given machine M, it is impossible to implement machines which have infinite storage capabilities. Every finite state machine therefore contains a finite number of memory devices, which store the information regarding to the past input. Here we concentrate only on past history of the machine which affects its future behaviour. So every finite state machine contains a finite number of memory elements (devices), which store the information regarding the past input history (internal states).

Note: No bound has been set on the duration for which a particular input may affect the future behaviour of machine.

Synchronous sequential machine. In general, a synchronous sequential machine is represented schematically by the circuit of Fig. 5.1.

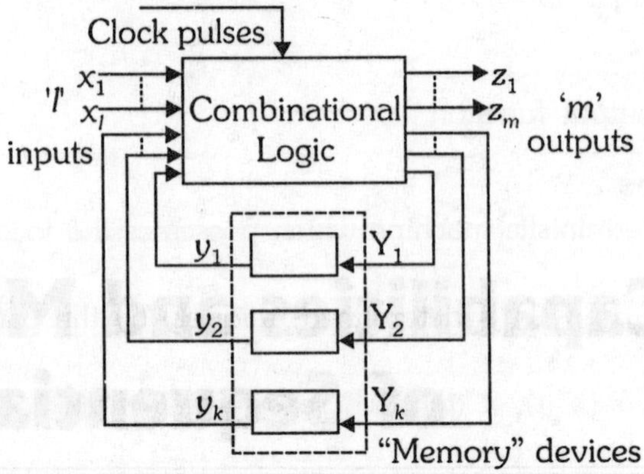

Fig. 5.1 Circuit representation of a synchronous sequential machine.

The circuit has a finite number of *l* input terminals. The signals entering the circuit via these terminals constitutes the set $\{x_1, x_2, x_3, ..., x_l\}$ of input variables, where x_j for all *j* may take on one of the two possible values of each input is either 0 (or) 1. There are an ordered *l-tuple* of 0's and 1's in an input configuration. The set of $p = 2^l$ distinct inputs is called the *'input alphabet'* I, and each configuration is referred to as a symbol of the alphabet. Thus the input is given by $I = \{I_1, I_2, ..., I_p\}$

For example, if a machine has two input variables, x_1 and x_2 then its input alphabet I consists of four symbols, that is I = {00, 01, 11, 10}.

Similarly, the circuit has a finite number '*m*' of output terminals which define the set $\{z_1, z_2, ..., z_m\}$ of output variables, where z_j for all *j* is a binary variable, (*i.e.*, its value is either 0 or 1). An ordered *m*-tuple of 0's and 1's is an output configuration (or simply output). The set of $q = 2^n$ ordered *m*-tuples is called the *output alphabet* and is given by

$$O = \{O_1, O_2, ..., O_q\} \text{ here } O_1, O_2, ... \text{ are the symbols of the output alphabet.}$$

The signal value at the output of each memory element is referred to as the *'state variable'* and the set $\{y_1, y_2, ..., y_k\}$ constitutes the set of state variables. The combination of values at the outputs of the *k* memory elements $y_1, y_2, ..., y_k$ defines the *'present internal state'* of the machine. The set $S = 2^k$ of *k* – tuples constitutes the entire set of states of the machine, where $S = \{s_1, s_2, ..., s_n)$ and $S_1, S_2, ...$ are the symbols of states.

The external inputs $x_1, x_2, ..., x_l$ and the values of the state variables $y_1, y_2, ..., y_k$ are supplied to the combinational circuit, which in turn produces the outputs $z_1, z_2, ..., z_m$ and the values $y_1, y_2, ..., y_k$. *The values of the y's, which appear at the outputs of the combinational circuit at time t, are identical with the values of the state variables at t+l,* and therefore they define the *'next state' of the machine.*

5.1.2 Deterministic Machine

The deterministic machines possess the property that the next state *s* (*t* + 1) is determined uniquely by the present state *s*(*t*) and present input *x*(*t*). Thus

$$S\ (t + 1) = \delta\ \{s\ (t),\ x\ (t)\} \qquad ...(1)$$

where δ is called the state transition function.

The value of the output $z(t)$ is, in the most general case, a function of the present state $s(t)$ and the inputs $x(t)$, *i.e.*,

$$Z(t) = \lambda \{s(t), x(t)\} \quad ...(2)$$

where λ is called the output function.

5.1.2.1 Mealy machine

A machine is a deterministic machine which possesses the following two properties known as Mealy machine.

1. The next state $S(t + 1)$ is determined uniquely by the present state $s(t)$ and present inputs $x(t)$.

Thus $S(t + 1) = \delta \{s(t), x(t)\}$ where δ = state transition function

2. The value of output function $Z(t)$ is a function of present state $s(t)$ and input $x(t)$

$\therefore$ $Z(t) = \lambda \{s(t), x(t)\}$ where λ = output function.

For example, consider the state table, shown in Table 5.1.

Table 5.1. State table

Present state	*Next state* x = 0	x = 1	*Z (output)* x = 0	x = 1
A	A	C	0	1
B	D	A	1	1
C	B	D	1	1
D	E	B	0	0
E	B	C	1	0

In the above state table, the next state depends upon present state and present input, and the output function Z(t) also depends upon present state and present input.

State diagram. The corresponding state diagram is shown in Fig. 5.2.

Here each node contains only the state and state transition depends on the present state and input. Each labelled arc indicates the input and output while the state transition.

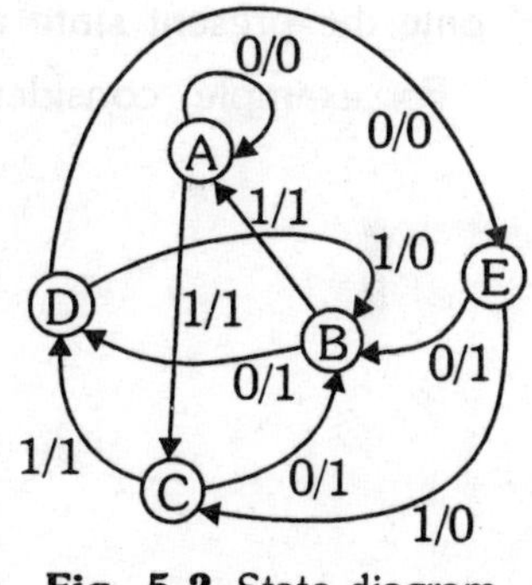

Fig. 5.2 State diagram.

Block diagram of Mealy machine. The Fig. 5.3 shows the simple block diagram of a Mealy machine.

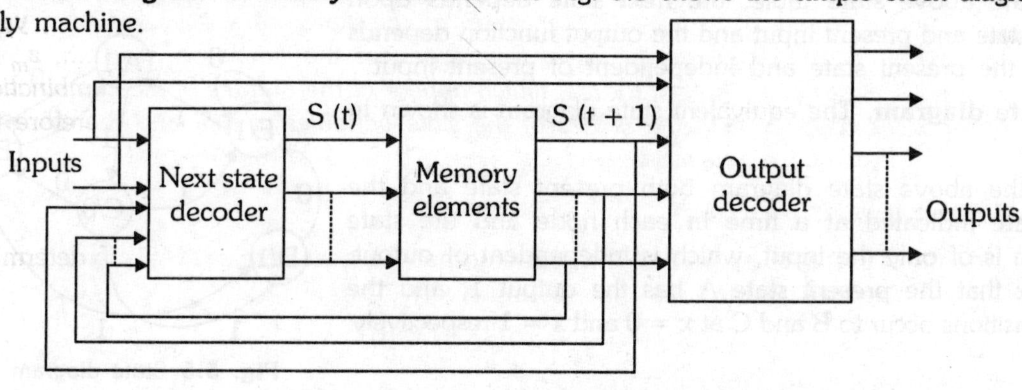

Fig. 5.3 Block diagram of Mealy machine.

The output Z(*t*) can change at any time *i.e.*, either the input (or) state changes, since Z(*t*) is a function of both. This gives rise to Z(*t*) unexpected output change at T_0 and T_3 as shown in Fig. 5.4. Thus in Mealy circuits, the unwanted gliches result.

Advantages. Since in Mealy circuits, the outputs are function of both inputs and the state, the designer has more flexibility in designing output and state transition functions.

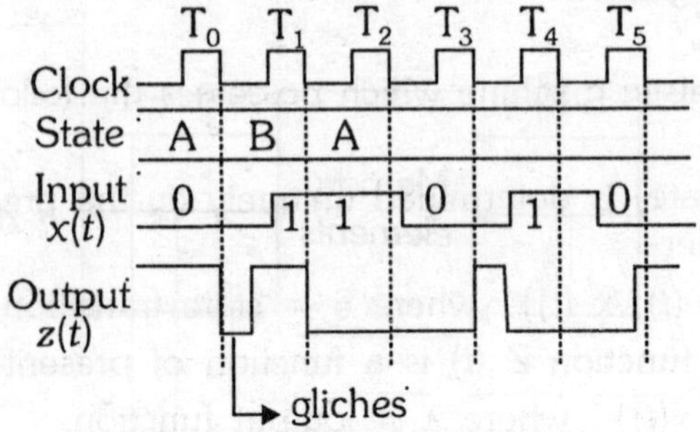

Fig. 5.4 Timing diagram.

Note: The output of a mealy machine is sampled only when the circuit has stabilised after an input change.

5.1.2.2 Moore machine

A machine possessing the properties that $s(t + 1) = \delta\{s(t), x(t)\}$ and $Z(t) = \lambda\{s(t)\}$ is generally known as Moore machine. In this, the output is a function of only the present state and is independent of the external input.

For example, consider the state table shown in Table 5.2.

Table 5.2. State table

Present state	*Next state* $x = 0$	$x = 1$	Output
A	B	C	1
B	A	D	0
C	B	E	0
D	E	C	1
E	E	A	1

In the above state table, the next state depends upon present state and present input and the output function depends only on the present state and independent of present input .

State diagram. The equivalent state diagram is shown in Fig. 5.5.

In the above state diagram both present state and the output are indicated at a time in each node and the state transition is of only the input, which is independent of output. Consider that the present state A has the output 1, and the state transitions occur to B and C at x = 0 and x = 1 respectively.

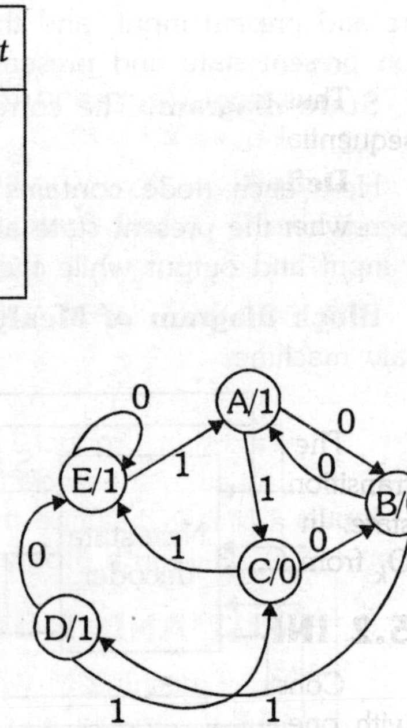

Fig. 5.5 State diagram.

Block diagram of Moore machine. It is shown in Fig. 5.6 in Moore circuits, the output changes are synchronized to the clock, since the output $z(t)$ is a function of the state only and can therefore change only when the state changes consequently, the output remains stable during any input changes as shown in Fig. 5.7. Unlike Mealy machine, the outputs of Moore model circuit typically exhibit better behaviour than those of Mealy circuits, *i.e.,* input changes do not result in unwanted gliches.

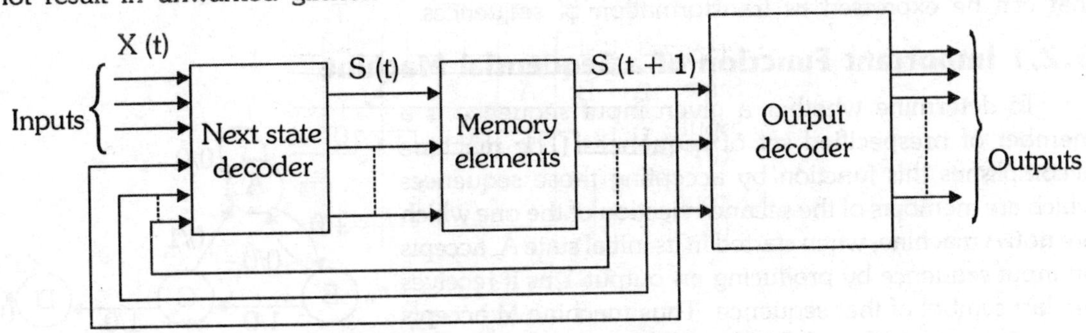

Fig. 5.6 Block diagram of Moore machine.

Timing diagram.

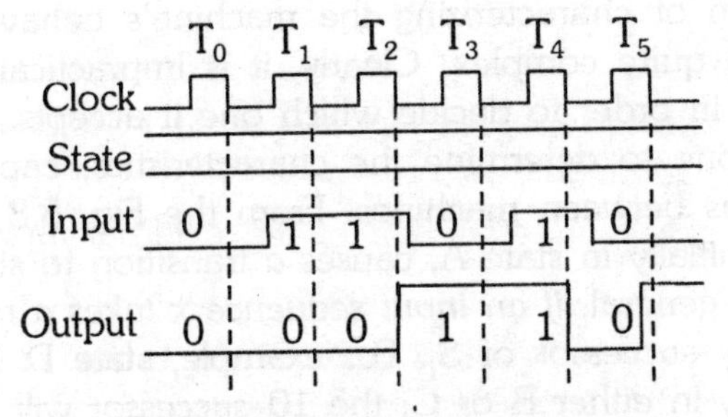

Fig. 5.7 Timing diagram.

Thus from Mealy and Moore machines, we arrive at the following formal definition of a sequential machine.

Definition. A synchronous sequential machine M is a quintuple. M = (I, O, S, δ, λ). where I, O, and S are finite, non-empty sets of inputs, outputs, and states, respectively;

$\delta : I \times S \rightarrow S$ is the state transition function;

λ is the output function such that

$\lambda : I \times S \rightarrow O$ for Mealy machines;

$\delta : s \rightarrow O$ for Moore machine.

The cartesian product I × S is the set containing all pairs of elements (I_i, S_j). The state transition function δ associates with each pair (I_i, S_j) an element S_k from S, is called the next state. In a Mealy machine the output function λ associates with each pair (I_i, S_j) an element O_k from O, while in a Moore machine a correspondence exists between states and the outputs.

5.2 INPUT AND OUTPUT TRANSFORMATIONS

Consider machine M whose state diagram is given in Fig. 5.8. It is a four state machine with one input variable and output variable for which

$$s = \{A, B, C, D\};\ I = \{0, 1\};\ O = \{O, 1\}.$$

Suppose that the initial state of M is A and the input sequence is 110; the machine will proceed through states B and C and return to state A, while producing the output sequence 001. Thus, for an initial state A, machine M transforms the input sequence 110 into 001. Similarly, for the same initial state, the input sequence 01100 is transformed into 00010. Since every computation involves some transformation of input to output sequences, a finite state machine is capable of performing a variety of computations and solving a number of problems that can be expressed as transformation of sequences.

5.2.1 Important Function of a Sequential Machine

To determine whether a given input sequence is a member of prespecified set of sequences. The machine accomplishes this function by accepting those sequences which are members of the set and rejection of the one which are not. A machine, when started in its initial state A, accepts an input sequence by producing an output 1 as it receives the last symbol of that sequence. Thus machine M accepts the sequences 110 and 0110 and rejects the sequence 01100, since its corresponding last output symbol is 0.

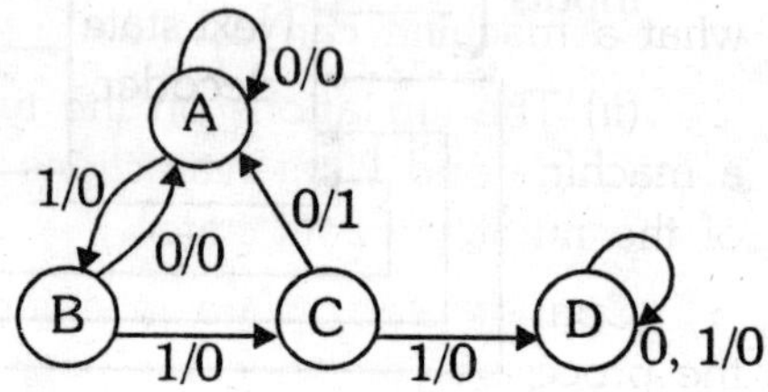

Fig. 5.8

The general problem of characterizing the machine's behaviour by observing it's input-output transformations is quite complex. Clearly, it is impractical to feed a machine with all possible input sequences in order to decide which one it accepts. In order to avoid this, some finite experiments are done to determine the characteristics, capabilities and limitations of a machine and the relations between machines. From the Fig. 5.8, the application of an input 1 to machine M, when initially in state A, causes *a* transition to state B. We thus say that B is the '*1–successor* of A'. In general, *if an input sequence x takes a machine from state S_i to state S_j then S_j is said to be x–successor* of S_i. For example, state D is the 111–successor of A. If M is known to be initially in either B or C, the 10–successor will be either state A (or) D. We say that (AD) is the 10–successor of (BC) to mean that A is the 10–successor of B and D is a 10-successor of C. From the state diagram it is evident that no input sequence exists which can take M out of state D, and thus D is said to be a *terminal state*. Generally, *a* state is called terminal state if either of the following is ture:

(1) The corresponding vertex in the state diagram is a *sink vertex i.e.,* no outgoing arcs which emanate from it and terminate in other vertices as shown in Fig. 5.9.

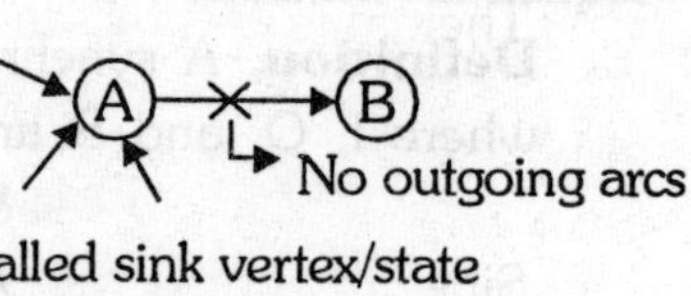

Fig. 5.9

(2) The corresponding vertex in the state diagram is a *source vertex/state i.e.,* no arcs which emanate from other vertices and terminate in it (see Fig. 5.10)

That is a source state/vertex is not accessible from any other state/vertex and no state/vertex is accessible from a sink vertex/state.

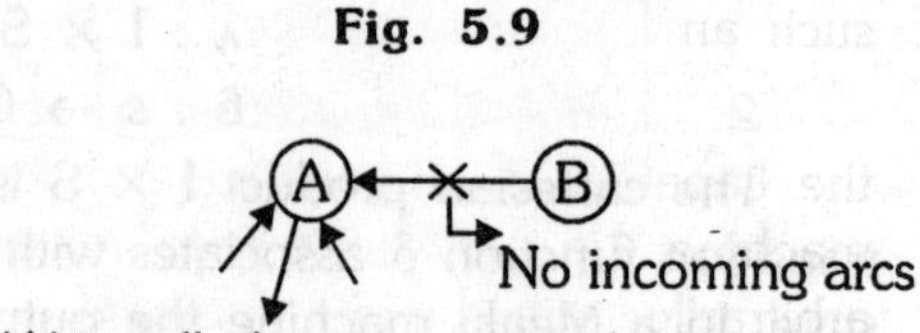

Fig. 5.10

The above two are extreme examples of situations, which limit the state transitions in a sequential machine.

In other cases, certain subsets of states may not be reachable from other subset of states, even machine does not contain any terminal state.

∴ *For every pair of states, S_i, S_j of a machine M, there exists an input sequence which takes M from S_i to S_j, the M is said to be strongly connected, clearly, any nontrivial machine which has terminal states is not strongly connected.*

5.3 CAPABILITIES AND LIMITATIONS OF FINITE-STATE MACHINES

The capabilities and limitations of finite state machines may determine the following:

(*i*) The behavioural properties and synthesis procedures for finite-state machine, that is what a machine can do.

(*ii*) The limitations on the type of input-output transformations that can be performed by a machine and what restrictions are imposed on the capabilities of the machine by finiteness of the number of its states.

Consider an *n*-state machine with the input of an arbitrarily long sequence of 1's. Then the process of the machine, starting from some initial state, through a succession of states, in accordance with its specified state of transitions. If the sequence be long enough so that it is larger than *n*, the machine must eventually arrive at its previous state. If $n = 5$, the maximum periodic sequence is 0, 1, 2, 3, 4, 0, 1, ... because the input remains the same, and must continue in a periodically repeating fashion. Thus, for an *n-state machine, the period cannot exceed n* and *could be smaller*. Moreover, the transient time until the output reaches its periodic pattern cannot exceed the number of states *n*.

The following two examples may demonstrate the capabilities and limitations of finite state machines.

1. Consider a finite state machine which can exhibit the limitations. If we want to design the machine which receives a long sequence of 1's and is to produce an output 1 when and only when the number of inputs that it has received so far is equal to $\frac{k(k+1)}{2}$, for $k = 1, 2, 3, \ldots\ldots$

Then, the desired input-output transformation has the following form :

Input = 1 1 1 1 1 1 1 1 1 1 1 1 1 1 1 1 ....

Output = 1 0 1 0 0 1 0 0 0 1 0 0 0 0 1 ....

Since the output does not become eventually periodic, no finite state machine can produce such an infinite sequence.

2. "*Serial–multiplication*" problem is another example for demonstrating the limitations on the capabilities of finite-state machines. This problem is not solvable by a fixed finite-state machine; *i.e.*, no finite-state machine with a fixed number of states that can multiply two arbitrarily large binary numbers.

Consider that there exists an *n*-state machine capable of serially multiplying any two binary numbers. Let us select the two numbers to be multiplied are $2^p \times 2^p = 2^{2p}$, where $p > n$. The inputs are fed serially into the machine, least significant digits first. 2^p is represented by a 1 followed by *p* number of 0's, and 2^{2p} is to be represented by a 1 followed by 2*p* number of 0's. The inputs are fed into the machine during the first $(p + 1)$ time units, *i.e.*, between

t_1 and t_{p+1} as shown below. During this period the machine produces 0's. At t_{p+1} the input stops, while the machine must go on producing p additional 0's followed by a 1.

t_{2p+1}	t_{2p}	t_{p+1}	t_p	t_2	t_1	= time
		1	0	0	0	= first number
		1	0	0	0	= second number
1	0	0	0	0	0	= product

At this time period between t_{p+1} and t_{2p}, the machine receives no input, but since $p>n$, it must have been twice during that time at one of the states. From this we are led to the conclusion, that its output must be periodic and the period is smaller than p. And therefore the machine will never produce the required 1 output.

Note: For any two finite numbers, we can find a machine which is capable of multiplying them. But the preceeding result demonstrates that, every finite-state machine capable of performing serial multiplication, can find such numbers which it could not multiply. The reason for this limitation is the *"limited memory available"* to the machine.

5.4 STATE EQUIVALENCE AND MACHINE MINIMIZATION

In implementing the logic diagram from state diagram (or table) of a finite-state machine, the diagram may contain redundant states, *i.e.*, states whose functions can be accomplished by other states. The number of memory elements required for the realization of the machine is directly related to the number of states. Consequently, the minimization of the number of states does in many cases, reduces the *complexity, number of Gates and flip-flops and finally the cost of* the final circuit.

Therefore it is necessary to determine the different techniques for transforming a given machine into another machine which has no redundant states, so that both the machines have the same terminal behaviour *i.e.*, same outputs and next states.

K-equivalence, minimization procedure and *machine equivalence* are the different techniques.

5.4.1 K-equivalence and K-distinguishable

Let S_i and S_j are the two states of a machine M, these states are distinguishable if and only if there exists at least one finite input sequence which, when applied to the machine M, causes different output sequences, depending on whether S_i (or) S_j is the initial state. The sequence which distinguishes these states is known as 'distinguishing sequence of the pair' (S_i, S_j). This sequence of the pair may conclude for the following :

(*a*) If there is uncertainty as to whether the state of M is S_i (or) S_j, the application of the corresponding distinguishing sequence yields an output sequence which is sufficient to determine uniquely the unknown state.

(*b*) If there exists a pair (S_i, S_j) of distinguishing sequence of length *k*, the states in (S_i, S_j) are said to be K-distinguishable. For example consider the machine M_1 whose state table as shown in Table 5.3.

Table 5.3

Present state	Nest state, Z	
	x = 0	x = 1
A	E, 0	D, 1
B	F, 0	D, 0
C	E, 0	B, 1
D	F, 0	B, 0
E	C, 0	F, 1
F	B, 0	C, 0

Consider the pair (A, B). The pair (A, B) is 1-distinguishable, since a 1 input applied to M_1 when initially in A yields an output 1, and an output 0 when it is initially in B.

On the other hand, the pair (A, E) is 3-distinguishable, since there is no input sequence of length 2 which distinguishes A from E. The only sequence of length 3 which is a distinguishing sequence for the pair (A, E) is X = 111, and the output sequences corresponding to initial states A and E are 100 and 101. Note that 1101 is also a sequence, which distinguishes A from E, but it is not the shortest.

K–equivalence. The concept of K-distinguishability leads directly to the definition of K-equivalence. States that are not K-distinguishable are said to be K-equivalent. For example, the states A and E of M_1 are 2–equivalent. This results the following definitions:

Definition. States S_i and S_j of machine M are said to be equivalent if and only if, for every possible input sequence, the same output sequence will be produced regardless of whether S_i (or) S_j is the initial state.

Thus S_i and S_j are equivalent (indicated by $S_i = S_j$) if there is no input sequence which distinguishes them. It will be subsequently shown that states are K-equivalent for all $k \leq n - 1$, where k is the length, and 'n' is the number of states equivalent.

Clearly, if $S_i = S_j$ and $S_j = S_k$,

then $S_i = S_k$

The set of states of the machine can be partitioned into different subsets, known as the 'equivalence classes', so that two states are in the same equivalence class if and only if they are equivalent, and are in different classes if and only if they are distinguishable.

Thus the procedure of determining the sets of equivalent states in a machine, *i.e.*, the equivalent classes ensures from the following property.

If S_i and S_j are equivalent states, their corresponding x–successors, for all x, are also equivalent, since otherwise it would be difficult to construct a distinguishing sequence for (S_i, S_j) by applying an input sequence that transfers the machine to the distinguishable successors of S_i and S_j.

Theorem 1. The equivalence partition is unique.

Proof. Let us assume that two equivalence partitions exist; these are P_A and P_B and such that $P_A \neq P_B$. This states that, there are two states S_i and S_j which are in the same block of one partition (P_A) and not in the same block of other partition (P_B). If S_i and S_j are in different blocks, there exists at least one input sequence which distinguishes S_i and S_j therefore they can be in the same block of P_A.

Theorem 2. If two states, S_i and S_j, of a machine M are distinguishable, then they are distinguishable by a sequence of length $(n - 1)$ or less, where n is number of states in machine M.

Proof. In the partition procedure, each next partition (P_{k+1}) must have at least one more block than the previous partition (p_k). Here always $p_{k+1} \geq p_k$. The maximum number of blocks are limited to number of states in the machine, M, is n, *i.e.*, at least one state in one block, at most $(n - 1)$ partitions can be generated in the minimization procedure, and thus we can say that if S_i and S_j are distinguishable, they are distinguishable by a sequence of length $(n - 1)$ or less.

5.4.2 The Minimization Procedure

The object of this section is to describe a procedure for determining the sets of equivalent states of a specified machine M. The result saught is a partition on the states of M such that two states are in the same block, if and only if they are equivalent. The following steps illustrate the minimization procedure.

Step 1. Let us partition the states of M into subsets, such that all states in the same subset are 1-equivalent. This is accomplished by placing states having identical outputs under all possible inputs in the same subset. Clearly, two states which are in different subsets are 1-distinguishable.

Consider a machine M_1 given in the Table 5.4.

Table 5.4

Present state	*Next state (Z)*	
	x = 0	*x = 1*
A	E, 0	D, 1
B	F, 0	D, 0
C	E, 0	B, 1
D	F, 0	B, 0
E	C, 0	F, 1
F	B, 0	C, 0

P_0 = (ABCDEF)
P_1 = (ACE) (BDF)
P_2 = (ACE) (BD)(F)
P_3 = (AC) (E)(BD)(F)
P_4 = (AC)(E)(BD)F

The first partition P_0 corresponds to 0-distinguishability and it is prior to the application of any input. P_1 is obtained simply by inspecting the table and placing *those states having the same outputs, under all inputs, in the same block*. Thus A, C and E are in the same block, since their outputs under 0 and 1 inputs are 0 and 1 respectively. A similar argument places B, D and F in the other block. Clearly P_1 establishes the sets of states which are 1-equivalent.

Step 2. Let us obtain the partition P_2 whose block consists of the sets of states, which are 2-equivalent, that is equivalent under any input sequences of length 2. This is accomplished by observing that two states are 2-equivalent if and only if they are 1-equivalent and their I_i successors for all possible I_i are also 1-equivalent. Consequently, two states are placed in the same block of P_2 if and only if they are in the same block of P_1, and for each possible I_i and their I_i – successors are also contained in block of P_1. This step is carried out by splitting blocks of P_1 whenever their successors are not contained in a common block of P_1.

The 0 – and 1– successors of (ACE) are (CE) and (BDF), respectively, and since both are contained in common blocks of P_1, the states in (ACE) are 2-equivalent, and therefore (ACE) constitutes a block in P_2. The 1-successor of (BDF) is (DBC), since (BD) and (C) are not contained in a single block of P_1, the block (BDF) must be split into (BD) and (F), so that the successors of the blocks in the required partition are 1-equivalent.

Step 3. In a similar manner P_3 is obtained by splitting the block (ACE) of P_2 into (AC) and (E) since the 1-successors of A, C, and E are D, B, and F, which are not 2-equivalent.

In general, the P_{k+1} partition is obtained from P_k by placing in the same block of P_{k+1} those states which are in the same block of P_k and whose I_i-successors for every possible I_i are also in a common block of P_k. This process places in the same block the states that are $(k + 1)$ equivalent and in different block states that are $(k + 1)$ distinguishable. One observation is that, no state can belong to more than one block, since this would make it distinguishable with respect to itself.

For example, for k, $P_{k+1} = P_k$, the process terminates and P_k defines the sets of equivalent states of the machine, *i.e.*, all states contained in the same block of P_k are equivalent, while states belonging to different blocks are distinguishable. P_k is thus called the "equivalence partition", and the foregoing procedure is rejected to as the Moore reduction procedure. For machine M_1, P_3 is the equivalence partition, and hence states A and C are equivalent and also B and D.

Note: (1) A partition P is said to be a *refinement* of a partition Q if P is smaller than Q.

(2) The equivalence partition is unique.

5.4.3 Machine Equivalence

Definition. The machines, M_1 and M_2 are said to be equivalent if and only if, for every state in M_1, there is corresponding equivalent state in M_2 and vice-versa.

The number of blocks in the equivalence partition of a machine M_1, defines the minimum no. of states that any machine equivalent must have. The machine which contains no equivalent states, and is equivalent to M_1 is called the *minimal (or) reduced form of* M_1.

Let us denote the blocks of the equivalence partition P_3 of M_1 by α, β, γ, and δ, corresponding respectively to (AC), (E), (BD) and (F), we obtain machine M_1* as shown in Table 5.5.

Table 5.5. Machine M_1*

Present state	*Next state (Z)*	
	x = 0	*x = 1*
(AC) ⇒ α	β, 0	γ, 1
(E) ⇒ β	α, 0	δ, 1
(BD) ⇒ γ	δ, 0	γ, 0
(F) ⇒ δ	γ, 0	α, 0

In constructing M_1*, we specify the 1-successor of α to be γ since the 1-successor of (AC) is (BD) and so on.

The selection of labels α, β, γ, and δ assigned to the blocks of P_3 is obviously arbitrary. A different assignment of labels would certainly have described a machine with the same behavioural properties. In general, if one machine can be obtained from the others by relabelling its states, they are said to be *isomorphic* to each other.

The foregoing results lead to the following basic conclusion:

> ⇒ To every machine M, there corresponds a minimal machine M* which is equivalent to M and is unique upto isomorsphism.

The detection of isomorphism is not always easy and is best accomplished by using a canonical representation for the machine. Such a representation is obtained by selecting a state and labelling it A. (Preferably the starting state if specified). The next labels are selected in such a way that when successive rows of the table, starting in A and going down through B, C etc. are read from left to right, the first occurrence of each new label will be in alphabetical order. Whenever a machine is given in this canonical representation, it is said to be in *standard form*.

Note: Clearly, when the starting state of a reduced machine is specified, its standard form is unique.

The transformation of a machine into its standard form will be illustrated by means of M_1*. Denoting 'α' by A implies that its 0-successor β must be denoted by B, because it is the first occurrence of a new label.

Similarly, its 1–successor γ, must be denoted by C. Row B (*i.e.*, β) must be relabelled next; its first entry is α, and since it is not a new label, make the second entry 'δ' denoted with a new label 'D'. The standard form of Table 5.6, results the machine M_1*.

Table 5.6. Standard form for M_1*

Present state	*Next state (Z)*		
	x = 0	x = 1	
α ⇒ A	B, 0	C, 1	α ⇒ A
β ⇒ B	A, 0	D, 1	β ⇒ B
γ ⇒ C	D, 0	C, 0	γ ⇒ C
δ ⇒ D	C, 0	A, 0	δ ⇒ D

The detection of isomorphism when the starting states are not specified is in general not as simple and not unique. When the number of states, however is not too large, isomorphism can be detected by inspecting the state diagrams of the machines. *The necessary and sufficient condition to be isomorphic to each other is that their state diagrams be identical except for the labelling of their vertices.*

5.5 TRANSITION DIAGRAMS

The Mealy machine and the Moore machine can be described by a transition table, shown in Table 5.7 (*a*) and 5.7 (*b*) respectively.

Table 5.7(a) Transition-table description of the Mealy machine

Present state S	Next state, Z I_1...	I_2...	I_l...
S_1	$\delta(S_1, I_1)\ \lambda(S_1, I_1)$	$\delta(S_1, I_2)\ \lambda(S_1, I_2)$ ...	$\delta(S_1, I_l)\ \lambda(S_1, I_l)$
S_2	$\delta(S_2, I_1)\ \lambda(S_2, I_1)$	$\delta(S_2, I_2)\ \lambda(S_2, I_2)$ ...	$\delta(S_2, I_l)\ \lambda(S_2, I_l)$
⋮	⋮	⋮	⋮
S_n	$\delta(S_n, I_1)\ \lambda(S_n, I_1)$	$\delta(S_n, I_2)\ \lambda(S_n, I_2)$ ...	$\delta(S_n, I_l)\ \lambda(S_n, I_l)$

Table 5.7(b) Transition-table for the Moore machine

Present state	Next state, Z I_1...	I_2...	I_l...
S_1	$\delta(S_1, I_1)\ \lambda(S_1, I_1)$	$\delta(S_1, I_2)\ \lambda(S_1, I_2)$	$\delta(S_1, I_l)\ \lambda(S_1, I_l)$
S_2	$\delta(S_2, I_1)\ \lambda(S_2, I_1)$	$\delta(S_2, I_2)\ \lambda(S_2, I_2)$	$\delta(S_2, I_l)\ \lambda(S_2, I_l)$
⋮	⋮	⋮	⋮
S_n	$\delta(S_n, I_1)\ \lambda(S_n, I_1)$	$\delta(S_n, I_2)\ \lambda(S_n, I_2)$	$\delta(S_n, I_l)\ \lambda(S_n, I_l)$

These transition tables can be explained with the following example.

Example. Consider a milk vendor machine which sells milk at the cost of 15 rupees per litre. The machine will accept 5, 10 and 20 rupees one at a time, and give the correct change. The coin return control unit M_1, which computes the amount of change to be returned to the customer, can be modeled by a sequential machine as follows.

$$M_1 = \{I_1, S_1, Z_1\ \delta_1, \lambda_1\}$$

where
$$I_1 = \{5, 10, 20\}$$
$$S_1 = \{S_0, S_5, S_{10}\}$$
$$Z_1 = \{0, 5, 10, 15, 20\}$$
$$\delta_1(S_0, 5) = S_5;\ \delta_1(S_5, 5) = S_{10};\ \delta_1(S_{10}, 5) = S_0$$
$$\delta_1(S_0, 10) = S_{10};\ \delta_1(S_5, 10) = S_0;\ \delta_1(S_{10}, 10) = S_0$$
$$\delta_1(S_0, 20) = S_0;\ \delta_1(S_5, 20) = S_0;\ \delta_1(S_{10}, 20) = S_0$$

where the terms S_0, S_5 and S_{10} denotes that, the machine has received 0, 5 and 10 rupees respectively.

However, in addition to the states, we add states S_{15}, S_{20}, S_{25}, S_{30} and S_{35} to denote that machine has received 15, 20, 25, 30 and 35 rupees respectively.

Thus
$$S_2 = \{S_0, S_5, S_{10}, S_{15}, S_{20}, S_{25}, S_{30}, S_{35})$$

and the output functions are defined as

$$\lambda_1 (S_0, 5) = 0;\ \lambda_1 (S_5, 5) = 0;\ \lambda_1 (S_{10}, 5) = 0$$
$$\lambda_1 (S_0, 10) = 0;\ \lambda_1 (S_5, 10) = 0;\ \lambda_1 (S_{10}, 10) = 5$$
$$\lambda_1 (S_0, 20) = 10;\ \lambda_1 (S_5, 20) = 15;\ \lambda_1 (S_{10}, 20) = 15$$

Similarly

$$\delta_2 (S_x, y) = \begin{cases} S_{x+y} & \text{if } x = 0, 5 \text{ or } 10 \\ S_y & \text{if } x = 15, 20, 25, 30 \text{ (or) } 35 \end{cases}$$

$$\lambda_2 (S_x) = \begin{cases} 0 & \text{if } x = 0, 5 \text{ or } 10 \\ (x - 15) & \text{if } x = 15, 20, 25, 30 \text{ (or) } 35 \end{cases}$$

Then the machine $M_2 = (I_2 = I_1, S_2, Z_2 = Z, \delta_2, x_2)$ is a Moore machine which is equivalent to the machine M_1. The two machines with same sets of input symbols and output symbols are equivalent if for any sequence of input symbols, they produce the identical sequence of output symbols.

Now we will see the transition tables for machine M_1, (Mealy machine) and machine M_2 (Moore machine) in Table 5.8 (*a*) and 5.8 (*b*)

Table 5.8(*a*) Mealy machine

Present state	Next state (Z) 5	10	20
S_0	S_5, 0	S_{10}, 0	S_0, 10
S_5	S_{10}, 0	S_0, 0	S_0, 15
S_{10}	S_0, 0	S_0, 5	S_0, 20

Table 5.8(*b*) Moore machine

Present state	Next state 5	10	20	Z
S_0	S_5	S_{10}	S_{20}	0
S_5	S_{10}	S_{15}	S_{25}	0
S_{10}	S_{15}	S_{20}	S_{30}	0
S_{15}	S_5	S_{10}	S_{35}	0
S_{20}	S_5	S_{10}	S_{25}	5
S_{25}	S_5	S_{10}	S_{25}	10
S_{30}	S_5	S_{10}	S_{25}	15
S_{35}	S_5	S_{10}	S_{25}	20

An alternative way of describing a sequential machine is to use the transition diagram. The transition diagram of machine M_1 and M_2 are given in the Fig. 5.11 (*a*) and 5.11(*b*) respectively.

It is very easy to convert the Moore machine into Mealy machine and this can be done by omitting the right most column in the Table 5.8 (*b*) and changing each next state $\delta(S_i\ I_j)$ where $i = 1, 2 \ldots n$ and $j = 1, 2 \ldots .l$ of the table by the next state and present output pair $[\delta (S_i, I_j), \lambda (\delta (S_i, I_j)]$. The resulting machine is a Mealy machine of machine M_2 (*i.e.*, machine M_1).

Note: The conventional symbol is input/output.

$S_0 \xrightarrow{10/0} S_{10}$ means that upon application of 10 rupees, the machine will go from state S_0 to S_{10} with an output of '0' rupees.

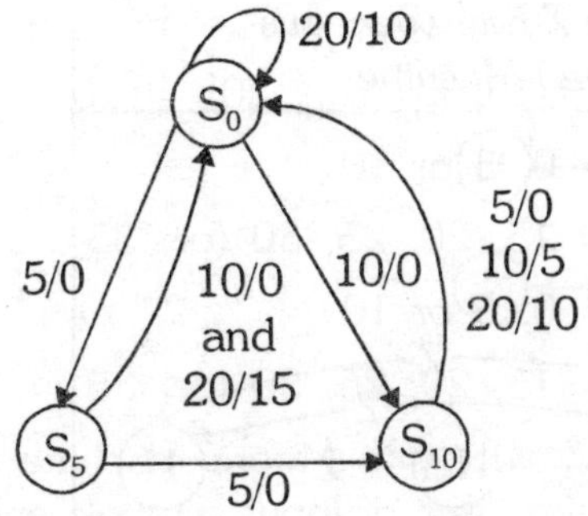

Fig. 5.11(*a*) Mealy machine M_1.

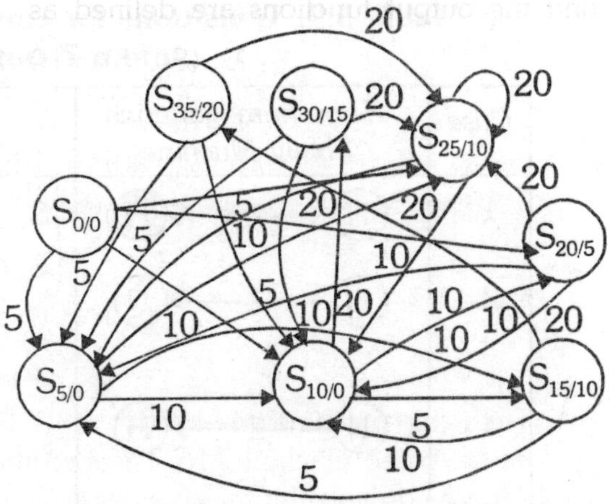

Fig. 5.11(*b*) Moore machine.

Note: The conventional is next state/output

S/z means that λ (s) = z.

To show that a Mealy machine can always be converted into a Moore machine we need the following definition :

Definition. A state of a Mealy machine is "Z-homogeneous" if the output associated with each transition line incident to that state is same otherwise, it is "*Z-non-homogeneous*". A Mealy machine is *Z*-homogeneous if every state of it is *Z*-homogeneous; otherwise it is a *Z*-nonhomogeneous.

Note: Transition line is an arc/line from one state to another in a transition diagram of a sequential machine.

It is clear that *Z*-homogeneous machine is a Moore machine. Now the problem is convertion of a Z-non-homogeneous machine into a Z-homogeneous. For this we have four possible cases of transitions, They are :

1. Transition from *a* Z-homogeneous state to a Z–homogeneous state.
2. Transition from a Z-homogeneous state to Z–non-homogeneous state.
3. Transition from a Z-non-homogeneous state to Z–homogeneous state.
4. Transition from a Z-non-homogeneous state to Z–non-homogeneous state.

One of the four basic types of conversions are used to convert any Z-non homogeneous Mealy machine into a Z-homogeneous (Moore) machine, which are shown in Table 5.9

Table 5.9. Transitions for converting Z–non homogeneous machine into a Z–homogeneous machine

Case	*Z-non homogeneous Mealy Machine*	*Equivalent Z-homogeneous (Moore) Machine*
1	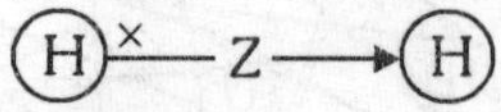	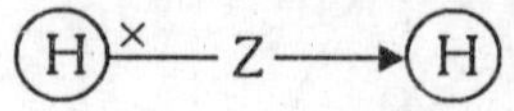
2	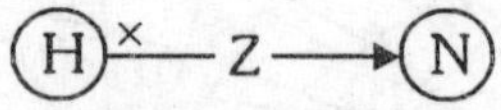	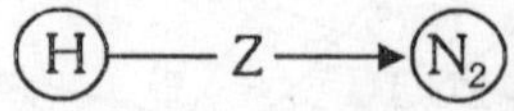
3	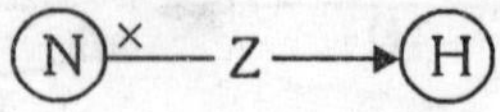	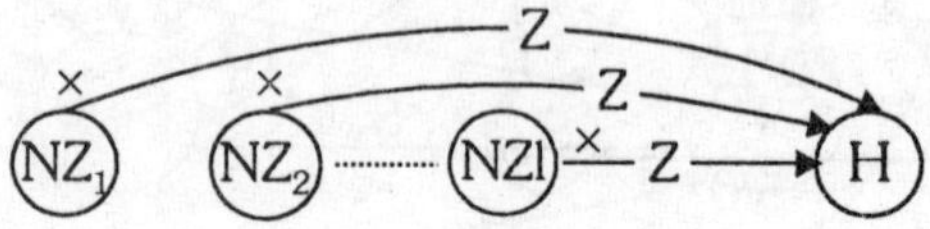
4	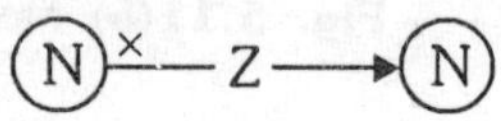	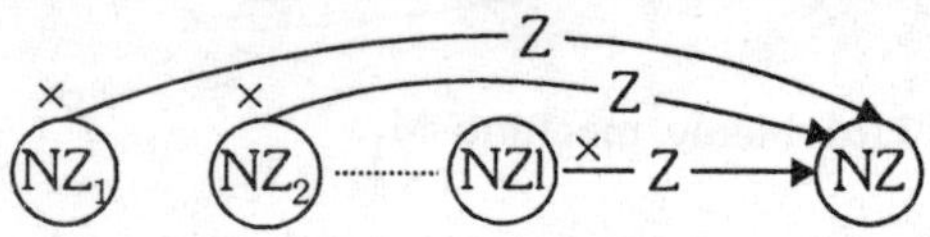

H : *a* Z–homogeneous state

N : *a* Z–non homogeneous state

NZ_1, NZ_2 ...NZ_l : equivalent Z–homogeneous states of N which receive output Z_1, Z_2 ... Z_l respectively.

By using these conversions M_1 can be converted into a Z-homogeneous machine M_3, which is shown in Fig. 5.12 (*a*). This Z-homogeneous machine is now readily to be drawn in the type of Moore machine is shown in Fig. 5.12(*b*), it contains states S_0, S_{20}, S_{25}, S_{30} and S_{35} represent states $S_0^{z=0}$, $S_0^{z=5}$, $S_0^{z=10}$, $S_0^{z=15}$ and $S_0^{z=20}$ respectively.

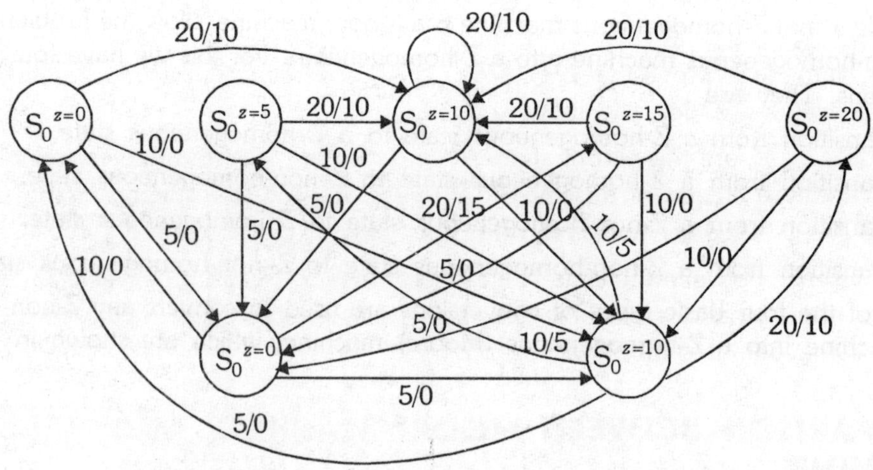

(*a*) Z–Homogeneous machine equivalent.

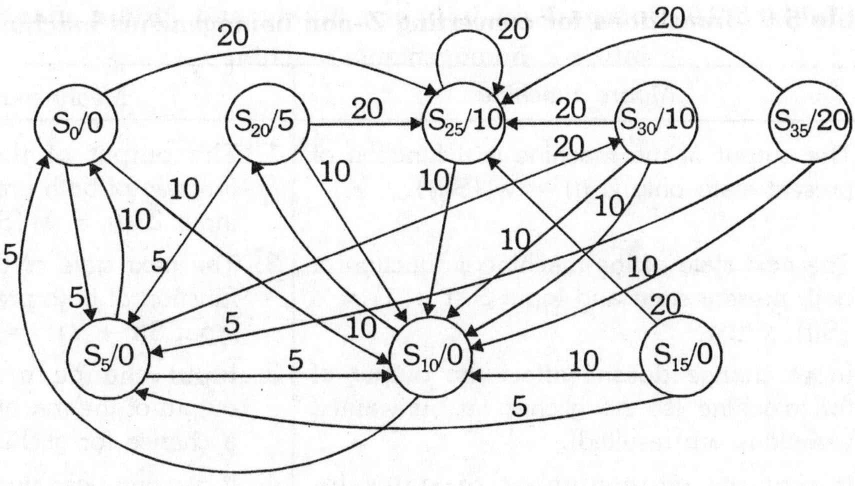

(*b*) Moore machine equivalent.

Fig. 5.12 Moore machine equivalent M_3 of Mealy machine M_1.

Theorem. Every Mealy machine can be converted into a Moore equivalent machine, and vice-versa.

Proof. To show that every Moore machine obtained from Mealy machine by using the conversions described in Table 5.4 is equivalent to be Mealy machine, it is necessary to show that:

1. All the new Z–homogeneous states split from a Z–non homogeneous state of the given machine are equivalent to each other.

2. They are equivalent to the original Z–non homogeneous state numbers.

Let NZ_1, NZ_2, ..., NZ_l, be *l* numbers Z–homogeneous states split from a Z–non homogeneous state N. To prove that NZ_1, NZ_2, ..., NZ_l are equivalent is to prove that they yield identical output sequences for any input sequence.

Since Z_1, Z_2,...., Z_l are the output symbols produced by the transitions incident to the state N, the transformations shown in the Table 5.9 guarantee that the splitting of the state N into state NZ_1, NZ_2, ..., NZ_l does not affect the other part of the machine.

This equivalence between the state N and the state NZ_1, NZ_2, ..., NZ_l, is again immediately seen from the transformations of Table 5.9, because when the same input symbol is applied to them, the same output symbol will be observed.

5.6 COMPARISON BETWEEN MOORE MACHINE AND MEALY MACHINE

The comparison between Mealy and Moore machines are shown in Table 5.10.

Table 5.10 Comparison between Mealy and Moore machine

Sl. No.	Moore machine	Mealy machine
1.	The output of the machine is a function of present state only Z (*t*) = λ {S(*t*)}.	1. The output of the machine is function of both present state and input Z (*t*) = λ {S(*t*), *x*(*t*)}.
2.	The next state of the machine is function of both present state and input S (*t* + 1) = δ {S(*t*), *x* (*t*)}.	2. The next state of the machine is function of both present state and input S(*t* + 1)} = δ {S(*t*), *x*(*t*)}.
3.	Input change doesn't affect the output of the machine (so no gliches *i.e.*, unwanted variations are resulted).	3. Input change may affect the output of the machine, so there is a chance for gliches.
4.	It requires more number of states for implementing.	4. It requires less number of states to implement the same circuit.
5.	Less flexible to design.	5. The designer has more flexibility, since the output and next state are functions of both input and present state.
6.	**Examples**	6. **Examples**
(*a*)	Block diagramDraw Fig. 5.6	Block diagram Draw Fig. 5.3
(*b*)	State table Table 5.2	State table Table 5.1
(*c*)	State diagram Fig. 5.5	State diagram Draw Fig. 5.2
(*d*)	Timing diagram Fig. 5.4	Timing diagram Fig. 5.4
(*e*)	Logic circuit.	

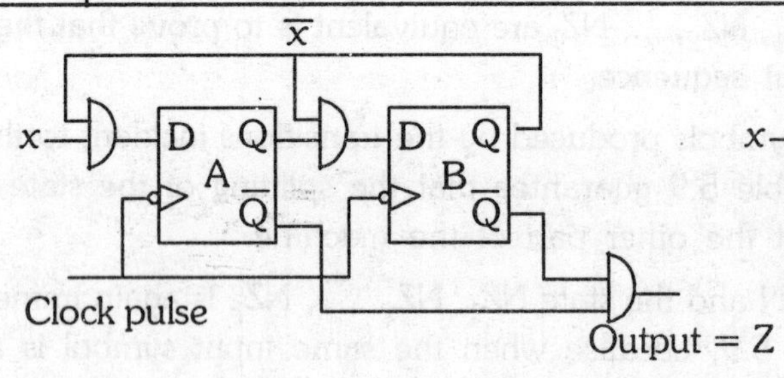

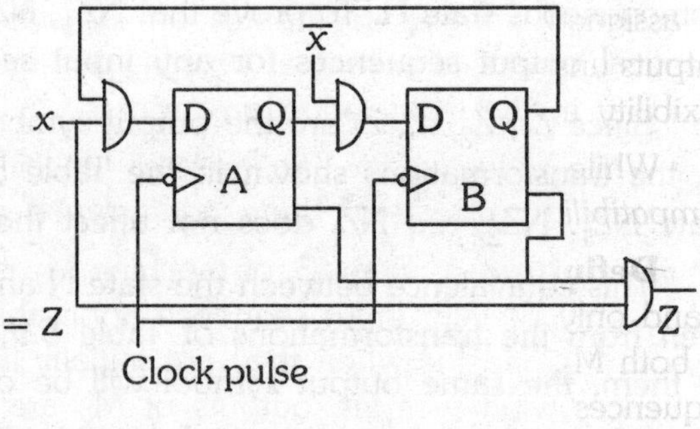

5.7 SIMPLIFICATION OF INCOMPLETELY SPECIFIED MACHINES

While designing some type combinational logic circuits, we encounter the condition where the truth table was incompletely specified *i.e.*, some terms are don't care. The same thing can happen in sequential circuits, either state transitions or output variables are not completely specified.

Such type of machines are said to be "incompletely specified machines". In this section, we see the properties and different methods for solving the incompletely specified machines.

When all state transitions are not specified, it is difficult to predict the machine's future behaviour. Such situation can be avoided as follows :

1. We shall apply the input sequence to the machine at any of its possible initial state, such that, no unspecified next state is encountered, except possibility at the final step. Such input sequence is said to be "*applicable*" to the initial starting state S_i of M.

Note: All outputs encountered need not be specified for a sequence to be applicable to S_i.

2. The specified behaviour of a machine with partially specified machine can be described by another machine whose state transitions are completely specified. This transformation is accomplished by replacing all the dashes (unspecified) in the next state entered by 'T' and adding a terminal state T whose output is unspecified.

This is illustrated in Table 5.11(*a*) and Table 5.11(*b*). Table 5.11(*a*) gives the incompletely specified sequential machine state table and Table 5.11(*b*) gives an equivalent description of state table where all transitions are specified *i.e.*, completely states specified state table.

Table 5.11(*a*) Incompletely specified machine

Present state	*Next state, Z*	
	x = 0	*x = 1*
A	A, 0	—
B	C, 1	—
C	— 1	D, 0
D	B, 1	—

Table 5.11(*b*) (Equivalent) transformation machine

Present state	*Next state, Z*	
	x = 0	*x = 1*
A	A, 0	T, –
B	C, 1	T, –
C	T, 1	D, 0
D	B, 1	T, –
T	T, –	T, –

Table 5.11(*b*) adds state T to describe specified behaviour of sequential machine, in which all state transitions are specified, and the outputs, are partially defined. Unspecified outputs can be assigned without any impact on the state sequence. However, it is advantageous to leave outputs unspecified as long as possible during the state reduction process. This provides additional flexibility in the state reduction process.

While reducing the incompletely specified state table we use terms 'state covers', '*state compatibility*' instead of '*state equivalence*'.

Definition 1. State S_i of machine M_1 is said to '*cover or contain*', state S_j of machine M_2 if and only if every input sequence applicable to S_i is also applicable to S_j, and its application to both M_1 and M_2 when they are initially in S_i and S_j, respectively results an identical output sequences whenever the outputs of M_2 are specified.

The covering concept can be extended to machines as follows :

Machine M_1 said to cover machine M_2 if and only if, for every state S_j in M_2, there is corresponding state S_i in M_1 such that S_i covers S_j.

Definition 2. (State Compatibility). The two states S_i and S_j of a machine M are said to be '*compatible*', if and only if, for every input sequence applicable to both S_i and S_j, the same output sequence will be produced whenever both outputs are specified, regardless of whether state S_i or S_j is the initial state.

So, S_i and S_j are compatible if and only if their outputs are identical when specified and their I_i–successors, for every I_i for which both are specified, are either the same are also compatible. In general, three or more states S_i, S_j, S_k, ... are compatible if and only if, for every applicable input sequence, with any one of the initial state may produce same outputs. Thus a set of states (S_i, S_j, ...,) are called compatible (*c*) if all of its members are compatible. A compatible C_i is said to be *larger* than or to cover, another compatible C_j if and only if every state contained in C_j is also contained in C_i. A compatible is maximum if it is not covered by any other compatible.

Consider the state Table 5.12 shown below. In this table there are following compatible states.

1. A and E are compatible, *i.e.*, A = E, if the output Z for E when x = 0 and x = 1 is '0' (instead of don't care).

2. States B and D are compatible, *i.e.*, B = D, if the output Z of states B and D are either 0 or 1 (instead of don't care).

Table 5.12. Incompletely specified state table

Present state	*Next state, Z*	
	x = 0	*x = 1*
A	A, 0	B, 0
B	A, –	C, 0
C	D, –	D, 1
D	A, –	C, 0
E	A, –	B, –

5.7.1 Merger Graphs

Merger graph is a state reducing tool used to reduce the states in the incompletely specified machines. It is defined as follows :

1. The number of vertices in the graph is equal to the number of states in the state table.

2. Each compatible state pair is indicated by a line (undirected arc) drawn between the two state vertices, whose *next states and output* are not conflecting.

3. If for a pair of states (S_i, S_j), the corresponding outputs underall inputs are same (not conflecting), but the next states are different, then those two states are connected by a broken line (arc), and the implied pairs are entered in the broken space.

The I_K successors S_i and S_j are S_p and S_q, respectively and outputs for any input sequence is same. If S_i and S_j is said to be a compatible pair, then S_p and S_q are said to be implied pair.

4. If the outputs and next states of the two states (S_i, S_j) are conflicting (or) only outputs are conflicting *i.e.*, two states are not compatible, no arc/line is drawn between the two states.

Consider an example shown in Table 5.13 and merger graph shown Fig. 5.13.

Table 5.13. State table

Present state	Next state, Z			
	x_1	x_2	x_3	x_4
A	–	C, 1	E, 1	B, 1
B	E, 0	–	–	–
C	F, 0	F, 1	–	–
D	–	–	B, 1	–
E	–	F, 0	A, 0	D, 1
F	C, 0	–	B, 0	C, 1

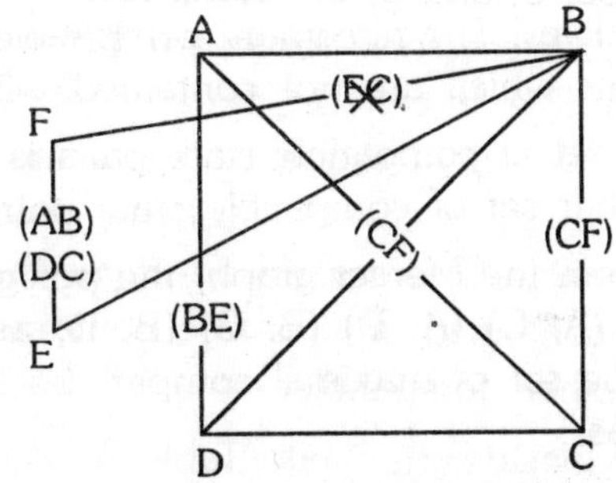

Fig. 5.13 Merger graph.

1. The states A and B are compatible, because the next states and outputs of both states for all inputs are same *i.e.*, not conflicting. Therefore, an arc/line is drawn between the states A and B.

Similarly BD, DC and BE are also compatible states.

2. The states A and C have non conflicting outputs, but the successors under input x_2 *are* C and F respectively. Therefore, states A and C are compatible only if implied states C and F are compatible. This is indicated by drawing a broken arc/line with CF written in between.

Similarly states A and D are compatible if implied states B and E are compatible.

Pairs B and C are compatible if implied pairs E and F are compatible.

Pairs B and F are compatible if implied pairs C and E are compatible.

Pairs F and E are compatible if implied pairs (A and B) and (C and D) are compatible.

3. No arc/ line is drawn between states A and E and A and F because the outputs under x_2, x_3 and x_3 respectively are conflecting *i.e.*, states A and E, and A and F are incompatible.

In the similar way, the Merger graph is drawn for all possible pair of states. From the Merger graph, we can see all possible compatible pairs and implied pairs, and we know that a pair of states is compatible if and only if the implied pair(s) is (are) compatible. Therefore, it is necessary to check whether the implied pairs are indeed compatibles. If implied states are incompatible *i.e.*, no line is drawn between those states, they are crossed and the corresponding line is ignored.

For example, the implied states C and E are incompatible, because no line is drawn *between* states C and E. So the states B and F are incompatible, hence that line is crossed/ ignored as shown in Fig. 5.13. Hence it is necessary to check whether the incompatibility of B and F does invalidate any other implied pair or not, if yes, that particular pair also comes under incompatible states list.

The broken lines which remain in the graph, after all the implied pairs have been verified to be compatible, are regarded as complete lines. After checking all possibilities of incompatibility in the merger graph, result the following nine compatible pairs

(A, B) (A, C) (A, D) (B, C) (B, D) (B, E) (C, D) (C, F) (E, F).

These compatible pairs are further checked for further compatibility. For example pairs (A, B) (A, C) and (B, C) are compatible, so (A, B, C) is also compatible. In this way, entire set of compatibles of the machine can be generated from its compatible pairs.

To find the minimal set of compatibles for state reduction, it is useful to find *maximal compatibles*. The maximum compatibles can be found by looking at merger graph for complete polygons which are not contained within any higher-order complete polygons.

A set of compatible state pairs is said to be maximum if it is not completely covered by any other set of compatible state pairs.

From the Merger graph, the polygon (A, B, C, D) is a higher order polygon and it covers (A, B) (A, C) (A, D) (B, C) (B, D) and (C, D) compatible pairs and these pairs are ignored from the set of maximal compatibles for the machine. The set of maximal compatibilities are given as :

{(A, B, C, D); (B, E); (C, F); (E, F)}

5.7.1.1 Closed set of compatibles

A set of compatibles for a machine M is said to be closed, if for every compatible contained in the set, all its implied compatibles are also contained in the set. A closed set of compatibilities which contains all the states of M is called a "*closed covering*".

Consider the set of compatibles {(ABCD) (EF)} of the above example, since this is the minimal number of compatibles covering all the states of the machine M, it defines *a lower bound* on the number of states in the minimal expression of machine.

The set containing all the maximal compatibles is clearly, a closed covering, since it covers all the states of the machine, and every implied compatible is contained in the set. The set of maximal compatibles places an "upper bound" on the number of states in the machine which covers the original one. For the above example, this bound is four.

Note: The upper bound is meaningless if the number of compatibles is larger than the number of states in the original machine.

5.7.2 Merger Table

When performing the state reduction of machines, having large number of states, it is difficult with the merger graph. It is more convenient to find the compatible pair and their implications using merger table, which is shown in Fig. 5.14. The state table of the machine is shown in the Table 5.14.

Table 5.14. State Table

Present state	*Next state, Z* x_1	x_2
A	E, 0	B, 0
B	F, 0	A, 0
C	E, –	C, 0
D	F, 1	D, 0
E	C, 1	C, 0
F	D, –	B, 0

B	EF				
C	BC	AC, EF			
D	×	×	EF		
E	×	×	√	CD, CF	
F	DE	AB, DE (crossed)	BC, DE	BD (crossed)	BC, CD
	A	B	C	D	E

Fig. 5.14 Merger table.

Procedure for plotting of merger table

1. Each cell of the merger table corresponds to the compatible pair represented by the intersection of row and column headings.

2. The incompatibility of two states is indicated by '×' mark in the corresponding cell and compatibility is indicated by '√' mark in the cell.

For example, states A and D, A and E are incompatible shown by × mark and C and E are compatible and the corresponding cell is placed with check mark.

3. The implied pair or pairs corresponding to compatible state pair are written within the cell.

For example, states A and C are compatible only when the implied states B and C are compatible, therefore pair BC is written in the cell corresponding to A and C. In a similar way, entire merger table is written. Now it is necessary to check whether implied pairs are compatible or not. This can be checked by observing the merger table. The implied states are incompatible if corresponding cells contain '×' mark.

For example, implied pair B D is incompatible, because cell BD contains '×' mark. Hence the states D and F are also incompatible. It is indicated by cross mark in Fig. 5.14.

Once the merger table is completed, the set of all maximal compatibles can be found by procedure as follows :

1. Start with right most column in the merger table and proceed to left until a column containing a compatible pair is encountered. List all the compatible pairs in that column. In our example, the compatible pair is EF.

2. Proceed left to the next column containing at least one compatible pair. If the state to which this column corresponds is compatible with all the states in the set of previously determined compatible states, then add this state to the set of compatible states to form a larger compatible. If the state is not compatible with all the states of previously determined set, but is compatible with some other state, form a new set of compatible states.

For example, state D is not compatible with both E and F which are previously determined set of states. However, it is compatible with state E. Therefore, it is not added in the previous set, but the new set is formed, DE. On the other hand, state C is compatible with both E and F which are previously determined set of states and hence it is included in that set to form a new set CEF. Similarly, C is also included in set DE to get a new set CDE.

3. Repeat step 2 until the left most column is reached.

After applying above procedure to the merger table shown in Fig. 5.14 gives the following set of maximal compatibles.

Column E : (EF)

Column D : (EF) (DE)

Column C : (EF) (DE) (CF) (CD) CE (C, D, E) (C, E, F)

Column B : (C, D, E) (C, E, F) (BC)

Column A : (C, D, E) (C, E, F) (BC), (AB) (AC) (AF) (C, D, E) (C, E, F)
(A, B, C) (A, C, F)

From the column C. It is evident that state C is compatible with states D, E and F and consequently the compatibilities generated previously are enlarged to include state C.

From the column B. It has a single compatible pair, which is added to the previously generated list.

From the column A. With rows B and C, we obtain the compatible (ABC), while with rows C and F, together with the previously available (CF), yield the compatible (ACF).

The final set of maximal compatibles clearly indicates that, the given machine can be covered by a four state machine. The minimum state machine, which covers the given machine contains only three states {(ABC) (CD) (EF)} and the Table 5.15 shows it.

Table 5.15. A state table of minimal machine

Present state	*Next state (Z)*	
	x_1	x_2
(ABC) → α	γ, 0	α, 0
(CD) → β	γ, 1	β, 0
(EF) → γ	β, 1	α, 0

5.8 SOLVED PROBLEMS

Problem 1. *(a) Determine the minimal state equivalent of the state table given in Table 5.16. (b) Find the minimum length sequence that distinguishes states A and F.*

Table 5.16

Present state	*Next state, Z*	
	x = 0	*x = 1*
1	*1, 0*	*1, 0*
2	*1, 1*	*6, 1*
3	*4, 0*	*5, 0*
4	*1, 1*	*7, 0*
5	*2, 1*	*3, 0*
6	*4, 1*	*5, 0*
7	*2, 0*	*3, 0*

Solution. Step 1. First partition P_0 corresponds to 0-distinguishability, it defines all initial states. Partition the states into subsets such that all states in the same subsets are 1-equivalent.

$$P_0 = (ABCDEFG)$$

The first partition P_1 can be obtained by placing those states having the same outputs under all inputs, in the same block. This partitioning gives two subsets.

(ABCDEFG) — x = 0: (ABCDFG) (E)
(ABCDEFG) — x = 1: (ABCDEFG)

$$\therefore \quad P_1 = (ABCDFG)\ (E)$$

Step 2. Partition the states into subsets that all states in the same subset are 2-equivalent.

This can be obtained by observing that two states are 2-equivalent if and only if they are 1-equivalent and their I_i-successor, for all possible I_i, are also 1-equivalent. (If two states are placed in the same block of partition P_2 if and only if they are in same block of P_1 and for each possible I_i those I_i-successors are in a block of P_1).

1. (a) The 0-successors of (ABCDFG) are (ECBGED), in which (CBGD) and (E) are not contained in the same block of P_1. Therefore block (ABCDFG) must be split into (BCDG) (AF).

(*b*) The 1-successors of (ABCDFG) are (CABADG), all successors group, that belong to the same group set. Set of P_1 belongs to (ABCDFG) is subset of P_1.

		next states	subsets
(ABCDFG)	x = 1	(CABADG)	(ABCD) (FG)
	x = 0	(ECBGED)	(AF) (BCDG)

2. The 0 and 1-successors of (F) are E and D.

Note: Whenever a subset is having only one state, then no need for partitioning. That itself is a separate subset.

$$\therefore \quad P_2 = \text{(AF) (BCDG) (E)}$$

Step 3. Partition the states into subsets such that all states in the same subset are 3-equivalent.

By applying the same procedure in step 2, we can split the block (BCDG) of P_2 into (BD) and (CG), since 1 successor of (BCDG) is (ABAG) which is not a simple block of P_2.

		next states	subsets
(BCDG)	x = 0	(CBGD)	(BC) (DG)
	x = 1	(ABAG)	(BD)(CG)

$$\therefore \quad P_3 = \text{(AF) (BD) (CG) (E)}$$

Step 4. Partition the states into subsets such that all states in the same subset are 4-equivalent.

By applying the same procedure, we can split the (AF) of P_3 into (A) and (F) since 1-successor of (AF) is (CD) which is not in the single block of P_3.

		next states	subsets
(AF)	x = 1	CD	(A)(F)
	x = 0	EE	

There is no split in (BD), (CG) since for 0 or 1-successor, the next states are same subset of P_3.

$$\therefore \quad P_4 = \text{(A) (F) (BD) (CG) (E).}$$

Step 5. Here no split in (BD) and (CG) takes place so $P_5 = P_4$

$\because$ $P_{k+1} = P_k$, hence the partitional process terminates and P_k defines the sets of equivalent states of the sequential machine.

The P_k thus called "equivalence partition" and this procedure is called "Moore reduction procedure".

$$\therefore \quad \boxed{P_4 = \text{(A) (F) (BD) (CG) (F)}}$$

The states in the common blocks of P_5 are equivalent and redundant. Therefore, state B = D and C = G are redundant states and can be eliminated by one state to give reduced minimised state table shown in Table 5.17 which contains only 5 states but in the original state table contains 7 states. With minimization two states have reduced.

In the B = D, 'B' is considered and in the C = G, C is considered.

The five states are A, B, C, E and F.

(*b*) States (A) and (F) separated in the P_4-partition. So A and F are 3-equivalent and 4-distinguishable. The distinguish sequence is "0011".

Table 5.17

Present state	Next state, Z x = 0	x = 1
A	E, 0	C, 0
BD = B	C, 0	A, 0
CG = C	B, 0	B, 0
E	F, 1	B, 0
F	B, 0	C, 0

Problem 2. *(a) Determine a minimal state table equivalent to the given state table (Table 5.18).*

(b) Find the minimum length sequence that distinguishes state 1 from 4.

Table 5.18

Present state	Next state, Z x = 0	x = 1
1	1, 0	1, 0
2	1, 1	6, 1
3	4, 0	5, 0
4	1, 1	7, 0
5	2, 1	3, 0
6	4, 1	5, 0
7	2, 0	3, 0

Solution.

Step 1. P_0 = (1, 2, 3, 4, 5, 6, 7)

(1, 2, 3, 4, 5, 6, 7) — x = 0: (1, 3, 5, 6, 7) (2, 4)
(1, 2, 3, 4, 5, 6, 7) — x = 1: (1, 3, 4, 5, 6, 7) (2)

∴ P_1 = (1, 3, 4, 5, 6, 7) (2)

Step 2.

		next states	subsets
(1, 3, 4, 5, 6, 7)	x = 0	(1, 4, 1, 2, 4, 2)	(1, 3, 4, 6) (5, 7)
	x = 1	(1, 5, 7, 3, 5, 3)	(1, 3, 4, 5, 6, 7)

∴ P_2 = (1, 3, 4, 6) (5, 7) (2)

Step 3.

		next states	subsets
(1, 3, 4, 6)	x = 0	(1, 4, 1, 4)	(1, 3, 4, 6)
	x = 1	(1, 5, 7, 5)	(1) (3, 4, 6)
(5, 7)	x = 0	(2, 2)	(5, 7)
	x = 1	(3, 3)	(5, 7)

∴ P_3 = (1) (3, 4, 6) (5, 7) (2)

Step 4.

(3, 4, 6) — x = 0: (4, 1, 4) ⟶ (3, 6) (4)
(3, 4, 6) — x = 1: (5, 7, 5)

No partitions are made in the subset (5, 7)

$\therefore$ $P_4 = (1)\ (2)\ (3, 6)\ (4)\ (5, 7).$

Step 5.

(3, 6) — $x = 0$: (4, 4) ⟶ (3, 6); $x = 1$: (5, 5) ⟶ (3, 6)

(5, 7) — $x = 0$: (2, 2) ⟶ (5, 7); $x = 1$: (3, 3) ⟶ (5, 7)

$\therefore$ $P_5 = (1)\ (2)\ (3, 6)\ (4)\ (5, 7)$

Here $P_5 = P_4.$

$\therefore$ Final minimal state equivalents are (1) (2) (3, 6) (4) (5, 7).

State 3 = 6 and 5 = 7 in the reduced state table containing only 5 states. In 3 = 6, 3 and 5 = 7, 5 are considered in the reduced state table. (see Table 5.19)

(*b*) The states 1 and 4 are distinguished in P_3 partition. Hence states 1 and 4 are 2- equivalent and are 3-distinguishable and the distinguished sequence is "101".

Table 5.19

Present state	*Next state, Z*	
	x = 0	*x = 1*
1	1, 0	1, 0
2	1, 0	3, 1
3	4, 0	5, 0
4	1, 1	5, 0
5	2, 0	3, 0

Problem 3. *Find the equivalence partition for the machine M_1 shown below. (see Table 5.20)*

Table 5.20

Present state	*Next state, Z*	
	x = 0	*x = 1*
A	B, 1	H, 1
B	F, 1	D, 1
C	D, 0	E, 1
D	C, 0	F, 1
E	D, 1	E, 1
F	C, 1	E, 1
G	C, 1	D, 1
H	C, 0	A, 1

(*i*) *Show the standard form of the corresponding reduced machine.*

(*ii*) *Find a minimum length sequence that distinguishes states A and B.*

Solution.

Step 1. First partition P_0 corresponds to 0-distinguishability.

$$P_0 = (A, B, C, D, E, F, G, H)$$

States having the same outputs under all input conditions can be grouped as

(ABCDEFGH) — $x = 0$ (ABEFG) (CDH)
(ABCDEFGH) — $x = 1$ (ABCDEF)

$$P_1 \overset{x=0}{=} \text{(ABEFG) (CDH)}$$

Step 2. 1-successors of (ABEFG) or 2-equivalent is

(ABEFG) — $x = 0$ (BFDCC) ⟶ (AB) (EFG)
(ABEFG) — $x = 1$ (HDEED) ⟶ (EF) (ABG)

and 1-successor of (CDH) is

(CDH) — $x = 0$ (DCC) ⟶ (C, D, H)
(CDH) — $x = 1$ (EFA) ⟶ (C, D, H)

$$\therefore \quad P_2 \overset{x=0}{=} \text{(AB) (EFG) (CDH)}$$

Step 3. The 1-successors of (AB), (EFG) and (CDH) or 3-equivalents are

(AB) — $x = 0$ (BF) ⟶ (A) (B)
(AB) — $x = 1$ (HD) ⟶ (AB)

(EFG) — $x = 0$ (DCC) ⟶ (EFG)
(EFG) — $x = 1$ (EED) ⟶ (EF) (G)

(CDH) — $x = 0$ (DCC) ⟶ (CDH)
(CDH) — $x = 1$ (EFA) ⟶ (CD) (H)

if we take $x = 1$, then

$$P_3 = \text{(A) (B) (EF) (G) (CD) (H)}$$

Step 4. 1-successors (or) 4-equivalents of (EF) and (CD) are

(EF) — $x = 0$ (DC) ⟶ (EF)
(EF) — $x = 1$ (EE) ⟶ (EF)

(CD) — $x = 0$ (DC) ⟶ (CD)
(CD) — $x = 1$ (EF) ⟶ (CD)

$$\therefore \quad P_4 \overset{x=0}{=} \text{(A) (B) (CD) (EF) (G) (H)}$$

(A) → S_1, (B) → S_2, (CD) → S_3, (EF) → S_4, (G) → S_5, (H) → S_6

Step 5. The equivalent states are C = D and E = F . The resultant state table contains only 6 states by labelling the 6 states namely S_1, S_2, S_3, S_4, S_5 and S_6.

The reduced machine M_1* is shown below. (see Table 5.21) The canonical form (or) standard form of a original machine can be obtained by the isomorphism.

Table 5.21

Present state	Next state, Z	
	x = 0	x = 1
S_1	S_2, 1	S_6, 1
S_2	S_4, 1	S_3, 1
S_3	S_3, 0	S_4, 1
S_4	S_3, 1	S_4, 1
S_5	S_3, 1	S_3, 1
S_6	S_3, 0	S_1, 1

Here rename (original machine state names) the 6 states by taking S_1 as initial state and name it as A (the state name of the machine M). The next labels are selected in such a way that moving row wise from left to right, starting with A and going down through B, C, etc. Whenever the new label appears in the reduced state table, assign the next state name (character) to that.

The standard form of machine M_1* is shown below. Here S_1 is assigned by A, its 0-successor is S_2 which is a new label, then assigned by next character/alphabet B. Similarly 1-successor S_1 is S_6, which is the first occurrence and is named as C.

In row B (S_2) the 0-successor is S_4, which is a new occurrence and is labelled as D, and 1-successor is S_3 labelled as E.

The next row is C (S_6) whose 0-successor and 1-successor are S_3 and S_1, which were already labelled as E and A. Next row is D (S_4) and so on. The standard form of M_1* is shown in Table 5.22.

(*b*) The minimum length sequence that distinguishes states A and B is 3 *i.e.*, states A and B are 2-equivalents and 3-distinguishable. The distinguishable sequence is "001"

Problem 4. *Find the equivalence partition for the machine M_2 whose state table is given below. (Table 5.23) Show the standard form of the corresponding reduced machine.*

Table 5.22

Present state	Next state, Z	
	x = 0	x = 1
$S_1 \to A$	B, 1	C, 1
$S_2 \to B$	D, 1	E, 1
$S_6 \to C$	E, 0	A, 1
$S_4 \to D$	E, 1	D, 1
$S_3 \to E$	E, 0	D, 1
$S_5 \to F$	E, 1	C, 1

Table 5.23

Present state	Next state, Z	
	x = 0	x = 1
A	D, 0	C, 1
B	E, 1	A, 0
C	H, 1	D, 1
D	D, 0	C, 1
E	B, 0	G, 1
F	H, 1	D, 1
G	A, 0	F, 1
H	C, 0	A, 1
I	G, 1	H, 1

Solution.

Step 1. First partition P_0 corresponds to 0-distinguishability.

$$P_0 = (A, B, C, D, E, F, G, H, I)$$

States having the same outputs under all input conditions can be grouped as

(ABCDEFGHI) — $x = 0$: (BCFI) (ADEGH)
$x = 1$: (ABCDEFGHI)

$\therefore \quad P_1 \overset{x=0}{=} \text{(ADEGH) (BCFI)}$

Step 2. 1-successors (or) 2-equivalents of (ADEGH) and (BCF) are

(ADEGH) — $x = 0$: (D, D, B, A, C) ⟶ (ADG) (EH)
$x = 1$: (C, C, G, F, H) ⟶ (ADG) (EH)

(BCFI) — $x = 0$: (E, H, H, G) ⟶ (BCF)(I)
$x = 1$: (A, D, D, H) ⟶ (BCF)

$\therefore \quad P_2 \xrightarrow{x=0} \text{(ADG) (EH) (BCF) (I)}$

Step 3. The 1-successors (or) 3-equivalents of (ADG) (EH) and (BCF) are

(ADG) — $x = 0$: (DDA) ⟶ (ADG)
$x = 1$: (CCF) ⟶ (ADG)

(BCF) — $x = 0$: (EHH) ⟶ (BCF)
$x = 1$: (ADD) ⟶ (BCF)

(EH) — $x = 0$: (BC) ⟶ (HE)
$x = 1$: (GA) ⟶ (EH)

$\therefore \quad P_3 \overset{x=0 \text{ (or) } x=1}{=} \text{(ADG) (BCF) (EH) (I)}$

Here $\quad P_3 = P_2$

The equivalent states are A = D = G, B = C = F and E = H. Hence the reduced state table of the machine M_2 contains only four states. All the above subsets are relabelled as

$$(ADG) \rightarrow S_1,\ (BCF) \rightarrow S_2,\ (HE) \rightarrow S_3 \text{ and } I \rightarrow S_4$$

and the reduced machine M_2^* state table is shown in Table 5.24.

Table 5.24

Present state	Next state, Z	
	$x = 0$	$x = 1$
S_1	S_1, 0	S_2, 1
S_2	S_3, 1	S_1, 1
S_3	S_2, 0	S_1, 1
S_4	S_1, 1	S_3, 1

To find the standard form of machine M_2^*, let us consider S_1 as initial state and name it as A. The 0-successor of A is S_1, which was already labelled and 1-successor is S_2, name it as B. Similarly S_3 as C and S_4 as D. The standard form of reduced machine M_2^* is shown in Table 5.25.

Table 5.25

Present state	Next state, Z	
	$x = 0$	$x = 1$
A (S_1)	A, 0	B, 1
B (S_2)	C, 1	A, 1
C (S_3)	B, 0	A, 1
D (S_4)	A, 1	C, 1

Problem 5. *Determine the minimal state equivalent of the Machine M_3 of the state table is shown in Table 5.26 and find its standard form.*

Table 5.26

Present state	Next state, Z x = 0	Next state, Z x = 1
q_0	q_0, 1	q_4, 0
q_1	q_0, 0	q_4, 0
q_2	q_1, 0	q_5, 0
q_3	q_1, 0	q_5, 0
q_4	q_2, 0	q_6, 1
q_5	q_2, 0	q_6, 1
q_6	q_3, 0	q_7, 1
q_7	q_3, 0	q_7, 1

Solution. The state of the machine M_3 is having 8 states, q_0 through q_7.

Step 1. The first partition P_0 has the 0-distinguishable states

$$P_0 = (q_0q_1q_2q_3q_4q_5q_6q_7)$$

The states having the same outputs under all input conditions are can be grouped as

$$q_0q_1q_2q_3q_4q_5q_6q_7 \begin{cases} x = 0 & (Q_0)\ (q_1q_2q_3q_4q_5q_6q_7) \\ x = 1 & (Q_0q_1q_2q_3\ (q_4q_5q_6q_7) \end{cases}$$

$$\therefore \quad P_1 \xrightarrow{x\,=\,0,\,1} (q_0)\,(q_1q_2q_3)\,(q_4q_5q_6q_7)$$

Step 2. The 1-successor (or) 1-equivalent of two above blocks are

		next states		subsets
$(q_1q_2q_3)$	x = 0	$(q_0q_1q_1)$	→	(q_1) (q_2q_3)
	x = 1	$(q_4q_5q_6q_7)$	→	$(q_1q_2q_3)$
$(Q_4q_5q_6q_7)$	x = 0	$(q_2q_2q_3q_3)$	→	$(q_4q_5q_6q_7)$
	x = 1	$(q_6q_6q_7q_7)$	→	$(q_4q_5q_6q_7)$

$$\therefore \quad P_2 \xrightarrow{x\,=\,0} (q_0)\,(q_1)\,(q_2q_3)\,(q_4q_5q_6q_7)$$

Step 3. The 1-successor (or) 2-equivalent of (q_2q_3) and $q_4q_5q_6q_7$ belong to the same block.

$$\therefore \quad P_3 = (q_0)\ (q_1)\ (q_2q_3)\ (q_4q_5q_6q_7)$$

Hence, the equivalent states are

$$q_2 = q_3 \text{ and } q_4 = q_5 = q_6 = q_7.$$

So the reduced machine, M_3^* of original machine M_3 contains only 4 states instead of 8 states and these blocks are renamed as

$$q_0 \rightarrow S_1$$
$$q_1 \rightarrow S_2$$
$$q_2q_3 \rightarrow S_3$$
$$(q_4q_5q_6q_7) \rightarrow S_4$$

The reduced machine state table is shown in Table 5.27.

Table 5.27

Present state	Next state, Z x = 0	Next state, Z x = 1
S_1	S_1, 1	S_4, 0
S_2	S_1, 0	S_4, 0
S_3	S_2, 0	S_4, 0
S_4	S_3, 0	S_4, 1

The standard form for the machine M_3, can be obtained by taking the S_1 as initial state and naming it q_0 or A (which is a original state name) and, $S_4 \rightarrow q_1$ or B, $S_3 \rightarrow q_2$ or C, $S_2 \rightarrow q_3$ or D, (see Table 5.27(*a*) and 5.27(*b*)).

Table 5.27(a)

Present state	Next state, Z	
	$x = 0$	$x = 1$
q_0 (S_1)	q_0, 1	q_1, 0
q_1 (S_4)	q_2, 0	q_1, 1
q_2 (S_3)	q_3, 0	q_1, 0
q_3 (S_2)	q_0, 0	q_1, 0

(Or)

Table 5.27(b)

Present state	Next state, Z	
	$x = 0$	$x = 1$
A (S_1)	A, 1	B, 1
B (S_4)	C, 0	B, 1
C (S_3)	D, 0	B, 0
D (S_2)	A, 0	B, 0

Problem 6. *Find the equivalence partition for the machine M_4 the state table is shown in Table 5.28. Show the standard form of the corresponding reduced machine.*

Table 5.28

Present state	Next state, Z	
	$x = 0$	$x = 1$
A	D, 0	H, 1
B	F, 1	C, 1
C	D, 0	F, 1
D	C, 0	E, 1
E	C, 1	D, 1
F	D, 1	D, 1
G	D, 1	C, 1
H	B, 1	A, 1

Solution. Step 1. The first partition P_0 is a 0-distinguishable state

$$P_0 = (A, B, C, D, E, F, G, H).$$

States having the same outputs under all input conditions can be grouped as

(ABCDEFGH) — $x = 0$: (A, C, D) (BEFGH)
(ABCDEFGH) — $x = 1$: (ABCDEFGH)

$\therefore$ $P_1 \xrightarrow{x=0}$ (ACD) (BEFGH)

Step 2. The 1-successor (or) 2-equivalent of (ACD) and (BEFGH) are

(ACD) — $x = 0$: (DDC) ⟶ (ACD)
(ACD) — $x = 1$: (HFE) ⟶ (ACD)

(BEFGH) — $x = 0$: (FCDDB) ⟶ (BH) (EFG)
(BEFGH) — $x = 1$: (CDDCA) ⟶ (BEFGH)

$\therefore$ $P_2 \xrightarrow{x=0}$ (ACD) (BH) (EFG)

Step 3. The 1-successor (or) 3-equivalent of (ACD), (BH) and (EFG) are

(ACD) — $x = 0$: (DDC) ⟶ (ACD)
(ACD) — $x = 1$: (HFE) ⟶ (A) (CD)

(BH) — $x = 0$: (FB) ⟶ (B) (H)
(BH) — $x = 1$: (CA) ⟶ (BH)

(EFG) — $x = 0$: (CDD) ⟶ (EFG)
(EFG) — $x = 1$: (DDC) ⟶ (EFG)

$$\therefore \quad P_3 \xrightarrow{x=0,1} (A)\ (B)\ (CD)\ (H)\ (EFG)$$

Step 4. The 1-successor (or) 4-equivalent of (CD) and (EFG) are

$$(CD) \begin{cases} x=0 & (DC) \longrightarrow (CD) \\ x=1 & (FE) \longrightarrow (CD) \end{cases}$$

$$(EFG) \begin{cases} x=0 & (CDD) \longrightarrow (EFG) \\ x=1 & (DCC) \longrightarrow (EFG) \end{cases}$$

$$\therefore \quad P_4 \longrightarrow (A)\ (B)\ (CD)\ (EFG)\ (H)$$

The equivalent states are C = D and E = F = G, hence the reduced machine contains only 5 states. The groups are relabelled as

$$A \longrightarrow S_1 \quad (CD) \longrightarrow S_3 \quad \text{and} \quad (H) \longrightarrow S_5$$
$$B \longrightarrow S_2 \quad (EFG) \longrightarrow S_4$$

The reduced/minimized state table for machine M_4 is M_4* as shown in Table 5.29

Table 5.29

Present state	*Next state, Z*	
	x = 0	*x = 1*
S_1	S_3, 1	S_5, 1
S_2	S_4, 1	S_3, 1
S_3	S_3, 1	S_4, 1
S_4	S_3, 1	S_3, 1
S_5	S_2, 1	S_1, 1

To represent in the standard form, let us consider S_1 as initial state and name it as A, and its 0-successor is S_3 and is named as B, similarly $S_5 \rightarrow$ C, $S_4 \rightarrow$ D and $S_2 \rightarrow$ E.

The standard form state table for M_4* is shown below (see Table 5.30)

Table 5.30

Present state	*Next state, Z*	
	x = 0	*x = 1*
A (S_1)	B, 0	C, 1
B (S_3)	B, 0	D, 1
C (S_5)	E, 1	A, 1
D (S_4)	B, 1	B, 1
E (S_2)	D, 1	B, 1

Problem 7. *For the machine given below, find a minimal state reduced machine using merger table. (see Table 5.31)*

Table 5.31

Present state	Next state, Z	
	I_1	I_2
A	E, 0	B, 0
B	F, 0	A, 0
C	E, –	C, 0
D	F, 1	D, 0
E	C, 1	C, 0
F	D, –	B, 0

Table 5.32

	A	B	C	D	E
B	~~EF AB~~				
C	BC	~~EF AC~~			
D	×	×	~~EF CD~~		
E	×	×	~~BC~~	DC FC	
F	~~BD~~	~~DF AB~~	~~ED BC~~	~~FD BD~~	~~CD BC~~

Solution. The merger table for this problem is shown in Table 5.32.

Step 1. First we check if the implied pairs are compatible or not starting with right most column E.

E and F are compatible if CD and BC are compatible, but from the merger table it is clear that 'CD' are not compatible hence EF are also not compatible.

Similarly, the incompatible pairs are (AD) (AE) (BD) (BE) (CD) (CE) and (EF). If any implied pair in the merger table belongs to that set, that corresponding pair is incompatible.

Because of EF and BD, CD as a implied to pair AB and DF and DE and BC are incompatible with these new four incompatible pairs,

(BF), (CF), (AF), (AC) are incompatible.

Finally it is concluded that no state is having another state to make a compatible pair. The given machine itself is a minimized machine.

5.9 REVIEW QUESTIONS

1. What are the limitations of a finite state machine ?
2. Prove the following related to sequential machine. "If two states, S_i and S_j, of machine M are distinguishable, then they are distinguishable by a sequence of length $(n - 1)$ or less, where '*n*' is a number of states in a machine M."
3. What are the capabilities and limitations of a finite state machine ?
4. What is a Moore machine ? What is its state equivalence ?
5. What is a Mealy machine ? What is its state equivalence ?
6. Write comparisons *between* Mealy and Moore machines.
7. Explain the state minimization procedure.
8. Explain the standard form of Mealy and Moore sequential circuits.
9. Explain the procedure of state minimization using merger graph and merger table.
10. Write short notes on
    (*a*) Compatibility (*b*) Equivalence (*c*) Distinguishing sequence

11. Define 'state equivalence' and 'machine equivalence' with reference to sequential machine.

12. What is the difference between Mealy and Moore sequential circuits ? Discuss their advantages and disadvantages.

5.10 EXERCISE

1. Find the equivalence partition and a corresponding reduced machine in standard form for the given three machines.

Present State	Next State, Z	
	x = 0	x = 1
A	B, 0	E, 0
B	E, 0	D, 0
C	D, 1	A, 0
D	C, 1	E, 0
E	B, 0	D, 0

(a)

Present State	Next State, Z	
	x = 0	x = 1
A	F, 0	B, 1
B	G, 0	A, 1
C	B, 0	C, 1
D	C, 0	B, 1
E	D, 0	A, 1
F	E, 1	F, 1
G	E, 1	G, 1

(b)

Present State	Next State, Z	
	x = 0	x = 1
A	A, 0	B, 0
B	C, 0	D, 1
C	E, 0	D, 0
D	F, 0	E, 1
E	G, 0	A, 0
F	G, 0	B, 1
G	C, 0	F, 0

(c)

2. For the machine given below, find a minimal state reduced machine using merger graph and merger table.

Present State	Next State, Z	
	x = 0	x = 1
A	A, –	B, –
B	–	C, 0
C	D, 0	D, 1
D	–	B, 1

3. An *n*-state machine is supplied with a periodic input sequence whose period is P.

Prove that the output sequence must eventually become periodic, and find a bound for the period.

4. Determine the conditions under which two equivalent machines are isomorphic.

5. Give the procedure that can be used to determine whether two incompletely specified machines M_1 and M_2, are related, so that either M_1 contains M_2 or vice-versa.

6. Find the minimal state reduced machine for the following

Present State	Next State, Z		
	I_1	I_2	I_3
A	C, 0	E, 1	—
B	C, 0	E, –	—
C	B, –	C, 0	A, –
D	B, 0	C, –	E, –
E	–	E, 0	A, –

(a)

Present State	Next State, Z	
	I_1	I_2
A	—	F, 0
B	B, 0	C, 0
C	E, 0	A, 1
D	B, 0	D, 0
E	F, 1	D, 0
F	A, 0	—

(b)

7. (*a*) Which of the following completely specified machines are minimal ?
(*b*) Minimize those which are not minimal.
(*c*) Which of them are equivalent machines ?

Present State	Next State, Z x = 0	x = 1
q_1	q_1, 0	q_2, 1
q_2	q_2, 0	q_4, 0
q_3	q_3, 1	q_2, 0
q_4	q_2, 1	q_3, 1

(*a*)

Present State	Next State, Z x = 0	x = 1
q_1	q_2, 0	q_5, 1
q_2	q_3, 1	q_4, 0
q_3	q_1, 1	q_4, 1
q_4	q_4, 1	q_1, 1
q_5	q_5, 0	q_3, 0

(*b*)

Present State	Next State, Z x = 0	x = 1
A	C, 1	D, 1
B	B, 0	C, 1
C	C, 1	A, 0
D	D, 0	C, 0
E	E, 0	C, 0
F	F, 0	C, 1

(*c*)

Present State	Next State, Z x = 0	x = 1
A	A, 0	G, 1
B	B, 0	D, 0
C	D, 1	E, 0
D	G, 1	E, 1
E	E, 0	G, 1
F	F, 0	D, 0
G	C, 0	F, 1

(*d*)

8. Minimize the following incompletely specified machines.

Present State	Next State x = 0	x = 1	Z
A	C	–	1
B	–	C	–
C	–	B	–
D	D	–	0
E	E	D	–

(*a*)

Present State	Next State, Z x = 0	x = 1
A	–, 0	B, 0
B	A, 1	—
C	F, –	A, 1
D	–, 1	E, 1
E	C, 0	D, –
F	D, 0	C, 1

(*b*)

9. Given a machine M (I, O, S, δ, λ), where |I| = 2, |O| = 3 and |S| = 12. The symbol | | denotes the number of elements of a set. Find an upper bound to the number of 1-equivalent classes. Derive the formula for it in terms of I, O and S.

10. Consider the machine M whose input space and state space are

$$|I| = \{0, 1\} \quad S = \{1, 2, \ldots, 9\}$$

The 4-equivalence partition P_4 is

$$P_4 = \{(1)\ (2, 3, 4, 5)\ (6, 7)\ (8)\ (9)\}.$$

It is also known that when 0 is applied to states 4 and 6, the next states are 4 and 6 respectively. Determine three–equivalence partition P_3 of the machine M.

11. If for every pair of states S_i and S_j of a machine M, there exists an input sequence that takes M from S_i to S_j then M is said to be "strongly connected". Show that a minimal machine M* of M is strongly connected if M is strongly connected.

12. The input/output strings of machine M are made up of the symbols 0, 1 and *x*. M reverses any string of 0's and 1's that lie between two *x*'s. For example,

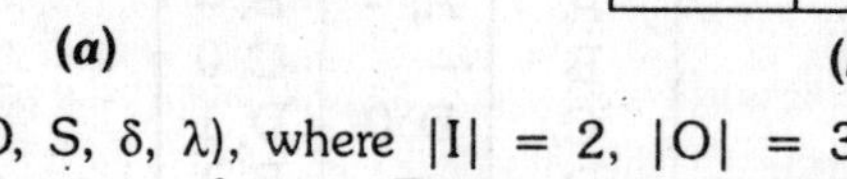

Could M be a sequential machine. Justify your answer.

13. Determine which of the following Mealy machines are readily convertible to Moore machine.

Present State	Next State, Z		
	I_1	I_2	I_3
A	B, 0	C, 1	A, 1
B	A, 1	B, 0	C, 0
C	C, 1	A, 0	B, 1

(*a*)

Present State	Next State, Z	
	$x = 0$	$x = 1$
A	B, 1	C, 0
B	D, 1	B, 0
C	A, 0	C, 0
D	B, 1	D, 1

(*b*)

Present State	Next State, Z		
	I_1	I_2	I_3
A	B, 0	C, 1	D, 1
B	A, 1	A, 0	A, 1
C	B, 1	B, 0	D, 1
D	B, 1	C, 0	C, 0
E	E, 0	B, 0	E, 0

(*c*)

5.11 MULTIPLE CHOICE QUESTIONS

1. A sequential circuit with 3 numbers of inputs produces an input alphabet contains ________ number of input sets.

(*a*) 3 (*b*) 2 (*c*) 7 (*d*) 8.

[Ans. (*d*)]

2. A sequential circuit with 4 numbers of outputs produces an output alphabet contains ________ number of output sets.

(*a*) 4 (*b*) 16 (*c*) 8 (*d*) 15.

[Ans. (*b*)]

3. The next state of a Mealy machine is a function of ________

(*a*) present state (*b*) present inputs (*c*) both (*d*) none.

[Ans. (*c*)]

4. The Mealy circuit output depends

(*a*) only on the present state of the flip-flop

(*b*) only on the next state of the flip-flop

(*c*) on both present state and next state of the flip-flop

(*d*) none.

[Ans. (*c*)]

5. The Moore circuit output depends

(*a*) only on the present state of the flip-flop

(*b*) only on the next state of the flip-flop

(*c*) on both present state and next state of the flip-flop

(*d*) none.

[Ans. (*a*)]

6. If every pair of states S_i, S_j of machine M, there exists an input sequence which takes M from S_i to S_j then the M is said to be

(*a*) strongly connected (*b*) loosely connected

(*c*) not connected (*d*) none.

[Ans. (*a*)]

7. If there exists for pair (S_i, S_j) a distinguish sequence of length K, then that states are said to be ________
(*a*) K-distinguishable (*b*) (K–1) distinguishable (*c*) (N–K) distinguishable (*d*) none.
[Ans. (*a*)]

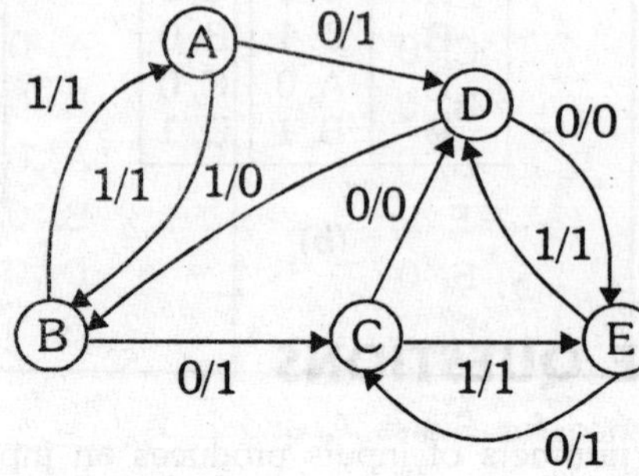

8. For the given state diagram with initial state A, the input sequence 110 transforms into ________ output sequence.
(*a*) 110 (*b*) 111 (*c*) 000 (*d*) 101.
[Ans. (*b*)]

9. With initial state C, and input sequence 0111 will transform into ________ output sequence.
(*a*) 0111 (*b*) 1110 (*c*) 1111 (*d*) 1010.
[Ans. (*b*)]

10. If an input sequence *y*-takes a machine from state A to B then B is said to be ________ of A.
(*a*) *y*-successor (*b*) *y*-equivalent (*c*) *y*-distinguishable (*d*) all.
[Ans. (*a*)]

11. From the given figure, D is ________ successor of A,
(*a*) 0 (*b*) 1 (*c*) 101 (*d*) 111.
[Ans. (*a*)]

12. From the given figure, E is ________ successor of A,
(*a*) 00 (*b*) 10 (*c*) 11 (*d*) 01.
[Ans. (*a*)]

13. From the given figure, (BD) is ________ successor of (CD).
(*a*) 1100 (*b*) 01 (*c*) 111 (*d*) 101.
[Ans. (*b*)]

14. Minimization of the number of states reduce the ________
(*a*) complexity (*b*) number of gates (*c*) cost (*d*) all.
[Ans. (*d*)]

15. If two states A and B, of a machine M are distinguishable, then they are distinguishable by a sequence of ________ length, (Assume n is number of states in the machine M).
(*a*) (*n* – 1) or less (*b*) *n* or less
(*c*) (*n* + 1) (*d*) does not depend on *n*.
[Ans. (*a*)]

From the given state table find the following :

Present state	Next state, X = 0	Output X = 1
A	E, 0	C, 0
B	C, 0	A, 0
C	B, 0	B, 0
D	G, 0	A, 0
E	F, 1	B, 0
F	E, 0	D, 0
G	D, 0	G, 0

16. The distinguishing sequence for States A and B is ________

(*a*) 00 (*b*) 01 (*c*) 11 (*d*) none.

[Ans. (*b*)**]**

17. The distinguishing sequence for States A and F is ________

(*a*) 000 (*b*) 0100 (*c*) 1011 (*d*) 01110.

[Ans. (*c*)**]**

18. The States G and F are ________ equivalence.

(*a*) 2 (*b*) 1 (*c*) 3 (*d*) none.

[Ans. (*b*)**]**

19. The distinguishing sequence for States B and C is ________

(*a*) 000 (*b*) 001 (*c*) 110 (*d*) 111.

[Ans. (*b*)**]**

20. The state reduction process in the incompletely specified machines can be done by ________.

(*a*) Merger graph (*b*) Merger table (*c*) both (*d*) not possible.

[Ans. (*c*)**]**

21. Two compatible state pairs in the merger graph are indicated by ________

(*a*) undirected arc (*b*) arc with implied pair (*c*) no arc (*d*) none.

[Ans. (*a*)**]**

22. The compatibility of two states in the merger table can be indicated by ________

(*a*) X-mark (*b*) tick mark (*c*) empty (*d*) implied pair.

[Ans. (*b*)**]**

23. For a 8-state machine if P_4 = (A) (B) (CD) (EFG) (H), then its P_3 partition may be ________.

(*a*) (CD) (A) (B) (EFG) (H) (*b*) (AB) (CD) (EF) (GH)

(*c*) (AB) (CD) (EH) (FG) (*d*) (ABC) (DEF) (GH).

[Ans. (*a*)**]**

24. For a 8-state machine if P_4 – (AB) (CD) (EFH) (G), then its P_3 partition may be________

(*a*) (ABG) (CD) (EFH) (*b*) (A) (B) (CD) (EFGH)

(*c*) (ABCD) (EF) (GH) (*d*) (A) (B) (C) (D) (EFGH).

[Ans. (*a*)**]**

5.11.1 True or False

1. Moore circuit requires more number of states for implementation.

[**Ans.** True]

2. Mealy circuit requires less number of states for implementation.

[**Ans.** True]

3. If an input sequence 'X' takes a machine from state S_i to S_j, then S_j is said to be the X-successor of S_i.

[**Ans.** True]

4. A state is called a terminal state. If the corresponding vertex in the state diagram is a sink vertex or source vertex.

[**Ans.** True]

5. State S_i and S_j of machine M are said to be equivalent if and only if every possible input sequence, the same output sequence will produce regardless of whether S_i or S_j is the initial state.

[**Ans.** True]

6. If A sequential circuit with 4 numbers of inputs produces an input alphabet contains 10 number of input sets.

[**Ans.** False]

7. A sequential circuit with 3 numbers of outputs produces an output alphabet contains 8 number of output sets.

[**Ans.** True]

8. If two states are said to be *k*-distinguishable, then they are also *k*-equivalent.

[**Ans.** False]

9. The output of the Mealy circuit depends on present states and inputs of the machine.

[**Ans.** True]

10. The next of the Mealy circuit depends on present states and inputs of the machine.

[**Ans.** True]

11. If states A = B, and B = C then A = C.

[**Ans.** True]

12. If two states are said to be equivalent, when there is no input sequence exists which to produce the different outputs which when applied or any one of the initial state.

[**Ans.** True]

13. States that are not *k*-distinguishable are said to be *k*-equivalent.

[**Ans.** True]

14. Two states A and B of a machine M are said to be equivalent if and only if, for every possible input sequence, the same output sequence will be produced regardless of whether A or B is the initial state.

[**Ans.** True]

15. A partition P is said to be a refinement of a partition Q if P is smaller than Q.

[**Ans.** True]

16. The number of vertices in the Merger graph is equal to number of states in the state table.

[**Ans.** True]

17. In the Merger graph an undirected are indicates each compatible pair.

[**Ans.** True]

18. In the Merger graph the outputs of two states are same but next states not same, this is represented by an arc with implied pair.

[**Ans.** True]

19. For large number of states, Merger table is suitable.

[**Ans.** True]

20. In the Merger table, the compatibility is indicated by tick mark.

[**Ans.** True]

5.11.2 Fill in the Blanks

1. States that are not K-distinguishable are said to be ________

[**Ans.** (K–1) equivalence]

2. The vertex in the state diagram is called as ________ if it does not have any outgoing arcs, which emanate from it and terminate in other verticles.

[**Ans.** sink vertex]

3. The vertex in the state diagram is called as ________ if no arcs which emanate from other vertices and terminate in it.

[**Ans.** source vertex]

4. For every pair of states S_i and S_j of a machine M, there exists an input sequence, which takes M from S_i to S_j, them M is said to be ________.

[**Ans.** strongly or fully connected]

5. A machine is said to be not strongly connected, it indicates that, it has at least one ________ vertex.

[**Ans.** terminal vertex (source or sink)]

6. The number of memory elements required for a realization of a machine is directly proportional to the ________.

[**Ans.** number of states in the state table]

7. The two states A and B of a machine M are said to be ________ if and only if there exists at least one finite input sequence which, when applied to the machine M, causes different output sequences, depending on whether A or B is the initial state.

[**Ans.** distinguishable]

8. The sequence, which distinguishes the two states of machine M, is called ________

[**Ans.** distinguishing sequence]

9. The distinguishing sequence of a pair A and B states is of length K, then the two states are said to be ________ distinguishable.

[**Ans.** K]

10. The distinguishing sequence of a pair A and B states is of length K, then the two states are said to be ________. equivalence.

[**Ans.** K–1]

11. The machines M_1 and M_2 are said to be ________ if every state in M_1, there is a corresponding equivalent state in M_2 and vice versa.

[**Ans.** equivalent]

12. In the sequential circuits, either state transitions or output variables are not completely specified, such machine are said to be ________.

[**Ans.** incompletely specified machines]

13. The two states A and B of a machine M are said to be ________, if and only if, for every input sequence applicable to both A and B, the same output will be produced, whenever both outputs are specified.

[**Ans.** compatible]

14. State A of machine M_1 is said to be ________, state B of machine M_2 if and only if every input sequence applicable to *b* is also applicable to A, and its application both M_1 and M_2 results the identical output sequences, when outputs are specified.

[Ans. cover (or) contains]

15. ________ Technique is used for state reduction process for incompletely specified machine.

[Ans. Merger graph (or) Merger table]

16. The two vertices in the Merger graph are not connected if their ________ are conflicting.

[Ans. outputs]

17. The incompatibility of two states is indicated with ________.

[Ans. X-mark]

6

Algorithmic State Machines

6.1 INTRODUCTION

The binary information stored in any digital circuit (system) can be classified into two ways. They are either data or control information. Data is an information, manipulated to perform various arithmetic and logical operations with the help of combinational and sequential circuits.

Control information gives command signals to supervise various sequence of operations to produce required result. On this basis, designing of any digital system can have two parts:

(1) *Data processor unit*—Design of digital circuits that perform the data-processing operations.

(2) *Control logic circuit*—Design of control circuit that supervises the operations and their sequence.

The relationship between the data processor and control logic in a digital system is shown in Fig. 6.1.

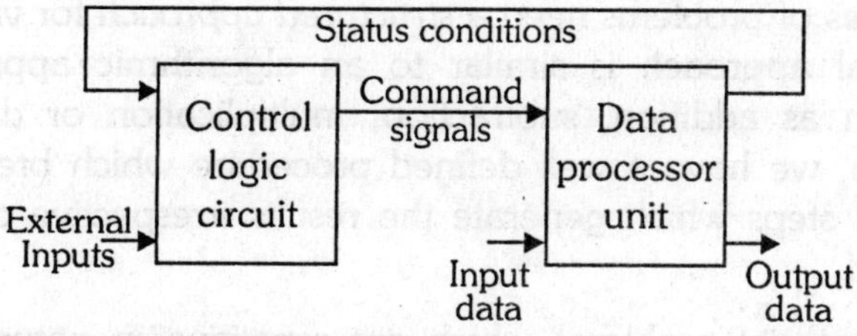

Fig. 6.1

The data processor unit contains registers to manipulate the data according to the system requirements. The control logic sends sequence commands to the data processor by using the status conditions from the data processor unit.

The data processing tests and control sequence of a digital system are specified by means of a hardware algorithm. An algorithm consists of a finite number of procedural steps that specify how to obtain a solution to a problem. The *flow chart* is the convenient way to specify the sequence of procedural steps and decision paths for an algorithm. A flow chart for a hardware algorithm translates the word statement into the information diagram which indicates the sequence of operations with necessary conditions for their execution, but in case of sequential circuits, it is necessary to define the timing relationship between sequence of operations and conditions required.

A special flow chart that has been developed specifically to define digital hardware algorithms is called an *algorithmic state machine (ASM) chart* (or) SM charts (or) state machine flow charts.

It is a special type of flow chart suitable for describing the special operations in step-by-step manner of a digital system to get the desired performance.

The MSI or LSI circuits are brought out by the designer only after he could translate the descriptive specifications of the problem into logical form *i.e.,* in truth table form in case of combinational logic and in state table or flow table form in case of synchronous sequential and asynchronous sequential circuits.

However the designer may sometimes find it very difficult to visualize the operation of the system from merely cumbersome descriptions. This may practically happen when the number of inputs to the system are large or when the system to be designed has large number of internal states or it may sometimes happen that the sequence of operations may be small but the descriptive specifications could be voluminous. Let us consider an example of traffic light controller.

"The north and south bound lights are in red and east and west bound lights are in green for 20 sec. Next, the east and west bound lights will turn yellow in preparation for changing the flow of traffic through intersection. This should be for 5 seconds. After 5 sec, the east and west bound lights will turn to red and north and south bound lights will turn to green for 20 seconds. Next, north and south bound lights will turn yellow for 5 seconds in preparation for another change in traffic flow. Finally, the north and south bound lights will turn to red and east and west bound lights will turn to green".

In the above problem, there is only small sequence of actions but one can see how voluminous it is and how difficult it may be to visualize the operation from the above descriptive specification. All such types of problems need a *structured approach* for visualizing the sequential operation. This structural approach is similar to an algorithmic approach in handling the arithmetic problems such as addition, subtraction, multiplication or division of numbers. In this algorithmic approach, we have a well defined procedure which breaks the total operation into a sequence of small steps which generate the results irrespective of the number of times the sequence is repeated.

Similarly, if the sequential problems which are repetitive in nature and when described verbally have a massive description can be broken down into small steps (states) and can be visualized in the form of a flow-chart as in case of mathematical algorithms. Such sequential descriptions will become easy to translate into hardware. This type of approach of handling the sequential problems is called as structural approach and the resulting circuits are called as ASM. The description of traffic light controller of Fig. 6.2(*a*) can be visualized if it is described by a diagram showing the series of steps involved as shown in Fig. 6.2(*b*).

Since these operations occur in sequence, *i.e.,* one after another, there is a flow of actions and the diagram is called flow-chart. Each step (state) of the sequential behaviour represents the individual action and whole sequential system consists of a set of such individual actions.

ASM chart can be implemented by two ways : (*a*) software program (*b*) using hardware.

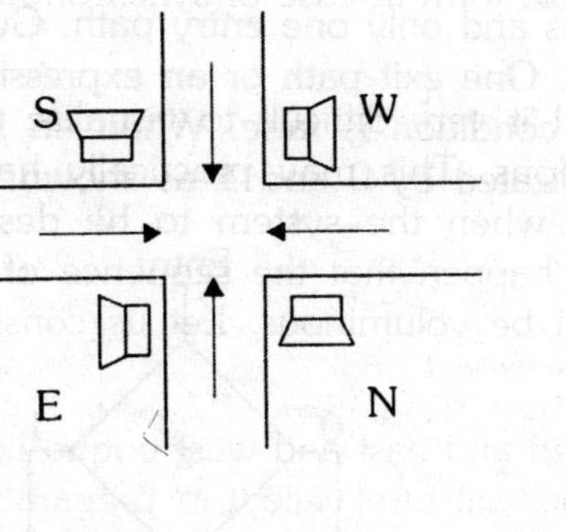

Fig. 6.2(*a*)

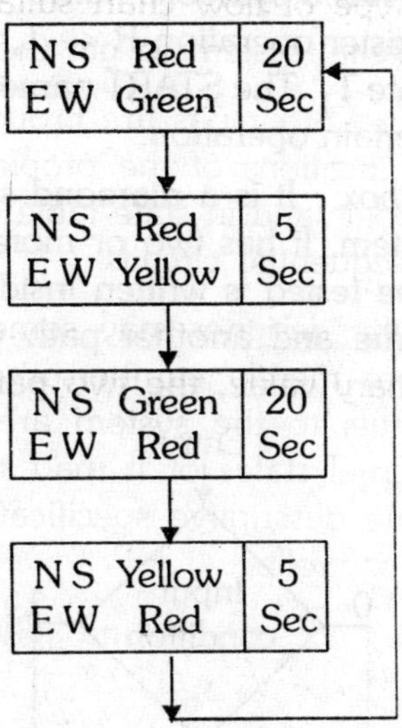

Fig. 6.2(*b*)

6.1.1. Difference between Conventional Flow Charts and ASM Charts

1. A conventional flow chart describes the sequence of procedural steps and decision paths for an algorithm without concern for their time relationship.

2. The ASM chart describes the sequence of events as well as the timing relationship between the states of a sequential controller and the events that occur while going from one state to another state.

6.2 ASM CHARTS

ASM chart is composed of three basic elements : State box, Decision box and Conditional box.

State box : A state of the system is indicated by a state box. The shape of the state box is a rectangle within which are written register operations or output signal names that control generates while being in this state. The state is given by a symbolic name, which is placed at the upper left corner of the box. The binary code assigned to the state is placed at the upper right corner, shown in Fig. 6.3(*a*). Each state box contains only one entry and one exit path. Fig. 6.3(*b*) shows a specific example of a state box.

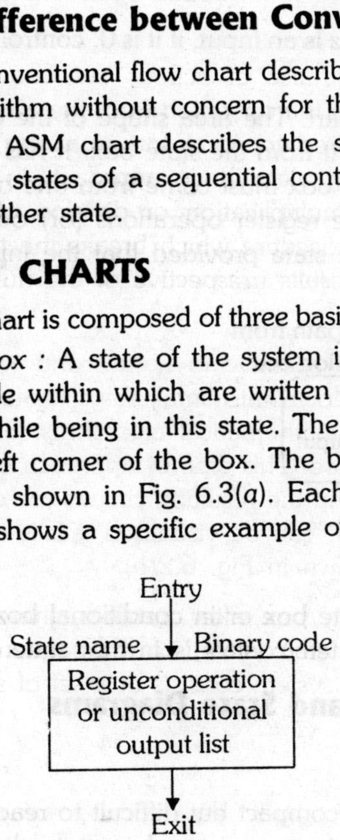

Fig. 6.3(*a*) General description.

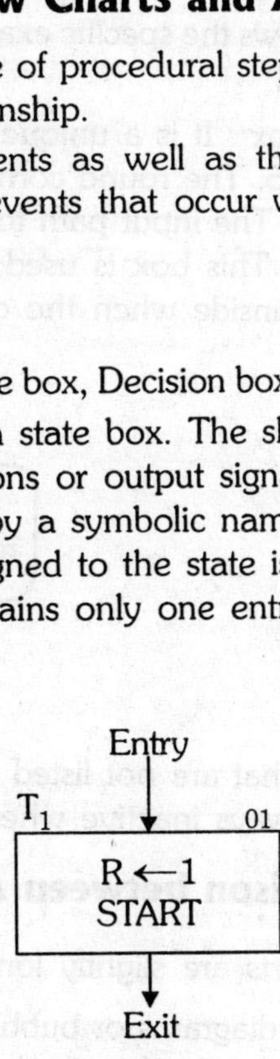

Fig. 6.3(*b*)

The state has a symbolic name T_1; and binary code assigned to it is 001. Inside the box is written the register operation R ← 1, which indicates that register R is initialized to 1, when the system is in state T_1. The START name inside the box may indicate, an unconditional output signal that starts a certain operation.

Decision box : It is a diamond shaped box used to describe the effect of an input on the control subsystem. It has two or more exit paths and only one entry path. Generally, the input condition to be tested is written inside the box. One exit path or an expression is taken if the condition is true and another path when the condition is false. When an input condition is assigned a binary value, the two parts are indicated by 0 and 1 as shown in Fig. 6.4(*a*).

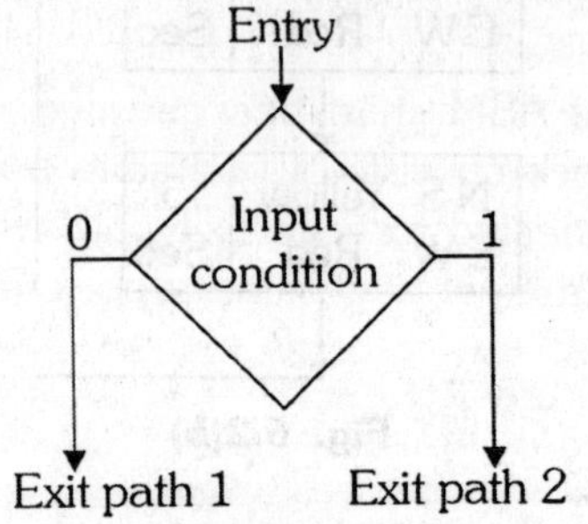

Fig. 6.4(*a*)

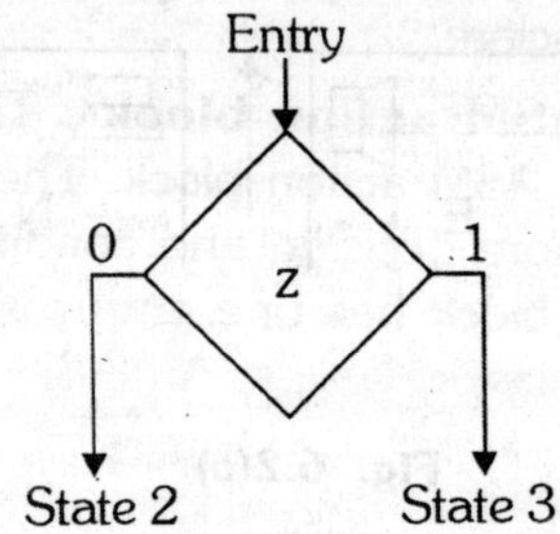

Fig. 6.4(*b*)

Fig. 6.4(*b*) shows the specific example. Here, the *z* is an input, if it is 0, control goes to state 2, else goes to state 3.

Conditional box : It is a unique box of ASM chart. The area shape of the conditional box is shown in Fig. 6.5. The round corners differentiate it from the state box. It has one entry path and one exit path. The input path to the conditional box must come from one of the exit paths of a decision box. This box is used to represent the register operations (or) output conditions which are written inside when the control is in that state provided that the input condition is satisfied.

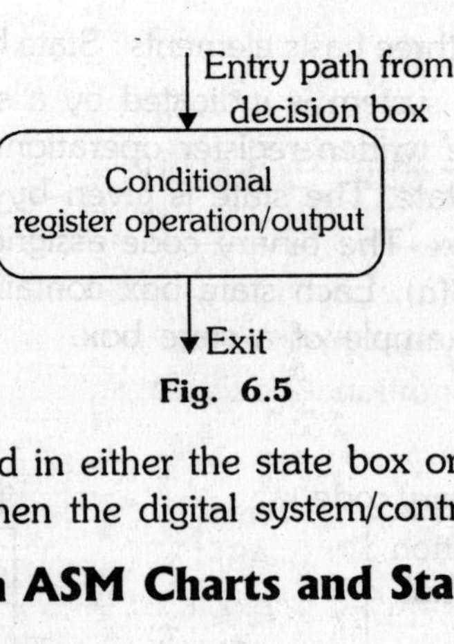

Fig. 6.5

The outputs that are not listed in either the state box or in conditional box in a particular ASM chart are always inactive when the digital system/control is in that state.

6.2.1 Comparison between ASM Charts and State Diagrams

(1) ASM charts are slightly longer.

(2) The state diagrams or bubble diagrams are compact but difficult to read *i.e.*, sequential operations can be easily described using ASM charts as compared to state diagrams.

(3) The construction of an ASM chart is more or less mechanical in nature. (Structured approach)

(4) Due to it's structured approach, the construction of ASM charts for complex systems will be easier than drawing a state diagram.

(5) The state box is equivalent to the state in a sequential circuit and the decision box is equivalent to the binary information written along the directed lines connected between two states.

In order to build insight into the sequential circuit specifications, we will have to develop the ASM charts for simple circuits to begin with. Before we proceed to develop the ASM charts, let us understand the definitions of some of the terms related to ASM chart coined by Mr. Chrisclare.

1. ASM action block : The portion of the ASM chart between two rectangular boxes is called an ASM action block. The ASM action block consists of one state box and may contain one or more decision and conditional boxes connected to each exit path as shown in Fig. 6.6. An ASM block has one entry path and any number of exit paths represented by the structure of the decision boxes. An ASM chart consists of one or more interconnected blocks.

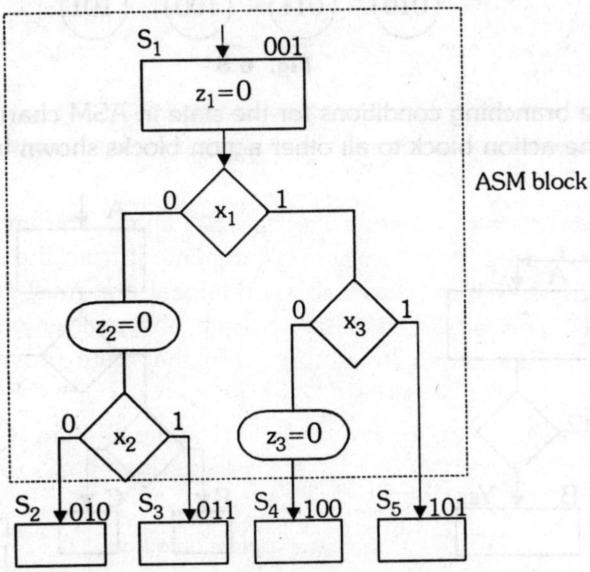

Fig. 6.6

In Fig. 6.6, state S_1 has three decision boxes and two conditional boxes. A state box without any decision or conditional boxes constitutes a simple ASM block as shown in Fig. 6.7.

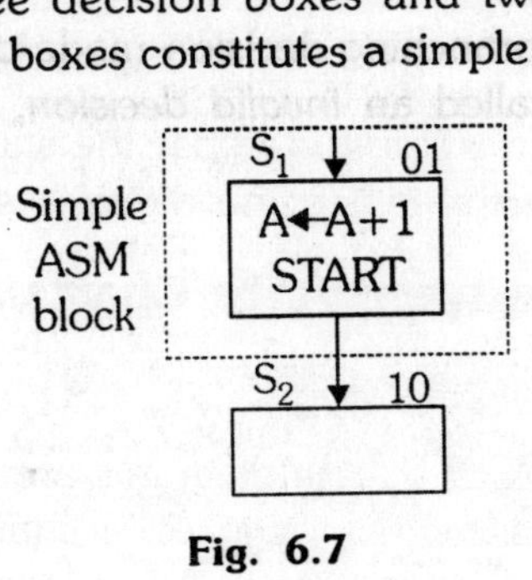

Fig. 6.7

Each block in the ASM chart describes the state of the system during one clock pulse interval. The operations within the state and conditional boxes in Fig. 6.6 are executed with a common clock pulse while the system is in state S_1. The same clock pulse transfers the system controller to one of the next states, S_2, S_3 or S_4, as directed by the binary values of x_1, x_2 and x_3.

Sometimes it is convenient to convert the ASM chart into a state diagram and then use the sequential circuit procedures to design the control logic. As an illustration, the ASM chart of Fig. 6.6 is drawn as a state diagram in Fig. 6.8. The five states are symbolized by circles, with their binary values written inside the circles. The directed lines indicate the conditions that determine the next state.

Note : The unconditional and conditional operations that must be performed are not indicated in the state diagram.

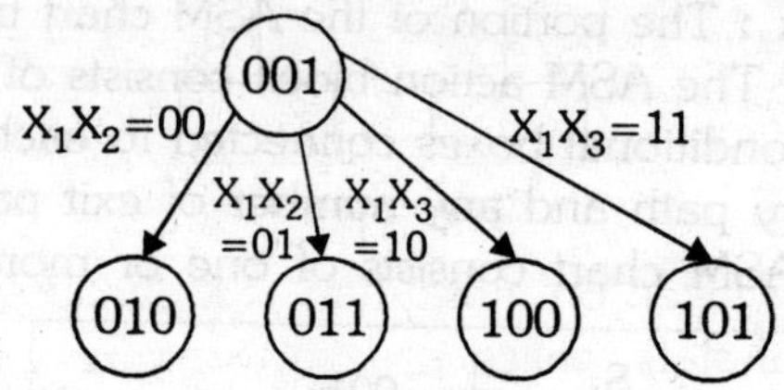

Fig. 6.8

2. Branch : The branching conditions for the state in ASM chart are derived by tracing the decision paths from the action block to all other action blocks shown in Fig. 6.9 and Fig. 6.10.

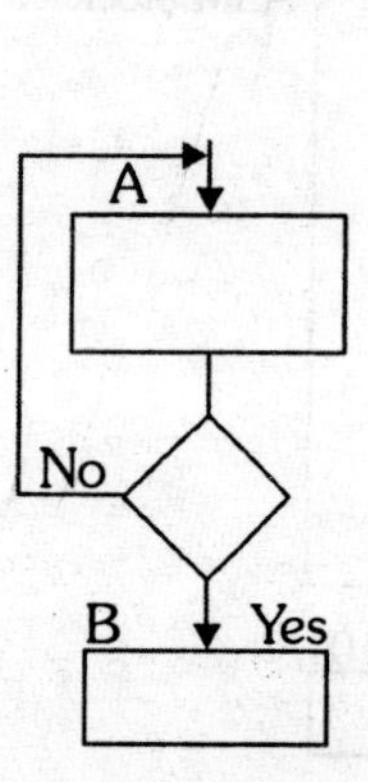

Fig. 6.9

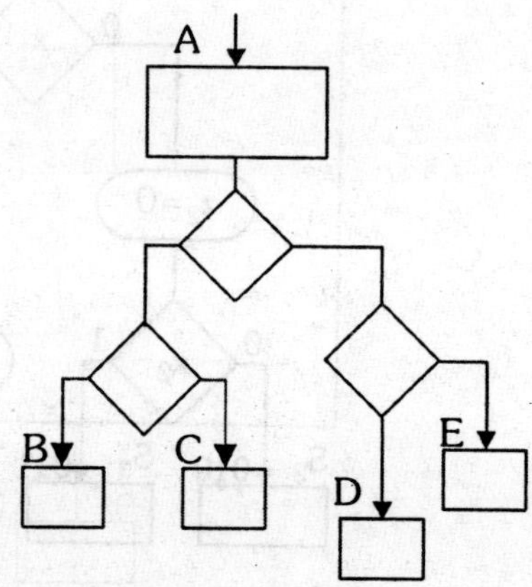

Fig. 6.10

3. Valid decision : The branching decision made on more than one purely asynchronous and independent variables is called an *invalid decision*, shown in Fig. 6.11.

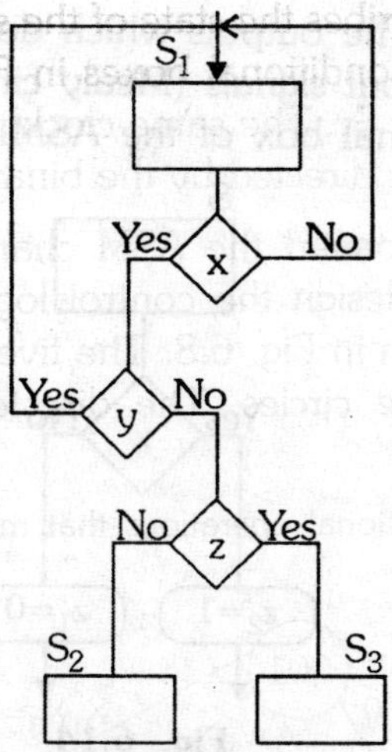

Fig. 6.11

While a branching decision taken on more than one purely asynchronous and independent variables by inserting additional state, called a *valid decision* is shown in Fig. 6.12.

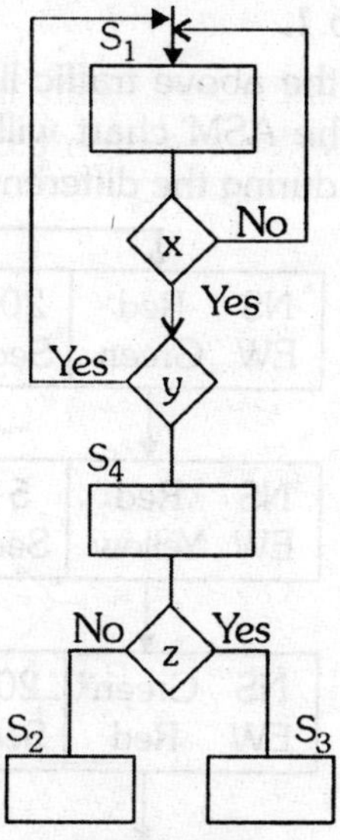

Fig. 6.12

4. Unconditional outputs : The outputs which are independent of the inputs and merely depend on the present state of the circuit (Moore circuit) are called unconditional outputs. These outputs are indicated in the rectangle-box as shown in Fig. 6.13. Here the unconditional output is *z* and its value = 1, when the controller is in state S.

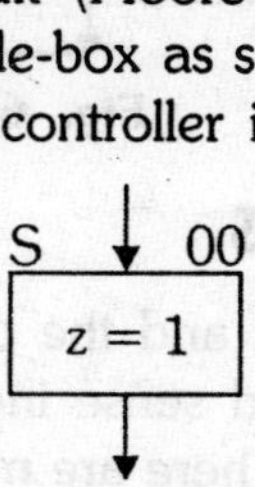

Fig. 6.13

5. Conditional outputs : The outputs which depend on the present state of the circuit as well as the condition of the input signals (Mealy circuits) are called as conditional outputs. These are represented in conditional box of the ASM chart as shown in Fig. 6.14.

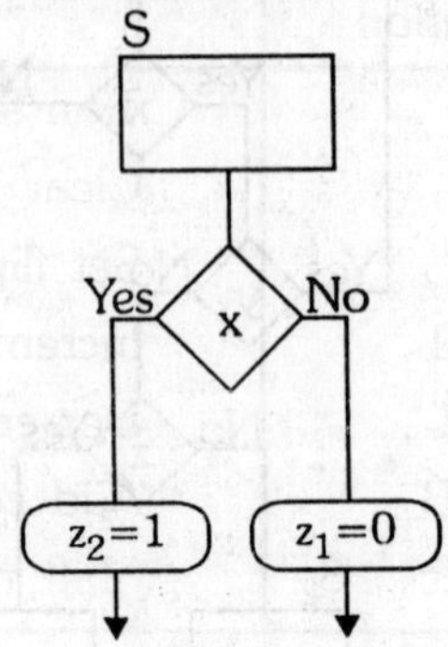

Fig. 6.14

Here z_1 and z_2 are two conditional outputs, their values are 0 and 1 respectively when the controller is in state S and input x is yes or no.

Example 6.1. *Develop an ASM chart for a traffic light controller with sequential specifications described in section 6.1.*

Solution. The ASM chart for the above traffic light controller is simple since there are no external inputs except the clock. The ASM chart will have just four rectangle-boxes and four different signals are to be generated during the different states of the circuit as shown in Fig. 6.15.

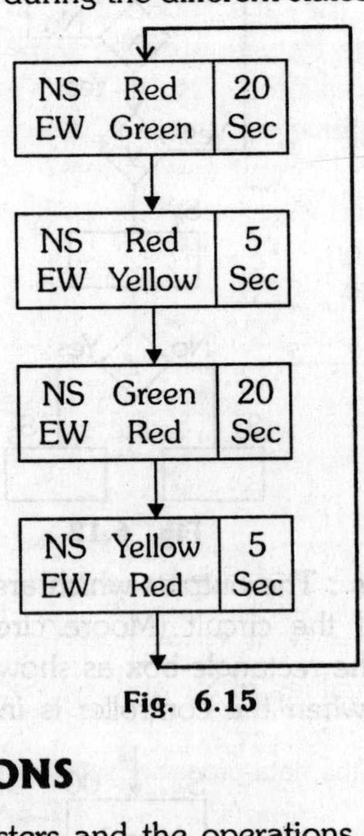

Fig. 6.15

6.3. REGISTER OPERATIONS

A digital system contains registers and the operations that are performed on the data are stored in them. A *register* in it's broad sense includes storage registers, shift registers, counter and single bit registers, *i.e.*, flip-flops. There are many types of operations that can be performed on the data stored in the registers, such as shift, increment, add, clear etc.

Table 6.1 gives examples of notation for some register operations.

Table 6.1

Symbolic notation	*Description*
A ← B	Transfer contents of register B into register A
R ← 0	Clear register R
E ← 1	Set flip-flop E to 1
R ← R + 1	Increment register R by 1 (count-up)
R ← R – 1	Decrement register R by 1 (count-down)
A ← A + B	Add contents of register B to register A
Shift right R	Shift right the contents of register R.

6.4 TIMING CONSIDERATIONS

The timing for all registers and flip-flop in a digital system is controlled by a master clock generator. The clock pulses are applied to the registers of the data processor and flip-flop in the control logic. All the inputs are also synchronised with the clock pulses, because they are normally generated as outputs of another circuit that uses the same clock signals. To simplify the design, we will assume that all inputs are synchronized with the clock and change of state takes place at edge transition of the clock pulse.

Consider Fig. 6.6, all the operations that are specified within the block must occur in synchronism during the edge transition of the same clock pulse, while the system changes from S_1 to next state. This is represented pictorially in Fig. 6.16. We assume negative edge triggering of all flipflops. The first negative transition of the clock transfers the control circuit into state S_1. While in state S_1, the control circuit checks the inputs x_1, x_2 and x_3 and generates appropriate signal accordingly. The following operations occur simultaneously during the next negative transition of the clock pulse.

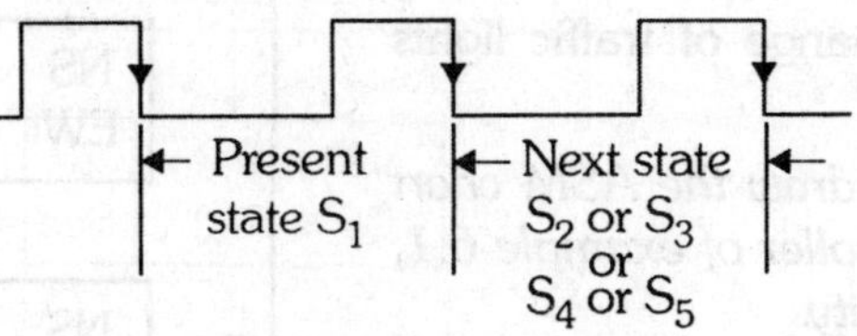

Fig. 6.16

1. Output $z_1 = 0$
2. If $x_1 = 0$, then $z_2 = 0$, if $x_1 = 1$ and $x_3 = 0$, the output $z_3 = 0$
3. Depending on the values of x_2 and x_3, control is transferred to next state S_2 or S_3 or S_4 or S_5.

Note : The operations in the data-processor subsystem and the change of state in the control logic occur at the same time.

Now, we will consider some examples for drawing the ASM charts for the given verbal problems.

Example 6.2. *Redevelop the ASM chart for the example 6.1, with the following additions.*

Consider the intersection where the NS street has major traffic which must not be stopped unnecessarily, whereas EW street has no traffic most of the time. Whenever the EW street is in use, the heavy traffic on NS street should be smoothly controlled.

Solution. One solution to this problem is to install traffic sensors in the pavement on EW street as shown in Fig. 6.17.

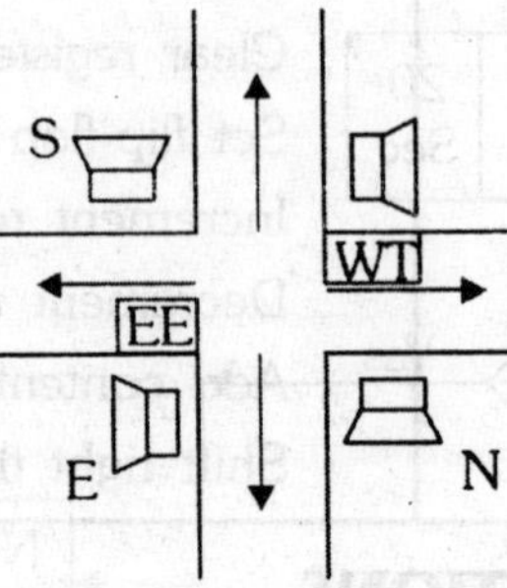

Fig. 6.17

Before switching OFF the green light for north and south, the circuit will check for the presence of vehicles on east and west roads from the signals generated by ET and WT sensors. If any of these signals are high, *i.e.*, there is a potential traffic on EW-street, than east-west light can be made green for the short period. *i.e.*, otherwise the NS-traffic may be permitted to continue. The ASM chart for this is shown in Fig. 6.18.

Here assuming that initially the NS lights are green and EW are red for the long period (20 sec), the change of traffic lights takes place smoothly.

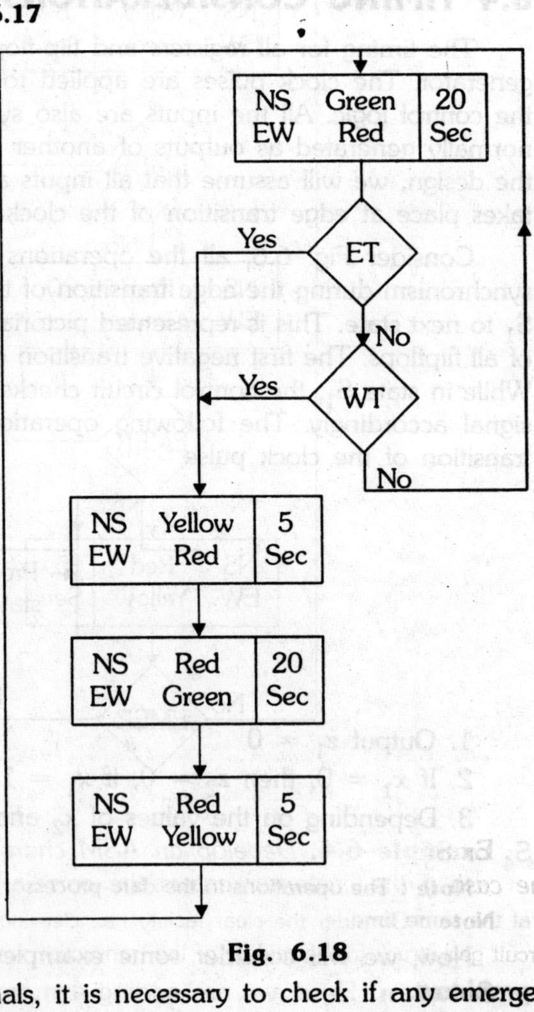

Fig. 6.18

Example 6.3. *Redraw the ASM chart for the traffic light controller of example 6.1, with the following facility.*

To allow an emergency vehicle to safely pass through the intersection by stopping traffic in all directions, the system should test for the presence of an emergency vehicle during the change in traffic lights.

Solution. Assume initially NS lights are in green and EW street lights are in red for 20 sec.

Whenever there is a change in traffic signals, it is necessary to check if any emergency vehicle is crossing the intersection from any direction or not. If so, keep all the traffic lights in red for 5 sec.

Here we are making one more assumption *i.e.*, after allowing the emergency vehicle, the traffic will go to initial condition, *i.e.*, NS street lights are green and EW street lights are red for 20 sec. The ASM chart for this is shown in Fig. 6.19

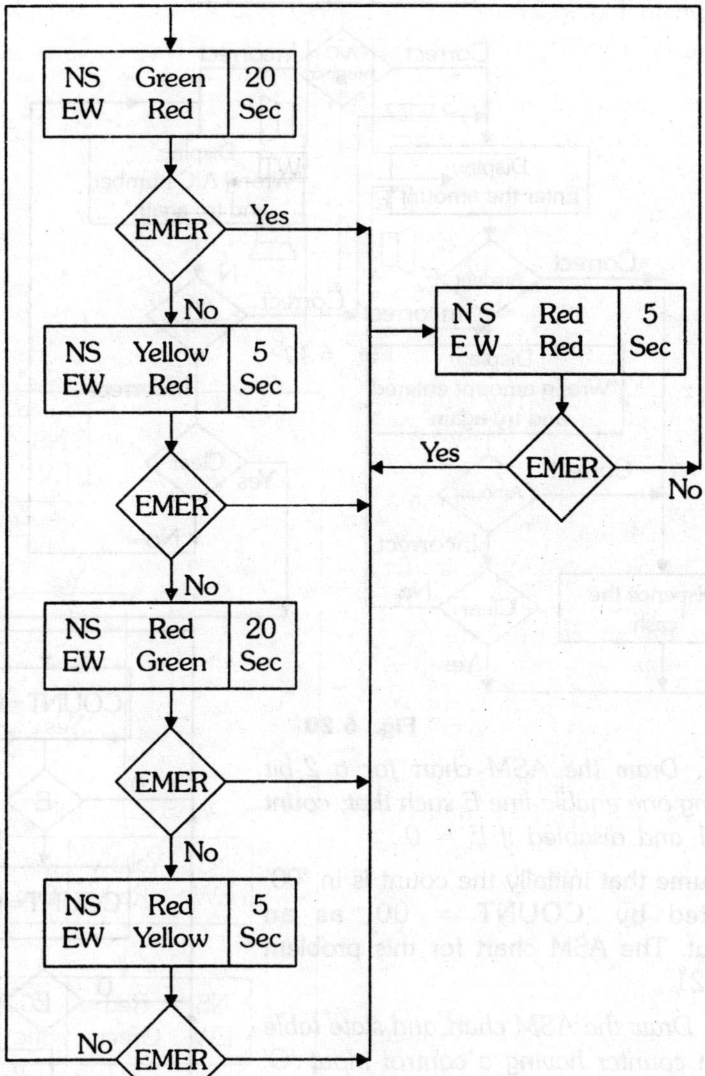

Fig. 6.19

Example 6.4. *Develop an ASM chart for an automated bank teller which will dispense the cash if the customer enters the correct account number and the correct amount.*

Note : Consider the clear facility *i.e.,* clearing of circuit (directing the circuit to the starting point) if the circuit gets struck in the middle of the transaction on account of customer's failure to complete the transaction.

Solution. Here, we are adding 4 more inputs for display to help the customer while he is in transaction. The four display inputs are "Enter the A/c number", "Wrong A/C number and try again" "Enter amount" and "Wrong amount entered and try again".

The ASM chart for this problem is shown in Fig. 6.20.

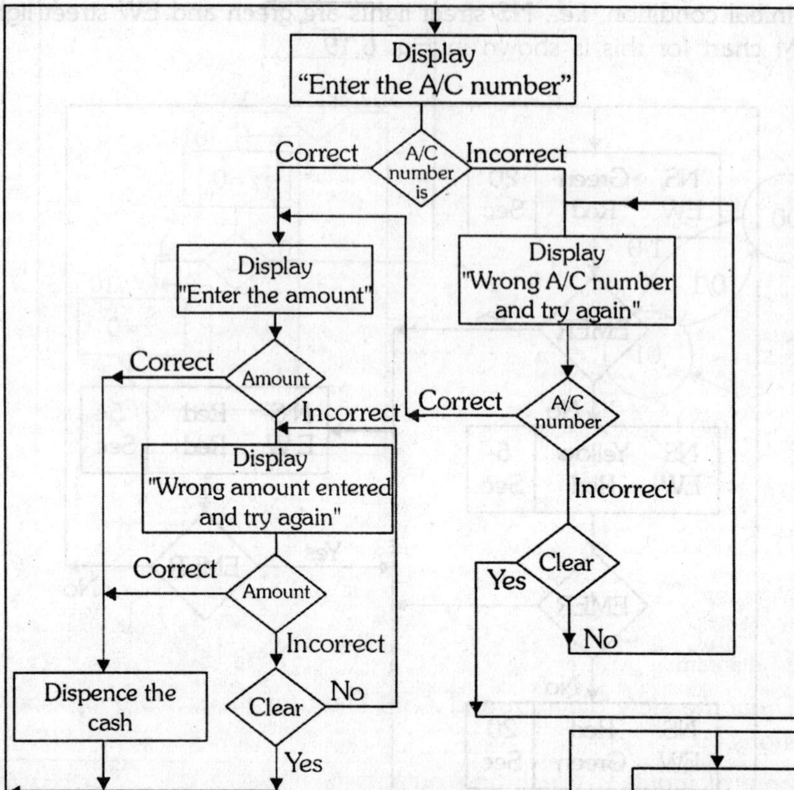

Fig. 6.20

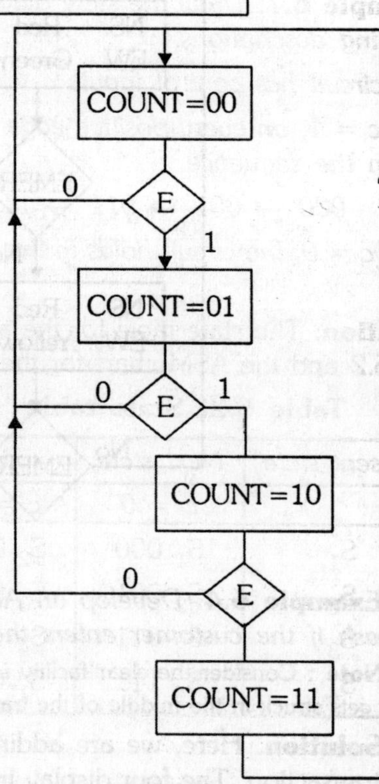

Fig. 6.21

Example 6.5. *Draw the ASM chart for a 2-bit binary counter having one enable line E such that, count is enabled if E = 1 and disabled if E = 0.*

Solution. Assume that initially the count is in '00' that is represented by 'COUNT = 00' as an unconditional output. The ASM chart for this problem is shown in Fig. 6.21.

Example 6.6. *Draw the ASM chart and state table for a 2-bit up/down counter having a control input 'C' in such a way that if C = 1, up counting and C = 0, down counting and the counter should generate an output 1, whenever count becomes minimum or maximum.*

Solution. The state diagram for this problem is shown in Fig. 6.22 and ASM chart is shown in Fig. 6.23.

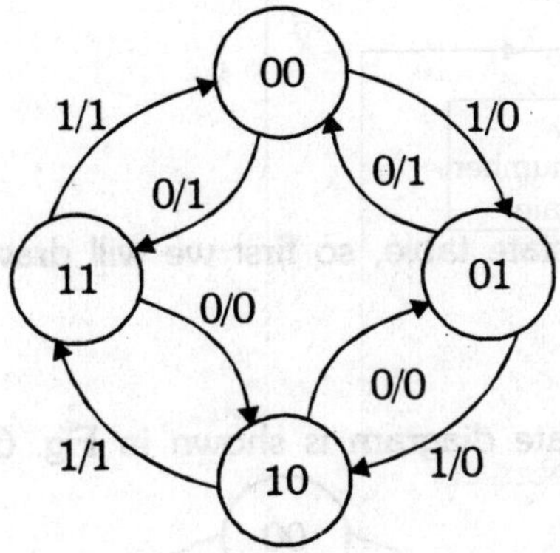

Fig. 6.22 State diagram.

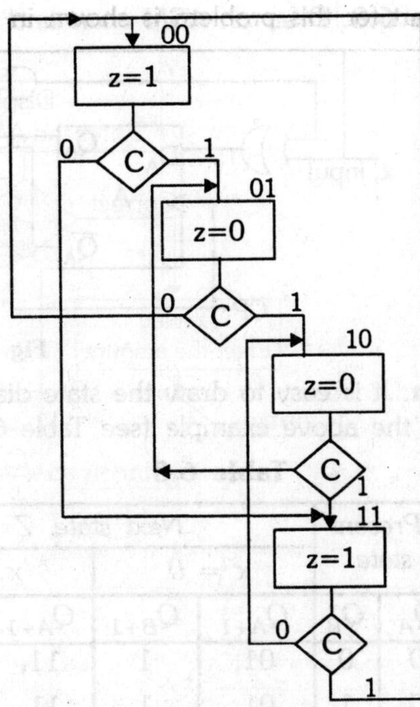

Fig. 6.23

Example 6.7. *Draw the state diagram and ASM chart for the synchronous circuit having the following descriptions.*

The circuit has control inputs c, clock and outputs a, b and c.

(i) If c = 1, on every positive edge of the clock, the output changes in the sequence.

000 → 001 → 011 → 111 → 000 and repeats.

(ii) If c = 0, the circuit holds in the present state (i.e., same state)

Solution. The state table for the above problem is shown in Table 6.2 and the ASM chart for the same example is 6.25.

Table 6.2. State table

Present state	*Next state*	*xyz*
	c = 0	*c = 1*
S_1	S_1 000	S_2 001
S_2	S_2 001	S_3 011
S_3	S_3 011	S_4 111
S_4	S_4 111	S_1 000

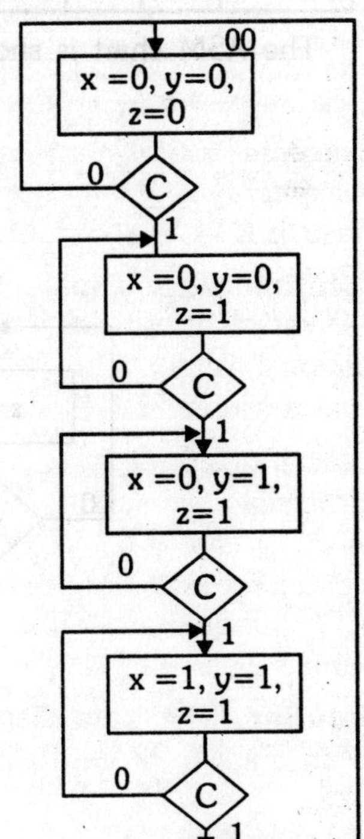

Fig. 6.24

Example 6.8. *Draw an ASM chart and state diagram for the circuit shown in Fig. 6.25.*

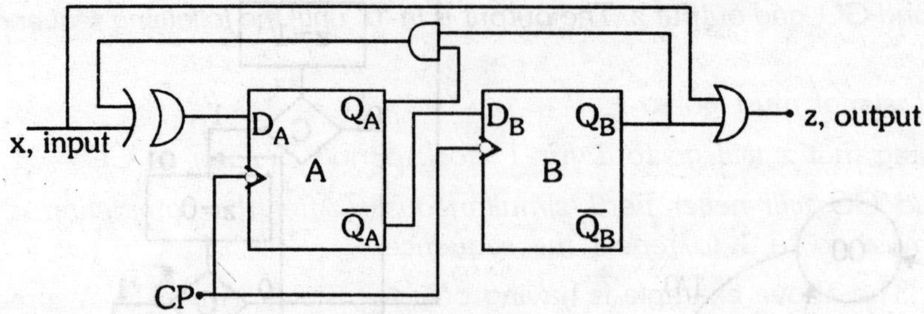

Fig. 6.25

Solution. It is easy to draw the state diagram from state table, so first we will draw the state table of the above example (see Table 6.3).

Table 6.3

Sl. No.	*Present state*		*Next state, Z*			
			x = 0		*x = 1*	
	Q_A	Q_B	Q_{A+1}	Q_{B+1}	Q_{A+1}	Q_{B+1}
1	0	0	01,	1	11,	1
2	0	1	01,	1	11,	1
3	1	0	00,	0	10,	1
4	1	1	10,	0	00,	1

State diagram is shown in Fig. 6.26.

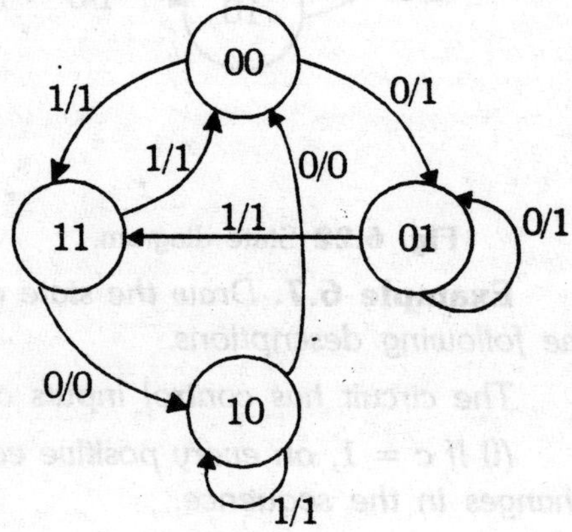

Fig. 6.26

The ASM chart is shown in Fig. 6.27.

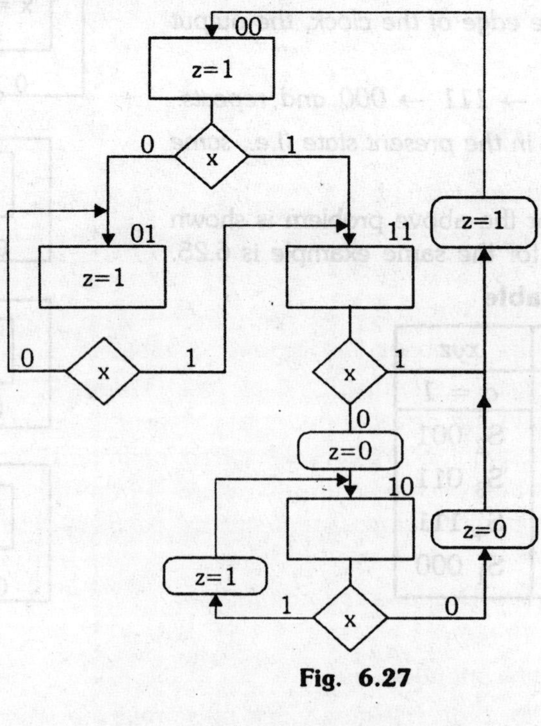

Fig. 6.27

Example 6.9. *Develop an ASM chart for a digital system that has been designed with two inputs, START and GO, and output z. The output is in '0' until the following sequence of events takes place.*

(i) START signal must go to 1.

(ii) Following that z will go to 1 for 1 clock period as soon as GO = 1.

START and GO will never be 1 simultaneously. After the completion of 1 entire cycle, START input must go to 1 to repeat the sequence.

Solution. The above example is having only two states, A and B. Assume that the circuit is initially at A and output is '0' in this state. If START input is 1, the circuit will go to the next state B, otherwise will remain in the same state A. When the circuit is in state B, if G = 1, then, circuit goes to state A with z = 1, otherwise will remain in the same state B. The ASM chart is shown in Fig. 6.28.

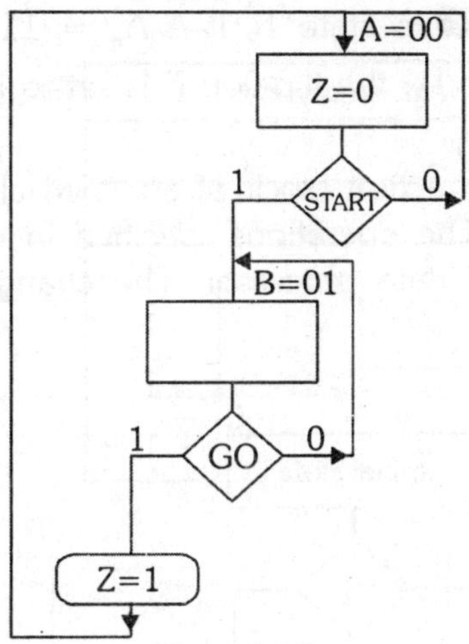

Fig. 6.28

6.5 CIRCUIT REALIZATION

The realization of a sequential circuit can be described by ASM chart which consists of three steps.

(1) To design the data-processor subsystem, the requirements are specified inside the state and conditional boxes.

(2) The control logic is determined from the decision boxes and the required state transitions.

(3) Combine both the circuits to obtain the required sequential circuit.

Example 6.10. *Design a digital system with a 4-bit binary counter, A and two flip-flops E and F. The individual flip-flops in A are denoted by A_4, A_3, A_2 and A_1 with A_1 holding the least significant bit of the count. A start signal 'S' initiates the system operation by clearing the counter A and flip-flop F. The counter is then incremented by 1 starting from the next clock pulse and continues to increment until the operation stop. Counter bits A_4 and A_3 determine the sequence of operations.*

(a) If A_3 = 0, E is cleared to 0 and count continues.

(b) If A_3 = 1, E is set to 1; then

(i) if A_4 = 0, the count continues, but

(ii) if A_4 = 1, F is set to 1 on the next clock pulse and the system stops counting.

Solution. Let us first see the ASM chart and consider the timing sequence of operations for system design.

(*i*) *ASM chart* : The ASM chart is shown in Fig. 6.29. The system is in the initial state T_0, when no operations are performed and waiting for start signal 'S'. When S = 1, the counter, A and flip-flop, F are cleared to 0 and the controller goes to state T_1. There is a conditional box that follows the decision box for S, which means that the counter and flip-flop will be cleared during T_0 if S = 1, and at the same time control transfers to state T_2.

The state T_1 has two decision boxes and two conditional boxes. As long as the control is in this state, the counter is incremented with every clock pulse. At the same time, one of the three possible operations occur during the same clock pulse transition :

(1) E = 0 and the control is in the same state T_1 (if A_3 = 0)

or (2) E = 1 and control may be in the same state T_1 if A_3A_4 = 10

or (3) E = 1 and control goes to state T_2 if A_3A_4 = 11.

When the control is in state T_2, the flip-flop F is set to 1 and the control goes back to the initial state T_0.

(*ii*) *Timing sequence* : Each action block of an ASM chart indicates the operations to be performed in one clock pulse. The operations specified in the state and conditional boxes in the block are performed in the data processor. The change from one state to next state is performed in the control logic.

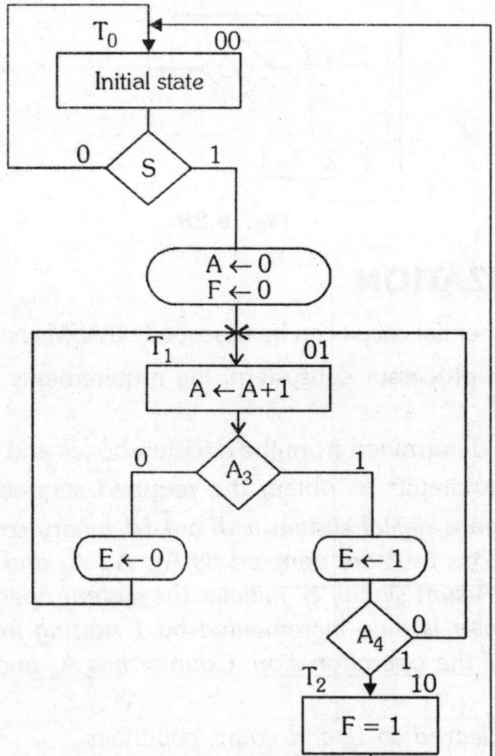

Fig. 6.29 ASM chart for design example.

Let us consider the step-by-step sequence of operations after each clock pulse from the time the start signal occurs until the system goes back to initial state. Table 6.3 shows the binary values of the counter and flip-flop after every clock pulse. The status of A_3 and A_4 as well as the present state of the controller are also indicated in the table. We start with T_1 after the start signal S = 1, *i.e.*, counter and flip-flop F are cleared. The value of E is assumed to be 1, because E = 1 at T_0 and it does not change during the state transition from T_0 to T_1. The system stays in state T_1 during the next three clock pulses. Each pulse increments the counter and either set or reset the flip-flop E. Note the time at which A_3 becomes 1 and the time at which E is set to 1.

When A = 0011, the next clock pulse increments the counter to 0100, but the same clock pulse sees the value of A_3 as 0, so E is cleared. In the next clock pulse, the counter changes from 0100 to 0101. Now, initially A_3 is equal to 1, so E = 1. Similarly, E is cleared when the count goes from 0111 to 1000, but when it goes from 1000 to 1001, which is when A_3 is 0 in the present value of the counter. When the count reaches 1100, both A_3 and A_4 are equal to 1. The next clock pulse increments the counter A by 1 and E sets to 11, and transfers the control to state T_2.

The control starts in T_2 for only one clock period. The clock pulse transition in state T_2 sets flip-flop, F to 1 and transfers control to state T_0, and the system starts in T_0 as long as S is equal to 0.

Table 6.4. Timing sequence operations for design example

Counter				*Flip-flops*		*Conditions*	*State*
A_4	A_3	A_2	A_1	E	F		
0	0	0	0	1	0		
0	0	0	1	0	0	$A_3 = 0, A_4 = 0$	T_1
0	0	1	0	0	0		
0	0	1	1	0	0		
0	1	0	0	0	0		
0	1	0	1	1	0	$A_3 = 1, A_4 = 0$	T_1
0	1	1	0	1	0		
0	1	1	1	1	0		
1	0	0	0	1	0		
1	0	0	1	0	0	$A_3 = 0, A_4 = 1$	T_1
1	0	1	0	0	0		
1	0	1	1	0	0		
1	1	0	0	0	0	$A_3 = 1, A_4 = 1$	T_1
1	1	0	1	1	0	$A_3 = 1, A_4 = 1$	T_2
1	1	0	1	1	1	$A_3 = 1, A_4 = 1$	T_0

From the observation of Table 6.3, it may be seen that the operations performed on flip-flop E are delayed by one clock pulse. The digital circuits in the control generate the signals for all the operations specified in the present block prior to the arrival of the next clock pulse.

The next clock transition executes all the operations in the registers and flip-flop including the flip-flops in the controller that determine the next state.

Design of data processor for the example 6.1.

Fig. 6.29 is the ASM chart of the problem, which has given necessary information in the state and conditions boxes to design the digital system. Fig. 6.30 shows the hardware requirement of the problem. Here, control subsystem is shown with only inputs and outputs.

The data processor consists of 4-bit binary counters, two flip-flops, E and F and a number of gates. The counter is incremented by 1 with every clock pulse when it is in state T_1. The counter is cleared when it is in T_1 and $S = 1$. This was represented by AND gate A_1 in the Fig. 6.30. The other two conditional operations *i.e.*, $E = 1$ when $A_3 = 1$ in T_1 state and $E = 0$ when $A_3 = 0$ in T_1 state use two other AND gates (A_2 and A_3) for setting or clearing flip-flop E.

The flip-flop F is set unconditionally during state T_2 and $F = 0$ when the system is in state T_0 with $S = 1$. All the flip-flops, Register of both data unit and control use a common clock pulse source.

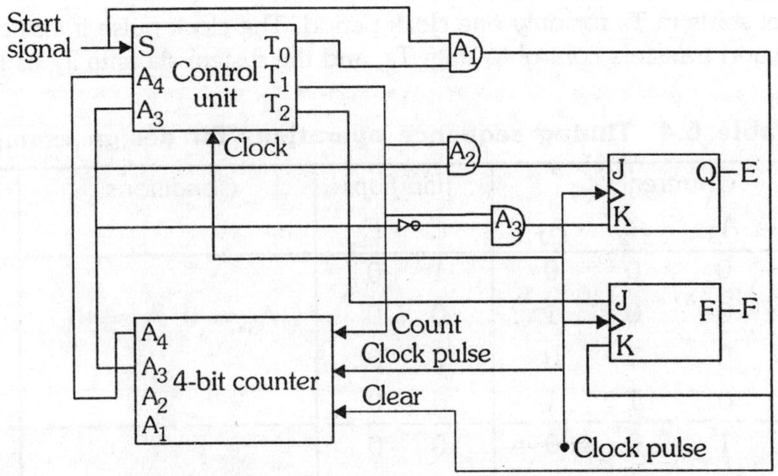

Fig. 6.30 Data processor for a design example.

6.6 CONTROL IMPLEMENTATION

The control section of a digital system is essentially a sequential circuit, that can be designed by using any one of the procedures such as using state tables and excitation tables. If a controller has a large number of states it may have more inputs and it is impractical to design. The digital designers use specialised methods for control logic design that may be considered as an extension of the classical sequential methods combined with other simplified assumptions.

Four different techniques are available for synthesizing the circuits using ASM charts.

(1) State table method

(2) Using multiplexers

(3) Using ROM and D flip-flop

(4) Using programmable devices such as PLAs or PALs.

The selection of any one of the techniques depends upon the requirements such as cost, reliability, speed of operation etc.

6.6.1 State Table Method

The ASM chart is first converted into state diagram, with each state box representing a state. The state diagram is then converted into a state table by which the controller circuit can be designed.

The state table for a controller is a list of present states and inputs and their corresponding next state outputs. Assign binary value to each state in the ASM chart. The number of flip-flops required is given by $n = [\log_2 N]$, where N = number of states in the controller

n = number of flip-flops required.

The ASM chart of the design example is shown in Fig. 6.29. The three states in the chart can be assigned by the following binary variables, $T_0 = 00$, $T_1 = 01$, $T_2 = 11$. The state '10' is not used and will be treated as don't care condition.

The state table for the ASM chart is shown in the Table 6.5.

Table 6.5. State table for the control circuit

Present state symbol	*Present state*		*Inputs*			*Next state*		*Outputs*		
	S_1	S_2	S	A_3	A_4	S_1	S_2	T_0	T_1	T_2
T_0	0	0	0	×	×	0	0	1	0	0
T_0	0	0	1	×	×	0	1	1	0	0
T_1	0	1	×	0	×	0	1	0	1	0
T_1	0	1	×	1	0	0	1	0	1	0
T_1	0	1	×	1	1	1	1	0	1	0
T_2	1	1	×	×	×	0	0	0	0	1

Two flip-flops are needed here and are labelled as S_1 and S_2. There are three inputs S, A_3 and A_4, taken from conditions in the decision box and three outputs T_0, T_1 and T_2. The outputs are equivalent to the present state of the control.

The state transition depends upon the condition, for example initial state 00 goes to 01 or stays in 00, depending on the value of 'S' input and other inputs are marked as don't care, X's. While the system is in state 00, the control provides an output labelled T_0 to initiate the required register operations.

The transition from binary state 01 depends on the values of A_3 and A_4. If the values of $A_3A_4 = 11$, then the system goes to state 11. Once the system is in 11 state, after the next clock pulse the system goes to 00 state independent of the input variables.

6.6.1.1 Logic diagram

Once state table is obtained, then the logic diagram of the controller can be drawn with the help of the excitation table. It contains 3 inputs and 2 present states. So, totally we need to use five-variable maps to simplify the input functions.

(*i*) *Using JK-flip-flop* : With the help of above procedure, the flip-flop functions for JK inputs of S_1 and S_2 flip-flops are given as

$$JS_1 = S_2A_3A_4 \qquad JS_2 = S$$
$$KS_1 = 1 \qquad KS_2 = S_1$$

The output functions are given as

$T_0 = \overline{S}_2$: From the state table S_2 values in the PS column are exactly opposite to the T_0 values in the output column, and similarly.

$$T_1 = \overline{S}_1S_2 \text{ and } T_2 = S_1.$$

The logic diagram of the control is shown in Fig. 6.31.

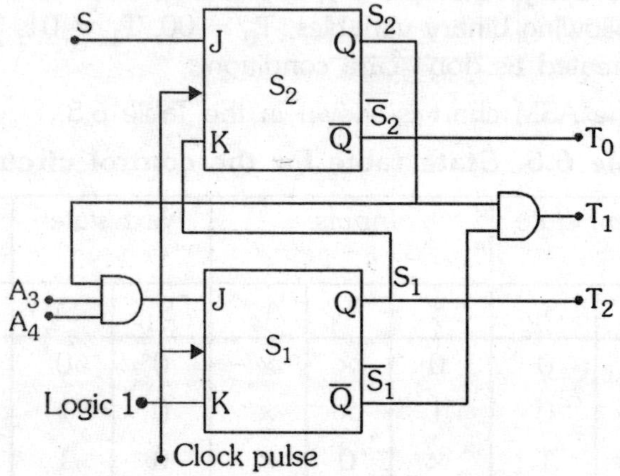

Fig. 6.31 Control circuit using JK flip-flop.

(*ii*) *D-flip-flops and decoder* : Whenever the number of flip-flops plus inputs in a state table is greater than 5, it is difficult to simplify them. One of the alternate methods for the design of controllers is D flip-flop with decoder.

It is very easy to design a sequential circuit using D flip-flop because the next state is same as the input requirement for D flip-flops. From the Table 6.4, we note that next state column of S_1 has a single 1 in the 5th row during state $T_1 = \overline{S}_2S_2$ when both inputs A_3 and A_4 are equal to 1.

Hence the D input of flip-flop $DS_1 = \overline{S}_1S_2A_3A_4$

Similarly, the next state column of S_2 has four 1's, and the condition for setting this flip-flop is

$$DS_2 = \overline{S}_1\overline{S}_2S + \overline{S}_1S_2 \qquad \because T_0 = \overline{S}_1\overline{S}_2$$

The decoder is used at the output of flip-flops to obtain the necessary three outputs, T_0, T_1 and T_2. Here the input of decoder is the output of flip-flops, hence we may express the input functions to the D flip-flops in terms of outputs as

$$DS_1 = A_3A_4S_1$$
$$DS_2 = ST_0 + T_1$$

The decoder with these two inputs produces four outputs, one outputs is don't care. These outputs are also used to determine the next state of each flip-flop. The logic diagram is shown in Fig. 6.32.

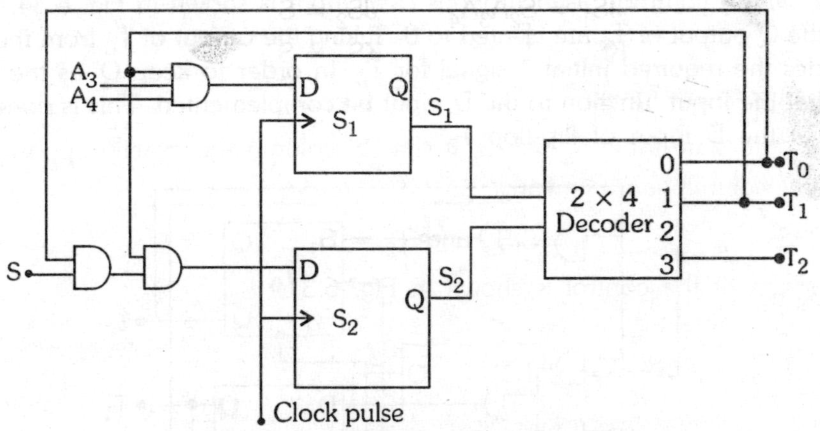

Fig. 6.32 Control circuit using D flip-flop and decoder.

(*iii*) *One flip-flop per state* : The next alternative method of control logic design is to use one flip-flop per state in the sequential circuit. Here, only one flip-flop is set at any particular time; all other flip-flops are cleared to 0. The single bit is made to propagate from one flip-flop to the other under the control of decision logic.

In this case, each flip-flop represents a state that is activated only when the control bit is transferred to it. A control organization that uses one flip-flop per state has the convenient characteristic that the circuit can be derived directly from the state diagram without the need of state or excitation tables. The equivalent state diagram from the ASM chart is shown in Fig. 6.33.

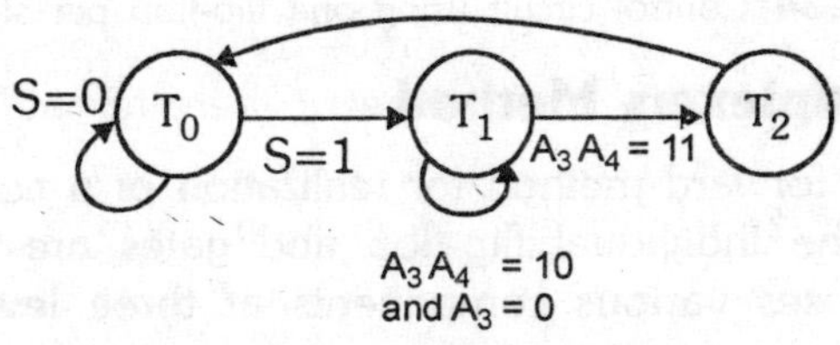

Fig. 6.33 State diagram of a controller.

The circuit has three states and needs three flip-flops and lable them as T_0, T_1 and T_2. It is easy to design the controller using D flip-flops. The Boolean function for setting the flip-flop is determined from the present state and inputs.

The flip-flop, T_0 is set with the next clock pulse if present state $T_2 = 1$ or present state $T_0 = 1$ and input, $\bar{S} = 0$. This is represented as input function of flip-flop T_0

$$DT_0 = T_2 + \bar{S}T_0$$

Similarly,

$$DT_1 = ST_0 + \bar{A}_3T_1 + A_3A'_4T_1 = ST_0 + (A_3A_4)'\,T_1$$

$$DT_2 = A_3A_4T_1$$

The logic diagram is shown in Fig. 6.34. It contains three D flip-flops T_0, T_1 and T_2 and the associated gates specified by the input functions listed before. Initially, the flip-flop T_0 must be set to 1 and all other flip-flops cleared to 0. Once started, one flip-flop per state controller will propagate itself from state to state in the proper manner.

The register with a common asynchronous clear input is shown in Fig. 6.34. Here, all flip-fiops including the Q output of T_0 are cleared to 0. Taking the output of T_0 from the complement output Q provides the required initial 1 signal for T_0. In order to keep Q' as the output of T_0, it is necessary that the input function to the D input be complemented. This is done by the extra inverter placed at the D input of flip-flop T_0.

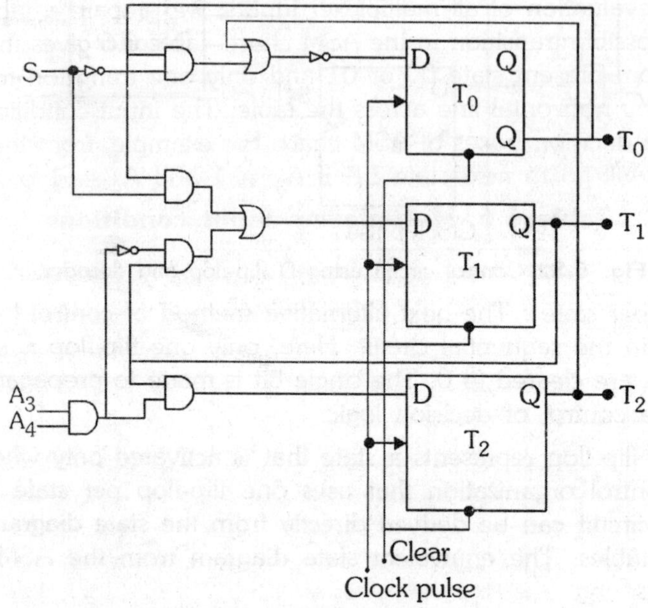

Fig. 6.34 Control circuit using one flip-flop per state.

6.6.2 Design with Multiplexers Method

It is simpler and straight forward method for realization of a combinational circuit for any controller. In this method, the individual flip-flop and gates are replaced by registers and multiplexers respectively. It uses various components at three levels. First level consists of multiplexers that determine the next state of the register.

The second level contains a register that holds the present binary state. The third level has the decoder that provides a separate output for each control state (sometimes a combinational circuit is used in place of decoder).

Consider, for example, the ASM chart of Fig. 6.29. It consists of 3 states and three contact inputs. The binary assignment for each state is indicated as T_0 = 00, T_1 = 01 and T_2 = 11.

The three level control implementation is shown in Fig. 6.35. It consists of two multiplexers, MUX1 and MUX2, a 2-bit register (2 flip-flops S_1 and S_2) and a 2 × 4 decoder. The outputs of register are applied to the decoder inputs and also to select the inputs of the multiplexers. In this way, the present state of the register is used to select one of the inputs from each multiplexer. The outputs of the multiplexers are then applied to the D flip-flop inputs (S_1 and S_2).

The purpose of each multiplexer is to produce an input to its corresponding flip-flop equal to the binary value of the next state.

The inputs of the multiplexers are determined from the state transition and decision boxes given in the ASM chart. For example, state 00 starts at 00 (or) goes to 01, depending on the input 'S'. Since next state of S_1 is '0' in either case, we place a signal equivalent to logic-0 in MUX_1 input 0. The next state of S_2 is 0 if S = 0 and 1 if S = 1. Since the next state of S_2 is equal to S, then we apply the control input 'S' to MUX_2 input 0.

To facilitate the evaluation of all multiplexer inputs, we prepare a table showing the input conditions for each possible transition in the ASM chart. Table 6.6 gives this information. There are two transitions from present state 00 or 01 and only one transition from present state 11. These are separated by horizontal line across the table. The input conditions listed in the table are obtained from the decision boxes of ASM chart. For example, from the Fig. 6.29, we note that present state 01 will go to next state 11, if A_3 = 1 and A_4 = 1 or to the next state 01,

Table 6.6. Multiplexers input conditions

Present state		*Next state*		*Input conditions*	*Multiplexer inputs*	
S_1	S_2	S_1	S_2		MUX_1	MUX_2
0	0	0	0	$\overline{S}$	0	S
0	0	0	1	S		
0	1	0	1	$\overline{A}_3$		$\overline{A}_3 + A_3\overline{A}_4$
0	1	0	1	$A_3\overline{A}_4$	A_3A_4	$+ A_3A_4$
0	1	1	1	A_3A_4		$= 1$
1	1	0	0	ϕ	ϕ	ϕ

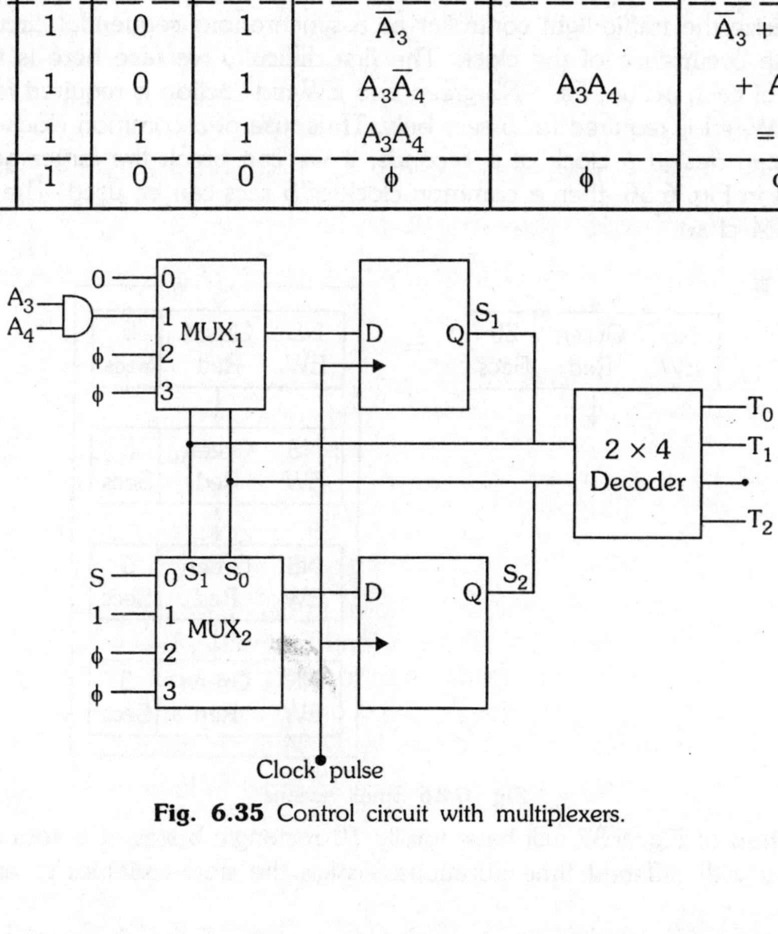

Fig. 6.35 Control circuit with multiplexers.

if $A_3 = 0$ (or) $A_3 = 1$ and $A_4 = 0$. In the table, we mark these conditions as A_3A_4 and 1 respectively. The two columns under 'multiplexer inputs' in the table specify the input values that must be applied to MUX1 and MUX2. The multiplexer input for each present state is determined from the input conditions when the next state of the flip-flop is equal to 1. Thus, after present state 01, the next state of S_2 is always equal to 1 and the next state of S_1 is equal to A_3A_4. Therefore, the input of MUX1 is made equal to A_3A_4 and that of MUX2 to 1, when the present state of the register is 01.

Another example is, after the present state 11, the next state is 00, irrespective of input condition. The inputs of MUX1 and MUX2 are made don't cares (ϕ). The remaining input 10 is not used and is given as don't care, ϕ.

6.6.3 Design Using ROM and D Flip-Flop

In ROM realization, don't care inputs/present state combinations are of no use; but synthesizing the circuit using ROMs is straight forward in the sense that ROM table can be obtained easily from the ASM chart.

The size of the ROM and number of the D flip-flops required depends upon the number of state variables used in the ASM chart. We will see this process with an example. For this, synthesize the traffic light controller as described by ASM chart of Fig. 6.15.

If we synthesize the traffic light controller as a synchronous sequential circuit, each action takes place at the occurrence of the clock. The first difficulty we face here is the unevenness of time duration of each action, *i.e.*, "NS-green and EW-red" action is required for 20 sec, while NS-yellow and EW-red is required for 5 secs only. Thus, use of a common clock will be difficult. However, if we can design a clock of 5 seconds, if we can break the larger action into small actions as shown in Fig. 6.36, then a common clock of 5 secs can be used. The Fig. 6.37 gives the modified ASM chart.

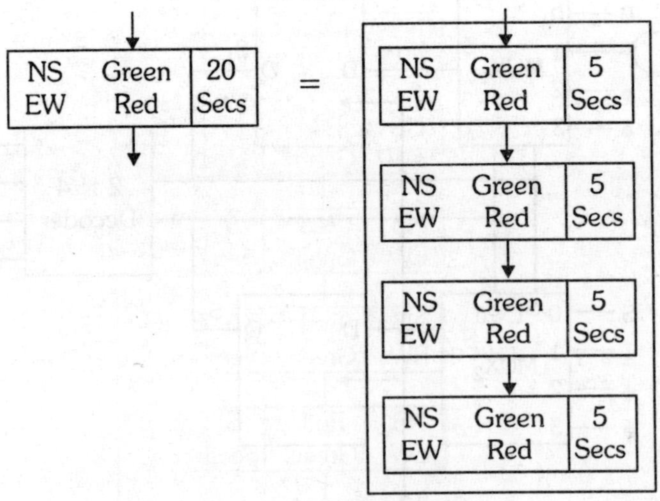

Fig. 6.36 Small actions.

The ASM chart of Fig. 6.37 will have totally 10 rectangle boxes of 5 secs each, instead of 4 rectangle-boxes with different time durations. Assign the state-variables to each state. Since there are 10 states, four variables are required. Thus, there will be 4-bits reserved at the input for present states and 4-bits at the output for the next states. The present state inputs can be called as internal inputs. Likewise, there are two external inputs WT and ET.

Thus, totally 6 inputs to the ROM. Now, let us decide the word length of the ROM. The different signals to be generated are :

(1) NS-Red
(2) NS-Green
(3) NW-Yellow
(4) EW-Red
(5) EW-Green
(6) EW-Yellow

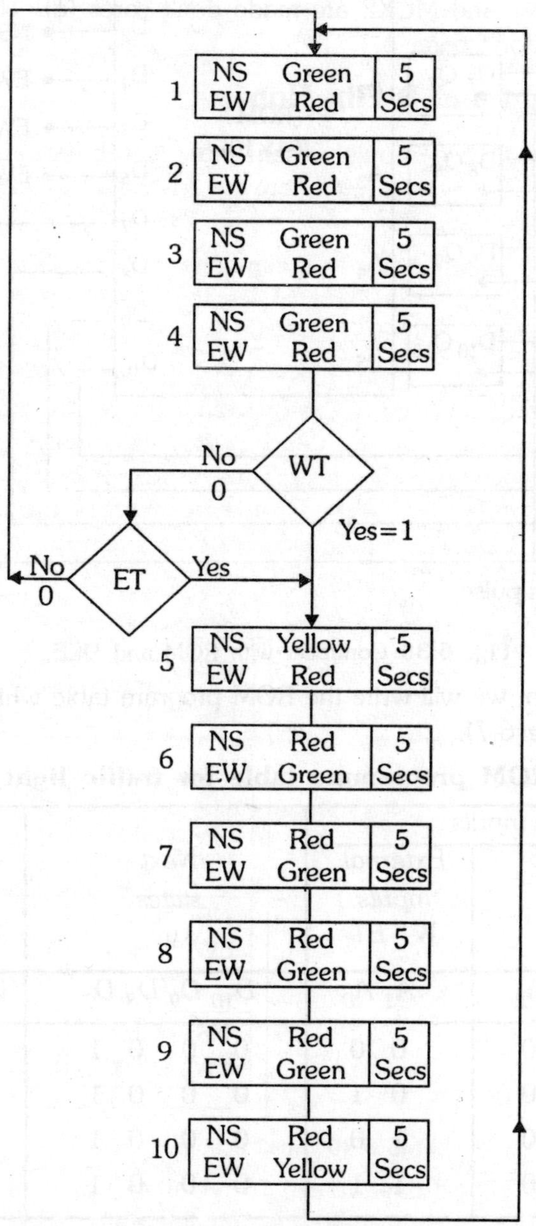

Fig. 6.37 Modified ASM chart.

The 6 output signals along with 4-bits reserved for next states constitute the word length of ROM as 10 bits. Thus, the size of the ROM will be $2^6 \times 10 = 64 \times 10$.

The block diagram of the traffic light controller using ROM and D flip-flops is shown in Fig. 6.38.

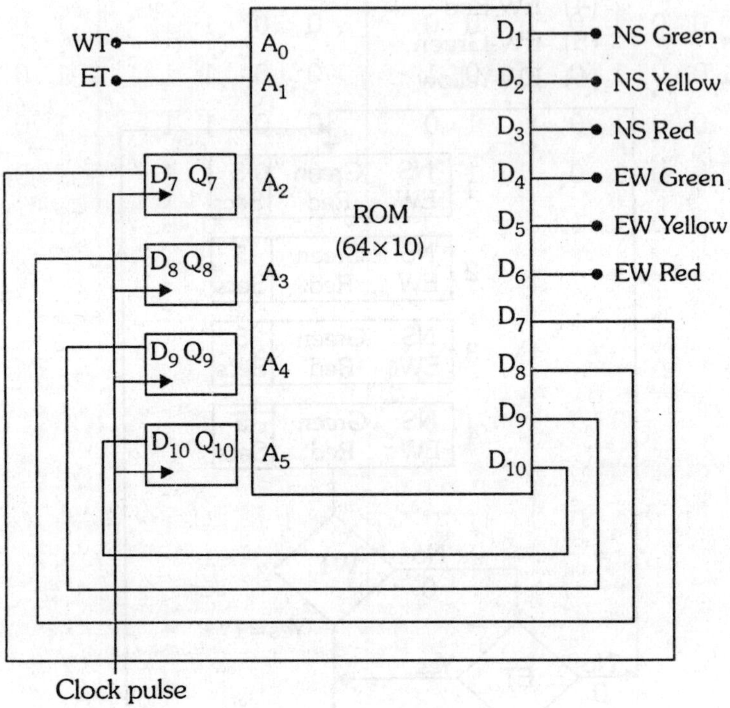

Fig. 6.38 Controller with ROM and DEF.

To complete the design, we will write the ROM program table which indicates the contents of each location (see Table 6.7).

Table 6.7. ROM programme table for traffic light controller

Decimal number	*Inputs*		*Next states* y_i	*Outputs actual outputs*
	Present states y_i	*External inputs WT ET*		
	$A_5\ A_4\ A_3\ A_2$	$A_1\ A_0$	$D_{10}\ D_9\ D_8\ D_7$	$D_6\ D_5\ D_4\ D_3\ D_2\ D_1$
0	0 0 0 0	0 0	0 0 0 1	1 0 0 0 0 1
1	0 0 0 0	0 1	0 0 0 1	1 0 0 0 0 1
2	0 0 0 0	1 0	0 0 0 1	1 0 0 0 0 1
3	0 0 0 0	1 1	0 0 0 1	1 0 0 0 0 1

4	0 0 0 1	0 0	0 0 1 0	1 0 0 0 0 1
5	0 0 0 1	0 1	0 0 1 0	1 0 0 0 0 1
6	0 0 0 1	1 0	0 0 1 0	1 0 0 0 0 1
7	0 0 0 1	1 1	0 0 1 0	1 0 0 0 0 1
8	0 0 1 0	0 0	0 0 1 1	1 0 0 0 0 1
9	0 0 1 0	0 1	0 0 1 1	1 0 0 0 0 1
10	0 0 1 0	1 0	0 0 1 1	1 0 0 0 0 1
11	0 0 1 0	1 1	0 0 1 1	1 0 0 0 0 1
12	0 0 1 1	0 0	0 0 0 0	1 0 0 0 0 1
13	0 0 1 1	0 1	0 1 0 0	1 0 0 0 1 0
14	0 0 1 1	1 0	0 1 0 0	1 0 0 0 1 0
15	0 0 1 1	1 1	0 1 0 0	1 0 0 0 1 0
16	0 1 0 0	0 0	0 1 0 1	0 0 1 1 0 0
17	0 1 0 0	0 1	0 1 0 1	0 0 1 1 0 0
18	0 1 0 0	1 0	0 1 0 1	0 0 1 1 0 0
19	0 1 0 0	1 1	0 1 0 1	0 0 1 1 0 0
20	0 1 0 1	0 0	0 1 1 0	0 0 1 1 0 0
21	0 1 0 1	0 1	0 1 1 0	0 0 1 1 0 0
22	0 1 0 1	1 0	0 1 1 0	0 0 1 1 0 0
23	0 1 0 1	1 1	0 1 1 0	0 0 1 1 0 0
24	0 1 1 0	0 0	0 1 1 1	0 0 1 1 0 0
25	0 1 1 0	0 1	0 1 1 1	0 0 1 1 0 0
26	0 1 1 0	1 0	0 1 1 1	0 0 1 1 0 0
27	0 1 1 0	1 1	0 1 1 1	0 0 1 1 0 0
28	0 1 1 1	0 0	1 0 0 0	0 0 1 1 0 0
29	0 1 1 1	0 1	1 0 0 0	0 0 1 1 0 0
30	0 1 1 1	1 0	1 0 0 0	0 0 1 1 0 0
31	0 1 1 1	1 1	1 0 0 0	0 0 1 1 0 0
32	1 0 0 0	0 0	1 0 0 1	0 1 1 0 0 0
33	1 0 0 0	0 1	1 0 0 1	0 1 1 0 0 0

34	1 0 0 0	1 0	1 0 0 1	0 1 1 0 0 0
35	1 0 0 0	1 1	1 0 0 1	0 1 1 0 0 0
36	1 0 0 1	0 0	0 0 0 0	1 0 0 0 0 1
37	1 0 0 1	0 1	0 0 0 0	1 0 0 0 0 1
38	1 0 0 1	1 0	0 0 0 0	1 0 0 0 0 1
39	1 0 0 1	1 1	0 0 0 0	1 0 0 0 0 1
40	1 0 1 0	0 0	0 0 0 0	0 0 0 0 0 0
41	1 0 1 0	0 1	0 0 0 0	0 0 0 0 0 0
42	1 0 1 0	1 0	0 0 0 0	0 0 0 0 0 0
43	1 0 1 0	1 1	0 0 0 0	0 0 0 0 0 0
44	1 0 1 1	0 0	0 0 0 0	0 0 0 0 0 0
45	1 0 1 1	0 1	0 0 0 0	0 0 0 0 0 0
46	1 0 1 1	1 0	0 0 0 0	0 0 0 0 0 0
47	1 0 1 1	1 1	0 0 0 0	0 0 0 0 0 0
48	1 1 0 0	0 0	0 0 0 0	0 0 0 0 0 0
49	1 1 0 0	0 1	0 0 0 0	0 0 0 0 0 0
50	1 1 0 0	1 0	0 0 0 0	0 0 0 0 0 0
51	1 1 0 0	1 1	0 0 0 0	0 0 0 0 0 0
52	1 1 0 1	1 0	0 0 0 0	0 0 0 0 0 0
53	1 1 0 1	0 1	0 0 0 0	0 0 0 0 0 0
54	1 1 0 1	1 0	0 0 0 0	0 0 0 0 0 0
55	1 1 0 1	1 1	0 0 0 0	0 0 0 0 0 0
56	1 1 1 0	0 0	0 0 0 0	0 0 0 0 0 0
57	1 1 1 0	0 1	0 0 0 0	0 0 0 0 0 0
58	1 1 1 0	1 0	0 0 0 0	0 0 0 0 0 0
59	1 1 1 0	1 1	0 0 0 0	0 0 0 0 0 0
60	1 1 1 1	0 0	0 0 0 0	0 0 0 0 0 0
61	1 1 1 1	0 1	0 0 0 0	0 0 0 0 0 0
62	1 1 1 1	1 0	0 0 0 0	0 0 0 0 0 0
63	1 1 1 1	1 1	0 0 0 0	0 0 0 0 0 0

ROM can be programmed as per the above table using ROM-programmer.

6.6.4 Design Using Programmable Devices

To implement sequential circuits, we may use memory elements *i.e.*, registers along with the combinational circuit. So, it is possible to design a control circuit with a register connected to a PLA. The combination of PLA and register used to implement the sequential circuit is known as 'PLA control'. Fig. 6.39 shows the block diagram of PLA control. The register has n flip-flops $f_1, f_2, \ldots, f_n$ which give the present state input for PLA-P_1, P_2, ..., P_n. The PLA has other inputs I_1, I_2, ..., I_n. The outputs of PLA from O_1 through O_n provides the values for the next state in the register and the control output variables.

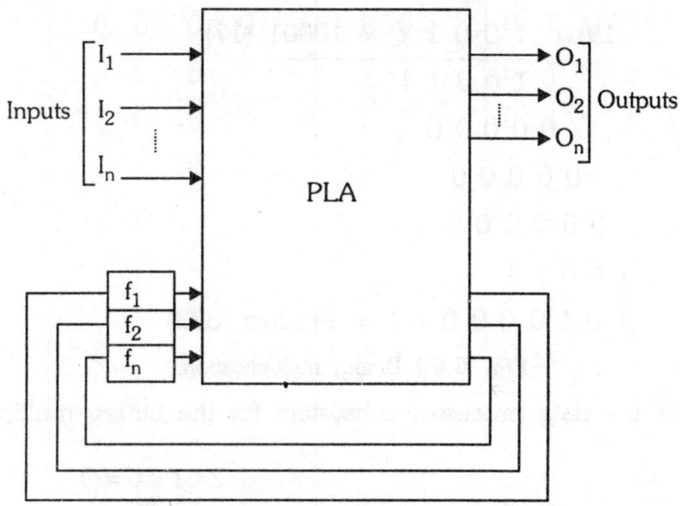

Fig. 6.39 Block diagram of PLA control.

The design of PLA control requires that we obtain the state table of the circuit. The PLA method should be used if the state table contains many don't care entries. Otherwise, it may be advantageous to use a ROM instead of a PLA.

The 'binary multiplier' design example may demonstrate the PLA control. This arithmetic circuit multiplies two unsigned binary numbers and produces their binary product.

Binary multiplier : Consider a numerical example for binary multiplication of two binary numbers, 10011 and 10001, as shown in Fig. 6.40. The decimal equivalents are indicated in brackets. During the multiplication process, if the multiplier is 1, the multiplicand is copied down, otherwise, 0's are copied down. The numbers copied down in successive lines are shifted one position to the left from the previous number. Finally, the numbers are added and their sum forms the product.

Note : The product obtained from the multiplication of two binary numbers of n-bits each, can be up to $2n$-bits long.

While implementing the above process with hardware, slight modification is made, that is :

(1) Instead of providing digital circuits to store and add simultaneously as many binary

numbers as there are 1's in the multiplier, it is convenient to provide circuits for the summation of only two binary numbers and successively accumulate the partial products in a register.

(2) Instead of shifting the multiplicand to left, the partial product is shifted to the right, which results that the multiplicand and partial product are in a required position for every bit multiplication.

(3) When the multiplier bit is 0, no need to add all 0's to the partial product, since this will not result in any change.

```
            Multiplicand  multiplier
     (19)    1 0 0 1 1 × 10001 (17)
            ---------------------------
             1 0 0 1 1
           0 0 0 0 0
         0 0 0 0 0
       0 0 0 0 0
     1 0 0 1 1
     ----------------------
     1 0 1 0 0 0 0 1 1 = Product (323)
```

Fig. 6.40 Binary multiplication.

Fig. 6.41 shows the data processor subsystem for the binary multiplier.

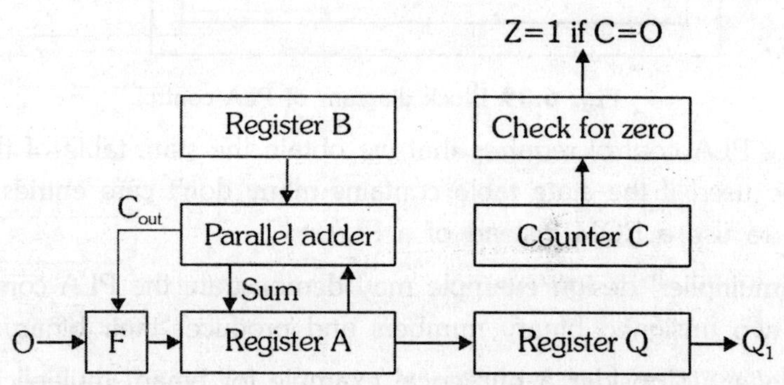

Fig. 6.41 Data processor for binary multiplication.

Here the multiplicand is stored in register B, the multiplier is stored in register Q, and the partial product is formed in register A.

Parallel adder is used to add the contents of the register B to register A, the resultant carry is stored in the flip-flop, F. The binary counter C is initially set to hold a binary number equal to the number of bits in the multiplier. This count value is decremented by one on the foundation of each partial product. When the content of the counter reaches zero, the resultant is (product) formed in the double register A and Q, and the process stops.

The control logic for this is :

The sum of A and B forms a partial product, which is transferred to A. The output carry from the addition is stored in E. After that EAQ are shifted right by one bit. The least significant bit of A is shifted to the most significant position of Q ; the carry from E is shifted to the most significant position of A and '0' is placed in E.

After this operation, one bit of the partial product is transferred into Q while the multiplier bits in Q are also shifted one position to the right. In this way, the right most bit of Q is Q_1, which indicates the multiplier bit that must be inspected next.

ASM chart for binary multiplier

Fig. 6.42 shows the ASM chart for the binary multiplier. The multiplication process is initiated with start signal 'S', when S = 1. The multiplicand and multiplier are stored in registers A and Q respectively. Initially, register A and flip-flop F are cleared and counter 'C' is set to 'n' where n = number of bits in the multiplier. This acts as a count to perform number of partial products. While forming partial product, the Q_1 bit is checked, if it is 1, the multiplicand in B is added with the partial product in A. The carry is stored in flip-flop F. If $Q_1 = 0$, no change in partial product. The counter value is decremented by 1 regardless of the value of Q_1. The combined registers of EAQ are shifted to right for every bit of Q_1, to obtain a new partial product.

The value in the counter 'C' is verified for every partial product. If it is zero, the control input $z = 0$ and the process is repeated to form a new partial product. Otherwise the process stops. The final product is available in A and Q registers, with MS bits holding in A and Q holding the least significant bits.

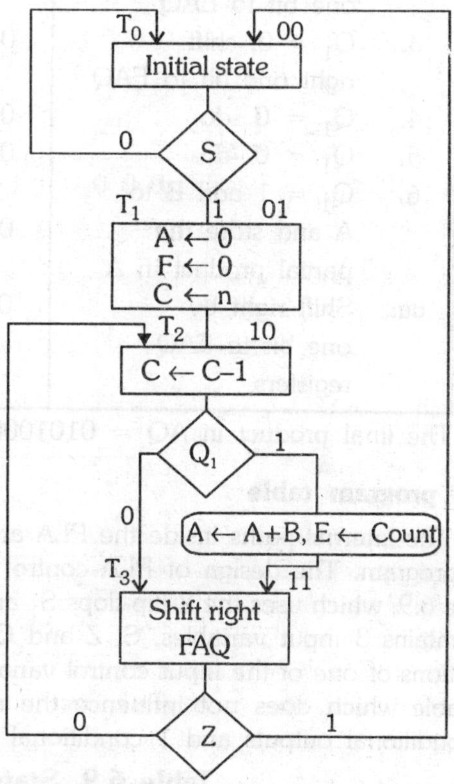

Fig. 6.42 ASM chart for binary multiplier

Table 6.8 illustrates the procedural steps involved in the ASM chart.

Table 6.8. Binary multiplication example

		Multiplicand = *B* = *10011* *Multiplier* = *Q* = *10001*			
		E	*A*	*Q*	*C*
1.	Initial values	0	00000	10001	101
2.	Multiplier in Q with Q_1 = 1, add B to A and store the partial product in A	0	10001 10001	First partial product	100
2*a*.	Shift right by one bit to EAQ	0	01000	11000	
3.	Q_1 = 0; shift right one bit to EAQ	0	00100	01100	011
4.	Q_1 = 0 -do-	0	00010	00110	010
5.	Q_1 = 0 -do-	0	00001	00011	001
6.	Q_1 = 1 add B to A and store the partial product in A	0	10011 10100	00011	000
6*a*.	Shift right by one bit to EAQ registers	0	01010	00011	

The final product in AQ = 0101000011

PLA program table

The internal paths inside the PLA are constructed according to the specifications given in the program. The design of PLA control requires the state table. The state table is shown in Table 6.9, which uses the 2 flip-flops S_1 and S_2 to represent 4 states T_0 to T_3 in the ASM chart. It contains 3 input variables, S, Z and Q, for control. The next states of S_1 and S_2 may be functions of one of the input control variables or may be independent of any inputs. The input variable which does not influence the next state is marked with don't care (×). It has 4 unconditional outputs and 1 conditional output. The conditional output affects only state T_2.

Table 6.9. State table for control circuit

Present state	*Inputs*	*Next state*	*Outputs*
$S_1\ S_2$	$S\ Z\ Q_1$	$S_1\ S_2$	$T_0\ T_1\ T_2\ D\ T_3$
0 0	0 × ×	0 0	1 0 0 0 0
0 0	1 × ×	0 1	1 0 0 0 0
0 1	× × ×	1 0	0 1 0 0 0
1 0	× × 0	1 1	0 0 1 0 0

1 0	× × 1	1 1	0 0 1 1 0
1 1	× 0 ×	1 0	0 0 0 0 1
1 1	× 1 ×	0 0	0 0 0 0 1

The PLA table can be obtained easily from state table without the need of simplification procedure. The PLA program table is shown in Table 6.10. It contains seven product terms, one for each row in the state table. The inputs (5 NO's) are marked with 0 (or) 1 and — (dash)

0 (or) 1 indicates that the particular inputs is present

— (dash) indicates that the particular input is not present (don't care).

i.e., '×' in the input column of state table is represented with dash (—). The zeros in the outputs and next states of state table are represented by — (dash) in the respective product term the program table and there is no change in the 1's place.

The inputs to the PLA table are same as the present state and inputs in the state table. The outputs of the PLA table are next state and outputs in the state table.

Table 6.10. PLA program table

Product term	*Inputs* 1	2	3	4	5	*Outputs* 1	2	3	4	5	6	7	*Comments*
1	0	0	0	–	–	–	–	1	–	–	–	–	$T_0 = 1, S = 0$
2	0	0	1	–	–	–	1	1	–	–	–	–	$T_0 = 1, S = 1$
3	0	1	–	–	–	1	–	–	1	–	–	–	$T_1 = 1$
4	1	0	–	–	0	1	1	–	–	1	–	–	$T_2 = 1$
5	1	0	–	–	1	1	1	–	–	1	1	–	$T_2 = 1, D = 1$
6	1	1	–	0	–	1	–	–	–	–	–	1	$T_3 = 1, Z = 0$
7	1	1	–	1	–	–	–	–	–	–	–	1	$T_3 = 1, Z = 1$

Control design using PLA

The controller for the binary multiplier has four states T_0, T_1, T_2 and T_3 and three inputs S, Z and Q_1. The design with PLA is shown in Fig. 6.43.

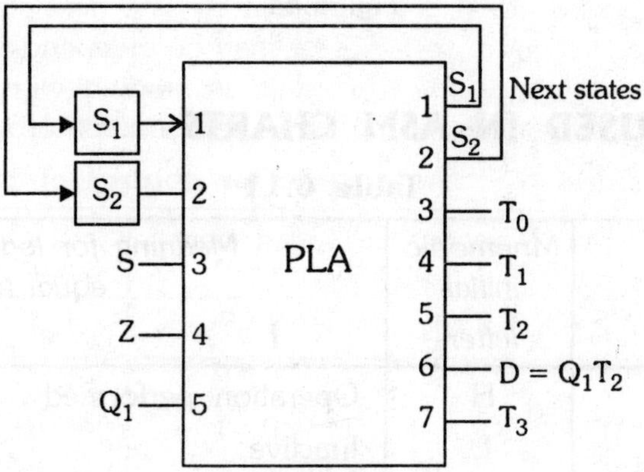

Fig. 6.43

The PLA connected with a register contains 2 flip-flops S_1S_2. The inputs to the PLA are the values of the present state of the register and three control inputs. The outputs of PLA are next states of the register and the 5 output variables. There is an output at each present state and one more conditional output at T_2 is $D = Q_1T_2$.

At any given time, the present state of the register together with the input conditions determine the output values and the next state for the register. Here, PLA acts as the combinational-circuit part of the sequential circuit to generate the control outputs and the next-state values for the register.

Example 6.11. *Draw the ASM chart for MOD-6 counter.*

Solution. It has 6 states namely *a, b, c, d, e* and *f* and there is a state transition per each clock pulse without any condition (see Fig. 6.44).

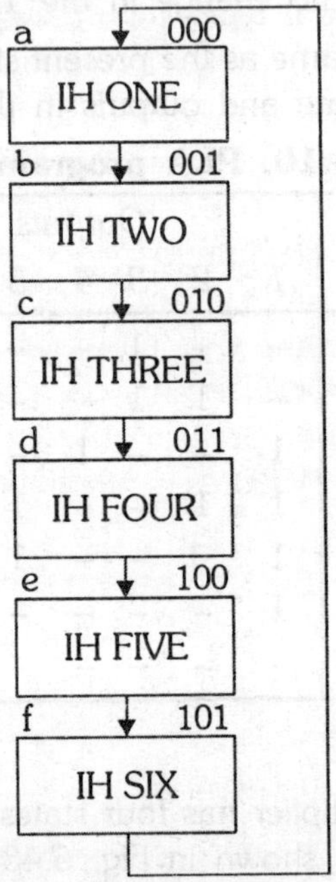

Fig. 6.44

6.7 MNEMONICS USED IN ASM CHARTS

Table 6.11

S. No.	*Type of logic terminal*	*Mnemonic initial letter*	*Meaning for logic terminal equal to* 1	0
1.	Output or Instruction	H	Operation performed	Inactive
		L	Inactive	Operation performed
2.	Inputs or qualifiers	Y	Statement is true	Statement is false

3.	Immediate	N I	Statement is false Proceeds either H or L to designate an immediate function	Statement is true

A system of mnemonics for defining input and output terminals and some semantics for defining the meaning associated with each terminal is represented in Table 6.11. The terminals that form the inputs and outputs of the sequential machine are given names, called "mnemonics", consisting of letters and numbers, which are used as memory aids in recalling the functions of operations associated with the logic levels on the terminals. The system fulfils four objectives, which are to provide :

(1) A single name for each common logic line.
(2) A consistent logic level throughout the machine.
(3) An indication of the terminal type (input, output)
(4) A means of identifying the logic interpretation.

The basic mnemonic is a group of three or four letters which is usually formed from the first letters of the words describing the operation or function of the related terminal. For example, an output terminal might be called OUT. An initial letter H, L, Y or N is used to signify the logic level and the type (Input or Output) of terminal. The prefixes H (high) and L (low) are used to signify the HIGH and LOW signal levels of an output; whereas Y (yes) and N (no) are used to signify the logic levels 1 and 0 of an input, respectively, if the test input is true. For example, $LOUT_1$ signifies that the OUT_1 occurs on a 0 (LOW) signal and YX means that the logic level is 1 if X is true. Another prefix, I, is useful in the interpretation of the output instruction terminals. I stands for IMMEDIATE function, which means a task that is completed in the present state time.

For example, $IHOUT_1$ signifies that the output OUT_1 occurs on a 1 (HIGH) signal in the present state time.

Example 6.12. *Draw the ASM chart for master slave D flip-flop.*

Solution. The next state function is realized by deriving the partial next state functions for each state variable. This is shown in Fig. 6.45.

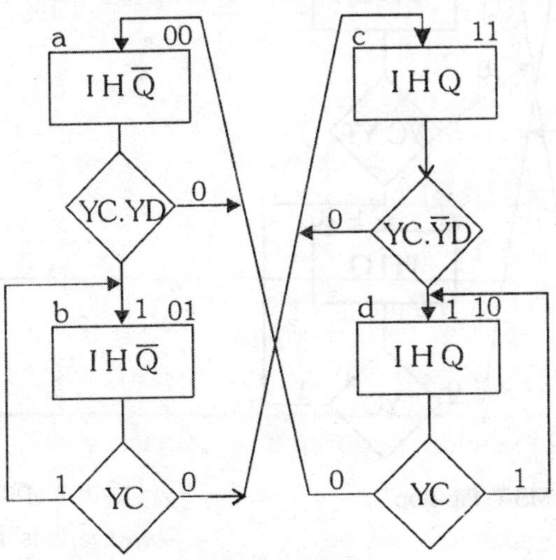

Fig. 6.45(a)

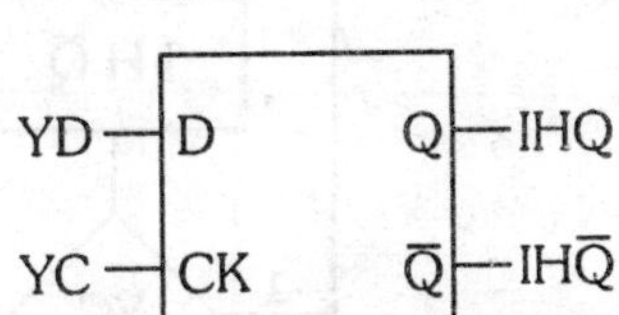

Fig. 6.45(b)

where IHQ = Output Q = 1

$\text{IH}\overline{\text{Q}} = \text{Output}/\overline{\text{Q}} = 1$

YC = Clock

YD = D Input

It has 4-states and need of two flip-flops namely A and B. The partial next state map for state variable A was found by observing all link paths for which the next state A was 1. From state '*a*' the next state A will be 1 if the quantifier YC.YD is true, so YCYD is entered in the state a location. Both exit paths of the decision box in the ASM block for state '*b*' lead to a next state of 1 for A, so a 1 is entered in location *b*.

For state 'c', the quantifier $\text{YC}.\overline{\text{YD}}$ must be false for the next state variable A to be a 1. Therefore $\overline{\text{YC}.\overline{\text{YD}}}$ is entered in location C. The map is completed with a '0' in the state '*d*' location, since no link path leads from '*d*' next state A is 1.

The partial next state for B is obtained similarly (see Fig. 6.46).

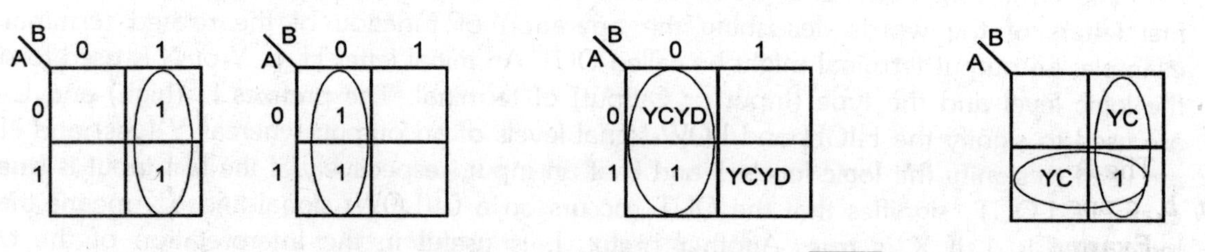

(*a*) IHQ = B (*b*) $\text{IH}\overline{\text{Q}} = \overline{\text{B}}$ (*c*) $g_A = \text{A}\overline{\text{B}} + \overline{\text{B}}\text{YC}\overline{\text{YD}} + \text{A}\overline{\text{YC}\overline{\text{YD}}}$ (*d*) $g_B = \text{A.B} + \text{A}\overline{\text{Y}}\text{C} + \text{BYC}$

Fig. 6.46

Example 6.13. *Draw the ASM chart for master slave T flip-flop.*

Solution.

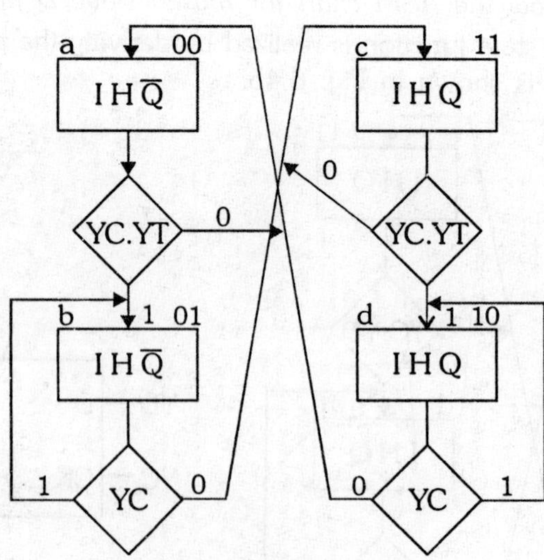

Fig. 6.47 MS-T flip-flop.

The ASM chart for the MS-T flip-flop is shown in Fig. 6.47 which contains 4 states *a*, *b*, *c* and *d*.

Example 6.14. *Draw the ASM chart for master slave JK flip-flop.*

Solution.

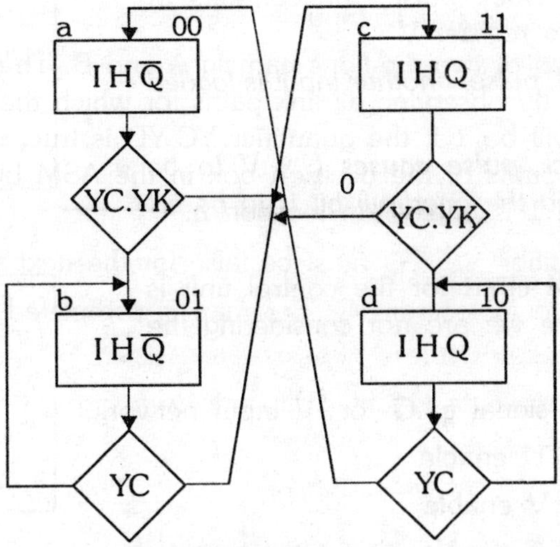

Fig. 6.48 MSJK flip-flop.

The ASM chart for master slave JK flip-flop is shown in Fig. 6.48. It has 4 states, namely *a*, *b*, *c* and *d*.

Example 6.15. *Draw the ASM chart for edge triggered JK flip-flop.*

Solution. The edge triggered-JK flip-flop allows the change in the input even while the clock is high, as long as the inputs are stable during the 1- to -0 transition of the clock is shown in Fig. 6.49.

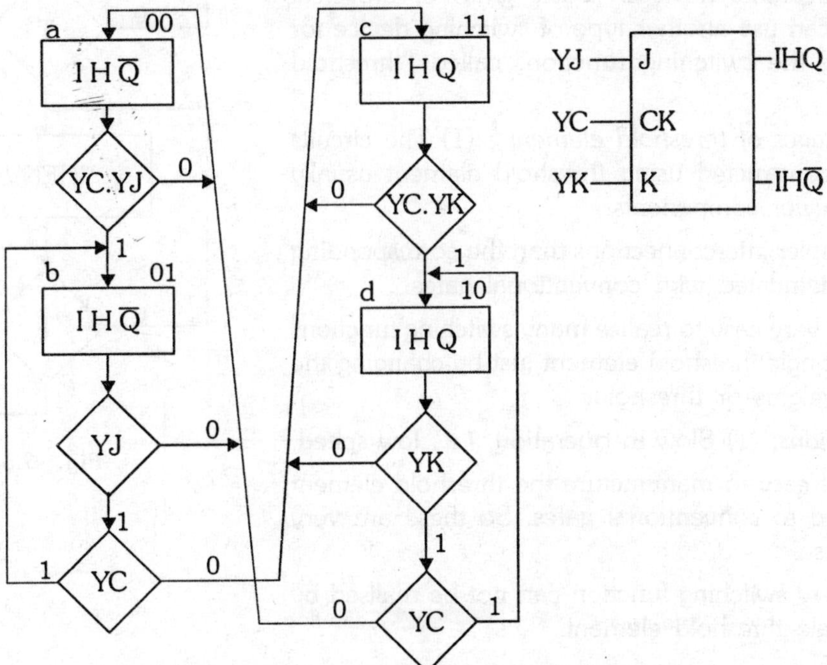

Fig. 6.49 ASM chart for edge triggered JK flip-flop.

Example 6.16. *Design a circuit which is to be operated as described below.*

(a) On the first clock pulse, the input X is loaded into the register R and the overflow bit is cleared.

(b) On the second clock pulse, the contents of register R are loaded into register V.

(c) On the third clock pulse, another input is loaded into register U.

(d) The fourth clock pulse causes U + V to be loaded into register V and the overflow bit is to be set if overflow occurs.

Solution. The ASM chart for the control unit is shown in Fig. 6.50. Here we are not considering the Data Processor Unit.

IHADD = control signal to O- or -V input network

IHENU = register U enable

IHENV = register V enable

IHENC = register C enable

T = clock pulse.

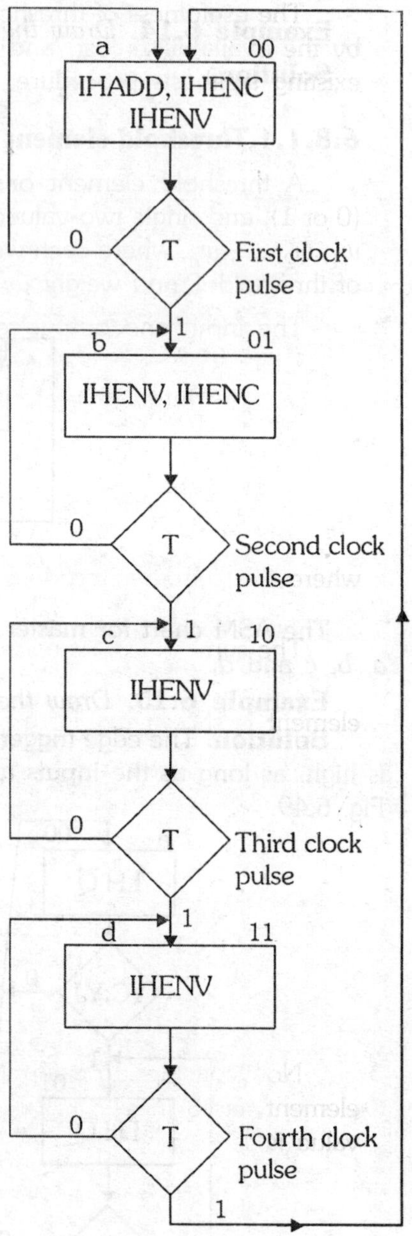

Fig. 6.50

6.8 THRESHOLD LOGIC

So far, we have constructed many switching circuits, such as sequential or combinational, the use of electronic gates *i.e.,* bilaterial devices. (Basic gates or universal gates). We can use another type of switching device for constructing the switching function, called "threshold element".

Advantages of threshold element : (1) The circuits which are constructed using threshold element usually consist of fewer components.

(2) Simpler interconnections than the corresponding circuits implemented with conventional gates.

(3) It is very easy to realize many switching functions by using a single threshold element just by changing the values of weights or threshold.

Limitations: (1) Slow in operation, *i.e.,* low speed.

(2) Not easy to manufacture the threshold element as compared to conventional gates. So these are very limited in use.

(3) Every switching function can not be realised by using a single threshold element.

6.8.1 Introduction

The usefulness of threshold logic, or of any new logic in digital system design, is determined by the availability, cost, and capabilities of the basic building block and also depends on the existing synthesis procedure.

6.8.1.1 Threshold element

A threshold element or gate, has 'n' number of two-valued inputs $x_1, x_2, x_3, \ldots, x_n$ (0 or 1), and single two-valued output y. It has two internal parameters, threshold T and weights, $w_1, w_2, \ldots, w_n$, where each weight w_i, is associated with a respective input variable x_i. The values of threshold T and weight w_i may be any real, finite, positive or negative numbers.

The input and output relationship of a threshold element is defined as follows :

$$\text{Output } y = 1, \text{ if and only if } \sum_{i=1}^{n} w_i x_i \geq T \qquad \ldots(6.1)$$

$$\text{Output } y = 0, \text{ if and only if } \sum_{i=1}^{n} w_i x_i < T \qquad \ldots(6.2)$$

where the sum and product operations are the conventional arithmetic ones.

The sum $\sum_{i=1}^{n} w_i x_i$ is called the "weighted sum" of the element. The symbol of the threshold element is shown in Fig. 6.51.

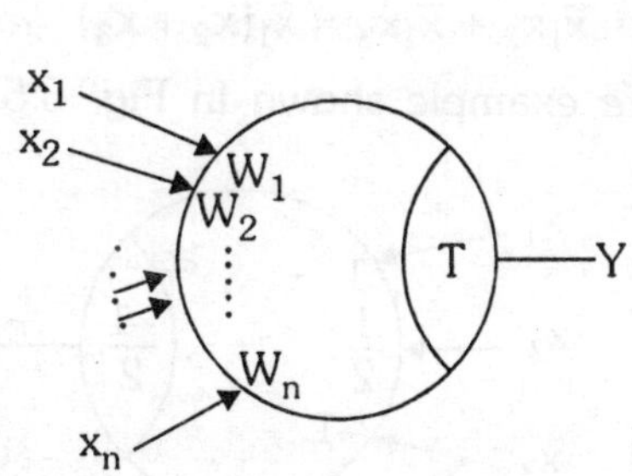

Fig. 6.51 Threshold element symbol.

Now consider an example of threshold element shown in Fig. 6.52. From this, the threshold element has 3 inputs, x_1, x_2, x_3 and their respective weights are – 2, 1 and 1. The threshold value is 1.

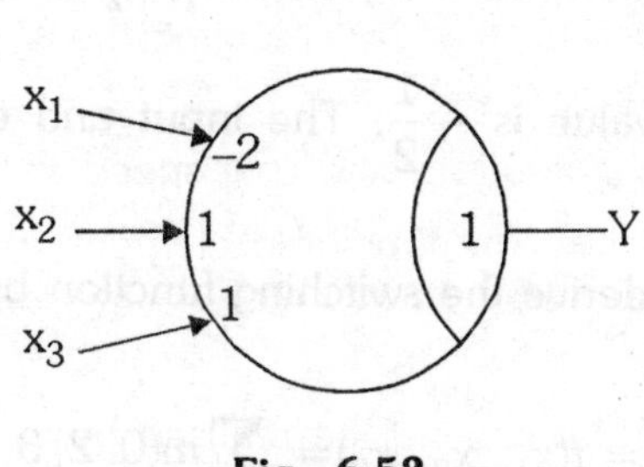

Fig. 6.52

The input output relation is shown in Table 6.12. In the output column of the table a logic 1 is entered when the weighted sum is greater than or equal to threshold value and '0' is entered elsewhere.

Table 6.12

Input variables			*Weighted sum*	*Output*
x_1	x_2	x_3	$-2x_1 + x_2 + x_3$	y
0	0	0	0	0
0	0	1	1	1
0	0	1	1	
0	1	1	2	1
1	0	0	− 2	0
1	0	1	− 1	0
1	1	0	− 1	0
1	1	1	0	0

From this, we can conclude that the threshold element can realise a switching function.

As per the truth table, output $y = 1$ for the minterms 1, 2 and 3.

$$\therefore \quad y = f(x_1, x_2, x_3) = \sum m(1, 2, 3)$$

$$= \bar{x}_1 x_2 + \bar{x}_1 x_3 = \bar{x}_1(x_2 + x_3)$$

Let us consider one more example shown in Fig. 6.53.

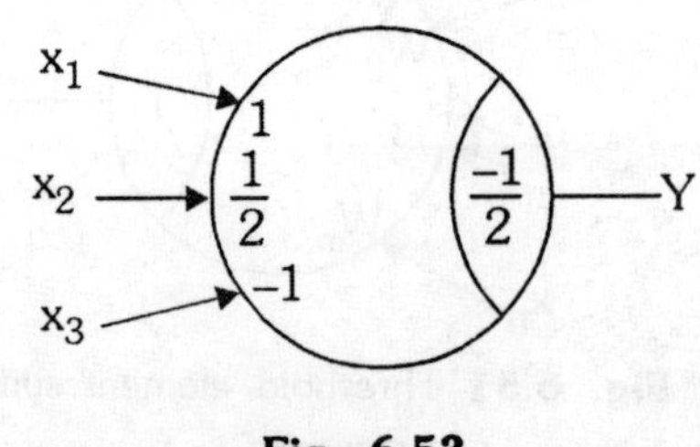

Fig. 6.53

Here, the threshold element has 3 inputs – x_1, x_2 and x_3 and respective weights are $1, \frac{1}{2}$ and -1. The threshold, T value is $-\frac{1}{2}$. The input and output relationship is shown in the Table 6.13.

From the table, we can derive the switching function by adding all minterms for which the output is 1.

$$\therefore \quad y = f(x_1, x_2, x_3) = \sum m(0, 2, 3, 4, 5, 6, 7)$$

$$= \Pi M(1)$$

Table 6.13

Input variable			*Weighted sum*	*Output*
x_1	x_2	x_3	$x_1 + \frac{1}{2}x_2 - x_3$	y
0	0	0	0	1
0	0	1	− 1	0
0	1	0	$+\frac{1}{2}$	1
0	1	1	$-\frac{1}{2}$	1
1	0	0	1	1
1	0	1	0	1
1	1	0	$1\frac{1}{2}$	1
1	1	1	$\frac{1}{2}$	1

6.8.2 Construction of Threshold Element

The threshold element/gate can be constructed by various ways. Two of them are :

(1) Register-transistor (gate) threshold element

(2) Magnetic-core threshold element

6.8.2.1 Register-transistor threshold element

The constructions of threshold element is shown in Fig. 6.54.

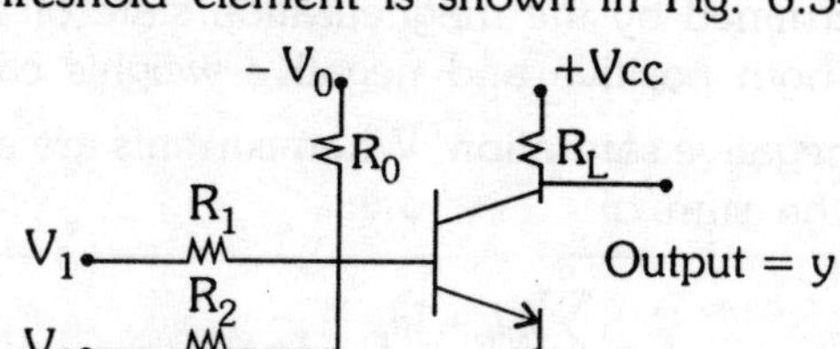

Fig. 6.54 Register-transistor logic threshold element.

Input of the above circuit is a linear summer which computes the weighted sum of the inputs.

The input voltages are V_1, V_2, ..., V_n and input resistances R_1, R_2, ..., R_n; replicate the inputs and weights of the threshold element.

The weighted sum of the circuit is a function of input voltages, input resistances and the particular combination of inputs whose logical value is 1.

The threshold value T, of the threshold element is determined by the register R_0 and voltage source V_0.

The output of the threshold element,

$$y = 1 \text{ when } V_1R_1 + V_2R_2 + \dots V_nR_n \geq V_0R_0 \qquad \dots(6.3)$$

$$\text{otherwise } y = 0$$

Limitation : Here all the resistor values are positive, hence this gate is capable of providing only positive weights.

6.8.2.2 Magnetic-core threshold element

The magnetic core threshold element is shown in Fig. 6.55. It has a magnetic core with $n+3$ windings.

The value of each input variable x_i is determined by the presence or absence of current (I) in the *i*th winding, the corresponding weight w_i is a function of the number of turns l_i. The threshold T is determined by the constant current I_t, and the number of turns l_t and their direction.

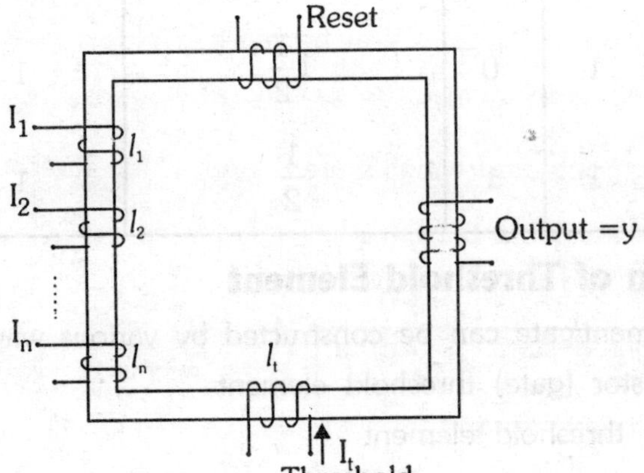

Fig. 6.55 Magnetic-core threshold element.

The value of y is determined by the magnetization state of the core. Depending upon the direction of the windings, both positive and negative weights can be obtained.

Initially, the core is in negative saturation. When currents are applied, it's saturation direction depends on the value of the sum of

$$\sum_{i=1}^{n} I_i l_i + I_t l_t \qquad \dots(6.4)$$

To determine the direction of solution, a reset pulse is applied sufficiently large to drive the core into negative saturation. (*a*) If at that time core is positively saturated, a pulse will appear on the output winding, which indicates that output is 1.

(*b*) If the core is negatively saturated when the reset pulse appears, no pulse will appear on the output winding, which indicates that output is '0'.

6.8.3 Capabilities and Limitations of Threshold Logic

(1) The single threshold element can realize a larger class of functions than is realizable by any one conventional gate.

(2) Any conventional gate can be realized by using a single threshold element. A two input NAND, for example, can be realized by a single threshold element with weights $-1, -1$, threshold $T = -1\frac{1}{2}$, as shown in Fig. 6.56.

Input x_1x_2	*Weighted sum* $-x_1-x_2$	*Output* y
0 0	0	1
0 1	− 1	1
1 0	− 1	1
1 1	− 2	0

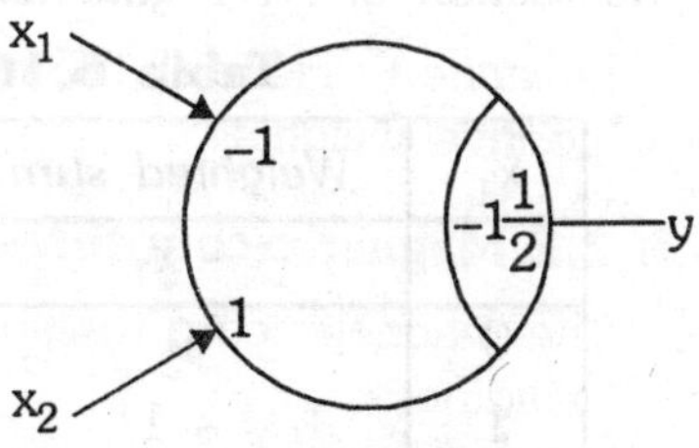

Fig. 6.56

Realization of various gates by using threshold element

All basic as well as universal gates can be realized by using a single threshold element.

(1) *Realization of OR gate using threshold element* : The realization of OR gate is shown in Fig. 6.57. It has two inputs and weights are 1 and 1, the threshold T is $\frac{1}{2}$. The input and output relation is shown in Table 6.14.

Table 6.14

Input x_1x_2	*Weighted sum* x_1+x_2	*Output* y
00	0	0
01	1	1
10	1	1
11	2	1

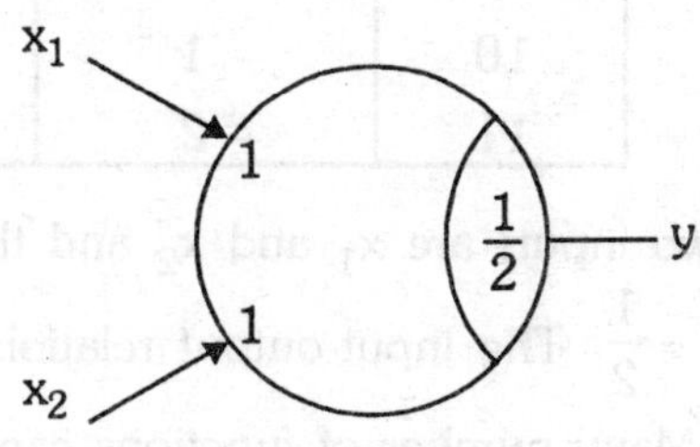

Fig. 6.57

(2) *Realization of AND gate using threshold element :*

Table 6.15

Input x_1x_2	*Weighted sum* x_1+x_2	*Output* = y
0 0	0	0
0 1	1	0
1 0	1	0
1 1	2	1

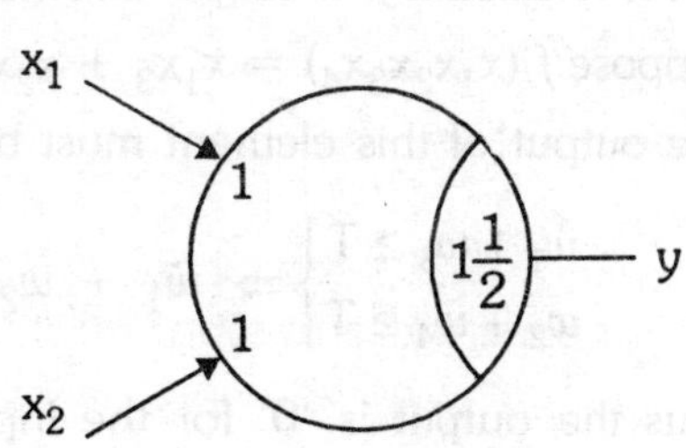

Fig. 6.58

The realization of AND gate is shown in Fig. 6.58. It has two inputs x_1 and x_2 with weights 1 and 1 respectively. The threshold value T is $1\frac{1}{2}$. The input-output relation of the gate is shown in Table 6.15.

Realization of NOT gate using threshold element

Table 6.16

x_1	*Weighted sum* $-x_1$	*output*
0	0	1
1	– 1	0

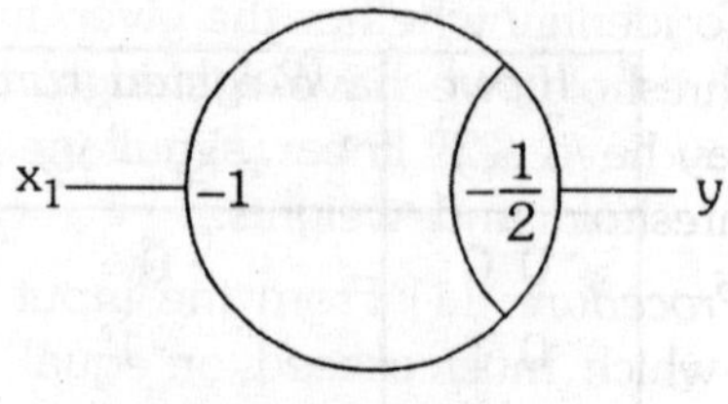

Fig. 6.59

It has one input x_1 and its weight is – 1. The threshold value $T = \frac{-1}{2}$. The input and output relationship of NOT gate is shown in Table 6.16.

Realization of NOR gate using threshold element.

Table 6.17

x_1x_2	$-x_1-x_2$	y
00	0	1
01	– 1	0
10	– 1	0
11	– 2	0

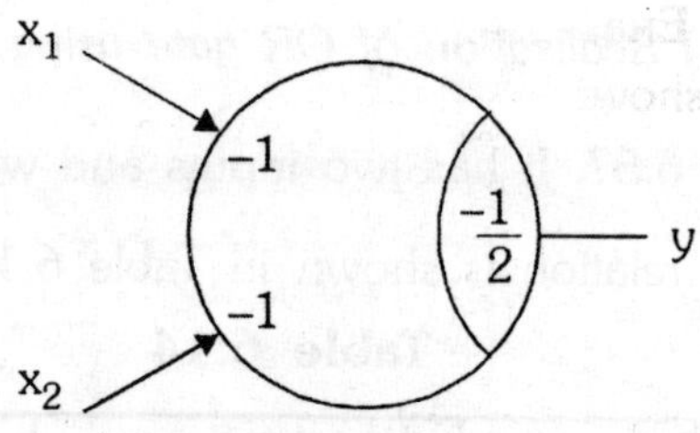

Fig. 6.60

If two inputs are x_1 and x_2 and their weights are – 1 and – 1 respectively, the threshold value is $-\frac{1}{2}$. The input-output relationship of NOR gate is shown in Table 6.17.

(3) Many number of functions can be realized by using a single threshold element just by changing weights or threshold and/or combinations. But every switching function cannot be realized by a single threshold element. Let us see one example of a switching function which cannot be realized by a single threshold element.

Suppose $f\,(x_1x_2x_3x_4) = x_1x_3 + x_2x_4$ and weights are w_1, w_2, w_3 and w_4 and threshold is T. The output of this element must be 1 for the input combinations $x_1x_3x'_2x'_4$ and $x_2x_4x'_1x'_3$.

$$\left.\begin{matrix} w_1 + w_3 \geq T \\ w_2 + w_4 \geq T \end{matrix}\right\} \Rightarrow w_1 + w_2 + w_3 + w_4 \geq 2T \qquad ...(6.5)$$

Thus the output is '0' for the input combinations $x'_1x'_3x_2x_4$ and $x'_2x'_4x_1x_3$. Thus

$$\left.\begin{matrix} w_2 + w_4 < T \\ w_1 + w_3 < T \end{matrix}\right\} \Rightarrow w_1 + w_2 + w_3 + w_4 < 2T. \qquad ...(6.6)$$

Clearly, the equations (6.5) and (6.6) indicate a conflicting in the requirements and no threshold value can satisfy them. So, $f = x_1x_3 + x_2x_4$ cannot be realized by a single threshold element.

Note : A switching function, $f(x_1, x_2, ..., x_n)$ can be realized by a single threshold element if it has appropriate weights and threshold.

A switching function that can be realized by a single threshold element is called "*threshold function*".

To identify whether the given function is a threshold function or not and to find the weights and threshold, we have a straight forward approach, called truth table. From the truth table, we may have a 2^n linear, simultaneous inequality equations and while solving them, we can get the threshold and weights.

Procedure: (1) From the input combinations for which $f = 1$, we derive all the weighted sums which must exceed or equal to the threshold T.

(2) From the input combinations for which $f = 0$, we derive all weighted sums which must be less than T.

(3) If a solution to the above inequalities exists, it provides the values for weights and threshold (The values are not unique).

(4) If no solution exists, then f is not a threshold function.

Example. Let $f(x_1, x_2, x_3) = \Sigma(0, 1, 3)$. The truth table and the corresponding inequalities are shown in Table 6.18.

Table 6.18. Truth table with linear inequalities for $f = \Sigma(0, 1, 3)$

Combination	*Inputs* $x_1x_2x_3$	*Output* f	*Inequalities*
0	000	1	$0 \geq T$
1	001	1	$w_3 \geq T$
2	010	0	$w_2 < T$
3	011	1	$w_2 + w_3 \geq T$
4	100	0	$w_1 < T$
5	101	0	$w_1 + w_3 < T$
6	110	0	$w_1 + w_2 < T$
7	111	0	$w_1 + w_2 + w_3 < T$

(1) From the combination 0, we observe that T must be negative.

(2) From the combinations 2 and 4, T is less than w_2 and w_1. So, weights w_1 and w_2 are also negative.

(3) From combinations 3 and 5, we conclude that w_2 must be greater than w_1.

(4) From combination 1, we conclude that w_3 is greater than or equal to T.

From the above conclusions, we can establish a relation $w_3 \geq T > w_2 > w_1$.

From the combinations 7, we conclude that $w_1 + w_2 + w_3 < T$.

From the above two equations, w_3 may be positive.

Let us choose $w_3 = 1$, $w_2 < w_3$ and w_2 is negative

$\therefore$ $w_1 = 1$

and $w_1 < w_2$ assume $w_1 = 2$

and we can choose $T = -\frac{1}{2}$ so with the $w_1 = -2$, $w_2 = -1$ and $w_3 = 1$ and $T = -\frac{1}{2}$ all inequalities are satisfied.

For an *n*-variable switching function, there are 2^n linear inequalities, some of which may usually be eliminated because they are implied by the others. (*e.g.*, if inequalities 2 and 4 are satisfied, since T is negative inequally 6 is automatically implied, and similarly, inequality 7 is implied by 2 and 5.)

The realization of a non-threshold switching function (whose corresponding AND-OR network may be quite complex) can often be accomplished with just a few threshold elements. Thus use of threshold element may result in a considerable reduction in the number of gates, inputs and components, as well as in the size of the final circuit.

6.8.3.1 Limitations

(1) The threshold logic is sensitive to various circuit parameters. *For example* : (*a*) the resistance values may change by as much as 20 per cent of their nominal value. (*b*) Input and supply voltages may also vary by an appreciable amount and consequently, the weighted sum for a particular combination, especially with a large number of inputs, it may deviate from its prescribed value so as to cause a circuit malfunction. Restrictions must therefore be imposed on the maximum allowable number of inputs and on the threshold value, T.

Care must be taken to increase the difference between the values of the weighted sum for which $f = 1$ and $f = 0$.

(2) *Lack of synthesis procedure* : This is the major limitation in usefulness of a threshold logic. The output of a switching function does not provide any indication of how it was expressed *i.e.*, the designer cannot specify the required number of input elements, their weight values and threshold and their interconnections.

(3) The weights and threshold values of a single switching function are not unique.

For example : $f(x_1x_2) = \Sigma(1, 2, 3)$ can be designed by using a threshold element with different weights and threshold as shown in Fig. 6.61.

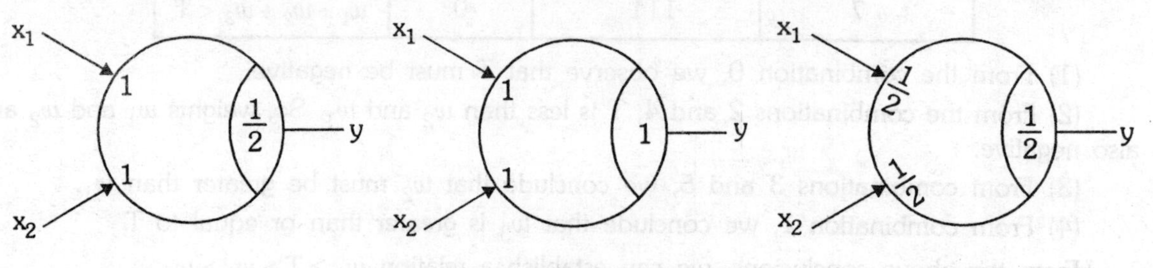

Fig. 6.61

6.8.4 Elementary Properties

Any threshold element will be specified by two parameters :

(a) Its input variables $\{x_1, x_2, ..., x_n\}$

(b) Weighted-threshold vector. $V = \{w_1, w_2, ..., w_{n_D} T\}$

Properties 1 : Consider the function, $f(x_1, x_2, ..., x_n)$, which is realized by a single threshold element whose inputs are $x_1, x_2\ x_i ..., x_n$ and weighted threshold vector $V = \{w_1, w_2, ..., w_i ...., w_{n_i} T\}$.

Statement : Now suppose one of the input, say x_i is complemented. The same function f is realizable by a single threshold element whose inputs are $x_1, x_2, ... x_i, ..., x_n$ and weight threshold vector is $V_2 = \{w_1, w_2, ..., w_i, ..., w_n: T - w_i\}$ and assume that the function is g.

From V_1 : The value of switching function, $f = 1$

When $$w_i x_i + \sum_{j \neq i} w_j x_j \geq T \qquad ...[6.7(a)]$$

and $$f = 0 \text{ when } w_i x_i + \sum_{j \neq i} w_j x_j < T \qquad ...[6.7(b)]$$

Similarly, from V_2, the function ouput $g = 1$ when

$$- w_i x_i + \sum_{j \neq i} w_j x_i \geq T - w_i \qquad ...[6.7(c)]$$

and $$g = 0 \text{ when } - w_i x_i + \sum_{j \neq i} w_j x_j < T - w_i \qquad ...[6.8(a)]$$

Proof : To prove that g and f are identical functions, let $x_i = 0$ so that $x'_i = 1$.

The equation (6.7) becomes

$$\left.\begin{aligned} \sum_{j \neq i} w_j x_j &\geq T \text{ when } f = 1 \\ &< T \text{ when } f = 0 \end{aligned}\right\} \qquad ...[6.8(b)]$$

The equation (6.8) becomes

$$- w_i.1 + \sum_{j \neq i} w_j x_j \geq T - w_i$$

$$< T - w_i$$

$$\therefore \qquad \left.\begin{aligned} \sum_{j \neq i} w_j x_j &\geq T \\ &< T \end{aligned}\right\} \qquad ...[6.8(c)]$$

Here equations [6.8(b)] and [6.8(c)] are identical and they are related to functions f and g respectively. So, both f and g assume for each input combination identical values, and thus are identical functions.

6.8.4.1. Conclusions

(1) If a function is realizable by a single threshold element, then by an appropriate selection of complemented and uncomplemented input variables, it is possible to obtain a realization by an element whose weights have any desired sign distribution. Therefore, if a function is realizable by a single threshold element, then it is realizable by an element with only positive weights and this assertion is valid only if the input variables are a variable in both forms.

Property 2 : A function, $f(x_1, x_2, ..., x_n)$ is realizable by a single threshold element whose weight threshold vector is $V_1 = \{w_1, w_2, ..., w_n; T\}$ and inputs are $(x_1, x_2, ..., x_n)$, then its complement $f'(x_1, x_2, ..., x_n)$ is realizable by a single threshold element whose weight threshold vector is $V_2 = \{-w_1, -w_2; .... -w_i; -T\}$.

From V_1, we can write

$$\left.\begin{array}{l}\sum_{i=1}^{n} w_i x_i \geq T \text{ where } f = 1 \\ \sum_{i=1}^{n} w_i x_i < T \text{ where } f = 0\end{array}\right\} \quad ...(6.9)$$

Multiplying both sides of equation (6.9) by – 1 yields

$$\left.\begin{array}{l}\sum_{i=1}^{n} (-w_i) x_i < -T \text{ whenever } f = 1 \text{ or } f' = 0 \\ \sum_{i=1}^{n} (-w_i) x_i > -T \text{ whenever } f = 0 \text{ or } f' = 1\end{array}\right\} \quad ...(6.10)$$

From equation (6.10), it is clearly demonstrated that f' is realizable by the threshold element whose weight threshold vector is V_2.

6.9. SYNTHESIS OF THRESHOLD NETWORKS

In this section, we see the development of methods for the identification and realization of threshold functions. Before proceeding for it, we shall present some more properties of threshold functions for simpler and more effective synthesis procedures.

6.9.1 Unate Functions

A function, $f(x_1, x_2, ..., x_n)$ which exists either in a disjunctive (SOP form) or conjunctive (POS form) expression, is said to be *positive* in a variable x_i, if the variable x_i appears only in uncomplement form in the given function.

Otherwise the function $f(x_1, x_2, ..., x_n)$ is said to be *negative* in x_i, if the variable x_i appears only in the complemented form.

If $f(x_1, x_2, ..., x_n)$ is either positive or negative in x_i, then the function is said to be *unate* in x_i.

Or

A given function is unate, if it can be represented by a SOP form or POS form, in which no variable appears in both it's complemented and uncomplemented forms.

Example. The function, $f = x'_1x_2 + x_2x_3 + x'_2x'_1$. The function f is positive in x_3, negative in x_1 but not unate in x_2, since x_3 appears only in uncomplement form and x_1 in complement form, whereas x_2 is appearing both in complement and uncomplement form.

Note : If a function is unate in each one of it's variables, then the function is called "*unate function*".

Example. The function, $f = x'_1x_2 + x_2x'_3$ is a unate function, because for each variable, f is either positive or negative. On the other hand, the function $f = x_1x'_2 + x_2x'_1$ is not unate in any variable, so it is a *non-unate function*.

If $f(x_1, x_2, ..., x_n)$ is positive in x_i, then it can be expressed as

$$f(x_1, x_2, ..., x_n) = x_i\, g_1\,(x_1....x_{i-1}, x_{i+1}...x_n) + h_1\,(x_1....x_{i-1}, x_{i+1}...x_n) \quad ...(6.11)$$

Similarly, if $f(x_1, x_2, ..., x_n)$ is negative in x_i, then it can be represented by

$$f(x_1, x_2, ..., x_n) = x'_i\, g_2\,(x_1....x_{i-1}, x_{i+1}...x_n) + h_2\,(x_1....x_{i-1}, x_{i+1}...x_n) \quad ...(6.12)$$

The function 'f' *can* be expressed by equation (6.11) [equation (6.12)], then it is positive (negative) in x_i.

Note : Hence, the existence of two such functions g_1 and h_1 (g_2 and h_2) is a necessary and sufficient condition for f to be positive (negative) in x_i.

6.9.2 Geometrical Representation of a Unate Function

The properties of unate functions are illustrated in a better way by a geometrical representation in the cube form.

A *n*-cube contains 2^n vertices, each of which represents one minterm. A line drawn between every pair of vertices indicates that those two vertices differ by only one variable and remaining all other variables appear in the same form. If two vertices are not connected by a line, it indicates that they differ by more than one variable.

The vertices corresponding to *true minterms* (true minterm is a minterm for which the function $f = 1$) are called '*true vertices*', while the vertices corresponding to the false minterms are called 'false vertices'.

Example. The 3-cube representation of a function, $f = xy + x'z$ is shown in Fig. 6.62. The thick line connecting the two pairs of true vertices [(1, 1, 1) and (1, 1, 0) and (0, 0, 0) and (0, 0, 1)] represents the subcubes xy and $\bar{x}z$.

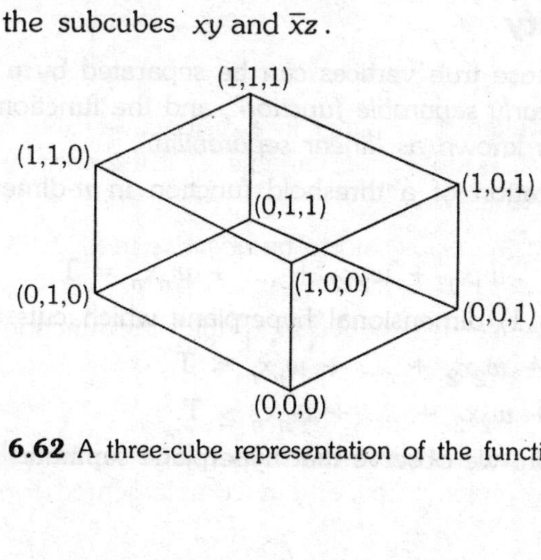

Fig. 6.62 A three-cube representation of the function f.

In the n-cube, the *partial-ordering* relation between the vertices can be defined as $(a_1, a_2, ..., a_n) \leq (b_1, b_2, ..., b_n)$ if and only if for all 'i', $a_i \leq b_i$. The least vertex is (0, 0, ..., 0) and the greatest vertex is (1, 1, ..., 1). In any partial ordering, some pairs of vertices may be incomparable, for example, (0, 0, ..., 0, 1) and (1, 0, ..., 0, 0).

While representing a switching function which is unate and is positive for all variables *i.e.*, every complemented variable in a unate function may be relabelled *e.g.*, $x'_i \rightarrow y_{ij}$ etc., then the resulting function is unate if and only if the original one is.

For example, the unate function $f = x_1x'_2 + x'_2x'_3$ may be converted to $g = x_1x_2 + x_2x_3$, using the relabelling $x'_2 \rightarrow x_2$ and $x'_3 \rightarrow x_3$. Here the function g is a unate function and is positive for all the variables.

Theorem 6.1. A switching function, $f(x_1, x_2, ..., x_n)$ is unate if and only if it is not a tautology and partial-ordering exists, so that for every pair of vertices $(a_1, a_2, ..., a_n)$ and $(b_1, b_2, ..., b_n)$, if $(a_1, a_2, ..., a_n)$ is a true vertex and $(b_1, b_2, ..., b_n) \geq (a_1, a_2, ..., a_n)$, then $(b_1, b_2, ..., b_n)$ is also a true vertex of f.

Note : A tautology is a function which is equal to 1 for all combinations of its variables. Let us consider an example to prove the above theorem.

Example. For the given unate function, $F = x_3x_4 + x_1x_2$, the minimal true vertices are S_1 = (1, 1, 0, 0) and S_2 = (0, 0, 1, 1) where 0 indicates that the variable is absent and 1 indicates that the variable is present in the term.

According to the theorem, any vertex (a_1, a_2, a_3, a_4) which is greater than S_1 or S_2 must be true vertex.

For example, (1, 1, 1, 0) and (0, 1, 1, 1) are true vertices,

Since (1, 1, 1, 0) > (1, 1, 0, 0) and

(0, 1, 1, 1) > (0, 0, 1, 1).

This indicates that these vertices corresponding product terms – $x_1\, x_2\, x_3$ and $x_2\, x_3\, x_4$ are covered by F.

Note : A true vertex S_i, is said to be *minimal* if for no other true vertex $S_j < S_i$.

6.9.3 Linear Separability

A switching function whose true vertices can be separated by a linear equation from it's false vertices is called a "*linearly separable function*', and the functional property which makes such a separation possible is known as '*linear separability*'.

In an n-cubic representation of a threshold function in n-dimensional space, the linear equation is

$$w_1x_1 + w_2x_2 + ... + w_nx_n = T \qquad ...[6.12(a)]$$

corresponds to an $(n - 1)$ dimensional hyperplane which cuts through the n-cube.

(i) $f = 0$ when $w_1x_1 + w_2x_2 + ... + w_nx_n < T$ ...[6.13(a)]

(ii) $f = 1$ when $w_1x_1 + w_2x_2 + ... + w_nx_n \geq T$ ...[6.13(b)]

From the above equation, we observe that hyperplane separates the true vertices from the false ones.

Conclusion : Every function whose true vertices are separated from false ones is called a threshold function (or) threshold function is a linearly separable one.

Let $f(x_1, x_2, ..., x_n)$ is a threshold function and is positive in variable x_i, it's weight threshold vector $V = \{w_1, w_2, ..., w_n; T\}$ since f is positive in variable x_i, there exists a set of values a_1, $a_2 .... a_{i-1}, a_{i+1} .... a_n$ for the input variables $x_1, x_2, ..., x_{i-1}, x_{i+1} .... x_n$, in such a way that

$$f(a_1, a_2, ... a_{i-1}, 1, a_{i+1}, ... a_n) = 1$$

and $$f(a_1, a_2, ... a_{i-1}, 0, a_{i+1}, ... a_n) = 0.$$

Hence, $$w_1a_1 + w_2a_2 +.... w_{i-1}a_{i-1} + w_i + w_{i+1}a_{i+1} + ... w_na_n > T \quad ...[6.14(a)]$$

and $$w_1a_1 + w_2a_2 +.... w_{i-1}a_{i-1} + w_{i+1}a_{i+1} + ... w_na_n < T. \quad ...[6.14(b)]$$

and $$w_i > 0.$$

The above equations may be applied to every x_i in $\{x_1, x_2, ..., x_n\}$. From this we conclude that "the weights associated with a threshold *function positive in all its variables are all positive*".

A threshold function which is positive (negative) in all its variables is called a *positive (negative) threshold function.*

Theorem 6.2. *Statement :* Every threshold function is unate.

Proof : Let $f(x_1, x_2, ..., x_n)$ be a threshold function whose true vertices can be separated from the false ones by the hyperplane $w_1x_1 + w_2x_2 + ... + w_nx_n = T$

(*i*) Suppose, f depends upon x_i and is positive, then $w_i > 0$; then for every combination $a_1, a_2 .... a_{i-1}, a_{i+1} .... a_n)$ of the variables $x_1, x_2 .... x_{i-1}, x_{i+1} .... x_n$, if vertex

$(a_1, a_2 .... a_{i-1}, 0, a_{i+1} .... a_n)$ is true, then

$(a_1, a_2 .... a_{i-1}, 1, a_{i+1} .... a_n)$ must also be true, because

$$w_1a_1 + w_2a_2 + ... w_i a_i + ... w_na_n > w_1a_1 + ... w_{i-1}a_{i-1} + w_{i+1}a_{i+1} + ... w_na_n \quad ...[6.15(a)]$$

Since f is not independent of x_i, vertex $(a_1, a_2, ..., a_{i-1}, 0, a_{i+1}, ..., a_n)$ must be false, proving that f is positive in x_i.

(*ii*) Now, consider the input variable x_i whose weight is negative *i.e.*, $w_i < 0$; then if vertex

$(a_1, a_2 .... a_{i-1}, 1, a_{i+1} .... a_n)$ is true, then

$(a_1, a_2 .... a_{i-1}, 0, a_{i+1} .... a_n)$ must be true, because

$$w_1a_1 + w_2a_2 + ...+ w_i + ... + w_na_n < w_1a_1 + w_2a_2 + ...+ w_{i-1}a_{i-1} + w_{i+1}a_{i+1} +....+ w_na_n \quad ...[6.15(b)]$$

Since f is not independent of x_i, vertex $(a_1, a_2 .... a_{i-1}, a_{i+1} .... a_n)$ must be false, proving that f is negative in x_i.

Conclusion : (*i*) f is either positive or negative in each of it's variables, and thus f is unate (or) unate function.

(*ii*) All unate functions are not linearly separable *e.g.*, $f = x_1x_2 + x_3x_4$.

6.10 IDENTIFICATION AND REALIZATION OF THRESHOLD FUNCTIONS

In this section, our aim is to present a procedure by taking the linearly separable property to determine.

(*i*) Whether the given switching function is a threshold function or not *i.e.*, to test whether or not there exists a hyperplane which separates the true vertices of the function from the false ones.

(*ii*) If it is a threshold function, then to find the values of weights and threshold.

Steps in the procedure : (1) Find the minimal form of the given function and test it for unateness. Since the unate function must have a unique minimal form, it is linearly separable.

(2) If it is unate, convert it into another function positive in all its variables.

For example, $f = x_1x_2x'_3x_4 + x_2x'_3x'_4$. The minimal form is $x_1x_2x'_3 + x'_3x_2x'_4$.

$\therefore \quad f = x_1x_2x'_3 + x_2x'_3x'_4$ is unate.

Now, convert this function into another function 'g' positive in all variables

$\therefore \quad g = x_1x_2x_3 + x_2x_3x_4$

(3) Now find the *minimal true* and maximal false vertices of g.

In the above example, there are two minimal true vertices, they are (1, 1, 1, 0) and (0, 1, 1, 1). Here '0' indicates that no variable and 1 indicates that variable is present.

The maximal false vertices are found by determining (*i*) all false vertices with just one variable whose value is 0, (*ii*) then all false vertices with two variables whose value is 0, and so on, (*iii*) leave all the vertices smaller than the ones already selected.

In our example, the maximal false vertices are (1, 1, 0, 1) (1, 0, 1, 1) and (0, 1, 1, 0).

Note : The list of minimal true vertices contains all the necessary information for the determination of the maximal false vertices.

(4) To determine whether 'g' is linearly separable or not, it is necessary to determine the co-efficients of the separating hyperplane, *i.e.*, to find the appropriate set of weights and threshold.

This can be solved by a system of '*pq-inequalities*', where p = minimal true vertices, and q = maximal false vertices.

For each pair of vertices A = $\{a_1, a_2, ..., a_n\}$ and B = $\{b_1, b_2, ..., b_n\}$, where A = minimal true vertices, B = maximal false vertices, the inequality is

$$a_1w_1 + a_2w_2 + ... + a_nw_n > b_1w_1 + b_2w_2 + ... + b_nw_n \quad ...(6.16)$$

In our example, $p = 2$ and $q = 3$. So, we find totally $pq = 6$ inequalities, which are

$$w_1 + w_2 + w_3 > w_1 + w_2 + w_4 \quad ...(i)$$
$$w_1 + w_2 + w_3 > w_1 + w_3 + w_4 \quad ...(ii)$$
$$w_1 + w_2 + w_3 > w_2 + w_3 \quad ...(iii)$$
$$w_2 + w_3 + w_4 > w_1 + w_2 + w_4 \quad ...(iv)$$
$$w_2 + w_3 + w_4 > w_1 + w_3 + w_4 \quad ...(v)$$
$$w_2 + w_3 + w_4 > w_2 + w_3 \quad ...(vi)$$

Since 'g' is a positive function, if it is linearly separable, the separating hyperplane will have positive coefficients.

Note : The hyperplane separating the minimal true vertices from the maximal false vertices separates all the true vertices from all false ones, *i.e.*, weight-threshold vector of '*g*'.

Solving the system inequalities from equations [(*i*) – (*vi*)], we observe the following necessary constraints to be satisfied

$w_3 > w_4$ From ...(*i*) $\quad$ $w_3 > w_1$ From ...(*iv*)

$w_2 > w_4$ From ...(*ii*) and $\quad$ $w_2 > w_1$ From ...(*v*)

$w_1 > 0$ From ...(*iii*) $\quad$ $w_4 > 0$ From ...(*vi*)

Let us assume $w_1 = w_4 = 1$, and $w_2 = w_3 = 2$.

Substitute these values in any inequality equation (or) in the minimal true vertices and maximal false vertices.

This should be less than minimal true vertex value and larger than maximal false vertex value. For the given example, T should be less than

$$w_1 + w_2 + w_3 \text{ or } w_2 + w_3 + w_4$$
$$= 1 + 2 + 2 \text{ (or) } 2 + 2 + 1$$
$$= 5$$

and larger than $\quad w_1 + w_2 + w_4 \text{ or } w_1 + w_3 + w_4 \text{ or } w_2 + w_3$

$$= 1 + 2 + 1 \text{ or } 1 + 2 + 1 \text{ or } 2 + 2$$
$$= 4.$$

$\therefore \quad 5 > T > 4.$

$\therefore \quad T = 9/2 = 4.5$

The weight-threshold vector for *g* is V = {1, 2, 2, 1; 4.5}

(5) It is necessary to convert this weight-threshold vector to one which corresponds to the original function, *f*. This conversion process is based on the property established in equation (6.8); where, if an input x_i is complemented in the original function, w_i must be changed to $-w_i$ and threshold T to T $- w_i$.

In our example, two inputs x_3 and x_4 appear in *f* in complemented form. Thus, in the new weight-threshold vector, the weights are '1, 2, – 2 and – 1' and the threshold is $\frac{9}{2} - 2 - 1$.

$\therefore$ The weight-threshold vector for the original function $f = \left\{1, 2, -2, -1; \frac{3}{2}\right\}$.

Example 6.17. *Determine whether the given function,* $f\ (x_1x_2x_3x_4) = \Sigma m$ *(0, 1, 3, 4, 5, 6, 7, 12, 13)*

(a) Is a threshold function or not

(b) If it is, find its weight-threshold vector.

Solution. Given function is

$$f(x_1x_2x_3x_4) = \Sigma m\ (0, 1, 3, 4, 5, 6, 7, 12, 13)$$

(1) The minimal expression of it can be obtained by using K-map. (see Fig. 6.63)

$$f = x'_1x'_3 + x_2x'_3 + x'_1x_4 + x'_1x_2$$

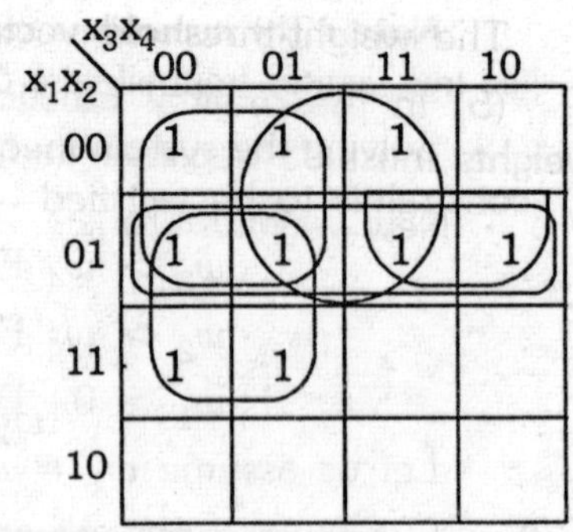

Fig. 6.63

(2) Here, the given function f is negative in x_1, positive in x_2, negative in x_3 and positive in x_4. So, the given function is a unate function. It's equivalent positive function is obtained by substituting $x'_1 \to x_1$, $x'_3 \to x_3$

$$g = x_1x_3 + x_2x_3 + x_1x_4 + x_1x_2$$

(3) The minimal true vertices are

(1, 0, 1, 0), (0, 1, 1, 0), (1, 0, 0, 1) and (1, 1, 0, 0)

The maximal false vertices are

(0, 1, 0, 1) (0, 0, 1, 1) and (1, 0, 0, 0)

(4) Here $p = 4$ and $q = 3$. So, totally we get 12-inequalities

$$\begin{Bmatrix} w_1 + w_3 \\ w_2 + w_3 \\ w_1 + w_4 \\ w_1 + w_2 \end{Bmatrix} > \begin{Bmatrix} w_2 + w_4 \\ w_3 + w_4 \\ w_1 \end{Bmatrix}$$

while solving, we will observe the following constraints to be satisfied.

$$w_1 > w_4 \quad w_2 > w_4 \quad w_2 > 0$$
$$w_1 > w_3 \quad w_3 > w_4 \quad w_3 > 0$$
$$w_1 > w_2 \qquad\qquad w_4 > 0$$

Let us assume that $w_4 = 1$, and $w_2 = w_3 = 2$ and $w_1 = 3$.

The T < minimal true vertex value

$$\min \{w_1 + w_3,\ w_1 + w_4,\ w_2 + w_3,\ w_1 + w_2\}$$
$$\min \{3 + 2,\ 3 + 1,\ 2 + 2,\ 3 + 2\} = \{5, 4, 4, 5\}$$
$$T < 4.$$

T must be larger than maximal false vertices.

$$T > \max \{w_2 + w_4,\ w_3 + w_4,\ w_1\}$$
$$> \max \{2 + 1,\ 2 + 1,\ 3\}$$
$$> \max \{3, 3, 3\}$$
$$T > 3$$
$$\therefore \quad 4 > T > 3$$
$$\therefore \quad T = \frac{7}{2}$$

The weight-threshold vector is = {3, 2, 2, 1; 7/2}.

(5) In the original function, the inputs x_1 and x_3 are in complemented form, so, their weights must be negative and threshold = T − w_1 − w_3

$$\therefore \text{ New weights are } \quad w_1 = -3,\quad w_2 = 2,\quad w_3 = -2,\quad w_4 = 1$$

and

$$\text{new threshold} = T = \frac{3}{2} - w_1 - w_3 = \frac{7}{2} - 3 - 2 = \frac{7}{2} - 4 = \frac{7-10}{2} = \frac{-3}{2}$$

$\therefore$ Weight-threshold vector = {− 3, 2, − 2, 1; − 3/2}

6.11 SYNTHESIS OF THRESHOLD NETWORKS USING MAP AS A TOOL

In this section, our aim is to synthesize the threshold network constructed of threshold elements to realize any arbitrary switching function. Here, we are concentrating on the non threshold functions, *i.e.*, a switching function can not be realized by a single threshold element. One approach is, decomposition of non threshold function into two or more patterns, in such a way that each part may be constructed by a single threshold element.

In the functions of three or four variables, the identification of patterns which can be realized by a single threshold element is obtained by using the map method. In the map, a pattern of 1 cells can be realized by a single threshold element, that pattern is called "*admissible pattern*".

The admissible pattern for the functions of 3 variables are shown in Fig. 6.64.

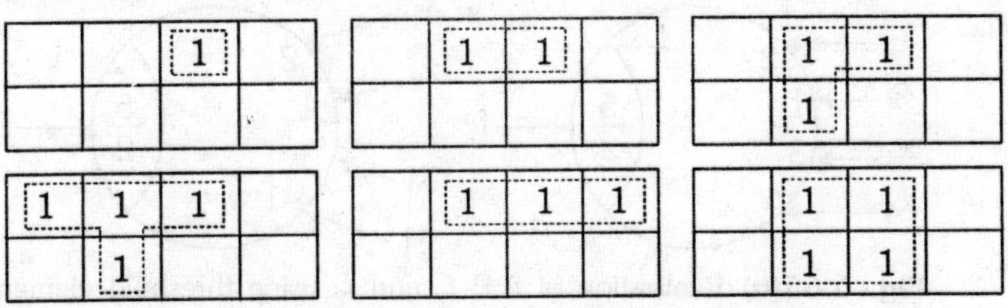

Fig. 6.64 Admissible patterns of 3-variables.

(1) Each admissible pattern may be in any position on the map, provided that it's basic topological structure is preserved.

(2) Any admissible pattern for functions of 3-variables is also an admissible pattern for functions of four variables.

(3) Since, the complement of a threshold function is also a threshold function, the patterns formed by 0 cells are admissible.

(4) While realizing a threshold function, which belongs to AND-OR or any network select the minimal number of admissible patterns, such that each 1 cell of the map is covered by at least one admissible pattern. This type of realization is called "minimal realization", *i.e.*, which contains less number of threshold elements.

The clear picture for the above synthesis procedure is illustrated in the below example.

Example 6.18. *Find the minimal threshold-logic realization for the function* $f(x_1, x_2, x_3, x_4) = \Sigma m$ *(2, 3, 6, 7, 10, 12, 14, 15).*

Solution. (1) Plot the K-map of the given function. This is shown in Fig. 6.65(*a*).

$f(x_1, x_2, x_3, x_4) = \Sigma m$ (2, 3, 6, 7, 10, 12, 14, 15).

Simplified expression is $f = x_1x_3x'_4 + x_2x_3x_4 + x'_1x'_2x_3 + x_1x_2x'_4$.

The given function, f is positive in x_3 and not unate in all variables *i.e.*, x_1, x_2 and x_4. So, the given switching function cannot be realized by a single threshold element. Now, our aim is to arrange all 1 cells into admissible patterns.

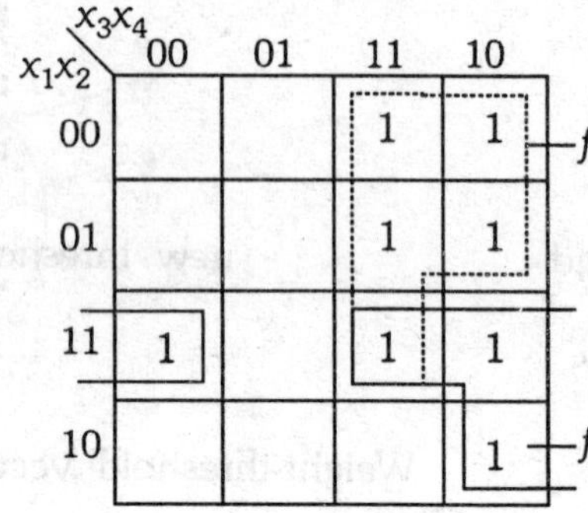

Fig. 6.65(*a*)

(2) From the map, it is clear that two admissible patterns are existing and each admissible pattern is realized by a single threshold element.

The two admissible patterns are f_1 and f_2,

where, $f_1(x_1, x_2, x_3, x_4) = \Sigma m$ (2, 3, 6, 7, 15) and

$f_2(x_1, x_2, x_3, x_4) = \Sigma m$ (10, 12, 14, 15)

By applying the previous procedures,

$$f_1 = x'_1x_3 + x_2x_3x_4 \text{ and it's positive function} = x_1x_3 + x_2x_3x_4$$

minimal true vertices are

(1, 0, 1, 0) and (0, 1, 1, 1) and so on.

The weight-threshold vector, Vf_1 = {– 2, 1, 3, 1; 5/2} and Vf_2 = {2, 1, 1, – 1; 5/2}

The realization of two threshold element for f_1 and f_2 independently is shown in Fig. 6.65(*b*).

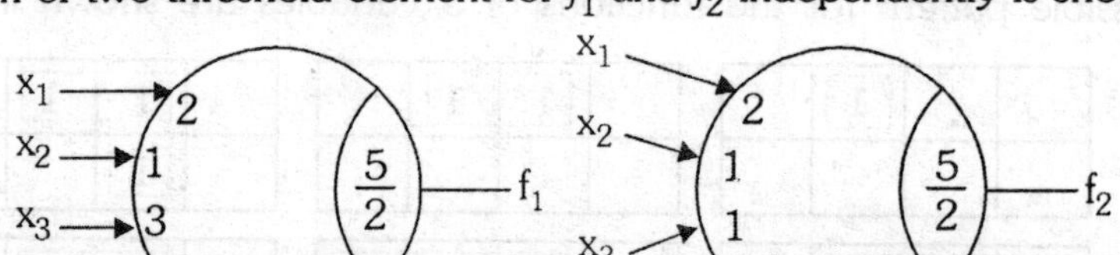

Fig. 6.65(*b*) Realization of A.P. f_1 and f_2 using threshold element.

We have selected a threshold element which realizes the function f_1 as the first element, then the second element must be such that it will realize f_2 and at the same time allow f_1 to propagate through it without any effect. The second element must, in addition to realizing f_2, act as an OR gate whose output is 1 if either g or h or both are equal to 1.

This is accomplished by providing it with five inputs as shown in Fig. 6.65(c). The five inputs are x_1, x_2, x_3, x_4 and output of the first threshold element *i.e.*, f_1. The corresponding weights are 2, 1, 1, – 1 and Wf_1. Now, it is necessary to determine the weight associated with the fifth input, f_1, Wf_1. This weight can be determined by computing the minimal weight sum that an occur in the second element when f_1 has the value 1. Since f must have value 1 whenever f_1 does, this minimal weighted sum must be larger than the threshold value of the second element.

In this case, minimal weighted sum Wf_1 must be larger than 5/2, therefore the value $Wf_1 = 3$ has been selected.

To simplify the computation of Wf_1, it can be set equal to (or larger than) the sum of the threshold and the absolute values of all negative weights of the second element. However, this will not always yield a minimal value for Wg.

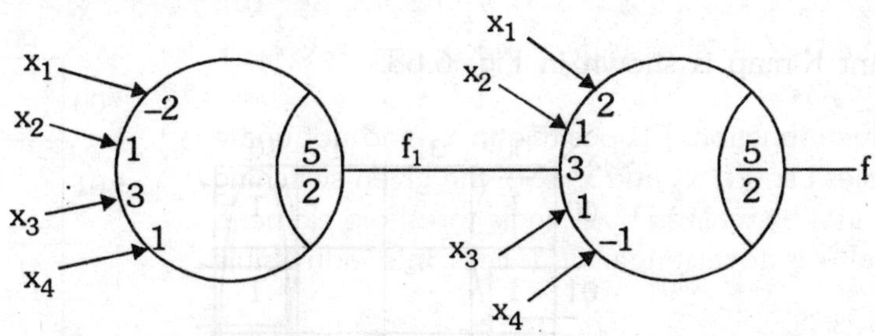

Fig. 6.65(c) Threshold-logic realization for the given switching function f.

This synthesis procedure is very useful when the number of admissible patterns are less. For the functions with five or more variables, the admissible patterns are determined by a tabulation method.

6.12. SOLVED PROBLEMS

Problem 1. *Find the function, $f(x_1, x_2, x_3, x_4)$ realized by each of the threshold networks shown below. Show the map of each function.*

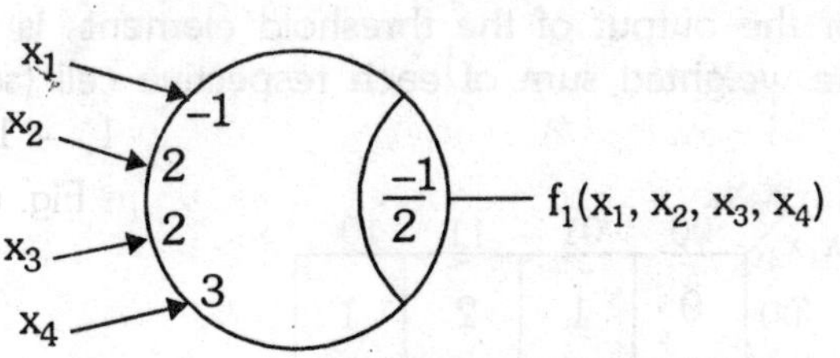

Fig. 6.66(a)

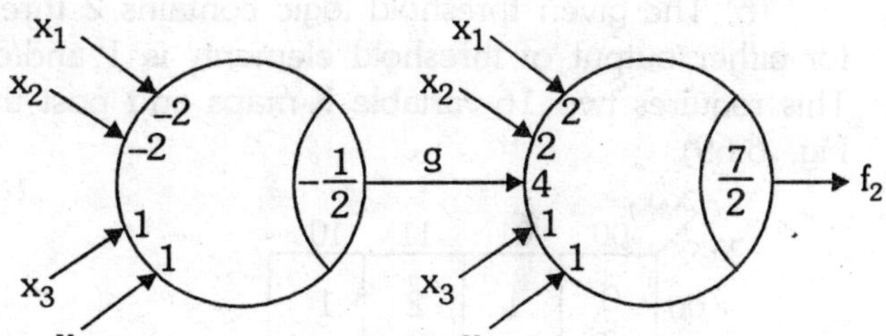

Fig. 6.66(b)

Solution. (*a*) First, plot the K-map of the given threshold element and post the value of the weighted sum in the corresponding cell (see Fig. 6.66(*a*)).

For example, if $x_1, x_2, x_3, x_4 = 1111$, then the weighted sum is $(1 \times -1) + (1 \times +2) + (1 \times 2) + (1 \times 3) = -1 + 2 + 2 + 3 = 0$ (see Table 6.19).

Table 6.19

Binary value $x_1x_2x_3x_4$		*Weighted sum* $-x_1 + 2x_2 + 2x_3 - 3x_4$		*B.V.*
0 0 0 0	—	0	1000	− 1
0 0 0 1	—	− 3	1001	− 4
0 0 1 0	—	2	1010	+ 1
0 0 1 1	—	− 1	1011	− 2
0 1 0 0	—	2	1100	+ 1

Weighted sum

x_1x_2 \ x_3x_4	00	01	11	10
00	0	−3	−1	2
01	2	−1	1	4
11	+1	−2	0	+3
10	−1	−4	−2	+1

Fig. 6.67

0 1 0 1	—	−1	1101	−2
0 1 1 0	—	4	1110	+3
0 1 1 1	—	1	1111	0

Since the threshold $T = -\frac{1}{2}$, place 1 in the cells whose value is greater than or equal to $\frac{-1}{2}$. The resultant K-map is shown in Fig. 6.68.

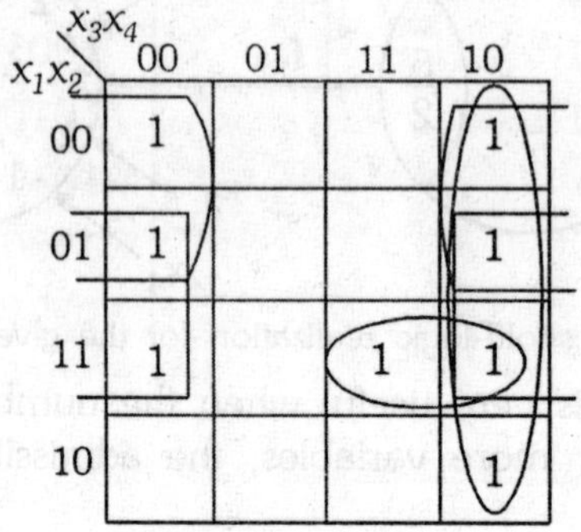

Fig. 6.68

The simplified expression from the K-map is

$$f = x_3x'_4 + x'_1x'_4 + x_2x'_4 + x_1x_2x_3$$

(*b*) The given threshold logic contains 2 threshold element in cascade. The output $f = 1$ for either output of threshold element$_1$ is 1 and/or the output of the threshold element$_2$ is 1. This requires two 16-variable K-maps and post the weighted sum of each respective cell (see Fig. 6.69).

x_1x_2 \ x_3x_4	00	01	11	10
00	0	1	2	1
01	−2	−1	0	−1
11	−4	−3	−2	−3
10	−2	−1	0	−1

(*a*)

x_1x_2 \ x_3x_4	00	01	11	10
00	0	1	2	1
01	2	3	4	3
11	4	5	6	5
10	2	3	4	3

(*b*)

Fig. 6.69

TE_1 **Table 6.20**(*a*)

Inputs	*Weighted sum*
$x_1x_2x_3x_4$	$-2x_1 - 2x_2 + x_3 + x_4$
0000	0
0001	1
0010	1

TE_2 **Table 6.20**(*b*)

Inputs	*Weighted sum*
$x_1x_2x_3x_4$	$2x_1 + 2x_2 + x_3 + x_4$
0000	0
0001	1
0010	1

0011	2
0100	– 2
0101	– 1
0110	– 1
0111	0
1000	–2
1001	–1
1010	–1
1011	0
1100	– 4
1101	– 3
1110	– 3
1111	– 2

0011	2
0100	2
0101	3
0110	3
0111	4
1000	2
1001	3
1010	3
1011	4
1100	4
1101	5
1110	5
1111	6

The threshold of TE_1 is $\frac{-1}{2}$ and TE_2 is $\frac{7}{2}$. Now place 1 in the corresponding cell whose value is greater than or equal to the respective threshold. The resultant K-maps are shown below (see Fig. 6.70).

x_1x_2 \ x_3x_4	00	01	11	10
00	1	1	1	1
01			1	
11				
10			1	

(a)

x_1x_2 \ x_3x_4	00	01	11	10
00				
01			1	
11	1	1	1	1
10			1	

(b)

Fig. 6.70

The resultant f output = 1 for both cases. The first K-map (see Fig. 6.71)

x_1x_2 \ x_3x_4	00	01	11	10
00	1	1	1	1
01			1	
11	1	1	1	1
10			1	

Fig. 6.71

The minimal expression of

$$f = x'_1x'_2 + x_1x_2 + x_3x_4.$$

Problem 2. *By examining the linear inequalities, determine which of the following function is a threshold function, and for each one that is, find the corresponding weight-threshold vector.*

(*a*) $f_1(x_1, x_2, x_3) = \Sigma m\ (1, 2, 3, 7)$

(*b*) $f_2(x_1, x_2, x_3) = \Sigma m\ (0, 2, 4, 5, 6)$

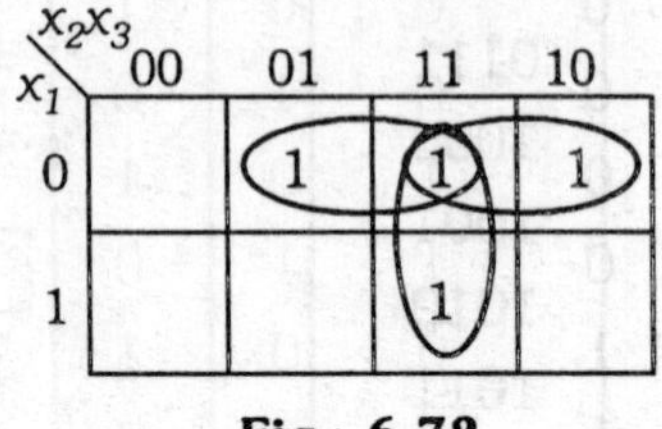

Fig. 6.72

Solution. (*a*) The given threshold function is,

$$f_1(x_1, x_2, x_3) = \Sigma m\ (1, 2, 3, 7)$$

The minimal expression for it is shown below by using K-map (see Fig. 6.72)

$$f_1 = x'_1x_3 + x'_1x_2 + x_2x_3$$

The given function is negative in x_1 and is positive in x_2 and x_3.

Hence, it is a unate function, *i.e.*, threshold function. It can be realized by single threshold element. The inequalities to be solved for finding weight threshold vector are (see Fig. 6.73)

– 1	1	1	1	
x_1	x_2	x_3	f'_1	
0	0	0	0	(*i*) $0 < T$
0	0	1	1	(*ii*) $w_3 \geq T$
0	1	0	1	(*iii*) $w_2 \geq T$
0	1	1	1	(*iv*) $w_2 + w_3 \geq T$
1	0	0	0	(*v*) $w_1 < T$
1	0	1	0	(*vi*) $w_1 + w_3 < T$
1	1	0	0	(*vii*) $w_1 + w_2 < T$
1	1	1	1	(*viii*) $w_1 + w_2 + w_3 \geq T.$

Fig. 6.73

From the above inequalities, it is clear that the weight w_1 and threshold T and w_2 and w_3 weights are positive.

Let us assume $\quad w_2 = w_3 = 1$ and $T = +1$

and $\quad w_1 + 2 \geq 1 \quad \boxed{\therefore w_1 = -1}$

$$w_1 \geq -1$$

The weight threshold vector is :

$$V = \{-1, 1, 1; +1\}.$$

(*b*) The given function is $f_2(x_1, x_2, x_3) = \Sigma m\ (0, 2, 4, 5, 6)$.

The minimal expression of this function is (from K-map) (see Fig. 6.74)

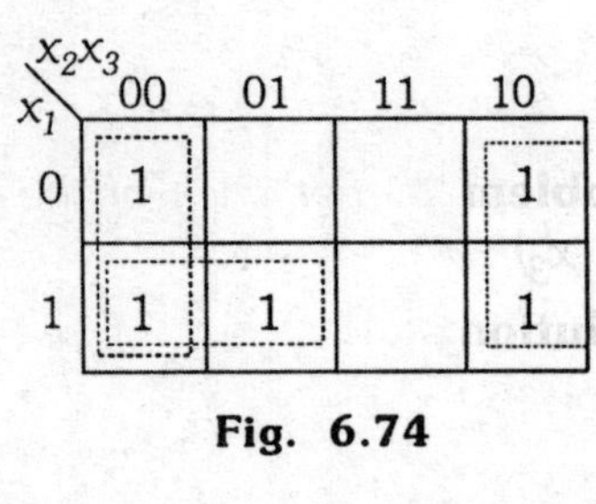

Fig. 6.74

$$\Rightarrow \qquad f_2 = x'_3 + x_1x'_2$$

The given function is positive in x_1 and negative in x_2 and x_3. So, it is a unate function *i.e.*, threshold function. It can be realized by using a single threshold element. To find its weight threshold vector, the following inequalities must be solved (see Fig. 6.75).

x_1	x_2	x_3	f_2	
0	0	0	1	(*i*) $0 \geq T$
0	0	1	0	(*ii*) $w_3 < T$
0	1	0	1	(*iii*) $w_2 \geq T$
0	1	1	0	(*iv*) $w_2 + w_3 < T$
1	0	0	1	(*v*) $w_1 \geq T$
1	0	1	1	(*vi*) $w_1 + w_3 \geq T$
1	1	0	1	(*vii*) $w_1 + w_2 \geq T$
1	1	1	0	(*viii*) $w_1 + w_2 + w_3 < T$

Fig. 6.75

From these inequalities it is clear that

(*i*) T is negative.

(*ii*) w_1 and w_2 are positive and negative but > T.

(*iii*) weight w_3 is negative.

Let us assume $\quad w_1 = 1$ and $T = -\frac{3}{2}$

$w_2 = -1 \quad$ from (*viii*) $w_3 = 1 - 1 + w_3 < -\frac{3}{2}$

$\therefore \quad w_3 < -\frac{3}{2} \quad (\therefore w_3 = -2)$.

The weight-threshold vector is $V = \left\{1, -1, -2; -\frac{3}{2}\right\}$ (see Fig. 6.76)

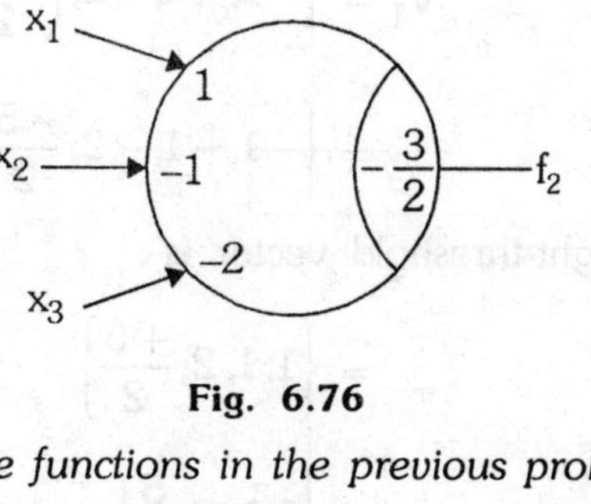

Fig. 6.76

Problem 3. *For each of the functions in the previous problem, find the realization for f' (x'_1, x_2, x_3)*

Solution. (*a*) $\quad f_1(x_1, x_2, x_3) = \Sigma m(1, 2, 3, 7)$

$= x'_1x_3 + x'_1x_2 + x_2x_3.$

$$f_1 f\ (x'_1, x_2, x_3) = x_1x_3 + x_1x_2 + x_2x_3.$$

$$V_f = \{-1, 1, 1; +1\}$$

The new weight threshold vector for $f_1(x'_1, x_2, x_3)$

$$V_1 = \{1, 1, 1; 1 + 1\} = \{1, 1, 1; 2\} \text{ (see Fig. 6.77)}$$

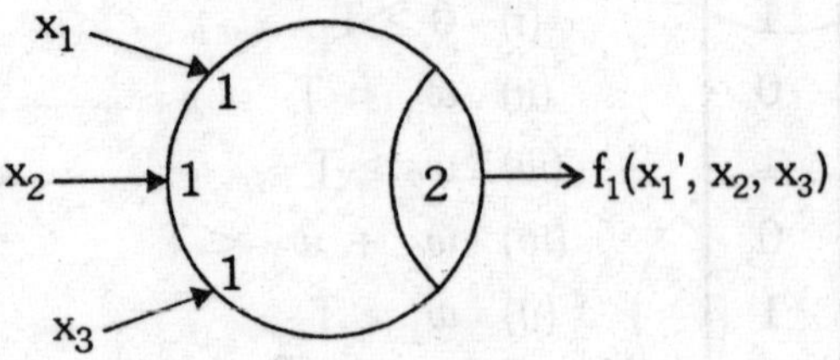

Fig. 6.77

The weight-threshold vector of

∴ $$f'_1(x'_1, x_2, x_3) = \{-1, -1, -1; -2\} \text{ (see Fig. 6.78)}$$

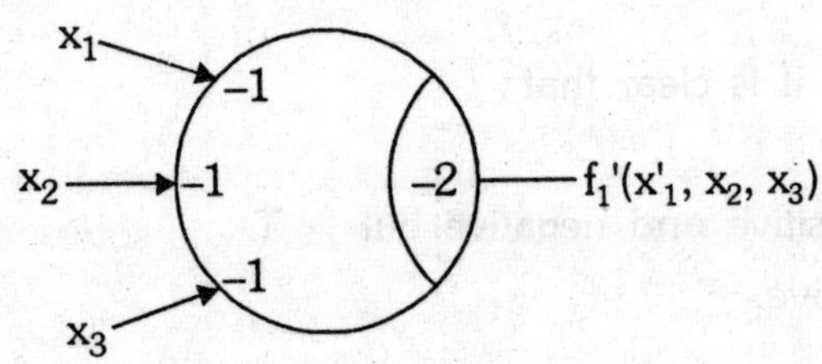

Fig. 6.78

(*b*) The given function $f_2\ (x_1, x_2, x_3) = \Sigma\ (0, 2, 4, 5, 6)$ $\quad f_2 = x'_3 + x'_1x'_2$

It's weight-threshold vector is

$$V = \left\{1, -1, -2; -\frac{3}{2}\right\}$$

$$f_2(x'_1, x_2, x_3) = x'_3 + x'_1x'_2$$

The new weight-threshold vector for $f_2(x'_1, x_2, x_3)$ is

$$V_1 = \left\{-1, -1, -2; \left(\frac{-3}{2} - 1\right)\right\}$$

$$= \left\{-1, -1, -2; \frac{-5}{2}\right\}$$

and for $f'_2(x'_1, x_2, x_3)$, the weight-threshold vector is

$$= \left\{1, 1, 2; \frac{+5}{2}\right\}$$

$$= \left\{1, 1, 2; \frac{5}{2}\right\}$$

Problem 4. *(a) Determine the function $f(x_1, x_2, x_3, x_4)$ realized by the network shown below (see Fig. 6.79 (a)).*

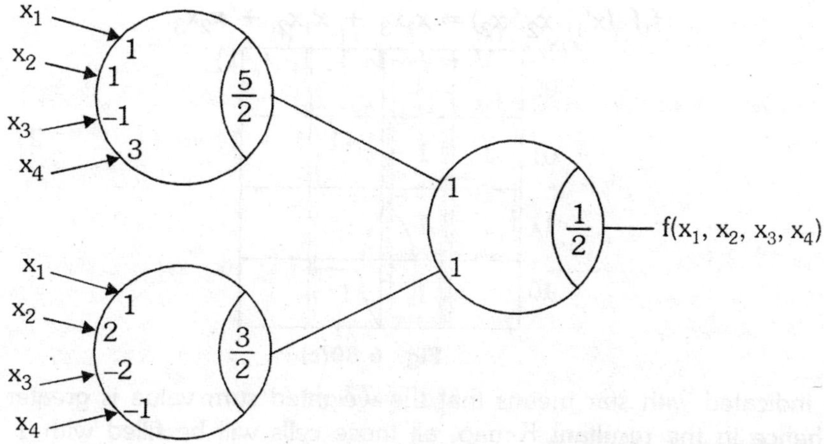

Fig. 6.79(*a*)

(b) Show that $f(x_1, x_2, x_3, x_4)$ can be realized by a single threshold element and find such realization.

Solution. (*a*) The given problem contains 3 threshold elements. We can name then A, B, C as shown below. To find the $f(x_1, x_2, x_3, x_4)$, C, first it is necessary to find output of A and B, for which we can plot the two individual K-maps with weighted sums in corresponding cell. (see Fig. 6.79(*b*))

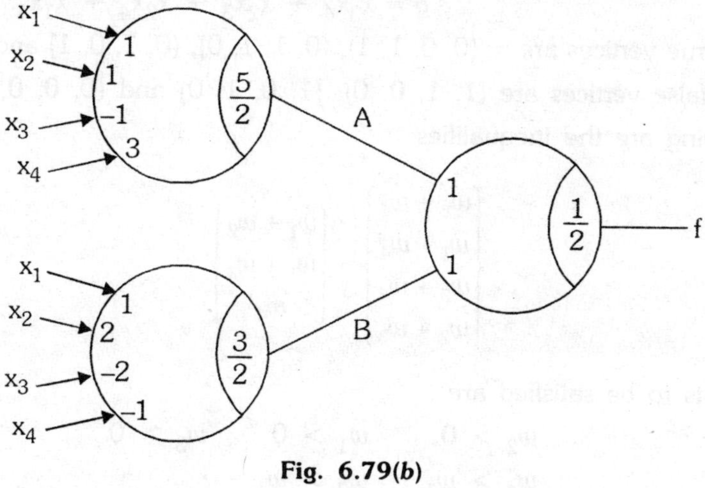

Fig. 6.79(*b*)

Weights ↓	x_1x_2 \ x_3x_4 (Weights →)	00 (0)	01 (3)	11 (2)	10 (–1)
0	00	0	3*	2	–1
1	01	1	4*	3*	0
2	11	2	5*	4*	1
1	10	1	4*	3*	0

Fig. 6.80(*a*) K-map for A.

Weights ↓	x_1x_2 \ x_3x_4 (Weights →)	00 (0)	01 (–1)	11 (–3)	10 (–2)
0	00	0	–1	–3	–2
2	01	2*	1	–1	0
3	11	3*	2*	0	1
1	10	1	0	–2	–1

Fig. 6.80(*b*) K-map for B.

$x_1x_2 \backslash x_3x_4$	00	01	11	10
00		1		
01	1	1	1	
11	1	1	1	
10		1	1	

Fig. 6.80(c)

The cells indicated with star means that the weighted sum value is greater than or equal to threshold, hence in the resultant K-map, all those cells will be filled with 1.

The simplified expression for $f(x_1, x_2, x_3, x_4)$

$$= x'_3x_4 + x_2x'_3 + x_2x_4 + x_1x_4$$

(*b*) Given, $\quad f(x_1, x_2, x_3, x_4) = x'_3x_4 + x_2x'_3 + x_2x_4 + x_1x_4$

So, the given function is positive in x_1, x_2 and x_4 negative in x_3. Hence, the given function is a unate function. It's equivalent positive function is obtained by replacing $x'_3 \rightarrow x_3$.

$$\therefore \quad g = x_3x_4 + x_2x_3 + x_2x_4 + x_1x_4.$$

The minimal true vertices are = (0, 0, 1, 1), (0, 1, 1, 0), (0, 1, 0, 1) and (1, 0, 0, 1) and the maximal false vertices are (1, 1, 0, 0), (1, 0, 1, 0) and (0, 0, 0, 1)

Hence, following are the inequalities

$$\begin{Bmatrix} w_2 + w_4 \\ w_1 + w_4 \\ w_3 + w_4 \\ w_2 + w_3 \end{Bmatrix} > \begin{Bmatrix} w_1 + w_2 \\ w_1 + w_3 \\ w_4 \end{Bmatrix}$$

The constraints to be satisfied are :

$$\begin{array}{lll} w_2 > 0 & w_1 > 0 & w_3 > 0 \\ w_4 > w_1 & w_3 > w_1 & \\ w_4 > w_2 & w_2 > w_1 & \\ w_4 > w_3 & & \end{array}$$

Here $\quad w_4 > \begin{Bmatrix} w_1 \\ w_2 \\ w_3 \end{Bmatrix}$ and $\begin{Bmatrix} w_3 \\ w_2 \end{Bmatrix} > w_1.$

Hence, let us assume that $w_1 = 1$, $w_2 = w_3 = 2$ and $w_4 = 3$.

The choice of T is less than min of $\{w_1 + w_4, w_2 + w_3, w_2 + w_4, w_2 + w_4\}$ and larger than max of $\{w_1+w_2, w_1+w_3, w_4\}$

$$\text{Min of } \{4, 4, 5, 5\} = \text{min value} = 4$$

$$\text{Max of } \{3, 3, 3\} = \text{max value} = 3$$

$$\therefore \quad 4 > T > 3$$

$$\therefore \quad T = 3.5 = \frac{7}{2}.$$

The weight-threshold vector is {1, 2, 2, 3; 7/2}.

In the original function, x_3 is present in complement form and it's corresponding weight $= -2$ and the new threshold $= \frac{7}{2} - 2 = \frac{7-4}{2} = \frac{3}{2}$

∴ The required weight-threshold vector is

$$V = \{1, 2, -2, 3; 3/2\}$$

The threshold element is shown below (Fig. 6.81)

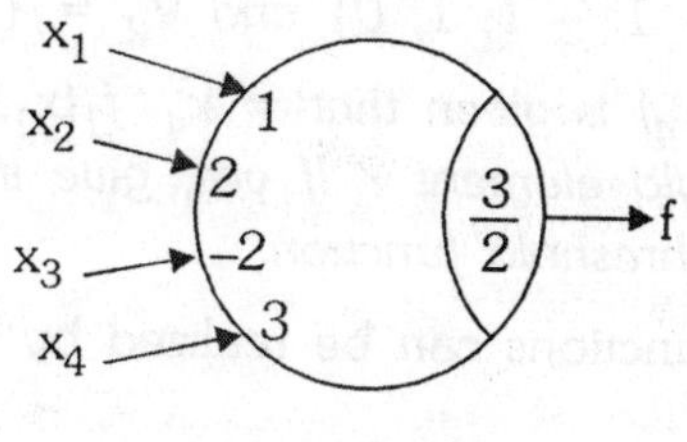

Fig. 6.81

Problem 5. *(a) If $f(x_1, x_2, ..., x_n)$ is a threshold function with weight-threshold vector $V_1 = \{w_1, w_2, ..., w_n; T\}$, then it's dual, $f_d(x_1, x_2, ..., x_n)$, is also a threshold function. Determine it's weight-threshold vector.*

(b) Prove that if f is a threshold function, then so

$g = x'_i f + x_i f_d$, where x_i may or may not be a member of the set $\{x_1, x_2, ..., x_n\}$. Find the weight-threshold vector of g.

Solution. (a) By definition of dual property, we know that

$$f_d(x_1, x_2, ..., x_n) = f'(x'_1, x'_2, ..., x'_n)$$

Let $g(x_1, x_2, ..., x_n) = f(x'_1, x'_2, ..., x'_n)$ then the weight-threshold vector for g is,

$$V_g = \{-w_1, -w_2, ..., -w_n; T - (w_1 + w_2 + ... w_n)\}$$

where $w_1, w_2, ..., w_n$ are respective weights of $x_1, x_2, ..., x_n$.

Now
$$f_d = g' = f'(x'_1, x'_2, ..., x'_n)$$
$$= \{w_1, w_2, ..., w_n; (w_1 + w_2 + ... w_n) - T\}.$$

Hence, the dual of a threshold function is also a threshold function.

(*b*) Let, V the weight-threshold vector for f, is

$$V = \{w_1, w_2, ..., w_n; T\}$$

(*i*) $$g = x'_i f + x_i f_d = 1 \text{ if and only if } x_i = 0 \text{ and } \sum_{j \neq i} x_j w_j > T.$$

(*ii*) $$x_i = 1 \text{ and } \sum_{j \neq i} w_j x_j > (w_1 + w_2 + ... + w_n) - T.$$

Establishing the same relation for the condition g = 0, we find the weight w_i of x_i

$$\therefore \quad w_i = (w_1 + w_2 + ... w_n) - 2T = \sum_{j \neq i} w_j - 2T.$$

$$\text{Thus, } V_g = \text{weight-threshold vector} = \left\{ w_1, w_2 .... \sum_j w_j - 2T, ..., w_n; T \right\}.$$

Problem 6. *The given two functions $f_1(x_1, x_2, x_3)$ and $f_2(x_1, x_2, x_3)$ are each realizable by a single threshold element. The weight-threshold vectors of these elements are,*

$$V_1 = \{-1, -1, 1; 0\} \text{ and } V_2 = \{1, 2, -2; 2\}$$

If the function $f(x_1, x_2, x_3, x_4)$ is given that = x_4 $f_1(x_1, x_2, x_3)$ x'_4 f_2 (x_1, x_2, x_3). Can it be realizable by a single threshold element ? If yes, give it's weight-threshold vector. If not, indicate clearly why it is not a threshold function.

Solution. The given two functions can be realized by using separate threshold elements as shown below. (see Fig. 6.82)

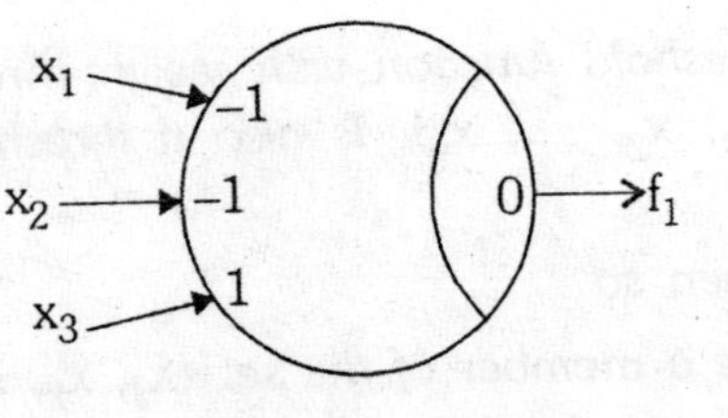

Fig. 6.82(*a*)

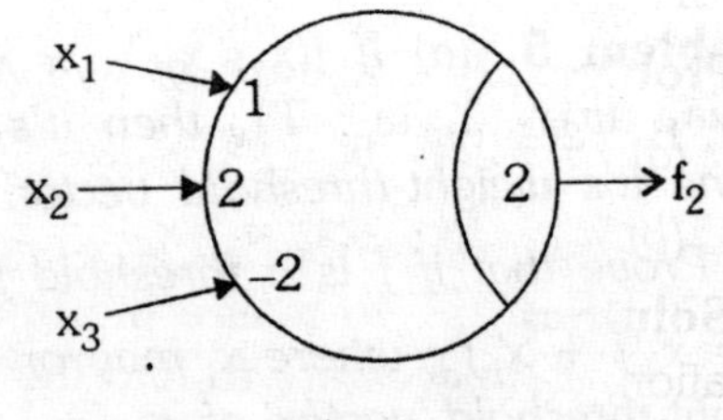

Fig. 6.82(*b*)

The given function, $f = x_4 . f_1(x_1, x_2, x_3) + x'_4 . f_2(x_1, x_2, x_3)$

if $$x_4 = 0 \Rightarrow f = f_2(x_1, x_2, x_3)$$

and if $$x_4 = 1 \Rightarrow f = f_1(x_1, x_2, x_3)$$

We can realize the functions from the threshold element by using K-maps (see Fig. 6.83)

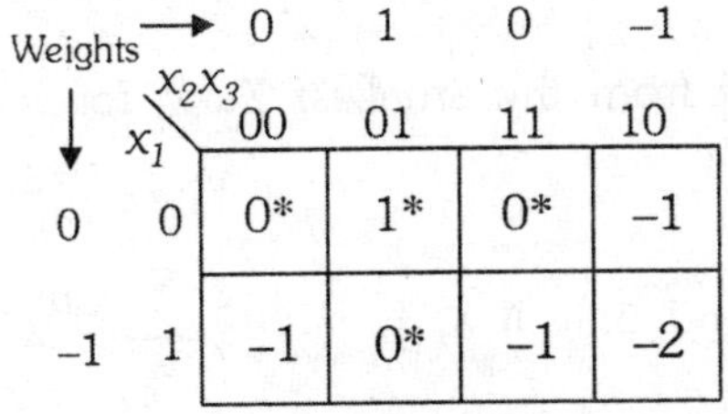

Fig. 6.83(*a*)

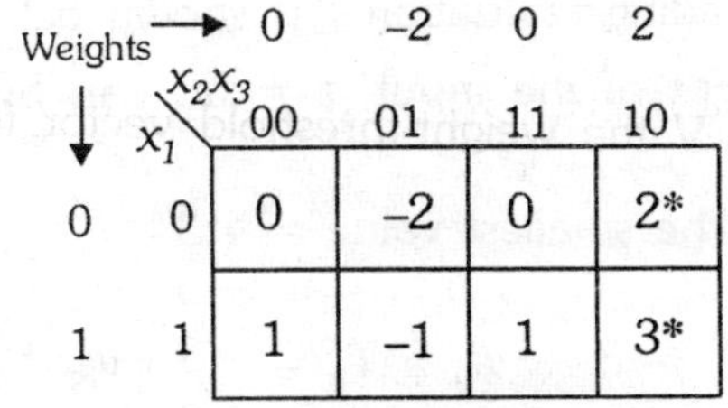

Fig. 6.83(*b*)

Here, $f_1 = \Sigma m(0, 1, 3, 5)$ and $f_2 = \Sigma m(2, 6)$

The * cells in the K-map indicate that the weighted sum is greater than or equal to the threshold value. So, placing in those cells the resultant K-maps are (see Fig. 6.84)

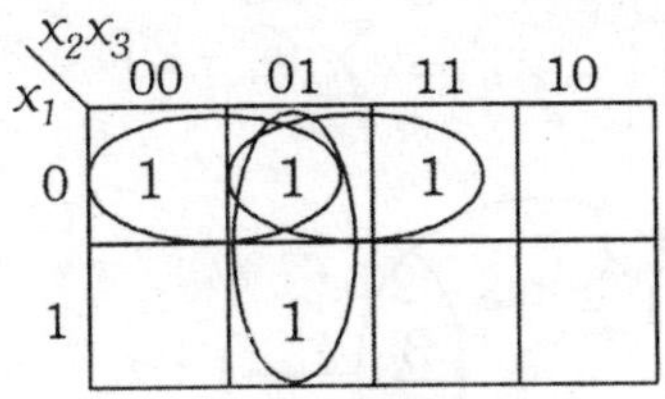

Fig. 6.84(*a*)

and

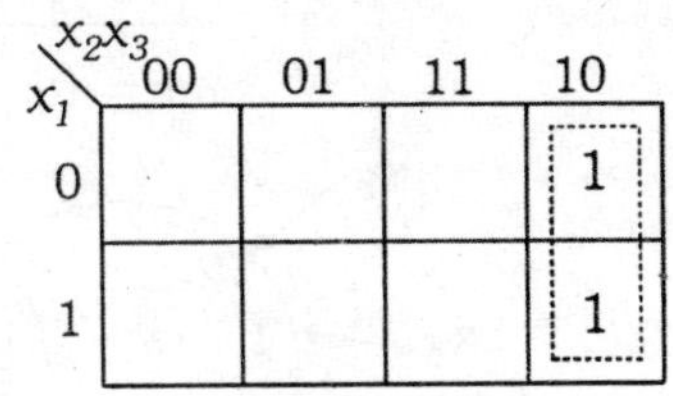

Fig. 6.84(*b*)

$$\therefore \quad f_1 = x'_1x'_2 + x'_2x_3 + x'_1x_3 \qquad f_2 = x_2x'_3.$$

$$\therefore \quad f = x_4.f_1 + x'_4.f_2$$

$$= x'_1x'_2x_4 + x'_2x_3x_4 + x'_1x_3x_4 + x'_4x_2x'_3$$

The given function is not a unate function since the variables x_2 and x_4 are appearing in both forms.

So, it cannot be realized by using single threshold element.

Note: $f = x_4f_1 + x'_4f_2$. Here x_4 is appearing in both forms. So, it is not a unate or threshold function. So, it can not be realized by a single threshold element.

Problem 7. *For each of the following functions, find a two-element cascade realization.*

(a) $f(x_1, x_2, x_3, x_4) = \Sigma(2, 3, 6, 7, 8, 9, 13, 15)$

(b) $f(x_1, x_2, x_3, x_4) = \Sigma m(0, 3, 4, 5, 6, 7, 8, 11, 12, 15)$.

Solution. (*a*) To realize the function with cascade realization, it is necessary to find the admissible patterns by using K-map plot. The K-map for this function is shown below, which contains two admissible patterns and need of cascading of two threshold elements.

$$g = \Sigma m(8, 9, 13) = x_1x'_2x'_3 + x_1x'_3x_4$$

$$h = \Sigma m(2, 3, 6, 7, 15) = x'_1x_3 + x_2x_3x_4$$

From the procedure 6.9.3

$$V_g = \{-2, 1, 3, 1; 5/2\} \text{ and}$$

$$V_h = \{-2, -1, -2, 1; 3/2\}$$

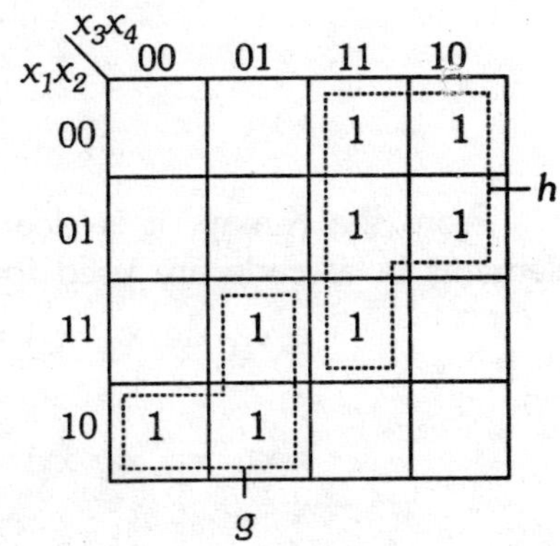

Fig. 6.85

The cascading realization f is shown below.

The weight of the input $g = w_g$ can be calculated from the smallest $\Sigma w_i x_i$ for h when $g = 1$.

$\therefore$ The smallest value = – 3

$$\therefore \quad -3 + w_g \geq T_h = -3 + w_g \geq \frac{3}{2}$$

$$\therefore \quad w_g \geq \frac{3}{2} + 3 = \frac{9}{2}$$

$\therefore w_g$ can be taken as 5 (see Fig. 6.86)

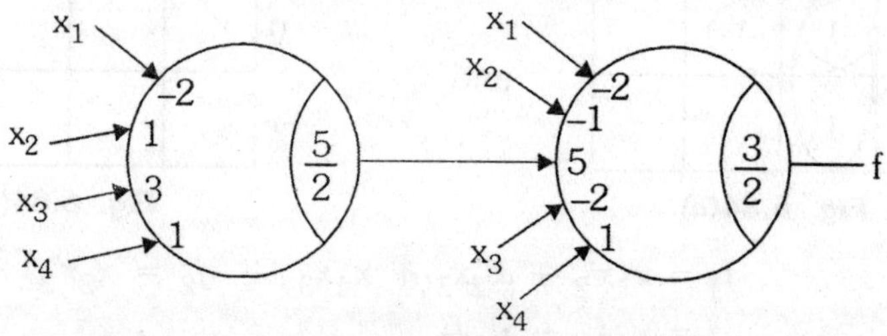

Fig. 6.86

(*b*) The given function is $f_2 = \Sigma(0, 3, 4, 5, 6, 7, 8, 11, 12, 15)$. The K-map for it is shown below (see Fig. 6.87).

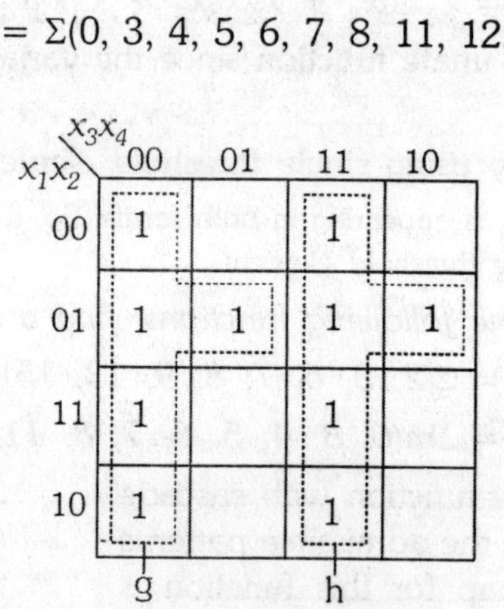

$x_1x_2 \backslash x_3x_4$	00	01	11	10
00	1		1	
01	1	1	1	1
11	1		1	
10	1		1	

Fig. 6.87

From the K-map, it is clear that two admissible patterns are existing and two threshold elements in cascade are used for realization.

$$g(x_1, x_2, x_3, x_4) = \Sigma m(0, 4, 5, 8, 12)$$
$$= x'_3x'_4 + x'_1x_2x'_3$$
$$h(x_1, x_2, x_3, x_4) = \Sigma m(3, 6, 7, 11, 15)$$
$$= x_3x_4 + x'_1x_2x_3$$

$\therefore$ $V_g = \{-1, 1, -3, -2; -3/2\}$. It is verified that

$$g(x_1, x_2, x'_3, x'_4) = h(x_1, x_2, x_3, x_4).$$

$$\therefore \quad V_h = \{-1, 1, +3, +2; 7/2\}.$$

The minimum weighted sum from the threshold element while realizes the

$$h\ \Sigma w_i x_i = -1$$

The 5th weight in the $TE_h = w_g$ *i.e.*, output of TE_g must be

$$-1 + w_g \geq \frac{7}{2}$$

$$\therefore \quad w_g \geq \frac{7}{2} + 1 = \frac{9}{2} \qquad \boxed{\therefore w_g \text{ can be taken as } 5}$$

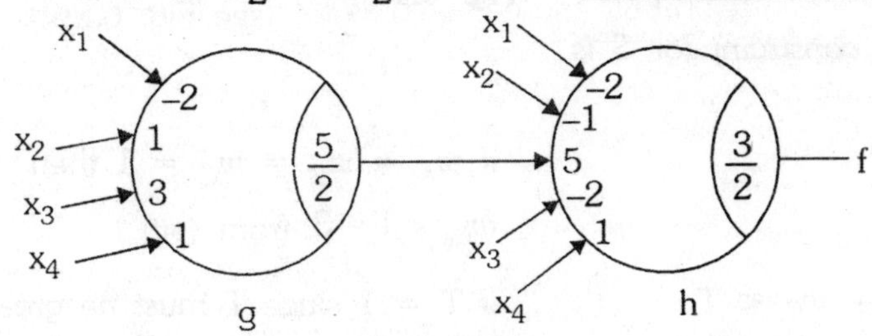

Fig. 6.88

Problem 8. *Show that a threshold-logic realization of a full adder requires only two threshold elements (Note : both the sum and carry-out must be generated).*

Solution. The truth table of a Full Adder is shown below (see Table 6.21).

Table 6.21

A	B	C_{in}	S	C_0
0	0	0	0	0
0	0	1	1	0
0	1	0	1	0
0	1	1	0	1
1	0	0	1	0
1	0	1	0	1
1	1	0	0	1
1	1	1	1	1

$$\therefore \quad S(x_1, x_2, x_3) = \Sigma m(1, 2, 4, 7)$$

$$C_0(x_1, x_2, x_3) = \Sigma m(5, 6, 7)$$

The K-map for S is shown in Fig. 6.89(*a*). The K-map for C_0 is shown in Fig. 6.89(*b*).

x_1 \ x_2x_3	00	01	11	10
0		1		1
1	1		1	

Fig. 6.89(*a*)

x_1 \ x_2x_3	00	01	11	10
0			1	
1		1	1	1

Fig. 6.89(*b*)

(*a*) $S = x'_1x'_2x_3 + x'_1x_2x'_3 + x_1x'_2x'_3 + x_1x_2x_3$ (*b*) $C_0 = x_1x_3 + x_2x_3 + x_1x_2$

The $V_{C_0} = \{1, 1, 1; 3/2\}$

$\therefore$ Threshold element for realization of C_0 is shown in Fig. 6.90.

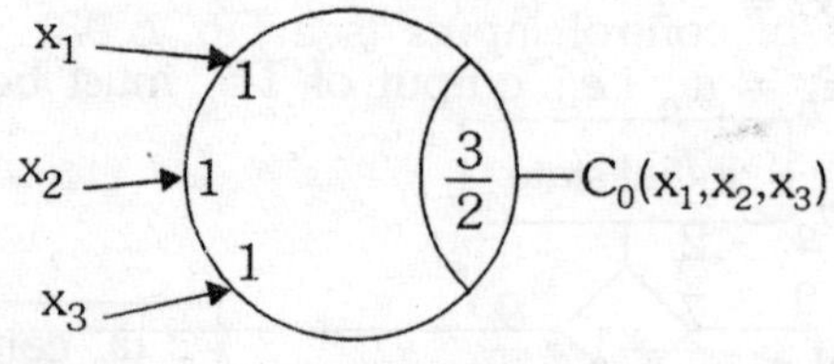

Fig. 6.90

The inequalities constraint for S is

(*i*) $T > 0$

(*ii*) $w_C \geq T$

(*iii*) $w_B \geq T$

(*iv*) $w_B + w_B + w_{C_0} < T$

(*v*) $w_A > T$

(*vi*) $w_A + w_{C_i} + w_{C_0} < T$

(*vii*) $w_A + w_B + w_C + w_{C_0} < T$

if $w_A = w_B = w_C = 1$ then

$w_{C_0} < T - 3$ from (*vii*)

if $T = 1$ since T must be greater than 0

$\therefore w_{C_0} < 1 - 3 = -2$

$\therefore w_{C_0} < -2$

assume $w_{C_0} = -3$

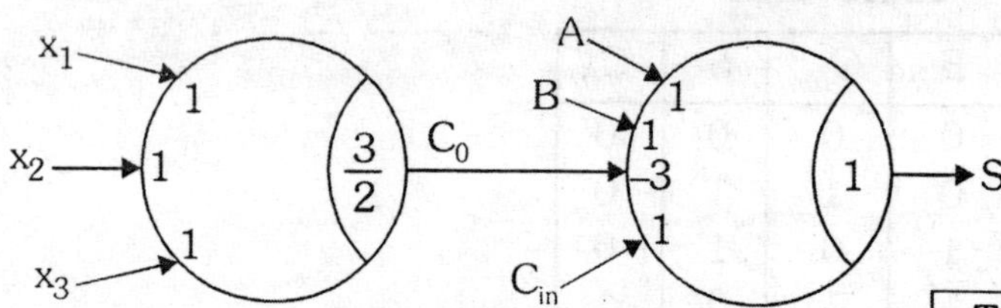

Fig. 6.91

Problem 9. *Draw the portion of an ASM chart that specifies a conditional operation to increment register R during state T_1 and transfer to state T_2 if control inputs z and y are equal to 1 and 0, respectively.*

Solution. Here assuming that the digital system is in initial system and uses a start signal to go to state T_1.

When the control is in state T_1, the register R is incremented and goes to state T_2 if control inputs z and y are 1 and 0 respectively. Otherwise, the control is in the same state. Once the controller is in state T_2, in the next clock pulse it goes to initial state T_0 (see Fig. 6.92).

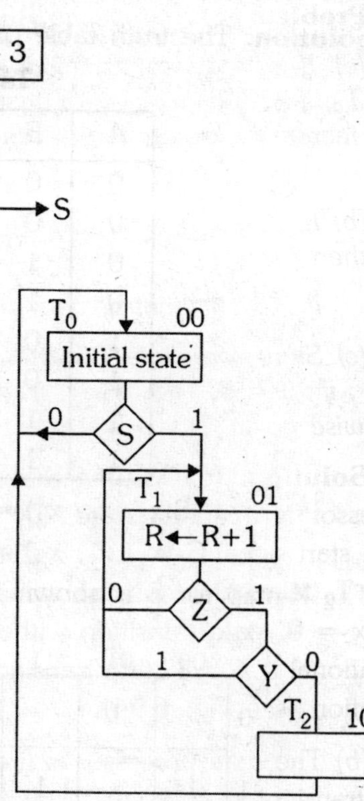

Fig. 6.92

Problem 10. *Show the eight exit paths in an ASM block emanating from the decision boxes that check the eight possible binary values of three control variables x, y and z.*

Solution. The given problem has three control inputs. So, we will get $2^3 = 8$ exit paths and assume that the digital circuit is initially in initial state and there are 8 different next states which depend on the values of control inputs (see Fig. 6.93).

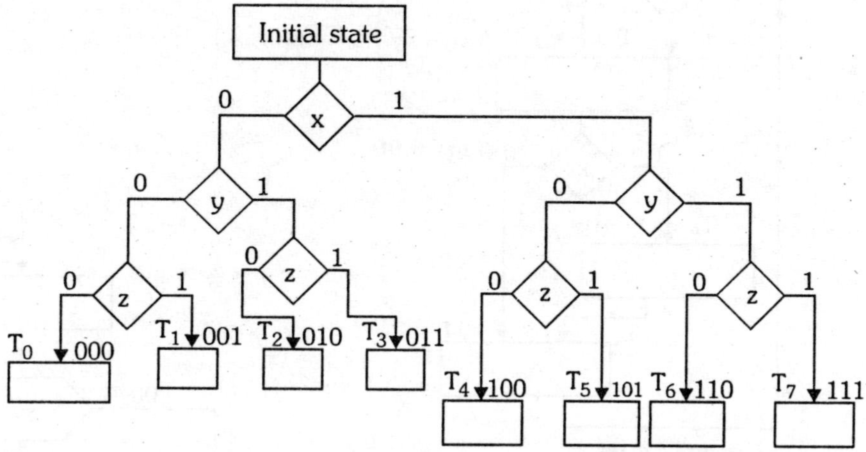

Fig. 6.93

Problem 11. *Draw the ASM charts for the following state transitions :*

(a) If x = 0, control goes from state T_1 to state T_2; if x = 1, generate a conditional operation (i.e., increment a counter A) and go from T_1 to T_2 ;

(b) If x = 1, control goes from state T_1 to T_2 and then to T_3 ;

If x = 0, control goes from T_1 to T_3.

(c) Start from state T_1 ; then ; if xy = 00, go to T_2 ; if xy = 01, go to T_3 ; if xy = 10 go to T_1; otherwise go to T_3.

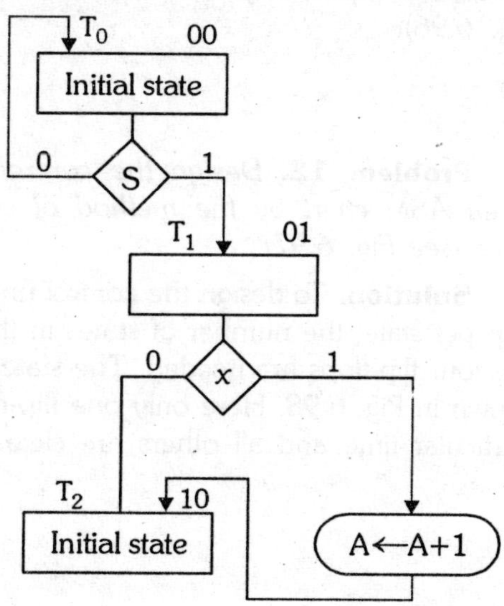

Fig. 6.94

Solution. (*a*) Assume that initially the data processor is in initial state, T_0 and goes to T_1, when start signal S = 1. The state transition from T_1 to T_2 takes place on the value of control input x. If x = 0, state transition without any condition operational if x = 1 state transition with condition operation (see Fig. 6.94).

(*b*) The data processor is first at initial state and goes to T_1 when start signal S = 1. The state transition from T_1 to T_2 takes place when control input *x* = 1 and if *x* = 0, state transition from T_1 to T_3 takes place.

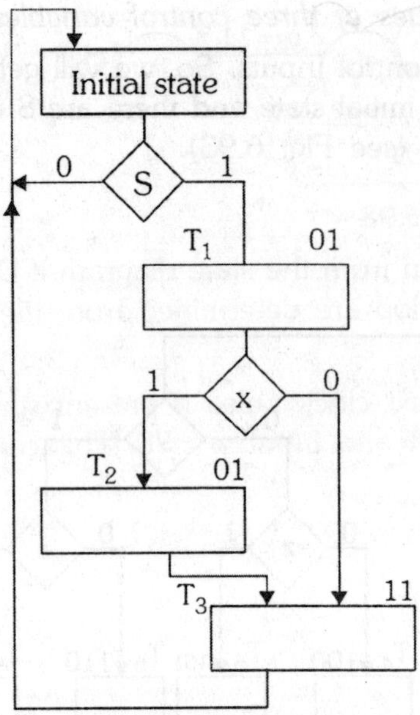

Fig. 6.95

(c) Here, the data processor is initially in state T_1. The next states are T_2, T_3, T_1 and T_3 if the two combined controlled inputs $x.y$ are 00, 01, 10 and 11 (see Fig. 6.96).

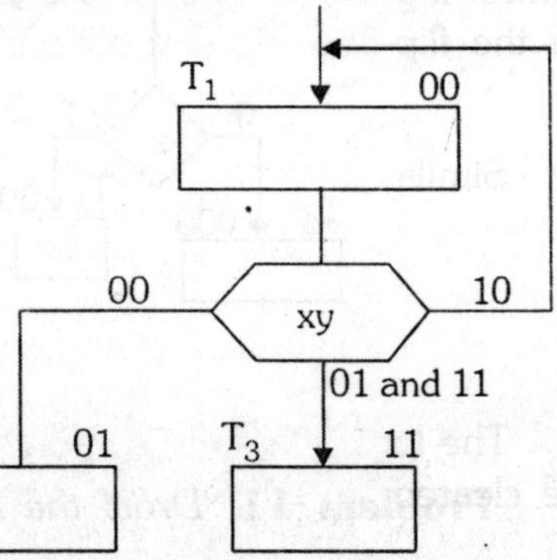

Fig. 6.96

Problem 12. *Design the control circuit for the given ASM chart by the method of one flip-flop per state (see Fig. 6.97).*

Solution. To design the control unit using one flip-flop per state, the number of states in the circuit is four. So, four flip-flops are needed. The state diagram of it is shown in Fig. 6.98. Here only one flip-flop is set at any particular time and all others are cleared to '0'.

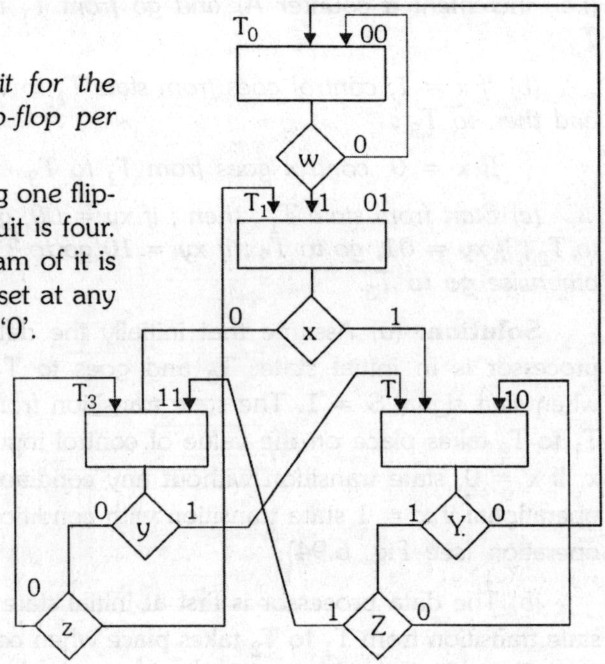

Fig. 6.97

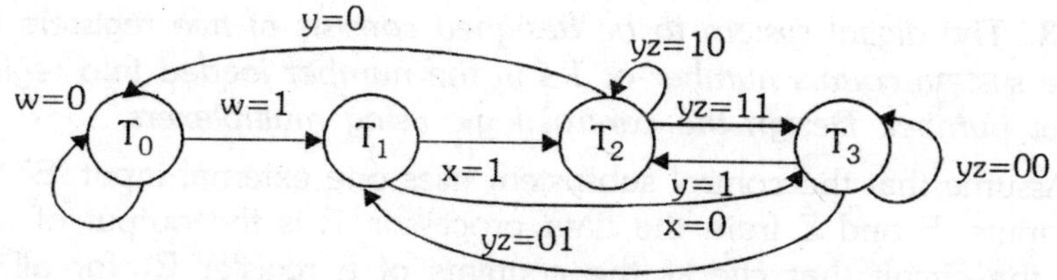

Fig. 6.98

The controller can be designed by inspection from the state diagram if D-type flip-flops are used. The Boolean function for setting the flip-flop are determined from the present state and the input conditions along the directed line.

For example, flip-flop T_0 is set with the next clock pulse if present state is $T_2 = 1$ with control input $y = 0$ or if present state is $T_0 = 1$ and input $w = 0$. This condition is specified by the flip-flop input function.

$DT_0 = T_2\, y' + T_0 w'$, where DT_0 designates the D input of flip-flop T_0.

Similarly,

$$DT_1 = T_0 w + T_3 y'z$$
$$DT_2 = T_1 x + T_2 yz' + T_3 y$$
$$DT_3 = T_3 y'z' + T_1 x' + T_2 yz$$

The logic diagram is shown below and initially flip-flop T_0 is set to 1 and all other flip-flops are cleared to '0'. So, the cleared inputs of all flip-flops are connected (see Fig. 6.99).

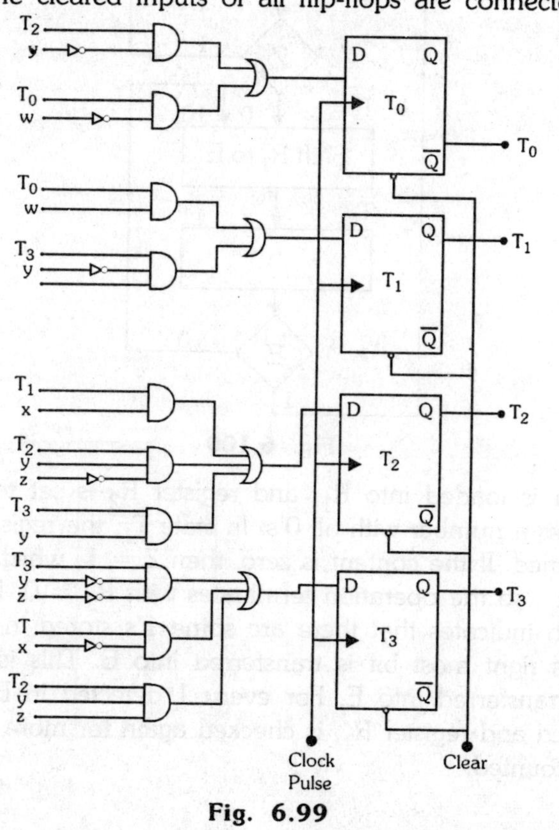

Fig. 6.99

Problem 13. *The digital system to be designed consists of two registers R_1 and R_2, and a flip-flop, E. The system counts number of 1's in the number loaded into register R_1 and sets register R_2 to that number. Design the control logic using multiplexers.*

Solution. Assume that the control subsystem uses one external input 'S' to the operation and two status inputs, E and Z from the data processor. E is the output of a flip-flop and Z is the output of the circuit that checks the contents of a register R_1 for all 0's. The circuit produces an output, Z = 1 when R_1 is equal to 0.

The ASM chart for the design example is shown below (see Fig. 6.100).

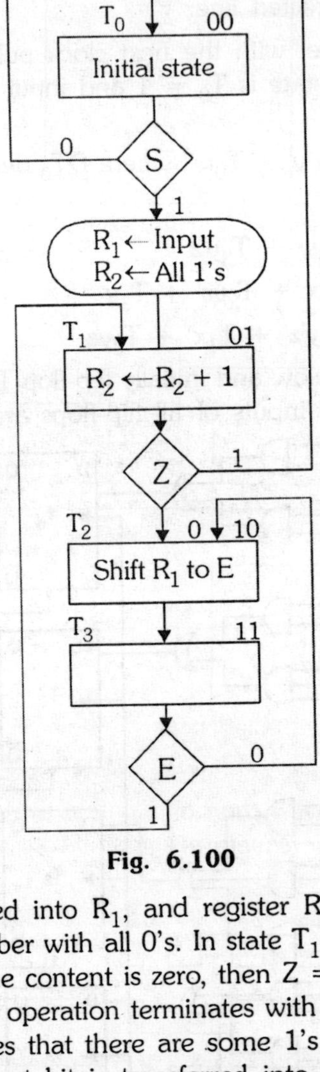

Fig. 6.100

The binary number is loaded into R_1, and register R_2 is set to an all 1's value. When incremented, R_2 produces a number with all 0's. In state T_1, the register R_2 is incremented and the content of R_1 is verified. If the content is zero, then Z = 1, which means that there are no 1's stored in the register ; so the operation terminates with R_2 = 0. If the content of R_1 is not zero, then Z = 0, which indicates that there are some 1's stored in the register. The number in R_1 is shifted and it's right most bit is transferred into E. This is done as many times as necessary until a 1 is transferred into E. For every 1 detected in E, in the next clock pulse register R_2 is incremented and register R_1, is checked again for more 1's. This loop is repeated until all 1's in R_1 are counted.

To implement the control logic using multiplexer, first we find the multiplexer input conditions from the state table. Here, 4 states are there. So, 2 flip-flops are necessary, namely S_1 and S_2. It has 3 control inputs S_1 Z and E and 4 outputs T_0, T_1, T_2 and T_3 (see Table 6.22).

Table 6.22

Present state S_1S_2	*Next state* S_1S_2	*Input conditions*	*Multiplex MUX1*	*Inputs MUX2*
0 0	0 0	S'		
0 0	0 1	S	0	S
0 1	0 0	Z		
0 1	1 0	Z'	Z'	0
1 0	1 1	ϕ	ϕ	ϕ
1 1	1 0	E'		
1 1	0 1	E	E'	E

From the above, it is clear that, it is a three level implementation with multiplexers in the first level as shown below (see Fig. 6.101).

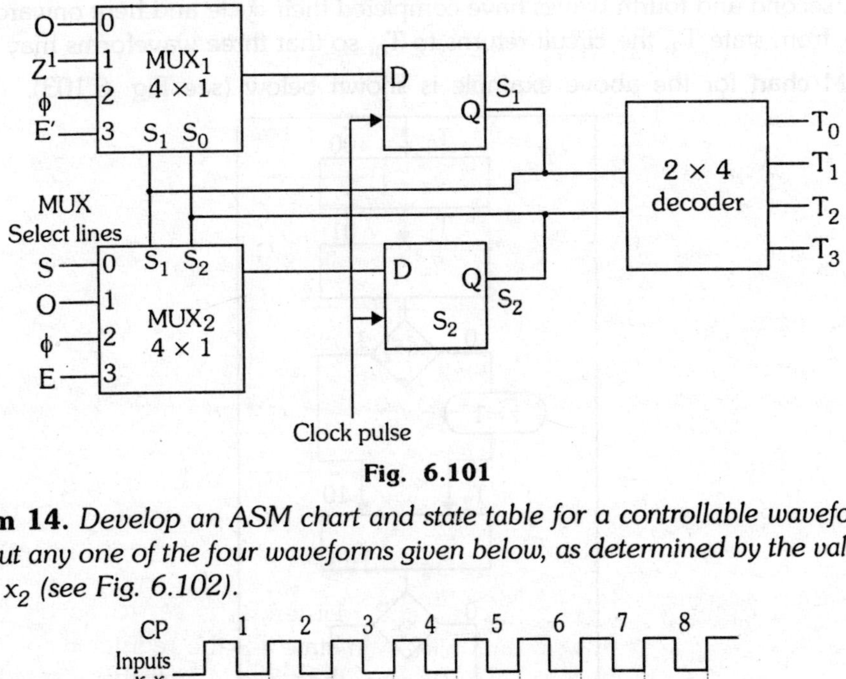

Fig. 6.101

Problem 14. *Develop an ASM chart and state table for a controllable waveform generator that will output any one of the four waveforms given below, as determined by the values of it's two input x_1 and x_2 (see Fig. 6.102).*

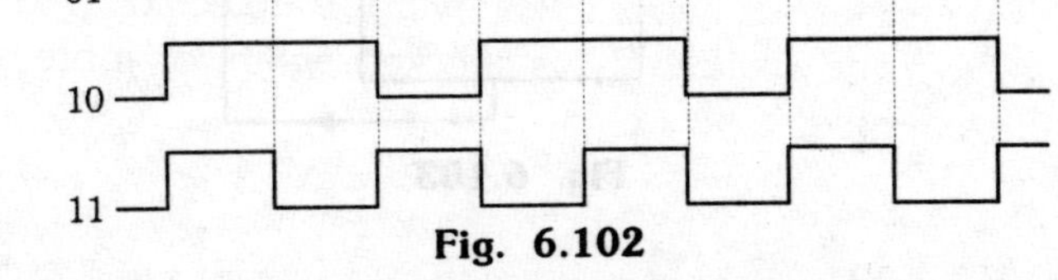

Fig. 6.102

Solution. The ASM chart can be drawn for the above waveform with four states, one for each clock pulse of the waveform with the longest period. For each state, the output will be conditional on the values of the inputs x_1 and x_2. Let us observe the different conditions at different states.

State T_0 : In the first state, T_0, *i.e.*, in the first clock pulse, all the waveforms are at logic 1. ∴ The output Z = 1, which was used in the state box T_0.

State T_1 : In the second state, T_1, *i.e.*, in the second clock pulse, the output is 1 except for the waveform whose input is x_1x_2 = 11. This condition is tested with the expression $x_1.x_2$ in the decision box. When the result of expression is 0, the input is other than x_1, x_2 = 11, and hence output Z = 1.

State T_2 : During the third state, T_2, *i.e.*, in the third clock pulse, the output is 1, if x_2 = 1 and irrespective of value of x_1. Hence, the condition of x_2 is checked in the decision box at state T_2.

In the third waveform, the new cycle was stated in this state. Therefore, when x_1x_2 = 10, the fourth state is not used and the line goes back to first state to start new cycle (T_0).

State T_3 : In the fourth state, T_3, *i.e.*, in the fourth clock pulse, the output is always zero. Because first, second and fourth waves have completed their cycle and here onwards, they repeat the cycle. So, from state T_3, the circuit returns to T_0, so that three waveforms may be repeated.

The ASM chart for the above example is shown below (see Fig. 6.103).

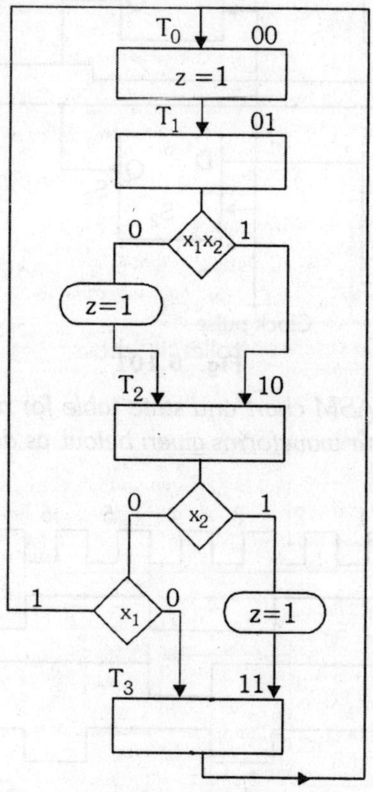

Fig. 6.103

Let us express the state table of the above example.

The state table contains the outputs, next states corresponding to each combination of input values. The state table is shown below (Table 6.23).

Table 6.23

Present state	*Next state Output* *Inputs→* 0 0	0 1	1 0	1 1
T_0	T_1, 1	T_1, 1	T_1, 1	T_1, 1
T_1	T_2, 1	T_2, 1	T_2, 1	T_2, 0
T_2	T_3, 0	T_3, 1	T_0, 0	T_3, 1
T_3	T_0, 0	T_0, 0	ϕ	T_0, 0

The control circuit consists of 4 states and two control inputs-x_1 and x_2. To design the control logic with multiplexer, first it is necessary to find the multiplexer inputs. It consists of 2 multiplexers-MUX_1 and MUX_2, a register with two flip-flops S_1 and S_2: and a combinational circuit (decoder) to determine the output. The inputs of the multiplexers are determined from the decision boxes and state transition given in the ASM charts (see Table 6.24).

Table 6.24

Present state		*Next state*		*Input conditions*	*Multiplexer inputs*	
S_1	S_2	S_1	S_2		MUX_1	MUX_2
0	0	0	1	ϕ	ϕ	1
0	1	1	0	$x'_1x'_2$	(0 → 0)	(0 → 1)
0	1	1	0	x_1x_2	1	0
1	0	1	1	x_2		
1	0	1	1	$x'_2x'_1$	x'_1+x_2	x'_1+x_2
1	0	0	0	$x_4x'_2$		
1	1	0	0	ϕ (1 → 0)	ϕ (1 → 0)	ϕ

The three level logic diagram for controller is shown below. (see Fig. 6.104)

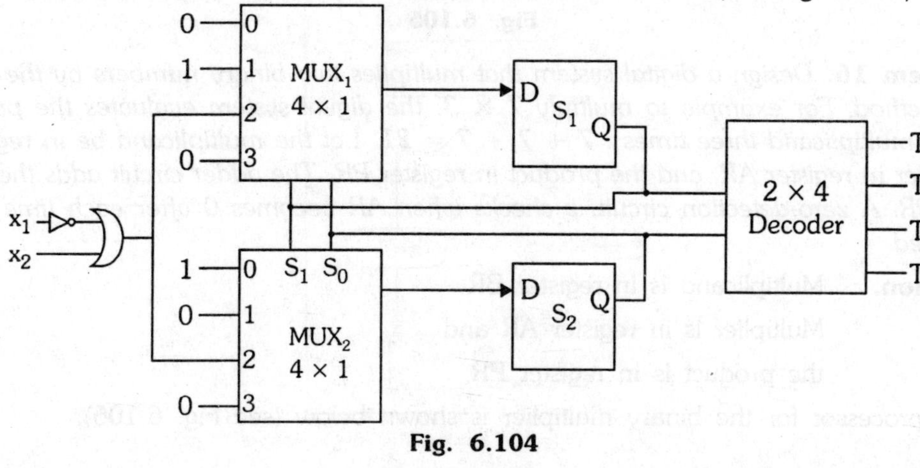

Fig. 6.104

Problem 15. *Draw the ASM chart for a digital system with three 4-bit registers-A, B and C, to perform the following operations :*

(a) Transfer two binary numbers (n, m) to A and B when a start signal is enabled.

(b) If A < B, shift left the contents of A and transfer the results to register C.

(c) If A > B, shift right the contents of B and transfer the results to register C.

(d) If A = B, transfer the number to register C unchanged.

Solution. Assume that the start signal used here is 'S' and the digital system is initially in initial state. The ASM chart for the above problem is drawn below (see Fig. 6.105).

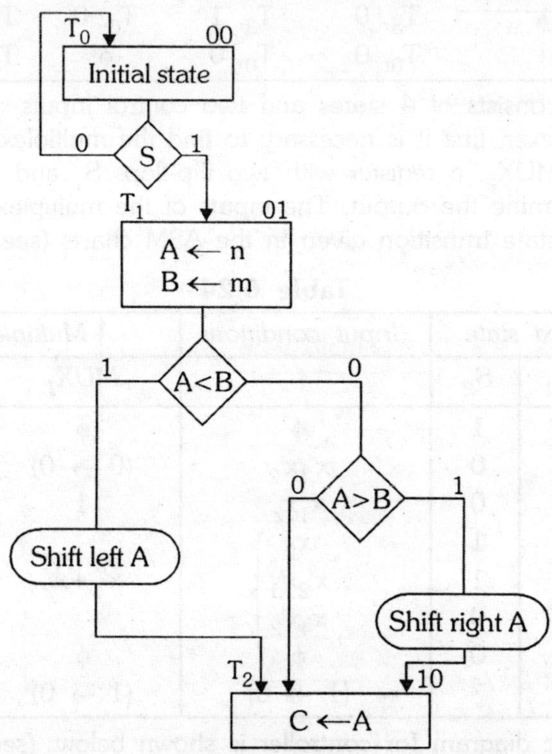

Fig. 6.105

Problem 16. *Design a digital system that multiplies two binary numbers by the repeated addition method. For example to multiply 7 × 3, the digital system evaluates the product by adding the multiplicand three times : 7 + 7 + 7 = 21. Let the multiplicand be in register BR, the multiplier in register AR, and the product in register PR. The adder circuit adds the contents of BR to PR. A zero-detection circuit, z checks when AR becomes 0 after each time that it is decremented.*

Solution. Multiplicand is in register BR

Multiplier is in register AR and

the product is in register PR.

Data processor for the binary multiplier is shown below (see Fig. 6.106).

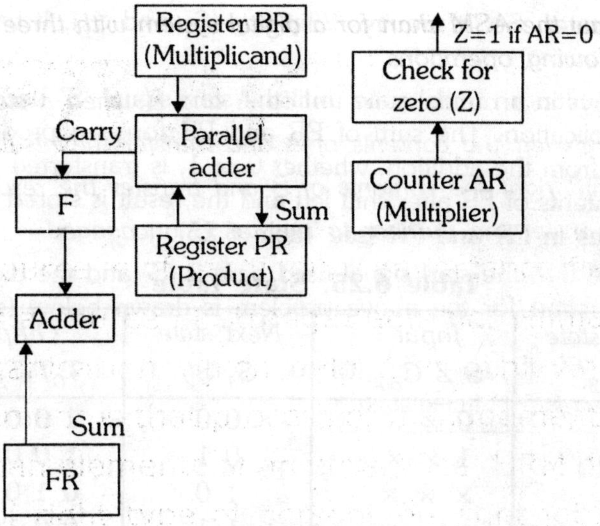

Fig. 6.106

The ASM chart for the binary multiplier is shown in Fig. 6.107.

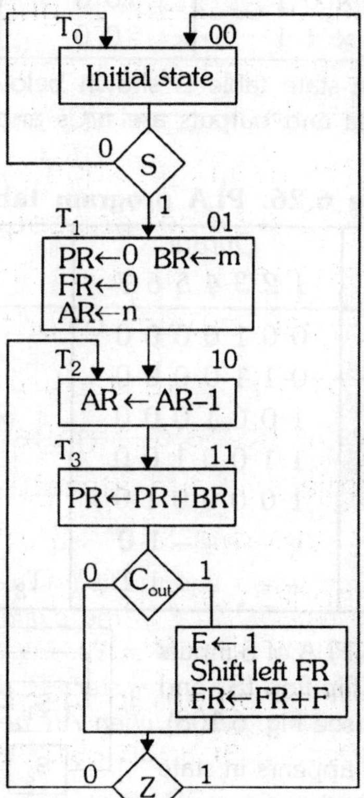

Fig. 6.107

Control circuit design by using PLA : To design the control circuit by using PLA, it is necessary to find PLA table. First, we will find the state table of the circuit.

The control has 4-states. So, it requires 2flip-flop-(S_1 and S_2). It has 3 control inputs S, C_{out} and Z.

The control logic stays in an initial state until the start signal 'S' becomes a 1. The system then performs the multiplication. The sum of BR and PR forms a product and is transferred to PR. The output carry from the addition, whether 0 or 1, is transferred to F. If C_{out} = 1, then carry is added to the contents of FR after shift left and the result is stored in FR. The final result is concatenation of values in FR and PR (see Table 6.25).

Table 6.25. State table

Present state $S_1\ S_2$	*Input* $S\ Z\ C_{out}$	*Next state* $S_1\ S_2$	*Output* $T_0T_1T_2T_3D$
0 0	0 × ×	0 0	1 0 0 0 0
0 0	1 × ×	0 1	1 0 0 0 0
0 1	× × ×	1 0	0 1 0 0 0
1 0	× × ×	1 1	0 0 1 0 0
1 1	× 0 0	1 0	0 0 0 1 0
1 1	× 0 1	0 0	0 0 0 1 0
1 1	× 1 1	0 0	0 0 0 1 1

The PLA table for the above state table is shown below. The inputs of PLA table are present state of flip-flop and inputs and outputs are n/Cs and outputs in the state table (see Table 6.26).

Table 6.26. PLA program table

Product terms	*Inputs* 1 2 3 4 5	*Output* 1 2 3 4 5 6 7	*Comments*
1	0 0 0 – –	0 0 1 0 0 0 0	T_0 = 1, S = 0
2	0 0 1– –	0 1 1 0 0 0 0	T_0 = 1, S = 1
3	0 1 – – –	1 0 0 1 0 0 0	T_1 = 1
4	1 0 – – –	1 1 0 0 1 0 0	T_2 = 1
5	1 1 – 0 0	1 0 0 0 0 1 0	T_3 = 1, C_{out} = 0, Z = 0
6	1 1 – 0 1	– – – – – 1 0	T_3 = 1, C_{out} = 0, Z = 1
7	1 1 – 1 1	– – – – – 1 1	T_3 = 1, C_{out} = 1, Z_1 = D = 1

To design the PLA, we need a PLA of 5 inputs and 7 outputs and register with 2 flip-flop (S_1 and S_2). The diagram is shown below (see Fig. 6.108).

where D is a control output which appears in state T_3 when C_{out} = 1

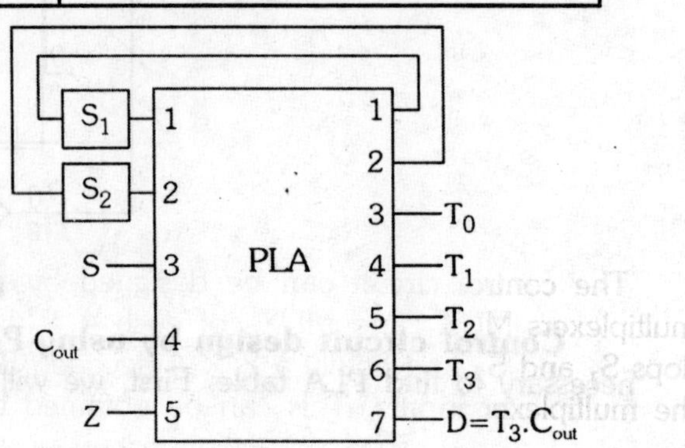

Fig. 6.108

Problem 17. *The state diagram of a control unit is shown in figure. It has four states and two inputs x and y.*

(a) Draw the equivalent ASM chart, leaving the state box empty.

(b) Design the control circuit with multiplexers (see Fig. 6.109).

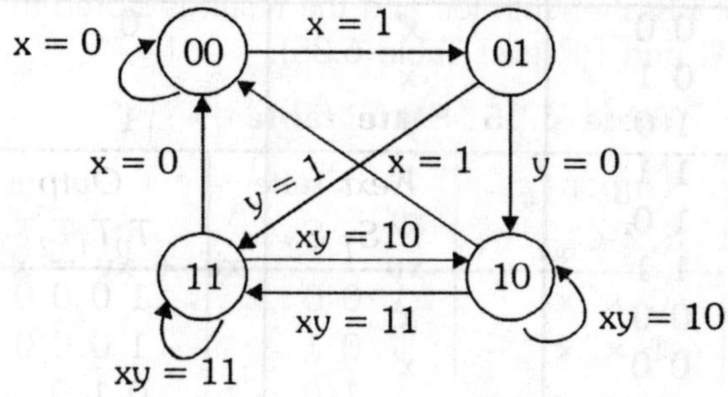

Fig. 6.109

Solution. (*a*) The ASM chart must have four states named T_0, T_1, T_2 and T_3 and two control inputs *x* and *y* (see Fig. 6.110).

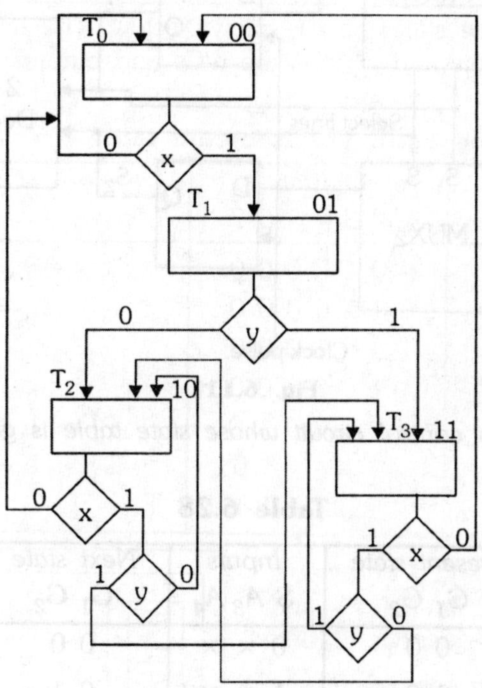

Fig. 6.110

The control circuit can be designed by three level implementation, which contains two multiplexers MUX_1 and MUX_2 of each 4 × 1 type (see Fig. 6.111). A register with two flip-flops S_1 and S_2 and finally a 2 × 4 decoder. Before designing the circuit, it is necessary to find the multiplexer inputs. This can be obtained by the following table : (see Table 6.27)

Table 6.27

Present state	Next state	Input* conditions	Multiplexer inputs	
S_1 S_2	S_1 S_2		MUX_1	MUX_2
0 0	0 0	x'	0	x
0 0	0 1	x		
0 1	1 0	y'	1	y
0 1	1 1	y		
1 0	1 0	xy'		
1 0	1 1	xy	$xy' + xy = x$	xy
1 0	0 0	x'		
1 1	0 0	x'		
1 1	1 0	xy'	$xy' + xy = x$	xy
1 1	1 1	xy		

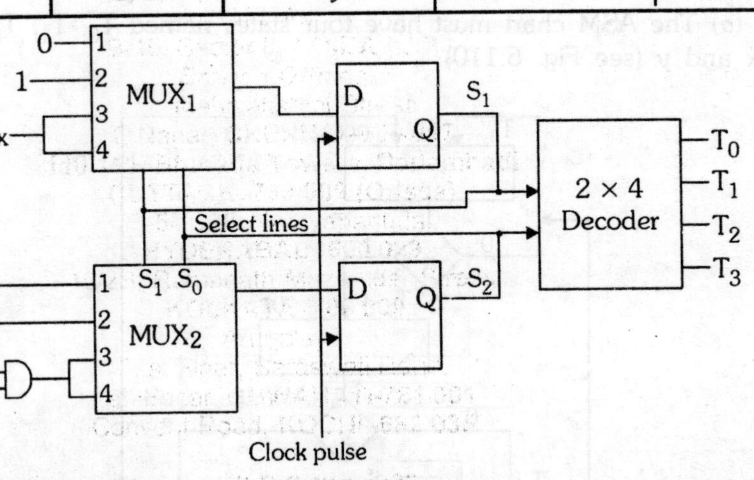

Fig. 6.111

Problem 18. *Design a control circuit whose state table is given below using the PLA method (see Table 6.28).*

Table 6.28

Present state symbol	Present state G_1 G_2	Inputs S A_3 A_4	Next state G_1 G_2	Outputs T_0 T_1 T_2
T_0	0 0	0 × ×	0 0	1 0 0
T_0	0 0	1 × ×	0 1	1 0 0
T_1	0 1	× 0 ×	0 1	0 1 0
T_1	0 1	× 1 0	0 1	0 1 0
T_1	0 1	× 1 1	1 1	0 1 0
T_2	1 1	× × ×	0 0	0 0 1

Solution. Let us convert the state table into PLA program table. The PLA table is shown below, which contains 5 inputs and 5 outputs (see Table 6.29).

Table 6.29

Product term	*Inputs* 1 2 3 4 5	*Outputs* 1 2 3 4 5	*Comments*
1	0 0 0 – –	– – 1 – –	$T_0 = 1, S = 0$
2	0 0 1 – –	– 1 1 – –	$T_0 = 1, S = 1$
3	0 1 – 0 –	– 1 – 1 –	$T_1 = 1, A_3 = 0$
4	0 1 – 1 0	– 1 – 1 –	$T_1 = 1, A_3A_4 = 10$
5	0 1 – 1 1	1 1 – 1 –	$T_1 = 1, A_3A_4 = 11$
6	1 1 – – –	– – – – 1	$T_2 = 1$

To design the logic diagram using PLA requires a PLA of 5 inputs and 5 outputs with a register containing 2 flip-flops (see Fig. 6.112).

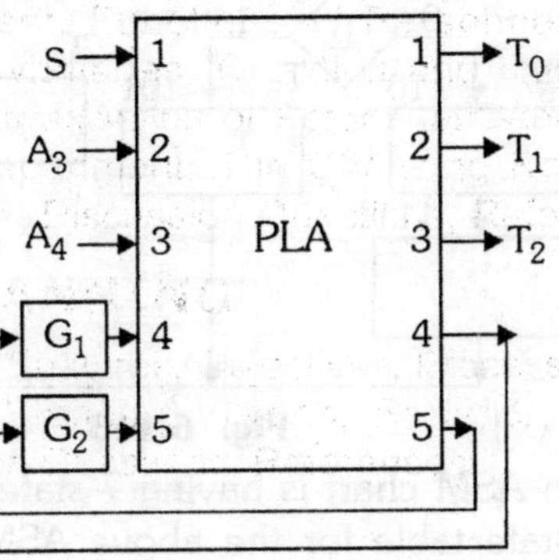

Fig. 6.112

Problem 19. *Consider the ASM chart given below. The register operations are not specified, (see Fig. 6.113)*

(a) Draw the equivalent state diagram.

(b) Design the controller with one flip-flop per state.

(c) Design the controller with three D flip-flops, a decoder and gates.

(d) Design the controller with three multiplexers, a register with three flip-flops and a 3 × 8 decoder.

(e) Design the control with a PLA and a register.

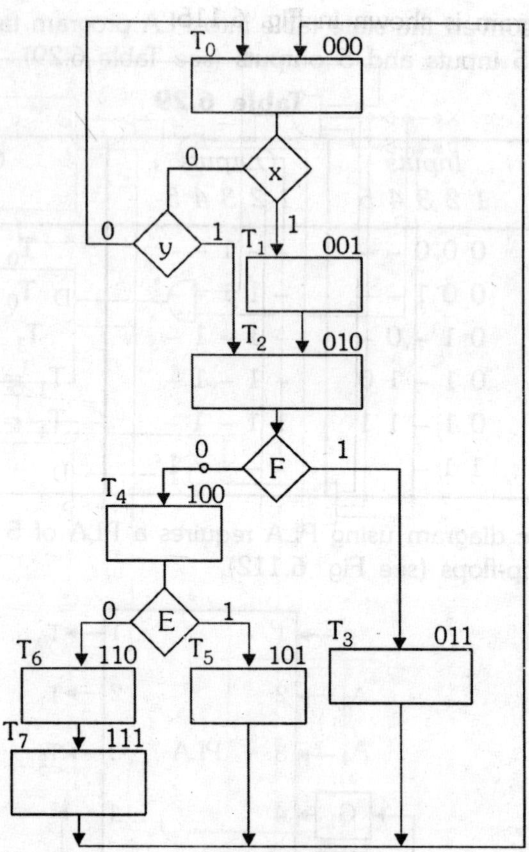

Fig. 6.113

Solution. (*a*) The given ASM chart is having 7 states from T_0 through T_7. It has 4 control inputs *x*, *y*, F and E. The state table for the above ASM chart is shown in Fig. 6.114.

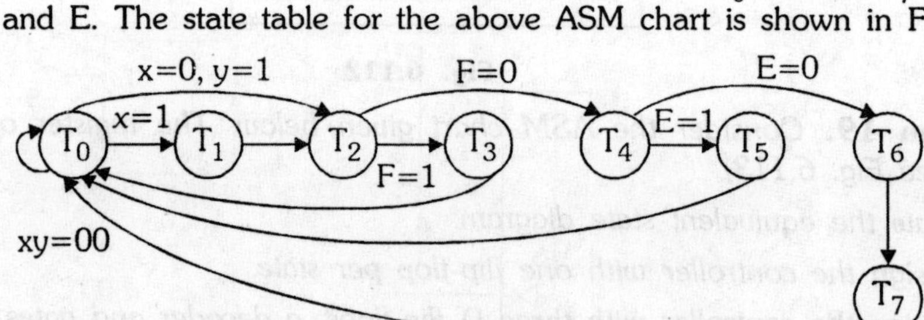

Fig. 6.114

(*b*) To design the controller with one flip-flop per state requires 7 D flip-flops. The Boolean functions for setting the flip-flop is determined from the present state and the input conditions along the directed lines. Now, we will derive the input conditions at which each flip-flop is in set condition.

$$DT_0 = T_0x'y' + T_7 + T_5 + T_3$$

$$DT_1 = T_0x \qquad DT_5 = T_4E$$

$$DT_2 = T_1 + T_0x'y \qquad DT_6 = T_4E'$$

$$DT_3 = T_2F \qquad DT_7 = T_6$$

$$DT_4 = T_2F'$$

The logic diagram is shown in Fig. 6.115.

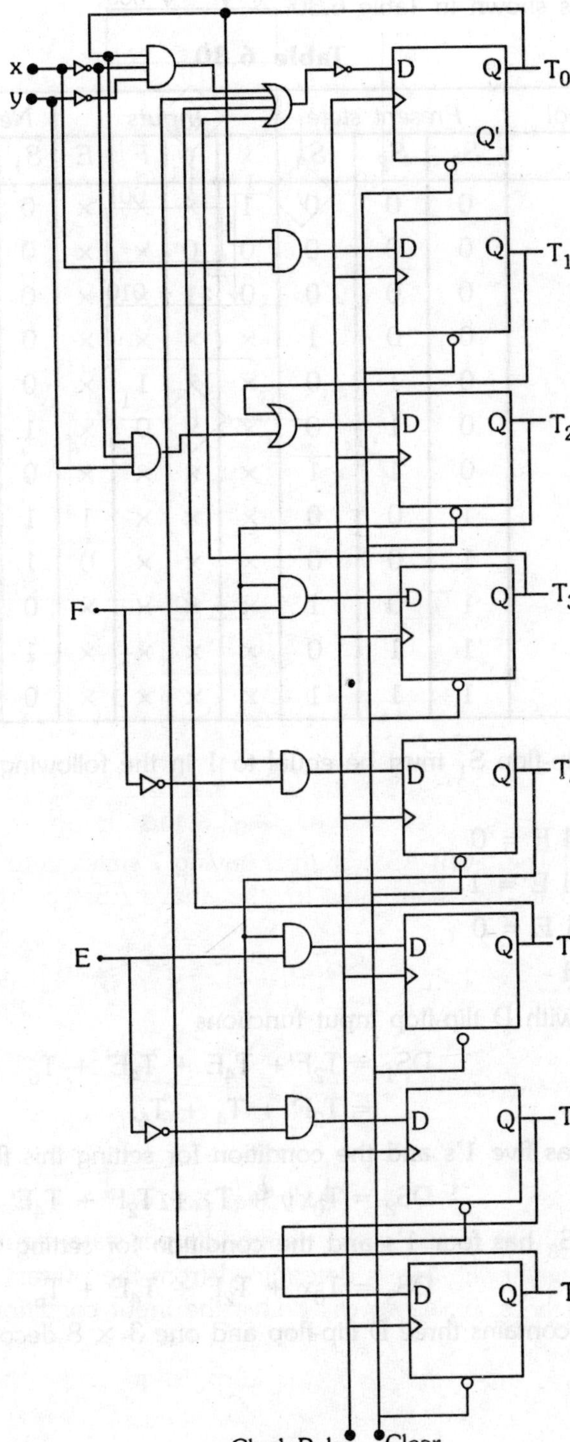

Fig. 6.115

(*c*) To design the controller with D flip-flop and decoder first we have to write the state table. The state table is shown in Table 6.30.

Table 6.30

Symbol	*Present state*			*Inputs*				*Next state*		
	S_1	S_2	S_3	x	y	F	E	S_1	S_2	S_3
T_0	0	0	0	1	×	×	×	0	0	1
T_0	0	0	0	0	0	×	×	0	0	0
T_0	0	0	0	0	1	×	×	0	1	0
T_1	0	0	1	×	×	×	×	0	1	0
T_2	0	1	0	×	×	1	×	0	1	1
T_2	0	1	0	×	×	0	×	1	0	0
T_3	0	1	1	×	×	×	×	0	0	0
T_4	1	0	0	×	×	×	1	1	0	1
T_4	1	0	0	×	×	×	0	1	1	0
T_5	1	0	1	×	×	×	×	0	0	0
T_6	1	1	0	×	×	×	×	1	1	1
T_7	1	1	1	×	×	×	×	0	0	0

The D inputs of flip-flop S_1 must be equal to 1 in the following present states and input conditions.

(*a*) at state T_2 and F = 0

(*b*) at state T_4 and E = 1

(*c*) at state T_4 and E = 0

(*d*) at state T_6 and

This is expressed with D flip-flop input functions

$$DS_1 = T_2F' + T_4E + T_4E' + T_6$$
$$= T_2F' + T_4 + T_6.$$

Similarly, the S_2 has five 1's and the condition for setting this flip-flop is

$$DS_2 = T_0x'y + T_1 + T_2F + T_4E' + T_6.$$

The next column S_3 has four 1's and the condition for setting this flip-flop is

$$DS_3 = T_0x + T_2F + T_4E + T_6$$

The logic diagram contains three D flip-flop and one 3 × 8 decoder as shown below (see Fig. 6.116).

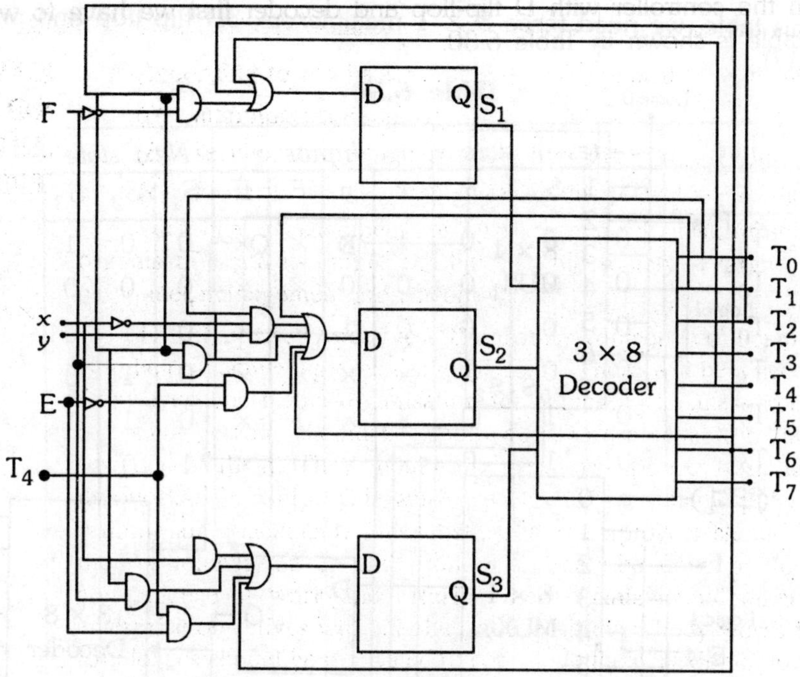

Fig. 6.116

(*d*) To design the controller with multiplexers first we have to find the multiplexer input conditions from the state table. This is shown in Table 6.31.

Table 6.31

Present state	*Next state*	*Input condition*	*Multiplexer inputs*		
S_1 S_2 S_3	S_1 S_2 S_3		MUX_1	MUX_2	MUX_3
0 0 0	0 0 0	$x'y'$			
0 0 0	0 0 1	x	0	$x'y$	x
0 0 0	0 1 0	$x'y$			
0 0 1	0 1 0	ϕ	0	1	0
0 1 0	0 1 1	F	F'	F	F
0 1 0	1 0 0	F'			
0 1 1	0 0 0	ϕ	0	0	0
1 0 0	1 0 1	E	1	E'	E
1 0 0	1 1 0	E'			
1 0 1	0 0 0	ϕ	0	0	0
1 1 0	1 1 1	ϕ	1	1	1
1 1 1	0 0 0	ϕ	0	0	0

From this table, we need three 8 × 1 multiplexers, 3D flip-flop and one 3 × 8 decoder (see Fig. 6.117).

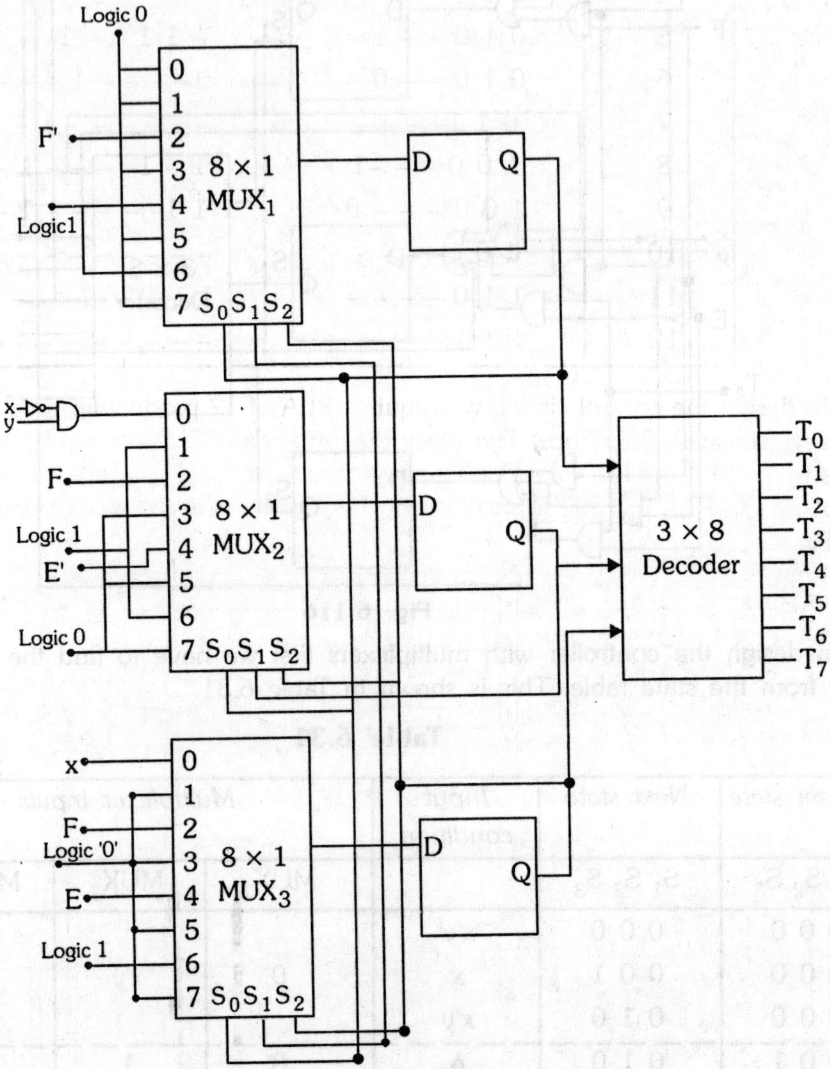

Fig. 6.117

(e) To design the controller with PLA and register, we need to have PLA program table. The PLA program table can be obtained directly from the state table without the need of simplification procedures. From the state table, the PLA table may contain twelve product terms, seven inputs and eleven outputs. The inputs and outputs are indicated by numbers (see Table 6.32).

Table 6.32

Product	*Inputs* *1 2 3 4 5 6 7*	*Outputs* *1 2 3 4 5 6 7 8 9 10 11*
1	0 0 0 1 – – –	– – 1 1 – – – – – – –
2	0 0 0 0 0 – –	– – –1– – – – – – – –

3	0 0 0 0 1– –	– 1 – 1 – – – – – – –
4	0 0 1 – – – –	– 1 – – 1 – – – – – –
5	0 1 0 – – 1 –	– 1 1 – – 1– – – – –
6	0 1 0 – – 0 –	1– – – 1 – – – –
7	0 1 1– – – –	– – – – – – – 1 – – – –
8	1 0 0 – – –1	1 – 1– – – – 1 – – –
9	1 0 0 – – – 0	1 1– – – – – 1 – – –
10	1 0 1– – – –	– – – – – – – – – 1 – –
11	1 1 0 – – – –	1 1 1 – – – – – – – 1 –
12	1 1 1 – – – –	– – – – – – – – – – – 1

To design the control circuit, we requires PLA of 12 product terms, 7 inputs and 11 outputs and a register with 3flip-flops. The block diagram of the PLA control is shown in the figure. The inputs to the PLA are the values of the present state of the register and the four control inputs. The outputs of the PLA provide the values for the next state in the register and the outputs variables (see Fig. 6.118).

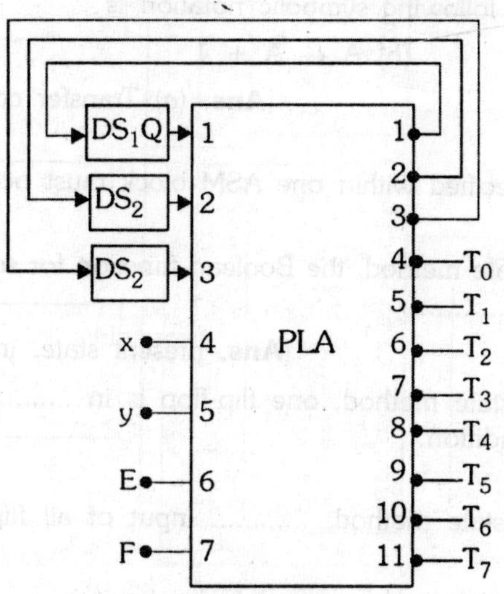

Fig. 6.118

6.13 SHORT ANSWER QUESTIONS

1. ASM is an acronym for ....

[Ans. asynchronous state machines.**]**

2. .... box is used to represent register operations.

[Ans. State**]**

3. .... box is used to represent conditional outputs.

[Ans. Decision**]**

4. .... box is used to represent unconditional outputs.

[Ans. State**]**

5. .... is written inside the decision box.

[**Ans.** Input conditions/functions]

6. The number paths from a decision box is ....

[**Ans.** two (or) more.]

7. The entry path to a conditional box is from the .... box.

[**Ans.** decision]

8. State name is written at .... place on the state box.

[**Ans.** left side top most]

9. The shape of the state box is ....

[**Ans.** rectangular.]

10. The register operations or outputs listed inside the conditional box are generated during the state provided that .... is satisfied.

[**Ans.** input condition]

11. The ASM block consists of ....

[**Ans.** one state box and one or more decision and conditional boxes.]

12. A simple ASM block contains only .... box.

[**Ans.** one state box]

13. The meaning for the following symbolic notation is

(*a*) A ← B (*b*) A ← A + 1

[**Ans.** (*a*) Transfer contents of register B into register A.
(*b*) Increment register A by 1.]

14. All the operations specified within one ASM block must occur in synchronism during ....

[**Ans.** same clock pulse.]

15. In one flip-flop per state method, the Boolean function for setting the flip-flop is determined from the .... and ....

[**Ans.** present state, input condition on the directed line.]

16. In one flip-flop per state method, one flip-flop is in .... condition and all other flip-flop are in .... condition.

[**Ans.** set, clear]

17. In one flip-flop per state method, ...input of all flip-flops are connected common.

[**Ans.** clear.]

18. The purpose of decoder in the control logic circuit is ....

[**Ans.** to generate control outputs.]

19. The design of control circuit with multiplexers is a .... level implementation.

[**Ans.** three]

20. The three levels in the multiplexer method are ....

[**Ans.** (1) first level is multiplexer (2) second level is a register (3) third level has the decoder.]

21. The purpose of multiplexer in control circuit design is ....

[**Ans.** to determine the next state of the register.]

22. The number of multiplexers required to design the control logic of a 4 state problem is ....

[**Ans.** two.]

23. The number of inputs of each multiplexer while designing control circuit depends on ....

[**Ans.** number of states.]

24. In PLA control design, the inputs of PLA are ....

[**Ans.** present states and control inputs.]

25. In PLA control design, the outputs of the PLA provide ....

[**Ans.** next states and control outputs.]

26. PLA program table contains ....

[**Ans.** inputs, outputs and product terms.]

6.14 REVIEW QUESTIONS

1. Write short notes on ASM chart.

2. Explain in detail the Mealey state diagram and ASM chart for it with an example.

3. Draw the state diagram for a serial adder and convert it into ASM chart.

4. Draw the ASM chart for a binary divider and realize it with T flip-flops and gates.

5. Draw the state diagram for a full adder and convert it to ASM chart and realize it with D flip-flops.

6. Explain the multiplexer method of implementing an ASM chart with an example.

7. Write short notes on implementation of control circuit using PLA.

8. The state diagram is shown in Fig. 6.119. Draw the equivalent ASM chart and implement it using JK-flip-flops.

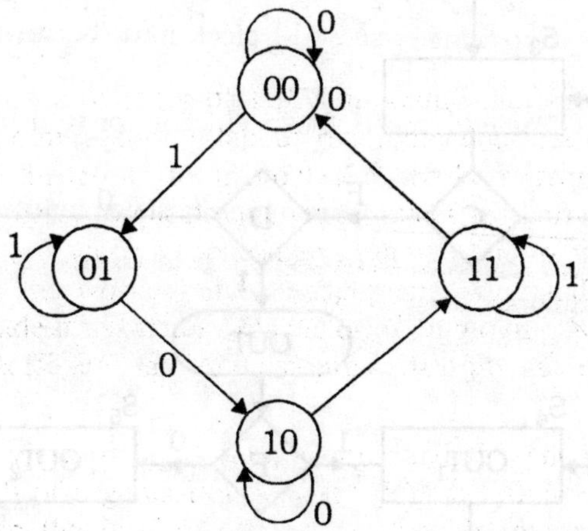

Fig. 6.119

9. Explain the following

(*a*) Unate function (*b*) Threshold logic.

10. Develop an ASM chart and a transition table for a controllable waveform generator that will output any one of the four waveforms given in the figure. The inputs are x_1 and x_2.

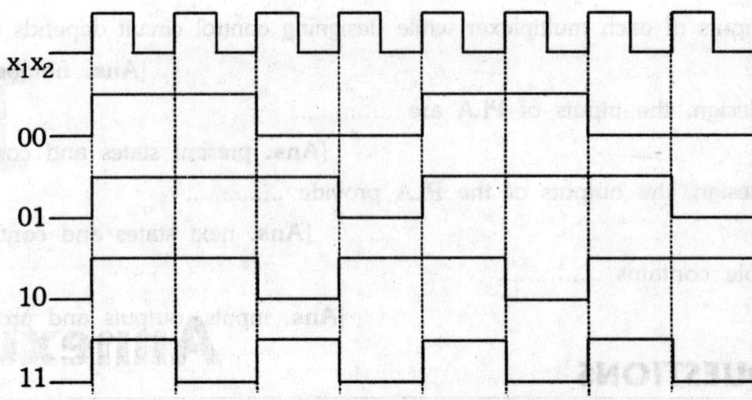

Fig. 6.120

11. Design a controller for the ASM chart given below.

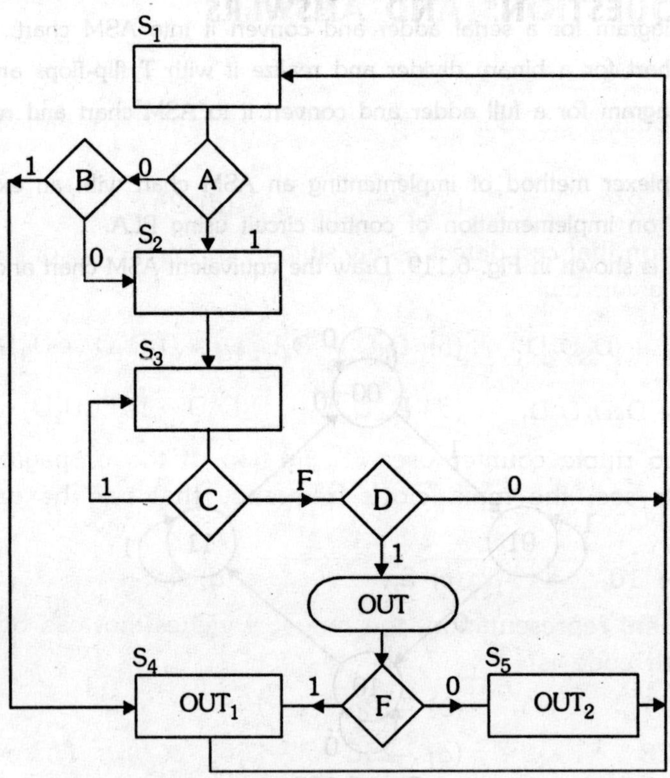

Fig. 6.121

1

Annexure

PREVIOUS 'GATE' QUESTIONS AND ANSWERS

GATE-1990

1. The number of boolean functions that can be generated by n-variables, is equal to

   (*a*) 2^n (*b*) 2^{2n} (*c*) 2^{n-1} (*d*) $2n$ **[Ans.** (*a*)]

2. The minimal function that can detect a "divisible by 3" BCD code digit (represented by $D_8D_4D_2D_1$) is given by

   (*a*) $D_8D_1 + D_4D_2 + D_8D_2D_1$ (*b*) $D_8D_1 + \overline{D}_4D_2D_1 + D_4D_2\overline{D}_1 + \overline{D}_8\overline{D}_4\overline{D}_2\overline{D}_1$

   (*c*) $D_4D_1 + D_4D_2 + D_8\overline{D}_4\overline{D}_2D_1$ (*d*) $D_4D_2\overline{D}_1 + \overline{D}_4D_2D_1 + D_8D_4D_2D_1$ **[Ans.** (*b*)]

3. A 4-bit modulo-16 ripple counter uses JK flip-flop. If the propagation delay of each FF is 50 m sec, the max. clock frequency that can be used is equal to—MHz

   (*a*) 20 (*b*) 10 (*c*) 5 (*d*) 4 **[Ans.** (*c*)]

4. With 2's complement representation, the range of values that can be represented for an 8-bit length code is

   (*a*) –128 to +127 (*b*) –128 to +128

   (*c*) –127 to +128 (*d*) –256 to +256 **[Ans.** (*a*)]

GATE-1991

1. Two-dimensional addressing of 256×8-bit ROM using 8 to 1 selectors requires _______ NAND gates. **[Ans.** 104]

2. An SR FF can be converted into a TFF by connecting _______ E_0Q and _______ $E_0\,\overline{Q}$. **[Ans.** R and S]

3. A sequential MUX is connected as shown in Fig. 1.1 Each time the multiplexer receives clock, it switches to the next channel (from 6 it goes to 1). If input signals are

$A = 5 \cos 2\pi\ (4 \times 10^3 t)$

$B = 2 \cos 2\pi\ (3.8\ Kt)$

$C = 6 \cos 2\pi\ (2.2\ Kt)$

and $D = 4 \cos 2\pi\ (1.7\ Kt)$

The minimum clock frequency should be ______ kHz. **[Ans.** 6**]**

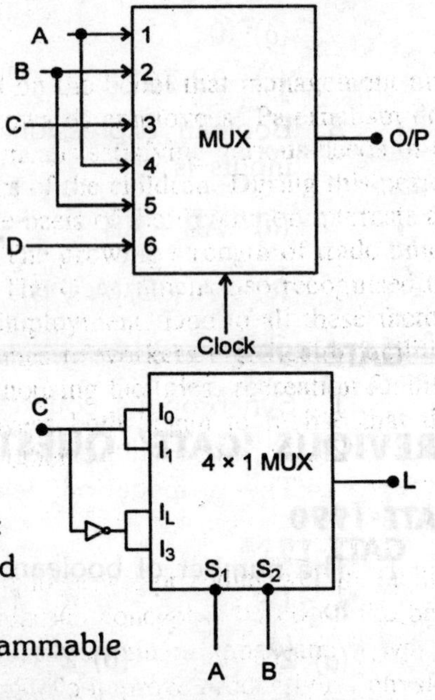

Fig. 1.1

GATE-1992

1. The logic realized by the circuit shown below:

(*a*) Y = AC (*b*) $Y = A \oplus C$

(*c*) Y = *b.c* (*d*) Y = B + C

[Ans. (*b*)**]**

2. Choose the correct statement from the following:

(*a*) PROM contains programmable AND array and a fixed OR array.

(*b*) PLA contains a fixed AND array and a programmable OR array.

(*c*) PROM contains fixed AND array and a programmable OR array.

(*d*) PLA contains a programmable AND array and a programmable OR array.

[Ans. (*c*) and (*d*)**]**

3. The initial controls of 40 bit serial-in parallel-out, righ shift shift register is shown in figure is 0110. After three clock pulses are applied, the controls of the right register will be

(*a*) 0000 (*b*) 0101

(*c*) 1010 (*d*) 1111 **[Ans.** (*c*)**]**

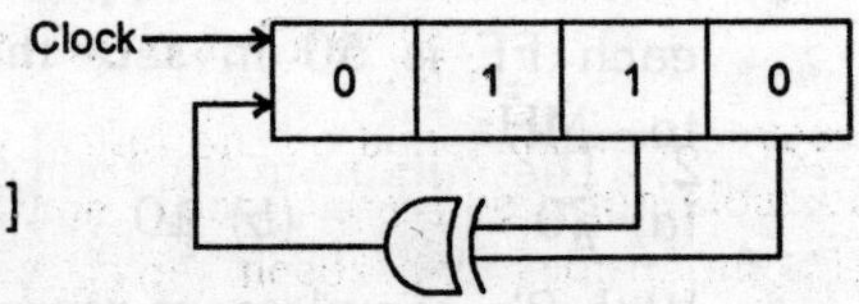

Fig. 1.2

GATE-1993

1. For the logic circuit shown in figure the out put Y is equal to

(*a*) $\overline{ABC}$ (*b*) $\overline{A} + \overline{B} + \overline{C}$

(*c*) $\overline{AB} + \overline{BC} + \overline{A} + \overline{C}$ (*d*) 1 **[Ans.** (*b*)**]**

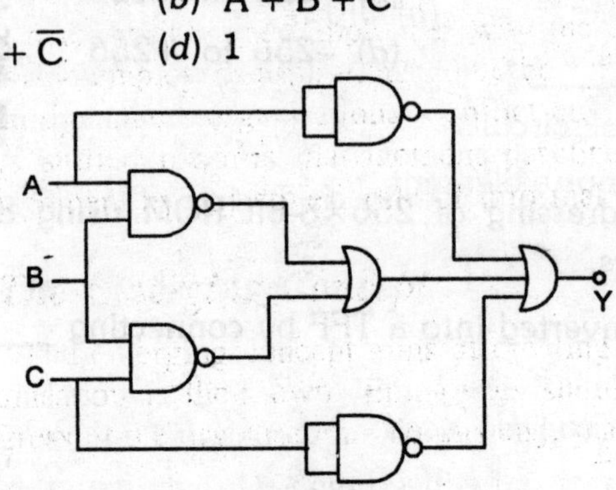

Fig. 1.3

2. The 2's complement representation of a 16-bit number (one sign bit and 15 magnitude bits) is FFFFH, the magnitude in decimal represents

   (*a*) 0 (*b*) 1

   (*c*) 32,767 (*d*) 65,535 **[Ans.** (*c*)**]**

3. Boolean expression for the output of XNOR (equivalent) gate with A and B inputs is

   (*a*) $A\overline{B} + \overline{A}B$ (*b*) $\overline{A}\overline{B} + AB$

   (*c*) $(\overline{A} + B)(A + \overline{B})$ (*d*) $(\overline{A} + \overline{B})(A + B)$ **[Ans.** (*b*)**]**

GATE-1994

1. Synchronous counters are—than the ripple counters. **[Ans.** Faster**]**
2. A ring oscillator consisting of 5 inverters is running at a frequency of 1.0 MHz. The propagation delay per gate is ________ n sec. **[Ans.** 200 n sec**]**

GATE-1995

1. The output of the circuit shown below is equal to

   (*a*) 0 (*b*) 1

   (*c*) $\overline{A}B + A\overline{B}$ (*d*) $\overline{\overline{A}.\overline{B}}.\overline{(\overline{A}+\overline{B})}$ **[Ans.** (*b*)**]**

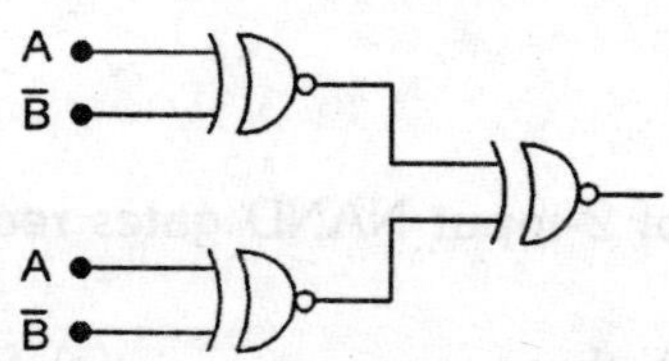

Fig. 1.4

2. The minimum number of NAND gates required to implement the Boolean function $A + A\overline{B} + A\overline{B}C$ is

   (*a*) 0 (*b*) 1 (*c*) 4 (*d*) 7 **[Ans.** (*a*)**]**

3. A switch-tailring counter is made by using a single DFF. The resulting circuit is

   (*a*) SRFF (*b*) JKFF (*c*) DFF (*d*) TFF **[Ans.** (*d*)**]**

4. RS latch is a ______

   (*a*) combinational circuit (*b*) synchronous sequential circuit

   (*c*) one bit memory element (*d*) one clock delay element.

   [Ans. (*c*)**]**

GATE-1996

1. Match the following

   (*a*) Shift register can be used (*i*) for code conversions

   (*b*) A multiplexer, can be used (*ii*) to generate memory chip select

(c) A decoder can be used (iii) for parallel to serial conversion
(iv) as a many-to-one switch
(v) for analog-to-digital conversion

Ans. (a) — (iii)
(b) — (iv)
(c) — (ii)

GATE-1997

1. A 2-bit binary multiplier can be implemented using ______
(a) 2 inputs AND gate only
(b) 2 input XOR gate and 4 input AND gate only
(c) two 2-input NOR and on XNOR gate
(d) XOR gates and shift register **[Ans. (a)]**

2. The out put of the logic gate is______
(a) 0 (b) 1
(c) A (d) F

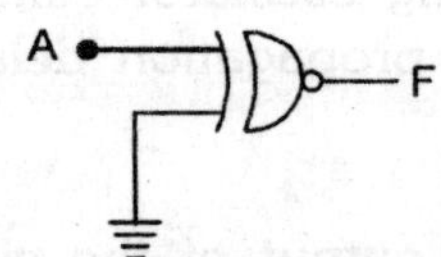

Fig. 1.5

3. The Boolean function A + BC is a reduced form of
(a) AB + BC (b) (A+B) (A+C) (c) $\overline{A}B + A\overline{B}C$ (d) (A+C)B
[Ans. (b)]

GATE-1998

1. The minimum number of 2-input NAND gates required to implement the Boolean function Z = $A\overline{B}C$ is
(a) two (b) three (c) five (d) six **[Ans. (c)]**

2. In the figure shown A = 1 and B = 1, the input B is now replaced by a sequence 101010...., the outputs x and y will be
(a) fixed at 0 and 1, respectively
(b) x = 1010... and y = 0101....
(c) x = 1010... and y = 1010....
(d) fixed at 1 and 0, respectively

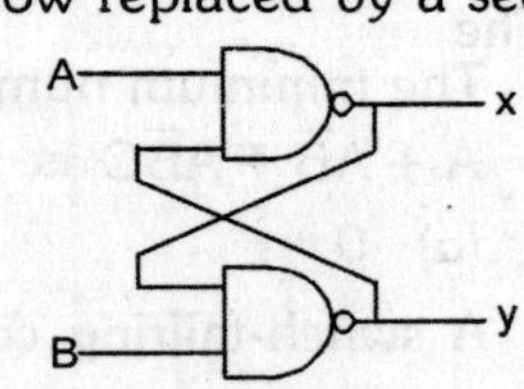

Fig. 1.6

3. An equivalent 2's complement representation of the 2's complement number 1101 bit form is
(a) 110100 (b) 001101 (c) 110111 (d) 111101
[Ans. (d)]

4. The 2's complement number having sign bits x and y are added and the sign bit of the result is Z. Then, the occurrence of overflow is indicated by the Boolean function
(a) xyz (b) $\overline{x}\,\overline{y}\,\overline{z}$ (c) $\overline{x}\,\overline{y}\,z + xy\overline{z}$ (d) xy + yz + zx
[Ans. (d)]

5. The identity $AB + \overline{A}C + BC = AB + \overline{A}C$ the duel form is

(*a*) $(A + B)(\overline{A} + C)(B + C) = (A + B)(\overline{A} + C)$

(*b*) $(A + B)(A + \overline{C})(\overline{B} + \overline{C}) = (A + B)(\overline{A} + \overline{C})$

(*c*) $(\overline{A} + \overline{B})(\overline{A} + C)(B + C) = (A + B)(\overline{A} + \overline{C})$

(*d*) $\overline{A}\,\overline{B} + A\overline{C} + \overline{B}\,\overline{C} = AB + \overline{A}C$ **[Ans.** (*a*)**]**

6. The K-map of the Boolean function is shown in Fig. 1.7. The number of essential prime implements for this function is

(*a*) 4 (*b*) 5

(*c*) 6 (*d*) 8 **[Ans.** (*a*)**]**

CD \ AB	00	01	11	10
00	1	1	0	1
01	0	0	0	1
11	1	0	0	0
10	1	0	0	1

Fig. 1.7

GATE-1999

1. The logical expression $y = A + \overline{A}B$ is equivalent to

(*a*) $y = AB$ (*b*) $y = \overline{A}B$

(*c*) $y = \overline{A} + B$ (*d*) $y = A + B$ **[Ans.** (*d*)**]**

2. The minimized form off the logical expression $\overline{ABC} + \overline{ABC} + \overline{A}BC + AB\overline{C}$ is

(*a*) $\overline{A}\,\overline{C} + B\overline{C} + \overline{A}B$ (*b*) $A\overline{C} + \overline{B}C + \overline{A}B$

(*c*) $\overline{A}C + \overline{B}C + \overline{A}B$ (*d*) $A\overline{C} + \overline{B}C + A\overline{B}$ **[Ans.** (*a*)**]**

3. For binary half-subtractor having two inputs A and B, the correct set of logical expression for the outputs D (A – B) and X (borrow) are

(*a*) $D = AB + \overline{A}\,\overline{B},\ X = \overline{A}B$ (*b*) $D = \overline{A}B + A\overline{B},\ X = A\overline{B}$

(*c*) $D = \overline{A}B + A\overline{B},\ X = \overline{A}B$ (*d*) $D = AB + \overline{A}\,\overline{B},\ X = A\overline{B}$ **[Ans.** (*c*)**]**

4. The ripple counter shown in figure works as a

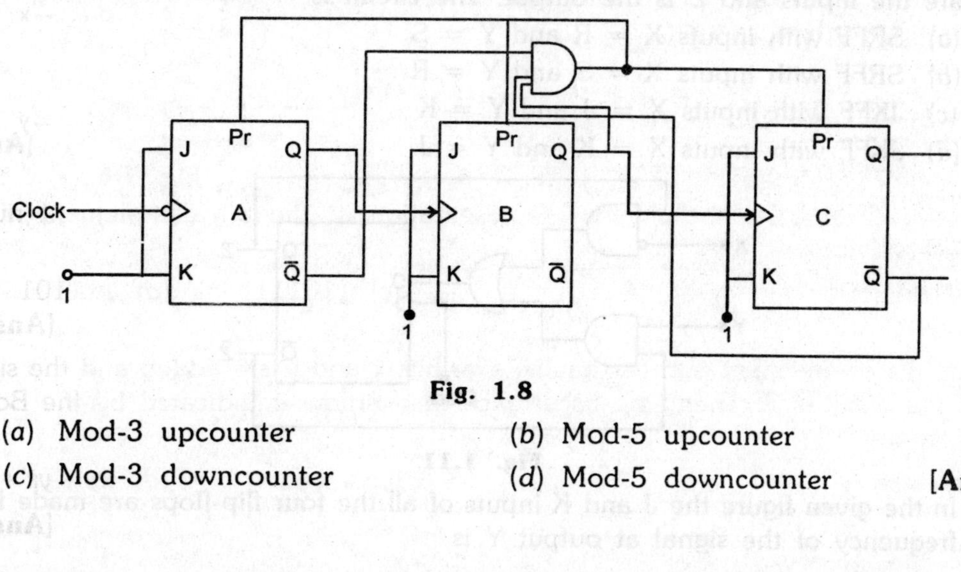

Fig. 1.8

(*a*) Mod-3 upcounter (*b*) Mod-5 upcounter

(*c*) Mod-3 downcounter (*d*) Mod-5 downcounter **[Ans.** (*d*)**]**

GATE-2000

1. For the logic circuit shown in figure the required input condition (A, B, C) to make the output X = 1 is

(*a*) 1, 0, 1 (*b*) 0, 0, 1 (*c*) 1, 1, 1 (*d*) 0, 1, 1 [**Ans.** (*d*)]

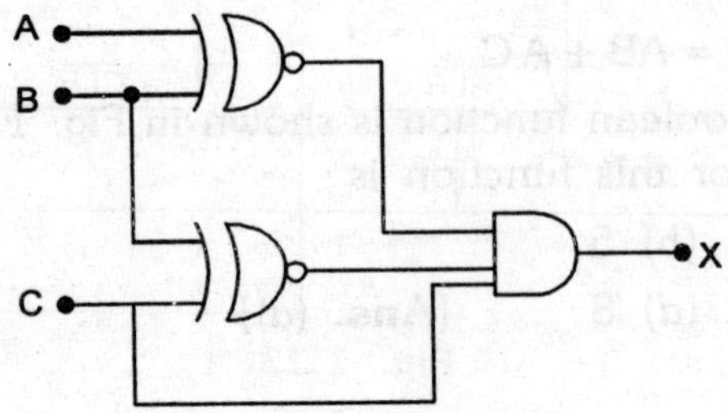

Fig. 1.9

2. For the logic circuit shown in figure the simplified Boolean expression for the output Y is

(*a*) A + B + C (*b*) A (*c*) B (*d*) C [**Ans.** (*c*)]

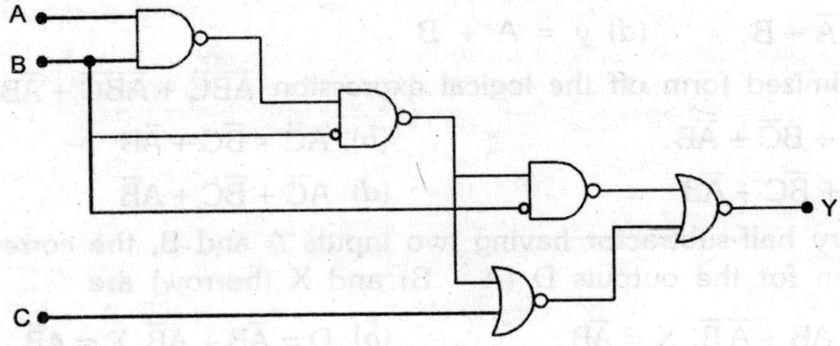

Fig. 1.10

3. A sequential circuit using DFF and logic gates is shown in figure where X and Y are the inputs and Z is the output. The circuit is

(*a*) SRFF with inputs X = R and Y = S

(*b*) SRFF with inputs X = S and Y = R

(*c*) JKFF with inputs X = J and Y = K

(*d*) JKFF with inputs X = K and Y = J [**Ans.** (*d*)]

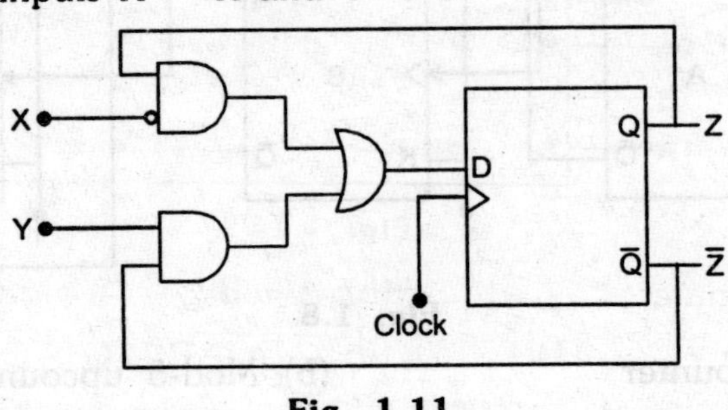

Fig. 1.11

4. In the given figure the J and K inputs of all the four flip-flops are made high the frequency of the signal at output Y is

(*a*) 0.833 kHz (*b*) 1.0 kHz (*c*) 0.91 kHz (*d*) 0.77 kHz

[Ans. (*a*)**]**

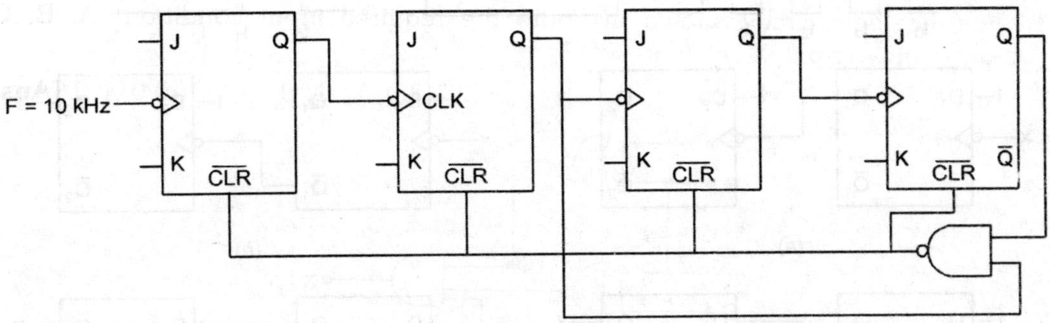

Fig. 1.12

GATE-2001

1. The 2's complement representation of –17 is

(*a*) 01110 (*b*) 01111 (*c*) 11110 (*d*) 10001

[Ans. (*b*)**]**

2. For the ring oscillator shown in figure the propagation delay of each inverter is 100 pico sec. What is the fundamental frequency of the oscillator output

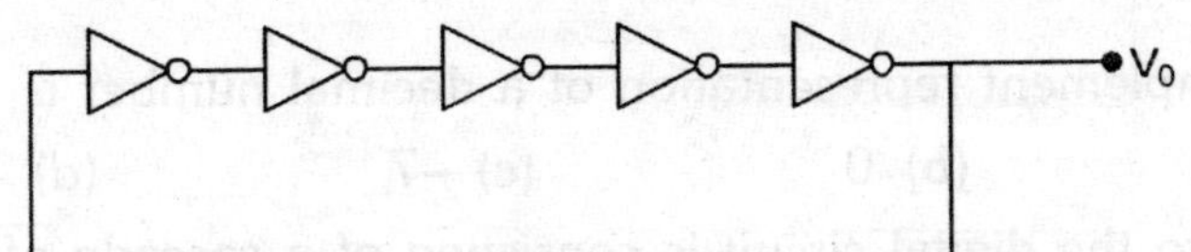

Fig. 1.13

(*a*) 10 MHz (*b*) 100 MHz (*c*) 1 GHz (*d*) 2 GHz

[Ans. (*c*)**]**

3. In the TTL circuit shown in figure S_2 to S_0 are select lines and x_7 to X_0 are input lines, S_0 and X_0 are LSB's. The output Y is equal to

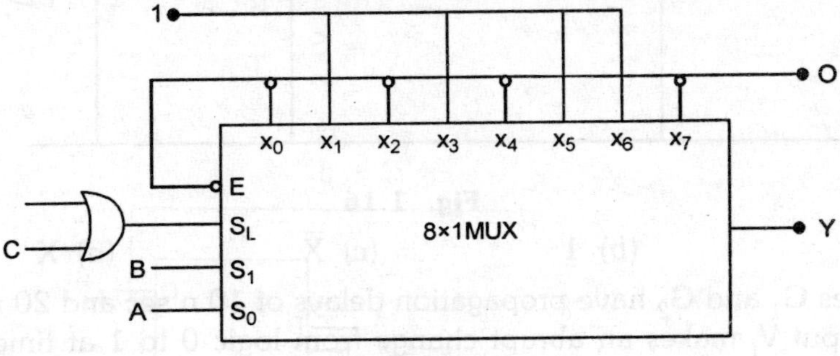

Fig. 1.14

(*a*) Indeterminate (*b*) $A \oplus B$

(*c*) $\overline{A \oplus B}$ (*d*) $\overline{C}\,(\overline{A \oplus B}) + C\,(A \oplus B)$ **[Ans.** (*d*)**]**

4. The digital block shown in figure is required using two positive edge triggered DFFs. Assume that for $t < t_0$, $Q_1 = Q_2 = 0$. The circuit in the digital block is given by:

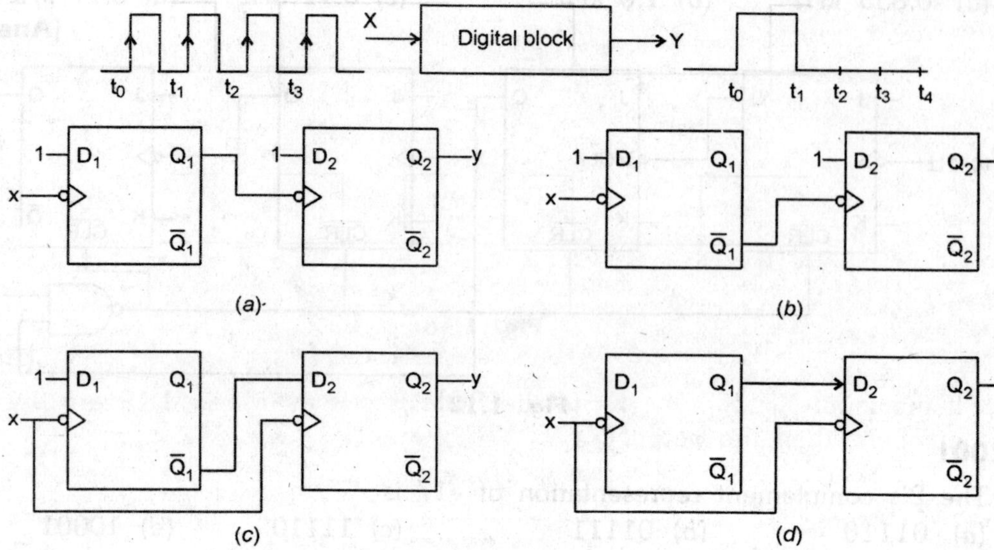

Fig. 1.15

[Ans. (*c*)**]**

GATE-2002

1. 4-bit 2's complement representation of a decimal number is 1000. The number is
   (*a*) +8 (*b*) 0 (*c*) –7 (*d*) –8
2. If the input to the digital circuit is consisting of a cascade of 20XOR–gates is 'X', then the output equal to

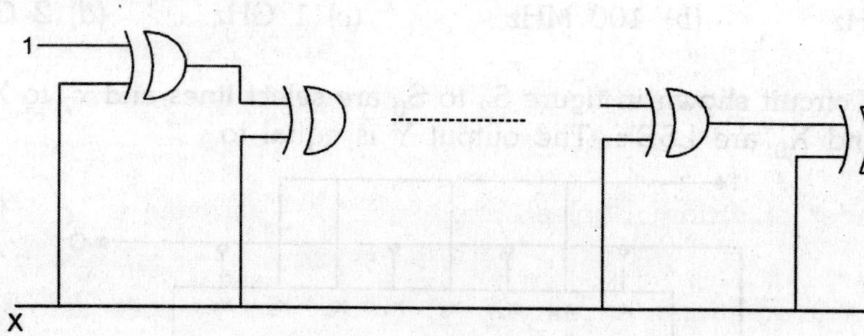

Fig. 1.16

(*a*) 0 (*b*) 1 (*c*) $\overline{X}$ (*d*) X **[Ans.** (*b*)**]**

3. The gates G_1 and G_2 have propagation delays of 10 n sec and 20 n sec respectively. If the input V_i makes an abrupt change from logic 0 to 1 at time $t = t_0$, then the output wave form is

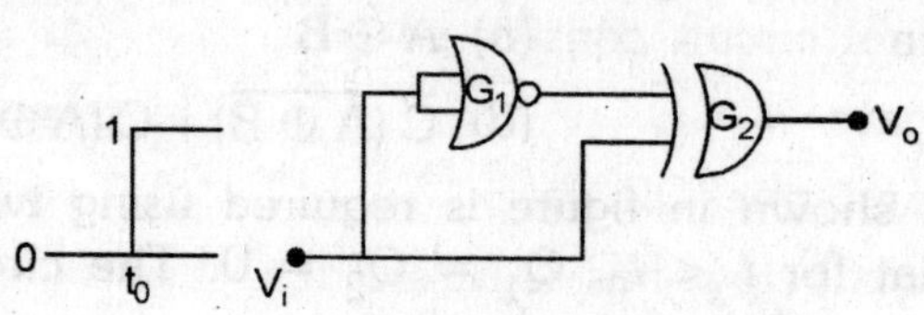

Fig. 1.17

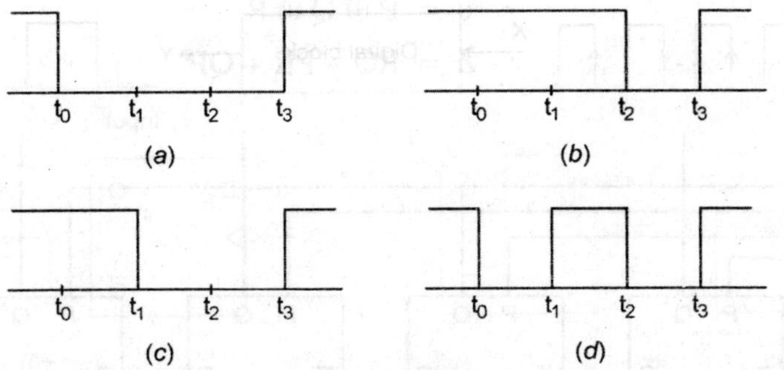

Fig. 1.18

[Ans. (*b*)**]**

4. If the input X_3, X_2, X_1, X_0 to the ROM is shown in Fig. 1.19 are 8-4-1 BCD numbers, then the output $Y_3\ Y_2\ Y_1\ Y_0$ are

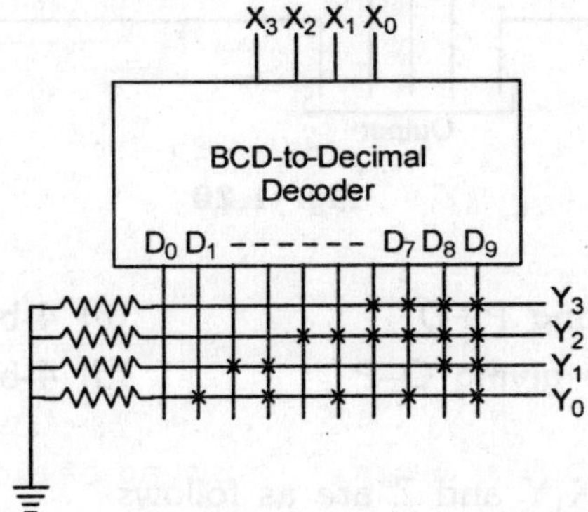

Fig. 1.19

(*a*) gray code numbers (*b*) 2-4-2-1 BCD numbers
(*c*) Ex-3 code numbers (*d*) none of the above **[Ans.** (*b*)**]**

GATE-2003

1. The number of distant Boolean expression of 4 variable is
(*a*) 16 (*b*) 256 (*c*) 1024 (*d*) 65536 **[Ans.** (*d*)**]**

2. Without any additional circuitry, an 8×1MUX can be used to obtain
(*a*) Some but not all Boolean functions of 3 variables
(*b*) All functions of 3 variable but none of 4 variables
(*c*) All functions of 3-variables and some but not all of 4-variables
(*d*) All the functions of 4-variables. **[And.** (*c*)**]**

3. A 0 to 6 counter consisting of 3FFs and a combinational circuit of 2 input gate(s). The combinational circuit consists of
(*a*) one AND gate (*b*) one OR gate
(*c*) one AND and one OR gate (*d*) Two AND gates. **[Ans.** (*a*)**]**

4. The circuit shown below has 4 boxes described by inputs P, Q, R, outputs y, z with

$$y = P \oplus Q \oplus R$$

$$Z = RQ + \bar{P}R + Q\bar{P}$$

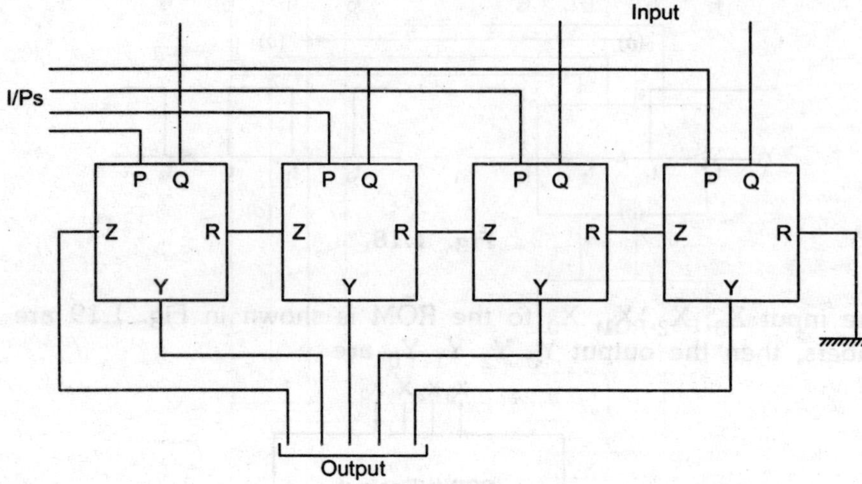

Fig. 1.20

The circuit act as a

(a) 4-bit adder giving P+Q (b) 4-bit subtractor giving P–Q

(c) 4-bit subtractor giving Q–P (d) 4-bit adder giving P+Q+R

[Ans. (b)**]**

5. If the functions W_1X_1Y and Z are as follows

$$W = R + \bar{P}Q + \bar{R}S$$

$$X = PQ\bar{R}\bar{S} + \bar{P}\bar{Q}\bar{R}\bar{S} + P\bar{Q}\bar{R}\bar{S}$$

$$Y = RS + \overline{PR + P\bar{Q} + \bar{P}.\bar{Q}}$$

$$Z = R + S + \overline{PQ + \bar{P}\bar{Q}\bar{R} + P\bar{Q}\bar{S}}$$

Then

(a) $W = Z, X = \bar{Z}$ (b) $W = Z, X = Y$

(c) $W = Y$ (d) $W = Y = \bar{Z}$ **[Ans.** (a)**]**

6. A 4-bit ripple counter and a 4-bit synchronous counter are made using flip-flops having a propagation delay of 10 n sec each. If the worst case delay in the ripple counter and the synchronous counter be R and S respectively, then

(a) R = 10 n sec, S = 40 n sec (b) R = 40 n sec, S = 10 n sec

(c) R = 10 n sec, S = 30 n sec (d) R = 30 n sec, S = 10 n sec

[Ans. (b)**]**

7. In the circuit shown below, A is a parallel-in, parallel-out 4-bit register, which loads at the rising edge of the clock C. The input lines are connected to a 4-bit bus, W. Its output acts as the input to a 16 × 4 ROM whose output is floating when the enable input E is 0. A partial table of the contents of the ROM is as follows

Address	0	2	4	6	8	10	11	14
Data	0011	1111	0100	1010	1011	1000	0010	1000

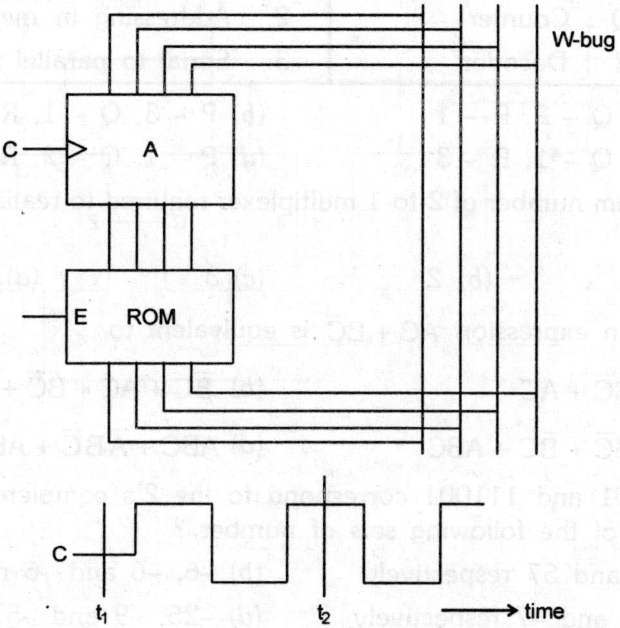

Fig. 1.21

The clock to the register is shown, and the data on the W bus at time t_1 is 0110. The data on the bus at t_2 is

(*a*) 1111 (*b*) 1011 (*c*) 1000 (*d*) 0010 [**Ans.** (*c*)]

GATE-2004

1. A MSFF has the characteristic that

   (*a*) change in the input immediately reflected in the output.

   (*b*) change in the output occurs when the state of the master is affected.

   (*c*) change in the output occurs when the state of the slave is affected.

   (*d*) both master and slake states are affected at the same time. [**Ans.** (*c*)]

2. The range of the signed decimal numbers that can be represented by 6-bit 1's complement number is

   (*a*) –31 to +32 (*b*) –63 to +64 (*c*) –64 to +63 (*d*) –32 to +31 [**Ans.** (*a*)]

3. A digital system is required to amplify a binary encoded audio signal. The user should be able to control the gain of the amplifier from a min. to a max in 100 increments. The min. number of bits required to encode, in straight binary is

   (*a*) 8 (*b*) 6 (*c*) 5 (*d*) 7 [**Ans.** (*d*)]

4. Choose the correct one from the among the alternatives A, B, C, D after matching an item from Group 1 with most appropriate item in Group 2.

Group 1	Group 2
P : Shift register	1. Frequency division
Q : Counter	2. Addressing in memory
R : Decoder	3. Serial to parallel to data

(*a*) P – 3, Q – 2, R – 1 (*b*) P – 3, Q – 1, R – 2
(*c*) P – 2, Q – 1, R – 3 (*d*) P – 1, Q – 2, R – 3 **[Ans.** (*b*)**]**

5. The minimum number of 2-to-1 multiplexer required to realize a 4-to-1 multiplexer is

(*a*) 1 (*b*) 2 (*c*) 3 (*d*) 4 **[Ans.** (*b*)**]**

6. The Boolean expression $AC + B\overline{C}$ is equivalent to

(*a*) $\overline{A}C + B\overline{C} + AC$ (*b*) $\overline{B}C + AC + B\overline{C} + \overline{A}C\overline{B}$
(*c*) $AC + B\overline{C} + \overline{B}C + ABC$ (*d*) $ABC + \overline{A}B\overline{C} + AB\overline{C} + AB\overline{C}$ **[Ans.** (*d*)**]**

7. 11001, 1001 and 111001 correspond to the 2's complement representation of which one of the following sets of number ?

(*a*) 25, 9 and 57 respectively (*b*) –6, –6 and –6 respectively
(*c*) –7, –7 and –7 respectively (*d*) –25, –9 and –57 respectively

[Ans. (*c*)**]**

8. In the modulo-6 ripple counter shown in Fig. 1.22 the output of the 2-input gate is used to clear the JK flip-flop

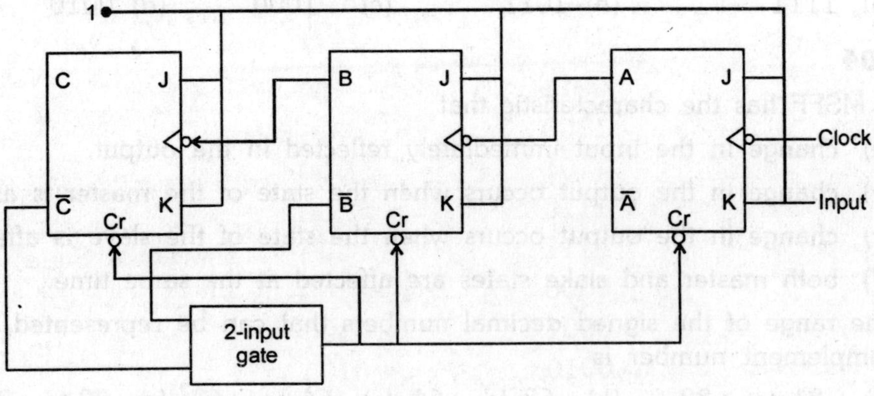

Fig. 1.22

The 2-input gate is ______ gate

(*a*) NAND (*b*) NOR
(*c*) OR (*d*) AND **[Ans.** (*c*)**]**

2

Annexure

GATE-1990

Q.1. *Design a 4-bit synchronous counter using j-K flip-flips to count in excess-3 BCD code (0011, 0100, 0101, 0110, 0111, 1000, 1001, 1010, 1011, 1100). Label the counter state Q_D Q_C Q_B Q_A with Q_A = LSB.*

Ans. Count excess – 3

0011	0100	
0100	0101	
0101	0110	Wanted State
0110	0111	
0111	1000	
1000	1001	
1001	1010	
1010	1011	
0011	1100	
1100	0011	
1101	0100	
1110	0100	
1111	0100	Unwanted State
0000	0100	
0001	0100	
0010	0100	

For J-K flip-flop transition table is

$Q(t)$	$Q(t+1)$	J	K
0	0	0	×
0	1	1	×
1	0	×	1
1	1	×	0

$Q_BQ_A \backslash Q_DQ_C$	00	01	11	10
00	0	1	1	1
01	×	×	×	×
11	×	×	×	×
10	0	1	×	1

J_0

$Q_BQ_C \backslash Q_DQ_A$	00	01	11	10
00	×	×	×	×
01	1	1	×	1
11	1	1	×	1
10	×	×	×	×

K_0

$Q_BQ_A \backslash Q_DQ_C$	00	01	11	10
00	0	0	1	0
01	0	1	×	1
11	×	×	×	×
10	1	×	×	×

J_1

$Q_BQ_A \backslash Q_DQ_C$	00	01	11	10
00	×	×	×	×
01	×	×	×	×
11	1	1	×	1
10	1	0	×	0

K_1

$Q_BQ_A \backslash Q_DQ_C$	00	01	11	10
00	1	×	×	0
01	0	×	×	0
11	1	×	×	1
10	1	×	×	0

J_2

$Q_BQ_A \backslash Q_DQ_C$	00	01	11	10
00	×	0	1	×
01	×	0	×	×
11	×	1	×	1
10	×	0	×	0

K_2

$Q_BQ_A \backslash Q_DQ_C$	00	01	11	10
00	0	0	×	×
01	0	0	×	×
11	0	1	×	×
10	0	0	×	×

J_3

$Q_BQ_A \backslash Q_DQ_C$	00	01	11	10
00	×	×	1	0
01	×	×	×	0
11	×	×	×	0
10	×	×	×	0

K_3

Fig. 2.1

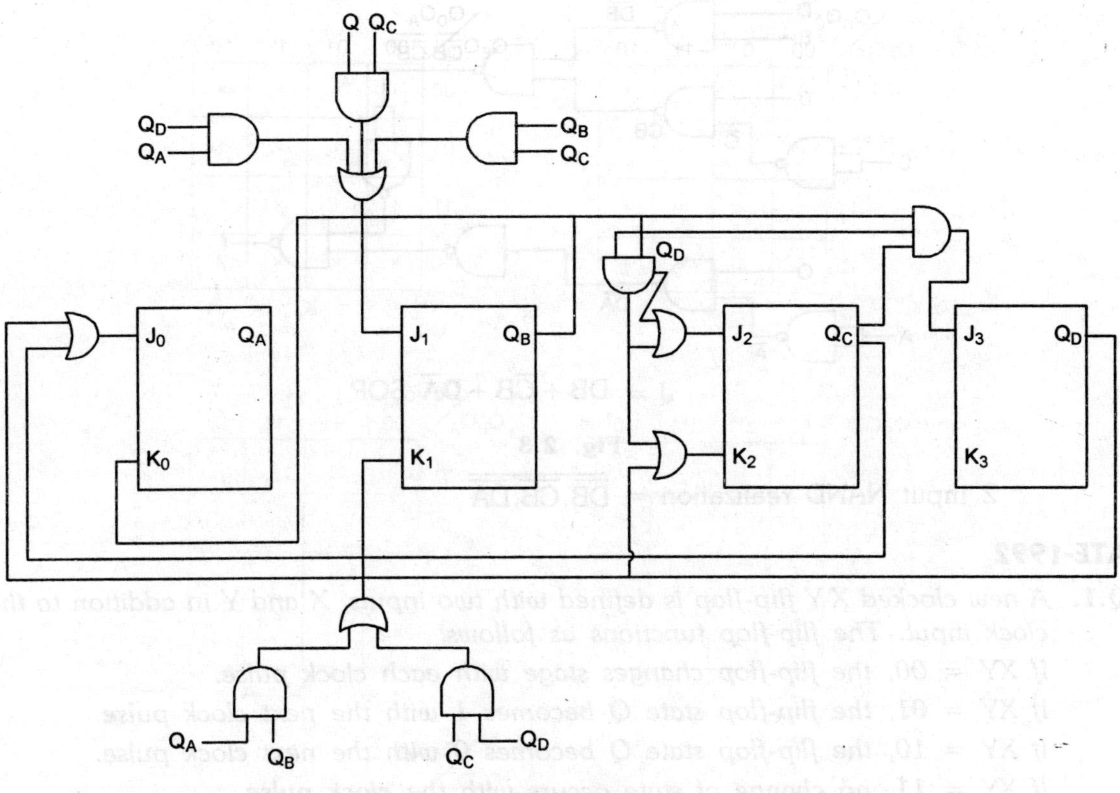

Fig. 2.2

$$J_0 = Q_C + Q_D; \qquad K_0 = 1$$
$$J_1 = Q_B + Q_CQ_A + Q_DQ_A + Q_BQ_A; \quad K_1 = Q_A + \overline{Q_D}\,\overline{Q_C}$$
$$J_2 = \overline{Q_A}\,\overline{Q_D} + Q_BQ_A; \qquad K_2 = Q_C + Q_BQ_A$$
$$J_3 = Q_BQ_AQ_C; \qquad K_3 = Q_C.$$

GATE-1991

Q.1. *The four variable function f is given in terms of minterms as:*

f(A, B, C, D) = Σm (2, 3, 8, 10, 11, 12, 14, 15)

Using the K-map minimize the function in the sum of products form. Also, give the realization using only two–input NAND gates.

Ans. f(A, B, C, D) = Σm (2, 3, 8, 10, 11, 12, 14, 15)

DC \ BA	00	01	11	10
00	(0) 0	(1) 0	(3) 1	(2) 1
01	(4) 0	(5) 0	(7) 0	(6) 0
11	(12) 1	(13) 0	(15) 1	(14) 1
10	(8) 1	(9) 0	(11) 1	(10) 1

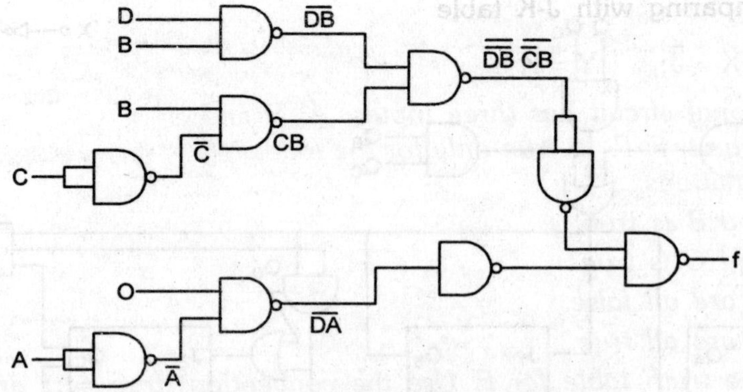

$$J = DB + \overline{C}B + D\overline{A} \quad SOP$$

Fig. 2.3

2 input NAND realization $= \overline{\overline{DB}.\overline{CB}.\overline{DA}}$

GATE-1992

Q.1. *A new clocked X-Y flip-flop is defined with two inputs, X and Y in addition to the clock input. The flip-flop functions as follows:*

If XY = 00, the flip-flop changes stage with each clock pulse.

If XY = 01, the flip-flop state Q becomes 1 with the next clock pulse.

If XY = 10, the flip-flop state Q becomes 0 with the next clock pulse.

If XY = 11, no change of state occurs with the clock pulse.

(a) Write the Truth Table for the X-Y flip-flop.

(b) Write the excitation table for the X-Y flip-flop.

(c) It is desired to convert a J-K flip-flop into the X-Y flip-flop by adding some external gates, if necessary. Draw a circuit to show how you will implement the X-Y flip-flop using a J-K flip-flop.

Ans. O.P. Implementation 2 Input NAND

$$F = \overline{A} + BC = \overline{AB.BC}$$

Clocked X-Y flip-flop

(*a*) CLK

X	Y	Q_{n+1}
0	0	$\overline{Q}_n$
0	1	1
1	0	0
1	1	Q_n

Fig. 2.4

(*b*) Excitation table for the X-Y flip-flop.

Q	Q_{n+1}	X	Y
0	0	1	×
0	1	0	×
1	0	×	0
1	1	×	1

(c) On comparing with J-K table

$$X = \bar{J}; \qquad Y = \bar{K}$$

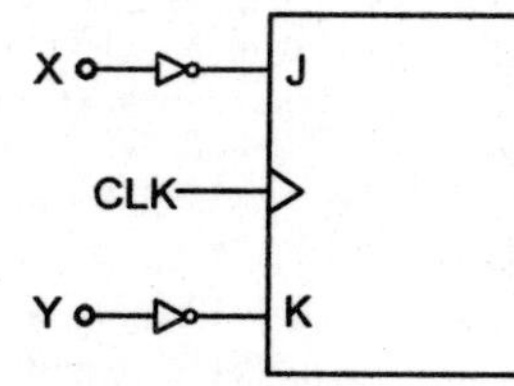

Fig. 2.5

Q.2. *A combinational circuit has three inputs, A, B and C and an output, F. F. is true only for the following input combinations:*
A is false and B is true.
A is false and C is true.
A, B and C are all false.
A, B and C are all true.
(a) Write the truth table for F. Use the convention, true = 1 and false = 0.
(b) Write the simplified expression for F as a sum of products.
(c) Write the simplified expression for F as a product of sums.
(d) Draw a logic circuits implementation of F using the minimum number of 2-input NAND gates only.

Ans.

	Inputs			*Output*
	A	*B*	*C*	*F*
1	0	0	0	1
2	0	0	1	1
3	0	1	0	1
4	0	1	1	1
5	1	0	0	0
6	1	0	1	0
7	1	1	0	0
8	1	1	1	1

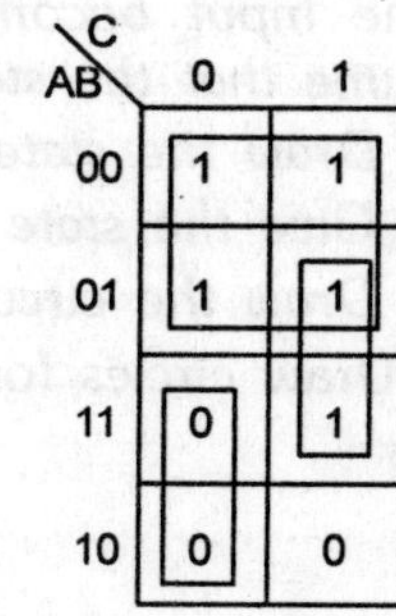

Fig. 2.6

GATE-1993

Q.1. *Signals A, B, C, D and D are available. Using a single 8 to 1 multiplexer and no other gate, implement the Boolean function* $f(A, B, C, D) = B.C + A.B.\bar{D} + \bar{A}.\bar{C}.D$

Ans. Signals A, B, C and D are available. A, B, C will be fed to the select lines S_2, S_1, S_0. A MSB and LSB, D will be the input to multiplexer I_0 to I_7.

$$I(A, B, C, D) = \bar{A}.B.C.\bar{D} + A.B.\bar{C}.\bar{D} + \bar{A}.\bar{B}.\bar{C}.D$$

0 1 1 0   1 1 0 0   0 0 0 1

(6)   (12)   (1)

	I_0	I_1	I_2	I_3	I_4	I_5	I_6	I_7
$\bar{D}$	0	2	4	(6)	8	10	(12)	14
D	(1)	3	5	7	9	11	13	15
	D	0	0	$\bar{D}$	0	0	$\bar{D}$	0

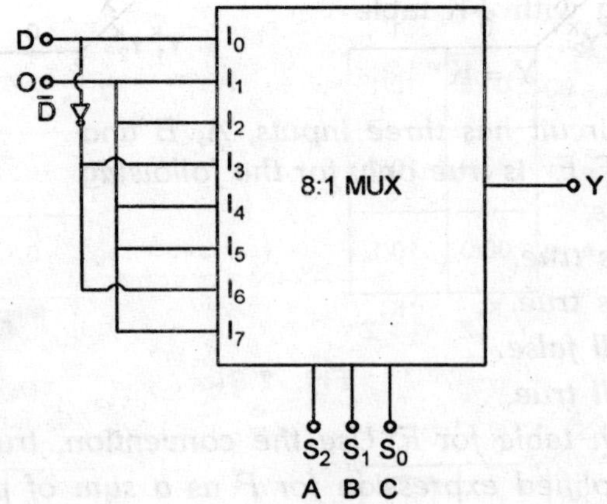

Fig. 2.7

Q.2. *A clocked sequential circuit has three states, A, B and C and one input X. As long as the input x is 0, the circuit alternates between the states A and B. If the input x becomes 1 (either in state A (or) in state B), the circuit goes to state C and remains in state C as long as X continues to be 1. The circuit returns to state A if the input becomes 0 once again and from then on repeats its behaviour. Assume that the state assignments are A = 00, B = 01, and C = 10.*

(a) Draw the state diagram of the circuit.

(b) Give the state table for the circuit.

(c) Draw the circuit using D-flip-flops.

Ans. (*a*) Draw circles for the 3 states A, B, C and write state assignments.

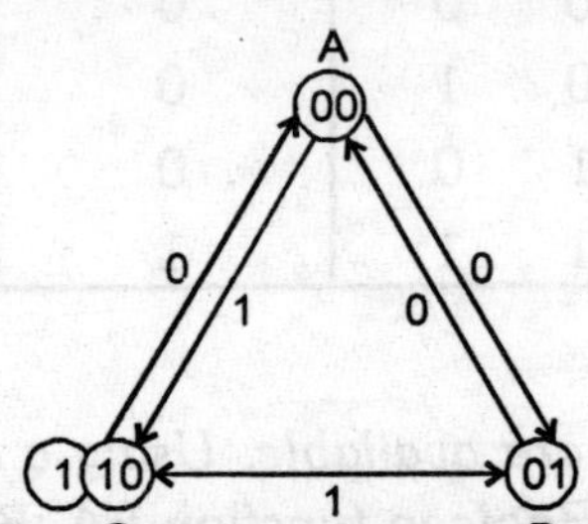

Fig. 2.8

Draw the directed lines to indicate the transition and mark the input that causes the change on the line.

(*b*) State Table

PS \ X	0	1
A	B/0	C/1
B	A/0	C/1
C	A/0	C/1

NS/Z

X–Input
PS–Present table
NS–Next State
Z–Output

Fig. 2.9

Given the state assignments A = 00, B = 01, C = 10.

$Y_1^K Y_2^K$ \ X	0	1
A = 00	01/0	10/1
B = 01	00/0	10/1
C = 10	00/0	10/1

$Y_1^{K_1} Z \quad Y_2^{K_1} Z$

$Y_1^K Y_2^K$ \ X	0	1
00	0	1
01	0	1
10	0	1

Map for Z

Fig. 2.9(*a*)

(*c*)

$Y_1 Y_2$ \ X	0	1
00	0	1
01	0	1
10	0	1

Map for D_1

$Y_1 Y_2$ \ X	0	1
00	1	0
01	0	0
10	0	0

Map for D_2

Fig. 2.10

GATE-1994

Q.1. *For the digital circuit shown in the Fig. 2.11. Explain what happens at the nodes, N_1, N_2, F and F, when*

(a) C_k = 1 and 'A' changes from '0' to '1'.

(b) A = 1 and 'C_k' changes from '1' to '0'.

(c) C_k = 0 and 'A' changes from '1' to '0'.

(d) Initially, C_k = 0 and 'A' changes from '0' to '1' and then C_k goes to '1'.

(e) What is the circuit performing ?

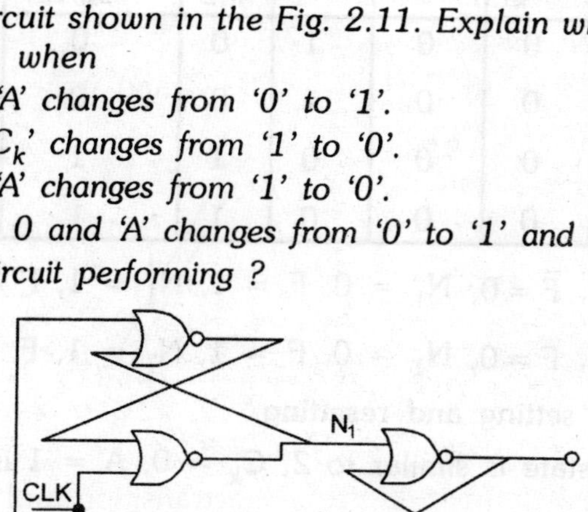

Fig. 2.11

Ans. (*a*) NOR Gates G_1 to G_4 are from bottom to top.

$N_{1(n)}$	$N_{2(n)}$	C_k	A	G_1	G_2	G_3	G_4	$N_{2(n+1)}$	$N_{1(n+1)}$
0	0	1	1	0	0	0	1	0	0
1	0	1	1	0	0	0	0	0	0
0	1	1	1	0	0	0	1	0	0
1	1	1	1	0	0	0	0	0	0

Initially if $N_1.N_2$ are 0, if remains at 0. If 1, they 90 to 0.

As $N_1.N_2$ are at 0. Initially if F is 1, $\overline{F}$ is 0 (or) initially if F = 0, $\overline{F}$ is 1.

That is F, $\overline{F}$ do not change their initial states.

(b)

$N_{1(n)}$	$N_{2(n)}$	C_k	A	G_1	G_2	$N_{2(n+1)}$	G_3	G_4	$N_{1(n+1)}$
0	0	0	1	0	1	1	0	1	0
1	0	0	1	0	1	1	1	0	1
0	1	0	1	0	1	1	0	1	0
1	1	0	1	0	1	1	1	0	1

N_2 continues to be 1 whatever be its initial state.

N_1 remains at 0 if initially 0, and at 1 if initially 1.

If F = 0, $\overline{F} = 0$.

If N_1 = 0, F will become 1 and $\overline{F} = 0$.

If N_1 = 1, F will be 0, and $\overline{F}$ also 0, which is a prohibited state.

(c)

$N_{1(n)}$	$N_{2(n)}$	C_k	A	G_1	G_2	$N_{2(n+1)}$	G_3	G_4	$N_{1(n+1)}$
0	0	0	0	1	0	0	1	0	1
1	0	0	0	1	0	0	1	0	1
0	1	0	0	0	1	1	0	1	0
1	1	0	0	0	1	1	1	0	1

N_2 = 0, F = 1, $\overline{F} = 0$, N_1 = 0, F = 1, N_1 = 1, F = 0, $\overline{F} = 1$.

N_2 = 1, F = 1, $\overline{F} = 0$, N_1 = 0, F = 1, N_1 = 1, F = 0, $\overline{F} = 1$.

Here it toggles, setting and resetting.

(d) The change of state is similar to 2, C_k = 0, A = 1 initially and finally as at 1, C_k = 1, A = 1.

(e) The circuit is functioning as a SR latch.

GATE-1995

Q.1. *A 'code converter' is to be designed to convert from the BCD (5421) to the normal BCD (8421). The input BCD combinations for each digit the given below. A block diagram of the converter is shown in figure.*

(a) Draw K-maps for outputs, W, X, Y and Z.

(b) Obtain minimized expression for the outputs W, X, Y and Z.

Decimal	BCD (5421)			
	A	*B*	*C*	*D*
0	0	0	0	0
1	0	0	0	1
2	0	0	1	0
3	0	0	1	1
4	0	1	0	0
5	1	0	0	0
6	1	0	0	1
7	1	0	1	0
8	1	0	1	1
9	1	1	0	0

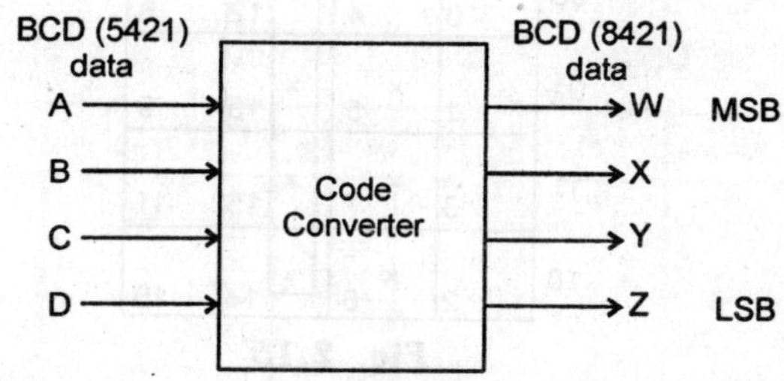

Fig. 2.12

Ans. Decimal

BCD 5421 | BCD

	A	*B*	*C*	*D*
0	0	0	0	0
1	0	0	0	1
2	0	0	1	0
3	0	0	1	1
4	0	1	0	0
5	1	0	0	0
6	1	0	0	1
7	1	0	1	0
8	1	0	1	1
9	1	1	0	0

	W	X	Y	Z
0	0	0	0	0
1	0	0	0	1
2	0	0	1	0
3	0	0	1	1
4	0	1	0	0
5	0	1	0	1
6	0	1	1	0
7	0	1	1	1
8	1	0	0	0
9	1	0	0	1

K map for Z | K map for Y

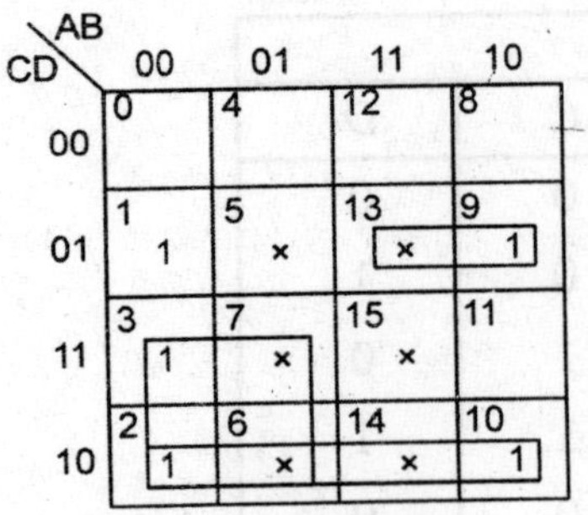

Fig. 2.13

$Z = A\overline{D} + \overline{A}D = A \oplus D$

5, 6, 7, 13, 14, 15,
are don't care conditions.
K-map for X.

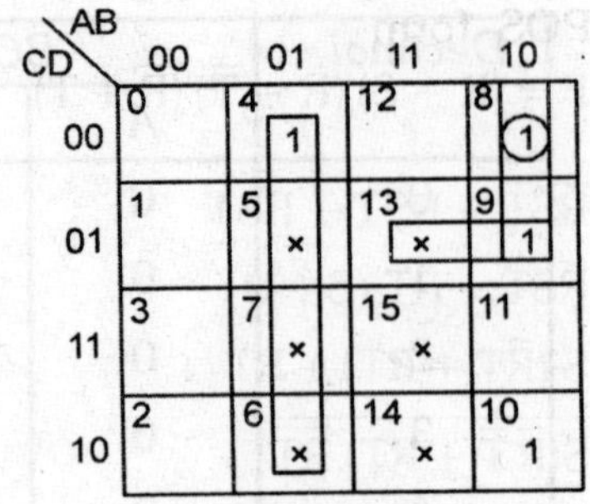

Fig. 2.14

$Y = \overline{A}C + C\overline{D} + A\overline{C}D$

$= \overline{C}\,(\overline{A}\,\overline{D}) + \overline{C}\,(AD) = C \oplus AD$

K-map for w.

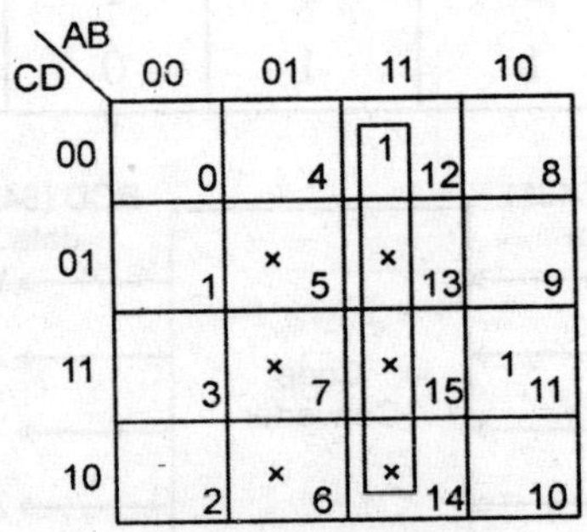

Fig. 2.15

$$X = \overline{A}B + A\overline{B}\,\overline{C} + A\overline{C}D + A\overline{B}\,\overline{D}$$

$$= \overline{A}B + A(\overline{B}\,\overline{C} + \overline{C}D + \overline{B}\,\overline{D})$$

$$W = AB + ACD = A\,(B + CD)$$

GATE-1996

Q.1. *Given the Boolean function F in the three variables R_1S_1 and T as*

$$F = \overline{R}S\overline{T} + R\overline{S}T + RST$$

(a) Express F in the minimum sum-of-products form.

(b) Express F in the minimum product-of-sums form.

(c) Assuming that both true and complement forms of the input variables are availables, draw a circuit to implement F using the minimum number of 2-input NAND gates only.

Ans. (a) SOP form

$$F = \overline{R}S\overline{T} + R\overline{S}T + RST = \overline{R}S\overline{T} + RT\,(S + \overline{S}) = \overline{R}S\overline{T} + RT$$

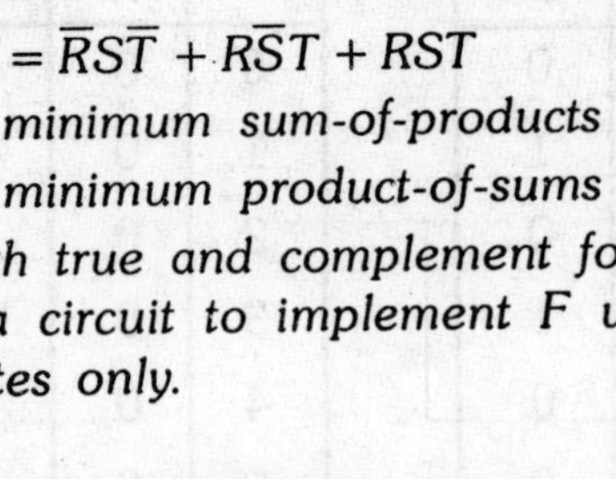

Fig. 2.16

(*b*) POS form

$$F = (R + S)(R + \bar{T})(\bar{R} + T)$$

(*c*)

$$\bar{R}S\bar{T} + R\bar{S}T + RST$$

$$\bar{R}S\bar{T} + RT\,(S+1)$$

$$S\,(\bar{R}T + RT) + RT$$

$$\overline{\overline{S(\bar{R}\,\bar{T} + RT).\overline{RT}}}$$

$$\overline{\left[\bar{S} + (\bar{R}\,\bar{T} + RT).\,\overline{RT}\right]}$$

$$\overline{[\bar{S} + (R+T)(\bar{R}+\bar{T})]\overline{RT}}$$

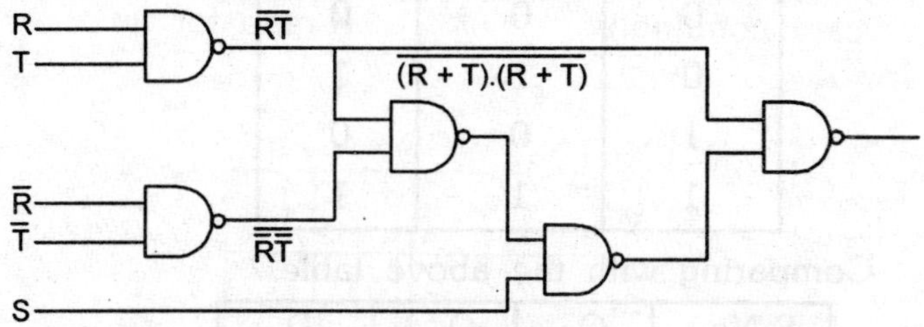

Fig. 2.17

Q.2. *A state machine is required to cycle through the following sequence of states.*

ABC

$$\rightarrow 000 \rightarrow 010 \rightarrow 111 \rightarrow 100 \rightarrow 011 \rightarrow 101$$

one possible implementation of the state machine is shown in Fig. 2.18 specify what signals should be applied to each of the multiplexer inputs.

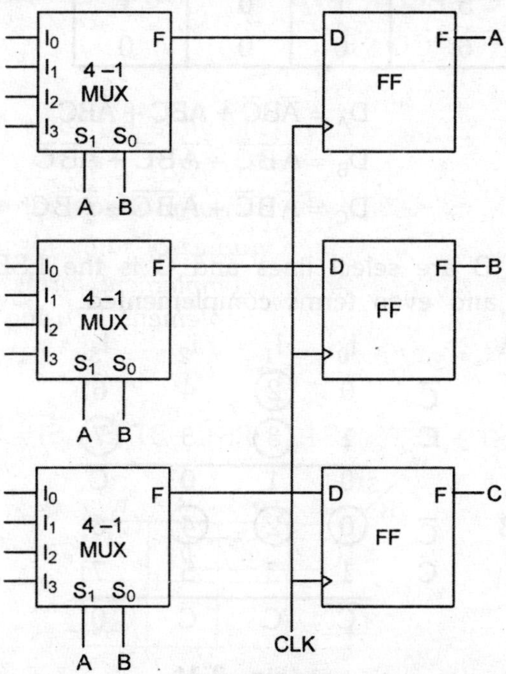

Fig. 2.18

Ans.

	Present count			Next count		
S.No.	Q_A	Q_B	Q_C	Q_A	Q_B	Q_C
1.	0	0	0	0	1	0
2.	0	1	0	1	1	1
3.	1	1	1	1	0	0
4.	1	0	0	0	1	1
5.	0	1	1	1	0	1
6.	1	0	1	0	0	0

Excitation table of D–FF

Q_n	Q_{n+1}	D_n
0	0	0
0	1	1
1	0	0
1	1	1

Comparing with the above table.

S.No.	Q_A	Q_B	D_C
1.	0	1	0
2.	1	1	1
3.	1	0	0
4.	0	1	1
5.	1	0	1
6.	0	0	0

$$D_A = \overline{A}B\overline{C} + ABC + \overline{A}BC$$
$$D_B = \overline{A}\,\overline{B}\,\overline{C} + \overline{A}B\overline{C} + A\overline{B}\,\overline{C}$$
$$D_C = \overline{A}B\overline{C} + A\overline{B}\,\overline{C} + \overline{A}BC$$

In the MUX, A, B are select lines and C is the LSB. In this odd terms are incomplemented and even terms complemented.

DA		I_0	I_1	I_2	I_3
	$\overline{C}$	0	(2)	4	6
	C	1	(3)	5	(7)
		0	1	0	C

DB					
	$\overline{C}$	(0)	(2)	(4)	6
	C	1	3	5	7
		$\overline{C}$	$\overline{C}$	$\overline{C}$	0

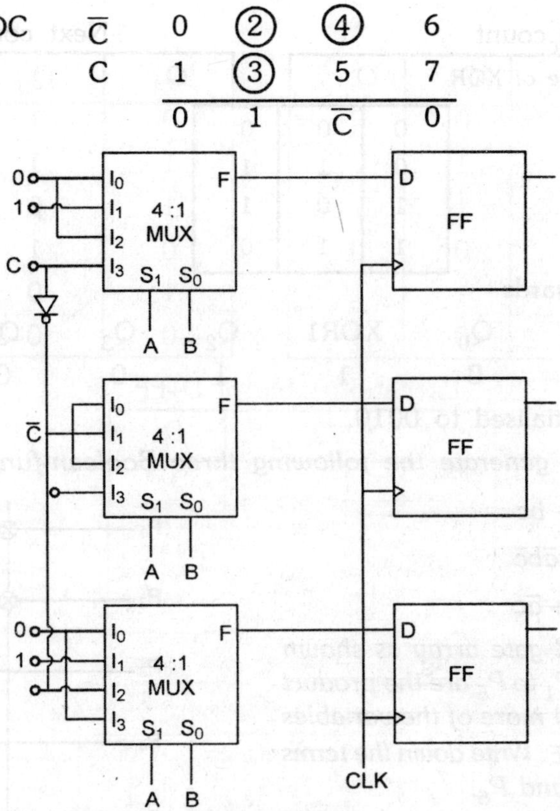

Fig. 2.19

Q.3. A 4-bit shift register, which shifts 1 bit to the right at every clock pulse, is initialised to values (1000) for (Q_0 Q_1 Q_2 Q_3). The D-input derived from Q_0, Q_2 and Q_3 through two XOR gates as shown in Fig. 2.20

(*a*) write the 4-bit values (Q_0 Q_1 Q_2 Q_3) after each clock pulse till the pattern (1000) reappears on (Q_0 Q_1 Q_2 Q_3).

(*b*) To what values should the shift register be initialised so that the pattern (1001) occurs after the first clock pulse ?

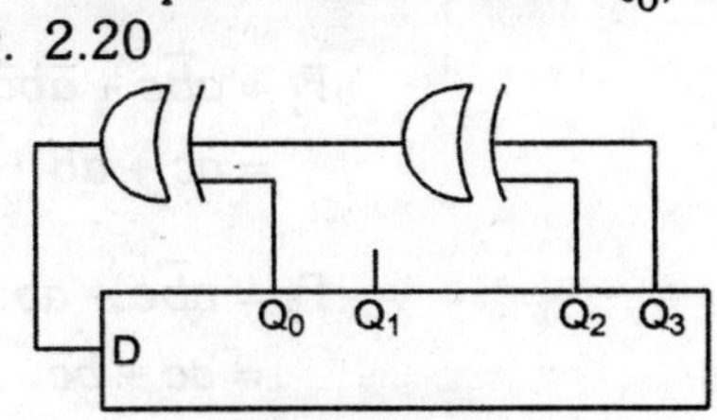

Fig. 2.20

Ans. (*a*)

Clock	D_{out}/Q_0	Q_1	Q_2	Q_3	*XOR* $(Q_2 - Q_3)$	*XOR2* $(Q_0 - XOR1)$	D_{in}
Initial	1	0	0	0	0	1	1
1st	1	1	0	0	0	1	1
2nd	1	1	1	0	1	0	0
3rd	0	1	1	1	0	0	0
4th	0	0	1	1	0	0	0
5th	0	0	0	1	1	1	1
6th	1	0	0	0	0	1	1

1000 reappears at the 6th clock pulse.

Note: (1) $D_{out} = Q_0$

(2) Truth table of XOR

0	0	0
0	1	1
1	0	1
1	1	0

(*b*) Working backwards

D_{in}	XOR2	Q_0	XOR1	Q_2	Q_3	Q_1
1	1	0	1	1	0	0

∴ It should be initialised to 0010.

Q.4. *It is desired to generate the following three Boolean-functions:*

$F_1 = a\bar{b}c + \bar{a}b\bar{c} + bc$

$F_2 = a\bar{b}c + ab + \bar{a}b\bar{c}$

$F_3 = \bar{a}\bar{b}\bar{c} + abc + \bar{a}c$

By using an OR gate array as shown in figure where P_1 to P_5 are the product terms in one (or) more of the variables $a, \bar{a}, b, \bar{b}, c$ and $\bar{c}$. Write down the terms P_1, P_2, P_3, P_4 and P_5.

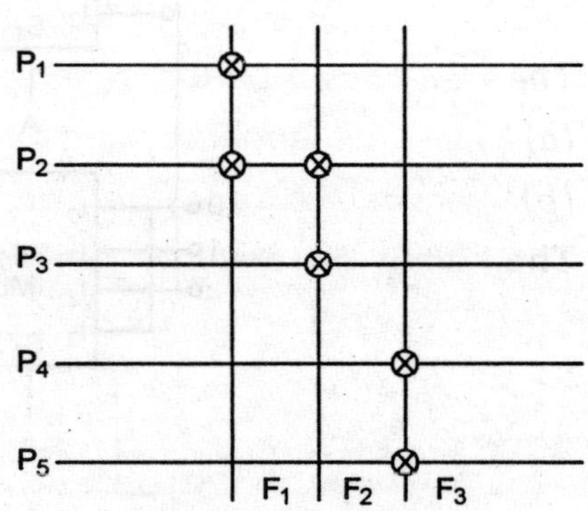

Fig. 2.21

Ans. P_1 to P_5 are AND gate outputs to OR arrays. First we bring each function to minimum SOP form.

$$F_1 = a\bar{b}c + \bar{a}b\bar{c} + bc$$
$$= ac + \bar{a}b$$

$$F_2 = a\bar{b}c + ab + \bar{a}b\bar{c}$$
$$= ac + b\bar{c}$$

$$F_3 = \bar{a}\bar{b}\bar{c} + abc + \bar{a}c$$
$$= \bar{a}\bar{b} + bc$$

∴

$P_1 = \bar{a}b$

$P_2 = a\bar{c}$ (This is common to F_2 also)

$P_3 = b\bar{c}$

$P_4 = \bar{a}\,\bar{b}$

$P_5 = bc$

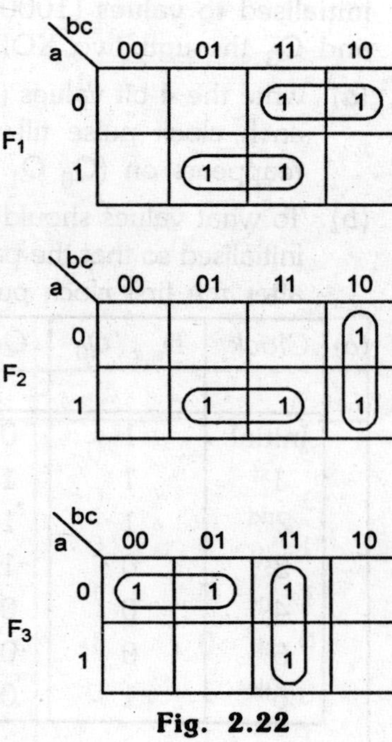

Fig. 2.22

GATE-1997

Q.1. *A sequence generator is shown in Fig. 2.23. The counter status (Q_0, Q_1, Q_2) is initialised to 010 using preset/clear inputs.*

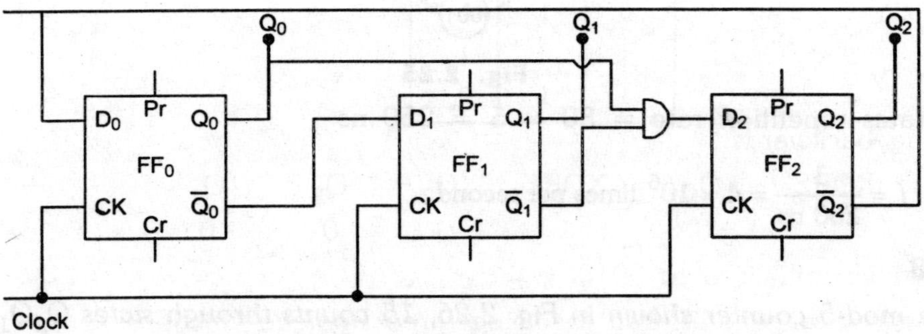

Fig. 2.23

The clock has a period of sons and transitions take place at the rising clock edge.

(a) Give the sequence generate at Q_0 till it repeats.

(b) What is the repetition rate of the generated sequence ?

Ans. The circuit of sequence generator is given as:

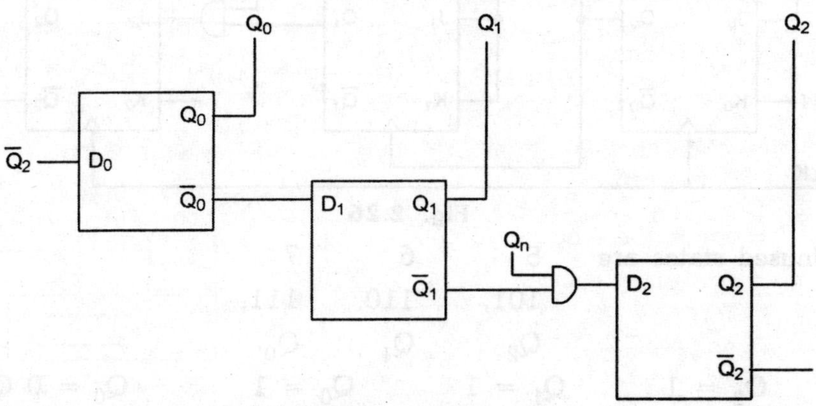

Fig. 2.24

			$\overline{Q_2}$	$\overline{Q_0}$	$\overline{Q_0}\,\overline{Q_1}$			
$Q_0{}'$	$Q_1{}'$	$Q_2{}'$	D_0	D_1	D_2	Q_0	Q_1	Q_2
0	1	0	1	1	0	1	1	0
1	1	0	1	0	0	1	0	0
1	0	0	1	0	1	1	0	1
1	0	1	0	0	1	0	0	1
0	0	1	0	1	0	0	1	0
0	1	0	1	1	0	1	1	0
1	1							

(*a*)

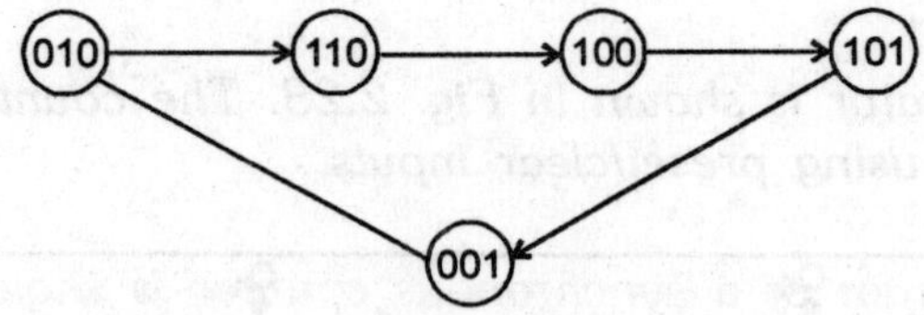

Fig. 2.25

(*b*) 5 states repetition rate = 50 × 5 = 250 ns

$\therefore \quad f = \dfrac{1}{250 \text{ ns}} = 4 \times 10^6$ times per second.

GATE-1998

Q.1. *The mod-5 counter shown in Fig. 2.26, 15 counts through states $Q_2Q_1Q_0$ = 000, 001, 010, 011 and 100.*

(*a*) *Will the counter lockout if it happen to be in any one of the unused states?*

(*b*) *Find the maximum rate at which the counter will operate satisfactorily. Assume the propagation delays of flip-flop and AND gate to be t_F and t_A respectively.*

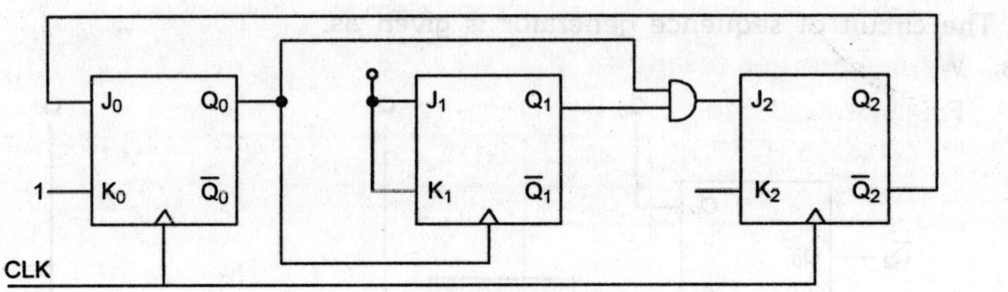

Fig. 2.26

Ans. (*a*) Unused states are 5 6 7

101, 110, 111,

Q_2 Q_1 Q_0

If $Q_2 = 1$ $Q_1 = 1$ $Q_0 = 1$ $Q_0 = 0$ $Q_1 = 0$

$\boxed{J_0 = 0}$ $\boxed{J_1 = 1}$ $\boxed{J_2 = 0}$ $Q_2 = 0$

$\therefore$ used state.

$K_0 = 1$ $K_1 = 1$ $K_2 = 1$ $Q_0 = 0, Q_1 = 0$

If $Q_2 = 1$ $Q_1 = 1$ $Q_0 = 0$ $Q_2 = 0$

$\boxed{J_0 = 0}$ $\boxed{J_1 = 1}$ $\boxed{J_2 = 0}$ $\therefore$ used state.

$K_0 = 1$ $K_1 = 1$ $K_2 = 1$ $Q_0 = 0, Q_1 = 0$

If $Q_2 = 1$ $Q_1 = 1$ $Q_0 = 1$ $Q_2 = 0$

$\boxed{J_0 = 0}$ $\boxed{J_1 = 1}$ $\boxed{J_2 = 0}$ $\therefore$ used state.

$K_0 = 1$ $K_1 = 1$ $K_2 = 1$

Therefore not locked.

(b) $N_F = \dfrac{1}{3t_F + t_A}$.

GATE-1999

Q.1. *The circuit diagram of a synchronous counter is shown in Fig. 2.27. Determine the sequence of states of the counter assuming that the initial state is '000'. Give your answer in a tabular form showing the present state $Q_{t(n)}$, $Q_{B(n)}$, $Q_{C(n)}$, J-K inputs (J_A, K_A, J_B, J_C, K) and the next state $Q_{A(n+1)}$, $Q_{B(n+1)}$, $Q_{C(n+1)}$, from the table, determine the modulus of the counter.*

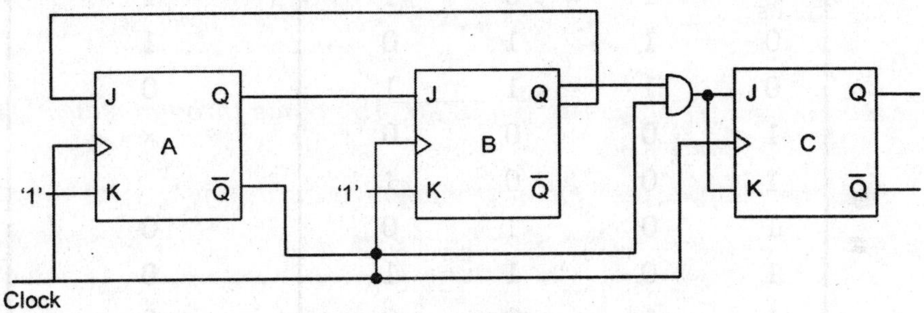

Fig. 2.27

Ans. We assume initial state as 000

For J-K flip-flop.

J	K	$Q_{(n+1)}$
0	0	Q_n
0	1	0
1	0	1
1	1	$\overline{Q}_n$

$Q_{A(n)}$	$Q_{B(n)}$	$Q_{C(n)}$	J_A	K_A	J_B	K_B	J_C	K_C	$Q_{A(n+1)}$	$Q_{B(n+1)}$	$Q_{C(n+1)}$
0	0	0	1	1	0	1	0	0	1	0	0
1	0	0	1	1	1	1	0	0	0	1	0
0	1	0	0	1	0	1	1	1	0	0	1
0	0	1	1	1	0	1	0	0	1	0	1
1	0	1	1	1	1	1	0	0	0	1	1
0	1	1	0	1	0	1	1	1	0	0	0

It is a Mod 6 counter.

Q.2. *In a certain application four inputs A, B, C, D (both true and complement forms available) are fed to logic circuit, producing an output F which operates a relay. The relay turns on when F(ABCD) = 1 for the following states of the inputs (ABCD): '0000', '0010', '0101', '0110', '1101' and '1110'. States '1000' and '1001' do not occur, and for the remaining states, the relay is off minimize F with the help of a Karnaugh-map and realize it using a minimum number of 3-input NAND gates.*

Ans.

Input				Relay output
A	*B*	*C*	*D*	*F*
0	0	0	0	1
0	0	0	1	0
0	0	1	0	1
0	0	1	1	0
0	1	0	0	0
0	1	0	1	1
0	1	1	0	1
0	1	1	1	0
1	0	0	0	×
1	0	0	1	
1	0	1	0	0
1	0	1	1	0
1	1	0	0	0
1	1	0	1	1
1	1	1	1	1
1	1	1	1	0

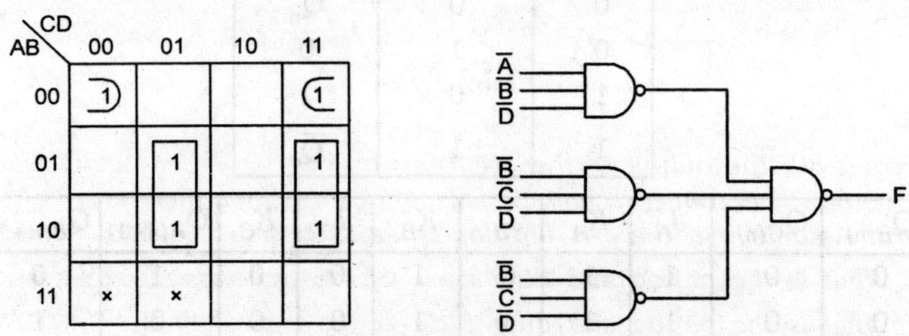

$$F = \overline{A}\,\overline{B}\,\overline{D} + B\overline{C}D + BC\overline{D}$$

Fig. 2.28

GATE-2000

Q.1 *A one-bit full adder is to be implemented using 8-to-1 multiplexers (MUX).*

(a) Write the truth table for sum (S) and carry to the next state (C_N) in terms of the two bits (A, B) and carry from the previous stage (C_p), the truth table should be in the ascending order of (A, B, C_p), i.e., (000, 001, 010, ... etc.).

(b) Implement S and C_N using 8-to-1 multiplexers.

Ans. (*a*) Truth table

	A	B	C_P	$Sum(s)$	$Carry\ (C_N)$
0	0	0	0	0	0
1	0	0	1	1	0
2	0	1	0	1	0
3	0	1	1	0	1
4	1	0	0	1	0
5	1	0	1	0	1
6	1	1	0	0	1
7	1	1	1	1	1

(*b*)

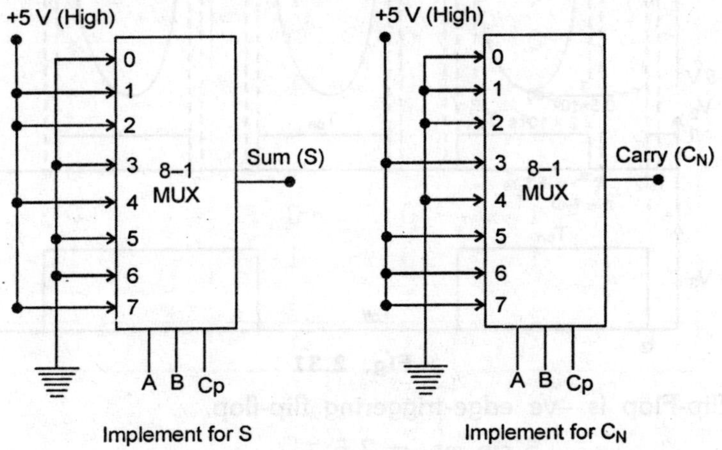

Fig. 2.29

GATE-2001

Q.1. *In Fig. 2.30, the output of the oscillator, V_1 has 10 V peak amplitude with zero DC value. The transfer characteristic of the Schmitt inverter is also shown in figure. Assume that the J-K flip-flop is reset at time t = 0.*

(*a*) *What is the period and duty cycle of the waveform V_2 ?*

(*b*) *What is the period and duty cycle of the waveform V_3 ?*

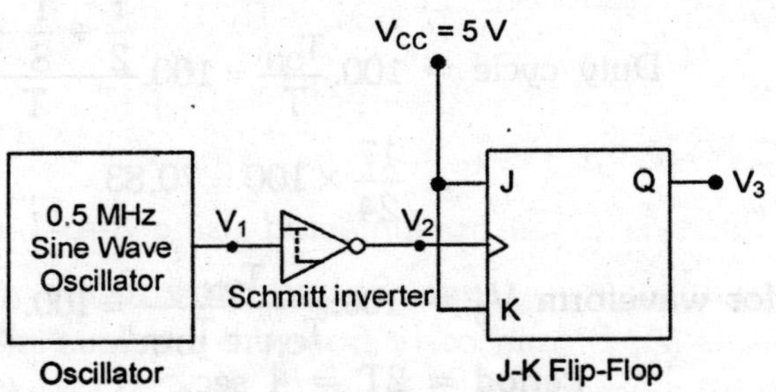

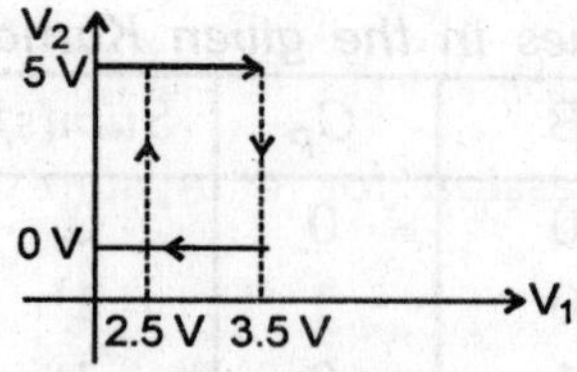

Fig. 2.30

Ans.

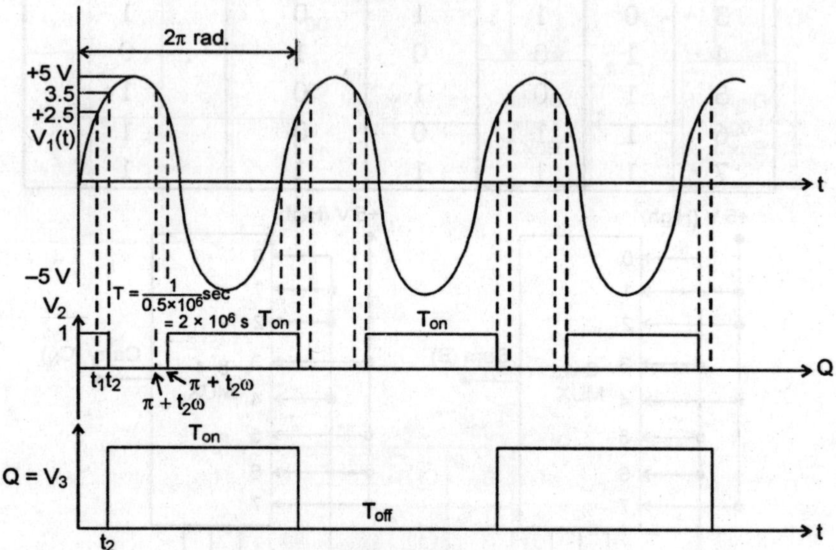

Fig. 2.31

Given J-K flip-Flop is –ve edge-triggering flip-flop.

$$5 \sin \omega t_1 = 2.5$$

$$t_1 = \frac{1}{\omega}\sin^{-1}\left(\frac{2.5}{5}\right) = \frac{T}{12}.$$

And $\quad 5 \sin \omega t_2 = 3.5, \quad t_2 = \frac{T}{8}$

For V_2,

(*a*) $\quad \text{Period} = \frac{1}{f} = T = 2\ \mu\ \text{sec.}$

$$\text{Duty cycle} = 100.\frac{T_{ON}}{T} = 100.\frac{\frac{T}{2}+\frac{T}{8}+\frac{T}{12}}{T}$$

$$= \frac{17}{24} \times 100 = 70.83$$

(*b*) Duty cycle for waveform $V_3 = 100.\frac{T_{ON}}{T_{ON}+T_{OFF}} = 100.\frac{T}{2T} = 50$

$$\text{Period} = 2T = 4\ \text{sec.}$$

Q.2. *For the digital block shown in Fig. 2.32(a), the output $y = f(S_3, S_2, S_1, S_0)$ where S_3 is MSB and S_0 is LSB, y is given in terms of minterms as $y = \Sigma m$ (1, 5, 6, 7, 11, 12, 13, 15) and its complement is $\bar{y} = \Sigma m(0, 2, 3, 4, 8, 9, 10, 14)$.*

(a) *Enter the logical values in the given Karnaugh map [Fig. 2.32 (b)] for the output y.*

(b) *Write down the expression for y in sum-of-products from using minimum number of terms.*

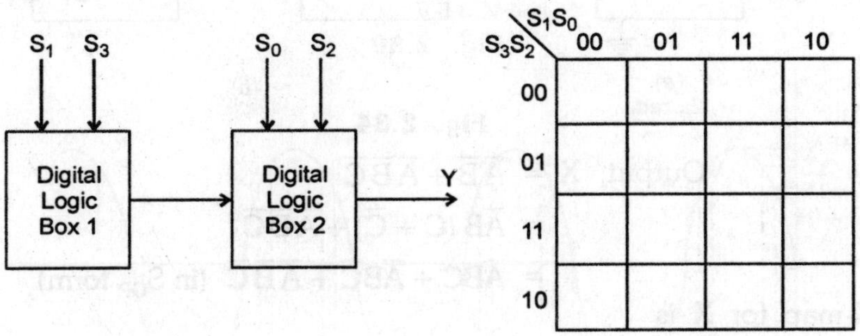

(a) (b)

Fig. 2.32

Ans. (a)

$S_3S_2 \backslash S_1S_0$	00	01	11	10
00	^{0} 0	^{1} (1)	^{3} 0	^{2} 0
01	^{4} 0	^{5} (1)	^{7} (1)	^{6} (1)
11	^{12} (1)	^{13} (1)	^{15} (1)	^{14} (1)
10	^{8} 0	^{9} 0	^{11} (1)	^{10} 0

Fig. 2.33

(b) The minimized expression will be

$$y = S_3S_2\overline{S}_1 + S_0S_1S_3 + \overline{S}_1S_0\overline{S}_3 + \overline{S}_3S_2S_1$$

$$= S_0\left[S_1S_3 + \overline{S}_1\overline{S}_3\right] + S_2\left[S_1\overline{S}_3 + \overline{S}_1S_3\right]$$

GATE-2002

Q.1. *The inputs to a digital circuit shown in Fig. (a) are the external signals A, B and C.*

($\overline{A}, \overline{B}$ and $\overline{C}$ are not availables). The +5 V power supply (logic 1) and the ground (logic 0) are also available. The output of the circuit is $X = \overline{A}B + \overline{A}\,\overline{B}\,\overline{C}$.

(a) *Write down the output function in its canonical S_{OP} and P_{OS} forms.*

(b) *Implement the circuit using only tow 2-1 multiplexers shown in Fig. (b), where S is the data-select line, D_0 and D_1 are the input data lines and y is the output line. The function table for the multiplexer is given in table.*

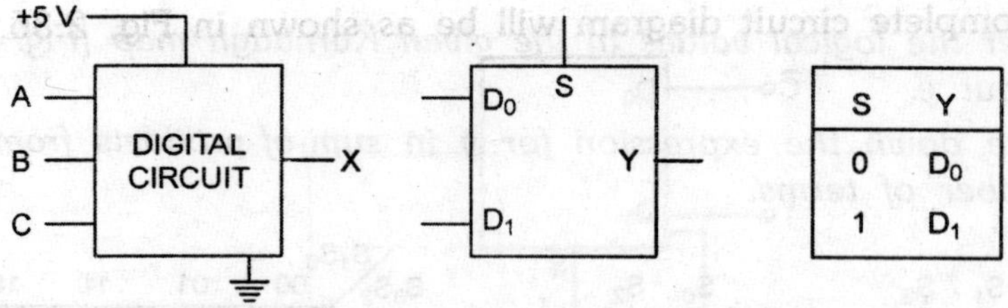

(*a*) (*b*)

Fig. 2.34

Ans. (*a*) Output, $X = \bar{A}\bar{B} + \bar{A}\bar{B}\bar{C}$

$= \bar{A}B(C + \bar{C}) + \bar{A}\bar{B}\bar{C}$

$= \bar{A}BC + \bar{A}B\bar{C} + \bar{A}\bar{B}\bar{C}$ (in S_{OP} form)

The K-map for X is

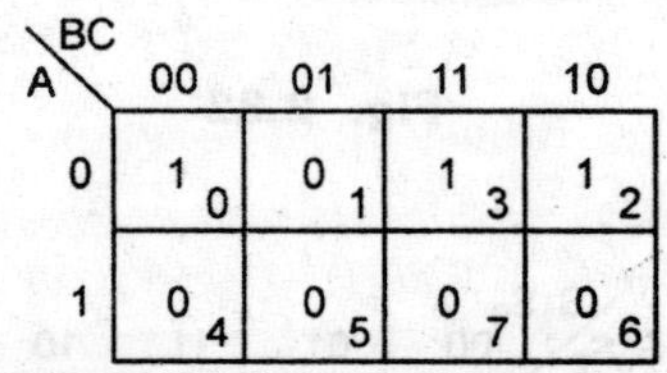

Fig. 2.35

From the K-map, output X is

$$x = \pi M\ (1, 4, 5, 6, 7)$$

$$(A + \bar{B} + C)\ (A + B + \bar{C})\ (A + \bar{B} + \bar{C})\ (\bar{A} + \bar{B} + \bar{C})\ (\bar{A} + B + \bar{C})$$

(*b*) Preparing Truth-table

Input			*Output*
A	*B*	*C*	*X*
0	0	0	1
0	0	1	0
0	1	0	1
0	1	1	1
1	0	0	0
1	0	1	0
1	1	0	0
1	1	1	0

Now, comparing input C and output X, and preparing Truth Table, we have

A	*B*	*X*
0	0	$\bar{C}$
0	1	1
1	0	0
1	1	0

Now, complete circuit diagram will be as shown in Fig. 2.36.

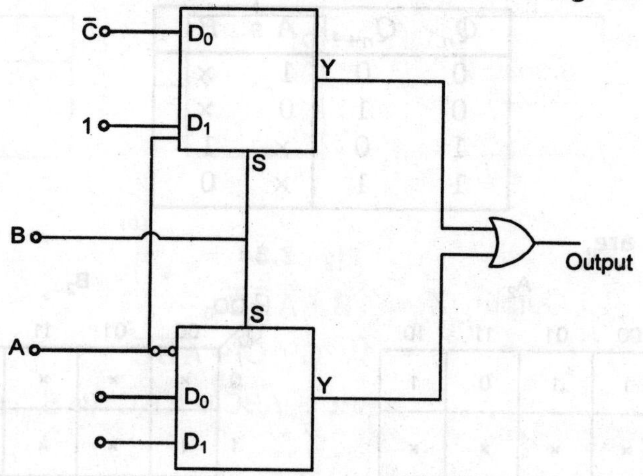

Fig. 2.36

Q.2. *It is required to design a binary mod-5 synchronous counter using AB flip-flops such that the output $Q_2Q_1Q_0$ changes as 000 → 001 → 010 ... and so on. The excitation table for the AB flip-flop is given in table.*

(a) Write down the state table for the mod-5 counter.

(b) Obtain simplified S_{OP} expressions for the inputs A_2, B_2, A_1, B_1, A_0 and B_0 in terms of Q_2, Q_1, Q_0 and their complements.

(c) Hence, complete the circuit diagram for the mod-5 counter given in Fig. using minimum number of 2-input NAND-gate only.

A	B	Q_{n+1}
0	0	1
0	1	$\overline{Q}_n$
1	0	Q_n
1	1	0

Table

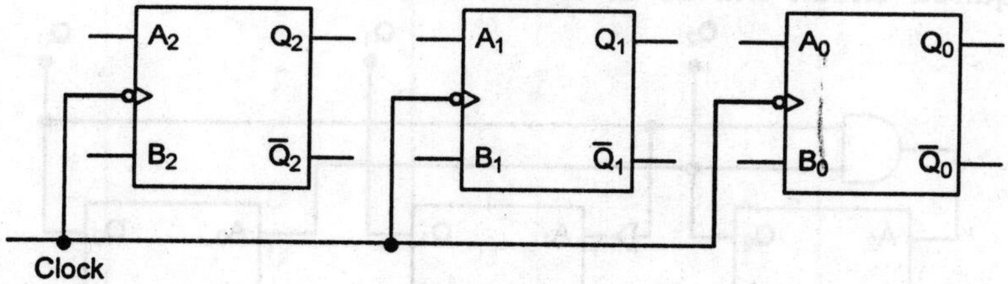

Fig. 2.37

Ans. (a) Count-sequence

Q_2	Q_1	Q_0	A_2	B_2	A_1	B_1	A_0	B_0
0	0	0	1	×	1	×	0	×
0	0	1	1	×	0	×	×	1
0	1	0	1	×	×	0	0	×
0	1	1	0	×	×	1	×	1
1	0	0	×	1	1	×	1	×
0	0	0						

Excitation-Table

Q_n	Q_{n+1}	A	B
0	0	1	×
0	1	0	×
1	0	×	1
1	1	×	0

(*b*) The K-maps are,

A_2

Q_2 \ QQ_0	00	01	11	10
0	1	1	0	1
1	×	×	×	×

$A_2 = \overline{Q.Q_0}$

B_2

Q_2 \ QQ_0	00	01	11	10
0	×	×	×	×
1	1	×	×	×

$B_2 = 1$

Q_2 \ QQ_0	00	01	11	10
0	1	0	×	×
1	1	×	×	×

$A_1 = \overline{Q_0}$

Q_2 \ Q_1Q_0	00	01	11	10
0	×	×	1	0
1	×	×	×	×

$B_2 = Q_0$

Q_2 \ Q_1Q_0	00	01	11	10
0	0	×	×	0
1	1	×	×	×

$A_0 = Q_2$

Q_2 \ Q_1Q_0	00	01	11	10
0	×	1	1	×
1	×	×	×	×

$B_0 = 1$

Fig. 2.38

(*c*) The required circuit will be as shown below:

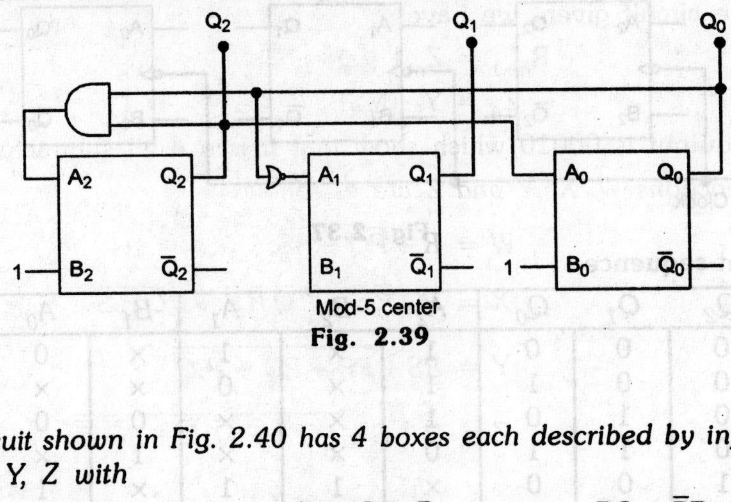

Mod-5 center

Fig. 2.39

GATE-2003

Q.1. *The circuit shown in Fig. 2.40 has 4 boxes each described by inputs P, Q, R and outputs Y, Z with*

$$y = P \oplus Q \oplus R \qquad z = RQ + \overline{P}R + Q\overline{P}$$

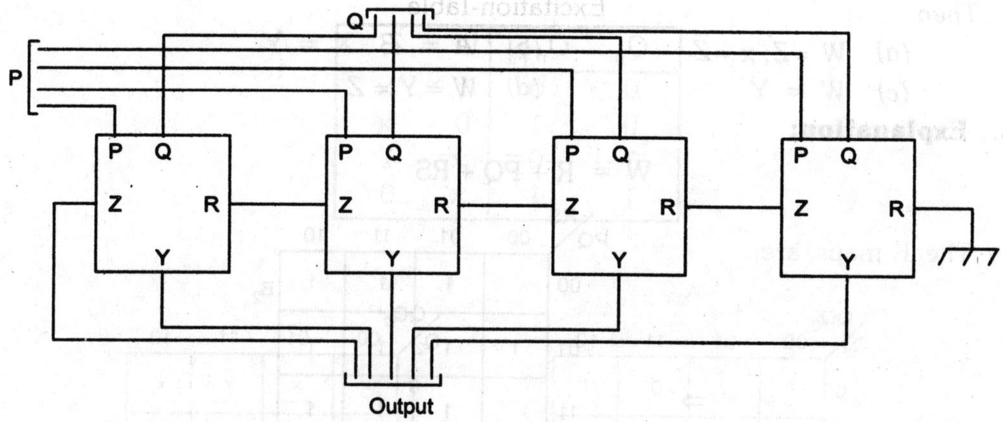

Fig. 2.40

The circuit acts as a

(a) 4 bit adder giving P+Q *(b) 4 bit subtractor giving P – Q*

(c) 4 bit subtractor giving Q – P *(d) 4 bit adder giving P + Q + R*

Ans. (b) Explanation: Let P = 1011 and Q = 1001

Given,
$$Y_n = P_n \oplus Q_n \oplus R_n$$
$$Z = R_nQ_n + \overline{P}_nR_n + Q_n\overline{P}_n$$

Now, constructing the Truth-table.

	P_n	Q_n	R_n	Z_n	Y_n
n = 1	1	1	0	0	0
n = 2	1	0	0	0	1
n = 3	0	0	0	0	0
n = 4	1	1	0	0	0
					0

From the circuit given, we have
$$R_{n+1} = Z_n, 1 \le n \le 3$$
and
$$Z_4 = Y_5 \text{ (MSB)}$$

Hence, output is 00010 which show that it is a 4-bit subtractor giving P – Q.

Q.2. *If the functions W, X, Y and Z are as follows*

$$W = R + \overline{P}Q + \overline{R}S$$
$$X = PQ\overline{R}\,\overline{S} + \overline{P}\,\overline{Q}\,\overline{R}\,\overline{S} + P\overline{Q}\,\overline{R}\,\overline{S}$$
$$Y = RS + \overline{PR + P\overline{Q} + \overline{P}.\overline{Q}}$$
$$Z = R + S + \overline{PQ + \overline{P}.\overline{Q}.\overline{R} + P\overline{Q}.\overline{S}}$$

Then

(a) $W = Z, x = \overline{Z}$ (b) $W = Z, \; X = Y$

(c) $W = Y$ (d) $W = Y = \overline{Z}$

Ans. Explanation:

$$W = R + \overline{P}Q + \overline{R}S$$

$\Rightarrow$

PQ \ RS	00	01	11	10
00		1	1	1
01	1	1	1	1
11		1	1	1
10		1	1	1

Fig. 2.41

$$X = PQ\overline{R}\,\overline{S} + \overline{P}\,\overline{Q}\,\overline{R}\,\overline{S} + P\overline{Q}\,\overline{R}\,\overline{S}$$

$\Rightarrow$

PQ \ RS	00	01	11	10
00	1			
01				
11	1			
10	1			

Fig. 2.42

$$Y = RS + \overline{PR + P\overline{Q} + \overline{P}\,\overline{Q}}$$

$$= RS + \overline{PR}\,.\,\overline{P\overline{Q}}\,.\,\overline{\overline{P}.\overline{Q}}$$

$$= RS + (\overline{P} + \overline{R})\,(\overline{P} + Q)\,(P + Q)$$

$$= RS + (\overline{P} + \overline{R}Q)\,(P + Q)$$

$$= RS + \overline{P}Q + \overline{R}PQ + Q\overline{R}$$

$\Rightarrow$

PQ \ RS	00	01	11	10
00			1	
01		1	1	1
11	1	1	1	
10			1	

Fig. 2.43

$$Z = R+S+\overline{PQ + \bar{P}.\bar{Q}.\bar{R} + P\bar{Q}\bar{S}}$$

$$= R+S+\overline{PQ}.\overline{\bar{P}\bar{Q}\bar{R}}.\overline{P\bar{Q}\bar{S}}$$

$$= R+S+(P+\bar{Q})(P+Q+R)(\bar{P}+Q+S)$$

$$= R+S+(\bar{P}+\bar{Q}(Q+S))(P+Q+R)$$

$$= R+S+(\bar{P}+\bar{Q}S)(P+Q+R)$$

$$= R+S+\bar{P}Q+\bar{P}R+\bar{Q}PS+\bar{Q}RS$$

$\Rightarrow$

PQ \ RS	00	01	11	10
00		1	1	1
01	1	1	1	1
11		1	1	1
10		1	1	1

Fig. 2.44

From Fig. 2.41 and Fig. 2.44, we see that W = Z.

From Fig. 2.42 and Fig. 2.44, we see that X = $\bar{Z}$.

Q.3. *A 4-bit ripple counter and a 4-bit synchronous counter are made using flip-flops having a propagation delay of 10 ns each. If the worst case delay in the ripple counter and the synchronous counter be R and S respectively, then*

(a) R = 10 ns, S = 40 ns

(b) R = 40 ns, S = 10 ns

(c) R = 10 ns, S = 30 ns

(d) R = 30 ns, S = 10 ns.

Ans. (b) Explanation: Propagation delay in 4-bit ripple counter = $4T_d$ = 40 ns and, since all the flip-flops in the synchronous counter are clocked simultaneously, then its worst delay will be equal to 10 ns.

Q.4. *In the circuit shown in Fig. 2.45 A is a parallel-in, parallel-out 4-bit register, which loads at the rising edge of the clock C. The input lines are connected to a 4-bit bus, W. Its output acts as the input to a 16×4 ROM whose output is floating when the enable input E is 0. A partial table of the contents of the ROM is as follows.*

Address	*0*	*2*	*4*	*6*	*8*	*10*	*11*	*14*
Data	*0011*	*1111*	*0100*	*1010*	*1011*	*1000*	*0010*	*1000*

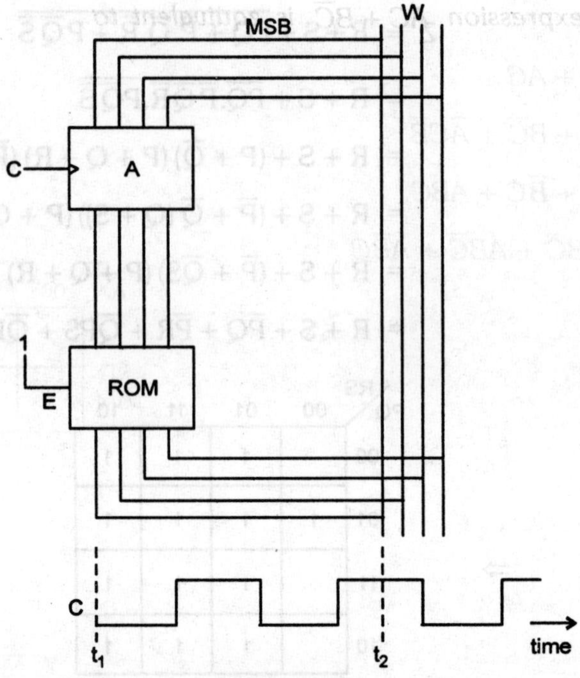

Fig. 2.45

The clock to the register is shown, and the data on the w bus at time t_1 is 0110. The data on the bus at time t_2 is

(a) 1111 *(b) 1011* *(c) 1000* *(d) 0010*

Ans. (c) Explanation: At the first rising edge of clock after $t = t_2$, output of shift register is 0110 = 6 in decimal. At the address 6 data 1010 is stored which is applied to the input of shift register and at the next rising edge, the output of register is 1010 = 10, at the address 10 of ROM, it contains 1000.

GATE-2004

Q.1. *The minimum number of 2-to-1 Multiplexers required to realize a 4-to-1 Multiplexer is*

(a) 1 *(b) 2* *(c) 3* *(d) 4*

Ans. (b)

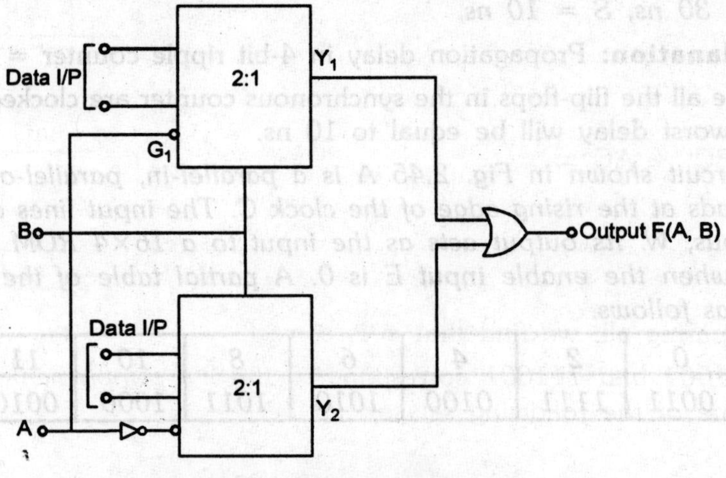

Fig. 2.46

Q.2. *The Boolean expression* $AC + B\overline{C}$ *is equivalent to*

(a) $\overline{A}C + B\overline{C} + AC$

(b) $\overline{B}C + AC + B\overline{C} + \overline{A}C\overline{B}$

(c) $AC + B\overline{C} + \overline{B}C + ABC$

(d) $ABC + \overline{A}B\overline{C} + AB\overline{C} + A\overline{B}C$

Ans. AC + BC′ ⇒

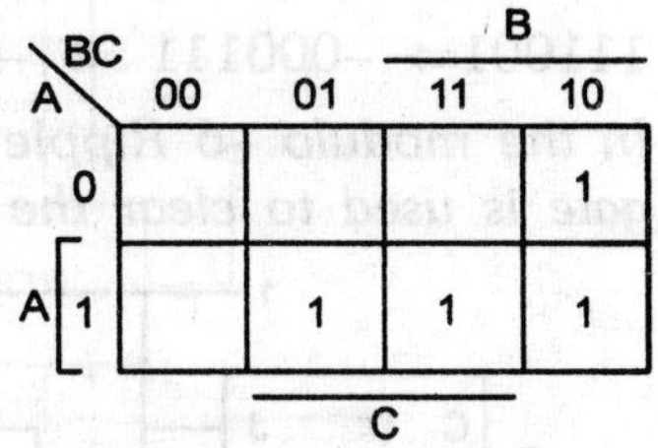

ABC + A′BC′ + ABC′ + ABC′

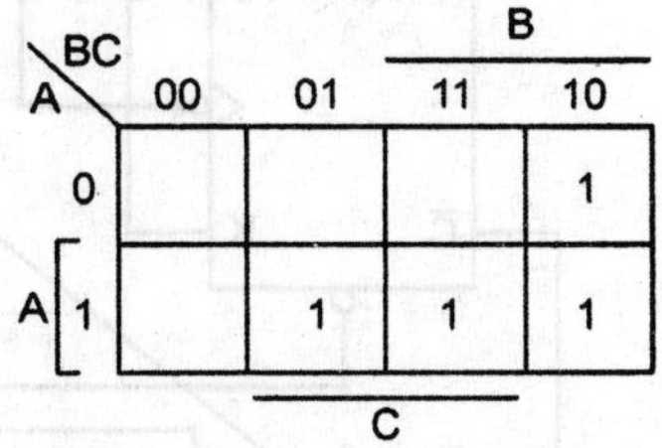

AC + BC′ + B′C + ABC ⇒

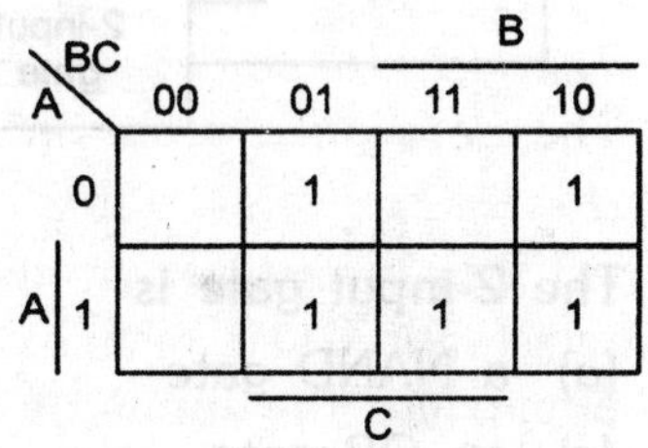

B′C + AC + BC′ + A′CB′ ⇒

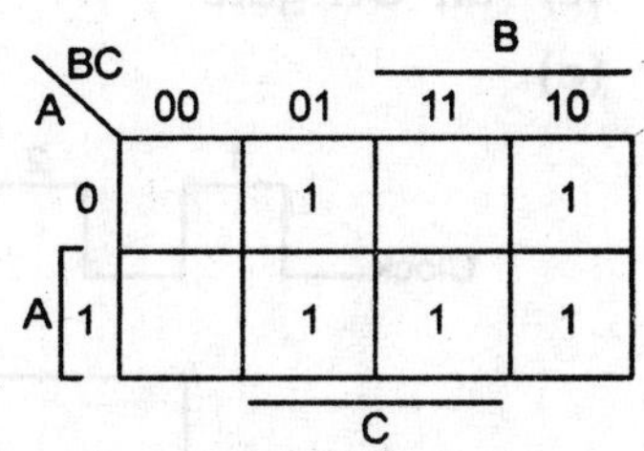

A′C + BC′ + AC ⇒

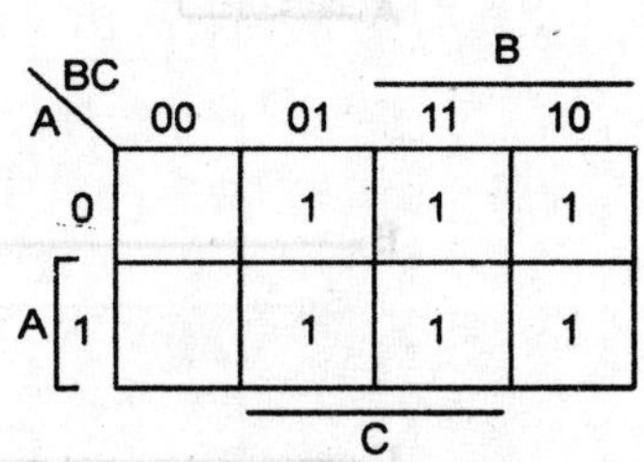

On comparing all, we find that ∞ is correct.

Fig. 2.47

Q.3. *11001, 1001 and 111001 correspond to the 2's complement representation of which one of the following sets of number?*

(*a*) *25, 9 and 57 respectively*

(*b*) *–6, –6 and –6 respectively*

(*c*) *–7, –7 and –7 respectively*

(*d*) *–25, –9 and –57 respectively.*

Ans. **(c)** 11001 ⇒ –00111 ⇒ –7

1001 ⇒ –0111 ⇒ –7

111001⇒ –000111 ⇒ –7

Q.4. *In the modulo –6 Ripple counter shown in Fig. 2.48, the output of the 2-input gate is used to clear the J-K flip-flops.*

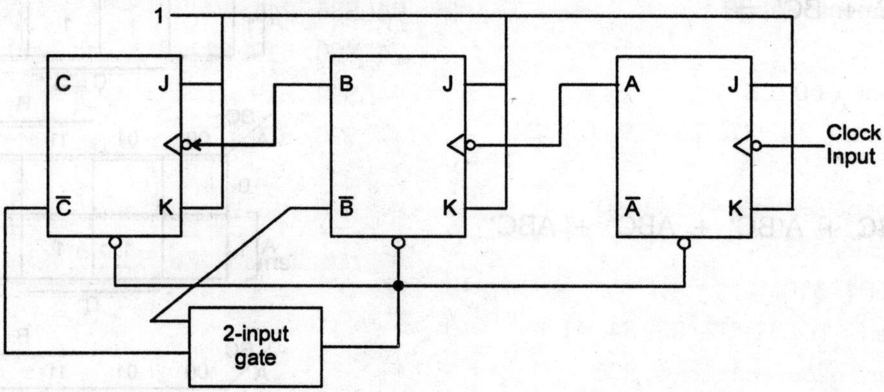

Fig. 2.48

The 2-input gate is

(*a*) a NAND gate (*b*) a NOR gate

(*c*) an OR gate (*d*) an AND gate.

Ans. **(c)**

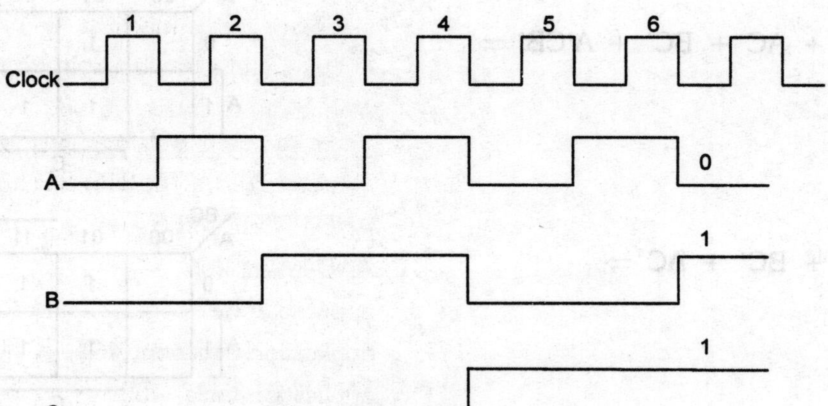

Fig. 2.49

Index

SYMBOL

(r-1)'s complement 26, 39
(XS3) 52
'2-out-of-5 code' 60
1's complement 26, 27, 28, 29, 30, 31, 32, 41
10's complement 26, 39, 40, 41, 43
10's Complement 38
1-successor 604, 610
2's complement 26, 31, 32, 33, 34, 41
2's complement adder-subtractor 306
2-out-of-5 code 60
3-to-8 decoder 325, 326, 327
4-bit comparator 320, 322
74×138 326
7402 239
74161 550
74162 550
74163 550
74190 550
74191 550
74192 550
74193 548, 550
74HC190 550
74HC191 550
74HC192 550
74HC193 548, 550
74HC4024 523
74HC93 519
74LS283 284
74LS290 522
74LS86 241
74LS90 519, 522
74LS92 519
74LS93 519
74XX151 341
8-to-3 priority encoder 336
9's complement 26, 38, 39, 40, 47

A

absorptive law 109
addition 231, 276
adjacent 143, 147, 148, 149, 150, 151, 152, 153, 159, 160
adjacent cells (pairs) 147
admissible patterns 695, 696, 697, 707, 708
algorithmic state machine 641
alpha numeric number system 6
alphanumeric codes 46
AND 227, 228, 229, 230, 231, 236, 237, 238, 241, 243, 244, 246, 247, 248
AND-INVERT 265
AND-invert 266
applicable 617
application equation 462
application table 452
arithmetic circuits 276
array logic 371
ASCII 48, 61, 62, 63, 67, 73, 74, 101

ASM action block 645
ASM chart 641, 643, 644, 645, 646, 648, 649, 650, 651, 652, 653, 654, 655, 656
ASM Chart 643, 644, 674
associative 109, 110
asynchronous counters 508, 509, 519, 523, 526, 532
asynchronous inputs 450, 451
asynchronous sequential circuits 430, 431
augend 277, 283, 286, 287

B

base 2, 3, 5, 6, 7, 8, 9, 11, 12, 13, 18, 19, 20, 21, 26,
BCD 48, 49, 50, 51, 52, 53, 59, 60, 78, 92, 98, 99, 100, 101, 102, 103, 106
BCD adder 290, 292, 293, 294
BCD addition 50, 51
BCD subtractor 296, 302, 303, 304
BCD-to-Excess 3 316, 317
bidirectional counter 540
bidirectional shift register 563
binary 429, 433, 437, 439, 461, 466, 468, 469, 473, 483, 486, 493
binary addition 23
binary arithmetic 22, 25
binary codes 46
binary coding 46
binary counter 508
binary division 36
binary multiplier 669, 671, 673
binary number system 3, 4, 6, 8, 9, 26
binary operators 109, 111
binary parallel adder 283, 284
binary subtraction 24, 26, 28, 29
binary subtractors 295
binary-coded-decimal 48
binary-to-gray code conversion 312
binary-to-octal converter 325
biquinary code 61
bistable multivibrator 432
Boolean algebra 108, 109, 110, 111, 114, 116, 131, 132, 133, 139, 140, 141
Boolean postulates 108
Boolean product 108
Boolean sum 108
borrow 24, 295, 296, 297, 298, 299, 301, 329
branching 646, 647
branching method 191
BUFFER 227, 235, 236
buffer 236, 260
buffer register 555
bushless circuit 484
byte 4

C

cannonical SOP 121, 122, 131, 145
capabilities and limitations 604, 605, 632
CARRY 257
carry 6, 11, 14, 21, 23, 28, 29, 30, 32, 33, 34, 39, 40, 41, 42, 43, 44
carry generate 287, 288, 289, 293
carry propagate 286, 287, 288, 289
carry propagation 286, 287
carry propagation delay 286, 287
cells 142, 143, 144, 145, 146, 147, 148, 149, 151, 152, 160
characteristic equation 437, 438, 440, 443, 462, 464
characteristic table 437, 439, 443
circulating shift register 543
clear 440, 450, 451, 468, 474, 481, 482
clock pulse 431, 435, 436, 439, 440, 442, 444, 445, 446, 447, 449, 451, 457
clocked sequential circuits 431
closed covering 620

closure 109
CMOS family 523, 550
CMOS ripple counter 523
code converter 307, 308, 309, 310, 312, 313, 314, 315, 316, 317, 331
coding circuits 161
combinational circuit 227, 253, 269, 270
combinational circuits 227, 269
commutative 109
comparator 317, 318, 319, 320, 321, 322, 323
complement 108, 109, 110, 111, 115, 116, 117, 119, 120, 122, 123, 124, 125
complementation 235, 266, 268
complementing 227, 266
complements 26, 31, 39, 41
conditional box 644, 645, 646, 655
conditional outputs 648, 672
conjective normal expression 122
conjunctive 688
control inputs 450
control logic 641, 649, 655, 656, 658, 661, 671
controlled buffer register 555
controlled shift register 562
correcting codes 48, 98
counter 508, 509, 510, 511, 512, 513, 514
counter code 61
cover 617, 618, 620, 622, 640
cyclic code 56, 92, 98, 106, 143, 152
cyclic PIC 192

D

D Flip-Flop 438, 456, 458
data 432, 439, 432
DATA Distributors 355
data processor unit 641
data selector 339, 340
De Morgan 115, 116, 117, 128, 130, 132, 160
decade ripple counter 515, 518
decimal adder 290
decimal number system 2, 3, 26, 41
decision box 643, 644, 645, 655, 656, 659, 663, 676
decision paths 641, 643, 646
decoder 323, 324, 325, 326, 327, 328, 329, 330, 331
delay flip-flop 439, 452
demultiplexer 354, 355, 356, 357
demultiplexer 354, 355, 357
demultiplexing 323, 331
deterministic machines 600
Dibble-dabble method 9
difference 24, 295, 296, 297, 298, 299, 306, 329
digital circuit 227, 228, 229, 232, 265, 273
digital codes 1, 46
digital multiplexer 339, 340
digital switch 339
digital systems 307, 323
digits 1, 2, 3, 4, 5, 6, 8, 26, 36, 38, 41, 47
diminished radix 26
direct feedback 543
direct reset 450
direct set 450
direct-inputs 450
disjunctive 688
disjunctive normal expression 121
distinguishing sequence 606, 607, 632, 637, 639
distributive law 109, 110, 111, 117, 131
dominates column 191
dominates row 190
don't care 161, 162, 163, 166, 169, 179, 195, 196, 197, 212
don't care conditions 161, 179

don't care outputs 161
down counters 523
duality theorem 117, 118
dual-line package 230
dynamic indicator 444

E

EAC 28, 29, 40, 44
EBCDIC 48, 63
edge-triggering 444
EEPROMs 364
electronic gate 228
enable 339, 340, 341, 342, 354, 355
encoder 331, 332, 333, 334, 335, 336, 337, 338
End-Around Carry 28
EPROMs 363
EPROMS 427
equivalence classes 607
equivalences 247
equivalent 617
error correcting code 48, 73, 103
error detecting 48, 73, 98, 103, 104
error detecting codes 48
essential prime implicants 171, 183, 185, 192, 222
essential row 176, 190
even parity 66, 67, 68, 74, 75, 76, 77, 99
Ex-3 48, 52, 53, 54, 55
EX-3 52
EX-3 subtractor 305
Excess-3 adder 294, 295
excess-3 code 52, 53, 98, 100
Excess-3 code 52, 98, 104
excitation table 453, 454, 455, 458, 459, 470, 471, 474, 475, 476, 478, 483
EXCLUSIVE-OR 227
Exclusive-OR 241
EX-NOR 242
EX-OR 241, 242
expansion theorem 116

F

falling edge 444
field-programmable PLA 366
finite state machine 599, 604, 605, 632
flip-flop 430, 431, 432, 433, 434, 435, 436, 437, 438, 439, 440, 441, 442, 443, 444
flow chart 641, 642, 643
fraction 2, 5, 6, 9, 16, 38
full subtractor 296, 298, 299, 300, 301, 329, 330

G

gated D-latch 439
gray code 55, 56, 57, 58, 59, 100, 102, 103

H

half adder 277, 279, 281
half subtractor 296, 297
Hamming 48, 74, 76, 77, 78, 104, 105, 107
Hamming code 48, 74, 76, 77, 78, 104, 107
hex dabble 16
hexadecimal number system 6, 7, 16
hinged 163
Hollerith code 48
horizontal quad 162
Huntington postulates 109, 110, 111

I

IC 230, 239

IC7432 233
IC counters 548
IC7400 237
IC7404 235
IC7408 230
IC74266 243
IC7483 284
IC7485 322
IC74LS147 336
IC74LS148 336
implementation table 347, 348, 349, 350, 351, 352, 353, 354, 406, 407, 409, 410, 411, 412, 413, 419
implicant 171, 172, 173, 175, 176, 177, 179, 185, 190, 191, 216, 217, 222
incompletely specified machines 616
indeterminate state 433, 436, 437, 438, 440, 441
index 173, 176, 178, 182, 183, 184, 185, 186, 187
input alphabet 600
integer 2, 4, 5, 6, 7, 8, 9, 10, 11, 12, 14, 15, 16, 20, 21, 26, 38
internal states 269, 429, 459, 466
invalid decision 646
inverse 109, 110, 127
inverse feedback 543
inversion 227, 235, 259
inverted inputs 244, 245, 246, 247
inverted logic 336
inverter 235, 242, 259, 260, 265, 266, 267, 268
INVERTER 235, 250
INVERT-OR 265
invert-OR 266
irreducible 142
irredundant 142
irredundant expression 171, 175
isomorphic 633

J

JK flip-flop 440, 441, 442, 443, 445, 446, 447, 448, 449, 450, 451, 453, 454, 456
Johnson 545

K

Karnaugh map 141, 142, 144
K-equivalent 607, 638
K-map 142, 143, 144, 145, 146, 147, 148

L

latch 432
least significant digit 2
level triggering 446
linear separability 690
linearly separable 692
linearly separable function 690
loading 550
lockout 484, 485
logic 228, 229, 232, 235, 269, 270
logic gate 227, 229, 248, 260, 262, 264, 265, 269, 276
logic level 228, 230, 235
logic symbol 229, 232, 236, 238, 241, 242, 245
logic variables 231
logical operators 227
longitudinal redundancy 68
look ahead carry 287, 289, 290
LRC 66, 68, 69
LSI 323

M

masked ROM 363
mask-programmable PLA 366
master 447

master-slave flip-flop 447
maximal false vertices 693, 694, 704
maxterm 120, 122, 128, 131, 141, 147, 160
Mealy circuits 602, 603
Mealy machine 601, 603, 611, 612, 613, 615, 632, 635
Merger graph 618, 619, 620, 637, 638, 640
Merger table 620, 621, 631, 632, 633, 637, 639, 640
microprogrammed control unit 364
minimal 690
minimal true 692, 693, 694, 696, 704
minimal true vertices 692, 693, 694, 696, 704
minimum' changes codes 56
minterm 120, 121, 128, 131, 140, 142, 143, 144, 145, 147, 149, 152
minuend 295, 296, 298, 301, 305
mirror image 163
mixed-base code 49
Mod-N counter 508
MOD-N counter 526
MOD-n counter 526
modulus counter 508
Moore machine 602, 603, 610, 611, 612, 613, 614, 615, 616, 632, 635
Moore model 603
MSI 289, 323, 336
multi-digit coding 49
multi-output functions 262
multiplexer 339, 340, 341, 342, 343, 344, 345, 346, 347, 354, 355, 356, 357, 428

N

NAND 227, 236, 237, 238, 241, 243, 244, 245, 246, 247, 254, 255, 256, 257, 258
negation 227
negative 649, 679, 682, 685, 686, 688, 689, 691
negative edge 444, 445, 447
negative level triggered flip-flop 446
negative levels 446
negative logic 228, 229
next state 429, 437, 439, 441, 442, 452, 453, 454, 455, 456, 457, 458, 459, 460
NEXT STATE 469
nibble 4
non-positional number system 1
non-unate function 689
non-weighted 46, 47
NOR 227, 238, 239, 240, 241, 242, 243, 244, 246, 247, 254, 258, 259, 261, 262
NOT 227, 234, 235, 236, 237, 238, 239, 241, 242, 243, 244, 247, 248, 249, 250
NOT-AND 236
NOT-XOR 242

O

octal dabble 14
octal-to-binary 331, 332, 333, 336
octet 151, 152, 153, 160
odd parity 66, 68, 76, 77
OR 227, 228, 231, 232, 233, 234, 235, 238, 239, 240, 241, 242, 243, 244, 245, 246
OR operation 231, 232, 233, 267, 268
output alphabet 600, 635

P

PAL 358, 371, 372, 373, 375, 428
parallel binary subtractor 301
parallel in-parallel out shift register 558
parallel in-serial out shift register 557
parity 48, 60, 63, 66, 67, 68, 74, 75, 76, 77, 78, 79, 98, 99
parity checkers 241
parity generators 241
partially specified machine 617

partial-ordering 690
perfect induction 113, 117
periodic 605, 633
permuting counter 517
PLA 364, 365, 366, 367, 368, 369, 370, 371, 372, 373, 374, 375, 423, 428
PLA control 669, 672
PLA program table 366, 367, 368, 370, 373, 375
PLD 358, 364
POS 117, 118, 121, 122, 123, 124, 125, 126, 127, 128, 129, 130, 131, 141, 144
positional number system 1, 2, 5, 6
positive 653, 679, 682, 685, 688, 689, 690, 691
positive edge 444, 445, 447
positive level 446
positive level triggered flip-flop 446
positive logic 228, 229
pq-inequalities 692
present state 429, 430, 431, 437, 439, 441, 452, 454, 455, 456, 457, 458, 459, 460
preset 450, 451
presettable 549
previous states 269
prime implicant 171, 172, 173, 175, 177, 179, 185, 191, 216, 217
prime implicant chart 175, 190, 191, 192, 224, 225
primed 120
priority encoder 333, 334, 336, 338
product-of-sums 121, 124, 131
programmable I/O PALs 372
programmable logic array 358, 364
programmable logic devices 358, 371
prohibited state 433
PROM 363, 375, 427
propagation delay 509, 513, 525, 531, 533, 596, 598
pulse transition 439, 444, 445
pulsed operation 230, 237, 240, 242, 244

Q

quad 149, 151, 152, 153, 160
quadruple multiplexer 341
Quine-McCluskey 170
Quine-Mccluskey 221

R

r's complement 26, 39
race-arround condition 446
radix 2, 3, 5, 6, 7, 18, 20, 22, 26
radix point 2
read only memory 358, 363, 364
reduced chart 190, 193
reduced dimension map 195, 225
redundancy 65, 66, 68, 69, 72, 73, 74, 75
reflected code 56
reflective 47
register 641, 643, 644, 648, 649, 658, 659, 662, 664, 669, 670, 671, 672, 674, 678
reset 432, 434, 436, 437, 440, 442, 449, 450, 451
RESET 450
ring counters 543
ripple carry 286
ripple counter 508, 509, 510, 511, 512, 513, 514, 515, 516, 517, 518, 520, 523, 524
ROM 358, 359, 360, 361, 362, 363, 364, 365, 366, 414, 415, 423, 427
roman numerical system 1

S

selective prime implicants 171, 172
self complementing code 47
self-dual function 118

sequence detector 493
sequence generator 487, 488, 491
sequential circuits 269
sequential circuits 429, 430, 431, 472
Sequential circuits 430
sequential codes 47
sequential data selector 340
sequential machine 599, 600, 603, 604, 611, 612, 613, 617, 623, 632, 633, 634
serial adder 285, 286
serial in-parallel out shift register 556
serial in-serial out shift register 556
set 430, 432, 433, 434, 436, 437, 440, 442, 448, 450, 452, 458
SET 450
Shannon's Theorem 115
shift counter 545
shift counters 543
shift left register 560
shift register 554
shift register 558
shift right register 558, 559, 597
shift-counter 60
signed magnitude format 25
signed numbers 25, 41
single input 234, 235, 262
single output 235, 262, 263, 264
sink vertex 604, 638, 639
six-variable K-map 166
skip states 514
slave 447
SOP 117, 118, 121, 122, 123, 124, 125, 126, 127, 128, 129, 131, 132, 140, 141
source vertex/state 604
spike or glitch 517
SR flip-flop 438, 439, 453, 454, 455
SR Flip-Flop 438, 452, 453, 455
SSI 323
standard form 131
standard package 230, 233, 235, 237, 239, 241, 243
standard product 120
STATE 469
state assignment 468
state box 643, 644, 645, 659
state compatibility 617
state covers 617
state diagram 461
state equation 462, 463
state equivalence 617, 633
state table 452, 454, 455, 459, 460, 461, 462, 463, 464, 465, 467, 468, 470
state transition 438, 461, 462
state variable 600
state-reduction 466
storage capacity 554
strobe 340
strongly connected 605, 634, 635, 639
structured approach 642, 645
subtrahend 295, 296, 298, 301, 302, 305
sum 232, 276
SUM 257
sum of minterms 121, 128, 143
sum-of-product 121, 131
switching algebra 114, 115, 117
switching expression 114, 118
switching function 114, 115, 116, 119, 120, 131, 141, 144, 145
switching variable 114, 115
synchronous binary counter 527, 528, 529
synchronous counter 508, 527, 528, 529, 530, 531, 532, 533
synchronous device 435
synchronous inputs 450
synchronous sequential circuit 430, 431, 505, 507
synchronous sequential machine 599, 600, 603

T

T flip-flop 443, 455, 459, 473, 474, 475
T Flip-Flop 443, 459
tabulation method 170, 181, 225
threshold element 678
threshold network 695, 697
toggle flip-flop 443
toggling 442, 473
transition diagrams 523
transition tables 611, 612
triggering 444, 445, 446, 447
trigonometric functions 397
true minterms 689
true vertices 689
truncated sequence 514
truth table 113, 116, 117, 119, 122, 124, 125, 126, 128, 130, 131, 142, 144, 145
truth values 114
TTL 230, 233, 235, 237, 239, 243
twisted ring counter 545
two-valued inputs 679
two-valued output 679

U

unary operators 108
unate 688, 689, 700, 691
unate function 689
unconditional 644, 646, 647, 652, 672
union law 109
unit distance code 56
universal flip-flop 442
universal gate 237, 239, 247, 262, 264, 265, 267
universal shift register 558, 563
unprimed 120
un-weighted 47
upcounters 523

V

valid decision 646, 647
valid-output indicator 333
Veitch diagram 142
VLSI 323
VRC 66, 68

W

weight 1, 2, 3, 4, 5, 8, 9, 15, 16, 18, 46, 47
weighted code 46, 47
weighted sum 675
word 2, 4, 31, 47

X

XNOR 242
XOR 241, 242
x-successor 604, 607

Z

Z-homogeneous 613
Z-non-homogeneous 613